Lecture Notes in Computer Science 7777

Commenced Publication in 1973
Founding and Former Series Editors:
Gerhard Goos, Juris Hartmanis, and Jan van Leeuwen

Editorial Board

David Hutchison
Lancaster University, UK
Takeo Kanade
Carnegie Mellon University, Pittsburgh, PA, USA
Josef Kittler
University of Surrey, Guildford, UK
Jon M. Kleinberg
Cornell University, Ithaca, NY, USA
Alfred Kobsa
University of California, Irvine, CA, USA
Friedemann Mattern
ETH Zurich, Switzerland
John C. Mitchell
Stanford University, CA, USA
Moni Naor
Weizmann Institute of Science, Rehovot, Israel
Oscar Nierstrasz
University of Bern, Switzerland
C. Pandu Rangan
Indian Institute of Technology, Madras, India
Bernhard Steffen
TU Dortmund University, Germany
Madhu Sudan
Microsoft Research, Cambridge, MA, USA
Demetri Terzopoulos
University of California, Los Angeles, CA, USA
Doug Tygar
University of California, Berkeley, CA, USA
Gerhard Weikum
Max Planck Institute for Informatics, Saarbruecken, Germany

Lecture Notes in Computer Science 7777

Harout Aydinian   Ferdinando Cicalese
Christian Deppe (Eds.)

Information Theory, Combinatorics, and Search Theory

In Memory of Rudolf Ahlswede

Springer

Volume Editors

Harout Aydinian
Christian Deppe
University of Bielefeld
Department of Mathematics
Universitätsstr. 25, 33615 Bielefeld, Germany
E-mail: {ayd, cdeppe}@math.uni-bielefeld.de

Ferdinando Cicalese
University of Salerno
Department of Computer Science
via Ponte don Melillo, 84084 Fisciano, Italy
E-mail: cicalese@dia.unisa.it

ISSN 0302-9743 e-ISSN 1611-3349
ISBN 978-3-642-36898-1 e-ISBN 978-3-642-36899-8
DOI 10.1007/978-3-642-36899-8
Springer Heidelberg Dordrecht London New York

Library of Congress Control Number: 2013933137

CR Subject Classification (1998): E.4, G.2.1-2, F.2.2

LNCS Sublibrary: SL 1 – Theoretical Computer Science and General Issues

Typesetting: Camera-ready by author, data conversion by Scientific Publishing Services, Chennai, India

Printed on acid-free paper

Springer is part of Springer Science+Business Media (www.springer.com)

Rudolf Ahlswede
1938 – 2010

Preface

The Center for Interdisciplinary Research (ZiF) at the University of Bielefeld hosted a cooperation group under the title "Search Methodologies", from October 1, 2010 to December 31, 2012. The main aim of the cooperation group "Search Methodologies" was to gain a deep understanding of the fundamental mathematical structures shared by different models of search encountered in many fields of the hard sciences and the natural sciences. The starting point was the discovery that certain mathematical structures and methodological approaches that we identify as typical of search problems are found in surprisingly diverse scenarios and areas that are, a priori, considered not connected to the field of combinatorial search.

One of our goals has been to create a common forum for theoreticians but also practitioners for comparing different approaches to search problems and different fields of application of search paradigms.

The ZiF cooperation group started in 2010 promoted and led by Rudolf Ahlswede and Ferdinando Cicalese. The seeds of this project had been planted in a preceding Dagstuhl seminar on "Search Methodologies", organized by R. Ahlswede, F. Cicalese, and U. Vaccaro, and held at the Leibniz Institut of Dagstuhl July 5–10, 2009. In fact, the opening conference of the ZiF cooperation group was named "Search Methodologies II", held on October 25–29, 2010.

As clearly stated by Rudolf Ahlswede in the opening lecture of this conference, the time was ripe for founding a comprehensive general theory of search.

Sadly, Rudi will not be able to see the full development of this plant he has so much and devotedly contributed to. On December 18, 2010, suddenly and unexpectedly, Rudolf Ahlswede passed away, but the project continued thanks to the effort of Christian Deppe (who took over as co-organizer), Harout Aydinian, and Vladimir Lebedev. This book is meant as a memorial and includes papers dedicated to Rudolf Ahlswede and results of the cooperation group "Search Methodologies".

On July 25 and 26, 2011 a memorial symposium for Rudolf Ahlswede took place at the ZiF at the University of Bielefeld. The symposium was organized in close cooperation with Rudolf Ahlswede's former students and partners of his many research projects. About 100 participants came to Bielefeld to remember the life and the work of Rudi, the great mathematician, the friend, the mentor, the man.

"Search Methodologies III" was the closing event of the cooperation group. The list of the four macro-topics of this workshop—theory of games and strategic planning, combinatorial group testing and database mining, computational biology and string matching, information coding and spreading and patrolling on networks—provides a comprehensive picture of the vision Rudolf Ahlswede had and put forward of a broad and systematic theory of search.

This book includes 36 thoroughly refereed research papers dedicated to the memory of Rudolf Ahlswede and the research areas in which he worked and that, in several cases, he contributed to shape. One third of the papers originated within the framework of the ZiF cooperation group "Search Methodologies". Furthermore three obituaries and several stories and anecdotes related to Rudolf Ahlswede's life are included.

We are deeply grateful to the members of the administration of the ZiF in Bielefeld for their invaluable and always extremely kind and patient cooperation.

November 2012

Harout Aydinian
Ferdinando Cicalese
Christian Deppe

Organization

Referees

Harout Aydinian
Bernhard Balkenhol
Igor Bjelaković
Vladimir Blinovsky
Holger Boche
Miklos Bona
Jérémie Bourdon
Minglai Cai
Ning Cai
Hong-Bin Chen
Ferdinando Cicalese
Éva Czabarka
Imre Csiszár
Christian Deppe
Christian Donninger
Arkadii D'yachkov
Konrad Engel
Peter Erdős
János Folláth
Nathan Gartner
Dániel Gerbner
Oleg Granichin
Katalin Gyarmati
Pawel Hitczenko
Thomas Kalinowski
Keiichi Kaneko
Gyula Katona
Kingo Kobayashi
Gerhard Kramer
Eyal Kushilevitz
Vladimir Lebedev
Uwe Leck
Robert Leonard
Linyuan Lu
Heinrich Matzinger
Franz Merkl
Balázs Patkós
Vyacheslav Rykov
Oriol Serra
Ulrich Tamm
Eberhard Triesch
Moritz Wiese
Andreas Winter
Jürg Wullschleger
Sergey Yekhanin
Zhen Zhang

Supporting Institutions

Fakultät für Mathematik (Department of Mathematics)
Universität Bielefeld
Postfach 100131
D-33501 Bielefeld
Germany

Zentrum für interdisziplinäre Forschung (ZiF)
(Center for Interdisciplinary Research)
Universität Bielefeld
Wellenberg 1
33615 Bielefeld
Germany

More than restoring
strings of symbols transmitted
means transfer today

What is information?

$\boxed{Cn \textit{ bits}}$ *in Shannon's fundamental theorem*
or $\boxed{\log Cn \textit{ bits}}$ *in our Theory of Identification ?*

Time is mature
for a general Theory of Search.

When we speak of General Theory of Information Transfer today
we mean it to include at its core
the theory of information transmission,
common randomness, identification and
its generalizations and applications,
but it goes far beyond it even outside communication theory
when we think about probabilistic algorithms with identification
as concepts of solution!

Close to the Shannon Island we can see
1.) Mathematical Statistics
2.) Communication Networks
3.) Computer Storage and Distributive Computing
4.) Memory Cells
Since those islands are close there is hope that they can be connected by dams.

There are various stimuli concerning the concepts of communication
and information from the sciences,
for instance from quantum theory in physics,
the theory of learning in psychology, theories in linguistics, etc

We live in a world vibrating with information
and in most cases we dont know
how the information is processed
or even what it is at the semantic and pragmatic levels.

It seems that the whole body of present day
Information Theory will undergo serious revisions
and some dramatic expansions.

Rudolf Ahlswede

1 Introduction

Above are some quotations directly taken from Rudolf Ahlswede's talks, presenting his scientific vision. This is probably the best way to start a volume dedicated to his invaluable contribution to our community. The scientific articles included in this volume are organized according to a broad classification of Ahlswede's interests into three main areas: Information Theory, Combinatorics, Search Theory. The boundaries of such classification are necessarily fuzzy and many papers might well fit into two or all of such major areas, very much like most of Ahlswede's work.

The papers were thoroughly refereed; in the following they are referred to by number, preceded by the letter B, to indicate the present book. We refer to two other lists of publications. They are labelled A and I, where

- **A** indicates the list of publications of Rudolf Ahlswede at the backmatter of the book
- **I** is used for papers mentioned in the introduction, in particular in these comments

The volume is concluded by several obituaries and anecdotes about Rudolf Ahlswede's life written by friends and coauthors.

Papers related to Rudolf Ahlswede's work

Here are some numerical facts about Rudolf Ahlswede's prolific work, with a classification of his 242 papers according to the main research area each paper addresses:

area	number of papers	percentage
Information Theory	**138**	**$\sim$ 57%**
- probabilistic models	56	$\sim$ 23%
- combinatorial models, coding theory	29	$\sim$ 12%
- related to search	21	$\sim$ 8%
- flows in networks	8	$\sim$ 3%
- cryptography as dual	9	$\sim$ 3%
- statistics	7	$\sim$ 3%
- memories	5	$\sim$ 2%
- quantum theoretical	7	$\sim$ 3%
Mainly Combinatorial Structures	**104**	**$\sim$ 43%**
-combinatorial extremal problems	64	$\sim$ 26%
-combinatorial number theory	16	$\sim$ 7%
-inequalities (combinatorial - but also probabilistic, number theoretic, analytical)	11	$\sim$ 5%
-computing and complexity (esp. communication complexity)	13	$\sim$ 5%

The complete list of Ahlswede's publications can be found at the end of this book. Here are listed the thirty papers with the highest number of citations (source: Google Scholar, Oct. 2012):

[A155]	[A12]	[A79]	[A18]	[A21]	[A122]	[A50]	[A161]	[A59]	[A30]	[A29]	[A36]	[A55]	[A131]	[A47]	
4529	473	437	232	215	149	142	137	130	121	106	103	101	91	89	
[A51]	[A27]	[A23]	[A156]	[A31]	[A132]	[A32]	[A60]	[A99]	[A17]	[A53]	[A114]	[A63]	[A44]	[A65]	[A3]
87	72	67	53	51	49	46	45	38	38	37	36	36	35	31	30

In fact, Ahlswede's h-index is 30, however, noticeably, the average number of citations of these papers is 259. Most of the co-authored papers were written with Ning Cai (49), Levon Khachatrian (45), Zhen Zhang (22), Harout Aydinian (22), András Sárközy (9), Christian Deppe (9), Vladimir Blinovsky (8), Mark Pinsker (8), Leonid Bassalygo (7), and Imre Csiszár (6). Altogether, Ahlswede co-authored papers with more than 70 researchers from all over the world.

In the backmatter one can read more about the importance and the impact of Rudolf Ahlswede's work.

I Information Theory

One of the main research goals of Rudolf Ahlswede's was the development of what he defined as a general theory of information transfer (GTIT) [A220], which includes Shannon's classical theory of transmission and his theory of identification as special cases.

A step beyond Shannon's celebrated theory of communication [I07], the foundations for a theory of identification in the presence of noise were laid together with Dueck and carried on by Verboven, van der Meulen, Zhang, Cai, Csiszár, Han, Verdú, Steinberg, Anantharam, Venkataram, Wei, Csibi, Yeong, Yang, Shamai, Merhav, Burnashev, Bassalygo, Narayan and many others.

To fix first ideas, the (classical) theory of information transmission is concerned with the question of how many different messages we can transmit over a noisy channel. Basically, in a (two-party) communication, the receiver tries to answer the question "What is the actual message sent from the set $\mathcal{M} = \{1, \ldots, M\}$?"

On the other hand in the theory of identification, the problem is about identifying rather than recognizing (or reconstructing) a message sent over a (noisy) channel. In short, one tries to answer the question "Is the actual message sent i?", where i is some fixed member of the set of possible messages $\mathcal{M} = \{1, 2, \ldots, N\}$.

This change of perspective extends the frontiers of information theory in several directions and has led to the discovery of new methods which are fruitful also for the classical theory of transmission, for instance in studies of robustness, arbitrarily varying channels, optimal coding procedures in case of complete feedback, novel approximation problems for output statistics and generation of common randomness, a key issue in cryptology.

In this volume, some aspects of the theory of identification are studied in [B01] by Christian Heup, one of Ahlswede's PhD-students. This paper is devoted to the proof of bounds for the expected running time and worst- case running time for identification process for sources.

[B02] introduces L-Identification as a generalization of identification for sources, and the concept of identification entropy of second order as a lower bound.

Shannon's paper about two-way communication was the starting point of multi-user information theory, an area in which Rudolf Ahlswede worked intensively, particularly in collaboration with Ning Cai, who got his PhD under the supervision of Ahlswede. Their life-long collaboration produced 49 papers altogether. Ning Cai's contribution to this volume is contained in the joint paper [B03] with Binyue Liu. The article studies the following channel problem: two senders would like to simultaneously transmit a message each to a receiver. This is done in two steps. First, the message pair is transmitted to n relays. In the second step, the n relays transmit an amplified version of what they received to the receiver. The receiver does not receive anything in the first step. All channels are subject to additive Gaussian noise. The senders have individual power constraints. For the relays two cases are considered. In the first case they have a sum power constraint. In the second case they are subject to individual power constraints. The problem treated in the paper is in each case to find the optimal amplify-and-forward scheme, i.e. the powers at which the relays forward the received signals to the receiver, and to find the corresponding rate region.

In the field of information technology starting around the year 2005, there was much interest in and discussion of Rudolf Ahlswedes information theoretical approach to the development of future communication systems. Because of this, Rudolf Ahlswede was often invited to information technology conferences as the plenary speaker. He lively exchanged developments with many engineers and this led him to attend research meetings at the TU Berlin and the Heinrich Hertz Institute, where he worked closely with Holger Boches group. The following topics were discussed extensively in his lectures and in joint seminars: information theoretic security, arbitrary varying classical and quantum channels, common randomness and de-randomization, resource theory for classical and quantum channels, multiuser systems, new approaches for frequency usage, and approaches to model channel uncertainty. Rudolf Ahlswede was also very much interested in the practical aspects of communication systems, and the resulting discussions led to many publications. This tradition of his close cooperation with communication engineers that had its beginnings in Berlin was to have been continued with his appointment to the Institute of Advanced Study at the TU München in 2011.

Moritz Wiese, one of Boche's group, and Holger Boche considered a multiuser model in [B04]. The multiple access channel with eavesdropper is considered with two main variations: with common randomness (between the transmitters) and with rate-limited conferencing. In addition, in the first variant, the senders not only have their own message but also have to transmit a common one. It comes to no surprise that the problem is complex, and the present authors don't completely solve it. But they use some well-established bits of machinery and quite some ingenuity to provide achievable regions.

This paper introduces us to another important research area to which Ahswede gave a significant contribution: cryptography. In this field, the article by Igor Bjelaković, Holger Boche, and Jochen Sommerfeld [B05] presents several results on AVCs and wiretapping: a lower bound on the random code secrecy capacity for the average error criterion and the strong secrecy criterion in the case where of the eavesdropper has the best channel; an upper bound on the deterministic code secrecy capacity in the general case, which results in a multi-letter expression for the secrecy capacity in the case of best channel for the eavesdropper.

A third paper on cryptography is by Csiszár (joint work with Rudolf Ahlswede). In [B06] the author shows new results on the so-called oblivious transfer (OT) capacity in the honest-but-curious model. OT is a fundamental concept in cryptography, see for example [I04]. The term has been used with different meanings, including a simple transmission over a binary erasure channel. Here OT means "1-2 oblivious string transfer" [I04]: Alice has two length-k binary strings K_0 and K_1 and Bob has a single bit Z as inputs; an OT protocol should let Bob learn K_Z while Alice remains ignorant of Z and Bob of $K_{\bar{Z}}$ ($\bar{Z} = 1 - Z$). The Shannon-theoretic approach is used, thus ignorance means negligible amount of information. In 1988 Rudolf Ahlswede received together with Imre Csiszár the Best Paper Award of the IEEE Information Theory Society for work in the area of the hypothesis testing.

Another major breakthrough brought by Rudolf Ahlswede to the area of information theory was the definition of network coding in [A155]. Network coding is by now a generally accepted concept and has started a revolution in communication networks! We quote from page 58 of [I03] "On the Internet and other shared networks, information currently gets relayed by routers – switches that operate at nodes where signaling pathways, or links, intersect. The routers shunt incoming messages to links heading toward the messages' final destinations. But if one wants efficiency, are routers the best devices for these intersections? Is switching even the right operation to perform?

Until seven years ago, few thought to ask such questions. But then Rudolf Ahlswede of the University of Bielefeld in Germany, along with Ning Cai, Shuo-Yen Robert Li and Raymond W. Yeung, all then at the Chinese University of Hong Kong, published groundbreaking work that introduced a new approach to distributing information across shared networks. In this approach, called network coding, routers are replaced by coders, which transmit evidence about messages instead of sending the messages themselves. When receivers collect the evidence, they deduce the original information from the assembled clues.

Although this method may sound counterintuitive, network coding, which is still under study, has the potential to dramatically speed up and improve the reliability of all manner of communications systems and may well spark the next revolution in the field. Investigators are, of course, also exploring additional avenues for improving efficiency; as far as we know, though, those other approaches generally extend existing methods."

In [B07] Anas Chaaban, Aydin Sezgin, and Daniela Tuninetti consider the symmetric butterfly network (BFN) with a full-duplex relay operating in a bi-directional fashion for feedback. This network is relevant for a variety of wireless networks, including cellular systems dealing with cell-edge users. Upper bounds on the capacity region of the general memoryless BFN with feedback are derived based on cut-set and cooperation arguments and then specialized to the linear deterministic BFN with really-source feedback.

In [B08] David Einstein and Lee Jones consider the case of a network with given total flows into and out of each of the sink and source nodes. It is useful to select uniformly at random an origin-destination (O-D) matrix for which the total in and out flows at sinks and sources (column and row sums) match the given data. The authors give an algorithm for small networks (less than 16 nodes) for sampling such O-D matrices with exactly the uniform distribution and apply it to traffic network analysis. This algorithm can also be used in the statistical analysis of contingency tables.

Ahlswede and Jones were together in Göttingen from 1974 to 1975 and in Bielefeld Summer semester 1991. They never published jointly. However they together discovered and provided several information-theoretic counterexamples for Imre Csiszár. In addition they found several very short original proofs of probability theorems (published by others in the mid 1970's) on partially ordered sets. And for a brief period they claimed an extension of Christofides traveling salesman inequality to the k-salesmen case (Daniel Kleitman even presented their proof at the Vancouver International 1974 Congress.) Shortly thereafter they discovered an error in the proof and showed with an example that the extension was impossible. Both agreed not to publish the counterexample as Ahlswede (at that time) believed that Science is better advanced with positive not negative results.

In 1996 Ahlswede started to work intensively on Quantum Information Theory. He was the supervisor of Andreas Winter who is now one of the main experts in the field. In the article [B09], Andreas Winter provides a survey of basic concepts, and main results in identification over quantum channels. This quantum counterpart of the classical channel identification was introduced in P. Löber's PhD thesis, under the supervision of Ahlswede. Andreas Winter's paper also includes a list of open problems in the area.

In [B10], Vladimir Blinovsky and Minglai Cai – the last student who started his PhD under the supervision of Ahlswede – derive a lower bound on the capacity of classical-quantum arbitrarily varying wiretap channel. The theory of the classical-quantum arbitrarily varying channel was developed by Ahlswede and Blinovsky. The complete characterization of the capacity of classical-quantum arbitrarily varying channel was given in [A216]. For the corresponding wiretap channel the situation is different. Only a lower bound is derived in the paper. The technique for the proof combines ideas from the theory of arbitrarily varying channel with ideas from the theory of the wiretap channel. Furthermore the authors also determine the capacity of the classical quantum arbitrarily varying wiretap channel with channel state information at the transmitter.

On quantum information theory, Rudolf Ahlswede had established a very active collaboration with Holger Boche and his group (see also above). This cooperation among the two groups still continues. Their contribution in [B11] deals with compound as well as arbitrarily varying classical-quantum channel models. The paper pulls together a whole set of previous results, improving them and simplifying their proofs. Also, the connection between zero-error capacity and certain arbitrarily varying channels is discussed.

At the crossroad of coding and information theory, the area of data compression is also represented in this volume. In [B12] Travis Gagie studies the power of multiple read/write streams compared to the single stream case for compression, constrained by polylog memory and passes is considered. The results provided are: universal compression is possible with one stream and one pass; grammar based compression cannot be achieved with one stream and, regarding streams, entropy-only bounds are achievable iff two streams are available.

The article by Matthias Löwe in [B13] provides a review of the area of scenery reconstruction, distinguishing sceneries and related topics. This article explains in a simple manner several of the main mechanisms of the algorithms which have been developed in this subfield. The author describes the model as an information theretic channel model and ask the question “Can we transmit information over such a channel at all?”.

Rudolf Ahlswede entertained a continuing and fruitfull collaboration with Armenian reasearchers at the Institute for Informatics and Automation Problems in Yerevan from beginning of 90's. Among his collaborators from this institute were Levon Khachatrian, Harout Aydinian, Evgueni Haroutunian and other reasearchers from the group of Rom Varshamov. Also Ahlswede was the supervisor of two PhD students, Marina Kyureghyan and Gohar Kyureghyan, from that institute. During the ZiF project General Theory of Information Transfer and Combinatorics (2001-2004) Ahlswede and Haroutunian investigated problems of hypothesis testing for arbitrarily varying sources and problems of hypothesis idendification. In [B14] Evgueni Haroutunian and Parandzem Hakobyan survey the investigations on optimal testing of multiple hypotheses, including the results of Ahlswede and Haroutunian, concerning various multiobject models. These studies show how useful are application of methods and techniques developed in Shannon Information Theory for solution of typical statistical problems.

II Combinatorics

When discussing Ahlswede and Combinatorics, we have to speak of his joint work with Levon Khachatrian. In 1991 Khachatrian visited the University of Bielefeld, as a guest of the Research Project SFB 343 (Diskrete Strukturen in der Mathematik). In Bielefeld Khachatrian began a very fruitful collaboration with Ahlswede, which continued for more than ten years, until the unexpected death of Khachatrian in early 2002. A series of deep results and solutions of long standing problems have been settled by Ahlswede and Khachatrian during this period. Among them two famous problems of Erdős were solved, the first one in number theory, the coprimality problem raised in 1962, and the second, in extremal set theory, the intersection problem raised by Erdős, Ko, and Rado in 1938. Combinatorial number theory was one of the favourite subjects of Ahlswede and Khachatrian. In the mid 90's Ahlswede and Khachatrian started a close cooperation with the well-known number theorist András Sárközy (the most frequent co-author of Erdős). A series of remarkable joint papers with Khachatrian and Sárközy, on divisibility (primitivity) of sequences of integers, have been written during this collaboration. Paper [B15] by Christian Mauduit and András Sárközy is dedicated to the memory of Ahlswede and Khachatrian.

[B15] gives a comprehensive survey on the results regarding the f-complexity of sequences. The notion of f-complexity (family-complexity) of sequences was first introduced and studied by Ahlswede, Khachatrian, Mauduit, and Sárközy in 2003. This quantitative measure of a property of families of sequences to have "rich" and "complex" structure plays an important role in cryptography. During the last decade several new papers on f-complexity and related problems have appeared. [B15] gives a nice overview of those results and related research problems. In the last section of the paper the authors answer the question asked by Csiszár and Gács, at the Ahlswede's memorial conference (at the ZiF), about the connection between f-complexity and VC-dimension.

In [B16] a shadow minimization problem under word-subword relation is considered for the restricted case. The authors give a complete solution to the problem. This problem was raised by Ahlswede, motivated by the study of capacity error function for q-ary error-correcting codes with feedback, the subject on which Rudolf Ahlswede, Christian Deppe, and Vladimir Lebedev were working on in late 2010. Vladimir Lebedev cooperated with Ahlswede's research group during the last years visiting regularly the University of Bielefeld.

In [B17] Harout Aydinian, Éva Czabarka, and Läszló Székely introduce and study k-dimensional M-part multifamilies which is a generalization of M-part Sperner families studied in the literature. BLYM type inequalities for these families are obtained and connections with multitransversals and mixed orthogonal arrays are established. The convex hull method is extended to k-dimensional M-part multifamilies which in turn provides new results for Sperner families. Harout Aydinian has worked for many years in the research group of Rudolf Ahlswede. His joint projects with Ahlswede include topics on extremal combinatorics, coding theory, and communication networks. Ahlswede and Aydinian, together with Khachatrian, introduced and studied a new type of extremal problems for finite sets: extremal problems under dimensional constraints. Many of them are still open.

Rudolf Ahlswede had good relations with many of his co-authors and their families. Thus, he kept good relations with David Daykin and his family, which is told in the obituary [B41] by Jacky Daykin, the daughter of David Daykin. David Daykin died in October 2010 and Rudolf Ahlswede gave a talk at his memorial. One of the last results of David Daykin is [B18]. In this paper, the authors revisit a well-known problem concerning the factorization of strings in a Lyndon-like manner. As the main result, they describe efficient sequential and parallel algorithms for factoring of words over some general classes of factorization families.

One of the last papers of Rudolf Ahlswede is related to a security problem for databases. In [B19] Sergei Bezrukov, Uwe Leck, and Victor Piotrowski consider an interesting combinatorial problem arising in the context of distributed databases. For a given configuration T, defined as a set of faulty nodes in an n-dimensional grid graph, the problem asks if (and how fast) is it possible to recover the nodes in T, where an allowed step is to recover all nodes along a line l, parallel to a coordinate axis, as long as l does not contain too many faulty nodes. The authors obtained nontrivial bounds for the size of completely recoverable configurations, for the numbers of steps needed to recover a fixed node or the full configuration. They also developed fast recovery algorithms for some variants of the recovery problem. Sergey Bezrukov, a co-author of Ahlswede, was a visiting researcher in Ahlswede's group in the early 90's.

In [B20] Carlos Hoppen, Yoshiharu Kohayakawa, and Hanno Lefmann look for an extremal problem for simple hypergraphs with unique extremal configuration but without stability: there are several almost optimal solutions with far apart structures. This phenomenon is known for non- simple graphs, however was not known for simple hypergraphs so far. The problem class studied in the paper is as follows: fix a family of forbidden hypergraphs F and for a fixed hypergraph H and integer k we are looking for all possible k-colorings such that no monochromatic copy of any element of F is present. The problem is to find the maximum of this number, over all choises of H. This problem is studied in details for $k = 4$ and it is shown that the extremal family is unique while the problem (of determining the maximum) is unstable. Hanno Lefmann was a student and later a colleague of Rudolf Ahlswede at the University of Bielefeld.

The paper [B21] by Ulrich Tamm is devoted to the communication complexity problems studied by Ahlswede and his co-authors. Ulrich Tamm was the last postdoc of Rudolf Ahlswede who did his habilitation at the University of Bielefeld. He was educated by Rudolf Ahlswede as an expert in communication complexity. Motivated by the communication complexity of the Hamming distance, Ahlswede raised several challenging combinatorial extremal problems (called Two Family Extremal Problems). Some of them are still open. In [B21] first the author surveys some important results on communication complexity of vector-valued and sum-type functions, for two-party model. In the second part he shows that some of these results can be extended to a multiparty model for vector-valued functions and the pairwise comparison scheme.

In [B22] Éva Czabarka, Matteo Marsili, and László Székely study the threshold function to make almost sure that no two bins contain the same number of balls, when n balls are put into k bins. Depending whether balls and bins are distinguishable/non-distinguishable, there are four combinations. The authors determined the threshold functions for all four scenarios. The non-distinguishable ball problems are essentially equivalent to the Erdős-Lehner asymptotic formula for the number of partitions of the integer n into k parts.

In [B23] Elena Konstantinova gives a survey on two problems of Cayley graphs on the symmetric group: the Star graphs and the Pancake graphs. The first problem deals with the characterization of efficient dominating sets (or perfect codes) and the second one on the cycle structure of these two families of graphs. The main contribution of the author is the conclusion that this description covers in fact all possible constructions of perfect codes in these families. Furthermore, the author gives a structural description of small cycles (of lengths between six and nine) in the Pancake graph.

III Search Theory

When the book [A243] by Rudolf Ahlswede was written in the late 70's during the preceding three decades theoretical and practical contributions classified as "search" were made in Operation Research, Information Theory, Medicine, Computer Science, Management Science, Optimisation Theory etc. Quite different tasks were lumped together under the same name and time was not mature to even try a unified theory. Instead, typical examples were treated and this is reflected in the choice of the title.

However, this was at the start of the aim to develop a theory in the long run. We quote "We want to use our concepts and classifications to make a contribution to working out the essential, common points of the various search problems. By contrasting the various search problems and the methods for their solution, we hope finally to improve the exchange of information between scientists in the various fields."

This book was divided into several parts each one of them dealing with a different class and model of search problems.

At the time of the cooperation group, the following canonical models of search could be shown to encompass all of the different classes considered in the book:

Combinatorial Model	Probabilistic Model
1. $\mathcal{X}$ set of objects.	1. $(\mathcal{X}, P)$ search space.
2. $\mathcal{T} \subset \mathcal{P}(\mathcal{X})$ set of tests.	2. $\mathcal{W}$ set of stochastical matrices $W : \mathcal{X} \to \mathcal{Y}$ as set of tests.

Familiar performance criteria are number of errors, error probability, costs, search duration, complexity, ... and expectations thereof (see also [A195]). Non-deterministic strategies are to be also considered.

Going through the fulminant work on search in the last years one notices a great emphasis on optimisation. If one wants to find the maximum of an integer-valued function one notices that this search problem is almost in the canonical form, but not quite: One has to replace objects by sets of objects (putting together functions with the same maximum value). After this slight generalisation optimisation is covered by our model and the main issue is to handle adaptiveness.

One of the main topics in the cooperation group was group testing. In combinatorial group testing, in an N-element set $\mathcal{U}$, the search space, there is a special subset D (the set of defectives of positives, usually $|D| \ll N$) which we want to discover by asking as few tests as possible. A test is any subset $\mathcal{A}$ of $\mathcal{U}$ and has two possible outcomes: 0 if $\mathcal{A} \cap D = \emptyset$ and 1 otherwise. Adaptive and non-adaptive strategies have been investigated and also both the cases where an exact estimate or only an upper bound is provided on the size of the set D.

Threshold group testing is a generalization of the basic group testing model introduced in [I02]. Let l and u be nonnegative integers with $l < u$, called the *lower* and *upper threshold*, respectively. Suppose that a group test for A says Yes if A contains at least u positives, and No if at most l positives are present. If the number of positives in A is between l and u, the test can give an arbitrary answer. We suppose that l and u are *constant* and previously known. The obvious questions are: What can we figure out about $\mathcal{D}$? How many tests and how much computation are needed? Can we do better in special cases? We call $g := u - l - 1$ the *gap* between the thresholds. The gap is 0 iff a sharp threshold separates Yes and No, so that all answers are determined. Obviously, the classical case of group testing is $l = 0$, $u = 1$.

Fig. 1. The photo shows Rudolf Ahlswede discussing with Anna Voronina (PhD of Arkadii D'yachkov) during the conference Search Methodologies II, 2010

In [B24] Rudolf Ahlswede, Christian Deppe, and Vladimir Lebedev consider non-adaptive threshold testing together with another generalization called majority group testing. They generalize and improve several earlier results by showing that appropriate codes are good test sets.

Christian Deppe was the last assitant of Rudolf Ahlswede. After Ahlswede received his emeritus status, Deppe worked together with him in several projects. Since 2011 he took the guidance of the remaining projects of Ahlswede. Together with Harout Aydinian, Vladimir Blinovsky, Ning Cai, Minglai Cai, Vladimir Lebedev, and Christian Wischmann he continues the work in Bielefeld started by Rudolf Ahlswede. He is thankful for the support of the Department of Mathematics in Bielefeld and for the help of Holger Boche and Igor Bjelaković.

In [B25] Arkadii D'yachkov, Vyacheslav Rykov, Christian Deppe, and Vladimir Lebedev study superimposed codes and their connections to two non-adaptive group testing models introduced by Erlich *et al.* (2010) and threshold group testing. The first part of this paper provides a survey on superimposed codes, which are ubiquitously used for group testing strategies.

D'yachkov first met Ahlswede in 1971 in Tsahkadzor (Armenia) during the Second International Information Theory Symposium, which was organized in the USSR. Ahlswede invited D'yachkov several times to Bielefeld.

In paper [B26] Rudolf Ahlswede and Harout Aydinian give a construction of error-tolerant pooling designs, associated with finite projective spaces. A new family of d^z-disjunct inclusion matrices obtained from packings in finite projective spaces is presented. In particular, a family of disjunct matrices with near optimal parameters is obtained.

In [B27] Dániel Gerbner, Balázs Kesegh, Dömötör Pálvölgyi, and Gabor Wiener consider a new model of group testing where the set of defectives D is assumed to be large, but we are only interested in finding at least m of them, where $m < |D|$. In this model, a test asks if the proportion of the defective elements in the subset tested is above a certain threshold. The paper gives lower and upper estimates on the minimum number of tests (in the worst case) in an adaptive algorithm, which are nearly sharp in different cases of the values of the parameters. The first results were presented in the first workshop of the cooperation group and motivated Ahlswede, Deppe, and Lebedev to work on it. Their results on this model were already published in [A237].

In [B28] Hong-Bin Chen and Hung-Lin Fu consider the case where more than one type of defectives are present in the search space and the result of a test can be affected by the simultaneous presence of different types of defectives. The authors provide lower and upper bounds on the total number of tests required, the number of stages needed to perform all tests and the decoding complexity. An asymptotically optimal 2-stage algorithm based on selectors is also given.

In [B29] a new variant of combinatorial group testing, called complete group testing problem is introduced: the number of defectives among N items is unknown and one tries to minimize the worst case number of tests needed if there happen to be at most D defectives. However, the case of more than D defectives is not excluded, and if more than D defectives are actually present, this must be detected as well and reported. The author proves that the optimal algorithm for the complete group testing problem needs exactly one additional test compared to the optimal algorithm for the basic model.

In [B30] a survey of results on randomized post-optimization for t-restrictions is given. These constructions can be used for many search problems, also for group testing algorithms.

The theory of screening experimental design (SED) can be located in applied mathematics in the border region of search and information theory ([I08]). It comprises models of so-called discrete search problems with randomly disturbed tests; these problems are equivalent to certain coding problems from information theory. In particular coding theory for so-called multiple-access channels finds application in practical problems here for the first time.

In many "processes" which are dependent on a large number of factors, it is natural, that one assumes a small number of "meaningful" factors, which really control the process, and considers the influence of the other factors as mere "experiment errors". Experiments to identify the meaningful factors are called selecting experiments.

In the middle of the seventies Mikhail Maljutov published a series of papers about special models that cleared up the connection to information theory ([I06],

[I05]). In [B31] Mikhail Maljutov gives a survey of the known results in SED focussing on the relations with the new flourishing area of compressed sensing.

Ahlswede was the first outstanding Western researcher to recognize the results of Maljutov. He endorsed its description in the Addendum to the Russian translation of his book "Search Problems". Ahlswede's diploma student Jochen Viemeister at Bielefeld University prepared a detailed survey 45 of results on Maljutov's results spanning around 200 pages in 1982.

Rudolf Ahlswede became interested in Combinatorics in the early seventies. After a short visit in Budapest he invited Gyula Katona to Göttingen. Their 6 month long joint work in 1974 there resulted in two joint papers. This was the beginning of Ahlswede's long and bright journey in the field of Combinatorics. They have not written any more joint papers, but they met many times in Bielefeld and Budapest and their cooperation in Combinatorics and Search Theory continued in a less intensive way as it can be traced in their publications.

In [B32] Gyula Katona and Krisztián Tichler consider an interesting combinatorial search problem with errors. In their model, the errors depend on the target in the following sense: every permitted question specifies two sets of elements B and a $A \subset B$ with $|A| \geq a$, where a is a parameter of the problem. The set A defines the border of reliability of the test, in the sense that if the object to be found is in A, then the answer can be arbitrarily YES or NO. The authors give optimal algorithms for the adaptive and for the non-adaptive case.

For the adaptive case, bounds on the size of the smallest size strategies had been previously given by Ahlswede in [A220].

In [B33] Christian Deppe and Christian Wischmann consider the 1-dimensional cutting problem, a search/optimization problem consisting in minimizing the cost of sawing steel tubes to lengths requested by customers starting from longer tubes. The solution given in this paper improves upon the previous results of Gilmore and Gomory.

To find an (approximately) optimal choice of batches, the graph theoretical problem of finding a minimal weighted partition for a weighted hypergraph is used. This is done by interpreting the lengths requested by customers as vertices and every possibility of combining customer's lengths to a batch as an edge. The weights of an edge correspond to the minimal cost which arise from combining all nodes on the edge into one initial length.

The authors show both how to calculate the edge weights, i.e. the minimal cost which arise from combining all nodes into one initial length by using Gilmore and Gomory's method, and how to faster calculate the (approximately) minimal cost by using a heuristic method for the knapsack problem. Furthermore, they derive a heuristic algorithm for the minimal weighted partition of a weighted hypergraph.

Tree search is one of the central algorithms of any game playing program. [B34] is an overview by Ingo Althöfer of the evolution of research on the topic of game tree search. The paper provides a succinct overview of the topic of game tree search, running all the way from Zermelo at the beginning of the 20th century to the present day. For the specialist in the early history of game theory, it provides useful informations and an overview of the central approaches and concepts in the development of computer-play of chess and Go.

Rudolf Ahlswede was Ingo Althöfer's teacher at Bielefeld University from 1983 to 1993. In his Ackknowledgement he describes how happy he was about the tolerant and unaffected research style of Rudolf Ahlswede. Ingo Althöfer introduced several students to do computer supported mathematics. Among them was Bernhard Balkenhol who initated a group working in data compression. Furthermore together with Ahlswede and Khachatrian he formulated the 3/4-conjecture for fix-free codes in 1996. This conjecture is still open and attracts a lot of papers.

A fix-free code is a code, which is prefix-free and suffix-free, i.e. any codeword of a fix-free code is neither a prefix, nor a suffix of another codeword. Fix-free codes were first introduced by Schützenberger (1956) and Gilbert and Moore (1959), where they were called never-self-synchronizing codes. [B35] establishes a collection of results regarding fix-free codes, i.e., variable length binary code where no codeword is either a prefix or a suffix of any other codeword. In particular the author shows that in any fix free code of Kraft sum 2/3 or less one of the shorter words can be replaced by a large number of longer words preserving both the fix free property and the Kraft sum. The author also presents some new families of complete thin fix free codes.

In [A61] and [A64] Ahlswede, Ye, and Zhang started a new area of research on creating order in sequence spaces with simple machines. On a very high level it was to understand how much "order" can be created in a "system" under constraints on the "knowledge about the system" and on the "actions which can be performed in the system". Rudolf Ahlswede was disappointed that this did not receive much attention. He pointed this out also during his Shannon Lecture.

In [B36]a "multi-user version" is discussed. Here, the output sequence is no longer a permutation of the input sequence, rather it is a permutation of a selected subsequence of the input sequence. The word of multi-user may not be precise for the problem studied, rather it is borrowed from multi-user information theory. The model studies a case that the rate of input is higher than the rate of output. Another major difference between this model and the original model is the measurement of the "efficiency" for a particular organization method. It no longer uses the entropy or cardinality of the output space, it uses the expected value of the time of getting the first 0. Although the model is quite different, it is an interesting mathematical model. Its solution is non-trivial and is related to other interesting mathematical problems. Like for the original creating order model, many variations of the model exist which are of interest.

Obituaries and Personal Memories

We refer to the prefaces of [I01] and [A246] for many biographical information on Rudolf Ahswede. These books appeared on the occasion of Ahlswede's 60th and six years later on the occasion of Ahlswede's ZiF Project "General Theory of Information Transfer and Combinatorics". In this part of the present volume we have collected obituaries written by Ahlswede's family and friends, his supervisor Konrad Jacobs, students and coauthors. We have also included some nice anecdotes. Rudolf Ahlswede liked to tell jokes and laugh about himself and his mischance. It was typical of him to interrupt a lecture to tell a funny story, a habit which was very much appreciated by his students.

It is not very easy to translate into written form such anecdotes with their emotional content, however we hope that these stories will help the reader to get yet another glimpse on Rudi's character and life.

November 2012

Ingo Althöfer, Harout Aydinian, Holger Boche, Ning Cai, Ferdinando Cicalese, Christian Deppe, Vladimir Lebedev, Ulrich Tamm

References

[I01] Althöfer, I., Cai, N., Dueck, G., Khachatrian, L.H., Pinsker, M., Sárközy, A., Wegener, I., Zhang, Z. (eds.): Numbers, Information and Complexity, Special volume in honour of R. Ahlswede on occasion of his 60th birthday. Kluwer Acad. Publ., Boston (2000)

[I02] Damaschke, P.: Threshold Group Testing. In: Ahlswede, R., Bäumer, L., Cai, N., Aydinian, H., Blinovsky, V., Deppe, C., Mashurian, H. (eds.) General Theory of Information Transfer and Combinatorics. LNCS, vol. 4123, pp. 707–718. Springer, Heidelberg (2006)

[I03] Effros, M., Kötter, R., Médard, M.: Breaking network logjams. Scientific American 296, 78–85 (2007)

[I04] Kilian, J.: Founding cryptography on oblivious transfer. In: Proc. STOC 1988, pp. 20–31 (1988)

[I05] Malyutov, M.B., Mateev, P.S.: Screening designs for non-symmetric response function. Mat. Zametki 27, 109–127 (1980)

[I06] Malyutov, M.B.: On the maximal rate of screening designs. Theory Probab. and Appl. 24, 655–657 (1979)

[I07] Shannon, C.E.: Mathematical theory of communication. Bell Syst. Techn. Journal 27, 339–425, 623–656 (1948)

[I08] Viemeister, J.: Die Theorie selektierender Versuche und MAC, Diplomarbeit. Fakultät für Mathematik. Universität Bielefeld (1982)

Search Methodologies I

organized by Rudolf Ahlswede, Ferdinando Cicalese, and Ugo Vaccaro
Dagstuhl, 05.07.09–10.07.09

The main purpose of this seminar was to provide a common forum for researchers interested in the mathematical, algorithmic, and practical aspects of the problem of efficient searching, as seen in its polymorphic incarnation in the areas of computer science, communication, bioinformatics, information theory, and related fields of the applied sciences. We believe that only the on site collaboration of a variety of established and young researchers engaged in different aspects of search theory might provide the necessary humus for the identification of the basic search problems at the conceptual underpinnings of the new scientific issues in the above mentioned areas.

Ingo Althöfer	Monte Carlo Search in Trees and Game Trees
Harout Aydinian	On Error-Tolerant Pooling Designs Based on Vector Spaces
Charles Colbourn	Locating an Interaction Fault
Peter Damaschke	Competitive Group Testing and Learning Hidden Vertex Covers with Minimum Adaptivity
Christian Deppe	Adaptive Coding Strategies and the Counting of Sequences with Forbidden Pattern
Travis Gagie	Minimax Trees in Linear Time with Applications
Mordecai Golin	A Generic Top-Down Dynamic-Programming Approach to Prefix-Free Coding
Gyula O.H. Katona	Finding at Least One Defective Element in Two Rounds
Kingo Kobayashi	Some Aspects of Finite State Channel related to Hidden Markov Process
Evangelos Kranakis	Memory/Time Tradeoffs for Rendezvous on a Ring
Anthony J. Macula	Covert Combinatorial DNA Taggant Signatures for Authentication, Tracking and Trace-Back
Martin Milanic	Competitive Evaluation of Threshold Functions and Game Trees in the Priced Information Model

Olgica Milenkovic	Iterative Algorithms for Low-Rank Matrix Completion
Ely Porat	The Beauty of Prime Numbers vs the Beauty of the Random
Søren Riis	Graph Entropy, Network Coding and Guessing Games
Eberhard Triesch	Searching for Defective Edges in Graphs and Hypergraphs
Gábor Wiener	Rounds in Combinatorial Search
Anatoly Zhigljavsky	Existence Theorems for Search Problems with Lies

The following researchers spoke in the problem session:

Rudolf Ahlswede, Ferdinando Cicalese, Gianluca De Marco, Vladimir Lebedev, Ugo Vaccaro, and Sören Werth.

Search Methodologies II

organized by Rudolf Ahlswede and Ferdinando Cicalese ZiF, Bielefeld, 25.10.10–29.10.10

This workshop was the opening event of the cooperation group "Search Methodologies" at the ZiF. We were interested in deepening our understanding of search problems and we envision that some unifying principles might be disclosed. At least we want to analyze methods and techniques of searching, simplify, compare and/or merge them. One day was devoted to these activities.

The opening lecture by Rudolf Ahlswede explained that time seems to be mature for a more systematic understanding with the goal to build general theories of search. Harout Aydinian and Vladimir Lebedev helped as fellows in this endeavor.

Harout Aydinian	On Generic Erasure Correcting Sets and Related Problems
Vladimir Blinovsky	Solutions of Some Problems from Extremal Combinatorics
Huilan Chang	Identication and Classication Problems on Pooling Designs for Inhibitor Complex Models
Hong-Bin Chen	Pooling Designs with Sensitive and Ecient Decoding
Ferdinando Cicalese	Superselectors: Efficient Constructions and Applications
Éva Czabarka	Full Transversals and Mixed Orthogonal Arrays
Peter Damaschke	Randomized and Competitive Group Testing in Stages
Annalisa De Bonis	Combinatorial Group Testing for Corruption Localizing Hashing
Christian Deppe	Threshold and Majority Group Testing
Arkadii D'yachkov	DNA Codes for Additive Stem Distance
Dániel Gerbner	Search with Density Tests
Tobias Jacobs	Searching in Trees
Sampath Kannan	Sampling from Constrained Multivariate Distributions
Gyula O. H. Katona	Average Length in q-ary Search with Restricted Sizes of Question Sets
Balázs Keszegh	Path-Search in a Pyramid and Other Graphs
Kingo Kobayashi	Some Structures of Tower of Hanoi with Four Pegs

Elena Konstantinova	Search Problems on Cayley Graphs
Vladimir Lebedev	Shadows Under the Word-Subword Relation
Ulf Lorenz	Searching for Solutions of Hard Problems
Anthony J. Macula	Combinatorial Method for Anomaly Detection
Martin Milanic	Haplotype Inference and Graphs of Small Separability
Ely Porat	An Extension of List Disjunct Matrices that can Correct Errors in Test Outcomes
K. Rüdiger Reischuk	Searching for Hidden Information
Søren Riis	A New Max-Flow Min-Cut Theorem for Information Flows
Vyacheslav V. Rykov	An Application of Superimposed Coding Theory to the Multiple Access OR Channel and Two-Stage Group Testing Algorithms
Christian Sohler	Streaming Algorithms for the Analysis of Massive Data Sets
László Székely	M-part Sperner families
Olivier Teytaud	Monte-Carlo Tree Search: a New Paradigm for Computational Intelligence
Eberhard Triesch	A Lower Bound for the Complexity of Monotone Graph Properties
Anna N. Voronina	DNA Codes for Non-Additive Stem Distance

Search Methodologies III

organized by Ferdinando Cicalese and Christian Deppe
ZiF, Bielefeld, 03.09.12–07.09.12

Search Methodologies III has been the last of three main events organized within this cooperation group. In this final workshop, the programme included four major topics: theory of games and strategic planning, combinatorial group testing and database mining, computational biology and string matching, coding, information spreading and patrolling on networks. Besides providing an opportunity for the dissemination of recent results, beyond the border of the restricted community, we pursued cross-fertilization via several RUMP sessions, namely, special open problems and discussion sessions where experts of diverse disciplines could cooperate on the definition of new models, problems or solutions to open problems. Further there were two tutorial talks on communication complexity and quantum computing.

Matthew Aldridge	Adaptive Group Testing: a Channel Coding Approach
Ingo Althöfer	Strange Phenomena in Monte-Carlo Game Tree Search
Harout Aydinian	Quantum Error Correction: An Introduction
Vladimir Balakirsky	Extracting Signicant Parameters from Noisy Observations and Their Use for Authentication
Bernhard Balkenhol	Fuzzy Self-Learning Search
Michael Bodewig	Completeness and Multiplication of Fix-Free Codes Regarding Ahlswedes 3/4-Conjecture
Éva Czabarka	k-Dimensional m-Part Sperner Families and Mixed Orthogonal Arrays
Peter Damaschke	On Optimal Strict Multistage Group Testing
Gianluca De Marco	Computing Majority with Triple Queries
Andreas Dress	Using Suffix Trees for Alignment-Free Sequence Comparison and Phylogenetic Reconstruction
Arkadii D'yachkov	Superimposed Codes and Non-Adaptive Threshold Group Testing
Leszek A. Gąsieniec	Communication-Less Agent Location Discovery
Dániel Gerbner	Majority and Plurality Problems
Gyula O. H. Katona	When the Lie Depends on the Target

Evangelos Kranakis	Boundary Patrolling by Mobile Agents
Vladimir Lebedev	Group Testing with Two Defectives
Zsuzsanna Lipták	Some Problems and Algorithms in Non-Standard String Matching
Mikhail Malyutov	On Connection with Compressed Sensing and Some Other New Developments in the Search Theory
Nikita Polyansky	Random Coding Bounds on the Rate for Non-Adaptive Threshold Group Testing
Ely Porat	Efficient Signature Scheme for Network Coding
Rüdiger Reischuk	The Smoothed Competitive Ratio of Online Caching
Søren Riis	Search Problems Related to Dispersion, Graph Entropy and Guessing Games
Atri Rudra	Group Testing and Coding Theory
Vyacheslav V. Rykov	On Superimposed Codes and Designs
László Székely	Some Constructions for the Diamond Problem
Ulrich Tamm	Multiparty Communication Complexity of Vector-Valued and Sum-Type Functions
Olivier Teytaud	Search with Partial Information
Eberhard Triesch	Upper and Lower Bounds for Competitive Group Testing
Gábor Wiener	On a Problem of Rényi and Katona

Symposium in Remembrance of Rudolf Ahlswede

ZiF, Bielefeld, 25.07.11–26.07.11

Organizers:

Harout Aydinian	Ingo Althöfer	Ning Cai
Ferdinando Cicalese	Christian Deppe	Gunter Dueck
Ulrich Tamm	Andreas Winter	

On July 25 and 26 a memorial symposium for Rudolf Ahlswede took place in the Center for Interdisciplinary Research (ZIF) at the University of Bielefeld. The symposium was organized by the ZIF in close cooperation with Rudolf Ahlswede's former students and partners of his many research projects. About 100 participants came to Bielefeld to remember life and work of Rudolf Ahlswede who had passed away suddenly and unexpectedly on December 18, 2010.

Outside the lecture hall there were a poster wall with pictures and permanent video presentations from Rudolf Ahlswede's Shannon lecture and from the memorial session at the ITA in San Diego. At the conference dinner further videos were presented by Vladimir Lebedev and Tatiana Dolgova about their last visit in Polle and by Rüdiger Reischuk from the conference on Rudolf Ahlswede's 60th birthday.

Further participants include among others - just to mention the information theorists - Arkadii D'yachkov, Peter Gács, Te Sun Han, Gerhard Kramer, Aydin Sezgin, Faina Soloveeva, and Frans Willems.

Before the official start of the symposium, a warm-up lecture by Charles Bennett on "Quantum information" already attracted a lot of interest.

The first day then was devoted to personal memories on Rudolf Ahlswede's life and academic career.

M. Egelhaaf	Commemorative address as representatives of the ZiF
E. Emmrich	Commemorative address as representative of the Department of Mathematics of the University of Bielefeld
A. Ahlswede and B. Ahlswede-Loghin	Joint life with Rudolf Ahlswede
K. Jacobs	Rudolf Ahlswede's time in Göttingen
E. van der Meulen	R. Ahlswede 1970-74
J. Daykin	R. Ahlwede's cooperation with David Daykin
M. Maljutov	R. Ahlswede and search theory
G. Dueck	Rudolf Ahlswede 1975-1985

I. Althöfer	Rudolf Ahlswede 1985-1995
U. Tamm	R. Ahlswede 1995-2000
G. Khachatrian	R. Ahlswede's cooperation with Levon Khachatrian
N. Cai	My cooperation with R. Ahlswede
B. Balkenhol	R. Ahlswede and the computer
K. Kobayashi	R. Ahlswede's cooperation with Japanese researchers
V. Blinovsky	R. Ahlswede's cooperation with Russian researchers
C. Deppe	R. Ahlswede 2000-2010
A. Winter	R. Ahlswede and quantum information theory
H. Aydinian	My cooperation with R. Ahlswede
C. Heup	The identification of the lucky-dog-entropy
F. Cicalese	My projects with R. Ahlswede
H. Boche	R. Ahlswede in Berlin

The second day of the symposium was devoted to scientific lectures on Ahlswede's fields of research.

I. Csiszár	Common randomness in information theory
P. Narayan	Common randomness and multiterminal secure computation
A. Winter	Quantum channels and identification theory
A. Sárközy	On the complexity of families of binary sequences and lattices
A. Orlitsky	String reconstruction from substring compositions
R. Reischuk	Stochastic search for locally clustered targets
G. Katona	The deep impact of Rudolf Ahlswede on combinatorics
H. Aydinian	AZ-identity for regular posets
N. Cai	Secure network coding
S. Riis	Information flows and bottle necks in dynamic communication networks
L. Tolhuizen	A generalisation of the Gilbert-Varshamov bound and its asymptotic evaluation
L. Székely	Higher order extremal problems
S. Bezrukov	Local-global principles in discrete extremal problems

Participants: H. Abels, A. Ahlswede, B. Ahlswede (Bielefeld), M. Ahlert (Halle a.d.Saale), I. Althöfer (Jena), H. Aydinian (Bielefeld), A. Bak (Bielefeld), B. Balkenhol (Bielefeld), C. Bennett (New York, USA), S. Bezrukov (Wisconsin, USA), I. Bjelaković (München), P. Blanchard (Bielefeld), V. Blinovsky (Moscow, RUS), H. Boche (München), G. Brinkmann (Gent, BE), J. Bültermann (Gütersloh), M. Cai (Bielefeld), N. Cai (Xian, China), H.-G. Carstens (Bielefeld) F. Cicalese (Salerno, I), I. Csiszár (Budapest, H), A. D'yachkov (Moscow, RUS), J. Daykin (London, UK), C. Deppe, E. Diemann (Bielefeld), T. Dolgova (Moscow, RUS), A. Dress (Bielefeld), G. Dueck (Mannheim), P. Eichelsbacher (Bochum), E. Emmrich (Bielefeld), P. Gács (Boston, USA), L. Galumyan, F. Götze, T. Grotendiek (Heilbronn), O. Gutjahr (Bielefeld), Te Sun Han (Tokyo, J), C. Heup (Grafschaft), A. Hilton (Reading, UK), W. Hoffmann (Bielefeld), K. Jacobs (Bamberg), G. Janßen (München), N. Kalus (Berlin), G. Katona (Budapest, H), A. Khachatrian (Bielefeld), G. Khachatrian (Yerevan, Armenia), C. Kleinewächter (Bielefeld), K. Kobayashi (Tokyo, J), K.-U. Koschnick (Dreieich), G. Kramer (München), H. Krause (Bielefeld), U. Krengel (Göttingen), G. Kyureghyan (Magdeburg), M. Kyureghyan (Eschborn), V. Lebedev (Moscow, RUS), H. Lefmann (Chemnitz), Z. Lipták (Salerno, I), P. Löber (Langen), M. Löwe (Münster), M. Malioutov (Boston, USA), M. Matz (Bielefeld), P. Narayan (Maryland, USA), J. Nötzel (München), J. Obbelode (Gütersloh), A. Orlitsky (San Diego, USA), T. Partner, C. Petersen, D. Poguntke, U. Rehmann (Bielefeld), K.R. Reischuk (Lübbeck), R. Reischuk (Bielefeld), S. Riis (London, UK), M. Rudolf (Halle), A. Sárközy (Budapest, H), A. Sezgin (Bochum), F. I. Solovyeva (Novosibirsk, RUS), N. Spenst (Kirchlengern), E. Steffen (Paderborn), H. Steidl (Bielefeld), L. A. Székely (Columbia, USA), U. Tamm (Istanbul, T), L. Tolhuizen (Eindhoven, NL), E. Van der Meulen (Leuven, Be), D. Voigt, H.M. Wallmeier, C. Wegener-Mürbe, W. Wenzel (Kassel), F. Willems (Eindhoven, NL), A. Winter (Bristol, UK)

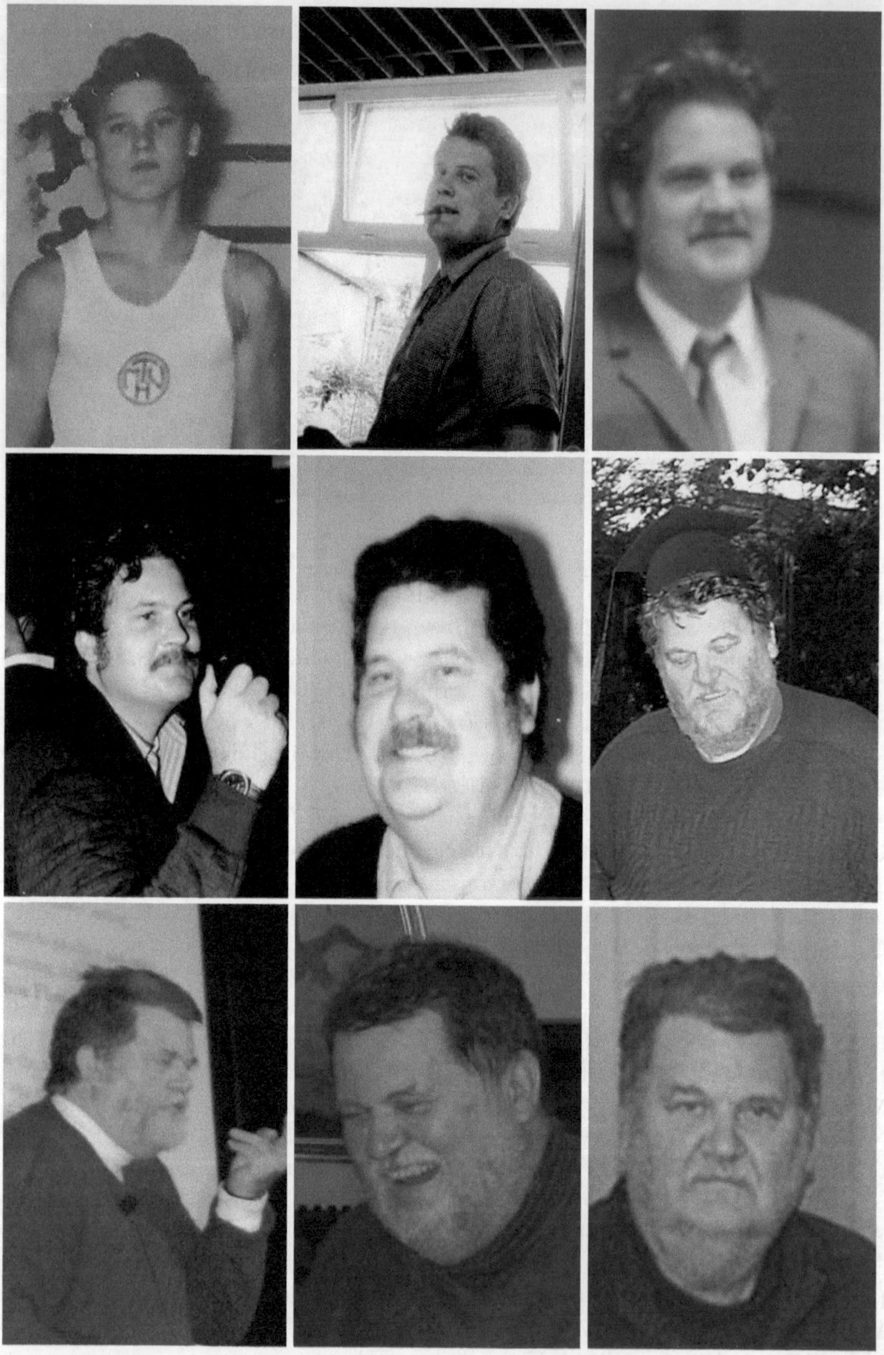

L. Khachatrian, P. Erdős, R. Ahlswede (Bielefeld 1995, check of Erdős for the solution of the $4m$-Conjecture)

R. Ahlswede, D.L. Neuhoff (Seattle 2006, Shannon Award)

C. Deppe, A. Ahlswede, R. Ahlswede, and H. Aydinian (Seattle 2006)

Table of Contents

I. Information Theory

II. Combinatorics

III. Search Theory

Obituaries and Personal Memories

Two New Results for Identification for Sources

Christian Heup

Universität Bielefeld, Fakultät für Mathematik,
Universitätsstraße 25, 33615 Bielefeld, Germany
christian.heup@gmail.com

Dedicated to the memory of Rudolf Ahlswede

Abstract. We provide two new results for identification for sources. The first result is about block codes. In [Ahlswede and Cai, IEEE-IT, 52(9), 4198-4207, 2006] it is proven that the q-ary identification entropy $H_{I,q}(P)$ is a lower bound for the average number $L(P,P)$ of expected checkings during the identification process. A necessary assumption for this proof is that the uniform distribution minimizes the symmetric running time $L_{\mathcal{C}}(P,P)$ for binary block codes $\mathcal{C} = \{0,1\}^k$. This assumption is proved in Sect. 2 not only for binary block codes but for any q-ary block code. The second result is about upper bounds for the worst-case running time. In [Ahlswede, Balkenhol and Kleinewchter, LNCS, 4123, 51-61, 2006] the authors proved in Theorem 3 that $L(P) < 3$ by an inductive code construction. We discover an alteration of their scheme which strengthens this upper bound significantly.

Keywords: identification, source coding.

1 Terminology

We consider a discrete source $(\mathcal{U}, P)$, where $\mathcal{U} := \{1, \ldots, N\}$ and P is a probability distribution on $\mathcal{U}$, together with a q-ary code $\mathcal{C}$ on $\mathcal{U}$. Additionally and in contrast to classical source coding we also introduce the so-called user space $\mathcal{V}$, with $|\mathcal{V}| = |\mathcal{U}|$. Let $f : \mathcal{V} \to \mathcal{U}$ be a bijective mapping. We encode the *users* v with the same code $\mathcal{C}$ as before. That is, we set $c_v = c_{f(v)}$. Without loss of generality we assume that $\mathcal{V} = \mathcal{U}$ and $f = id_{\mathcal{U}}$.

The task of identification is to decide for any user $v \in \mathcal{U}$ and every output $u \in \mathcal{U}$ whether or not $v = u$. To achieve this goal we compare step by step the first, second, third etc. digit of c_v with the corresponding digits of c_u. This process halts with a negative answer if a digit occurs which is different for c_v and c_u. It halts with a positive answer if all digits coincide. The number of steps until the process halts is called the *identification running time* for $(u,v) \in \mathcal{U} \times \mathcal{U}$ and is denoted by $L_{\mathcal{C}}(u,v)$. Obviously, it holds that $L_{\mathcal{C}}(u,v) = lcp(c_u, c_v) + 1$, where $lcp(c_u, c_v)$ is the longest common prefix of the two codewords c_u and c_v.

The goal of this article is to analyze the expected length of the identification running time, also called the *average running time*, for a given user $v \in \mathcal{U}$

$$L_{\mathcal{C}}(P,v) = \sum_{u \in \mathcal{U}} P_u L_{\mathcal{C}}(u,v) \ .$$

H. Aydinian, F. Cicalese, and C. Deppe (Eds.): Ahlswede Festschrift, LNCS 7777, pp. 1–10, 2013.

This can be done in different ways. In the first scenario we assume that also user v is randomly chosen according to a probability distribution Q on $\mathcal{U}$. We are now interested in the *expected average running time* or shortly the *expected running time*

$$L_{\mathcal{C}}(P,Q) = \sum_{v\in\mathcal{U}} Q(\{v\})L_{\mathcal{C}}(P,v)$$

where we focus on the special case $Q = P$

$$L_{\mathcal{C}}(P,P) = \sum_{v\in\mathcal{U}} p_v L_{\mathcal{C}}(P,v) = \sum_{(u,v)\in\mathcal{U}^2} p_u p_v L_{\mathcal{C}}(u,v)$$

We call $L_{\mathcal{C}}(P,P)$ the *symmetric running time* for a given code $\mathcal{C}$ and show that it is minimized by the uniform distribution if $\mathcal{C}$ is a saturated block code.

In the second scenario we are interested in the *worst-case average running time*

$$L_{\mathcal{C}}(P) = \max_{v\in\mathcal{U}} L_{\mathcal{C}}(P,v) \ ,$$

which we shortly call the *worst-case running time.* We construct a specific code whose running time yields an upper bound for the *optimal worst-case running time*

$$L(P) = \min_{\mathcal{C}} L_{\mathcal{C}}(P) \ .$$

2 On the Optimality of the Uniform Distribution for Identification on Saturated Block Codes

A saturated q-ary block code of depth n is a code of size q^n where all codewords have length n. It is denoted by $\mathcal{C}_{q^n}$. In order to show that the uniform distribution is optimal for identification on those codes we modify any given probability distribution step by step until we reach the uniform distribution without increasing $L_{\mathcal{C}_{q^n}}(P,P)$. It turns out that not only the uniform distribution is optimal. In fact, all distributions $P = (p_1, \dots, p_{q^n})$ are optimal with the property that for all $i \in \{1, \dots, q^{n-1}\}$

$$\sum_{u=(i-1)q+1}^{iq} p_u = \frac{1}{q^{n-1}} \ .$$

This is due to the fact that the running time regarding v is the same for all u whose codewords c_u coincide with c_v in all but the last digit. The individual steps of modification and their monotone decreasing property are formalized in

Lemma 1. *Let $n \in \mathbb{N}$, $q \in \mathbb{N}_{\geq 2}$, $k \in \{0, \dots, n-1\}$ and $t \in \{0, \dots, q^{n-k-1}-1\}$. Further, let $P = (p_1, \dots, p_{q^n})$ and $\tilde{P} = (\tilde{p}_1, \dots, \tilde{p}_{q^n})$ be probability distributions on $\{1, \dots, q^n\}$ with*

$$P = (p_1, \dots, p_{tq^{k+1}}, \underbrace{r_1, \dots, r_1}_{q^k}, \underbrace{r_2, \dots, r_2}_{q^k}, \dots, \underbrace{r_q, \dots, r_q}_{q^k}, p_{(t+1)q^{k+1}+1}, \dots, p_{q^n})$$

and

$$\tilde{P} = (p_1, \dots, p_{tq^{k+1}}, \underbrace{\frac{1}{q}\sum_{i=1}^{q} r_i, \dots, \frac{1}{q}\sum_{i=1}^{q} r_i}_{q^{k+1}}, p_{(t+1)q^{k+1}+1}, \dots, p_{q^n}) \ .$$

Then, it holds

$$L_{\mathcal{C}_{q^n}}(P,P) - L_{\mathcal{C}_{q^n}}(\tilde{P},\tilde{P}) = \frac{q^k(q^k-1)}{2(q-1)} \sum_{i,j=1}^{q} (r_i - r_j)^2 \geq 0 \ .$$

The inequality holds with equality if and only if either $k = 0$ or $r_i = r_j$ for all $i, j \in \{1, \dots, q\}$.

Proof. Without loss of generality we assume that $t = 0$, such that

$$P = (p_1, \dots, p_{q^n}) = (r_1, \dots, r_1, r_2, \dots, r_2, \dots, r_q, \dots, r_q, p_{q^{k+1}+1}, \dots, p_{q^n})$$

and

$$\tilde{P} = (\tilde{p}_1, \dots, \tilde{p}_{q^n}) = (\frac{1}{q}\sum_{i=1}^{q} r_i, \dots, \frac{1}{q}\sum_{i=1}^{q} r_i, p_{q^{k+1}+1}, \dots, p_{q^n}) \ .$$

Also, we abbreviate $\Delta := L_{\mathcal{C}_{q^n}}(P,P) - L_{\mathcal{C}_{q^n}}(\tilde{P},\tilde{P})$, $L_{u,v} := L_{\mathcal{C}_{q^n}}(u,v)$ and $\alpha_{u,v} := (p_u p_v - \tilde{p}_u \tilde{p}_v) L_{u,v}$ so that we have

$$\Delta = \sum_{u,v=1}^{q^n} \alpha_{u,v} \ .$$

It is clear that $L_{u,v} = L_{v,u}$ and hence $\alpha_{u,v} = \alpha_{v,u}$. Also, $\alpha_{u,v} = 0$ if both u and v are in $\{q^{k+1}+1, \dots, q^n\}$. This yields

$$\Delta = \sum_{u,v=1}^{q^{k+1}} \alpha_{u,v} + 2 \sum_{u=1}^{q^{k+1}} \sum_{v=q^{k+1}+1}^{q^n} \alpha_{u,v} \ .$$

Furthermore, for $u \in \{1, \dots, q^{k+1}\}$ and $v \in \{q^{k+1}+1, \dots, q^n\}$ we have

i) $p_v = \tilde{p}_v$,
ii) $L_{u,v} = L_{1,v}$, which we denote by L_v, and
iii) $\sum_{u=1}^{q^{k+1}} p_u = \sum_{u=1}^{q^{k+1}} \tilde{p}_u$.

The above observations yield

$$\sum_{u=1}^{q^{k+1}} \sum_{v=q^{k+1}+1}^{q^n} \alpha_{u,v} \overset{i),ii)}{=} \sum_{v=q^{k+1}+1}^{q^n} p_v L_v \sum_{u=1}^{q^{k+1}} (p_u - \tilde{p}_u) \overset{iii)}{=} 0$$

and hence

$$\begin{aligned}\Delta &= \sum_{u,v=1}^{q^{k+1}} \left[p_u p_v - \tfrac{1}{q^2} \left(\sum_{i=1}^{q} r_i \right)^2 \right] L_{u,v} \\ &= \sum_{j,m=1}^{q} \left[r_j r_m - \tfrac{1}{q^2} \left(\sum_{i=1}^{q} r_i \right)^2 \right] \sum_{u=(j-1)q^k+1}^{jq^k} \sum_{v=(m-1)q^k+1}^{mq^k} L_{u,v} \ .\end{aligned}$$

Here, the first equality follows from the definition of $\tilde{P}$. The second equality is due to the definition of P. We now take a look at $L_{u,v}$ and see that for $u \in \{(j-1)q^k + 1, \ldots, jq^k\}$ and $v \in \{(m-1)q^k + 1, \ldots, mq^k\}$ we have

$$L_{u,v} = \begin{cases} n-k & \text{, if } j \neq m \\ n-k+L_{\mathcal{C}_{q^k}}(u,v) & \text{, if } j = m \ , \end{cases}$$

where $\mathcal{C}_{q^k}$ is the code consisting for all $u \in \{1, \ldots, q^k\}$ of the codewords $\mathring{c}_u$ which we obtain by deleting the leading $n-k$ digits of c_u. With this and the additional fact that $\sum_{u,v=(j-1)q^k+1}^{jq^k} L_{u,v}$ is independent of the choice of $j \in \{1, \ldots, q\}$ we get

$$\begin{aligned}\Delta = \quad & (n-k)q^{2k} \sum_{j,m=1}^{q} \left[r_j r_m - \tfrac{1}{q^2} \left(\sum_{i=1}^{q} r_i \right)^2 \right] \\ & + \sum_{j=1}^{q} \left[r_j^2 - \tfrac{1}{q^2} \left(\sum_{i=1}^{q} r_i \right)^2 \right] \sum_{u,v=1}^{q^k} L_{\mathcal{C}_{q^k}}(u,v) \ .\end{aligned}$$

Next we see that

$$\sum_{j,m=1}^{q} \left[r_j r_m - \frac{1}{q^2} \left(\sum_{i=1}^{q} r_i \right)^2 \right] = \sum_{j,m=1}^{q} r_j r_m - \left(\sum_{i=1}^{q} r_i \right)^2 = 0$$

so that

$$\Delta = \sum_{j=1}^{q} \left[r_j^2 - \frac{1}{q^2} \left(\sum_{i=1}^{q} r_i \right)^2 \right] \sum_{u,v=1}^{q^k} L_{\mathcal{C}_{q^k}}(u,v) \ .$$

We now focus on the running times and see that

$$\sum_{u,v=1}^{q^k} L_{\mathcal{C}_{q^k}}(u,v) = \sum_{v=1}^{q^k} \sum_{l=1}^{k} l \cdot \#\{u \in \{1, \ldots, q^k\} \mid L_{\mathcal{C}_{q^k}}(u,v) = l\} \ .$$

For $l = 1, \ldots, k-1$ the codeword of each element in the above sets has to coincide with $\mathring{c}_v$ in the first $l-1$ digits. Those are q^{k-l+1} many. Furthermore, each one of those codewords has to differ from $\mathring{c}_v$ in the l-th digit. These are $q-1$ out of q. We end up with $(q-1)q^{k-l}$ elements. If $l = k$, also v itself is contained in the corresponding set. Again it is easy to see that this does not depend on the choice of $v \in \{1, \ldots, q^k\}$. It follows that

$$\sum_{u,v=1}^{q^k} L_{C_{q^k}}(u,v) = q^k \left[\sum_{l=1}^{k-1} l(q-1)q^{k-l} + kq\right] = q^k \left[(q-1)q^k \sum_{l=1}^{k} lq^{-l} + k\right] .$$

The partial sum behavior of the geometric series yields

$$\sum_{l=1}^{k} lq^{-l} = \frac{q^{k+1} - (k+1)q + k}{q^k(q-1)^2} = \frac{q(q^k-1) - k(q-1)}{q^k(q-1)^2} .$$

It follows that

$$\sum_{u,v=1}^{q^k} L_{C_{q^k}}(u,v) = \frac{q}{q-1} q^k (q^k - 1)$$

and thus

$$\Delta = \frac{q}{q-1} q^k (q^k - 1) \sum_{j=1}^{q} \left[r_j^2 - \frac{1}{q^2} \left(\sum_{i=1}^{q} r_i \right)^2 \right] .$$

Finally, since

$$\sum_{j=1}^{q} \left[r_j^2 - \frac{1}{q^2} \left(\sum_{i=1}^{q} r_i \right)^2 \right] = \frac{1}{2q} \sum_{i,j=1}^{q} (r_i - r_j)^2 ,$$

we obtain

$$\Delta = \frac{q^k(q^k-1)}{2(q-1)} \sum_{i,j=1}^{q} (r_i - r_j)^2 .$$

Obviously $\Delta \geq 0$ with equality if and only if either $k = 0$ or $r_i = r_j$ for all $i, j \in \{1, \ldots, q\}$. □

Lemma 1 provides a way to transform any given distribution P step-by-step to the uniform distribution without increasing the symmetric running time. In the first step of the first round ($t = 0$, $k = 0$) we level out the probabilities $p_1, \ldots, p_q$. In the second step ($t = 1$, $k = 0$) we level out $p_{q+1}, \ldots, p_{2q}$ and so on until the last block of probabilities $p_{q^n-q+1}, \ldots, p_{q^n}$ is leveled out in the last step of the first round ($t = q^{n-1} - 1$, $k = 0$). Since $k = 0$ we have not changed the symmetric identification running time. Furthermore, since within every block of q subsequent elements the probabilities are now identical we have constructed a probability distribution which enables us to apply Lemma 1 again.

In round 2 ($k = 1$) we level out all q^{n-1} blocks of q identical probabilities. Lemma 1 guarantees that during these actions the symmetric identification running time does not increase. In fact, the only way it does not decrease but stays the same is when probabilities within all subsequent blocks of q^2 elements had already been the same. Again we end up with a distribution which allows us to apply Lemma 1 in round three and so on. Finally, in the last round $k = n - 1$ we level out the q blocks of q^{n-1} identical probabilities and end up with the uniform distribution. We have proven the following

Theorem 1. *Let $n \in \mathbb{N}$ and $q \in \mathbb{N}_{\geq 2}$. Further, let $\mathcal{C} = \mathcal{C}_{q^n}$. Then it holds for all probability distributions P on $\{1, \ldots, q^n\}$ that*

$$L_{\mathcal{C}}(\bar{P}, \bar{P}) \leq L_{\mathcal{C}}(P, P) ,$$

where $\bar{P}$ denotes the uniform distribution on $\{1, \ldots, q^n\}$. The inequality holds with equality if and only if it holds for all $i \in \{1, \ldots, q^{n-1}\}$ that

$$\sum_{u=(i-1)q+1}^{iq} p_u = \frac{1}{q^{n-1}} .$$

3 An Improved Upper Bound for the Worst-Case Running Time for Binary Codes

In Sect. 4 of [2] the authors proved in Theorem 3 by an inductive code construction that $L(P) < 3$ in the case of binary codes ($q = 2$). They assumed that without loss of generality $p_1 \geq p_2 \geq \cdots \geq p_N$. In the first step $\mathcal{U}$ is partitioned into $\mathcal{U}_0 = \{1, \ldots, t\}$ and $\mathcal{U}_1 = \{t + 1, \ldots, N\}$ such that $\sum_{i=1}^{t} p_i$ is as close as possible to $1/2$. Then, they inductively construct codes on $\mathcal{U}_0$ (resp. $\mathcal{U}_1$) and prefix the codewords for all elements in $\mathcal{U}_0$ (resp. $\mathcal{U}_1$) by **0** (resp. **1**). The proof of the theorem is given by analyzing several cases. The upper bound is given by the case where $\sum_{i=1}^{t} p_i < 1/2$ and the user $v_{\max}$ which maximizes $L_{\mathcal{C}}(P, v)$ is in $\mathcal{U}_1$.[1]

In order to optimize this worst case we take up to a certain number of additional outputs from $\mathcal{U}_1$ and put them into $\mathcal{U}_0$. As we will show this significantly speeds up the identification process.

Lemma 2. *We define $\mathcal{U}_{\max} = \{u \in \mathcal{U} \mid c_{u,1} = c_{v_{\max},1}\}$ and $p_{\max} = \sum_{u \in \mathcal{U}_{\max}} p_u$. Further, let $P_{\max}$ be a probability distribution on $\mathcal{U}_{\max}$ defined for all $u \in \mathcal{U}_{\max}$ by*

$$P_{\max,u} = \frac{p_u}{p_{\max}}$$

and $\mathcal{C}_{\max}$ be the code on $\mathcal{U}_{\max}$ which we obtain by deleting the leading bit of all c_u's. Then, it holds that

$$L_{\mathcal{C}}(P) \leq 1 + p_{\max} \cdot L_{\mathcal{C}_{\max}}(P_{\max}) .$$

[1] $v_{\max}$ may not be unique, but if there are more than one, it does not matter which one is chosen.

Proof. We have

$$\begin{aligned}
L_{\mathcal{C}}(P) &= \sum_{u\in\mathcal{U}} p_u \cdot L_{\mathcal{C}}(u, v_{\max}) \\
&= 1 + \sum_{u\in\mathcal{U}_{\max}} p_u \cdot L_{\mathcal{C}_{\max}}(u, v_{\max}) \\
&= 1 + p_{\max} \cdot L_{\mathcal{C}_{\max}}(P_{\max}, v_{\max}) \\
&\leq 1 + p_{\max} \cdot L_{\mathcal{C}_{\max}}(P_{\max}) \ .
\end{aligned}$$

The first equality is due to the definition of $L_{\mathcal{C}}(P)$. The second holds since we have to check the first bit for all elements but the next bit only for those which coincided with $v_{\max}$ during this check. The third equality holds because of the definition of the average running time and the final inequality is a direct consequence of the definition of both the average and the worst-case running time. □

The above lemma provides the induction step for the proof of

Theorem 2. *It holds for all probability distributions P on $\mathcal{U}$ that the worst-case running time for binary identification is upper bounded by*

$$L(P) < \frac{5}{2} \ .$$

Proof. Without loss of generality we assume that $p_1 \geq p_2 \geq \cdots \geq p_N$. For the induction bases $N = 1, 2$ we have that $L(P) = 1 < 5/2$ for all P. Now let $N > 2$ and we distinguish between the following cases.

Case 1: $\mathbf{p_1 \geq \frac{1}{2}}$

In this case we assign $c_1 = \mathbf{0}$ and $\mathcal{U}_1 = \{2, \ldots, N\}$. Inductively we construct a code $\mathcal{C}' = \{c'_u \mid u = 2, \ldots, N\}$ on $\mathcal{U}_1$ and we extend this code to a code on $\mathcal{U}$ by setting $c_u = \mathbf{1}c'_u$ for $u \in \mathcal{U}_1$.

It is clear that $v_{\max} \neq 1$ because in this case $L(P)$ would equal 1. This is a contradiction since $N > 2$ and thereby we have more than one output whose codeword begins with $\mathbf{1}$ and each of these outputs results in a running time strictly greater than 1.

Thus, the maximum is assumed on the "right" side. This yields $p_{\max} \leq 1/2$. Further, by Lemma 2 and the induction hypothesis we have that

$$L_{\mathcal{C}}(P) < 1 + \frac{1}{2} \cdot \frac{5}{2} = \frac{9}{4} < \frac{5}{2} \ .$$

Case 2: $\mathbf{p_1 < \frac{1}{2}}$

In this case we choose t such that $|1/2 - \sum_{u=1}^{t} p_u|$ is minimized. Now we distinguish again between two subcases.

Case 2.1: t = 1

In this case we set $\mathcal{U}_0 = \{1,2\}$ and $\mathcal{U}_1 = \{3,\dots,N\}$. Again by we inductively construct $\mathcal{C}' = \{c'_u \mid u = 3,\dots,N\}$. And we obtain $\mathcal{C}$ by setting $c_1 = \mathbf{00}$, $c_2 = \mathbf{01}$ and $c_u = \mathbf{1}c'_u$ for $u = 3,\dots,N$.

If $v_{\max} \in \mathcal{U}_0$, we have that $p_{\max} = p_1 + p_2$ and $\mathcal{C}_{\max} = \{\mathbf{0},\mathbf{1}\}$. Again by Lemma 2 we obtain

$$L_{\mathcal{C}}(P) \le 1 + (p_1 + p_2) L_{\mathcal{C}_{\max}}(P_{\max}) \le 2 < \frac{5}{2} \ .$$

Otherwise it follows from the definition of t that $p_1 + p_2 > 1/2$. By this we get $p_{\max} < 1/2$ and $\mathcal{C}_{\max} = \mathcal{C}'$. By induction and Lemma 2 this yields

$$L_{\mathcal{C}}(P) < 1 + \frac{1}{2} \cdot \frac{5}{2} = \frac{9}{4} < \frac{5}{2} \ .$$

Case 2.2: t ≥ 2

We now set $\mathcal{U}_0 = \{1,\dots,t\}$ and $\mathcal{U}_1 = \{t+1,\dots,N\}$ and construct inductively codes $\mathcal{C}' = \{c'_u \mid u = 1,\dots,t\}$ and $\mathcal{C}'' = \{c''_u \mid u = t+1,\dots,N\}$. We obtain a code $\mathcal{C}$ on $\mathcal{U}$ by setting

$$c_u = \begin{cases} \mathbf{0}c'_u & \text{for } u = 1,\dots,t \\ \mathbf{1}c''_u & \text{for } u = t+1,\dots,N \ . \end{cases}$$

Case 2.2.1: $\mathbf{v}_{\max} \in \mathcal{U}_0$

It follows that $p_{\max} = \sum_{u=1}^{t} p_u$. If $\sum_{u=1}^{t} p_u \le 1/2$, we get again by induction and Lemma 2 that

$$L_{\mathcal{C}}(P) < 1 + \frac{1}{2} \cdot \frac{5}{2} = \frac{9}{4} < \frac{5}{2} \ .$$

In the case that $\sum_{u=1}^{t} p_u > 1/2$ we have by the definition of t that

$$\sum_{u=1}^{t} p_u - \frac{1}{2} \le \frac{1}{2} - \sum_{u=1}^{t-1} p_u \ .$$

It follows $\sum_{u=1}^{t} p_u \le (p_t + 1)/2$. We also have $p_t \le p_{t-1} < 1/(2(t-1))$ because otherwise $\sum_{u=1}^{t-1} p_u \ge 1/2$. This would be a contradiction to the definition of t. This together implies

$$p_{\max} = \sum_{u=1}^{t} p_u < \frac{1 + 2(t-1)}{4(t-1)} \ . \tag{1}$$

If $t = 2$, we obtain for the same reasons as in Case 2.1 that

$$L_{\mathcal{C}}(P) < \frac{5}{2} \ .$$

If $t = 3$, we get that $\mathcal{C}_{\max} = \mathcal{C}' = \{c'_1, c'_2, c'_3\}$, with $c'_1 = \mathbf{0}$, $c'_2 = \mathbf{10}$ and $c'_3 = \mathbf{11}$. Further, $p_{\max} = p_1 + p_2 + p_3$ and $P_{\max} = (p_1/p_{\max}, p_2/p_{\max}, p_3/p_{\max})$. Since $p_1 \geq p_2 \geq p_3$ it follows that

$$\frac{p_2 + p_3}{p_{\max}} \leq \frac{2}{3} .$$

This yields

$$L_{\mathcal{C}_{\max}}(P_{\max}) = 1 + \frac{p_2 + p_3}{p_{\max}} \leq \frac{5}{3}.$$

It now follows from Lemma 2 and (1) that

$$L_{\mathcal{C}}(P) \leq 1 + \frac{5}{3} p_{\max} < 1 + \frac{5}{3} \cdot \frac{5}{8} = \frac{49}{24} < \frac{5}{2} .$$

For $t \geq 4$ the induction hypothesis and (1) yield

$$L_{\mathcal{C}}(P) < 1 + \frac{1 + 2(t-1)}{4(t-1)} \cdot \frac{5}{2} \leq 1 + \frac{7}{12} \cdot \frac{5}{2} = \frac{59}{24} < \frac{5}{2} .$$

Case 2.2.2: $\mathbf{v}_{\max} \in \mathcal{U}_1$

We get that $p_{\max} = \sum_{u=t+1}^{N} p_u$. If $\sum_{u=t+1}^{N} p_u \leq 1/2$, we get like before

$$L_{\mathcal{C}}(P) < 1 + \frac{1}{2} \cdot \frac{5}{2} = \frac{9}{4} < \frac{5}{2} .$$

If $\sum_{u=t+1}^{N} p_u > 1/2$, it follows that

$$\sum_{u=1}^{t} p_u \geq \frac{1}{2} - \frac{1}{2} p_{t+1} .$$

Since $p_{t+1} \leq \left(\sum_{u=1}^{t} p_u \right) / t$, we further obtain

$$\sum_{u=1}^{t} p_u \geq \frac{t}{2t+1} \geq \frac{2}{5} . \tag{2}$$

Since $p_{\max} = 1 - \sum_{u=1}^{t} p_u$, we finally get by induction and (2) that

$$L_{\mathcal{C}}(P) < 1 + \frac{3}{5} \cdot \frac{5}{2} = \frac{5}{2} .$$

□

The following corollary combines the lower bound from Theorem 2 in [1] and the upper bound from above Theorem 2 as follows

Corollary 1. *It holds for all probability distributions P on $\mathcal{U}$ that*

$$2 \left(1 - \sum_{u \in \mathcal{U}} p_u^2 \right) \leq L(P, P) \leq L(P) < \frac{5}{2} ,$$

where $L(P, P) = \min_{\mathcal{C}} L_{\mathcal{C}}(P, P)$.

References

1. Ahlswede, R.: Identification Entropy. In: Ahlswede, R., Bäumer, L., Cai, N., Aydinian, H., Blinovsky, V., Deppe, C., Mashurian, H. (eds.) Information Transfer and Combinatorics. LNCS, vol. 4123, pp. 595–613. Springer, Heidelberg (2006)
2. Ahlswede, R., Balkenhol, B., Kleinewächter, C.: Identification for Sources. In: Ahlswede, R., Bäumer, L., Cai, N., Aydinian, H., Blinovsky, V., Deppe, C., Mashurian, H. (eds.) Information Transfer and Combinatorics. LNCS, vol. 4123, pp. 51–61. Springer, Heidelberg (2006)
3. Ahlswede, R., Cai, N.: An interpretation of identification entropy. IEEE Trans. Inf. Theory 52(9), 4198–4207 (2006)

L-Identification for Uniformly Distributed Sources and the q-ary Identification Entropy of Second Order

Christian Heup

Universität Bielefeld, Fakultät für Mathematik,
Universitätsstraße 25, 33615 Bielefeld, Germany
christian.heup@gmail.com

Dedicated to the memory of Rudolf Ahlswede

Abstract. In this article we generalize the concept of identification for sources, which was introduced by Ahlswede, to the concept of L-identification for sources. This means that we do not only consider a discrete source but a discrete memoryless source (DMS) with L outputs. The task of L-identification is now to check for any previously given output whether it is part of the L outputs of the DMS. We establish a counting lemma and use it to show that, if the source is uniformly distributed, the L-identification symmetric running time asymptotically equals the rational number

$$K_{L,q} = -\sum_{l=1}^{L}(-1)^l \binom{L}{l} \frac{q^l}{q^l-1} \quad .$$

We then turn to general distributions and aim to establish a lower bound for the symmetric 2-identification running time. In order to use the above asymptotic result we first concatenate a given code sufficiently many times and show that for 2-identification the uniform distribution is optimal, thus yielding a first lower bound. This lower bound contains the symmetric (1-)identification running time negatively signed so that (1-)identification entropy cannot be applied immediately. However, using the fact that the (1-)identification entropy is attained iff the probability distribution consists only of q-powers, we can show that our lower bound is in this case also exactly met for 2-identification. We then prove that the obtained expression is in general a lower bound for the symmetric 2-identification running time and that it obeys fundamental properties of entropy functions. Hence, the following expression is called the *q-ary identification entropy of second order*

$$H_{\mathrm{ID}}^{2,q}(P) = 2\frac{q}{q-1}\left(1-\sum_{u\in\mathcal{U}} p_u^2\right) - \frac{q^2}{q^2-1}\left(1-\sum_{u\in\mathcal{U}} p_u^3\right) \quad .$$

Keywords: identification, source coding, entropy.

H. Aydinian, F. Cicalese, and C. Deppe (Eds.): Ahlswede Festschrift, LNCS 7777, pp. 11–43, 2013.

1 Introduction and Notations

We consider the discrete memoryless source $(\mathcal{U}^L, P^L)$, where $\mathcal{U} := \{1, \ldots, N\}$ and P is a probability distribution on $\mathcal{U}$, its individual probabilities denoted by p_u. Then P^L is a probability distribution on $\mathcal{U}^L$ induced by P and is defined by $P^L_{u^L} := p_{u_1} \cdot \cdots \cdot p_{u_L}$ for $u^L = (u_1, \ldots, u_L)$. Together with this, $\mathcal{C}$ is a q-ary prefix code on $\mathcal{U}$. The codeword of user $u \in \mathcal{U}$ is denoted by $c_u = (c_{u,1}, ..., c_{u,\|c_u\|})$, while $c_u^k = (c_{u,1}, ..., c_{u,k})$ denotes the prefix of length k of this codeword. We also introduce the so-called user space $\mathcal{V}$ and as in [7] without loss of generality we assume that $\mathcal{V} = \mathcal{U}$.

In contrast to the classical identification task introduced in [2], the task of L-identification is to decide for every user $v \in \mathcal{U}$ not only if it coincides with a single output $u \in \mathcal{U}$ but whether for a given output tuple $u^L = (u_1, .., u_L) \in \mathcal{U}^L$ there exists at least one $l \in \{1, \ldots, L\}$ such that $v = u_l$. To achieve this goal we compare step by step the first, second, third etc. q-bit of c_v with the corresponding q-bits of $c_{u_1}, ..., c_{u_L}$. After each step i all u_l with $c_{u_l,i} \neq c_{v,i}$ are eliminated from the set of possible candidates. We continue with step $i+1$ comparing only those u_l which still are candidates. If at some point during this procedure the last possible candidate is eliminated, the L-identification process stops with a negative answer. On the other hand, if there are still candidates left after the comparison of the last q-bit of c_v, the L-identification returns a positive answer. The number of steps until the process halts is called the *L-identification running time* for $(u^L, v) \in \mathcal{U}^L \times \mathcal{U}$.

Algorithm LID in Fig. 1 accomplishes L-identification. As its input serve the codewords $c_{u_1}, ..., c_{u_L}$ and c_v and it returns the triple $(A, s, \mathcal{S})$. Here, A is a boolean variable which is "TRUE" if v is contained in u^L and "FALSE" if not. The second component s equals the number of steps until the algorithm halted and the third component returns the set of positions of the output vector u^L which coincide with the user v. This means that if there exist one or more components of u^L which coincide with v, we also know their exact number and positions. This is not a requirement to L-identification but an extra feature. It follows from the fact that up to the last comparison of q-bits still all possible candidates may not coincide with v.

Formally, we define the L-identification running time for given $(u^L, v) \in \mathcal{U}^{L+1}$ and q-ary code $\mathcal{C}$ by

$$\mathcal{L}^{L,q}_{\mathcal{C}}(u^L, v) = \mathtt{LID}_2(c_{u_1}, ..., c_{u_L}, c_v) \ ,$$

where $\mathtt{LID}_2(c_{u_1}, ..., c_{u_L}, c_v)$ is the second component of the triple returned by algorithm LID. We will analyze the expected length of the L-identification running time, also called the *average running time*, for a given user $v \in \mathcal{U}$

$$\mathcal{L}^{L,q}_{\mathcal{C}}(P, v) = \sum_{u^L \in \mathcal{U}^L} P_{u^L} \mathcal{L}^{L,q}_{\mathcal{C}}(u^L, v) \ .$$

```
LID {
  S_0 := {1,...,L};
  for i from 0 to ||c_v|| - 1 do {
    if (∀ l ∈ S_i :  c_{u_l,i} ≠ c_{v,i}) then {
      return (''FALSE'',i,∅);
    }
    else {
      set S_{i+1} := {l ∈ S_i  :  c_{u_l,i} = c_{v,i}};
      if (i = ||c_v||) then {
        return (''TRUE'',||c_v||,S_{||c_v||});
      }
    }
  }
}
```

Fig. 1. The L-identification algorithm

This can be done in different ways. The first is the worst-case scenario where we are interested in the *worst-case average running time*, which we shortly call the *worst-case running time*, and which is defined by

$$\mathcal{L}_{\mathcal{C}}^{L,q}(P) = \max_{v \in \mathcal{U}} \mathcal{L}_{\mathcal{C}}^{L,q}(P, v) \ .$$

Let us assume that also user v is chosen at random according to a probability distribution Q on $\mathcal{U}$. We are now interested in the *expected average running time* or shortly the *expected running time*

$$\mathcal{L}_{\mathcal{C}}^{L,q}(P, Q) = \sum_{v \in \mathcal{U}} Q(\{v\}) \mathcal{L}_{\mathcal{C}}^{L,q}(P, v) \tag{1}$$

and in particular in the *optimal expected running time*

$$\mathcal{L}^{L,q}(P, Q) = \min_{\mathcal{C}} \mathcal{L}_{\mathcal{C}}^{L,q}(P, Q). \tag{2}$$

A special case is $Q = P$ so that (1) and (2) become

$$\mathcal{L}_{\mathcal{C}}^{L,q}(P, P) = \sum_{v \in \mathcal{U}} p_v \mathcal{L}_{\mathcal{C}}^{L,q}(P, v) = \sum_{(u^L, v) \in \mathcal{U}^{L+1}} P_{u^L} p_v \mathcal{L}_{\mathcal{C}}^{L,q}(u^L, v)$$

and

$$\mathcal{L}^{L,q}(P, P) = \min_{\mathcal{C}} \mathcal{L}_{\mathcal{C}}^{L,q}(P, P) \ .$$

We call $\mathcal{L}_{\mathcal{C}}^{L,q}(P, P)$ the *symmetric running time* for a given code $\mathcal{C}$ and $\mathcal{L}^{L,q}(P, P)$ the *optimal symmetric running time*. All the above values also depend on $N = |\mathcal{U}|$. We do not state this fact explicitly since it is contained implicitly in both P and $\mathcal{C}$.

2 L-Identification for Uniformly Distributed Sources

We first focus on the symmetric running time when P is the uniform distribution and we will establish a formula for its asymptotic behavior when $N = |\mathcal{U}|$ tends to infinity, namely that it asymptotically equals the rational number

$$K_{L,q} = -\sum_{l=1}^{L}(-1)^l \binom{L}{l} \frac{q^l}{q^l - 1} \ .$$

2.1 A Detour to Balanced Huffman Codes

We assume familiarity with the concept of Huffman coding (e.g. see [8]) and shall start by recalling the concept of balanced Huffman codes, which was introduced in [2]. Let $N = q^{n-1} + d$, where $0 \leq d \leq (q-1)q^{n-1} - 1$. The q-ary Huffman coding for the uniform distribution of size N yields a code where some codewords have length n and the other codewords have length $n-1$. More precisely, if $0 \leq d < q^{n-1}$, then $q^{n-1} - d$ codewords have length $n-1$ and $2d$ codewords have length n, while in the case $q^{n-1} \leq d \leq (q-1)q^{n-1} - 1$ all codewords have length n. It is well-known that for data compression all Huffman codes are optimal. This, however, is not the case for identification.

In [2] it is shown (for $q = 2$) that for identification it is crucial which codewords have length n or, in terms of code trees, where in the code tree these longer codewords lie. Moreover, those Huffman codes have a shorter expected and worst-case running time for which the longer codewords are distributed along the code tree in such a way that for every inner node the difference between the number of leaves of its left side and the number of leaves of its right side is at most one. Huffman trees satisfying this property were called *balanced.*

Analogously, we say that a q-ary Huffman code is *balanced* if its corresponding q-ary code tree $\mathcal{H}$ obeys the property that for every inner node the difference between the number of leaves of each two of its branches is at most one. We further denote by $\mathcal{H}_{q,N}$ the set of all q-ary balanced Huffman trees with N leaves and the corresponding set of q-ary balanced Huffman codes of size N is denoted by $\mathcal{C}_{q,N}$. If $N = q^n$, there exists a single balanced Huffman code, namely the code $\mathcal{C}_{q^n}$ where all codewords have length n. We denote the balanced Huffman tree which corresponds to $\mathcal{C}_{q^n}$ by $\mathcal{H}_{q^n}$. For every (inner or outer) node x of a balanced Huffman tree $\mathcal{H} \in \mathcal{H}_{q,N}$ we denote the set of all leaves of $\mathcal{H}$ which have x as prefix by $\mathcal{H}_x$. An easy consequence of the balancing property is the following

Lemma 1. *Let $q^{n-1} < N \leq q^n$, $\mathcal{H} \in \mathcal{H}_{q,N}$ and x be a node of $\mathcal{H}$, then it follows*

$$\left\lfloor \frac{N}{q^{\|x\|}} \right\rfloor \leq |\mathcal{H}_x| \leq \left\lceil \frac{N}{q^{\|x\|}} \right\rceil \ .$$

The inequality holds with equality for all x if and only if $N = q^n$. Moreover, it simplifies to

$$|\mathcal{H}_x| = q^{n-\|x\|} \ .$$

In identification what is relevant is not the length of a codeword but the length of the maximal common prefix of two or more different codewords. This is why a balanced Huffman code is better for identification than an unbalanced one. It is easy to see by the pigeonhole principle that if we consider Huffman codes with codewords of lengths $n-1$ and n, a balanced Huffman code is optimal for the worst-case running time and we will see in the proof of Theorem 1 that the balancing property is also crucial for the symmetric running time of L-identification.

The q-ary Shannon-Fano coding procedure [6] constructs codes where for every inner node the difference between the sum of the normalized probabilities within its individual branches is as close as possible to $1/q$. It is an easy observation that if we are dealing with uniform distributions, a code is a Shannon-Fano code if and only if it is a balanced Huffman code.

2.2 Counting Leaves

Our goal is to analyze the asymptotic behavior of

$$\mathcal{L}_{\mathcal{C}}^{L,q}\left((\frac{1}{N},\dots,\frac{1}{N}),(\frac{1}{N},\dots,\frac{1}{N})\right)=\frac{1}{N^{L+1}}\sum_{u_1,\dots,u_L,v=1}^{N}\mathcal{L}_{\mathcal{C}}^{L,q}(u^L,v),$$

with $\mathcal{C}\in\mathcal{C}_{q,N}$. This will be done by applying a different counting method. The above equation suggests to calculate $\mathcal{L}_{\mathcal{C}}^{L,q}(u^L,v)$ for all pairs (u^L,v) individually. Instead we merge all u^L having the same running time regarding some v into sets

$$\mathcal{R}_{\mathcal{C}}^{L,q}(k,v)=\left\{u^L\in\mathcal{U}^L \mid \mathcal{L}_{\mathcal{C}}^{L,q}(u^L,v)=k\right\}$$

for $k\in\{1,\dots,\|c_v\|\}$. Again, the dependency on N of the above defined sets is contained implicitly within $\mathcal{C}$. The above equation now becomes

$$\mathcal{L}_{\mathcal{C}}^{L,q}\left((\frac{1}{N},\dots,\frac{1}{N}),(\frac{1}{N},\dots,\frac{1}{N})\right)=\frac{1}{N^{L+1}}\sum_{v=1}^{N}\sum_{k=1}^{\|c_v\|}k|\mathcal{R}_{\mathcal{C}}^{L,q}(k,v)|. \quad (3)$$

In order to apply this equation we need to know upper and lower bounds on the cardinalities of these sets. Corollary 1 below provides such bounds and exact values for the case when N is a q-power. The base for this corollary is the following

Lemma 2. *Let $q^{n-1}<N\le q^n$, $\mathcal{C}\in\mathcal{C}_{q,N}$, $\mathcal{H}\in\mathcal{C}_{q,N}$ be its corresponding code tree and $v\in\mathcal{U}$. Then, it holds for $k\in\{1,\dots,\|c_v\|-1\}$ that*

$$|\mathcal{R}_{\mathcal{C}}^{L,q}(k,v)|=\sum_{m=1}^{L}\binom{L}{m}|\mathcal{H}_{c_v^{k-1}}\backslash\mathcal{H}_{c_v^k}|^m\left(N-|\mathcal{H}_{c_v^{k-1}}|\right)^{L-m}$$

and

$$|\mathcal{R}_{\mathcal{C}}^{L,q}(\|c_v\|,v)|=\sum_{m=1}^{L}\binom{L}{m}|\mathcal{H}_{c_v^{\|c_v\|-1}}|^m\left(N-|\mathcal{H}_{c_v^{\|c_v\|-1}}|\right)^{L-m}.$$

Proof. In order to simplify notation we shall write $\mathcal{R}(k,v)$ for $\mathcal{R}_{\mathcal{C}}^{L,q}(k,v)$.

Case 1: k = 1
The L-identification algorithm terminates after the first step if and only if the codewords of all components of u^L differ already in the first q-bit from $c_{v,1}$. This gives us

$$\mathcal{R}(1,v) = \{u^L \in \mathcal{U}^L \mid c_{u_i} \in \mathcal{H}\backslash\mathcal{H}_{c_{v,1}} \ \forall\, i \in \{1,\ldots,L\}\}$$

and thus

$$|\mathcal{R}(1,v)| = |\mathcal{H}\backslash\mathcal{H}_{c_{v,1}}|^L = (N - |\mathcal{H}_{c_{v,1}}|)^L .$$

This coincides with the first equation of Lemma 2 since all summands vanish except for $m = L$.

Case 2: k = 2, ..., $\|\mathbf{c_v}\| - 1$
The identification time of u^L and v equals k if and only if $c_{u_i}^k \neq c_v^k$ holds for all $i \in \{1,\ldots,L\}$ and there exists at least one $i \in \{1,\ldots,L\}$ such that $c_{u_i}^{k-1} = c_v^{k-1}$. This consideration yields

$$\mathcal{R}(k,v) = \{u^L \in \mathcal{U}^L \mid \exists\, i \in \{1,\ldots,L\} \text{ with } c_{u_i} \in \mathcal{H}_{c_v^{k-1}}\backslash\mathcal{H}_{c_v^k}$$
$$\text{and } c_{u_i} \notin \mathcal{H}_{c_v^k} \ \forall\, i \in \{1,\ldots,L\}\} .$$

In order to count the elements we partition $\mathcal{R}(k,v)$ into L subsets $S_{k,1}, ..., S_{k,L}$, where

$$S_{k,m} = \Big\{u^L \in \mathcal{U}^L \mid \exists\, i_1,...,i_m \in \{1,\ldots,L\} \text{ with } c_{u_{i_1}},...,c_{u_{i_m}} \in \mathcal{H}_{c_v^{k-1}}\backslash\mathcal{H}_{c_v^k}$$
$$\text{and } c_{u_i} \in \mathcal{H}\backslash\mathcal{H}_{c_v^{k-1}} \ \forall\, i \in \{1,\ldots,L\}\backslash\{i_1,...,i_m\}\Big\} .$$

If we fix the positions $i_1, ..., i_m$, we see that the number of possible vectors is

$$|\mathcal{H}_{c_v^{k-1}}\backslash\mathcal{H}_{c_v^k}|^m (N - |\mathcal{H}_{c_v^{k-1}}|)^{L-m} .$$

Since we have no restrictions for these positions, it follows that

$$|S_{k,m}| = \binom{L}{m} |\mathcal{H}_{c_v^{k-1}}\backslash\mathcal{H}_{c_v^k}|^m \left(N - |\mathcal{H}_{c_v^{k-1}}|\right)^{L-m} .$$

Altogether we obtain

$$|\mathcal{R}(k,v)| = |\bigcup_{m=1}^{L} S_{k,m}| = \sum_{m=1}^{L} |S_{k,m}|$$
$$= \sum_{m=1}^{L} \binom{L}{m} |\mathcal{H}_{c_v^{k-1}}\backslash\mathcal{H}_{c_v^k}|^m \left(N - |\mathcal{H}_{c_v^{k-1}}|\right)^{L-m} .$$

Case 3: k = $\|\mathbf{c_v}\|$
In this case also c_v itself may be one of the components of u^L. This yields

$$\mathcal{R}(n,v) = \{u^L \in \mathcal{U}^L \mid \exists\, i \in \{1,\ldots,L\} \text{ with } c_{u_i} \in \mathcal{H}_{c_v^{\|c_v\|-1}}\} .$$

According to this we adjust the subsets $S_{n,1}, ..., S_{n,L}$, such that

$$S_{n,m} = \{u^L \in \mathcal{U}^L \mid \exists\, i_1, ..., i_m \in \{1, \dots, L\} \text{ with } c_{u_{i_1}}, ..., c_{u_{i_m}} \in \mathcal{H}_{c_v^{\|c_v\|-1}}$$
$$\text{and } c_{u_i} \in \mathcal{H} \backslash \mathcal{H}_{c_v^{\|c_v\|-1}} \; \forall\, i \in \{1, \dots, L\} \backslash \{i_1, ..., i_m\}\}.$$

Of course, these sets partition $\mathcal{R}(n,1)$ and since

$$|S_{n,m}| = \binom{L}{m} |\mathcal{H}_{c_v^{\|c_v\|-1}}|^m (N - |\mathcal{H}_{c_v^{\|c_v\|-1}}|)^{L-m} \;,$$

for all $m \in \{1, \dots, L\}$, we obtain the desired result for $|\mathcal{R}(n,v)|$. □

If we combine Lemma 1 and Lemma 2, we obtain

Corollary 1. *Let $q^{n-1} < N \leq q^n$, $\mathcal{C} \in \mathcal{C}_{q,N}$ and $v \in \mathcal{U}$. Then, it holds for $k \in \{1, \dots, \|c_v\| - 1\}$ that*

$$|\mathcal{R}_{\mathcal{C}}^{L,q}(k,v)| \leq \sum_{m=1}^{L} \binom{L}{m} \left(\left\lceil \frac{N}{q^{k-1}} \right\rceil - \left\lfloor \frac{N}{q^k} \right\rfloor \right)^m \left(N - \left\lfloor \frac{N}{q^{k-1}} \right\rfloor \right)^{L-m}$$

and

$$|\mathcal{R}_{\mathcal{C}}^{L,q}(\|c_v\|,v)| \leq \sum_{m=1}^{L} \binom{L}{m} \left\lceil \frac{N}{q^{\|c_v\|-1}} \right\rceil^m \left(N - \left\lfloor \frac{N}{q^{\|c_v\|-1}} \right\rfloor \right)^{L-m} .$$

Additionally, we get lower bounds for $k \in \{1, \dots, \|c_v\| - 1\}$

$$|\mathcal{R}_{\mathcal{C}}^{L,q}(k,v)| \geq \sum_{m=1}^{L} \binom{L}{m} \left(\left\lfloor \frac{N}{q^{k-1}} \right\rfloor - \left\lceil \frac{N}{q^k} \right\rceil \right)^m \left(N - \left\lceil \frac{N}{q^{k-1}} \right\rceil \right)^{L-m}$$

and

$$|\mathcal{R}_{\mathcal{C}}^{L,q}(\|c_v\|,v)| \geq \sum_{m=1}^{L} \binom{L}{m} \left\lfloor \frac{N}{q^{\|c_v\|-1}} \right\rfloor^m \left(N - \left\lceil \frac{N}{q^{\|c_v\|-1}} \right\rceil \right)^{L-m} .$$

The above inequalities hold with equality for all $v \in \mathcal{U}$ if and only if $N = q^n$. Moreover, they simplify for all $k \in \{1, \dots, n-1\}$ to

$$|\mathcal{R}_{\mathcal{C}}^{L,q}(k,v)| = q^{nL} \sum_{m=1}^{L} \binom{L}{m} q^{-km} (q-1)^m (1 - q^{-k+1})^{L-m}$$

and

$$|\mathcal{R}_{\mathcal{C}}^{L,q}(n,v)| = \sum_{m=1}^{L} \binom{L}{m} q^m (q^n - q)^{L-m} .$$

2.3 The Asymptotic Theorem for Uniform Distributions

With the above estimates we are now ready to prove the asymptotic theorem for uniform distributions. If we consider the uniform distribution and use a balanced Huffman code for the encoding, the symmetric L-identification running time asymptotically equals a rational number $K_{L,q}$.

Theorem 1. *Let $L, n \in \mathbb{N}$, $q \in \mathbb{N}_{\geq 2}$, $q^{n-1} < N \leq q^n$, $\mathcal{C} \in \mathcal{C}_{q,N}$ and P be the uniform distribution on $[N]$. Then it holds that*

$$\lim_{N\to\infty} \mathcal{L}_{\mathcal{C}}^{L,q}(P,P) = K_{L,q} = -\sum_{l=1}^{L}(-1)^l\binom{L}{l}\frac{q^l}{q^l-1} \ .$$

Proof. **Case 1: $\mathbf{N = q^n}$**
It follows from Corollary 1 and (3) that

$$\mathcal{L}_{\mathcal{C}}^{L,q}(P,P) = \tfrac{1}{q^{nL}}\left[\sum_{k=1}^{n-1} kq^{nL}\sum_{m=1}^{L}\tbinom{L}{m}q^{-km}(q-1)^m(1-q^{-k+1})^{L-m}\right.$$
$$\left. +\ n\sum_{m=1}^{L}\tbinom{L}{m}q^m(q^n-q)^{L-m}\right] .$$

It holds that

$$\frac{n}{q^{nL}}\sum_{m=1}^{L}\binom{L}{m}q^m(q^n-q)^{L-m} = \sum_{m=1}^{L}\binom{L}{m}nq^{-m(n-1)}(1-q^{-n+1})^{L-m} \overset{n\to\infty}{\to} 0$$

since $nq^{-m(n-1)} \overset{n\to\infty}{\to} 0$ and $(1-q^{-n+1})^{L-m} \overset{n\to\infty}{\to} 1$. Thus, we get

$$\lim_{n\to\infty}\mathcal{L}_{\mathcal{C}}^{L,q}(P,P) = \sum_{k=1}^{\infty} k\sum_{m=1}^{L}\tbinom{L}{m}q^{-km}(q-1)^m(1-q^{-k+1})^{L-m}$$
$$= \sum_{m=1}^{L}\sum_{t=0}^{L-m}(-q)^t\tbinom{L}{m}\tbinom{L-m}{t}(q-1)^m\sum_{k=1}^{\infty}kq^{-k(m+t)} \qquad (4)$$
$$= \sum_{m=1}^{L}\sum_{t=0}^{L-m}(-q)^t\tbinom{L}{m}\tbinom{L-m}{t}(q-1)^m\tfrac{q^{m+t}}{(q^{m+t}-1)^2} \ .$$

The second equality follows from

$$(1-q^{-k+1})^{L-m} = \sum_{t=0}^{L-m}\binom{L-m}{t}(-q)^tq^{-tk} \ ,$$

while the last equality is a consequence of the geometric series. In the following we set $x_{m,t} = (-q)^t\binom{L}{m}\binom{L-m}{t}(q-1)^m$ as well as $z_l = q^l/(q^l-1)^2$ and change the order of summation. This yields

$$\lim_{n\to\infty} \mathcal{L}_{\mathcal{C}}^{L,q}(P,P) = \sum_{m=1}^{L}\sum_{t=0}^{L-m} x_{m,t}z_{m+t} = \sum_{l=1}^{L} z_l \sum_{t=0}^{l-1} x_{l-t,t}$$

$$= \sum_{l=1}^{L} \frac{q^l}{(q^l-1)^2} \sum_{t=0}^{l-1} (-q)^t \binom{L}{l-t}\binom{L-l+t}{t}(q-1)^{l-t}$$

$$= \sum_{l=1}^{L} \binom{L}{l} \frac{q^l}{(q^l-1)^2} \sum_{t=0}^{l-1} \binom{l}{t}(-q)^t(q-1)^{l-t}$$

$$= \sum_{l=1}^{L} \binom{L}{l} \frac{q^l}{(q^l-1)^2} \left((-1)^l - (-q)^l\right)$$

$$= -\sum_{l=1}^{L} (-1)^l \binom{L}{l} \frac{q^l}{q^l-1} \ .$$

Case 2: $\mathbf{q^{n-1} < N < q^n}$
For this case we obtain

$$\begin{aligned}
&\mathcal{L}_{\mathcal{C}}^{L,q}(P,P) \\
&\le \frac{1}{N^{L+1}} \sum_{v=1}^{N} \left[\sum_{k=1}^{\|c_v\|-1} k \sum_{m=1}^{L} \binom{L}{m} \left(\lceil \tfrac{N}{q^{k-1}} \rceil - \lfloor \tfrac{N}{q^k} \rfloor\right)^m \left(N - \lfloor \tfrac{N}{q^{k-1}} \rfloor\right)^{L-m} \right] \\
&+ \frac{1}{N^{L+1}} \sum_{v=1}^{N} \left[\|c_v\| \sum_{m=1}^{L} \binom{L}{m} \lceil \tfrac{N}{q^{\|c_v\|-1}} \rceil^m \left(N - \lfloor \tfrac{N}{q^{\|c_v\|-1}} \rfloor\right)^{L-m} \right] \qquad (5) \\
&\le \frac{1}{N} \sum_{v=1}^{N} \left[\sum_{k=1}^{\|c_v\|-1} k \sum_{m=1}^{L} \binom{L}{m} (q^{-k+1} - q^{-k} + \tfrac{2}{N})^m (1 - q^{-k+1} + \tfrac{1}{N})^{L-m} \right] \\
&+ \frac{1}{N} \sum_{v=1}^{N} \left[\|c_v\| \sum_{m=1}^{L} \binom{L}{m} (q^{-\|c_v\|+1} + \tfrac{1}{N})^m (1 - q^{-\|c_v\|+1} + \tfrac{1}{N})^{L-m} \right] .
\end{aligned}$$

The first inequality is obtained by the insertion of the upper bound in Corollary 1 into (3). $\lceil N/q^k \rceil \le N/q^k + 1$ and $\lfloor N/q^k \rfloor \ge N/q^k - 1$ yield the second inequality. We now divide this case into two subcases.

Case 2.1: $\mathbf{2q^{n-1} \le N < q^n}$
In this case all codewords have length n. Hence, (5) reduces to

$$\begin{aligned}
\mathcal{L}_{\mathcal{C}}^{L,q}(P,P) \le &\sum_{k=1}^{n-1} k \sum_{m=1}^{L} \binom{L}{m} (q^{-k+1} - q^{-k} + \tfrac{2}{N})^m (1 - q^{-k+1} + \tfrac{1}{N})^{L-m} \\
&+ n \sum_{m=1}^{L} \binom{L}{m} (q^{-n+1} + \tfrac{1}{N})^m (1 - q^{-n+1} + \tfrac{1}{N})^{L-m} \ .
\end{aligned} \qquad (6)$$

As in the case $N = q^n$ the second summand goes to zero as N goes to infinity. Thus, we only have to consider the first summand. In fact, we can reduce this case to the previous one by applying the binomial theorem. We obtain

$$\left(q^{-k}(q-1)+\frac{2}{N}\right)^m = \left(q^{-k}(q-1)\right)^m + \sum_{t=0}^{m-1}\binom{m}{t}(q^{-k}(q-1))^t\left(\frac{2}{N}\right)^{m-t}$$

and

$$\left(1-q^{-k+1}+\frac{1}{N}\right)^{L-m} = \left(1-q^{-k+1}\right)^{L-m} + \sum_{s=0}^{L-m-1}\binom{L-m}{s}\frac{(1-q^{-k+1})^s}{N^{L-m-s}} .$$

In the following we use

$$A = \sum_{t=0}^{m-1}\binom{m}{t}(q^{-k}(q-1))^t\left(\frac{2}{N}\right)^{m-t}$$

and

$$B = \sum_{s=0}^{L-m-1}\binom{L-m}{s}\frac{(1-q^{-k+1})^s}{N^{L-m-s}} .$$

With this notation the right hand side of (6) asymptotically equals

$$\sum_{k=1}^{n-1} k \sum_{m=1}^{L}\binom{L}{m}\Big[\left(q^{-k}(q-1)\right)^m\left(1-q^{-k+1}\right)^{L-m} + \left(q^{-k}(q-1)\right)^m B + \left(1-q^{-k+1}\right)^{L-m}A + AB\Big] . \tag{7}$$

If we focus on the second summand in the square brackets, we see that

$$\begin{aligned}
&\sum_{k=1}^{n-1} k \sum_{m=1}^{L}\binom{L}{m}\left(q^{-k}(q-1)\right)^m B \\
&= \sum_{m=1}^{L}\sum_{s=0}^{L-m-1}\binom{L-m}{s}\binom{L}{m}\frac{(q-1)^m}{N^{L-m-s}}\sum_{k=1}^{n-1} kq^{-km}\left(1-q^{-k+1}\right)^{L-m} \\
&= \sum_{m=1}^{L}\sum_{s=0}^{L-m-1}\sum_{r=0}^{L-m}(-1)^r\binom{L-m}{r}\binom{L-m}{s}\binom{L}{m}\frac{q^r(q-1)^m}{N^{L-m-s}}\sum_{k=1}^{n-1} kq^{-k(m+r)} \\
&= \sum_{m=1}^{L}\sum_{s=0}^{L-m-1}\sum_{r=0}^{L-m}\alpha(m,s,r)\frac{1}{N^{L-m-s}}\frac{1}{(q^{m+r}-1)^2}\left(q^{m+r}-\frac{(q^{m+r}-1)n+q^{m+r}}{q^{n(m+r)}}\right) ,
\end{aligned}$$

where $\alpha(m,s,r) = (-1)^r\binom{L-m}{r}\binom{L-m}{s}\binom{L}{m}q^r(q-1)^m$. The last equality follows from the partial sum behavior of the geometric series. This expression tends to zero as N (resp. $n \approx \log_q N$) goes to infinity because $L-m-s \geq 1$. In the same way it can be shown that the third and the fourth summand of (7) also tend to zero. Thus, we end up with exactly the same expression like (4). This proves the upper bound for this case. By using the same arguments and the lower estimates in Corollary 1 one can easily show the matching lower bound.

Case 2.2: $\mathbf{q^{n-1} < N < 2q^{n-1}}$
In this case $N = q^{n-1} + d$, with $0 < d < q^{n-1}$, and there exist exactly $q^{n-1} - d$ codewords of length $n-1$ and $2d$ codewords of length n. Then, (5) becomes

$$\begin{aligned}
&\mathcal{L}_C^{L,q}(P,P) \\
&\leq \tfrac{q^{n-1}-d}{N}\Bigg[\sum_{k=1}^{n-2} k \sum_{m=1}^{L} \binom{L}{m}(q^{-k+1} - q^{-k} + \tfrac{2}{N})^m(1 - q^{-k+1} + \tfrac{1}{N})^{L-m} \\
&\qquad + (n-1)\sum_{m=1}^{L} \binom{L}{m}(q^{-n+2} + \tfrac{1}{N})^m(1 - q^{-n+2} + \tfrac{1}{N})^{L-m}\Bigg] \\
&+\tfrac{2d}{N}\Bigg[\sum_{k=1}^{n-1} k \sum_{m=1}^{L} \binom{L}{m}(q^{-k+1} - q^{-k} + \tfrac{2}{N})^m(1 - q^{-k+1} + \tfrac{1}{N})^{L-m} \\
&\qquad + n\sum_{m=1}^{L} \binom{L}{m}(q^{-n+1} + \tfrac{1}{N})^m(1 - q^{-n+1} + \tfrac{1}{N})^{L-m}\Bigg] \\
&= \sum_{k=1}^{n-2} k \sum_{m=1}^{L} \binom{L}{m}(q^{-k+1} - q^{-k} + \tfrac{2}{N})^m(1 - q^{-k+1} + \tfrac{1}{N})^{L-m} \\
&+\tfrac{(q^{n-1}-d)(n-1)}{N}\sum_{m=1}^{L} \binom{L}{m}(q^{-n+2} + \tfrac{1}{N})^m(1 - q^{-n+2} + \tfrac{1}{N})^{L-m} \\
&+\tfrac{2d(n-1)}{N}\sum_{m=1}^{L} \binom{L}{m}(q^{-n+2} - q^{-n+1} + \tfrac{2}{N})^m(1 - q^{-n+2} + \tfrac{1}{N})^{L-m} \\
&+\tfrac{2dn}{N}\sum_{m=1}^{L} \binom{L}{m}(q^{-n+1} + \tfrac{1}{N})^m(1 - q^{-n+1} + \tfrac{1}{N})^{L-m} \; .
\end{aligned}$$

For the same reason as in the preceding cases the last three summands tend to zero as $N \to \infty$ and since the first summand asymptotically equals the first summand of (6), the upper bound also in this last case is settled. Omitting the details we limit ourselves to remark that also in this case the matching lower bound can be easily obtained by analogous arguments. □

3 2-Identification for General Distributions

We now focus on the case $L = 2$ and consider general distributions. Thus, we consider the discrete memoryless source $(\mathcal{U}^2, P^2)$, where $\mathcal{U} := \{1, \ldots, N\}$ and P is a probability distribution on $\mathcal{U}$, its individual probabilities denoted by p_u.

Then, P^2 is a probability distribution on $\mathcal{U}^2$ induced by P and is defined by $P^2_{u^2} := p_{u_1} \cdot p_{u_2}$ for $u^2 = (u_1, u_2)$. The task of 2-identification is to decide for every $v \in \mathcal{U}$ whether for a given output pair $(u_1, u_2) \in \mathcal{U}^2$ either $v = u_1$ or $v = u_2$. We establish the *q-ary identification entropy of second order*

$$H^{2,q}_{\mathrm{ID}}(P) = 2\frac{q}{q-1}\left(1 - \sum_{u\in\mathcal{U}} p_u^2\right) - \frac{q^2}{q^2-1}\left(1 - \sum_{u\in\mathcal{U}} p_u^3\right)$$

and show that it obeys important desiderata for entropy functions. Furthermore, we show that it serves as a lower bound for the optimal symmetric 2-identification running time:

$$\mathcal{L}^{2,q}(P,P) \geq H^{2,q}_{\mathrm{ID}}(P) .$$

3.1 Concatenated Codes and Asymptotic Calculations

Let us focus on the case $N = q^n$, $P = (1/q^n, ..., 1/q^n)$ and $\mathcal{C} = \mathcal{C}_{q^n}$, namely a block code of length n with q^n codewords. It is clear that $\mathcal{C}_{q^n}$ can be constructed by concatenating the minimal code $\mathcal{C}_q$ appropriately. Every q-ary comparison which is done during 2-identification for u^2 and v is itself an identification within $\mathcal{C}_q$. It is either a 1-identification if only one element of u^2 still remains as possible candidate or it is a 2-identification if both elements u_1 and u_2 are still left. The running time of each of those "small" identifications within $\mathcal{C}_q$ always equals 1. In fact, we have applied up to n "small" identifications within the code $\mathcal{C}_q$ in order to perform the original 2-identification within $\mathcal{C}_{q^n}$.

Let now $r_{t,l}$ be the probability that after the t-th comparison there are still l possible candidates left. We can now calculate 2-identification running time within $\mathcal{C}_{q^n}$ by

$$\begin{aligned}
&\mathcal{L}^{2,q}_{\mathcal{C}_{q^n}}\left(\left(\tfrac{1}{q^n}, ..., \tfrac{1}{q^n}\right), \left(\tfrac{1}{q^n}, ..., \tfrac{1}{q^n}\right)\right)\\
&= 1 + \sum_{t=2}^{n}\sum_{l=1}^{2} \binom{2}{l} r_{t,l}\mathcal{L}^{l,q}_{\mathcal{C}_q}\left(\left(\tfrac{1}{q}, ..., \tfrac{1}{q}\right), \left(\tfrac{1}{q}, ..., \tfrac{1}{q}\right)\right)\\
&= 1 + 2\sum_{t=2}^{n} r_{t,1} + \sum_{t=2}^{n} r_{t,2} .
\end{aligned}$$

Here, the binomial coefficient in the first equality occurs since in the case $l = 1$ either u_1 or u_2 is the leftover candidate. As stated before l-identification running time within $\mathcal{C}_q$ always equals 1 which explains the second equality.

The above observations lead us to the attempt of doing the same for any given source code $\mathcal{C}$. Namely, to consider the discrete memoryless source $((\mathcal{U}^n)^2, (P^n)^2)$ together with the concatenated code $\mathcal{C}^n$ and try to establish a connection between the 2-identification running time within $\mathcal{C}^n$ and the l-identification running times within $\mathcal{C}$, which we call the *basic code.* This relation is the content of

Lemma 3. *Let $(\mathcal{U}, P)$ be a discrete source and $\mathcal{C}$ be a q-ary prefix code with $q \in \mathbb{N}_{\geq 2}$. Further, let $\mathcal{C}^n$ be the concatenated code corresponding to $\mathcal{C}$. It holds that*

$$\mathcal{L}^{2,q}_{\mathcal{C}^n}(P^n, P^n) = \mathcal{L}^{2,q}_{\mathcal{C}}(P,P)\left(1 + \sum_{t=1}^{n-1}\left(\sum_{u\in\mathcal{U}} p_u^3\right)^t\right)$$

$$+2\mathcal{L}^{1,q}_{\mathcal{C}}(P,P)\left(\sum_{t=1}^{n-1}\left(\sum_{u\in\mathcal{U}} p_u^2\right)^t - \sum_{t=1}^{n-1}\left(\sum_{u\in\mathcal{U}} p_u^3\right)^t\right).$$

Proof. It is clear that while we are in the first basic code we have to apply 2-identification and there are three possibilities of what might happen.

1. Both elements u_1^n and u_2^n do not coincide with v^n.
   The reason would be that their first components $u_{1,1}, u_{2,1}$ do not coincide with v_1. This stops the identification process.
2. Only one element, e.g. u_1^n, coincides with v^n.
   This would be because $u_{1,1} = v_1$ and $u_{2,1} \neq v_1$. Then, we continue with applying (1-)identification in the next code.
3. Both elements coincide with v^n.
   In this case also in the next code 2-identification would have to be applied.

The main idea now is to exploit the fact that the symmetric 2-identification running time is an expectation. Therefore we introduce X_{t+1} as the random variable which indicates how many components of (U_1^n, U_2^n) are still candidates at step t. For all $t \in \{1, ..., n-1\}$ we define

$$X_{t+1} = \begin{cases} 0 \text{ if } U_1^t \neq V^t \text{ and } U_2^t \neq V^t \\ 1 \text{ if } (U_1^t = V^t \text{ and } U_2^t \neq V^t) \text{ or } (U_1^t \neq V^t \text{ and } U_2^t = V^t) \\ 2 \text{ if } U_1^t = U_2^t = V^t \end{cases}$$

and we set $X_1 = 2$. In order to calculate the corresponding probabilities we use the facts that U_1, U_2 and V are independently identically distributed. With this we get

$$Prob(X_{t+1} = 2) = Prob(U_1^t = U_2^t = V^t)$$

$$= \sum_{u^t\in\mathcal{U}^t} p_{u^t}^3 = \sum_{u_1,\dots,u_t\in\mathcal{U}} (p_{u_1}\cdots p_{u_t})^3 = \left(\sum_{u\in\mathcal{U}} p_u^3\right)^t$$

and

$$Prob(X_{t+1} = 1) = 2Prob(U_1^t = V^t \text{ and } U_2^t \neq V^t) = 2\sum_{u^t\in\mathcal{U}^t} p_{u^t}^2(1 - p_{u^t})$$

$$= 2\left[\sum_{u_1,\dots,u_t\in\mathcal{U}} (p_{u_1}\cdots p_{u_t})^2 - \sum_{u_1,\dots,u_t\in\mathcal{U}} (p_{u_1}\cdots p_{u_t})^3\right]$$

$$= 2\left[\left(\sum_{u\in\mathcal{U}} p_u^2\right)^t - \left(\sum_{u\in\mathcal{U}} p_u^3\right)^t\right].$$

As stated before the symmetric 2-identification running time is an expectation. Since for the first time step $X_1 = 2$ and for all other time steps the case $X_t = 0$ leads to the termination of the identification process before time step t, we obtain

$$\mathcal{L}_{\mathcal{C}^n}^{2,q}(P^n, P^n) = \sum_{t=1}^{n} \mathbf{E}\left[\mathcal{L}_{\mathcal{C}}^{X_t,q}(P,P)\right] = \sum_{t=0}^{n-1} \mathbf{E}\left[\mathcal{L}_{\mathcal{C}}^{X_{t+1},q}(P,P)\right]$$

$$= \mathcal{L}_{\mathcal{C}}^{2,q}(P,P) + \sum_{t=1}^{n-1} Prob(X_{t+1} = 1)\mathcal{L}_{\mathcal{C}}^{1,q}(P,P)$$

$$+ \sum_{t=1}^{n-1} Prob(X_{t+1} = 2)\mathcal{L}_{\mathcal{C}}^{2,q}(P,P)$$

$$= \mathcal{L}_{\mathcal{C}}^{2,q}(P,P)\left(1 + \sum_{t=1}^{n-1}\left(\sum_{u\in\mathcal{U}} p_u^3\right)^t\right)$$

$$+ 2\mathcal{L}_{\mathcal{C}}^{1,q}(P,P)\left(\sum_{t=1}^{n-1}\left(\sum_{u\in\mathcal{U}} p_u^2\right)^t - \sum_{t=1}^{n-1}\left(\sum_{u\in\mathcal{U}} p_u^3\right)^t\right).$$

□

If we now establish the limit for n going to infinity and apply the geometric series for $k = 2, 3$ we obtain

$$\sum_{t=1}^{\infty}\left(\sum_{u\in\mathcal{U}} p_u^k\right)^t = \frac{1}{1 - \sum_{u\in\mathcal{U}} p_u^k} - 1$$

and thus,

$$\lim_{n\to\infty} \mathcal{L}_{\mathcal{C}^n}^{2,q}(P^n, P^n) = \frac{\mathcal{L}_{\mathcal{C}}^{2,q}(P,P)}{1 - \sum_{u\in\mathcal{U}} p_u^3} + 2\left(\frac{1}{1 - \sum_{u\in\mathcal{U}} p_u^2} - \frac{1}{1 - \sum_{u\in\mathcal{U}} p_u^3}\right)\mathcal{L}_{\mathcal{C}}^{1,q}(P,P).$$

This proves

Corollary 2. *Let $(\mathcal{U}, P)$ be a discrete source and $\mathcal{C}$ be q-ary prefix code with $q \in \mathbb{N}_{\geq 2}$. It then holds that*

$$\mathcal{L}_{\mathcal{C}}^{2,q}(P,P) = (1 - \sum_{u\in\mathcal{U}} p_u^3)\lim_{n\to\infty} \mathcal{L}_{\mathcal{C}^n}^{2,q}(P^n, P^n) - 2\left(\frac{1 - \sum_{u\in\mathcal{U}} p_u^3}{1 - \sum_{u\in\mathcal{U}} p_u^2} - 1\right)\mathcal{L}_{\mathcal{C}}^{1,q}(P,P).$$

We try now to lower bound the expression $\lim_{n\to\infty} \mathcal{L}_{\mathcal{C}^n}^{2,q}(P^n, P^n)$. In Theorem 1 we have shown how L-identification and in particular 2-identification behaves asymptotically on block codes if we consider the uniform distribution. To use this result we have to show that for 2-identification uniform distribution is optimal for block codes. The following lemma provides a way for coming from any probability distribution to the uniform distribution without increasing the symmetric identification running time. Its rather long and technical proof can be found in the appendix. It is not necessary to follow it in order to understand the remaining part of this paper. It suffices to note that it uses the same ideas as Lemma 1 in [7].

Lemma 4. *Let $(\mathcal{U}, P)$ and $(\mathcal{U}, \tilde{P})$ be two discrete sources with $\mathcal{U} = \{1, \ldots, q^n\}$, for $n \in \mathbb{N}$ and $q \in \mathbb{N}_{\geq 2}$, and*

$$P = (p_1, \ldots, p_{tq^{k+1}}, \underbrace{r_1, \ldots, r_1}_{q^k}, \underbrace{r_2, \ldots, r_2}_{q^k}, \ldots, \underbrace{r_q, \ldots, r_q}_{q^k}, p_{(t+1)q^{k+1}+1}, \ldots, p_{q^n})$$

and

$$\tilde{P} = (p_1, \ldots, p_{tq^{k+1}}, \underbrace{\frac{1}{q}\sum_{i=1}^{q} r_i, \ldots, \frac{1}{q}\sum_{i=1}^{q} r_i}_{q^{k+1}}, p_{(t+1)q^{k+1}+1}, \ldots, p_{q^n}) \ .$$

Further, let $k \in \{0, \ldots, n-1\}$ and $t \in \{0, \ldots, q^{n-k-1} - 1\}$. It then holds that

$$\mathcal{L}^{2,q}_{\mathcal{C}_{q^n}}(P, P) - \mathcal{L}^{2,q}_{\mathcal{C}_{q^n}}(\tilde{P}, \tilde{P}) \geq 0 \ .$$

The inequality holds with equality if and only if either $k = 0$ or $r_i = r_j$ for all $i, j \in \{1, \ldots, q\}$.

By applying Lemma 4 in the same iterative way as Lemma 1 in [7] we obtain

Corollary 3. *Let $(\mathcal{U}, P)$ be a discrete source with $\mathcal{U} = \{1, \ldots, q^n\}$ for some $n \in \mathbb{N}$ and $q \in \mathbb{N}_{\geq 2}$. Further, let $\mathcal{C} = \mathcal{C}_{q^n}$ and $T = T_{\mathcal{C}}$. It then holds that*

$$\mathcal{L}^{2,q}_{\mathcal{C}}(P, P) \geq \mathcal{L}^{2,q}_{\mathcal{C}}\left((\frac{1}{q^n}, \ldots, \frac{1}{q^n}), (\frac{1}{q^n}, \ldots, \frac{1}{q^n})\right) \ .$$

The inequality holds with equality if and only if $P(T_x) = q^{-\|x\|}$ for all inner nodes $x \in \mathring{\mathcal{N}}(T)$.

We now make a short detour to *δ-typical sequences.* These are defined e.g. in [4] Definition 2.8 (p. 33). We will change some of the notation of this definition in order to harmonize it with the notation used in this paper:

"For any distribution P on $\mathcal{U}$, a sequence $u^n \in \mathcal{U}^n$ is called *P-typical* with constant δ if

$$\left| \frac{1}{n} < u^n | a > - p_a \right| \leq \delta \tag{8}$$

for every $a \in \mathcal{U}$ and, in addition, no $a \in \mathcal{U}$ with $p_a = 0$ occurs in u^n. The set of such sequences will be denoted by $\mathcal{T}^n_{P,\delta}$."

Here, the value of $< u^n | a >$ is the number of appearances of a as a component of u^n. In words, a sequence $u^n \in \mathcal{U}^n$ is called *P-typical* with constant δ if for all $a \in \mathcal{U}$ the difference between the relative frequency of a in u^n and the actual probability of a with respect to P is at most δ. Lemma 2.12 in [4] and its subsequent remark state that

$$P^n(\mathcal{T}^n_{P,\delta}) \geq 1 - \frac{|\mathcal{U}|}{4n\delta^2} . \tag{9}$$

Further, it follows from (8) for all $u^n \in \mathcal{T}^n_{P,\delta}$ that

$$P^n_{u^n} = \prod_{a\in\mathcal{U}} p_a^{<u^n|a>} \leq \prod_{a\in supp(P)} p_a^{n(p_a-\delta)} = 2^{-n\left(H(P)+\delta \sum\limits_{a\in supp(P)} \log p_a\right)} . \tag{10}$$

Here, $H(P) = -\sum_{a\in supp(P)} p_a \log p_a$ is Shannon's classical entropy. In the following we use $M_P = -\sum_{a\in supp(P)} \log p_a$. It holds that $0 \leq M_P < \infty$ with equality on the left hand side if and only if $supp(P) = 1$. In our further analysis we assume that $supp(P) > 1$. It follows that for all $\epsilon > 0$ exists $\delta > 0$ such that on the one hand it holds that

$$P^n((\mathcal{T}^n_{P,\delta})^c) \leq \frac{|\mathcal{U}|M_P}{4n\epsilon^2} . \tag{11}$$

On the other hand it holds for all $u^n \in \mathcal{T}^n_{P,\delta}$ that

$$P^n_{u^n} \leq 2^{-n(H(P)-\epsilon)} . \tag{12}$$

To see this choose $\delta = \epsilon/M_P$ and apply (9) and (10). Things are now settled to prove

Lemma 5. *Let $(\mathcal{U}, P)$ be a discrete source with $|supp(P)| > 1$. For all $\epsilon > 0$ and all q-ary prefix codes $\mathcal{C}$ over $\mathcal{U}$ there exist sequences $\alpha_n(\epsilon) = \alpha_n \to 0$ and $K_n(\epsilon) = K_n \to \infty$ such that*

$$\mathcal{L}^{2,q}_{\mathcal{C}^n}(P^n, P^n) \geq (1-\alpha_n)^3 \mathcal{L}^{2,q}_{\mathcal{C}_{q^{K_n}}} \left(\left(\frac{1}{q^{K_n}}, \ldots, \frac{1}{q^{K_n}}\right), \left(\frac{1}{q^{K_n}}, \ldots, \frac{1}{q^{K_n}}\right) \right)$$

holds for all sufficiently large n.

Proof. The proof of this theorem follows the same guidelines as the proof of Lemma 3 in [3]. However, we changed some of its steps in order to obtain a more explanatory proof. We begin the proof without explicitly specifying K_n and α_n. This will be done later. We partition $\mathcal{U}^n$ according to the given code $\mathcal{C}^n$ into $\mathcal{U}^n_1 = \{u^n \in \mathcal{U}^n : \|c_{u^n}\| \leq K_n\}$ and $\mathcal{U}^n_2 = \mathcal{U}^n \backslash \mathcal{U}^n_1$. Since $\mathcal{C}^n$ is a q-ary prefix code, we have that

$$|\mathcal{U}^n_1| \leq q^{K_n} . \tag{13}$$

For $\epsilon > 0$ we choose $\delta = \epsilon/M_P$ and obtain

$$\begin{aligned} P^n(\mathcal{U}^n_1) &= P^n(\mathcal{U}^n_1 \cap \mathcal{T}^n_{P,\delta}) + P^n(\mathcal{U}^n_1 \cap (\mathcal{T}^n_{P,\delta})^c) \\ &\leq |\,\mathcal{U}^n_1 \cap \mathcal{T}^n_{P,\delta}| 2^{-n(H(P)-\epsilon)} + P^n((\mathcal{T}^n_{P,\delta})^c) \\ &\leq q^{K_n} 2^{-n(H(P)-\epsilon)} + \tfrac{|\mathcal{U}|M_P}{4n\epsilon^2} . \end{aligned}$$

The first inequality follows by (12). Equations (11) and (13) yield the second inequality. We now set $K_n = \left\lfloor \frac{n(H(P)-2\epsilon)}{\log q} \right\rfloor$ as well as $\alpha_n = 2^{-n\epsilon} + \frac{|\mathcal{U}|M_P}{4n\epsilon^2}$ and obtain

$$P^n(\mathcal{U}_1^n) \leq \alpha_n$$

and thus

$$P^n(\mathcal{U}_2^n) \geq 1 - \alpha_n \ . \tag{14}$$

We will now construct a new source code by cutting all codewords in $\mathcal{U}_2^n$ back to length K_n. Formally, we define the new source $\tilde{\mathcal{U}} = \tilde{\mathcal{U}}_1 \cup \tilde{\mathcal{U}}_2$, where $\tilde{\mathcal{U}}_1 = \mathcal{U}_1^n$ and $\tilde{\mathcal{U}}_2$ is defined as follows. Let $\cong$ be an equivalence relation on $\mathcal{U}_2^n$ with

$$u^n \cong v^n :\Leftrightarrow c_{u^n}^{K_n} = c_{v^n}^{K_n}$$

and let $\mathcal{E}_1, ..., \mathcal{E}_m$ be the equivalence classes. Further, we associate with every equivalence class $\mathcal{E}_i$ the object e_i and define $\tilde{\mathcal{U}}_2 = \{e_1, ..., e_m\}$. Moreover, we define a probability distribution $\tilde{P}$ on $\tilde{\mathcal{U}}$ by $\tilde{P}(u^n) = P(u^n)$ for all $u^n \in \tilde{\mathcal{U}}_1$ and $\tilde{P}(e_k) = \sum_{u^n \in \mathcal{E}_k} P(u^n)$ for $k \in \{1, \ldots, m\}$. Finally, we obtain a new code $\tilde{\mathcal{C}} : \tilde{\mathcal{U}} \to \mathcal{Q}^*$ by $\tilde{c}_{u^n} = c_{u^n}$ if $u^n \in \tilde{\mathcal{U}}_1$ and $\tilde{c}_{e_k}$ will be the common prefix of length K_n of the objects in $\mathcal{E}_k$. It follows that

$$\mathcal{L}_{\mathcal{C}^n}^{2,q}(P^n, P^n) \geq \mathcal{L}_{\tilde{\mathcal{C}}}^{2,q}(\tilde{P}, \tilde{P}) \ . \tag{15}$$

The next step is to focus only on the $\tilde{\mathcal{U}}_2$-part of $\tilde{\mathcal{U}}$. Again we operate without increasing the symmetric 2-identification running time since

$$\begin{aligned}
\mathcal{L}_{\tilde{\mathcal{C}}}^{2,q}(\tilde{P}, \tilde{P}) &= \sum_{\tilde{u}_1, \tilde{u}_2, \tilde{v} \in \tilde{\mathcal{U}}} \tilde{P}(\tilde{u}_1)\tilde{P}(\tilde{u}_2)\tilde{P}(\tilde{v})\mathcal{L}_{\tilde{\mathcal{C}}}^{2,q}((\tilde{u}_1, \tilde{u}_2), \tilde{v}) \\
&\geq \sum_{\tilde{u}_1, \tilde{u}_2, \tilde{v} \in \tilde{\mathcal{U}}_2} \tilde{P}(\tilde{u}_1)\tilde{P}(\tilde{u}_2)\tilde{P}(\tilde{v})\mathcal{L}_{\tilde{\mathcal{C}}}^{2,q}((\tilde{u}_1, \tilde{u}_2), \tilde{v}) \\
&= \sum_{i_1, i_2, j=1}^{m} \tilde{P}(e_{i_1})\tilde{P}(e_{i_2})\tilde{P}(e_j)\mathcal{L}_{\tilde{\mathcal{C}}}^{2,q}((e_{i_1}, e_{i_2}), e_j) \\
&= \left(\sum_{k=1}^{m} \tilde{P}(e_k)\right)^3 \sum_{i_1, i_2, j=1}^{m} \tilde{P}_2(e_{i_1})\tilde{P}_2(e_{i_2})\tilde{P}_2(e_j)\mathcal{L}_{\tilde{\mathcal{C}}_2}^{2,q}((e_{i_1}, e_{i_2}), e_j) \ .
\end{aligned}$$

Here, $\tilde{P}_2$ is a probability distribution on $\tilde{\mathcal{U}}_2$ defined for $j \in \{1, \ldots, m\}$ by

$$\tilde{P}_2(e_j) = \tilde{P}(e_j) / \sum_{k=1}^{m} \tilde{P}(e_k) \ .$$

Further, $\tilde{\mathcal{C}}_2$ is the restriction of $\tilde{\mathcal{C}}$ to $\tilde{\mathcal{U}}_2$. Since

$$\sum_{k=1}^{m} \tilde{P}(e_k) = \sum_{k=1}^{m} \sum_{u^n \in \mathcal{E}_k} P^n(u^n) = P^n(\mathcal{U}_2^n) \ ,$$

we obtain by (14) that

$$\mathcal{L}^{2,q}_{\tilde{\mathcal{C}}}(\tilde{P},\tilde{P}) \geq (1-\alpha_n)^3 \mathcal{L}^{2,q}_{\tilde{\mathcal{C}}_2}(\tilde{P}_2,\tilde{P}_2) \ . \tag{16}$$

Although $\tilde{\mathcal{C}}_2$ is a block code with codewords of length K_n it might not be saturated. To achieve this property we extend $\tilde{\mathcal{U}}_2$ to a set of cardinality q^{K_n}, assign zero probabilities to the additional elements and use for them codewords from $\mathcal{Q}^{K_n}\backslash\tilde{\mathcal{C}}_2$. We now obey the conditions of Corollary 3 by which we obtain

$$\mathcal{L}^{2,q}_{\tilde{\mathcal{C}}_2}(\tilde{P}_2,\tilde{P}_2) \geq \mathcal{L}^{2,q}_{\mathcal{C}_{q^{K_n}}}((\frac{1}{q^{K_n}},...,\frac{1}{q^{K_n}}),(\frac{1}{q^{K_n}},...,\frac{1}{q^{K_n}})) \ . \tag{17}$$

Inequalities (15), (16) and (17) finally yield the statement of the lemma. □

By applying Theorem 1 and Lemma 5 to Corollary 2 we obtain

Corollary 4. *Let $(\mathcal{U},P)$ be a discrete source with $|supp(P)|>1$, $q\in\mathbb{N}_{\geq 2}$ and $\mathcal{C}$ be a q-ary prefix code. It then holds that*

$$\mathcal{L}^{2,q}_{\mathcal{C}}(P,P) \geq (1-\sum_{u\in\mathcal{U}} p_u^3)\left(2\frac{q}{q-1}-\frac{q^2}{q^2-1}\right)-2\left(\frac{1-\sum_{u\in\mathcal{U}} p_u^3}{1-\sum_{u\in\mathcal{U}} p_u^2}-1\right)\mathcal{L}^{1,q}_{\mathcal{C}}(P,P) \ .$$

3.2 The q-ary Identification Entropy of Second Order

Unfortunately, in Corollary 4 the symmetric (1-)identification running time appears negatively signed so that we cannot immediately apply its lower bound $\mathcal{L}^{1,q}_{\mathcal{C}}(P,P) \geq H^{1,q}_{\mathrm{ID}}(P)$, which has been proven in [3]. In the same work, however, it has been shown that this lower bound can be attained if P consists only of q-powers. Using this fact we can prove the following important

Proposition 1. *Let $(\mathcal{U},P)$ be a discrete source, where P consists only of q-powers and $\mathcal{C}$ be a q-ary prefix code, where $\|c_u\| = -\log_q p_u$ for all $u\in\mathcal{U}$. It then holds that*

$$\mathcal{L}^{2,q}_{\mathcal{C}}(P,P) = 2\frac{q}{q-1}\left(1-\sum_{u\in\mathcal{U}} p_u^2\right)-\frac{q^2}{q^2-1}\left(1-\sum_{u\in\mathcal{U}} p_u^3\right) \ .$$

Proof. It is an immediate consequence from the condition $\|c_u\| = -\log_q p_u$ for all $u\in\mathcal{U}$ that

$$P(T_x) = q^{-\|x\|} \tag{18}$$

holds for all $x\in\mathcal{N}(T)$, where $T=T_{\mathcal{C}}$. We now introduce for all $v\in\mathcal{U}$ and $k=1,...,\|c_v\|$ the set

$$\bar{\mathcal{R}}^{1,q}_{\mathcal{C}}(k,v) = \mathcal{R}^{1,q}_{\mathcal{C}}(1,v) \ \dot{\cup} \ ... \ \dot{\cup} \ \mathcal{R}^{1,q}_{\mathcal{C}}(k-1,v) \ . \tag{19}$$

Proceeding as in the proof of Theorem 1 we obtain

$$\mathcal{L}_{\mathcal{C}}^{L,q}(P,P) = \sum_{v\in\mathcal{U}} p_v \sum_{k=1}^{\|c_v\|} k \sum_{(u_1,u_2)\in\mathcal{R}_{\mathcal{C}}^{2,q}(k,v)} p_{u_1}p_{u_2} \ .$$

In the following we use $S_{k,v} = \sum_{(u_1,u_2)\in\mathcal{R}_{\mathcal{C}}^{2,q}(k,v)} p_{u_1}p_{u_2}$. With the notation of (19) it holds that

$$S_{k,v} = 2 \sum_{u_1\in\mathcal{R}_{\mathcal{C}}^{1,q}(k,v)} \sum_{u_2\in\bar{\mathcal{R}}_{\mathcal{C}}^{1,q}(k,v)} p_{u_1}p_{u_2} + \sum_{u_1,u_2\in\mathcal{R}_{\mathcal{C}}^{1,q}(k,v)} p_{u_1}p_{u_2} \ .$$

The above equality holds because there are two cases to consider. The first is that there exists one component for which (1-)identification against v takes exactly k steps and the other yields a (1-)identification running time regarding v of at most $k-1$. In the second case both components have a (1-)identification time regarding v of k.

Case 1: $\mathbf{k} = \mathbf{1}, ..., \|\mathbf{c_v}\| - \mathbf{1}$

Here we have that $\mathcal{R}_{\mathcal{C}}^{1,q}(k,v) = \bar{T}_{c_v^{k-1}}\backslash\bar{T}_{c_v^k}$ and $\bar{\mathcal{R}}_{\mathcal{C}}^{1,q}(k,v) = \mathcal{U}\backslash\bar{T}_{c_v^{k-1}}$. Together with (18) this yields

$$\sum_{u\in\mathcal{R}_{\mathcal{C}}^{1,q}(k,v)} p_u = P(T_{c_v^{k-1}}) - P(T_{c_v^k}) = q^{-k+1} - q^{-k} = q^{-k}(q-1)$$

and

$$\sum_{u\in\bar{\mathcal{R}}_{\mathcal{C}}^{1,q}(k,v)} p_u = 1 - P(T_{c_v^{k-1}}) = 1 - q^{-k+1} \ .$$

Thus,

$$\begin{aligned} S_{k,v} &= 2q^{-k}(q-1)(1-q^{-k+1}) + q^{-2k}(q-1)^2 \\ &= (1-q^{-k})^2 - (1-q^{-k+1})^2 \ . \end{aligned}$$

Case 2: $\mathbf{k} = \|\mathbf{c_v}\|$

We have that $\mathcal{R}_{\mathcal{C}}^{1,q}(\|c_v\|,v) = \bar{T}_{c_v^{\|c_v\|-1}}$ and $\bar{\mathcal{R}}_{\mathcal{C}}^{1,q}(\|c_v\|,v) = \mathcal{U}\backslash\bar{T}_{c_v^{\|c_v\|-1}}$. Equation (18) yields

$$\sum_{u\in\mathcal{R}_{\mathcal{C}}^{1,q}(\|c_v\|,v)} p_u = P(T_{c_v^{\|c_v\|-1}}) = q^{-\|c_v\|+1}$$

and

$$\sum_{u\in\bar{\mathcal{R}}_{\mathcal{C}}^{1,q}(\|c_v\|,v)} p_u = 1 - P(T_{c_v^{\|c_v\|-1}}) = 1 - q^{-\|c_v\|+1} \ .$$

Thus, we obtain

$$S_{\|c_v\|,v} = 2q^{-\|c_v\|+1}(1-q^{-\|c_v\|+1}) + q^{-2(\|c_v\|-1)} = 1 - (1-q^{-\|c_v\|+1})^2 \ .$$

Together, the above two cases yield

$$\sum_{k=1}^{\|c_v\|} kS_{k,v}$$

$$= \sum_{k=1}^{\|c_v\|-1} k\left[(1-q^{-k})^2 - (1-q^{-k+1})^2\right] + \|c_v\|\left[1-(1-q^{-\|c_v\|+1})^2\right]$$

$$= \sum_{k=1}^{\|c_v\|-1} k(1-q^{-k})^2 + \|c_v\| - \sum_{k=1}^{\|c_v\|} k(1-q^{-k+1})^2 \ .$$

If we take a look at the first sum plus $\|c_v\|$, we see that

$$\sum_{k=1}^{\|c_v\|-1} k(1-q^{-k})^2 + \|c_v\| = \sum_{k=1}^{\|c_v\|-1} k(1-2q^{-k}+q^{-2k}) + \|c_v\|$$

$$= \sum_{k=1}^{\|c_v\|} k - 2\sum_{k=1}^{\|c_v\|-1} kq^{-k} + \sum_{k=1}^{\|c_v\|-1} kq^{-2k} \ .$$

Further, we obtain

$$\sum_{k=1}^{\|c_v\|} k(1-q^{-k+1})^2 = \sum_{k=1}^{\|c_v\|} k(1-2q^{-k+1}+q^{-2k+2})$$

$$= \sum_{k=1}^{\|c_v\|} k - 2\sum_{k=1}^{\|c_v\|} kq^{-k+1} + \sum_{k=1}^{\|c_v\|} kq^{-2k+2} \ .$$

Subtracting the second from the first result we get

$$\sum_{k=1}^{\|c_v\|} kS_{k,v} = 2(q-1)\sum_{k=1}^{\|c_v\|} kq^{-k} - (q^2-1)\sum_{k=1}^{\|c_v\|} kq^{-2k}$$

$$+\|c_v\|q^{-\|c_v\|}(2-q^{-\|c_v\|})$$

$$= 2\tfrac{q}{q-1}(1-p_v) - 2\|c_v\|p_v - \tfrac{q^2}{q^2-1}(1-p_v^2) + \|c_v\|p_v^2$$

$$+\|c_v\|p_v(2-p_v)$$

$$= 2\tfrac{q}{q-1}(1-p_v) - \tfrac{q^2}{q^2-1}(1-p_v^2) \ .$$

Here, the first equality follows from the previously calculated sums. The second equality holds since by assumption $q^{-\|c_v\|} = p_v$ for all $v \in \mathcal{U}$ and since we have for $j = 1, 2$ that

$$\sum_{k=1}^{\|c_v\|} kq^{-jk} = \tfrac{1}{(q^j-1)^2}[q^j - (q^j(\|c_v\|+1) - \|c_v\|)q^{-j\|c_v\|}]$$

$$= \tfrac{q^l}{(q^l-1)^2}(1-p_v^l) - \tfrac{\|c_v\|}{q^l-1}p_v^l \ .$$

Finally the above calculations yield

$$\mathcal{L}_{\mathcal{C}}^{L,q}(P,P) = \sum_{v\in\mathcal{U}} p_v \sum_{k=1}^{\|c_v\|} kS_{k,v}$$

$$= 2\tfrac{q}{q-1}\left(1-\sum_{v\in\mathcal{U}} p_v^2\right) - \tfrac{q^2}{q^2-1}\left(1-\sum_{v\in\mathcal{U}} p_v^3\right) .$$

□

This result encourages us in the belief that the right side of the equation in Proposition 1 is in general a lower bound for 2-identification. As we will see in Theorem 2 it obeys fundamental properties for entropy functions. We therefore introduce

Definition 1. *Let $(\mathcal{U},P)$ be a discrete source and $q \in \mathbb{N}_{\geq 2}$. The q-ary identification entropy of second order $H_{ID}^{2,q}$ is then defined by*

$$H_{ID}^{2,q}(P) := 2\frac{q}{q-1}\left(1-\sum_{u\in\mathcal{U}} p_u^2\right) - \frac{q^2}{q^2-1}\left(1-\sum_{u\in\mathcal{U}} p_u^3\right) .$$

Before we will show that this entropy function serves as a general lower bound for the symmetric 2-identification running time we will first analyze its functional properties. A list of desiderata for entropy functions can be found in [1], pp. 50. As we will now see our entropy function $H_{\mathrm{ID}}^{2,q}$ obeys some important ones. Those are stated in

Theorem 2. *Let $(\mathcal{U},P)$ be a discrete source and $q \in \mathbb{N}_{\geq 2}$. The following properties then hold for $H_{ID}^{2,q}(P)$:*

1. *Symmetry:*
$$H_{ID}^{2,q}(p_1,...,p_N) = H_{ID}^{2,q}(p_{\pi(1)},...,p_{\pi(N)}) ,$$
where π is a permutation on $\{1,\ldots,N\}$.
2. *Expansibility:*
$$H_{ID}^{2,q}(p_1,...,p_N) = H_{ID}^{2,q}(p_1,...,p_N,0) .$$
3. *Decisiveness:*
$$H_{ID}^{2,q}(1,0,...,0) = 0 .$$
4. *Normalization:*
$$H_{ID}^{2,q}\left(\frac{1}{q},...,\frac{1}{q}\right) = 1 .$$
5. *Bounds:*
$$H_{ID}^{2,q}(1,0,...,0) \leq H_{ID}^{2,q}(P) \leq H_{ID}^{2,q}\left(\frac{1}{N},...,\frac{1}{N}\right) .$$

6. *Additive Grouping Behavior:*
For $m \leq N$ let $\mathcal{U}_1, \mathcal{U}_2, ..., \mathcal{U}_m$ be a partition of $\mathcal{U}$ of non-empty sets, $Q = (Q_1, ..., Q_m)$ be the probability distribution on $\{1, \ldots, m\}$ defined by

$$Q_i := \sum_{u \in \mathcal{U}_i} p_u$$

and P_i be the probability distribution on $\mathcal{U}_i$ defined by

$$p_{i,u} := p_u / Q_i$$

for all $i \in \{1, \ldots, m\}$ and $u \in \mathcal{U}_i$. It then holds that

$$H^{2,q}_{ID}(P) = H^{2,q}_{ID}(Q) + \sum_{i=1}^{m} \left[2Q_i^2(1 - Q_i)H^{1,q}_{ID}(P_i) + Q_i^3 H^{2,q}_{ID}(P_i)\right] .$$

Proof. 1. Symmetry, 2. Expansibility, 3. Decisiveness, 4. Normalization:

Symmetry, expansibility and decisiveness follow directly from the definition of $H^{2,q}_{\mathrm{ID}}$. Further, the normalization property follows from

$$H^{2,q}_{\mathrm{ID}}\left(\frac{1}{q}, ..., \frac{1}{q}\right) = 2\frac{q}{q-1}\left(1 - \frac{1}{q}\right) - \frac{q^2}{q^2-1}\left(1 - \frac{1}{q^2}\right) = 1 .$$

5. Bounds:

Let $f(p_1, ..., p_{N-1}) = H^{2,q}_{\mathrm{ID}}(p_1, ..., p_{N-1}, 1 - \sum_{i=1}^{N-1} p_i)$. We will show that the gradient $\nabla f(p_1, ..., p_{N-1}) = \mathbf{0}$ if and only if $(p_1, ..., p_{N-1}) = (1/N, ..., 1/N)$. For that we set $p_N = 1 - \sum_{i=1}^{N-1} p_i$ and obtain that

$$\frac{\delta}{\delta p_j} f(p_1, ..., p_{N-1}) = -4\frac{q}{q-1}(p_j - p_N) + 3\frac{q^2}{q^2-1}(p_j^2 - p_N^2)$$

holds for all $j \in \{1, \ldots, N-1\}$. It follows directly that $\nabla f(1/N, ..., 1/N) = \mathbf{0}$. Assume now that for any $P' \neq (1/N, ..., 1/N)$ it holds that $\nabla f(P') = \mathbf{0}$. It follows that there exists $j \in \{1, \ldots, N-1\}$ such that $p_j \neq p_N$. If we now take a look at $\frac{\delta}{\delta p_j} f(P')$, we see that

$$\frac{\delta}{\delta p_j} f(P') = 0 \Leftrightarrow 3\frac{q}{q+1}(p_j + p_N) = 4 .$$

This is a contradiction because $\frac{q}{q+1}(p_j + p_N) < 1$. In order to ensure that $(1/N, ..., 1/N)$ is indeed a maximum we show that the Hessian is negative definite. In fact, we will obtain a stronger result namely that all second derivatives $\frac{\delta^2}{\delta p_k \delta p_j} f(1/N, ..., 1/N)$ are strictly negative.

$$\frac{\delta^2}{\delta p_k \delta p_j} f\left(\frac{1}{N}, ..., \frac{1}{N}\right) = \begin{cases} 4\frac{q}{q-1}\left(\frac{3q}{N(q+1)} - 2\right) & \text{if } k = j \\ 2\frac{q}{q-1}\left(\frac{3q}{N(q+1)} - 2\right) & \text{if } k \neq j \end{cases} .$$

From $q \geq 2$ now follows that $\frac{3q}{N(q+1)} - 2 < 0$ if $N \geq 2$. And for $N = 1$ we are in the trivial case, where $H_{\text{ID}}^{2,q}(1) = 0$.

6. Additive Grouping Behavior:

We use

$$S_i = 2Q_i^2(1 - Q_i)H_{\text{ID}}^{1,q}(P_i) + Q_i^3 H_{\text{ID}}^{2,q}(P_i) \ ,$$

for all $i \in \{1, \dots, m\}$ and observe that

$$S_i = 2Q_i^2(1 - Q_i)\tfrac{q}{q-1}(1 - \tfrac{1}{Q_i^2} \sum_{u \in \mathcal{U}_i} p_u^2)$$

$$+Q_i^3 \left[2\tfrac{q}{q-1}(1 - \tfrac{1}{Q_i^2} \sum_{u \in \mathcal{U}_i} p_u^2) - \tfrac{q^2}{q^2-1}(1 - \tfrac{1}{Q_i^3} \sum_{u \in \mathcal{U}_i} p_u^3) \right]$$

$$= 2\tfrac{q}{q-1}(Q_i^2 - \sum_{u \in \mathcal{U}_i} p_u^2) - \tfrac{q^2}{q^2-1}(Q_i^3 - \sum_{u \in \mathcal{U}_i} p_u^3) \ .$$

By summing the S_i's up we obtain

$$\sum_{i=1}^m S_i = 2\frac{q}{q-1}(\sum_{i=1}^m Q_i^2 - \sum_{u \in \mathcal{U}} p_u^2) - \frac{q^2}{q^2-1}(\sum_{i=1}^m Q_i^3 - \sum_{u \in \mathcal{U}} p_u^3)$$

and thus

$$H_{\text{ID}}^{2,q}(Q) + \sum_{i=1}^m S_i = 2\tfrac{q}{q-1}(1 - \sum_{i=1}^m Q_i^2) - \tfrac{q^2}{q^2-1}(1 - \sum_{i=1}^m Q_i^3)$$

$$+2\tfrac{q}{q-1}(\sum_{i=1}^m Q_i^2 - \sum_{u \in \mathcal{U}} p_u^2) - \tfrac{q^2}{q^2-1}(\sum_{i=1}^m Q_i^3 - \sum_{u \in \mathcal{U}} p_u^3)$$

$$= 2\tfrac{q}{q-1}(1 - \sum_{u \in \mathcal{U}} p_u^2) - \tfrac{q^2}{q^2-1}(1 - \sum_{u \in \mathcal{U}} p_u^3)$$

$$= H_{\text{ID}}^{2,q}(P) \ .$$

□

We will provide an inductive proof in order to show that $H_{\text{ID}}^{2,q}$ is a general lower bound for the symmetric 2-identification running time. Therefore, we will partition $\mathcal{U}$ into several smaller sets for which we can use the additive grouping behavior of our entropy function. In order to link this to the 2-identification process we also need a decomposition formula for the symmetric 2-identification running time. It turns out that such a decomposition exists and that it mainly behaves in the same way as the additive grouping behavior. We prove this formula for general $L \in \mathbb{N}$ in

Lemma 6. *Let $(\mathcal{U}, P)$ be a discrete source, $q \in \mathbb{N}_{\geq 2}$ and $\mathcal{C}$ be a q-ary prefix code. Further, we define for all $i \in \{0, \dots, q-1\}$:*

1. $\mathcal{U}_i := \{u \in \mathcal{U} \ : \ c_{u,1} = i\}$
2. $Q_i := \sum_{u \in \mathcal{U}_i} p_u$
3. *P_i to be a probability distribution on $\mathcal{U}_i$ defined by*
$$p_{i,u} := \frac{p_u}{Q_i}$$
for all $u \in \mathcal{U}_i$.
4. *$\mathcal{C}^{(i)} : \mathcal{U}_i \to \mathcal{Q}^*$ to be a code on $\mathcal{U}_i$ defined by*
$$c_u^{(i)} := c_{u,2} c_{u,3} \cdots c_{u,\|c_u\|}$$
for all $u \in \mathcal{U}_i$.

Then it holds that

$$\mathcal{L}_{\mathcal{C}}^{L,q}(P,P) = 1 + \sum_{i \in \mathcal{Q}} \sum_{l=1}^{2} \binom{L}{l} Q_i^{l+1} (1-Q_i)^{L-l} \mathcal{L}_{\mathcal{C}^{(i)}}^{l,q}(P_i, P_i) \ .$$

For $L = 2$ this becomes

$$\mathcal{L}_{\mathcal{C}}^{2,q}(P,P) = 1 + \sum_{i \in \mathcal{Q}} \left[2Q_i^2(1-Q_i)\mathcal{L}_{\mathcal{C}^{(i)}}^{1,q}(P_i,P_i) + Q_i^3 \mathcal{L}_{\mathcal{C}^{(i)}}^{2,q}(P_i,P_i) \right] \ .$$

Proof. First, we observe that

$$\mathcal{L}_{\mathcal{C}}^{L,q}(P,P) = \sum_{u^2 \in \mathcal{U}^2} \sum_{v \in \mathcal{U}} P_{u^2}^2 p_v \mathcal{L}_{\mathcal{C}}^{L,q}(u^2, v)$$

$$= \sum_{i \in \mathcal{Q}} \sum_{v \in \mathcal{U}_i} \sum_{u^2 \in \mathcal{U}^2} P_{u^2}^2 p_v \mathcal{L}_{\mathcal{C}}^{L,q}(u^2, v) \ .$$

Since $\mathcal{L}_{\mathcal{C}}^{L,q}(u^2,v) = \mathcal{L}_{\mathcal{C}}^{L,q}((u_1,\dots,u_L),v) = \mathcal{L}_{\mathcal{C}}^{L,q}((u_{\pi(1)},\dots,u_{\pi(L)}),v)$ for all permutations π on $\{1,\dots,L\}$, we get for all $i \in \mathcal{Q}$

$$\sum_{v \in \mathcal{U}_i} \sum_{u^2 \in \mathcal{U}^2} P_{u^2}^2 p_v \mathcal{L}_{\mathcal{C}}^{L,q}(u^2, v)$$

$$= \sum_{l=0}^{2} \tbinom{L}{l} \sum_{v \in \mathcal{U}_i} \sum_{u_1,\dots,u_l \in \mathcal{U}_i} \sum_{u_{l+1},\dots,u_L \in \mathcal{U} \setminus \mathcal{U}_i} P_{u^2}^2 p_v \mathcal{L}_{\mathcal{C}}^{L,q}(u^2, v)$$

$$= \sum_{l=0}^{2} \tbinom{L}{l} (1-Q_i)^{L-l} \sum_{u_1,\dots,u_l,v \in \mathcal{U}_i} p_{u_1} \cdots p_{u_l} p_v (1 + \mathcal{L}_{\mathcal{C}^{(i)}}^{l,q}((u_1,\dots,u_l),v))$$

$$= Q_i \sum_{l=0}^{2} \tbinom{L}{l} Q_i^l (1-Q_i)^{L-l} + \sum_{l=1}^{2} \tbinom{L}{l} (1-Q_i)^{L-l} Q_i^{l+1} \mathcal{L}_{\mathcal{C}^{(i)}}^{l,q}(P_i,P_i)$$

$$= Q_i + \sum_{l=1}^{2} \tbinom{L}{l} (1-Q_i)^{L-l} Q_i^{l+1} \mathcal{L}_{\mathcal{C}^{(i)}}^{l,q}(P_i,P_i) \ .$$

The second equality follows since $\mathcal{L}_{\mathcal{C}}^{L,q}(u^2, v) = 1 + \mathcal{L}_{\mathcal{C}(i)}^{l,q}((u_1, ..., u_l), v)$ holds if $u_1, ..., u_l, v \in \mathcal{U}_i$ and $u_{l+1}, ..., u_L \in \mathcal{U} \backslash \mathcal{U}_i$. Adding this up for $i \in \mathcal{Q}$ we obtain the desired result. □

As one can see there is a strong relation between the above decomposition formula for 2-identification and the additive grouping behavior of the identification entropy of second order. This fact can be used in the induction step of the proof of the following theorem which expresses the role of the q-ary identification entropy of second order as a general lower bound for 2-identification.

Theorem 3. *Let $(\mathcal{U}, P)$ be a discrete source, $q \in \mathbb{N}_{\geq 2}$ and $\mathcal{C}$ be a q-ary prefix code. It then holds that*

$$\mathcal{L}_{\mathcal{C}}^{2,q}(P, P) \geq H_{ID}^{2,q}(P) \ ,$$

where equality is attained if and only if P consists only of q-powers, and $\mathcal{C}$ is a prefix code, with $\|c_u\| = -\log_q p_u$ for all $u \in \mathcal{U}$.

Proof. In the following proof by induction we have to consider all the cases $N = 1, .., q$ as induction base for N. Since we have $\mathcal{L}_{\mathcal{C}}^{2,q}(P, P) = 1$ for $N \leq q$, we have to show that $H_{\mathrm{ID}}^{2,q}(P) \leq 1$. It follows by the expansibility property that we only have to consider the case $N = q$. Further, the maximality of the uniform distribution and the normalization property yield

$$H_{\mathrm{ID}}^{2,q}(p_1, ..., p_q) \leq H_{\mathrm{ID}}^{2,q}\left(\frac{1}{q}, ..., \frac{1}{q}\right) = 1 \ .$$

In order to apply the induction step we use the same notation as in Lemma 6. That is, we partition $\mathcal{U}$ into sets $\mathcal{U}_i$ and set $Q = (Q_0, ..., Q_{q-1})$. The inequality of Theorem 3 now follows directly since

$$\begin{aligned} \mathcal{L}_{\mathcal{C}}^{2,q}(P, P) &= 1 + \sum_{i \in \mathcal{Q}} \left[2Q_i^2(1 - Q_i)\mathcal{L}_{\mathcal{C}(i)}^{1,q}(P_i, P_i) + Q_i^3 \mathcal{L}_{\mathcal{C}(i)}^{2,q}(P_i, P_i)\right] \\ &\geq H_{\mathrm{ID}}^{2,q}(Q) + \sum_{i \in \mathcal{Q}} \left[2Q_i^2(1 - Q_i)H_{\mathrm{ID}}^{1,q}(P_i) + Q_i^3 H_{\mathrm{ID}}^{2,q}(P_i)\right] \qquad (20) \\ &= H_{\mathrm{ID}}^{2,q}(P) \ . \end{aligned}$$

Here, the first equality follows from Lemma 6. The inequality is a consequence of the induction step together with the normalization property and the established bounds of $H_{\mathrm{ID}}^{2,q}$. Finally, the additive grouping behavior of $H_{\mathrm{ID}}^{2,q}$ yields the second equality.

The fact that this lower bound is attained for every q-ary prefix code $\mathcal{C}$ for which (18) holds has already been proven in Proposition 1. If instead the inequality of Theorem 3 holds with equality, then also the inequality of (20) is in fact an equality and thus

i) $H_{\mathrm{ID}}^{2,q}(Q) = 1$

ii) $H_{\mathrm{ID}}^{1,q}(P_i) = \mathcal{L}_{\mathcal{C}_i}^{1,q}(P_i, P_i)$

iii) $H_{\mathrm{ID}}^{2,q}(P_i) = \mathcal{L}_{\mathcal{C}_i}^{2,q}(P_i, P_i)$.

We have seen in the proof of the bounds of the entropy function that the uniform distribution is the only point where the first derivative of the identification entropy function equals zero and thus $(1/q, ..., 1/q)$ is the only point for which $H_{\mathrm{ID}}^{2,q}(Q) = 1$. Together with i) this means that we get

$$Q_i = \frac{1}{q} \tag{21}$$

for all $i \in \mathcal{Q}$. The crucial part is now ii). For all $i \in \mathcal{Q}$ we obtain from (21) and the definitions of P_i and $\mathcal{C}^{(i)}$ (see Lemma 6) that we have for any $u \in \mathcal{U}_i$ that

$$p_u = Q_i p_{i,u} = \frac{p_{i,u}}{q} \tag{22}$$

and

$$\|c_u\| = \|c_u^{(i)}\| + 1 \ . \tag{23}$$

Moreover, Theorem 1 in [3] stated that for (1-)identification an equality between the running time and identification entropy is only attained if and only if the probability distribution consists only of q-powers and the lengths of the codewords equal the negative logarithm of the probability of their corresponding elements. Thus it follows from ii) that all the $p_{i,u}$'s are q-powers and that $\|c_u^{(i)}\| = -\log_q p_{i,u}$. Together with (22) and (23) we finally obtain that P consists only of q-powers and that

$$\|c_u\| = -\log_q p_{i,u} + 1 = -\log_q \frac{p_{i,u}}{q} = -\log_q p_u \ .$$

□

Since Theorem 3 applies to any prefix code $\mathcal{C}$ we immediately get

Corollary 5. *Let $(\mathcal{U}, P)$ be a discrete source and $q \in \mathbb{N}_{\geq 2}$. It then holds that*

$$\mathcal{L}^{2,q}(P, P) \geq H_{ID}^{2,q}(P) \ .$$

References

1. Aczél, J., Daróczy, Z.: On Measures of Information and Their Characterizations. Mathematics in Science and Engineering, vol. 115 (1975)
2. Ahlswede, R.: Identification Entropy. In: Ahlswede, R., Bäumer, L., Cai, N., Aydinian, H., Blinovsky, V., Deppe, C., Mashurian, H. (eds.) Information Transfer and Combinatorics. LNCS, vol. 4123, pp. 595–613. Springer, Heidelberg (2006)
3. Ahlswede, R., Cai, N.: An interpretation of identification entropy. IEEE Trans. Inf. Theory 52(9), 4198–4207 (2006)
4. Csiszár, I., Körner, J.: Information Theory: Coding Theorems for Discrete Memoryless Systems. Academic Press (1981)

5. Cover, T.M., Thomas, J.A.: Elements of Information Theory. Wiley-Interscience (1991)
6. Fano, R.M.: Transmission of Information. MIT Press, Wiley, Cambridge, MA (1961)
7. Heup, C.: Two New Results for Identification for Sources. In: Aydinian, H., Cicalese, F., Deppe, C. (eds.) Ahlswede Festschrift. LNCS, vol. 7777, pp. 1–10. Springer, Heidelberg (2013)
8. Huffman, D.A.: A method for the construction of minimum-redundancy codes. Proceedings of the I.R.E. 40, 1098–1101 (1952)

Appendix

A The Proof of Lemma 4

Without loss of generality we assume that $t = 0$ such that for $i \in \{1, \dots, q\}$

$$p_{(i-1)q^k+1} = p_{(i-1)q^k+2} = \dots = p_{iq^k} = r_i \ .$$

Also, we use for simplicity the abbreviations $L_{u_1u_2,v} = \mathcal{L}^{2,q}_{\mathcal{C}_{q^n}}((u_1,u_2),v)$ and $\alpha_{u_1u_2,v} = (p_{u_1}p_{u_2}p_v - \tilde{p}_{u_1}\tilde{p}_{u_2}\tilde{p}_v)L_{u_1u_2,v}$. With this notation we obtain

$$\begin{aligned}
&\mathcal{L}^{L,q}_{\mathcal{C}_{q^n}}(P,P) - \mathcal{L}^{L,q}_{\mathcal{C}_{q^n}}(\tilde{P},\tilde{P}) \\
&= \sum_{u_1,u_2,v=1}^{q^n} \alpha_{u_1u_2,v} \\
&= \sum_{v=1}^{q^n}\left[\sum_{u_1,u_2=1}^{q^{k+1}} \alpha_{u_1u_2,v} + 2\sum_{u_1=1}^{q^{k+1}}\sum_{u_2=q^{k+1}+1}^{q^n} \alpha_{u_1u_2,v} + \sum_{u_1,u_2=q^{k+1}+1}^{q^n} \alpha_{u_1u_2,v}\right] \\
&= \sum_{i=1}^{6} R_i \ ,
\end{aligned}$$

where the second equality comes from the fact that $L_{u_1u_2,v} = L_{u_2u_1,v}$ and where

$$R_1 = \sum_{u_1,u_2,v=1}^{q^{k+1}} \alpha_{u_1u_2,v} \qquad R_2 = \sum_{u_1,u_2=1}^{q^{k+1}} \sum_{v=q^{k+1}+1}^{q^n} \alpha_{u_1u_2,v}$$

$$R_3 = 2\sum_{u_1,v=1}^{q^{k+1}} \sum_{u_2=q^{k+1}+1}^{q^n} \alpha_{u_1u_2,v} \quad R_4 = 2\sum_{u_1=1}^{q^{k+1}} \sum_{u_2,v=q^{k+1}+1}^{q^n} \alpha_{u_1u_2,v}$$

$$R_5 = \sum_{u_1,u_2=q^{k+1}+1}^{q^n} \sum_{v=1}^{q^{k+1}} \alpha_{u_1u_2,v} \quad R_6 = \sum_{u_1,u_2,v=q^{k+1}+1}^{q^n} \alpha_{u_1u_2,v} \ .$$

As one might expect the above summands disappear, except for R_1 and R_3. This is obvious for R_6 since $p_u = \tilde{p}_u$ for all $u \in \{q^{k+1}+1, \dots, q^n\}$. If $u_1, u_2 \in \{q^{k+1}+1, \dots, q^n\}$, we have on the one hand that $L_{u_1u_2,v} = L_{u_1u_2,1}$ for all

$v \in \{1, \ldots, q^{k+1}\}$. We denote this by $L_{u_1u_2}$. On the other hand $p_{u_i} = \tilde{p}_{u_i}$ for $i = 1, 2$. This yields

$$R_5 = \sum_{u_1,u_2=q^{k+1}+1}^{q^n} \sum_{v=1}^{q^{k+1}} L_{u_1u_2}p_{u_1}p_{u_2}\left[p_v - \frac{1}{q}\sum_{i=1}^{q} r_i\right]$$

$$= \sum_{u_1,u_2=q^{k+1}+1}^{q^n} L_{u_1u_2}p_{u_1}p_{u_2}\left[\sum_{v=1}^{q^{k+1}} p_v - q^k\sum_{i=1}^{q} r_i\right] = 0 .$$

Here, the final equality follows from $\sum_{v=1}^{q^{k+1}} p_v = \sum_{i=1}^{q} q^k r_i$.

If $u_2, v \in \{q^{k+1}+1, \ldots, q^n\}$ and $u_1 \in \{1, \ldots, q^{k+1}\}$, we see that $L_{u_1u_2,v} = L_{1u_2,v}$ and $p_{u_2} = \tilde{p}_{u_2}$ as well as $p_v = \tilde{p}_v$. Thus, proceeding as before we have that $R_4 = 0$. If $u_1, u_2 \in \{1, \ldots, q^{k+1}\}$ and $v \in \{q^{k+1}+1, \ldots, q^n\}$, it follows that $L_{u_1u_2,v} = L_{11,v}$, which is denoted by L_v, and $p_v = \tilde{p}_v$. With this we get

$$R_2 = \sum_{u_1,u_2=1}^{q^{k+1}} \sum_{v=q^{k+1}+1}^{q^n} L_v p_v\left[p_{u_1}p_{u_2} - \frac{1}{q^2}\left(\sum_{i=1}^{q} r_i\right)^2\right]$$

$$= \sum_{v=q^{k+1}+1}^{q^n} L_v p_v\left[\sum_{u_1,u_2=1}^{q^{k+1}} p_{u_1}p_{u_2} - q^{2k}\left(\sum_{i=1}^{q} r_i\right)^2\right] = 0 .$$

Here, $\sum_{u_1,u_2=1}^{q^{k+1}} p_{u_1}p_{u_2} = \left(\sum_{i=1}^{q} q^k r_i\right)^2$ yields the final equality. Altogether we end up with

$$\mathcal{L}^{L,q}_{\mathcal{C}_{q^n}}(P,P) - \mathcal{L}^{L,q}_{\mathcal{C}_{q^n}}(\tilde{P},\tilde{P}) = R_1 + R_3 .$$

We begin with R_3. Similar as before we get $L_{u_1u_2,v} = L_{u_11,v}$, which we denote by $L_{u_1,v}$, and $p_{u_2} = \tilde{p}_{u_2}$ if $u_1, v \in \{1, \ldots, q^{k+1}\}$ and $u_2 \in \{q^{k+1}+1, \ldots, q^n\}$. We obtain

$$\frac{1}{2}R_3 = \sum_{u_1,v=1}^{q^{k+1}} \sum_{u_2=q^{k+1}+1}^{q^n} L_{u_1,v}p_{u_2}\left[p_{u_1}p_v - \frac{1}{q^2}\left(\sum_{i=1}^{q} r_i\right)^2\right]$$

$$= \sum_{u_1,v=1}^{q^{k+1}} L_{u_1,v}\left[p_{u_1}p_v - \frac{1}{q^2}\left(\sum_{i=1}^{q} r_i\right)^2\right] \sum_{u_2=q^{k+1}+1}^{q^n} p_{u_2}$$

$$= (1 - q^k\sum_{i=1}^{q} r_i)\sum_{u,v=1}^{q^{k+1}} L_{u,v}\left[p_u p_v - \frac{1}{q^2}\left(\sum_{i=1}^{q} r_i\right)^2\right] .$$

We set $A = \sum_{u,v=1}^{q^{k+1}} L_{u,v}\left[p_u p_v - \frac{1}{q^2}\left(\sum_{i=1}^{q} r_i\right)^2\right]$ and separate the different areas in which u and v can occur. We get

$$A = \sum_{s,t=1}^{q} \sum_{u=(s-1)q^k+1}^{sq^k} \sum_{v=(t-1)q^k+1}^{tq^k} L_{u,v}\left[r_s r_t - \frac{1}{q^2}\left(\sum_{i=1}^{q} r_i\right)^2\right] .$$

It holds for $s,t \in \{1,\dots,q\}$, $u \in \{(s-1)q^k+1,\dots,sq^k\}$ and $v \in \{(t-1)q^k+1,\dots,tq^k\}$ that

$$L_{u,v} = \begin{cases} n-k & \text{if } s \neq t \\ n-k+\mathcal{L}^{1,q}_{\mathcal{C}_{q^k}}(u,v) & \text{if } s = t \ . \end{cases}$$

Thus, the above equation becomes

$$A = (n-k)\left[\sum_{s,t=1}^{q} q^{2k} r_s r_t - q^{2k}\left(\sum_{i=1}^{q} r_i\right)^2\right]$$

$$+\sum_{s=1}^{q}\sum_{u,v=1}^{q^k} \mathcal{L}^{1,q}_{\mathcal{C}_{q^k}}(u,v)\left[r_s^2 - \tfrac{1}{q^2}\left(\sum_{i=1}^{q} r_i\right)^2\right]$$

$$= \tfrac{1}{2q}\sum_{i,j=1}^{q}(r_i-r_j)^2 \sum_{u,v=1}^{q^k} \mathcal{L}^{1,q}_{\mathcal{C}_{q^k}}(u,v) \ .$$

The second equation follows on the one hand from $\sum_{s,t=1}^{q} r_s r_t = \left(\sum_{i=1}^{q} r_i\right)^2$. From this follows that the first summand is 0. On the other hand

$$\sum_{s=1}^{q} r_s^2 - \frac{1}{q}\left(\sum_{i=1}^{q} r_i\right)^2 = \frac{1}{2q}\sum_{i,j=1}^{q}(r_i-r_j)^2 \ .$$

By applying Corollary 1 we obtain

$$\sum_{u,v=1}^{q^k} \mathcal{L}^{1,q}_{\mathcal{C}_{q^k}}(u,v) = q^k \sum_{l=1}^{k} l|\mathcal{R}^{1,q}_{\mathcal{C}_{q^k}}(q^k,l,1)|$$

$$= q^k\left[\sum_{l=1}^{k-1} l q^{k-l}(q-1) + kq\right]$$

$$= q^k\left[q^k(q-1)\sum_{l=1}^{k} l q^{-l} + k\right]$$

$$= q^k\left[q^k(q-1)\tfrac{q(q^k-1)-k(q-1)}{q^k(q-1)^2} + k\right]$$

$$= \tfrac{q}{q-1} q^k(q^k-1) \ .$$

Putting all this together we get

$$R_3 = \frac{1}{q-1} q^k(q^k-1)(1-q^k\sum_{i=1}^{q} r_i)\sum_{i,j=1}^{q}(r_i-r_j)^2 \geq 0 \ .$$

This equals 0 if and only if either $k=0$ or $r_i = r_j$ for all $i,j \in \{1,\dots,q\}$ or $\sum_{i=1}^{q} r_i = q^{-k}$. The last condition is equivalent to $p_i = 0$ for all $i \in \{q^{k+1}+1,\dots,q^n\}$.

We now turn to R_1. With the same notation as before we have

$$R_1 = \sum_{u_1,u_2,v=1}^{q^{k+1}} L_{u_1u_2,v}\left[p_{u_1}p_{u_2}p_v - \tfrac{1}{q^3}\left(\sum_{i=1}^{q} r_i\right)^3\right]$$

$$= \sum_{s_1,s_2,t=1}^{q}\sum_{r=1}^{2}\sum_{u_r=(s_r-1)q^k+1}^{s_rq^k}\sum_{v=(t-1)q^k+1}^{tq^k} L_{u_1u_2,v}\left[r_{s_1}r_{s_2}r_t - \tfrac{1}{q^3}\left(\sum_{i=1}^{q} r_i\right)^3\right]$$

$$= \sum_{s_1,s_2,t=1}^{q}\left[r_{s_1}r_{s_2}r_t - \tfrac{1}{q^3}\left(\sum_{i=1}^{q} r_i\right)^3\right]\sum_{r=1}^{2}\sum_{u_r=(s_r-1)q^k+1}^{s_rq^k}\sum_{v=(t-1)q^k+1}^{tq^k} L_{u_1u_2,v}\ .$$

For $u_r \in \{(s_r-1)q^k+1,\ldots,s_rq^k\}$ and $v \in \{(t-1)q^k+1,\ldots,tq^k\}$ it holds that

$$L_{u_1u_2,v} = \begin{cases} n-k & \text{if } s_1 \neq t \text{ and } s_2 \neq t \\ n-k+\mathcal{L}^{1,q}_{\mathcal{C}_{q^k}}(u_1,v) & \text{if } s_1 = t \text{ and } s_2 \neq t \\ n-k+\mathcal{L}^{1,q}_{\mathcal{C}_{q^k}}(u_2,v) & \text{if } s_1 \neq t \text{ and } s_2 = t \\ n-k+\mathcal{L}^{2,q}_{\mathcal{C}_{q^k}}((u_1,u_2),v) & \text{if } s_1 = s_2 = t\ . \end{cases}$$

If we insert the above equations into R_1, we get

$$R_1 = (n-k)\left[\sum_{s_1,s_2,t=1}^{q} q^{3k}r_{s_1}r_{s_2}r_t - q^{3k}\left(\sum_{i=1}^{q} r_i\right)^3\right]$$

$$+\sum_{s_1=1}^{q}\sum_{s_2=1,s_2\neq s_1}^{q} q^k \sum_{u_1,v=1}^{q^k}\mathcal{L}^{1,q}_{\mathcal{C}_{q^k}}(u_1,v)\left[r_{s_1}^2 r_{s_2} - \tfrac{1}{q^3}\left(\sum_{i=1}^{q} r_i\right)^3\right]$$

$$+\sum_{s_2=1}^{q}\sum_{s_1=1,s_1\neq s_2}^{q} q^k \sum_{u_2,v=1}^{q^k}\mathcal{L}^{1,q}_{\mathcal{C}_{q^k}}(u_2,v)\left[r_{s_1} r_{s_2}^2 - \tfrac{1}{q^3}\left(\sum_{i=1}^{q} r_i\right)^3\right]$$

$$+\sum_{s=1}^{q}\sum_{u_1,u_2,v=1}^{q^k}\mathcal{L}^{2,q}_{\mathcal{C}_{q^k}}((u_1,u_2),v)\left[r_s^3 - \tfrac{1}{q^3}\left(\sum_{i=1}^{q} r_i\right)^3\right]$$

$$= 2q^k\left[\sum_{s=1}^{q}\sum_{t=1,t\neq s}^{q} r_s^2 r_t - \tfrac{q-1}{q^2}\left(\sum_{i=1}^{q} r_i\right)^3\right]\sum_{u,v=1}^{q^k}\mathcal{L}^{1,q}_{\mathcal{C}_{q^k}}(u,v)$$

$$+\left[\sum_{s=1}^{q} r_s^3 - \tfrac{1}{q^2}\left(\sum_{i=1}^{q} r_i\right)^3\right]\sum_{u_1,u_2,v=1}^{q^k}\mathcal{L}^{2,q}_{\mathcal{C}_{q^k}}((u_1,u_2),v)\ .$$

If all r_i's are zero, we obtain $R_1 = 0$. We exclude this case and normalize the probabilities $r_1, ..., r_q$ by setting $\bar{r}_i = r_i / \sum_{j=1}^{q} r_j$ for $i \in \{1, \ldots, q\}$. This yields

$$R_1 = \left(\sum_{i=1}^{q} r_i\right)^3 \left[2q^k \left(\sum_s \sum_{t \neq s} \bar{r}_s^2 \bar{r}_t - \tfrac{q-1}{q^2}\right) \sum_{u,v=1}^{q^k} \mathcal{L}^{1,q}_{\mathcal{C}_{q^k}}(u,v) \right.$$
$$\left. + \left(\sum_s \bar{r}_s^3 - \tfrac{1}{q^2}\right) \sum_{u_1,u_2,v=1}^{q^k} \mathcal{L}^{2,q}_{\mathcal{C}_{q^k}}((u_1,u_2),v) \right] .$$

We have already seen during the calculations of R_3 that

$$\sum_{u,v=1}^{q^k} \mathcal{L}^{1,q}_{\mathcal{C}_{q^k}}(u,v) = \frac{q}{q-1} q^k (q^k - 1) .$$

By applying Corollary 1 we further get that

$$\sum_{u_1,u_2,v=1}^{q^k} \mathcal{L}^{2,q}_{\mathcal{C}_{q^k}}((u_1,u_2),v) = q^k \sum_{l=1}^{k} l |\mathcal{R}^{2,q}_{\mathcal{C}_{q^k}}(q^k, l, 1)|$$

$$= q^k \left[\sum_{l=1}^{k-1} l q^{2k} \left(2q^{-l}(q-1)(1-q^{-l+1}) + q^{-2l}(q-1)^2\right) \right]$$
$$+ q^k k \left(2q(q^k - q) + q^2\right)$$

$$= q^k \left[\sum_{l=1}^{k} l q^{2k} \left(2q^{-l}(q-1)(1-q^{-l+1}) + q^{-2l}(q-1)^2\right) \right]$$
$$+ q^k k (2q^k - 1)$$

$$= (q-1) q^{3k} \left[2 \sum_{l=1}^{k} l q^{-l} - (q+1) \sum_{l=1}^{k} l q^{-2l} \right]$$
$$+ k q^k (2q^k - 1)$$

$$= (q-1) q^{3k} \left[2 \tfrac{q(q^k-1) - k(q-1)}{q^k (q-1)^2} - (q+1) \tfrac{q^2(q^{2k}-1) - k(q^2-1)}{q^{2k}(q^2-1)^2} \right]$$
$$+ k q^k (2q^k - 1)$$

$$= 2 \tfrac{q}{q-1} q^{2k} (q^k - 1) - \tfrac{q^2}{q^2-1} q^k (q^{2k} - 1)$$

$$= \tfrac{q}{q-1} q^k (q^k - 1) \tfrac{(q+2)q^k - q}{q+1} .$$

Applying this result we obtain

$$R_1 = \left(\sum_{i=1}^{q} r_i\right)^3 \tfrac{q}{q-1} q^k(q^k-1)\left[2q^k\left(\sum_s \bar{r}_s^2 - \sum_s \bar{r}_s^3 - \tfrac{q-1}{q^2}\right)\right.$$
$$\left. + \tfrac{(q+2)q^k-q}{q+1}\left(\sum_s \bar{r}_s^3 - \tfrac{1}{q^2}\right)\right]$$
$$= -\left(\sum_{i=1}^{q} r_i\right)^3 \tfrac{q}{q-1} q^k(q^k-1)\left[\tfrac{q}{q+1}(q^k+1)\sum_s \bar{r}_s^3 - 2q^k\sum_s \bar{r}_s^2\right.$$
$$\left. + \tfrac{(2q+1)q^k-1}{q(q+1)}\right] .$$

It remains to show that

$$\frac{q}{q+1}(q^k+1)\sum_s \bar{r}_s^3 - 2q^k\sum_s \bar{r}_s^2 + \frac{(2q+1)q^k-1}{q(q+1)} \le 0 .$$

The left hand side obviously equals 0 if $\bar{r}_1 = ... = \bar{r}_q = 1/q$, i.e. $r_1 = ... = r_q$. Let us define $f : \Delta_{q-1} \to \mathbb{R}$ by

$$f(x_1, ..., x_{q-1}) = a_1\left[\sum_{s=1}^{q-1} x_s^3 + \left(1 - \sum_{s=1}^{q-1} x_s\right)^3\right] - a_2\left[\sum_{s=1}^{q-1} x_s^2 + \left(1 - \sum_{s=1}^{q-1} x_s\right)^2\right] ,$$

where $a_1 = q(q^k+1)/(q+1)$ and $a_2 = 2q^k$. We will show that $(1/q, ..., 1/q)$ is the only extremal point of f in Γ_q and that it is a local maximum. The first partial derivative for $j \in \{1, \ldots, q-1\}$ is

$$\tfrac{\delta}{\delta x_j} f(x_1, ..., x_{q-1}) = 3a_1\left(x_j^2 - (1 - \sum_{i=1}^{q-1} x_i)^2\right) - 2a_2\left(x_j - (1 - \sum_{i=1}^{q-1} x_i)\right)$$
$$= \left(x_j - (1 - \sum_{i=1}^{q-1} x_i)\right)\left[3a_1\left(x_j + 1 - \sum_{i=1}^{q-1} x_i\right) - 2a_2\right] .$$

It follows that the gradient $\nabla f = \mathbf{0}$ if and only if either $x_j = 1 - \sum_{i=1}^{q-1} x_i$ for all $j \in \{1, \ldots, q-1\}$, which yields $x_1 = ... = x_{q-1} = 1/q$, or $3a_1(x_j + 1 - \sum_{i=1}^{q-1} x_i) - 2a_2 = 0$ for all $j \in \{1, \ldots, q-1\}$. Since

$$3a_1(1 - \sum_{i=1, i\neq j}^{q-1} x_i) - 2a_2 \le 3\frac{q}{q+1}(q^k+1) - 4q^k < -q^k + 3 \le 0 ,$$

the latter is impossible. We conclude that the only extremal point of f is $(1/q, .., 1/q)$. Further, the second partial derivatives are

$$\frac{\delta^2}{\delta x_k \delta x_j} f(x_1, ..., x_{q-1}) = \begin{cases} 6a_1(1 - \sum_{i=1}^{q-1} x_i) - 2a_2 & \text{if } k \neq j \\ 6a_1(1 - \sum_{i=1, i\neq j}^{q-1} x_i) - 4a_2 & \text{if } k = j \end{cases}$$

such that

$$\frac{\delta^2}{\delta x_k \delta x_j} f\left(\frac{1}{q}, .., \frac{1}{q}\right) = \begin{cases} \frac{6a_1}{q} - 2a_2 & \text{if } k \neq j \\ \frac{12a_1}{q} - 4a_2 & \text{if } k = j \end{cases} .$$

Since $(6a_1/q) - 2a_2 = [6(q^k + 1)/(q + 1)] - 4q^k \leq -2(q^k - 1) < 0$, we see that $(1/q, .., 1/q)$ is a global maximum. With this we obtain that $R_1 \geq 0$, with equality if and only if either $k = 0$ or $r_i = r_j$ for all $i, j \in \{1, \ldots, q\}$. Remember that $R_3 \geq 0$. It equals zero if and only if either $k = 0$ or $r_i = r_j$ for all $i, j \in \{1, \ldots, q\}$ or $p_i = 0$ for $i \in \{q^{k+1} + 1, \ldots, q^n\}$. Further, $\mathcal{L}^{2,q}_{\mathcal{C}_{q^n}}(P, P) - \mathcal{L}^{2,q}_{\mathcal{C}_{q^n}}(\tilde{P}, \tilde{P}) = R_1 + R_3$. It follows that this difference is not negative. Moreover, it equals 0 if and only if either $k = 0$ or $r_i = r_j$ for all $i, j \in \{1, \ldots, q\}$. This concludes the proof. □

Optimal Rate Region of Two-Hop Multiple Access Channel via Amplify-and-Forward Scheme

Binyue Liu and Ning Cai

State Key Lab. of ISN, Xidian University, Xi'an 710071, China
{liuby,caining}@mail.xidian.edu.cn

Dedicated to the memory of Rudolf Ahlswede

Abstract. We study a two-hop multiple access channel (MAC), where two source nodes communicate with the destination node via a set of amplify-and-forward (AF) relays. To characterize the optimal rate region, we focus on deriving the boundary points of it, which is formulated as a weighted sum rate maximization problem. In the first part, we are concerned with the scenario that all relays are under a sum power constraint. Although the optimal AF rate region for the case has been obtained, we revisit the results by an alternative method. The first step is to investigate the algebraic structures of the three SNR functions in the rate set of the two-hop MAC with a specific AF scheme. Then an equivalent optimization problem is established for deriving each boundary point of the optimal rate region. From the geometric perspective, the problem has a simple solution by optimizing a one-dimensional problem *without* constraint. In the second part, the optimal rate region of a two-hop MAC under the individual power constraints is discussed, which is still an open problem. An algorithm is proposed to compute the maximum individual and sum rates along with the corresponding AF schemes.

Keywords: multiple access channel, amplify-and-forward, achievable rate region.

1 Introduction

Since the introduction of the amplify-and-forward (AF) relay scheme, it has been studied in the context of cooperative communication [5,6,11,17]. It is an interesting technique from the practical standpoint because the complexity and cost of relaying, always an issue in designing cooperative networks, is minimal for AF relay networks. Furthermore, as the simplest coding scheme, the optimal AF rate can be viewed as a lower bound to the network capacity. In addition to its simplicity, AF is known to be the optimal relay strategy in many interesting cases [7,8,16].

H. Aydinian, F. Cicalese, and C. Deppe (Eds.): Ahlswede Festschrift, LNCS 7777, pp. 44–70, 2013.

Finding the optimal performance of the AF relay scheme has recently received tremendous attention from the research community. For a two-hop parallel relay network, Marić and Yates [14] have found the optimal AF relay scheme in closed-form along with the maximum achievable rate under the sum power constraint. To generalize the result, Gomadam and Jafar [9] consider the case when the relay nodes introduce the correlated Gaussian noises. With a similar approach, they have found the optimal AF relay scheme for this scenario and have pointed out the influence of the correlation between noises on the end-to-end performance. In [15], Marić *et al.* have studied a multi-hop AF scheme for a layered relay network under the individual power constraints in the high-SNR regime. An asymptotical optimal multi-hop AF scheme is obtained with each relay node amplifying the received signal to the maximum possible value. Later, Liu and Cai [12] characterize an asymptotical behavior of another suboptimal multi-hop AF scheme in the generalized high-SNR regime. In the proposed scheme, not all the relays transmit with the maximum possible powers. The results have shown that the achievable rate of the new scheme can approach the upper bound in the generalized high-SNR regime. Recently, Agnihotri *et al.* [2] showed that by optimizing the sum rate between the adjacent two layers, the optimal multi-hop AF scheme for a layered relay network in general SNR regime can be obtained. However, to derive the maximum sum rate itself is computational intractable. Therefore, it is no surprising that the maximum achievable AF rate for general unicast networks is still an open problem. Many works have focused on this issue [1] and [13]. For a two-hop relay network with a single source-destination pair, Agnihotri *et al.* [3] obtained the optimal AF scheme via an iterative algorithm. The problem can also be solved by the approach proposed in this paper, but the method is completely different from the one in [3]. Specifically, in each iteration cycle of the algorithm [3], the solution should be updated via solving a sequence of non-linear equations until the resulting AF scheme is feasible. However, in the algebraic approach proposed in our paper, the solution candidates are obtained via solving a system of *linear* equations. Furthermore, the approach [3] cannot be directly extended to solve the problem considered in this paper.

Employing AF relay scheme in a multiuser scenario is also an interesting topic in recent research community. The optimal AF rate region of a two-hop multiple access channel (MAC) has been studied in [10]. With a specific AF scheme, the two-hop MAC is equivalent to the conventional MAC, which has first been studied in a celebrated work by Ahlswede [4]. Owing to his outstanding contribution to the MAC capacity [4], the achievable rate set corresponding to a specific AF scheme of the two-hop MAC can be directly obtained. Under a sum power constraint, Jafar *et al.* [10] fully characterized the optimal rate region by obtaining all the boundary points via solving a weighted sum rate maximization problem. The Lagrange's method is used in deriving the results, however, it may be hard to explore the key insight of the problem. Moreover, the individual power constraint case is not considered.

Motivated by this work, we also study a two-hop MAC with AF relays either under sum or individual power constraint. The main contributions are given as follows.

In the first part, we revisit the optimal rate region of the two-hop MAC with sum power constraint. It is first formulated as a weighted sum rate maximization problem. We propose a novel technique to solve it, which is motivated by our previous work [13]. The algebraic structure of the object function provides an opportunity to simplify the form of the problem. The key step is to convert the constraint optimization problem into an *unconstraint* one and the dimension of the problem is reduced from n to one. Then it can be easily solved. The method also gives an intuitive interpretation of each AF scheme from the geometric perspective. In particular, the maximum sum rate and individual rates are obtained in closed-form. The other boundary points can be determined by solving an equation numerically according to different cases.

In the second part, we study the AF rate region under individual power constraints. In our previous work [13], an outer bound has been obtained. In two special scenarios, either the upper layer noise or the lower layer noise dominates the total noises received at the destination node, we have proposed two mixed AF schemes, with which two inner bounds are obtained. The results have shown that each of the maximum sum rate and individual rates can achieve the outer bound within half a bit. However, to the best of our knowledge, the optimal rate region have not been fully characterized so far. We work progressively toward this goal and provide an algorithm to compute the AF schemes achieving the maximum sum and individual rates.

Notation: Scalars are denoted by lower-case letters, e.g., x, and bold-face lower-case letters are used for vectors, e.g., $\mathbf{x}$, and bold-face upper-case letters for matrices, e.g., $\mathbf{X}$. In addition, $\mathbf{X}^T$, $\mathbf{X}^{-1}$ and $\mathrm{tr}(\mathbf{X})$ denote the transpose, inverse and trace of $\mathbf{X}$, respectively, and $\mathrm{diag}(\mathrm{x}_1, \cdots, \mathrm{x}_\mathrm{n})$ denotes a block-diagonal square matrix with $x_1, \cdots, x_n$ as the diagonal elements. $||\mathbf{x}||$ denotes the Euclidean norm of a vector $\mathbf{x}$. $E[\cdot]$ is the expectation operation. $\log(\cdot)$ denotes the logarithm in the base 2 and $\ln(\cdot)$ denotes the natural logarithm.

2 Network Model

We study the optimal rate region of a wireless two-hop MAC with AF relays. A two-source case is considered in the paper, which is depicted in Fig. 1. The two source nodes S_1 and S_2 transmit the signals with fixed powers P_{S_1} and P_{S_2} respectively. Assume no direct path from the source nodes to the destination node appears. So, the destination node D can only receive the signals from the relays. With the help of n AF relays, the two sources communicate with the destination. Each relay node works in a full duplex mode. It is also assumed that each non-source node introduces an i.i.d. Gaussian noise with unit variance and zero mean. In the scope of this paper, all channel gains are real-valued numbers and remain constant during the transmission. Perfect channel state information are known through the network. Furthermore, the channel gains between the

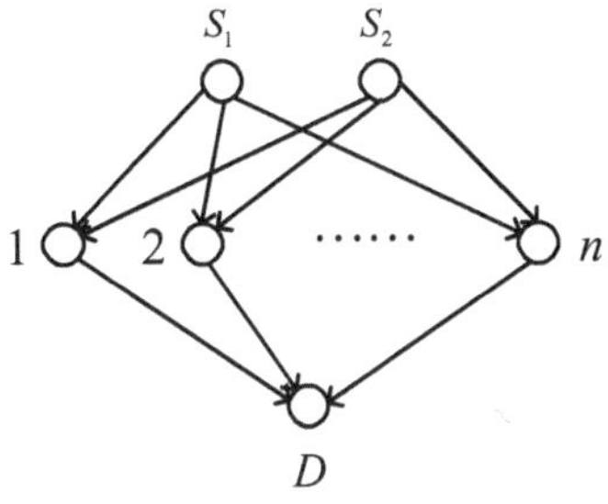

Fig. 1. Two-source two-hop multiple access channel

relays and the destination and the channel gains between the source nodes and the relays are supposed to be positive and non-negative respectively. Note that if there is a channel gain between the relay and the destination node equal to zero, it implies that the relay cannot transmit anything to the destination node. So, we can omit this relay in the network. Therefore, the non-zero assumption is without loss of generality in our system model. Note that the assumption of non-negative channel gains is only used to simplify the discussion. The techniques proposed below directly apply to the general case as well.

Denote by a_k the amplitude of the transmitting signal at relay node k, $k \in \{1, \cdots, n\} \stackrel{\Delta}{=} \mathcal{V}$, which is also a real-valued number. Thus relay node k retransmits the received noisy signal with power $a_k^2 = P_k$. $\mathbf{a} = [a_1 \cdots a_k \cdots a_n]^T$ is referred to as a specific AF scheme. We consider two different assumptions on the relay power constraint in the paper. In the first scenario, we assume that all the relays are subject to a sum power constraint P_{sum}, i.e.,

$$\sum_{k \in \mathcal{V}} a_k^2 = \sum_{k \in \mathcal{V}} P_k \leq P_{sum}. \tag{1}$$

It is equivalent to a power allocation scheme among relays. Therefore we do not distinguish them in the sequel. In the second scenario, we assume that each relay k is subject to an individual power constraint $P_{k,max}$, i.e.,

$$a_k^2 \leq P_{k,max}, k \in \mathcal{V}. \tag{2}$$

In this scenario, each relay selects the transmitting power individually. Due to the channel assumption, we only consider the non-negative part of the amplitude constraint according to the corresponding power constraint. The constraint may be changed to the non-positive part if we get rid of the non-negative assumption.

The notation $\{\mathbf{a}\}$ is used to indicate all feasible AF schemes that are allowed in both scenarios. With the superposition property of the wireless channel, the received signal at each relay is the linear combination of the signals from the two sources and the Gaussian noise. Therefore, the signal received at node k can be expressed as

$$y_k = h_{S_1,k} x_{S_1} + h_{S_2,k} x_{S_2} + z_k, k \in \mathcal{V}, \tag{3}$$

where x_{S_i}, $i = 1, 2$, is the transmitting signal from source node S_i, $i = 1, 2$, $h_{S_i,k}$, $i = 1, 2$ denotes the channel gain from S_i to relay k and z_k denotes the

Gaussian noise introduced by relay k. With a specific AF scheme $\mathbf{a}$, the signal received at the destination node can be expressed as,

$$\begin{aligned} y_D &= \sum_{k \in \mathcal{V}} \gamma_k a_k \left(h_{S_1,k} x_{S_1} + h_{S_2,k} x_{S_2} + z_k \right) + z_D \\ &= \mathbf{a}^T \mathbf{\Gamma} \left(\mathbf{h}_{11} x_{S_1} + \mathbf{h}_{12} x_{S_2} + \mathbf{z} \right) + z_D, \end{aligned} \quad (4)$$

where $\mathbf{\Gamma} = \text{diag}\{\gamma_1, \cdots, \gamma_n\}$, $\gamma_k = h_{k,D}\sqrt{1/(1 + h_{S_1,k}^2 P_{S_1} + h_{S_2,k}^2 P_{S_2})}$ is a constant related to the network settings, $h_{k,D}$ denotes the channel gain from the relay node k to the destination node, $\mathbf{h}_{1i} = [h_{S_i,1}, \cdots, h_{S_i,n}]^T$, $i = 1, 2$, and $\mathbf{z} = [z_1, \cdots, z_n]^T$. From (4), the two-hop MAC can be considered as a conventional Gaussian MAC with $x_{S_1,eq} = \mathbf{a}^T \mathbf{\Gamma} \mathbf{h}_{11} x_{S_1}$, $x_{S_2,eq} = \mathbf{a}^T \mathbf{\Gamma} \mathbf{h}_{12} x_{S_2}$, and $z_{eq} = \mathbf{a}^T \mathbf{\Gamma} \mathbf{z} + z_D$ as the equivalent signals of the two source nodes and Gaussian noise respectively. To distinguish them, we denote the two-hop MAC with respect to the AF scheme $\mathbf{a}$ as MAC($\mathbf{a}$). As the well-known result of the Gaussian MAC, the source nodes adopt Gaussian codebooks. The independent random variables used to generate the codebooks are $x_{S_1} \sim \mathcal{N}[0, P_{S_1}]$ and $x_{S_2} \sim \mathcal{N}[0, P_{S_2}]$. The codebooks consist of $\lceil 2^{nR_1} \rceil$ and $\lceil 2^{nR_2} \rceil$ codewords of length n respectively, and the decoding error probability tends to zero as $n \to \infty$.

The achievable rate set of MAC($\mathbf{a}$) is shown as follows.

$$\begin{aligned} \mathcal{R}(\mathbf{a}) = \Bigg\{ (R_1, R_2) : R_1 \le \mathcal{C}\left(\frac{(\mathbf{a}^T \mathbf{\Gamma} \mathbf{h}_{11})^2 P_{S_1}}{\|\mathbf{a}^T \mathbf{\Gamma}\|^2 + 1} \right), R_2 \le \mathcal{C}\left(\frac{(\mathbf{a}^T \mathbf{\Gamma} \mathbf{h}_{12})^2 P_{S_2}}{\|\mathbf{a}^T \mathbf{\Gamma}\|^2 + 1} \right), \\ R_1 + R_2 \le \mathcal{C}\left(\frac{(\mathbf{a}^T \mathbf{\Gamma} \mathbf{h}_{11})^2 P_{S_1} + (\mathbf{a}^T \mathbf{\Gamma} \mathbf{h}_{12})^2 P_{S_2}}{\|\mathbf{a}^T \mathbf{\Gamma}\|^2 + 1} \right) \Bigg\} \end{aligned} \quad (5)$$

For notation brevity, $\mathcal{C}(x) = 0.5 \log (1 + x)$ is used to denote the Gaussian capacity formula with SNR x. The optimal rate region is then obtained as the union of all rate regions and denoted by $\mathcal{R}(\{\mathbf{a}\})$.

A commonly used method to characterize different rate-tuples on the boundary of a multiuser capacity region is via solving a sequence of weighted sum rate maximization problems, each for a different nonnegative rate weight vector of two sources (see [13] and also [10]). That is to solve an optimization problem

$$\max \left(\mu_1 R_1 + \mu_2 R_2 \right), \; \mu_1, \mu_2 \ge 0, \quad (6)$$

under the power constraint given by (1) or (2). It is clear that the rate pair $(R_1, R_2) \in \mathbf{R}_+^2$ on the boundary of the union region must be on the boundary of some rate region $\mathcal{R}(\mathbf{a})$ (5). Hence, the problem can be separated into several subproblems. Without loss of generality, assume that μ_1 and μ_2 are normalized in $[0, 1]$, and that $\mu_1 + \mu_2 = 1$. Let $\mu = \mu_1$, $\bar{\mu} = 1 - \mu_1$. Therefore, the maximization can be recast as $\max \mu R_1 + \bar{\mu} R_2$, $\mu \in [0, 1]$. Given a specific AF scheme $\mathbf{a}$, it is easy to check that

- **case 1:** when $\mu = 1/2$, the rate pairs that maximize the sum rate in rate set $\mathcal{R}(\mathbf{a})$ are the solutions of the optimization problem, which satisfy

$$R_1(\mathbf{a}) + R_2(\mathbf{a}) = \mathcal{C}\left(\frac{\left(\mathbf{a}^T\mathbf{\Gamma}\mathbf{h}_{11}\right)^2 P_{S_1} + \left(\mathbf{a}^T\mathbf{\Gamma}\mathbf{h}_{12}\right)^2 P_{S_2}}{\|\mathbf{a}^T\mathbf{\Gamma}\|^2+1}\right). \quad (7)$$

- **case 2:** when $\mu = 1$, the rate pairs that maximize the individual rate R_1 in rate set $\mathcal{R}(\mathbf{a})$ are the solutions of the optimization problem, which satisfy

$$\begin{aligned} R_1(\mathbf{a}) &= \mathcal{C}\left(\frac{\left(\mathbf{a}^T\mathbf{\Gamma}\mathbf{h}_{11}\right)^2 P_{S_1}}{\|\mathbf{a}^T\mathbf{\Gamma}\|^2+1}\right), \\ R_2(\mathbf{a}) &\le \mathcal{C}\left(\frac{\left(\mathbf{a}^T\mathbf{\Gamma}\mathbf{h}_{11}\right)^2 P_{S_1} + \left(\mathbf{a}^T\mathbf{\Gamma}\mathbf{h}_{12}\right)^2 P_{S_2}}{\|\mathbf{a}^T\mathbf{\Gamma}\|^2+1}\right) - \mathcal{C}\left(\frac{\left(\mathbf{a}^T\mathbf{\Gamma}\mathbf{h}_{11}\right)^2 P_{S_1}}{\|\mathbf{a}^T\mathbf{\Gamma}\|^2+1}\right). \end{aligned} \quad (8)$$

- **case 3:** when $\mu = 0$, the rate pairs that maximize the individual rate R_2 in rate set $\mathcal{R}(\mathbf{a})$ are the solutions of the optimization problem, which satisfy

$$\begin{aligned} R_1(\mathbf{a}) &\le \mathcal{C}\left(\frac{\left(\mathbf{a}^T\mathbf{\Gamma}\mathbf{h}_{11}\right)^2 P_{S_1} + \left(\mathbf{a}^T\mathbf{\Gamma}\mathbf{h}_{12}\right)^2 P_{S_2}}{\|\mathbf{a}^T\mathbf{\Gamma}\|^2+1}\right) - \mathcal{C}\left(\frac{\left(\mathbf{a}^T\mathbf{\Gamma}\mathbf{h}_{12}\right)^2 P_{S_2}}{\|\mathbf{a}^T\mathbf{\Gamma}\|^2+1}\right), \\ R_2(\mathbf{a}) &= \mathcal{C}\left(\frac{\left(\mathbf{a}^T\mathbf{\Gamma}\mathbf{h}_{12}\right)^2 P_{S_2}}{\|\mathbf{a}^T\mathbf{\Gamma}\|^2+1}\right). \end{aligned} \quad (9)$$

- **case 4:** when $\mu \in \left(0, \frac{1}{2}\right)$, the rate pair at the "upper-diagonal" corner point of the pentagon region in rate set $\mathcal{R}(\mathbf{a})$ is the solution of the optimization problem, i.e.,

$$\begin{aligned} R_1(\mathbf{a}) &= \mathcal{C}\left(\frac{\left(\mathbf{a}^T\mathbf{\Gamma}\mathbf{h}_{11}\right)^2 P_{S_1} + \left(\mathbf{a}^T\mathbf{\Gamma}\mathbf{h}_{12}\right)^2 P_{S_2}}{\|\mathbf{a}^T\mathbf{\Gamma}\|^2+1}\right) - \mathcal{C}\left(\frac{\left(\mathbf{a}^T\mathbf{\Gamma}\mathbf{h}_{12}\right)^2 P_{S_2}}{\|\mathbf{a}^T\mathbf{\Gamma}\|^2+1}\right), \\ R_2(\mathbf{a}) &= \mathcal{C}\left(\frac{\left(\mathbf{a}^T\mathbf{\Gamma}\mathbf{h}_{12}\right)^2 P_{S_2}}{\|\mathbf{a}^T\mathbf{\Gamma}\|^2+1}\right). \end{aligned} \quad (10)$$

- **case 5:** when $\mu \in \left(\frac{1}{2}, 1\right)$, the rate pair at the llower-diagonalcorner point of the pentagon region in rate set $\mathcal{R}(\mathbf{a})$ is the solution of the optimization problem, i.e.,

$$\begin{aligned} R_1(\mathbf{a}) &= \mathcal{C}\left(\frac{\left(\mathbf{a}^T\mathbf{\Gamma}\mathbf{h}_{11}\right)^2 P_{S_1}}{\|\mathbf{a}^T\mathbf{\Gamma}\|^2+1}\right), \\ R_2(\mathbf{a}) &= \mathcal{C}\left(\frac{\left(\mathbf{a}^T\mathbf{\Gamma}\mathbf{h}_{11}\right)^2 P_{S_1} + \left(\mathbf{a}^T\mathbf{\Gamma}\mathbf{h}_{12}\right)^2 P_{S_2}}{\|\mathbf{a}^T\mathbf{\Gamma}\|^2+1}\right) - \mathcal{C}\left(\frac{\left(\mathbf{a}^T\mathbf{\Gamma}\mathbf{h}_{11}\right)^2 P_{S_1}}{\|\mathbf{a}^T\mathbf{\Gamma}\|^2+1}\right). \end{aligned} \quad (11)$$

Substituting the above results into the object function of (6) according to different μ, a set of subproblems are formulated, e.g., if $\mu = \frac{1}{2}$, then the problem is cast as $\max_{\{\mathbf{a}\}} \frac{1}{2}(R_1(\mathbf{a}) + R_2(\mathbf{a}))$, where $(R_1(\mathbf{a}), R_2(\mathbf{a}))$ satisfies (7) and the feasible region $\{\mathbf{a}\}$ are determined by the power constraint either (1) or (2). By standard Lagrange's method, the set of subproblems have been solved for the sum power constraint scenario as given in [10]. However, it is observed that the derivative operations make the problem complicated to be handled. For the individual constraints case, the issue has not been solved yet but an asymptotical analysis can be found in [13]. In the rest of the paper, the focus is put on finding an alternative method to solve the subproblems, which is based on the algebraic structures of the object functions of them. With such method, the complicated derivative operations as used in Lagrange's method are avoided.

3 Optimal Rate Region of Two-Hop MAC via AF

In this section, we propose a novel technique to investigate the optimal rate region of the two-hop MAC via AF scheme. The basic idea is as follows. We first relax the original optimization problem by removing some of the constraints, which makes the problem easy to be handled. Then a set of solutions for the relaxed problem is found. Finally, it follows that there exists a specific solution in the set satisfying all the constraints of the original problem. Therefore, the optimal solution of the original problem is obtained.

3.1 Sum Power Constraint

Now let us come to the first topic, that is the relays of the two-hop MAC are under a sum power constraint. The weighted sum maximization problem is given as

$$\begin{array}{ll} \max & \mu R_1(\mathbf{a}) + \bar{\mu} R_2(\mathbf{a}) \\ \text{s.t.} & \sum\limits_{k\in\mathcal{V}} a_k^2 \le P_{sum} \end{array} . \tag{12}$$

We first claim that to obtain the optimal rate region, the power constraint (1) should always take the equality. This can be proved as follows. For an AF scheme $\mathbf{a}_0 = [a_{01}, \cdots, a_{0n}]^T$ such that $\sum_{k\in\mathcal{V}} a_{0k}^2 < P_{sum}$, it can be found a constant $c = \sqrt{P_{sum}/\sum_{k\in\mathcal{V}} a_{0k}^2} > 1$ such that $\mathbf{a} = c\mathbf{a}_0$ is a new scheme with $c^2 \sum_{k\in\mathcal{V}} a_{0k}^2 = P_{sum}$. From (5), it is easy to verify that $\mathcal{R}(\mathbf{a}_0) \subset \mathcal{R}(\mathbf{a})$. Therefore, the boundary points of the union region cannot be in the rate set $\mathcal{R}(\mathbf{a}_0)$. Hence, we only consider the AF schemes with an equality constraint in (1) in the sequel, i.e., $\sum_{k\in\mathcal{V}} a_k^2 = P_{sum}$. So, the optimization problem turns to be

$$\begin{array}{ll} \max & \mu R_1(\mathbf{a}) + \bar{\mu} R_2(\mathbf{a}) \\ \text{s.t.} & \sum\limits_{k\in\mathcal{V}} a_k^2 = P_{sum} \end{array} . \tag{13}$$

With a little abuse of notation, we define a rate set $\mathcal{R}(\mathbf{x})$ given as follows.

$$\mathcal{R}(\mathbf{x}) = \left\{ (R_1, R_2) : R_1 \le \mathcal{C}\left(\frac{\mathbf{x}^T\mathbf{d}_1\mathbf{d}_1^T\mathbf{x}P_{S_1}}{\mathbf{x}^T\mathbf{x}}\right), R_2 \le \mathcal{C}\left(\frac{\mathbf{x}^T\mathbf{d}_2\mathbf{d}_2^T\mathbf{x}P_{S_2}}{\mathbf{x}^T\mathbf{x}}\right), \right.$$
$$\left. R_1 + R_2 \le \mathcal{C}\left(\frac{\mathbf{x}^T\mathbf{d}_1\mathbf{d}_1^T\mathbf{x}P_{S_1} + \mathbf{x}^T\mathbf{d}_2\mathbf{d}_2^T\mathbf{x}P_{S_2}}{\mathbf{x}^T\mathbf{x}}\right) \right\}, \tag{14}$$

where $\mathbf{\Lambda} = \text{diag}\left\{\sqrt{\gamma_1^2 + \frac{1}{P_{sum}}}, \cdots, \sqrt{\gamma_n^2 + \frac{1}{P_{sum}}}\right\}$, $\mathbf{x} = \mathbf{\Lambda a}$, $\mathbf{a} \in \mathbf{R}^n$, and $\mathbf{d}_i = \mathbf{\Lambda}^{-1}\mathbf{\Gamma h}_{1i}$, $i = 1, 2$. Similarly, the notation $\mathcal{R}(\{\mathbf{x}\})$ is used to represent the union of all sets.

We first claim that $\mathcal{R}(\mathbf{a}) = \mathcal{R}(\mathbf{x})$. It is clear that $\mathcal{R}(\{\mathbf{a}\}) \subset \mathcal{R}(\{\mathbf{x}\})$ since there is no constraint on $\mathbf{x}$ in rate set (14). Then we need to show that $\mathcal{R}(\{\mathbf{x}\}) \subset \mathcal{R}(\{\mathbf{a}\})$ also holds. From (14), it follows that for any constant $c \neq 0$, $\mathcal{R}(c\mathbf{x}) = \mathcal{R}(\mathbf{x})$. Therefore, for any given $\mathbf{x}$, we can always find an AF scheme $\mathbf{a} = c_0\mathbf{\Lambda}^{-1}\mathbf{x}$ where the constant c_0 is chosen such that $\|\mathbf{a}\|^2 = P_{sum}$. Then, we can conclude that $\mathcal{R}(\mathbf{a}) = \mathcal{R}(c_0\mathbf{x})$ and thus $\mathcal{R}(\mathbf{a}) = \mathcal{R}(\mathbf{x})$. It suffices to show that $\mathcal{R}(\{\mathbf{x}\}) \subset \mathcal{R}(\{\mathbf{a}\})$ and thus $\mathcal{R}(\{\mathbf{a}\}) = \mathcal{R}(\{\mathbf{x}\})$. Consequently, the rate set $\mathcal{R}(\mathbf{a})$ can be replaced by $\mathcal{R}(\mathbf{x})$.

It can be observed that the SNR functions in (14) are all in the form of generalized Rayleigh quotient. With the algebraic properties of the generalized Rayleigh quotient, the problem (13) can be easily solved. It should be pointed out that the technique used below was first proposed in [14], including the lemmas. We find it very useful to obtain the desired results. To make the paper self-contained, we provide the detailed procedures and give the proofs of the lemmas in the appendices.

Lemma 1. *To obtain the boundary points of* $\mathcal{R}(\{\mathbf{x}\})$ *of the two-hop MAC with AF relays, it is sufficient to take* $\mathbf{x}$ *in the linear subspace* $\text{span}\{\mathbf{d}_1, \mathbf{d}_2\}$.

Proof. The proof is given in the appendix.

It has been proved that $\mathbf{x}$ and $c\mathbf{x}$, $c \neq 0$, correspond to the same rate set (14), i.e., $\mathcal{R}(c\mathbf{x}) = \mathcal{R}(\mathbf{x})$. Therefore, we can always assume that $\mathbf{x}$ is a normalized vector. By absorbing $\|\mathbf{d}_1\|^2$ and $\|\mathbf{d}_2\|^2$ into P_{S_1} and P_{S_2} respectively, it is assumed without loss of generality, $\mathbf{d}_i$, $i = 1, 2$ are normalized vectors as well. To emphasize the differences, they are denoted by $\mathbf{d}_{i0}$, $i = 1, 2$ and $\mathbf{x}_0$ respectively. Moreover, the results for the case when $\mathbf{d}_i$, $i = 1, 2$ are linearly dependent can always be considered as an immediate consequence of those for the independent case. Therefore, the independent assumption is given in the sequel. Choose an orthonormal basis of linear space $\text{span}\{\mathbf{d}_1, \mathbf{d}_2\}$, denoted by $(\mathbf{u}_1, \mathbf{u}_2)$, such that

$$\mathbf{x}_0 = \cos\theta\mathbf{u}_1 + \sin\theta\mathbf{u}_2, \tag{15}$$

$$\mathbf{d}_{10} = \cos\alpha\mathbf{u}_1 + \sin\alpha\mathbf{u}_2, \tag{16}$$

$$\mathbf{d}_{20} = \cos\beta\mathbf{u}_1 + \sin\beta\mathbf{u}_2, \tag{17}$$

where $\cos\theta = \mathbf{x}_0^T \mathbf{u}_1$, $\cos\alpha = \mathbf{d}_{10}^T \mathbf{u}_1$ and $\cos\beta = \mathbf{d}_{20}^T \mathbf{u}_1$. Since all the channel gains are non-negative, it is clear that $\mathbf{d}_{10}^T \mathbf{d}_{20} = \cos(\alpha - \beta) \geq 0$ and thus $|\alpha - \beta| \leq \frac{\pi}{2}$. So, it is convenient to choose $(\mathbf{u}_1, \mathbf{u}_2)$ such that $\alpha, \beta \in \left[0, \frac{\pi}{2}\right]$, which simplifies the following discussion.

By substituting (15)-(17) into (14), the rate set $\mathcal{R}(\mathbf{x})$ can be recast as

$$\begin{aligned} \{(R_1, R_2) : R_1 &\leq \mathcal{C}\left(\phi_1\left(\theta\right)\right), \\ R_2 &\leq \mathcal{C}\left(\phi_2\left(\theta\right)\right), R_1 + R_2 \leq \mathcal{C}(\phi\left(\theta\right))\}, \end{aligned} \quad (18)$$

where $\theta \in [-\pi, \pi]$, and

$$\begin{aligned} \phi_1\left(\theta\right) &= P_{S_1} \cos^2\left(\theta - \alpha\right), \\ \phi_2\left(\theta\right) &= P_{S_2} \cos^2\left(\theta - \beta\right), \\ \phi\left(\theta\right) &= P_{S_1} \cos^2\left(\theta - \alpha\right) + P_{S_2} \cos^2\left(\theta - \beta\right). \end{aligned}$$

To emphasize the expression obtained in (18), later we use $\mathcal{R}\left(\theta\right)$ to denote a specific rate set and $\mathcal{R}\left(\{\theta\}\right)$ to denote the union of them. Note that to fully characterize $\mathcal{R}\left(\{\theta\}\right)$, we can always consider $\alpha, \beta \in \left[0, \frac{\pi}{2}\right]$ without the assumption that all the channel gains are non-negative. Intuitively, this can be interpreted that for any channel vector, we can always find one with all channel gains non-negative such that the rate regions of them are equal. From the geometric perspective, the reason is given as follows. If we drop the assumption, we may have $|\alpha - \beta| > \frac{\pi}{2}$. But it can be observed from (18) that the rate region $\mathcal{R}\left(\theta\right)$ will not be changed if we rotate α (or β) by π and then redefine it as α (or β). Then it follows that $|\alpha - \beta| \leq \frac{\pi}{2}$. Further we can rotate α and β by an angle δ simultaneously such that both α and β are in $\left[0, \frac{\pi}{2}\right]$ and keep the union of the rate region $\mathcal{R}\left(\{\theta\}\right)$ unchanged. So, to characterize the union rate region, the assumption that all the channel gains are non-negative and thus $\alpha, \beta \in \left[0, \frac{\pi}{2}\right]$ is only given for simplification of the proof and is without loss of generality. So far, we have reduced an n-dimensional constraint optimization problem (12) to the following 1-dimensional one *without* constraint.

$$\max_{\theta \in [-\pi, \pi]} \left(\mu R_1(\theta) + \bar{\mu} R_2(\theta)\right). \quad (19)$$

Furthermore, following the previous arguments, the optimal values of the two problems (13) and (19) for each μ are identical. Moreover, we can reconstruct an optimal solution for (13) from the optimal solution for (19). For example, if $\theta^*(\mu)$ is the optimal solution of (19), then the corresponding AF scheme is given as $\mathbf{a}^* = c_0 \mathbf{\Lambda}^{-1}[\mathbf{u}_1 \cos\theta^*(\mu) + \mathbf{u}_2 \sin\theta^*(\mu)]$, where the parameters are given in (14) and (15). Before solving the problem, we further investigate the relationship between different rate sets and find an interesting result with which the scope of θ to be considered is further narrowed. Assume, without loss of generality, $0 \leq \alpha \leq \beta \leq \pi/2$.

Lemma 2. *To obtain the boundary points of $\mathcal{R}\left(\{\theta\}\right)$ of the two-hop MAC with AF relays, it is sufficient to take $\theta \in [\alpha, \beta]$.*

Proof. The proof is given in the appendix.

Then for a fixed μ, the boundary points of $\mathcal{R}(\{\theta\})$ can be derived by solving the problem

$$\max_{\theta\in[\alpha,\beta]} \left(\mu R_1(\theta) + \bar{\mu} R_2(\theta)\right), \tag{20}$$

and the optimal solution is denoted by $\theta^*(\mu)$. Then we work towards the ultimate goal to solve the optimization problem. The results obtained in (7)-(11) can be directly applied here by substituting the Gaussian capacities given in (18). The rate pairs on the boundary which maximize the sum rate and the individual rates of $\mathcal{R}(\{\theta\})$ are first obtained in closed-form. Then two equations will be presented, whose solutions maximize $\mu R_1(\theta) + \bar{\mu} R_2(\theta)$ for $\mu \in \left(0, \frac{1}{2}\right)$ and $\mu \in \left(\frac{1}{2}, 1\right)$ respectively.

Let us start with the maximum sum rate of $\mathcal{R}(\{\theta\})$. The following theorem is established.

Theorem 1. *The maximum value of* R_1+R_2 *of* $\mathcal{R}(\{\theta\})$ *is given by* $\mathcal{C}\left(\phi\left(\theta^*(\frac{1}{2})\right)\right)$*, where* $\theta^*(\frac{1}{2})$ *is shown as follows.*

$$\theta^*(\frac{1}{2}) = \begin{cases} \frac{1}{2}\arctan(x), x \geq 0 \\ \frac{1}{2}\arctan(\pi + x), x < 0 \end{cases} \tag{21}$$

where $x = \dfrac{P_{S_1}\sin 2\alpha + P_{S_2}\sin 2\beta}{P_{S_1}\cos 2\alpha + P_{S_2}\cos 2\beta}$.

Proof. From (18), for any θ, the maximum sum rate $R_1(\theta) + R_2(\theta)$ is upper bounded by $\mathcal{C}(\phi(\theta))$. Since $\mathcal{C}(x)$ is a monotonically increasing function of x, to obtain the maximum of $\mathcal{C}(\phi(\theta))$ is to obtain the maximum of $\phi(\theta)$. By setting the derivative $\phi'(\theta)$ to zero, we have

$$\phi'(\theta) = -P_{S_1}\sin 2(\theta - \alpha) - P_{S_2}\sin 2(\theta - \beta) = 0. \tag{22}$$

Then, it follows that the solution θ_{opt} of the above equation satisfies

$$\tan 2\theta_{opt} = \frac{P_{S_1}\sin 2\alpha + P_{S_2}\sin 2\beta}{P_{S_1}\cos 2\alpha + P_{S_2}\cos 2\beta}. \tag{23}$$

Then we claim that the points that maximize the sum rate are on the boundary of the union rate region. It is easy to see that $\max_{\theta\in[\alpha,\beta]} R_1(\theta)+R_2(\theta)$ and problem (20) with $\mu = 1/2$ has the same optimal solution. Therefore, $\theta^*(\frac{1}{2}) = \theta_{opt}$ and the rate pairs satisfying $R_1 + R_2 = \mathcal{C}\left(\phi\left(\theta^*(\frac{1}{2})\right)\right)$ in rate set $\mathcal{R}(\theta^*(\frac{1}{2}))$ are on the boundary of the union rate region as shown in *case 1* (7) in section 2.

Then we complete the proof.

From (14) and (15), the AF scheme corresponding to $\theta^*(\frac{1}{2})$ is shown as follows.

$$\mathbf{a}^{(10)} = c^{(10)}\mathbf{\Lambda}^{-1}\left(\mathbf{u}_1\cos\theta^*\left(\frac{1}{2}\right) + \mathbf{u}_2\sin\theta^*\left(\frac{1}{2}\right)\right), \tag{24}$$

where the constant $c^{(10)}$ is chosen such that $\|\mathbf{a}^{(10)}\|^2 = P_{sum}$.

Actually, we have also found an alternative approach as observed in [13] to obtain the maximum sum rate in terms of the amplification gains rather than the parameters used in Theorem 1. From (14), the maximum SNR value of the sum rate is equal to the maximum eigenvalue of the matrix $\mathbf{d}_1\mathbf{d}_1^T P_{S_1} + \mathbf{d}_2\mathbf{d}_2^T P_{S_2}$. We first consider the case when $\mathbf{d}_1$ and $\mathbf{d}_2$ are linearly independent, i.e., $\mathbf{d}_1\mathbf{d}_1^T P_{S_1} + \mathbf{d}_2\mathbf{d}_2^T P_{S_2}$ has two non-zero eigenvalues. By Sylvester theorem, the matrix $\mathbf{A} = [\mathbf{d}_1, \mathbf{d}_2]^T[\mathbf{d}_1, \mathbf{d}_2]\mathrm{diag}(\mathrm{P}_{\mathrm{S}_1}, \mathrm{P}_{\mathrm{S}_2})$ has the same non-zero eigenvalues as $\mathbf{d}_1\mathbf{d}_1^T P_{S_1} + \mathbf{d}_2\mathbf{d}_2^T P_{S_2}$, which are denoted by λ_1 and λ_2. Then by the property of the eigenvalue, we have $\lambda_1 + \lambda_2 = \mathrm{tr}(\mathbf{A})$ and $\lambda_1\lambda_2 = \det(\mathbf{A})$. The maximum eigenvalue can be easily solved and the maximum sum rate is given as follows.

$$\mathcal{C}\left(\theta^*\left(\frac{1}{2}\right)\right) = \frac{1}{2}\log\left(1 + \frac{c_2 + \sqrt{c_2^2 - 4c_1}}{2}\right), \tag{25}$$

where $c_1 = P_{S_1}P_{S_2}\left[\|\mathbf{d}_1\|^2\|\mathbf{d}_2\|^2 - (\mathbf{d}_1^T\mathbf{d}_2)^2\right]$ and $c_2 = \left(\|\mathbf{d}_1\|^2 P_{S_1} + \|\mathbf{d}_2\|^2 P_{S_2}\right)$.

Now, let us consider the case when $\mathbf{d}_1$ and $\mathbf{d}_2$ are linearly dependent. The maximum SNR value of the sum rate can be easily obtained since there is only one non-zero eigenvalue of $\mathbf{A}$, denoted by λ. Therefore, $\lambda = \mathrm{tr}(\mathbf{A}) = \|\mathbf{d}_1\|^2 \mathrm{P}_{\mathrm{S}_1} + \|\mathbf{d}_2\|^2 \mathrm{P}_{\mathrm{S}_2}$. From (25), it is easy to see that when $\mathbf{d}_1$ and $\mathbf{d}_2$ are linearly dependent, $c_1 = 0$. So, we conclude that (25) can represent the result for both the cases.

We find the maximum sum rate has exactly the same expression as in [10, Theorem 5]. Then we consider the maximum individual rates of $\mathcal{R}(\{\theta\})$.

Theorem 2. *The maximum value of individual rates R_i, $i = 1, 2$ of $\mathcal{R}(\{\theta\})$ are given by $\mathcal{C}(\phi_1(\alpha))$ and $\mathcal{C}(\phi_2(\beta))$.*

Proof. For any θ, the maximum individual rates $R_1(\theta)$ and $R_2(\theta)$ are upper bounded by $\mathcal{C}(\phi_1(\theta))$ and $\mathcal{C}(\phi_2(\theta))$ respectively. By setting the derivative $\phi_1'(\theta)$ to zero, we have

$$\phi_1'(\theta) = -P_{S_1}\sin 2(\theta - \alpha) = 0. \tag{26}$$

The solution of the above equation is $\theta_{opt} = \alpha$.

We observe that the problem of finding the maximum rate R_1 is equivalent to the maximization problem (20) with $\mu = 1$. Therefore, $\theta^*(1) = \alpha$ and the maximum individual rate $R_1 = \mathcal{C}(\phi_1(\alpha))$. The rate pairs that satisfy (8) in rate set $\mathcal{R}(\theta^*(1))$ are on the boundary of the optimal rate region as shown in *case 2* in section 2.

Similarly, by setting the derivative $\phi_2'(\theta)$ to zero, we have

$$\phi_2'(\theta) = -P_{S_2}\sin 2(\theta - \beta) = 0. \tag{27}$$

The solution of the above equation is $\theta_{opt} = \beta$.

We observe that the problem of finding the maximum rate R_2 is equivalent to the maximization problem (20) with $\mu = 0$. Therefore, $\theta^*(0) = \beta$ and the maximum individual rate $R_2 = \mathcal{C}(\phi_2(\beta))$. The rate pairs that satisfy (9) in rate set $\mathcal{R}(\theta^*(0))$ are on the boundary of the optimal rate region as shown in *case 3* in section 2.

Then we complete the proof.

From Theorem 2, the AF scheme corresponding to α is denoted by

$$\begin{aligned}\mathbf{a}^{(11)} &= c^{(11)}\mathbf{\Lambda}^{-1}\left(\mathbf{u}_1\cos\alpha + \mathbf{u}_2\sin\alpha\right) \\ &= c^{(11)}\mathbf{\Lambda}^{-1}\mathbf{d}_{10},\end{aligned} \tag{28}$$

where the constant $c^{(11)} = \sqrt{\frac{P_{sum}}{\| \mathbf{\Lambda}^{-1}\mathbf{d}_{10} \|^2}}$. To compare with the result obtained in [16], we replace P_{S_1} in (18) by $P_{S_1}\| \mathbf{d}_1 \|^2$. Therefore, the corresponding maximum rate of R_1 can be explicitly expressed as follows.

$$\mathcal{C}(\phi_1(\alpha)) = \frac{1}{2}\log\left(1 + P_{sum}P_{S_1}\sum_{k=1}^{n}\frac{h_{S_1,k}^2 h_{k,D}^2}{P_{sum}h_{k,D}^2 + h_{S_1,k}^2 P_{S_1} + h_{S_2,k}^2 P_{S_2} + 1}\right). \tag{29}$$

Similarly, from Theorem 2, the AF scheme corresponding to β is denoted by

$$\begin{aligned}\mathbf{a}^{(12)} &= c^{(12)}\mathbf{\Lambda}^{-1}\left(\mathbf{u}_1\cos\beta + \mathbf{u}_2\sin\beta\right), \\ &= c^{(12)}\mathbf{\Lambda}^{-1}\mathbf{d}_{20},\end{aligned} \tag{30}$$

where the constant $c^{(12)} = \sqrt{\frac{P_{sum}}{\| \mathbf{\Lambda}^{-1}\mathbf{d}_{20} \|^2}}$. To compare with the result obtained in [16], we replace P_{S_2} in (18) by $P_{S_2}\| \mathbf{d}_2 \|^2$. Therefore, the corresponding maximum rate of R_2 is given as follows.

$$\mathcal{C}(\phi_2(\beta)) = \frac{1}{2}\log\left(1 + P_{sum}P_{S_2}\sum_{k=1}^{n}\frac{h_{S_2,k}^2 h_{k,D}^2}{P_{sum}h_{k,D}^2 + h_{S_2,k}^2 P_{S_2} + h_{S_1,k}^2 P_{S_1} + 1}\right). \tag{31}$$

It is not surprising that the maximum individual rates given above are exactly the same as the ones obtained in [10, Theorem 4].

Theorem 3. *The weighted sum rate* $\mu R_1(\theta) + \bar{\mu}R_2(\theta)$ *is maximized by* $\theta^*(\mu)$ *satisfying (32) for* $\mu \in \left(0, \frac{1}{2}\right)$*, and is maximized by* $\theta^*(\mu)$ *satisfying (33) for* $\mu \in \left(\frac{1}{2}, 1\right)$.

$$\begin{aligned}&\mu\left[P_{S_1}P_{S_2}\sin 2(\beta-\theta)\cos^2(\theta-\alpha) + P_{S_1}\sin 2(\theta-\alpha)\left(1 + P_{S_2}\cos^2(\theta-\beta)\right)\right] \\ &\quad = \bar{\mu}\left[P_{S_1}P_{S_2}\sin 2(\beta-\theta)\cos^2(\theta-\alpha) + P_{S_2}\sin 2(\beta-\theta)\left(1 + P_{S_2}\cos^2(\theta-\beta)\right)\right]\end{aligned} \tag{32}$$

$$\begin{aligned}&\mu\left[P_{S_1}P_{S_2}\sin 2(\theta-\alpha)\cos^2(\theta-\beta) + P_{S_1}\sin 2(\theta-\alpha)\left(1 + P_{S_1}\cos^2(\theta-\alpha)\right)\right] \\ &\quad = \bar{\mu}\left[P_{S_1}P_{S_2}\sin 2(\theta-\alpha)\cos^2(\theta-\beta) + P_{S_2}\sin 2(\beta-\theta)\left(1 + P_{S_1}\cos^2(\theta-\alpha)\right)\right]\end{aligned} \tag{33}$$

Proof. Since for $\mu \in \left(0, \frac{1}{2}\right)$, the rate pair (R_1, R_2) on the boundary can be determined by solving the problem $\max \mu \left(\mathcal{C}\left(\phi\left(\theta\right)\right) - \mathcal{C}\left(\phi_2\left(\theta\right)\right)\right) + \bar{\mu}\mathcal{C}\left(\phi_2\left(\theta\right)\right)$ as shown in *case 4* in section 2. By setting the derivative of the object function to zero, we obtain (32). Then we argue that the function has a solution for each $\mu \in \left(0, \frac{1}{2}\right)$. Denote by

$$f\left(\theta\right) = \frac{P_{S_1}P_{S_2}\sin 2\left(\beta-\theta\right)\cos^2\left(\theta-\alpha\right) + P_{S_2}\sin 2\left(\beta-\theta\right)\left(1+P_{S_2}\cos^2\left(\theta-\beta\right)\right)}{P_{S_1}P_{S_2}\sin 2\left(\beta-\theta\right)\cos^2\left(\theta-\alpha\right) + P_{S_1}\sin 2\left(\theta-\alpha\right)\left(1+P_{S_2}\cos^2\left(\theta-\beta\right)\right)}, \tag{34}$$

where $\theta \in \left[\theta^*\left(\frac{1}{2}\right), \beta\right]$. We observe that $f\left(\theta\right)$ is continuous in $\left[\theta^*\left(\frac{1}{2}\right), \beta\right]$. By (22), it follows that $f\left(\theta^*\left(\frac{1}{2}\right)\right) = 1$. It is easy to see that $f\left(\beta\right) = 0$. Fix a $\mu \in [0, 1/2]$, then $\mu/\bar{\mu} \in (0, 1)$. By mean value theorem, there exists a $\theta^*(\mu) \in \left[\theta^*\left(\frac{1}{2}\right), \beta\right]$ such that $f\left(\theta^*(\mu)\right) = \mu/\bar{\mu}$. Then $\theta^*(\mu)$ is the solution corresponding to μ.

By symmetry, the proof of the second part is exactly the same as the first part.

In the proof of Theorem 3, the complicated derivative operations as observed in [10, Appendix I] are avoided. It shows that with the help of the expression given in (18) the equations (32) and (33) used to determine the optimal solutions are easier to be established here. However, as the result obtained in [10, Theorem 3], the optimal solution obtained above cannot be expressed in closed-form either. So the results can only be compared via numerical results.

So far, we have fully obtained the optimal rate region by characterizing all the boundary points of it for a two-hop MAC under the sum power constraint.

3.2 Individual Power Constraints

Now let us come to another topic. Assume the relays of the two-hop MAC are subject to individual power constraints. The problem of finding the optimal rate region becomes even more complicated than it was by changing the constraint. To the best of our knowledge, the optimal AF rate region has not been fully characterized in general SNR regime so far. In this section, we make progress towards this goal and give some interesting results. Let us start with the weighted sum maximization problem given as follows.

$$\begin{array}{ll} \max & \mu R_1(\mathbf{a}) + \bar{\mu} R_2(\mathbf{a}) \\ \text{s.t.} & 0 \le a_k \le \sqrt{P_{k,max}},\ k \in \mathcal{V} \end{array}. \tag{35}$$

It is clear that the rate pairs on the boundary of the optimal rate region are also the boundary points of some rate set $\mathcal{R}(\mathbf{a})$. So we focus on deriving these points and the corresponding AF schemes. It is first observed that in such AF schemes at least one of the relays transmits the received noisy signal with the maximum possible power. To prove this statement, consider a scheme $\mathbf{a}_0 = [a_{01}, \cdots, a_{0n}]^T$, where all the transmitting powers are strictly less than the corresponding upper bounds. Choose a constant $c = \min\left\{\frac{\sqrt{P_{k,\max}}}{a_{0k}}, k \in \mathcal{V}\right\}$. It is clear that $\mathbf{a} = c\mathbf{a}_0$ is also a feasible scheme. According to the assumption, it is clear that $c > 1$. Consequently, the corresponding rate region $\mathcal{R}(\mathbf{a})$ given in (5) is strictly larger than $\mathcal{R}(\mathbf{a}_0)$. Therefore, to obtain the optimal rate region for a two-hop MAC, we only consider the AF schemes with $a_k = \sqrt{P_{k,\max}}, k \in \mathcal{M}$, where $\mathcal{M}$ is a nonempty subset of relay nodes $\mathcal{V}$. Then given a μ, for each $\mathcal{M}$, an optimization problem can be formulated as follows.

$$\begin{array}{ll} \max & \mu R_1(\mathbf{a}) + \bar{\mu} R_2(\mathbf{a}) \\ \text{s.t.} & \quad a_k = \sqrt{P_{k,max}},\ k \in \mathcal{M} \\ & 0 \le a_j < \sqrt{P_{j,max}},\ j \in \mathcal{V}\backslash\mathcal{M} \end{array} . \tag{36}$$

We first consider a trivial case when $\mathcal{M} = \mathcal{V}$. The feasible region of the corresponding optimization problem (36) only contains a single point. Therefore, the resulting AF scheme is a fixed one with all the elements equal to their corresponding upper bounds. Besides this one, there are altogether $2^n - 2$ optimization problems for each μ, which have nontrivial solutions. Furthermore, the optimal solution of (35) should be achieved at one of the solutions of the above problems by the previous statement. For general μ, problem (36) is computational intractable. We put emphasis on several special cases, i.e., $\mu = 1$, 0, and 1/2 in the sequel, which correspond to the maximum individual rates $R_{1,max}$ and $R_{2,max}$, and maximum sum rate of the optimal region respectively.

For each $\mathcal{M}$, two sets are defined as follows.

$$\mathcal{A}_{\mathcal{M}}^{(1)} = \left\{\mathbf{a}|\mathbf{a} \in \mathbf{R}^n, a_k = \sqrt{P_{k,\max}}, k \in \mathcal{M}, 0 \le a_j < \sqrt{P_{j,\max}}, j \in \mathcal{V}\backslash\mathcal{M}\right\}, \tag{37}$$

$$\mathcal{A}_{\mathcal{M}}^{(2)} = \left\{\mathbf{a}|\mathbf{a} \in \mathbf{R}^n, a_k = \sqrt{P_{k,\max}}, k \in \mathcal{M}\right\}. \tag{38}$$

It is easy to see that $\mathcal{A}_{\mathcal{M}}^{(1)} \subset \mathcal{A}_{\mathcal{M}}^{(2)}$.

Maximum Individual Rates. Let us start with the maximum individual rates. The main idea we used to solve problem (35) with $\mu = 1$ and 0 is described as follows. A relaxed problem of (36) is first considered by relaxing the feasible region from $\mathcal{A}_{\mathcal{M}}^{(1)}$ to $\mathcal{A}_{\mathcal{M}}^{(2)}$. Then a set of solutions of the relaxed problem is obtained corresponding to each $\mathcal{M}$. Finally, we claim that the AF scheme that maximizes the individual rate is in the solution set.

In [3], the optimal AF scheme for a two-hop relay network is obtained via an algorithm. In each iteration cycle of the algorithm, the solution should be updated via solving a sequence of non-linear equations until the resulting AF scheme is feasible. However, the procedure may be exhausted when deriving the first partial derivatives and solving the system of non-linear equations. We propose a novel algebraic approach in the following lemma. By solving a system of linear equations, each relaxed problem can be easily solved.

Lemma 3. *For a given $\mathcal{M}$, the AF scheme in $\mathcal{A}_{\mathcal{M}}^{(2)}$ that maximizes the individual rate R_i, $i = 1, 2$, is given as*

$$\mathbf{a}_{\mathcal{M}}^{(i)} = c_{\mathcal{M}}^{(i)} \left(\mathbf{\Gamma}^2 + \mathbf{G}_{\mathcal{M}}^{(i)}\right)^{-1} \mathbf{\Gamma} \mathbf{h}_{1i}, \; i = 1, 2, \tag{39}$$

where $\mathbf{\Gamma} = \operatorname{diag}(\gamma_1, \cdots \gamma_n)$, $\mathbf{h}_{1i} = [h_{S_i,1}, \cdots, h_{S_i,n}]^T$, *and* $\mathbf{G}_{\mathcal{M}}^{(i)} = \operatorname{diag}\{g_1, \cdots, g_n\}$, *where* $g_k = g_k^{(i)}$, $k \in \mathcal{M}$ *given as follows, and* $g_j = 0$, $j \in \mathcal{V} \backslash \mathcal{M}$,

$$c_{\mathcal{M}}^{(i)} = \frac{1 + \sum\limits_{j \in \mathcal{M}} \gamma_j^2 P_{j,\max}}{\sum\limits_{j \in \mathcal{M}} \gamma_j h_{S_i,j} \sqrt{P_{j,\max}}},$$

$$g_k^{(i)} = c_{\mathcal{M}}^{(i)} \frac{\gamma_k h_{S_i,k}}{\sqrt{P_{k,\max}}} - \gamma_k^2, \; k \in \mathcal{M}.$$

Proof. The SNR function of R_i, $i = 1, 2$ in (5) is first rewritten as

$$SNR^{(1)}(\mathbf{a}) = \frac{\left(\mathbf{a}^T \mathbf{\Gamma} \mathbf{h}_{1i}\right)^2}{\mathbf{a}^T \mathbf{\Gamma}^2 \mathbf{a} + 1} P_{S_i}. \tag{40}$$

Then we define a matrix $\mathbf{G}_{\mathcal{M}} = \operatorname{diag}\{g_1, \cdots, g_n\}$, such that $g_j = 0$, $j \in \mathcal{V} \backslash \mathcal{M}$ and $\sum\limits_{k \in \mathcal{M}} g_k P_{k,\max} = 1$. Then for each $\mathbf{G}_{\mathcal{M}}$ consider the SNR function given as follows,

$$SNR_{\mathbf{G}_{\mathcal{M}}}^{(2)}(\mathbf{a}) = \frac{\left(\mathbf{a}^T \mathbf{\Gamma} \mathbf{h}_{1i}\right)^2}{\mathbf{a}^T \left(\mathbf{\Gamma}^2 + \mathbf{G}_{\mathcal{M}}\right) \mathbf{a}} P_{S_i}. \tag{41}$$

It follows that $SNR^{(1)}(\mathbf{a}) = SNR_{\mathbf{G}_{\mathcal{M}}}^{(2)}(\mathbf{a})$ for $\mathbf{a} \in \mathcal{A}_{\mathcal{M}}^{(2)}$. If $\mathbf{\Gamma}^2 + \mathbf{G}_{\mathcal{M}}$ is a positive definite matrix, $SNR_{\mathbf{G}_{\mathcal{M}}}^{(2)}(\mathbf{a})$ is a generalized Rayleigh quotient. Therefore, it is easy to obtain the global maximizers of it in $\mathbf{R}^n$, which is denoted by

$$\mathbf{a}_{\mathbf{G}_{\mathcal{M}}} = c \left(\mathbf{\Gamma}^2 + \mathbf{G}_{\mathcal{M}}\right)^{-1} \mathbf{\Gamma} \mathbf{h}_{1i}, c \neq 0. \tag{42}$$

Obviously, given a matrix $\mathbf{G}_{\mathcal{M}}$, if for some constant c, there exists a specific $\mathbf{a}_{\mathbf{G}_{\mathcal{M}}} \in \mathcal{A}_{\mathcal{M}}^{(2)}$, it should be the maximum point of $SNR_{\mathbf{G}_{\mathcal{M}}}^{(2)}(\mathbf{a})$ in $\mathcal{A}_{\mathcal{M}}^{(2)}$. Combining the observation that $SNR^{(1)}(\mathbf{a}) = SNR_{\mathbf{G}_{\mathcal{M}}}^{(2)}(\mathbf{a})$ for $\mathbf{a} \in \mathcal{A}_{\mathcal{M}}^{(2)}$, it should be the maximum point of $SNR^{(1)}(\mathbf{a})$ in $\mathcal{A}_{\mathcal{M}}^{(2)}$ as well. To find such parameters, we establish the following system of equations with respect to c and g_k, $k \in \mathcal{M}$.

$$\sum_{k \in \mathcal{M}} g_k P_{k,\max} = 1, \tag{43}$$

$$\gamma_k h_{S_i,k} c = \sqrt{P_{k,\max}} \left(\gamma_k^2 + g_k\right), k \in \mathcal{M}. \tag{44}$$

The solution of the above linear equations is easy to derive which is given as follows.

$$c_{\mathcal{M}}^{(i)} = \frac{1 + \sum\limits_{j\in\mathcal{M}} \gamma_j^2 P_{j,\max}}{\sum\limits_{j\in\mathcal{M}} \gamma_j h_{S_i,j}\sqrt{P_{j,\max}}}, \tag{45}$$

$$g_k^{(i)} = c_{\mathcal{M}}^{(i)} \frac{\gamma_k h_{S_i,k}}{\sqrt{P_{k,\max}}} - \gamma_k^2, k \in \mathcal{M}, \tag{46}$$

Since $\mathcal{C}(\cdot)$ is a monotonically increasing function of the SNR value, the maximum point of $SNR^{(1)}(\mathbf{a})$ in $\mathcal{A}_{\mathcal{M}}^{(2)}$ also maximizes R_i. From (42), (45), and (46), the AF scheme in $\mathcal{A}_{\mathcal{M}}^{(2)}$ that achieves the maximum rate R_i is given by $\mathbf{a}_{\mathcal{M}}^{(i)} = c_{\mathcal{M}}^{(i)}\left(\mathbf{\Gamma}^2 + \mathbf{G}_{\mathcal{M}}^{(i)}\right)^{-1}\mathbf{\Gamma}\mathbf{h}_{1i}$.

Then we complete the proof.

The set of the relaxed solutions can be derived via the above approach. Since the discussion of the maximum individual rates are exactly the same, we do not distinguish them in the sequel. We use the notation $\mathbf{a}_{\mathcal{M}}$ to represent the relaxed solution and drop the subscript and superscript of all parameters used to indicate the two sources. As it is only required that $\mathbf{a}_{\mathcal{M}} \in \mathcal{A}_{\mathcal{M}}^{(2)}$, some elements in $\mathbf{a}_{\mathcal{M}}$ may violate the corresponding upper bounds given in $\mathcal{A}_{\mathcal{M}}^{(1)}$. We define a collection of nonempty subsets of $\mathcal{V}$ as

$$\mathcal{S}_0 = \left\{\mathcal{M}|\mathbf{a}_{\mathcal{M}} \in \mathcal{A}_{\mathcal{M}}^{(1)}, \mathcal{M} \subset \mathcal{V}, \mathcal{M} \neq \emptyset\right\}. \tag{47}$$

Consequently, only the relaxed solutions in $\{\mathbf{a}_{\mathcal{M}}|\mathcal{M} \in \mathcal{S}_0\}$ are feasible AF schemes with respect to the power constraint. Without considering the other relaxed solutions, we doubt whether it will lead to a loss of optimality. Fortunately, it is convenient to verify that the optimal solution of (35) can be found in the set $\{\mathbf{a}_{\mathcal{M}}|\mathcal{M} \in \mathcal{S}_0\}$ in the following lemma.

Lemma 4. *Let $\mathbf{a}^*$ be the AF scheme that achieves the maximum individual rate. Then it should be the optimal solution of a relaxed optimization problem, i.e., $\mathbf{a}^* \in \{\mathbf{a}_{\mathcal{M}}|\mathcal{M} \in \mathcal{S}_0\}$.*

Proof. As claimed before, if $\mathbf{a}^*$ is the desired AF scheme, it should be in some $\mathcal{A}_{\mathcal{M}}^{(1)}$ *such that* $a_k = \sqrt{P_{k,max}}$, $k \in \mathcal{M}$ *and* $0 \leq a_j < \sqrt{P_{j,max}}$, $j \in \mathcal{V}\backslash\mathcal{M}$. It is assumed without loss of generality that $\mathbf{a}^* \neq \mathbf{0}$, since it leads to a global minimum of the SNR function.

Case 1: Assume $\mathcal{M} \in \mathcal{S}_0$ then $\mathbf{a}_{\mathcal{M}} \in \mathcal{A}_{\mathcal{M}}^{(1)}$. It is clear that we cannot find another AF scheme in $\mathcal{A}_{\mathcal{M}}^{(1)}$ with a strictly larger SNR value.

Case 2: Assume $\mathcal{M} \notin \mathcal{S}_0$ then $\mathbf{a}_{\mathcal{M}} \notin \mathcal{A}_{\mathcal{M}}^{(1)}$. Then we need to show that any scheme in $\mathcal{A}_{\mathcal{M}}^{(1)}$ should not be the optimal one. Then it follows that we should saturate an amplitude in $\mathcal{V}\backslash\mathcal{M}$ to its corresponding upper bound. By doing so repetitively, case 2 can always reduced to case 1, i.e., for a larger subset $\mathcal{M}'$,

$\mathcal{M} \subset \mathcal{M}'$ such that $\mathbf{a}_{\mathcal{M}'} \in \mathcal{A}^{(1)}_{\mathcal{M}'}$. It implies that to obtain the optimal solution in terms of maximum individual rate, we can omit the subset $\mathcal{M} \notin \mathcal{S}_0$ and thus prove the lemma. The detailed procedure is given as follows.

Define a matrix

$$\mathbf{D}_{\mathcal{M}} = \mathrm{diag}\left\{\sqrt{\gamma_1^2 + g_1}, \cdots, \sqrt{\gamma_n^2 + g_n}\right\}, \tag{48}$$

where g_k, $k \in \mathcal{M}$ is obtained in (46) and $g_j = 0$, $j \in \mathcal{V}\backslash\mathcal{M}$. Then $SNR^{(2)}_{\mathbf{G}_{\mathcal{M}}}(\mathbf{a})$ can be recast as

$$SNR^{(3)}_{\mathbf{G}_{\mathcal{M}}}(\mathbf{x}) = \frac{\mathbf{x}^T\mathbf{v}\mathbf{v}^T\mathbf{x}}{\mathbf{x}^T\mathbf{x}} P_S, \tag{49}$$

where $\mathbf{v} = \mathbf{D}_{\mathcal{M}}^{-1}\mathbf{\Gamma}\mathbf{h}_1$ and $\mathbf{x} = \mathbf{D}_{\mathcal{M}}\mathbf{a}$. Since it only has a variable substitution, $SNR^{(2)}_{\mathbf{G}_{\mathcal{M}}}(\mathbf{a}) = SNR^{(3)}_{\mathbf{G}_{\mathcal{M}}}(\mathbf{x})$ when $\mathbf{x} = \mathbf{D}_{\mathcal{M}}\mathbf{a}$. Denote by $\mathbf{x}_{\mathcal{M}} = \mathbf{D}_{\mathcal{M}}\mathbf{a}_{\mathcal{M}}$. Let $\mathrm{span}\{\mathbf{x}_{\mathcal{M}}\}$ and $\mathrm{span}^{\perp}\{\mathbf{x}_{\mathcal{M}}\}$ represent the linear subspace spanned by $\mathbf{x}_{\mathcal{M}}$ and its orthogonal space respectively. Then any $\mathbf{x}$ can be decomposed into $\mathrm{span}\{\mathbf{x}_{\mathcal{M}}\}$ and $\mathrm{span}^{\perp}\{\mathbf{x}_{\mathcal{M}}\}$, that is,

$$\mathbf{x} = c\mathbf{x}_{\mathcal{M}} + \mathbf{x}^{\perp}_{\mathcal{M}}, \tag{50}$$

where $\mathbf{x}^{\perp}_{\mathcal{M}} \in \mathrm{span}^{\perp}\{\mathbf{x}_{\mathcal{M}}\}$ and $c = \mathbf{x}^T\mathbf{x}_{\mathcal{M}}/\|\mathbf{x}_{\mathcal{M}}\| > 0$. By Lemma 3, it follows that $\mathrm{span}\{\mathbf{x}_{\mathcal{M}}\} = \mathrm{span}\{\mathbf{v}\}$. It can be verified that for each $\mathbf{a} \in \mathcal{A}^{(1)}_{\mathcal{M}}$, there exits an $\mathbf{a}_0 = \delta(\mathbf{a}_{\mathcal{M}} - \mathbf{a}) + \mathbf{a} \in \mathcal{A}^{(1)}_{\mathcal{M}}$ for some sufficiently small positive δ. Denote by $\mathbf{x}_0 = \mathbf{D}_{\mathcal{M}}\mathbf{a}_0$. To compare the corresponding SNR values of $\mathbf{a}_0$ and $\mathbf{a}$, we have

$$\begin{aligned}
SNR^{(1)}(\mathbf{a}_0) &\overset{(a)}{=} SNR^{(2)}_{\mathbf{G}_{\mathcal{M}}}(\mathbf{a}_0) = SNR^{(3)}_{\mathbf{G}_{\mathcal{M}}}(\mathbf{x}_0) = \frac{\mathbf{x}_0^T\mathbf{v}\mathbf{v}^T\mathbf{x}_0}{\mathbf{x}_0^T\mathbf{x}_0} P_S \\
&= \frac{\left[\left((1-\delta)\, c\mathbf{x}_{\mathcal{M}} + \delta\mathbf{x}_{\mathcal{M}} + (1-\delta)\,\mathbf{x}^{\perp}_{\mathcal{M}}\right)^T \mathbf{v}\right]^2}{\|(1-\delta)\, c\mathbf{x}_{\mathcal{M}} + \delta\mathbf{x}_{\mathcal{M}} + (1-\delta)\,\mathbf{x}^{\perp}_{\mathcal{M}}\|^2} P_S \\
&= \frac{\left(\mathbf{x}_{\mathcal{M}}^T\mathbf{v}\right)^2}{\|\mathbf{x}_{\mathcal{M}}\|^2 + \left(c + \frac{\delta}{1-\delta}\right)^{-2} \|\mathbf{x}^{\perp}_{\mathcal{M}}\|^2} P_S \\
&\overset{(b)}{\geq} \frac{\left(\mathbf{x}_{\mathcal{M}}^T\mathbf{v}\right)^2}{\|\mathbf{x}_{\mathcal{M}}\|^2 + c^{-2}\|\mathbf{x}^{\perp}_{\mathcal{M}}\|^2} P_S \\
&= SNR^{(3)}_{\mathbf{G}_{\mathcal{M}}}(\mathbf{x}) = SNR^{(2)}_{\mathbf{G}_{\mathcal{M}}}(\mathbf{a}) \overset{(c)}{=} SNR^{(1)}(\mathbf{a}),
\end{aligned} \tag{51}$$

where (a) and (c) follow from that both $\mathbf{a}_0$ and $\mathbf{a}$ are in $\mathcal{A}^{(1)}_{\mathcal{M}}$, and (b) follows from the fact that $c^{-2} \geq (c + \frac{\delta}{1-\delta})^{-2}$ and the equality holds only when $\mathbf{x}^{\perp}_{\mathcal{M}} = 0$, which implies that $\mathbf{a} = \mathbf{a}_{\mathcal{M}}$, yielding a contradiction to the assumption that $\mathbf{a}_{\mathcal{M}} \notin \mathcal{A}^{(1)}_{\mathcal{M}}$. It concludes that $\mathbf{a}$ should not be the AF scheme that achieves the maximum value of the individual rate.

Then we complete the proof.

It is observed that if $\mathcal{M}_1 \subset \mathcal{M}_2$, then by the definition (38), $\mathcal{A}^{(2)}_{\mathcal{M}_2} \subset \mathcal{A}^{(2)}_{\mathcal{M}_1}$, and thus $SNR^{(2)}_{\mathbf{G}_{\mathcal{M}_1}}(\mathbf{a}_{\mathcal{M}_1}) \geq SNR^{(2)}_{\mathbf{G}_{\mathcal{M}_2}}(\mathbf{a}_{\mathcal{M}_2})$. Recall that the subsets of a finite set form a partial order set with order $\subset$, so-called the Boolean lattice. A mapping can be established from the subset $\mathcal{M}$ to the SNR value $SNR^{(2)}_{\mathbf{G}_{\mathcal{M}}}(\mathbf{a}_{\mathcal{M}})$. The function is monotonically decreasing with respect to the partial order. Combining the result obtained in Lemma 4, the maximum SNR value is achieved at some minimal element in $\mathcal{S}_0$. So far, to find out the feasible AF scheme maximizing the individual rate for the two-hop MAC, an algorithm can be proposed as follows. We use the notation $\mathcal{M}_{i,j}$ to denote the nonempty subsets with cardinality equal to i. Since there are altogether $n_i = \binom{n}{i}$ such subsets, j can take values from 1 to n_i.

Algorithm 1

```
Initialization: S = ∅, a* = 0.
for i = 1; i++; i ≤ n; do
  j = 1, n_i = (n choose i);
  while j ≤ n_i; do
    if ∃ M_0 ∈ S, s.t. M_0 ⊂ M_{i,j} then
      j = j + 1;
    else
      Compute a_{M_{i,j}} by the technique given in Lemma 3.
      if a_{M_{i,j}} ∉ A^(1)_{M_{i,j}} then
        j = j + 1;
      else
        Let S = S ∪ {M_{i,j}} and set a* = a_{M_{i,j}} if SNR^(1)(a*) < SNR^(1)(a_{M_{i,j}});
        j = j + 1;
```

Note that in Algorithm 1, the set $\mathcal{S}$ denotes the collection of minimal elements in $\mathcal{S}_0$. Therefore, we only compute the relaxed solutions corresponding to the subsets in $\mathcal{S}$ by the algorithm. Therefore, denote by

$$\mathcal{M}^* = \arg\max_{\mathcal{M} \in \mathcal{S}} SNR^{(1)}(\mathbf{a}_{\mathcal{M}}), \tag{52}$$

and the corresponding AF scheme is denoted by $\mathbf{a}^*$. We conclude the result in the following theorem.

Theorem 4. *The AF scheme that maximizes the individual rate R_i, $i = 1, 2$ for the two-hop MAC with individual power constraints can be computed by Algorithm 1 and the analytical solution is given as follows.*

$$\mathbf{a}^* = c^{(i)}_{\mathcal{M}^*} \left(\mathbf{\Gamma}^2 + \mathbf{G}^{(i)}_{\mathcal{M}^*}\right)^{-1} \mathbf{\Gamma} \mathbf{h}_{1i}, \ i = 1, 2. \tag{53}$$

Maximum Sum Rate. To obtain the maximum sum rate, the main idea is quite similar to the one described above. That is, we first solve the relaxed problem of (36) with $\mu = 1/2$ in a similar way. Then after proving that the AF scheme maximizing the sum rate is in the relaxed solution set, we propose an algorithm to solve the problem (35). However, the procedure is more complicated than the previous case.

The SNR function of the sum rate $R_1 + R_2$ is first recast as follows. With a little abuse of notation, we still denote the SNR function by $SNR^{(i)}(\cdot)$, $i = 1$, 2, and 3.

$$SNR^{(1)}(\mathbf{a}) = \frac{\left(\mathbf{a}^T \mathbf{\Gamma} \mathbf{h}_{11}\right)^2 P_{S_1} + \left(\mathbf{a}^T \mathbf{\Gamma} \mathbf{h}_{12}\right)^2 P_{S_2}}{\mathbf{a}^T \mathbf{\Gamma}^2 \mathbf{a} + 1}. \tag{54}$$

Assume $\mathbf{h}_{11}$ and $\mathbf{h}_{12}$ are linearly independent, otherwise the problem can be solved by exactly the same approach that is used to solve the maximum individual rate problem. Given a subset $\mathcal{M}$, define a diagonal matrix

$$\mathbf{D}_{\mathcal{M}} = \text{diag}\{d_1, \cdots, d_n\}, \tag{55}$$

where $d_j = \gamma_j$, $j \in \mathcal{V} \backslash \mathcal{M}$, and $d_k \neq 0$, $k \in \mathcal{M}$, such that

$$\sum_{k \in \mathcal{M}} \left(d_k^2 - \gamma_k^2\right) P_{k,\max} = 1. \tag{56}$$

Consider the following SNR function,

$$\begin{aligned} SNR^{(2)}_{\mathbf{D}_{\mathcal{M}}}(\mathbf{a}) &= \frac{\mathbf{a}^T \mathbf{\Gamma} [\mathbf{h}_{11}, \mathbf{h}_{12}] \text{diag}\{P_{S_1}, P_{S_2}\} [\mathbf{h}_{11}, \mathbf{h}_{12}]^T \mathbf{\Gamma} \mathbf{a}}{\mathbf{a}^T \mathbf{D}^2_{\mathcal{M}} \mathbf{a}} \\ &= \frac{\|[\mathbf{v}_1, \mathbf{v}_2]^T \mathbf{D}_{\mathcal{M}} \mathbf{a}\|^2}{\|\mathbf{D}_{\mathcal{M}} \mathbf{a}\|^2}, \end{aligned} \tag{57}$$

where $\mathbf{v}_i = \sqrt{P_{S_i}} \mathbf{D}^{-1}_{\mathcal{M}} \mathbf{\Gamma} \mathbf{h}_{1i}$, $i = 1, 2$.

It is easy to see that $SNR^{(1)}(\mathbf{a}) = SNR^{(2)}_{\mathbf{D}_{\mathcal{M}}}(\mathbf{a})$ for $\mathbf{a} \in \mathcal{A}^{(2)}_{\mathcal{M}}$. Denote by $\mathbf{v} = \mathbf{D}_{\mathcal{M}} \mathbf{a}$. With the variable substitution, $SNR^{(2)}_{\mathbf{D}_{\mathcal{M}}}$ can be recast as follows.

$$SNR^{(3)}_{\mathbf{D}_{\mathcal{M}}}(\mathbf{v}) = \frac{\mathbf{v}^T [\mathbf{v}_1, \mathbf{v}_2][\mathbf{v}_1, \mathbf{v}_2]^T \mathbf{v}}{\mathbf{v}^T \mathbf{v}}, \tag{58}$$

which forms a Rayleigh quotient and thus the maximum value of $SNR^{(3)}_{\mathbf{D}_{\mathcal{M}}}$ is achieved at the eigenvector $\mathbf{v}_{\mathcal{M}}$ corresponding to the maximum eigenvalue of matrix $[\mathbf{v}_1, \mathbf{v}_2][\mathbf{v}_1, \mathbf{v}_2]^T$.

Lemma 5. *For each $\mathcal{M}$, the solution in $\mathcal{A}^{(2)}_{\mathcal{M}}$ that maximizes the sum rate is given by $\mathbf{a}_{\mathcal{M}} = \mathbf{D}^{-1}_{\mathcal{M}}(c_1 \mathbf{v}_1 + c_2 \mathbf{v}_2)$, where c_1, c_2 and d_k, $k \in \mathcal{M}$ are given in (61)-(63).*

Proof. By Lemma 1, it follows that $\mathbf{v}_{\mathcal{M}}$ is in the linear subspace span$\{\mathbf{v}_1, \mathbf{v}_2\}$, i.e., $\mathbf{v}_{\mathcal{M}} = c_1\mathbf{v}_1 + c_2\mathbf{v}_2$, where c_1 and c_2 are chosen such that $\mathbf{a}_{\mathcal{M}} = \mathbf{D}_{\mathcal{M}}^{-1}(c_1\mathbf{v}_1 + c_2\mathbf{v}_2) \in \mathcal{A}_{\mathcal{M}}^{(2)}$, which implies that

$$c_1 \frac{\gamma_k h_{S_1,k}\sqrt{P_{S_1}}}{d_k^2} + c_2 \frac{\gamma_k h_{S_2,k}\sqrt{P_{S_2}}}{d_k^2} = \sqrt{P_{k,\max}},\ k \in \mathcal{M}. \tag{59}$$

Combining with (56), there are altogether $|\mathcal{M}|+1$ linear equations with respect to d_k^2, $k \in \mathcal{M}$, c_1 and c_2. Furthermore, we observe that c_1/c_2 should be a constant because adding the original point, the eigenvectors form a rank-1 linear subspace. Denote by c_0 the desired ratio of c_1 and c_2. Then, with this additional condition, a system of equations is established as follows. For notation brevity, assume without loss of generality, $\mathcal{M} = \{1, \cdots, |\mathcal{M}|\}$.

$$\mathbf{A}\mathbf{x} = \mathbf{b}, \tag{60}$$

where

$$\mathbf{A} = \begin{bmatrix} P_{1,\max} & \cdots & P_{|\mathcal{M}|,\max} & 0 & 0 \\ -\sqrt{P_{1,\max}} & & 0 & \gamma_1 h_{S_1,1}\sqrt{P_{S_1}} & \gamma_1 h_{S_2,1}\sqrt{P_{S_2}} \\ & \ddots & & \vdots & \vdots \\ 0 & & -\sqrt{P_{|\mathcal{M}|,\max}} & \gamma_{|\mathcal{M}|} h_{S_1,|\mathcal{M}|}\sqrt{P_{S_1}} & \gamma_{|\mathcal{M}|} h_{S_2,|\mathcal{M}|}\sqrt{P_{S_2}} \\ 0 & \cdots & 0 & 1 & -c_0 \end{bmatrix},$$

$\mathbf{x} = \left[d_1^2, \cdots, d_{|\mathcal{M}|}^2, c_1, c_2\right]^T$ and $\mathbf{b} = \left[1 + \sum_{k\in\mathcal{M}} \gamma_k^2 P_{k,\max}, 0, \cdots, 0\right]^T$.

Solving it, we have

$$d_k^2 = \frac{1 + \sum_{m\in\mathcal{M}} \gamma_m^2 P_{m,\max}}{\sqrt{P_{k,\max}}} \times \frac{\gamma_k h_{S_2,k}\sqrt{P_{S_2}} + c_0\gamma_k h_{S_1,k}\sqrt{P_{S_1}}}{\sqrt{P_{S_2}}\sum_{m\in\mathcal{M}} \gamma_m h_{S_2,m}\sqrt{P_{m,\max}} + c_0\sqrt{P_{S_1}}\sum_{m\in\mathcal{M}} \gamma_m h_{S_1,m}\sqrt{P_{m,\max}}},\ k \in \mathcal{M}, \tag{61}$$

$$c_1 = \frac{c_0(1 + \sum_{m\in\mathcal{M}} \gamma_m^2 P_{m,\max})}{\sqrt{P_{S_2}}\sum_{m\in\mathcal{M}} \gamma_m h_{S_2,m}\sqrt{P_{m,\max}} + c_0\sqrt{P_{S_1}}\sum_{m\in\mathcal{M}} \gamma_m h_{S_1,m}\sqrt{P_{m,\max}}}, \tag{62}$$

$$c_2 = \frac{1 + \sum_{m\in\mathcal{M}} \gamma_m^2 P_{m,\max}}{\sqrt{P_{S_2}}\sum_{m\in\mathcal{M}} \gamma_m h_{S_2,m}\sqrt{P_{m,\max}} + c_0\sqrt{P_{S_1}}\sum_{m\in\mathcal{M}} \gamma_m h_{S_1,m}\sqrt{P_{m,\max}}}. \tag{63}$$

It is easy to see that if $c_0 \geq 0$ then $d_k^2 > 0$, which implies that $\mathbf{D}_{\mathcal{M}}$ is positive definite.

Finally, we shall check whether there exists a non-negative c_0, such that

$$[\mathbf{v}_1, \mathbf{v}_2][\mathbf{v}_1, \mathbf{v}_2]^T \mathbf{v}_{\mathcal{M}} = \lambda_{\max} \mathbf{v}_{\mathcal{M}}, \tag{64}$$

where $\lambda_{\max}$ is the maximum eigenvalue of $[\mathbf{v}_1, \mathbf{v}_2][\mathbf{v}_1, \mathbf{v}_2]^T$. For any eigenvalue λ of $[\mathbf{v}_1, \mathbf{v}_2][\mathbf{v}_1, \mathbf{v}_2]^T$, we have

$$[\mathbf{v}_1, \mathbf{v}_2][\mathbf{v}_1, \mathbf{v}_2]^T \mathbf{v} = \lambda \mathbf{v}, \tag{65}$$

where $\mathbf{v}$ is the eigenvector corresponding to λ. Substituting $\mathbf{v} = c_1\mathbf{v}_1 + c_2\mathbf{v}_2$ and $c_1/c_2 = c_0$ into (65), we have

$$\left(c_0\mathbf{v}_1^T\mathbf{v}_1 + \mathbf{v}_1^T\mathbf{v}_2 - c_0\lambda\right)\mathbf{v}_1 + \left(c_0\mathbf{v}_1^T\mathbf{v}_2 + \mathbf{v}_2^T\mathbf{v}_2 - \lambda\right)\mathbf{v}_2 = 0. \tag{66}$$

As assumed that $\mathbf{h}_{11}$ and $\mathbf{h}_{12}$ are linearly independent, then it follows that $\mathbf{v}_1$ and $\mathbf{v}_2$ are linearly independent as well. So (66) holds only when both the coefficients are equal to zero, which yields

$$\begin{cases} c_0\mathbf{v}_1^T\mathbf{v}_1 + \mathbf{v}_1^T\mathbf{v}_2 - c_0\lambda = 0 \\ c_0\mathbf{v}_1^T\mathbf{v}_2 + \mathbf{v}_2^T\mathbf{v}_2 - \lambda = 0 \end{cases}. \tag{67}$$

Then c_0 and λ can be expressed in terms of $\mathbf{v}_i$, $i = 1, 2$.

$$\begin{cases} c_0 = \dfrac{-b + \sqrt{b^2+4}}{2} \\ \lambda = \dfrac{\|\mathbf{v}_1\|^2 + \|\mathbf{v}_2\|^2 + \sqrt{\left(\|\mathbf{v}_1\|^2 - \|\mathbf{v}_2\|^2\right)^2 + 4\left(\mathbf{v}_1^T\mathbf{v}_2\right)^2}}{2} \end{cases}, \tag{68}$$

and

$$\begin{cases} c_0 = \dfrac{-b - \sqrt{b^2+4}}{2} \\ \lambda = \dfrac{\|\mathbf{v}_1\|^2 + \|\mathbf{v}_2\|^2 - \sqrt{\left(\|\mathbf{v}_1\|^2 - \|\mathbf{v}_2\|^2\right)^2 + 4\left(\mathbf{v}_1^T\mathbf{v}_2\right)^2}}{2} \end{cases}, \tag{69}$$

where $b = \dfrac{\mathbf{v}_2^T\mathbf{v}_2 - \mathbf{v}_1^T\mathbf{v}_1}{\mathbf{v}_1^T\mathbf{v}_2}$.

From (68) and (69), it is clear that $c_0 = \frac{1}{2}\left(-b + \sqrt{b^2+4}\right)$ corresponds to the maximum eigenvalue. Recall that the elements in $\mathbf{v}_1$ and $\mathbf{v}_2$ are also functions of c_0. We consider the following function of c_0.

$$f(c_0) = c_0 - \frac{-b + \sqrt{b^2+4}}{2}. \tag{70}$$

It is easy to see that the solution of $f(c_0) = 0$ is also the solution of (67) with respect to c_0. Consider the following two cases.

1). $c_0 = 0$, $f(0) < 0$.
2). $c_0 \to +\infty$, $\lim\limits_{c_0 \to +\infty} f(c_0) \to +\infty$, because from (61),

$$\lim_{c_0\to+\infty} d_k^2 = \frac{\gamma_k h_{S_1,k}\left(1+\sum\limits_{m\in\mathcal{M}}\gamma_m^2 P_{m,\max}\right)}{\sqrt{P_{k,\max}}\sum\limits_{m\in\mathcal{M}}\gamma_m h_{S_1,m}\sqrt{P_{m,\max}}},\ k\in\mathcal{M} \text{ is a finite constant and}$$

thus b.

Since $f(c_0)$ is a continuous function of c_0. Then by the intermediate value theorem, there exists a $c_0^* > 0$ such that $f(c_0^*) = 0$ holds. Substituting it into (61)-(63), we obtain the solution $\mathbf{a}_{\mathcal{M}} = \mathbf{D}_{\mathcal{M}}^{-1}(c_1\mathbf{v}_1 + c_2\mathbf{v}_2)$ that maximizes the sum rate in $\mathcal{A}_{\mathcal{M}}^{(2)}$.

Then we complete the proof.

Again, we use the notation $\mathcal{S}_0$ to denote the collection of subsets of $\mathcal{V}$ as defined in (47). Similarly, we shall prove that the AF scheme maximizing the sum rate is in $\{\mathbf{a}_{\mathcal{M}}|\mathcal{M}\in\mathcal{S}_0\}$.

Lemma 6. *Let* $\mathbf{a}^*$ *be the AF scheme that achieves the maximum sum rate* $R_1 + R_2$. *Then it should be the optimal solution of a relaxed optimization problem, i.e.,* $\mathbf{a}^*\in\{\mathbf{a}_{\mathcal{M}}|\mathcal{M}\in\mathcal{S}_0\}$.

Proof. The proof is similar to Lemma 4. If $\mathbf{a}^*$ is the maximum point of the SNR function of the sum rate, it should be in some set $\mathcal{A}_{\mathcal{M}}^{(1)}$. Assume $\mathbf{a}^*\notin\{\mathbf{a}_{\mathcal{M}}|\mathcal{M}\in\mathcal{S}_0\}$, then $\mathbf{a}^*\neq\mathbf{a}_{\mathcal{M}}$ for all $\mathcal{M}\in\mathcal{S}_0$. Then we can find another scheme given as $\mathbf{a}_0 = \delta(\mathbf{a}_{\mathcal{M}}-\mathbf{a}^*)+\mathbf{a}^*$, where $\mathbf{a}_{\mathcal{M}}\in\mathcal{A}_{\mathcal{M}}^{(2)}$ is the relaxed solution obtained in Lemma 5. It is clear that if $\delta<1$ is a sufficiently small positive value, then $\mathbf{a}_0\in\mathcal{A}_{\mathcal{M}}^{(1)}$. Denote by $\mathbf{v}^* = \mathbf{D}_{\mathcal{M}}\mathbf{a}^*$, $\mathbf{v}_0 = \mathbf{D}_{\mathcal{M}}\mathbf{a}_0$, and $\mathbf{v}_{\mathcal{M}} = \mathbf{D}_{\mathcal{M}}\mathbf{a}_{\mathcal{M}}$. We decompose $\mathbf{v}^*$ into $\text{span}\{\mathbf{v}_{\mathcal{M}}\}$ and $\text{span}^{\perp}\{\mathbf{v}_{\mathcal{M}}\}$, i.e., $\mathbf{v}^* = c\mathbf{v}_{\mathcal{M}} + \mathbf{v}_{\mathcal{M}}^{\perp}$, where $c>0$ and $\mathbf{v}_{\mathcal{M}}^{\perp}\in\text{span}^{\perp}\{\mathbf{v}_{\mathcal{M}}\}$. Then we have the following result.

$$\begin{aligned}
SNR^{(1)}(\mathbf{a}_0) &\overset{(a)}{=} SNR^{(2)}_{\mathbf{D}_{\mathcal{M}}}(\mathbf{a}_0) = SNR^{(3)}_{\mathbf{D}_{\mathcal{M}}}(\mathbf{v}_0) = \frac{\mathbf{v}_0^T[\mathbf{v}_1,\mathbf{v}_2][\mathbf{v}_1,\mathbf{v}_2]^T\mathbf{v}_0}{\mathbf{v}_0^T\mathbf{v}_0}\\
&= \frac{\left(\left[(c(1-\delta)+\delta)\mathbf{v}_{\mathcal{M}} + (1-\delta)\mathbf{v}_{\mathcal{M}}^{\perp}\right]^T[\mathbf{v}_1,\mathbf{v}_2]\right)^2}{\|(c(1-\delta)+\delta)\mathbf{v}_{\mathcal{M}}+(1-\delta)\mathbf{v}_{\mathcal{M}}^{\perp}\|^2}\\
&= \frac{\lambda_{\max}\|\mathbf{v}_{\mathcal{M}}\|^2 + c_3\|[\mathbf{v}_1,\mathbf{v}_2]^T\mathbf{v}_{\mathcal{M}}^{\perp}\|^2}{\|\mathbf{v}_{\mathcal{M}}\|^2 + c_3\|\mathbf{v}_{\mathcal{M}}^{\perp}\|^2}\\
&\overset{(b)}{\geq} \frac{\lambda_{\max}\|\mathbf{v}_{\mathcal{M}}\|^2 + c^{-2}\|[\mathbf{v}_1,\mathbf{v}_2]^T\mathbf{v}_{\mathcal{M}}^{\perp}\|^2}{\|\mathbf{v}_{\mathcal{M}}\|^2 + c^{-2}\|\mathbf{v}_{\mathcal{M}}^{\perp}\|^2}\\
&= SNR^{(2)}_{\mathbf{D}_{\mathcal{M}}}(\mathbf{v}^*) = SNR^{(2)}_{\mathbf{D}_{\mathcal{M}}}(\mathbf{a}^*) \overset{(c)}{=} SNR^{(1)}(\mathbf{a}^*),
\end{aligned} \tag{71}$$

where $c_3 = \left(c+\frac{\delta}{1-\delta}\right)^{-2}$, (a) and (c) follow from that both $\mathbf{a}_0$ and $\mathbf{a}^*$ are in $\mathcal{A}_{\mathcal{M}}^{(1)}$, and (b) follows from $c_3 < c^{-2}$ and $\frac{\|[\mathbf{v}_1,\mathbf{v}_2]^T\mathbf{v}_{\mathcal{M}}^{\perp}\|^2}{\|\mathbf{v}_{\mathcal{M}}^{\perp}\|^2} < \lambda_{\max}$. Since

$\mathbf{v}^* \neq \mathbf{v}_{\mathcal{M}}$, the equality cannot hold. It concludes that $\mathbf{a}^*$ should not be the AF scheme that maximizes the sum rate.

Then the proof is completed.

Since the same notations are used in the sum rate case, by Lemmas 5 and 6, Algorithm 1 also applies to compute the AF scheme that maximizes the sum rate of the two-hop MAC.

Discussion. We first discuss the complexity of the algorithm. Note that to obtain the AF scheme that achieves the maximum individual or sum rate, the computing step in Algorithm 1 may repeat $2^n - 2$ times in the worst case. It seems that the algorithm is infeasible for n too large in practice. Fortunately, in our previous work [13, Theorem 6], we have derived a result which shows that there exists a simple mixed-AF scheme consisting of three AF schemes that can achieve the optimal rate region within half a bit with respect to sum and individual rates. The condition under which the scheme works is that the upper layer noise dominates the total noises received at the destination node [13, Definition 6]. It is clear that in our system model, as the number of relays sufficiently large, the condition can always be satisfied. Furthermore, it is generally the case that the power of the upper layer noise increases as the number of relays becomes large. So, we can first determine whether the dominative condition holds or not before running the algorithm. Therefore, we conjecture that our algorithm combined with the asymptotical optimal AF scheme obtained in [13, Theorem 6] is efficient enough to obtain a good AF scheme from the practice perspective. However, to obtain the other boundary points and the corresponding AF schemes is not such straightforward.

Then we should point out that, the non-negative assumption of the channel gains simplifies the discussion of the issue. It does not loose generality for the maximum individual rate case. From (5), we can see that even if the channel gain takes negative value, by properly choosing a_k the elements in the summation in the numerator of the SNR function can always have the same sign. However, if we relax this assumption for the sum rate case, we find the problem becomes much more complicated. For example, if $h_{S_1,1}h_{S_2,1} < 0$, and $h_{S_1,k}h_{S_2,k} > 0$, $k \in \mathcal{V}\backslash\{1\}$, we cannot decide whether to choose $a_1 \in [0, \sqrt{P_{1,\max}}]$ or to choose $a_1 \in [-\sqrt{P_{1,\max}}, 0]$. It makes the problem intractable by the current approach.

4 Conclusion

In this paper, we have studied the optimal rate region of a two-hop MAC both under the sum and individual power constraints. We obtain the results for the sum power constraint and propose a novel technique to solve it. The new method provides a geometric interpretation of the optimal AF rate region. However, for the individual power constraints scenario, there are still several problems unsolved. First, the algorithm may not be efficient enough to compute the exact AF scheme that achieves maximum individual rate when n is too large from the practical perspective. Secondly, the general boundary points and the corresponding AF schemes are not found. Thirdly, if the non-negative assumption of the channel gains is removed, the current approach is inadequate to solve the maximum sum rate problem. To solve these problems is the goal of our future work.

Acknowledgments. This work is supported by grants from the National Natural Science Foundation of China (60832001 and 61271174).

References

1. Agnihotri, S., Jaggi, S., Chen, M.: Amplify-and-Forward in Wireless Relay Networks. In: Proc. IEEE Inf. Theory Workshop 2011, Paraty, Brazil, pp. 311–315 (2011)
2. Agnihotri, S., Jaggi, S., Chen, M.: Analog Network Coding in General SNR Regime. In: IEEE Int. Symp. Inf. Theory 2012, Cambridge, MA, pp. 2052–2056 (2012)
3. Agnihotri, S., Jaggi, S., Chen, M.: Analog network coding in general SNR regime: Performance of a greedy scheme. In: Proc. 2012 Int. Symp. NetCod, Cambridge, MA, pp. 137–142 (2012)
4. Ahlswede, R.: Multi-way communication channels. In: Proc. 2nd. Int. Symp. Inf. Theory, Tsahkadsor, Armenia, pp. 23–52 (1971)
5. Azarian, K., Gamal, H.E., Schniter, P.: On the Achievable Diversity-Multiplexing Tradeoff in Half-Duplex Cooperative Channels. IEEE Trans. Inf. Theory 51(12), 4152–4172 (2005)
6. Borade, S., Zheng, L., Gallager, R.: Amplify-and-forward in wireless relay networks: Rate, diversity, and network size. IEEE Trans. Inf. Theory 53(10), 3302–3318 (2007)
7. Dana, A.F., Gowaikar, R., Hassibi, B., Effros, M., Medard, M.: Should we break a wireless network into subnetworks? In: Allerton Conf. Commun., Contr. Comput. (2003)
8. Gastpar, M., Vetterli, M.: On the capacity of large Gaussian relay networks. IEEE Trans. Inf. Theory 51(3), 765–779 (2005)
9. Gomadam, K.S., Jafar, S.A.: The Effect of Noise Correlation in Amplify-and-Forward Relay Networks. IEEE Trans. Inf. Theory 55(2), 731–745 (2009)
10. Jafar, S.A., Gomadam, K.S., Huang, C.: Duality and Rate Optimization for Multiple Access and Broadcast Channels with Amplify-and-Forward Relays. IEEE Trans. Inf. Theory 53(10), 3350–3370 (2007)
11. Laneman, J.N., Tse, D.N.C., Wornell, G.W.: Cooperative diversity in wireless networks: efficient protocols and outage behavior. IEEE Trans. Inf. Theory 50(12), 3062–3080 (2004)
12. Liu, B., Cai, N.: Analog Network Coding in the Generalized High-SNR Regime. In: Proc. IEEE Int. Symp. Inf. Theory 2011, St. Pertersburg, Russia, pp. 74–78 (2011)
13. Liu, B., Cai, N.: Multi-hop Analog Network Coding: An Amplify-and-forward Approach. Submitted to IEEE Trans. Inf. Theory (2012), `http://arxiv.org/abs/1203.4867`

14. Marić, I., Yates, R.D.: Bandwidth and power allocation for cooperative strategies in Gaussian relay networks. IEEE Trans. Inf. Theory 56(4), 1880–1889 (2010)
15. Marić, I., Goldsmith, A., Médard, M.: Multihop Analog Network Coding via Amplify-and-Forward: The High SNR Regime. IEEE Trans. Inf. Theory 58(2), 793–803 (2012)
16. Zahedi, S., Mohseni, M., Gamal, A.E.: On the capacity of AWGN relay channels with linear relaying functions. In: Proc. IEEE Int. Symp. Inf. Theory, p. 399 (2004)
17. Zhao, Y., Adve, R., Lim, T.J.: Improving amplify-and-forward relay networks: optimal power allocation versus selection. IEEE Trans. Wireless Comm., 3114–3123 (2007)

Appendix

A Proof of Lemma 1

Proof. To prove the lemma, we justify the following proposition. To have each rate set $\mathcal{R}(\mathbf{x})$, it is sufficient to take $\mathbf{x}$ in the linear subspace span $\{\mathbf{d}_i, i = 1, 2\}$. Then the lemma can be viewed as an direct corollary of such proposition.

On one hand, any n-dimensional vector $\mathbf{x}$ can be decomposed as

$$\mathbf{x} = \mathbf{x}_1 + \mathbf{x}_2, \tag{72}$$

where $\mathbf{x}_1 \in \text{span}\{\mathbf{d}_i, i = 1, 2\}$, $\mathbf{x}_2 \in \text{span}^{\perp}\{\mathbf{d}_i, i = 1, 2\}$, and thus $\mathbf{x}_1^T\mathbf{x}_2 = 0$. On the other hand, $\text{span}^{\perp}\{\mathbf{d}_i, i = 1, 2\}$ is the solution space of the homogenous linear equations $[\mathbf{d}_1, \mathbf{d}_2]^T\mathbf{x} = \mathbf{0}$. Therefore, it follows that $\mathbf{d}_i^T\mathbf{x}_2 = 0$, $i = 1, 2$.

Given any AF scheme $\mathbf{a}$ such that $\| \mathbf{a} \|^2 = P_{sum}$, we have $\mathbf{x} = \Lambda\mathbf{a}$. Suppose $\mathbf{x} \notin \text{span}\{\mathbf{d}_i, i = 1, 2\}$, otherwise, there is nothing to prove. It can be found another scheme $\mathbf{a}_1 = c\Lambda^{-1}\mathbf{x}_1$, where c is chosen such that $\| \mathbf{a}_1 \|^2 = P_{sum}$. Then from (14), it is clear that $\mathcal{R}(\mathbf{a}) = \mathcal{R}(\mathbf{x}) \subseteq \mathcal{R}(c\mathbf{x}_1) = \mathcal{R}(\mathbf{a}_1)$, which implies that the points in rate set $\mathcal{R}(\mathbf{x})$ cannot be on the boundary of $\mathcal{R}(\{\mathbf{x}\})$.

Then we complete the proof.

B Proof of Lemma 2

Proof. The lemma can be proved by a consequence of careful calculation. Several basic conditions are described first. In the previous section, all the channel gains are assumed to be positive values, and the angles α and β are both in $\left[0, \frac{\pi}{2}\right]$ as shown in Fig. 2. Without loss of generality, assume $0 \le \alpha \le \beta \le \frac{\pi}{2}$ hence $0 \le \beta - \alpha \le \frac{\pi}{2}$. Although θ can take all values in $[-\pi, \pi]$, it is easy to find that $\mathbf{x}^T\mathbf{d}_{10} = \cos^2(\theta - \alpha) = \cos^2(\theta - \alpha \pm \pi)$ and the same argument applies to the inner product between $\mathbf{x}$ and $\mathbf{d}_{20}$. This implies that the value of θ can be limited to $\left[-\frac{\pi}{2}, \frac{\pi}{2}\right]$. Then the proof can be completed by the following discussions.

– Case 1. When $\theta \in \left[\beta, \frac{\pi}{2}\right]$,

$$\mathbf{x}^T\mathbf{d}_{10} = \cos^2(\theta - \alpha) \overset{(a)}{\le} \cos^2(\beta - \alpha), \tag{73}$$

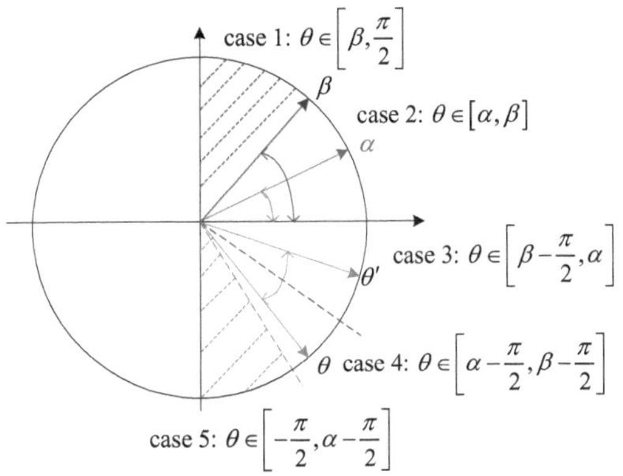

Fig. 2. Geometric Interpretation

where (a) follows by $0 \leq \beta - \alpha \leq \theta - \alpha \leq \frac{\pi}{2}$, and

$$\mathbf{x}^T \mathbf{d}_{20} = \cos^2(\theta - \beta) \leq \cos^2(\beta - \beta) = 1. \tag{74}$$

Therefore, by taking $\theta^* = \beta$, $\mathcal{R}(\theta) \subset \mathcal{R}(\theta^*)$ holds for all $\theta \in \left[\beta, \frac{\pi}{2}\right]$.

- Case 2. When $\theta \in [\alpha, \beta]$, there is nothing to prove.
- Case 3. When $\theta \in \left[\beta - \frac{\pi}{2}, \alpha\right]$,

$$\mathbf{x}^T \mathbf{d}_{10} = \cos^2(\theta - \alpha) \leq \cos^2(\alpha - \alpha) = 1, \tag{75}$$

and

$$\mathbf{x}^T \mathbf{d}_{20} = \cos^2(\beta - \theta) \overset{(a)}{\leq} \cos^2(\beta - \alpha), \tag{76}$$

where (a) follows by $0 \leq \beta - \alpha \leq \beta - \theta \leq \frac{\pi}{2}$.
Therefore, by taking $\theta^* = \alpha$, $\mathcal{R}(\theta) \subset \mathcal{R}(\theta^*)$ holds for all $\theta \in \left[\beta - \frac{\pi}{2}, \alpha\right]$.

- Case 4. When $\theta \in \left[\alpha - \frac{\pi}{2}, \beta - \frac{\pi}{2}\right]$,
let $\delta = \beta - \frac{\pi}{2} - \theta \in [0, \beta - \alpha]$ and $\theta' = \beta - \frac{\pi}{2} + \delta \in \left[\beta - \frac{\pi}{2}, 2\beta - \alpha - \frac{\pi}{2}\right] \subset \left[\beta - \frac{\pi}{2}, \beta\right]$,

$$\begin{aligned}\mathbf{x}^T \mathbf{d}_{20} &= \cos^2(\beta - \theta) = \cos^2\left(\frac{\pi}{2} + \delta\right) \\ &= \cos^2\left(\frac{\pi}{2} - \delta\right) = \cos^2(\beta - \theta')\end{aligned} \tag{77}$$

1). If $\theta' \leq \alpha$, then

$$\begin{aligned}\mathbf{x}^T \mathbf{d}_{10} &= \cos^2(\alpha - \theta) = \cos^2\left(\alpha - \beta + \frac{\pi}{2} + \delta\right) \\ &\overset{(a)}{\leq} \cos^2\left(\alpha - \beta + \frac{\pi}{2} - \delta\right) = \cos^2(\alpha - \theta'),\end{aligned} \tag{78}$$

where (a) follows from $0 \leq \alpha - \theta' = \alpha - \beta + \frac{\pi}{2} - \delta \leq \alpha - \beta + \frac{\pi}{2} + \delta = \alpha - \theta \leq \frac{\pi}{2}$.

Therefore, in such case, for any θ there always exists a $\theta' \in \left[\beta - \frac{\pi}{2}, \alpha\right]$ that $\mathcal{R}(\theta) \subset \mathcal{R}(\theta')$ holds. Hence from the result of case 3, by taking $\theta^* = \alpha$, $\mathcal{R}(\theta) \subset \mathcal{R}(\theta^*)$ holds.
2). If $\alpha < \theta' \leq \beta$, then

$$\mathbf{x}^T \mathbf{d}_{10} = \cos^2(\alpha - \theta) = \cos^2\left(\alpha - \beta + \frac{\pi}{2} + \delta\right) \\ \overset{(a)}{\leq} \cos^2\left(\beta - \alpha - \frac{\pi}{2} + \delta\right) = \cos^2(\theta' - \alpha), \tag{79}$$

where (a) follows from $0 \leq \alpha - \theta' = \alpha - \beta + \frac{\pi}{2} - \delta \leq \alpha - \beta + \frac{\pi}{2} + \delta = \alpha - \theta \leq \frac{\pi}{2}$.

Therefore, in such case, for any θ there always exists a $\theta' \in [\alpha, \beta]$ that $\mathcal{R}(\theta) \subset \mathcal{R}(\theta')$ holds.

- Case 5. When $\theta \in \left[-\frac{\pi}{2}, \alpha - \frac{\pi}{2}\right]$,

$$\mathbf{x}^T \mathbf{d}_{10} = \cos^2(\theta - \alpha) = \cos^2(\theta - \alpha + \pi) \\ \overset{(a)}{\leq} \cos^2(\beta - \alpha), \tag{80}$$

where (a) follows from $\frac{\pi}{2} - \beta \leq \frac{\pi}{2} - \alpha \leq \theta - \alpha + \pi \leq \frac{\pi}{2}$, and

$$\mathbf{x}^T \mathbf{d}_{20} = \cos^2(\theta - \beta) \leq \cos^2(\beta - \beta) = 1. \tag{81}$$

Therefore, by taking $\theta^* = \beta$, $\mathcal{R}(\theta) \subset \mathcal{R}(\theta^*)$ holds for all $\theta \in \left[-\frac{\pi}{2}, \alpha - \frac{\pi}{2}\right]$.

Then we complete the proof.

Strong Secrecy for Multiple Access Channels

Moritz Wiese and Holger Boche

Technische Universität München, Lehrstuhl für Theoretische Informationstechnik,
Theresienstr. 90, 80333 München, Germany
{wiese,boche}@tum.de

Dedicated to the memory of Rudolf Ahlswede

Abstract. We show strongly secret achievable rate regions for two different wiretap multiple-access channel coding problems. In the first problem, each encoder has a private message and both together have a common message to transmit. The encoders have entropy-limited access to common randomness. If no common randomness is available, then the achievable region derived here does not allow for the secret transmission of a common message. The second coding problem assumes that the encoders do not have a common message nor access to common randomness. However, they may have a conferencing link over which they may iteratively exchange rate-limited information. This can be used to form a common message and common randomness to reduce the second coding problem to the first one. We give the example of a channel where the achievable region equals zero without conferencing or common randomness and where conferencing establishes the possibility of secret message transmission. Both coding problems describe practically relevant networks which need to be secured against eavesdropping attacks.

Keywords: wiretap multiple access channel, information-theoretic security, strong secrecy, common randomness.

1 Introduction

The wiretap Multiple-Access Channel (MAC) combines two areas where Rudolf Ahlswede has made major contributions. In the area of multi-user information theory, he [9] and Liao [25] independently gave one of the first complete characterizations of the capacity region of a multi-user channel – the MAC with one message per sender. Later, Dueck [18] proved the strong converse for the MAC and Ahlswede [10] gave an elementary proof immediately afterwards. Slepian and Wolf generalized the results from [9] and [25] to the case where the senders additionally have a common message [32]. Willems used Slepian and Wolf's result to derive the capacity region of the MAC with conferencing encoders. This is a MAC without common message, but the encoders can exchange rate-limited information about their messages in an interactive conferencing protocol [38, 39]. The results of Slepian and Wolf as well as Willems' result were only recently generalized to general compound MACs with partial channel state

H. Aydinian, F. Cicalese, and C. Deppe (Eds.): Ahlswede Festschrift, LNCS 7777, pp. 71–122, 2013.

information in [37], arbitrarily varying MACs with conferencing encoders were treated in [36]. The latter paper made substantial use of techniques developed by Ahlswede for single-sender arbitrarily varying channels in [1–3] and also of his and Cai's contribution to arbitrarily varying MACs [5].

The other area of Ahlswede's interest which plays a role in this paper is secrecy and common randomness. Among other problems, he considered together with Csiszár in [6, 7] how a secret key can be shared at distant terminals in the presence of an eavesdropper. Work on secret key sharing aided by public communication goes back to Maurer [28]. The first paper which exploits the statistics of a discrete memoryless channel to establish secret communication is due to Wyner [40]. He considers the wiretap channel, the simplest model of a communication scenario where secrecy is relevant: a sender would like to transmit a message to a receiver over a discrete memoryless channel and transmission is overheard by a second receiver who should be kept ignorant of the message. It was noted by Wyner that a secret key shared at both legitimate terminals is not necessary to establish secret transmission – if the channel statistics are taken into consideration, it is sufficient that the sender randomizes his inputs in order to secure transmission.

Since Wyner discovered this fact, information-theoretic secrecy for message transmission without a key shared between sender and legitimate receiver has been generalized in various directions. The first paper on multi-user information-theoretic security is due to Csiszár and Körner [15]. Here, the second receiver only is a partial eavesdropper: there is a common message intended for both receivers, but as in the original wiretap channel, an additional private message intended for the first receiver must be kept secret from the second. We come to multiple-access models below. An overview over the area is given in [24].

The original secrecy criterion used in [40] and [15] and in most of the subsequent work until today has become known as the "weak secrecy criterion". Given a code, it measures the mutual information normalized by the code blocklength between the randomly chosen message and the eavesdropper's output corresponding to the application of the code and transmission over the channel. Maurer introduced the "strong secrecy criterion" in [29] by omitting the normalization. The advantage of this criterion was revealed in [11]: it can be given an operational meaning, i.e. one can specify the attacks it withstands. It is possible to show that if transmission obeys the strong secrecy criterion, then the eavesdropper's average error tends to one for any decoder it might apply. Translated into practical secrecy schemes, this means that no matter how large the computing power of a possible eavesdropper might be, it will not succeed in breaking the security of this scheme. For the weak criterion, there are still only heuristic argumentations as to why it should be secret. Further secrecy metrics are presented in [12], but without giving them an operational meaning, strong secrecy remains the strongest of these metrics. To our knowledge, there are three different approaches to establishing strong secrecy in a wiretap channel so far [14, 17, 28]. In fact, the last of these approaches also applies to classical-quantum wiretap channels [17] and also was used to give an achievable rate for the classical compound wiretap channel [11].

There exist many MAC models where secrecy is an issue. This may even be the case when there is no eavesdropper, as each encoder might have access to noisy observations of the other sender's codeword but wants to protect its own message from decoding at the other sender [19, 23, 27]. The case where the encoders have access to generalized feedback but only keep their messages secret from an external eavesdropper is considered in [33]. In the cognitive MAC, only one encoder has a private message, and together, the encoders have a common message. There are again two cases: In the case without an eavesdropper, the encoder without a private message has access to the codeword sent by the other encoder through a noisy channel and must be kept ignorant of the other encoder's private message [26]. In [31], the cognitive MAC without feedback was investigated where the messages must be kept secret from an eavesdropper and the encoders have unrestricted access to common randomness. All of these papers use the weak secrecy criterion.

The first part of this article generalizes and strengthens the achievability result from [20] where multi-letter characterizations of an achievable region and of an outer bound on the capacity region of a MAC without common message and with an external eavesdropper under the weak secrecy criterion are given. The channel needs to satisfy certain relatively strong conditions for the bounds to work. Extensions to the Gaussian case can be found in [20, 21, 34].

We consider two senders Alice_1 and Alice_2. Each has a private message and together they have a common message. This message triple must be transmitted to Bob over a discrete memoryless MAC in such a way that Eve who obtains a version of the sent codewords through another discrete memoryless MAC cannot decode the messages. We apply the strong secrecy criterion. In order to find a code which satisfies this criterion, we use Devetak's approach [17], which in the quantum case builds on the Ahlswede-Winter lemma [8] and classically on a Chernoff bound. It is similar to the approach taken in [13]. As the senders have a common message and as the second part of the paper deals with the wiretap MAC with conferencing encoders, we assume that the encoders have access to a restricted amount of common randomness. Common randomness for encoding has so far only been used in [31], but without setting any limitations on its amount. Note that this use of common randomness in order to establish secrecy differs from the use made in [6, 7]. We only obtain an achievable region. In this achievable region it is not possible to transmit a common message if no common randomness is available. Further it is notable that we use random coding and have to apply time-sharing before derandomizing.

The wiretap MAC with common message and common randomness is also needed in the second part of this paper about the wiretap MAC with conferencing encoders. Conferencing was introduced by Willems in [38, 39] and is an iterative protocol for the senders of a MAC to exchange information about their messages. One assumes that the amount of information that is exchanged is rate-limited because otherwise one would obtain a single-encoder wiretap MAC. Willems already used the coding theorem for the MAC with common message to deduce an achievable region for the conferencing MAC. The same can be done for the wiretap MAC with conferencing encoders. More precisely, aside from the senders' private messages, there are no further messages to be transmitted, and no common randomness is available. However, conferencing is used to produce both a common message and common randomness, which allows the reduction. A consequence of the fact that no common message can be transmitted by the wiretap MAC with common message if there is no common randomness is that one has to use conferencing to establish some common randomness if this is supposed to enlarge the achievable region compared to what would be achievable without conferencing. Again, this consequence presumes that the achievable region equals the capacity region even though we cannot prove this.

Information-theoretic security has far-reaching practical consequences. As digital communication replaces more and more of the classical paper-based ways of communication even for the transmission of sensible data, the problem of securing these data becomes increasingly important. Information-theoretic secrecy provides an alternative to the traditional cryptographic approach which bases on the assumption of limited computing power. However, as information-theoretic security uses the imperfections of the channels to secure data, its models must be sufficiently complex to describe realistic scenarios. Our article shows how encoder cooperation can be utilized to secure data. The cooperation of base stations in mobile networks is included in future wireless network standards, and our work can be seen as a contribution to the theoretical analysis of how it fares when it comes to security. But already Csiszár and Körner's paper on the broadcast channel with confidential messages shows how messages with different secrecy requirements can be combined in one transmission. A more recent example which also applies the strong secrecy criterion is given in [41].

Organization of the Paper: The next section introduces the general model of a wiretap MAC and also presents the Willems conferencing protocol. Section 3 contains the two achievability theorems for the wiretap MAC with common message and the wiretap MAC with conferencing encoders.

The common message theorem is treated in the rather long Section 4. First, the regions we claim to be achievable are decomposed into regions whose achievability can be shown more easily. Following Devetak, it is shown that it is sufficient to make Eve's output probability given a message triple almost independent of this triple in terms of variation distance. Then, in the mathematical core of the paper, we derive lower bounds on the randomness necessary to achieve strong secrecy using probabilistic concentration results. Here we also follow Devetak. Having derived these bounds, we finally find a realization of the random codes which defines a good wiretap code.

Section 5 gives the proof of the achievability theorem for the wiretap MAC with conferencing encoders. We again have to decompose the claimed regions into regions whose achievability can be shown more easily. Then we can reduce the problem of achieving a certain rate pair with conferencing to the problem of achieving a certain rate triple by the wiretap MAC with common message more or less in the same way as done by Willems in the non-wiretap situation. Finally, Section 6 shows that conferencing may help in situations where no secret transmission is possible without and that for our approach it is necessary to do the time-sharing within the random coding.

Notation: For sets $\{1, \ldots, M\}$, where M is a positive integer, we use the combinatorial shorthand $[M]$. For a real number x we define $[x]_+ := \max\{x, 0\}$.

For any set $\mathcal{X}$ and subset $A \subset \mathcal{X}$ we write $A^c := \mathcal{X} \setminus A$. We let $1_A : \mathcal{X} \to \{0, 1\}$ be the indicator function of A which takes on the value 1 at $x \in \mathcal{X}$ if and only if $x \in A$. Given a probability space $(\Omega, \mathcal{A}, \mathbb{P})$ we write $\mathbb{E}$ for the expectation corresponding to $\mathbb{P}$ and for $A \in \mathcal{A}$ and a real-valued random variable X we write $\mathbb{E}[X; A] := E[X 1_A]$.

The space of probability distributions on the finite set $\mathcal{X}$ is denoted by $\mathcal{P}(\mathcal{X})$. In particular, it contains for every $x \in \mathcal{X}$ the probability measure δ_x defined by $\delta_x(x) = 1$. The product of two probability distributions P and Q is denoted by $P \otimes Q$. A stochastic matrix with input alphabet $\mathcal{X}$ and output alphabet $\mathcal{Z}$ is written as a mapping $W : \mathcal{X} \to \mathcal{P}(\mathcal{Z})$. The n-fold memoryless extension of a channel $W : \mathcal{X} \to \mathcal{P}(\mathcal{Z})$ is denoted by $W^{\otimes n}$, so that for $\mathbf{x} = (x_1, \ldots, x_n) \in \mathcal{X}^n$ and $\mathbf{z} = (z_1, \ldots, z_n) \in \mathcal{Z}^n$,

$$W^{\otimes n}(\mathbf{z}|\mathbf{x}) = \prod_{i=1}^{n} W(z_i|x_i).$$

We also define for $P \in \mathcal{P}(\mathcal{X})$ and $W : \mathcal{X} \to \mathcal{P}(\mathcal{Z})$ the probability distribution $P \otimes W \in \mathcal{P}(\mathcal{X} \times \mathcal{Z})$ by $(P \otimes W)(x, z) = P(x)W(z|x)$.

Every measure μ on the finite set $\mathcal{X}$ can be identified with a unique function $\mu : \mathcal{X} \to [0, \infty]$. Then for any subset $A \subset \mathcal{X}$ we have $\mu(A) = \sum_{x \in A} \mu(x)$. On the set of measures on $\mathcal{X}$, we define the total variation distance by

$$\|\mu_1 - \mu_2\| := \sum_{x \in \mathcal{X}} |\mu_1(x) - \mu_2(x)|.$$

Given a random variable X living on $\mathcal{X}$ and a $P \in \mathcal{P}(\mathcal{X})$, we mean by $X \sim P$ that P is the distribution of X. Given a pair of random variables (X, Y) taking values in the finite set $\mathcal{X} \times \mathbf{y}$, we write $P_X \in \mathcal{P}(\mathcal{X})$ for the distribution of X and $P_{X|Y}$ for the conditional distribution of X given Y. We also write $T^n_{X,\delta} \subset \mathcal{X}^n$ for the subset of δ-typical sequences with respect to X and $T^n_{X|Y,\delta}(\mathbf{y}) \subset \mathbf{y}^n$ for the subset of conditionally δ-typical sequences with respect to $P_{X|Y}$ given $\mathbf{y} \in \mathbf{y}^n$. Given a sequence $\mathbf{x} \in \mathcal{X}^n$ and an $x \in \mathcal{X}$, we let $N(x|\mathbf{x})$ be the number of coordinates of $\mathbf{x}$ equal to x.

For random variables X, Y, Z we write $H(X)$ for the entropy of X, $H(X|Y)$ for the conditional entropy of X given Y, $I(X \wedge Y)$ for the mutual information of X and Y and $I(X \wedge Y|Z)$ for the conditional mutual information of X and Y given Z.

2 The Wiretap Multiple-Access Channel

The wiretap Multiple-Access Channel (MAC) is described by a stochastic matrix

$$W : \mathcal{X} \times \mathcal{Y} \to \mathcal{T} \times \mathcal{Z},$$

where $\mathcal{X}, \mathcal{Y}, \mathcal{T}, \mathcal{Z}$ are finite sets. We write W_b and W_e for the marginal channels to $\mathcal{T}$ and $\mathcal{Z}$, so e.g.

$$W_b(t|x,y) := \sum_{z\in\mathcal{Z}} W(t,z|x,y).$$

$\mathcal{X}$ and $\mathcal{Y}$ are the finite alphabets of Alice$_1$ and Alice$_2$, respectively. $\mathcal{T}$ is the finite alphabet of the receiver called Bob and the outputs received by the eavesdropper Eve are elements of the finite alphabet $\mathcal{Z}$.

2.1 With Common Message

Let H_C be a nonnegative real number. A wiretap MAC code with common message and blocklength n satisfying the common randomness bound H_C consists of a stochastic matrix

$$G : [K_0] \times [K_1] \times [K_2] \to \mathcal{P}(\mathcal{X}^n \times \mathcal{Y}^n)$$

and a decoding function

$$\phi : \mathcal{T}^n \to [K_0] \times [K_1] \times [K_2].$$

G is required to have the form

$$G(\mathbf{x}, \mathbf{y}|k_0, k_1, k_2) = \sum_{j\in\mathcal{J}} G_0(j|k_0)G_1(\mathbf{x}|k_0, k_1, j)G_2(\mathbf{y}|k_0, k_2, j),$$

where $\mathcal{J}$ is some finite set and

$$G_0 : [K_0] \to \mathcal{P}(\mathcal{J}),$$
$$G_1 : [K_0] \times [K_1] \times \mathcal{J} \to \mathcal{P}(\mathcal{X}),$$
$$G_2 : [K_0] \times [K_2] \times \mathcal{J} \to \mathcal{P}(\mathcal{Y}).$$

Further, G_0 has to satisfy that $H(J|M_0) \leq nH_C$ for M_0 uniformly distributed on $[K_0]$ and $P_{J|M_0} = G_0$. $[K_0]$ is called the set of common messages, $[K_1]$ is the set of Alice$_1$'s private messages and $[K_2]$ the set of Alice$_2$'s private messages.

Let M_0, M_1, M_2 be independent random variables uniformly distributed on $[K_0]$, $[K_1]$ and $[K_2]$, respectively. Further, let X^n, Y^n, T^n, Z^n be random variables such that for $(\mathbf{x}, \mathbf{y}, \mathbf{t}, \mathbf{z}) \in \mathcal{X}^n \times \mathcal{Y}^n \times \mathcal{T}^n \times \mathcal{Z}^n$

$$P_{X^nY^n|M_0M_1M_2}(\mathbf{x}, \mathbf{y}|k_0, k_1, k_2) = G(\mathbf{x}, \mathbf{y}|k_0, k_1, k_2),$$
$$P_{T^nZ^n|X^nY^nM_0M_1M_2}(\mathbf{t}, \mathbf{z}|\mathbf{x}, \mathbf{y}, k_0, k_1, k_2) = W^{\otimes n}(\mathbf{t}, \mathbf{z}|\mathbf{x}, \mathbf{y}).$$

Then the average error of the code defined above equals

$$\mathbb{P}[\phi(T^n) \neq (M_0, M_1, M_2)].$$

Definition 1. *A rate pair $(R_0, R_1, R_2) \in \mathbb{R}^3_{\geq 0}$ is achievable by the wiretap MAC with common message under the common randomness bound $H_C \geq 0$ if for every $\eta > 0$ and every $\varepsilon \in (0,1)$ and n large there exists a wiretap MAC code with common message and blocklength n satisfying the common randomness bound H_C and*

$$\frac{1}{n} \log K_\nu \geq R_\nu - \eta \qquad (\nu = 0, 1, 2),$$
$$\mathbb{P}[\phi(T^n) \neq (M_0, M_1, M_2)] \leq \varepsilon,$$
$$I(Z^n \wedge M_0M_1M_2) \leq \varepsilon.$$

Remark 1. *It was shown in [11] that no matter how Eve tries to decode the messages from the Alices, the average error must tend to one. More precisely, assume that a wiretap code with common message and blocklength n is given, and assume that Eve has a decoding function*

$$\chi : \mathcal{Z}^n \to [K_0] \times [K_1] \times [K_2].$$

Then

$$\mathbb{P}[\chi(Z^n) \neq (M_0, M_1, M_2)] \geq 1 - \varepsilon'$$

for some ε' which tends to zero as ε tends to zero. If ε tends to zero exponentially fast and K_0, K_1, K_2 grow exponentially, then ε' tends to zero at exponential speed.

More generally assume that $f : [K_0] \times [K_1] \times [K_2] \to [K']$ is a function satisfying $\mathbb{P}[f(M_0, M_1, M_2) = k'] = 1/K'$ for all $k' \in [K']$. Then with the same argument as in [11] one can show that for every function $g : \mathcal{Z}^n \to [K']$, one has $\mathbb{P}[f(M) \neq g(Z^n)] \geq 1 - 1/K' - \varepsilon'$ for the same ε' as above. That is, even for $K' = 2$, blind guessing is the best way for Eve to estimate $f(M)$. In particular, no subset of the message random variables, like M_0 or (M_1, M_2), can be reliably decoded by Eve.

2.2 With Conferencing Encoders

In the wiretap MAC with conferencing encoders, Alice$_1$ and Alice$_2$ do not have a common message nor common randomness. However before forming their codewords, they may exchange some information about their private messages according to an iterative and randomized "conferencing" protocol whose deterministic form was introduced by Willems [38, 39]. If the respective message sets are $[K_1]$ and $[K_2]$, such a stochastic Willems conference can be described as follows. Let finite sets $\mathcal{J}_1$ and $\mathcal{J}_2$ be given which can be written as products

$$\mathcal{J}_\nu = \mathcal{J}_{\nu,1} \times \ldots \times \mathcal{J}_{\nu,I} \qquad (\nu = 1, 2)$$

for some positive integer I which does not depend on ν. A Willems conferencing stochastic matrix c completely describing such a conference is determined in an iterative manner via sequences of stochastic matrices $c_{1,1}, \ldots, c_{1,I}$ and $c_{2,1}, \ldots, c_{2,I}$. $c_{1,i}$ describes the probability distribution of what Alice$_1$ tells Alice$_2$ in the i-th conferencing iteration given the knowledge accumulated so far at Alice$_1$. Thus in general, using the notation

$$\bar{\nu} := \begin{cases} 1 & \text{if } \nu = 2, \\ 2 & \text{if } \nu = 1, \end{cases}$$

these stochastic matrices satisfy for $\nu = 1, 2$ and $i = 2, \ldots, I$,

$$\begin{aligned} c_{\nu,1} &: [K_\nu] \to \mathcal{P}(\mathcal{J}_{\nu,1}), \\ c_{\nu,i} &: [K_\nu] \times \mathcal{J}_{\bar{\nu},1} \times \ldots \times \mathcal{J}_{\bar{\nu},i-1} \to \mathcal{P}(\mathcal{J}_{\nu,i}). \end{aligned}$$

The stochastic matrix $c : [K_1] \times [K_2] \to \mathcal{P}(\mathcal{J}_1 \times \mathcal{J}_2)$ is obtained by setting

$$\begin{aligned} &c(j_{1,1}, \ldots, j_{1,I}, j_{2,1}, \ldots, j_{2,I} | k_1, k_2) \\ &:= c_{1,1}(j_{1,1}|k_1)\, c_{2,1}(j_{2,1}|k_2) \cdots \\ &\quad \cdots c_{1,I}(j_{1,I}|k_1, j_{2,1}, \ldots, j_{2,I-1})\, c_{2,I}(j_{2,I}|k_2, j_{1,1}, \ldots, j_{1,I-1}). \end{aligned}$$

We denote the $\mathcal{J}_1$- and $\mathcal{J}_2$-marginals of this stochastic matrix by c_1 and c_2, so $c_1(j_{1,1}, \ldots, j_{1,I}|k_1, k_2)$ is obtained by summing over $j_{2,1}, \ldots, j_{2,I}$ and c_2 is obtained analogously.

Now we define a wiretap MAC code with conferencing encoders. It consists of a Willems conferencing stochastic matrix $c : [K_1] \times [K_2] \to \mathcal{P}(\mathcal{J}_1 \times \mathcal{J}_2)$ as above together with encoding stochastic matrices

$$\begin{aligned} G_1 &: [K_1] \times \mathcal{J}_2 \to \mathcal{X}^n, \\ G_2 &: [K_2] \times \mathcal{J}_1 \to \mathcal{Y}^n \end{aligned}$$

and a decoding function

$$\phi : \mathcal{T}^n \to [K_1] \times [K_2].$$

$[K_1]$ is the set of Alice$_1$'s messages and $[K_2]$ is the set of Alice$_2$'s messages. A pair $(k_1, k_2) \in [K_1] \times [K_2]$ is encoded into the codeword pair $(\mathbf{x}, \mathbf{y}) \in \mathcal{X}^n \times \mathcal{Y}^n$ with probability

$$\sum_{(j_1,j_2) \in \mathcal{J}_1 \times \mathcal{J}_2} c(j_1, j_2|k_1, k_2)\, G_1(\mathbf{x}|k_1, j_2)\, G_2(\mathbf{y}|k_2, j_1). \qquad (1)$$

In particular, conferencing generates common randomness. As both c_1 and c_2 may depend on both encoders' messages, the codewords may as well depend on both messages. Thus if conferencing were unrestricted, this would transform the MAC into a single-user wiretap channel with input alphabet $\mathcal{X} \times \mathcal{Y}$. However, Willems introduces a restriction in terms of the blocklength of the code which is used for transmission. For conferencing under conferencing capacities $C_1, C_2 \geq 0$, he requires that for a blocklength-n code, $|\mathcal{J}_1|$ and $|\mathcal{J}_2|$ satisfy

$$\frac{1}{n}\log|\mathcal{J}_\nu| \leq C_\nu \qquad (\nu = 1,2). \tag{2}$$

We also impose this constraint and define a wiretap MAC code with conferencing capacities $C_1, C_2 \geq 0$ to be a wiretap MAC code with conferencing encoders satisfying (2).

Let a wiretap MAC code with conferencing encoders be given and let M_1, M_2 be independent random variables uniformly distributed on $[K_1]$ and $[K_2]$, respectively. Let X^n, Y^n, T^n, Z^n be random variables such that conditional on (M_1, M_2), the distribution of (X^n, Y^n) is given by (1) and such that

$$P_{T^n Z^n | X^n Y^n M_1 M_2} = W^{\otimes n}.$$

Then the average error of the code defined above equals

$$\mathbb{P}[\phi(T^n) \neq (M_1, M_2)].$$

Definition 2. *A rate pair $(R_1, R_2) \in \mathbb{R}^3_{\geq 0}$ is achievable by the wiretap MAC with conferencing encoders at conferencing capacities $C_1, C_2 > 0$ if for every $\eta > 0$ and every $\varepsilon \in (0,1)$ and for n large there exists a wiretap MAC code with conferencing capacities C_1, C_2 and blocklength n satisfying*

$$\begin{aligned} \frac{1}{n}\log K_\nu &\geq R_\nu - \eta \qquad (\nu = 1,2), \\ \mathbb{P}[\phi(T^n) \neq (M_1, M_2)] &\leq \varepsilon, \\ I(Z^n \wedge M_1 M_2) &\leq \varepsilon. \end{aligned}$$

Remark 2. *Here again, as in Remark 1, the average decoding error for any decoder Eve might apply tends to 1 if the security criterion is satisfied.*

3 Coding Theorems

3.1 For the Wiretap MAC with Common Message

Let $H_C \geq 0$ be the common randomness bound. The rate region whose achievability we are about to claim in Theorem 1 can be written as the closure of the convex hull of the union of certain rate sets which are parametrized by the elements of a subset Π_{H_C} of the set Π which is defined as follows. Π contains all probability distributions p of random vectors $(U, V_1, V_2, X, Y, T, Z)$ living on sets $\mathcal{U} \times \mathcal{V}_1 \times \mathcal{V}_2 \times \mathcal{X} \times \mathcal{Y} \times \mathcal{T} \times \mathcal{Z}$, where $\mathcal{U}, \mathcal{V}_1, \mathcal{V}_2$ are finite subsets of the integers and where p has the form

$$p = P_U \otimes P_{V_1|U} \otimes P_{V_2|U} \otimes P_{X|V_1} \otimes P_{Y|V_2} \otimes W.$$

Next we define Π_{H_C}. There are four cases altogether, numbered Case 0 to Case 3. Case 0 corresponds to $H_C = 0$ and if $H_C > 0$, then Π_{H_C} has the form $\Pi_{H_C} = \Pi_{H_C}^{(1)} \cup \Pi_{H_C}^{(2)} \cup \Pi_{H_C}^{(3)}$, and each of these subsets corresponds to one of these cases. The one condition all cases have in common is that $I(Z \wedge V_1V_2) \leq I(T \wedge V_1V_2)$.

Case 0: If $H_C = 0$ define the set Π_0 as the set of those $p \in \Pi$ where V_1 and V_2 are independent of U (so we can omit U in this case and V_1 and V_2 are independent) and where p satisfies the inequalities

$$I(Z \wedge V_1) \leq I(T \wedge V_1|V_2), \tag{3}$$
$$I(Z \wedge V_2) \leq I(T \wedge V_2|V_1). \tag{4}$$

For $p \in \Pi_0$ define the set $\mathcal{R}^{(0)}(p)$ to be the set of nonnegative triples (R_0, R_1, R_2) satisfying

$$\begin{aligned}
R_0 &= 0,\\
R_1 &\leq I(T \wedge V_1|V_2) - I(Z \wedge V_1) - [I(Z \wedge V_2|V_1) - I(T \wedge V_2|V_1)]_+,\\
R_2 &\leq I(T \wedge V_2|V_1) - I(Z \wedge V_2) - [I(Z \wedge V_1|V_2) - I(T \wedge V_1|V_2)]_+,\\
R_1 + R_2 &\leq I(T \wedge V_1V_2) - I(Z \wedge V_1V_2).
\end{aligned}$$

Case 1: $\Pi_{H_C}^{(1)}$ is the set of those $p \in \Pi$ which satisfy $I(Z \wedge U) < H_C$ and

$$I(Z \wedge V_1|U) \leq I(T \wedge V_1|V_2U), \tag{5}$$
$$I(Z \wedge V_2|U) \leq I(T \wedge V_2|V_1U), \tag{6}$$
$$I(Z \wedge V_1V_2|U) \leq I(T \wedge V_1|V_2U) + I(T \wedge V_2|V_1U). \tag{7}$$

Then we denote by $\mathcal{R}^{(1)}(p)$ the set of nonnegative real triples (R_0, R_1, R_2) satisfying

$$\begin{aligned}
R_1 &\leq I(T \wedge V_1|V_2U) - I(Z \wedge V_1|U)\\
&\qquad - [I(Z \wedge V_2|V_1U) - I(T \wedge V_2|V_1U)]_+,\\
R_2 &\leq I(T \wedge V_2|V_1U) - I(Z \wedge V_2|U)\\
&\qquad - [I(Z \wedge V_1|V_2U) - I(T \wedge V_1|V_2U)]_+,\\
R_1 + R_2 &\leq I(T \wedge V_1V_2|U) - I(Z \wedge V_1V_2|U),\\
R_0 + R_1 + R_2 &\leq I(T \wedge V_1V_2) - I(Z \wedge V_1V_2).
\end{aligned}$$

Case 2: The conditions for p to be contained in $\Pi_{H_C}^{(2)}$ cannot be phrased as simply as for $\Pi_{H_C}^{(1)}$. Generally, if $p \in \Pi_{H_C}^{(2)}$ then

$$\min\{I(Z \wedge V_1U), I(Z \wedge V_2U)\} < H_C \leq I(Z \wedge V_1V_2).$$

This is sufficient if $I(Z \wedge V_1|V_2U) = I(Z \wedge V_2|V_1U)$. If $I(Z \wedge V_1|V_2U) > I(Z \wedge V_2|V_1U)$ then we additionally require that

$$\alpha_0^{(2)} := \max\left(\frac{I(Z \wedge V_1U) - H_C}{I(Z \wedge V_1|V_2U) - I(Z \wedge V_2|V_1U)}, 1 - \frac{I(T \wedge V_2|V_1U)}{I(Z \wedge V_2|V_1U)}, 0\right)$$
$$\leq \alpha_1^{(2)} := \min\left(\frac{I(T \wedge V_1|V_2U)}{I(Z \wedge V_1|V_2U)}, \frac{I(T \wedge V_1V_2|U) - I(Z \wedge V_2|V_1U)}{I(Z \wedge V_1|V_2U) - I(Z \wedge V_2|V_1U)}, 1\right)$$

whereas if $I(Z \wedge V_1|V_2U) < I(Z \wedge V_2|V_1U)$ then we need

$$\alpha_0^{(2)} := \max\left(1 - \frac{I(T \wedge V_2|V_1U)}{I(Z \wedge V_2|V_1U)}, \frac{I(T \wedge V_1V_2|U) - I(Z \wedge V_2|V_1U)}{I(Z \wedge V_1|V_2U) - I(Z \wedge V_2|V_1U)}, 0\right)$$
$$\leq \alpha_1^{(2)} := \min\left(\frac{H_C - I(Z \wedge V_1U)}{I(Z \wedge V_2|V_1U) - I(Z \wedge V_1|V_2U)}, \frac{I(T \wedge V_1|V_2U)}{I(Z \wedge V_1|V_2U)}, 1\right).$$

In the case of equality, i.e. if $I(Z \wedge V_1|V_2U) = I(Z \wedge V_2|V_1U)$, we define $\mathcal{R}^{(2)}(p)$ as

$$\begin{aligned} R_1 &\leq I(T \wedge V_1|V_2U), \\ R_2 &\leq I(T \wedge V_2|V_1U), \\ R_1 + R_2 &\leq I(T \wedge V_1V_2|U) - I(Z \wedge V_1|V_2U), \\ R_0 + R_1 + R_2 &\leq I(T \wedge V_1V_2) - I(Z \wedge V_1V_2). \end{aligned}$$

If $I(Z \wedge V_1|V_2U) > I(Z \wedge V_2|V_1U)$, we define $\mathcal{R}^{(2)}(p)$ by

$$\begin{aligned} R_1 &\leq I(T \wedge V_1|V_2U) - \alpha_0^{(2)} I(Z \wedge V_1|V_2U), \\ R_2 &\leq I(T \wedge V_2|V_1U) - (1 - \alpha_1^{(2)}) I(Z \wedge V_2|V_1U), \\ R_1 + R_2 &\leq I(T \wedge V_1V_2|U) - \alpha_0^{(2)} I(Z \wedge V_1|V_2U) \qquad (8) \\ &\quad - (1 - \alpha_0^{(2)}) I(Z \wedge V_2|V_1U), \\ R_1 + \frac{I(Z \wedge V_2|V_1U)}{I(Z \wedge V_1|V_2U)} R_2 &\leq I(T \wedge V_2|V_1U) \qquad (9) \\ &\quad + \left(\frac{I(T \wedge V_1|U)}{I(Z \wedge V_1|V_2U)} - 1\right) I(Z \wedge V_2|V_1U), \\ R_0 + R_1 + R_2 &\leq I(T \wedge V_1V_2) - I(Z \wedge V_1V_2). \end{aligned}$$

The bound (8) on $R_1 + R_2$ can be reformulated as

$$\begin{aligned} R_1 + R_2 \leq\ & I(T \wedge V_1V_2|U) - I(Z \wedge V_1V_2|U) \\ & + \min\Big\{ H_C - I(Z \wedge U), I(Z \wedge V_1|U), \\ & \quad I(T \wedge V_1|V_2U)\left(\frac{I(Z \wedge V_2|V_1U)}{I(Z \wedge V_1|V_2U)} - 1\right) + I(Z \wedge V_1|U) \Big\}, \end{aligned}$$

and if $I(Z \wedge V_2|V_1U) > 0$, we can give the weighted sum bound (9) the almost symmetric form

$$\frac{R_1}{I(Z \wedge V_1|V_2U)} + \frac{R_2}{I(Z \wedge V_2|V_1U)} \leq \frac{I(T \wedge V_1|U)}{I(Z \wedge V_1|V_2U)} + \frac{I(T \wedge V_2|V_1U)}{I(Z \wedge V_2|V_1U)} - 1.$$

For the case that $I(Z\wedge V_1|V_2U) < I(Z\wedge V_2|V_1U)$, we define $\mathcal{R}^{(2)}(p)$ by exchanging the roles of V_1 and V_2.

Case 3: We define $\Pi_{H_C}^{(3)}$ to be the set of those $p \in \Pi$ with $I(Z \wedge V_1V_2) < H_C$ and for such a p let $\mathcal{R}^{(3)}(p)$ equal

$$\begin{aligned} R_1 &\leq I(T \wedge V_1|V_2U), \\ R_2 &\leq I(T \wedge V_2|V_1U), \\ R_1 + R_2 &\leq I(T \wedge V_1V_2|U), \\ R_0 + R_1 + R_2 &\leq I(T \wedge V_1V_2) - I(Z \wedge V_1V_2). \end{aligned}$$

Theorem 1. *For the common randomness bound $H_C = 0$, the wiretap MAC W with common message achieves the set*

$$\mathit{closure}\left(\mathit{conv}\left(\bigcup_{p\in\Pi_0} \mathcal{R}^{(0)}(p)\right)\right). \tag{10}$$

If $H_C > 0$, then the closure of the convex hull of the set

$$\bigcup_{p\in\Pi_{H_C}^{(1)}} \mathcal{R}^{(1)}(p) \cup \bigcup_{p\in\Pi_{H_C}^{(2)}} \mathcal{R}^{(2)}(p) \cup \bigcup_{p\in\Pi_{H_C}^{(3)}} \mathcal{R}^{(3)}(p)$$

is achievable.

Remark 3. *Using the standard Carathéodory-Fenchel technique, one can show that one may without loss of generality assume $|\mathcal{U}| \leq |\mathcal{X}||\mathcal{Y}| + 5$. However, $|\mathcal{V}_1|$ and $|\mathcal{V}_2|$ cannot be bounded in this way, as the application of the Carathéodory-Fenchel theorem does not preserve the conditional independence of V_1 and V_2. Thus a characterization of the above achievable region involving sets with upper-bounded cardinality is currently not available. As it would be important for an efficient calculation of the achievable region, it still requires further consideration.*

Remark 4. *If no common randomness is available, then no common message can be transmitted.*

Remark 5. *We have $\mathcal{R}^{(1)}(p) \subset \mathcal{R}^{(2)}(p) \subset \mathcal{R}^{(3)}(p)$. This can be seen directly at the beginning of the proof in Subsection 4.1 where we decompose the regions $\mathcal{R}^{(\nu)}(p)$ for $\nu = 1, 2$ into a union of simpler regions.*

In particular, if H_C is larger than the capacity of the single-sender discrete memoryless channel W_e with input alphabet $\mathcal{X} \times \mathcal{Y}$ and output alphabet $\mathcal{Z}$, then $\Pi_{H_C}^{(3)} = \Pi$ and the achievable set equals

$$\mathit{closure}\left(\mathit{conv}\left(\bigcup_{p\in\Pi} \mathcal{R}^{(3)}(p)\right)\right).$$

In this case the maximal sum rate equals

$$\mathcal{C} := \max_{p\in\Pi}\big(I(T\wedge V_1V_2) - I(Z\wedge V_1V_2)\big). \tag{11}$$

This equals the secrecy capacity of the single-sender wiretap channel when Alice$_1$ *and Alice*$_2$ *together are considered as one single sender. In order to see this, we have to show that for any pair* (V_1', V_2') *of random variables on any Cartesian product* $\mathcal{V}_1\times\mathcal{V}_2$ *of finite sets one can find random variables* (V_1, V_2, U) *satisfying* $P_{UV_1V_2} = P_U\otimes(P_{V_1|U}\otimes P_{V_2|U})$ *and* $P_{V_1V_2} = P_{V_1'V_2'}$*. Given such arbitrary* (V_1', V_2') *as above, just define* $U = V_1'$ *and* $P_{V_1|U} =$ *the identity on* $\mathcal{V}_1$ *and* $P_{V_2|U} = P_{V_2'|V_1'}$*. Then a simple calculation shows that the above conditions are satisfied. Thus* (11) *equals the secrecy capacity of the single-sender wiretap channel with Alice*$_1$ *and Alice*$_2$ *combined into a single sender. The remaining conditions on* R_1 *and* R_2 *formulated in the definition of* $\mathcal{R}^{(3)}(p)$ *are not concerned with* W_e*, they are required by the non-wiretap MAC coding theorem applied to* W_b*.*

3.2 For the Wiretap MAC with Conferencing Encoders

For conferencing capacities $C_1, C_2 > 0$, the achievable rate region is parametrized by the members of $\Pi_{C_1+C_2}$. We have Cases 1-3 from the common message part.

Case 1: For $p\in\Pi^{(1)}_{C_1+C_2}$ we define $\mathcal{R}^{(1)}(p, C_1, C_2)$ by

$$\begin{aligned}
R_1 &\le I(T\wedge V_1|V_2U) - I(Z\wedge V_1|U)\\
&\quad -[I(Z\wedge V_2|V_1U) - I(T\wedge V_2|V_1U)]_+ + C_1 - [I(Z\wedge U) - C_2]_+,\\
R_2 &\le I(T\wedge V_2|V_1U) - I(Z\wedge V_2|U)\\
&\quad -[I(Z\wedge V_1|V_2U) - I(T\wedge V_1|V_2U)]_+ + C_2 - [I(Z\wedge U) - C_1]_+,\\
R_1 + R_2 &\le \min\{I(T\wedge V_1V_2|U) + C_1 + C_2, I(T\wedge V_1V_2))\} - I(Z\wedge V_1V_2)).
\end{aligned}$$

Case 2: For $p\in\Pi^{(2)}_{C_1+C_2}$, we set $J_0^{(\alpha)} := \alpha I(Z\wedge V_2U) + (1-\alpha)I(Z\wedge V_1U)$. For $\alpha\in[\alpha_0^{(2)}, \alpha_1^{(2)}]$ define the set $\mathcal{R}^{(2)}_\alpha(p, C_1, C_2)$ by

$$\begin{aligned}
R_1 &\le I(T\wedge V_1|V_2U) - \alpha I(Z\wedge V_1|V_2U) + C_1 - [J_0^{(\alpha)} - C_2]_+,\\
R_2 &\le I(T\wedge V_2|V_1U) - (1-\alpha)I(Z\wedge V_2|V_1U) + C_2 - [J_0^{(\alpha)} - C_1]_+,\\
R_1 + R_2 &\le \min\{I(T\wedge V_1V_2|U) + C_1 + C_2, I(T\wedge V_1V_2)\} - I(Z\wedge V_1V_2).
\end{aligned}$$

Then we set

$$\mathcal{R}^{(2)}(p, C_1, C_2) := \bigcup_{\alpha_0^{(2)}\le\alpha\le\alpha_1^{(2)}} \mathcal{R}^{(2)}_\alpha(p, C_1, C_2).$$

Case 3: For $p\in\Pi^{(3)}_{C_1+C_2}$ we define $\mathcal{R}^{(3)}(p, C_1, C_2)$ by

$$\begin{aligned}
R_1 &\le I(T\wedge V_1|V_2U) + C_1 - [I(Z\wedge V_1V_2) - C_2]_+,\\
R_2 &\le I(T\wedge V_2|V_1U) + C_2 - [I(Z\wedge V_1V_2) - C_1]_+,\\
R_1 + R_2 &\le \min\{I(T\wedge V_1V_2|U) + C_1 + C_2, I(T\wedge V_1V_2)\} - I(Z\wedge V_1V_2).
\end{aligned}$$

Theorem 2. *For the conferencing capacities $C_1, C_2 > 0$, the wiretap MAC W with conferencing encoders achieves the closure of the convex hull of the set*

$$\bigcup_{p\in\Pi_{HC}^{(1)}} \mathcal{R}^{(1)}(p, C_1, C_2) \cup \bigcup_{p\in\Pi_{HC}^{(2)}} \mathcal{R}^{(2)}(p, C_1, C_2) \cup \bigcup_{p\in\Pi_{HC}^{(3)}} \mathcal{R}^{(3)}(p, C_1, C_2).$$

Remark 6. *Remark 3 applies here, too.*

Remark 7. *The stochastic conferencing protocols employed to achieve the sets in Theorem 2 are non-iterative. That means that the c we use in the proof have the form*

$$c(v_1, v_2|k_1, k_2) = c_1(v_1|k_1)c_2(v_2|k_2).$$

Remark 8. *If $C_1 = C_2 = 0$, then the maximal rate set whose achievability we can show is* (10). *Conferencing only enlarges this set in the presence of a wiretapper if it is used to establish common randomness between the encoders. At least this is true for the achievable region we can show, it cannot be verified in general as long as one does not have a converse. The reason is that conferencing generates a common message shared by Alice$_1$ and Alice$_2$. As noted in Remark 4, a common message can only be kept secret if common randomness is available. As the Alices do not have common randomness a priori, this also has to be generated by conferencing, so the Willems conferencing protocol has to be stochastic.*

Remark 9. *With the coding method we apply, conferencing may enable secure transmission if this is not possible without. That means that there are wiretap MACs where the achievable region without conferencing as derived in Theorem 1 only contains the rate pair* $(0, 0)$ *whereas it contains non-trivial rate pairs with* $C_1, C_2 > 0$. *See Section 6 for an example.*

Remark 10. *If C_1, C_2 are sufficiently large, then the maximal achievable sum rate equals the secrecy capacity $\mathcal{C}$ of the single-sender wiretap channel with input alphabet $\mathcal{X} \times \mathcal{Y}$ and channel matrix W, see* (11). *In fact, this happens if*

1) *C_1+C_2 is strictly larger than the capacity of the single-sender discrete memoryless channel W_e with input alphabet $\mathcal{X} \times \mathcal{Y}$ and output alphabet $\mathcal{Z}$,*
2) *$C_1+C_2 \geq \min_{p\in\Pi^*} I(T \wedge U)$, where Π^* contains those $p \in \Pi$ which achieve $\mathcal{C}$.*

Condition 1) is sufficient to guarantee that $\mathcal{C}$ is achievable by an element of $\Pi_{C_1+C_2}^{(3)}$ which then equals Π, see Remark 5. In particular Π^* is nonempty, and 2) ensures that the maximum over Π of the sum rate bounds from $\mathcal{R}^{(3)}(p, C_1, C_2)$ equals $\mathcal{C}$.

4 Proof of Theorem 1

4.1 Elementary Rate Regions

For Cases 0, 1 and 2 we first show the achievability of certain rate regions whose union or convex combination then yields the achievable regions claimed in the theorem.

For Case 0 and 1: We only consider Case 1, Case 0 is analogous. The considerations hold for $I(Z \wedge V_1|U) < I(Z \wedge V_1|V_2U)$ which is equivalent to $I(Z \wedge V_2|U) < I(Z \wedge V_2|V_1U)$. In the case of equality we can prove the achievability of $\mathcal{R}(p)$ directly. Define

$$\alpha_0^{(1)} := \left[\frac{I(T \wedge V_2|V_1U) - I(Z \wedge V_2|V_1U)}{I(Z \wedge V_2|U) - I(Z \wedge V_2|V_1U)}\right]_+,$$
$$\alpha_1^{(1)} := \min\left\{\frac{I(T \wedge V_1|V_2U) - I(Z \wedge V_1|U)}{I(Z \wedge V_1|V_2U) - I(Z \wedge V_1|U)}, 1\right\}.$$

Note that conditions (5)-(7) are equivalent to $\alpha_0^{(1)} \leq \alpha_1^{(1)}$. For $\alpha \in [\alpha_0^{(1)}, \alpha_1^{(1)}]$ we define a rate region $\mathcal{R}_\alpha^{(1)}(p)$ by the bounds

$$\begin{aligned}
R_1 &\leq I(T \wedge V_1|V_2U) - \alpha I(Z \wedge V_1|V_2U) - (1-\alpha)I(Z \wedge V_1|U),\\
R_2 &\leq I(T \wedge V_2|V_1U) - \alpha I(Z \wedge V_2|U) - (1-\alpha)I(Z \wedge V_2|V_1U),\\
R_1 + R_2 &\leq I(T \wedge V_1V_2|U) - I(Z \wedge V_1V_2|U),\\
R_0 + R_1 + R_2 &\leq I(T \wedge V_1V_2) - I(Z \wedge V_1V_2).
\end{aligned}$$

Lemma 1. *We have*

$$\mathcal{R}^{(1)}(p) = \bigcup_{\alpha_0^{(1)} \leq \alpha \leq \alpha_1^{(1)}} \mathcal{R}_\alpha^{(1)}(p).$$

Thus if $\mathcal{R}_\alpha^{(1)}(p)$ is an achievable rate region for every $\alpha \in [\alpha_0^{(1)}, \alpha_1^{(1)}]$, then $\mathcal{R}^{(1)}(p)$ is achievable.

For the proof we use the following lemma which is proved in the appendix.

Lemma 2. *Assume that $a_1, a_2, b_1, b_2, c, d, r_1, r_2, r_{12}, r_{012}$ are nonnegative reals satisfying*

$$a_1 > b_1, \quad a_2 < b_2, \quad a_1 + a_2 = b_1 + b_2 = c, \quad r_1 + r_2 \geq r_{12}.$$

Let $0 \leq \alpha_0 \leq \alpha_1 \leq 1$. For every $\alpha \in [\alpha_0, \alpha_1]$, let a three-dimensional convex subset $\mathcal{K}_\alpha$ of $\mathbb{R}^3_{\geq 0}$ be defined by

$$\begin{aligned}
R_1 &\leq r_1 - \alpha a_1 - (1-\alpha)b_1,\\
R_2 &\leq r_2 - \alpha a_2 - (1-\alpha)b_2,\\
R_1 + R_2 &\leq r_{12} - c,\\
R_0 + R_1 + R_2 &\leq r_{012} - d
\end{aligned}$$

and assume that $\mathcal{K}_\alpha \neq \varnothing$ for every α. Then

$$\bigcup_{\alpha_0 \leq \alpha \leq \alpha_1} \mathcal{K}_\alpha = \mathcal{K}, \tag{12}$$

where $\mathcal{K}$ is defined by

$$R_1 \leq r_1 - \alpha_0 a_1 - (1-\alpha_0) b_1, \tag{13}$$
$$R_2 \leq r_2 - \alpha_1 a_2 - (1-\alpha_1) b_2, \tag{14}$$
$$R_1 + R_2 \leq r_{12} - c, \tag{15}$$
$$R_0 + R_1 + R_2 \leq r_{012} - d. \tag{16}$$

Proof (Lemma 1). The proof is a direct application of Lemma 2 by setting

$$\begin{aligned} r_1 &= I(T \wedge V_1|V_2U), & r_2 &= I(T \wedge V_2|V_1U), \\ r_{12} &= I(T \wedge V_1V_2|U), & r_{012} &= I(T \wedge V_1V_2), \\ a_1 &= I(Z \wedge V_1|V_2U), & a_2 &= I(Z \wedge V_2|U), \\ b_1 &= I(Z \wedge V_1|U), & b_2 &= I(Z \wedge V_2|V_1U), \\ \alpha_0 &= \alpha_0^{(1)}, & \alpha_1 &= \alpha_1^{(1)}. \end{aligned}$$

We just need to show that the bounds (13) and (14) coincide with those from the definition of $\mathcal{R}^{(1)}(p)$. This is easy for the case $\alpha_0^{(1)} = 0$ because in that case we have $I(T \wedge V_2|V_1U) \geq I(Z \wedge V_2|V_1U)$ and the positive part in the bound on R_1 in the definition of $\mathcal{R}^{(1)}(p)$ vanishes. Similarly $\alpha_1^{(1)} = 1$ implies $I(T \wedge V_1|V_2U) \geq I(Z \wedge V_1|V_2U)$ and the positive part in the bound on R_2 in the definition of $\mathcal{R}^{(1)}(p)$ vanishes. Now assume that $\alpha_0^{(1)} > 0$. This assumption implies $I(Z \wedge V_2|V_1U) > I(T \wedge V_2|V_1U)$. Thus we obtain for the equivalent of (13)

$$\begin{aligned} &I(T \wedge V_1|V_2U) - I(Z \wedge V_1|U) \\ &\quad - \frac{I(T \wedge V_2|V_1U) - I(Z \wedge V_2|V_1U)}{I(Z \wedge V_2|U) - I(Z \wedge V_2|V_1U)} (I(Z \wedge V_1|V_2U) - I(Z \wedge V_1|U)) \\ &= I(T \wedge V_1|V_2U) - I(Z \wedge V_1|U) \\ &\quad - \frac{I(T \wedge V_2|V_1U) - I(Z \wedge V_2|V_1U)}{I(Z \wedge V_2|U) - I(Z \wedge V_2|V_1U)} (I(Z \wedge V_2|V_1U) - I(Z \wedge V_2|U)) \\ &= I(T \wedge V_1|V_2U) + I(T \wedge V_2|V_1U) - I(Z \wedge V_1V_2|U) \\ &= I(T \wedge V_1|V_2U) - I(Z \wedge V_1|U) - [I(Z \wedge V_2|V_1U) - I(T \wedge V_2|V_1U)]_+. \end{aligned}$$

If $\alpha_1^{(1)} < 1$, we obtain the analog for the bound on R_2. This shows with Lemma 2 that $\mathcal{R}^{(1)}(p)$ can be represented as the union of the sets $\mathcal{R}_\alpha^{(1)}(p)$ for $\alpha_0^{(1)} \leq \alpha \leq \alpha_1^{(1)}$. □

For Case 2: Here we assume that $I(Z \wedge V_1|V_2U) \neq I(Z \wedge V_2|V_1U)$ which is equivalent to $I(Z \wedge V_1U) \neq I(Z \wedge V_2U)$. In the case of equality, the achievability of $\mathcal{R}^{(2)}(p)$ can be shown directly. Define for $\alpha \in [\alpha_0^{(2)}, \alpha_1^{(2)}]$ the rate set $\mathcal{R}_\alpha^{(2)}(p)$ by the conditions

$$
\begin{aligned}
R_1 &\leq I(T \wedge V_1|V_2U) - \alpha I(Z \wedge V_1|V_2U),\\
R_2 &\leq I(T \wedge V_2|V_1U) - (1-\alpha)I(Z \wedge V_2|V_1U),\\
R_1 + R_2 &\leq I(T \wedge V_1V_2|U) - \alpha I(Z \wedge V_1|V_2U) - (1-\alpha)I(Z \wedge V_2|V_1U),\\
R_0 + R_1 + R_2 &\leq I(T \wedge V_1V_2) - I(Z \wedge V_1V_2).
\end{aligned}
$$

Lemma 3. *We have that*

$$\mathcal{R}^{(2)}(p) = \bigcup_{\alpha_0^{(2)} \leq \alpha \leq \alpha_1^{(2)}} \mathcal{R}_\alpha^{(2)}(p).$$

In particular, if $\mathcal{R}_\alpha^{(2)}(p)$ is achievable for every $\alpha \in [\alpha_0^{(2)}, \alpha_1^{(2)}]$, then so is $\mathcal{R}^{(2)}(p)$.

Remark 11. *The similarity between the rate regions for Case 1 and Case 2 becomes clear in these decompositions. The description for Case 2 is more complex because $\alpha_0^{(2)}$ and $\alpha_1^{(2)}$ are defined through three minima/maxima. This is due to the fact that the sum $\alpha I(Z \wedge V_1|V_2U) + (1-\alpha)I(Z \wedge V_2|V_1U)$ is not constant in α. Hence the conditions for $\alpha_0^{(2)} \leq \alpha_1^{(2)}$ cannot be reformulated into simple conditions on the corresponding p.*

One obtains Lemma 3 from the next lemma by making the following replacements:

$$
\begin{aligned}
r_1 &= I(T \wedge V_1|V_2U), & r_2 &= I(T \wedge V_2|V_1U),\\
r_{12} &= I(T \wedge V_1V_2|U), & r_{012} &= I(T \wedge V_1V_2),\\
a &= I(Z \wedge V_1|V_2U), & b &= I(Z \wedge V_2|V_1U),\\
c &= I(Z \wedge V_1V_2), & &\\
\alpha_0 &= \alpha_0^{(2)}, & \alpha_1 &= \alpha_1^{(2)}.
\end{aligned}
$$

Lemma 4. *Let $r_1, r_2, r_{12}, r_{012}, a, b, c$ be nonnegative reals with $\max(r_1, r_2) \leq r_{12} \leq r_1 + r_2$. Let $\alpha_0, \alpha_1 \in [0,1]$ be given such that for every $\alpha \in [\alpha_0, \alpha_1]$ the set $\mathcal{K}_\alpha$ defined by*

$$
\begin{aligned}
R_1 &\leq r_1 - \alpha a,\\
R_2 &\leq r_2 - (1-\alpha)b,\\
R_1 + R_2 &\leq r_{12} - \alpha a - (1-\alpha)b,\\
R_0 + R_1 + R_2 &\leq r_{012} - c
\end{aligned}
$$

is nonempty. If $a \leq b$, the convex hull of the union of these sets is given by the set $\mathcal{K}$ which is characterized by

$$0 \leq R_1 \leq r_1 - \alpha_0 a, \tag{17}$$

$$0 \leq R_2 \leq r_2 - (1-\alpha_1)b, \tag{18}$$

$$R_1 + R_2 \leq r_{12} - \alpha_1 a - (1-\alpha_1)b, \tag{19}$$

$$bR_1 + aR_2 \leq r_{12}a + r_1(b-a) - ab, \tag{20}$$

$$R_0 + R_1 + R_2 \leq r_{012} - c. \tag{21}$$

If $a > b$, the convex hull of the union of the sets $\mathcal{K}_\alpha$ is given by analogous bounds where a and b are exchanged in (20).

The proof of Lemma 4 can be found in the appendix.

4.2 How to Prove Secrecy

Proving secrecy using Chernoff-type concentration inequalities (see Subsection 4.3) is the core of Devetak's approach to the wiretap channel [17]. Due to the multi-user structure of the inputs of the wiretap MAC, we need several such Chernoff-type inequalities basing on each other compared to the one used by Devetak (actually an application of the Ahlswede-Winter lemma). However, once these are established, the way of obtaining secrecy is exactly the same as presented by Devetak. With the help of the inequalities one obtains a code with stochastic encoding and a measure ϑ (not necessarily a probability measure!) such that for all k_0, k_1, k_2

$$\|P_{Z^n|M_0=k_0,M_1=k_1,M_2=k_2} - \vartheta\| \leq \frac{\varepsilon}{2}. \tag{22}$$

Given this, we now derive an upper bound on $I(Z^n \wedge M_0M_1M_2)$, where the random triple (M_0, M_1, M_2) is uniformly distributed on the possible input message triples and Z^n represents the output received by Eve. Observe that

$$\begin{aligned}
&I(Z^n \wedge M_0M_1M_2)\\
&= \frac{1}{K_0K_1K_2}\sum_{k_0,k_1,k_2}(H(Z^n) - H(Z^n|M_0 = k_0, M_1 = k_1, M_2 = k_2)).
\end{aligned} \tag{23}$$

By [16, Lemma 2.7], every summand on the right-hand side is upper-bounded by $\varepsilon_{k_0k_1k_2}\log(|\mathcal{Z}|^n/\varepsilon_{k_0k_1k_2})$ if

$$\varepsilon_{k_0k_1k_2} := \|P_{Z^n} - P_{Z^n|M_0=k_0,M_1=k_1,M_2=k_2}\| \leq \frac{1}{2}.$$

But due to (22),

$$\begin{aligned}
&\|P_{Z^n} - P_{Z^n|M_0=k_0,M_1=k_1,M_2=k_2}\|\\
&\leq \|P_{Z^n} - \vartheta\| + \|\vartheta - P_{Z^n|M_0=k_0,M_1=k_1,M_2=k_2}\|\\
&\leq \frac{1}{K_0K_1K_2}\sum_{\tilde{k}_0,\tilde{k}_1,\tilde{k}_2}\|P_{Z^n|M_0=\tilde{k}_0,M_1=\tilde{k}_1,M_2=\tilde{k}_2} - \vartheta\| + \frac{\varepsilon}{2}\\
&\leq \varepsilon.
\end{aligned}$$

Thus if ε tends to zero exponentially in blocklength, then (23) is upper-bounded by $\varepsilon\log(|\mathcal{Z}|^n/\varepsilon)$ which tends to zero in n.

4.3 Probabilistic Bounds for Secrecy

In this subsection we define the random variables from which we will build a stochastic wiretap code in Subsection 4.5. For this family of random variables we prove several Chernoff-type estimates which will serve to find a code satisfying (22). For Case 3, two such estimates are sufficient, Case 0 and 2 require three and Case 1 requires four. Within each case, one deals with the joint typicality of the inputs at Alice$_1$ and Alice$_2$, and the other estimates base on each other. This is due to the complex structure of our family of random variables. Still, all the cases are nothing but a generalization of Devetak's approach taken in [17]. For each case, we first show the probabilistic bounds in one paragraph and then in another paragraph how to achieve (22) from those bounds.

Let $p = P_U \otimes P_{X|U} \otimes P_{Y|U} \otimes W \in \Pi$, i.e. p is the distribution of a random vector (U, X, Y, T, Z). The auxiliary random variables V_1 and V_2 will be introduced later in the usual way of prefixing a channel as a means of additional randomization. Let $\delta > 0$ and define for any n

$$P_U^n(\mathbf{u}) := \frac{P_U^{\otimes n}(\mathbf{u})}{P_U^{\otimes n}(T_{U,\delta}^n)} \qquad (\mathbf{u} \in T_{U,\delta}^n),$$

$$P_{X|U}^n(\mathbf{x}|\mathbf{u}) := \frac{P_{X|U}^{\otimes n}(\mathbf{x}|\mathbf{u})}{P_{X|U}^{\otimes n}(T_{X|U,\delta}^n(\mathbf{u})|\mathbf{u})} \qquad (\mathbf{x} \in T_{X|U,\delta}^n(\mathbf{u}), \mathbf{u} \in T_{U,\delta}^n),$$

$$P_{Y|U}^n(\mathbf{y}|\mathbf{u}) := \frac{P_{Y|U}^{\otimes n}(\mathbf{y}|\mathbf{u})}{P_{Y|U}^{\otimes n}(T_{Y|U,\delta}^n(\mathbf{u})|\mathbf{u})} \qquad (\mathbf{y} \in T_{Y|U,\delta}^n(\mathbf{u}), \mathbf{u} \in T_{U,\delta}^n).$$

Let L_0, L_1, L_2 be positive integers. We define L_0 independent families of random variables $(U^{l_0}, \mathcal{F}_{l_0})$ as follows. U^{l_0} is distributed according to P_U^n. We let $\mathcal{F}_{l_0} := \{X^{l_0 l_1}, Y^{l_0 l_2} : l_1 \in [L_1], l_2 \in [L_2]\}$ be a set of random variables which are independent given U^{l_0} and which satisfy $X^{l_0 l_1} \sim P_{X|U}^n(\,\cdot\,|U^{l_0})$ and $Y^{l_0 l_2} \sim P_{Y|U}^n(\,\cdot\,|U^{l_0})$. Finally we define

$$\mathcal{F} := \bigcup_{l_0 \in [L_0]} (U^{l_0}, \mathcal{F}_{l_0}). \tag{24}$$

Throughout the section, let a small $\varepsilon > 0$ be fixed. The core of the proofs of all the lemmas of this subsection is the following Chernoff bound, see e.g. [4].

Lemma 5. *Let $b > 0$ and $0 < \varepsilon < 1/2$. For an independent sequence of random variables $Z_1, \dots, Z_L$ with values in $[0, b]$ with $\mu_l := \mathbb{E}[X_l]$ and with $\mu := \frac{1}{L}\sum_l \mu_l$ one has*

$$\mathbb{P}\left[\frac{1}{L}\sum_{l=1}^{L} Z_l > (1+\varepsilon)\mu\right] \le \exp\left(-L \cdot \frac{\varepsilon^2 \mu}{2b \ln 2}\right)$$

and

$$\mathbb{P}\left[\frac{1}{L}\sum_{l=1}^{L} Z_l < (1-\varepsilon)\mu\right] \le \exp\left(-L \cdot \frac{\varepsilon^2 \mu}{2b \ln 2}\right).$$

In order to obtain useful bounds in the following we collect here some well-known estimates concerning typical sets, see e.g. [16, Lemma 17.8]. Let (A, B) be a random pair on the finite Cartesian product $\mathcal{A} \times \mathcal{B}$. Let $\xi, \zeta > 0$. Then there exists a $\tilde{c} = \tilde{c}(|\mathcal{A}||\mathcal{B}|) > 0$ such that for sufficiently large n

$$P_{B|A}^{\otimes n}(T_{B|A,\zeta}^n(\mathbf{a})^c|\mathbf{a}) \leq 2^{-n\tilde{c}\zeta^2}. \tag{25}$$

Further there is a $\tau = \tau(P_{AB}, \xi, \zeta)$ with $\tau \to 0$ as $\xi, \zeta \to 0$ such that

$$P_A^{\otimes n}(\mathbf{a}) \leq 2^{-n(H(A)-\tau)} \qquad \text{if } \mathbf{a} \in T_{A,\xi}^n, \tag{26}$$

$$P_{B|A}^{\otimes n}(\mathbf{b}|\mathbf{a}) \leq 2^{-n(H(B|A)-\tau)} \qquad \text{if } \mathbf{a} \in T_{A,\xi}^n, \mathbf{b} \in T_{B|A,\zeta}^n(\mathbf{a}), \tag{27}$$

and that for n sufficiently large,

$$|T_{A,\xi}^n| \leq 2^{n(H(A)+\tau)}, \tag{28}$$

$$|T_{B|A,\zeta}^n(\mathbf{a})| \leq 2^{n(H(B|A)+\tau)} \qquad \text{if } \mathbf{a} \in T_{A,\xi}^n. \tag{29}$$

We set

$$c := \tilde{c}(|\mathcal{U}||\mathcal{X}||\mathcal{Y}||\mathcal{Z}|),$$

this is the minimal $\tilde{c}$ we will need in the following.

Bounds for Case 0 and 1: Let L_0, L_1, L_2 be arbitrary. Due to their conditional independence, the $X^{l_0 l_1}$ and $Y^{l_0 l_2}$ cannot be required to be jointly conditionally typical given U^{l_0}. However, the next lemma shows that most of them are jointly conditionally typical with high probability.

Lemma 6. *For $(l_0, l_2) \in [L_0] \times [L_2]$, let the event $A_*^{(1)}(l_0, l_2)$ be defined by*

$$A_*^{(1)}(l_0, l_2) := \{|\{l_1 \in [L_1] : X^{l_0 l_1} \in T_{X|YU,\delta}^n(Y^{l_0 l_2}, U^{l_0})\}| \geq (1-\varepsilon)(1 - 2 \cdot 2^{-nc\delta^2})L_1\}.$$

Then

$$\mathbb{P}[A_*^{(1)}(l_0, l_2)^c] \leq \exp\left(-L_1 \cdot \frac{\varepsilon^2(1 - 2 \cdot 2^{-nc\delta^2})}{2 \ln 2}\right).$$

Proof. Let $\mathbf{u} \in T_{U,\delta}^n$ and $\mathbf{y} \in T_{Y|U,\delta}^n(\mathbf{u})$. We first condition on the event $\{Y^{l_0 l_2} = \mathbf{y}, U^{l_0} = \mathbf{u}\}$. Due to (25), we have

$$\begin{aligned}
&\mathbb{P}[X^{11} \notin T_{X|YU,\delta}^n(\mathbf{y}, \mathbf{u})|Y^{l_0 l_2} = \mathbf{y}, U^1 = \mathbf{u}] \\
&= \frac{1}{P_{X|U}^{\otimes n}(T_{X|U,\delta}^n(\mathbf{u})|\mathbf{u})} \sum_{\mathbf{x} \in T_{X|U,\delta}^n(\mathbf{u}) \setminus T_{X|YU,\delta}^n(\mathbf{y},\mathbf{u})} P_{X|U}^{\otimes n}(\mathbf{x}|\mathbf{u}) \\
&\leq \frac{1}{P_{X|U}^{\otimes n}(T_{X|U,\delta}^n(\mathbf{u})|\mathbf{u})} \sum_{\mathbf{x} \notin T_{X|YU,\delta}^n(\mathbf{y},\mathbf{u})} P_{X|YU}^{\otimes n}(\mathbf{x}|\mathbf{y}, \mathbf{u}) \\
&\leq \frac{2^{-nc\delta^2}}{1 - 2^{-nc\delta^2}}.
\end{aligned}$$

In particular,

$$\mu := \mathbb{P}[X^{11} \in T^n_{X|YU,\delta}(\mathbf{y},\mathbf{u})|Y^{11} = \mathbf{y}, U^1 = \mathbf{u}] \geq 1 - 2 \cdot 2^{-nc\delta^2}.$$

Therefore

$$\mathbb{P}[A_*^{(1)}(l_0,l_2)^c|Y^{l_0l_2} = \mathbf{y}, U^{l_0} = \mathbf{u}]$$
$$\leq \mathbb{P}\left[\sum_{l_1} 1_{T^n_{X|YU,\delta}(\mathbf{y},\mathbf{u})}(X^{l_0l_1}) \leq (1-\varepsilon)\mu\, L_1 \middle| Y^{l_0l_2} = \mathbf{y}, U^{l_0} = \mathbf{u}\right],$$

which by Lemma 5 can be bounded by

$$\exp\left(-L_1 \cdot \frac{\varepsilon^2\mu}{2\ln 2}\right) \leq \exp\left(-L_1 \cdot \frac{\varepsilon^2(1 - 2 \cdot 2^{-nc\delta^2})}{2\ln 2}\right).$$

This completes the proof as this bound is independent of $(\mathbf{y},\mathbf{u})$. □

Lemma 6 is not needed for a single sender. As we cannot guarantee the joint conditional typicality of both senders' inputs, we need to introduce an explicit bound on the channel transition probabilities. This is done in the set $E_1^{(1)}$. Then we prove three lemmas each of which exploits one of the three types of independence contained in $\mathcal{F}$. Altogether these lemmas provide lower bounds on L_0, L_1, L_2 which if satisfied allow the construction of a wiretap code satisfying (22). Let

$$E_1^{(1)}(\mathbf{u},\mathbf{x},\mathbf{y}) := \{\mathbf{z} \in T^n_{Z|YU,2|\mathcal{X}|\delta}(\mathbf{y},\mathbf{u}) : W_e^{\otimes n}(\mathbf{z}|\mathbf{x},\mathbf{y}) \leq 2^{-n(H(Z|XY)-f_2(\delta))}\},$$

where $f_2(\delta) = \tau(P_{UXYZ}, 3\delta, \delta)$ (see (27)). Let

$$\vartheta^{(1)}_{\mathbf{uy}}(\mathbf{z}) := \mathbb{E}[W_e^{\otimes n}(\mathbf{z}|X^{11},\mathbf{y})1_{E_1^{(1)}(\mathbf{u},X^{11},\mathbf{y})}(\mathbf{z})|U^1 = \mathbf{u}]$$

and for

$$F_1^{(1)}(\mathbf{u},\mathbf{y}) := \{\mathbf{z} \in T^n_{Z|YU,2|\mathcal{X}|\delta}(\mathbf{y},\mathbf{u}) : \vartheta^{(1)}_{\mathbf{uy}}(\mathbf{z}) \geq \varepsilon|T^n_{Z|YU,2|\mathcal{X}|\delta}(\mathbf{y},\mathbf{u})|^{-1}\}$$

define

$$\hat{\vartheta}^{(1)}_{\mathbf{uy}} := \vartheta^{(1)}_{\mathbf{uy}} \cdot 1_{F_1^{(1)}(\mathbf{u},\mathbf{y})}, \quad E_2^{(1)}(\mathbf{u},\mathbf{x},\mathbf{y}) := E_1^{(1)}(\mathbf{u},\mathbf{x},\mathbf{y}) \cap F_1^{(1)}(\mathbf{u},\mathbf{y}).$$

Lemma 7. *For every $\mathbf{z} \in \mathcal{Z}^n$ and $(l_0,l_2) \in [L_0] \times [L_2]$, let $A_1^{(1)}(l_0,l_2,\mathbf{z})$ be the event that*

$$\frac{1}{L_1}\sum_{l_1} W_e^{\otimes n}(\mathbf{z}|X^{l_0l_1},Y^{l_0l_2})1_{E_2^{(1)}(U^{l_0},X^{l_0l_1},Y^{l_0l_2})}(\mathbf{z}) \in [(1\pm\varepsilon)\hat{\vartheta}^{(1)}_{U^{l_0}Y^{l_0l_2}}(\mathbf{z})].$$

Then

$$\mathbb{P}[A_1^{(1)}(l_0,l_2,\mathbf{z})^c] \leq 2\exp\left(-L_1 \cdot \frac{\varepsilon^3 2^{-n(I(Z\wedge X|YU)+f_1(\delta)+f_2(\delta))}}{2\ln 2}\right)$$

for $f_1(\delta) = \tau(P_{UYZ}, 2\delta, 2|\mathcal{X}|\delta)$ and n sufficiently large.

Proof. For $\mathbf{u} \in T^n_{U,\delta}$ and $\mathbf{y} \in T^n_{Y|U,\delta}(\mathbf{u})$ we condition on the event $\{Y^{l_0 l_2} = \mathbf{y}, U^{l_0} = \mathbf{u}\}$. The conditional expectation of the bounded conditionally i.i.d. random variables

$$W_e^{\otimes n}(\mathbf{z}|X^{l_0 l_1}, \mathbf{y}) 1_{E_2^{(1)}(\mathbf{u}, X^{l_0 l_1}, \mathbf{y})}(\mathbf{z}) \le 2^{-n(H(Z|XY) - f_2(\delta))} \qquad (l_1 \in [L_1])$$

is $\hat{\vartheta}^{(1)}_{\mathbf{uy}}(\mathbf{z})$. We use Lemma 5, the definition of $F_1^{(1)}(\mathbf{u}, \mathbf{y})$, and (29) to obtain for n sufficiently large

$$\begin{aligned}
&\mathbb{P}[A_1^{(1)}(l_0, l_2, \mathbf{z})^c | Y^{l_0 l_2} = \mathbf{y}, U^{l_0} = \mathbf{u}] \\
&\le 2 \exp\left(-L_1 \cdot \frac{\varepsilon^2 \hat{\vartheta}^{(1)}_{\mathbf{uy}}(\mathbf{z}) 2^{n(H(Z|XY) - f_2(\delta))}}{2 \ln 2} \right) \\
&\le 2 \exp\left(-L_1 \cdot \frac{\varepsilon^3 2^{-n(I(Z \wedge X|YU) + f_1(\delta) + f_2(\delta))}}{2 \ln 2} \right).
\end{aligned}$$

This bound is uniform in $\mathbf{u}$ and $\mathbf{y}$, so the proof is complete. □

For the next lemma, define

$$\vartheta^{(1)}_{\mathbf{u}}(\mathbf{z}) := \mathbb{E}[W_e^{\otimes n}(\mathbf{z}|X^{11}, Y^{11}) 1_{E_2^{(1)}(\mathbf{u}, X^{11}, Y^{11})}(\mathbf{z}) | U^1 = \mathbf{u}].$$

Further let

$$F_2^{(1)}(\mathbf{u}) := \{\mathbf{z} \in T^n_{Z|U, 3|\mathcal{Y}||\mathcal{X}|\delta}(\mathbf{u}) : \vartheta^{(1)}_{\mathbf{u}}(\mathbf{z}) \ge \varepsilon |T^n_{Z|U,3|\mathcal{Y}||\mathcal{X}|\delta}(\mathbf{u})|^{-1}\}$$

and

$$\hat{\vartheta}^{(1)}_{\mathbf{u}} = \vartheta^{(1)}_{\mathbf{u}} \cdot 1_{F_2^{(1)}(\mathbf{u})}, \quad E_0^{(1)}(\mathbf{u}, \mathbf{x}, \mathbf{y}) := E_2^{(1)}(\mathbf{u}, \mathbf{x}, \mathbf{y}) \cap F_2^{(1)}(\mathbf{u}, \mathbf{y}).$$

Lemma 8. *For every $\mathbf{z} \in \mathcal{Z}^n$ and $l_0 \in [L_0]$, let $A_2^{(1)}(l_0, \mathbf{z})$ be the event*

$$\frac{1}{L_1 L_2} \sum_{l_1 l_2} W_e^{\otimes n}(\mathbf{z}|X^{l_0 l_1}, Y^{l_0 l_2}) 1_{E_0^{(1)}(U^{l_0}, X^{l_0 l_1}, Y^{l_0 l_2})}(\mathbf{z}) \in [(1 \pm 3\varepsilon) \hat{\vartheta}^{(1)}_{U^{l_0}}(\mathbf{z})].$$

Then for ε sufficiently small and n sufficiently large,

$$\begin{aligned}
\mathbb{P}[A_2^{(1)}(l_0, \mathbf{z})^c] \le 2|\mathbf{Y}|^n \exp&\left(-L_1 \cdot \frac{\varepsilon^3 2^{-n(I(Z \wedge X|YU) + f_1(\delta) + f_2(\delta))}}{2 \ln 2} \right) \\
&+ 2 \exp\left(-L_2 \cdot \frac{\varepsilon^3 2^{-n(I(Z \wedge Y|U) + f_1(\delta) + f_4(\delta))}}{4 \ln 2} \right),
\end{aligned}$$

where $f_4(\delta) = \tau(P_{UZ}, \delta, 3|\mathbf{Y}||\mathcal{X}|\delta)$.

Proof. Let $\mathbf{u} \in T^n_{U,\delta}$. We define the set $B_{\mathbf{u}} \subset (T^n_{X|U,\delta}(\mathbf{u}))^{L_1}$ as

$$\bigcap_{\mathbf{y} \in T^n_{Y|U,\delta}(\mathbf{u})} \Big\{ (\mathbf{x}^1, \ldots, \mathbf{x}^{L_1}) \in (T^n_{X|U,\delta}(\mathbf{u}))^{L_1} : \\
\frac{1}{L_1} \sum_{l_1} W_e^{\otimes n}(\mathbf{z}|\mathbf{x}^{l_0 l_1}, \mathbf{y}) 1_{E_0^{(1)}(\mathbf{u}, X^{l_0 l_1}, \mathbf{y})}(\mathbf{z}) \in [(1 \pm \varepsilon) \hat{\vartheta}^{(1)}_{\mathbf{uy}}(\mathbf{z})] \Big\}.$$

One has

$$\begin{aligned}
&\mathbb{P}[A_2^{(1)}(l_0,\mathbf{z})^c|U^{l_0}=\mathbf{u}]\\
&\leq \mathbb{P}\big[\{(X^{l_01},\dots,X^{l_0L_1})\notin B_{\mathbf{u}}\}|U^{l_0}=\mathbf{u}\big]\\
&\quad+\sum_{(\mathbf{x}^1,\dots,\mathbf{x}^{L_1})\in B_{\mathbf{u}}}\mathbb{P}\big[A_2^{(1)}(l_0,\mathbf{z})^c|X^{l_01}=\mathbf{x}^1,\dots,X^{l_0L_1}=\mathbf{x}^{L_1},U^{l_0}=\mathbf{u}\big]\cdot\\
&\qquad\qquad\cdot\mathbb{P}[X^{l_01}=\mathbf{x}^1,\dots,X^{l_0L_1}=\mathbf{x}^{L_1}|U^{l_0}=\mathbf{u}\big].
\end{aligned}$$

From the proof of Lemma 7 it follows that

$$\mathbb{P}[\{(X^{l_01},\dots,X^{l_0L_1})\notin B_{\mathbf{u}}\}|U^{l_0}=\mathbf{u}]\leq 2|\mathbf{Y}|^n\exp\left(-L_1\cdot\frac{\varepsilon^3 2^{-n(I(Z\wedge X|YU)+f_1(\delta)+f_2(\delta))}}{2\ln 2}\right),\tag{30}$$

which gives a bound independent of $\mathbf{u}$. Now let $(\mathbf{x}^1,\dots,\mathbf{x}^{L_1})\in B_{\mathbf{u}}$. By (25) and (27),

$$\begin{aligned}
\hat{\vartheta}_{\mathbf{uy}}^{(1)}(\mathbf{z})&=\mathbb{E}[W_e^{\otimes n}(\mathbf{z}|X^{11},\mathbf{y})1_{E_2^{(1)}(\mathbf{u},X^{11},\mathbf{y})}(\mathbf{z})|U^1=\mathbf{u}]\\
&\leq\mathbb{E}[W_e^{\otimes n}(\mathbf{z}|X^{11},\mathbf{y})|U^1=\mathbf{u}]\\
&\leq\frac{1}{P_{X|U}^{\otimes n}(T_{X|U,\delta}^n(\mathbf{u})|\mathbf{u})}(P_{Z|YU})^{\otimes n}(\mathbf{z}|\mathbf{y},\mathbf{u})\\
&\leq(1-2^{-nc\delta^2})^{-1}2^{-n(H(Z|YU)-f_1(\delta))}.
\end{aligned}$$

Hence the random variables

$$\tilde{W}_{\mathbf{uz}}^{(1)}(l_0,l_2):=\frac{1}{L_1}\sum_{l_1}W_e^{\otimes n}(\mathbf{z}|\mathbf{x}^{l_1},Y^{l_0l_2})1_{E_0^{(1)}(\mathbf{u},\mathbf{x}^{l_1},Y^{l_0l_2})}(\mathbf{z})\qquad(l_2\in[L_2]),$$

which are independent conditional on $\{U^{l_0}=\mathbf{u}\}$, are upper-bounded by

$$\frac{(1+\varepsilon)}{(1-2^{-nc\delta^2})}\cdot 2^{-n(H(Z|YU)-f_1(\delta))}.$$

For their conditional expectation we have

$$\mu_{l_0l_2}:=\mathbb{E}[\tilde{W}_{\mathbf{uz}}^{(1)}(l_0,l_2)|U^{l_0}=\mathbf{u}]\in[(1\pm\varepsilon)\mathbb{E}[\hat{\vartheta}_{\mathbf{u}Y^{l_0l_2}}^{(1)}(\mathbf{z})|U^1=\mathbf{u}]]=[(1\pm\varepsilon)\hat{\vartheta}_{\mathbf{u}}^{(1)}(\mathbf{z})].$$

Thus their arithmetic mean $\bar{\mu}=(1/L_2)\sum_{l_2}\mu_{l_0l_2}$ must also be contained in $[(1\pm\varepsilon)\hat{\vartheta}_{\mathbf{u}}^{(1)}(\mathbf{z})]$. Applying Lemma 5, we conclude

$$\begin{aligned}
&\mathbb{P}\big[A_2^{(1)}(l_0,\mathbf{z})^c|X^{l_01}=\mathbf{x}^1,\dots,X^{l_0L_1}=\mathbf{x}^{L_1},U^{l_0}=\mathbf{u}\big]\\
&=\mathbb{P}\left[\frac{1}{L_2}\sum_{l_2}\tilde{W}_{\mathbf{uz}}^{(1)}(l_0,l_2)\notin[(1\pm 3\varepsilon)\hat{\vartheta}_{\mathbf{u}}^{(1)}(\mathbf{z})]\middle|U^{l_0}=\mathbf{u}\right]\\
&\leq\mathbb{P}\left[\frac{1}{L_2}\sum_{l_2}\tilde{W}_{\mathbf{uz}}^{(1)}(l_0,l_2)\notin[(1\pm\varepsilon)\bar{\mu}]\middle|U^{l_0}=\mathbf{u}\right]\\
&\leq 2\exp\left(-L_2\cdot\frac{\varepsilon^2(1-2^{-nc\delta^2})2^{n(H(Z|YU)-f_1(\delta))}(1-\varepsilon)\hat{\vartheta}_{\mathbf{u}}^{(1)}(\mathbf{z})}{2(1+\varepsilon)\ln 2}\right).
\end{aligned}$$

Due to the definition of $F_2^{(1)}(\mathbf{u})$ and to (29), this is smaller than

$$2\exp\left(-L_2\cdot\frac{\varepsilon^3 2^{-n(I(Z\wedge Y|U)+f_1(\delta)+f_4(\delta))}}{4\ln 2}\right) \tag{31}$$

if ε is sufficiently small and n is sufficiently large, giving a bound independent of $\mathbf{u}$ and $\mathbf{x}^1,\dots,\mathbf{x}^{L_1}$. Adding the bounds (30) and (31) concludes the proof. □

The next lemma is only needed in Case 1. Let $A_2^{(1)}(\mathbf{z}) := A_2^{(1)}(1,\mathbf{z})\cap\dots\cap A_2^{(1)}(L_0,\mathbf{z})$. For every $\mathbf{z}$, we then define a new probability measure by $\hat{\mathbb{P}}_{\mathbf{z}}^{(1)} := \mathbb{P}[\cdot|A_2^{(1)}(\mathbf{z})]$. With $\vartheta^{(1)}(\mathbf{z}) := \hat{\mathbb{E}}_{\mathbf{z}}^{(1)}[\hat{\vartheta}_{U^1}^{(1)}(\mathbf{z})]$ define

$$F_0^{(1)} := \{\mathbf{z}\in T^n_{Z,4|\mathcal{Y}||\mathcal{X}||\mathcal{U}|\delta} : \vartheta^{(1)}(\mathbf{z}) \geq |T^n_{Z,4|\mathcal{Y}||\mathcal{X}||\mathcal{U}|\delta}|^{-1}\}$$

and $\hat{\vartheta}^{(1)} := \vartheta^{(1)}\cdot 1_{F_0^{(1)}}$.

Lemma 9. *Let $\mathbf{z}\in F_0^{(1)}$ and let $A_0^{(1)}(\mathbf{z})$ be the event that*

$$\frac{1}{L_0L_1L_2}\sum_{l_0,l_1,l_2} W_e^{\otimes n}(\mathbf{z}|X^{l_0l_1},Y^{l_0l_2})1_{E_0^{(1)}(U^{l_0},X^{l_0l_1},Y^{l_0l_2})}(\mathbf{z}) \in [(1\pm 5\varepsilon)\hat{\vartheta}^{(1)}(\mathbf{z})].$$

Then for $f_6(\delta) = \tau(P_Z, 4|\mathcal{Y}||\mathcal{X}||\mathcal{U}|\delta,\delta)$, sufficiently small ε and n sufficiently large,

$$\begin{aligned}
&\mathbb{P}[A_0^{(1)}(\mathbf{z})^c]\\
&\leq 2L_0|\mathbf{Y}|^n\exp\left(-L_1\cdot\frac{\varepsilon^3 2^{-n(I(Z\wedge X|YU)+f_1(\delta)+f_2(\delta))}}{2\ln 2}\right)\\
&\quad+2L_0\exp\left(-L_2\cdot\frac{\varepsilon^3 2^{-n(I(Z\wedge Y|U)+f_1(\delta)+f_4(\delta))}}{4\ln 2}\right)\\
&\quad+2\exp\left(-L_0\cdot\frac{\varepsilon^3 2^{-n(I(Z\wedge U)+f_4(\delta)+f_6(\delta))}}{4\ln 2}\right).
\end{aligned}$$

Proof. We have

$$\mathbb{P}[A_0^{(1)}(\mathbf{z})^c] \leq \hat{\mathbb{P}}_{\mathbf{z}}^{(1)}[A_0^{(1)}(\mathbf{z})^c] + \mathbb{P}[A_2^{(1)}(\mathbf{z})^c]. \tag{32}$$

By Lemma 8, for ε sufficiently small and n sufficiently large,

$$\begin{aligned}
\mathbb{P}[A_2^{(1)}(\mathbf{z})^c] &\leq 2L_0|\mathbf{y}|^n\exp\left(-L_1\cdot\frac{\varepsilon^3 2^{-n(I(Z\wedge X|YU)+f_1(\delta)+f_2(\delta))}}{2\ln 2}\right)\\
&\quad+2L_0\exp\left(-L_2\cdot\frac{\varepsilon^3 2^{-n(I(Z\wedge Y|U)+f_1(\delta)+f_4(\delta))}}{4\ln 2}\right).
\end{aligned} \tag{33}$$

In order to bound $\hat{\mathbb{P}}_{\mathbf{z}}^{(1)}[A_0^{(1)}(\mathbf{z})^c]$, note that the sets $A_2^{(1)}(1,\mathbf{z}),\dots,A_2^{(1)}(L_0,\mathbf{z})$ are independent with respect to $\mathbb{P}$. Thus under $\hat{\mathbb{P}}_{\mathbf{z}}^{(1)}$, the random variables

$$\tilde{W}_{\mathbf{z}}^{(1)}(l_0) := \frac{1}{L_1L_2}\sum_{l_1,l_2} W_e^{\otimes n}(\mathbf{z}|X^{l_0l_1},Y^{l_0l_2})1_{E_0^{(1)}(U^{l_0},X^{l_0l_1},Y^{l_0l_2})}(\mathbf{z}) \quad (l_0\in[L_0])$$

retain their independence and are upper-bounded by

$$(1+3\varepsilon)\max_{\mathbf{u}\in T^n_{U,\delta}} \hat{\vartheta}^{(1)}_{\mathbf{u}}(\mathbf{z}).$$

We can further bound this last term as follows: for $\mathbf{u} \in T^n_{U,\delta}$, applying (25) and (27),

$$\begin{aligned}
\hat{\vartheta}^{(1)}_{\mathbf{u}}(\mathbf{z}) &= \mathbb{E}[W_e^{\otimes n}(\mathbf{z}|X^{11},Y^{11})1_{E_0^{(1)}(\mathbf{u},X^{11},Y^{11})}(\mathbf{z})|U^1=\mathbf{u}] \\
&\le \mathbb{E}[W_e^{\otimes n}(\mathbf{z}|X^{11},Y^{11})|U^1=\mathbf{u}] \\
&\le \frac{1}{P_1^{\otimes n}(T^n_{X|U,\delta}(\mathbf{u})|\mathbf{u})P_2^{\otimes n}(T^n_{Y|U,\delta}(\mathbf{u})|\mathbf{u})} P^{\otimes n}_{Z|U}(\mathbf{z}|\mathbf{u}) \\
&\le (1-2^{-nc_1\delta^2})^{-2} 2^{-n(H(Z|U)-f_4(\delta))}.
\end{aligned}$$

Observing that $\hat{\mathbb{E}}^{(1)}_{\mathbf{z}}[\tilde{W}^{(1)}_{\mathbf{z}}(1)] \in [(1\pm 3\varepsilon)\hat{\vartheta}^{(1)}(\mathbf{z})]$ and applying Lemma 5 and (29) in the usual way yields

$$\begin{aligned}
\hat{\mathbb{P}}^{(1)}_{\mathbf{z}}[A_0^{(1)}(\mathbf{z})^c] &\le 2\exp\left(-L_0 \cdot \frac{\varepsilon^2(1-2^{-nc\delta^2})^2\, 2^{n(H(Z|U)-f_4(\delta))}(1-3\varepsilon)\,\hat{\vartheta}^{(1)}(\mathbf{z})}{2(1+3\varepsilon)\ln 2}\right) \\
&\le 2\exp\left(-L_0 \cdot \frac{\varepsilon^3 2^{-n(I(Z\wedge U)+f_4(\delta)+f_6(\delta))}}{4\ln 2}\right)
\end{aligned}$$

if ε is sufficiently small and n sufficiently large. Inserting this and (33) in (32) completes the proof. □

We finally note that results analogous to Lemma 6-9 hold where the roles of X and Y are exchanged. We denote the corresponding events by $A_*^{(1)}(l_0,l_2)'$ and $A_1^{(1)}(l_0,l_2,\mathbf{z})'$, $A_2^{(1)}(l_0,\mathbf{z})'$, $A_0^{(1)}(\mathbf{z})'$.

Secrecy for Case 0 and 1: The following lemma links the above probabilistic bounds to secrecy. In the next subsection, roughly speaking, we will associate a family $\mathcal{F}$ to every message triple (k_0,k_1,k_2). If L_0, L_1, L_2 are large enough, the bounds of Lemma 10 are satisfied for every such $\mathcal{F}$ with high probability. Hence there is a joint realization of the $\mathcal{F}$ such that the statement of the lemma is satisfied for every message triple. By an appropriate choice of random code one then obtains (22).

Lemma 10. *Denote by $p^{(1)}$ the bound on $\mathbb{P}[A_2^{(1)}(l_0,\mathbf{z})^c]$ derived in Lemma 8. Let $\{\mathbf{u}^{l_0},\mathbf{x}^{l_0l_1},\mathbf{y}^{l_0l_2} : (l_0,l_1,l_2)\in[L_0]\times[L_1]\times[L_2]\}$ be a realization of $\mathcal{F}$ satisfying the conditions of*

$$\bigcap_{l_0,l_2} A_*^{(1)}(l_0,l_2), \tag{34}$$

$$\bigcap_{l_0,l_2}\bigcap_{\mathbf{z}\in\mathcal{Z}^n} A_1^{(1)}(l_0,l_2,\mathbf{z}), \tag{35}$$

$$\bigcap_{l_0}\bigcap_{\mathbf{z}\in\mathcal{Z}^n} A_2^{(1)}(l_0,\mathbf{z}), \tag{36}$$

$$\bigcap_{\mathbf{z}\in F_0^{(1)}} A_0^{(1)}(\mathbf{z}). \tag{37}$$

Then

$$\|\hat{\vartheta}^{(1)} - \frac{1}{L_0L_1L_2}\sum_{l_0,l_1,l_2} W_e^{\otimes n}(\cdot|\mathbf{x}^{l_0l_1},\mathbf{y}^{l_0l_2})\| \leq 20\varepsilon + 9\cdot 2^{-nc\delta^2} + L_0|\mathcal{Z}|^n p^{(1)}.$$

The same inequality is true if we require conditions (34′)-(37′) which contain the primed equivalents of (34)-(37) defined at the end of the previous paragraph. If $L_0 = 1$, then (37) and (37′) do not have to hold.

We now prove the above lemma. We have

$$\begin{aligned}
&\|\hat{\vartheta}^{(1)} - \frac{1}{L_0L_1L_2}\sum_{l_0,l_1,l_2} W_e^{\otimes n}(\cdot|\mathbf{x}^{l_0l_1},\mathbf{y}^{l_0l_2})\| \\
&\leq \|\hat{\vartheta}^{(1)} - \frac{1}{L_0L_1L_2}\sum_{l_0,l_1,l_2} W_e^{\otimes n}(\cdot|\mathbf{x}^{l_0l_1},\mathbf{y}^{l_0l_2})1_{E_0^{(1)}(\mathbf{u}^{l_0},\mathbf{x}^{l_0l_1},\mathbf{y}^{l_0l_2})}1_{F_0^{(1)}}\| &(38)\\
&+\|\frac{1}{L_0L_1L_2}\sum_{l_0,l_1,l_2} W_e^{\otimes n}(\cdot|\mathbf{x}^{l_0l_1},\mathbf{y}^{l_0l_2})1_{E_0^{(1)}(\mathbf{u}^{l_0},\mathbf{x}^{l_0l_1},\mathbf{y}^{l_0l_2})}(1-1_{F_0^{(1)}})\| &(39)\\
&+\|\frac{1}{L_0L_1L_2}\sum_{l_0,l_1,l_2} W_e^{\otimes n}(\cdot|\mathbf{x}^{l_0l_1},\mathbf{y}^{l_0l_2})1_{E_2^{(1)}(\mathbf{u}^{l_0},\mathbf{x}^{l_0l_1},\mathbf{y}^{l_0l_2})}(1-1_{F_2^{(1)}(\mathbf{u}^{l_0})})\| &(40)\\
&+\|\frac{1}{L_0L_1L_2}\sum_{l_0,l_1,l_2} W_e^{\otimes n}(\cdot|\mathbf{x}^{l_0l_1},\mathbf{y}^{l_0l_2})1_{E_1^{(1)}(\mathbf{u}^{l_0},\mathbf{x}^{l_0l_1},\mathbf{y}^{l_0l_2})}(1-1_{F_1^{(1)}(\mathbf{u}^{l_0},\mathbf{y}^{l_0l_2})})\| &(41)\\
&+\|\frac{1}{L_0L_1L_2}\sum_{l_0,l_1,l_2} W_e^{\otimes n}(\cdot|\mathbf{x}^{l_0l_1},\mathbf{y}^{l_0l_2})(1-1_{E_1^{(1)}(\mathbf{u}^{l_0},\mathbf{x}^{l_0l_1},\mathbf{y}^{l_0l_2})})\|. &(42)
\end{aligned}$$

Due to (37), we know that (38) $\leq 5\varepsilon$.

Next we consider (41). Due to (35) we have

$$\begin{aligned}
&(41)\\
&\leq 1 - \frac{1}{L_0L_1L_2}\sum_{l_0,l_1,l_2} W_e^{\otimes n}(E_2^{(1)}(\mathbf{u}^{l_0},\mathbf{x}^{l_0l_1},\mathbf{y}^{l_0l_2})|\mathbf{x}^{l_0l_1},\mathbf{y}^{l_0l_2})\\
&\leq 1 - \frac{1-\varepsilon}{L_0L_2}\sum_{l_0,l_2}\hat{\vartheta}^{(1)}_{\mathbf{u}^{l_0}\mathbf{y}^{l_0l_2}}(\mathcal{Z}^n)
\end{aligned}$$

(we defined the general measure of a set in the notation section at the beginning of the paper). The support of $\vartheta^{(1)}_{\mathbf{u}^{l_0}\mathbf{y}^{l_0 l_2}}$ is contained in $T^n_{Z|YU,2|\mathcal{X}|\delta}(\mathcal{Y}^{l_0 l_2}, \mathbf{u}^{l_0})$, so by the definition of $F_1^{(1)}(\mathbf{u}^{l_0}, \mathbf{y}^{l_0 l_2})$ we obtain

$$\hat{\vartheta}^{(1)}_{\mathbf{u}^{l_0}\mathbf{y}^{l_0 l_2}}(\mathcal{Z}^n) \geq \vartheta^{(1)}_{\mathbf{u}^{l_0}\mathbf{y}^{l_0 l_2}}(\mathcal{Z}^n) - \varepsilon. \tag{43}$$

Lemma 11. *If* $\mathbf{u} \in T^n_{U,\delta}$ *and* $\mathbf{y} \in T^n_{Y|U,\delta}(\mathbf{u})$, *then*

$$\vartheta^{(1)}_{\mathbf{uy}}(\mathcal{Z}^n) \geq 1 - 2 \cdot 2^{-nc\delta^2}.$$

Proof. First of all note that

$$\begin{aligned}
&\vartheta^{(1)}_{\mathbf{uy}}(\mathcal{Z}^n)\\
&= \mathbb{E}[W_e^{\otimes n}(E_1^{(1)}(\mathbf{u}, X^{11}, \mathbf{y})|X^{11}, \mathbf{y})|U^1 = \mathbf{u}]\\
&\geq \mathbb{E}[W_e^{\otimes n}(E_1^{(1)}(\mathbf{u}, X^{11}, \mathbf{y})|X^{11}, \mathbf{y}); X^{11} \in T^n_{X|YU,\delta}(\mathbf{y}, \mathbf{u})|U^1 = \mathbf{u}].
\end{aligned} \tag{44}$$

Now we claim that for $\mathbf{x} \in T^n_{X|YU,\delta}(\mathbf{y}, \mathbf{u})$

$$T^n_{Z|YXU,\delta}(\mathbf{y}, \mathbf{x}, \mathbf{u}) \subset T^n_{Z|YU,2|\mathcal{X}|\delta}(\mathbf{y}, \mathbf{u}). \tag{45}$$

To verify this, let $(z, y, u) \in \mathcal{Z} \times \mathbf{y} \times \mathcal{U}$ and $\mathbf{z} \in T^n_{Z|YXU,\delta}(\mathbf{y}, \mathbf{x}, \mathbf{u})$. Then

$$\begin{aligned}
&\left|\frac{1}{n}N(z, y, u|\mathbf{z}, \mathbf{y}, \mathbf{u}) - P_{Z|YU}(z|y, u)\frac{1}{n}N(y, u|\mathbf{y}, \mathbf{u})\right|\\
&\leq \sum_x \left|\frac{1}{n}N(z, y, x, u|\mathbf{z}, \mathbf{y}, \mathbf{x}, \mathbf{u}) - W(z|x, y)\frac{1}{n}N(y, x, u|\mathbf{y}, \mathbf{x}, \mathbf{u})\right|\\
&\quad + \sum_x W(z|x, y)\left|\frac{1}{n}N(y, x, u|\mathbf{y}, \mathbf{x}, \mathbf{u}) - P_{X|YU}(x|y, u)\frac{1}{n}N(y, u|\mathbf{y}, \mathbf{u})\right|\\
&\leq 2|\mathcal{X}|\delta.
\end{aligned}$$

This proves (45). Due to the choice of $f_2(\delta)$ and to (27), we thus see that $T^n_{Z|YXU,\delta}(\mathbf{y}, \mathbf{x}, \mathbf{u})$ is contained in $E_1^{(1)}(\mathbf{u}, \mathbf{x}, \mathbf{y})$ for $\mathbf{x} \in T^n_{X|YU,\delta}(\mathbf{y}, \mathbf{u})$, and we have that (44) is lower-bounded by

$$\mathbb{E}[W_e^{\otimes n}(T^n_{Z|YXU,\delta}(\mathbf{y}, X^{11}, \mathbf{u})|X^{11}, \mathbf{y}); X^{11} \in T^n_{X|YU,\delta}(\mathbf{y}, \mathbf{u})|U^1 = \mathbf{u}]. \tag{46}$$

Further, as in the proof of Lemma 6 one sees that

$$\mathbb{P}[X^{11} \in T^n_{X|YU,\delta}(\mathbf{y}, \mathbf{u})|U^1 = \mathbf{u}] \geq 1 - \frac{2^{-nc\delta^2}}{1 - 2^{-nc\delta^2}}. \tag{47}$$

Due to (47) and (25), we can lower-bound (46) for sufficiently large n by

$$(1 - 2^{-nc\delta^2}) \cdot \left(1 - \frac{2^{-nc\delta^2}}{1 - 2^{-nc\delta^2}}\right) \geq 1 - 2 \cdot 2^{-nc\delta^2},$$

which proves Lemma 11. □

Using (43) and Lemma 11 we can conclude that

$$(41) \leq 2(\varepsilon + 2^{-nc\delta^2}).$$

One starts similarly for (40). We have by (36)

$$\begin{aligned}(40) &\leq 1 - \frac{1}{L_0 L_1 L_2} \sum_{l_0,l_1,l_2} W_e^{\otimes n}(E_0^{(1)}(\mathbf{u}^{l_0}, \mathbf{x}^{l_0 l_1}, \mathbf{y}^{l_0 l_2})|\mathbf{x}^{l_0 l_1}, \mathbf{y}^{l_0 l_2}) \\ &\leq 1 - \frac{(1-3\varepsilon)}{L_0} \sum_{l_0} \hat{\vartheta}_{\mathbf{u}^{l_0}}^{(1)}(\mathcal{Z}^n).\end{aligned}$$

As the support of $\vartheta_{\mathbf{u}^{l_0}}^{(1)}$ is contained in $T_{Z|U,3|\mathcal{Y}||\mathcal{X}|\delta}^n(\mathbf{u}^{l_0})$, we can lower-bound $\hat{\vartheta}_{\mathbf{u}^{l_0}}^{(1)}(\mathcal{Z}^n)$ by $\vartheta_{\mathbf{u}^{l_0}}^{(1)}(\mathcal{Z}^n) - \varepsilon$. Using (43) and Lemma 11, we have

$$\vartheta_{\mathbf{u}^{l_0}}^{(1)}(\mathcal{Z}^n) = \mathbb{E}[\hat{\vartheta}_{\mathbf{u}^{l_0} Y^{11}}^{(1)}(\mathcal{Z}^n)|U^1 = \mathbf{u}^{l_0}] \geq 1 - 2 \cdot 2^{-nc\delta^2} - \varepsilon, \tag{48}$$

so we conclude

$$(40) \leq 5\varepsilon + 2 \cdot 2^{-nc\delta^2}.$$

For (39), one has by (37)

$$\begin{aligned}(39) &\leq 1 - \frac{1}{L_0 L_1 L_2} \sum_{l_0,l_1,l_2} W_e^{\otimes n}(E_0^{(1)}(\mathbf{u}^{l_0}, \mathbf{x}^{l_0 l_1}, \mathbf{y}^{l_0 l_2}) \cap F_0^{(1)}|\mathbf{x}^{l_0 l_1}, \mathbf{y}^{l_0 l_2}) \\ &\leq 1 - (1-5\varepsilon)\hat{\vartheta}^{(1)}(F_0^{(1)}).\end{aligned}$$

It remains to lower-bound $\hat{\vartheta}^{(1)}(F_0^{(1)})$. Observe that the support of $\vartheta^{(1)}$ is restricted to $T_{Z,4|\mathbf{y}||\mathcal{X}||\mathcal{U}|\delta}^n$, so due to the definition of $F_0^{(1)}$, one has $\hat{\vartheta}^{(1)}(F_0^{(1)}) = \vartheta^{(1)}(F_0^{(1)}) \geq \vartheta^{(1)}(\mathcal{Z}^n) - \varepsilon$. Further,

$$\begin{aligned}\vartheta^{(1)}(\mathcal{Z}^n) &= \sum_{\mathbf{z} \in \mathcal{Z}^n} \hat{\mathbb{E}}_{\mathbf{z}}^{(1)}[\hat{\vartheta}_{U^1}^{(1)}(\mathbf{z})] \\ &\geq \mathbb{E}[\vartheta_{U^1}^{(1)}(\mathcal{Z}^n)] - \sum_{\mathbf{z} \in \mathcal{Z}^n} \mathbb{P}[A_2^{(1)}(\mathbf{z})^c] \\ &= \mathbb{E}[\vartheta_{U^1}^{(1)}(\mathcal{Z}^n)] - L_0|\mathcal{Z}|^n p^{(1)}.\end{aligned}$$

In (48), the integrand of $\mathbb{E}[\vartheta_{U^1}^{(1)}(\mathcal{Z}^n)]$ was lower-bounded by $1 - 2 \cdot 2^{-nc\delta^2} - \varepsilon$. We conclude

$$(39) \leq 7\varepsilon + 2 \cdot 2^{-nc\delta^2} + L_0|\mathcal{Z}|^n p^{(1)}.$$

Finally, we use condition (34) to bound (42). We have

$$(42) \tag{49}$$
$$= \frac{1}{L_0L_1L_2}\sum_{l_0,l_1,l_2} W_e^{\otimes n}(E_1^{(1)}(\mathbf{u}^{l_0},\mathbf{x}^{l_0l_1},\mathbf{y}^{l_0l_2})^c|\mathbf{x}^{l_0l_1},\mathbf{y}^{l_0l_2})$$
$$= \frac{1}{L_0L_2}\sum_{l_0,l_2}\Bigg($$
$$\frac{1}{L_1}\sum_{l_1:\mathbf{x}^{l_0l_1}\in T^n_{X|YU,\delta}(\mathbf{y}^{l_0l_2},\mathbf{u}^{l_0})} W_e^{\otimes n}(E_1^{(1)}(\mathbf{u}^{l_0},\mathbf{x}^{l_0l_1},\mathbf{y}^{l_0l_2})^c|\mathbf{x}^{l_0l_1},\mathbf{y}^{l_0l_2}) \tag{50}$$
$$+\frac{1}{L_1}\sum_{l_1:\mathbf{x}^{l_0l_1}\notin T^n_{X|YU,\delta}(\mathbf{y}^{l_0l_2},\mathbf{u}^{l_0})} W_e^{\otimes n}(E_1^{(1)}(\mathbf{u}^{l_0},\mathbf{x}^{l_0l_1},\mathbf{y}^{l_0l_2})^c|\mathbf{x}^{l_0l_1},\mathbf{y}^{l_0l_2})\Bigg). \tag{51}$$

For every (l_0,l_2), we use $T^n_{Z|YXU,\delta}(\mathbf{y},\mathbf{x},\mathbf{u})\subset E_1^{(1)}(\mathbf{u},\mathbf{x},\mathbf{y})$ for $(\mathbf{u},\mathbf{x},\mathbf{y})\in T^n_{U,\delta}\times T^n_{Y|U,\delta}(\mathbf{u})\times T^n_{X|YU,\delta}(\mathbf{y},\mathbf{u})$ as shown in the proof of Lemma 11 to upper-bound the term in (50) by $2^{-nc\delta^2}$. For (51), we know from assumption (34) that it is at most $1-(1-\varepsilon)(1-2\cdot 2^{-nc\delta^2})$. Thus

$$(42)\le 2^{-nc\delta^2}+(1-\varepsilon)(1-2\cdot 2^{-nc\delta^2})\le\varepsilon+3\cdot 2^{-nc\delta^2}.$$

Collecting the bounds on (38)-(42), we obtain a total upper bound of

$$20\varepsilon+9\cdot 2^{-nc\delta^2}+L_0|\mathcal{Z}|^n p^{(1)}.$$

This finishes the proof of Lemma 10.

Bounds for Case 2: Now we specialize to the case that $L_2=1$, but L_0 and L_1 arbitrary. This reduces the number of Chernoff-type estimates needed by one. Lemma 7 carries over, Lemma 8 is not needed, but Lemma 9 changes. We write $Y^{l_01}=: Y^{l_0}$. The definitions of $E_1^{(1)}(\mathbf{u},\mathbf{x},\mathbf{y}), F_1^{(1)}(\mathbf{u},\mathbf{y})$ and $\vartheta^{(1)}_{\mathbf{uy}}$ carry over to this case, we just call them $E_1^{(2)}(\mathbf{u},\mathbf{x},\mathbf{y}), F_1^{(2)}(\mathbf{u},\mathbf{y})$ and $\vartheta^{(2)}_{\mathbf{uy}}$. Further we define

$$E_0^{(2)}(\mathbf{u},\mathbf{x},\mathbf{y}):=E_1^{(2)}(\mathbf{u},\mathbf{x},\mathbf{y})\cap F_1^{(2)}(\mathbf{u},\mathbf{y}).$$

For every l_0, let $A_1^{(2)}(l_0,\mathbf{z}):=A_1^{(1)}(l_0,1,\mathbf{z})$ and we set $A_1^{(2)}(\mathbf{z}):=A_1^{(2)}(1,\mathbf{z})\cap\ldots\cap A_1^{(2)}(L_0,\mathbf{z})$. We define for every $\mathbf{z}$ a new probability measure by $\hat{\mathbb{P}}^{(2)}_{\mathbf{z}}:=\mathbb{P}[\cdot|A_1^{(2)}(\mathbf{z})]$. Let

$$\vartheta^{(2)}(\mathbf{z}):=\hat{\mathbb{E}}^{(2)}_{\mathbf{z}}[\hat{\vartheta}^{(2)}_{U^1Y^1}(\mathbf{z})].$$

Further let

$$F_0^{(2)}:=\{\mathbf{z}\in T^n_{Z,4|\mathbf{y}||\mathcal{X}||\mathcal{U}|\delta}:\vartheta^{(2)}(\mathbf{z})\ge\varepsilon|T^n_{Z,\delta}|^{-1}\}$$

and

$$\hat{\vartheta}^{(2)}=\vartheta^{(2)}\cdot 1_{F_0^{(2)}}.$$

Lemma 12. *Let* $\mathbf{z} \in F_0^{(2)}$. *Let* $A_0^{(2)}(\mathbf{z})$ *be the event*

$$\frac{1}{L_0 L_1} \sum_{l_0,l_1} W_e^{\otimes n}(\mathbf{z}|X^{l_0 l_1}, Y^{l_0}) 1_{E_0^{(2)}(U^{l_0}, X^{l_0 l_1}, Y^{l_0})}(\mathbf{z}) \in [(1 \pm 3\varepsilon)\hat{\vartheta}^{(2)}(\mathbf{z})].$$

Then for ε *sufficiently small and* n *sufficiently large,*

$$\mathbb{P}[A_0^{(2)}(\mathbf{z})^c] \leq 2L_0 \exp\left(-L_1 \cdot \frac{\varepsilon^3 2^{-n(I(Z\wedge X|YU)+f_1(\delta)+f_2(\delta))}}{2\ln 2}\right) + 2\exp\left(-L_0 \cdot \frac{\varepsilon^3 2^{-n(I(Z\wedge YU)+f_1(\delta)+f_6(\delta))}}{4\ln 2}\right).$$

Proof. We have

$$\mathbb{P}[A_0^{(2)}(\mathbf{z})^c] \leq \hat{\mathbb{P}}_{\mathbf{z}}^{(2)}[A_0^{(2)}(\mathbf{z})^c] + \mathbb{P}[A_1^{(2)}(\mathbf{z})^c]. \tag{52}$$

By Lemma 7, we know that

$$\mathbb{P}[A_1^{(2)}(\mathbf{z})^c] \leq 2L_0 \exp\left(-L_1 \cdot \frac{\varepsilon^3 2^{-n(I(Z\wedge X|YU)+f_1(\delta)+f_2(\delta))}}{2\ln 2}\right). \tag{53}$$

In order to bound $\hat{\mathbb{P}}_{\mathbf{z}}^{(2)}[A_0^{(2)}(\mathbf{z})]$, note that the sets $A_1^{(2)}(1,\mathbf{z}),\ldots,A_1^{(2)}(L_0,\mathbf{z})$ are independent with respect to $\mathbb{P}$. Thus under $\hat{\mathbb{P}}_{\mathbf{z}}^{(2)}$, the random variables

$$\tilde{W}_{\mathbf{z}}^{(2)}(l_0) := \frac{1}{L_1} \sum_{l_1} W_e^{\otimes n}(\mathbf{z}|X^{l_0 l_1}, Y^{l_0}) 1_{E_0^{(2)}(U^{l_0}, X^{l_0 l_1}, Y^{l_0})}(\mathbf{z}) \qquad (l_0 \in [L_0])$$

retain their independence and are upper-bounded by

$$(1+\varepsilon) \max_{\mathbf{u}\in T_{U,\delta}^n} \max_{\mathbf{y}\in T_{Y|U,\delta}^n(\mathbf{u})} \hat{\vartheta}_{\mathbf{uy}}^{(2)}(\mathbf{z}).$$

We can further bound this last term as follows: for $\mathbf{u} \in T_{U,\delta}^n$ and $\mathbf{y} \in T_{Y|U,\delta}^n(\mathbf{u})$ one obtains by (25) and (27)

$$\begin{aligned}\hat{\vartheta}_{\mathbf{uy}}^{(2)}(\mathbf{z}) &\leq \mathbb{E}[W_e^{\otimes n}(\mathbf{z}|X^{11}, \mathbf{y})|U^{l_0} = \mathbf{u}] \\ &\leq \frac{1}{1-2^{-nc\delta^2}} P_{Z|YU}^{\otimes n}(\mathbf{z}|\mathbf{y},\mathbf{u}) \\ &\leq \frac{1}{1-2^{-nc\delta^2}} 2^{-n(H(Z|YU)-f_1(\delta))}.\end{aligned}$$

Observing that $\hat{\mathbb{E}}_{\mathbf{z}}^{(2)}[\tilde{W}_{\mathbf{z}}^{(2)}(1)] \in [(1\pm\varepsilon)\hat{\vartheta}^{(2)}(\mathbf{z})]$ and applying Lemma 5 in the usual way yields

$$\begin{aligned}\hat{\mathbb{P}}_{\mathbf{z}}^{(2)}[A_0^{(2)}(\mathbf{z})] &\leq 2\exp\left(-L_0 \cdot \frac{\varepsilon^2(1-2^{-nc\delta^2})2^{n(H(Z|YU)-f_1(\delta))}(1-\varepsilon)\hat{\vartheta}^{(2)}(\mathbf{z})}{2(1+\varepsilon)\ln 2}\right) \\ &\leq 2\exp\left(-L_0 \cdot \frac{\varepsilon^3 2^{-n(I(Z\wedge YU)+f_1(\delta)+f_6(\delta))}}{4\ln 2}\right)\end{aligned}$$

if ε is sufficiently small and n sufficiently large. Inserting this and (53) in (52) completes the proof. □

Again we note that a result analogous to Lemma 12 holds where the roles of X and Y are exchanged. Setting $A_*^{(2)}(l_0) := A_*^{(1)}(l_0, 1)$, we denote the events corresponding to such an exchange by $A_*^{(2)}(l_0)'$ and $A_1^{(2)}(l_0, \mathbf{z})', A_0^{(2)}(\mathbf{z})'$.

Secrecy for Case 2:

Lemma 13. *Denote by $p^{(2)}$ the bound on $\mathbb{P}[A_1^{(2)}(l_0, \mathbf{z})^c]$ derived in Lemma 7. Let $\{\mathbf{u}^{l_0}, \mathbf{x}^{l_0 l_1}, \mathbf{y}^{l_0} : (l_0, l_1, l_2) \in [L_0] \times [L_1] \times [L_2]\}$ be a realization of $\mathcal{F}$ satisfying the conditions of*

$$\bigcap_{l_0} A_*^{(2)}(l_0), \tag{54}$$

$$\bigcap_{l_0} \bigcap_{\mathbf{z} \in \mathcal{Z}^n} A_1^{(2)}(l_0, \mathbf{z}), \tag{55}$$

$$\bigcap_{\mathbf{z} \in F_0^{(2)}} A_0^{(2)}(\mathbf{z}). \tag{56}$$

Then

$$\|\hat{\vartheta}^{(2)} - \frac{1}{L_0 L_1} \sum_{l_0, l_1} W_e^{\otimes n}(\cdot|\mathbf{x}^{l_0 l_1}, \mathbf{y}^{l_0})\| \leq 9\varepsilon + 7 \cdot 2^{-nc\delta^2} + L_0 |\mathcal{Z}|^n p^{(2)}.$$

The same inequality is true if we require conditions (54′)-(56′) which contain the primed equivalents of (54)-(56) defined at the end of the previous paragraph.

We now prove the above lemma. We have

$$\|\hat{\vartheta}^{(2)} - \frac{1}{L_0 L_1} \sum_{l_0, l_1} W_e^{\otimes n}(\cdot|\mathbf{x}^{l_0 l_1}, \mathbf{y}^{l_0})\|$$

$$\leq \|\hat{\vartheta}^{(2)} - \frac{1}{L_0 L_1} \sum_{l_0, l_1} W_e^{\otimes n}(\cdot|\mathbf{x}^{l_0 l_1}, \mathbf{y}^{l_0}) 1_{E_0^{(2)}(\mathbf{u}^{l_0}, \mathbf{x}^{l_0 l_1}, \mathbf{y}^{l_0})} 1_{F_0^{(2)}}\| \tag{57}$$

$$+ \|\frac{1}{L_0 L_1} \sum_{l_0, l_1} W_e^{\otimes n}(\cdot|\mathbf{x}^{l_0 l_1}, \mathbf{y}^{l_0}) 1_{E_0^{(2)}(\mathbf{u}^{l_0}, \mathbf{x}^{l_0 l_1}, \mathbf{y}^{l_0})} (1 - 1_{F_0^{(2)}})\| \tag{58}$$

$$+ \|\frac{1}{L_0 L_1} \sum_{l_0, l_1} W_e^{\otimes n}(\cdot|\mathbf{x}^{l_0 l_1}, \mathbf{y}^{l_0}) 1_{E_1^{(2)}(\mathbf{u}^{l_0}, \mathbf{x}^{l_0 l_1}, \mathbf{y}^{l_0})} (1 - 1_{F_1^{(2)}(\mathbf{u}^{l_0}, \mathbf{y}^{l_0})})\| \tag{59}$$

$$+ \|\frac{1}{L_0 L_1} \sum_{l_0, l_1} W_e^{\otimes n}(\cdot|\mathbf{x}^{l_0 l_1}, \mathbf{y}^{l_0}) (1 - 1_{E_1^{(2)}(\mathbf{u}^{l_0}, \mathbf{x}^{l_0 l_1}, \mathbf{y}^{l_0})})\|. \tag{60}$$

Due to (56), we know that (57) $\leq \varepsilon$.

Next we consider (59). Due to (55), we have

$$(59) \leq 1 - \frac{1}{L_0 L_1} \sum_{l_0, l_1} W_e^{\otimes n}(E_0^{(2)}(\mathbf{u}^{l_0}, \mathbf{x}^{l_0 l_1}, \mathbf{y}^{l_0}) | \mathbf{x}^{l_0 l_1}, \mathbf{y}^{l_0})$$

$$\leq 1 - \frac{1 - \varepsilon}{L_0} \sum_{l_0} \hat{\vartheta}^{(2)}_{\mathbf{u}^{l_0} \mathbf{y}^{l_0}}(\mathcal{Z}^n).$$

As for Case 1, one lower-bounds $\hat{\vartheta}^{(2)}_{\mathbf{u}^{l_0}\mathbf{y}^{l_0}}(\mathcal{Z}^n) \geq \vartheta^{(2)}_{\mathbf{u}^{l_0}\mathbf{y}^{l_0}}(\mathcal{Z}^n) - \varepsilon$ by $1 - 2 \cdot 2^{-nc\delta^2} - \varepsilon$. Thus we can conclude that

$$(59) \leq 2(\varepsilon + 2^{-nc\delta^2}).$$

For (58), we have by (55)

$$\begin{aligned}(58) &\leq 1 - \frac{1}{L_0L_1}\sum_{l_0,l_1} W_e^{\otimes n}(E_0^{(2)}(\mathbf{u}^{l_0}, \mathbf{x}^{l_0l_1}, \mathbf{y}^{l_0}) \cap F_0^{(2)} | \mathbf{x}^{l_0l_1}, \mathbf{y}^{l_0}) \\ &\leq 1 - (1 - 3\varepsilon)\hat{\vartheta}^{(2)}(F_0^{(2)}).\end{aligned}$$

It remains to lower-bound $\hat{\vartheta}^{(2)}(F_0^{(2)}) \geq \vartheta^{(2)}(\mathcal{Z}^n) - \varepsilon$. As in the lower bound on $\vartheta^{(1)}(\mathcal{Z}^n)$ above, one obtains the bound

$$\vartheta^{(2)}(\mathcal{Z}^n) \geq 1 - 2 \cdot 2^{-nc\delta^2} - \varepsilon.$$

Thus we conclude

$$(58) \leq 5\varepsilon + 2 \cdot 2^{-nc\delta^2}.$$

Finally, we use condition (54) to bound (60). We have

$$\begin{aligned}(60) &= \frac{1}{L_0L_1}\sum_{l_0,l_1} W_e^{\otimes n}(E_1^{(2)}(\mathbf{u}^{l_0}, \mathbf{x}^{l_0l_1}, \mathbf{y}^{l_0}) | \mathbf{x}^{l_0l_1}, \mathbf{y}^{l_0}) \\ &= \frac{1}{L_0}\sum_{l_0}\Biggl(\frac{1}{L_1}\sum_{l_1:\mathbf{x}^{l_0l_1}\in T^n_{X|YU,\delta}(\mathbf{y}^{l_0},\mathbf{u}^{l_0})} W_e^{\otimes n}(E_1^{(2)}(\mathbf{u}^{l_0}, \mathbf{x}^{l_0l_1}, \mathbf{y}^{l_0}) | \mathbf{x}^{l_0l_1}, \mathbf{y}^{l_0}) \qquad (61) \\ &\quad + \frac{1}{L_1}\sum_{l_1:\mathbf{x}^{l_0l_1}\notin T^n_{X|YU,\delta}(\mathbf{y}^{l_0},\mathbf{u}^{l_0})} W_e^{\otimes n}(E_1^{(2)}(\mathbf{u}^{l_0}, \mathbf{x}^{l_0l_1}, \mathbf{y}^{l_0}) | \mathbf{x}^{l_0l_1}, \mathbf{y}^{l_0})\Biggr). \qquad (62)\end{aligned}$$

For every l_0, the summand appearing in (61) can be upper-bounded by $2^{-nc\delta^2}$. By assumption (54), (62) is upper-bounded by $1 - (1 - \varepsilon)(1 - 2 \cdot 2^{-nc\delta^2})$. Thus

$$(60) \leq \varepsilon + 3 \cdot 2^{-nc\delta^2}.$$

Collecting the bounds for (57)-(60), we obtain a total upper bound of

$$9\varepsilon + 7 \cdot 2^{-nc\delta^2} + L_0|\mathcal{Z}|^n p^{(2)}.$$

This finishes the proof of Lemma 13.

Bounds for Case 3: Now we treat the case $L_1 = L_2 = 1$. Lemma 14 is the analog of Lemma 6, the proofs are analogous.

Lemma 14. *Let the event $A_*^{(3)}$ be defined by*

$$A_*^{(3)} := \{|\{l_0 \in [L_0] : X^{l_0} \in T^n_{X|YU,\delta}(Y^{l_0}, U^{l_0})\}| \geq (1 - \varepsilon)(1 - 2 \cdot 2^{-nc_1\delta^2})L_0\}.$$

Then

$$\mathbb{P}[(A_*^{(3)})^c] \leq \exp\left(-L_0 \cdot \frac{\varepsilon^2(1 - 2 \cdot 2^{-nc_1\delta^2})}{2\ln 2}\right).$$

Let

$$E^{(3)}(\mathbf{x},\mathbf{y}) := \{\mathbf{z} \in T^n_{Z,4|\mathcal{Y}||\mathcal{X}||\mathcal{U}|\delta} : W_e^{\otimes n}(\mathbf{z}|\mathbf{x},\mathbf{y}) \leq 2^{-n(H(Z|XY)-f_2(\delta))}\},$$

where $f_2(\delta) = \tau(P_{XYZ}, 3\delta, \delta)$. Let

$$\vartheta^{(3)}(\mathbf{z}) := \mathbb{E}[W_e^{\otimes n}(\mathbf{z}|X^1, Y^1)1_{E_1(X^1,Y^1)}(\mathbf{z})]$$

and for

$$F^{(3)} := \{\mathbf{z} \in T^n_{Z,4|\mathcal{Y}||\mathcal{X}||\mathcal{U}|\delta} : \vartheta(\mathbf{z}) \geq \varepsilon |T^n_{Z,\delta}|^{-1}\}$$

define the measure

$$\hat{\vartheta}^{(3)} := \hat{\vartheta}^{(3)} \cdot 1_{F^{(3)}}.$$

Lemma 15. *Let $\mathbf{z} \in F^{(3)}$. Let $A^{(3)}(\mathbf{z})$ be the event that*

$$\frac{1}{L_0}\sum_{l_0} W_e^{\otimes n}(\mathbf{z}|X^{l_0}, Y^{l_0})1_{E^{(3)}(X^{l_0},Y^{l_0})}(\mathbf{z}) \in [(1 \pm \varepsilon)\hat{\vartheta}^{(3)}(\mathbf{z})].$$

Then for $f_1(\delta) = \tau(P_{UYZ}, 4|\mathcal{Y}||\mathcal{X}||\mathcal{U}|\delta, \delta)$,

$$\mathbb{P}[A^{(3)}(\mathbf{z})^c] \leq 2\exp\left(-L_0 \cdot \frac{\varepsilon^3 2^{-n(I(Z\wedge XY)+f_1(\delta)+f_2(\delta))}}{2\ln 2}\right).$$

The proof of this lemma is analogous to that of Lemma 7.

Secrecy for Case 3:

Lemma 16. *Let $\{(\mathbf{u}^{l_0}, \mathbf{x}^{l_0}, \mathbf{y}^{l_0})\}$ be a realization of $\mathcal{F}$ satisfying the conditions of*

$$A_*^{(3)}, \tag{63}$$

$$\bigcap_{\mathbf{z}\in F^{(3)}} A^{(3)}(\mathbf{z}). \tag{64}$$

Then for sufficiently large n,

$$\|\hat{\vartheta}^{(3)} - \frac{1}{L_0}\sum_{l_0} W_e^{\otimes n}(\cdot|\mathbf{x}^{l_0}, \mathbf{y}^{l_0})\| \leq 4\varepsilon + 5 \cdot 2^{-nc\delta^2}. \tag{65}$$

We now prove the above lemma. We have

$$\|\hat{\vartheta}^{(3)} - \frac{1}{L_0}\sum_{l_0} W_e^{\otimes n}(\cdot|\mathbf{x}^{l_0}, \mathbf{y}^{l_0})\|$$

$$\leq \|\hat{\vartheta}^{(3)} - \frac{1}{L_0}\sum_{l_0} W_e^{\otimes n}(\cdot|\mathbf{x}^{l_0}, \mathbf{y}^{l_0})1_{E^{(3)}(\mathbf{x}^{l_0},\mathbf{y}^{l_0})}1_{F^{(3)}}\| \tag{66}$$

$$+ \|\frac{1}{L_0}\sum_{l_0} W_e^{\otimes n}(\cdot|\mathbf{x}^{l_0}, \mathbf{y}^{l_0})1_{E^{(3)}(\mathbf{x}^{l_0},\mathbf{y}^{l_0})}(1 - 1_{F^{(3)}})\| \tag{67}$$

$$+ \|\frac{1}{L_0}\sum_{l_0} W_e^{\otimes n}(\cdot|\mathbf{x}^{l_0}, \mathbf{y}^{l_0})(1 - 1_{E^{(3)}(\mathbf{x}^{l_0},\mathbf{y}^{l_0})})\|. \tag{68}$$

Due to (64) we have (66) $\leq \varepsilon$.

Next we bound (67). Again using (64),

$$(67) \leq 1 - \frac{1}{L_0} \sum_{l_0} W_e^{\otimes n}(E^{(3)}(\mathbf{x}^{l_0}, \mathbf{y}^{l_0}) \cap F^{(3)} | \mathbf{x}^{l_0}, \mathbf{y}^{l_0})$$

$$\leq 1 - (1-\varepsilon)\hat{\vartheta}^{(3)}(F^{(3)}). \tag{69}$$

As in Case 1 and 2, $\hat{\vartheta}^{(3)}(F^{(3)})$ can be lower-bounded by $1 - 2 \cdot 2^{-nc\delta^2} - \varepsilon$, so

$$(67) \leq 1 - (1-\varepsilon)(1 - 2 \cdot 2^{-nc\delta^2} - \varepsilon) \leq 2(\varepsilon + 2^{-nc\delta^2}).$$

Finally, the third term (68) equals

$$\frac{1}{L_0} \sum_{l_0} W_e^{\otimes n}(E^{(3)}(\mathbf{x}^{l_0}, \mathbf{y}^{l_0})^c | \mathbf{x}^{l_0}, \mathbf{y}^{l_0})$$

$$= \frac{1}{L_0} \sum_{l_0 : \mathbf{x}^{l_0} \in T_{X|YU,\delta}(\mathbf{y}^{l_0}, \mathbf{u}^{l_0})} W_e^{\otimes n}(E^{(3)}(\mathbf{x}^{l_0}, \mathbf{y}^{l_0})^c | \mathbf{x}^{l_0}, \mathbf{y}^{l_0}) \tag{70}$$

$$+ \frac{1}{L_0} \sum_{l_0 : \mathbf{x}^{l_0} \notin T_{X|YU,\delta}(\mathbf{y}^{l_0}, \mathbf{u}^{l_0})} W_e^{\otimes n}(E^{(3)}(\mathbf{x}^{l_0}, \mathbf{y}^{l_0})^c | \mathbf{x}^{l_0}, \mathbf{y}^{l_0}). \tag{71}$$

and is lower-bounded by

$$(68) \leq 2^{-nc\delta^2} + (1-\varepsilon)(1 - 2 \cdot 2^{-nc\delta^2}) \leq \varepsilon + 3 \cdot 2^{-nc\delta^2}.$$

Combining the above bounds, we can conclude that

$$(66) + (67) + (68) \leq 4\varepsilon + 5 \cdot 2^{-nc\delta^2},$$

which completes the proof of Lemma 16.

4.4 Random Coding for the Non-wiretap MAC with Common Message

Assume we are given another family of random variables

$$\mathcal{F}' := \bigcup_{l_0 \in [L_0]} (U^{l'_0}, \mathcal{F}'_{l'_0})$$

with $\mathcal{F}'_{l'_0} = \{X^{l'_0 l'_1}, Y^{l'_0 l'_2} : l'_1, l'_2 \in [L'_1] \times [L'_2]\}$ for other positive integers L'_0, L'_1, L'_2 with blocklength n' which is independent of $\mathcal{F}$, but which has the same structure as $\mathcal{F}$ and whose distribution is defined according to the same p as $\mathcal{F}$. Define the rate set $\tilde{\mathcal{R}}(p)$ by the bounds

$$\begin{aligned} \tilde{R}_1 &\leq I(T \wedge X|YU), \\ \tilde{R}_2 &\leq I(T \wedge Y|XU), \\ \tilde{R}_1 + \tilde{R}_2 &\leq I(T \wedge XY|U), \\ \tilde{R}_0 + \tilde{R}_1 + \tilde{R}_2 &\leq I(T \wedge XY). \end{aligned}$$

Assume that for some $0 < \eta < I_* := \min\{I_\nu > 0 : \nu = 1, 2, 3, 4\}$ we have

$$\frac{n \log L_1 + n' \log L_1'}{n + n'} \leq [I(T \wedge X|YU) - \eta]_+,$$
$$\frac{n \log L_2 + n' \log L_2'}{n + n'} \leq [I(T \wedge Y|XU) - \eta]_+,$$
$$\frac{n \log(L_1 L_2) + n' \log(L_1' L_2')}{n + n'} \leq [I(T \wedge XY|U) - \eta]_+,$$
$$\frac{n \log(L_0 L_1 L_2) + n' \log(L_0' L_1' L_2')}{n + n'} \leq [I(T \wedge XY) - \eta]_+.$$

Define a new family of random vectors

$$\mathcal{F} \circ \mathcal{F}' := \{\tilde{U}^{l_0 l_0'}, \tilde{X}^{l_0 l_0' l_1 l_1'}, \tilde{Y}^{l_0 l_0' l_2 l_2'}\}$$

by concatenating the corresponding elements of $\mathcal{F}$ and $\mathcal{F}'$, so e.g. $\tilde{U}^{l_0 l_0'} = (U^{l_0}, U^{l_0'}) \in \mathcal{U}^{n+n'}$, $\tilde{X}^{l_0 l_0' l_1 l_1'} = (X^{l_0 l_1}, X^{l_0' l_1'}) \in \mathcal{X}^{n+n'}$.

Lemma 17. *For any $\delta, \eta > 0$ there are $\zeta_1, \zeta_2 = \zeta_1(\eta, \delta), \zeta_2(\eta, \delta) > 0$ such that the probability of the event A_{MAC} that the family*

$$\{\tilde{X}^{l_0 l_0' l_1 l_1'}, \tilde{Y}^{l_0 l_0' l_2 l_2'} : (l_0, l_0', l_1, l_1', l_2, l_2')\}$$

is the codeword set of a deterministic MAC code with average error at most $\exp(-(n + n')\zeta_1)$ is lower-bounded by $1 - \exp(-(n + n')\zeta_2)$. The same result is true if it is formulated only for $\mathcal{F}$ or $\mathcal{F}'$ without concatenation.

Proof. The difference to standard random coding proofs is that the random variables from $\mathcal{F}$ and $\mathcal{F}'$ are conditioned on typicality. Using the random sets

$$E^{l_0 l_0' l_1 l_1' l_2 l_2'} := \{\mathbf{t} \in \mathcal{T}^n : (\tilde{U}^{l_0 l_0'}, \tilde{X}^{l_0 l_0' l_1 l_1'}, \tilde{Y}^{l_0 l_0' l_2 l_2'}, \mathbf{t}) \in T^n_{UXYT,\delta}\},$$

we define the decoding sets $F^{l_0 l_0' l_1 l_1' l_2 l_2'}$ by deciding for $(l_0, l_0', l_1, l_1', l_2, l_2')$ if the output is contained in $E^{l_0 l_0' l_1 l_1' l_2 l_2'}$ and if at the same time it is not contained in any $E^{\tilde{l}_0 \tilde{l}_0' \tilde{l}_1 \tilde{l}_1' \tilde{l}_2 \tilde{l}_2'}$ for a different message tuple $(\tilde{l}_0, \tilde{l}_0', \tilde{l}_1, \tilde{l}_1', \tilde{l}_2, \tilde{l}_2')$. This decoder is known to be the right decoder in the case where the codewords have the standard i.i.d. structure, i.e. for a family of random variables

$$\{\hat{U}^{l_0 l_0'}, \hat{X}^{l_0 l_0' l_1 l_1'}, Y^{l_0 l_0' l_2 l_2'}\}$$

where $\hat{U}^{l_0 l_0'} \sim P_U^{\otimes (n+n')}$ and where conditional on $\hat{U}^{l_0 l_0'}$, the $\hat{X}^{l_0 l_0' l_1 l_1'}$ and $\hat{Y}^{l_0 l_0' l_2 l_2'}$ are independent with $\hat{X}^{l_0 l_0' l_1 l_1'} \sim P_{X|U}^{\otimes (n+n')}$ and $\hat{Y}^{l_0 l_0' l_2 l_2'} \sim P_{Y|U}^{\otimes (n+n')}$. It is easily seen that

$$\begin{aligned} &\mathbb{E}[W^{\otimes n}((F^{l_0 l_0' l_1 l_1' l_2 l_2'})^c | \tilde{X}^{l_0 l_0' l_1 l_1'}, \tilde{Y}^{l_0 l_0' l_2 l_2'})] \\ &\leq (1 - 2^{-nc\delta^2})^3 (1 - 2^{-n' c\delta^2})^3 \mathbb{E}[W^{\otimes n}((F^{l_0 l_0' l_1 l_1' l_2 l_2'})^c | \hat{X}^{l_0 l_0' l_1 l_1'}, \hat{Y}^{l_0 l_0' l_2 l_2'})]. \end{aligned}$$

Then the standard random coding proof technique yields the result. The specialization for the case that only $\mathcal{F}$ or $\mathcal{F}'$ is treated is obvious. □

4.5 Coding

In this subsection we show the achievability of the rate sets $\mathcal{R}^{(\nu)}(p)$ for $\nu = 0, 1, 2, 3$ and appropriate p. For the cases where we showed that $\mathcal{R}^{(\nu)}(p)$ can be written as the union over certain α of rate sets $\mathcal{R}_\alpha^{(\nu)}(p)$, we show the achievability of the latter for every α.

Throughout this section fix a common randomness bound $H_C \geq 0$. Let $\delta > 0$ which will be specified later and n a blocklength which will have to be large enough. Every p considered in this section has the form $p = P_U \otimes (P_{X|U} \otimes P_{Y|U}) \otimes W$. Without loss of generality we may assume that $I(Z \wedge XY) < I(T \wedge XY)$, in particular, $I(T \wedge XY) > 0$. Letting

$$K_0, K_1, K_2, L_0, L_1, L_2, n, \qquad K_0', K_1', K_2', L_0', L_1', L_2', n' \tag{72}$$

be arbitrary positive integers, we define two independent families $\mathcal{G}, \mathcal{G}'$ of random vectors. $\mathcal{G}$ has the same form as $\mathcal{F}$ with the parameters L_0, L_1, L_2 replaced by K_0L_0, K_1L_1, K_2L_2. $\mathcal{G}'$ is defined analogously with the parameters on the left-hand side of (72) replaced by those on its right-hand side. Every choice of (k_0, k_1, k_2) induces a subfamily $\mathcal{F}$ of $\mathcal{G}$ which has the same parameters as the $\mathcal{F}$ treated above, every subfamily of $\mathcal{G}'$ corresponding to any (k_0', k_1', k_2') induces an $\mathcal{F}'$ with parameters L_0', L_1', L_2', n'. Further recall the notation $\mathcal{G} \circ \mathcal{G}'$ as the family of concatenated words from $\mathcal{G}$ and $\mathcal{G}'$.

Case 0 and 1: Let $p \in \Pi_0$ or $p \in \Pi_{H_C}^{(1)}$. Note that $\alpha_0^{(1)} \leq \alpha_1^{(1)}$ if and only if the vector $(J_0^{(\alpha)}, J_1^{(\alpha)}, J_2^{(\alpha)})$ whose components are given by

$$\begin{aligned}
J_0^{(\alpha)} &= I(Z \wedge U), \\
J_1^{(\alpha)} &= \alpha I(Z \wedge X|YU) + (1-\alpha) I(Z \wedge X|U), \\
J_2^{(\alpha)} &= \alpha I(Z \wedge Y|U) + (1-\alpha) I(Z \wedge Y|XU)
\end{aligned}$$

is contained in $\tilde{\mathcal{R}}(p)$. We first consider Case 1. Let a rate vector (R_0, R_1, R_2) with positive components be given such that $(\tilde{R}_0, \tilde{R}_1, \tilde{R}_2) := (R_0, R_1, R_2) + (J_0^{(\alpha)}, J_1^{(\alpha)}, J_2^{(\alpha)}) \in \tilde{\mathcal{R}}(p)$, which means that $(R_0, R_1, R_2) \in \mathcal{R}_\alpha(p)$. We now define a wiretap code whose rates approximate (R_0, R_1, R_2). If $\alpha = 0$, we only need $\mathcal{G}'$, if $\alpha = 1$, we only need $\mathcal{G}$. Otherwise we do time-sharing in the following way: choose for a small $0 < \gamma < \min\{\alpha, 1-\alpha\}$ blocklengths n and n' with $n/(n+n') \in (\alpha - \gamma, \alpha + \gamma)$. For some $0 < 2\eta < \min\{R_0, R_1, R_2\}$ and every $\nu = 0, 1, 2$ let

$$\tilde{R}_\nu - \eta \leq \frac{\log(K_\nu L_\nu) + \log(K_\nu' L_\nu')}{n + n'} \leq \tilde{R}_\nu - \frac{\eta}{2}$$

(and this modifies accordingly for $\alpha \in \{0, 1\}$). By Lemma 17 we know that with probability exponentially close to 1, the random variables $\tilde{X}_{k_0 k_0' k_1 k_1'}^{l_0 l_0' l_1 l_1'}$ and $\tilde{Y}_{k_0 k_0' k_2 k_2'}^{l_0 l_0' l_2 l_2'}$ form the codewords of a code for the non-wiretap MAC given by W_b

with an average error at most $\exp(-(n+n')\zeta_1)$ for some $\zeta_1 > 0$. We denote Bob's corresponding random decoder by Φ. Now let

$$\frac{\log L_1 + \log L_1'}{n+n'} \in J_1^{(\alpha)} + \big(f_1(\delta) + (\alpha f_2(\delta) + (1-\alpha) f_4(\delta))\big) \cdot [2,3],$$
$$\frac{\log L_2 + \log L_2'}{n+n'} \in J_2^{(\alpha)} + \big(f_1(\delta) + (\alpha f_4(\delta) + (1-\alpha) f_2(\delta))\big) \cdot [2,3],$$
$$\frac{\log L_0 + \log L_0'}{n+n'} \in J_0^{(\alpha)} + \big(f_4(\delta) + f_6(\delta)\big) \cdot [2,3].$$

This is possible if $4(f_1(\delta) + f_2(\delta) + f_4(\delta)) \le \min\{\eta, H_C - J_0^{(\alpha)}\}$. If additionally ε is chosen according to

$$-\frac{1}{n}\log\varepsilon = \frac{1}{4}\min\{4\zeta_1, f_1(\delta) + f_2(\delta) + f_4(\delta) + f_6(\delta)\},$$

then for every $(k_0, k_1, k_2) \in [K_0] \times [K_1] \times [K_2]$, the corresponding subfamily $\mathcal{F}$ of $\mathcal{G}$ satisfies (34)-(37) with probability exponentially close to 1, and for every $(k_0', k_1', k_2') \in [K_0'] \times [K_1'] \times [K_2']$, the corresponding subfamily $\mathcal{F}'$ of $\mathcal{G}'$ satisfies (34′)-(37′) with probability exponentially close to 1. Thus we can choose a realization of $\mathcal{G} \circ \mathcal{G}'$ which has all these properties and use it to define a stochastic wiretap code. We define independent encoders G and G' by setting

$$G_0(l_0|k_0) = \frac{1}{L_0}, \qquad (k_0 \in [K_0], l_0 \in [L_0]),$$
$$G_1(\mathbf{x}|k_0, k_1, l_0) = \frac{1}{L_1}\sum_{l_1} \delta_{\mathbf{x}_{k_0 k_1}^{l_0 l_1}}(\mathbf{x}), \quad (\mathbf{x} \in \mathcal{X}^n, k_1 \in [K_1], k_0 \in [K_0], l_0 \in [L_0]),$$
$$G_2(\mathbf{y}|k_0, k_2, l_0) = \frac{1}{L_2}\sum_{l_2} \delta_{\mathbf{y}_{k_0 k_2}^{l_0 l_2}}(\mathbf{y}), \quad (\mathbf{y} \in \mathcal{Y}^n, k_2 \in [K_2], k_0 \in [K_0], l_0 \in [L_0]),$$

and defining G' analogously. G_0 and G_0' satisfy the common randomness constraint. We choose the decoder ϕ to be the realization of Φ corresponding to the chosen realization of $\mathcal{G} \circ \mathcal{G}'$. The average error of the stochastic encoding code equals the average error of the deterministic MAC code for W_b determined by the realization of $\mathcal{G} \circ \mathcal{G}'$, in particular it is bounded by ε. Due to the choice of δ the rates of this code satisfy

$$\frac{\log K_\nu + \log K_\nu'}{n+n'} \ge R_\nu - 2\eta \qquad (\nu = 0, 1, 2,).$$

Finally if we let M_ν be uniformly distributed on $[K_\nu]$ and M_ν' on $[K_\nu']$, then it follows from Lemma 10 and (22) together with the fact that ε is exponentially small that the strong secrecy criterion is satisfied. Thus the rate triple (R_0, R_1, R_2) is achievable. So far, this excludes (R_0, R_1, R_2) where one component equals zero, but as δ and η may be arbitrarily close to 0 and the achievable region of W is closed by definition, we can conclude that the whole region $\mathcal{R}_\alpha(p)$ is achievable.

For Case 0, everything goes through if one sets $K_0 = K_0' = L_0 = L_0' = 1$ and $R_0 = 0$. The difference to Case 1 is that even if $J_0^{(\alpha)} = 0$, one needs a little bit more common randomness than that in order to protect a common message, as can be seen in the choice of L_0 and L_0' above. Thus the transmission of a common message is impossible if common randomness is not available.

Case 2: Let $p \in \Pi_{HC}^{(2)}$. In this case we generally need both a $\mathcal{G}$ and a $\mathcal{G}'$, where $\mathcal{G}$ has $L_2 = 1$ and $\mathcal{G}'$ has $L_1 = 1$. We define the vector $(J_0^{(\alpha)}, J_1^{(\alpha)}, J_2^{(\alpha)})$ by

$$J_0^{(\alpha)} = \alpha I(Z \wedge YU) + (1-\alpha) I(Z \wedge XU),$$
$$J_1^{(\alpha)} = \alpha I(Z \wedge X|YU),$$
$$J_2^{(\alpha)} = (1-\alpha) I(Z \wedge Y|XU)$$

As it should always be clear which case we are treating, this should not lead to confusion with case 1. Note that $\alpha_0^{(2)} \leq \alpha \leq \alpha_1^{(2)}$ if and only if $(J_0^{(\alpha)}, J_1^{(\alpha)}, J_2^{(\alpha)})$ is contained in $\tilde{\mathcal{R}}(p)$ and satisfies $J_0^{(\alpha)} < H_C$. Let a rate vector (R_0, R_1, R_2) be given whose ν-th component may only vanish if $L_\nu = L_\nu' = 1$. Further we require that $(\tilde{R}_0, \tilde{R}_1, \tilde{R}_2) = (R_0, R_1, R_2) + (J_0^{(\alpha)}, J_1^{(\alpha)}, J_2^{(\alpha)})$ is contained in $\tilde{\mathcal{R}}(p)$. If $\alpha = 0$, we only need $\mathcal{G}'$, if $\alpha = 1$, we only need $\mathcal{G}$. Otherwise, let $0 < \gamma < \min\{\alpha, 1-\alpha\}$ be small and let n and n' be large enough such that $n/(n+n') \in (\alpha - \gamma, \alpha + \gamma)$. Further for some $0 < 2\eta < \min\{R_\nu : \nu = 0, 1, 2, R_\nu > 0\}$ let

$$[\tilde{R}_\nu - \eta]_+ \leq \frac{\log(K_\nu L_\nu) + \log(K_\nu' L_\nu')}{n+n'} \leq [\tilde{R}_\nu - \frac{\eta}{2}]_+,$$

and modify this accordingly for $\alpha \in \{0, 1\}$. By Lemma 17 we know that with probability exponentially close to 1, the random variables $\tilde{X}_{k_0 k_0' k_1 k_1'}^{l_0 l_0' l_1 l_1'}$ and $\tilde{Y}_{k_0 k_0' k_2 k_2'}^{l_0 l_0' l_2 l_2'}$ form the codewords of a code for the non-wiretap MAC given by W_b with an average error at most $\exp(-(n+n')\zeta_1)$ for some $\zeta_1 > 0$. We denote the corresponding random decoder by Φ. We define $(j_1^1, j_1^2) = (j_2^1, j_2^2) = (1, 2)$ and $(j_0^1, j_0^2) = (1, 6)$. Then let for $\nu = 0, 1, 2$

$$J_\nu^{(\alpha)} + 2(f_{j_\nu^1}(\delta) + f_{j_\nu^2}(\delta)) \leq \frac{\log L_\nu + \log L_\nu'}{n+n'} \leq J_\nu^{(\alpha)} + 3(f_{j_\nu^1}(\delta) + f_{j_\nu^2}(\delta)),$$

which is possible if $4(f_{j_\nu^1}(\delta) + f_{j_\nu^2}(\delta)) \leq \min\{\eta, H_C - J_0^{(\alpha)}\}$ for all ν. If additionally ε is chosen according to

$$-\frac{1}{n} \log \varepsilon = \frac{1}{4} \min\{4\zeta_1, f_1(\delta) + f_2(\delta), f_1(\delta) + f_6(\delta)\},$$

then for every $(k_0, k_1, k_2) \in [K_0] \times [K_1] \times [K_2]$, the corresponding subfamily $\mathcal{F}$ of $\mathcal{G}$ satisfies (54)-(56) with probability exponentially close to 1, and for every $(k_0', k_1', k_2') \in [K_0'] \times [K_1'] \times [K_2']$, the corresponding subfamily $\mathcal{F}'$ of $\mathcal{G}'$ satisfies (54′)-(56′) with probability exponentially close to 1. Thus we can choose a realization of $\mathcal{G} \circ \mathcal{G}'$ which has all these properties plus those defining A_{MAC} and use it to define a stochastic wiretap code. We define independent encoders G and G' by setting

$$G_0(l_0|k_0) = \frac{1}{L_0}, \qquad (l_0 \in [L_0], k_0 \in [K_0]),$$

$$G_1(\mathbf{x}|k_0, k_1, l_0) = \frac{1}{L_1} \sum_{l_1} \delta_{\mathbf{x}_{k_0 k_1}^{l_0 l_1}}(\mathbf{x}), \quad (\mathbf{x} \in \mathcal{X}^n, k_1 \in [K_1], k_0 \in [K_0], l_0 \in [L_0]),$$

$$G_2(\mathbf{y}|k_0, k_2, l_0) = \delta_{\mathbf{y}_{k_0 k_2}^{l_0}}(\mathbf{y}), \qquad (\mathbf{y} \in \mathcal{Y}^n, k_2 \in [K_2], k_0 \in [K_0], l_0 \in [L_0]),$$

and defining G' analogously. The decoder ϕ is the realization of Φ corresponding to the chosen realization of $\mathcal{G} \circ \mathcal{G}'$. G_0 and G_0' satisfy the common randomness constraint. Due to the simple form of G and G', the average error of the stochastic encoding code equals the average error of the deterministic MAC code for W_b determined by the realization of $\mathcal{G} \circ \mathcal{G}'$, in particular it is bounded by ε. Due to the choice of δ, the rates of this code satisfy

$$\frac{\log K_\nu + \log K_\nu'}{n + n'} \geq R_\nu - 2\eta \qquad (\nu = 0, 1, 2,).$$

Finally if we let M_ν be uniformly distributed on $[K_\nu]$ and M_ν' on $[K_\nu']$, then it follows from Lemma 10 and (22) together with the fact that ε is exponentially small that the strong secrecy criterion is satisfied. Thus the rate triple (R_0, R_1, R_2) is achievable. So far, this may exclude rate triples (R_0, R_1, R_2) where one component equals zero, but as δ and η may be arbitrarily close to 0 and the achievable region of W is closed by definition, we can conclude that the whole region $\mathcal{R}_\alpha(p)$ is achievable.

Case 3: In this case we only need $\mathcal{G}$ with $L_1 = L_2 = 1$. Let $R_0 > 0$ and assume that the rate vector $(\tilde{R}_0, \tilde{R}_1, \tilde{R}_2) := (R_0 + I(Z \wedge XY), R_1, R_2)$ is contained in $\tilde{\mathcal{R}}(p)$. Further for some $0 < 2\eta < \min\{R_\nu : \nu = 0, 1, 2, R_\nu > 0\}$ let

$$[\tilde{R}_\nu - \eta]_+ \leq \frac{1}{n} \log(K_\nu L_\nu) \leq [\tilde{R}_\nu - \frac{\eta}{2}]_+.$$

$\mathcal{G}$ satisfies A_{MAC} with probability exponentially close to 1, so the $X_{k_0 k_1}^{l_0 l_1}$ and $Y_{k_0 k_2}^{l_0 l_2}$ form the codewords of a deterministic non-wiretap MAC code whose average error for transmission over W_b is bounded by $\exp(-n\zeta_1)$ for some $\zeta_1 > 0$. We denote the corresponding random decoder by Φ. Now let

$$I(Z \wedge XY) + 2(f_1(\delta) + f_2(\delta)) \leq \frac{1}{n} \log L_0 \leq I(Z \wedge XY) + 3(f_1(\delta) + f_2(\delta))$$

for δ so small that $4(f_1(\delta) + f_2(\delta)) \leq \min(\eta, H_C - I(Z \wedge XY))$ and choose ε such that

$$-\frac{1}{n} \log \varepsilon = \frac{1}{4} \min\{4\zeta_1, f_1(\delta) + f_2(\delta)\}.$$

Then for every (k_0, k_1, k_2) the corresponding family $\mathcal{F}$ satisfies the conditions (63) and (64) with probability exponentially close to 1. We choose a realization $\{(\mathbf{u}_{k_0}^{l_0}, \mathbf{x}_{k_0 k_1}^{l_0}, \mathbf{y}_{k_0 k_2}^{l_0})\}$ which satisfies the conditions of (63) and (64) and which determines a deterministic non-wiretap code for W_b with decoder ϕ. Now we can define a wiretap code whose decoder is ϕ and whose stochastic encoder G is given by

$$G_0(l_0|k_0) = \frac{1}{L_0}, \qquad (k_0 \in [K_0], l_0 \in [L_0]),$$

$$G_1(\mathbf{x}|k_0, k_1, l_0) = \delta_{\mathbf{x}_{k_0 k_1}^{l_0}}(\mathbf{x}), \qquad (\mathbf{x} \in \mathcal{X}^n, k_1 \in [K_1], k_0 \in [K_0], l_0 \in [L_0]),$$

$$G_2(\mathbf{y}|k_0, k_2, l_0) = \delta_{\mathbf{y}_{k_0 k_2}^{l_0}}(\mathbf{y}), \qquad (\mathbf{y} \in \mathcal{Y}^n, k_2 \in [K_2], k_0 \in [K_0], l_0 \in [L_0]).$$

Note that G_0 satisfies the common randomness constraint. Due to the uniform distribution of G_0, its average error is identical to that of the deterministic MAC code determined by the $\mathbf{x}_{k_0 k_1}^{l_0}$ and the $\mathbf{y}_{k_0 k_2}^{l_0}$, in particular, it is exponentially small with rate at most ε. We have for $\nu = 0, 1, 2$

$$\frac{1}{n} \log K_\nu \geq R_\nu - 2\eta.$$

due to the choice of δ. Finally if we let M_ν be uniformly distributed on $[K_\nu]$, then it follows from Lemma 16 and (22) together with the fact that ε is exponentially small that the strong secrecy criterion is satisfied. Thus the rate triple (R_0, R_1, R_2), and hence $\mathcal{R}(p)$, is achievable.

4.6 Concluding Steps

We can reduce coding for a general p which is the distribution of a random vector $(U, V_1, V_2, X, Y, T, Z)$ to the case treated above by constructing a new wiretap MAC as follows: its input alphabets are $\mathcal{V}_1$ and $\mathcal{V}_2$, its output alphabets still are $\mathcal{T}$ and $\mathcal{Z}$. The transition probability for inputs (v_1, v_2) and outputs (t, z) is given by

$$\tilde{W}(t, z|v_1, v_2) := \sum_{(x,y) \in \mathcal{X} \times \mathbf{y}} W(t, z|x, y) P_{X|V_1}(x|v_1) P_{Y|V_2}(y|v_2).$$

For this channel we do the same construction as above considering the joint distribution of random variables (U, V_1, V_2, T, Z) which we denote by $\tilde{p}$. In this way we also construct a wiretap code for the original channel W because the additional randomness $P_{V_1 V_2|U}$ can be integrated into the stochastic encoders G_1 and G_2. G_0 remains unchanged, so the additional randomness in the encoders does not increase the common randomness needed to do the encoding.

On the other hand, we need to show that the rate regions thus obtained are those appearing in the statement of Theorem 1. As the sets $\Pi_0, \Pi_{HC}^{(1)}, \ldots, \Pi_{HC}^{(3)}$ depend on the channel, we write $\Pi_0(W)$, $\Pi_0(\tilde{W})$, $\Pi_{HC}^{(1)}(W), \ldots, \Pi_{HC}^{(3)}(\tilde{W})$. Note that $\tilde{p}$ is contained in $\Pi_0(\tilde{W})$ or $\Pi_{HC}^{(\nu)}(\tilde{W})$ for some $\nu = 1, 2, 3$ if and only if p is contained in the corresponding $\Pi_0(W)$ or $\Pi_{HC}^{(\nu)}(W)$. This immediately implies that the rate regions also coincide.

5 Proof of Theorem 2

5.1 Elementary Rate Regions

As for the wiretap MAC with common message we show that we can write the claimed achievable regions as unions of simpler sets whose achievability will be show in the next step.

For Case 1: Define

$$\beta_0^{(1)} := [1 - \frac{C_2}{I(Z \wedge U)}]_+, \qquad \beta_1^{(1)} := \min\{\frac{C_1}{I(Z \wedge U)}, 1\}.$$

We have $\beta_0^{(1)} \leq \beta_1^{(1)}$ because $I(Z \wedge U) < C_1 + C_2$.

Lemma 18. *For $\beta_0^{(1)} \leq \beta \leq \beta_1^{(1)}$, let $\mathcal{R}_\beta^{(1)}(p, C_1, C_2)$ be the set of those real pairs (R_1, R_2) satisfying*

$$\begin{aligned}
R_1 &\leq I(T \wedge V_1|V_2U) - I(Z \wedge V_1|U) \\
&\qquad - [I(Z \wedge V_2|V_1U) - I(T \wedge V_2|V_1U)]_+ - \beta I(Z \wedge U) + C_1, \\
R_2 &\leq I(T \wedge V_2|V_1U) - I(Z \wedge V_2|U) \\
&\qquad - [I(Z \wedge V_1|V_2U) - I(T \wedge V_1|V_2U)]_+ - (1-\beta) I(Z \wedge U) + C_2, \\
R_1 + R_2 &\leq \min\{I(T \wedge V_1V_2|U) - I(Z \wedge V_1V_2|U) - I(Z \wedge U) + C_1 + C_2, \\
&\qquad I(T \wedge V_1V_2) - I(Z \wedge V_1V_2)\}.
\end{aligned}$$

Then

$$\mathcal{R}^{(1)}(p, C_1, C_2) = \bigcup_{\beta_0^{(1)} \leq \beta \leq \beta_1^{(1)}} \mathcal{R}_\beta^{(1)}(p, C_1, C_2).$$

Thus it is sufficient to show the achievability of $\mathcal{R}_\beta^{(1)}(p, C_1, C_2)$ for every β. For the proof one uses Lemma 2.

For Case 2: Recall the vector $(J_0^{(\alpha)}, J_1^{(\alpha)}, J_2^{(\alpha)})$ defined as in Case 2 from the common message part. Define

$$\beta_0^{(2,\alpha)} := [1 - \frac{C_2}{J_0^{(\alpha)}}]_+, \qquad \beta_1^{(2,\alpha)} := \min\{\frac{C_1}{J_0^{(\alpha)}}, 1\}.$$

We show that every $\mathcal{R}_\alpha^{(2)}(p, C_1, C_2)$ can be represented as the union of sets $\mathcal{R}_{\alpha,\beta}^{(2)}(p, C_1, C_2)$ for $\beta_0^{(2,\alpha)} \leq \beta \leq \beta_1^{(2,\alpha)}$. Define $\mathcal{R}_{\alpha,\beta}^{(2)}(p, C_1, C_2)$ by

$$\begin{aligned}
R_1 &\leq I(T \wedge V_1|V_2U) - \alpha I(Z \wedge V_1|V_2U) + C_1 - \beta J_0^{(\alpha)}, \\
R_2 &\leq I(T \wedge V_2|V_1U) - (1-\alpha) I(Z \wedge V_2|V_1U) + C_2 - (1-\beta) J_0^{(\alpha)}, \\
R_1 + R_2 &\leq I(T \wedge V_1V_2|U) - \alpha I(Z \wedge V_1|V_2U) - (1-\alpha) I(Z \wedge V_2|V_1U) \\
&\qquad + C_1 + C_2 - J_0^{(\alpha)}, \\
R_1 + R_2 &\leq I(T \wedge V_1V_2) - I(Z \wedge V_1V_2).
\end{aligned}$$

Lemma 19. *We have for every* $\alpha \in [\alpha_0^{(2)}, \alpha_1^{(2)}]$

$$\mathcal{R}_\alpha^{(2)}(p, C_1, C_2) = \bigcup_{\beta_0^{(2,\alpha)} \leq \beta \leq \beta_1^{(2,\alpha)}} \mathcal{R}_{\alpha,\beta}^{(2)}(p, C_1, C_2).$$

This is seen immediately using Lemma 2.

For Case 3: Define

$$\beta_0^{(1)} := [1 - \frac{C_2}{I(Z \wedge V_1 V_2)}]_+, \qquad \beta_1^{(1)} := \min\{\frac{C_1}{I(Z \wedge V_1 V_2)}, 1\}.$$

We have $\beta_0^{(1)} \leq \beta_1^{(1)}$ because $I(Z \wedge V_1 V_2) < C_1 + C_2$.

Lemma 20. *For* $\beta_0^{(3)} \leq \beta \leq \beta_1^{(3)}$, *let* $\mathcal{R}_\beta^{(3)}(p, C_1, C_2)$ *be the set of those real pairs* (R_1, R_2) *satisfying*

$$\begin{aligned} R_1 &\leq I(T \wedge V_1 | V_2 U_0) + C_1 - \beta I(Z \wedge V_1 V_2), \\ R_2 &\leq I(T \wedge V_2 | V_1 U_0) + C_2 - (1 - \beta) I(Z \wedge V_1 V_2), \\ R_1 + R_2 &\leq \min\{I(T \wedge V_1 V_2 | U) + C_1 + C_2 - I(Z \wedge V_1 V_2), \\ &\qquad I(T \wedge V_1 V_2) - I(Z \wedge V_1 V_2)\}. \end{aligned}$$

Then

$$\mathcal{R}^{(1)}(p, C_1, C_2) = \bigcup_{\beta_0^{(1)} \leq \beta \leq \beta_1^{(1)}} \mathcal{R}_\beta^{(1)}(p, C_1, C_2).$$

Thus it is sufficient to show the achievability of $\mathcal{R}_\beta^{(3)}(p, C_1, C_2)$ for every β. For the proof one uses Lemma 2.

5.2 Coding

Let $C_1, C_2 > 0$ and let $p \in \Pi_{C_1+C_2}$. Further let $(R_1, R_2) \in \mathcal{R}(p, C_1, C_2)$. In Case 1 we then know that there is a $\beta \in [\beta_0^{(1)}, \beta_1^{(1)}]$ such that $(R_1, R_2) \in \mathcal{R}_\beta^{(1)}(p, C_1, C_2)$, in Case 2 we have an $\alpha \in [\alpha_0^{(2)}, \alpha_1^{(2)}]$ and a $\beta \in [\beta_0^{(2,\alpha)}, \beta_1^{(2,\alpha)}]$ with $(R_1, R_2) \in \mathcal{R}_{\alpha,\beta}^{(2)}(p, C_1, C_2)$. For Case 3, there is a $\beta \in [\beta_0^{(3)}, \beta_1^{(3)}]$ with $(R_1, R_2) \in \mathcal{R}_\beta^{(3)}(p, C_1, C_2)$. Recall the notation

$$J_0^{(\alpha)} = \begin{cases} I(Z \wedge U) & \text{in Case 1,} \\ \alpha I(Z \wedge V_2 U) + (1 - \alpha) I(Z \wedge V_1 U) & \text{in Case 2,} \\ I(Z \wedge V_1 V_2) & \text{in Case 3.} \end{cases}$$

We set

$$\tilde{R}_0^{(1)} := R_1 \wedge (C_1 - \beta J_0^{(\alpha)}), \quad \tilde{R}_0^{(2)} := R_2 \wedge (C_2 - (1 - \beta) J_0^{(\alpha)})$$

and

$$\tilde{R}_\nu := R_\nu - \tilde{R}_0^{(\nu)} \qquad (\nu = 1, 2).$$

Then setting

$$\tilde{R}_0 := \tilde{R}_0^{(1)} + \tilde{R}_0^{(2)},$$

we conclude that

$$(\tilde{R}_0, \tilde{R}_1, \tilde{R}_2) \in \begin{cases} \mathcal{R}_\beta^{(1)}(p) & \text{in Case 1,} \\ \mathcal{R}_{\alpha,\beta}^{(2)}(p) & \text{in Case 2,} \\ \mathcal{R}_\beta^{(3)}(p) & \text{in Case 3.} \end{cases}$$

In particular, $(\tilde{R}_0, \tilde{R}_1, \tilde{R}_2)$ is achievable by the wiretap MAC W with common message under the common randomness bound C_1+C_2. That means that for any $\eta, \varepsilon > 0$ and for sufficiently large n, there is a common-message blocklength-n code which has the form

$$\tilde{G} : [\tilde{K}_0] \times [\tilde{K}_1] \times [\tilde{K}_2] \to \mathcal{P}(\mathcal{X}^n \times \mathbf{y}^n),$$
$$\phi : \mathcal{T}^n \to [\tilde{K}_0] \times [\tilde{K}_1] \times [\tilde{K}_2],$$

and the proof of Theorem 1 shows that we may assume that $\tilde{G}$ is given by

$$\tilde{G}(\mathbf{x}, \mathbf{y} | \tilde{k}_0, \tilde{k}_1, \tilde{k}_2) = \frac{1}{\tilde{L}_0} \sum_{l_0=1}^{\tilde{L}_0} \tilde{G}_1(\mathbf{x} | \tilde{k}_0, \tilde{k}_1, l_0) \tilde{G}_2(\mathbf{y} | \tilde{k}_0, \tilde{k}_2, l_0)$$

for two stochastic matrices $\tilde{G}_1, \tilde{G}_2$. For $\tilde{L}_0$ we have the bounds

$$J_0^{(\alpha)} + \frac{\eta}{4} \leq \frac{1}{n} \log \tilde{L}_0 \leq J_0^{(\alpha)} + \frac{\eta}{2}.$$

Without loss of generality we may additionally assume that $\tilde{L}_0^{(1)} := \tilde{L}_0^\beta$ and $\tilde{L}_0^{(2)} := \tilde{L}_0^{(1-\beta)}$ are integers. If $0 < 2\eta < \min\{\tilde{R}_\nu : \nu = 0, 1, 2, \tilde{R}_\nu > 0\}$, the codelength triple $(\tilde{K}_0, \tilde{K}_1, \tilde{K}_2)$ may be assumed to satisfy

$$[\tilde{R}_\nu - 2\eta]_+ \leq \frac{1}{n} \log \tilde{K}_\nu \leq [\tilde{R}_\nu - \eta]_+, \qquad (\nu = 0, 1, 2), \tag{73}$$

and both the average error as well as $I(\tilde{M}_0 \tilde{M}_1 \tilde{M}_2 \wedge Z^n)$ are upper-bounded by ε, where $(\tilde{M}_0, \tilde{M}_1, \tilde{M}_2)$ is distributed uniformly on $[\tilde{K}_0] \times [\tilde{K}_1] \times [\tilde{K}_2]$ and Z^n is Eve's corresponding output random variable. The definitions imply that

$$\frac{1}{n} \log \tilde{K}_0 \tilde{L}_0 \leq C_1 + C_2.$$

We can find $\tilde{K}_0', \tilde{K}_0^{(1)}, \tilde{K}_0^{(2)}$ such that $\tilde{K}_0' = \tilde{K}_0^{(1)} \tilde{K}_0^{(2)}$ and $\tilde{K}_0' \leq \tilde{K}_0$ and satisfying

$$[\tilde{R}_0^{(\nu)} - 2\eta]_+ \leq \frac{1}{n} \log \tilde{K}_0^{(\nu)} \leq [\tilde{R}_0^{(\nu)} - \frac{\eta}{2}]_+, \tag{74}$$

$$[\tilde{R}_0 - 2\eta]_+ \leq \frac{1}{n} \log \tilde{K}_0'. \tag{75}$$

Thus one obtains a natural embedding

$$[\tilde{K}_0^{(\nu)}] \times [L_0^{(\nu)}] \subset [\lfloor 2^{nC_\nu} \rfloor] \qquad (\nu = 1, 2). \tag{76}$$

We now construct a wiretap code with conferencing encoders. Let

$$K_\nu := \tilde{K}_0^{(\nu)} \tilde{K}_\nu \qquad (\nu = 1, 2).$$

Thus every $k_\nu \in [K_\nu]$ has the form $(a_\nu(k_\nu), b_\nu(k_\nu))$ with $a_\nu(k_\nu) \in [\tilde{K}_0^{(\nu)}]$ and $b_\nu(k_\nu) \in [\tilde{K}_\nu]$. We then define a stochastic one-shot Willems conferencing protocol

$$c_1 : [K_1] \to \mathcal{P}([\lfloor 2^{nC_1} \rfloor]), \qquad c_2 : [K_2] \to \mathcal{P}([\lfloor 2^{nC_2} \rfloor])$$

which is used to generate both a common message as well as common randomness. Given a message $k_\nu \in [K_\nu]$, Alice$_\nu$ chooses an l_ν uniformly at random from the set $[L_0^{(\nu)}]$ and then maps the pair (k_ν, l_ν) to $(a_\nu(k_\nu), l_\nu)$, so $c_\nu(k_\nu, l_\nu) = (a_\nu(k_\nu), l_\nu)$.

Next we define stochastic encoders G_1, G_2 as in the definition of a code with conferencing encoders by setting

$$\mathcal{J} := [\lfloor 2^{nC_1} \rfloor] \times [\lfloor 2^{nC_2} \rfloor]$$

and, using the embedding (76),

$$G_1(\mathbf{x}|k_1, j) = \tilde{G}_1(\mathbf{x}|(a_1(k_1), k_0^{(2)}), b_1(k_1), (l_1, l_2)) .$$

if $j = ((a_1(k_1), l_1), (k_0^{(2)}, l_2))$ and letting $G_1(\mathbf{x}|k_1, j)$ be arbitrary else; G_2 is defined analogously. For decoding, one takes the decoder from the common message code and lets it combine the messages it receives into elements of $[K_1]$ and $[K_2]$. By (73), the numbers K_1 and K_2 satisfy

$$\frac{1}{n} \log K_1 \geq R_1 - 3\eta,$$
$$\frac{1}{n} \log K_2 \geq R_2 - 3\eta.$$

Thus depending on the case we are in, every rate pair (R_1, R_2) contained in $\mathcal{R}_\beta^{(1)}(p, C_1, C_2)$ or $\mathcal{R}_{\alpha,\beta}^{(2)}(p, C_1, C_2)$ or $\mathcal{R}_\beta^{(3)}(p, C_1, C_2)$ is achievable.

6 Discussion

6.1 Conferencing and Secret Transmission

This subsection is devoted to the comparison of the wiretap MAC without conferencing nor common randomness and the wiretap MAC if conferencing is allowed. As our focus is on conferencing, we assume that common randomness can only be established by conferencing. We show that there exists a wiretap MAC where the only rate pair contained in the region (10) achievable without conferencing is $(0,0)$, whereas if conferencing is enabled with arbitrarily small $C_1, C_2 > 0$, then the corresponding achievable region contains positive rates. Note that this does not mean that there are cases where conferencing is necessary to establish secret transmission as we do not have a converse. This restriction limits the use of this discussion and should be kept in mind.

Our goal is to find multiple access channels W_b and W_e such that for every Markov chain $((V_1,V_2),(X,Y),(T,Z))$ where $P_{T|XY} = W_b$ and $P_{Z|XY} = W_e$ and where V_1 and V_2 are independent one has

$$I(T \wedge V_1V_2) \leq I(Z \wedge V_1V_2). \tag{77}$$

We noted in Remark 8 that (10) is the achievable region without conferencing and it is easy to see that condition (77) is an equivalent condition for this region to equal $\{(0,0)\}$. Thus the only rate pair which is achievable according to our above considerations is $(R_1,R_2) = (0,0)$. At the same time, there should be a Markov chain $(U,(X,Y),(T,Z))$ for the same pair of channels W_b and W_e such that

$$I(T \wedge XY) > I(Z \wedge XY).$$

This would prove the existence of a rate pair (R_1,R_2) with positive components for arbitrary $C_1, C_2 > 0$.

We recall one concept of comparison for single-sender discrete memoryless channels (DMCs) introduced by Körner and Marton [22].

Definition 3. *A DMC $W_e : \mathcal{X} \to \mathcal{P}(\mathcal{Z})$ is less noisy than a DMC $W_b : \mathcal{X} \to \mathcal{P}(\mathcal{T})$ if for every Markov chain $(U,X,(T,Z))$ with $P_{T|X} = W_b$ and $P_{Z|X} = W_e$ one has*

$$I(Z \wedge U) \geq I(T \wedge U).$$

It was observed by van Dijk [35] that this is nothing but saying that the function

$$P_X \mapsto I(Z \wedge X) - I(T \wedge X), \qquad P_X \in \mathcal{P}(\mathcal{X})$$

is concave. Now we generalize this to the MAC case to obtain an equivalent condition for (77).

Lemma 21. (77) *holds for every Markov chain $((V_1,V_2),(X,Y),(T,Z))$ with independent V_1,V_2 and X independent of V_2 and Y independent of V_1 and $P_{T|XY} = W_b$ and $P_{Z|XY} = W_e$ if and only if the function*

$$(P_X,P_Y) \mapsto I(Z \wedge XY) - I(T \wedge XY), \quad X,Y \text{ independent r.v.s on } \mathcal{X} \times \mathcal{Y}$$

is concave in each of its components.

Proof. Let a Markov chain be given as required in the lemma. One has

$$\begin{aligned} &I(Z \wedge V_1V_2) - I(T \wedge V_1V_2) \qquad (78)\\ &= \big(I(Z \wedge XY) - I(T \wedge XY)\big) - \big(I(Z \wedge XY|V_1V_2) - I(T \wedge XY|V_1V_2)\big). \end{aligned}$$

Now note that the rightmost bracket equals

$$\sum_{v_1}\sum_{v_2} P_{V_1}(v_1)P_{V_2}(v_2)\big(I(Z\wedge XY|V_1=v_1,V_2=v_2)-I(T\wedge XY|V_1=v_1,V_2=v_2)\big),$$

so it is clear that the nonnegativity of (78) is equivalent to the concavity in each component of the function from the lemma statement. □

We now define the channels W_b and W_e which will provide the desired example. Let N_1, N_2 be i.i.d. random variables uniformly distributed on $\{0,1\}$. The input alphabets are $\mathcal{X}=\mathcal{Y}=\{0,1\}$. The output alphabet of W_b is $GF(3)$ and the output alphabet of W_e is $\{-2,\dots,3\}$. The outputs t of W_b are given by

$$t = x + y + N_1,$$

those of W_e by

$$z = 2x - 2y + N_2.$$

The intuition is that in W_e, one can exactly determine through the output whether or not the inputs were equal and if they were unequal, which input was 0 and which was 1. For W_b, however, there are for every output at least two input possibilities, so it is reasonable that an independent choice of the inputs makes W_e better than W_b. However, if one may choose the inputs with some correlation, one may choose the inputs to be equal. Then the output of W_e is only noise, whereas one can still extract some information about the input from W_b.

As the entries of the corresponding stochastic matrices of both channels are only 1/2 or 0, the conditional output entropy is independent of the input distribution and equals 1. Further any pair of independent random variables on $\mathcal{X}$ and $\mathcal{Y}$ is given by parameters $q, r \in [0,1]$ such that

$$\mathbb{P}[X^{(q)}=0]=q, \qquad \mathbb{P}[Y^{(r)}=0]=r.$$

Thus in order to determine whether (77) holds, it is enough to consider the function $H(Z^{(q,r)})-H(T^{(q,r)})$ for $T^{(q,r)}, Z^{(q,r)}$ being the outputs of W_b and W_e, respectively, corresponding to the pair $(X^{(q)}, Y^{(r)})$. One has

$$\begin{aligned} f_Z(q,r) := H(Z^{(q,r)}) &= -q(1-r)\log(q(1-r)/2)\\ &\quad -(qr+(1-q)(1-r))\log((qr+(1-q)(1-r))/2)\\ &\quad -(1-q)r\log((1-q)r/2) \end{aligned}$$

and

$$\begin{aligned} f_T(q,r) &:= H(T^{(q,r)})\\ &= -\frac{1}{2}(qr+(1-q)(1-r))\log((qr+(1-q)(1-r))/2)\\ &\quad -\frac{1}{2}(qr+q(1-r)+(1-q)r)\log((qr+q(1-r)+(1-q)r)/2)\\ &\quad -\frac{1}{2}(q(1-r)+(1-q)r+(1-q)(1-r))\cdot\\ &\qquad \cdot\log((q(1-r)+(1-q)r+(1-q)(1-r))/2). \end{aligned}$$

Both entropies are symmetric in q and r and continuous on $[0,1]^2$ and differentiable on $(0,1)^2$, so by Lemma 21 it suffices to find the second derivatives in q of both of them and to compare.

We have

$$\begin{aligned}\frac{\partial f_Z}{\partial q}(q,r) = &-(1-r)\log(q(1-r)/2)\\ &-(2r-1)\log((qr+(1-q)(1-r))/2)\\ &+r\log((1-q)r/2)\end{aligned}$$

and

$$\begin{aligned}\frac{\partial f_T}{\partial q}(q,r) = &-\frac{1}{2}(2r-1)\log((qr+(1-q)(1-r))/2)\\ &-\frac{1}{2}(1-r)\log((qr+q(1-r)+(1-q)r)/2)\\ &+\frac{r}{2}\log((q(1-r)+(1-q)r+(1-q)(1-r))/2).\end{aligned}$$

Thus

$$\frac{\partial^2 f_Z}{\partial q^2}(q,r) = -\frac{1-r}{q} - \frac{(2r-1)^2}{qr+(1-q)(1-r)} - \frac{r}{1-q}$$

and

$$\begin{aligned}\frac{\partial^2 f_T}{\partial q^2}(q,r) = &-\frac{(2r-1)^2}{2(qr+(1-q)(1-r))}\\ &-\frac{(1-r)^2}{2(qr+q(1-r)+(1-q)r)}\\ &-\frac{r^2}{2(q(1-r)+(1-q)r+(1-q)(1-r))}.\end{aligned}$$

After some algebra, it turns out that for $q,r \in (0,1)$,

$$\begin{aligned}\frac{\partial^2 f_Z}{\partial q^2}(q,r) - \frac{\partial^2 f_T}{\partial q^2}(q,r) &= -\frac{1-r}{2q}\cdot\frac{q+2r-qr}{q+r-qr}\\ &\quad -\frac{(2r-1)^2}{2(qr+(1-q)(1-r))}\\ &\quad -\frac{r}{2(1-q)}\cdot\frac{2-r-qr}{1-qr}\\ &< 0.\end{aligned}$$

Thus $f_Z - f_T$ is concave and (77) is true for W_b, W_e.

Now we show that there exists an input distribution with $I(T \wedge XY) > I(Z \wedge XY)$. Of course, X and Y cannot be independent any more in this case. Every probability distribution p on $\{0,1\}$ induces a probability distribution p^2 on $\{0,1\}^2$ via $p^2(x,x) = p(x)$. Let the pair (X,Y) be distributed according to p. It is immediate from the definition of W_e that $I(Z \wedge XY) = 0$. On the other hand, P_T can be described by the vector $(1/2)(1, p(0), p(1))$. Thus one sees easily that this is maximized for $p(0) = p(1) = 1/2$, resulting in

$$I(T \wedge XY) = \frac{1}{2}.$$

p^2 is identified as an element of Π by setting $\mathcal{U} = \{0,1\}$, $P_U = P_X$, and $P_{X|U} = P_{Y|U} = \delta_U$. Note that $I(Z \wedge U) = 0$, so secret transmission is possible with arbitrarily small conferencing capacities $C_1, C_2 > 0$.

6.2 Necessity of Time-Sharing in Random Coding

We show here that doing time-sharing during random coding is necessary for our method to work. This only serves to justify the effort we had to make in coding. We concentrate on Case 0 and 1. Then we have to show that it may happen that $\alpha_0^{(1)} > 0$ or $\alpha_1^{(1)} < 1$. Let $\mathcal{X} = \mathcal{Y} = \mathcal{T} = \mathcal{Z} = \{0,1\}$ and let $W_b, W_e : \{0,1\}^2 \to \mathcal{P}(\{0,1\})$ be defined by

$$W_b = \begin{pmatrix} 0.6178 & 0.3822 \\ 0.0624 & 0.9376 \\ 0.9350 & 0.0650 \\ 0.2353 & 0.7647 \end{pmatrix}, \qquad W_e = \begin{pmatrix} 0.0729 & 0.9271 \\ 0.7264 & 0.2736 \\ 0.3662 & 0.6338 \\ 0.4643 & 0.5357 \end{pmatrix},$$

where the output distribution for the input pair (x, y) is given in row number $2x + y$ for each matrix. With $q = 0.6933$ and $r = 0.3151$, let $p = p^{(q)} \otimes p^{(r)} \in \mathcal{P}(\mathcal{X} \times \mathcal{Y})$ be the product measure with the marginals

$$p^{(q)} = (q, 1-q), \qquad p^{(r)} = (r, 1-r).$$

Note that $p \in \Pi_0$. One obtains the following entropies:

$$\begin{aligned} H(T|XY) &\approx 0.5685, & H(Z|XY) &\approx 0.7851, \\ H(T|X) &\approx 0.8532, & H(Z|X) &\approx 0.9952, \\ H(T|Y) &\approx 0.6251, & H(Z|Y) &\approx 0.8442, \\ H(T) &\approx 0.8866, & H(Z) &\approx 0.9999. \end{aligned}$$

Calculating with the above values returns

$$\begin{aligned} I(T \wedge XY) &= 0.3181, & I(Z \wedge XY) &= 0.2147, \\ I(T \wedge X|Y) &= 0.0566, & I(Z \wedge X|Y) &= 0.0590, \\ I(T \wedge Y|X) &= 0.2847, & I(Z \wedge Y|X) &= 0.2101, \\ & & I(Z \wedge X) &= 0.0047, \\ & & I(Z \wedge Y) &= 0.1557. \end{aligned}$$

Thus the conditions (3) and (4) are satisfied. If $H_C < \min\{I(Z \wedge X|Y), I(Z \wedge Y|X)\} = 0.0590$, then we can only show that $\mathcal{R}^{(0)}(p)$ or $\mathcal{R}^{(1)}(p)$ is achievable and might have to use time-sharing during random coding to do so. In fact, this is necessary as

$$I(Z \wedge X|Y) > I(T \wedge X|Y),$$

whereas

$$I(Z \wedge Y|X) < I(T \wedge Y|X).$$

Hence $\alpha_0^{(1)} > 0$, but $\alpha_1^{(1)} = 1$. This example was found by a brute-force search using the computer.

Acknowledgment. We would like to thank A. J. Pierrot for bringing the papers [30] and [42] to our attention. They consider strong secrecy problems in multi-user settings with the help of resolvability theory. In particular, in [42], an achievable region for the wiretap MAC without common message or conferencing is derived.

References

1. Ahlswede, R.: Elimination of correlation in random codes for arbitrarily varying channels. Z. Wahrscheinlichkeitstheorie verw. Gebiete 44, 159–175 (1978)
2. Ahlswede, R.: Coloring hypergraphs: A new approach to multi-user source coding—II. J. Comb. Inform. Syst. Sci. 5(3), 220–268 (1980)
3. Ahlswede, R.: Arbitrarily varying channels with states sequence known to the sender. IEEE Trans. Inf. Theory IT 32(5), 621–629 (1986)
4. Ahlswede, R.: On Concepts of Performance Parameters for Channels. In: Ahlswede, R., Bäumer, L., Cai, N., Aydinian, H., Blinovsky, V., Deppe, C., Mashurian, H. (eds.) Information Transfer and Combinatorics. LNCS, vol. 4123, pp. 639–663. Springer, Heidelberg (2006)
5. Ahlswede, R., Cai, N.: Arbitrarily varying multiple-access channels part I—Ericson's symmetrizability is adequate, Gubner's conjecture is true. IEEE Trans. Inf. Theory 45(2), 742–749 (1999)
6. Ahlswede, R., Csiszár, I.: Common randomness in information theory and cryptography—part I: Secret sharing. IEEE Trans. Inf. Theory 39(4) (1993)
7. Ahlswede, R., Csiszár, I.: Common randomness in information theory and cryptography—part II: CR capacity. IEEE Trans. Inf. Theory 44(1) (1998)
8. Ahlswede, R., Winter, A.: Strong converse for identification vie quantum channels. IEEE Trans. Inf. Theory 48(3), 569–579 (2002)
9. Ahlswede, R.: Multi-way communication channels. In: Proceedings of 2nd International Symposium on Information Theory, Tsahkadsor, Armenian SSR, Akadémiai Kiadó, Budapest, pp. 23–52 (1971)
10. Ahlswede, R.: An elementary proof of the strong converse theorem for the multiple-access channel. J. Comb. Inf. Syst. Sci. 7, 216–230 (1982)
11. Bjelaković, I., Boche, H., Sommerfeld, J.: Secrecy results for compound wiretap channels, `http://arxiv.org/abs/1106.2013`
12. Bloch, M.R., Laneman, J.N.: Secrecy from resolvability. Submitted to IEEE Trans. Inf. Theory (2011)
13. Cai, N., Winter, A., Yeung, R.W.: Quantum privacy and quantum wiretap channels. Problems of Information Transmission 40(4), 318–336 (2004)
14. Csiszár, I.: Almost independence and secrecy capacity. Problems of Information Transmission 32(1), 40–47 (1996)
15. Csiszár, I., Körner, J.: Broadcast channels with confidential messages. IEEE Trans. Inf. Theory IT-24(3), 339–348 (1978)

16. Csiszár, I., Körner, J.: Information Theory: Coding Theorems for Discrete Memoryless Systems, 2nd edn. Cambridge University Press, Cambridge (2011)
17. Devetak, I.: The private classical capacity and quantum capacity of a quantum channel. IEEE Trans. Inf. Theory 51(1), 44–55 (2005)
18. Dueck, G.: The strong converse of the coding theorem for the multiple-access channel. J. Comb. Inf. Syst. Sci. 6, 187–196 (1981)
19. Ekrem, E., Ulukus, S.: Effects of cooperation on the secrecy of multiple access channels with generalized feedback. In: Proc. Conf. on Inf. Sciences and Systems (CISS), Princeton, NJ, pp. 791–796 (2008)
20. Ekrem, E., Ulukus, S.: On the secrecy of multiple access wiretap channel. In: Proc. Allerton Conference, Allerton House, UIUC, IL, pp. 1014–1021 (2008)
21. He, X., Yener, A.: Mimo wiretap channel with arbitrarily varying eavesdropper channel states. Submitted to IEEE Trans. Inf. Theory (2010), `http://arxiv.org/abs/1007.4801`
22. Körner, J., Marton, K.: The comparison of two noisy channels. In: Csiszár, I., Elias, P. (eds.) Topics in Information Theory . Coll. Math. Soc. J. Bolyai, vol. 16. North Holland, Amsterdam (1977)
23. Liang, Y., Poor, H.V.: Multiple-access channels with confidential messages. IEEE Trans. Inf. Theory 54(3), 976–1002 (2008)
24. Liang, Y., Poor, H.V., Shamai, S.: Information theoretic security. Found. Trends Commun. Inf. Theory 5(4-5), 355–580 (2008)
25. Liao, H.J.: Multiple access channels. PhD thesis, Dept. of Electrical Engineering, University of Hawaii, Honolulu (1972)
26. Liu, R., Liang, Y., Poor, H.V.: Fading cognitive multiple-access channels with confidential messages. Submitted to IEEE Trans. Inf. Theory (2009), `http://arxiv.org/abs/0910.4613`
27. Liu, R., Marić, I., Yates, R., Spasojević, P.: The discrete memoryless multiple-access channel with confidential messages. In: Proc. Int. Symp. Inf. Theory, Seattle, pp. 957–961 (2006)
28. Maurer, U.M.: Secret key agreement by public discussion from common information. IEEE Trans. Inf. Theory 39(3), 733–742 (1993)
29. Maurer, U.M.: The strong secret key rate of discrete random triples. In: Blahut, R. (ed.) Communication and Cryptography — Two Sides of One Tapestry, pp. 271–285. Kluwer Academic Publishers (1994)
30. Pierrot, A.J., Bloch, M.R.: Strongly secure communications over the two-way wiretap channel. IEEE Trans. Inf. Forensics Secur. 6(3) (2011)
31. Simeone, O., Yener, A.: The cognitive multiple access wire-tap channel. In: Proc. Conf. on Inf. Sciences and Systems (CISS), Baltimore, NJ (2009)
32. Slepian, D., Wolf, K.: A coding theorem for multiple access channels with correlated sources. Bell Sytem Techn. J. 52(7), 1037–1076 (1973)
33. Tang, X., Liu, R., Spasojević, P., Poor, H.V.: Multiple acess channels with generalized feedback and confidential messages. In: Proc. Inf. Theory Workshop, Lake Tahoe, CA, pp. 608–613 (2007)
34. Tekin, E., Yener, A.: The gaussian multiple access wire-tap channel. IEEE Trans. Inf. Theory 54(12), 5747–5755 (2008)
35. van Dijk, M.: On a special class of broadcast channels with confidential messages. IEEE Trans. Inf. Theory 43(2), 712–714 (1997)
36. Wiese, M., Boche, H.: The arbitrarily varying multiple-access channel with conferencing encoders. Submitted to IEEE Trans. Inf. Theory (2011), `http://arxiv.org/abs/1105.0319`

37. Wiese, M., Boche, H., Bjelaković, I., Jungnickel, V.: The compound multiple access channel with partially cooperating encoders. IEEE Trans. Inf. Theory 57(5), 3045–3066 (2011)
38. Willems, F.M.J.: Informationtheoretical results for the discrete memoryless multiple access channel. PhD thesis, Katholieke Universiteit Leuven (1982)
39. Willems, F.M.J.: The discrete memoryless multiple access channel with partially cooperating encoders. IEEE Trans. Inf. Theory IT-29(3), 441–445 (1983)
40. Wyner, A.: The wire-tap channel. The Bell System Tech. J. 54(8), 1355–1387 (1975)
41. Wyrembelski, R.F., Wiese, M., Boche, H.: Strong secrecy in bidirectional relay networks. In: Proc. Asilomar Conference on Signals, Systems and Computers (ACSSC 2011), Pacific Grove, CA (2011)
42. Yassaee, M.H., Aref, M.R.: Multiple access wiretap channels with strong secrecy. In: Proc. IEEE Information Theoy Workshop (ITW 2010), Dublin (2010)

Appendix

A Proof of Lemma 2

The direction "$\subset$" in (12) is obvious. For the other direction, let $(R_0, R_1, R_2) \in \mathcal{K}$. We may assume that for some $0 \leq \beta \leq 1$,

$$\begin{aligned} R_1 &= r_1 - \beta(\alpha_1 a_1 + (1-\alpha_1)b_1) - (1-\beta)(\alpha_0 a_1 + (1-\alpha_0)b_1) \\ &= r_1 - (\beta\alpha_1 + (1-\beta)\alpha_0)a_1 - (\beta(1-\alpha_1) + (1-\beta)(1-\alpha_0))b_1 \end{aligned}$$

because the claim is obvious for $R_1 \leq r_1 - \alpha_1 a_1 - (1-\alpha_1)b_1$. We show that $(R_0, R_1, R_2) \in \mathcal{K}_{\beta\alpha_1+(1-\beta)\alpha_0}$. The R_1-bound is satisfied due to our assumption. Further due to the bound on $R_1 + R_2$,

$$\begin{aligned} &R_2 \\ &\leq r_{12} - c - r_1 + (\beta\alpha_1 + (1-\beta)\alpha_0)a_1 + (\beta(1-\alpha_1) + (1-\beta)(1-\alpha_0))b_1 \\ &\leq r_2 - (\beta\alpha_1 + (1-\beta)\alpha_0)a_2 - (\beta(1-\alpha_1) + (1-\beta)(1-\alpha_0))b_2, \end{aligned}$$

so R_2 also satisfies the necessary upper bound. The sum constraints are independent of α. Hence all upper bounds in the definition of $\mathcal{K}_{\beta\alpha_1+(1-\beta)\alpha_0}$ are satisfied, and Lemma 2 is proved.

B Proof of Lemma 4

For $\alpha \in [\alpha_0, \alpha_1]$, the set $\mathcal{K}_\alpha$ is contained in the convex hull of $\mathcal{K}_{\alpha_0} \cup \mathcal{K}_{\alpha_1}$. Thus we only have to prove that $\mathcal{K} = conv(\mathcal{K}_{\alpha_0} \cup \mathcal{K}_{\alpha_1})$. Without loss of generality we assume that $b > a$.

We first prove $conv(\mathcal{K}_{\alpha_0} \cup \mathcal{K}_{\alpha_1}) \subset \mathcal{K}$. Let $(R_0, R_1, R_2) \in conv(\mathcal{K}_{\alpha_0} \cup \mathcal{K}_{\alpha_1})$. Using the convexity of $\mathcal{K}_{\alpha_0}$ and $\mathcal{K}_{\alpha_1}$ we infer that there is a $(R_0^{(0)}, R_1^{(0)}, R_2^{(0)}) \in \mathcal{K}_{\alpha_0}$ and a $(R_0^{(1)}, R_1^{(1)}, R_2^{(1)}) \in \mathcal{K}_{\alpha_1}$ and a $\beta \in [0, 1]$ such that

$$(R_0, R_1, R_2) = \beta(R_0^{(0)}, R_1^{(0)}, R_2^{(0)}) + (1-\beta)(R_0^{(1)} R_1^{(1)}, R_2^{(1)}).$$

One sees immediately that (R_0, R_1, R_2) satisfies the bounds (17)-(19) and (21). It is sufficient to check that (20) is satisfied by the triples $(R_0^{(0)}, R_1^{(0)}, R_2^{(0)})$ and $(R_0^{(1)}, R_1^{(1)}, R_2^{(1)})$. For $(R_0^{(0)}, R_1^{(0)}, R_2^{(0)})$ we assume that

$$R_1^{(0)} = \gamma(r_1 - \alpha_0 a)$$

for some $\gamma \in [0,1]$. After some calculations this yields

$$\begin{aligned} bR_1^{(0)} + aR_2^{(0)} &\leq (b-a)r_1 + ar_{12} - ab - (1-\gamma)(b-a)(r_1 - \alpha_0 a) \\ &\leq (b-a)r_1 + ar_{12} - ab. \end{aligned}$$

One proceeds analogously for $(R_0^{(1)}, R_1^{(1)}, R_2^{(1)})$.

Next we have to check that $\mathcal{K} \subset conv(\mathcal{K}_{\alpha_0} \cup \mathcal{K}_{\alpha_1})$. It is sufficient to check whether those points (R_0, R_1, R_2) are contained in $conv(\mathcal{K}_{\alpha_0} \cup \mathcal{K}_{\alpha_1})$ that satisfy both (20) and one of (17)-(19) with equality. So assume that

$$bR_1 + aR_2 = r_{12}a + r_1(b-a) - ab. \tag{79}$$

First we also assume that

$$R_1 + R_2 = r_{12} - \alpha_0 a - (1-\alpha_1)b.$$

Then

$$R_2 = r_{12} - \alpha_0 a - (1-\alpha_1)b - R_1$$

and using (79) we obtain

$$R_1 = r_1 - \frac{\alpha_1 b - \alpha_0 a}{b-a} a \leq r_1 - \alpha_1 a.$$

For R_2 this gives

$$R_2 = r_{12} - r_1 - \left(\alpha_0 + \frac{\alpha_1 b - \alpha_0 a}{b-a}\right) a - (1-\alpha_1)b \leq r_2 - (1-\alpha_1)b,$$

so $(R_1, R_2) \in \mathcal{K}_{\alpha_1}$.

Now we assume

$$R_1 = r_1 - \alpha_0 a.$$

Then inserting this in (79) one obtains

$$R_2 \leq r_2 - (1-\alpha_0)b,$$

so $(R_1, R_2) \in \mathcal{K}_{\alpha_0}$.

Finally for

$$R_2 = r_2 - (1-\alpha_1)b$$

we obtain

$$R_1 \leq r_1 - \alpha_1 a,$$

so $(R_1, R_2) \in \mathcal{K}_{\alpha_1}$. This proves the lemma.

Capacity Results for Arbitrarily Varying Wiretap Channels

Igor Bjelaković, Holger Boche, and Jochen Sommerfeld

Lehrstuhl für theoretische Informationstechnik,
Technische Universität München, 80290 München, Germany
{igor.bjelakovic,boche,jochen.sommerfeld}@tum.de

Dedicated to the memory of Rudolf Ahlswede

Abstract. In this work the arbitrarily varying wiretap channel AVWC is studied. We derive a lower bound on the random code secrecy capacity for the average error criterion and the strong secrecy criterion in the case of a best channel to the eavesdropper by using Ahlswede's robustification technique for ordinary AVCs. We show that in the case of a non-symmetrisable channel to the legitimate receiver the deterministic code secrecy capacity equals the random code secrecy capacity, a result similar to Ahlswede's dichotomy result for ordinary AVCs. Using this we can derive that the lower bound is also valid for the deterministic code capacity of the AVWC. The proof of the dichotomy result is based on the elimination technique introduced by Ahlswede for ordinary AVCs. We further prove upper bounds on the deterministic code secrecy capacity in the general case, which results in a multi-letter expression for the secrecy capacity in the case of a best channel to the eavesdropper. Using techniques of Ahlswede, developed to guarantee the validity of a reliability criterion, the main contribution of this work is to integrate the strong secrecy criterion into these techniques.

Keywords: arbitrarily varying wiretap channels, strong secrecy, classes of attacks, jamming, active wiretapper.

1 Introduction

Models of communication systems taking into account both the requirement of security against a potential eavesdropper and reliable information transmission to legitimate receivers which suffer from channel uncertainty, have received much interest in current research. One of the simplest communication models with channel uncertainty are compound channels, where the channel realisation remains fixed during the whole transmission of a codeword. Compound wiretap channels were the topic of previous work of the authors [7], [8] and for example of [13], [9]. In the model of an arbitrarily varying wiretap channel AVWC the channel state to both the legitimate receiver and the eavesdropper varies from symbol to symbol in an unknown and arbitrary manner. Thus apart from eavesdropping the model takes into account an active adversarial jamming situation in

H. Aydinian, F. Cicalese, and C. Deppe (Eds.): Ahlswede Festschrift, LNCS 7777, pp. 123–144, 2013.

which the jammer chooses the states at her/his will. Then reliable transmission to the legitimate receiver must be guaranteed in the presence of the jammer.

In this paper we consider families of pairs of channels $\mathfrak{W} = \{(W_{s^n}, V_{s^n}) : s^n \in S^n\}$ with common input alphabets and possibly different output alphabets, where $s^n \in S^n$ denotes the state sequence during the transmission of a codeword. The legitimate users are connected via W_{s^n} and the eavesdropper observes the output of V_{s^n}. In our communication scenario the legitimate users have no channel state information. We derive capacity results for the AVWC $\mathfrak{W}$ under the average error probability criterion and a strong secrecy criterion. The investigation of the corresponding problem concerning the maximum error criterion is left as a subject of future investigations. However, we should emphasize that there is no full capacity result for an ordinary (i.e. without secrecy constraints) AVC for the maximum error criterion. Together with Wolfowitz in [6] Ahlswede determined the capacity for AVCs with binary output alphabets under this criterion. In [1] he showed that the general solution is connected to Shannon's zero error capacity problem [15].

Two fundamental techniques, called *elimination* and *robustification technique* discovered by Ahlswede will play a crucial role in this paper. In [2] he developed the *elimination technique* to derive the deterministic code capacity for AVCs under the average error probability criterion, which is either zero or equals its random code capacity, a result, which is called Ahlswede's dichotomy for single user AVCs. With the so-called *robustification technique* [3] in turn he could link random codes for the AVC to deterministic codes for compound channels. Further in the papers [4], [5] on *common randomness* in information theory Ahlswede together with Csiszár studied, inter alia, problems of information theoretic security by considering a model which enables secret sharing of a random key, in particular in the presence of a wiretapper. Because the arbitrarily varying wiretap channel AVWC combines both the wiretap channel and the AVC it is not surprising that we can use the aforementioned techniques to derive capacity results for the AVWC. The actual challenge of our work was to integrate the strong secrecy criterion in both the *elimination* and the *robustification technique*, approaches, both were developed to guarantee a reliability criterion. As it was shown in [8], compared with weaker secrecy criteria, the strong secrecy criterion ensures that the average error probability of every decoding strategy of the eavesdropper in the limit tends to one.

In Section 3.2 we give a lower bound on the random code secrecy capacity in the special case of a "best" channel to the eavesdropper. The proof is based on the *robustification technique* by Ahlswede [3] combined with results for compound wiretap channels given by the authors in [8].

In Section 3.3 we use the *elimination technique* [2], which is composed of the *random code reduction* and the *elimination of randomness* [10], to show that, provided that the channel to the legitimate receiver is non-symmetrisable, the deterministic code secrecy capacity equals the random code secrecy capacity and to give a condition when it is greater than zero. Thus we establish a result for the AVWC that is similar to that of Ahlswede's dichotomy result for ordinary AVCs.

As a consequence the above-mentioned lower bound on the random code secrecy capacity can be achieved by a deterministic code under the same assumptions.

In Section 3.4 we give a single-letter upper bound on the deterministic code secrecy capacity, which corresponds to the upper bound of the secrecy capacity of a compound wiretap channel. Moreover, by establishing an multi-letter upper bound on the secrecy capacity we can conclude to a multi-letter expression of the secrecy capacity of the AVWC in the special case of a best channel to the eavesdropper.

The lower bound on the secrecy capacity as well as other results were given earlier in [14] for a weaker secrecy criterion, but the proof techniques for the stronger secrecy criterion differ significantly, especially in the achievability part for the random codes.

2 Arbitrarily Varying Wiretap Channels

2.1 Definitions

Let A, B, C be finite sets and consider a non-necessarily finite family of channels $W_s : A \to \mathcal{P}(B)$[1], where $s \in S$ denotes the state of the channel. Now, given $s^n = (s_1, s_2, \ldots, s_n) \in S^n$ we define the stochastic matrix

$$W^n(y^n|x^n, s^n) := \prod_{i=1}^{n} W(y_i|x_i, s_i) := \prod_{i=1}^{n} W_{s_i}(y_i|x_i) \tag{1}$$

for all $y^n = (y_1, \ldots, y_n) \in B^n$ and $x^n = (x_1, \ldots, x_n) \in A^n$. An arbitrarily varying channel is then defined as the sequence $\{\mathcal{W}^n\}_{n=1}^{\infty}$ of the family of channels $\mathcal{W}^n = \{W^n(\cdot|\cdot, s^n) : s^n \in S^n\}$. Now let $\mathcal{W}^n$ represent the communication link to a legitimate receiver to which the transmitter wants to send a private message, such that a possible second receiver should be kept as ignorant of that message as possible. We call this receiver the eavesdropper, which observes the output of a second family of channels $\mathcal{V}^n = \{V^n(\cdot|\cdot, s^n) : s^n \in S^n\}$ with an analogue definition of $V^n(\cdot|\cdot, s^n)$ as in (1) for $V_s : A \to \mathcal{P}(C)$, $s \in S$. Then we denote the set of the two families of channels with common input by $\mathfrak{W} = \{(W_{s^n}, V_{s^n}) : s^n \in S^n\}$ and call it the arbitrarily varying wiretap channel. In addition, we assume that the state sequence s^n is unknown to the legitimate receiver, whereas the eavesdropper always knows which channel is in use.

A (n, J_n) code $\mathcal{C}_n$ for the arbitrarily varying wiretap channel $\mathfrak{W}$ consists of a stochastic encoder $E : \mathcal{J}_n \to \mathcal{P}(A^n)$ (a stochastic matrix) with a message set $\mathcal{J}_n := \{1, \ldots, J_n\}$ and a collection of mutually disjoint decoding sets $\{D_j \subset B^n : j \in \mathcal{J}_n\}$. The average error probability of a code $\mathcal{C}_n$ is given by

$$e(\mathcal{C}_n) := \max_{s^n \in S^n} \frac{1}{J_n} \sum_{j=1}^{J_n} \sum_{x^n \in A^n} E(x^n|j) W^n_{s^n}(D_j^c|x^n) \ . \tag{2}$$

[1] $\mathcal{P}(B)$ denotes the set of probability distributions on B.

A *correlated random* (n, J_n, Γ, μ) *code* $\mathcal{C}_n^{\text{ran}}$ for the arbitrarily varying wiretap channel is given by a family of wiretap codes $\{\mathcal{C}_n(\gamma)\}_{\gamma\in\Gamma}$ together with a random experiment choosing γ according to a distribution μ on Γ. The mean average error probability of a random $(n.J_n, \Gamma, \mu)$ code $\mathcal{C}_n^{\text{ran}}$ is defined analogously to the ordinary one but with respect to the random experiment choosing γ by

$$\bar{e}(\mathcal{C}_n^{\text{ran}}) := \max_{s^n\in S^n} \frac{1}{J_n} \sum_{j=1}^{J_n} \sum_{\gamma\in\Gamma} \sum_{x^n\in A^n} E^\gamma(x^n|j) W_{s^n}^n((D_j^\gamma)^c|x^n)\mu(\gamma) .$$

Definition 1. *A non-negative number R_S is an achievable secrecy rate for the AVWC $\mathfrak{W}$, if there is a sequence $(\mathcal{C}_n)_{n\in\mathbb{N}}$ of (n, J_n) codes such that*

$$\lim_{n\to\infty} e(\mathcal{C}_n) = 0 ,$$

$$\liminf_{n\to\infty} \frac{1}{n} \log J_n \geq R_S ,$$

and

$$\lim_{n\to\infty} \max_{s^n\in S^n} I(p_J; V_{s^n}^n) = 0 , \tag{3}$$

where J is a uniformly distributed random variable taking values in $\mathcal{J}_n$ and $I(p_J; V_{s^n}^n)$ is the mutual information of J and the output variable Z^n of the eavesdropper's channel $V_{s^n}^n$. The secrecy capacity then is given as the supremum of all achievable secrecy rates R_S and is denoted by $C_S(\mathfrak{W})$.

Analogously we define the secrecy rates and the secrecy capacity for random codes $C_{S,\text{ran}}(\mathfrak{W})$, if we replace $\mathcal{C}_n$ by $\mathcal{C}_n^{\text{ran}}$ in the above definition.

Definition 2. *A non-negative number R_S is an achievable secrecy rate for correlated random codes for the AVWC $\mathfrak{W}$, if there is a sequence $(\mathcal{C}_n^{\text{ran}})_{n\in\mathbb{N}}$ of (n, J_n, Γ, μ) codes such that*

$$\lim_{n\to\infty} \bar{e}(\mathcal{C}_n^{\text{ran}}) = 0 ,$$

$$\liminf_{n\to\infty} \frac{1}{n} \log J_n \geq R_S ,$$

and

$$\lim_{n\to\infty} \max_{s^n\in S^n} \sum_{\gamma\in\Gamma} I(p_J, V_{s^n}^n; \mathcal{C}(\gamma))\mu(\gamma) = 0 , \tag{4}$$

where $I(p_J, V_{s^n}^n; \mathcal{C}(\gamma))$ is the mutual information according to the code $\mathcal{C}_(\gamma)$, $\gamma \in \Gamma$ chosen according to the distribution μ. The secrecy capacity then is given as the supremum of all achievable secrecy rates R_S and is denoted by $C_{S,\text{ran}}(\mathfrak{W})$.

3 Capacity Results

3.1 Preliminaries

In what follows we use the notation as well as some properties of *typical* and *conditionally typical* sequences from [10]. For $p \in \mathcal{P}(A)$, $W : A \to \mathcal{P}(B)$, $x^n \in A^n$, and $\delta > 0$ we denote by $\mathcal{T}^n_{p,\delta}$ the set of typical sequences and by $\mathcal{T}^n_{W,\delta}(x^n)$ the set of conditionally typical sequences given x^n in the sense of [10].

The basic properties of these sets that are needed in the sequel are summarised in the following three lemmata.

Lemma 1. *Fixing $\delta > 0$, for every $p \in \mathcal{P}(A)$ and $W : A \to \mathcal{P}(B)$ we have*

$$p^{\otimes n}(\mathcal{T}^n_{p,\delta}) \geq 1 - (n+1)^{|A|} 2^{-nc\delta^2}$$
$$W^{\otimes n}(\mathcal{T}^n_{W,\delta}(x^n)|x^n) \geq 1 - (n+1)^{|A||B|} 2^{-nc\delta^2}$$

for all $x^n \in A^n$ with $c = 1/(2\ln 2)$. In particular, there is $n_0 \in \mathbb{N}$ such that for each $\delta > 0$ and $p \in \mathcal{P}(A)$, $W : A \to \mathcal{P}(B)$

$$p^{\otimes n}(\mathcal{T}^n_{p,\delta}) \geq 1 - 2^{-nc'\delta^2}$$
$$W^{\otimes n}(\mathcal{T}^n_{W,\delta}(x^n)|x^n) \geq 1 - 2^{-nc'\delta^2}$$

holds with $c' = \frac{c}{2}$.

Proof. Standard Bernstein-Sanov trick using the properties of types from [10] and Pinsker's inequality. The details can be found in [16] and references therein for example. □

Recall that for $p \in \mathcal{P}(A)$ and $W : A \to \mathcal{P}(B)$, $pW \in \mathcal{P}(B)$ denotes the output distribution generated by p and W and that $x^n \in \mathcal{T}^n_{p,\delta}$ and $y^n \in \mathcal{T}^n_{W,\delta}(x^n)$ imply that $y^n \in \mathcal{T}^n_{pW,2|A|\delta}$.

Lemma 2. *Let $x^n \in \mathcal{T}^n_{p,\delta}$, then for $V : A \to \mathcal{P}(C)$*

$$|\mathcal{T}^n_{pV,2|A|\delta}| \leq \alpha^{-1}$$
$$V^n(z^n|x^n) \leq \beta \quad \text{for all} \quad z^n \in \mathcal{T}^n_{V,\delta}(x^n)$$

hold, where

$$\alpha = 2^{-n(H(pV)+f_1(\delta))} \tag{5}$$
$$\beta = 2^{-n(H(V|p)-f_2(\delta))} \tag{6}$$

with universal $f_1(\delta), f_2(\delta) > 0$ satisfying $\lim_{\delta\to\infty} f_1(\delta) = 0 = \lim_{\delta\to\infty} f_2(\delta)$.

Proof. Cf. [10].

The next lemma is a standard result from large deviation theory.

Lemma 3. *(Chernoff bounds) Let $Z_1, \ldots, Z_L$ be i.i.d. random variables with values in $[0,1]$ and expectation $\mathbb{E}Z_i = \mu$, and $0 < \epsilon < \frac{1}{2}$. Then it follows that*

$$Pr\left\{\frac{1}{L}\sum_{i=1}^{L} Z_i \notin [(1 \pm \epsilon)\mu]\right\} \leq 2\exp\left(-L \cdot \frac{\epsilon^2 \mu}{3}\right),$$

where $[(1 \pm \epsilon)\mu]$ denotes the interval $[(1-\epsilon)\mu, (1+\epsilon)\mu]$.

For the optimal random coding strategy of the AVWC we need the *robustification technique* by Ahlswede [3] which is formulated as a further lemma. Therefore let Σ_n be the group of permutations acting on $(1, 2, \ldots, n)$. Then every permutation $\sigma \in \Sigma_n$ induces a bijection $\pi \in \Pi_n$ defined by $\pi : \mathcal{S}^n \to \mathcal{S}^n$ with $\pi(s^n) = (s_{\sigma(1)}, \ldots, s_{\sigma(n)})$ for all $s^n = (s_1, \ldots, s_n) \in \mathcal{S}^n$ and Π_n denotes the group of these bijections.

Lemma 4. *(Robustification technique) If a function $f : \mathcal{S}^n \to [0,1]$ satisfies*

$$\sum_{s^n \in \mathcal{S}^n} f(s^n) q(s_1) \cdot \ldots \cdot q(s_n) \geq 1 - \gamma \tag{7}$$

for all $q \in \mathcal{P}_0(n, \mathcal{S})$ and some $\gamma \in [0,1]$, then

$$\frac{1}{n!}\sum_{\pi \in \Pi_n} f(\pi(s^n)) \geq 1 - 3 \cdot (n+1)^{|\mathcal{S}|} \cdot \gamma \quad \forall s^n \in \mathcal{S}^n \ . \tag{8}$$

Proof. The proof is given in [3].

To reduce the random code for the AVWC $\mathfrak{W}$ to a deterministic code we need the concept of symmetrisability, which was established for ordinary AVCs in the following representation by [12], [11].

Definition 3. *[11] An AVC is symmetrisable if for some channel $U : A \to S$*

$$\sum_{s \in S} W(y|x,s)U(s|x') = \sum_{s \in S} W(y|x',s)U(s|x) \tag{9}$$

for all $x, x' \in A$, $y \in B$.

A new channel defined by (9) then would be symmetric with respect to all $x, x' \in A$. The authors of [11] proved the following theorem which is a concretion of Ahlswede's dichotomy result for single-user AVC, which states that the deterministic code capacity C is either $C = 0$ or equals the random code capacity.

Theorem 1. *[11] $C > 0$ if and only if the AVC is non-symmetrisable. If $C > 0$, then*

$$C = \max_{p \in \mathcal{P}(A)} \min_{W \in \bar{\mathcal{W}}} I(p, W) \tag{10}$$

Here the RHS gives the random code capacity and $\bar{\mathcal{W}}$ denotes the convex closure of all channels W_s with $s \in S$, S finite or countable.

3.2 Random Code Construction

First let us define the convex hull of the set of channels $\{W_s : s \in S\}$ by the set of channels $\{W_q : q \in \mathcal{P}(S)\}$, where W_q is defined by

$$W_q(y|x) = \sum_{s \in S} W(y|x,s)q(s), \tag{11}$$

for all possible distributions $q \in \mathcal{P}(S)$. Accordingly we define V_q and its convex hull $\{V_q : q \in \mathcal{P}(S)\}$. Then we denote the convex closure of the set of channels $\{(W_s, V_s) : s \in S\}$ by $\overline{\mathfrak{W}} := \{(W_q, V_q) : q \in \mathcal{P}(\tilde{S}), \tilde{S} \subseteq S, \tilde{S} \text{ is finite}\}$. Occasionally, we restrict q to be from the set of all types $\mathcal{P}_0(n, S)$ of state sequences $s^n \in S^n$.

Lemma 5. *The secrecy capacity $C_S(\mathfrak{W})$ of the arbitrarily varying wiretap channel AVWC $\mathfrak{W}$ equals the secrecy capacity of the arbitrarily varying wiretap channel $\overline{\mathfrak{W}}$.*

Proof. The proof was given for an ordinary arbitrarily varying channel AVC without secrecy criterion in [10] and for an AVWC under the weak secrecy criterion in [14]. Let $\tilde{W}_1, \ldots, \tilde{W}_n$ be averaged channels as defined in (11) and a channel $W_{\tilde{q}}^n : A^n \to \mathcal{P}(B^n)$ with $\tilde{q} = \prod_{i=1}^n q_i$, $\tilde{q} \in \mathcal{P}(S^n)$, $q_i \in \mathcal{P}(S)$ defined by

$$W_{\tilde{q}}^n(y^n|x^n) = \prod_{i=1}^n \tilde{W}_i(y_i|x_i) = \prod_{i=1}^n W_{q_i}(y_i|x_i) = \sum_{s^n \in S^n} W^n(y^n|x^n, s^n)\tilde{q}(s^n)$$

If we now use the same (n, J_n) code $\mathcal{C}_n$ defined by the same pair of encoder and decoding sets as for the AVWC $\mathfrak{W}$ the error probability for transmission of a single codeword by the channel $W_{\tilde{q}}^n$ is given by

$$\sum_{x^n \in A^n} E(x^n|j) W_{\tilde{q}}^n(D_j^c|x^n) = \sum_{s^n \in S^n} \tilde{q}(s^n) \sum_{x^n \in A^n} E(x^n|j) W_{s^n}^n(D_j^c|x^n)$$

and we can bound the average error probability by

$$\begin{aligned} &\frac{1}{J_n} \sum_{j=1}^{J_n} \sum_{x^n \in A^n} E(x^n|j) W_{\tilde{q}}^n(D_j^c|x^n) \\ &\leq \max_{s^n \in S^n} \frac{1}{J_n} \sum_{j=1}^{J_n} \sum_{x^n \in A^n} E(x^n|j) W_{s^n}^n(D_j^c|x^n) = e(\mathcal{C}_n) \ . \end{aligned}$$

Otherwise, because $\mathfrak{W}$ is a subset of $\overline{\mathfrak{W}}^n$, which is the closure of the set of channels $(W_{\tilde{q}}^n, V_{\tilde{q}}^n)$, the opposite inequality holds for the channel $W_{\tilde{q}}^n$ that maximizes the error probability. Because $V_{\tilde{q}}^n$ is defined analogously to $W_{\tilde{q}}^n$, we can define for the (n, J_n) code

$$\hat{V}(z^n|j) := \sum_{x^n \in A^n} E(x^n|j) V_{\tilde{q}}^n(z^n|x^n) \tag{12}$$

for all $z^n \in C^n$, $j \in \mathcal{J}_n$. Then

$$\hat{V}(z^n|j) = \sum_{s^n \in S^n} \tilde{q}(s^n) \sum_{x^n \in A^n} E(x^n|j) V^n_{s^n}(z^n|x^n) = \sum_{s^n \in S^n} \tilde{q}(s^n) \hat{V}^n_{s^n}(z^n|j) \quad (13)$$

and because of the convexity of the mutual information in the channel $\hat{V}$ and (13) it holds that

$$I(J, Z^n_{\tilde{q}}) \leq \sum_{s^n \in S^n} \tilde{q}(s^n) I(J; Z^n_{s^n}) \leq \sup_{s^n} I(J, Z^n_{s^n}). \quad (14)$$

Now because $\{\hat{V}^n_{s^n}(z^n|j) : s^n \in S^n\}$ is a subset of $\{\hat{V}(z^n|j) : \tilde{q} \in \mathcal{P}(S^n)\}$ we end in

$$\sup_{\tilde{q} \in \mathcal{P}(S^n)} I(J, Z^n_{\tilde{q}}) = \sup_{s^n} I(J, Z^n_{s^n}) \ .$$

□

Now we can proceed in the construction of the random code of the AVWC $\mathfrak{W}$.

Definition 4. *We call a channel to the eavesdropper a best channel if there exist a channel $V_{q^*} \in \{V_q : q \in \mathcal{P}(S)\}$ such that all other channels from $\{V_q : q \in \mathcal{P}(S)\}$ are degraded versions of V_{q^*}. If we denote the output of any channel V_q, $q \in \mathcal{P}(S)$ by Z_q it holds that*

$$X \to Z_{q^*} \to Z_q, \quad \forall q \in \mathcal{P}(S). \quad (15)$$

Proposition 1. *Provided that there exist a best channel to the eavesdropper, for the random code secrecy capacity $C_{S,ran}(\mathfrak{W})$ of the AVWC $\mathfrak{W}$ it holds that*

$$C_{S,ran}(\mathfrak{W}) \geq \max_{p \in \mathcal{P}(A)} (\min_{q \in \mathcal{P}(S)} I(p, W_q) - \max_{q \in \mathcal{P}(S)} I(p, V_q)). \quad (16)$$

Proof. The proof is based on Ahlswedes *robustification technique* [3] and is divided in two parts:
step 1): The set

$$\overline{\mathcal{W}} := \{(W^n_q, V^n_q) : q \in \mathcal{P}(S)\}$$

corresponds to a compound wiretap channel indexed by the set of all possible distributions $q \in \mathcal{P}(S)$ on the set of states S. First we show, that there exist a deterministic code for the compound wiretap channel $\overline{\mathcal{W}}$ that achieves the lower bound on the random code secrecy capacity of the AVWC $\mathfrak{W}$ given in (16).

In [8] it was shown that for a compound wiretap channel $\{(W_t, V_t) : t \in \theta\}$ without channel state information at the legitimate receivers the secrecy capacity is bounded from below by

$$C_{S,\text{comp}} \geq \max_{p \in \mathcal{P}(A)} (\min_{t \in \theta} I(p, W_s) - \max_{t \in \theta} I(p, V_s)). \quad (17)$$

In accordance with the proof of (17) in [8] we define a set of i.i.d. random variables $\{X_{jl}\}_{j \in [J_n], l \in [L_n]}$ each according to the distribution $p' \in \mathcal{P}(A^n)$ with

$$p'(x^n) := \begin{cases} \frac{p^{\otimes n}(x^n)}{p^{\otimes n}(\mathcal{T}^n_{p,\delta})} & \text{if } x^n \in \mathcal{T}^n_{p,\delta}, \\ 0 & \text{otherwise}, \end{cases} \quad (18)$$

for any $p \in \mathcal{P}(A)$, and where J_n and L_n are chosen as

$$J_n = \lfloor 2^{n[\inf_{q\in\mathcal{P}(S)} I(p,W_q) - \sup_{q\in\mathcal{P}(S)} I(p,V_q) - \tau]} \rfloor \tag{19}$$
$$L_n = \lfloor 2^{n[\sup_{q\in\mathcal{P}(S)} I(p,V_q) + \frac{\tau}{4}]} \rfloor \tag{20}$$

with $\tau > 0$. Now we assume that there exist a best channel to the eavesdropper V_{q^*} in contrast to the proof in [8]. Hence by the definition of V_{q^*} in (15) and because the mutual Information $I(p, V)$ is convex in V and every member of $\{V_q\}_{q\in\mathcal{P}(S)}$ is a convex combination of the set $\{V_s\}_{s\in S}$, it holds that

$$I(p, V_{q^*}) = \sup_s I(p, V_s) = \sup_{q\in\mathcal{P}(S)} I(p, V_q) \tag{21}$$

for all $p \in \mathcal{P}(A)$. Note that because of (21) for $|S| < \infty$ $V_{q^*} \in \{V_s : s \in S\}$, which means that q^* is a one-point distribution.

By the definition of the compound channel $\overline{\mathcal{W}}$ the channels to the eavesdropper are of the form

$$V_q^n(z^n|x^n) := \prod_{i=1}^n V_q(z_i|x_i) \tag{22}$$

for all $q \in \mathcal{P}(S)$. Then following the same approach as in the proof in [8] we define

$$\tilde{Q}_{q,x^n}(z^n) = V_q^n(z^n|x^n) \cdot \mathbf{1}_{\mathcal{T}^n_{V_q,\delta}(x^n)}(z^n),$$

and

$$\Theta'_q(z^n) = \sum_{x^n\in\mathcal{T}^n_{p,\delta}} p'(x^n)\tilde{Q}_{q,x^n}(z^n). \tag{23}$$

for all $z^n \in C^n$. Now let $\mathcal{B} := \{z^n \in C^n : \Theta'_q(z^n) \geq \epsilon\alpha_q\}$ where $\epsilon = 2^{-nc'\delta^2}$ (cf. Lemma 1) and α_q is from (5) in Lemma 2 computed with respect to p and V_q. By Lemma 2 the support of Θ'_q has cardinality $\leq \alpha_q^{-1}$ since for each $x^n \in \mathcal{T}^n_{p,\delta}$ it holds that $\mathcal{T}^n_{V_q,\delta}(x^n) \subset \mathcal{T}^n_{pV_q,2|A|\delta}$, which implies that $\sum_{z^n\in\mathcal{B}} \Theta_q(z^n) \geq 1 - 2\epsilon$, if

$$\Theta_q(z^n) = \Theta'_q(z^n) \cdot \mathbf{1}_{\mathcal{B}}(z^n) \quad \text{and}$$
$$Q_{q,x^n}(z^n) = \tilde{Q}_{q,x^n}(z^n) \cdot \mathbf{1}_{\mathcal{B}}(z^n). \tag{24}$$

Now it is obvious from (23) and the definition of the set $\mathcal{B}$ that for any $z^n \in \mathcal{B}$ $\Theta_q(z^n) = \mathbb{E}Q_{q,X_{jl}}(z^n) \geq \epsilon\alpha_q$ if $\mathbb{E}$ is the expectation value with respect to the distribution p'. Let β_q defined as in (6) with respect to V_q. For the random variables $\beta_q^{-1}Q_{q,X_{jl}}(z^n)$ define the event

$$\iota_j(q) = \bigcap_{z^n\in C^n} \left\{ \frac{1}{L_n}\sum_{l=1}^{L_n} Q_{q,X_{jl}}(z^n) \in [(1\pm\epsilon)\Theta_q(z^n)] \right\}, \tag{25}$$

and keeping in mind that $\Theta_q(z^n) \geq \epsilon\alpha_q$ for all $z^n \in \mathcal{B}$. Then it follows that for all $j \in [J_n]$ and for all $s \in S$

$$\Pr\{(\iota_j(q))^c\} \leq 2|C|^n \exp\Big(- L_n \frac{2^{-n[I(p,V_q)+g(\delta)]}}{3}\Big) \tag{26}$$

by Lemma 3, Lemma 2, and our choice $\epsilon = 2^{-nc'\delta^2}$ with $g(\delta) := f_1(\delta) + f_2(\delta) + 3c'\delta^2$. Making $\delta > 0$ sufficiently small we have for all sufficiently large $n \in \mathbb{N}$

$$L_n 2^{-n[I(p,V_q)+g(\delta)]} \geq 2^{n\frac{\tau}{8}}.$$

Thus, for this choice of δ the RHS of (26) is double exponential in n uniformly in $q \in \mathcal{P}(S)$ and can be made smaller than ϵJ_n^{-1} for all $j \in [J_n]$ and all sufficiently large $n \in \mathbb{N}$. I.e.

$$\Pr\{(\iota_j(q))^c\} \leq \epsilon J_n^{-1} \quad \forall q \in \mathcal{P}(S) \tag{27}$$

Now we will show that we can achieve reliable transmission to the legitimate receiver governed by $\{(W_q^n : q \in \mathcal{P}(S)\}$ for all messages $j \in [J_n]$ when randomising over the index $l \in L_n$ but without the need of decoding $l \in [L_n]$. To this end define $\mathcal{X} = \{X_{jl}\}_{j\in[J_n],l\in[L_n]}$ to be the set of random variables with X_{jl} are i.i.d. according to p' defined in (18). Define now the random decoder $\{D_j(\mathcal{X})\}_{j\in[J_n]} \subseteq B^n$ analogously as in [8], [7]. Then it was shown by the authors, that there exist a sequence of (n, J_n) codes for the compound wiretap channel in the particular case without CSI with arbitrarily small mean average error

$$\mathbb{E}_{\mathcal{X}}(\lambda_n^{(q)}(\mathcal{X})) \leq 2^{-na}$$

for all $q \in \mathcal{P}(S)$ and sufficiently large $n \in \mathbb{N}$. Additionally we define for each $q \in \mathcal{P}(S)$

$$\iota_0(q) = \{\lambda_n^{(q)}(\mathcal{X})) \leq 2^{-n\frac{a}{2}}\} \tag{28}$$

and set

$$\iota := \bigcap_{q\in\mathcal{P}_0(n,S)} \bigcap_{j=0}^{J_n} \iota_j(q) \tag{29}$$

Then with (27), (28) and applying the union bound we obtain

$$\Pr\{\iota^c\} \leq 2^{-nc}$$

for a suitable positive constant $c > 0$ and all sufficiently large $n \in \mathbb{N}$ (Cf. [8]).

Hence, we have shown that there exist realisations $\{x_{jl}\}$ of $\{X_{jl}^n\}_{j\in[J_n],l\in[L_n]}$ such that $x_{jl} \in \iota$ for all $j \in [J_n]$ and $l \in [L_n]$. Now following the same argumentation as in [8], [7] we obtain that there is a sequence of (n, J_n) codes that for all codewords $\{x_{jl}\}$ it follows by construction that

$$\frac{1}{J_n}\sum_{j\in[J_n]}\frac{1}{L_n}\sum_{l\in[L_n]} W_q^n(D_j^c|x_{jl}) \leq 2^{-na'} \tag{30}$$

is fulfilled for $n \in \mathbb{N}$ sufficiently large and for all $q \in \mathcal{P}(S)$ with $a' > 0$. So we have found a (n, J_n) code with average error probability upper bounded by (30). Further, for the given code and a random variable J uniformly distributed on the message set $\{1, \ldots, J_n\}$ it holds that

$$I(p_J; V_q^n) \leq \epsilon' \tag{31}$$

uniformly in $q \in \mathcal{P}(S)$. Both (30) and (31) ensure that in the scenario of the compound wiretap channel $\overline{\mathcal{W}}$ the legitimate receiver can identify each message j from the message set $\{1, \ldots, J_n\}$ with high probability, while at the same time the eavesdropper receives almost no information about it. That is, that all numbers R_S with

$$R_S \leq \inf_{q \in \mathcal{P}(S)} I(p, W_q) - \sup_{q \in \mathcal{P}(S)} I(p, V_q) \tag{32}$$

are achievable secrecy rates of the compound wiretap channel $\overline{\mathcal{W}}$.

step 2): *Robustification*: In the second step we derive from the deterministic (n, J_n) code for the above mentioned compound wiretap channel $\overline{\mathcal{W}}$ a (n, J_n) random code $\mathcal{C}_n^{\text{ran}}$ for the AVWC $\mathfrak{W}$, which achieves the same secrecy rates. We note first that by (21) and (31)

$$\max_{s^n \in S^n} I(p_J, V_{s^n}) = I(p_J, V_{q^*}^n) \leq \epsilon', \tag{33}$$

which means, that, due to the assumption of a best channel to the eavesdropper, the code achieving the secrecy rate for the best channel to the eavesdropper fulfills the secrecy criterion for a channel with any state sequence $s^n \in S^n$. Now, as already mentioned we use the robustification technique (cf. Lemma 4) to derive from the deterministic code $\mathcal{C}_{\overline{\mathcal{W}}} = \{x_{jl}, D_j : j \in [J_n], l \in [L_n]\}$ of the compound wiretap channel $\overline{\mathcal{W}}$ the random code for the AVWC $\mathfrak{W}$. Therefore, for now let S to be finite. With (30) it holds that

$$\frac{1}{J_n} \sum_{j \in [J_n]} \frac{1}{L_n} \sum_{l \in [L_n]} \sum_{s^n \in S^n} W^n(D_j | x_{jl}, s^n) q^{\otimes n}(s^n) \geq 1 - 2^{-na'} \tag{34}$$

for all $q^{\otimes n} = \prod_{i=1}^n q$ and in particular for all $q \in \mathcal{P}_0(n, S)$. Now let $\pi \in \Pi_n$ be the bijection on S^n induced by the permutation $\sigma \in \Sigma_n$. Since (7) is fulfilled with

$$f(s^n) = \frac{1}{J_n} \sum_{j \in [J_n]} \frac{1}{L_n} \sum_{l \in [L_n]} W^n(D_j | x_{jl}, s^n) \tag{35}$$

it follows from (8) that

$$\frac{1}{n!} \sum_{\pi \in \Pi_n} \frac{1}{J_n} \sum_{j \in [J_n]} \frac{1}{L_n} \sum_{l \in [L_n]} W^n(D_j | x_{jl}, \pi(s^n)) \geq 1 - (n+1)^{|S|} 2^{-na'} \tag{36}$$

for all $s^n \in S^n$. Hence by defining $\mathcal{C}^\pi := \{\pi^{-1}(x_{jl}^n), \pi^{-1}(D_j)\}$ as a member of a family of codes $\{\mathcal{C}^\pi\}_{\pi \in \Pi_n}$ together with a random variable K distributed according to μ as the uniform distribution on Π_n, (36) is equivalent to

$$\mathbb{E}_\mu(\bar{\lambda}_n(\mathcal{C}^K, W_{s^n}^n)) \leq (n+1)^{|S|} 2^{-na'} =: \lambda_n \tag{37}$$

with $\bar{\lambda}_n(\mathcal{C}^\pi, W_{s^n}^n)$ as the respective average error probability for $K = \pi$ and it holds for all $s^n \in S^n$. Thus we have shown that

$$\mathcal{C}_n^{\text{ran}} := \{(\pi^{-1}(x_{jl}), \pi^{-1}(D_j)) : j \in [J_n], l \in [L_n], \pi \in \Pi_n, \mu\} \tag{38}$$

is a (n, J_n, Π_n, μ) random code for the AVC channel $\mathcal{W}^n = \{W_{s^n} : s^n \in S^n\}$ with the mean average error probability $\mathbb{E}_\mu(\bar{\lambda}_n(\mathcal{C}^K, W^n_{s^n}))$ upper bounded by λ_n as in (37).

Now it is easily seen that

$$p^{\mathcal{C}^\pi}_{JZ^n_{q^*}}(j, z^n) = \frac{1}{J_n}\frac{1}{L_n}\sum_{l=1}^{L_n} V^n_{q^*}(\pi^{-1}(z^n)|\pi^{-1}(x_{jl})) = p_{JZ^n_{q^*}} . \tag{39}$$

Actually, it still holds that

$$p^{\mathcal{C}^r}_{JZ^n_{q^*}}(j, z^n) = \frac{1}{n!}\sum_{\pi\in\Pi_n} p^{\mathcal{C}^\pi}_{JZ^n_{q^*}}(j, z^n) = p_{JZ^n_{q^*}} . \tag{40}$$

With (39) and the representation of the mutual information by the information divergence we obtain from (33)

$$\begin{aligned}\mathbb{E}_\mu(D(p^{\mathcal{C}^K}_{JZ^n_{q^*}}\|p_J \otimes p^{\mathcal{C}^K}_{Z^n_{q^*}})) &= \frac{1}{n!}\sum_{\pi\in\Pi_n} D(p^{\mathcal{C}^\pi}_{JZ^n_{q^*}}\|p_J \otimes p^{\mathcal{C}^\pi}_{Z^n_{q^*}}) \\ &= \frac{1}{n!}\sum_{\pi\in\Pi_n} D(p_{JZ^n_{q^*}}\|p_J \otimes p_{Z^n_{q^*}}) = I(p_J, V^n_{q^*}) \le \epsilon' .\end{aligned} \tag{41}$$

Thus we have constructed a random (n, J_n, Γ, μ) code $\mathcal{C}^{\text{ran}}_n$ with mean average error probability bounded for all $s^n \in S^n$ as in (37) and which fulfills the strong secrecy criterion almost surely, provided that there exist a best channel to the eavesdropper. By the construction of the random code it follows that the secrecy rates given by (32) for the compound wiretap channel $\overline{\mathcal{W}}$ achieved by the deterministic code $\mathcal{C}_{\overline{\mathcal{W}}}$ are achievable secrecy rates for the AVWC $\mathfrak{W}$ with random code $\mathcal{C}^{\text{ran}}_n$. That is, we have shown that all rates R_S with

$$R_S \le \max_{p\in\mathcal{P}(A)}\left(\min_{q\in\mathcal{P}(S)} I(p, W_q) - \max_{q\in\mathcal{P}(S)} I(p, V_q)\right) . \tag{42}$$

are achievable secrecy rates of the arbitrarily varying wiretap channel AVWC with random code $\mathcal{C}^{\text{ran}}_n$. □

3.3 Deterministic Code Construction

Because the code $\mathcal{C}^\pi$ that is used for the transmission of a single message is subjected to a random selection, reliable transmission can only be guaranteed if the outcome of the random experiment can be shared by both the transmitter and the receiver. One way to inform the receiver about the code that is chosen is to add a short prefix to the actual codeword. Provided that the number of codes is small enough, the transmission of these additional prefixes causes no essential loss in rate. In the following we use the *elimination technique* by Ahlswede [2] which has introduced the above approach to derive deterministic codes from random codes for determining capacity of arbitrarily varying channels. Temporarily we drop the requirement of a best channel to the eavesdropper and state the following theorem.

Theorem 2. *1. Assume that for the AVWC $\mathfrak{W}$ it holds that $C_{S,\text{ran}}(\mathfrak{W}) > 0$. Then the secrecy capacity $C_S(\mathfrak{W})$ equals its random code secrecy capacity $C_{S,\text{ran}}(\mathfrak{W})$,*

$$C_S(\mathfrak{W}) = C_{S,\text{ran}}(\mathfrak{W}), \tag{43}$$

if and only if the channel to the legitimate receiver is non-symmetrisable.
2. If $C_{S,\text{ran}}(\mathfrak{W}) = 0$ it always holds that $C_S(\mathfrak{W}) = 0$.

First, if the channel to the legitimate receiver is symmetrisable then the deterministic code capacity of the channel to the legitimate receiver equals zero by Theorem 1 and no reliable transmission of messages is possible. Hence the deterministic code secrecy capacity of the arbitrarily varying wiretap channel also equals zero although the random code secrecy capacity could be greater than zero. So we can restrict to the case in which the channel to the legitimate receiver is non-symmetrisable. If $C_S(\mathfrak{W}) = C_{S,\text{ran}}(\mathfrak{W}) > 0$, then the channel to the legitimate receiver must be nonsymmetrisable. For the other direction, because the secrecy capacity of the AVWC $\mathfrak{W}$ cannot be greater than the random code secrecy capacity it suffices to show that $C(\{W_{s^n}\}) > 0$ implies that $C_S(\mathfrak{W}) \geq C_{S,\text{ran}}(\mathfrak{W})$. Here $C(\{W_{s^n}\})$ denotes the capacity of the arbitrarily varying channels to the legitimate receiver without secrecy. The proof is given in the two paragraphs *Random code reduction* and *Elimination of randomness.*

Random Code Reduction. We first reduce the random code $\mathcal{C}^{\text{ran}}$ to a new random code selecting only a small number of deterministic codes from the former, and averaging over this codes gives a new random code with a constant small mean average error probability, which additionally fulfills the secrecy criterion.

Lemma 6. *(Random Code Reduction) Let $\mathcal{C}(\mathcal{Z})$ be a random code for the AVWC $\mathfrak{W}$ consisting of a family $\{\mathcal{C}(\gamma)\}_{\gamma\in\Gamma}$ of wiretap codes where γ is chosen according to the distribution μ of $\mathcal{Z}$. Then let*

$$\bar{e}(\mathcal{C}_n^{\text{ran}}) = \max_{s^n} \mathbb{E}_\mu e(s^n|\mathcal{C}(\mathcal{Z})) \leq \lambda_n \quad \textit{and,} \quad \max_{s^n} \mathbb{E}_\mu I(p_J, V_{s^n}; \mathcal{C}(\mathcal{Z})) \leq \epsilon_n' \ . \tag{44}$$

Then for any ϵ and K satisfying

$$\epsilon > 4\max\{\lambda_n, \epsilon_n'\} \quad \textit{and} \quad K > \frac{2n\log|A|}{\epsilon}(1 + n\log|S|) \tag{45}$$

there exist K deterministic codes $\mathcal{C}_i$, $i = 1, \ldots, K$ chosen from the random code by random selection such that

$$\frac{1}{K}\sum_{i=1}^{K} e(s^n|\mathcal{C}_i) \leq \epsilon \quad \textit{and} \quad \frac{1}{K}\sum_{i=1}^{K} I(p_J, V_{s^n}; \mathcal{C}_i) \leq \epsilon \tag{46}$$

for all $s^n \in S^n$.

Proof. The proof is analogue to the proof of Lemma 6.8 [10], where a similar assertion in terms of the maximal probability of error for single user AVCs without secrecy criterion is established. Cf. also [2]. Let $\mathcal{Z}$ be the random variable distributed according to μ on Γ for the (n, J_n, Γ, μ) random code. Now consider K independent repetitions of the random experiment of code selections according to μ and call the according random variables $\mathcal{Z}_i$, $i \in \{1, \ldots, K\}$. Then for any $s^n \in S^n$ it holds that

$$\begin{aligned}
&\Pr\Big\{\frac{1}{K}\sum_{i=1}^{K} e(s^n|\mathcal{C}(\mathcal{Z}_i)) \geq \epsilon \quad \text{or} \quad \frac{1}{K}\sum_{i=1}^{K} I(p_J, V_{s^n}; \mathcal{C}(\mathcal{Z}_i)) \geq \epsilon\Big\} \\
&\leq \Pr\Big\{\exp\sum_{i=1}^{K}\frac{e(s^n|\mathcal{C}(\mathcal{Z}_i))}{n\log|A|} \geq \exp\frac{K\epsilon}{n\log|A|}\Big\} \\
&\qquad + \Pr\Big\{\exp\sum_{i=1}^{K}\frac{I(p_J, V_{s^n}; \mathcal{C}(\mathcal{Z}_i))}{n\log|A|} \geq \exp\frac{K\epsilon}{n\log|A|}\Big\},
\end{aligned}$$

and by Markov's inequality

$$\begin{aligned}
&\Pr\Big\{\frac{1}{K}\sum_{i=1}^{K} e(s^n|\mathcal{C}(\mathcal{Z}_i)) \geq \epsilon \quad \text{or} \quad \frac{1}{K}\sum_{i=1}^{K} I(p_J, V_{s^n}; \mathcal{C}(\mathcal{Z}_i)) \geq \epsilon\Big\} \\
&\leq \exp\Big(-\frac{K\epsilon}{n\log|A|}\Big)\mathbb{E}\exp\sum_{i=1}^{K}\frac{e(s^n|\mathcal{C}(\mathcal{Z}_i))}{n\log|A|} \\
&\qquad + \exp\Big(-\frac{K\epsilon}{n\log|A|}\Big)\mathbb{E}\exp\sum_{i=1}^{K}\frac{I(p_J, V_{s^n}; \mathcal{C}(\mathcal{Z}_i))}{n\log|A|}.
\end{aligned}$$

Now because of the independency of the random variables $\mathcal{Z}_i$ and because all $\mathcal{Z}_i$ are distributed as $\mathcal{Z}$ and we have $\exp t \leq 1 + t$, for $0 \leq t \leq 1$ (exp to the base 2), we can give the following upper bounds

$$\Big(\mathbb{E}\exp\frac{e(s^n|\mathcal{C}(\mathcal{Z}))}{n\log|A|}\Big)^K \leq \Big(1+\mathbb{E}\frac{e(s^n|\mathcal{C}(\mathcal{Z}))}{n\log|A|}\Big)^K \leq \Big(1+\frac{\lambda_n}{n\log|A|}\Big)^K \tag{47}$$

and

$$\Big(\mathbb{E}\exp\frac{I(p_J, V_{s^n}; \mathcal{C}(\mathcal{Z}))}{n\log|A|}\Big)^K \leq \Big(1+\mathbb{E}\frac{I(p_J, V_{s^n}; \mathcal{C}(\mathcal{Z}))}{n\log|A|}\Big)^K \leq \Big(1+\frac{\epsilon_n'}{n\log|A|}\Big)^K. \tag{48}$$

Hence we obtain for any $s^n \in S^n$

$$\begin{aligned}
&\Pr\Big\{\frac{1}{K}\sum_{i=1}^{K} e(s^n|\mathcal{C}(\mathcal{Z}_i)) \geq \epsilon \quad \text{or} \quad \frac{1}{K}\sum_{i=1}^{K} I(p_J, V_{s^n}; \mathcal{C}(\mathcal{Z}_i)) \geq \epsilon\}\\
&\leq \exp\Big[-K\Big(\frac{\epsilon}{n\log|A|} - \log(1+\frac{\lambda_n}{n\log|A|})\Big)\Big]\\
&\qquad\qquad + \exp\Big[\Big(-K(\frac{\epsilon}{n\log|A|} - \log(1+\frac{\epsilon'_n}{n\log|A|})\Big)\Big]\\
&\leq 2\exp\Big[-K\Big(\frac{\epsilon}{n\log|A|} - \log(1+\max\{\frac{\lambda_n}{n\log|A|}, \frac{\epsilon'_n}{n\log|A|}\})\Big)\Big] \ .
\end{aligned}$$

Then

$$\begin{aligned}
&\Pr\Big\{\frac{1}{K}\sum_{i=1}^{K} e(s^n|\mathcal{C}(\mathcal{Z}_i)) \leq \epsilon \text{ and } \frac{1}{K}\sum_{i=1}^{K} I(p_J, V_{s^n}; \mathcal{C}(\mathcal{Z}_i)) \leq \epsilon, \forall s^n \in S^n\Big\}\\
&\geq 1 - 2|S|^n \exp\Big[-K\Big(\frac{\epsilon}{n\log|A|} - \log(1+\max\{\frac{\lambda_n}{n\log|A|}, \frac{\epsilon'_n}{n\log|A|}\})\Big)\Big],
\end{aligned} \tag{49}$$

which is strictly positive, if we choose

$$\epsilon \geq 2n\log|A|\log(1+\max\{\frac{\lambda_n}{n\log|A|}, \frac{\epsilon'_n}{n\log|A|}\})$$

and

$$K \geq \frac{2\log|A|}{\epsilon}(n + n^2\log|S|) \ . \tag{50}$$

Now because for $0 \leq t \leq 1$ and log to the base 2 it holds that

$$t \leq \log(1+t) \leq 2t \ ,$$

we increase the lower bound for choosing ϵ if

$$\epsilon \geq 4\max\{\lambda_n, \epsilon'_n\} \ .$$

and with (50) the assertion of (49) still holds. Hence, we have shown that there exist K realisations $\mathcal{C}_i := \mathcal{C}(\mathcal{Z}_i = \gamma_i)$, $\gamma_i \in \Gamma$, $i \in \{1, \ldots, K\}$ of the random code, which build a new reduced random code with uniform distribution on these codes with mean average error probability and mean secrecy criterion fulfilled by (46). □

Now, if we assume that the channel to the legitimate receiver is non-symmetrisable, which means that $C(\{W_{s^n}\}) > 0$, and that there exist a random code $\mathcal{C}_n^{\text{ran}}$ that achieves the random code capacity $C_{S,\text{ran}}(\mathfrak{W}) > 0$, then there exist a sequence of random (n, J_n) codes with

$$\lim_{n\to\infty}\max_{s^n\in S^n}\frac{1}{J_n}\sum_{j=1}^{J_n}\sum_{\gamma\in\Gamma}\sum_{x^n\in A^n} E^{\gamma}(x^n|j)\cdot W^n_{s_n}((D^{\gamma}_j)^c|x^n)\mu(\gamma) = 0 \ ,$$

$$\liminf_{n\to\infty} \frac{1}{n} \log J_n \to C_{S,\text{ran}}(\mathfrak{W}) > 0,$$

and

$$\lim_{n\to\infty} \max_{s^n\in S^n} \sum_{\gamma\in\Gamma} I(p_J; V^n_{s^n}; \mathcal{C}(\gamma))\mu(\gamma) = 0. \tag{51}$$

Then on account of the random code reduction lemma there exist a sequence of random (n, J_n) codes consisting only of n^3 deterministic codes (cf. (45)) chosen from the former random code, and it holds for any $\epsilon > 0$ and sufficiently large n that

$$\max_{s^n\in S^n} \frac{1}{J_n} \sum_{j=1}^{J_n} \frac{1}{n^3} \sum_{i=1}^{n^3} \sum_{x^n\in A^n} E^i(x^n|j) W^n_{s_n}((D^i_j)^c|x^n) \le \epsilon \tag{52}$$

and

$$\max_{s^n\in S^n} \frac{1}{n^3} \sum_{i=1}^{n^3} I(p_J; V^n_{s^n}; \mathcal{C}_i) \le \epsilon, \tag{53}$$

where $\mathcal{C}_i = \{(E^i_j, D^i_j), j \in \mathcal{J}_n\}$, $i = 1, \ldots, n^3$, and E^i is the stochastic encoder of the deterministic wiretap code. Then the reduced random code consists of the family of codes $\{\mathcal{C}_i\}_{i\in\{1,\ldots,n^3\}}$ together with the uniform distribution $\mu'(i) = \frac{1}{n^3}$ for all $i \in \{1, \ldots, n^3\}$.

Elimination of Randomness. (Cf. Theorem 6.11 in [10])
Now if there exist a deterministic code and $C(\{W_{s^n}\}) > 0$ then there exist a code

$$\{x^{k_n}_i, F_i \subset B^{k_n} : i = 1, \ldots n^3\} \tag{54}$$

where $x^{k_n}_i$ is chosen according to an encoding function $f_i : \{1, \ldots, n^3\} \to A^{k_n}$ with $\frac{k_n}{n} \to 0$ as $n \to \infty$ with error probability

$$\frac{1}{n^3} \sum_{i=1}^{n^3} W^{k_n}(F^c_i | x^{k_n}_i, s^{k_n}) \le \epsilon \tag{55}$$

for any $\epsilon > 0$ and sufficiently large n (cf. (52)) for all $s^{k_n} \in S^{k_n}$. If we now compose a new deterministic code for the AVWC $\mathfrak{W}$ by prefixing the codewords of each $\mathcal{C}_i$

$$\{f_i E^i_j, F_i \times D^i_j : i = 1, \ldots, n^3, j \in [J_n]\} =: \mathcal{C}\ , \tag{56}$$

the decoder is informed of which encoder E^i is in use for the actual message j if he identifies the prefix correctly. Note that for the transmission of the prefix only the reliability is of interest, because it contains no information about the message $j \in \mathcal{J}_n$ to be sent. Now the new codewords has a length of $k_n + n$, transmit a message from $\{1, \ldots, n^3\} \times \mathcal{J}_n$, where the channel which is determined by the state sequence $s^{k_n+n} \in S^{k_n+n}$ yields an average error probability of

$$\begin{aligned}\bar{\lambda}_n(\mathcal{C}, W^{(k_n+n)}_{s^{k_n+n}}) &\leq \frac{1}{n^3 J_n} \sum_{i=1}^{n^3} \sum_{j\in[J_n]} (\lambda_i + \lambda_j(i)) \\ &\leq \frac{1}{n^3}\sum_{i=1}^{n^3} \lambda_i + \frac{1}{n^3}\sum_{i=1}^{n^3} e_n(s^n, \mathcal{C}_i) \leq 2\epsilon.\end{aligned} \tag{57}$$

Here, for each $s^{k_n} \in S^{k_n}$ λ_i means the error probability for transmitting i from $\{1, \dots, n^3\}$ encoded in $x_i^{k_n}$ by $W^{k_n}_{s^{k_n}}$ followed by the transmission of j, where the codeword is chosen according to the stochastic encoder E^i_j, over the last n channel realisations determined by s^n with error probability $\lambda_j(i)$. This construction is possible due to the memorylessness of the channel.

Now if we turn to the security part of the transmission problem it is easily seen that

$$\begin{aligned}p^{\mathcal{C}}_{JZ^{k_n+n}_{s^{k_n+n}}}(j, z^{k_n+n}) &= \frac{1}{J_n}\frac{1}{n^3}\sum_{i=1}^{n^3} V^{k_n}_{s^{k_n}}(\hat{z}^{k_n}|x_i^{k_n}) \sum_{x^n} E^i(x^n|j) V^n_{s^n}(z^n|x^n) \\ &= \frac{1}{n^3}\sum_{i=1}^{n^3} V^{k_n}_{s^{k_n}}(\hat{z}^{k_n}|x_i^{k_n}) \cdot p^{\mathcal{C}_i}_{JZ^n_{s^n}} ,\end{aligned} \tag{58}$$

where $\hat{z}^{k_n}$ are the first k_n components of z^{k_n+n}. With (58) and the representation of the mutual information by the information divergence we obtain that

$$\begin{aligned}&D(p^{\mathcal{C}}_{JZ^{k_n+n}_{s^{k_n+n}}} \| p_J \otimes p^{\mathcal{C}}_{Z^{k_n+n}_{s^{k_n+n}}}) \\ &= D\Big(\frac{1}{n^3}\sum_{i=1}^{n^3} V^{k_n}_{s^{k_n}}(\hat{z}^{k_n}|x_i^{k_n}) p^{\mathcal{C}_i}_{JZ^n_{s^n}} \Big\| \frac{1}{n^3}\sum_{i=1}^{n^3} V^{k_n}_{s^{k_n}}(\hat{z}^{k_n}|x_i^{k_n}) p_J \otimes p^{\mathcal{C}_i}_{Z^n_{s^n}}\Big) \\ &\leq \frac{1}{n^3}\sum_{i=1}^{n^3} D(V^{k_n}_{s^{k_n}}(\hat{z}^{k_n}|x_i^{k_n}) p^{\mathcal{C}_i}_{JZ^n_{s^n}} \| V^{k_n}_{s^{k_n}}(\hat{z}^{k_n}|x_i^{k_n}) p_J \otimes p^{\mathcal{C}_i}_{Z^n_{s^n}}) \\ &= \frac{1}{n^3}\sum_{i=1}^{n^3} D(p^{\mathcal{C}_i}_{JZ^n_{s^n}} \| p_J \otimes p^{\mathcal{C}_i}_{Z^n_{s^n}}) = \frac{1}{n^3}\sum_{i=1}^{n^3} I(p_J, V^n_{s^n}; \mathcal{C}_i) \leq \epsilon\end{aligned} \tag{59}$$

for all $s^n \in S^n$ and $n \in \mathbb{N}$ sufficiently large, where the first inequality follows because for two probability distributions p, q the relative entropy $D(p\|q)$ is a convex function in the pair (p, q) and the last inequality follows by the random code reduction lemma.

Because $\frac{k_n}{n} \to 0$ as $n \to \infty$

$$\lim_{n\to\infty} \frac{1}{k_n + n} \log(n^3 J_n) = \lim_{n\to\infty} (\frac{1}{n}\log J_n + \frac{1}{n}\log(n^3)) = \lim_{n\to\infty} \frac{1}{n}\log J_n , \tag{60}$$

$\mathcal{C}_n$ is a deterministic (n, J_n) code which achieves the same rates as the random code $\mathcal{C}_n^{\text{ran}}$ and so the random code capacity $C_{S,\text{ran}}$ as given in (51), provided that the channel to the legitimate receiver is non-symmetrisable.

Thus, with $\{1,\ldots,J_n\}$ as the message set, $\mathcal{C}_n$ is a deterministic $(n+o(n), n^3\cdot J_n)$ code with average error probability bounded for all $s^{k_n+n}\in S^{k_n+n}$ as in (57) and which fulfills the strong secrecy criterion as in (59), and which achieves the random code secrecy capacity $C_{S,\text{ran}}$ of the arbitrarily varying wiretap channels AVWC $\mathcal{W}$ which implies that $C_S = C_{S,\text{ran}}$. This concludes the proof.

Note that in the case in which the channel to the legitimate receiver is non-symmetrisable and we know that the deterministic code secrecy capacity $C_S(\mathfrak{W})$ equals zero we can conclude that the random code secrecy capacity $C_{S,\text{ran}}(\mathfrak{W})$ equals zero. As a consequence of the theorem we can state the following assertion.

Corollary 1. *The deterministic code secrecy capacity of the arbitrarily varying wiretap channel $\mathfrak{W}$, provided that there exists a best channel to the eavesdropper and under the assumption that the channel to the legitimate receiver is non-symmetrisable, is lower bounded by*

$$C_S(\mathfrak{W}) \geq \max_{p\in\mathcal{P}(A)}\Big(\min_{q\in\mathcal{P}(S)} I(p,W_q) - \max_{q\in\mathcal{P}(S)} I(p,V_q)\Big)\ .$$

Proof. Combine the assertions of Proposition 1 and Theorem 2. □

3.4 Upper Bound on the Capacity of the AVWC $\mathfrak{W}$ and a Multi-letter Coding Theorem

In this section we give an upper bound on the secrecy capacity of the AVWC $\mathfrak{W}$ which corresponds to the bound for the compound wiretap channel built by the same family of channels. In addition we give the proof of the multi-letter converse of the AVWC $\mathfrak{W}$.

Theorem 3. *The secrecy capacity of the arbitrarily varying wiretap channel AVWC $\mathfrak{W}$ is upper bounded,*

$$C_S(\mathfrak{W}) \leq \min_{q\in\mathcal{P}(S)} \max_{U\to X\to(YZ)_q} \big(I(U,Y_q) - I(U,Z_q)\big)\ . \tag{61}$$

Proof. By Lemma 5 the capacity of the AVWC $\mathfrak{W}$ equals the capacity of the AVWC $\overline{\mathfrak{W}}$. Obviously, the set $\overline{\mathcal{W}} = \{(W_q^{\otimes n}, V_q^{\otimes n}) : q\in\mathcal{P}(S)\}$ which describes a compound wiretap channel is a subset of $\overline{\mathfrak{W}}^n = \{(W_{\tilde{q}}^n, V_{\tilde{q}}^n) : \tilde{q}\in\mathcal{P}(S^n), \tilde{q} = \prod_{i=1}^n q_i\}$. Now, because we can upper bound the secrecy capacity of the AVWC $\mathfrak{W}$ by the secrecy capacity of the worst wiretap channel in the family $\overline{\mathfrak{W}}^n$, together with the foregoing we can upper bound it by the capacity of the worst channel of the compound channel $\overline{\mathcal{W}}$. Hence,

$$\begin{aligned} C_S(\mathfrak{W}) = C_S(\overline{\mathfrak{W}}) &\leq \inf_{\tilde{q}} C_S((W_{\tilde{q}}^n, V_{\tilde{q}}^n)) \\ &\leq \inf_q C_S((W_q^n, V_q^n)) = \inf_q C_S(W_q, V_q)\ , \end{aligned}$$

The minimum is attained because of the continuity of $C_S(W_q, V_q)$ on the compact set $\overline{\mathfrak{W}}$. □

Remark 1. *Consider the special case of an AVWC* $\mathfrak{W} = \{(W_{s^n}, V_{r^n}) : s^n \in S_1^n,\ r^n \in S_2^n\}$, *where both the state of the main channel* $s \in S_1$ *and the state of the eavesdropper's channel* $r \in S_2$ *in every time step can be chosen independently. In addition let us assume that there exist a channel* $W_{q_1^*} \in \{W_{q_1} : q_1 \in \mathcal{P}(S_1)\}$, *which is a degraded version of all other channels from* $\{W_{q_1} : q_1 \in \mathcal{P}(S_1)\}$, *and a best channel to the eavesdropper* $V_{q_2^*}$ *from the set* $\{V_{q_2} : q_2 \in \mathcal{P}(S_2)\}$ *(cf. Definition 4). Then in accordance with Section* 3.5 *of [8] the lower bound on the secrecy capacity given in Corollary 1 matches the upper bound from Theorem 3. Thus we can conclude that under the assumption, that the channel to legitimate receiver is non-symmetrisable, the capacity of the AVWC* $\mathfrak{W}$ *is given by*

$$C_S(\mathfrak{W}) = \max_{U \to X \to (Y_{q_1^*} Z_{q_2^*})} (I(U, Y_{q_1^*}) - I(U, Y_{q_2^*})) \ .$$

Now in addition to Theorem 3 we give a multi-letter formula of the upper bound of the secrecy rates. Therefore we need the following lemma used in analogy to Lemma 3.7 in [8].

Lemma 7. *For the arbitrarily varying wiretap channel AVWC* $\mathfrak{W}^n$ *the limit*

$$\lim_{n\to\infty} \frac{1}{n} \max_{U \to X^n \to (Y^n Z^n)_{\tilde{q}}} \Big(\inf_{\tilde{q} \in \mathcal{P}(S^n)} I(U, Y_{\tilde{q}}^n) - \sup_{\tilde{q} \in \mathcal{P}(S^n)} I(U, Z_{\tilde{q}}^n) \Big)$$

exists.

The proof is carried out in analogy to Lemma 3.7 in [8] and therefore omitted.

Theorem 4. *The secrecy capacity of the arbitrarily varying wiretap channel AVWC* $\mathfrak{W}$ *is upper bounded by*

$$C_S(\mathfrak{W}) \le \lim_{n\to\infty} \frac{1}{n} \max_{U \to X^n \to (Y^n Z^n)_{\tilde{q}}} \Big(\inf_{\tilde{q} \in \mathcal{P}(S^n)} I(U, Y_{\tilde{q}}^n) - \sup_{\tilde{q} \in \mathcal{P}(S^n)} I(U, Z_{\tilde{q}}^n) \Big) \ , \qquad (62)$$

where $\tilde{q} = \prod_{i=1}^n q_i$, $q_i \in \mathcal{P}(S)$ *and* $Y_{\tilde{q}}^n, Z_{\tilde{q}}^n$ *are the outputs of the channels* $W_{\tilde{q}}^n$ *and* $V_{\tilde{q}}^n$ *respective.*

Proof. Let $(\mathcal{C}_n)_{n\in\mathbb{N}}$ be any sequence of (n, J_n) codes such that with

$$\sup_{s^n \in S^n} \frac{1}{J_n} \sum_{j=1}^{J_n} \sum_{x^n \in A^n} E(x^n|j) W_{s_n}^n(D_j^c|x^n) =: \varepsilon_{1,n} \text{ and, } \sup_{s^n \in S^n} I(J, Z_{s^n}^n) =: \varepsilon_{2,n}$$

it holds that $\lim_{n\to\infty} \varepsilon_{1,n} 0 =$ and $\lim_{n\to\infty} \varepsilon_{2,n}$, where J denotes the random variable which is uniformly distributed on the message set $\mathcal{J}_n$. Because of Lemma 5 we obtain that for the same sequences of (n, J_n) codes

$$\lim_{n\to\infty} \sup_{\tilde{q} \in \mathcal{P}(S^n)} \frac{1}{J_n} \sum_{j=1}^{J_n} \sum_{x^n \in A^n} E(x^n|j) W_{\tilde{q}}^n(D_j^c|x^n) = \lim_{n\to\infty} \varepsilon_{1,n} = 0 \qquad (63)$$

and

$$\lim_{n\to\infty} \sup_{\tilde{q}\in\mathcal{P}(S^n)} I(J, Z^n_{\tilde{q}}) = \lim_{n\to\infty} \varepsilon_{2,n} = 0 \ . \tag{64}$$

Now let us denote another random variable by $\hat{J}$ with values in $\mathcal{J}_n$ determined by the Markov chain $J \to X^n \to Y^n_{\tilde{q}} \to \hat{J}$, where the first transition is governed by E, the second by $W^n_{\tilde{q}}$, and the last by the decoding rule. Now the proof is analogue to the proof of Proposition 3.8 in [8]. For any $\tilde{q} \in \mathcal{P}(S^n)$ we have from data processing and Fano's inequality

$$(1-\varepsilon_{1,n}) \log J_n \le I(J, Y^n_{\tilde{q}}) + 1.$$

We then use the validity of the secrecy criterion (64) to derive

$$(1-\varepsilon_{1,n}) \log J_n \le I(J, Y^n_{\tilde{q}}) - \sup_{\tilde{q}} I(J, Z^n_{\tilde{q}}) + \varepsilon_{2,n} + 1$$

for any $\tilde{q} \in \mathcal{P}(S^n)$. Since the LHS does not depend on $\tilde{q}$ we end in

$$(1-\varepsilon_{1,n}) \log J_n \le \max_{U\to X^n\to Y^n_{\tilde{q}} Z^n_{\tilde{q}}} (\inf_{\tilde{q}} I(U, Y^n_{\tilde{q}}) - \sup_{\tilde{q}} I(U, Z^n_{\tilde{q}})) + \varepsilon_{2,n} + 1 \ .$$

Dividing by $n \in \mathbb{N}$ and taking $\overline{\lim}$ concludes the proof. □

Now if we consider the set $\overline{\mathcal{W}} = \{(W^{\otimes n}_q, V^{\otimes n}_q) : q \in \mathcal{P}(S)\}$ as a subset of $\overline{\mathfrak{W}}^n = \{(W^n_{\tilde{q}}, V^n_{\tilde{q}}) : \tilde{q} \in \mathcal{P}(S^n), \tilde{q} = \prod_{i=1}^n q_i\}$ and the same sequence $(\mathcal{C}_n)_{n\in\mathbb{N}}$ of (n, J_n) codes for the AVWC $\mathfrak{W}$ for which (63) and (64) holds, we can conclude that

$$\lim_{n\to\infty} \sup_{q\in\mathcal{P}(S)} \frac{1}{J_n} \sum_{j=1}^{J_n} \sum_{x^n\in A^n} E(x^n|j) W^{\otimes n}_q (D^c_j|x^n) \le \lim_{n\to\infty} \varepsilon_{1,n} \tag{65}$$

and

$$\lim_{n\to\infty} \sup_{q\in\mathcal{P}(S)} I(J, Z^n_q) \le \lim_{n\to\infty} \varepsilon_{2,n} \ , \tag{66}$$

with $\varepsilon_{1,n}$ and $\varepsilon_{2,n}$ as above. Then we can conclude with the same argumentation as in the previous proof,

Corollary 2. *The secrecy capacity of the arbitrarily varying wiretap channel AVWC $\mathfrak{W}$ is upper bounded by*

$$C_S(\mathfrak{W}) \le \lim_{n\to\infty} \frac{1}{n} \max_{U\to X^n\to(Y^n Z^n)_q} (\inf_{q\in\mathcal{P}(S)} I(U, Y^n_q) - \sup_{q\in\mathcal{P}(S)} I(U, Z^n_q)) \ ,$$

where $q \in \mathcal{P}(S)$ and Y^n_q, Z^n_q are the outputs of the channels $W^{\otimes n}_q$ and $V^{\otimes n}_q$ respective.

Now, using standard arguments concerning the use of the channels defined by $P_{Y_q|U} = W_q \cdot P_{X|U}$ and $P_{Z_q|U} = V_q \cdot P_{X|U}$ instead of W_q and V_q and applying the assertion of Corollary 1 to the n-fold product of channels W_q and V_q, we are able to give the coding theorem for the multi-letter case of the AVWC with a best channel to the eavesdropper.

Theorem 5. *Provided that there exist a best channel to the eavesdropper, the multi-letter expression for the secrecy capacity $C_S(\mathfrak{W})$ of the AVWC $\mathfrak{W}$ is given by*

$$C_S(\mathfrak{W}) = \lim_{n\to\infty} \frac{1}{n} \max_{U\to X^n\to(Y^n Z^n)_q} \left(\inf_{q\in\mathcal{P}(S)} I(U, Y_q^n) - \sup_{q\in\mathcal{P}(S)} I(U, Z_q^n) \right) ,$$

if the channel to the legitimate receiver is non-symmetrisable, and is zero otherwise.

Acknowledgment. Support by the Deutsche Forschungsgemeinschaft (DFG) via projects BO 1734/16-1, BO 1734/20-1, and by the Bundesministerium für Bildung und Forschung (BMBF) via grant 01BQ1050 is gratefully acknowledged.

References

1. Ahlswede, R.: A note on the existence of the weak capacity for channels with arbitrarily varying channel probability functions and its relation to shannon's zero error capacity. The Annals of Mathematical Statistics 41(3), 1027–1033 (1970)
2. Ahlswede, R.: Elimination of correlation in random codes for arbitrarily varying channels. Zeitschrift für Wahrscheinlichkeitstheorie und verwandte Gebiete 44, 159–175 (1978)
3. Ahlswede, R.: Arbitrarily varying channels with states sequence known to the sender. IEEE Trans. on Inf. Theory 32(5), 621–629 (1986)
4. Ahlswede, R., Csiszár, I.: Common randomness in information theory and cryptography-part I: Secret sharing. IEEE Trans. on Inf. Theory 39(4), 1121–1132 (1993)
5. Ahlswede, R., Csiszár, I.: Common randomness in information theory and cryptography-part ii: Cr capacity. IEEE Trans. on Inf. Theory 44(1), 225–240 (1998)
6. Ahlswede, R., Wolfowitz, J.: The capacity of a channel with arbitrarily varying channel probability functions and binary output alphabet. Z. Wahrscheinlichkeitstheorie verw. Gebiete 15, 186–194 (1970)
7. Bjelacović, I., Boche, H., Sommerfeld, J.: Capacity results for compound wiretap channels. In: Proc. IEEE Information Theory Workshop, pp. 60–64 (2011)
8. Bjelacović, I., Boche, H., Sommerfeld, J.: Secrecy results for compound wiretap channels. Submitted to Problems of Information Transmission (2011), `http://arxiv.org/abs/1106.2013v1`
9. Bloch, M., Laneman, J.: On the secrecy capacity of arbitrary wiretap channel. In: Forty-Sixth Annual Allerton Conference, Allerton House, Illinois, USA (2008)
10. Csiszár, I., Körner, J.: Information Theory: Coding Theorems for Discrete Memoryless Systems. Akademiai Kiado (1981)
11. Csiszár, I., Narayan, P.: The capacity of the arbitrarily varying channel revisited: positivity, constraints. IEEE Trans. on Inf. Theory 34(2), 181–193 (1988)
12. Ericson, T.: Exponential error bounds for random codes in the arbitrarily varying channel. IEEE Trans. on Inf. Theory 31(1), 42–48 (1985)

13. Liang, Y., Kramer, G., Poor, H., Shamai, S.: Compound Wiretap Channels. EURASIP Journal on Wireless Communications and Networking (2008)
14. MolavianJazi, E.: Secure Communications over Arbitrarily Varying Wiretap Channels. Master's thesis, Graduate School of the University of Notre Dame (2009)
15. Shannon, C.: The zero error capacity of a noisy channel. IRE Trans. Information Theory IT-2, 8–19 (1956)
16. Wyrembelski, R.F., Bjelaković, I., Oechtering, T.J., Boche, H.: Optimal coding strategies for bidirectional broadcast channels under channel uncertainty. IEEE Trans. on Communications 58(10), 2984–2994 (2010)

On Oblivious Transfer Capacity

Rudolf Ahlswede* and Imre Csiszár[1,]**

[1] Rényi Institute of Mathematics, P.O. Box 127, Budapest, Hungary
csiszar.imre@renyi.mta.hu

Abstract. Upper and lower bounds to the oblivious transfer (OT) capacity of discrete memoryless channels and multiple sources are obtained, for 1 of 2 strings OT with honest but curious participants. The upper bounds hold also for one-string OT. The results provide the exact value of OT capacity for a specified class of models, and the necessary and sufficient condition of its positivity, in general.

Keywords: entropy difference bound, generalized erasure channel, honest but curious, oblivios transfer, one of two strings, secret key, wiretap channel.

This paper is based on the ISIT-07 contribution [2]. The authors did intend to write up a full version and devoted substantial amount of work to that project, but abandoned it as other obligations delayed completion and the elapsed time caused loss of novelty. Still, the second author considers it proper to publish this paper in this volume, paying tribute to the memory of Rudolph Ahlswede. The results in [2] are completed by some previously unpublished ones which originated from the authors' discussions during their work towards a full version of [2].

1 Introduction

Oblivious transfer (OT) is a fundamental concept in cryptography, see for example [9]. The term has been used with different meanings, including a simple transmission over a binary erasure channel. In this paper, unless stated otherwise, OT means "1 out of 2 oblivious string transfer" [9]. Two parties are involved, commonly called Alice and Bob. Alice is initially given two binary strings K_0, K_1 of length k, and Bob is given a single bit Z. An OT protocol performed by Alice and Bob is supposed to let Bob learn K_Z while he remains ignorant of $K_{\overline{Z}}$ ($\overline{Z} = 1 - Z$) and Alice remains ignorant of Z. The Shannon-theoretic approach is used, thus ignorance means negligible amount of information. Formal definitions are in Section 2.

* Rudolf Ahlswede, one of the world's top information theorists at the University of Bielefeld, died December 18, 2010.

** Supported by Hungarian National Foundation for Scientific Research, Grant 76088.

H. Aydinian, F. Cicalese, and C. Deppe (Eds.): Ahlswede Festschrift, LNCS 7777, pp. 145–166, 2013.

Throughout this paper, it is assumed that Alice and Bob may use the following resources for free: (i) unlimited computing power (ii) local randomness provided by random experiments they may perform, independently of each other (iii) a noiseless public channel, available for unlimited communication in any number of rounds. These free resources alone are not sufficient for OT. In this paper, two kinds of models will be considered which involve an additional (non-free) resource, either a discrete memoryless multiple source (DMMS) or a noisy discrete memoryless channel (DMC).

A *source model* is determined by a DMMS with two component sources, i.e., a sequence of i.i.d. repetitions (X_i, Y_i), $i = 1, 2, \ldots$ of a pair (X, Y) of "generic" random variables (RVs) taking values in finite sets $\mathcal{X}, \mathcal{Y}$ called source alphabets. At the ith access to this DMMS, Alice observes X_i and Bob Y_i. A *channel model* is determined by a DMC whose (finite) input and output alphabets are denoted by $\mathcal{X}, \mathcal{Y}$, and the conditional probability of Bob receiving $y \in \mathcal{Y}$ when Alice sends $x \in \mathcal{X}$ is denoted by $W(y|x)$. At the ith access to this DMC, Alice selects an input X_i and Bob observes the corresponding output Y_i. In either model, the cost of one access to the DMMS resp. DMC is one unit. Thus the cost of an OT protocol is the number of accesses to the DMMS resp. DMC.

The OT capacity C_{OT} of a DMMS or DMC is the limit as $n \to \infty$ of $1/n$ times the largest k for which OT is possible with cost n. This concept has been introduced by Nascimento and Winter [11,12] who also proved $C_{\mathrm{OT}} > 0$ under a natural condition. See also Imai et al. [7] who for the binary erasure channel with erasure probability $1/2$ proved $C_{\mathrm{OT}} = 1/2$. For previous results showing that a DMMS or DMC makes OT possible for any k (but not that k/n may be bounded away from 0 while the conditions (1)-(3) below are satisfied) see the references in [12]. A related concept of commitment capacity has been introduced and characterized in [15].

In the literature of OT much of the effort is devoted to designing protocols that prevent a malicious Alice from learning Bob's bit Z or a malicious Bob from obtaining information also about $K_{\overline{Z}}$. This issue is not entered here, we assume following [11,12] that Alice and Bob are "honest but curious". This means that they honestly follow the protocol but do not discard any information they get access to in the process, and may use all of it to infer what they are supposed to remain ignorant about. Nevertheless, we will point out that a modification of the basic protocol does provide some protection against cheating, while not decreasing OT capacity.

2 Preliminaries

The basic notation of the book [6] is used, except that source and channel alphabets are denoted by script rather than boldface capitals. In particular, log denotes logarithm to base 2, and a DMC with matrix $W = \{W(y|x), x \in \mathcal{X}, y \in \mathcal{Y}\}$ is referred to as DMC $\{W : \mathcal{X} \to \mathcal{Y}\}$ or just $\{W\}$. In order to define admissible OT protocols for source and channel models, general two-party protocols are described first.

A *noiseless protocol*, assuming Alice and Bob have initial knowledge or *view* U and V, is described as follows; here U and V are not necessarily independent RVs. At the beginning of the protocol, both Alice and Bob perform a random experiment to generate RVs M resp. N, where M, N and (U, V) are independent. Then Alice sends Bob over the noiseless public channel a message F_1 which is a function of U and M, and Bob returns Alice a message F_2, a function of V, N and F_1. The formal role of the RVs M, N is to model possible randomization in Alice's choice of F_1 and Bob's choice of F_2, as well as in their actions later on. In following rounds (as many as desired) Alice and Bob alternatingly send messages $F_3, F_4, \ldots, F_{2t}$ which are functions of their instantenous views. In other words, F_i is a function of U, M and $\{F_j,\ j < i\}$ if i is odd, and of V, N and $\{F_j,\ j < i\}$ is i is even (here the messages F_j with j of the same parity as i are redundant). At the end of the protocol, Alice's view will be $(U, M, \mathbf{F})$ and Bob's $(V, N, \mathbf{F})$, where $\mathbf{F} = F_1 \ldots, F_{2t}$.

A *noisy protocol* with n accesses to the DMC $\{W\}$ is described as follows. Alice and Bob, whose initial views are represented by RVs U and V, start the protocol by generating RVs M, N as above. Then Alice selects the DMC input X_1 as a function of U and M, and Bob observes the corresponding output Y_1. After this, in a first session of public communication, they may exchange messages according to a noiseless protocol in which the role of their initial views is played by (U, M) and (V, N, Y_1), respectively; X_1 need not be indicated as part of Alice's view for it is a function of (U, M). In this public communication session, and in subsequent ones, Alice and Bob need not generate new RVs for randomization, the original M and N may be assumed to contain all randomness needed for that purpose.

Next, DMC accesses and public communication sessions alternate. Denote the total public communication in the first i sessions by F^i. Before the i'th access to the DMC, Alice's view is (U, M, F^{i-1}). She selects the DMC input X_i as a function of that view, and Bob observes the corresponding output Y_i. Formally, on the condition that $X_i = x$, the RV Y_i is conditionally independent of $U, V, M, N, Y^{i-1}, F^{i-1}$, and its conditional distribution is $W(\cdot|x)$. Then, in the i'th session of public communication, Alice and Bob perform a noiseless protocol in which their original views are (U, M, F^{i-1}) resp. (V, N, Y^i, F^{i-1}). The protocol ends with the n'th public session, and Alice's and Bob's final views are $(U, M, \mathbf{F})$ and $(V, N, Y^n, \mathbf{F})$ where $\mathbf{F} = F^n$. Alice's knowledge of $X^n = X_1, \ldots, X_n$ need not be indicated for X^n is a function of $(U, M, \mathbf{F})$.

Using the above general concepts, admissible protocols for cost-n oblivious transfer of length-k messages, or briefly (n, k) protocols for OT, are described as follows. Below, $X^n = (X_1, \ldots, X_n)$ and $Y^n = (Y_1, \ldots . Y_n)$ denote, in case of source models, the source output sequences observed by Alice and Bob, and in case of channel models, the sequences of DMC inputs and outputs selected by Alice resp. observed by Bob.

In case of a source model, Alice and Bob may perform any noiseless protocol in which their initial views are $U = (K_0, K_1, X^n)$ and $V = (Z, Y^n)$. Here K_0 and K_1, representing the two binary strings given to Alice, are uniformly distributed on $\{0,1\}^k$, the RV Z, representing the bit given to Bob, is uniformly distributed on $\{0,1\}$, and $K_0, K_1, Z, (X^n, Y^n)$ are mutually independent. In case of a channel model, Alice and Bob may perform any noisy protocol with n accesses to the DMC, in which their initial views are $U = (K_0, K_1)$ and $V = Z$ with K_0, K_1, Z independent and uniformly distributed on $\{0,1\}^k$ resp. $\{0,1\}$. In both cases, upon completing the protocol, Bob produces an estimate $\hat{K}_Z$ of K_Z as a function of his view $(Z, N, Y^n, \mathbf{F})$.

Of course, such an (n,k) protocol is suitable for OT only if it meets the goals stated in the Introduction. These are formalized, in the limit $n \to \infty$, by conditions (1)-(3) below in which the dependence on n of the RVs involved is suppressed to keep the notation transparent. Condition (1) means that Bob learns K_Z with negligible probability of error. Conditions (2) and (3) mean that Alice remains ignorant of Z and Bob of $K_{\overline{Z}}$, in the sense of obtaining negligible amount of information about Z resp. $K_{\overline{Z}}$. In exceptional cases when these conditions hold with equality rather than merely convergence to 0, one speaks of perfect OT.

Definition 1. *A positive number R is an achievable OT rate for a given DMMS or DMC if for $n \to \infty$ there exist (n,k) protocols with $\frac{k}{n} \to R$ such that*

$$\Pr\{\hat{K}_Z \neq K_Z\} \to 0 \tag{1}$$

$$I(K_0 K_1 M X^n \mathbf{F} \wedge Z) \to 0 \tag{2}$$

$$I(Z N Y^n \mathbf{F} \wedge K_{\overline{Z}}) \to 0. \tag{3}$$

The OT capacity of a DMMS or DMC is the supremum of achievable OT rates, or 0 if no $R > 0$ is achievable.

Note that since $I(Z \wedge K_{\overline{Z}}) = 0$, condition (3) is equivalent to

$$I(NY^n\mathbf{F} \wedge K_1 | Z = 0) \to 0; \quad I(NY^n\mathbf{F} \wedge K_0 | Z = 1) \to 0. \tag{4}$$

Remark 1. *An alternative definition of achievable OT rates reqiures exponentially fast convergence to 0 in (1)-(3) as $n \to \infty$. Another alternative relaxes (3) to $\frac{1}{n} I(ZNY^n\mathbf{F} \wedge K_{\overline{Z}}) \to 0$. The results in this paper hold under either definition. Note that Definition 1 admits arbitrarily complex protocols. This is necessary for the generality of our upper bound to OT capacity (Theorem 1). On the other hand, for our achievability results (lower bounds to OT capacity) rather simple protocols will suffice. See also Remark 2.*

Given any DMC $\{W : \mathcal{X} \to \mathcal{Y}\}$ and distribution P on $\mathcal{X}$ (referred to as an input distribution), consider a DMMS with generic RVs X, Y whose joint distribution is given by $P(x)W(y|x)$. The OT capacity of this DMMS will be denoted by $C_{\mathrm{OT}}(P, W)$, while the OT capacity of the DMC $\{W\}$ is denoted by $C_{\mathrm{OT}}(W)$.

Lemma 1. *For each DMC $\{W\}$ and input distribution P*

$$C_{\mathrm{OT}}(W) \geq C_{\mathrm{OT}}(P, W).$$

Proof. Let R be an achievable OT rate for the source model given by the DMMS with generic RVs X, Y as above. Then (n, k) protocols achieving OT rate R for the source model give rise to OT protocols for the channel model achieving the same OT rate, simply as follows. In the first stage Alice selects i.i.d. repetitions of X as DMC inputs $X_1, \ldots, X_n$, and Bob observes the corresponding outputs $Y_1, \ldots, Y_n$; in this stage the public channel is not used, thus the first $n-1$ public sessions are empty. Upon completing this stage, Alice and Bob have views as their initial views would be in the source model. Then they perform the given source model protocol.

Remark 2. *Lemma 1 may be applied to the DMC $\{W^l\colon \mathcal{X}^l \to \mathcal{Y}^l\}$ defined by*

$$W^l(y_1, \ldots, y_l | x_1, \ldots, x_l) = \prod_{i=1}^{l} W(y_i|x_i),$$

whose OT capacity clearly equals $lC_{\mathrm{OT}}(W)$. This gives

$$C_{\mathrm{OT}}(W) \geq \frac{1}{l} C_{\mathrm{OT}}(P^{(l)}, W^l), \qquad \text{for every distribution } P^{(l)} \text{ on } \mathcal{X}^l.$$

In this paper, for channel models only protocols as in the proof of Lemma 1 will be used, in effect employing the DMC merely to emulate a DMMS (with alphabets $\mathcal{X}$, $\mathcal{Y}$ or $\mathcal{X}^2$, $\mathcal{Y}^2$; we will not use $l > 2$). For DMCs with the property that in Lemma 1 some input distribution P attains the equality, or at least that $\frac{1}{l} C_{\mathrm{OT}}(P^{(l)}, W^l) \to C_{\mathrm{OT}}(W)$ for suitable distributions $P^{(l)}$ on $\mathcal{X}^l$, the OT capacity can be attained via source model emulating protocols. It remains open whether every DMC has that property.

Let us briefly mention also a more general concept of OT, where Alice is initially given m strings $K_0, \ldots, K_{m-1}$, and Bob may be interested in any subset $\{K_j, j \in J\}$ of those, with index set J in a specified family $\mathcal{J}$ of subsets of $\{0, \ldots, m-1\}$. Formally, Bob is given a RV Z with $|\mathcal{J}|$ possible values, and an OT protocol is supposed to let him learn all K_j with index j in the set $J \in \mathcal{J}$ specified by the value of Z, while keeping him ignorant of the remaining strings. At the same time, Alice has to remain ignorant of Z, i.e., of which strings of her has Bob chosen to learn. This general OT concept will not be addressed but its simplest special case $m = 1$, $\mathcal{J} = \{\{0\}, \varnothing\}$ will. In that case, referred to below as *one-string OT*, Alice is given only one string K_0, and Bob one bit Z. He is supposed to learn K_0 if $Z = 0$ and remain ignorant of K_0 if $Z = 1$, while Alice should remain ignorant of Z.

The concepts of (n, k) protocol and OT capacity immediately extend to the above general version of OT, and in particular to one-string OT. For the latter case, the analogues of the conditions (1)-(3) in Definition 1 are

$$\Pr\{\hat{K}_0 \neq K_0 | Z = 0\} \to 0 \tag{5}$$

$$I(K_0 M X^n \mathbf{F} \wedge Z) \to 0 \tag{6}$$

$$I(N Y^n \mathbf{F} \wedge K_0 | Z = 1) \to 0. \tag{7}$$

3 Statement of Results

Theorem 1. *The OT capacity of a DMMS with generic RVs X, Y or of a DMC $\{W\}$ is bounded above by*

$$\min\left[I(X \wedge Y), H(X|Y)\right],$$

respectively by the maximum of this expression for RVs X, Y connected by the channel, i.e., satisfying $P_{Y|X} = W$. The same upper bounds hold for one-string OT, as well.

A first example that the upper bound in Theorem 1 may be achievable is provided by the *binary erasure channel* (BEC). A BEC with erasure probability $0 < p < 1$ is a DMC with input alphabet $\{0,1\}$, output alphabet $\{0,1,2\}$, and $W(0|0) = W(1|1) = 1-p$, $W(2|0) = W(2|1) = p$. It has been shown in [7] that a BEC with erasure probability $1/2$ has OT capacity $1/2$.

Theorem 2. *If $\{W\}$ is a BEC with erasure probability p, and P is any distribution on $\{0,1\}$, then*

$$C_{\mathrm{OT}}(W) = \min(p, 1-p), \quad C_{\mathrm{OT}}(P, W) = H(P)\min(p, 1-p).$$

The next theorem addresses a larger class of channels than BECs.

Definition 2. *A generalized erasure channel (GEC) is a DMC $\{W : \mathcal{X} \to \mathcal{Y}\}$ such that for some nonempty $\mathcal{Y}_1 \subset \mathcal{Y}$ the probabilities $W(y|x), y \in \mathcal{Y}_1$ do not depend on $x \in \mathcal{X}$.*

As outputs $y \in \mathcal{Y}_1$ carry no information about the input, they are interpreted as erasures. The BEC is a special case with $\mathcal{X} = \{0,1\}$, $\mathcal{Y} = \{0,1,2\}$, $\mathcal{Y}_1 = \{2\}$. The *erasure probabability* of a GEC is $p = \sum_{y \in \mathcal{Y}_1} W(y|x)$ which does not depend on $x \in \mathcal{X}$.

Theorem 3. *If $\{W : \mathcal{X} \to \mathcal{Y}\}$ is a GEC with erasure probability p, and P is any distribution on $\mathcal{X}$, then*

$$C_{\mathrm{OT}}(W) = C(W), \quad C_{\mathrm{OT}}(P, W) = I(P, W) \qquad \textit{if } p \geq 1/2$$
$$C_{\mathrm{OT}}(W) \geq \frac{p}{1-p} C(W), \quad C_{\mathrm{OT}}(P, W) \geq \frac{p}{1-p} I(P, W) \quad \textit{if } p < 1/2.$$

Here $C(W) = \max_P I(P, W)$ is the *Shannon capacity* of the DMC $\{W\}$, and $I(P, W)$ denotes the mutual information of RVs X, Y with joint distribution given by $P(x)W(y|x)$.

The proof technique of the lower bounds in Theorem 3 works beyond the class of GECs. It provides lower bounds to OT capacity for the larger class of DMCs that can be represented as a mixture of two channels with identical input alphabet $\mathcal{X}$ and disjoint output alphabets $\mathcal{Y}_0$ and $\mathcal{Y}_1$, namely as

$$W(y|x) = \begin{cases} (1-p)W_0(y|x), & x \in \mathcal{X}, y \in \mathcal{Y}_0 \\ pW_1(y|x), & x \in \mathcal{X}, y \in \mathcal{Y}_1. \end{cases} \tag{8}$$

Note that if the matrix W_1 has identical rows then (8) gives a GEC.

The following result is not contained in [2]. The auxiliary RV U in its second assertion is unrelated to U appearing in the Preliminaries.

Theorem 4. *For a DMC $\{W\}$ of form (8) and any distribution P on $\mathcal{X}$*

$$C_{\mathrm{OT}}(P,W) \geq [I(P,W_0) - I(P,W_1)] \min(p, 1-p).$$

A possibly better bound is

$$C_{\mathrm{OT}}(P,W) \geq \left[I(U \wedge Y^{(0)}) - I(U \wedge Y^{(1)})\right] \min(p, 1-p),$$

where U is any RV and $X, Y^{(0)}, Y^{(1)}$ are RVs with $P_{XY^{(j)}}(x,y) = P(x)W_j(y|x)$, $j = 0, 1$, such that

$$U \to X \to (Y^{(0)}, Y^{(1)}) \text{ is a Markov chain.} \tag{9}$$

Consequently, $C_{\mathrm{OT}}(W)$ is bounded below by $\min(p, 1-p)$ times the secrecy capacity of the wiretap channel with component channels W_0, W_1.

The model called wiretap channel with component channels W_0, W_1 has been introduced by Wyner [17] assuming a special relationship between W_0, W_1 and by Csiszár and Körner [5] for any W_0, W_1 with the same input alphabet. In this model, Alice selects the inputs, Bob observes the W_0-outputs and an eavesdropper Eve the W_1-outputs. The *secrecy capacity* is the supremum of rates at which Alice can reliably send Bob messages in such a way that Eve remains ignorant about them. According to [5], it equals the maximum of $I(U \wedge Y^{(0)}) - I(U \wedge Y^{(1)})$ for RV's satisfying (9), with X and $Y^{(j)}$ connected by the channel W_j, $j = 0, 1$. Hence the second assertion of Theorem 4 implies the last one by Lemma 1.

Remark 3. *In (8) the indices 0 and 1 can be exchanged if simultaneously p and $1-p$ are exchanged. Hence the bounds in Theorem 4 hold also with the reversed order of W_0 and W_1.*

Theorems 1 and 3 admit to give a necessary and sufficient condition for the positivity of OT capacity.

Theorem 5. *A DMC $\{W : \mathcal{X} \to \mathcal{Y}\}$ has positive OT capacity iff there exist x', x'' in $\mathcal{X}$ such that the corresponding rows of the matrix W are not identical, and $W(y|x')W(y|x'') > 0$ for some $y \in \mathcal{Y}$. Further, $C_{\mathrm{OT}}(P,W) > 0$ for an input distribution P iff x', x'' as above exist with $P(x')P(x'') > 0$.*

Remark 4. *A similar result appears in [11,12], but there a stronger condition is claimed necessary and sufficient for $C_{\mathrm{OT}}(W) > 0$; it can be equivalently stated by adding to the requirements on x' and x'' in Theorem 5 that neither of the corresponding rows of W is a convex combination of other rows. That additional requirement, however, is not necessary in the "honest but curious" framework, see Example 3 for a counterexample and additional discussion. Nevertheless, the proof of Theorem 5 uses an idea as [11,12], simplified by the availability of Theorem 3.*

4 Proofs

Proof of Theorem 1. It suffices to prove the claimed bounds for one-string OT capacity. Indeed, (n,k) protocols satisfying (1)-(3) trivially give rise to (n,k) protocols for one-string OT satisfying (5)-(7), just letting the pair of RVs K_1, M in the former protocols play the role of M in the latter. Below, attention is restricted to channel models since the proof for source models is similar but simpler. In the proof, instead of condition (7) only its relaxation

$$I(NY^n\mathbf{F} \wedge K_0|Z=1) = o(n) \tag{10}$$

will be used, see Remark 1 after Definition 1.

Now, given a DMC $\{W : \mathcal{X} \to \mathcal{Y}\}$, consider (n,k) protocols for one-string OT that satisfy (5), (6) and (10). By Lemma 3 in Appendix A, the condition (6) implies

$$H(K_0|X^n\mathbf{F}, Z=0) - H(K_0|X^n\mathbf{F}, Z=1) = o(n) \tag{11}$$

as well as

$$H(K_0|\mathbf{F}, Z=0) - H(K_0|\mathbf{F}, Z=1) = o(n). \tag{12}$$

Since $H(K_0|Z=0) = H(K_0|Z=1) = k$, equation (12) is equivalent to

$$I(K_0 \wedge \mathbf{F}|Z=0) = I(K_0 \wedge \mathbf{F}|Z=1) + o(n)$$

and hence (10) implies

$$I(K_0 \wedge \mathbf{F}|Z=0) = o(n). \tag{13}$$

The conditions (5),(13) are similar to those defining a secret key for Alice and Bob, with (weak sense) security from an eavesdropper who observes their public communication $\mathbf{F}$. If (5),(13) held without the conditioning on $Z=0$ then K_0 would be, by definition, such a secret key, see [10],[1]. Then by these references

$$k = H(K_0) \le \sum_{t=1}^{n} I(X_t \wedge Y_t) + o(n) \tag{14}$$

would hold. Actually, (14) holds also in the present case. Indeed, the conditioning on $Z=0$ affects the mentioned result only by changing the terms $I(X_t \wedge Y_t)$ to $I(X_t \wedge Y_t|Z=0)$. This has a negligible effect if n is large, because (6) implies that $\max_t I(X_t \wedge Z) \to 0$, and hence the conditional distribution of X_t on the condition $Z=0$ differs negligibly from the unconditional one, uniformly in t.

To derive another bound on k, we use that K_0 and NY^nZ are conditionally independent given $X^n\mathbf{F}$. For a formal proof of this, see Lemma 6 in Appendix B. It follows using (5) and Fano's inequality that

$$H(K_0|X^n\mathbf{F}, Z=0) \le H(K_0|NY^n\mathbf{F}, Z=0) \le H(K_0|\hat{K}_0, Z=0) + o(n),$$

whence by (11) also

$$H(K_0|X^n\mathbf{F}, Z=1) = o(n). \tag{15}$$

Using (10) and (15) we obtain

$$\begin{aligned}k &= H(K_0|Z=1) = H(K_0|NY^n\mathbf{F},\, Z=1) + o(n)\\ &\le H(K_0X^n|NY^n\mathbf{F},\, Z=1) + o(n) = H(X^n|NY^n\mathbf{F},\, Z=1) + o(n)\\ &\le H(X^n|Y^n,\, Z=1) + o(n) \le \sum_{t=1}^{n} H(X_t|Y_t,\, Z=1) + o(n).\end{aligned}$$

In the last sum, the conditioning on $Z = 1$ may be omitted with negligible effect as before. Thus we have shown that

$$k \le \sum_{t=1}^{n} H(X_t|Y_t) + o(n). \tag{16}$$

Finally, the sums in (14) and (16) may be written as $nI(X_T \wedge Y_T|T)$ and $nH(X_T|Y_T,\, T)$, respectively, where T is a RV uniformly distributed on $\{1, \dots, n\}$ and independent of (X^n, Y^n). The RVs X_T and Y_T are connected by the channel W and satisfy

$$I(X_T \wedge Y_T|T) \le I(X_T \wedge Y_T), \quad H(X_T|Y_T, T) \le H(X_T|Y_T).$$

The proof of Theorem 1 is complete.

Proof of Theorem 2. If X and Y are RVs connected by a BEC with erasure probability p then

$$H(X|Y=0) = H(X|Y=1) = 0, \quad H(X|Y=2) = H(X),$$

hence

$$H(X|Y) = pH(X), \quad I(X \wedge Y) = H(X) - H(X|Y) = (1-p)H(X).$$

It follows by Theorem 1 that

$$C_{\mathrm{OT}}(P, W) \le H(P)\min(p, 1-p), \quad C_{\mathrm{OT}}(W) \le \min(p, 1-p).$$

It remains to show that these upper bounds are achievable.

By Lemma 1, it suffices to show that each $R < H(X)\min(p, 1-p)$ is an achievable OT rate for the source model defined by a DMMS with generic RVs X, Y as above. To this end, an OT protocol will be described for this source model. It will involve only two messages sent over the public noiseless channel, the first by Bob and the second by Alice; formally, Alice's message F_1 and Bob's message F_4 will be empty.

Upon observing $Y^n = (Y_1, \dots, Y_n)$, Bob first determines two subsets G and B of $\{1, \dots, n\}$, called the good and bad sets, both of size about $n\min(p, 1-p)$. If $p \ge 1/2$ then Bob takes for G the set of all indices i with $Y_i \ne 2$, and he assigns the indices i with $Y_i = 2$ to B with probability $(1-p)/p$, independently of each other. If $p < 1/2$ then Bob takes for B the set of all indices with $Y_i = 2$, and he assigns the indices with $Y_i \neq 2$ to G with probability $p/(1-p)$, independently of each other. Formally, in order to comply with the description of protocols in Section 2, Bob may be assumed to use a RV N generated at the outset, when he has to assign indices i to B or to G in a randomized manner. E.g., when $p > 1/2$, this N may consist of n independent bits, each equal to 0 with probability $(1-p)/p$, and an index i with $Y_i = 2$ is assigned to B if the i'th bit of N is 0.

Bob's next action is to send Alice a message telling her the sets G and B but not which is which: he lets her learn two sets S_0, S_1 where $S_0 = G, S_1 = B$ if $Z = 0$, and $S_0 = B, S_1 = G$ if $Z = 1$. Note that the pair of random sets G, B is independent of X^n, the events $\{i \in G\}$, $i = 1, \dots, n$ are independent and have probability $\min(p, 1-p)$, and the same holds for the events $\{i \in B\}$. This implies, in particular, that Bob's message gives Alice no information about Z.

Consider first the case when X is uniformly distributed on $\{0,1\}$. Suppose Alice's strings K_0, K_1 are of length[1] $k = nr$ where $r < \min(p, 1-p)$ is arbitrarily fixed. If $|G| \geq nr$ and $|B| \geq nr$, which holds with probability going to 1 exponentially fast as $n \to \infty$, let S_0' and S_1' denote the subsets of S_0 resp. S_1 consisting of their first nr elements. Then Alice encrypts K_0 and K_1 with the "keys" $\{X_i, i \in S_0'\}$ resp. $\{X_i, i \in S_1'\}$, and sends Bob the "cryptograms" $K_j + \{X_i, i \in S_j'\}$, $j = 0, 1$, where $+$ means componentwise addition mod 2. If $|G| < nr$ or $|B| < nr$ then she sends nothing. Except for the latter case of negligible probability, Bob can decrypt K_Z since $S_Z = G$ implies that he knows $\{X_i, i \in S_Z'\} = \{Y_i, i \in S_Z'\}$. On the other hand, Bob remains fully ignorant of $K_{\overline{Z}}$, since the "key" $\{X_i, i \in S_{\overline{Z}}'\}$ is uniformly distributed on $\{0,1\}^{nr}$ and $S_{\overline{Z}} = B$ implies that Bob has 0 information about it. Note that this already suffices for the proof of $C_{\mathrm{OT}}(W) = \min(p, 1-p)$.

If X is not uniformly distributed on $\{0,1\}$, the strings $\{X_i, i \in S_j'\}$, $j = 0, 1$ are not directly suitable as encryption keys, they have to be transformed to binary strings of length $k < rn$ whose distribution is nearly uniform on $\{0,1\}^k$. It is well-known that given any $\delta > 0$, in the case of large n there exists a mapping $\kappa : \{0,1\}^n \to \{0,1\}^k$ with $k = n(H(X) - \delta)$ such that $k - H(\kappa(X^n))$ is exponentially small (in later proofs we will need a stronger result, Proposition 1). Applying this replacing n by rn, there exists a mapping $\kappa : \{0,1\}^{nr} \to \{0,1\}^k$ with $k = nr(H(X) - \delta)$ such that $\kappa_j = \kappa(\{X_i, i \in S_j'\})$, $j = 0, 1$ are nearly uniformly distributed, in the sense that their entropy differs from k only by an exponentially small amount.

To complete the proof, assume Alice's strings K_0, K_1 are of length $k = nr(H(X) - \delta)$. She encrypts them by the keys κ_0, κ_1, and sends Bob the strings $K_j + \kappa_j$, $j = 0, 1$. Again, Bob can decipher K_Z, and he remains ignorant of $K_{\overline{Z}}$ in the sense that he has an exponentially small amount of information about $K_{\overline{Z}}$, see, e.g. [6, Proposition 17.1].

Remark 5. *The protocol in the above proof achieves more than required in Definition 1: Eve's amount of information about Z is not only asymptotically but*

[1] Here and later on, if a specified length of sequences is not an integer, the next integer is meant.

exactly 0*, and in the case when* X *is uniformly distributed on* $\{0,1\}$*, Bob's information about* $K_{\overline{Z}}$ *is also* 0*. The latter need not hold for the described protocol when* X *is not uniformly distributed, but can be achieved also in that case by a slightly modified protocol. As* $k-H(\kappa_j)$ *equals the* I*-divergence of the distribution of* κ_j *from the uniform distribution on* $\{0,1\}^k$*, its exponential smallness implies that of the variation distance of these distributions. Hence Alice can generate RVs* $\overline{\kappa}_j$ *uniformly distributed on* $\{0,1\}^k$ *with* $\Pr\{\kappa_j \neq \overline{\kappa}_j\}$ *exponentially small,* $j=0,1$*, and send Bob* $K_j+\overline{\kappa}_j$ *rather than* $K_j+\kappa_j$*,* $j=0,1$*. Then Bob can still reconstruct* K_Z *with exponentially small probability of error (an error occurring when* $\kappa_Z \neq \overline{\kappa}_Z$*), and he has* 0 *information about* $K_{\overline{Z}}$*.*

Proof of Theorem 3. Let $\{W\}$ be a GEC. Then (8) holds with $\mathcal{Y}_0=\mathcal{Y}\setminus\mathcal{Y}_1$, $W_0(y|x)=\frac{1}{1-p}W(y|x)$ $(y\in\mathcal{Y}_0)$ and with $W_1(y|x)$ $(y\in\mathcal{Y}_1)$ not depending on $x\in\mathcal{X}$. Hence by Lemma 7 in Appendix B,

$$I(P,W)=(1-p)I(P,W_0)\ . \tag{17}$$

On account of Theorem 1, Lemma 1 and (17), it suffices to prove that if $\{W\}$ is a GEC then $C_{\mathrm{OT}}(P,W)\geq I(P,W_0)\min(p,1-p)$. This is a special case of the first assertion of Theorem 4, and the proof of that more general result is not really more difficult. Below we proceed directly with the latter.

The following basic proposition about generating a secret key will be used.

Proposition 1. *([10,1]) Let* (X_i,Y_i) $i=1,\ldots,n$ *and* $(\tilde{X}_i,T_i)$ $i=1,\ldots,n$ *be i.i.d. repetitions of pairs of RVs* (X,Y) *resp.* (X,T)*. For any* $\delta>0$ *and* $n\to\infty$ *there exist functions* κ *and* f *on* $\mathcal{X}^n$*, where the range of* κ *is* $\{0,1\}^k$ *with*

$$k=n(I(X\wedge Y)-I(X\wedge T)-\delta) \tag{18}$$

such that $\kappa(X^n)$ *is recoverable from* $f(X^n)$ *and* Y^n *with exponentially small probability of error, and*

$$k-H(\kappa(\tilde{X}^n|f(\tilde{X}^n),T^n)\to 0 \quad \textit{exponentially fast.} \tag{19}$$

Such functions κ *and* f *also exist with*

$$k=n(I(U\wedge Y)-I(U\wedge T)-\delta)\ , \tag{20}$$

for any RV U *satisfying the Markov condition* $U\to X\to(Y,T)$*.*

Remark 6. *In the usual setting, Alice and Bob have to generate a secret key assuming Alice observes* X^n*, Bob observes* Y^n*, only Alice is permitted to send Bob a public message, and the key has to be concealed from Eve who observes Alice's message and has side information* T^n*. This setting is formally less general than that in Proposition 1, for it regards the sequences* X^n *and* $\tilde{X}^n$ *identical rather than only identically distributed. Mathematically, however, this makes no difference, and the stated form of Proposition 1 is more convenient for the* *purpose of this paper. Note that originally weak secrecy had been addressed, i.e., the difference in (19) was shown to be* $o(n)$ *rather than to approach* 0 *(in [10] for (18) and in [1] also for (20); in [1] the largest key rate* k/n *asymptotically achievable with unidirectional public communication is also determined). Still, the "strong" version with (19) is also well-known, see, e.g. [6, Theorem 17.21].*

Proof of Theorem 4. Let $\{W : \mathcal{X} \to \mathcal{Y}\}$ with $\mathcal{Y} = \mathcal{Y}_0 \cup \mathcal{Y}_1$ be a DMC of form (8), and consider a DMMS with generic RVs X, Y whose joint distribution is given by $P(x)W(y|x)$, $x \in \mathcal{X}, y \in \mathcal{Y}$. To prove the claimed bounds on $C_{\mathrm{OT}}(P, W)$, protocols for the corresponding source model similar to those in the proof of Theorem 2 will be used.

Upon observing $Y^n = (Y_1, \ldots, Y_n)$, Bob first determines a "good set" G and a "bad set" B as in the proof of Theorem 2, with the only modification that the criteria $Y_i \neq 2$ resp. $Y_i = 2$ are replaced by $Y_i \in \mathcal{Y}_0$ resp. $Y_i \in \mathcal{Y}_1$. As there, the pair of random sets G, B is independent of $X^n = (X_1, \ldots, X_n)$, the events $\{i \in G\}$, $i = 1, \ldots, n$ have probability $\min(p, 1-p)$ and are independent of each other and X^n, and the same holds also for the events $\{i \in B\}$. Then Bob sends Alice a message telling her two sets S_0, S_1 where $S_0 = G, S_1 = B$ if $Z = 0$, and $S_0 = B, S_1 = G$ if $Z = 1$. Thereby Alice receives 0 information about Z.

The i.i.d. pairs (X_i, Y_i) are conditionally independent conditioned on the value of Z and the sets S_0, S_1, moreover, those with $i \in S_0$ as well as those with $i \in S_1$ are conditionally i.i.d. If $i \in S_0$ resp. $i \in S_1$, the conditional distribution of (X_i, Y_i) is given by $P(x)W_0(y|x)$ resp. $P(x)W_1(y|x)$ if $Z = 0$, and by $P(x)W_1(y|x)$ resp. $P(x)W_0(y|x)$ if $Z = 1$. To verify this, suppose first that $Z = 0$. Then $i \in S_0$ means $i \in G$, which implies $Y_i \in \mathcal{Y}_0$, and for $x \in \mathcal{X}, y \in \mathcal{Y}_0$ the conditional probability $\Pr\{X_i = x, Y_i = y | S_0, S_1, Z = 0\} = \Pr\{X_i = x, Y_i = y | G, B\}$ is equal to

$$\Pr\{X_i = x, Y_i = y | i \in G\} = \frac{\Pr\{X_i = x, Y_i = y, i \in G\}}{\Pr\{i \in G\}} = P(x)W_0(y|x) \ ;$$

here the second equality holds because, by the construction of G, the probability in the numerator is equal to $P(x)W(y|x)$ if $p \geq 1/2$ and to $P(x)W(y|x)\frac{p}{1-p}$ if $p < 1/2$, where $W(y|x) = (1-p)W_0(y|x)$ by (8), while the probability in the denominator equals $\min(p, 1-p)$. For $i \in S_1$ the calculation is similar. In the case $Z = 1$ the roles of S_0 and S_1 are simply reversed.

The proof of the first assertion of Theorem 4 will be completed by showing that, for any $r < \min(p, 1-p)$, if Alice's strings K_0, K_1 have length

$$k = rn(I(P, W_0) - I(P, W_1) - \delta)$$

then she, knowing S_0, S_1, can send Bob a message that enables him to recover K_Z while keeping him ignorant of $K_{\overline{Z}}$.

Apply the first assertion of Proposition 1 with rn in the role of n, taking $\{P(x)W_0(y|x), x \in \mathcal{X}, y \in \mathcal{Y}_0\}$ resp. $\{P(x)W_1(y|x), x \in \mathcal{X}, y \in \mathcal{Y}_1\}$ for the joint distribution of X, Y resp. X, T. Let f and κ denote the corresponding functions on $\mathcal{X}^{rn}$ where the range of κ is $\{0,1\}^k$ with the above k, see (18). Supposing

$$|S_0| \geq rn, \quad |S_1| \geq rn \ , \tag{21}$$

denote by S'_0 and S'_1 the sets of the first rn elements of S_0 resp. S_1. Let Alice compute $f_j = f(\{X_i, i \in S'_j\})$ and $\kappa_j = \kappa(\{X_i, i \in S'_j\})$, $j = 0, 1$, and send Bob a message consisting of f_0, f_1 and the "cryptograms" $K_0 + \kappa_0, K_1 + \kappa_1$; if (21) does not hold then she sends nothing.

Consider first the case $Z = 0$. Then, conditioned on Z and S_0, S_1 satisfying (21), the pairs (X_i, Y_i), $i \in S'_0$ are conditionally i.i.d. with distribution $P(x)W_0(y|x)$. Hence, due to the choice of the mappings f and κ, Bob can recover κ_0 from f_0 and $\{X_i, i \in S'_0\}$ with exponentially small (conditional) probability of error, enabling him to recover K_0. As this always holds when (21) does, the probability of error in recovering K_0 conditioned only on $Z = 0$ is also exponentially small. Further, the pairs (X_i, Y_i), $i \in S'_1$ are conditionally i.i.d. with distribution $P(x)W_1(y|x)$. Hence the choice of κ and f implies that f_1 and $\{Y_i, i \in S'_1\}$ give a negligible amount of information about κ_1; in turn, since κ_1 is nearly uniformly distributed, Bob's amount of information about K_1 provided by $f_1, \{Y_i, i \in S'_1\}$ and $K_1 + \kappa_1$ is also negligible: $I(K_1 \wedge f_1, K_1 + \kappa_1, \{Y_i, i \in S'_1\}|S_0, S_1, Z = 0)$ is exponentially small. To formally verify that the last conditional mutual information coincides with that in the first condition in (4), assuming the RV N has been generated and used by Bob as in the proof of Theorem 2, note that the total communication is now $\mathbf{F} = (S_0, S_1, f_0, K_0 + \kappa_0, f_1, K_1 + \kappa_1)$, and K_1 is independent of (N, S_0, S_1, Z). Hence

$$I(NY^n\mathbf{F} \wedge K_1|Z = 0) = I(Y^n, f_0, K_0 + \kappa_0, f_1, K_1 + \kappa_1 \wedge K_1|N, S_0, S_1, Z = 0) \ .$$

Here, N in the condition may be omitted. It remains to show that

$$I(\{Y_i, i \notin S'_0\}, f_0, K_0 + \kappa_0 \wedge K_1|S_0, S_1, f_1, K_1 + \kappa_1, Z = 0) = 0 \ .$$

This follows because $(X_i, Y_i), i = 1, \ldots, n$ are conditionally independent given $S_0, S_1, Z = 0$, and f_j and κ_j are functions of K_j and $\{(X_i, Y_i), i \in S'_j\}$, $j = 0, 1$.

In the case $Z = 1$ it follows similarly that Bob can recover K_1 and he remains ignorant of K_0. This completes the proof of the first assertion of Theorem 4.

The second assertion follows in the same way, applying this time the second assertion of Proposition 1. The third assertion follows from the second one as noted in the passage following Theorem 4.

Remark 7. *Another suitable protocol is obtained by modifying the choice of the sets G and B as follows. According as $p \geq 1/2$ or $p < 1/2$, let G resp. B contain all indices i with Y_i in $\mathcal{Y}_0$ resp. in $\mathcal{Y}_1$ as before, and let the other indices i be assigned to G or B with probabilities $(\pi, 1 - \pi)$. Here π is chosen to make sure that $\Pr\{i \in G\} = \Pr\{i \in B\} = 1/2$, thus π equals $1 - 1/2p$ if $p \geq 1/2$ and $1/2(1 - p)$ if $p < 1/2$. Consider first the case $p \geq 1/2$. Then, by similar calculation as in the proof of Theorem 4,*

$$\Pr\{X_i = x, Y_i = y|i \in G\} = \begin{cases} 2(1-p)P(x)W_0(y|x), & x \in \mathcal{X}, y \in \mathcal{Y}_0 \\ (2p-1)P(x)W_1(y|x), & x \in \mathcal{X}, y \in \mathcal{Y}_1 \end{cases} ,$$

$$\Pr\{X_i = x, Y_i = y|i \in B\} = P(x)W_1(y|x), \ x \in \mathcal{X}, y \in \mathcal{Y}_1 \ .$$

It follows, in turn, that the conditional mutual information $I(X_i \wedge Y_i|G,B)$ is equal to $2(1-p)I(P,W_0)+(2p-1)I(P,W_1)$ if $i \in G$ (using Lemma 7) and to $I(P,W_1)$ if $i \in B$. This implies via Proposition 1, again as in the proof of Theorem 4, that with this modified protocol one can achieve OT rate

$$1/2\,[2(1-p)I(P,W_0)+(2p-1)I(P,W_1)-I(P,W_1)]\ ,$$

the same as with the original protocol. In the case $p<1/2$ the situation is similar. It follows similarly that OT rates in the second assertion of Theorem 4 can also be achieved with protocols in which G and B are selected as above.

To the proof of Theorem 5 a simple fact is sent forward.

Lemma 2. *If a DMC $\{W'\}$ is obtained from $\{W : \mathcal{X} \to \mathcal{Y}\}$ by restricting the input alphabet $\mathcal{X}$ to a subset $\mathcal{X}'$ then $C_{\mathrm{OT}}(W') \le C_{\mathrm{OT}}(W)$.*

The proof is obvious but depends on the "honest but curious" assumption. Were Alice allowed to deviate from the agreed-upon protocol, a larger input alphabet would give her more room for deviations undetectable for Bob and letting her gain information about Bob's bit Z; this might decrease OT capacity.

Proof of Theorem 5. (i) Necessity. Given a DMC $\{W : \mathcal{X} \to \mathcal{Y}\}$, let $\mathcal{X}'$ be a maximal subset of $\mathcal{X}$ such that the rows of the matrix W corresponding to input symbols $x' \in \mathcal{X}'$ are all distinct; let W' be the matrix that has these distinct rows. Clearly $C_{\mathrm{OT}}(W)=C_{\mathrm{OT}}(W')$. If $C_{\mathrm{OT}}(W)>0$ then $C_{\mathrm{OT}}(W')>0$ implies by Theorem 1 that the outputs of W' do not unambiguously determine the inputs. In other words, for some $y \in \mathcal{Y}$ there exist x' and x'' in $\mathcal{X}'$ such that $W(y|x')W(y|x'')>0$; this proves necessity for channel models. For source models the proof is similar, this time using that $C_{\mathrm{OT}}(P,W)=C_{\mathrm{OT}}(P',W')$ where $P'(x')$, $x' \in \mathcal{X}'$ equals the sum of $P(x)$ for all $x \in \mathcal{X}$ such that the rows of W corresponding to x and x' are equal.

(ii) Sufficiency. Let $\{W\}$ be a DMC satisfying the conditions in Theorem 5. Consider an auxiliary DMC $\{\widetilde{W}\}$, restricting the input alphabet $\mathcal{X}\times\mathcal{X}$ of W^2 (see Remark 2) to the pairs (x',x''), (x'',x'), where x',x'' as in Theorem 5 are fixed. Formally, $\{\widetilde{W} : ((x',x''),(x'',x')) \to \mathcal{Y}\times\mathcal{Y}\}$ is defined by

$$\widetilde{W}(y_1,y_2|x',x'')=W(y_1|x')W(y_2|x''),\ \widetilde{W}(y_1,y_2|x'',x')=W(y_1|x'')W(y_2|x'). \tag{22}$$

This auxiliary DMC is a GEC, the role of $\mathcal{Y}_1$ in Definition 2 being played by the subset $\{(y,y) : y \in \mathcal{Y}\}$ of $\mathcal{Y}\times\mathcal{Y}$; hence Theorem 3 implies $C_{\mathrm{OT}}(\widetilde{W})>0$. On account of Lemma 2, this proves the positivity of $C_{\mathrm{OT}}(W)=\frac{1}{2}C_{\mathrm{OT}}(W^2)$.

Consider next a source model defined by a DMMS with generic RVs X,Y whose joint distribution $P_{XY}(x,y)=P(x)W(y|x)$ satisfies the condition in Theorem 5. Fixing x',x'' as there, for $2n$ i.i.d. repetitions of X, viz. $X^{2n}=(X_1,\dots,X_{2n})$ let J denote the set of indices $i \in \{1,\dots,n\}$ for which (X_{2i-1},X_{2i}) equals either (x',x'') or (x'',x'). The tuples $\{(X_{2i-1},X_{2i}),(Y_{2i-1},Y_{2i}), i \in J\}$ are conditionally i.i.d. given J, their (conditional) distribution is equal to $P_{\tilde{X}\tilde{Y}}$

where $P_{\tilde{X}}$ is the uniform distribution on $\{(x', x''), (x'', x')\}$ and $P_{\tilde{Y}|\tilde{X}}$ equals $\widetilde{W}$ in (22). Consider an auxiliary DMMS with generic RVs $\tilde{X}, \tilde{Y}$ as above. Since $\Pr\{i \in J\} = 2P(x')P(x'')$, the size of J exceeds $\ell = nP(x')P(x'')$ with probability approaching 1 exponentially fast as $n \to \infty$. It follows that each (ℓ, k) protocol for the auxiliary DMMS gives rise to a $(2n, k)$ protocol for the original one: Alice tells Bob the set J in her first message, then Alice and Bob perform the given (ℓ, k) protocol using only the first $\ell = nP(x')P(x'')$ tuples $(X_{2i-1}, X_{2i}), (Y_{2i-1}, Y_{2i})$ with $i \in J$. Since the auxiliary DMMS has positive OT capacity by Theorem 3, this completes the proof of Theorem 5.

5 Examples

Example 1 (Binary symmetric channel). *A DMC $\{W : \{0,1\} \to \{0,1\}\}$ is a binary symmetric channel (BSC) with crossover probability $p \neq 1/2$ if $W(1|0) = W(0|1) = p$. To obtain a lower bound to its OT capacity, consider as in the proof of Theorem 5 an auxiliary channel $\{\widetilde{W} : \{(0,1),(1,0)\} \to \{0,1\}^2\}$, see (22) with $x' = 0, x'' = 1$, i.e.,*

$$\widetilde{W}(0,1|\,0,1) = \widetilde{W}(1,0|1,0) = (1-p)^2, \qquad \widetilde{W}(1,0|0,1) = \widetilde{W}(0,1|\,1,0) = p^2,$$
$$\widetilde{W}(0,0|\,0,1) = \widetilde{W}(1,1|0,1) = \widetilde{W}(0,0|1,0) = \widetilde{W}(1,1|1,0) = p(1-p).$$

This $\{\widetilde{W}\}$ is a GEC with erasure probability $\tilde{p} = 2p(1-p) < 1/2$. The role of the set $\mathcal{Y}_1$ in Definition 2 is played by $\{(0,0),(1,1)\}$, and that of $\{W_0\}$ in (8) is played by a channel $\{\widetilde{W}_0\}$ with input and output alphabets equal to $\{(0,1),(1,0)\}$ which is a BSC with crossover probability $\frac{p^2}{1-\tilde{p}} = \frac{p^2}{p^2+(1-p)^2}$.

By Theorem 3 and (17), $C_{\mathrm{OT}}(\widetilde{W}) \geq \frac{\tilde{p}}{1-\tilde{p}} C(\widetilde{W}) = \tilde{p} C(\widetilde{W}_0)$. Finally, since Lemma 2 implies $C_{\mathrm{OT}}(\widetilde{W}) \leq C_{\mathrm{OT}}(W^2) = 2C_{\mathrm{OT}}(W)$, we obtain

$$C_{\mathrm{OT}}(W) \geq \frac{1}{2} C_{\mathrm{OT}}(\widetilde{W}) \geq \frac{1}{2} \tilde{p} C(\widetilde{W}_0) = p(1-p)\left[1 - h\left(\frac{p^2}{p^2+(1-p)^2}\right)\right] .$$

Example 2 (*Z* channel). *A Z channel is a DMC $\{W : \{0,1\} \to \{0,1\}\}$ with $W(0|0) = 1$, $W(0|1) = p$, $W(1|1) = 1-p$. To bound its OT capacity from below, consider an auxiliary channel $\{\widetilde{W} : \{(0,1),(1,0)\} \to \{0,1\}^2\}$ as in Example 1, where this time*

$$\widetilde{W}(0,1|0,1) = \widetilde{W}(1,0|1,0) = 1-p, \quad \widetilde{W}(1,1|0,1) = \widetilde{W}(1,1|1,0) = p,$$

and the other entries of the matrix $\widetilde{W}$ are 0. This auxiliary channel is a BEC with erasure probability p, hence $C_{\mathrm{OT}}(\widetilde{W}) = \min(p, 1-p)$ by Theorem 2. It follows that

$$C_{\mathrm{OT}}(W) \geq \frac{1}{2} C_{\mathrm{OT}}(\widetilde{W}) = \frac{1}{2}\min(p, 1-p) .$$

Example 3. *The DMC* $\{W: \{0,1,2\} \to \{0,1\}\}$ *with*

$$W(0|0) = W(1|1) = 1,\ W(0|2) = p,\ W(1|2) = 1 - p$$

is, in a sense, a reversed BEC. By Lemma 2, its OT capacity is not smaller than that of the Z channel in Example 2, hence $C_{\mathrm{OT}}(W) \geq \frac{1}{2}\min(p, 1-p)$*. Note that while this channel satisfies the condition for* $C_{\mathrm{OT}}(W) > 0$ *in Theorem 5, it fails to satisfy the stronger condition mentioned in Remark 4. Recall that in the proof of Theorem 5 we have used the fact that the OT capacity of a DMC* $\{W\}$ *is not changed by a reduction of the input alphabet that keeps only the distinct rows of* W*. In [11,12] the same is claimed for a further reduction that removes also those rows of* W *which are convex combinations of others, but that claim is valid only in a "malicious" setting. In the "honest but curious" setting the above DMC is a counterexample, it has positive OT capacity but if the input symbol 2 were removed, the OT capacity would become 0.*

The lower bounds to OT capacity in the above examples are smaller than the upper bound in Theorem 1, and the exact value of OT capacity remains an open problem. The next example shows that the upper bound in Theorem 1 may be tight even if the channel is not a GEC. The authors have found this example unaware of the work of Wolf and Wullschleger [16] in which the channel below plays a key role and, in particular, another simple $(1,1)$ protocol for perfect OT of 1 bit is given.

Example 4. *For* $\mathcal{X} = \mathcal{Y} = \{0,1,2,3\}$*, let* $\{W : \mathcal{X} \to \mathcal{Y}\}$ *be a channel with additive noise such that the RVs* X, Y *are connected by it if* $Y = X + N \pmod 4$ *for a RV* N *uniformly distributed on* $\{0,1\}$*, independent of* X*. Theorem 1 gives* $C_{\mathrm{OT}}(W) \leq 1$*, and* $C_{\mathrm{OT}}(P,W) \leq 1$ *if* P *is the uniform distribution on* $\mathcal{X}$*. These upper bounds are tight; indeed, the next* $(1,1)$ *protocol achieves perfect OT for the source model with generic RVs* X, Y *as above and* X *uniformly distributed on* $\mathcal{X}$*. Now, Alice has two bits* K_0*,* K_1*, Bob one bit* Z*, independent of each other and* (X, Y)*, and uniformly distributed; Alice observes* X *and Bob* Y*. First, let Bob tell Alice the parity of* $Y+Z$*, sending her* $\phi = 0$ *or* $\phi = 1$ *according as* $Y+Z$ *is even or odd; this gives Alice no information about* Z*. Then Alice reports Bob the* mod 2 *sums* $K_0 + i_\phi(X)$ *and* $K_1 + i_{1-\phi}(X)$ *where* i_0 *and* i_1 *are the indicator functions of the sets* $\{1,2\}$ *resp.* $\{2,3\}$*. Note that Bob knowing* Y *also knows either the bit* $i_0(X)$ *(if* Y *is even) or* $i_1(X)$ *(if* Y *is odd), but he is fully ignorant of the other bit, in both cases. It follows that Bob can unambiguously determine* K_Z *but remains fully ignorant of* $K_{\overline{Z}}$*.*

6 Discussion

Oblivious transfer has been approached from an information theoretic point of view, addressing OT capacity for (discrete memoryless) source and channel models, concentrating on 1 of 2 strings OT.

A general upper bound to OT capacity has been derived, with essential use of inequalities for information measures, see Appendix A. Let us call attention to an improved bound on the difference of conditional entropies via variation distance (Lemma 5), included for its own sake, though a weaker previous bound would also suffice. A remarkable feature of our upper bound to OT capacity is its validity for one-string OT, as well. It remains open whether this is a coincidence caused by the weakness of our method, or perhaps the rate of one-string OT can never exceed the optimal rate of 1 of 2 strings OT.

Our achievability results (lower bounds to OT capacity) rely on rather simple protocols, still they shed light on relationships of OT and other problems of information theoretic security, such as secret key agreement using public discussion [10,1] and secure transmission over insecure channels [17,5]. It remains open whether the OT capacity of channel models can always be attained via source model emulating protocols, as in those cases when we were able to determine OT capacity. These cases are the binary erasure channels with any erasure probability p, and generalized erasure channels (introduced here) with $p \geq 1/2$. An additional such channel appears in Example 4; it remains open whether this is exceptional, or perhaps a member of another "good" class.

Throughout this paper, only models with "honest but curious" participants are studied. Still, let us briefly address some issues arising in "malicious" settings. In case of a BEC or GEC, with agreed-upon protocol as in the proofs of Theorems 2 and 3, a malicious Alice has no opportunity to learn about Bobs bit Z if he follows the protocol. In Examples 1-2, however, a malicious Alice can well gain information about Z if she deviates from using DMC input pairs $(0, 1)$ and $(1, 0)$ only. In Example 3, the malicious model admits no OT at all, see [11,12]. Indeed, Eve may send instead of DMC input 2 always 0 or 1, with probabilities $(p, 1-p)$; this cheating is undetectable to Bob, and reduces any protocol, in effect, to one for a noiseless channel.

Even the BEC and GEC models are vulnerable to cheating by Bob, who may gain illegitimate information by deviating from the agreed-upon protocol, maliciously selecting the set B. Suppose $p \leq 1/2$, when the protocol requires Bob to take for B the set of indices i with $Y_i = 2$ (or $Y_i \in \mathcal{Y}_1$). He may instead chose B as follows, not modifying the choice of G. If $p \leq 1/3$, he may take B to consist only of indices with $Y_i \neq 2$ (or $Y_i \in \mathcal{Y}_0$), assigning each such index with the same probability $p/(1-p)$ to B as to G. If $1/3 < p < 1/2$, Bob may assign to B all indices with $Y_i \neq 2$ (or $Y_i \in \mathcal{Y}_0$) not assigned to G, and assign to B the remaining indices with probability $(3p-1)/p$. If Bob uses this fake B in giving Alice the sets S_0, S_1, she has no way to detect cheating; in case $p \leq 1/3$ Bob will learn both of Alice's strings, and also when $1/3 < p < 1/2$, he will get nonzero information about $K_{\overline{Z}}$, in addition to learning K_Z.

Note, however, that if $p = 1/2$ then the sets G and B provided by the agreed-upon protocol are complements of each other, thus no deviation in selecting B is possible without one in selecting G. This amounts to a kind of limited protection against Bob's cheating: while a malicious Bob can still gain information about both of Alice's strings, to do so he has to give up his goal of fully learning K_Z (the situation is similar if $p > 1/2$). Recall that protocols as in the proof of Theorem 4 can always be modified to protocols of equal power that use complementary sets G and B, see Remark 7. It is plausible that for a BEC or GEC, modified protocols of this kind provide limited protection as above against Bob's cheating also when $p < 1/2$.

This issue is not pursued here any further, since by a recent result of Pinto et al. [13] the OT capacity of a GEC, determined in this paper, is actually achievable also in the "malicious" model. Another recent work, Ishai et al. [8], regarded Alice's pair of strings (K_0, K_1) as a sequence of k pairs (K_{0i}, K_{1i}), $i = 1, \ldots, k$. Bob selects one component of each pair he wants to learn, this selection is specified by a k-bit string $Z = Z_1, \ldots, Z_k$. Then an (n, k) protocol is supposed to let Bob learn $K_{Z_1 1}, \ldots, K_{Z_k k}$ and keep him ignorant of $K_{\overline{Z}_1 1}, \ldots, K_{\overline{Z}_k k}$, while Eve remains ignorant of Z. Ishai et al. show that this goal is achievable with k/n bounded away from 0, see [8] for details. Finally, the reader's attention is called to recent works that address more general problems via similar techniques, and also contain results relevant for OT capacity, as pointed out by an anonymous referee. See Prabhakaran and Prabhakaran [14] and references there.

References

1. Ahlswede, R., Csiszár, I.: Common randomness in information theory and cryptography, part I. IEEE Trans. Inf. Theory 39, 1121–1132 (1993)
2. Ahlswede, R., Csiszár, I.: On oblivious transfer capacity. In: Proc. ISIT 2007, Nice, pp. 2061–2064 (2007)
3. Alicki, R., Fannes, M.: Continuity of quantum conditional information. J. Phys. A: Math. Gen. 37, L55–L57 (2004)
4. Audenaert, K.M.R.: A sharp Fannes-type inequality for the von Neumann entropy. J. Phys. A. 40, 8127–8136 (2007)
5. Csiszár, I., Körner, J.: Broadcast channels with confidential messages. IEEE Trans. Inf. Theory 24, 339–348 (1978)
6. Csiszár, I., Körner, J.: Information Theory: Coding Theorems for Discrete Memoryless Systems, 2nd edn. Cambridge University Press (2011)
7. Imai, H., Nascimento, A., Morozov, K.: On the oblivious transfer capacity of the erasure channel. In: Proc. ISIT 2006, Seattle, pp. 1428–1431 (2006)
8. Ishai, Y., Kushilevitz, E., Ostrovsky, R., Prabhakaran, M., Sahai, A., Wullschleger, J.: Constant-Rate Oblivious Transfer from Noisy Channels. In: Rogaway, P. (ed.) CRYPTO 2011. LNCS, vol. 6841, pp. 667–684. Springer, Heidelberg (2011)
9. Kilian, J.: Founding cryptography on oblivious transfer. In: Proc. STOC 1988, pp. 20–31 (1988)
10. Maurer, U.: Secret key agreement by public discussion. IEEE Trans. Inf. Theory 39, 733–742 (1993)
11. Nascimento, A., Winter, A.: On the oblivious transfer capacity of noisy correlations. In: Proc. ISIT 2006, Seattle, pp. 1871–1875 (2006)
12. Nascimento, A., Winter, A.: On the oblivious transfer capacity of noisy resources. IEEE Trans. Inf. Theory 54, 2572–2581 (2008)
13. Pinto, A., Dowsley, R., Morozov, K., Nascimento, A.: Achieving oblivious transfer apacity of generalized erasure channels in the malicious model. IEEE Trans. Inf. Theory 57, 5566–5571 (2011)
14. Prabhakaran, V., Prabhakaran, M.: Assisted common information with an application to secure two-party sampling. arXiv:1206.1282v1 [cs.IT] (2012)

15. Winter, A., Nascimento, A.C.A., Imai, H.: Commitment Capacity of Discrete Memoryless Channels. In: Paterson, K.G. (ed.) Cryptography and Coding 2003. LNCS, vol. 2898, pp. 35–51. Springer, Heidelberg (2003)
16. Wolf, S., Wullschleger, J.: Oblivious Transfer Is Symmetric. In: Vaudenay, S. (ed.) EUROCRYPT 2006. LNCS, vol. 4004, pp. 222–232. Springer, Heidelberg (2006)
17. Wyner, A.: The wiretap channel. Bell System Tech. J. 54, 1355–1387 (1975)
18. Zhang, Z.: Estimating mutual information via Kolmogorov distance. IEEE Trans. Inf. Theory 53, 3280–3283 (2007)

Appendix A

Let U, V, Z denote RVs with values in finite sets $\mathcal{U}, \mathcal{V}, \mathcal{Z}$. Suppose $z_1, z_2 \in \mathcal{Z}$ with $\Pr\{Z = z_1\} = p > 0$, $\Pr\{Z = z_2\} = q > 0$.

Lemma 3

$$|H(U|V, Z = z_1) - H(U|V, Z = z_2)| \le 3\sqrt{\frac{(p+q)\ln 2}{2pq} I(UV \wedge Z) \log |\mathcal{U}| + 1} \ .$$

Remark 8. *It will be clear from the proof that the constant term* $+1$ *could be replaced by a term that goes to* 0 *as* $I(UV \wedge Z)$ *does, which may be relevant for some purposes but not here.*

The proof of Lemma 3 will rely on two auxiliary lemmas. The variation distance of probability distributions P and Q on the same finite set, say $\mathcal{S}$, is

$$|P - Q| = \sum_{s \in \mathcal{S}} |P(s) - Q(s)| \ .$$

Lemma 4. *The variation distance of the conditional distributions of* U *on the conditions* $Z = z_1$ *resp.* $Z = z_2$ *is bounded as*

$$\left|P_{U|Z=z_1} - P_{U|Z=z_2}\right| \le \sqrt{\frac{2(p+q)\ln 2}{pq} I(U \wedge Z)} \ .$$

Proof.

$$\begin{aligned} I(U \wedge Z) &= \sum_{z \in \mathcal{Z}} \Pr\{Z = z\} D(P_{U|Z=z} \| P_U) \\ &\ge pD(P_{U|Z=z_1} \| P_U) + qD(P_{U|Z=z_2} \| P_U) \\ &\ge \frac{p|P_{U|Z=z_1} - P_U|^2}{2\ln 2} + \frac{q|P_{U|Z=z_2} - P_U|^2}{2\ln 2} \ ; \end{aligned}$$

the last step is by Pinsker inequality. Since

$$|P_{U|Z=z_1} - P_U| + |P_{U|Z=z_2} - P_U| \ge |P_{U|Z=z_1} - P_{U|Z=z_2}| \ ,$$

it follows by the easily checked inequality $pa^2 + qb^2 \ge \frac{pq}{p+q}(a+b)^2$ that $I(U \wedge Z)$ is further bounded below by

$$\frac{pq}{2(p+q)\ln 2} |P_{U|Z=z_1} - P_{U|Z=z_2}|^2 \ .$$

Lemma 5. *For RVs U_1, U_2 with values in $\mathcal{U}$, and V_1, V_2 with values in $\mathcal{V}$,*

$$|H(U_1|V_1) - H(U_2|V_2)| \leq \left[\frac{1}{2}|P_{U_1V_1} - P_{U_2V_2}| + |P_{V_1} - P_{V_2}|\right] \log|\mathcal{U}|$$
$$+h\left(\frac{1}{2}\min\left[1, |P_{U_1V_1} - P_{U_2V_2}| + |P_{V_1} - P_{V_2}|\right]\right)$$
$$\leq \frac{3}{2}|P_{U_1V_1} - P_{U_2V_2}| \log|\mathcal{U}| + h\left(\min\left[\frac{1}{2}, |P_{U_1V_1} - P_{U_2V_2}|\right]\right),$$

where $h(t) = -t\log t - (1-t)\log(1-t)$, $0 \leq t \leq 1$.

Remark 9. *The main feature of this lemma, for our purposes, is that it does not involve the cardinality of $\mathcal{V}$, only that of $\mathcal{U}$. A previous bound of this kind to the difference of conditional entropies, due to Alicki and Fannes [3], would also suffice for the proof of Theorem 1. but we preferred to sharpen it to obtain Lemma 3 in the stated form.*

Proof. The following bound for the entropy difference of two distributions on $\mathcal{U}$ will be used:

$$|H(P) - H(Q)| \leq \frac{1}{2}|P - Q| \log|\mathcal{U}| + h\left(\frac{1}{2}|P - Q|\right). \tag{23}$$

This sharpening of a more familiar weaker bound is rather recent [4,18]. Let us recall its simple proof: Let X and Y be RVs with $P_X = P$, $P_Y = Q$ such that $\Pr\{X \neq Y\}$ is smallest possible subject to these conditions, thus $\Pr\{X \neq Y\} = \frac{1}{2}|P - Q|$. Then, as $H(P) - H(Q) \leq H(X|Y)$ and $H(Q) - H(P) \leq H(Y|X)$, (23) follows from Fano's inequality.

Now,

$$H(U_1|V_1) - H(U_2|V_2) = \sum_{v\in\mathcal{V}} \left[P_{V_1}(v)H(P_{U_1|V_1=v}) - P_{V_2}(v)H(P_{U_2|V_2=v})\right]$$
$$\leq \sum_{v\in\mathcal{V}} P_{V_1}(v)\left[H(P_{U_1|V_1=v}) - H(P_{U_2|V_2=v})\right]$$
$$+ \sum_{v: P_{V_1}(v) > P_{V_2}(v)} \left[P_{V_1}(v) - P_{V_2}(v)\right] H(P_{U_2|V_2=v}).$$

Bounding the first sum via (23), and the entropies in the second sum by $\log|\mathcal{U}|$, this can be continued as

$$\leq \frac{1}{2}\sum_{v\in\mathcal{V}} P_{V_1}(v)|P_{U_1|V_1=v} - P_{U_2|V_2=v}| \log|\mathcal{U}|$$
$$+ \sum_{v\in\mathcal{V}} P_{V_1}(v) h\left(\frac{1}{2}|P_{U_1|V_1=v} - P_{U_2|V_2=v}|\right) + \frac{1}{2}|P_{V_1} - P_{V_2}| \log|\mathcal{U}|.$$

Let U_3 be an auxiliary RV such that $P_{U_3V_1}(u,v) = P_{V_1}(v)P_{U_2|V_2=v}(u)$. Then

$$\begin{aligned}&\sum_{v\in\mathcal{V}} P_{V_1}(v)|P_{U_1|V_1=v} - P_{U_2|V_2=v}| = |P_{U_1V_1} - P_{U_3V_1}|\\ &\le |P_{U_1V_1} - P_{U_2V_2}| + |P_{U_3V_1} - P_{U_2V_2}|\\ &= |P_{U_1V_1} - P_{U_2V_2}| + |P_{V_1} - P_{V_2}| \le 2|P_{U_1V_1} - P_{U_2V_2}| \ .\end{aligned}$$

Using this, and that the concave function $h(t)$ is increasing in $[0,1/2]$, and noting that the above arguments hold also with the roles of (U_1,V_1) and (U_2,V_2) interchanged, Lemma 5 follows.

Proof of Lemma 3. Apply Lemma 5 to RVs U_1,V_1 with joint distribution $P_{U_1V_1} = P_{UV|Z=z_1}$ and U_2,V_2 with $P_{U_2V_2} = P_{UV|Z=z_2}$, replacing the $h()$ term by its upper bound 1. This gives

$$|H(U|V,Z=z_1) - H(U|V,Z=z_2)| \le \frac{3}{2}|P_{UV|Z=z_1} - P_{UV|Z=z_2}|\log|\mathcal{U}| + 1 \ .$$

Combining this with Lemma 4 completes the proof of Lemma 3.

Appendix B

Lemma 6. *With the notation in the proof of Theorem 1,*

$$I(K_0M \wedge NY^nZ|X^n\mathbf{F}) = 0 \ .$$

Proof. Recall that $\mathbf{F} = F^n$ where F^t denotes the total public communication in the first t sessions. For each $1 \le t \le n$ we have

$$\begin{aligned}I(K_0M \wedge NY^tZ|X^tF^t) &\le I(K_0M \wedge NY^tZ|X^tF^{t-1})\\ = I(K_0M \wedge NY^{t-1}Z|X^tF^{t-1}) &\le I(K_0MX_t \wedge NY^{t-1}Z|X^{t-1}F^{t-1})\\ &= I(K_0M \wedge NY^{t-1}Z|X^{t-1}F^{t-1}) \ .\end{aligned}$$

Here the first inequality holds by [6, Lemma 17.18] (or previous similar results in [10,1]), the next equality holds because $I(K_0M \wedge Y_t|X^tF^{t-1}NY^{t-1}Z) = 0$ due to the conditional independence of Y_t given X_t from the other RVs, and the last equality holds since X_t is a function of K_0, M and F^{t-1}. The lemma follows since $I(K_0M \wedge NY^{t-1}Z|X^{t-1}F^{t-1}) = 0$ trivially holds for $t=1$.

Lemma 7. *For $\{W : \mathcal{X} \to \mathcal{Y}_0 \cup \mathcal{Y}_1\}$ as in (8), the identity*

$$I(P,W) = (1-p)I(p,W_0) + pI(P,W_1)$$

holds for each input distribution P.

Proof. Let X and Y have joint distribution $P(x)W(y|x)$. Define $T = j$ if $Y \in \mathcal{Y}_j$, $j = 0, 1$, then $P_T = (1-p, p)$ and T is independent of X. The claimed identity follows since

$$I(P, W) = I(X \wedge Y) = I(X \wedge YT) = I(X \wedge Y|T),$$

and for each $x \in \mathcal{X}$ and $y \in \mathcal{Y}_j$, $j = 0, 1$,

$$\Pr\{X = x, Y = y|T = j\} = \frac{\Pr\{X = x, Y = y\}}{\Pr\{T = j\}} = P(x)W_j(y|x) \ .$$

Achieving Net Feedback Gain in the Linear-Deterministic Butterfly Network with a Full-Duplex Relay

Anas Chaaban[1], Aydin Sezgin[1], and Daniela Tuninetti[2]

[1] Ruhr-University of Bochum, 44801 Bochum Germany
{anas.chaaban,aydin.sezgin}@rub.de
[2] University of Illinois at Chicago, Chicago, IL 60607 USA
danielat@uic.edu

Dedicated to the memory of Rudolf Ahlswede

Abstract. A symmetric butterfly network (BFN) with a full-duplex relay operating in a bi-directional fashion for feedback is considered. This network is relevant for a variety of wireless networks, including cellular systems dealing with cell-edge users. Upper bounds on the capacity region of the general memoryless BFN with feedback are derived based on cut-set and cooperation arguments and then specialized to the linear deterministic BFN with relay-source feedback. It is shown that the upper bounds are achievable using combinations of the compute-forward strategy and the classical decode-and-forward strategy, thus fully characterizing the capacity region. It is shown that net rate gains are possible in certain parameter regimes.

Keywords: butterfly network, interference relay channel with feedback, capacity, inner bound, outer bound.

1 Introduction

Ahlswede [1] introduced the Interference Channel (IC) as an information theoretic model to capture scenarios where simultaneous transmission of dedicated messages by multiple sources to their respective destination takes place on a shared channel. Such a channel is important, for instance, in cellular networks with cell edge users that suffer from interference caused by base stations in neighboring cells. The phenomenon of interference is not limited to cellular networks and occurs in many other networks such as ad-hoc wireless networks. In the most extreme case, there might be no direct communication link between the transmitting node and its intended receiver due to large obstructing objects. In these cases simply increasing the power level at the transmitting base stations will not resolve the problem. A possible solution is to use dedicated relay stations to enable communication among source-destination pairs. Such a network was studied by Avestimehr *et al.* in [4] under the assumption that the relay

H. Aydinian, F. Cicalese, and C. Deppe (Eds.): Ahlswede Festschrift, LNCS 7777, pp. 167–208, 2013.

nodes are half-duplex; their channel model is known as *the butterfly network (BFN) with a half-duplex relay.*[1] In [4] the authors exploited network coding ideas in order to design transmission strategies that were shown to be optimal for the linear deterministic approximation of the Gaussian noise BFN at high SNR[2] and to achieve capacity to within 1.95 bits per channel use at any finite SNR. Note that the BFN is a special case of the interference relay channel (IRC) [17,13,7] shown in Fig. 1 obtained by setting the direct links to zero. In this paper, we consider a BFN in which the nodes are full-duplex and where a dedicated feedback channel exists from the relay to the sources. From a slightly different perspective, the resulting setup can be considered as an IC utilizing a bi-directional relay for interference management to achieve higher data rates.

1.1 Contributions

The main contribution of this paper is the characterization of the capacity region of the full-duplex linear deterministic BFN with relay-source feedback.

First, we introduce the general memoryless IRC with Feedback (IRCF) where each node is full-duplex and has both an input to and an output from the channel. For such an IRCF, depicted in Fig. 2, we provide upper bounds on the achievable rates based on the cut-set bound [9, Thm.15.10.1] and based on an upper bound recently derived for the general cooperative IC [18]. We then specialize these upper bounds to the linear deterministic BFN with relay-source feedback depicted in Fig. 3 for which we provide a complete characterization of the capacity region.

Our achievable strategies aim to establish cooperation among the source nodes and the relay and to exploit the feedback from the relay to the source nodes. The relay participates in the delivery of the messages, since clearly in the setup of Fig. 3 communications is only possible via the relay. We develop transmission strategies where both the relay-destination links and the feedback links are used to deliver messages from the sources to the destination. We use the following main ingredients:

- **Decode-forward (DF):** Each source sends a "D-signal" to be decoded and forwarded by the relay using classical DF [8].
- **Compute-forward (CF):** Each source sends a "C-signal". The relay decodes a function (in our specific case the sum) of the C-signals and forwards it to the destinations. Since the processing at the relay does not involve decoding each C-signal separately, but "computing" their sum, the strategy is refered to as compute-forward [15]. This strategy is designed in such a way that each destination can decode both the interfering C-signal and the forwarded sum of C-signals. Backward decoding is used at the destinations

[1] Note that the classical butterfly network with multicast message was used by Ahlswede *et al.* in [2] to demonstrate the capabilities of network coding.

[2] The deterministic approximation of a Gaussian noise network is a deterministic model where the Gaussian additive noises are neglected so as to focus on the interaction of users' signals [3].

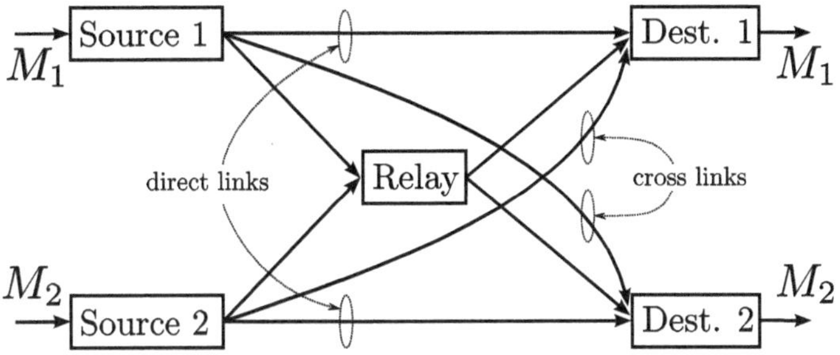

Fig. 1. The Interference Relay Channel (IRC)

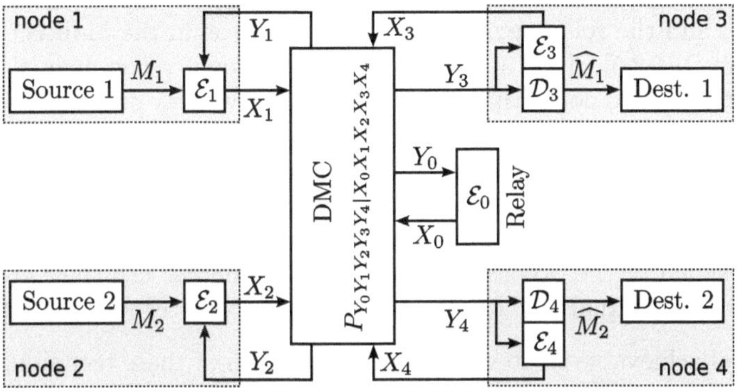

Fig. 2. The general memoryless Interference Relay Channel with Feedback (IRCF)

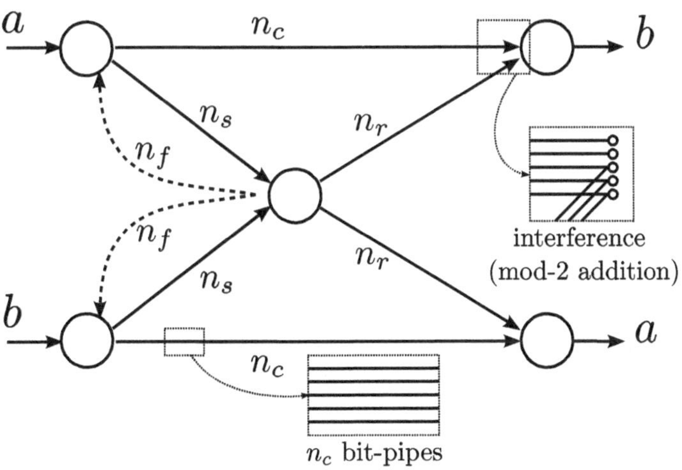

Fig. 3. The linear deterministic butterfly network with relay-source feedback

to recover the desired C-signal. A similar strategy was used in [7] for the IRC, and a half-duplex variant was also used in [4] for the half-duplex BFN.

- **Cooperative Neutralization (CN):** Each source sends two "N-signals": a "present N-signal" and a "future N-signal". The future N-signal is intended for the relay only, which computes the sum of the future N-signals. The relay then forwards this sum in the next channel use (note that the "future N-signals" of the i-th channel use are "present N-signals" in channel use $i+1$). This strategy is designed as follows. The forwarded N-signal sum from the relay and the interfering N-signal from the cross link interfere at the destination in such a way that neutralizes interference (on the fly) leaving the desired N-signal interference free. A similar strategy was used for the interference channel with cooperation in [19] and for the half-duplex BFN in [4].
- **Feedback (F):** Each source sends an "F-signal" to the relay. The two sources and the relay operate on the F-signals as in the bi-directional relay channel [16,12,5]. In a nutshell, the bi-directional relay channel is a setup consisting of two nodes that want to establish two way communications via a relay node, where each node is a transmitter and a receiver at the same time. In the BFN with feedback, the relay-source feedback channels together with the source-relay forward channels establish such bi-directional relay channel. Therefore, as in the bi-directional relay channel, each source is able to obtain the F-signal of the other source. Then, the sources use their cross link to deliver the F-signal of the other source node to its respective destination.

Our general achievable strategy uses a combination of these techniques depending on the channel parameters. The following give a rational as of why certain schemes should be used for a specific scenarios:

- If the source-relay channel is stronger than the source-destination (cross) channel, then the sources can pass some future information to the relay without the destinations noticing (below their noise floor). This future information is to be used in the next channel use for interference neutralization. If the source-relay channel is weaker than the source-destination channel, the CN strategy should be avoided since the transmission of future information to the relay disturbs the destinations in this case.
- On the other hand, the F strategy is to be used when the source-destination (cross) channel is stronger than the relay-destination channel. In this case, the sources can send the signal acquired via feedback to the destinations, which is received by the destination at a higher SNR than the relay signal. This allows the destination to decode this signal, strip it, and then proceed with decoding the relay signal. Otherwise, if the cross channel is weaker than the relay-destination channel, then such transmission would disturb the relay transmission and should be avoided.
- In the CF strategy, each destination has to decode two observations of the C-signals in each channel use (the interfering C-signal and the sum of the C-signals), whereas the relay has to decode only one observation (the C-signal sum). Therefore, this scheme requires more levels at the destinations than

at the relay. For this reason, the CF strategy is to be used by the relay if the source-relay channel is weaker than either the relay-destination channel or the source-destination (cross) channel (as in [7]).
- The DF strategy can be always used to achieve asymmetric rate points.

By using this intuition, we design achievable strategies for different parameter regimes that meet the derived outer bounds for the linear deterministic BFN with relay-source feedback, thus characterizing its capacity region completely.

1.2 Paper Organization

The general flow of the paper is as follows. We define the general memoryless IRCF in Sect. 2 where we also provide upper bounds. The linear deterministic BFN with relay-source feedback is defined in Sect. 3 and its upper bounds are derived in Sect. 4. The coding strategies (DF, CF, CN, and F) that constitute the basic building block of our achievable schemes are described in Sect. 5. The capacity achieving scheme is described and analyzed in Sect. 6 and 7, for the two regimes where relay-source feedback does not and does, respectively, increase the capacity with respect to the non-feedback case. We discuss the net-gain due to feedback in Sect. 8. Sect. 9 concludes the paper.

1.3 Notation

We use X^N to denote the length-N sequence $(X_1, X_2, \dots, X_N)$, $(x)^+ := \max\{0, x\}$ for $x \in \mathbb{R}$, and 0_ℓ to denote the all-zero vector of length $\ell \in \mathbb{N}$. For a vector $x(i)$ given as

$$x(i) = \begin{bmatrix} x^{[1]}(i) \\ x^{[2]}(i) \\ \vdots \\ x^{[K]}(i) \end{bmatrix},$$

i denotes the time index, and $x^{[k]}(i)$ is the k-th component of $x(i)$, which can be scalar or vector depending on the context. x^T is the transpose of the vector x.

2 The Memoryless IRC with Relay-Source Feedback: Channel Model and Outer Bounds

In Section 2.1 we introduce the memoryless IRC with general feedback even though in the rest of the paper we will be analyzing the case of relay-source feedback only. The reason for doing so is that the general feedback model allows us to easily describe the proposed outer bounds for the relay-source feedback model in Section 2.2.

2.1 The Memoryless IRC with General Feedback

A memoryless IRC with general feedback is a five node network with a relay (node 0), two sources (nodes 1 and 2), and two destinations (nodes 3 and 4) sharing the same channel, as shown in Fig. 2. All nodes are full-duplex and causal. Node j, $j \in \{1,2\}$, has an independent message $M_j \in \{1,\ldots,2^{NR_j}\}$, where $N \in \mathbb{N}$ is the code-length and $R_j \in \mathbb{R}_+$ the rate in bits per channel use, to be sent to node $j+2$. The operations performed at each node can be described in general as follows:

- Node 0 receives Y_0 and sends X_0, where the i-th symbol of X_0^N is constructed from Y_0^{i-1} using an encoding function $\mathcal{E}_{0,i}$, i.e., $X_{0,i} = \mathcal{E}_{0,i}(Y_0^{i-1})$.
- Node 1 receives feedback information Y_1 and sends X_1, where $X_{1,i}$ is constructed from the message M_1 and from Y_1^{i-1} using an encoding function $\mathcal{E}_{1,i}$, i.e., $X_{1,i} = \mathcal{E}_{1,i}(M_1, Y_1^{i-1})$.
- Node 2 operates similarly to node 1, i.e., $X_{2,i} = \mathcal{E}_{2,i}(M_2, Y_2^{i-1})$.
- Node 3 receives Y_3 and sends X_3, where $X_{3,i}$ is constructed from Y_3^{i-1} using an encoding function $\mathcal{E}_{3,i}$, i.e., $X_{3,i} = \mathcal{E}_{3,i}(Y_3^{i-1})$. After N channel uses, node 3/destination 1 tries to obtain M_1 from Y_3^N using a decoding function $\mathcal{D}_3$, i.e., $\widehat{M}_1 = \mathcal{D}_3(Y_3^N)$. An error occurs if $M_1 \neq \widehat{M}_1$.
- Node 4 operates similarly to node 3/destination 1, i.e., $X_{4,i} = \mathcal{E}_{4,i}(Y_4^{i-1})$ and $\widehat{M}_2 = \mathcal{D}_4(Y_4^N)$. An error occurs if $M_2 \neq \widehat{M}_2$.

The channel has transition probability $P_{Y_0,Y_1,Y_2,Y_3,Y_4|X_0,X_1,X_2,X_3,X_4}$ and is assumed to be memoryless, that is, for all $i \in \mathbb{N}$ the following Markov chain holds

$$(W_1, W_2, X_0^{i-1}, X_1^{i-1}, X_2^{i-1}, X_3^{i-1}, X_4^{i-1}, Y_0^{i-1}, Y_1^{i-1}, Y_2^{i-1}, Y_3^{i-1}, Y_4^{i-1})$$
$$\to (X_{0,i}, X_{1,i}, X_{2,i}, X_{3,i}, X_{4,i}) \to (Y_{0,i}, Y_{1,i}, Y_{2,i}, Y_{3,i}, Y_{4,i}).$$

We use the standard information theoretic definition of a code, probability of error and achievable rates [9]. We aim to characterize the capacity defined as the convex closure of the set of non-negative rate pairs (R_1, R_2) such that $\max_{j\in\{1,2\}} \mathbb{P}[M_j \neq \widehat{M}_j] \to 0$ as $N \to \infty$.

This model generalizes various well studied channel models. For instance, it models the classical IC [6] (for $Y_1 = Y_2 = Y_0 = X_0 = X_3 = X_4 = \varnothing$), the IC with cooperation [18] (for $Y_0 = X_0 = \varnothing$), the classical IRC [13,7] (for $Y_1 = Y_2 = X_3 = X_4 = \varnothing$), etc.

2.2 Upper Bounds for the Memoryless IRC with Relay-Source Feedback

The memoryless IRC with relay-source feedback is obtained from the model in Section 2.1 by setting $X_3 = X_4 = \varnothing$. We next derive several upper bounds on achievable rate pairs for the general memoryless IRC with relay-source feedback. We note that the described techniques apply to the general IRCF and do not require necessarily $X_3 = X_4 = \varnothing$. We start with the cut-set bound [9] and then we adapt upper bounds for the general memoryless IC with cooperation given in [18] to our channel model.

Cut-Set Bounds: The cut-set bound [9] applied to a general network with independent messages at each node states that an achievable rate vector must satisfy

$$R(\mathcal{S} \to \mathcal{S}^c) \leq I(X(\mathcal{S}); Y(\mathcal{S}^c) | X(\mathcal{S}^c)), \tag{1}$$

for some joint distribution on the inputs, where $\mathcal{S}$ is a subset of the nodes in the network, $\mathcal{S}^c$ is the complement of $\mathcal{S}$, and $R(\mathcal{S} \to \mathcal{S}^c)$ indicates the sum of the rates from the source nodes in $\mathcal{S}$ to the destination nodes in $\mathcal{S}^c$.

For the IRCF, by using (1), the rate R_1 can be bounded as

$$R_1 \leq I(X_1; Y_0, Y_2, Y_3 | X_0, X_2) \tag{2a}$$
$$R_1 \leq I(X_1, X_2; Y_0, Y_3 | X_0) \tag{2b}$$
$$R_1 \leq I(X_0, X_1; Y_2, Y_3 | X_2) \tag{2c}$$
$$R_1 \leq I(X_0, X_1, X_2; Y_3), \tag{2d}$$

for some input distribution P_{X_0,X_1,X_2}.

Similarly, we can bound R_2 by replacing the subscripts 1, 2, and 3 with 2, 1, and 4, respectively, in (2).

The sum-rate can be bounded as

$$R_1 + R_2 \leq I(X_1, X_2; Y_0, Y_3, Y_4 | X_0) \tag{3a}$$
$$R_1 + R_2 \leq I(X_0, X_1, X_2; Y_3, Y_4), \tag{3b}$$

for some input probability distribution P_{X_0,X_1,X_2}.

Cooperation Upper Bounds: As mentioned earlier, the IC with general cooperation is a special case of the IRCF obtained by setting $Y_0 = X_0 = \varnothing$. An upper bound for the sum-capacity of the IC with general cooperation is [18]

$$R_1 + R_2 \leq I(X_1; Y_3, Y_2 | Y_4, X_2, X_3, X_4) + I(X_1, X_2, X_3; Y_4 | X_4), \tag{4a}$$
$$R_1 + R_2 \leq I(X_2; Y_4, Y_1 | Y_3, X_1, X_3, X_4) + I(X_1, X_2, X_4; Y_3 | X_3). \tag{4b}$$

for some P_{X_1,X_2,X_3,X_4}.

In the interference relay channel with feedback, if we let the relay perfectly cooperate with one of the other nodes in the network, then the model again reduces to an IC with general cooperation in which one of the nodes has an enhanced input and output. Since cooperation cannot decrease capacity, any outer bound for the IC with general cooperation is an upper bound to the capacity of the interference relay channel with feedback. In particular, if node j, $j \in \{1, 2, 3, 4\}$, cooperates with the relay (node 0), then in (4) we replace X_j with (X_j, X_0) and Y_j with (Y_j, Y_0). Moreover, since we do not consider feedback from the destinations in this paper, we set $X_3 = X_4 = \varnothing$ after this substitution. This yields the following upper bounds:

1. Full cooperation between node 1 and node 0, giving an IC with bi-directional cooperation between nodes 1 and 2 where node 1 sends (X_1, X_0) and receives (Y_1, Y_0):

$$R_1 + R_2 \leq I(X_1, X_0; Y_3, Y_2 | Y_4, X_2) + I(X_1, X_0, X_2; Y_4), \tag{5a}$$
$$R_1 + R_2 \leq I(X_2; Y_4, Y_1, Y_0 | Y_3, X_1, X_0) + I(X_1, X_0, X_2; Y_3). \tag{5b}$$

2. Full cooperation between node 2 and node 0, giving an IC with bi-directional cooperation between nodes 1 and 2 where node 2 sends (X_2, X_0) and receives (Y_2, Y_0):

$$R_1 + R_2 \leq I(X_1; Y_3, Y_2, Y_0 | Y_4, X_2, X_0) + I(X_1, X_2, X_0; Y_4), \tag{5c}$$
$$R_1 + R_2 \leq I(X_2, X_0; Y_4, Y_1 | Y_3, X_1) + I(X_1, X_2, X_0; Y_3). \tag{5d}$$

3. Full cooperation between node 3 and node 0, giving an IC with uni-directional cooperation between node 3 and 4 and with feedback from node 3 to nodes 1 and 2, where node 3 sends X_0 and receives (Y_3, Y_0):

$$R_1 + R_2 \leq I(X_1; Y_3, Y_0, Y_2 | Y_4, X_2, X_0) + I(X_1, X_2, X_0; Y_4), \tag{5e}$$
$$R_1 + R_2 \leq I(X_2; Y_4, Y_1 | Y_3, Y_0, X_1, X_0) + I(X_1, X_2; Y_3, Y_0 | X_0). \tag{5f}$$

4. Finally, full cooperation between node 4 and node 0, giving an IC with uni-directional cooperation between node 4 and 3 and with feedback from node 4 to nodes 1 and 2, where node 4 sends X_0 and receives (Y_4, Y_0):

$$R_1 + R_2 \leq I(X_1; Y_3, Y_2 | Y_4, Y_0, X_2, X_0) + I(X_1, X_2; Y_4, Y_0 | X_0), \tag{5g}$$
$$R_1 + R_2 \leq I(X_2; Y_4, Y_0, Y_1 | Y_3, X_1, X_0) + I(X_1, X_2, X_0; Y_3). \tag{5h}$$

These upper bounds will be used next to upper bound the capacity region of the butterfly network with relay-source feedback. As it turns out, these bounds suffice to characterize the capacity of the symmetric linear deterministic butterfly network.

3 The Linear Deterministic Butterfly Network with Feedback

We consider here a special case for the IRC with relay-source feedback described in the previous section, namely the linear deterministic channel that is by now customarily used to approximate a Gaussian noise network at high SNR as originally proposed by [3].

We assume a dedicated out-of-band feedback channel between node 0 on one side, and nodes 1 and 2 on the other side. For this reason, we write X_0 as (X_r, X_f) where X_r is the in-band relay signal to the destinations and X_f is the out-of-band feedback signal to the sources. The input-output relations of this linear deterministic IRCF with out-of-band feedback from the relay to the sources is

$$Y_0 = \mathbf{S}^{q-n_{10}} X_1 + \mathbf{S}^{q-n_{20}} X_2, \tag{6a}$$
$$Y_1 = \mathbf{S}^{q-n_{01}} X_f, \tag{6b}$$
$$Y_2 = \mathbf{S}^{q-n_{02}} X_f, \tag{6c}$$
$$Y_3 = \mathbf{S}^{q-n_{13}} X_1 + \mathbf{S}^{q-n_{23}} X_2 + \mathbf{S}^{q-n_{03}} X_r, \tag{6d}$$
$$Y_4 = \mathbf{S}^{q-n_{14}} X_1 + \mathbf{S}^{q-n_{24}} X_2 + \mathbf{S}^{q-n_{04}} X_r, \tag{6e}$$

where Y_0 is the channel output at relay, Y_1 and Y_2 are the received feedback signal at the sources, and Y_3 and Y_4 are the received signals at the destinations. Here $q := \max\{n_{jk}\}$, with $n_{jk} \in \mathbb{N}$ for $j \in \{0, 1, 2\}$ and $k \in \{0, 1, 2, 3, 4\}$ and $\mathbf{S}$ is the $q \times q$ shift matrix

$$\mathbf{S} := \begin{bmatrix} 0\,0\,0\ldots \\ 1\,0\,0\ldots \\ 0\,1\,0\ldots \\ \vdots\;\vdots\;\vdots\;\ddots \end{bmatrix}.$$

All signals are binary vectors of length q and addition is the component-wise addition over the binary field.

As the number of parameters in the general channel model in (6) is large, we resort to a symmetric setup for simplicity of exposition. This simplification reduces the number of parameters, and thus leads to complete analytical, clean, and insightful capacity region characterization. In the *symmetric* scenario the channel model in (6) has the following parameters

$$\begin{aligned}
n_{13} &= n_{24} = 0 && \text{(direct channel)},\\
n_{14} &= n_{23} = n_c && \text{(cross channel)},\\
n_{03} &= n_{04} = n_r && \text{(relay-destination channel)},\\
n_{10} &= n_{20} = n_s && \text{(source-relay channel)},\\
n_{01} &= n_{02} = n_f && \text{(feedback channel)}.
\end{aligned}$$

Thus, the symmetric linear deterministic BFN with feedback shown in Fig. 3 has the following input-output relationship

$$Y_0 = \mathbf{S}^{q-n_s} (X_1 + X_2), \tag{7a}$$
$$Y_1 = \mathbf{S}^{q-n_f} X_f, \tag{7b}$$
$$Y_2 = \mathbf{S}^{q-n_f} X_f, \tag{7c}$$
$$Y_3 = \mathbf{S}^{q-n_c} X_2 + \mathbf{S}^{q-n_r} X_r, \tag{7d}$$
$$Y_4 = \mathbf{S}^{q-n_c} X_1 + \mathbf{S}^{q-n_r} X_r. \tag{7e}$$

The main focus of the rest of the paper is to determine the capacity region of the network described by (7). In the following section, we provide matching upper and lower bounds for the linear deterministic BFN with feedback thereby completely characterizing the capacity region.

4 Upper Bounds for the Linear Deterministic BFN with Feedback

In this section we specialize the general bounds given in Section 2.2 to the linear deterministic BFN described in Section 3. Our main result is as follows.

Theorem 1. *The capacity region of the linear deterministic BFN with source-relay feedback is contained in the set of rate pairs* (R_1, R_2) *such that*

$$0 \leq R_1 \leq \min\{n_s, n_r + n_f, \max\{n_c, n_r\}\} \tag{8a}$$
$$0 \leq R_2 \leq \min\{n_s, n_r + n_f, \max\{n_c, n_r\}\} \tag{8b}$$
$$R_1 + R_2 \leq \max\{n_r, n_c\} + n_c \tag{8c}$$
$$R_1 + R_2 \leq \max\{n_r, n_c\} + (n_s - n_c)^+ \tag{8d}$$
$$R_1 + R_2 \leq n_s + n_c. \tag{8e}$$

The details of the proof can be found in the Appendix.

An intuitive explanation of the single-rate bounds in Thm. 1 is as follows. Since communications is only possible via the relay, source 1 can not send more bits per channel use than the relay can receive; thus, we have the bound $R_1 \leq n_s$ in (8a). Now assume that the channel to the relay is very strong (say of infinite capacity); in this case, the rate achieved by a source can not exceed the capacity of the outgoing channels from the relay, i.e., $n_r + n_f$ in (8a). Finally, the rate R_1 can not exceed the amount of information that can be received by node 3/destination 1, which is given by $\max\{n_c, n_r\}$, and hence the bound $R_1 \leq \max\{n_c, n_r\}$ in (8a). Similar reasoning holds for the bound in (8b).

Interestingly, the sum-rate bounds in (8c)-(8e) do not depend on the feedback parameter n_f. As we shall see in the following sections, given $n_r > n_c$, the region in (8) is as for $n_f = 0$, i.e., no gain from the availability of a dedicated relay-source feedback channel. In this case, the relay-destination link is so strong that the relay can help the destinations resolve their signals without the need of source cooperation. On the other hand, when $n_r < n_c$, relaying can be improved upon by source cooperation enabled by the presence of feedback; in this case, we can have a 'net-gain' from feedback that is larger than the 'cost' of feedback. We will expand on this idea after we proved the achievability of the outer bound in Thm. 2.

5 Achievable Strategies

The main result of this section is as follows:

Theorem 2. *The outer bound region in Thm. 1 is achievable.*

Before we prove the achievability of the outer bound in Thm. 1, we describe the different coding strategies that we will use in the achievability proof. Each strategy is discussed separately in the rest of this section. The proof of Thm. 2 is a careful combination of these strategies for different parameter regimes. The actual proof of Thm. 2, due to its length, is split between Section 6 and Section 7.

5.1 Cooperative Interference Neutralization

We propose a signaling scheme which we call *cooperative interference neutralization*, or CN for short. The main idea of CN is to allow the relay to know some information about future source transmissions, in order to facilitate interference neutralization. This is done as follows. Each source sends two N-signals in the i-th channel use, which we call $u_{j,n}(i)$ and $u_{j,n}(i+1)$, where $j \in \{1,2\}$ is the source index, the subscript n is used to denote N-signals, and where i is the channel use index. $u_{j,n}(i)$ is the N-signal to be decoded by the destination in the i-th channel use, while $u_{j,n}(i+1)$ is to be decoded in the next channel use $i+1$. Therefore, the source sends the present and the future N-signals. The future one, $u_{j,n}(i+1)$ is intended for the relay, and is not decoded at the destinations. The relay attempts to decode $u_{1,n}(i+1) \oplus u_{2,n}(i+1)$ in the i-th channel use. This sum is then sent in the next channel use $i+1$, on the same levels where $u_{2,n}(i+1)$ is observed at node 3/destination 1 (note that $u_{2,n}(i+1)$ is interference from node 3/destination 1's perspective), resulting in interference neutralization since $u_{1,n}(i+1) \oplus u_{2,n}(i+1) \oplus u_{2,n}(i+1) = u_{1,n}(i+1)$. This allows node 3/destination 1 to decode its desired N-signal in channel use $i+1$.

In Fig. 4, as well as in similar figures in the following, the vertical bars represent bit vectors and the circles inside them represent bits. On the left we represent the bits of the sources (node 1 on top and node 2 at the bottom) and on the right the bits of the destinations (node 4/destination 2 on top and node 3/destination 1 at the bottom); the relay is represented in the middle. Lines connecting circles represent bit-pipes, and when a level (circle) receives 2 bit-pipes (lines), the modulo-2 sum of the bits is observed (valid at the relay and the destinations). The in-band channel is drawn in black, while the out-of-band feedback channel is drawn in red. When the red channel is not shown, this means that either $n_f = 0$ (no feedback channel to the sources) or the feedback channel is not used.

An illustrative example for CN is given in Fig. 4. In Fig. 4 destination 1/node 3 receives on its the second level $u_{2,n}(i) \oplus u_{1,n}(i) \oplus u_{2,n}(i) = u_{1,n}(i)$, that is, thanks to CN, the signal $u_{1,n}(i)$ is received interference free. Similarly, destination 2/node 4 obtains $u_{2,n}(i)$ interference free. Note how the sources pass the future N-signals to the relay without disturbing the destinations.

From Fig. 4 we remark that by using CN each source can send R_n bits per channel use over R_n levels at the destination while using $2R_n$ levels at the relay. Due to this fact, this strategy is preferable when n_s is larger than n_c.

To realize the CN strategy we use block Markov coding. Each source sends N signals in $N+1$ channel uses. Starting with an initialization step, the sources send $u_{j,n}(1)$ in channel use $i = 0$ while the relay remains silent. Then, each source sends both $u_{j,n}(i)$ and $u_{j,n}(i+1)$ in the i-th channel use for $i = 1, \ldots, N-1$ while the relay sends $u_{1,n}(i) \oplus u_{2,n}(i)$. Finally, in the N-th channel use, each source sends $u_{j,n}(N)$ only and the relay sends $u_{1,n}(N) \oplus u_{2,n}(N)$. Each destination decodes its desired N-signal starting from $i = 1$ till $i = N$. Thus, assuming that $u_{j,n}$ is a binary vector of length R_n, each source is able to successfully deliver NR_n bits over the span of $N + 1$ channel uses. Hence, the rate per channel use would be $\frac{N}{N+1}R_n$ which approaches R_n for large N. This factor $\frac{N}{N+1}$ will be ignored from now on, as we always choose N to be large.

A strategy similar to the CN strategy was also used in the interference channel with generalized feedback in [19], where the sources exchanges bits below the noise floor of the receivers, which are then used in the next slot to 'zero force' the interference. A half-duplex variant of this scheme was also used for the half-duplex BFN in [4].

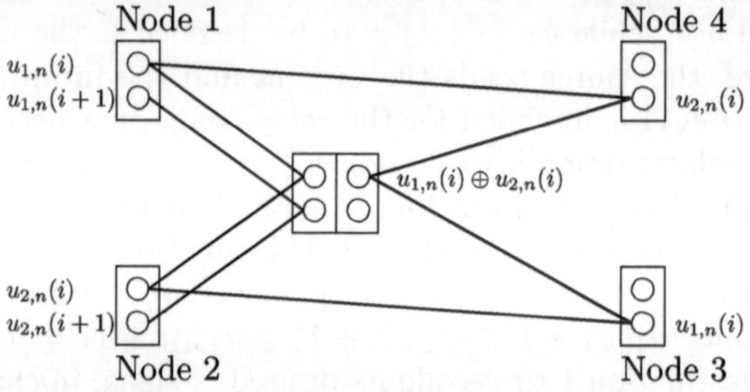

Fig. 4. A graphical illustration of the CN strategy. Due to interference neutralization, node 3/destination 1 receives $u_{2,n}(i) \oplus u_{1,n}(i) \oplus u_{2,n}(i) = u_{1,n}(i)$ interference free at the second level. Similarly, node 4/destination 2 obtains $u_{2,n}(i)$. Using this strategy in this setup, each source can send 1 bit per channel use. Note how the sources pass the future N-signals to the relay without disturbing the destinations.

5.2 Compute-Forward

We use compute-forward at the relay [15] (CF) to deliver both source messages to both destinations. The CF strategy works as follows (see Fig. 5 for an example). Each source sends a signal $u_{j,c}(i)$ in the i-th channel use, $i = 1, \ldots, N$ and where we use the subscript c to indicate C-signals. The relay decodes the function/sum $u_{1,c}(i) \oplus u_{2,c}(i)$ in the i-th channel use and sends it in the next channel use on a different level at the destinations. This process is repeated from $i = 1$ till $i = N + 1$, where the sources are active in channel uses $i = 1, \ldots, N$ and the relay is active in channel uses $i = 2, \ldots, N + 1$.

Thus, node 3/destination 1 for instance receives $u_{2,c}(i)$ and $u_{1,c}(i-1) \oplus u_{2,c}(i-1)$ in the i-th channel use, $i = 2, \ldots, N$. In the first channel use, it only receives $u_{2,c}(1)$ since the relay has no information to send in this channel use. In channel use $N + 1$, it only receives $u_{1,c}(N) \oplus u_{2,c}(N)$ from the relay since the sources do not send in this channel use. Decoding is performed backwards starting from

$i = N+1$, where only the relay is active and thus $u_{1,c}(N) \oplus u_{2,c}(N)$ is decoded. In the N-th channel use, destination 1 decodes $u_{1,c}(N-1) \oplus u_{2,c}(N-1)$ and $u_{2,c}(N)$. Then, it adds the two observations of the signals with time index N, i.e., $u_{1,c}(N) \oplus u_{2,c}(N)$ and $u_{2,c}(N)$ to obtain its desired signal $u_{1,c}(N)$. Similar decoding is performed at node 4/destination 2. Decoding proceeds backwards till $i = 1$ is reached. If the signals $u_{j,c}$ are binary vectors of length R_c, then each source achieves R_c bits per channel use for large N using this strategy. The signals sent using this strategy are "public", in the sense of the Han and Kobayashi's achievable region for the classical IC [11], i.e., each destination decode both C-signals from source 1 and 2.

An example of CF strategy is given in Fig. 5. Here node 3/destination 1 decodes $u_{1,c}(i-1) \oplus u_{2,c}(i-1)$ and $u_{2,c}(i)$ in the i-th channel use. By backward decoding, it can add $u_{1,c}(i-1) \oplus u_{2,c}(i-1)$ (decoded in the i-th channel use) and $u_{2,c}(i-1)$ (decoded in channel use $i-1$) to obtain its desired signal $u_{1,c}(i-1)$.

Notice from Fig. 5 that the CF strategy allows the sources to send R_c bits each while using R_c levels at the relay and $2R_c$ levels at the destinations. For this reason, this strategy is preferable when the number of levels at the destinations $\max\{n_c, n_r\}$ is larger than n_s.

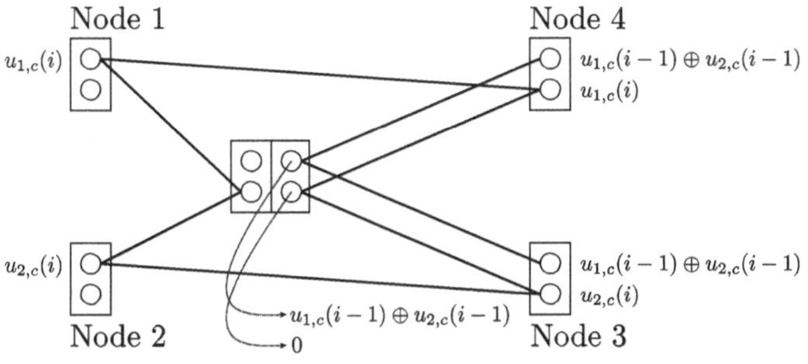

Fig. 5. A graphical illustration of the CF strategy. Node 3/destination 1 decodes $u_{1,c}(i-1) \oplus u_{2,c}(i-1)$ and $u_{2,c}(i)$ in the i-th channel use. By backward decoding, it can add $u_{1,c}(i-1) \oplus u_{2,c}(i-1)$ (decoded in the i-th channel use and $u_{2,c}(i-1)$ (decoded in channel use $i-1$) to obtain its desired CP signal $u_{1,c}(i-1)$. Using this strategy in this setup, each source can send 1 bit per channel use.

5.3 Feedback

Symmetric: Here both sources use the same strategy. This strategy exploits the feedback channel between the relay and the sources to establish cooperation between the sources. It is similar to the scheme used in the linear deterministic bi-directional relay channel in [14]. Each source j, $j \in \{1,2\}$, sends a feedback (F) signal $u_{j,f}(i)$ in the i-th channel use, where the subscript f is used to indicate F-signals. The relay decodes the sum $u_{1,f}(i) \oplus u_{2,f}(i)$ in the i-th channel use and feeds it back to the sources in channel use $i+1$. In channel use $i+1$, source 1 for instance decodes $u_{1,f}(i) \oplus u_{2,f}(i)$ from the feedback channel, and extracts

$u_{2,f}(i)$, having its own signal $u_{1,f}(i)$ as "side information". Then, it sends this information to destination 2 using its cross channel in channel use $i+2$ (see Fig. 6). A similar procedure is done at the second source.

Note that this scheme incurs a delay of 2 channel uses. Each source sends N F-signals from the first channel use till channel use $i = N$. The relay feeds these signals back in the channel uses $i = 2, \ldots, N+1$. Finally, nodes 1 and 2 send the F-signals to their respective destinations in the channel uses $i = 3, \ldots, N+2$. If the F-signals $u_{j,f}$ are vectors of length R_f, then each source can successfully deliver NR_f bits in $N+2$ channel uses. Thus the rate that each source can achieve per channel use approaches R_f for large N.

An illustrative example for symmetric F strategy is shown in Fig. 6. The sources send $u_{j,f}(i)$ to the relay in the i-th channel use, which decodes the sum $u_{1,f}(i) \oplus u_{1,f}(i)$. In the same channel use, the relay feeds the signal $u_{1,f}(i-1) \oplus u_{1,f}(i-1)$ (decoded in channel use $i-1$) back to the sources. Nodes 1 and 2 use this sum to extract $u_{2,f}(i-1)$ and $u_{1,f}(i-1)$, respectively. Nodes 1 and 2 also send $u_{2,f}(i-2)$ and $u_{1,f}(i-2)$ (decoded in channel use $i-1$), respectively, to their respective destinations via the cross link in the i-th channel use. Nodes 3 and 4 decode $u_{1,f}(i-2)$ and $u_{2,f}(i-2)$, respectively, in channel use i. Note that nodes 1 and 2 always send information to the relay which renders some levels at the sources always occupied. Thus, the sources have to use *other levels* for sending the F-signals to the respective destinations. In general, for each F-signal, the symmetric F strategy uses 2 levels at the sources and 1 level for feedback.

Notice from Fig. 6 that we have sent the F-signals on levels that could have also been used by the relay to send the same amount of bits (using CN or CF). As we shall see, this symmetric F strategy does not increase the capacity if $n_c \le n_r$. The F strategy would increase the capacity if n_c is larger than n_r, in which case the sources would send the F-signals to their respective destinations over levels that are not accessible by the relay, thus not disturbing the relay transmission while doing so.

Asymmetric: The symmetric F strategy achieves symmetric rates for the F-signals, i.e., the rate achieved by source 1 is equal to that of source 2. We can also use the F strategy in an asymmetric fashion as follows. Node 1 sends $u_{1,f}(i)$ to the relay in the i-th channel use, the relay decodes this signal and feeds it back to node 2 in channel use $i+1$, which sends it to node 3/destination 1 in the channel use $i+2$ *on the same level used by node 1.* This causes the signals $u_{1,f}(i)$ and $u_{1,f}(i-2)$ to interfere at the relay. However, the relay can always resolve this interference since it decoded $u_{1,f}(i-2)$ in channel use $i-2$. If the vector $u_{1,f}(i)$ has length R_f, then this strategy achieves the rate point $(R_f, 0)$.

An illustrative example for the asymmetric F strategy is given in Fig. 7 where source 1 can send 1 bit per channel use to destination 1, achieving the rate pair $(1,0)$. Note that the same rate pair can be achieved using the symmetric F strategy (Fig. 6) by setting $u_{2,f} = 0$. But this would be inefficient since it consumes 2 levels at the relay for reception. The same rate pair can be achieved using the asymmetric F strategy while using only 1 level at the relay as shown

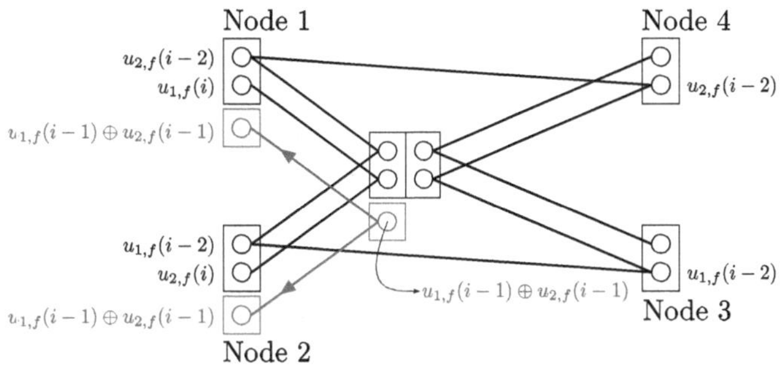

Fig. 6. A symmetric feedback strategy. Node 1 sends its own F-signal $u_{1,f}(i)$ to the relay to be fed back to node 2 in the next channel use. At the same time, node 1 sends node 2's F-signal $u_{2,f}(i-2)$, acquired via feedback, to node 4/destination 2. Node 2 performs similar operations. Notice the bi-directional relay channel formed by nodes 1 and 2 and the relay.

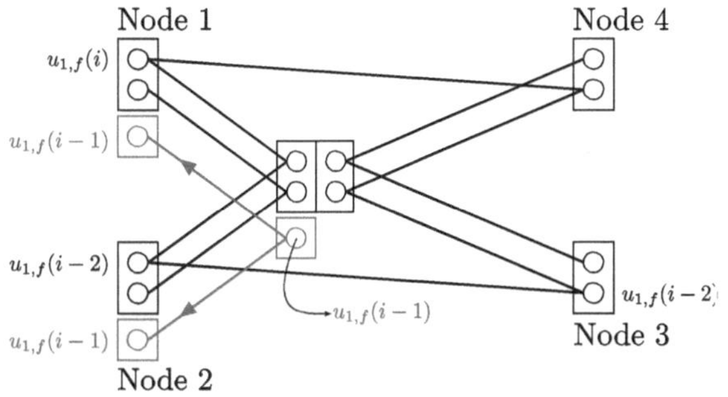

Fig. 7. An asymmetric feedback strategy. Node 1 sends the F-signal $u_{1,f}(i)$ to the relay. The relay decodes this signal and feeds it back to node 2 in the next channel use. Node 2 in its turn sends node 1's F-signal $u_{1,f}(i-2)$, acquired via feedback, to node 3/destination 1.

in Fig. 7. This leaves one level at the relay unused, providing more flexibility to combine the F strategy with other strategies. Since our aim is to characterize the capacity region of the linear deterministic BFN with feedback, we are going to need strategies which achieve asymmetric rates efficiently. Both the symmetric and the asymmetric F strategies will be used in the sequel.

5.4 Decode-Forward

The last strategy we describe in this section is the decode-forward (DF). Although this strategy is well known[8], we describe it here to draw the reader's attention to a convention we will adopt in the following. In classical DF, each source sends a public signal $u_{j,d}(i)$ in the i-th channel use, $i = 1, \ldots, N$ and where the subscript d is used to denote D-signals, the relay decodes both $u_{1,d}(i)$ and $u_{2,d}(i)$ in the i-th channel use, maps them to $u_{r,d}(i)$ which it forwards in channel use $i+1$ (see Fig. 9).

For convenience, this operation is represented as follows (see Fig. 8). Let the D-signal of source 1 in channel use i, $u_{1,d}(i)$, be a vector of length $R_{1d}+R_{2d}$ where the lower-most R_{2d} positions of $u_{1,d}(i)$ are zeros. Similarly, let $u_{2,d}(i)$ be of length $R_{1d}+R_{2d}$ with zeros in the top-most R_{1d} positions. Then, source 1 sends $u_{1,d}(i)$ and source 2 sends $u_{2,d}(i)$. The relay then decodes $u_{1,d}(i) \oplus u_{2,d}(i)$, a process which is equivalent to decoding both $u_{1,d}(i)$ and $u_{2,d}(i)$ separately due to the zero padding. The relay then forwards $u_{r,d}(i) = u_{1,d}(i) \oplus u_{2,d}(i)$ in channel use $i+1$ (see Fig. 9).

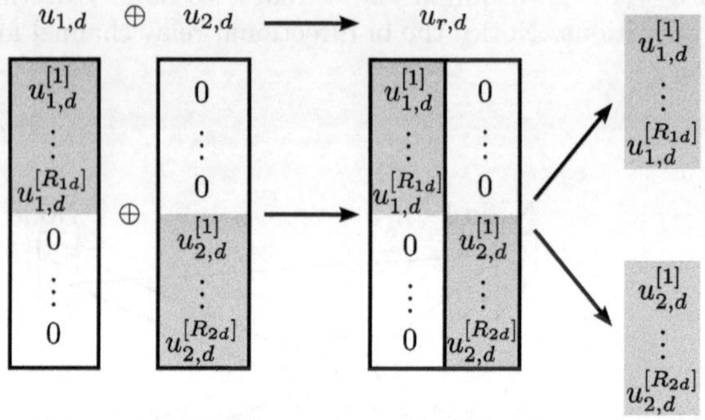

Fig. 8. A graphical illustration of the structure of the D-signals. Notice how $u_{1,d}$ and $u_{2,d}$ are zero-padded. Notice also that decoding the sum $u_{1,d} \oplus u_{2,d}$ is equivalent to decoding the D-signals separately. In the sequel, we will use these colored bars to represent the D-signals.

The destinations start decoding from channel use $N+1$ where only the relay is active, and they both decode $u_{r,d}(N)$, which allows them to obtain both $u_{1,d}(N)$ and $u_{2,d}(N)$. Decoding proceeds backward to the N-th channel use. In the N-th channel use, the destinations start by removing $u_{j,d}(N)$ from the received signal (which they know from channel use $N+1$). Then, they decode $u_{r,d}(N-1)$ to obtain $u_{1,d}(N-1)$ and $u_{2,d}(N-1)$. In this way, the destinations obtain their desired D-signals, R_{1d} bits from source 1 and R_{2f} bits from source 2. Decoding proceeds backwards till the first channel use is reached. Thus, source 1 and source 2 achieve R_{1d} and R_{2d} bits per channel use, respectively, for large N. Notice that the D-signals are public since they are decoded at both destinations.

5.5 Remark on the Use of the Different Strategies

At this point, a remark about the DF strategy as compared to the CN strategy in Sect. 5.1 is in order. Due to backward decoding, the interference caused by the D-signal, $u_{1,d}(i)$ at node 4/destination 2 for instance, is not harmful since it can be removed as long as the decoding of the D-signals was successful in

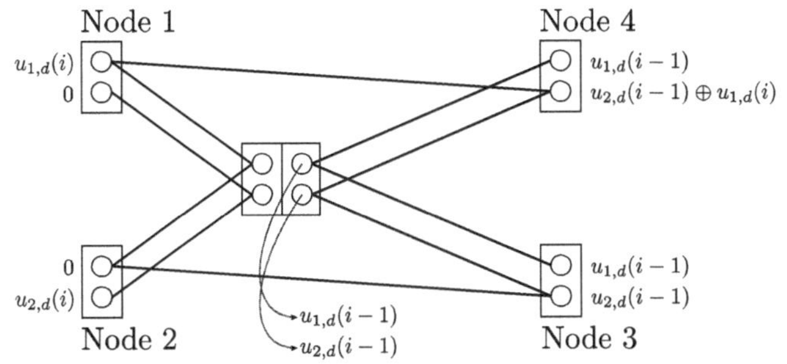

Fig. 9. A graphical illustration of the decode-forward strategy. Node 4/destination 2 starts be removing $u_{1,d}(i)$ (known from the decoding process in channel use $i+1$) from its received signal. Then it decodes the D-signal with time index $i-1$. Using this strategy in this setup, each source can send 1 bit per channel use.

channel use $i+1$. This is the reason why the relay and the sources can send over the same levels at the destinations (as in Fig. 9), in contrast to CN, CF, and F where separate levels have to be allocated to the source and the relay signals. We summarize this point by saying that *the D-signals $u_{j,d}(i)$ (from the sources) should be received 'clean' at the relay but not necessarily so at the destinations.* In fact, the D-signals arriving from the sources do not have to be received at all at the destinations since they are decoded from the relay signal.

Now consider the N-signals where an opposite statement holds. Since the relay decodes in a forward fashion, and since the sources send 'present' and 'future' N-signals, i.e., $u_{j,n}(i)$ and $u_{j,n}(i+1)$ in the i-th channel use, then interference from the $u_{j,n}(i)$ is not harmful at the relay. This is true since this interference is known from the decoding in channel use $i-1$ at the relay, and hence can be removed. The 'present' N-signal is however important at the destinations, since it is the signal that participates in interference neutralization. We summarize this statement by saying that *the 'present' N-signal must be received 'clean' at the destinations but not necessarily so at the relay.* Additionally, *the 'future' N-signal must be received 'clean' at the relay, but does not have to be received at all at the destinations.*

Combining these properties, we can construct a hybrid scheme where both CN and DF are used, and where the N-signals and the D-signals overlap at the relay and the destinations in a not harmful way, as illustrated in Fig. 10. Here, node 1 allows its present N-signal $u_{1,n}(i) = [u_{1,n}^{[1]T}(i),\ u_{1,n}^{[2]T}(i)]^T$ to overlap with the D-signal $u_{1,d}(i)$. And thus these signals also overlap at the destination nodes. Nevertheless, the relay is still able to decode the necessary information and forward it to the destinations which can still recover their desired information. This overlap allows a more efficient exploitation of the channel levels.

In the following sections, we develop capacity achieving schemes for the linear deterministic BFN with feedback which are based on combinations of the four strategies explained above.

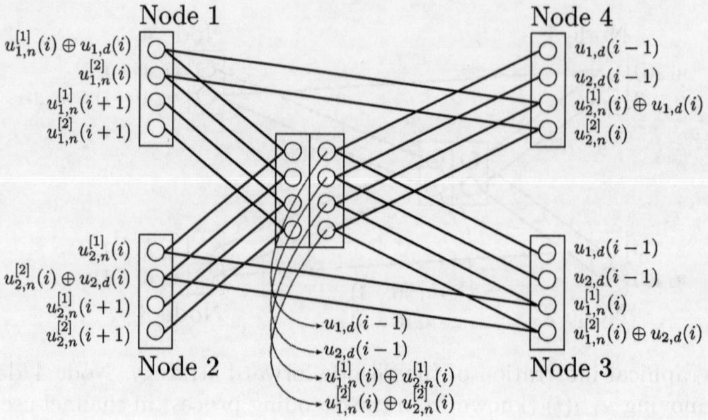

Fig. 10. A graphical illustration of the combination of DF and CN. The relay can obtain $u_{1,d}(i)$ and $u_{2,d}(i)$ in the i-th channel use after removing $u_{1,n}^{[1]}(i) \oplus u_{2,n}^{[1]}(i)$ and $u_{1,n}^{[2]}(i) \oplus u_{2,n}^{[2]}(i)$ which it has decoded in channel use $i-1$. Thus, this interference between the N-signal and the D-signals at the relay is not harmful. In the i-th channel use, node 3/destination 1 starts by removing $u_{2,d}(i)$ (known from the decoding process in channel use $i+1$) from its received signal. Then it decodes $u_{1,d}(i-1)$, $u_{2,d}(i-1)$, $u_{1,n}^{[1]}(i)$, and $u_{1,n}^{[2]}(i)$. Using this strategy each source can send 3 bit per channel use which achieves the sum-capacity upper bound (cf. Thm. 2).

6 Achievability for $n_c \leq n_r$

In this section, we show that the outer bound region given in Thm. 1 is achievable for the case $n_c \leq n_r$. First, we notice that if $n_c \leq n_r$, then the feedback channel n_f does not have a contribution to the upper bounds in Thm. 1, which reduces to

$$\begin{aligned} 0 &\leq R_1 \leq \min\{n_s, n_r\} \\ 0 &\leq R_2 \leq \min\{n_s, n_r\} \\ R_1 + R_2 &\leq n_r + n_c \\ R_1 + R_2 &\leq n_r + (n_s - n_c)^+ \\ R_1 + R_2 &\leq n_s + n_c. \end{aligned}$$

In this case feedback does not increase the capacity of the BFN with respect to the non-feedback case. The outer bound can be achieved without exploiting the feedback link n_f, and thus without using the F strategy, as per the discussion at the end of Sect. 4. Hence, in this section we only use the strategies that do not exploit feedback, i.e., CF, CN, and DF.

6.1 Case $n_s \leq \min\{n_c, n_r\} = n_c \leq n_r$:

Lemma 1. *In the linear deterministic BFN with feedback with $n_s \leq n_c \leq n_r$ the following region is achievable*

$$
\begin{aligned}
0 \le R_1 &\le n_s \\
0 \le R_2 &\le n_s \\
R_1 + R_2 &\le n_r,
\end{aligned}
$$

This achievable rate region coincides with the outer bound given in Thm. 1. Thus, the achievability of this region characterizes the capacity region of the linear deterministic BFN with feedback with $n_s \le n_c \le n_r$.

The rest of this subsection is devoted for the proof of this Lemma. In this case $\max\{n_c, n_r\} \ge n_s$ and thus we use the CF strategy according to the discussion in Sect. 5.2. We also use DF for achieving asymmetric rate tuples. Moreover, since $n_s \le n_c$ we do not use the CN strategy following the discussion in Sect. 5.1.

Encoding: Let us construct $x_1(i)$ in the i-th channel use as follows

$$
x_1(i) = \begin{bmatrix} u_{1,c}(i) \\ u_{1,d}(i) \\ 0_{n_c - R_c - R_{1d} - R_{2d}} \\ 0_{q-n_c} \end{bmatrix}.
$$

The signal $u_{1,d}$ is a vector of length $R_{1d} + R_{2d}$ with the lower R_{2d} components equal to zero as described in Sect. 5.4. Thus, it contains R_{1d} information bits. The signal $u_{1,c}$ is a vector of length R_c. We construct $x_2(i)$ similarly, with $u_{2,d}$ and $u_{2,c}$ being $(R_{1d}+R_{2d}) \times 1$ and $R_c \times 1$ binary vectors, respectively, where the first R_{1d} components of $u_{2,d}$ are zeros. The rates of the C-signal of both sources are chosen to be equal.

Relay Processing: In the i-th channel use, the relay observes the top-most n_s bits of $x_1(i) \oplus x_2(i)$. Under the following condition

$$
R_c + R_{1d} + R_{2d} \le n_s, \tag{9}
$$

the relay is able to observe both $u_{1,c}(i) \oplus u_{2,c}(i)$ and $u_{1,d}(i) \oplus u_{2,d}(i)$ and hence to decode them.

Since in this case $n_r \ge n_c$, the relay can access levels at the destinations above those that can be accessed by the sources. Then, the signal $u_{1,c}(i) \oplus u_{2,c}(i)$ to be forwarded by the relay is split into two parts as follows

$$
u_{1,c}(i) \oplus u_{2,c}(i) = \begin{bmatrix} u_{1,c}^{[1]}(i) \oplus u_{2,c}^{[1]}(i) \\ u_{1,c}^{[2]}(i) \oplus u_{2,c}^{[2]}(i) \end{bmatrix},
$$

where the upper part of length $R_c^{[1]}$ is sent such that it arrives on top of the signals from the sources, and the lower part of length $R_c^{[2]}$ is sent below, with $R_c = R_c^{[1]} + R_c^{[2]}$. Fig. 11 shows the transmit signal of node 2 ($x_2(i)$) and the relay ($x_r(i)$) and received signal of node 3 ($y_3(i)$). For clarity, from this point on we drop the labels of signals that do not undergo any change from left to right in this type of pictorial illustration. Thus, the relay forwards $x_r(i+1)$ in the next channel use where

$$x_r(i+1) = \begin{bmatrix} 0_{n_r-n_c-R_c^{[1]}} \\ u_{1,c}^{[1]}(i) \oplus u_{2,c}^{[1]}(i) \\ 0_{R_c} \\ u_{1,d}(i) \oplus u_{2,d}(i) \\ u_{1,c}^{[2]}(i) \oplus u_{2,c}^{[2]}(i) \\ 0_{n_c-R_c^{[1]}-2R_c^{[2]}-R_{1d}-R_{2d}} \\ 0_{q-n_r} \end{bmatrix}.$$

This construction requires

$$R_c^{[1]} + 2R_c^{[2]} + R_{1d} + R_{2d} \leq n_c, \tag{10}$$

$$R_c^{[1]} \leq n_r - n_c. \tag{11}$$

Decoding at the Destinations: Node 3/destination 1 waits until the end of channel use $N+1$ where only the relay is active, and it receives the top-most n_r bits of $x_r(N+1)$. Then, it decodes $u_{1,c}^{[1]}(N) \oplus u_{2,c}^{[1]}(N)$, $u_{1,c}^{[2]}(N) \oplus u_{2,c}^{[2]}(N)$, and $u_{1,d}(N) \oplus u_{2,d}(N)$. Similarly at the second receiver. At this point, both receivers have obtained both D-signals $u_{1,d}(N)$ and $u_{2,d}(N)$ which they extract from $u_{1,d}(N) \oplus u_{2,d}(N)$ (recall our discussion in Sect. 5.4).

The destinations proceed to the N-th channel use. The received signal at node 3/destination 1 can be written as (see Fig. 11 or 12)

$$y_3(N) = \begin{bmatrix} 0_{q-n_r} \\ 0_{n_r-n_c-R_c^{[1]}} \\ u_{1,c}^{[1]}(N-1) \oplus u_{2,c}^{[1]}(N-1) \\ u_{2,c}(N) \\ u_{2,d}(N) \oplus u_{1,d}(N-1) \oplus u_{2,d}(N-1) \\ u_{1,c}^{[2]}(N-1) \oplus u_{2,c}^{[2]}(N-1) \\ 0_{n_c-R_c^{[1]}-2R_c^{[2]}-R_{1d}-R_{2d}} \end{bmatrix}.$$

Since node 3/destination 1 knows $u_{2,d}(N)$, it can remove it from the received signal (see Fig. 12). Then it proceeds with decoding

$$u_{1,c}^{[1]}(N-1) \oplus u_{2,c}^{[1]}(N-1), \quad u_{1,c}^{[2]}(N-1) \oplus u_{2,c}^{[2]}(N-1), \quad u_{2,c}(N),$$

$$\text{and} \quad u_{1,d}(N-1) \oplus u_{2,d}(N-1)$$

Having $u_{2,c}(N)$ allows node 3/destination 1 to obtain $u_{1,c}(N)$ as $u_{1,c}(N) \oplus u_{2,c}(N) \oplus u_{2,c}(N) = u_{1,c}(N)$. Additionally, node 3/destination 1 obtains $u_{1,d}(N-1)$ which is a desired signals. Furthermore, $u_{2,d}(N-1)$ and $u_{1,c}(N-1) \oplus u_{2,c}(N-1)$ are obtained which are used in the decoding process in channel use $N-1$.

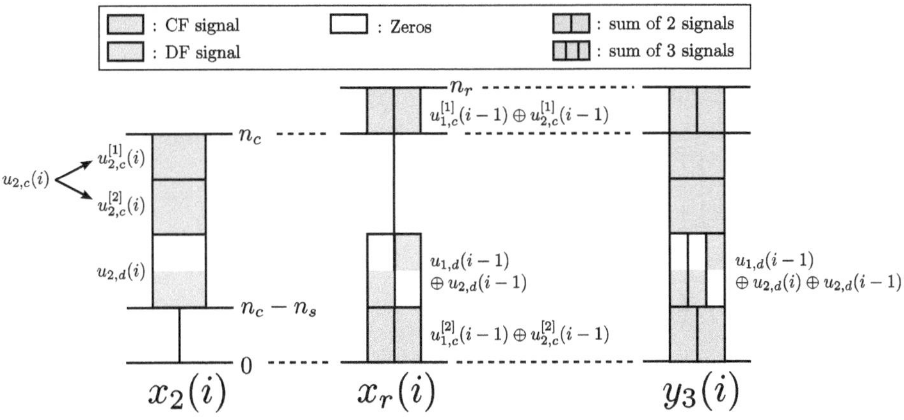

Fig. 11. A graphical illustration of the transmit signal of node 2, i.e., $x_2(i)$, the transmit signal of the relay, i.e., $x_r(i)$, and the received signal at node 3/destination 1, i.e., $y_3(i)$, using the capacity achieving scheme of the linear deterministic BFN with feedback with $n_s \leq n_c \leq n_r$. The color legend is shown on top.

In this process, node 3/destination 1 was able to recover its C and its D-signals comprising of R_c and R_{1d} bits, respectively. Node 4 performs similar operations. The receivers proceed backwards till channel use 1 is reached.

Achievable Region: The rates achieved by source 1 and 2 are $R_1 = R_c + R_{1d}$ and $R_2 = R_c + R_{2d}$, respectively. Collecting the rate constraints (9), (10), and (11), we get the following constraints on the non-negative rates $R_c^{[1]}$, $R_c^{[2]}$, R_{1d}, and R_{2d}:

$$\begin{aligned} R_c^{[1]} + R_c^{[2]} + R_{1d} + R_{2d} &\leq n_s \\ R_c^{[1]} + 2R_c^{[2]} + R_{1d} + R_{2d} &\leq n_c \\ R_c^{[1]} &\leq n_r - n_c. \end{aligned}$$

Using Fourier-Motzkin's elimination [10, Appendix D] we get the following achievable region

$$\begin{aligned} 0 \leq R_1 &\leq n_s \\ 0 \leq R_2 &\leq n_s \\ R_1 + R_2 &\leq n_r, \end{aligned}$$

which proves Lemma 1.

6.2 Case $n_s > \min\{n_c, n_r\}$ or Equivalently $n_c \leq \min\{n_s, n_r\}$:

Now we consider the case where $n_c \leq \min\{n_s, n_r\}$, i.e., the cross channel is weaker than both the source-relay channel and the relay-destination channel. As

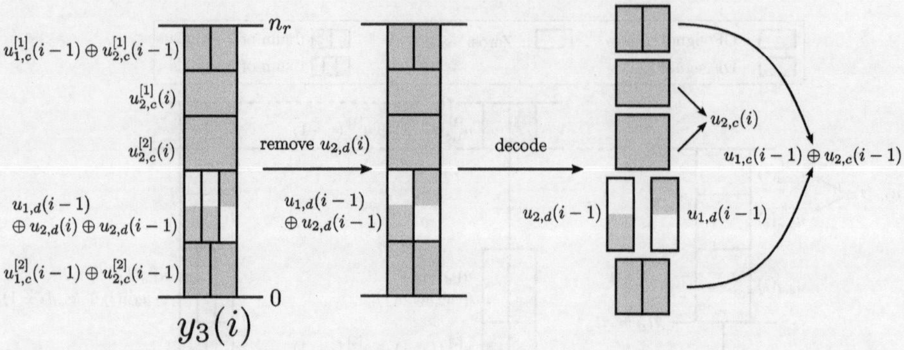

Fig. 12. The decoding steps at the destination. Due to backward decoding, node 3/destination 1 knows $u_{2,d}(i)$ and $u_{1,c}(i) \oplus u_{2,c}(i)$ when decoding $y_3(i)$ (decoded in channel use $i+1$). It starts by removing $u_{2,d}(i)$ from $y_3(i)$. Then it decodes the C-signal sum $u_{1,c}(i-1) \oplus u_{2,c}(i-1)$, the C-signal interference $u_{2,c}(i)$, and the D-signals $u_{1,d}(i-1)$ and $u_{2,d}(i-1)$. Finally, it uses the CF sum $u_{1,c}(i) \oplus u_{2,c}(i)$ and $u_{2,c}(i)$ to extract $u_{1,c}(i)$.

we have mentioned earlier, if $n_s \geq n_c$, then we can pass some future information to the relay without the destinations noticing by using the CN strategy. Thus, we use CN in addition to CF and DF. For this case, we have the following lemma.

Lemma 2. *The rate region defined by the following rate constraints*

$$\begin{aligned} 0 \leq R_1 &\leq \min\{n_s, n_r\} \\ 0 \leq R_2 &\leq \min\{n_s, n_r\} \\ R_1 + R_2 &\leq n_s + n_c \\ R_1 + R_2 &\leq n_r + n_c \\ R_1 + R_2 &\leq n_r + n_s - n_c, \end{aligned}$$

is achievable in the linear deterministic BFN with feedback with $n_c \leq \min\{n_s, n_r\}$.

This region coincides with the outer bound given in Thm. 1. Thus, the scheme which achieves this region achieves the capacity of the linear deterministic BFN with feedback with $n_c \leq \min\{n_s, n_r\}$. We provide this capacity achieving scheme in the rest of this subsection.

Encoding: At time instant i, node 1 sends the following signal

$$x_1(i) = \begin{bmatrix} 0_{n_c - R_c - R_{1d} - R_{2d} - R_n} \\ u_{1,c}(i) \\ u_{1,d}(i) \\ \begin{bmatrix} u_{1,n}(i) \\ 0_{n_s - n_c - R_n} \end{bmatrix} \oplus \begin{bmatrix} 0_{n_s - n_c - \overline{R}_{1d} - \overline{R}_{2d}} \\ \overline{u}_{1,d}(i) \end{bmatrix} \\ u_{1,n}(i+1) \\ 0_{q-n_s} \end{bmatrix},$$

where $u_{1,n}(i+1)$ is the future information passed to the relay. The signals $u_{1,c}$, $u_{1,d}$, $u_{1,n}$, and $\overline{u}_{1,d}$ are vectors of length R_c, $R_{1d} + R_{2d}$, R_n, and $\overline{R}_{1d} + \overline{R}_{2d}$, respectively. Notice that this construction requires that

$$R_c + R_{1d} + R_{2d} + R_n \leq n_c \tag{12}$$

$$\overline{R}_{1d} + \overline{R}_{2d} \leq n_s - n_c \tag{13}$$

$$R_n \leq n_s - n_c. \tag{14}$$

Using this construction, there can be an overlap between $u_{1,n}(i)$ and $\overline{u}_{1,d}(i)$ at the relay and at the destinations (see Fig. 13). However, this overlap is not harmful (similar to the one discussed in Sect. 5.4). The overlapping D-signal is marked with an overline to distinguish it from $u_{1,d}$ which does not overlap with any signal at the relay.

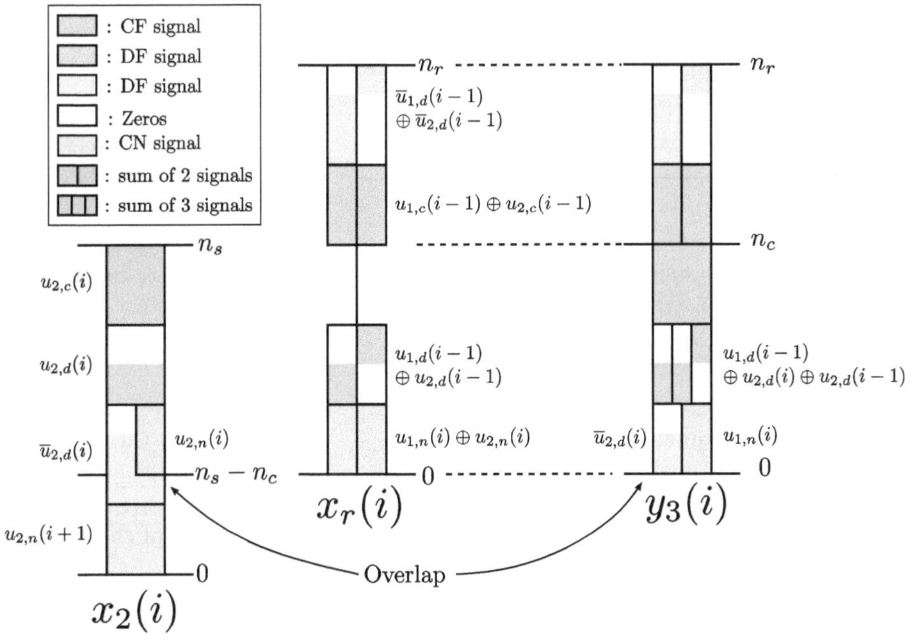

Fig. 13. A graphical illustration of the $x_2(i)$, $x_r(i)$, and $y_3(i)$ for the capacity achieving scheme of the linear deterministic BFN with feedback with $n_c \leq \min\{n_s, n_r\}$. Notice the overlap of $\overline{u}_{j,d}$ and $u_{j,n}$.

A similar construction is employed by the second source. As we show next, this construction allows us to achieve the capacity region of the linear deterministic BFN with feedback in this case. The task now is to find the conditions that R_{1d}, R_{2d}, $\overline{R}_{1d}$. $\overline{R}_{2d}$ R_c, and R_n should satisfy in order to guarantee reliable decoding.

Relay Processing: The received signal at the relay consists of the top n_s bits of $x_1(i) \oplus x_2(i)$. Let us write $y_0(i)$ as follows

$$y_0(i) = \begin{bmatrix} 0_{q-n_s} \\ 0_{n_c-R_c-R_{1d}-R_{2d}-R_n} \\ u_{1,c}(i) \oplus u_{2,c}(i) \\ u_{1,d}(i) \oplus u_{2,d}(i) \\ \begin{bmatrix} u_{1,n}(i) \oplus u_{2,n}(i) \\ 0_{n_s-n_c-R_n} \end{bmatrix} \oplus \begin{bmatrix} 0_{n_s-n_c-\overline{R}_{1d}-\overline{R}_{2d}} \\ \overline{u}_{1,d}(i) \oplus \overline{u}_{2,d}(i) \end{bmatrix} \\ u_{1,n}(i+1) \oplus u_{2,n}(i+1) \end{bmatrix}.$$

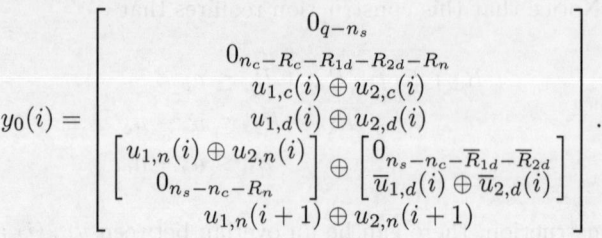

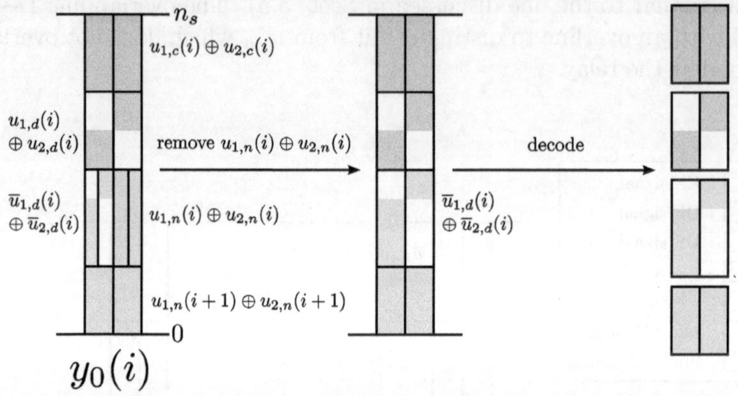

Fig. 14. The relay receives the superposition of $x_1(i)$ and $x_2(i)$ in the i-th channel use. First, it removes $u_{2,n}(i) \oplus u_{1,n}(i)$ which it knows from the decoding process in channel use $i-1$. Next, it decodes $u_{1,n}(i+1) \oplus u_{2,n}(i+1)$, $u_{1,c}(i) \oplus u_{2,c}(i)$, $u_{1,d}(i) \oplus u_{2,d}(i)$, and $\overline{u}_{1,d}(i) \oplus \overline{u}_{2,d}(i)$ which it forwards in channel use $i+1$ as shown in Fig. 13.

In the i-th channel use, the relay knows $u_{1,n}(i) \oplus u_{2,n}(i)$ from the decoding process in the channel use $i-1$. This allows it to remove $u_{1,n}(i) \oplus u_{2,n}(i)$ from $y_0(i)$ (see Fig. 14). Then, the relay can decode $u_{1,c}(i) \oplus u_{2,c}(i)$, $u_{1,d}(i) \oplus u_{2,d}(i)$, $\overline{u}_{1,d}(i) \oplus \overline{u}_{2,d}(i)$, and finally $u_{1,n}(i+1) \oplus u_{2,n}(i+1)$. At the end of channel use i, the relay constructs the following signal

$$x_r(i+1) = \begin{bmatrix} 0_{n_r-\overline{R}_{1d}-\overline{R}_{2d}-2R_c-R_{1d}-R_{2d}-R_n} \\ \overline{u}_{1,d}(i) \oplus \overline{u}_{2,d}(i) \\ u_{1,c}(i) \oplus u_{2,c}(i) \\ 0_{R_c} \\ u_{1,d}(i) \oplus u_{2,d}(i) \\ u_{1,n}(i+1) \oplus u_{2,n}(i+1) \\ 0_{q-n_r} \end{bmatrix} \quad (15)$$

and sends it in channel use $i+1$. The constituent signals of $x_r(i+1)$ in (15) fit in an interval of size n_r if

$$\overline{R}_{1d} + \overline{R}_{2d} + 2R_c + R_{1d} + R_{2d} + R_n \le n_r. \quad (16)$$

Decoding at the Destinations: In the following, consider the processing at node 3/destination 1 (the processing at node 4/destination 2 follows similar lines). Node 3/destination 1 will observe the top-most n_c bits of $x_2(i)$ plus the top-most n_r bits of $x_r(i)$ (modulo-2) at the i-th channel use. That is, we can write the received signal at node 3/destination 1 as

$$y_3(i) = \begin{bmatrix} 0_{q-n_r} \\ 0_{n_r-\overline{R}_{1d}-\overline{R}_{2d}-2R_c-R_{1d}-R_{2d}-R_n} \\ \overline{u}_{1,d}(i-1) \oplus \overline{u}_{2,d}(i-1) \\ u_{1,c}(i-1) \oplus u_{2,c}(i-1) \\ u_{2,c}(i) \\ u_{1,d}(i-1) \oplus u_{2,d}(i-1) \oplus u_{2,d}(i) \\ u_{1,n}(i) \oplus \overline{u}^{u}_{2,d}(i) \end{bmatrix}$$

where we used $\overline{u}^{u}_{2,d}(i)$ to denote the top-most $(n_c - n_s + \overline{R}_{1d} + \overline{R}_{2d} + R_n)^+$ bits of $\overline{u}_{2,d}(i)$. Notice the effect of CN: node 3/destination 1 receives $u_{1,n}(i)$ interference free (except the interference caused by the previously decoded $\overline{u}^{u}_{2,d}(i)$ which is not harmful) on the lowest R_n levels.

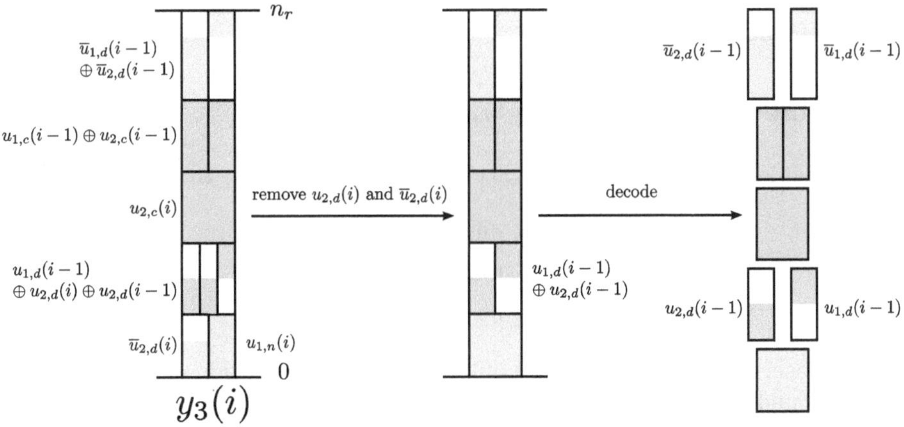

Fig. 15. The decoding process at node 3/destination 1. First, the receiver removes $u_{2,d}(i)$ and $\overline{u}_{2,d}(i)$ from $y_3(i)$ which it knows from the decoding process in channel use $i+1$ (backward decoding). Then, it decodes the signals $\overline{u}_{1,d}(i-1) \oplus \overline{u}_{2,d}(i-1)$, $u_{1,c}(i-1) \oplus u_{2,c}(i-1)$, followed by $u_{2,c}(i)$, $u_{1,d}(i-1) \oplus u_{2,d}(i-1)$ and $u_{1,n}(i)$. Notice how cooperative neutralization CN allows node 3/destination 1 to decode $u_{1,n}(i)$ interference free. Finally, $u_{1,c}(i)$ is extracted from $u_{2,c}(i)$ and $u_{1,c}(i) \oplus u_{2,c}(i)$ (known from the decoding process in block $i+1$).

Decoding at the receivers proceeds backwards. As shown in Fig. 15, node 3/destination 1 starts with removing $\overline{u}^{u}_{2,d}(i)$ and $u_{2,d}(i)$. Then, it decodes $\overline{u}_{1,d}(i-1)$, $\overline{u}_{2,d}(i-1)$, $u_{1,c}(i-1) \oplus u_{2,c}(i-1)$, $u_{2,c}(i)$, $u_{1,d}(i-1)$, $u_{2,d}(i-1)$, and $u_{1,n}(i)$. Decoding then proceeds backwards till $i = 1$.

Achievable Region: Collecting the rate constraints (12)-(14) and (16), we conclude that the non-negative rates $\overline{R}_{1,d}$, $\overline{R}_{2,d}$, $R_{1,d}$, $R_{2,d}$, R_c, and R_n can be achieved if they satisfy

$$\begin{aligned} R_c + R_{1d} + R_{2d} + R_n &\leq n_c \\ \overline{R}_{1d} + \overline{R}_{2d} &\leq n_s - n_c \\ R_n &\leq n_s - n_c \\ \overline{R}_{1d} + \overline{R}_{2d} + 2R_c + R_{1d} + R_{2d} + R_n &\leq n_r. \end{aligned}$$

Using Fourier Motzkin's elimination with $R_1 = R_{1d} + \overline{R}_{1d} + R_c + R_n$ and $R_2 = R_{2d} + \overline{R}_{2d} + R_c + R_n$, we can show that the following region is achievable

$$\begin{aligned} 0 \leq R_1 &\leq \min\{n_s, n_r\} \\ 0 \leq R_2 &\leq \min\{n_s, n_r\} \\ R_1 + R_2 &\leq n_s + n_c \\ R_1 + R_2 &\leq n_r + n_c \\ R_1 + R_2 &\leq n_r + n_s - n_c, \end{aligned}$$

which proves Lemma 2. At this point, we have finished the proof of the achievability of Thm. 2 for $n_c \leq n_r$.

7 Achievability for $n_c > n_r$

In this section, we prove Thm. 2 for the BFN with $n_c > n_r$, which reduces to

$$\begin{aligned} R_1 &\leq \min\{n_s, n_r + n_f, n_c\} \\ R_2 &\leq \min\{n_s, n_r + n_f, n_c\} \\ R_1 + R_2 &\leq n_c + [n_s - n_c]^+. \end{aligned}$$

In this case, n_f contributes to the outer bounds. If the region defined by these upper bounds is achievable, then feedback has a positive impact on the BFN. This is what we shall prove next. That is, we show that this region is in fact achievable, and hence that relay-source feedback increases the capacity of the network if $n_c > n_r$ when compared to the case $n_f = 0$.

We first show a toy example to explain the main ingredients of the achievable scheme. Then, depending on the relation between n_s and n_c, we split the proof of the achievability of Thm. 1 for $n_c > n_r$ to two cases: $\max\{n_r, n_s\} < n_c$ and $n_r < n_c \leq n_s$.

7.1 A Toy Example: Feedback Enlarges the Capacity Region

Consider a linear deterministic BFN with feedback where $(n_c, n_s, n_r) = (2, 3, 1)$. According to the upper bounds above, the capacity of this setup is outer bounded by

$$R_1 \leq 1$$
$$R_2 \leq 1$$

if $n_f = 0$ (please refer to Fig. 16). This capacity region is achieved using CN as shown in Fig. 17.

Now assume that this network has a feedback channel from the relay to the sources with capacity $n_f = 1$. The capacity in this case is outer bounded by

$$\begin{aligned} R_1 &\leq 2 \\ R_2 &\leq 2 \\ R_1 + R_2 &\leq 3 \end{aligned}$$

as shown in Fig. 16. In this case, the two sources can use the F strategy in Sect. 5.3 to exchange messages among each other. Then, the sources can use their cross channels n_c to send some more bits and achieve higher rates. This idea is illustrated and described in the caption of Fig. 18, where we show how to achieve the corner point $(2, 1)$ in Fig. 16. The other corner point can be achieved similarly by swapping the roles of the sources. The corner points $(2, 0)$ and $(0, 2)$ can be achieved by setting $u_{2,n}$ and $u_{1,n}$ to zero, respectively. The whole region is achievable by time sharing between corner points, and hence this scheme is optimal.

7.2 Case $\max\{n_r, n_s\} < n_c$

We start by stating the achievable region described in this subsection in the following lemma.

Lemma 3. *The rate region defined by the following inequalities*

$$\begin{aligned} 0 \leq R_1 &\leq \min\{n_s, n_r + n_f\} \\ 0 \leq R_2 &\leq \min\{n_s, n_r + n_f\} \\ R_1 + R_2 &\leq n_c, \end{aligned}$$

is achievable in the linear deterministic BFN with feedback with $\max\{n_r, n_s\} < n_c$.

Notice that this region matches the outer bound given in Thm. 1. Therefore, the achievability of this region proves the achievability of Thm. 2 for $\max\{n_r, n_s\} < n_c$. We prove this lemma in the rest of this subsection.

Since $n_c < n_r$ the sources can use the upper $n_c - n_r$ levels at the destinations which are not accessible by the relay to send feedback information to the destinations. Thus, in this case we use the F strategy. Since $\max\{n_c, n_r\} > n_s$ in this case, we also use the CF strategy following the intuition in Sect. 5.2. Furthermore, we use DF to achieve asymmetric rate pairs. The capacity achieving scheme is described next.

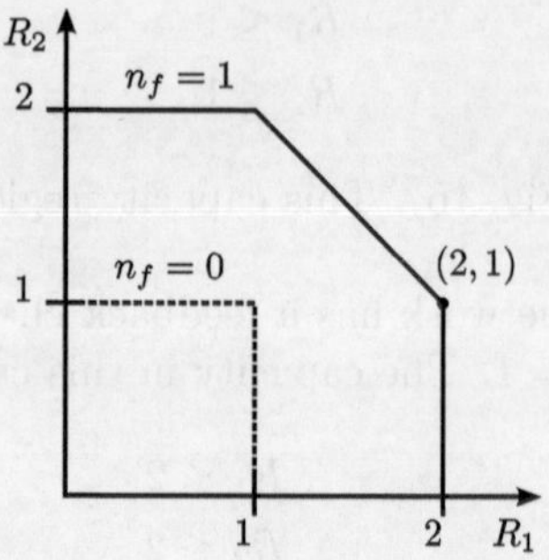

Fig. 16. Capacity regions for the deterministic BFN with $(n_c, n_s, n_r) = (2, 3, 1)$. Dotted line: without feedback ($n_f = 0$); solid line: with feedback with $n_f = 1$.

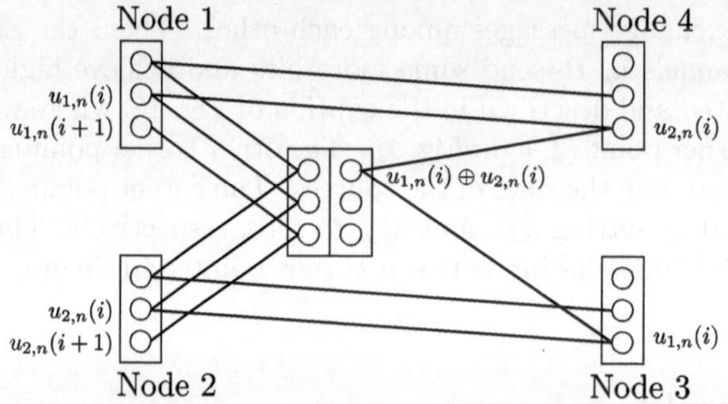

Fig. 17. The deterministic BFN with $(n_c, n_s, n_r) = (2, 3, 1)$ and $n_f = 0$. The given scheme achieves the corner point $(1, 1)$ of the capacity region in Fig. 16.

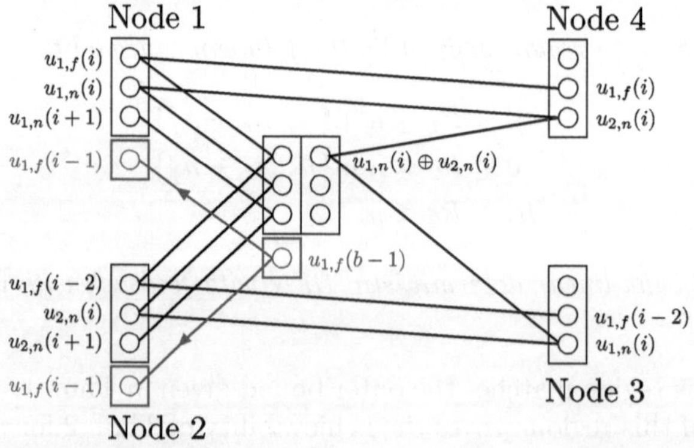

Fig. 18. The deterministic BFN with $(n_c, n_s, n_r) = (2, 3, 1)$ with $n_f = 1$. The sources use the same scheme as in Fig. 17 to achieve 1 bit each, and source 1 uses feedback to achieve one additional bit per channel use, thus enlarging the acievable region and achieving the optimal corner point $(2, 1)$ in Fig. 16. In the i-th channel use, node 1 sends $u_{1,f}(i)$ to the relay, the relay feeds back $u_{1,f}(i-1)$, which it decoded in channel use $i-1$, and node 2 sends $u_{1,f}(i-2)$ to node 3/destination 1.

Encoding: In the i-th channel use, node 1 sends the following signal (as shown in Fig. 19)

$$x_1(i) = \begin{bmatrix} u_{1,c}(i) \\ u_{1,d}(i) \\ u_{1,f}(i) \\ u_{2,f}(i-2) \\ \overline{u}_{1,f}(i) \\ \overline{u}_{2,f}(i-2) \\ 0_{n_c-R_c-R_{1d}-R_{2d}-R_{1f}-R_{2f}-2\overline{R}_f} \\ 0_{q-n_c} \end{bmatrix}. \tag{17}$$

Here, the signals $u_{1,f}$ is the signals used to establish the asymmetric F strategy, which is a vector of length R_{1f}. Similarly, $u_{2,f}(i-2)$ is the F-signal of node 2 of length R_{2f}, and is available at node 1 via feedback. The signal $\overline{u}_{1,f}$ is the signal used in the symmetric F strategy, and is a vector of length $\overline{R}_f$. We use both symmetric feedback and asymmetric feedback to achieve all points on the closure of the region given in Lemma 3. The C-signal $u_{1,c}$ has length R_c, and the D-signal $u_{1,d}$ has length $R_{1d}+R_{2d}$, containing information in the upper R_{1d} bits and zeros in the lower R_{2d} bits as described in Sect. 5.4. Node 2 sends a similar signal, with $u_{2,d}$ having zeros in the upper R_{1d} positions, and with $u_{1,f}(i)$ and $u_{2,f}(i-2)$ replaced by $u_{1,f}(i-2)$ and $u_{2,f}(i)$, respectively. The given construction works if

$$R_c + R_{1d} + R_{2d} + R_{1f} + R_{2f} + 2\overline{R}_f \leq n_c. \tag{18}$$

Relay Processing: The relay observes the top-most n_s bits of $x_1(i) \oplus x_2(i)$. We want the relay to be able to observe $u_{1,c}(i) \oplus u_{2,c}(i)$, $u_{1,d}(i) \oplus u_{2,d}(i)$, $u_{1,f}(i) \oplus u_{1,f}(i-2)$, $u_{2,f}(i-2) \oplus u_{2,f}(i)$, and $\overline{u}_{1,f}(i) \oplus \overline{u}_{2,f}(i)$. This is possible if we choose

$$R_c + R_{1d} + R_{2d} + R_{1f} + R_{2f} + \overline{R}_f \leq n_s. \tag{19}$$

Given this condition is satisfied, the relay starts by removing $u_{1,f}(i-2)$ and $u_{2,f}(i-2)$ (known from past decoding) from $y_0(i)$ as shown in Fig. 20. Next, it decodes the sum of the C-signals $u_{1,c}(i) \oplus u_{2,c}(i)$, the sum of the D-signals $u_{1,d}(i) \oplus u_{2,d}(i)$, the F-signals $u_{1,f}(i)$ and $u_{2,f}(i)$ and $\overline{u}_{1,f}(i) \oplus \overline{u}_{2,f}(i)$. Then it sends the following signal

$$x_r(i+1) = \begin{bmatrix} 0_{n_r-R_c-R_{1d}-R_{2d}} \\ u_{1,d}(i) \oplus u_{2,d}(i) \\ u_{1,c}(i) \oplus u_{2,c}(i) \\ 0_{q-n_r} \end{bmatrix},$$

over the forward channel (relay-destination channel) in channel use $i+1$, which requires

$$R_c + R_{1d} + R_{2d} \leq n_r. \tag{20}$$

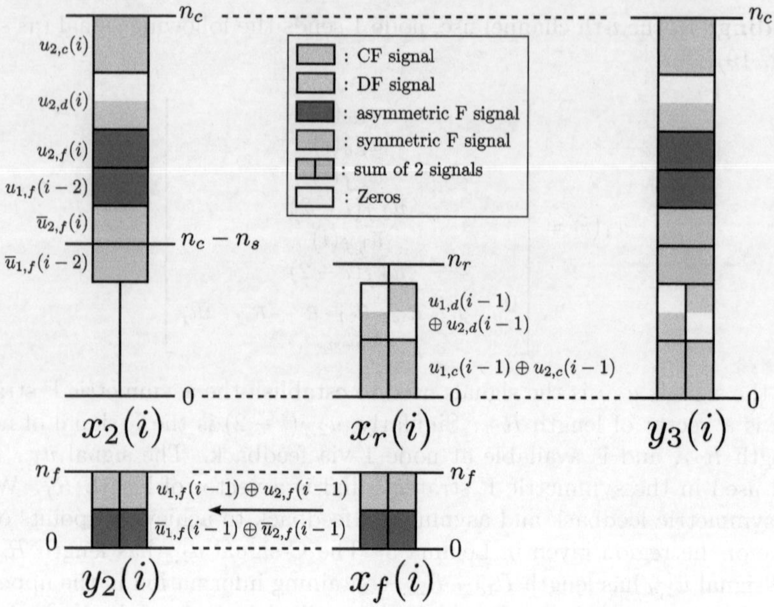

Fig. 19. A graphical illustration of the transmit and received signal at node 2, i.e., $x_2(i)$ and $y_2(i)$, the relay signal $x_r(i)$, the feedback signal $x_f(i)$, and the received signal at node 3/destination 1 $y_3(i)$, for the capacity achieving scheme of the linear deterministic BFN with feedback with $\max\{n_r, n_s\} \leq n_c$ with $n_f > 0$. Node 2 uses the received feedback signals to extract $u_{1,f}(i-1)$ and $\overline{u}_{1,f}(i-1)$ which it sends to node 3/destination 1 in channel use $i+1$. Node 3/destination 1 uses $u_{1,c}(i) \oplus u_{2,c}(i)$ (decoded in time channel use $i+1$) and $u_{2,c}(i)$ to extract $u_{1,c}(i)$.

It also sends the feedback signal $x_f(i+1)$ on the backward channel (feedback channel), where

$$x_f(i+1) = \begin{bmatrix} u_{1,f}(i) \oplus u_{2,f}(i) \\ \overline{u}_{1,f}(i) \oplus \overline{u}_{2,f}(i) \\ 0_{q-R_f} \end{bmatrix},$$

in channel use $i+1$. The signal x_f represents feedback to nodes 1 and 2. Note that we feed back the signal $u_{1,f} \oplus u_{2,f}(i)$ instead of separately sending $u_{1,f}$ and $u_{2,f}(i)$. This allows a more efficient use of the feedback channel. If the vectors $u_{1,f}$ and $u_{2,f}(i)$ have different lengths, the shorter is zero padded till they have equal length.

The construction of the feedback signal $x_f(i+1)$ requires

$$R_{1f} + \overline{R}_f \leq n_f \tag{21}$$

$$R_{2f} + \overline{R}_f \leq n_f. \tag{22}$$

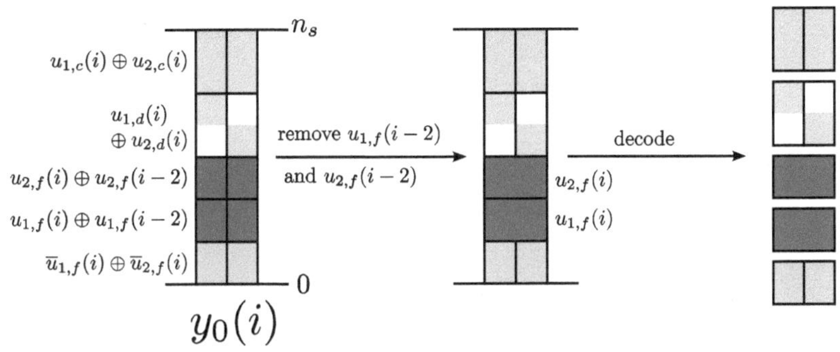

Fig. 20. The decoding process at the relay. In the i-th channel use, the relay starts with removing the known signals $u_{1,f}(i-2)$ and $u_{2,f}(i-2)$ (which it decoded in channel use $i-2$). Then, it decodes $u_{1,c}(i) \oplus u_{2,c}(i)$, $u_{1,d}(i) \oplus u_{2,d}(i)$ and $\overline{u}_{1,f}(i) \oplus \overline{u}_{2,f}(i)$ as well as $u_{2,f}(i)$ and $u_{1,f}(i)$. The F-signals are then sent back to the sources, and the DF and C-signals to the destinations as illustrated in Fig. 19.

Processing Feedback at the Sources: Consider node 1 at time instant $i+1$. Node 1 receives the feedback signal given by

$$y_1(i+1) = \mathbf{S}^{q-n_f} x_f(i+1) = \begin{bmatrix} 0_{q-n_f} \\ u_{1,f}(i) \oplus u_{2,f}(i) \\ \overline{u}_{1,f}(i) \oplus \overline{u}_{2,f}(i) \\ 0_{n_f - R_f} \end{bmatrix}.$$

Node 1 decodes $u_{1,f}(i) \oplus u_{2,f}(i)$ and $\overline{u}_{1,f}(i) \oplus \overline{u}_{2,f}(i)$. Since node 1 knows its own F-signal $u_{1,f}(i)$, then it can extract $u_{2,f}(i)$ from this feedback information. Similarly, it can extract $\overline{u}_{2,f}(i)$. Therefore, in channel use $i+2$, node 1 knows the F-signals of node 2 which are $u_{2,f}(i)$ and $\overline{u}_{2,f}(i)$ which justifies the transmission of $u_{2,f}(i-2)$ and $\overline{u}_{2,f}(i-2)$ in $x_1(i)$ in (17). After processing this feedback, node 1 is able to send node 2's F-signals to node 4/destination 2. A similar processing is performed at node 2, which sends the F-signals of node 1 to node 3/destination 1.

Decoding at the Destinations: Assume that

$$2R_c + 2R_{1d} + 2R_{2d} + R_{1f} + R_{2f} + 2\overline{R}_f \leq n_c. \tag{23}$$

In this case, node 3/destination 1 for instance is able to observe all the signals sent by node 2 and the relay. The received signal $y_3(i)$ is then

$$y_3(i) = \begin{bmatrix} 0_{q-n_c} \\ u_{2,c}(i) \\ u_{2,d}(i) \\ u_{2,f}(i) \\ u_{1,f}(i-2) \\ \overline{u}_{2,f}(i) \\ \overline{u}_{1,f}(i-2) \\ 0_{n_c-2R_c-2R_{1d}-2R_{2d}-R_{1f}-R_{2f}-2\overline{R}_f} \\ u_{1,d}(i-1) \oplus u_{2,d}(i-1) \\ u_{1,c}(i-1) \oplus u_{2,c}(i-1) \end{bmatrix}.$$

Node 3/destination 1 decodes backwards starting with $i = N+2$ where the desired F-signals $u_{1,f}(N)$ and $\overline{u}_{1,f}(N)$ are decoded. In channel use $N+1$, node 3/destination 1 decodes the desired F-signals $u_{1,f}(N-1)$ and $\overline{u}_{1,f}(N-1)$, in addition to its desired D-signal $u_{1,d}(N)$ (obtained from $u_{1,d}(N) \oplus u_{2,d}(N)$) and the C-signal sum $u_{1,c}(N) \oplus u_{2,c}(N)$. Next, in the N-th channel use, it decodes $u_{2,c}(N)$, $u_{1,f}(N-2)$, $\overline{u}_{1,f}(N-2)$, $u_{1,d}(N-1)$, and $u_{1,c}(N-1) \oplus u_{2,c}(N-1)$. Then it adds $u_{1,c}(N) \oplus u_{2,c}(N)$ to $u_{2,c}(N)$ to obtain the desired C-signal $u_{1,c}(N)$. Decoding proceeds backwards till channel use $i = 1$. Similar processing is performed by node 4/destination 2. The number of bits recovered by node 3/destination 1 is $R_1 = R_c + R_{1d} + R_{1f} + \overline{R}_f$, and similarly node 4/destination 2 obtains $R_2 = R_c + R_{2d} + R_{2f} + \overline{R}_f$.

Achievable Region: Collecting the bounds (18), (19), (20), (21), (22), and (23), we see that a pair (R_1, R_2) with $R_1 = R_c + R_{1d} + R_{1f} + \overline{R}_f$ and $R_2 = R_c + R_{2d} + R_{2f} + \overline{R}_f$, where the rates R_c, R_{1d}, R_{2d}, R_{1f}, R_{2f}, $\overline{R}_f$ are non-negative, is achievable if

$$\begin{aligned} R_c + R_{1d} + R_{2d} + R_{1f} + R_{2f} + \overline{R}_f &\leq n_s \\ R_c + R_{1d} + R_{2d} &\leq n_r \\ R_{1f} + \overline{R}_f &\leq n_f \\ R_{2f} + \overline{R}_f &\leq n_f \\ 2R_c + 2R_{1d} + 2R_{2d} + R_{1f} + R_{2f} + 2\overline{R}_f &\leq n_c. \end{aligned}$$

Solving this set of linear inequalities using the Fourier Motzkin elimination, we get the achievable region given by

$$0 \leq R_1 \leq \min\{n_s, n_r + n_f\} \tag{24}$$

$$0 \leq R_2 \leq \min\{n_s, n_r + n_f\} \tag{25}$$

$$R_1 + R_2 \leq n_c, \tag{26}$$

which proves Lemma 3. This also proves Thm. 2 for the case $\max\{n_r, n_s\} < n_c$.

7.3 Case $n_r < n_c \leq n_s$

In this case, the relay observes more bits than the destinations since $n_s \geq n_c$. Thus, the sources can exploit the additional $n_s - n_c$ bits by using the CN strategy of Sect. 5.1. Additionally, we use the F strategy for feedback, and the DF strategy to achieve asymmetric rates. In the rest of this subsection, we prove the following lemma.

Lemma 4. *The region defined by*

$$\begin{aligned} 0 \leq R_1 &\leq \min\{n_r + n_f, n_c\} \\ 0 \leq R_2 &\leq \min\{n_r + n_f, n_c\} \\ R_1 + R_2 &\leq n_s, \end{aligned}$$

is achievable in the linear deterministic BFN with feedback with $n_r < n_c \leq n_s$.

This lemma proves Thm. 2 for the given case since the achievable region of this lemma matches the outer bound given in Thm. 1. Next, we describe the scheme which achieves the region in Lemma 4. The transmit signals of node 2 and the relay, and the received signals at node 2 and node 3/destination 1 for the capacity achieving scheme are depicted graphically in Fig. 21.

Encoding: In this case, node 1 sends a D-signal vector $u_{1,d}(i)$ of length $R_{1d} + R_{2d}$ (zero padded as explained in Sect. 5.4), two N-signal vectors $u_{1,n}(i)$ and $u_{1,n}(i+1)$ of length R_n each, two F-signal vectors $u_{1,f}(i)$ (asymmetric) and $\overline{u}_{1,f}(i)$ (symmetric) of length R_{1f} and $\overline{R}_f$, respectively. Additionally, it sends the F-signals of node 2 (acquired through feedback) $u_{2,f}(i-2)$ and $\overline{u}_{2,f}(i-2)$ of length R_{2f} and $\overline{R}_f$, respectively, as shown if Fig. 21.

Notice that out of these signals, two do not have to be observed at the destinations, namely $u_{1,n}(i+1)$ and $\overline{u}_{1,f}(i)$. These two signals have to be decoded at the relay to establish the F and the CN strategies. Thus, these signals can be sent below the noise floor of the destinations, i.e., in the lower $n_s - n_c$ levels observed at the relay. Assume that these signals do not fit in this interval of length $n_s - n_c$, i.e., $R_n + \overline{R}_f > n_s - n_c$. In this case, a part of these signals is sent below the noise floor, and a part above it. For this reason, we split these signals to two parts:

$$\overline{u}_{1,f}(i) = \begin{bmatrix} \overline{u}_{1,f}^{[1]}(i) \\ \overline{u}_{1,f}^{[2]}(i) \end{bmatrix}, \quad u_{1,n}(i) = \begin{bmatrix} u_{1,n}^{[1]}(i) \\ u_{1,n}^{[2]}(i) \end{bmatrix},$$

where $\overline{u}_{1,f}^{[m]}$ has length $\overline{R}_f^{[m]}$ and $u_{1,n}^{[m]}$ has length $R_n^{[m]}$, $m = 1, 2$, such that $\overline{R}_f^{[1]} + \overline{R}_f^{[2]} = \overline{R}_f$ and $R_n^{[1]} + R_n^{[2]} = R_n$ (this split is not shown in Fig. 21 for clarity). As a result, node 1 sends

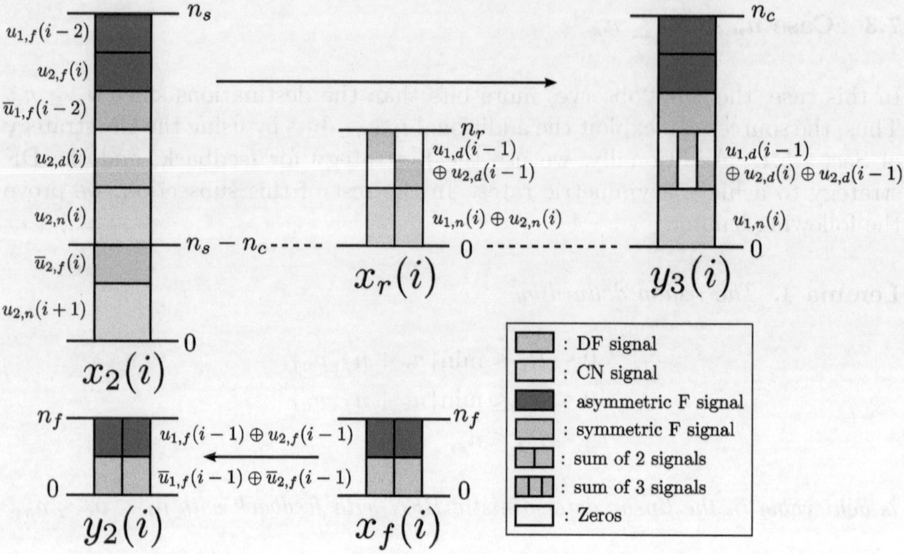

Fig. 21. The transmit signal and received signal of node 2, the transmit signals of the relay, and the received signal of node 3/destination 1 for the capacity achieving scheme of the linear deterministic BFN with feedback with $n_r < n_c \leq n_s$ and $n_f > 0$. Node 2 makes use of the feedback signals $u_{1,f}(i-1) \oplus u_{2,f}(i-1)$ and $\overline{u}_{1,f}(i-1) \oplus \overline{u}_{2,f}(i-1)$ to extract $u_{1,f}(i-1)$ and $\overline{u}_{1,f}(i-1)$ which are sent to node 3/destination 1 in channel use $i+1$. Node 3/destination 1 decodes $u_{1,f}(i-2)$, $\overline{u}_{1,f}(i-2)$, $u_{1,d}(i-1)$, $u_{2,d}(i-1)$, and $u_{1,n}(i)$ in channel use i.

$$x_1(i) = \begin{bmatrix} 0_{n_c - R_{1f} - R_{2f} - 2\overline{R}_f^{[1]} - \overline{R}_f^{[2]} - R_{1d} - R_{2d} - 2R_n^{[1]} - R_n^{[2]}} \\ u_{1,f}(i) \\ u_{2,f}(i-2) \\ \overline{u}_{2,f}^{[1]}(i-2) \\ \overline{u}_{2,f}^{[2]}(i-2) \\ \overline{u}_{1,f}^{[1]}(i) \\ u_{1,n}^{[1]}(i+1) \\ u_{1,d}(i) \\ u_{1,n}^{[1]}(i) \\ u_{1,n}^{[2]}(i) \\ \overline{u}_{1,f}^{[2]}(i) \\ u_{1,n}^{[2]}(i+1) \\ 0_{n_s - n_c - \overline{R}_f^{[2]} - R_n^{[2]}} \\ 0_{q-n_s} \end{bmatrix}. \quad (27)$$

The vectors $\overline{u}_{1,f}^{[1]}(i)$ and $u_{1,n}^{[1]}(i+1)$ are sent above $u_{1,d}(i)$, $u_{1,n}^{[1]}(i)$, and $u_{1,n}^{[2]}(i)$ since that latter signals have to align with the signals sent from the relay (see Sect. 5.1 and 5.3), where the relay can only access lower levels since $n_r < n_c$ in this case. The transmit signal of node 2, $x_2(i)$, is constructed similarly, by replacing $u_{1,f}(i)$ and $u_{2,f}(i-2)$ with $u_{1,f}(i-2)$ and $u_{2,f}(i)$, respectively, and replacing the user index of the other signals with 2. This construction requires

$$R_{1f} + R_{2f} + 2\overline{R}_f^{[1]} + \overline{R}_f^{[2]} + R_{1d} + R_{2d} + 2R_n^{[1]} + R_n^{[2]} \leq n_c \tag{28}$$

$$\overline{R}_f^{[2]} + R_n^{[2]} \leq n_s - n_c. \tag{29}$$

Relay Processing: The relay receives the top-most n_s bits of $x_1(i) \oplus x_2(i)$. We write $y_0(i)$ as

$$y_0(i) = \begin{bmatrix} 0_{q-n_s} \\ 0_{n_c - R_{1f} - R_{2f} - 2\overline{R}_f^{[1]} - \overline{R}_f^{[2]} - R_{1d} - R_{2d} - 2R_n^{[1]} - R_n^{[2]}} \\ u_{1,f}(i) \oplus u_{1,f}(i-2) \\ u_{2,f}(i) \oplus u_{2,f}(i-2) \\ \overline{u}_{1,f}^{[1]}(i-2) \oplus \overline{u}_{2,f}^{[1]}(i-2) \\ \overline{u}_{1,f}^{[2]}(i-2) \oplus \overline{u}_{2,f}^{[2]}(i-2) \\ \overline{u}_{1,f}^{[1]}(i) \oplus \overline{u}_{2,f}^{[1]}(i) \\ u_{1,n}^{[1]}(i+1) \oplus u_{2,f}^{[1]}(i+1) \\ u_{1,d}(i) \oplus u_{2,d}(i) \\ u_{1,n}^{[1]}(i) \oplus u_{2,f}^{[1]}(i) \\ u_{1,n}^{[2]}(i) \oplus u_{2,f}^{[2]}(i) \\ \overline{u}_{1,f}^{[2]}(i) \oplus \overline{u}_{2,f}^{[2]}(i) \\ u_{1,n}^{[2]}(i+1) \oplus u_{2,f}^{[2]}(i+1) \\ 0_{n_s - R_{1f} - R_{2f} - 2\overline{R}_f - R_{1d} - R_{2d} - 2R_n} \end{bmatrix}$$

The relay starts processing this signal by removing the past F-signals $u_{1,f}(i-2)$ and $u_{2,f}(i-2)$ (decoded in channel use $i-2$) from $y_0(i)$. Then it decodes the remaining signals as shown in Fig. 22.

Then, the relay forwards

$$x_r(i+1) = \begin{bmatrix} 0_{n_r - R_{1d} - R_{2d} - R_n^{[1]} - R_n^{[2]}} \\ u_{1,d}(i) \oplus u_{2,d}(i) \\ u_{1,n}^{[1]}(i+1) \oplus u_{2,f}^{[1]}(i+1) \\ u_{1,n}^{[2]}(i+1) \oplus u_{2,f}^{[2]}(i+1) \\ 0_{q-n_r} \end{bmatrix},$$

in channel use $i+1$. The given signals fit in the interval of length n_r if

$$R_{1d} + R_{2d} + R_n^{[1]} + R_n^{[2]} \leq n_r. \tag{30}$$

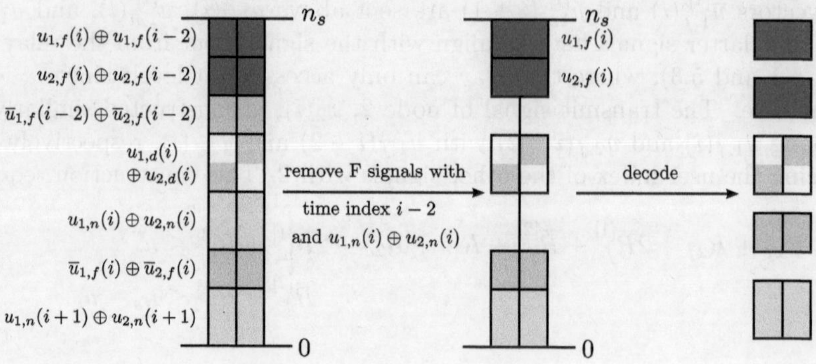

Fig. 22. The processing steps at the relay. The relay starts by removing its past-decoded signals. Then it decodes the signals $u_{1,f}(i)$, $u_{2,f}(i)$, $u_{1,d}(i) \oplus u_{2,d}(i)$, $\overline{u}_{1,f}(i) \oplus \overline{u}_{2,f}(i)$, and $u_{1,n}(i+1) \oplus u_{2,n}(i+1)$ in channel use i. In channel use $i+1$, the decoded F-signals are fed back to the sources, and the N-signal and D-signals are forwarded to the destinations.

The relay also sends a feedback signal $x_f(i+1)$ in channel use $i+1$ to node 1 and node 2 on the backward channel, where

$$x_f(i+1) = \begin{bmatrix} u_{1,f}(i) \oplus u_{2,f}(i) \\ \overline{u}_{1,f}^{[1]}(i) \oplus \overline{u}_{2,f}^{[1]}(i) \\ \overline{u}_{1,f}^{[2]}(i) \oplus \overline{u}_{2,f}^{[2]}(i) \\ 0_{n_f - \max\{R_{1f}, R_{2f}\} - \overline{R}_f^{[1]} - \overline{R}_f^{[2]}} \\ 0_{q-n_f} \end{bmatrix}.$$

For efficient use of the feedback channel, the relay adds the signals $u_{1,f}(i)$ and $u_{2,f}(i)$ together, and feeds the sum back. If these signals do not have the same length, the shorter is zero padded at the relay till both signals have the same length. This construction requires

$$R_{1f} + \overline{R}_f^{[1]} + \overline{R}_f^{[2]} \leq n_f \tag{31}$$

$$R_{2f} + \overline{R}_f^{[1]} + \overline{R}_f^{[2]} \leq n_f. \tag{32}$$

Processing Feedback at the Sources: Consider node 1 at channel use $i+1$. Node 1 receives the feedback signal

$$y_1(i+1) = \begin{bmatrix} 0_{q-n_f} \\ u_{1,f}(i) \oplus u_{2,f}(i) \\ \overline{u}_{1,f}^{[1]}(i) \oplus \overline{u}_{2,f}^{[1]}(i) \\ \overline{u}_{1,f}^{[2]}(i) \oplus \overline{u}_{2,f}^{[2]}(i) \\ 0_{n_f - \max\{R_{1f}, R_{2f}\} - \overline{R}_f^{[1]} - \overline{R}_f^{[2]}} \end{bmatrix}.$$

Node 1 then subtracts its own F-signals from $y_1(i+1)$, and obtains the F-signals of node 2, i.e., $u_{2,f}(i)$, $\overline{u}_{2,f}^{[1]}(i)$, and $\overline{u}_{2,f}^{[2]}(i)$. These signals are sent in channel use $i+2$ as seen in (27).

Decoding at the Destinations: In the i-th channel use, node 3/destination 1 observes

$$y_3(i) = \begin{bmatrix} 0_{q-n_c} \\ 0_{n_c-R_{1f}-R_{2f}-2\overline{R}_f^{[1]}-\overline{R}_f^{[2]}-R_{1d}-R_{2d}-2R_n^{[1]}-R_n^{[2]}} \\ u_{1,f}(i-2) \\ u_{2,f}(i) \\ \overline{u}_{1,f}^{[1]}(i-2) \\ \overline{u}_{1,f}^{[2]}(i-2) \\ \overline{u}_{2,f}^{[1]}(i) \\ u_{2,n}^{[1]}(i+1) \\ u_{2,d}(i) \oplus u_{1,d}(i-1) \oplus u_{2,d}(i-1) \\ u_{1,n}^{[1]}(i) \\ u_{1,n}^{[2]}(i) \end{bmatrix}.$$

Decoding at node 3/destination 1 is done in a backward fashion. In the i-th channel use, it starts with removing the already known D-signal $u_{2,d}(i)$ (decoded in channel use $i+1$). Then it proceeds with decoding each of

$$u_{1,f}(i-2),\quad \overline{u}_{1,f}^{[1]}(i-2),\quad \overline{u}_{1,f}^{[2]}(i-2),\quad u_{1,d}(i-1),\quad u_{2,d}(i-1),\quad u_{1,n}^{[1]}(i),\quad u_{1,n}^{[2]}(i).$$

It recovers its desired signals for a total rate of $R_1 = R_{1f} + \overline{R}_f^{[1]} + \overline{R}_f^{[2]} + R_{1d} + R_n^{[1]} + R_n^{[2]}$. Similarly, node 4/destination 2 recovers $R_2 = R_{2f} + \overline{R}_f^{[1]} + \overline{R}_f^{[2]} + R_{2d} + R_n^{[1]} + R_n^{[2]}$ bits per channel use.

Achievable Region: Collecting the bounds (28), (29), (30), (31), and (32) we get

$$\begin{aligned} R_{1f} + R_{2f} + 2\overline{R}_f^{[1]} + \overline{R}_f^{[2]} + R_{1d} + R_{2d} + 2R_n^{[1]} + R_n^{[2]} &\le n_c \\ \overline{R}_f^{[2]} + R_n^{[2]} &\le n_s - n_c \\ R_{1d} + R_{2d} + R_n^{[1]} + R_n^{[2]} &\le n_r \\ R_{1f} + \overline{R}_f^{[1]} + \overline{R}_f^{[2]} &\le n_f \\ R_{2f} + \overline{R}_f^{[1]} + \overline{R}_f^{[2]} &\le n_f, \end{aligned}$$

where the rates R_{1f}, R_{2f}, $\overline{R}_f^{[1]}$, $\overline{R}_f^{[2]}$, R_{1d}, R_{2d}, $R_n^{[1]}$, and $R_n^{[2]}$ are non-negative. Solving this set in linear inequalities using the Fourier Motzkin elimination with

$R_1 = R_{1f} + \overline{R}_f^{[1]} + \overline{R}_f^{[2]} + R_{1d} + R_n^{[1]} + R_n^{[2]}$ and $R_2 = R_{2f} + \overline{R}_f^{[1]} + \overline{R}_f^{[2]} + R_{2d} + R_n^{[1]} + R_n^{[2]}$ yields the following achievable rate region

$$0 \leq R_1 \leq \min\{n_r + n_f, n_c\} \tag{33}$$
$$0 \leq R_2 \leq \min\{n_r + n_f, n_c\} \tag{34}$$
$$R_1 + R_2 \leq n_s, \tag{35}$$

which proves Lemma 4. By the end of this section, we finish the proof of Thm. 2.

8 Net Feedback Gain

At this point, it is clear that relay-source feedback link can increases the capacity of the BFN with respect to the non-feedback case. However, is this feedback efficient? In other words, is there a *net-gain* when using feedback? In this section, we discuss the net-gain attained by exploiting feedback and we answer the question above in the affirmative.

First, let us define what we mean by net-gain. Let C_0 be the sum-capacity of a BFN without feedback ($n_f = 0$), and let C_{n_f} be the sum-capacity with feedback ($n_f \neq 0$), which is achieved by feeding back r_f bits per channel use through the feedback channel. Let η be defined as the ratio

$$\eta = \frac{C_{n_f} - C_0}{r_f}.$$

We say that we have a net-gain if the ratio of the sum-capacity increase to the number of feedback bits is larger than 1, i.e., $\eta > 1$. Otherwise, if $\eta \leq 1$, then we have no net-gain because then $C_{n_f} - C_0 \leq r_f$, i.e., the gain is less than the number of bits sent over the feedback channel.

Note that if $n_c \leq n_r$, then there is no feedback gain at all, since in this case, the capacity region in Thm. 2 is the same as $n_f = 0$.

Now, consider for sake of example the case $n_c > n_f$ with a BFN with $(n_c, n_s, n_r) = (6, 3, 1)$. The capacity region of this BFN without feedback is shown in Fig. 23. The no-feedback sum-capacity of this network is $C_0 = 2$ bits per channel use corresponding to the rate pair $(R_1, R_2) = (1, 1)$. This rate pair is achieved by using the CF strategy, where node 1 sends $x_1(i) = [u_{1,c}(i),\ 0_5^T]^T$ and node 2 sends $x_2(i) = [u_{2,c}(i),\ 0_5^T]^T$, and the relay sends $x_r(i) = [u_{1,c}(i-1) \oplus u_{2,c}(i-1),\ 0_5]^T$. Now consider the case with $n_f = 1$. In this case, the sum-capacity is $C_1 = 4$ bits per channel use corresponding to the corner point of the capacity region $(R_1, R_2) = (2, 2)$ as shown in Fig. 23. To achieve this, the sources use the same CF strategy used for $n_f = 0$, which achieves $R_1 = R_2 = 1$ bit per channel use. Additionally each source sends a feedback bit $u_{j,f}(i)$ to the other source via the relay using the symmetric F strategy. This way, each source acquires the F-signal of the other source, which it forwards then to the respective destination. This F strategy requires feeding back only $r_f = 1$ bit, namely $u_{1,f}(i) \oplus u_{2,f}(i)$. With this we have

$$\eta = \frac{C_1 - C_0}{r_f} = \frac{4-2}{1} = 2,$$

i.e., a net-gain: *for each feedback bit, we gain 2 bits in the sum-capacity.*

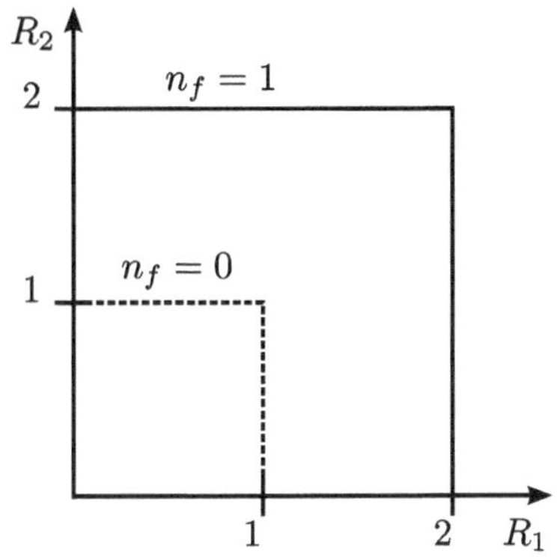

Fig. 23. The capacity region of the deterministic BFN with $(n_c, n_s, n_r) = (6, 3, 1)$ with $(n_f = 1)$ and without $(n_f = 0)$ feedback

9 Summary

We have studied the butterfly network with relay-source feedback and examined the benefit of feedback for this network. We have derived capacity upper bounds, and proposed transmission schemes that exploit the feedback channel. The result was a characterization of the capacity region of the network. While feedback does not affect the capacity of the network in some cases, it does enlarge its capacity region in other cases. Moreover, the proposed feedback scheme which is based on bi-directional relaying is an efficient form of feedback, it provides a net-gain in the regimes where feedback helps. It turns out that the increase in the sum-capacity of the network is twice the number of feedback bits.

Acknowledgement. The work of A. Chaaban and A. Segzin is supported by the German Research Foundation, Deutsche Forschungsgemeinschaft (DFG), Germany, under grant SE 1697/7.

The work of Dr. D. Tuninetti was partially funded by NSF under award number 0643954; the contents of this article are solely the responsibility of the author and do not necessarily represent the official views of the NSF. The work of Dr. D. Tuninetti was possible thanks to the generous support of Telecom-ParisTech, Paris France, while the author was on a sabbatical leave at the same institution.

References

1. Ahlswede, R.: Multi-way communication channels. In: Proc. of 2nd International Symposium on Info. Theory, Tsahkadsor, Armenian S.S.R., pp. 23–52 (1971)
2. Ahlswede, R., Cai, N., Li, S.Y.R., Yeung, R.W.: Network information flow. IEEE Trans. on Inf. Theory 46(4), 1204–1216 (2000)
3. Avestimehr, A.S., Diggavi, S., Tse, D.: A deterministic approach to wireless relay networks. In: Proc. of Allerton Conference (2007)
4. Avestimehr, A.S., Ho, T.: Approximate capacity of the symmetric half-duplex Gaussian butterfly network. In: Proc. of the IEEE Information Theory Workshop (ITW), pp. 311–315 (2009)
5. Avestimehr, A.S., Sezgin, A., Tse, D.: Capacity of the two-way relay channel within a constant gap. European Trans. in Telecommunications (2009)
6. Carleial, A.B.: Interference channels. IEEE Trans. on Inf. Theory 24(1), 60–70 (1978)
7. Chaaban, A., Sezgin, A.: Achievable rates and upper bounds for the Gaussian interference relay channel. IEEE Trans. on Inf. Theory 58(7), 4432–4461 (2012)
8. Cover, T.M., El-Gamal, A.: Capacity theorems for the relay channel. IEEE Trans. on Inf. Theory IT-25(5), 572–584 (1979)
9. Cover, T., Thomas, J.: Elements of Information Theory, 2nd edn. John Wiley and Sons, Inc. (2006)
10. Gamal, A.E., Kim, Y.H.: Network Information Theory. Cambridge University Press (2011)
11. Han, T.S., Kobayashi, K.: A new achievable rate region for the interference channel. IEEE Trans. on Inf. Theory IT-27(1), 49–60 (1981)
12. Kim, S., Devroye, N., Mitran, P., Tarokh, V.: Comparisons of bi-directional relaying protocols. In: Proc. of the IEEE Sarnoff Symposium, Princeton, NJ (2008)
13. Mariç, I., Dabora, R., Goldsmith, A.J.: Relaying in the presence of interference: achievable rates, interference forwarding, and outer bounds. IEEE Trans. on Info. Theory 58(7), 4342–4354 (2012)
14. Narayanan, K., Wilson, M.P., Sprintson, A.: Joint physical layer coding and network coding for bi-directional relaying. In: Proc. of the Forty-Fifth Allerton Conference, Illinois (2007)
15. Nazer, B., Gastpar, M.: Compute-and-forward: harnessing interference through structured codes. IEEE Trans. on Inf. Theory 57(10), 6463–6486 (2011)
16. Rankov, B., Wittneben, A.: Spectral efficient signaling for half-duplex relay channels. In: Proc. of the Asilomar Conference on Signals, Systems, and Computers, Pacific Grove, CA (2005)
17. Sahin, O., Erkip, E.: Achievable rates for the Gaussian interference relay channel. In: Proc. of 2007 GLOBECOM Communication Theory Symposium, Washington D.C (2007)
18. Tuninetti, D.: An outer bound for the memoryless two-user interference channel with general cooperation. In: Proc. of the IEEE Information Theory Workshop (ITW), Lausanne (2012)
19. Yang, E., Tuninetti, D.: Interference channels with source cooperation in the strong cooperation regime: symmetric capacity to within 2 bits/s/Hz with dirty paper coding. In: Proc. of 42nd Asilomar Conference on Signals, Systems and Computers, Pacific Grove, CA (2011)

Appendix

A Proof of Thm. 2

We set $X_0 = (X_r, X_f)$ and use the output definition in (7) in the outer bounds in Section 2.2.

From the cut-set bound in (2a) we have

$$\begin{aligned} R_1 &\leq I(X_1; Y_0, Y_2, Y_3 | X_0, X_2) \\ &= I(X_1; Y_0, Y_2, Y_3 | X_r, X_f, X_2) \\ &= H(Y_0, Y_2, Y_3 | X_r, X_f, X_2) - H(Y_0, Y_2, Y_3 | X_r, X_f, X_2, X_1) \\ &= H(\mathbf{S}^{q-n_s} X_1 | X_r, X_f, X_2) \\ &\leq H(\mathbf{S}^{q-n_s} X_1) \\ &\leq n_s, \end{aligned}$$

Similarly, the cut-set bounds in (2c) and (2d) reduce to

$$\begin{aligned} R_1 &\leq n_r + n_f, \\ R_1 &\leq \max\{n_c, n_r\}, \end{aligned}$$

respectively. These bounds combined give (8a). Similarly, the bound in (8b) for R_2 follows by the symmetry in the network.

The sum-rate cut-set bound in (3b) becomes

$$\begin{aligned} R_1 + R_2 &\leq I(X_0, X_1, X_2; Y_3, Y_4) \\ &= I(X_r, X_f, X_1, X_2; Y_3, Y_4) \\ &= H(Y_3, Y_4) \\ &= H(Y_3) + H(Y_4 | Y_3), \end{aligned}$$

which leads to

$$R_1 + R_2 \leq \max\{n_r, n_c\} + n_c. \tag{36}$$

These are the neccessary cut-set upper bounds for our problem. The remaining cut-set bounds are redundant given the cooperation bounds that we derive next, and are thus omitted.

Next, we evaluate the cooperation bounds in (5). In the symmetric case, bounds (5c), (5d), (5g), and (5h) are equivalent to bounds (5b), (5a), (5f), and (5e), respectively. Notice also that due to symmetry, the bounds (5b) and (5e) are similar. Thus, we need only to specialize the bounds (5a), (5b), and (5f) to the linear deterministic BFN with feedback. It turns out that the bound (5a) for the linear deterministic BFN with feedback is redundant given (36). Thus, we omit its derivation.

Next, we consider the bound in (5b), which yields

$$\begin{aligned}R_1 + R_2 &\leq I(X_2; Y_4, Y_1, Y_0|Y_3, X_1, X_0) + I(X_1, X_0, X_2; Y_3)\\&= H(Y_4, Y_1, Y_0|Y_3, X_1, X_r, X_f) - H(Y_4, Y_1, Y_0|Y_3, X_1, X_r, X_f, X_2)\\&\quad + H(Y_3) - H(Y_3|X_1, X_0, X_2)\\&= H(\mathbf{S}^{q-n_s}X_2|\mathbf{S}^{q-n_c}X_2, X_1, X_r, X_f) + H(\mathbf{S}^{q-n_c}X_2 + \mathbf{S}^{q-n_r}X_r)\\&\leq (n_s - n_c)^+ + \max\{n_c, n_r\}.\end{aligned}$$

Notice that this bound can be tighter than the sum-rate cut-set bound in (36) and is equal to (8d).

Finally, the bound in (5f) becomes

$$\begin{aligned}R_1 + R_2 &\leq I(X_2; Y_4, Y_1|Y_3, Y_0, X_1, X_0) + I(X_1, X_2; Y_3, Y_0|X_0)\\&= H(Y_4, Y_1|Y_3, Y_0, X_1, X_r, X_f) - H(Y_4, Y_1|Y_3, Y_0, X_1, X_r, X_f, X_2)\\&\quad + H(Y_3, Y_0|X_r, X_f) - H(Y_3, Y_0|X_r, X_f, X_1, X_2)\\&= H(\mathbf{S}^{q-n_c}X_2, \mathbf{S}^{q-n_s}X_1 + \mathbf{S}^{q-n_s}X_2|X_r, X_f)\\&\leq n_s + n_c.\end{aligned}$$

This bound yields (8e).

Uniformly Generating Origin Destination Tables

David M. Einstein and Lee K. Jones*

Department of Mathematical Sciences, University of Massachusetts Lowell
deinst@gmail.com, Lee_Jones@uml.edu

Dedicated in memory of Prof. Rudolf Ahlswede for his valuable contributions to combinatorics, probability and statistics, computer science and information theory

Abstract. Given a network and the total flows into and out of each of the sink and source nodes, it is useful to select uniformly at random an origin-destination (O-D) matrix for which the total in and out flows at sinks and sources (column and row sums) matches the given data. We give an algorithm for small networks (less than 16 nodes) for sampling such O-D matrices with exactly the uniform distribution and apply it to traffic network analysis. This algorithm also can be applied to communication networks and used in the statistical analysis of contingency tables.

Keywords: p-quantile, robust traffic control, O-D matrix, importance sampling, rejection method.

1 Introduction

For a traffic network or sub-network, only partial information concerning the flow (in vehicles/min) is available. In many cases, noisy measurements of flow on the links of the network allow only source and sink total out and in flows to be fairly accurately estimated, but path flows from a given source to a given sink are not accurately known. We assume here that each source is also a possible sink so that the O-D matrix is square but that there are no loops from a source to itself so that the matrix has zeros on the diagonal. Traffic planners desire a full O-D matrix of flows from each source to each sink so that they may find equilibrium and system optimal routes for travelers. Total cost may then be determined for existing or planned networks or sub-networks. Knowledge of the O-D matrix would then further allow traffic planners to minimize cost by optimizing over signal settings.

Several approaches exist for network analysis under the O-D uncertainty. In [Xie et al.(2010)] a maximum entropy estimate of the O-D matrix is derived using available link flow measurements. In [Jones et al.(2012)] the unknown O-D matrices are characterized by the discrete manifold of all non-negative integer

* Lee K. Jones was supported by a grant from the New England University Traffic Center.

H. Aydinian, F. Cicalese, and C. Deppe (Eds.): Ahlswede Festschrift, LNCS 7777, pp. 209–216, 2013.

valued matrices with row and column sums fixed corresponding to the given total out and in flows. Probability distributions on the manifold are then proposed based on additional data to represent the uncertainty of actual O-D values. Using these distributions Bayesian and minimax strategies are developed for optimal traffic signal controls.

The uniform distribution on the set of possible O-D tables plays an important role in the analysis; as in [Jones et al.(2012)], one may require cost estimates assuming a conditional discrete Gaussian distribution on the manifold. Uniform generation of tables from the manifold can be used via the accept-reject method to generate the conditional Gaussian and hence the cost estimate for use in finding optimal controls.

We may also consider the uniform distribution as least informative on the manifold and calculate, for fixed controls and fixed traveler routing, the p-quantile cost for this distribution (that cost which exceeds exactly $100p\%$ of the costs under the uniform assumption). This criterion can then be optimized appropriately. We estimate p-quantiles here for a simple network.

The problem of generating such matrices (allowing nonzero diagonals) is also important in statistical analysis [Diaconis and Efron(1985)] for testing independence of row-column effects after sampling two properties of individuals in a population.

Jones et al [Jones et al.(2012)] have analyzed traffic systems by uniformly generating origin destination tables given the flows into and out of each sink and source node. The algorithm for generating such matrices is not specified in the paper, and is given here.

There are a number of ways to generate origin destination matrices. We can consider the polytope formed by the constraints (sums of flows in, sums of flows out, zeros on the diagonal) and use the Hit and Run algorithm [Dyer et al.(1995)] [Kannan et al.(1997)] [Lovasz and Simonovits(1993)]. Unfortunately, although these algorithms seem to converge quickly, it is difficult to get precise estimates of the mixing times.

The Sequential Importance Sampling (SIS) algorithm of [Chen et al.(2005)] can also be used. It has been modified by Chen [Chen(2007)] for tables that have zeros in specified locations. Unfortunately, it is known to behave badly in certain situations [Bezáková et al.(2006)].

We modify the "ordinary" method of [Holmes and Jones(1996)] (described below) to generate tables with exactly the uniform distribution where the diagonal elements must be zero. Consider $I \times J$ tables with column sums c_j and row sums r_i. We usually arrange the row sums in increasing order to get maximum efficiency of the algorithm but this is not necessary for the algorithm to work. We present the H-J algorithm for general $I \times J$ matrices and the modified algorithm for $I = J$.

2 Exact Sampling

In the following rejection algorithms, we generate the tables by uniformly sampling one row at a time, abandoning the attempt, and starting over from the first row if any of the column sums become too large. We start by restating the rejection algorithms from [Holmes and Jones(1996)].

2.1 Ordinary Naive Rejection Algorithm

For the naive rejection algorithm, we generate rows uniformly constrained only by the row sum. This is easy to do, but often inefficient as a single entry may be larger than the column sum.

1. Choose the first row vector uniformly among all those with component sum r_1. Such vectors are generated from the uniform generation of 0-1 sequences with r_1 1s and $J-1$ 0s. (Vector components are the number of consecutive 1s before the first 0, after the first 0 and before the second 0, ... after the last 0. There are $\binom{r_1+J-1}{J-1}$ such sequences.)
2. Check that the jth entry does not exceed c_j. If it does reject and restart at 1. Otherwise generate a second vector with component sum equal to r_2 as in step 1. Again check that, for each j, the sum of the jth entries of vectors constructed do not exceed c_j. If not reject and restart at 1.
3. Continue until $I-1$ vectors have been generated without rejection. An Ith vector can now be formed uniquely to yield an admissible table. The set of such tables so generated are equally likely since the set of tables, formed by generating I -1 vectors as above but without rejections, are equally likely so are also the ones in the algorithm with no rejections occurring.

Modification for Tables with 0s on the Diagonal ($I = J$). To adapt the naive algorithm to tables with zeros on the diagonal we proceed as above, only generating the rows with zeros on the diagonal, until we reach the last generated row (the $(n-1)$st.) For the $(n-1)$st row, we know the value of the last element (as the last element of the last row is 0.) So we randomly sample the last row with its last element fixed, and if the column sums are not violated accept the result with probability proportional to the the number of rows with the specified last entry and row sum divided by the maximum over all last enries of the number of rows with specified last entry and row sum.

We perform the above steps except that the vectors have 0 on the diagonal and a sum of r_i for the remaining coordinates (We now choose at random from the $\binom{r_i+J-2}{J-2}$ sequences of $J-2$ 0s and r_i 1s to generate them.) But we stop after generating $I-2$ vectors (instead of $I-1$).

Assuming we have $I-2$ vectors without rejection, the possibilities are equally likely (for the same reasons as before). We now have only one choice for the values in the $I-1$, I and I, $I-1$ positions. We now generate a vector of length $J-2$ for the $I-1$st row. However the number of possible vectors $n(x) = \binom{r_{I-1}-x+J-3}{J-3}$ varies with x , the table value in position $I-1$, J. We have $x < \min\{r_{I-1}, c_I\}$. So we accept the vector of length $J-2$ with probability proportional to $n(x)$, $p(x) = n(x)/\max n(y)$. This makes all of the accepted values equally likely so far. Then we reject if column sums are not possible and complete to the desired table and achieve the uniform distribution for the accepted tables.

2.2 Ordinary Revised Rejection Algorithm

We can, at the cost of some precomputation of polynomials, choose the rows uniformly given that each entry is less than its column sum. This gives us the revised algorithm.

Here the vectors generated at each step i are uniformly chosen from those with component sum r_i but with the additional condition that the kth component bounded by c_k. This is achieved in [Holmes and Jones(1996)] : Precompute φ_q^j = coefficient of z^q in the polynomial $(1+z+\cdots+z^{c_1})(1+z+\cdots+z^{c_2})\cdots(1+z+\cdots+z^{c_j})$ for $j = 1,2,\ldots,J-1$ and $q = 0,1,\ldots,N_{\max}$ = maximum row sum among first$I-1$ rows. It is then demonstrated in [Holmes and Jones(1996)] how, using these coefficients, to generate uniformly J- dimensional vectors with component sum $N < N_{\max}$ and component k bounded by c_k. We reproduce the method here so that our treatment is self contained: To generate a J-dimensional nonnegative integer vector $\mathbf{v}$ with component sum N, we initially define a probability function on the set

$$\{0,1,2,\ldots,\min\{c_J,N\}\}$$

whose value at t is proportional to ϕ_{N-t}^{J-1} and choose v_J at random from this distribution, and then successively choose v_{J-k} from the distribution on

$$\{0,1,2,\ldots,\min\{c_{J-k},N-(v_J+\cdots+v_{J-k+1})\}\}$$

whose probability at t is proportional to

$$\varphi_{N-(v_J+\cdots+v_{J-k+1})-t}^{J-k-1}.$$

For $k = J-1$ choose $v_1 = N-(v_J+\cdots+v_2)$.

Modification of Revised Algorithm for Tables with 0s on the Diagonal ($I = J$). Same as in naive case but we need more precomputing. For $i = 1,\ldots,I-2$ we need to compute coefficients ${}_i\varphi_q^j$ of z^q in

$$\prod_{k=1,\ldots,j\ k\neq i} (1+z+\cdots+z^{c_j})$$

for $j = 1,\ldots,J-1$ for $q = 0,1,\ldots,r_i$. Use Holmes-Jones generation for each i as described above. This gets us the first $I-2$ rows. For the $I-1$st row we need the coefficients ρ_q^j of z^q in

$$\prod_{k=1,\ldots,j} (1+z+\cdots+z^{c_k})$$

for $j = 1,2,\ldots,I-2$ and $q = 0,1,\ldots,r_{I-1}$. We now have $n(x) = \rho_q^{I-2}$ for $q = r_{I-1}-x$. This row is generated using Holmes-Jones (given above) with the r_q^j. Acceptance is using the same method applied to the $n(x)$ for this revised case.

2.3 Experiments

A number of experiments were performed by generating random matrices by placing $10n(n-1)$ items into an $n \times n$ matrix with zeros on the diagonal, computing the row and column sums, and then generating random matrices with those row and column sums.

The modified revised algorithm takes about 10-50 thousand tries to find a 10×10 matrix, depending on the matrix. This is near the limit of practicality. A 13×13 matrix with row sums 121, 106, 124, 99, 140, 124, 120, 134, 122, 112, 102, 130, 126, and column sums 123, 114, 117, 106, 122, 136, 145, 111, 118, 128, 116, 112, 112 generates a suitable matrix about every million tries. A random matrix failed to be generated for a 15×15 matrix after 20 million tries.

So, for larger tables, modifications are necessary.

2.4 Splitting

Most of the failures occur when the matrix is almost completed, but we can uniformly sample half a matrix with a small probability of failure (less than 0.5). We can use this idea for accelerating. The process is to uniformly sample two "halves" of the $(n-1) \times n$ matrix, and search for matches that allow the last row to be filled.

1. Generate a set of k $\lfloor n/2 \rfloor \times n$ matrices whose row sums match the first $\lfloor n/2 \rfloor$ row sums of the original matrix and whose column sums are less than or equal to the column sums of the original matrix.
2. Generate a set of k $\lfloor (n-1)/2 \rfloor \times n$ matrices whose row sums match the row sums of rows $\lfloor n/2 \rfloor + 1$ to $n-1$, and whose column sums are less than or equal to the column sums of the original matrix.
3. Collect the pairs of matrices where the first $\lfloor n/2 \rfloor$ rows come from the first set, and the second $\lfloor (n-1)/2 \rfloor$ rows come from the second set, the sum of both column sums are all less than or equal to the column sums of the original matrix, and the column sum of the last column equals the column sum of the last column of the original matrix because the cell in the last row and last column is 0. Randomly choose one of the matching pairs.

The last step can be performed by first filtering the pairs so that the last column sums match (as we know the last element of the last row is zero), and then using a divide and conquer method on each column by excluding all pairs where the column sum of each half matrix is greater than half of the column sum of the corresponding column of the original matrix.

If we assume that all the time is spent in generating the candidate half matrices (which is true to a first approximation) and that each of the k^2 pairs of top and bottom half matrices has an equal and independent probability of extending to a complete matrix (they are definitely not independent but they should be close enough), then we can treat the number of candidates generated as a Poisson random variable where $\lambda \propto k^2$. Since we only use one candidate from each run, the expected number of samples generated per unit work is proportional to

$$\frac{(1 - e^{-\theta^2})}{\theta}$$

where θ is proportional to k. By taking the derivative of this quantity, we find that the maximum is achieved where θ is the positive root of $e^{x^2} - 2x^2 - 1$. This gives us a probability of getting at least one canditate of $1 - e^{-\theta^2} \approx 0.71$ for each run. As the constant of proportionality between θ and k is unknown, we choose k by trial and error, finding a value where the probability of choosing a table is about 0.71. For a 15×15 matrix with row and column sums about 150, this means a value of k about 25000.

3 Experimental Results

As a simple proxy for a transportation network 12 nodes which are both sources and sinks are arranged in a ring, with a central node that is neither a origin or destination, and each node has a link to and from its adjacent node and the central node. All traffic is assumed to take the shortest path as follows: travel between adjacent nodes goes along the rim, and all other traffic uses the central node. The cost of traversing a link was assumed to be the square of the flow along the link (The actual cost functions in practical networks are more complicated but this quadratic approximation should suffice to demonstrate the procedure of obtaining p-quantile cost.) The trips per five minute interval originating from the nodes are assumed to be 100, 130, 96, 117, 107, 124, 101, 107, 127, 85, 119, and 107 and the trips per five minute interval destined for each node are assumed to be 110, 99, 113, 92, 113, 109, 114, 110, 117, 123, 113, and 107. See figure 1.

Ten thousand uniformly distributed tables were generated ($n = 10000$), and the mean cost was determined to be 210162 ± 141. Using ten thousand tables generated by Sequential Importance Sampling [Chen(2007)] gives us the mean cost of 208633 ± 1871. The standard deviation for the SIS sample is the sample standard deviation divided by the square root of the effective sample size(ESS), defined below. This comparison is not completely fair as sequential importance sampling generates the 12×12 tables about 50 times as quickly as uniform sampling. Using the Effective Sample Size rule of thumb

$$\text{ESS} = \frac{n}{1 + \text{cv}^2}$$

from [Liu(2008)] section 2.5.3, or [Chen et al.(2005)] section 2, we get a value of $\text{cv}^2 = 131.2672$ so the effective sample size of a million SIS samples is approximately 7560, slightly less than the 10000 uniform samples. From the ESS we can estimate the variance of the estimate of our SIS sample by dividing the variance of the sample by $\frac{n}{1+\text{cv}^2}$. Generating a million tables using sequential importance sampling gives an estimate of 210076 ± 177. We also estimated the 0.90 and 0.95

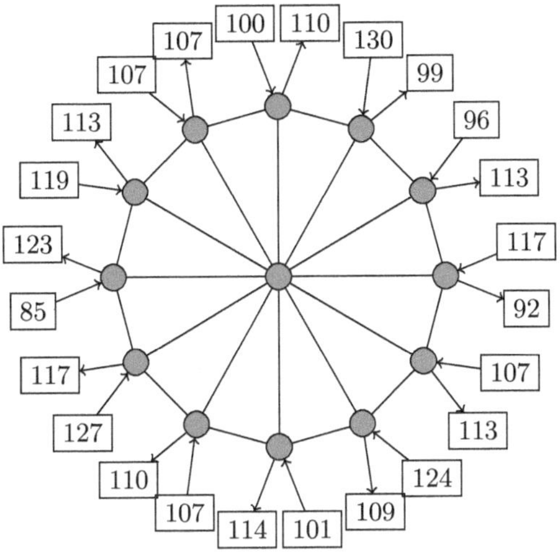

Fig. 1. Network Flows

quantiles. We can use order statistics to get a 95% confidence interval for the quantiles with the uniform sample [Hogg et al.(2005)] section 5.2.2. This gives us an estimate of 227429 for the 0.9 quantile with a 95% confidence interval running from 226938 to 227763. The SIS estimate for the 0.9 quantile is 227250 which lies well inside the confidence interval. For the 0.95 quantile the estimate was 232110 with a confidence interval that runs from 231532 to 232723, and the SIS estimate was 231713.

So for 12 × 12 tables uniform sampling seems to perform slightly better than Sequential Importance Sampling for the same amount of work without suffering from the problems in detecting convergence that are discussed in [Bezáková et al.(2006)]. A uniform sampling also makes the statistical analysis much easier.

Code is available on request.

We have already mentioned how uniform generation of O-D tables is important for generating O-D tables with conditional normal distributions so that cost estimates may be calculated for determining optimal control strategies. It would be interesting to obtain tight bounds on p-quantile costs under the conditional normality assumptions of [Jones et al.(2012)] in terms of p-quantile costs for the uniform distributions.This could yield both more efficient optimization algorithms and more robust control strategies.

References

[Bezáková et al.(2006)] Bezáková, I., Sinclair, A., Štefankovič, D., Vigoda, E.: Negative Examples for Sequential Importance Sampling of Binary Contingency Tables. In: Azar, Y., Erlebach, T. (eds.) ESA 2006. LNCS, vol. 4168, pp. 136–147. Springer, Heidelberg (2006)

[Chen(2007)] Chen, Y.: Conditional inference on tables with structural zeros. J. Comput. Graph. Statist. 16(2), 445–467 (2007)

[Chen et al.(2005)] Chen, Y., Diaconis, P., Holmes, S.P., Liu, J.S.: Sequential Monte Carlo methods for statistical analysis of tables. J. Amer. Statist. Assoc. 100, 109–120 (2005)

[Diaconis and Efron(1985)] Diaconis, P., Efron, B.: Testing for independence in a two-way table: new interpretations of the chi-square statistic. Annals of Statistics 13(3), 845–874 (1985)

[Dyer et al.(1995)] Dyer, M., Kannan, R., Mount, J.: Sampling contingency tables. Random Structures & Algorithms 10, 487–506 (1995)

[Hogg et al.(2005)] Hogg, R.V., McKean, J.W., Craig, A.T.: Introduction to mathematical statistics. Pearson education international. Pearson Education (2005) ISBN 9780130085078, `http://books.google.com/books?id=dX4pAQAAMAAJ`

[Holmes and Jones(1996)] Holmes, R.B., Jones, L.K.: On uniform generation of two-way tables with fixed margins and the conditional volume test of diaconis and efron. Annals of Statistics 24(1), 64–68 (1996)

[Jones et al.(2012)] Jones, L.K., Deshpande, R., Gartner, N.H., Stamatiadis, C., Zou, F.: Robust controls for traffic networks: the near-bayes near-minimax strategy. Transportation Research Part C (in press, 2012)

[Kannan et al.(1997)] Kannan, R., Lovász, L., Simonovits, M.: Random walks and an $o(n^5)$ volume algorithm for convex bodies. Random Structures & Algorithms 11(1), 1–50 (1997)

[Liu(2008)] Liu, J.S.: Monte Carlo Strategies in Scientific Computing. Springer Series in Statistics. Springer (2008) ISBN 9780387763699, `http://books.google.com/books?id=R8E-yHaKCGUC`

[Lovasz and Simonovits(1993)] Lovasz, L., Simonovits, M.: Random walks in a convex body and an improved volume algorithm. Random Structures & Algorithms 4(4), 359–412 (1993)

[Xie et al.(2010)] Xie, C., Kockelman, K.M., Waller, S.T.: Maximum entropy method for subnetwork origin-destination trip matrix estimation. Transportation Research Record: Journal of the Transportation Research Board 2196(4), 111–119 (2010)

Identification via Quantum Channels

Andreas Winter[1,2,3]

[1] ICREA – Institució Catalana de Recerca i Estudis Avançats,
Pg. Lluis Companys 23, ES-08010 Barcelona, Spain,
Física Teòrica: Informació i Fenomens Quàntics,
Universitat Autònoma de Barcelona, ES-08193 Bellaterra (Barcelona), Spain
[2] Department of Mathematics, University of Bristol, Bristol BS8 1TW, U.K.
[3] Centre for Quantum Technologies, National University of Singapore,
2 Science Drive 3, Singapore 117542, Singapore
der.winter@gmail.com

Dem Andenken an Rudolf Ahlswede (15/9/1938—18/12/2010)

Abstract. We review the development of the quantum version of Ahlswede and Dueck's theory of identification via channels. As is often the case in quantum probability, there is not just one but several quantizations: we know at least two different concepts of identification of classical information via quantum channels, and three different identification capacities for quantum information.

In the present summary overview we concentrate on conceptual points and open problems, referring the reader to the small set of original articles for details.

0 Quantum and Classical Channels

Our communication model is the quantum channel, also known as completely positive and trace preserving (cptp) linear map between quantum systems,

$$\mathcal{N} : \mathcal{L}(A) \longrightarrow \mathcal{L}(B).$$

Here, as in the rest of the paper, we assume that A, B, etc, are finite dimensional (complex) Hilbert spaces and $\mathcal{L}(A)$ is the set of linear operators (matrices) over A.

The cptp condition is necessary and sufficient for $\mathcal{N}$ mapping states on A, i.e. density operators $\rho \geq 0$ with $\mathrm{Tr}\rho = 1$, whose set we denote as $\mathcal{S}(A)$, to states on B, and the same for $\mathcal{N}\otimes\mathrm{id}_C$ for arbitrary systems C. Thus, the class of cptp maps is closed under composition, tensor products and taking convex combinations. One of the most useful characterizations of cptp maps is in terms of the Stinespring dilation [29]: namely, $\mathcal{N}$ is cptp if and only if there exists an ancilla (environment) system E and an isometry $V : A \hookrightarrow B\otimes E$ such that $\mathcal{N}(\rho) = \mathrm{Tr}_E V\rho V^\dagger$. The isometry V is essentially unique, up to unitary equivalence of E; hence it makes sense to define, for a chosen dilation V, the *complementary channel*

$$\widehat{\mathcal{N}} : \mathcal{L}(A) \longrightarrow \mathcal{L}(E),$$

H. Aydinian, F. Cicalese, and C. Deppe (Eds.): Ahlswede Festschrift, LNCS 7777, pp. 217–233, 2013.

by $\widehat{\mathcal{N}}(\rho) := \mathrm{Tr}_B V\rho V^\dagger$.

For a given channel $\mathcal{N}$, we are interested in the asymptotic performance of many iid copies, $\mathcal{N}^{\otimes n}$. One can also consider more complicated channel models (such as with feedback, or with pre-shared correlations), but here we will restrict ourselves to the simple forward channel – see however [33] and [1].

Classical channels are of course transition probability kernels $N : \mathcal{X} \rightarrow \mathcal{Y}$ (with finite input and output alphabets $\mathcal{X}$ and $\mathcal{Y}$, respectively). Such a channel may be identified with the cptp map

$$\begin{aligned} \mathcal{N} : \mathcal{L}(\mathbb{C}^{\mathcal{X}}) &\longrightarrow \mathcal{L}(\mathbb{C}^{\mathcal{Y}}) \\ \rho &\longmapsto \sum_{xy} N(y|x)|y\rangle\langle y|x\rho|x\rangle\langle x|y, \end{aligned}$$

while a probability distribution P on $\mathcal{X}$ is identified with the state $\sum_x P(x)|x\rangle\langle x|$.

Two special classes of channels we will have occasion to consider are the following, either whose input or whose output is classical: A *cq-channel* $\mathcal{N} : \mathcal{X} \longrightarrow \mathcal{S}(B)$ is a cptp map of the form

$$\mathcal{N}(\xi) = \sum_x \langle x|\xi|x\rangle \rho_x,$$

with states ρ_x on B. A *qc-channel* $\mathcal{M} : \mathcal{S}(A) \longrightarrow \mathcal{Y}$ instead is given by a quantum measurement, i.e. a positive operator values measure (POVM) $(M_y)_{y\in\mathcal{Y}}$ such that $M_y \geq 0$ and $\sum_y M_y = \mathbb{1}$. The channel then has the form

$$\mathcal{M}(\rho) = \sum_y \mathrm{Tr}\rho M_y |y\rangle\langle y|.$$

We refer the reader to the excellent text [31] for more details on quantum and classical channels, and the various transmission capacities associated with them, including their history. Here we need only two, the classical and the quantum capacity of a channel, $C(\mathcal{N})$ and $Q(\mathcal{N})$, respectively, defined as the maximum rates of asymptotically faithful transmission of classical bits and qubits, respectively, over many iid copies of the channel. They can be expressed as regularizations of entropic information quantities, based on the von Neumann entropy $S(\rho) = -\mathrm{Tr}\rho\log\rho$ of a quantum state ρ. They are given by the formulas

$$\begin{aligned} C(\mathcal{N}) &= \lim_{n\to\infty} \frac{1}{n} C^{(1)}(\mathcal{N}^{\otimes n}), \text{ with} \\ C^{(1)}(\mathcal{N}) &= \max_{\{p_x,\rho_x\}} S\left(\sum_x p_x \mathcal{N}(\rho_x)\right) - \sum_x p_x S(\mathcal{N}(\rho_x)), \end{aligned} \tag{1}$$

and

$$
\begin{aligned}
Q(\mathcal{N}) &= \lim_{n\to\infty} \frac{1}{n} Q^{(1)}(\mathcal{N}^{\otimes n}), \text{ with} \\
Q^{(1)}(\mathcal{N}) &= \max_{\rho\in\mathcal{S}(A)} S\big(\mathcal{N}(\rho)\big) - S\big(\widehat{\mathcal{N}}(\rho)\big),
\end{aligned}
\tag{2}
$$

both of which represent the culmination of concerted efforts of several researchers in the 1990s and early 2000s (Holevo-Schumacher-Westmoreland and Schumacher & Lloyd-Shor-Devetak, respectively). The classical capacity generalizes Shannon's channel capacity for classical channels N, for which $C(N) = C^{(1)}(N)$ reduces to the famous formula in terms of the mutual information [28].

The structure of the rest of the paper is as follows: In Section 1 we present the definitions for identification of classical information via quantum channels, after Löber [25], generalizing the model of Ahlswede and Dueck [8,9]. In Section 2 we move to identification of quantum information; Section 3 presents the recently developed theoretical underpinning to prove capacity formulas for two of the three quantum models. In Section 4 we show how the quantum identification results imply new lower bounds on classical identification capacities, which we illustrate with several examples, shedding new light also on Löber's founding work [25]. Finally, Section 5 is devoted to an outlook on open questions and possible conjectures.

1 Classical Identification

Ahlswede and Dueck [8,9] introduced identification by noting that while Shannon's theory of transmission presumes that the receiver wants to know everything about the message, in reality he may be interested only in certain aspects of it. In other words, the receiver may want to compute a function of the message. The most extreme case is that of identification: for sent message m and an arbitrary message m', the receiver would like to be able to answer the question "Is $m = m'$?" as accurately as possible.

Definition 1 (Löber [25]). *A* classical identification code for the channel $\mathcal{N}$ *with* error probability λ_1 of first, and λ_2 of second kind *is a set* $\{(\rho_i, D_i) : i = 1,\dots,N\}$ *of states* ρ_i *on* A *and operators* D_i *on* B *with* $0 \le D_i \le \mathbb{1}$, *i.e. the pair* $(D_i, \mathbb{1} - D_i)$ *forms a measurement, such that*

$$
\begin{aligned}
\forall i \quad & \operatorname{Tr}\big(\mathcal{N}(\rho_i) D_i\big) \ge 1 - \lambda_1, \\
\forall i \ne j \quad & \operatorname{Tr}\big(\mathcal{N}(\rho_i) D_j\big) \le \lambda_2.
\end{aligned}
$$

For the special case of memoryless channels $\mathcal{N}^{\otimes n}$, *we speak of an* $(n, \lambda_1, \lambda_2)$-ID code, *and denote the largest size* N *of such a code* $N(n, \lambda_1, \lambda_2)$.

An identification code as above is called simultaneous *if all the* D_i *are coexistent: this means that there exists a positive operator valued measure (POVM)* $(E_t)_{t=1}^T$ *and pairwise disjoint sets* $\mathcal{D}_i \subset \{1,\dots,T\}$ *such that* $D_i = \sum_{t\in\mathcal{D}_i} E_t$. *The largest size of a simultaneous* $(n, \lambda_1, \lambda_2)$*-ID code is denoted* $N_{\text{sim}}(n, \lambda_1, \lambda_2)$.

Note that $N_{\mathrm{sim}}(n, \lambda_1, \lambda_2) = N(n, \lambda_1, \lambda_2) = \infty$ *if* $\lambda_1 + \lambda_2 \geq 1$, *hence to avoid this triviality one has to assume* $\lambda_1 + \lambda_2 < 1$.

It is straightforward to verify that in the case of a classical channel, this definition reduces to the famous one of Ahlswede and Dueck [8], in particular all codes are without loss of generality automatically simultaneous. It was in fact Löber [25] in his PhD thesis who noticed that in the quantum case we have to make a choice – whether the receiver should be able to answer *all* or *any one* of the "Is the message $= m'$?" questions. It was the original realization of Ahlswede and Dueck [8] that $N(n, \lambda_1, \lambda_2)$ grows doubly exponential in n, hence the following definition of the (classical) identification capacity:

Definition 2. *The* (simultaneous) classical ID-capacity *of a quantum channel* $\mathcal{N}$ *is given by*

$$C_{\mathrm{ID}}(\mathcal{N}) = \inf_{\lambda>0} \liminf_{n\to\infty} \frac{1}{n} \log\log N(n, \lambda, \lambda),$$
$$C_{\mathrm{ID}}^{\mathrm{sim}}(\mathcal{N}) = \inf_{\lambda>0} \liminf_{n\to\infty} \frac{1}{n} \log\log N_{\mathrm{sim}}(n, \lambda, \lambda),$$

respectively. We say that the strong converse *holds for the identification capacity if for all* $\lambda_1 + \lambda_2 < 1$,

$$\lim_{n\to\infty} \frac{1}{n} \log\log N(n, \lambda_1, \lambda_2) = C_{\mathrm{ID}}(\mathcal{N}),$$

and similarly for $C_{\mathrm{ID}}^{\mathrm{sim}}$.

Theorem 3 (Ahlswede/Dueck [8], Han/Verdú [18,19], Ahlswede [4]). *For a classical channel* N *and any* $\lambda_1, \lambda_2 > 0$ *with* $\lambda_1 + \lambda_2 < 1$,

$$\lim_{n\to\infty} \frac{1}{n} \log\log N(n, \lambda_1, \lambda_2) = C(N),$$

in particular, $C_{\mathrm{ID}}^{\mathrm{sim}}(N) = C_{\mathrm{ID}}(N) = C(N)$. □

The direct part of the above theorem, due to Ahlswede and Dueck [8], can be seen by concatenating a sufficiently good Shannon channel code with an identification code for the ideal bit channel. For the latter, [8] contains a combinatorial construction showing that by k-bit encodings, one can identify $\geq 2^{\Omega(2^k)}$ messages. Using this, the direct part of the following result is immediate:

Theorem 4 (Löber [25], Ahlswede/Winter [10]). *For quantum channel* $\mathcal{N}$,

$$C_{\mathrm{ID}}(\mathcal{N}) \geq C_{\mathrm{ID}}^{\mathrm{sim}}(\mathcal{N}) \geq C(\mathcal{N}).$$

The simultaneous ID-capacity obeys a strong converse under the additional restriction that the signal states ρ_i *are from a set that is the convex hull of* $\leq 2^{2^{o(n)}}$

quantum states on A^n. *I.e., denoting the maximum number of messages under this constraint by* $\underline{N}_{\text{sim}}(n, \lambda_1, \lambda_2)$,

$$\overline{\lim_{n\to\infty}} \frac{1}{n} \log\log \underline{N}_{\text{sim}}(n, \lambda_1, \lambda_2) \leq C(\mathcal{N}),$$

for $\lambda_1 + \lambda_2 < 1$. *(For instance, the* ρ_i *could be restricted to be – approximately – separable states.)*

For cq-channels, the constraint is w.l.o.g. satisfied since there are only $|\mathcal{X}|^n$ *classical input symbols, so for these channels the simultaneous ID-capacity obeys a strong converse, with* $C_{\text{ID}}^{\text{sim}}(\mathcal{N}) = C(\mathcal{N}) = C^{(1)}(\mathcal{N})$.

Indeed, in the case of cq-channels, the strong converse holds even without the simultaneity constraint:

$$\lim_{n\to\infty} \frac{1}{n} \log\log N(n, \lambda_1, \lambda_2) = C(\mathcal{N}) = C^{(1)}(\mathcal{N}),$$

for $\lambda_1 + \lambda_2 < 1$. □

[To be precise, Löber's results are in the framework of Han and Verdú [18,19], of "arbitrary" sequences of channels and using information spectrum methods. As we are focusing on the iid case here, we stated only a special case of his theorem.]

The last, non-simultaneous part of the Theorem is the main identification result of [10], which was proved by developing a theory of tail bounds for sums of random matrices, extending classical Hoeffding bounds, and inspired by Ahlswede's strong converse for the ID-capacity of classical channels [4]. The simplest, and most useful, version is as follows.

Lemma 5 (Ahlswede/Winter [10]). *For i.i.d. random variables* X_i *in* $d \times d$ *Hermitian matrices and* $0 \leq X_i \leq \mathbb{1}$, *such that* $\mathbb{E}X_i = \mu\mathbb{1}$. *Then, for* $\mu \leq \alpha \leq 1$ *and* $0 \leq \alpha \leq \mu$, *respectively,*

$$\Pr\left\{\frac{1}{n}\sum_{i=1}^{n} X_i \not\leq \alpha\mathbb{1}\right\} \leq d\, e^{-nD(\alpha\|\mu)},$$

$$\Pr\left\{\frac{1}{n}\sum_{i=1}^{n} X_i \not\geq \alpha\mathbb{1}\right\} \leq d\, e^{-nD(\alpha\|\mu)},$$

with the binary relative entropy $D(\alpha\|\mu) = \alpha \ln \frac{\alpha}{\mu} + (1-\alpha)\ln\frac{1-\alpha}{1-\mu}$.

As a consequence, for all $0 \leq \epsilon \leq \frac{1}{2}$,

$$\Pr\left\{\frac{1}{n}\sum_{i=1}^{n} X_i \not\in [(1-\epsilon)\mu\mathbb{1}, (1=\epsilon)\mu\mathbb{1}]\right\} \leq 2d\, e^{-\frac{1}{4}n\mu\epsilon^2}.$$

using elementary estimates for the relative entropy. □

The power of this Lemma is in its giving explicit and simple tail bounds, useful already for finite n and d, whereas general abstract large deviation theory –

which applies, see [7] for a version in infinite dimension – often incurs complex finite n behaviour, only yielding clear asymptotic statements. The proof of the Lemma is simple, too: it requires generalizing the elementary Markov-Chebyshev inequalities and the Bernstein trick from real random variables to matrices. It has since found countless applications in quantum information theory and beyond: The first proofs of some core results such as the quantum channel capacity, remote state preparation or decoupling heavily relied on it, cf. [31], as did the structurally simple proof of the Alon-Roichman theorem and matrix versions of compressed sensing, cf. [30] and references therein. The latter also presents far-reaching generalizations of the above bounds.

It is not known whether simultaneous and non-simultaneous ID-capacity coincide or not for general quantum channels. In any case, going beyond simultaneity seems to provide major freedom:

Example 6. *Buhrman* et al. *[15] found that in the space of n qubits, whilst the largest number of orthogonal pure state vectors is clearly the dimension of the Hilbert space, 2^n, there are $N \geq 2^{\Omega(2^n)}$* pairwise almost orthogonal *pure states, i.e. $|\langle\psi_i|\psi_j\rangle| \leq \epsilon$ for $i \neq j$.*

They dubbed this "fingerprinting" because a verifier who gets a copy of each $|\psi_i\rangle$ and $|\psi_j\rangle$ can efficiently determine whether $i = j$ or not. In particular, the set of these vectors forms a (non-simultaneous) ID-code, with $\rho_i = D_i = |\psi_i\rangle\langle\psi_i|$.

One can obtain a set of such vectors also by turning the probability distributions on n bits form [8] into superpositions – cf. [32] for details.

Fingerprinting ID-codes use quantum superpositions in a nontrivial way, albeit the almost-orthogonality is somewhat analogous to the way the classical distributions in [8] do not overlap too much. However, they only use pure states, whereas the power of classical identification comes from randomization. Hence it is natural to ask whether mixed states offer any improvement. As the classical capacity of a noiseless qubit channel is 1, the following result came as a bit of a surprise. It was proved using powerful geometric measure concentration techniques – cf. [21,11] for other applications in quantum information theory.

Theorem 7 (Winter [32]). *For the noiseless qubit channel* $\mathrm{id}_2 = \mathrm{id}_{\mathbb{C}^2}$*, and* $0 < \lambda_1, \lambda_2, \lambda_1 + \lambda_2 < 1$,

$$2^{\Omega(2^{2n})} \leq N(n, \lambda_1, \lambda_2) \leq 2^{O(2^{2n})}.$$

As a consequence, $C_{\mathrm{ID}}(\mathrm{id}_2) = 2$ *and the strong converse holds. If the encodings are restricted to pure states, the capacity is only* 1. □

[In [32] (Remark 13; the technical argument there has been elaborated in [17]) it was heuristically argued that one would expect $C_{\mathrm{ID}}^{\mathrm{sim}}(\mathrm{id}_2)$ to be 1 rather than 2.]

To appreciate why this result was so surprising, we need to go back to the insights from the original identification papers [8,9]: It was understood that what determines identification capacity of a communication system is its ability to establish common randomness (cf. [6]), as long as some sublinear amount of actual communication is available. But the common randomness capacity of a noiseless qubit channel is 1. However, a noiseless qubit channel can also establish entanglement (ebits) at rate 1. And indeed, in [33,1] it was found that k EPR pairs shared between sender and receiver, together with $o(n)$ bits of communication are sufficient to identify $2^{\Omega(2^{2k})}$ messages. In this respect, it may be interesting to draw attention to the following:

Proposition 8 (Winter [32]). *Given an ID-code of rate C and common randomness of rate R, one can construct an ID-code of rate $C + R - o(1)$ which uses the signal states of the first code and correlations with the common randomness.* □

In other words: Whatever your communication system, its identification capacity is increased by 1 by each bit of common randomness. This was used in [32] to derive a lower bound on the ID-capacity of a quantum channel: If $\mathcal{N}$ permits simultaneous transmission of classical bits and qubits at rates C and Q, respectively, then $C_{\mathrm{ID}}(\mathcal{N}) \geq C+2Q$. Thus the results of [16] become applicable, where the Q-C capacity region was determined. As we saw above, this bound, can be strictly larger than the classical capacity $C(\mathcal{N})$ of the channel, marking a decisive departure from the behaviour of classical channels.

Beyond these bounds and a few special examples in [32], the ID-capacity (simultaneous or not) of a general quantum channel remains elusive. However, in Section 4 below we shall present a new lower bound.

2 How to Identify Quantum States?

So far the only quantum element in the discussion pertained to the channel model. However, there is a natural way in which even the task of identification can be extended from classical to quantum information. This has been promoted in [32] and further in the more recent [22]. In the following, $\mathcal{P}(A) \subset \mathcal{S}(A)$ denotes the set of pure quantum states on a system A.

Definition 9 (Winter [32]). *A* quantum ID-code *for the channel $\mathcal{N}$ with error ϵ, for the Hilbert space K, is a pair of maps $\mathcal{E} : \mathcal{P}(K) \longrightarrow \mathcal{S}(A)$ and $\mathcal{D} : \mathcal{P}(K) \longrightarrow \mathcal{L}(B)$ with $0 \leq \mathcal{D}_\varphi \leq \mathbb{1}$ for all $\varphi = |\varphi\rangle\langle\varphi| \in \mathcal{P}(K)$, such that for all pure states/rank-one projectors $\psi, \varphi \in \mathcal{P}(K)$,*

$$\left|\mathrm{Tr}\psi\varphi - \mathrm{Tr}\mathcal{N}\big(\mathcal{E}(\psi)\big)\mathcal{D}_\varphi\right| \leq \epsilon.$$

If the encoding $\mathcal{E}$ is cptp we speak of a blind *code, in general and to contrast it with the former, we call it* visible.

For the case of an iid channel $\mathcal{N}^{\otimes n}$, we denote the maximum dimension of a blind (visible) quantum ID-code by $M(n, \epsilon)$ ($M_v(n, \epsilon)$).

This notion can be motivated as follows: In quantum transmission, the objective for the receiver is to recover the state ψ by means of a suitable decoding (cptp) map $\widetilde{\mathcal{D}} : \mathcal{L}(B) \longrightarrow \mathcal{L}(K)$, with high accuracy. Of course then the receiver can perform any measurement on the decoded state, effectively simulating an arbitrary measurement on the original input state, in the sense that for any state ρ and POVM $M = (M_i)_i$ on K, there exists another POVM $M' = (M_i')_i$ on B such that the measurement statistics of ρ under M is approximately that of $\mathcal{N}(\mathcal{E}(\rho))$ under M'. (M' can be written down directly via the adjoint $\widetilde{\mathcal{D}}^\dagger : \mathcal{L}(K) \longrightarrow \mathcal{L}(B)$ of the decoding map, which maps measurement POVMs on K to POVMs on B: $M_i' = \widetilde{\mathcal{D}}^\dagger(M_i)$.) The converse is also true: If the receiver can simulate sufficiently general measurements on the input state by suitable measurements on the channel output, then he can actually decode the state by a cptp map $\widetilde{\mathcal{D}}$ [26].

This allows us to relax the task of quantum information transmission to requiring only that the receiver be able to simulate the statistics of certain restricted measurements. In the case of quantum identification, these are $(\varphi, \mathbb{1} - \varphi)$ for arbitrary rank-one projectors $\varphi = |\varphi\rangle\langle\varphi| \in \mathcal{P}(K)$. They are the measurements which allow the receiver to ask the (quantum) question: "Is the state equal to φ or orthogonal to it?" Obviously, in quantum theory this question cannot be answered with certainty, but for each test state it yields a characteristic distribution. The quantum-ID task above is about reproducing this distribution.

Note that we can always concatenate a blind or visible quantum ID-code for the Hilbert space K with a fingerprinting set of pure states in K, to obtain a classical ID-code in the sense of Definition 1. This is because in fingerprinting the encodings are pure states ψ_i and the tests precisely the POVMs $(\psi_i, \mathbb{1} - \psi_i)$. Hence, as the cardinality of the fingerprinting set is exponential in the dimension $|K|$, $M(n, \epsilon)$ and $M_v(n, \epsilon)$ can be at most exponential in n.

Definition 10. *For a quantum channel* $\mathcal{N}$, *the* blind, respectively visible, quantum ID-capacity *is defined as*

$$Q_{\mathrm{ID}}(\mathcal{N}) := \inf_{\epsilon>0} \liminf_{n\to\infty} \frac{1}{n} \log M(n, \epsilon),$$

$$Q_{\mathrm{ID,v}}(\mathcal{N}) := \inf_{\epsilon>0} \liminf_{n\to\infty} \frac{1}{n} \log M_v(n, \epsilon).$$

If we leave out the qualifier, the quantum ID-capacity is by default the blind variety.

Note that by definition and the above remark,

$$Q_{\mathrm{ID}}(\mathcal{N}) \leq Q_{\mathrm{ID,v}}(\mathcal{N}) \leq C_{\mathrm{ID}}(\mathcal{N}). \tag{3}$$

The first quantum ID-capacity that had been determined was for the ideal qubit channel:

Theorem 11 (Winter [32]). *For the noiseless channel* id_A *on Hilbert space* A, *there exists a (blind) quantum ID-code with error* ϵ *and encoding a space* K *of dimension* $|K| \geq C(\epsilon)|A|^2$, *for some universal function* $C(\epsilon) > 0$.

As a consequence, $Q_{\mathrm{ID}}(\mathrm{id}_2) = Q_{\mathrm{ID,v}}(\mathrm{id}_2) = 2$, *twice the quantum transmission capacity.* □

In view of this theorem, we gain at least 2 in capacity for each noiseless qubit we use additionally to the given channel. This motivates the following definition.

Definition 12 (Hayden/Winter [22]). *For a quantum channel* $\mathcal{N}$, *the* amortized (blind/visible) quantum ID-capacity *is defined as*

$$Q_{\rm ID}^{\rm am}(\mathcal{N}) := \sup_k \; Q_{\rm ID}(\mathcal{N}\otimes{\rm id}_k) - 2\log k,$$
$$Q_{\rm ID,v}^{\rm am}(\mathcal{N}) := \sup_k \; Q_{\rm ID,v}(\mathcal{N}\otimes{\rm id}_k) - 2\log k,$$

respectively.

The blind quantum ID-capacities are among the best understood, thanks to recently made conceptual progress, which we review in the next section. We will then also ask the question *how much* amortization is required. This is formalized in the usual way: Namely, for a rate $Q \leq Q_{\rm ID}^{\rm am}(\mathcal{N})$, we say that A is an *achievable amortization rate* if there exist k_n for all n, such that

$$\liminf_{n\to\infty} \frac{1}{n}\left(Q_{\rm ID}(\mathcal{N}^{\otimes n}\otimes{\rm id}_{k_n}) - 2\log k_n\right) \geq Q \quad\text{and}\quad \overline{\lim_{n\to\infty}}\; \frac{1}{n}\log k_n \leq A,$$

giving rise to an achievable quantum ID-rate/amortization region, *viz.* a tradeoff between Q and A. Similarly of course for the visible variant.

3 Weak Decoupling Duality

The fundamental insight about quantum information transmission, which allowed an understanding of the quantum capacity as we have it today, is the *decoupling principle*: for a channel to permit (approximate) error correction it is necessary and sufficient that it leaks (almost) no information to the environment in the sense that the complementary channel $\widehat{\mathcal{N}}$ is close to constant. To be precise, ${\rm id}_{A'}\otimes\widehat{\mathcal{N}}$ should map an entangled test state $\Phi^{A'A}$ to $\approx \Phi^{A'}\otimes\sigma^E$, where the approximation is with respect to the trace norm on density operators. In practice, to define capacities it is enough to demand this for the maximally entangled test state between the code space and a reference system [27].

This condition is compactly expressed as saying that $\widehat{\mathcal{N}}$ is $\approx [\sigma^E]$ in the so-called diamond norm, the completely bounded version of the naive superoperator norm. Here, $[\sigma^E]$ denotes the constant channel mapping every input to σ^E. Because of this connection to completely bounded norms, we call channels with the above property *completely forgetful* or *decoupling.*

Indeed, it is well-known that this is a much stronger condition than $\widehat{\mathcal{N}}(\rho) \approx \sigma^E$ for all input states ρ on A. Cf. [20] for some instances of this effect relevant to quantum information processing. There, it is shown how to construct channels that are only (approximately) *forgetful* (or *weakly decoupling*), but far from completely forgetful.

To state the following conceptual points about blind(!) quantum ID-codes, it is useful to fix an encoding cptp map $\mathcal{E} : \mathcal{L}(K) \longrightarrow \mathcal{L}(A)$ and to combine it with the noisy channel, $\mathcal{N}' = \mathcal{N} \circ \mathcal{E}$, for which we choose a Stinespring dilation $V : K \hookrightarrow B \otimes F$. The quantum ID-code is now the entire input space K of this effective new channel, together with the previously given operators D_φ on B. The next result states that just as quantum error correctability of $\mathcal{N}'$ is equivalent to $\widehat{\mathcal{N}'}$ being decoupling [27,24], quantum identification is essentially equivalent to weak decoupling from the environment:

Theorem 13 (Hayden/Winter [22]). *If K is a ϵ-quantum ID-code for the channel $\mathcal{N}'$ with Stinespring dilation $V : K \hookrightarrow B \otimes F$, then the complementary channel $\widehat{\mathcal{N}'}$ is approximately forgetful:*

$$\forall |\varphi\rangle, |\psi\rangle \in K \quad \frac{1}{2} \left\| \widehat{\mathcal{N}'}(\varphi) - \widehat{\mathcal{N}'}(\psi) \right\|_1 \leq \delta := 7\sqrt[4]{\epsilon}.$$

Conversely, if $\widehat{\mathcal{N}'}$ is approximately forgetful with error δ, then the trace-norm geometry is approximately preserved by $\mathcal{N}'$:

$$\forall |\varphi\rangle, |\psi\rangle \in S \quad 0 \leq \left\| \varphi - \psi \right\|_1 - \left\| \mathcal{N}'(\varphi) - \mathcal{N}'(\psi) \right\|_1 \leq \epsilon := 4\sqrt{2\delta}.$$

If, in addition, the nonzero eigenvalues of the environment's states $\widehat{\mathcal{N}'}(\varphi)$ lie in the interval $[\mu, \lambda]$ for all $|\varphi\rangle \in K$, then one can construct an η-quantum ID-code for $\mathcal{N}'$ (i.e. a set of operators D_φ for all $|\varphi\rangle \in K$ as in Definition 9), with $\eta := 7\delta^{1/8}\sqrt{\lambda/\mu}$. □

Remark 14. *While it would be desirable to eliminate the eigenvalue condition at the end of the theorem, the condition is fairly natural in this context. If the environment's states $\widehat{\mathcal{N}'}(\varphi)$ are very close to a single state σ^F for all $|\varphi\rangle \in K$, then all the $V|\varphi\rangle$ are very close to being purifications of σ^F, meaning that they differ from one another only by a unitary plus a small perturbation. If σ^F is the maximally mixed state or close to it, then the assumption will be satisfied. In the asymptotic iid setting we are looking at this turns to be the case.*

This characterization of quantum ID-codes (albeit "only" blind ones) allows the determination of capacities by a random coding argument, for which only the weak decoupling has to be verified. The above duality theorem is not only the basis for the direct but also for the converse part(s) of the following capacity theorem.

Theorem 15 (Hayden/Winter [22]). *For a quantum channel $\mathcal{N}$, its (blind) quantum ID-capacity is given by*

$$Q_{\text{ID}}(\mathcal{N}) = \lim_{n\to\infty} \frac{1}{n} Q_{\text{ID}}^{(1)}(\mathcal{N}^{\otimes n}), \text{ where}$$
$$Q_{\text{ID}}^{(1)}(\mathcal{N}) = \sup_{|\phi\rangle} \{ I(A:B)_\rho \text{ s.t. } I(A\rangle B)_\rho > 0 \},$$

where $|\phi\rangle$ is the purification of an input state to $\mathcal{N}$, $\rho^{AB} = (\text{id} \otimes \mathcal{N})\phi$ and $I(A : B)_\rho = S(\rho^A) + S(\rho^B) - S(\rho^{AB})$ is the mutual information, and $I(A\rangle B)_\rho =$

$S(\rho^B) - S(\rho^{AB})$ the coherent information (which already appeared in eq. (2)). We declare the sup *to be* 0 *if the set above is empty. In particular, $Q_{\mathrm{ID}}(\mathcal{N}) = 0$ if and only if $Q(\mathcal{N}) = 0$.*

Furthermore, the amortized quantum ID-capacity equals

$$Q_{\mathrm{ID}}^{\mathrm{am}}(\mathcal{N}) = \sup_{|\phi\rangle} I(A:B)_\rho = C_E(\mathcal{N}),$$

the entanglement-assisted classical capacity of $\mathcal{N}$ [12]. □

Remark 16. *Let us say that a channel $\mathcal{N}$ has "sufficiently low noise" if for an input state $|\phi\rangle$ maximizing $I(A:B)_\rho$, $\rho = (\mathrm{id}\otimes\mathcal{N})\phi$, it holds that $I(A\rangle B)_\rho > 0$. This is motivated by the fact that in this case the channel has positive quantum capacity. Also, for any channel, $\mathcal{N}\otimes\mathrm{id}_k$ has sufficiently low noise if k is chosen large enough; likewise $p\mathcal{N} + (1-p)\mathrm{id}$ if $p > 0$ is small enough.*

In that case, the above tells us $Q_{\mathrm{ID}(\mathcal{N})} = Q_{\mathrm{ID}}^{\mathrm{am}}(\mathcal{N}) = \sup_{|\phi\rangle} I(A:B)_\rho$, which is an additive, single-letter formula.

This theorem also shows that amortized and non-amortized quantum ID-capacities are different – indeed, any channel $\mathcal{N}$ with vanishing quantum capacity also has $Q_{\mathrm{ID}}(\mathcal{N}) = 0$, whereas $Q_{\mathrm{ID}}^{\mathrm{am}}(\mathcal{N}) = 0$ only for trivial channels. In particular this implies that Q_{ID} is not additive. In [22] it is in fact proven that certain channels require a positive rate of amortization to attain or even to approximate $Q_{\mathrm{ID}}^{\mathrm{am}}$. The example analyzed there is the qubit erasure channel

$$\begin{aligned}\mathcal{E}_q : \mathcal{L}(\mathbb{C}^2) &\longrightarrow \mathcal{L}(\mathbb{C}^3)\\ \rho &\longmapsto (1-q)\rho \oplus q|*\rangle\langle *|,\end{aligned}$$

which will serve us again in the following section. To be precise, for $0 \le q < \frac{1}{2}$, the channel has sufficiently low noise and no amortization is required. For $\frac{1}{2} \le q \le 1$ instead, an amortized rate of at least $2q - 1$ qubits per channel use are necessary.

On the other hand, for all *symmetric* channels, i.e. those with $E = B$ in the Stinespring representation and $\mathcal{N} = \widetilde{\mathcal{N}}$, whereas quantum capacity and hence Q_{ID} are zero, only a vanishing rate of amortization is necessary to attain $Q_{\mathrm{ID}}^{\mathrm{am}}$. This is because they have $I(A\rangle B)_\rho = 0$ for every input state, so arbitrarily little is required to make the coherent information positive.

This includes qc-channels with rank-one POVM $(M_y)_{y\in\mathcal{Y}}$, and the noiseless classical bit channel

$$\overline{\mathrm{id}}_2 : \rho \longmapsto \sum_{b=0,1} |b\rangle\langle b|\rho|b\rangle\langle b|.$$

The latter implies that also cq-channels $\mathcal{N}$ only require a vanishing rate of amortization to attain $Q_{\mathrm{ID}}^{\mathrm{am}}(\mathcal{N}) = C(\mathcal{N}) = C^{(1)}(\mathcal{N})$: This is because we can use $n \gg 1$ copies of $\mathcal{N}$, with appropriate encoding and decoding, to simulate $\big(C(\mathcal{N}) - o(1)\big)n$ almost noiseless classical bits. This also shows that the rate $C(\mathcal{N})$ is attainable for all channels $\mathcal{N}$ as an amortized quantum ID-rate, with vanishing rate of amortization.

In fact, inspection of the proof of the direct part of Theorem 15 (Thm. 12 in [22]) reveals that for the noiseless classical channel $\overline{\mathrm{id}}_2$, a constant amount of amortization is enough, hence the same for all cq-channels, and also for certain rank-one POVM qc-channels, namely those for which the outputs $\mathcal{N}(\tau)$ and $\widetilde{\mathcal{N}}(\tau)$ for the maximally mixed input state τ_A are themselves maximally mixed. Because then the typicality arguments in the proof, which deal with eigenvalue fluctuations around the inverse exponential of the entropy, are unnecessary.

Remark 17. *The previous observations show that the amortized quantum ID-capacity of a cq-channel $\mathcal{N}$ (which equals its classical capacity) can be achieved by visible, non-amortized codes:*

$$Q_{\mathrm{ID,v}}(\mathcal{N}) = Q_{\mathrm{ID}}^{\mathrm{am}}(\mathcal{N}) = C(\mathcal{N}) = C^{(1)}(\mathcal{N}).$$

Indeed, choose a sequence of amortized quantum ID-codes for n uses of the channel, with amortized noiseless communication of a system of dimension $t = o(\log n)$. Then, whatever the code produces as the input state $\omega = \mathcal{E}(\psi)$ to the channel $\mathcal{N}^{\otimes n} \otimes \mathrm{id}_t$, the effect is the same if we first dephase the input to $\mathcal{N}^{\otimes n}$ as the channel is cq, so w.l.o.g.

$$\omega = \sum_{x^n} p_{x^n} |x^n\rangle\langle x^n| \otimes \omega_{x^n},$$

with states $\omega_{x^n} \in \mathcal{S}(\mathbb{C}^t)$. The latter can be described classically to good approximation using $o(n)$ bits [21,32], which can be communicated by $o(n)$ uses of the channel (if we exclude the trivial case of zero capacity).

This then is the visible scheme: the encoding of state ψ is to sample from the distribution p_{x^n} and sending $|x^n\rangle\langle x^n|$ through $\mathcal{N}^{\otimes n}$, and to send a classical description of ωx^n via $\mathcal{N}^{\otimes o(n)}$. The receiver creates then ω_{x^n} in addition to the other channel output, and otherwise uses the measurement D_φ from the amortized ID-code.

That the capacity cannot be larger than $C(\mathcal{N})$ follows from eq. (3) and Theorem 4.

On the other hand, $Q_{\mathrm{ID}}(\mathcal{N}) = 0$ by Theorem 15, so we obtain a separation between blind and visible quantum ID-capacity, a question left open in [32]. □

We close this section by pointing out that $Q_{\mathrm{ID}}^{\mathrm{am}}$ is one of only two fully understood identification capacities so far: it has a single letter formula which can be evaluated efficiently and it is additive. The other one is the classical ID-capacity of a quantum channel with "coherent feedback" (meaning that in each use of the channel, the environment of the Stinespring isometry ends up with the sender), which we did not discuss here; the interested reader is referred to [33].

4 From Q_{ID} to C_{ID}

As pointed out in Section 2, concatenating a quantum ID-code (blind or visible) with the fingerprinting construction (Example 6), yields a classical ID-code of asymptotically the same rate. Hence,

$$C_{\text{ID}}(\mathcal{N}) \geq Q_{\text{ID,v}}(\mathcal{N}) \geq \begin{cases} Q_{\text{ID}}(\mathcal{N}), \\ C(\mathcal{N}), \end{cases}$$

$$C_{\text{ID}}^{\text{am}}(\mathcal{N}) \geq Q_{\text{ID}}^{\text{am}}(\mathcal{N}) = C_E(\mathcal{N}),$$

where the amortized classical ID-capacity is defined analogously to the quantum variant.

Perhaps we do not find the amortized classical ID-capacity that interesting, but at least we get lots of channels for which $C_{\text{ID}}(\mathcal{N}) \geq C_E(\mathcal{N})$, namely all sufficiently low noise channels and of course all cq-channels. This bound improves on the earlier best bound

$$C_{\text{ID}}(\mathcal{N}) \geq \max\{C + 2Q : (Q, C) \text{ jointly achievable}\},$$

the right hand side of which is always $\leq C_E(\mathcal{N})$. For example for the erasure channel $\mathcal{E}_q$, the quantum-classical-capacity region is known [16] to be

$$\text{conv}\left\{(0,0),\ (0,1-q),\ \left((1-2q)_+,0\right)\right\},$$

so the above maximization yields

$$C_{\text{ID}}(\mathcal{E}_q) \geq \begin{cases} 2-4q & \text{for } 0 \leq q \leq \frac{1}{3}, \\ 1-q & \text{for } \frac{1}{3} \leq q \leq 1. \end{cases}$$

Our new bound instead is

$$C_{\text{ID}}(\mathcal{E}_q) \geq \begin{cases} 2-2q & \text{for } 0 \leq q < \frac{1}{2}, \\ 1-q & \text{for } \frac{1}{2} \leq q \leq 1, \end{cases}$$

which is strictly better in the interval $[0, \frac{1}{2})$.

5 Conclusion and Open Questions

As it should have become clear from the above exposition, identification theory in the quantum setting is an enormously fruitful area, much more so even than the classical version, if only because we have at least five natural capacities. And we did not yet even touch upon a general theory of information transfer in quantum information, or rather how quantum information would fit into this far-reaching vision [2,3], these aspects still awaiting development.

At the same time the subject of identification via quantum channels is wide open, with most of the questions implied in the original papers [25,10,32] remaining unsolved, despite significant progress over the last decade. In particular, it turned out that the quantum identification task lent itself much more easily to the currently available techniques, and that the recent progress satisfyingly shed a fresh light on the older, and seemingly more elementary classical identification task. The following seven broad open problems are recommended to the reader's attention.

1. Surely the biggest open problem is to determine the classical ID-capacity $C_{\rm ID}(\mathcal{N})$ of a general quantum channel, and to study its properties, such as additivity etc. Even obtaining non-trivial upper bounds would be a worthy goal. Note that practically all transmission capacities of a channel are upper bounded by its entanglement-assisted capacity, by way of the Quantum Reverse Shannon Theorem [12,13,14] through simulation of the channel by noiseless communication and unlimited shared entanglement. This argument is not available here since entanglement or even common randomness have an impact on the ID-capacities.

   In fact, the few cases for which $C_{\rm ID}$ is known are consistent with the idea that it is always equal to the entanglement-assisted classical capacity of the channel [32]. One might speculate that $C_{\rm ID}(\mathcal{N}) \geq Q_{\rm ID,v}(\mathcal{N}) \geq C_E(\mathcal{N})$ be true for all channels, seeing that for sufficiently low noise we can prove it, and that it is true for the *amortized* classical ID-capacity. The erasure channel $\mathcal{E}_q$ discussed in Section 4 is already an excellent test case for this idea.
2. Is there a deeper, operational, reason why the amortized quantum ID-capacity equals the entanglement-assisted classical(!) capacity of a channel? In the derivation of [22] this comes out naturally as a result of the analysis, but almost as an accident, and it seems difficult to connect it to [12]...
3. Is the simultaneous ID-capacity $C_{\rm ID}^{\rm sim}(\mathcal{N})$ equal to the non-simultaneous version $C_{\rm ID}(\mathcal{N})$? I suspect that they are different, possibly even for the noiseless qubit channel (see Theorem 7 and subsequent remarks). In such a case we face another problem to determine $C_{\rm ID}^{\rm sim}(\mathcal{N})$. When studying simultaneous ID-codes, Löber's technical condition in Theorem 4 deserves special attention, as it precludes using the entire input state space of the iid channels. A very interesting case to study will be (rank-one POVM) qc-channels as there any identification code is *per se* simultaneous. For these channels we know the amortized quantum ID-capacity (it is the entanglement-assisted classical capacity, which evaluates to $\log|A|$), and that amortization rate 0 is sufficient to achieve it, in some cases even a constant amount. In fact, it would be interesting to know whether the visible quantum ID-capacity for these channels is the same – cf. the case of cq-channels discussed in Remark 17 –; this would evidently prove $C_{\rm ID}(\mathcal{N}) \geq \log|A|$ for all these channels $\mathcal{N}$, whereas it is known that the classical capacity $C(\mathcal{N})$ for many of them is much smaller.
4. The role of amortization is extremely interesting: For the quantum ID-capacity it makes for a quasi-superactivation effect, since a vanishing rate of it (i.e. an arbitrary small rate of noiseless communication) can turn a capacity 0 channel into one of positive capacity. It is possible that vanishing rate of amortization likewise has an impact on classical ID-capacities – see the example of qc-channels discussed in the previous point.
   Finally, in [22] only the non-triviality of amortization (and only for the erasure channels) was proved. How to characterize the quantum ID-rate vs. amortization rate tradeoff?
5. We have seen that the visible quantum ID-capacity can be larger than the blind variant, indeed the former can be positive while the latter is 0 for

cq-channels. Let us note that the distinction visible/blind can also be made in the quantum transmission game, and there it is far from clear whether there will be a difference in capacities, see [32].

6. We did not comment much on the role of shared correlations in the identification game, indeed referring the reader to [32,33], where also the impact of feedback is discussed. However, in Proposition 8 we saw that not only is the classical ID-capacity of common randomness (in the presence of negligible communication) equal to 1 per bit, but it increases the rate of any given ID-code by 1 per bit. We also know that the classical ID-capacity of shared entanglement is 2 per ebit, but it is open whether we can augment a given ID-code with entanglement to increase its rate by 2 per ebit.
7. Finally: All the known upper bounds on classical ID-capacities are in fact strong converses. Does the strong converse also hold for (visible, blind, amortized, etc) quantum ID-capacities? This question seems to require new techniques to be answered.

Acknowledgements. I have thought about identification in quantum information theory for quite some time, going back all the way to the days of my PhD, from which period date my hugely enjoyable mathematical interactions with Peter Löber, and of course with Rudolf Ahlswede, a sometimes terrifying but always, and ever newly, inspiring *Doktorvater*. In an article such as the present one it may be permitted to say that I miss him. Why, he keeps influencing my work even now!

Fortunately, later on others began to share my enthusiasm for quantum identification, most importantly Patrick Hayden. Much of the present review was only motivated by our recent collaboration.

I am or have been supported by a U.K. EPSRC Advanced Research Fellowship, the European Commission (STREP "QCS" and Integrated Project "QESSENCE"), the ERC Advanced Grant "IRQUAT", a Royal Society Wolfson Merit Award and a Philip Leverhulme Prize.

References

1. Abeyesinghe, A., Hayden, P., Smith, G., Winter, A.: Optimal superdense coding of entangled states. IEEE Trans. Inf. Theory 52(8), 3635–3641 (2006), arXiv:quant-ph/0407061
2. Ahlswede, R.: General theory of information transfer. Elec. Notes Discr. Math. 21, 181–184 (2005)
3. Ahlswede, R.: General theory of information transfer: updated. Discr. Applied Math. 156(9), 1348–1388 (2008)
4. Ahlswede, R.: On Concepts of Performance Parameters for Channels. In: Ahlswede, R., Bäumer, L., Cai, N., Aydinian, H., Blinovsky, V., Deppe, C., Mashurian, H. (eds.) General Theory of Information Transfer and Combinatorics. LNCS, vol. 4123, pp. 639–663. Springer, Heidelberg (2006)
5. Ahlswede, R., Balakirsky, V.B.: Identification under random processes. Probl. Inf. Transm. 32(1), 123–138 (1996)

6. Ahlswede, R.: Introduction. In: Ahlswede, R., Bäumer, L., Cai, N., Aydinian, H., Blinovsky, V., Deppe, C., Mashurian, H. (eds.) Information Transfer and Combinatorics. LNCS, vol. 4123, pp. 1–44. Springer, Heidelberg (2006)
7. Ahlswede, R., Blinovsky, V.M.: Large deviations in quantum information theory. Probl. Inf. Transm. 39(4), 373–379 (2003); Translated form Problemy Peredachi Informatsii 39(4), 63–70 (2003)
8. Ahlswede, R., Dueck, G.: Identification via channels. IEEE Trans. Inf. Theory 35(1), 15–29 (1989)
9. Ahlswede, R., Dueck, G.: Identification in the presence of feedback – a discovery of new capacity formulas. IEEE Trans. Inf. Theory 35(1), 30–36 (1989)
10. Ahlswede, R., Winter, A.: Strong converse for identification via quantum channels. IEEE Trans. Inf. Theory 48(3), 569–579 (2002), arXiv:quant-ph/0012127
11. Aubrun, G., Szarek, S., Werner, E.: Hastings' additivity counterexample via Dvoretzky's theorem. Commun. Math. Phys. 305(1), 85–97 (2011)
12. Bennett, C.H., Shor, P.W., Smolin, J.A., Thapliyal, A.V.: Entanglement-assisted capacity of a quantum channel and the reverse Shannon theorem. IEEE Trans. Inf. Theory 48(10), 2637–2655 (2002)
13. Bennett, C.H., Devetak, I., Harrow, A.W., Shor, P.W., Winter, A.: The quantum reverse Shannon theorem and resource tradeoffs for simulating quantum channels (2012), arXiv[quant-ph]:0912.5537v2
14. Berta, M., Christandl, M., Renner, R.: The quantum reverse Shannon theorem based on one-shot information theory. Commun. Math. Phys. 306(3), 579–615 (2011), arXiv[quant-ph]:0912.3805
15. Buhrman, H., Cleve, R., Watrous, J., de Wolf, R.: Quantum fingerprinting. Phys. Rev. Lett. 87, 167902 (2001), arXiv:quant-ph/0102001
16. Devetak, I., Shor, P.W.: The Capacity of a quantum channel for simultaneous transmission of classical and quantum information. Commun. Math. Phys. 256(2), 287–303 (2005)
17. Dupuis, F., Florjanczyk, J., Hayden, P., Leung, D.: Locking classical information (2010), arXiv[quant-ph]:1011.1612
18. Han, T.-S., Verdú, S.: New results in the theory of identification via channels. IEEE Trans. Inf. Theory 38(1), 14–25 (1992)
19. Han, T.-S., Verdú, S.: Approximation theory of output statistics. IEEE Trans. Inf. Theory 39(3), 752–772 (1993)
20. Hayden, P., Leung, D., Shor, P.W., Winter, A.: Randomizing quantum states: constructions and applications. Commun. Math. Phys. 250, 371–391 (2004), arXiv:quant-ph/0307104
21. Hayden, P., Leung, D., Winter, A.: Aspects of generic entanglement. Commun. Math. Phys. 265, 95–117 (2006), arXiv:quant-ph/0407049
22. Hayden, P., Winter, A.: Weak decoupling duality and quantum identification. IEEE Trans. Inf. Theory 58(7), 4914–4929 (2012), arXiv[quant-ph]:1003.4994
23. Harrow, A.W., Montanaro, A., Short, A.J.: Limitations on Quantum Dimensionality Reduction. In: Aceto, L., Henzinger, M., Sgall, J. (eds.) ICALP 2011. LNCS, vol. 6755, pp. 86–97. Springer, Heidelberg (2011)
24. Kretschmann, D., Schlingemann, D., Werner, R.F.: The information-disturbance tradeoff and the continuity of Stinespring's representation. IEEE Trans. Inf. Theory 54(4), 1708–1717 (2008)
25. Löber, P.: Quantum channels and simultaneous ID coding. PhD thesis, Universität Bielefeld, Bielefeld (Germany) (1999), `http://katalog.ub.uni-bielefeld.de/title/1789755`

26. Renes, J.M.: Approximate quantum error correction via complementary observables (2010), arXiv[quant-ph]:1003.1150
27. Schumacher, B., Westmoreland, M.D.: Approximate quantum error correction. Quantum Inf. Proc. 1(1/2), 5–12 (2002), arXiv:quant-ph/0112106
28. Shannon, C.E.: A mathematical theory of communication. Bell Syst. Tech. J. 27, 379–423 & 623–656 (1948)
29. Stinespring, W.F.: Positive functions on C^*-algebras. Proc. Amer. Math. Soc. 6(2), 211–216 (1955)
30. Tropp, J.A.: User-friendly tail bounds for sums of random matrices. Found. Comput. Math. 12(4), 389–434 (2012), arXiv[math.PR]:1004.4389 (2010)
31. Wilde, M.M.: From Classical to Quantum Shannon Theory (2011), arXiv[quant-ph]:1106.1445
32. Winter, A.: Quantum and classical message identification via quantum channels. In: Hirota, O. (ed.) Festschrift "A. S. Holevo 60", pp. 171–188. Rinton Press (2004), Reprinted in Quantum Inf. Comput. 4(6&7), 563–578 (2004), arXiv:quant-ph/0401060
33. Winter, A.: Identification Via Quantum Channels in the Presence of Prior Correlation and Feedback. In: Ahlswede, R., Bäumer, L., Cai, N., Aydinian, H., Blinovsky, V., Deppe, C., Mashurian, H. (eds.) General Theory of Information Transfer and Combinatorics. LNCS, vol. 4123, pp. 486–504. Springer, Heidelberg (2006)

Classical-Quantum Arbitrarily Varying Wiretap Channel

Vladimir Blinovsky[1,⋆] and Minglai Cai[2]

[1] Institute of Information Transmission Problems, Russian Academy of Sciences, B. Karetnii Per 19, Moscow 127 994, Russia
vblinovs@yandex.ru

[2] Fakultät für Mathematik Universität Bielefeld, Postfach 100131, 33501 Bielefeld
mlcai@math.uni-bielefeld.de

Dedicated to the memory of Rudolf Ahlswede

Abstract. We derive a lower bound on the secrecy capacity of classical-quantum arbitrarily varying wiretap channel for both the case with and without channel state information at the transmitter.

Keywords: quantum channel, arbitrarily varying channel, wiretap channel.

1 Introduction

The arbitrarily varying channel models transmission over a channel with an state that can change over time. We may interpret it as a channel with an evil jammer. The arbitrarily varying channel was first introduced by Blackwell, Breiman, and Thomasian in [8]. The wiretap channel models communication with security. It was first introduced by Wyner in [12]. We may interpret it as a channel with an evil eavesdropper. The arbitrarily varying wiretap channel models transmission with both a jammer and an eavesdropper. Its capacity has been determined by Bjelaković, Boche, and Sommerfeld in [3].

A quantum channel is a channel which can transmit both classical and quantum information. In this paper, we consider the capacity of quantum channels to carry classical information, or equivalently, the capacity of a classical quantum channels. The classical capacity of quantum channels has been determined by Holevo in [9]. A classical-quantum channel with a jammer is called a classical-quantum arbitrarily varying channel, its capacity has been determined by Ahlswede and Blinovsky in [1]. Bjelaković, Boche, Janßen, and Nötzel gave an alternative proof and a proof of the strong converse in [2]. A classical-quantum channel with an eavesdropper is called a classical-quantum wiretap channel, its capacity has been determined by Devetak in [7], and by N. Cai, Winter, and Yeung in [5].

⋆ Project "Informationstheorie des Quanten-Repeaters" supported by the Federal Ministry of Education and Research (Ref. No. 01BQ1052).

H. Aydinian, F. Cicalese, and C. Deppe (Eds.): Ahlswede Festschrift, LNCS 7777, pp. 234–246, 2013.

A classical-quantum channel with both a jammer and an eavesdropper is called a classical-quantum wiretap channel, it is defined as a pair of double indexed finite set of density operators $\{(\rho_{x,t}, \sigma_{x,t}) : x \in \mathcal{X}, t \in \Theta\}$ with common input alphabet $\mathcal{X}$ connecting a sender with two receivers, one legal and one wiretapper, where t is called a state of the channel pair. The legitimate receiver accesses the output of the first channel $\rho_{x,t}$ in the pair $(\rho_{x,t}, \sigma_{x,t})$, and the wiretapper observes the output of the second part $\sigma_{x,t}$ in the pair $(\rho_{x,t}, \sigma_{x,t})$, respectively, when a state t, which varies from symbol to symbol in an arbitrary manner, governs both the legitimate receiver's channel and the wiretap channel. A code for the channel conveys information to the legal receiver such that the wiretapper knows nothing about the transmitted information. This is a generalization of model of classical-quantum compound wiretap channels in [4] to the case when the channel states are not stationary, but can change over the time.

We will be dealing with two communication scenarios. In the first one only the transmitter is informed about the index t (channel state information, or simply CSI, at the transmitter), while in the second, the legitimate users have no information about that index at all (no CSI).

2 Definitions

Let $\mathcal{X}$ be a finite set (the set of code symbols). Let $\Theta := \{1, \cdots, T\}$ be finite set (the set of channel states). Denote the set of the (classical) messages by $\{1, \cdots, J_n\}$. Define the classical-quantum arbitrarily varying wiretap channel by a pair of double indexed finite set of density operators $\{(\rho_{x,t}, \sigma_{x,t}) : x \in \mathcal{X}, t \in \Theta\}$ on $\mathbb{C}^d$. Here the first family represents the communication link to the legitimate receiver while the output of the latter is under control of the wiretapper.

One important notation in [1] is the symmetrizable classical-quantum arbitrarily varying channel. We say $\{\rho_{x,t} : x \in \mathcal{X}, t \in \Theta\}$ is symmetrizable if there exists a parameterized set of distributions $\{U(t|x) : x \in \mathcal{X}\}$ on Θ such that for all $x, x' \in \mathcal{X}$ the following equalities are valid:

$$\sum_{t \in \Theta} U(t|x)\rho_{x',t} = \sum_{t \in \Theta} U(t|x')\rho_{x,t}$$

For any probability distribution $P \in \mathcal{P}$ and positive δ denote $\mathcal{T}^n_{P,\delta}$ the δ-typical set in sense of [6].

For a state ρ, the von Neumann entropy is defined as

$$S(\rho) := -\mathrm{tr}(\rho \log \rho) \ .$$

Let P be a probability distribution over a finite set J, and $\Phi := \{\rho(x) : x \in J\}$ be a set of states labeled by elements of J. Then the Holevo χ quantity is defined as

$$\chi(P, \Phi) := S\left(\sum_{x \in J} P(x)\rho(x)\right) - \sum_{x \in J} P(x) S\left(\rho(x)\right) \ .$$

A (deterministic) quantum code C of cardinality J_n and length n is a set of pairs $\{(c_j^n, D_j) : j = 1, \cdots J_n\}$, where $c_j^n = (c_{j,1}, c_{j,2}, \cdots, c_{j,n}) \in \mathcal{X}^n$, and $\{D_j : j=1, \cdots J_n\}$ is a collection of positive semi-definite operators which is a resolution of the identity in $(\mathbb{C}^d)^{\otimes n}$, i.e. $\sum_{j=1}^{J_n} D_j = id_{(\mathbb{C}^d)^{\otimes n}}$.

A non-negative number R is an achievable secrecy rate for the classical-quantum arbitrarily varying wiretap channel if for every $\epsilon > 0, \delta > 0, \zeta > 0$ and sufficiently large n there exist a code $C = \{(c_j^n, D_j) : j = 1, \cdots J_n\}$ such that

$$\frac{\log J_n}{n} > R - \delta \ ,$$

$$\max_{t^n \in \Theta^n} P_e(C, t^n) < \epsilon \ ,$$

$$\max_{t^n \in \Theta^n} \frac{1}{n} \chi \left(W, \{\sigma_{c_j^n, t^n} : j = 1, \cdots, J_n\} \right) < \zeta ,$$

where W is an uniformly distributed random variable with values in $\{1, \cdots J_n\}$. Here $P_e(C, t^n)$ (the average probability of the decoding error of a deterministic code C, when the state (sequence of states) of the classical-quantum arbitrarily varying wiretap channels is $t^n = (t_1, t_2, \cdots, t_n)$) is defined as follows

$$P_e(C, t^n) := 1 - \frac{1}{J_n} \sum_{j=1}^{J_n} \mathrm{tr}(\rho_{c_j^n, t^n} D_j) \ ,$$

where $\rho_{c_j^n, t^n} := \rho_{c_{j,1}, t_1} \otimes \rho_{c_{j,2}, t_2} \otimes \cdots \otimes \rho_{c_{j,n}, t_n}$.

A non-negative number R is an achievable secrecy rate for the classical-quantum arbitrarily varying wiretap channel with channel state information (CSI) at the transmitter if for every $\epsilon > 0$, $\delta > 0$, $\zeta > 0$ and sufficiently large n there exist for every t^n a code $C^{t^n} = \{(c_{j,t^n}^n, D_j) : j = 1, \cdots J_n\}$, where $c_{j,t^n}^n = (c_{j,1,t^n}, c_{j,2,t^n}, \cdots, c_{j,n,t^n}) \in \mathcal{X}^n$, such that

$$\frac{\log J_n}{n} > R - \delta \ ,$$

$$\max_{t^n \in \Theta^n} P_e^{CSI}(C^{t^n}, t^n) < \epsilon \ ,$$

$$\max_{t^n \in \Theta^n} \frac{1}{n} \chi \left(W, \{\sigma_{c_{j,t^n}^n, t^n} : j = 1, \cdots, J_n\} \right) < \zeta \ .$$

Here $P_e^{CSI}(C^{t^n}, t^n)$ is defined as follows:

$$P_e^{CSI}(C^{t^n}, t^n) := 1 - \frac{1}{J_n} \sum_{j=1}^{J_n} \mathrm{tr}(\rho_{c_{j,t^n}^n, t^n} D_j) \ ,$$

where $\rho_{c_j^{t^n}, t^n} := \rho_{c_{j,t^n}^n, t_1} \otimes \rho_{c_{j,1,t^n}, t_2} \otimes \cdots \otimes \rho_{c_{j,n,t^n}, t_n}$.

One tool we will use is the random quantum code, which we will define now. Let $\Lambda = \left(\mathcal{X}^n \times \mathcal{B}((\mathbb{C}^d)^{\otimes n})\right)^{J_n}$. A random quantum code $(\{C^\gamma : \gamma \in \Lambda\}, G)$

consists of the family of sets of J_n pairs $C^\gamma = \{(c_j^{n,\gamma}, D_j^\gamma) : j = 1, \cdots, J_n\}_{\gamma \in \Lambda}$, where $c_j^{n,\gamma} = (c_{j,1}^{n,\gamma} \cdots c_{j,n}^{n,\gamma}) \in \mathcal{X}^n$ and $\sum_{j=1}^{J_n} D_j^\gamma = id_{(\mathbb{C}^d)^{\otimes n}}$, together with a distribution G on Λ. The average probability of the decoding error is defined as follows

$$P_{er} := \inf_G \max_{t^n \in \Theta^n} \int_\Lambda P_e(C^\gamma, t^n) dG(\gamma) \ .$$

A non-negative number R is an achievable secrecy rate for the classical-quantum arbitrarily varying wiretap channel under random quantum coding if for every $\delta > 0$, $\zeta > 0$, and $\epsilon > 0$, if n is sufficiently large, we can find a J_n such that

$$\frac{\log J_n}{n} > R - \delta \ ,$$

$$P_{er} < \epsilon \ ,$$

$$\max_{t^n \in \Theta^n} \max_{\gamma \in \Lambda} \frac{1}{n} \chi\left(W, \{\sigma_{c_j^{n,\gamma}, t^n} : j = 1, \cdots, J_n\}\right) < \zeta \ .$$

Denote $P^n(x^n) := P(x_1)P(x_2) \cdots P(x_n)$. The following facts hold: (cf. [10])

Let $\mathcal{X}'$ be a finite set and for any $x \in \mathcal{X}'$, ς_x be a density operator on $\mathbb{C}^d$. For any distribution P on $\mathcal{X}'$ and $x^n \in \mathcal{T}_P^n$ let $\varsigma_{x^n} := \varsigma_{x_1} \otimes \varsigma_{x_2} \otimes \cdots \otimes \varsigma_{x_n}$. Let $\sum_k l_k |e_k\rangle\langle e_k|$ be a spectral decomposition of $P\varsigma := \sum_{x^n \in \mathcal{X}'^n} P^n(x^n) \varsigma_{x^n}$, where $l_k \in \mathbb{R}^+$, $\sum_k l_k = 1$. For $\alpha > 0$ denote $G_\alpha := \{k : 2^{-n[S(\sum_{x^n \in \mathcal{X}'^n} P^n(x^n)\varsigma_{x^n}) - \alpha]} \leq l_k \leq 2^{-n[S(\sum_{x^n \in \mathcal{X}'^n} P^n(x^n)\varsigma_{x^n}) + \alpha]}\}$. Denote $\Pi_{\varsigma,\alpha} := \sum_{k \in G_\alpha} |e_k\rangle\langle e_k|$. Then $\Pi_{\varsigma,\alpha}$ commuting with $P\varsigma$ and satisfying

$$\mathrm{tr}\,(P\varsigma \Pi_{P\varsigma,\alpha}) \geq 1 - \frac{d}{4n\alpha^2} \ ,$$

$$\mathrm{tr}\,(\Pi_{\rho,\alpha}) \leq 2^{S(\sum_{x^n \in \mathcal{X}'^n} P(x^n)\varsigma_{x^n}) + Kd\alpha\sqrt{n}} \ ,$$

$$\Pi_{P\varsigma,\alpha} \cdot P\varsigma \cdot \Pi_{P\varsigma,\alpha} \leq 2^{-S(\sum_{x^n \in \mathcal{X}'^n} P(x^n)\varsigma_{x^n}) + Kd\alpha\sqrt{n}} \Pi_{P\varsigma,\alpha} \ ,$$

$$\mathrm{tr}\,\left(P\varsigma \cdot \Pi_{P\varsigma,\alpha\sqrt{a}}\right) \geq 1 - \frac{ad}{4n\alpha^2} \ ,$$

where $a := \#\mathcal{X}'$ and K is a positive constant.

Let $\sum_k l_{x^n,k} |e_{x^n,j}\rangle\langle e_{x^n,k}|$ be a spectral decomposition of ς_{x^n}, where $l_{x^n,k} \in \mathbb{R}^+$, $\sum_k l_{x^n,k} = 1$. For $\alpha > 0$ denote $G_{x^n,\alpha} := \{k : 2^{-n[S(\varsigma_{x^n}) - \alpha]} \leq l_{x^n,k} \leq 2^{-n[S(\varsigma_{x^n}) + \alpha]}\}$, and $\Pi_{\varsigma_{x^n},\alpha} := \sum_{k \in G_{x^n,\alpha}} |e_{x^n,k}\rangle\langle e_{x^n,k}|$.

The subspace projector $\Pi_{\varsigma_{x^n},\alpha}$ commutes with ς_{x^n} and satisfies:

$$\mathrm{tr}\,(\varsigma_{x^n} \Pi_{\varsigma_{x^n},\alpha}) \geq 1 - \frac{ad}{4n\alpha^2} \ ,$$

$$\mathrm{tr}\,(\Pi_{\varsigma_{x^n},\alpha}) \leq 2^{\sum_{x^n \in \mathcal{X}'^n} P(x^n) S(\varsigma_{x^n}) + Kad\alpha\sqrt{n}} \ ,$$

$$\Pi_{\varsigma_{x^n},\alpha} \cdot \varsigma_{x^n} \cdot \Pi_{\varsigma_{x^n},\alpha} \leq 2^{-\sum_{x^n \in \mathcal{X}'^n} P(x^n) S(\varsigma_{x^n}) + Kad\alpha\sqrt{n}} \Pi_{\varsigma_{x^n},\alpha} \ ,$$

where K is a positive constant.

3 Main Result

Theorem 1. *Let $\mathcal{W} := \{(\rho_{x,t}, \sigma_{x,t}) : x \in \mathcal{X}, t \in \Theta\}$ be a classical-quantum arbitrarily varying wiretap channel, if for all $t \in \Theta$ it holds: $\{\rho_{x,t}, x \in \mathcal{X}\}$ is not symmetrizable, then the largest achievable secrecy rate, called secrecy capacity, of $\mathcal{W}$, is bounded as follow,*

$$C(\mathcal{W}) \geq \max_{P\in\mathcal{P}} \left(\min_{Q\in\mathcal{Q}} \chi\left(P, \{\rho_x^Q : x \in \mathcal{X}\}\right) - \lim_{n\to\infty} \max_{t^n\in\Theta^n} \frac{1}{n}\chi(P^n, \{\sigma_{x^n,t^n} : x^n \in \mathcal{X}^n\}) \right), \tag{1}$$

where $\mathcal{P}$ are distributions on $\mathcal{X}$, $\mathcal{Q}$ are distributions on Θ, and $\rho_x^Q = \sum_{t\in\Theta} Q(t)\rho_{x,t}$ for $Q \in \mathcal{Q}$.

If $\{\rho_{x,t}, \in \mathcal{X}, t \in \Theta\}$ is not symmetrizable, then the secrecy capacity of $\mathcal{W}$ with CSI at the transmitter is bounded as follow

$$C_{CSI}(\mathcal{W}) \geq \min_{Q\in\mathcal{Q}, t^n\in\Theta^n} \max_{P\in\mathcal{P}} \left(\chi(P, \{\rho_x^Q : x \in \mathcal{X}\}) - \lim_{n\to\infty} \frac{1}{n}\chi(P^n, \{\sigma_{x^n,t^n} : x^n \in \mathcal{X}^n\}) \right). \tag{2}$$

Is $\{\rho_{x,t}, x \in \mathcal{X}\}$ symmetrizable for some $t \in \Theta$, then we have:

$$C(\mathcal{W}) = C_{CSI}(\mathcal{W}) = 0. \tag{3}$$

Proof. At first, we are going to prove (1).

For $P \in \mathcal{P}$ denote $\rho^{P,Q} := \sum_{x\in\mathcal{X}} P(x)\rho_x^Q$. Let

$$J_n = \left\lfloor 2^{n \min_Q \chi(P,\{\rho_x^Q : x\in\mathcal{X}\}) - \max_{t^n\in\Theta^n} \chi(P^n,\{\sigma_{x^n,t^n} : x^n\in\mathcal{X}^n\} - 2n\eta)} \right\rfloor,$$

$$L_n = \left\lfloor 2^{\max_{t^n\in\Theta^n} \chi(P^n,\{\sigma_{x^n,t^n} : x^n\in\mathcal{X}^n\} + n\eta)} \right\rfloor,$$

where η is a positive constant.

Let $P'(x^n) := \begin{cases} \frac{P^n(x^n)}{P^n(\mathcal{T}^n_{P,\delta})} & \text{if } x^n \in \mathcal{T}^n_{P,\delta} \\ 0 & \text{else} \end{cases}$, and $X^n := (X^n_{j,l})_{j\in\{1,\cdots,J_n\}, l\in\{1,\cdots,L_n\}}$ be a family of random matrices such that their entries are i.i.d. according to P'.

Fix $P \in \mathcal{P}$. Denote $Q^n = (Q_1, \cdots, Q_n) \in \mathcal{Q}$. Let $\rho^{P,Q^n} := \rho^{P,Q_1} \otimes \rho^{P,Q_2} \otimes \cdots \otimes \rho^{P,Q_n}$, and $\sum_{j^n} \lambda_{j^n}^{P,Q^n} |e_{j^n}^{P,Q^n}\rangle\langle e_{j^n}^{P,Q^n}|$ be a spectral decomposition of ρ^{P,Q^n}, where $\lambda_{j^n}^{P,Q^n} \in \mathbb{R}^+$, $\sum_{j^n} \lambda_{j^n}^{P,Q^n} = 1$.

For $\delta > 0$ denote $F_{\delta,Q^n} := \{j^n : 2^{-\sum_{i=1}^n H(\rho^{P,Q_i}) - n\delta} \leq \lambda_{j^n}^{P,Q^n} \leq 2^{-\sum_{i=1}^n H(\rho^{P,Q_i}) + n\delta}\}$, and $\Pi_\delta^{P,Q^n} := \sum_{j^n\in F_{\delta,Q^n}} |e_j^{P,Q^n}\rangle\langle e_j^{P,Q^n}|$.

For any $x^n \in \mathcal{X}^n$ let $\rho_{x^n}^{Q^n} := \rho_{x_1}^{Q_1} \otimes \rho_{x_2}^{Q_2} \otimes \cdots \otimes \rho_{x_n}^{Q_n}$. Let $\sum_{j^n} \lambda_{x^n,j^n}^{Q^n} |e_{x^n,j^n}^{Q^n}\rangle\langle e_{x^n,j^n}^{Q^n}|$ be a spectral decomposition of $\rho_{x^n}^{Q^n}$, where $\lambda_{x^n,j^n}^{Q^n} \in \mathbb{R}^+$, $\sum_{j^n} \lambda_{x^n,j^n}^{Q^n} = 1$.

For $\delta > 0$ denote $F_{x^n,Q^n,\delta} := \{j_n : 2^{-\sum_{i=1}^n \sum_{x\in\mathcal{X}} P(x)H(\rho_x^{Q_i})-n\delta} \leq \lambda_{x^n,j}^{Q^n} \leq 2^{-\sum_{i=1}^n \sum_{x\in\mathcal{X}} P(x)H(\rho_x^{Q_i})+n\delta}\}$, and $\Pi_{x^n,\delta}^{Q^n} := \sum_{j_n \in F_{x^n,Q^n,\delta}} |e_{x^n,j}^{Q^n}\rangle\langle e_{x^n,j}^{Q^n}|$.

Our proof bases on the following two lemmas. The first lemma is due to Rudolf Ahlswede and Vladimir Blinovsky, the second one (the Covering Lemma) is due to Rudolf Ahlswede and Andreas Winter.

Lemma 1 (cf. [1]). *Let $\{\varrho_{x,t}, x \in \mathcal{X}, t \in \Theta\}$ be a classical-quantum arbitrarily varying wiretap channel, defined in sense of [1], where $\mathcal{X}$ is the set of code symbols and Θ is the set of states of the classical-quantum arbitrarily varying wiretap channel.*

For $\{c_i^n : i = 1, \cdots, N\} \subset \mathcal{X}^n$ and distribution $Q^n = (Q_1, Q_2, \cdots Q_n)$ on Θ^n define

$$D_i^{Q^n} := \left(\sum_{j=1}^N \Pi_\delta^{P,Q^n} \Pi_{c_j^n,\delta}^{P,Q^n} \Pi_\delta^{P,Q^n}\right)^{-1/2} \Pi_\delta^{P,Q^n} \Pi_{c_i^n,\delta}^{P,Q^n} \Pi_\delta^{P,Q^n} \left(\sum_{j=1}^N \Pi_\delta^{P,Q^n} \Pi_{c_j^n,\delta}^{P,Q^n} \Pi_\delta^{P,Q^n}\right)^{-1/2} . \tag{4}$$

Define the the set of the quantum codes

$$\mathcal{C} := \left\{C^{Q^n} = \{(c_i^n, D_i^{Q^n}) : i = 1, \cdots, N\} : c_i^n \in \mathcal{X}^n \forall i, Q^n \text{ is a distribution on} \Theta^n\right\} , \tag{5}$$

If $\frac{\log N}{n} < \min_Q \chi\left(P, \{\varrho_x^Q : x \in \mathcal{X}\}\right) - \delta$, where δ is a positive constant, and assume $\{\varrho_{x,t}, x \in \mathcal{X}, t \in \Theta\}$ is not symmetrizable. then following holds. For any $\epsilon > 0$, if n is large enough, then there exist a distribution G on $\mathcal{C}$ such that

$$\max_{t^n \in \Theta^n} \sum_{C \in \mathcal{C}} P_e(C, t^n) G(C) < \epsilon . \tag{6}$$

Lemma 2 (Covering Lemma, cf. [10]). *Suppose we are given a finite set $\mathcal{Y}$, an ensemble $\{\sigma_y : y \in \mathcal{Y}\}$ with probability distribution $p_\mathcal{Y}$ on $\mathcal{Y}$. Suppose there exist a total subspace projector Π and codeword subspace projectors $\{\Pi_y\}_{y\in\mathcal{Y}}$, they project onto subspaces of the Hilbert space in which the states $\{\sigma_y\}$ exist, and these projectors and the ensemble satisfy the following conditions:*

$$\mathrm{tr}\,(\sigma_y \Pi) \geq 1 - \epsilon$$
$$\mathrm{tr}\,(\sigma_y \Pi_y) \geq 1 - \epsilon$$
$$\mathrm{tr}\,(\Pi) \leq c$$
$$\Pi_y \sigma_y \Pi_y \leq \frac{1}{d} \Pi_y$$

Suppose that $\mathcal{M} \subset \mathcal{Y}$ is a set of size $|\mathcal{M}|$ with elements $\{m\}$, $C = \{C_m\}_{m\in\mathcal{M}}$ is a random code where the codewords C_m are chosen according to the distribution $p_{\mathcal{Y}}(y)$, and an ensemble $\{\sigma_{C_m} : m \in \mathcal{M}\}$ with uniform distribution on $\mathcal{M}$, then

$$Pr\left\{\left\|\sum_{y\in\mathcal{Y}} p_{\mathcal{Y}}(y)\sigma_y - \frac{1}{|\mathcal{M}|}\sum_{m\in\mathcal{M}} \sigma_{C_m}\right\|_1 \le \epsilon + 4\sqrt{\epsilon} + 24\sqrt[4]{\epsilon}\right\} \ge 1 - 2c\exp\left(-\frac{\epsilon^3|\mathcal{M}|d}{2\log 2c}\right). \tag{7}$$

Let $\{X^n_{j,l}\}_{j\in\{1,\cdots,J_n\},l\in\{1,\cdots,L_n\}}$ be a family of random matrices such that the entries are i.i.d. according to P'^n.

For every realization $\left(x^n_{j,l}\right)_{j\in\{1,\cdots,J_n\},l\in\{1,\cdots,L_n\}}$ of $\left(X^n_{j,l}\right)_{j\in\{1,\cdots,J_n\},l\in\{1,\cdots,L_n\}}$ and distribution Q^n on Θ^n define

$$D^{Q^n}_{x^n_{j,l}} := \left(\sum_{j'=1}^{J_n}\sum_{l'=1}^{L_n} \Pi^{P,Q^n}_{\delta}\, \Pi^{P,Q^n}_{x^n_{j',l'},\delta} \Pi^{P,Q^n}_{\delta}\right)^{-1/2}$$

$$\Pi^{P,Q^n}_{\delta}\, \Pi^{P,Q^n}_{x^n_{j,l},\delta} \Pi^{P,Q^n}_{\delta} \left(\sum_{j'=1}^{J_n}\sum_{l'=1}^{L_n} \Pi^{P,Q^n}_{\delta}\, \Pi^{P,Q^n}_{x^n_{j',l'},\delta} \Pi^{P,Q^n}_{\delta}\right)^{-1/2}. \tag{8}$$

Let $\sum_k \lambda^{t^n}_k |e^{t^n}_k\rangle\langle e^{t^n}_k|$ be a spectral decomposition of $P\rho_{t^n} := \sum_{x^n\in\mathcal{X}^n} P^n(x^n)\rho_{x^n,t^n}$ We denote

$$G^{t^n}_\alpha := \{k : 2^{-n[H(\sum_{x^n\in\mathcal{X}^n} P^n(x^n)\rho_{x^n,t^n})-\alpha]} \le \lambda^{t^n}_k \le 2^{-n[H(\sum_{x^n\in\mathcal{X}^n} P^n(x^n)\rho_{x^n,t^n})+\alpha]}\}$$

and $\Pi_{\rho_{t^n},\alpha} := \sum_{k\in G^{t^n}_\alpha} |e^{t^n}_j\rangle\langle e^{t^n}_k|$, where $\rho_{x^n,t^n} := \rho_{x_1,t_1}\otimes\cdots\otimes\rho_{x_n,t_n}$. Let $\sum_k \lambda^{t^n}_{x^n,k}|e^{t^n}_{x^n,k}\rangle\langle e^{t^n}_{x^n,k}|$ be a spectral decomposition of ρ_{x^n,t^n}. Denote $G^{t^n}_{x^n,\alpha} := \{k : 2^{-n[H(\rho_{x^n,t^n})-\alpha]} \le \lambda^{t^n}_{x^n,k} \le 2^{-n[H(\rho_{x^n,t^n})+\alpha]}\}$, and $\Pi_{\rho_{x^n,t^n},\alpha} := \sum_{k\in G^{t^n}_{x^n,\alpha}} |e^{t^n}_{x^n,k}\rangle\langle e^{t^n}_{x^n,k}|$.

Define

$$\overline{\sigma}_{x^n,t^n} := \Pi_{P\rho_{t^n},\alpha\sqrt{a}}\Pi_{\rho_{x^n,t^n},\alpha}\cdot\sigma_{x^n,t^n}\cdot\Pi_{\rho_{x^n,t^n},\alpha}\Pi_{P\rho_{t^n},\alpha\sqrt{a}}\,. \tag{9}$$

By

Lemma 3 (cf. [11]). *Let ρ be a state and X be a positive operator with $X \le id$ (the identity matrix) and $1 - \mathrm{tr}(\rho X) \le \lambda \le 1$. Then*

$$\|\rho - \sqrt{X}\rho\sqrt{X}\|_1 \le \sqrt{8\lambda}\,.$$

and the fact that $\Pi_{P\rho_{t^n},\alpha\sqrt{a}}$ and $\Pi_{\rho_{x^n,t^n},\alpha}$ are both projection matrices, for any t^n and x^n it holds:

$$\|\overline{\sigma}_{x^n,t^n} - \sigma_{x^n,t^n}\|_1 \leq \sqrt{\frac{2(ad+d)}{n\alpha^2}} \ ,$$

Thus for any positive α and any positive η if n is large enough

$$\|\overline{\sigma}_{x^n,t^n} - \sigma_{x^n,t^n}\|_1 \leq \eta \ . \tag{10}$$

Since

$$\operatorname{tr}\left(\Pi_{P\rho_{t^n},\alpha\sqrt{a}}\right) \leq 2^{S(\sum_{x^n\in\mathcal{X}^n} P^n(x^n)\rho_{x^n,t^n})} \ ,$$

$$\Pi_{\rho_{x^n,t^n},\alpha} \cdot \sigma_{x^n,t^n} \cdot \Pi_{\rho_{x^n,t^n},\alpha} \leq 2^{-\sum_{x^n\in\mathcal{X}^n} P^n(x^n)S(\rho_{x^n,t^n})}\Pi_{\rho_{x^n,t^n},\alpha} \ ,$$

and

$$L_n \geq 2^{\chi\left(P^n,\{\sigma_{x^n,t^n}:x^n\in\mathcal{X}^n\}\right)+2n\delta} = \frac{2^{\sum_{x^n} P^n(x^n)S(\rho_{x^n,t^n})+n\delta}}{2^{S\left(\sum_{x^n} P^n(x^n)\rho_{x^n,t^n}\right)-n\delta}} \ ,$$

by applying covering lemma, for every t^n and $j' \in \{1,\cdots,J_n\}$ there is a positive constant c_1' such that for any $\nu > 0$,

$$Pr\left\{\left\|\frac{1}{L_n}\frac{1}{J_n}\sum_{l=1}^{L_n}\sum_{j=1}^{J_n}\overline{\sigma}_{X^n_{j,l},t^n} - \frac{1}{L_n}\sum_{l=1}^{L_n}\overline{\sigma}_{X^n_{j',l},t^n}\right\|_1 < \nu\right\} \geq 1-2^{-\nu^3 2^{nc_1'}} \ . \tag{11}$$

Since $|\Theta^n| = O(2^n)$, and $J_n \ll 2^{\nu^3 2^{nc_1'}}$, there is a positive constant c_1 such that for any $\nu > 0$,

$$Pr\left\{\left\|\frac{1}{L_n}\frac{1}{J_n}\sum_{l=1}^{L_n}\sum_{j=1}^{J_n}\overline{\sigma}_{X^n_{j,l},t^n} - \frac{1}{L_n}\sum_{l=1}^{L_n}\overline{\sigma}_{X^n_{j',l},t^n}\right\|_1 < \nu \ \forall j' \in \{1,\cdots,J_n\}\forall t^n \in \Theta^n\right\} \geq 1-2^{-\nu^3 nc_1} \ . \tag{12}$$

Denote the set of all codes $\left\{\left(x^{\prime n}_{j,l}, D^{Q^n}_{x^n_{j,l}}\right) : j=1,\cdots,J_n, l=1,\cdots,L_n\right\}$, where $(x_{j,l})_{j=1,\cdots,J_n,l=1,\cdots,L_n}$ are realizations of $(X_{j,l})_{j=1,\cdots,J_n,l=1,\cdots,L_n}$, such that

$$\left\|\frac{1}{L_n}\frac{1}{J_n}\sum_{l=1}^{L_n}\sum_{j=1}^{J_n}\overline{\sigma}_{x^n_{j,l},t^n} - \frac{1}{L_n}\sum_{l=1}^{L_n}\overline{\sigma}_{x^n_{j',l},t^n}\right\|_1 < \nu \ \forall j' \in \{1,\cdots,J_n\}\forall t^n \in \Theta^n$$

by $\mathcal{C}'_\nu$.

Now we want to show the following alternative result to Lemma 1.

If n is large enough then for any any positive ν, there exist a distribution G on $\mathcal{C}'_\nu$ such that

$$\max_{t^n\in\Theta^n} \sum_{C\in\mathcal{C}'_\nu} P_e(C,t^n)G(C) < \epsilon \ . \tag{13}$$

In [1], following inequality is shown. There is a positive constant c_2 such that for any positive ν

$$Pr\left\{1-\frac{1}{L_n}\frac{1}{J_n}\sum_{l=1}^{L_n}\sum_{j=1}^{J_n}\mathrm{tr}\left(D_i^{Q^n}\rho_{X_{j,l}^n}^{Q^n}\right)\leq\nu\right\}\geq 1-J_nL_n\cdot 2^{n[\min_Q(H(\rho^{P,Q})-\sum_x P(x)H(\rho_x^Q))-c_2]}, \tag{14}$$

where c_2 is some positive constant. Since

$$J_nL_n\leq 2^{n\left[\min_{t^n\in\Theta^n}\chi(P,\{\rho_{x^n,t^n}:x^n\in\mathcal{X}^n\})-\eta\right]},$$

There is a positive constant c_3 such that if n is large enough then

$$J_nL_n\cdot 2^{n[\min_Q(H(\rho^{P,Q})-\sum_x P(x)H(\rho_x^Q))-c_2]}\leq 2^{-nc_3}$$

Denote the set of all codes $\left\{\left(x_{j,l}^{\prime n},D_{x_{j,l}^n}^{Q^n}\right):j=1,\cdots,J_n,l=1,\cdots,L_n\right\}$ such that

$$1-\frac{1}{L_n}\frac{1}{J_n}\sum_{l=1}^{L_n}\sum_{j=1}^{J_n}\mathrm{tr}\left(D_i^{Q^n}\rho_{x_{j,l}^n}^{Q^n}\right)\leq\nu$$

by $\mathcal{C}_\nu''$.

We have

$$Pr(\mathcal{C}_\nu'\cap\mathcal{C}_\nu'')\geq 1-2^{-n\nu c_1}-2^{-nc_3},$$

therefore if n is large enough, $\mathcal{C}_\nu'\cap\mathcal{C}_\nu''$ is not empty. This means if is large enough, then for any positive ν and for each set of distributions $T^n=(T_1,\cdots T_n)$ on Θ^n, there exists a $C^{X_{T^n}}\in\mathcal{C}_\nu'$ with a positive probability such that,

$$\sum_{t^n\in\Theta^n}T^n(t^n)P_e(C^{X_{T^n}},t^n)\leq\nu,$$

where $T^n(t^n)=T_1(t_1)T_2(t_2)\cdots T_n(t_n)$.

Let us denote the set of distributions on $\mathcal{C}_\nu'$ by $\Omega_{\mathcal{C}_\nu'}$. By applying the minimax theorem for mixed strategies (cf. [1]), we have

$$\max_{T^n}\min_{G\in\Omega_{\mathcal{C}_\nu'}}\sum_{t^n\in\Theta^n,C\in\mathcal{C}_\nu'}T^n(t^n)G(C)P_e(C,t^n)=\min_{G\in\Omega_{\mathcal{C}_\nu'}}\max_{T^n}\sum_{t^n\in\Theta^n,C\in\mathcal{C}_\nu'}T^n(t^n)G(C)P_e(C,t^n).$$

Therefore (13) holds.

Now we are going to use the derandomization technique in [1] to build a deterministic code.

Consider now n^2 independent and identically distributed random variables $Z_1,Z_2,\cdots,Z_{n^2}$ with values in $\mathcal{C}_\nu'$ such that $P(Z_i=C)=G(C)$ for all $C\in\mathcal{C}_\nu'$

and for all $i \in \{1, \cdots, n^2\}$. Then for given $t^n \in \Theta^n$

$$G\left(\sum_{i=1}^{n^2} P_e(Z_i, t^n) > \lambda n^2\right) < e^{-\lambda n^2} , \tag{15}$$

where $\lambda := \log(\nu \cdot e^2 + 1)$. If n is large enough then $1 - e^{-\lambda n^2}$ is positive, this means

$$\left\{C^{z_i} \text{ is a realization of } Z_i : \sum_{i=1}^{n^2} P_e(C^{z_i}, t^n) < \lambda n^2\right\}$$

is not the empty set, since $G(\varnothing) = 0$ by the definition of distribution.

In [1], it is shown that if $\left\{C^{z_i} \text{ is a realization of } Z_i : \sum_{i=1}^{n^2} P_e(C^{z_i}, t^n) < \lambda n^2\right\}$ is not the empty set, there exist codes $C_1, C_2, \cdots, C_{n^2} \in \mathcal{C}'_\nu$, where we denote

$$C_i = \left\{\left(x_{j,l}^{(i),n}, D^{Q^n}_{x_{j,l}^{(i),n}}\right) : j = 1, \cdots, J_n, l = 1, \cdots, L_n\right\}$$

for $i \in \{1, \cdots, n^2\}$ with a positive probability such that

$$\frac{1}{n^2}\sum_{i=1}^{n^2} P_e(C_i, t^n) < \lambda . \tag{16}$$

Following fact is trivial. There is a code $\left\{(c_i^{\mu(n)}, D_i) : i = 1, \cdots, n^2\right\}$ of length $\mu(n)$, where $\mu(n) = o(n)$ (this code does not need to be secure against the wiretapper, i.e. we allow the wiretapper to have the full knowledge of i), such that for any positive ϑ if n is large enough then

$$\min_{t^n \in \Theta^n} \frac{1}{n^2}\sum_{i=1}^{n^2} \mathrm{tr}(\rho_{c_i^{\mu(n)}, t^n} D_i) \geq 1 - \vartheta .$$

By (16) we can construct a code of length $\mu(n) + n$ (cf. [1])

$$C^{det} = \left\{\left(c_i^{\mu(n)} \otimes x_{j,l}^{(i),n}, D_i \otimes D^{Q^n}_{x_{j,l}^{(i),n}}\right) : i = 1, \cdots, n^2, j = 1, \cdots, J_n, l = 1, \cdots, L_n\right\} ,$$

which is a juxtaposition of words of the code $\{(c_i^{\mu(n)}, D_i) : i = 1, \cdots, n^2\}$ and the words of code $C_i = \{(x_{j,l}^{(i),n} D^{Q^n}_{x_{j,l}^{(i),n}}) : j = 1, \cdots, J_n, l = 1, \cdots, L_n\}$, with following feature. C^{det} is a deterministic code with $n^2 J_n L_n$ codewords such that for any positive ϵ if n is large enough then

$$\max_{t^n \in \Theta^n} P_e(C^{det}, t^n) < \epsilon . \tag{17}$$

Furthermore, since $C_1, C_2, \cdots, C_{n^2} \in \mathcal{C}'_\nu$, for all $i \in \{1, \cdots, n^2\}$, $t^n \in \Theta^n$, and $j' \in \{1, \cdots, J_n\}$ we have

$$\|\frac{1}{L_n}\sum_{l=1}^{L_n} \sigma_{x^{(i),n}_{j',l}, t^n} - \frac{1}{L_n}\frac{1}{J_n}\sum_{l=1}^{L_n}\sum_{j=1}^{J_n} \sigma_{x^{(i),n}_{j,l}, t^n}\|_1 < 3\eta \,. \tag{18}$$

Lemma 4 (Fannes inequality, cf. [11]). *Let $\mathfrak{X}$ and $\mathfrak{Y}$ be two states in a d-dimensional complex Hilbert space and $\|\mathfrak{X} - \mathfrak{Y}\|_1 \le \mu < \frac{1}{e}$, then*

$$|S(\mathfrak{X}) - S(\mathfrak{Y})| \le \mu \log d - \mu \log \mu \,.$$

Let W be a random variable uniformly distributed on $\{1, \cdots, J_n\}$, by Lemma 4, for all $t^n \in \Theta^n$ and all $i \in \{1, \cdots, n^2\}$

$$\begin{aligned}
&\chi\left(W, \{\frac{1}{L_n}\sum_{l=1}^{L_n} \sigma_{x^{(i),n}_{j,l}, t^n} : j = 1 \cdots J_n\}\right) \\
&= S\left(\frac{1}{J_n}\frac{1}{L_n}\sum_{j=1}^{J_n}\sum_{l=1}^{L_n} \sigma_{x^{(i),n}_{j,l}, t^n}\right) - \frac{1}{J_n}\sum_{j=1}^{J_n} S\left(\frac{1}{L_n}\sum_{l=1}^{L_n} \sigma_{x^{(i),n}_{j,l}, t^n}\right) \\
&= \frac{1}{J_n}\sum_{j'=1}^{J_n}\left[S\left(\frac{1}{J_n}\frac{1}{L_n}\sum_{j=1}^{J_n}\sum_{l=1}^{L_n} \sigma_{x^{(i),n}_{j,l}, t^n}\right) - S\left(\frac{1}{L_n}\sum_{l=1}^{L_n} \sigma_{x^{(i),n}_{j',l}, t^n}\right)\right] \\
&\le 3\eta \log d - 3\eta \log 3\eta \,.
\end{aligned}$$

Therefore for any $\zeta > 0$ we can choose such η that for all $i \in \{1, \cdots, n^2\}$ (i.e. even when the wiretapper has the full knowledge of i), for all $t^n \in \Theta^n$,

$$\chi\left(W, \left\{\left[\frac{1}{L_n}\sum_{l=1}^{L_n} \sigma_{x^{(i),n}_{j,l}, t^n}\right] : j = 1 \cdots J_n\right\}\right) < \zeta \,. \tag{19}$$

By (17) and (19), we see that for any distribution P on $\mathcal{X}$ and any positive δ, we can find a (n, ϵ)-code with secrecy rate $\min_Q \chi\left(P, \{\rho_x^Q : x \in \mathcal{X}\}\right) - \frac{1}{n}\max_{t^n\in\Theta^n} \chi\left(P^n, \{\sigma_{x^n,t^n} : x^n \in \mathcal{X}^n\}\right) - \delta$. Therefore (1) follows.

We are going to prove (2).

Fix P and let

$$J_n = \left\lfloor 2^{\min_{Q\in\mathcal{Q}, t^n\in\Theta^n}\left[n\chi(P,\{\rho_x^Q : x\in\mathcal{X}\}) - \chi(P^n,\{\sigma_{x^n,t^n} : x^n\in\mathcal{X}^n\})\right] - 2n\eta}\right\rfloor ,$$

$L_{t^n} = \left\lfloor 2^{\chi(P,\{\sigma_{x^n,t^n} : x^n\in\mathcal{X}^n\}) + n\eta}\right\rfloor$, where η is a positive constant. For any $t^n \in \Theta^n$ let $X^{(t^n)} := \{X^{(t^n)}_{j,l}\}_{j\in\{1,\cdots,J_n\}, l\in\{1,\cdots,L_{t^n}\}}$ be a family of random matrices whose components are i.i.d. according to P'.

For any realization $(x^{(t^n)}_{j,l})_{j\in\{1,\cdots,J_n\}, l\in\{1,\cdots,L_{t^n}\}}$ of $(X^{(t^n)}_{j,l})_{j\in\{1,\cdots,J_n\}, l\in\{1,\cdots,L_{t^n}\}}$ and distribution Q^n on Θ^n define

$$D^{Q^n,t^n}_{x^{(t^n)}_{j,l}} := \left(\sum_{j'=1}^{J_n}\sum_{l'=1}^{L_{t^n}} \Pi_{\delta}^{P,Q^n}\Pi^{P,Q^n}_{x^{(t^n)}_{j',l'},\delta}\Pi_{\delta}^{P,Q^n}\right)^{-1/2} \cdot \Pi_{\delta}^{P,Q^n}\Pi^{P,Q^n}_{x^{(t^n)}_{j,l},\delta}\Pi_{\delta}^{P,Q^n} \cdot \left(\sum_{j'=1}^{J_n}\sum_{l'=1}^{L_{t^n}} \Pi_{\delta}^{P,Q^n}\Pi^{P,Q^n}_{x^{(t^n)}_{j',l'},\delta}\Pi_{\delta}^{P,Q^n}\right)^{-1/2} . \tag{20}$$

Since $L_{t^n} \geq 2^{\chi(P^n,\{\sigma_{x^n,t^n}:x^n\in\mathcal{X}^n\})+2n\delta} = \frac{2^{\sum_{x^n}P^n(x^n)S(\rho_{x^n,t^n})+n\delta}}{2^{S\left(\sum_{x^n}P^n(x^n)\rho_{x^n,t^n}\right)-n\delta}}$, by applying covering lemma, there is a positive c_1' such that for any positive η we have:

$$Pr\left\{\left\|\frac{1}{L_{t^n}}\frac{1}{J_n}\sum_{l=1}^{L_{t^n}}\sum_{j=1}^{J_n}\overline{\sigma}_{X^{(t^n)}_{j,l},t^n} - \frac{1}{L_{t^n}}\sum_{l=1}^{L_{t^n}}\overline{\sigma}_{X^{(t^n)}_{j',l},t^n}\right\|_1 < \eta\ \forall j'\in\{1,\cdots,J_n\}\right\} = 1-2^{-\nu^3 c_1''} \tag{21}$$

Denote all $\{(x^{(t^n)}_{j,l}, D^{Q^n}_{x^{(t^n)}_{j,l}}) : j=1,\cdots,J_n, l=1,\cdots,L_{t^n}\}$ such that

$$\left\|\frac{1}{L_{t^n}}\frac{1}{J_n}\sum_{l=1}^{L_{t^n}}\sum_{j=1}^{J_n}\overline{\sigma}_{x^{(t^n)}_{j,l},t^n} - \frac{1}{L_{t^n}}\sum_{l=1}^{L_{t^n}}\overline{\sigma}_{x^{(t^n)}_{j',l},t^n}\right\|_1 < \eta \forall j'\in\{1,\cdots,J_n\} ,$$

by $\mathcal{C}_\nu^{t^n}$.

Since $\frac{1}{n}\log(J_n \cdot L_{t^n}) \leq \min_Q \chi\left(P,\{\rho_x^Q : x\in\mathcal{X}\}\right) - 2\delta$, analogue to our proof for (1), if n is large enough then there are n^2 codes$\left\{\{(x^{(i),(t^n)}_{j,l}, D^{Q^n}_{x^{(i),(t^n)}_{j,l}}) : j=1,\cdots,J_n, l=1,\cdots,L_{t^n}\} : i=1,\cdots,n^2\right\}$ $\in \mathcal{C}_\nu^{t^n}$ such that we can construct a code $C^{det}_{t^n}$ which is a juxtaposition of words of the code $\{(c_i^{\mu(n)}, D_i) : i=1,\cdots,n^2\}$, defined as in our proof for (1) above, and words of the code $\{(x^{(i),(t^n)}_{j,l}, D^{Q^n}_{x^{(i),(t^n)}_{j,l}}) : j=1,\cdots,J_n, l=1,\cdots,L_{t^n}\}$, with following property

$$P_e(C^{det}_{t^n}, t^n) < \epsilon . \tag{22}$$

and for all $i\in\{1,\cdots,n^2\}$ and all $j'\in\{1,\cdots,J_n\}$:

$$\left\|\frac{1}{L_{t^n}}\sum_{l=1}^{L_{t^n}}\sigma_{x^{(i),(t^n)}_{j',l},t^n} - \frac{1}{J_n}\sum_{j=1}^{J_n}\frac{1}{L_{t^n}}\sum_{l=1}^{L_{t^n}}\sigma_{x^{(i),(t^n)}_{j,l},t^n}\right\|_1 < 3\eta . \tag{23}$$

By Lemma 4, we have for all $t^n \in \Theta^n$ and all $i \in \{1, \cdots, n^2\}$

$$\chi\left(W; \left\{\left[\frac{1}{L_{t^n}} \sum_{l=1}^{L_{t^n}} \sigma_{x_{j,l}^{(i),(t^n)},t^n}\right] : j = 1, \cdots, J_n\right\}\right) \leq 3\eta \log d - 3\eta \log 3\eta \, .$$

Therefore for any $\zeta > 0$ we can choose such η that for all $i \in \{1, \cdots, n^2\}$, for all $t^n \in \Theta^n$

$$\chi\left(W, \left\{\left[\frac{1}{L_{t^n}} \sum_{l=1}^{L_{t^n}} \sigma_{x_{j,l}^{(i),(t^n)},t^n}\right] : j = 1, \cdots, J_n\right\}\right) < \zeta \, . \tag{24}$$

By (22) and (24), we see that for any distribution P on $\mathcal{X}$ and any positive δ, we can find a (n, ϵ)-code with secrecy rate $\min_{Q \in \mathcal{Q}, t^n \in \Theta^n} \left[\chi(P, \{\rho_x^Q : x \in \mathcal{X}\}) - \frac{1}{n}\chi(P^n, \{\sigma_{x^n,t^n} : x^n \in \mathcal{X}^n\})\right] - \delta$. Therefore (2) holds.

Is $\{\rho_{x,t}, x \in \mathcal{X}\}$ symmetrizable, then by [1], even in the case without wiretapper (we have only the arbitrarily varying channel $\{\rho_{x,t} : x \in \mathcal{X}, t \in \Theta\}$ instead of the pairs $\{(\rho_{x,t}, \sigma_{x,t}) : x \in \mathcal{X}, t \in \Theta\}$), the capacity is equal to 0. Since we cannot exceed the secrecy capacity of the worst wiretap channel, (3) holds.

References

1. Ahlswede, R., Blinovsky, V.: Classical capacity of classical-quantum arbitrarily varying channels. IEEE Trans. Inform. Theory 53(2), 526–533 (2007)
2. Bjelaković, I., Boche, H., Janßen, G., Nötzel, J.: Arbitrarily varying and compound classical-quantum channels and a note on quantum zero-error capacities, arXiv:1209.6325 (2012)
3. Bjelaković, I., Boche, H., Sommerfeld, J.: Capacity results for arbitrarily varying wiretap channels, arXiv:1209.5213 (2012)
4. Cai, M., Cai, N., Deppe, C.: Capacities of classical compound quantum wiretap and classical quantum compound wiretap channels, arXiv:1202.0773v1 (2012)
5. Cai, N., Winter, A., Yeung, R.W.: Quantum privacy and quantum wiretap channels. Problems of Information Transmission 40(4), 318–336 (2004)
6. Csiszár, I., Körner, J.: Information Theory: Coding Theorems for Discrete Memoryless Systems. Academic Press, New York (1981)
7. Devetak, I.: The private classical information capacity and quantum information capacity of a quantum channel. IEEE Trans. Inf. Theory 51(1), 44–55 (2005)
8. Blackwell, D., Breiman, L., Thomasian, A.J.: The capacities of a certain channel classes under random coding. Ann. Math. Statist. 31(3), 558–567 (1960)
9. Holevo, S.: Statistical problems in quantum physics. In: Maruyama, G., Prokhorov, J.V. (eds.) Proceedings of the Second Japan-USSR Symposium on Probability Theory. Lecture Notes in Mathematics, vol. 330, pp. 104–119. Springer, Berlin (1973)
10. Wilde, M.: From Classical to Quantum Shannon Theory, arXiv:1106-1445 (2011)
11. Winter, A.: Coding theorem and strong converse for quantum channels. IEEE Trans. Inf. Theory 45(7), 2481–2485 (1999)
12. Wyner, A.D.: The wiretap channel. Bell System Technical Journal 54(8), 1355–1387 (1975)

Arbitrarily Varying and Compound Classical-Quantum Channels and a Note on Quantum Zero-Error Capacities

Igor Bjelaković, Holger Boche, Gisbert Janßen, and Janis Nötzel

Lehrstuhl für Theoretische Informationstechnik,
Technische Universität München, 80290 München, Germany
{igor.bjelakovic,boche,gisbert.janssen,janis.noetzel}@tum.de

Dedicated to the memory of Rudolf Ahlswede

Abstract. We consider compound as well as arbitrarily varying classical-quantum channel models. For classical-quantum compound channels, we give an elementary proof of the direct part of the coding theorem. A weak converse under average error criterion to this statement is also established. We use this result together with the robustification and elimination technique developed by Ahlswede in order to give an alternative proof of the direct part of the coding theorem for a finite classical-quantum arbitrarily varying channels with the criterion of success being average error probability. Moreover we provide a proof of the strong converse to the random coding capacity in this setting.

The notion of symmetrizability for the maximal error probability is defined and it is shown to be both necessary and sufficient for the capacity for message transmission with maximal error probability criterion to equal zero.

Finally, it is shown that the connection between zero-error capacity and certain arbitrarily varying channels is, just like in the case of quantum channels, only partially valid for classical-quantum channels.

Keywords: arbitrarily varying classical-quantum channels, compound classical-quantum channels, zero error capacity, Ahlswedes dichotomy, weak converse, strong converse.

1 Introduction

Channel uncertainty is omnipresent and mostly unavoidable in real-world applications and one of the major technological challenges is the design of communication protocols that are robust against it. The incarnation of that challenge on the theoretical side delivers a plethora of interesting structural and methodological problems for Information Theory. Despite these facts it happened only recently that this range of problems received the necessary attention in Quantum Information Theory and especially in Quantum Shannon Theory [7], [15], [9], [11], [6]. In this paper we revisit two basic models for communication under

H. Aydinian, F. Cicalese, and C. Deppe (Eds.): Ahlswede Festschrift, LNCS 7777, pp. 247–283, 2013.

channel uncertainty, the compound and arbitrarily varying channels with classical input and quantum output and give essentially self-contained derivations of coding theorems for them. These results were originally obtained in [7] and [9].

The contributions of the paper and the difference to existing work are the following. First, in [9] a capacity result with strong converse for compound channels with a classical input and quantum output (compound cq-channel for short) under the maximum error criterion has been derived. However, the achievability proof given there lacks transparency and does not show that good codes with the uniformly bounded exponentially decreasing maximal error exist. Indeed, in [9] it is merely shown that good codes exist with uniformly super-polynomially decreasing maximal error probability. Here we prove that sharper result for the average error criterion and, at the same time, give a significantly simpler proof of the achievability part of the coding theorem based on a universal hypothesis testing result which is a generalization of the technique developed by Hayashi and Ogawa in [25]. The passage to the maximal error criterion can be carried out via a standard argument which can be found in [9].

It is interesting to compare this result with related work of Hayashi [21] and Datta and Hsieh [16]. The works [21] and [16] aim at showing the existence of codes depending on the input distribution and a prescribed rate only and achieving an exponential but *channel dependent* decay of error probability for all cq-channels whose Holevo information is strictly larger than that prescribed rate. The good codes in our approach depend on the input distribution and the set of cq-channels generating the compound cq-channel. Additionally we obtain a uniform exponential bound on error probabilities, a property that seems highly desirable in case that the channel is unknown.

Moreover, we prove the weak converse to the coding theorem under average error criterion by a reduction to the strong converse for the maximal error via a lemma of Ahlswede and Wolfowitz from [2].

Second, once we have the achievability result for compound cq-channels we can obtain the corresponding results for arbitrarily varying cq-channels (AVcqC) in a straight-forward fashion via Ahlswede's powerful elimination [4] and robustification [5] techniques. This way, we obtain an alternative approach to the coding theorem for AVcqCs which was originally proven by Ahlswede and Blinovsky in [7].

Finally, we show that a naive quantum analog of Ahlswede's beautiful relation [3] between Shannon's zero-error capacity [27] and the capacity of arbitrarily varying channels subject to maximal error criterion does hold neither for AVcqCs when employing the maximal error criterion nor for the strong subspace transmission over arbitrarily varying quantum channels. The latter communication scenario is widely acknowledged as a fully quantum counterpart to message transmission subject to the maximal error criterion.

2 Notation and Conventions

All Hilbert spaces are assumed to have finite dimension and are over the field $\mathbb{C}$. The set of linear operators from $\mathcal{H}$ to $\mathcal{H}$ is denoted $\mathcal{B}(\mathcal{H})$. The adjoint of

$b \in \mathcal{B}(\mathcal{H})$ is marked by a star and written b^*. The notation $\langle \cdot, \cdot \rangle_{HS}$ is reserved for the Hilbert-Schmidt inner product on $\mathcal{B}(\mathcal{H})$.

$\mathcal{S}(\mathcal{H})$ is the set of states, i.e. positive semi-definite operators with trace 1 acting on the Hilbert space $\mathcal{H}$. Pure states are given by projections onto one-dimensional subspaces. A vector $x \in \mathcal{H}$ of unit length spanning such a subspace will therefore be referred to as a state vector, the corresponding state will be written $|x\rangle\langle x|$. For a finite set $\mathbf{X}$ the notation $\mathfrak{P}(\mathbf{X})$ is reserved for the set of probability distributions on $\mathbf{X}$, and $|\mathbf{X}|$ denotes its cardinality. For any $l \in \mathbb{N}$, we define $\mathbf{X}^l := \{(x_1, \ldots, x_l) : x_i \in \mathbf{S}\ \forall i \in \{1, \ldots, l\}\}$, we also write x^l for the elements of $\mathbf{X}^l$. For any natural number N, we define $[N]$ to be the shortcut for the set $\{1, ..., N\}$.

The set of classical-quantum channels (cq-channels) mapping a finite alphabet $\mathbf{X}$ to a Hilbert space $\mathcal{H}$ is denoted $CQ(\mathbf{X}, \mathcal{H})$. Since $CQ(\mathbf{X}, \mathcal{H})$ is the set of functions $W : \mathbf{X} \to \mathcal{S}(\mathcal{H})$. It is naturally equipped with the norm $\| \cdot \|_{cq}$ (which is inherited from the usual one-norm $\| \cdot \|_1$ on operators) and is defined by

$$\|W\|_{cq} := \max_{x \in \mathbf{X}} \|W(x)\|_1 \qquad (W \in CQ(\mathbf{X}, \mathcal{H})).$$

It is common, to embed the set $\mathfrak{P}(\mathbf{X})$ of probability distributions into $\mathcal{B}(\mathbb{C}^{|\mathbf{X}|})$, i.e. to fix an orthonormal basis $\{e_x\}_{x \in \mathbf{X}}$ in $\mathbb{C}^{|\mathbf{X}|}$ and assign to every p in $\mathfrak{P}(\mathbf{X})$ an element of $\mathcal{B}(\mathbb{C}^{|\mathbf{X}|})$ which is diagonal in this basis. For a channel $W \in CQ(\mathbf{X}, \mathcal{H})$ and a given input probability distribution $p \in \mathfrak{P}(\mathbf{X})$ one defines the corresponding state on $\mathbb{C}^{|\mathbf{X}|} \otimes \mathcal{H}$ by

$$\rho := \sum_{x \in \mathbf{X}} p(x) |e_x\rangle\langle e_x| \otimes W(x). \tag{1}$$

The set of measurements with $N \in \mathbb{N}$ different outcomes is written $\mathcal{M}_N(\mathcal{H}) := \{(D_1, \ldots, D_N) : \sum_{i=1}^N D_i \leq \mathbb{1}_{\mathcal{H}} \text{ and } D_i \geq 0\ \forall i \in [N]\}$. To every $(D_1, \ldots, D_N) \in \mathcal{M}_N(\mathcal{H})$ there corresponds a unique operator defined by $D_0 := \mathbb{1}_{\mathcal{H}} - \sum_{i=1}^N D_i$.

The von Neumann entropy of a state $\rho \in \mathcal{S}(\mathcal{H})$ is given by

$$S(\rho) := -\mathrm{tr}(\rho \log \rho), \tag{2}$$

where $\log(\cdot)$ denotes the base two logarithm which is used throughout the paper (accordingly, $\exp(\cdot)$ is reserved for the base two exponential). For two states $\rho, \sigma \in \mathcal{S}(\mathcal{H})$, the quantum relative entropy is defined by

$$D(\rho||\sigma) := \begin{cases} \mathrm{tr}(\rho \log \rho - \rho \log \sigma) & \text{if } \ker \sigma \subseteq \ker \rho \\ +\infty & \text{else.} \end{cases} \tag{3}$$

The Holevo information is for a given channel $W \in CQ(\mathbf{X}, \mathcal{H})$ and input probability distribution $p \in \mathfrak{P}(\mathbf{X})$ defined by

$$\chi(p, W) := S(\overline{W}) - \sum_{x \in \mathbf{X}} p(x) S(W(x)) = \sum_{x \in \mathbf{X}} p(x) D(W(x)||\overline{W}), \tag{4}$$

where $\overline{W}$ is defined by $\overline{W} := \sum_{x \in \mathbf{X}} p(x)W(x)$. This quantity is concave w.r.t. the input probability distribution and convex w.r.t. the channel. Its concavity property follows directly from the concavity of the von Neumann entropy, its convexity in the channel is by joint convexity of the quantum relative entropy. For an arbitrary set $\mathcal{W} \subset CQ(\mathbf{X}, \mathcal{H})$ we denote its convex hull by $\mathrm{conv}(\mathcal{W})$ (for the definition of the convex hull, [28] is a useful reference). In fact, for a set $\mathcal{W} := \{W_s\}_{s \in \mathbf{S}}$

$$\mathrm{conv}(\mathcal{W}) = \left\{ W_q \in CQ(\mathbf{X}, \mathcal{H}) : W_q = \sum_{s \in \mathbf{S}} q(s) W_s,\ q \in \mathfrak{P}(\mathbf{S}), |\mathrm{supp}(q)| < \infty \right\}, \tag{5}$$

because of Carathéodory's Theorem.

3 Definitions

3.1 The Compound Classical-Quantum Channel

Let $\mathcal{W} \subset CQ(\mathbf{X}, \mathcal{H})$. The memoryless compound cq-channel associated with $\mathcal{W}$ is given by the family $\{W^{\otimes l}\}_{l \in \mathbb{N}, W \in \mathcal{W}}$. With slight abuse of notation it will be denoted $\mathcal{W}$ or, if necessary, 'the compound cq-channel $\mathcal{W}$' for short. In the remainder, using arbitrary index sets T, we will often write $\mathcal{W} = \{W_t\}_{t \in T}$ to enhance readability. Before we continue, let us put a brief remark in order to explain why this subsection contains no definition of random codes (while subsection 3.2 does):

Remark 1. *We abstain from defining random codes for compound cq-channels, the reason for this being that they do offer no increase in capacity. For the reader interested in the topic, we briefly outline one way of arriving at this conclusion.*

First, the capacity of compound channels, seen as a function from the power set of the set of channels with given input and output systems to the reals, is continuous (this can fact can be proven by an argument very similar to the one given for compound quantum channels in Sect. 8 of [11] together with continuity of the single channel classical capacity, cf. [23]). This allows for an arbitrarily good (speaking in terms of their capacity) approximation of infinite compound cq-channels by finite ones, so that we can restrict our discussion to finite compound cq-channels.

Second, given such a finite compound cq-channel $\{W_t\}_{t \in T}$ and a sequence of random codes which achieve a given rate r with asymptotically vanishing average error, we may simply use it for the memoryless cq-channel $\overline{W} := \frac{1}{|T|} \sum_{t \in T} W_t$. Since the average error is a convex function of the channel, this implies the existence of a sequence of deterministic codes at the same asymptotic rate with vanishing average error for $\overline{W}$.

Using affinity of the average error criterion once more, we see that the very same sequence of deterministic codes also has vanishing average error for the cq-compound channel $\{W_t\}_{t \in T}$, only with a slightly slower convergence. As in the definition of $\overline{W}$, the assumption that $|T| < \infty$ holds is crucial at this point of the *argument. This shows that random codes cannot have higher asymptotic rates than deterministic ones, if one insists on asymptotically vanishing average error.*

For the maximal error criterion, it is enough to note that both the random and the deterministic capacity for transmission of messages over a compound cq-channel using that criterion are upper bounded by the respective capacities for the average error criterion.

Definition 1. *An* (l, M_l)-code *for message transmission over a compound cq-channel* $\mathcal{W} \subset CQ(\mathbf{X}, \mathcal{H})$ *is a family* $(x^l_m, D^l_m)_{m=1}^{M_l}$*, where* $x^l_1, \dots, x^l_{M_l} \in \mathbf{X}^l$ *and* $(D^l_1, \dots, D^l_{M_l}) \in \mathcal{M}_{M_l}(\mathcal{H}^{\otimes l})$.

Definition 2. *For* $\lambda \in [0, 1)$*, a non-negative number* R *is called a* λ-achievable rate *for transmission of messages over the compound cq-channel* $\mathcal{W} = \{W_t\}_{t\in T}$ *using the average error criterion if there is a sequence* $\{(u^l_m, D^l_m)_{m=1}^{M_l}\}_{l\in\mathbb{N}}$ *of* (l, M_l)*-codes with*

$$\liminf_{l\to\infty} \frac{1}{l} \log M_l \geq R \qquad \textit{and}$$

$$\overline{\lim_{l\to\infty}} \sup_{t\in T} \frac{1}{M_l} \sum_{m=1}^{M_l} \operatorname{tr}(W_t^{\otimes l}(u^l_m)(\mathbb{1}_{\mathcal{H}^{\otimes l}} - D^l_m)) \leq \lambda.$$

Definition 3. *For* $\lambda \in [0, 1)$*, a non-negative number* R *is called a* λ-achievable rate *for transmission of messages over the compound cq-channel* $\mathcal{W} = \{W_t\}_{t\in T}$ *using the maximal error criterion if there is a sequence* $\{(u^l_m, D^l_m)_{m=1}^{M_l}\}_{l\in\mathbb{N}}$ *of* (l, M_l)*-codes with*

$$\liminf_{l\to\infty} \frac{1}{l} \log M_l \geq R \qquad \textit{and}$$

$$\overline{\lim_{l\to\infty}} \sup_{t\in T} \max_{m\in[M_l]} \operatorname{tr}(W_t^{\otimes l}(u^l_m)(\mathbb{1}_{\mathcal{H}^{\otimes l}} - D^l_m)) \leq \lambda.$$

Definition 4. *For* $\lambda \in [0, 1)$*, the* λ-capacity *for message transmission using the average error criterion of a compound cq-channel* $\mathcal{W}$ *is given by*

$$\overline{C}_C(\mathcal{W}, \lambda) := \sup \left\{ R : \begin{array}{l} R \textit{ is a } \lambda\textit{-achievable rate for} \\ \textit{transmission of messages over } \mathcal{W} \\ \textit{using the average error probability criterion} \end{array} \right\}. \tag{6}$$

The number $\overline{C}_C(\mathcal{W}, 0)$ *is called the* weak capacity *for message transmission using the average error criterion of* $\mathcal{W}$ *and abbreviated* $\overline{C}_C(\mathcal{W})$.

Definition 5. *For* $\lambda \in [0, 1)$*, the* λ-capacity *for message transmission using the maximal error criterion of a compound cq-channel* $\mathcal{W}$ *is given by*

$$C_C(\mathcal{W}, \lambda) := \sup \left\{ R : \begin{array}{l} R \textit{ is a } \lambda\textit{-achievable rate for transmission} \\ \textit{of messages over } \mathcal{W} \\ \textit{using the maximal error probability criterion} \end{array} \right\}. \tag{7}$$

The number $C_C(\mathcal{W}, 0)$ *is called the* weak capacity *for message transmission using the maximal error criterion of* $\mathcal{W}$ *and abbreviated* $C_C(\mathcal{W})$.

3.2 The Arbitrarily Varying Classical-Quantum Channel

Let $\mathcal{A} \subset CQ(\mathbf{X}, \mathcal{H})$. In the remainder we will write $\mathcal{A} = \{A_s\}_{s\in\mathbf{S}}$, where $\mathbf{S}$ denotes an index set, in order to enhance readability. We also set

$$A_{s^l} := \otimes_{i=1}^{l} A_{s_i}. \tag{8}$$

The arbitrarily varying classical-quantum channel associated with $\mathcal{A}$ is given by the family $\{A_{s^l}\}_{s^l\in\mathbf{S}^l, l\in\mathbb{N}}$. Again, with slight abuse of notation it will be denoted $\mathcal{A}$ or, if necessary, 'the AVcqC $\mathcal{A}$' for short.

In this work, we will always consider the set $\mathbf{S}$ to be finite. Generalizations of our results to the case of arbitrary sets can be done by standard techniques (see [6]). We will now define random codes and the random capacity emerging from them. In order to do so, we have to clarify a few things.

A code for an AVcqC $\mathcal{A}$ will, for some choice of $l, N \in \mathbb{N}$, be given by a probability measure μ_l on the set $((\mathbf{X}^l)^N \times \mathcal{M}_N(\mathcal{H}^{\otimes l}), \Sigma_l)$, where Σ_l is a suitably chosen sigma-algebra.

It has to be taken care that a function f defined by

$$((x_1^l, \dots, x_N^l), (D_1^l, \dots, D_N^l)) \mapsto \min_{s^l\in\mathbf{S}^l} \frac{1}{N} \sum_{i=1}^{N} \mathrm{tr}\{W_{s^l}(x_i^l) D_i^l\}$$

is measurable w.r.t. Σ_l. Also, in order to define deterministic codes later, Σ_l has to contain all the singleton sets. In the remainder, we shall assume that such a choice is always made.

An explicit example of such a sigma-algebra is given by the Borel sigma-algebra defined using the topology induced by the metric $((x, D), (x', D')) \mapsto (1 - \delta(x, x')) + \|D - D'\|_2$ where $\delta(x, x) = 1\ \forall x \in \mathbf{X}$ and equal to zero else, and for sake of simplicity, we set $l = N = 1$. Finally, we note that the function f mentioned above is continuous w.r.t. to that metric.

In the following definitions, let $\lambda \in [0, 1)$.

Definition 6. *An* (l, M_l)-random code *for message transmission over* $\mathcal{A} = \{A_s\}_{s\in\mathbf{S}}$ *is a probability measure* μ_l *on* $((X^l)^{M_l} \times \mathcal{M}_N(\mathcal{H}^{\otimes l}), \Sigma_l)$. *In order to shorten our notation, we write elements of* $(X^l)^{M_l} \times \mathcal{M}_N(\mathcal{H}^{\otimes l})$ *in the form* $(x_i^l, D_i^l)_{i=1}^{M_l}$.

Definition 7. *An* (l, M_l)-deterministic code *for message transmission over* $\mathcal{A} = \{A_s\}_{s\in\mathbf{S}}$ *is given by a random code for message transmission over* $\mathcal{A}$ *with* μ_l *assigning probability one to a singleton set.*

Definition 8. *A non-negative number* R *is called* λ-achievable *for transmission of messages over the AVcqC* $\mathcal{A} = \{A_s\}_{s\in\mathbf{S}}$ *with random codes using the average error criterion if there is a sequence* $(\mu_l)_{l\in\mathbb{N}}$ *of* (l, M_l)*-random codes such that the following two lines are true:*

$$\liminf_{l\to\infty} \frac{1}{l} \log M_l \geq R \tag{9}$$

$$\overline{\lim_{l\to\infty}} \max_{s^l\in\mathbf{S}^l} \int \frac{1}{M_l}\sum_{i=1}^{M_l} \operatorname{tr}\left(A_{s^l}(x_i^l)(\mathbb{1}_{\mathcal{H}^{\otimes l}} - D_i^l)\right)\, d\mu_l((u_i^l, D_i^l)_{i=1}^{M_l}) \leq \lambda. \tag{10}$$

Definition 9. *A non-negative number R is called* λ-achievable *for transmission of messages over the AVcqC $\mathcal{A} = \{A_s\}_{s\in\mathbf{S}}$ with deterministic codes using the average error criterion if it is λ-achievable with random codes by a sequence $(\mu_l)_{l\in\mathbb{N}}$ which are deterministic codes.*

Definition 10. *The* λ-capacity *for message transmission using random codes and the average error criterion of an AVcqC $\mathcal{A}$ is given by*

$$\overline{C}_{\mathrm{A,r}}(\mathcal{A},\lambda) := \sup\left\{ R : \begin{array}{l} R \textit{ is a } \lambda\textit{-achievable rate for transmission of} \\ \textit{messages over } \mathcal{A} \textit{ with random codes} \\ \textit{using the average error probability criterion} \end{array}\right\}. \tag{11}$$

The number $\overline{C}_{\mathrm{A,r}}(\mathcal{A},0)$ is called the weak capacity *for message transmission using random codes and the average error criterion of $\mathcal{A}$ and abbreviated $\overline{C}_{\mathrm{A,r}}(\mathcal{A})$.*

Definition 11. *The* λ-capacity *for message transmission using deterministic codes and the average error criterion of an AVcqC $\mathcal{A}$ is given by*

$$\overline{C}_{\mathrm{A,d}}(\mathcal{A},\lambda) := \sup\left\{ R : \begin{array}{l} R \textit{ is a } \lambda\textit{-achievable rate for transmission of} \\ \textit{messages over } \mathcal{A} \textit{ with deterministic codes} \\ \textit{using the average error probability criterion} \end{array}\right\}. \tag{12}$$

The number $\overline{C}_{\mathrm{A,d}}(\mathcal{A},0)$ is called the weak capacity *for message transmission using deterministic codes and the average error criterion of $\mathcal{A}$ and abbreviated $\overline{C}_{\mathrm{A,d}}(\mathcal{A})$.*

Definition 12. *A non-negative number R is called* λ-achievable *for transmission of messages over the AVcqC $\mathcal{A} = \{A_s\}_{s\in\mathbf{S}}$ with deterministic codes using the maximal error probability criterion if there is a sequence of (l, M_l)-random codes with each μ_l being a deterministic code such that the following two lines are true:*

$$\liminf_{l\to\infty} \frac{1}{l}\log M_l \geq R \tag{13}$$

$$\overline{\lim_{l\to\infty}} \max_{s^l\in\mathbf{S}^l} \max_{i=1,\ldots,M_l} \int \operatorname{tr}\left(A_{s^l}(x_i^l)(\mathbb{1}_{\mathcal{H}^{\otimes l}} - D_i^l)\right)\, d\mu_l((u_i^l, D_i^l)_{i=1}^{M_l}) \leq \lambda. \tag{14}$$

Definition 13. *The* λ-capacity *for message transmission using deterministic codes and the maximal error probability criterion of an AVcqC $\mathcal{A}$ is given by*

$$C_{\mathrm{A,d}}(\mathcal{A},\lambda) := \sup\left\{ R : \begin{array}{l} R \textit{ is a } \lambda\textit{-achievable rate for transmission} \\ \textit{of messages over } \mathcal{A} \textit{ with deterministic codes} \\ \textit{using the maximal error probability criterion} \end{array}\right\}. \tag{15}$$

The number $C_{\mathrm{A,d}}(\mathcal{A},0)$ is called the weak capacity *for message transmission using deterministic codes and the maximal error criterion of $\mathcal{A}$ and abbreviated $C_{\mathrm{A,d}}(\mathcal{A})$.*

The following definition will turn out to be useful to decide whether a given $AVcqC$ has nonzero capacity for transmission of messages using average error criterion and deterministic codes.

Definition 14. *Let* $\mathcal{A} = \{A_s\}_{s\in\mathbf{S}} \subset CQ(\mathbf{X},\mathcal{H})$ *be an AVcqC. If, for every* $x, x' \in \mathbf{X}$*, we have*

$$\operatorname{conv}(\{\mathcal{A}_s(x)\}_{s\in\mathbf{S}}) \cap \operatorname{conv}(\{\mathcal{A}_s(x')\}_{s\in\mathbf{S}}) \neq \varnothing, \tag{16}$$

then $\mathcal{A}$ *is called* m-symmetrizable.

3.3 Zero-Error Capacity

Definition 15. *An* (l, M_l) zero-error code *for a stationary memoryless cq-channel defined by* $V \in CQ(\mathbf{X},\mathcal{H})$ *is given by a family* $(x_i^l, D_i^l)_{i=1}^{M_l}$*, where* $x_1^l, \ldots, x_{M_l}^l \in \mathbf{X}^l$ *and* $(D_1^l, \ldots, D_{M_l}^l) \in \mathcal{M}_{M_l}(\mathcal{H}^{\otimes l})$ *satisfy* $\operatorname{tr}(V^{\otimes l}(x_i^l)D_i^l) = 1$ *for every* $i \in [M_l]$.

Definition 16. *The* zero-error capacity *for message transmission over the cq-channel* $V \in CQ(\mathbf{X},\mathcal{H})$ *is given by*

$$C_0(V) := \lim_{l\to\infty} \frac{1}{l} \log\max\{M_l : \exists\ (l, M_l) \text{ zero}-\text{error code for } V\}. \tag{17}$$

4 Main Results

We now enlist the main results contained in this work. We will not state the results obtained in Subsection 6.3. These evolve around the relation between zero-error capacities and arbitrarily varying channels. They include both message transmission and entanglement transmission. Rather than stating a positive result, in this section we argue that certain straightforward quantum analogues of results that are valid in the classical theory do not hold. As always, this is a delicate task that involves much more than just embedding a commutative subalgebra into a non-commutative one. We therefore encourage the reader to consider this last subsection as something that should be read separately and in one piece.

Our first result is the following.

Theorem 1 (cq Compound Coding Theorem). *For every compound cq-channel* $\mathcal{W} \in CQ(\mathbf{X},\mathcal{H})$ *it holds*

$$\overline{C}_C(\mathcal{W}) = \max_{p\in\mathfrak{P}(\mathbf{X})} \inf_{W\in\mathcal{W}} \chi(p, W). \tag{18}$$

In subsection 6.1, an analogue of the Ahlswede dichotomy from [4] for arbitrarily varying classical-quantum channels will be derived. This statement has originally been obtained by Ahlswede and Blinovsky in [7]. The precise mathematical formulation reads as follows.

Theorem 2 (Ahlswede-Dichotomy for AVcqCs). *Let* $\mathcal{A} = \{A_s\}_{s \in \mathbf{S}} \subset CQ(\mathbf{X}, \mathcal{H})$ *be an AVcqC. Then*

$$1) \qquad \overline{C}_{A,r}(\mathcal{A}) = \overline{C}_C(\mathrm{conv}(\mathcal{A})) \tag{19}$$

$$2) \qquad \text{If } \overline{C}_{A,d}(\mathcal{A}) > 0, \text{ then } \overline{C}_{A,d}(\mathcal{A}) = \overline{C}_{A,r}(\mathcal{A}). \tag{20}$$

Also, this section contains the following statement, which asserts, that every sequence of random codes whith error strictly smaller than 1 for all but finitely many blocklenghts will not achieve rates higher than the rightmost term in (19).

Theorem 3 (Strong converse). *Let* $\mathcal{A} := \{A_s\}_{s \in \mathbf{S}}$ *be an AVcqC. For every* $\lambda \in [0, 1)$

$$\overline{C}_{A,r}(\mathcal{A}, \lambda) \leq \overline{C}_C(\mathrm{conv}(\mathcal{A})) \tag{21}$$

holds.

Remark 2. *The result can be gained for arbitrary (infinite) AVcqCs with only trivial modifications of the proof given below.*

In the next subsection 6.2, we show that the capacity for message transmission over an AVcqC using deterministic codes and the maximal error probability criterion is zero if and only if the AVcqC is $m - symmetrizable$.

This is an analog of [22, Theorem 1]. It can be formulated as follows.

Theorem 4. *Let* $\mathcal{A} = \{A_s\}_{s \in \mathbf{S}} \subset CQ(\mathbf{X}, \mathcal{H})$ *be an AVcqC. Then* $C_{A,d}(\mathcal{A})$ *is equal to zero if and only if* $\mathcal{A}$ *is m-symmetrizable.*

5 Compound cq-Channels

In this section, we consider compound cq-channels and give a rigourous proof for the achievability part of the coding theorem under the average error criterion together with a weak converse. The channel coding problem for compound cq-channels was treated, restricted to achievability, by Datta and Hsieh [16] for a certain class of compound channels, and Hayashi [21]. In our proof, we exploit the close relationship between channel coding and hypothesis testing which was utilized by Hayashi and Nagaoka [20] before. With focus set on the maximal error criterion, the compound cq channel coding theorem was proven in [9] already where also a strong converse theorem was proven for this setting.

For orientation of the reader we sketch the contents of this section. In Lemma 1 we reduce the problem of finding good channel codes for a finite compound channel to the problem of finding good hypothesis tests for certain quantum states generated by this channel. The existence of hypothesis tests with a performance sufficient for our purposes is shown in Lemma 5. In order to establish the coding theorem for arbitrary compound channels, we recall some approximation results in Lemma 6. With these preparations, we are able to prove the direct part of the coding theorem. Additionally, we give a proof of the weak converse (for which we utilize the strong converse result for the maximal error criterion given in [9] in Theorem 1). A strong converse for coding under the average error criterion does not hold in general for compound cq-channels (for further information, see Remark 4).

We consider a compound channel $\mathcal{W} := \{W_t\}_{t\in T} \subset CQ(\mathbf{X}, \mathcal{H})$ where T is a finite index set. We fix an orthonormal basis $\{e_x\}_{x\in\mathbf{X}}$ in $\mathbb{C}^{|\mathbf{X}|}$. For $\mathcal{W}$ and a given input probability distribution $p \in \mathfrak{P}(\mathbf{X})$ we define for every $t \in T$ states

$$\rho_t := \sum_{x\in\mathbf{X}} p(x)|e_x\rangle\langle e_x| \otimes W_t(x), \quad \text{and} \quad \hat{\sigma}_t := p \otimes \sigma_t, \tag{22}$$

on $\mathbb{C}^{|\mathbf{X}|} \otimes \mathcal{H}$, where p and σ_t are defined by

$$p := \sum_{x\in\mathbf{X}} p(x)|e_x\rangle\langle e_x|, \quad \text{and} \quad \sigma_t := \sum_{x\in\mathbf{X}} p(x)W_t(x). \tag{23}$$

With some abuse of notation, we use the letter p for the probability distribution as well as for the according quantum state defined above. Moreover, we define for every $l \in \mathbb{N}$ states

$$\rho_l := \frac{1}{|T|}\sum_{t\in T} v_l \rho_t^{\otimes l} v_l^* \tag{24}$$

$$\tau_l := \frac{1}{|T|}\sum_{t\in T} v_l \hat{\sigma}_t^{\otimes l} v_l^* = p^{\otimes l} \otimes \frac{1}{|T|}\sum_{t\in T} \sigma_t^{\otimes l} \tag{25}$$

where $v_l : (\mathbb{C}^{|\mathbf{X}|} \otimes \mathcal{H})^{\otimes l} \to (\mathbb{C}^{|\mathbf{X}|})^{\otimes l} \otimes \mathcal{H}^{\otimes l}$ is the ismorphism permuting the tensor factors. The next lemma is a variant of a result by Hayashi and Nagaoka in [20], which states that good hypothesis tests imply good message transmission codes for the average error criterion. Here it is formulated and proven for the states ρ_l and τ_l.

Lemma 1. *Let $\mathcal{W} := \{W_t\}_{t\in T} \subset CQ(\mathbf{X}, \mathcal{H})$ be a compound cq-channel with $|T| < \infty$, $p \in \mathfrak{P}(\mathbf{X})$, and $l \in \mathbb{N}$. Let further ρ_l, τ_l be the states associated to $\mathcal{W}$,p as defined in (24) and (25). If for $\lambda \in [0, 1]$, and $a > 0$ exists a projection $q_l \in \mathcal{B}((\mathbb{C}^{|\mathbf{X}|})^{\otimes l} \otimes \mathcal{H}^{\otimes l})$ which fulfills the conditions*

1. $\mathrm{tr}(q_l \rho_l) \geq 1 - \lambda$
2. $\mathrm{tr}(q_l \tau_l) \leq 2^{-la}$,

then for any γ with $a \geq \gamma > 0$ and $M_l := \lfloor 2^{l(a-\gamma)} \rfloor$ there is an (l, M_l)-code $(x_m^l, D_m^l)_{m\in[M_l]}$ with

$$\max_{t\in T} \frac{1}{M_l}\sum_{m=1}^{M_l} \mathrm{tr}(W_t^{\otimes l}(x_m^l)(\mathbb{1}_{\mathcal{H}^{\otimes l}} - D_m^l)) \leq |T|(2\lambda + 4 \cdot 2^{-l\gamma}) \tag{26}$$

The following operator inequality is a crucial ingredient in the proof of the lemma above, it was given in a more general form by Hayashi and Nagaoka in [20].

Lemma 2. *Let $a, b \in \mathcal{B}(\mathcal{H})$ be operators on $\mathcal{H}$ with $0 \leq a \leq 1$ and $b \geq 0$. Then*

$$\mathbb{1}_{\mathcal{H}} - (a+b)^{-\frac{1}{2}} a (a+b)^{-\frac{1}{2}} \leq 2(\mathbb{1}_{\mathcal{H}} - a) + 4b, \tag{27}$$

where $(\cdot)^{-1}$ denotes the generalized inverse.

Proof. See Lemma 2 in [20]. □

Proof (of Lemma 1). Let $l \in \mathbb{N}$, q_l a projection such that the assumptions of the lemma are fulfilled, and γ a number with $0 < \gamma \leq a$. According to the assumptions, q_l takes the form

$$q_l = \sum_{x^l \in \mathbf{X}^l} |e_{x^l}\rangle\langle e_{x^l}| \otimes q_{x^l}, \tag{28}$$

where $q_{x^l} \in \mathcal{B}(\mathcal{H}^{\otimes l})$ is a projection for every $x^l \in \mathbf{X}^l$. Set $M_l := \lfloor 2^{l(a-\gamma)} \rfloor$, and let $U_1, ..., U_{M_l}$ be i.i.d. random variables with values in $\mathbf{X}^l$, each distributed according to the l-fold product $p^{\otimes l}$ of the given distribution p. We define a random operator

$$D_m := \left(\sum_{n=1}^{M_l} q_{U_n} \right)^{-\frac{1}{2}} q_{U_m} \left(\sum_{n=1}^{M_l} q_{U_n} \right)^{-\frac{1}{2}} \tag{29}$$

for every $m \in [M_l]$ (we omit the superscript l here), where again generalized inverses are taken. The particular form of the decoding operators $D_1, ..., D_{M_l}$ in eq. (29) guarantees, that

$$\sum_{m=1}^{M_l} D_m \leq \mathbb{1}_{\mathcal{H}^{\otimes l}}$$

holds for every outcome of $U_1, ..., U_{M_l}$, and therefore $(U_m, D_m)_{m \in [M_l]}$ is a random code of size M_l. The remaining task is to bound the expectation value of the average error of this random code. We introduce an abbreviation for the average of the channels in $\mathcal{W}$ by

$$\overline{W}^l(\cdot) := \frac{1}{T} \sum_{t=1}^{T} W_t^{\otimes l}(\cdot).$$

The error probability of the random code is bounded as follows. By virtue of Lemma 2,

$$\begin{aligned} \mathbb{E}\left[\operatorname{tr}\left(\overline{W}^l(U_m)(\mathbb{1}_{\mathcal{H}^{\otimes l}} - D_m)\right)\right] \leq & 2\, \mathbb{E}\left[\operatorname{tr}\left(\overline{W}^l(U_m)(\mathbb{1}_{\mathcal{H}^{\otimes l}} - q_{U_m})\right)\right] \\ & + 4 \cdot \sum_{\substack{m \in [M_l]: \\ n \neq m}} \mathbb{E}\left[\operatorname{tr}\left(\overline{W}^l(U_m) q_{U_n}\right)\right] \end{aligned} \tag{30}$$

holds. The calculation of the expectation values on the r.h.s. of the above equation is straightforward, we obtain for every $m \in [M_l]$

$$\mathbb{E}[\mathrm{tr}(\overline{W}^l(U_m)(\mathbb{1}_{\mathcal{H}^{\otimes l}} - q_{U_m}))] = \mathrm{tr}(\rho_l(\mathbb{1}_{\mathcal{H}^{\otimes l}} - q_l)), \tag{31}$$

and, for $n \neq m$,

$$\mathbb{E}\left[\mathrm{tr}\left(\overline{W}^l(U_m)q_{U_n}\right)\right] = \mathrm{tr}(\tau_l q_l). \tag{32}$$

Together with the assumptions of the lemma, eqns. (31) and (32) imply

$$\mathbb{E}\left[\mathrm{tr}\left(\frac{1}{|T|}\sum_{t\in T} W_t^{\otimes l}(U_m)(\mathbb{1}_{\mathcal{H}^{\otimes l}} - D_m)\right)\right] \leq 2\lambda + 4\cdot M_l \cdot 2^{-la}$$
$$\leq 2\lambda + 4\cdot 2^{-l\gamma}$$

Because this error measure is an affine function of the channel we conclude, that there exists a cq-code $(x_m^l, D_m)_{m=1}^{M_l}$ for $\mathcal{W}$ with average error bounded by

$$\frac{1}{M_l}\sum_{m=1}^{M_l} \mathrm{tr}(W_t^{\otimes l}(x_m^l)(\mathbb{1}_{\mathcal{H}^{\otimes l}} - D_m)) \leq |T|(2\lambda + 4\cdot 2^{-l\gamma}) \tag{33}$$

for every $t \in T$, which is what we aimed to prove. □

The next two lemmata contain facts which are important for later considerations. The first lemma presents a bound on the cardinality of the spectrum of operators on a tensor product space which are invariant under permutations of the tensor factors. The group S_l of permutations on $[l]$ is, on $\mathcal{H}^{\otimes l}$, represented by defining (with slight abuse of notation) for each $\sigma \in S_l$ the unitary operator $\sigma \in \mathcal{B}(\mathcal{H}^{\otimes l})$

$$\sigma(v_1 \otimes ... \otimes v_l) := v_{\sigma^{-1}(1)} \otimes ... \otimes v_{\sigma^{-1}(l)}. \tag{34}$$

for all product vectors $v_1 \otimes ... \otimes v_l \in \mathbb{C}^l$ and linear extension to the whole space $\mathbb{C}^{\otimes l}$.

Lemma 3. *Let $Y \in \mathcal{B}(\mathcal{H}^{\otimes l})$ ($d := \dim \mathcal{H} \geq 2$) satisfy $\sigma Y = Y\sigma$ for every permutation $\sigma \in S_l$. Then*

$$|\,\mathrm{spec}(Y)| \leq (l+1)^{d^2}. \tag{35}$$

Proof. It is clear that, under the action of S_l, $\mathcal{H}^{\otimes l}$ decomposes into a finite direct sum $\mathcal{H}^{\otimes l} = \oplus_{i=1}^M \oplus_{j=1}^{m_i} \mathcal{H}_{i,j}$, where the $\mathcal{H}_{i,j}$ are irreducible subspaces of S_l, $m_i \in \mathbb{N}$ their multiplicity and $M \in \mathbb{N}$. Moreover, $\mathcal{H}_{i,j} \simeq \mathcal{H}_{i,k}$ f.a. $i \in [M]$, $j, k \in [m_i]$ and to every such choice of indices there exists a linear operator $Q_{i,j,k} : \mathcal{H}_{i,k} \mapsto \mathcal{H}_{i,j}$ such that $\sigma Q_{i,j,k} = Q_{i,j,k}\sigma$ f.a. $\sigma \in S_l$.

Let us write $Y = \sum_{i,j} Y_{i,m,j,n}$, where $Y_{i,m,j,n} : \mathcal{H}_{j,n} \mapsto \mathcal{H}_{i,m}$. Then according to Schur's lemma, $Y_{i,m,j,n} = 0$, $(i \neq j)$ and $Y_{i,m,i,n} = c_{i,m,n}Q_{i,m,n}$ for all valid choices of indices and unique complex numbers $c_{i,m,n} \in \mathbb{C}$.

Thus, defining the self-adjoint operators $Y_i := \sum_{m,n=1}^{m_i} c_{i,m,n} Q_{i,m,n}$, we see that

$$Y = \sum_{i=1}^{M} Y_i \tag{36}$$

holds. Obviously, $Y_{i,m,i,m} = \mathbb{1}_{\mathcal{H}_{i,m}}$. Thus, with an appropriate choice of bases in every single one of the $\mathcal{H}_{i,m}$ and defining the matrices C_i by $(C_i)_{mn} := c_{i,m,n}$, we can write a matrix representation $\tilde{Y}_i$ of Y_i as $\tilde{Y}_i = C_i \otimes \mathbb{1}_{\mathbb{C}^{\dim(\mathcal{H}_{i,1})}}$.

Clearly then, each of the Y_i can have no more than m_i different eigenvalues. Since $\operatorname{supp}(Y_i) \perp \operatorname{supp}(Y_j)$ $(i \neq j)$, we get

$$|\operatorname{spec}(Y)| \leq \sum_{i=1}^{M} m_i. \tag{37}$$

Now, taking a look at [13], equation (1.22), we see that $m_i \leq (l+1)^{d^2/2}$ holds. The number M is the number of different Young tableaux occuring in the representation of S_l on $\mathcal{H}^{\otimes l}$ and obeys the bound $M \leq N_T([d]^l)$, where $N_T([d]^l)$ is the number of different types on $[d]^l$, that itself obeys $N_T([d]^l) \leq (l+1)^d$ (Lemma 2.2 in [14]). For $d \geq 2$ we thus have

$$|\operatorname{spec}(Y)| \leq \sum_{i=1}^{M} m_i \leq (l+1)^{d^2/2}(l+1)^d \leq (l+1)^{d^2}. \tag{38}$$

□

Lemma 5 provides the result which will, together with Lemma 1, imply the existence of optimal codes for $\mathcal{W}$. We give a proof which is based on an idea of Ogawa and Hayashi which originally appeared in [25]. An important ingredience of their proof is the operator inequality stated in the following lemma.

Lemma 4 ([19]). *Let χ be a state on on a Hilbert space $\mathcal{K}$, and $\mathcal{M} := \{P_k\}_{k=1}^{K} \subset \mathcal{B}(\mathcal{K})$ be a collection of projections on $\mathcal{K}$ with $\sum_{k=1}^{K} P_k = \mathbb{1}_{\mathcal{K}}$. Then the operator inequality*

$$\chi \leq K \cdot \sum_{k=1}^{K} P_k \chi P_k \tag{39}$$

holds.

Lemma 5. *For every $\delta > 0$, finite compound cq-channel $\mathcal{W} := \{W_t\}_{t \in T} \subset CQ(\mathbf{X}, \mathcal{H})$ and $p \in \mathfrak{P}(\mathbf{X})$ there exists a constant $\tilde{c}$, such that for every sufficiently large $l \in \mathbb{N}$ there exists a projection $q_{l,\delta} \in \mathcal{B}((\mathbb{C}^{|\mathbf{X}|})^{\otimes l} \otimes \mathcal{H}^{\otimes l})$ which fulfills*

1. $\operatorname{tr}(q_{l,\delta}\rho_l) \geq 1 - |T| \cdot 2^{-l\tilde{c}}$, and
2. $\operatorname{tr}(q_{l,\delta}\tau_l) \leq 2^{-l(a-\delta)}$

where ρ_l, τ_l are the states belonging to $\mathcal{W}, p$ according to (24) and (25), and a is defined by $a := \min_{t\in[T]} D(\rho_t||p \otimes \sigma_t)$.

Proof. Let $\delta > 0$ be fixed, for $l \in \mathbb{N}$, we have $\mathrm{ran}(\rho_l) \subseteq \mathrm{ran}(\tau_l) := \mathcal{H}_l$, which allows us to restrict ourselves to $\mathcal{H}_l$, where τ_l is invertible. For every $\varepsilon \in (0,1)$, we define a regularized version $\rho_{l,\varepsilon}$ to ρ_l by

$$\rho_{l,\varepsilon} := (1-\varepsilon)\rho_l + \varepsilon\tau_l. \tag{40}$$

These operators are invertible on $\mathcal{H}_l$ and approximate ρ_l, i.e.

$$\|\rho_{l,\varepsilon} - \rho_l\|_1 \leq 2\varepsilon. \tag{41}$$

holds for every $\epsilon > 0$. We also define an operator

$$\overline{\rho}_{l,\varepsilon} := \sum_{\lambda\in\mathrm{spec}(\tau_l)\setminus\{0\}} E_\lambda \rho_{l,\varepsilon} E_\lambda, \tag{42}$$

which is the pinching of $\rho_{l,\varepsilon}$ to the eigenspaces of τ_l (here E_λ is the projection which projects onto the eigenspace belonging to the eigenvalue λ for every $\lambda \in \mathrm{spec}(\tau_l)$). This definition guarantees

$$\tau_l \overline{\rho}_{l,\varepsilon} = \overline{\rho}_{l,\varepsilon}\tau_l. \tag{43}$$

With a as assumed in the lemma, we define the operator

$$T_\varepsilon := \overline{\rho}_{l,\varepsilon} - 2^{l(a-\delta)}\tau_l \tag{44}$$

with spectral decomposition

$$T_\varepsilon = \sum_{\mu\in\mathrm{spec}(T_\varepsilon)} \mu P_\mu. \tag{45}$$

The projection $q_{l,\delta}$ onto the nonnegative part of T_ε, defined by

$$q_{l,\delta} := \sum_{\mu\in\mathrm{spec}(T_\varepsilon):\mu\geq 0} P_\mu. \tag{46}$$

will now be shown to suffice the bounds stated in the lemma. Clearly, $q_{l,\delta}T_\varepsilon q_{l,\delta}$ is a positive semidefinite operator, therefore, with (44) the inequality

$$q_{l,\delta}\tau_l q_{l,\delta} \leq 2^{-l(a-\delta)} q_{l,\delta}\overline{\rho}_{l,\varepsilon} q_{l,\delta}. \tag{47}$$

is valid. Taking traces in (47) yields

$$\mathrm{tr}(q_{l,\delta}\tau_l) \leq 2^{-l(a-\delta)}\mathrm{tr}(q_{l,\delta}\overline{\rho}_{l,\varepsilon}) \tag{48}$$

$$\leq 2^{-l(a-\delta)} \tag{49}$$

which shows, that $q_{l,\delta}$ fulfills the second bound in the lemma. We shall now prove, that $q_{l,\delta}$ for l large enough actually also suffices the first one. To this end we derive an upper bound on $\mathrm{tr}((\mathbb{1}-q_{l,\delta})\rho_{l,\varepsilon})$ for any given $\varepsilon>0$, which implies (together with (41)) a bound on $\mathrm{tr}((\mathbb{1}-q_{l,\delta})\rho_l)$. In fact it is sufficient to find an upper bound on $\mathrm{tr}((\mathbb{1}-q_{l,\delta})\overline{\rho}_{l,\varepsilon})$, which can be seen as follows. Because $\overline{\rho}_{l,\varepsilon}$ and τ_l commute by construction (see eq. (43)), T_ε and τ_l commute as well. This in turn implies that $q_{l,\delta}$ commutes with the operators $E_1,\dots,E_{|\,\mathrm{spec}(\tau_l)|}$ in the spectral decomposition of τ_l which eventually ensures us, that

$$\mathrm{tr}((\mathbb{1}_{\mathcal{H}^{\otimes l}}-q_{l,\delta})\overline{\rho}_{l,\varepsilon})=\mathrm{tr}((\mathbb{1}_{\mathcal{H}^{\otimes l}}-q_{l,\delta})\rho_{l,\varepsilon}) \tag{50}$$

holds. For an arbitrary but fixed number $s\in[0,1]$ we have

$$\begin{aligned}\mathrm{tr}((\mathbb{1}_{\mathcal{H}^{\otimes l}}-q_{l,\delta})\overline{\rho}_{l,\varepsilon}) &= \mathrm{tr}(\overline{\rho}_{l,\varepsilon}^{(1-s)}\overline{\rho}_{l,\varepsilon}^{s}(\mathbb{1}_{\mathcal{H}^{\otimes l}}-q_{l,\delta})) &(51)\\ &\le 2^{-ls(a-\delta)}\mathrm{tr}(\overline{\rho}_{l,\varepsilon}^{(1-s)}\tau_l^{s}(\mathbb{1}_{\mathcal{H}^{\otimes l}}-q_{l,\delta})) &(52)\\ &\le 2^{-ls(a-\delta)}\mathrm{tr}(\overline{\rho}_{l,\varepsilon}^{(1-s)}\tau_l^{s}). &(53)\end{aligned}$$

The inequality in (52) is justified by the following argument. Since $\overline{\rho}_{\varepsilon,l}$ and τ_l commute, they are both diagonal in the same orthonormal basis $\{g_i\}_{i=1}^d$, i.e. they have spectral decompositions of the form

$$\overline{\rho}_{l,\varepsilon}=\sum_{i=1}^{d}\chi_i|g_i\rangle\langle g_i|,\qquad \text{and}\quad \tau_l=\sum_{i=1}^{d}\theta_i|g_i\rangle\langle g_i|. \tag{54}$$

Because $q_{l,\delta}$ projects onto the eigenspaces corresponding to nonnegative eigenvalues of T_ε, we have

$$\mathbb{1}_{\mathcal{H}^{\otimes l}}-q_{l,\delta}=\sum_{i\in N}|g_i\rangle\langle g_i|, \tag{55}$$

where the set N is defined by $N:=\{i\in[d]:\chi_i-2^{l(a-\delta)}\theta_i<0\}$. It follows

$$\chi_i^s\le 2^{ls(a-\delta)}\theta_i^s \tag{56}$$

for all $i\in N$ and $s\in[0,1]$. This in turn implies, via (43) and (54),

$$\overline{\rho}_{l,\varepsilon}^{s}(\mathbb{1}_{\mathcal{H}^{\otimes l}}-q_{l,\delta})\le 2^{ls(a-\delta)}\tau_l^{s}(\mathbb{1}_{\mathcal{H}^{\otimes l}}-q_{l,\delta}), \tag{57}$$

which shows (52). Combining eqns. (50) and (53) we obtain

$$\begin{aligned}\mathrm{tr}((\mathbb{1}_{\mathcal{H}^{\otimes l}}-q_{l,\delta})\rho_{l,\varepsilon}) &\le 2^{ls(a-\delta)}\mathrm{tr}(\overline{\rho}_{l,\varepsilon}^{(1-s)}\tau_l^{s})\\ &= 2^{ls(a-\delta)}\mathrm{tr}(\overline{\rho}_{l,\varepsilon}\tau_l^{\frac{s}{2}}\overline{\rho}_{l,\varepsilon}^{-s}\tau_l^{\frac{s}{2}})\\ &= 2^{ls(a-\delta)}\mathrm{tr}(\rho_{l,\varepsilon}\tau_l^{\frac{s}{2}}\overline{\rho}_{l,\varepsilon}^{-s}\tau_l^{\frac{s}{2}}). &(58)\end{aligned}$$

Here we used the fact, that $\overline{\rho}_{l,\varepsilon}$ and τ_l commute in the first equality. Eq. (58) is justified, because the eigenprojections of τ_l wich appear in the definition of $\overline{\rho}_{l,\varepsilon}$

are absorbed by $\tau_l^{\frac{1}{2}}$. We can further upper bound the above expressions in the following way. Note, that

$$\rho_{l,\varepsilon} \leq |\operatorname{spec}(\tau_l)|\overline{\rho}_{l,\varepsilon}. \tag{59}$$

holds by Lemma 4. Because $-(\cdot)^{-s}$ is an operator monotone function for every $s \in [0,1]$ (see e.g. [8]), (59) implies

$$\overline{\rho}_{l,\varepsilon}^{-s} \leq |\operatorname{spec}(\tau_l)|^s \rho_{l,\varepsilon}^{-s}.$$

Using the above relation, one obtains

$$\operatorname{tr}(\rho_{l,\varepsilon}\tau_l^{\frac{s}{2}}\overline{\rho}_{l,\varepsilon}^{-s}\tau_l^{\frac{s}{2}}) \leq |\operatorname{spec}(\tau_l)|^s \operatorname{tr}(\rho_{l,\varepsilon}\tau^{\frac{s}{2}}\rho_{l,\varepsilon}^{-s}\tau_l^{\frac{s}{2}}).$$

By combination with (58) this leads to

$$\begin{aligned}
\operatorname{tr}((\mathbb{1}_{\mathcal{H}^{\otimes l}} - q_{l,\delta})\rho_{l,\varepsilon}) &\leq |\operatorname{spec}(\tau_l)|^s 2^{ls(a-\delta)} \operatorname{tr}(\rho_{l,\varepsilon}\tau^{\frac{s}{2}}\rho_{l,\varepsilon}^{-s}\tau_l^{\frac{s}{2}}) &(60)\\
&\leq (l+1)^{d^2} \exp\{l[(a-\delta)s - \tfrac{1}{l}\psi_{l,\varepsilon}(s)]\} &(61)\\
&= \exp\{l[(a-\delta)s - \tfrac{1}{l}\psi_{l,\varepsilon}(s) + w(l)]\}, &(62)
\end{aligned}$$

where $d := \dim\mathcal{H}$. In (61), we used the definition

$$\psi_{l,\varepsilon}(s) := -\log\operatorname{tr}(\rho_{l,\varepsilon}\tau_l^{\frac{s}{2}}\rho_{l,\varepsilon}^{-s}\tau_l^{\frac{s}{2}}), \tag{63}$$

in the last line we introduced the function w defined by $w(l) := \frac{d^2}{l}\log(l+1)$ for every $l \in \mathbb{N}$. Notice, that we also used the bound $|\operatorname{spec}(\tau_l)| \leq (l+1)^{d^2}$ on the spectrum of τ_l which is justified by Lemma 3. In fact, by observation of (25), it is easy to see, that for every σ in the tensor product representation of S_l on $\mathcal{H}^{\otimes l}$ (see (34)),

$$(\mathbb{1}_{\mathbb{C}^{|\mathbf{X}|}}^{\otimes l} \otimes \sigma)\tau_l = \tau_l(\mathbb{1}_{\mathbb{C}^{|\mathbf{X}|}}^{\otimes l} \otimes \sigma) \tag{64}$$

holds. We will now show, that the argument of the exponential in (62) becomes strictly negative for a suitable choice of s, sufficiently small ε and large enough l. We define

$$f_{l,\varepsilon}(s) := (a-\delta)s - \frac{1}{l}\psi_{l,\varepsilon}(s). \tag{65}$$

By the mean value theorem it suffices to show that $f'_{l,\varepsilon}(0) < 0$ for small enough $\varepsilon > 0$. For the derivative, we have

$$f'_{l,\varepsilon}(0) = a - \delta - \frac{1}{l}D(\rho_{l,\varepsilon}||\tau_l). \tag{66}$$

The relative entropy term in (66) can be lower bounded as follows. It holds

$$\begin{aligned}
D(\rho_{l,\varepsilon}||\tau_l) &= -S(\rho_{l,\varepsilon}) - \operatorname{tr}(\rho_{l,\varepsilon}\log\tau_l)\\
&= -S(\rho_{l,\varepsilon}) + lS(p) + S\left(\frac{1}{|T|}\sum_{t\in T}\sigma_t^{\otimes l}\right) &(67)\\
&\geq -S(\rho_{l,\varepsilon}) + lS(p) + \frac{1}{|T|}\sum_{t\in T} lS(\sigma_t). &(68)
\end{aligned}$$

Notice that the equality in (67) indeed holds, because the marginals on $(\mathbb{C}^{|\mathbf{X}|})^{\otimes l}$ and $\mathcal{H}^{\otimes l}$ of ρ_l and τ_l are equal and therefore equal to the marginals of $\rho_{l,\epsilon}$ by definition for each $\epsilon \in (0,1)$. The inequality in (68) is valid due to concavity of the von Neumann entropy. Because (41) holds,

$$\begin{aligned} S(\rho_{l,\varepsilon}) &\leq S(\rho_l) + 2\varepsilon \log \frac{\dim(\mathcal{H}_l)}{2\varepsilon} \\ &\leq S(\rho_l) + 2\varepsilon l \log \frac{d}{2\varepsilon} \end{aligned} \tag{69}$$

is valid for $\epsilon < \frac{1}{2e}$, since for two states $\rho, \sigma \in \mathcal{S}(\mathcal{H})$ with $\|\rho - \sigma\|_1 \leq \varepsilon \leq \frac{1}{e}$, Fannes' inequality [18],

$$|S(\rho) - S(\sigma)| \leq \varepsilon \log \frac{\dim \mathcal{H}}{\varepsilon}, \tag{70}$$

is valid. Together with (69), (68) implies

$$\begin{aligned} D(\rho_{l,\varepsilon}||\tau_l) &\geq -S(\rho_l) - 2\varepsilon l \log \frac{d}{2\varepsilon} + lS(p) + \frac{1}{|T|}\sum_{t\in T} lS(\sigma_t) \\ &\geq -\frac{1}{|T|}\sum_{t\in T} lS(\rho_t) - \log|T| - 2\varepsilon l \log \frac{d}{2\varepsilon} + lS(p) + \frac{1}{|T|}\sum_{t\in T} lS(\sigma_t) \qquad (71) \\ &= \frac{l}{|T|}\sum_{t\in T} D(\rho_t||p\otimes\sigma_t) - \log|T| - 2\varepsilon l \log \frac{d}{2\varepsilon}. \qquad (72) \end{aligned}$$

The inequality in (71) results from the fact, that the von Neumann entropy is an almost convex function, i.e.

$$S(\rho) \leq \sum_{i=1}^{N} p_i S(\rho_i) + \log(N) \tag{73}$$

for any mixture $\rho = \sum_{i=1}^{N} p_i \rho_i$ of states. Inserting (72) in (66) gives

$$\begin{aligned} f'_{l,\varepsilon}(0) &\leq \min_{t\in T} D(\rho_t||p\otimes\sigma_t) - \delta - \frac{1}{|T|}\sum_{t\in T} D(\rho_t||p\otimes\sigma_t) + 2\varepsilon\log\frac{d}{2\varepsilon} + \frac{1}{l}\log|T| \\ &< -\frac{\delta}{2} + \frac{1}{l}\log|T|, \end{aligned} \tag{74}$$

provided that $0 < \varepsilon < \varepsilon_0(\delta)$ where ε_0 is small enough to ensure $2\varepsilon \log \frac{d}{2\varepsilon} < \frac{\delta}{2}$. The mean value theorem shows that for $s \in (0,1]$

$$f_{l,\varepsilon}(s) = f_{l,\varepsilon}(0) + f'_{l,\varepsilon}(s')\cdot s$$

holds for some $s' \in (0,s)$. Since $f_{l,\varepsilon}(0) = 0$, (74) shows that we can guarantee

$$f_{l,\varepsilon}(s) < \left(-\frac{\delta}{2} + \frac{1}{l}\log|T|\right) s \tag{75}$$

for small enough s. By (62) and (75) we obtain for $\varepsilon < \varepsilon_0(\delta)$ and l large enough to make $w(l) < \frac{\delta s}{8}$ valid,

$$\begin{aligned} \mathrm{tr}((\mathbb{1}_{\mathcal{H}^{\otimes l}} - q_{l,\delta})\rho_{l,\varepsilon}) &\leq \exp\{l[f_{l,\varepsilon}(s) + w(l)]\} \\ &\leq \exp\left\{-l\left(\frac{\delta}{4}s - w(l)\right)\right\} \\ &\leq |T| \cdot \exp\{-l\frac{\delta}{8}s\}. \end{aligned} \tag{76}$$

Using (41), we have (with $\varepsilon < \varepsilon_0$)

$$\begin{aligned} \mathrm{tr}((\mathbb{1}_{\mathcal{H}^{\otimes l}} - q_{l,\delta})\rho_l) &\leq \|\rho_{l,\varepsilon} - \rho_l\|_1 + \mathrm{tr}((\mathbb{1}_{\mathcal{H}^{\otimes l}} - q_{l,\delta})\rho_{l,\varepsilon}) \\ &\leq 2\varepsilon + |T| \exp\{-l\frac{\delta}{8}s\}. \end{aligned}$$

We can in fact, choose the parameter ϵ dependent on l in a way that $(\varepsilon_l)_{l=1}^{\infty}$ decreases exponentially in l, which proves the second claim of the lemma. □

In order to prove the direct part of the coding theorem for general sets of channels we have to approximate arbitrary sets of channels by finite ones. For $\alpha > 0$, an α-net in $CQ(\mathbf{X}, \mathcal{H})$ is a finite set $\mathcal{N}_\alpha := \{W_i\}_{i=1}^{N_\alpha} \subset CQ(\mathbf{X}, \mathcal{H})$ with the property, that for every channel $W \in CQ(\mathbf{X}, \mathcal{H})$ there exists an index $i \in [N_\alpha]$ such that

$$\|W - W_i\|_{cq} < \alpha \tag{77}$$

holds. For a given set $\mathcal{W} \subset CQ(\mathbf{X}, \mathcal{H})$ an α-net $\mathcal{N}_\alpha$ in $CQ(\mathbf{X}, \mathcal{H})$ generates an approximating set $\widetilde{\mathcal{W}}_\alpha$ defined by

$$\widetilde{\mathcal{W}}_\alpha := \{W_i \in \mathcal{N}_\alpha : B_{cq}(\alpha, W_i) \cap \mathcal{W} \neq \varnothing\}. \tag{78}$$

where $B_{cq}(\alpha, A)$ denotes the α-ball with center A regarding the norm $\|\cdot\|_{cq}$. The above definition does not guarantee, that $\widetilde{\mathcal{W}}_\alpha$ is a subset of $\mathcal{W}$ but each $\widetilde{\mathcal{W}}_\alpha$ clearly generates a set $\mathcal{W}_{2\alpha} \subset \mathcal{W}$ of at most the same cardinality, such that for every $W \in \mathcal{W}$ exists an index $i \in [N_\alpha]$ with

$$\|W - W_i\|_{cq} < 2\alpha. \tag{79}$$

The next lemma states that we find good approximations of arbitrary compound cq-channels among such sets as defined above. The proof can be given by minor variations of the corresponding results in [9], [10], and we omit it here.

Lemma 6. *Let $\mathcal{W} := \{W_t\}_{t\in T} \subset CQ(\mathbf{X}, \mathcal{H})$ and $\alpha \in (0, \frac{1}{e})$. There exists a set $T_\alpha \subseteq T$ which fulfills the following conditions*

1. $|T_\alpha| < \left(\frac{6}{\alpha}\right)^{2|\mathbf{X}|d^2}$,
2. *given any $l \in \mathbb{N}$, to every $t \in T$ one finds an index $t' \in T_\alpha$ such that*

$$\|W_t^{\otimes l}(x^l) - W_{t'}^{\otimes l}(x^l)\|_1 < 2l\alpha. \tag{80}$$

holds for every $x^l \in \mathbf{X}^l$. Moreover,

3. for every $p \in \mathfrak{P}(\mathbf{X})$,

$$\left| \min_{t' \in T_\alpha} \chi(p, W_{t'}) - \inf_{t \in T} \chi(p, W_t) \right| \leq 2\alpha \log \frac{d}{2\alpha} \tag{81}$$

holds.

The following lemma is from [2] and will be used to establish the weak converse in Theorem 1. It states that codes which have small average error probability for a finite compound cq-channel contain subcodes with good maximal error probability of not substantially smaller size.

Lemma 7 (cf. [2], Lemma 1). *Let* $\mathcal{W} : \{W_t\}_{t \in T} \subset CQ(\mathbf{X}, \mathcal{H})$ *be a compound channel with* $|T| < \infty$ *and* $l \in \mathbb{N}$. *If* $(u_i^l, D_i^l)_{i=1}^{M_l}$ *is an* (l, M_l)*-code with*

$$\max_{t \in T} \frac{1}{M_l} \sum_{i=1}^{M_l} \operatorname{tr}(W_t^{\otimes l}(u_i^l)(\mathbb{1}_{\mathcal{H}} - D_i^l)) \leq \overline{\lambda}. \tag{82}$$

Then there exists for every $\epsilon > 0$ *a subcode* $(u_{i_j}^l, D_{i_j}^l)_{j=1}^{M_{l,\epsilon}}$ *of size* $M_{l,\epsilon} = \lfloor \frac{\epsilon}{1-\epsilon} M_l \rfloor$ *with*

$$\max_{t \in T} \max_{j \in [M_{l,\epsilon}]} \operatorname{tr}(W_t^{\otimes l}(u_{i_j}^l)(\mathbb{1}_{\mathcal{H}} - D_{i_j}^l)) \leq |T|(\overline{\lambda} + \epsilon) \tag{83}$$

Finally, we have gathered all the prerequisites to prove Theorem 1:

Proof (of Theorem 1). The direct part (i.e. the assertion that the r.h.s. lower-bounds the l.h.s. in (18)) is proven by combining Lemma 1 with Lemma 5. Let $p = \operatorname{argmax}_{p' \in \mathfrak{P}(\mathbf{X})} \inf_{t \in T} \chi(p', W_t)$. We show that for any $\delta > 0$,

$$\inf_{t \in T} \chi(p, W_t) - \delta \tag{84}$$

is an achievable rate. We can restrict ourselves to the case, where $\inf_{t \in T} \chi(p, W_t) > \delta > 0$ holds, because otherwise the above statement is trivially fulfilled. The above mentioned lemmata consider finite sets of channels, therefore we choose an approximating set $\mathcal{W}_{\alpha_l}$ (of cardinality T_{α_l}) according to Lemma 6 for every $l \in \mathbb{N}$, where we leave the sequence $\alpha_1, \alpha_2, \ldots$ initially unspecified. For every $l \in \mathbb{N}$ and $t' \in T_{\alpha_l}$, let $\rho_{t'}$, $\sigma_{t'}$ be defined according to eq. (22) and (23), and further define states

$$\rho_l := \frac{1}{|T_{\alpha_l}|} \sum_{t' \in T_{\alpha_l}} v_l \rho_{t'}^{\otimes l} v_l^* \tag{85}$$

and

$$\tau_l := p^{\otimes l} \otimes \frac{1}{|T_{\alpha_l}|} \sum_{t' \in T_{\alpha_l}} \sigma_{t'}^{\otimes l}. \tag{86}$$

For a given number η with $0 < \eta < a_l$, Lemma 5 guarantees (for large enough l), with a suitable constant $\tilde{c} > 0$, the existence of a projection $q_{l,\eta} \in \mathcal{B}((\mathbb{C}^{|\mathbf{X}|})^{\otimes l} \otimes \mathcal{H}^{\otimes l})$ with

$$\mathrm{tr}(q_{l,\eta}\rho_l) \geq 1 - |T_{\alpha_l}| \cdot 2^{-l\tilde{c}} \tag{87}$$

and

$$\mathrm{tr}(q_{l,\eta}\tau_l) \leq 2^{-l(a_l-\eta)} \tag{88}$$

where we defined $a_l := \min_{t' \in T_{\alpha_l}} D(\rho_{t'}||p \otimes \sigma_{t'})$. This by virtue of Lemma 1 implies for every $\gamma > 0$ such that $\eta + \gamma < a_l$ the existence of a cq-code $(x^l_m, D^l_m)_{m \in [M_l]}$ of size

$$M_l = \lfloor 2^{l(a_l-\eta-\gamma)} \rfloor \tag{89}$$

and average error bounded by

$$\max_{t' \in T_{\alpha_l}} \frac{1}{M_l} \sum_{m=1}^{M_l} \mathrm{tr}(W^{\otimes l}_{t'}(u^l_m)(\mathbb{1}_{\mathcal{H}^{\otimes l}} - D^l_m)) \leq 2|T_{\alpha_l}|^2 2^{-l\tilde{c}} + 4 \cdot |T_{\alpha_l}| 2^{-l\gamma}. \tag{90}$$

Notice, that for other positive numbers γ, δ, trivial codes have $M_l = 1 \geq \lfloor 2^{l(a_l-\eta-\gamma)} \rfloor$. Using (89) we obtain,

$$\frac{1}{l} \log M_l \geq \min_{t' \in T_{\alpha_l}} \chi(p, W_{t'}) - \eta - \gamma \tag{91}$$

$$\geq \inf_{t \in T} \chi(p, W_t) - \eta - \gamma - 4\alpha_l \log \frac{d}{2\alpha_l}, \tag{92}$$

where the second inequality follows from Lemma 6. For the average error, it holds,

$$\sup_{t \in T} \frac{1}{M_l} \sum_{m \in [M_l]} \mathrm{tr}\left(W^{\otimes l}_t(u^l_m)(\mathbb{1}_{\mathcal{H}^{\otimes l}} - D^l_m)\right) \tag{93}$$

$$\leq \max_{t' \in T_{\alpha_l}} \frac{1}{M_l} \sum_{m \in [M_l]} \mathrm{tr}\left(W^{\otimes l}_{t'}(u^l_m)(\mathbb{1}_{\mathcal{H}^{\otimes l}} - D^l_m)\right) + 2l\alpha_l \tag{94}$$

$$\leq 2|T_{\alpha_l}|^2 2^{-l\tilde{c}} + 4|T_{\alpha_l}| 2^{-l\gamma} + 2l\alpha_l. \tag{95}$$

The first of the above inequalities follows from Lemma 6, the second one is by (90). Because we chose the approximating sets according to Lemma 6,

$$|T_{\alpha_l}| \leq \left(\frac{6}{\alpha_l}\right)^{2|\mathbf{X}|d^2} \tag{96}$$

holds. In fact, if we specify α_l to be $\alpha_l := 2^{-l\hat{c}}$ for every $l \in \mathbb{N}$, where $\hat{c}$ is a constant with $0 < \hat{c} < \min\left\{\frac{\tilde{c}}{4|\mathbf{X}|d^2}, \frac{\eta}{2|\mathbf{X}|d^2}\right\}$, the r.h.s of (95) decreases exponentially for $l \to \infty$. If we additionally choose η and γ, small enough to validate $\delta > \eta + \gamma + 2\alpha_l \log \frac{d}{2\alpha_l}$ for sufficiently large l, the rate defined in (84) is shown to be achievable by (95) and (92). Since δ was arbitrary, the direct statement follows.

It remains to prove the converse statement. For the proof, we will construct a good code for transmission under the maximal error criterion and invoke the strong converse result given in [9] (see Remark 3). We show, that for any $\delta > 0$,

$$\overline{C}_C(\mathcal{W}) < \max_{p \in \mathfrak{P}(\mathbf{X})} \inf_{t \in T} \chi(p, W_t) + \delta. \tag{97}$$

Let $\delta > 0$ and assume that for some fixed $l \in \mathbb{N}$, $\mathcal{C}_l := (u_m^l, D_m^l)_{m=1}^{M_l}$ is an (l, M_l)-code with

$$\sup_{t \in T} \frac{1}{M_l} \sum_{m=1}^{M_l} \operatorname{tr}(W_t^{\otimes l}(u_m^l)(\mathbb{1}_{\mathcal{H}^{\otimes l}} - D_m^l)) \le \overline{\lambda}_l. \tag{98}$$

We always can find a finite subset $\hat{T} \subset T$ such that

$$\left| \max_{p \in \mathfrak{P}(\mathbf{X})} \inf_{t \in T} \chi(p, W_t) - \max_{p \in \mathfrak{P}(\mathbf{X})} \min_{t \in \hat{T}} \chi(p, W_t) \right| \le \frac{\delta}{2} \tag{99}$$

holds (e.g. a set T_α as in Lemma 6 for suitable α). We set $\epsilon := \frac{1}{2|\hat{T}|}$. By virtue of Lemma 7 we find a subcode $(u_{i_j}^l, D_{i_j}^l)_{j=1}^{M_{l,\epsilon}} \subseteq \mathcal{C}_l$ of $\mathcal{C}_l$ which has size

$$M_{l,\epsilon} := \left\lfloor \frac{\epsilon}{1-\epsilon} M_l \right\rfloor \tag{100}$$

and maximal error bounded by

$$\max_{t \in \hat{T}} \max_{j \in M_{l,\epsilon}} \operatorname{tr}\left(W_t^{\otimes l}(u_{i_j}^l)(\mathbb{1}_{\mathcal{H}^{\otimes l}} - D_{i_j}^l) \right) \le \overline{\lambda}_l |\hat{T}| + \frac{1}{2}. \tag{101}$$

$$\tag{102}$$

If l is sufficiently large, the r.h.s. is strictly smaller than one. Therefore, by the strong converse theorem for coding under the maximal error criterion (see [9], Theorem 5.13), we have (with some constant $K > 0$)

$$\frac{1}{l} \log M_{l,\epsilon} \le \max_{p \in \mathfrak{P}(\mathbf{X})} \min_{t \in \hat{T}} \chi(p, W_t) + K \frac{1}{\sqrt{l}} \tag{103}$$

$$\le \max_{p \in \mathfrak{P}(\mathbf{X})} \inf_{t \in T} \chi(p, W_t) + \frac{\delta}{2} + K \frac{1}{\sqrt{l}}. \tag{104}$$

The second line above follows from (99). On the other hand, by (100), we have

$$\log M_l \le \log M_{l,\epsilon} + \log \left(\frac{\varepsilon}{2(1-\varepsilon)} \right). \tag{105}$$

Dividing both sides of (105) by l and combinig the result with (104) shows that for sufficiently large l

$$\frac{1}{l}\log M_l \leq \max_{p\in\mathfrak{P}(\mathbf{X})}\inf_{t\in T}\chi(p,W_t)+\frac{\delta}{2}+K\frac{1}{\sqrt{l}}+\frac{1}{l}\log\left(\frac{\varepsilon}{2(1-\varepsilon)}\right) \tag{106}$$

$$\leq \max_{p\in\mathfrak{P}(\mathbf{X})}\inf_{t\in T}\chi(p,W_t)+\delta \tag{107}$$

holds, which shows (97). Since δ was an arbitrary positive number, we are done. □

Remark 3. *While the achievability part for cq-compound channels regarding the maximal error criterion given in [9] required technical effort, the strong converse proof was rather uncomplicated. It was given there by a combination of Wolfowitz' proof technique for the strong converse in case of classical compound channels and a lemma from [29].*

Remark 4. *We remark here, that a general strong converse does not hold for the capacity of compound cq-channels if the average error is considered as criterion for reliability of the message transmission. This can be seen by a counterexample given by Ahlswede in [1] (Example 1) regarding classical compound channels. However, we will see in the proof of Theorem 3, that in certain situations (especially, where $\mathcal{W}$ is a convex set) a strong converse proof can be established.*

As a corollary to the achievability part of Theorem 1 above, we immediately obtain a direct coding theorem for the capacity of a finite cq-compound channel under the maximal error criterion.

Corollary 1. *For a finite compound cq-channel $\mathcal{W}:=\{W_t\}_{t\in T}\subset CQ(\mathbf{X},\mathcal{H})$ we have*

$$C_C(\mathcal{W})\geq \max_{p\in\mathfrak{P}(\mathbf{X})}\min_{t\in T}\chi(p,W_t) \tag{108}$$

Proof. For an arbitrary number $\delta>0$, we show, that

$$\max_{p\in\mathfrak{P}(\mathbf{X})}\min_{t\in T}\chi(p,W_t)-\delta \tag{109}$$

is an achievable rate. Let $\{\mathcal{C}_l\}_{l\in\mathbb{N}}$, $\mathcal{C}_l:=(u_m^l,D_m^l)_{m=1}^{M_l}\forall l\in\mathbb{N}$, be a sequence of (l,M_l)-codes with

$$\liminf_{l\to\infty}\frac{1}{l}\log M_l\geq \max_{p\in\mathfrak{P}(\mathbf{X})}\min_{t\in T}\chi(p,W_t)-\frac{1}{\delta}. \tag{110}$$

and

$$\max_{t\in T}\frac{1}{M_l}\sum_{m=1}^{M_l}\operatorname{tr}\left(W_t^{\otimes l}(u_m^l)(\mathbb{1}_{\mathcal{H}^{\otimes l}}-D_m^l)\right)\leq\lambda_l \tag{111}$$

for every $l \in \mathbb{N}$, where $\lim_{l\to\infty} \lambda_l = 0$. Such codes exist by virtue of Theorem 1. Because of Lemma 7, we find for each $l \in \mathbb{N}$ a subcode $\tilde{\mathcal{C}}_l := (u^l_{m_i}, D^l_{m_i})_{i\in[\tilde{M}_l]} \subseteq \mathcal{C}_l$ of size $\widetilde{M}_l := \lfloor \frac{\epsilon_l}{1-\epsilon_l} M_l \rfloor$ and maximal error

$$\max_{t\in T} \max_{i\in[\widetilde{M}_l]} \operatorname{tr}\left(W_t^{\otimes l}(u^l_{m_i})(\mathbb{1}_{\mathcal{H}^{\otimes l}} - D^l_{m_i})\right) \leq (\lambda_l + \epsilon_l)|T|. \tag{112}$$

with the sequence $(\epsilon_l)_{l=1}^{\infty}$ defined by $\epsilon_l := 2^{-l\frac{\delta}{3}}$ f.a. $l \in \mathbb{N}$, it is clear that we find a sequence of $(l, \tilde{M}_l)$-subcodes $\{\tilde{\mathcal{C}}_l\}_{l\in\mathbb{N}}$, where $\widetilde{\mathcal{C}}_l := (u^l_{m_i}, D^l_{m_i})_{i=1}^{\widetilde{M}_l}$ f.a. $l \in \mathbb{N}$, which fulfills

$$\lim_{l\to\infty} \max_{t\in T} \max_{i\in[\widetilde{M}_l]} \operatorname{tr}\left(W_t^{\otimes l}(u^l_{m_i})(\mathbb{1}_{\mathcal{H}^{\otimes l}} - D^l_{m_i})\right) = 0 \tag{113}$$

and

$$\liminf_{l\to\infty} \frac{1}{l} \log \tilde{M}_l = \liminf_{l\to\infty} \frac{1}{l} \log M_l \geq \max_{p\in\mathfrak{P}(\mathbf{X})} \min_{t\in T} \chi(p, W_t) - \delta. \tag{114}$$

□

Remark 5. *The above corollary, although proven here for finite sets, can be extended to arbitrary compound sets by approximation arguments, as carried out in [9]. Moreover, an inspection of the proofs in this section shows that the speed of convergence of the errors remains exponential.*

6 AVCQC

6.1 The Ahlswede-Dichotomy for AVcqCs

In this section, we prove Theorem 2 and Theorem 3. The proof of Theorem 2 is carried out via robustification of codes for a suitably chosen compound cq-channel. More specifically, to a given AVcqC $\mathcal{A}$ we take a sequence of codes for the compound channel $\mathcal{W} := \operatorname{conv}(\mathcal{A})$ that operates close to the capacity of $\mathcal{W}$. Thanks to Theorem 1, we know that there exist codes for $\mathcal{W}$ that, additionally, have an exponentially fast decrease of average error probability. The robustification technique then produces a sequence of random codes for $\mathcal{A}$ that have a discrete, but super-exponentially large support and, again, an exponentially fast decrease of average error probability.

An intermediate result here is the (tight) lower bound on $\overline{C}_{A,r}(\mathcal{A})$.

A variant of the elimination technique of [4] is proven that is adapted to AVcqCs and reduces the amount of randomness from super-exponential to polynomial, while slowing down the speed of convergence of the average error probability from exponential to polynomial at the same time.

Then, under the assumption that $C_{A,d}(\mathcal{A}) > 0$ holds, the sender can send the required amount of subexponentially many messages in order to establish sufficiently much common randomness. After that, sender and receiver simply use the random code for $\mathcal{A}$.

We now start out on our predescribed way. The following Theorem 5 and Lemma 8 will be put to good use, but are far from being new so we simply state them without proof.

Let, for each $l \in \mathbb{N}$, Perm_l denote the set of permutations acting on $\{1, \dots, l\}$. Let us further suppose that we are given a finite set $\mathbf{S}$. We use the natural action of Perm_l on $\mathbf{S}^l$ given by $\sigma : \mathbf{S}^l \to \mathbf{S}^l$, $\sigma(s^l)_i := s_{\sigma^{-1}(i)}$.

Let $T(l, \mathbf{S})$ denote the set of types on $\mathbf{S}$ induced by the elements of $\mathbf{S}^l$, i.e. the set of empirical distributions on $\mathbf{S}$ generated by sequences in $\mathbf{S}^l$. Then Ahlswede's robustification can be stated as follows.

Theorem 5 (Robustification Technique, cf. Theorem 6 in [5]) *Let* $\mathbf{S}$ *be a set with* $|\mathbf{S}| < \infty$ *and* $l \in \mathbb{N}$. *If a function* $f : \mathbf{S}^l \to [0, 1]$ *satisfies*

$$\sum_{s^l \in \mathbf{S}^l} f(s^l) q(s_1) \cdot \ldots \cdot q(s_l) \geq 1 - \gamma \tag{115}$$

for all $q \in T(l, \mathbf{S})$ *and some* $\gamma \in [0, 1]$, *then*

$$\frac{1}{l!} \sum_{\sigma \in \mathrm{Perm}_l} f(\sigma(s^l)) \geq 1 - (l+1)^{|\mathbf{S}|} \cdot \gamma \qquad \forall s^l \in \mathbf{S}^l. \tag{116}$$

The original theorem can, together with its proof, be found in [5]. A proof of Theorem 5 can be found in [6]. The following Lemma is borrowed from [4].

Lemma 8. *Let* $K \in \mathbb{N}$ *and real numbers* $a_1, \dots, a_K, b_1, \dots, b_K \in [0, 1]$ *be given. Assume that*

$$\frac{1}{K} \sum_{i=1}^{K} a_i \geq 1 - \varepsilon \qquad \text{and} \qquad \frac{1}{K} \sum_{i=1}^{K} b_i \geq 1 - \varepsilon, \tag{117}$$

hold. Then

$$\frac{1}{K} \sum_{i=1}^{K} a_i b_i \geq 1 - 2\varepsilon. \tag{118}$$

We now come to the promised application of the robustification technique to AVcqCs.

Lemma 9. *Let* $\mathcal{A} = \{A_s\}_{s \in \mathbf{S}}$ *be an AVcqC. For every* $\eta > 0$ *there is a sequence of* (l, M_l)*-codes for the compound channel* $\mathcal{W} := \mathrm{conv}(\mathcal{A})$ *and an* $l_0 \in \mathbb{N}$ *such that the following two statements are true.*

$$\liminf_{l \to \infty} \frac{1}{l} \log M_l \geq \overline{C}_{\mathrm{C}}(\mathcal{W}) - \eta \tag{119}$$

$$\min_{s^l \in \mathbf{S}^l} \frac{1}{l!} \sum_{\sigma \in \mathrm{Perm}_l} \frac{1}{M_l} \sum_{i=1}^{M_l} \mathrm{tr}(A_{s^l}(\sigma^{-1}(x_i^l)) \sigma^{-1}(D_i^l)) \geq 1 - (l+1)^{|\mathbf{S}|} \cdot 2^{-lc} \qquad \forall l \geq l_0 \tag{120}$$

with a positive number $c = c(|\mathbf{X}|, \dim \mathcal{H}, \mathcal{A}, \eta)$.

Remark 6. *The above result can be gained for arbitrary, non-finite sets* $\mathbf{S}$ *as well. A central idea then is the approximation of* $\mathrm{conv}(\mathcal{A})$ *from the outside by a convex polytope. Since* $CQ(\mathbf{X},\mathcal{H})$ *is not a polytope itself (except for trivial cases), an additional step consists of applying a depolarizing channel* $\mathcal{N}_p$ *and approximate* $\mathcal{N}_p(\mathrm{conv}(\mathcal{A}))$*, a set which does not touch the boundary of* $CQ(\mathbf{X},\mathcal{H})$*, instead of* $\mathrm{conv}(\mathcal{A})$*.*

This step can then be absorbed into the measurement operators, i.e. one uses operators $\mathcal{N}_p^*(D_i^l)$ *instead of the original* D_i^l $(i=1,\ldots,M_l)$*.*

A thorough application of this idea can be found in [6], where the robustification technique gets applied in the case of entanglement transmission over arbitrarily varying quantum channels.

Proof. According to Lemma 1 there is a sequence of (l,M_l) codes for the compound channel $\mathrm{conv}(\mathcal{A})=\{W_q : W_q=\sum_{s\in\mathbf{S}} q(s)A_s,\ q\in\mathfrak{P}(\mathbf{S})\}$ fulfilling

$$\liminf_{l\to\infty}\frac{1}{l}\log M_l \geq \overline{C}_{\mathrm{C}}(\mathrm{conv}(\mathcal{A}))-\eta \tag{121}$$

and

$$\exists l_0\in\mathbb{N}:\ \inf_{W\in\mathrm{conv}(\mathcal{A})}\frac{1}{M_l}\sum_{i=1}^{M_l}\mathrm{tr}(W^{\otimes l}(x_i^l)D_i^l)\geq 1-2^{-lc}\ \forall l\geq l_0. \tag{122}$$

The idea is to apply Theorem 5. Let us, for the moment, fix an $\mathbb{N}\ni l\geq l_0$ and define a function $f_l:\mathbf{S}^l\to[0,1]$ by

$$f_l(s^l):=\frac{1}{M_l}\sum_{i=1}^{M_l}\mathrm{tr}(A_{s^l}(x_i^l)D_i^l). \tag{123}$$

Then for every $q\in\mathfrak{P}(\mathbf{S})$ we have

$$\sum_{s^l\in\mathbf{S}^l} f_l(s^l)\prod_{i=1}^{l}q(s_i)=\frac{1}{M_l}\sum_{i=1}^{M_l}\mathrm{tr}(W_q^{\otimes l}(x_i^l)D_i^l)\geq 1-2^{-lc}. \tag{124}$$

It follows from Theorem 5, that

$$1-(l+1)^{|\mathbf{S}|}\cdot 2^{-lc}\leq\frac{1}{l!}\sum_{\sigma\in\mathrm{Perm}_l} f_l(\sigma(s^l)) \tag{125}$$

$$=\frac{1}{l!}\sum_{\sigma\in\mathrm{Perm}_l}\frac{1}{M_l}\sum_{i=1}^{M_l}\mathrm{tr}(A_{s^l}(\sigma^{-1}(x_i^l))\sigma^{-1}(D_i^l))\qquad\forall s^l\in\mathbf{S}^l \tag{126}$$

holds, where

$$\sigma(B_1\otimes\ldots\otimes B_l):=B_{\sigma^{-1}(1)}\otimes\ldots\otimes B_{\sigma^{-1}(l)}\qquad\forall\ B_1,\ldots,B_l\in\mathcal{B}(\mathcal{H}) \tag{127}$$

defines, by linear extension, the usual representation of Perm_l on $\mathcal{B}(\mathcal{H})^{\otimes l}$ and the action of Perm_l on $\mathbf{X}^l$ is analogous to that on $\mathbf{S}^l$. □

It is easily seen from the above Lemma 9 and Theorem 1, that the following theorem holds.

Theorem 6. *For every AVcqC $\mathcal{A}$,*

$$\overline{C}_{A,r}(\mathcal{A}) \geq \overline{C}_C(\mathrm{conv}(\mathcal{A})) = \max_{p\in\mathfrak{P}(\mathbf{X})} \inf_{A\in\mathrm{conv}(\mathcal{A})} \chi(p, A). \tag{128}$$

In the following we give a proof of the remaining inequality in (19). In fact, we prove the stronger statement Theorem 3:

Proof (of Theorem 3:). We define $\mathcal{W} := \mathrm{conv}(\mathcal{A})$. Since $|\mathbf{S}|$ is finite, this set is compact. The function $\chi(\cdot,\cdot)$ is a concave-convex function (see eq. (4)), therefore by the Minimax Theorem,

$$\max_{p\in\mathfrak{P}(\mathbf{X})} \min_{W\in\mathcal{W}} \chi(p, W) = \min_{W\in\mathcal{W}} \max_{p\in\mathfrak{P}(\mathbf{X})} \chi(p, W) \tag{129}$$

holds. Both sides of the equality are well defined, because we are dealing with a compact set. Let an arbitrary $W_q \in \mathcal{W}$ be given by the formula

$$W_q = \sum_{s\in\mathbf{S}} q(s)A_s, \tag{130}$$

where $q \in \mathfrak{P}(\mathbf{X})$. Set, for every $l \in \mathbb{N}$, $q^{\otimes l}(s^l) := \prod_{i=1}^{l} q(s_i)$. Let $\lambda \in [0, 1)$, $\delta > 0$ and $(\mu_l)_{l\in\mathbb{N}}$ be a sequence of (l, M_l)-random codes such that both

$$\liminf_{l\to\infty} \frac{1}{l} \log M_l = \overline{C}_{A,r}(\mathcal{A}, \lambda) - \delta \tag{131}$$

and

$$\liminf_{l\to\infty} \min_{s^l\in\mathbf{S}^l} \frac{1}{M_l} \sum_{i=1}^{M_l} \mathrm{tr}(A_{s^l}(u_i^l)D_i^l) d\mu_l((u_i^l, D_i^l)_{i=1}^{M_l}) \geq 1 - \lambda. \tag{132}$$

For every $l \in \mathbb{N}$ it holds that

$$\int \sum_{i=1}^{M_l} \mathrm{tr}(W_q^{\otimes l}(u_i^l)D_i^l)\ d\mu_l((u_i^l, D_i^l)_{i=1}^{M_l}) \tag{133}$$

$$= \sum_{s^l\in\mathbf{S}^l} q^{\otimes l}(s^l) \int \sum_{i=1}^{M_l} \mathrm{tr}(A_{s^l}(u_i^l)D_i^l)\ d\mu_l((u_i^l, D_i^l)_{i=1}^{M_l}) \tag{134}$$

$$\geq \min_{s^l\in\mathbf{S}^l} \int \sum_{i=1}^{M_l} \mathrm{tr}(A_{s^l}(u_i^l)D_i^l)\ d\mu_l((u_i^l, D_i^l)_{i=1}^{M_l}), \tag{135}$$

which shows, that

$$\liminf_{l\to\infty} \int \frac{1}{M_l} \sum_{i=1}^{M_l} \mathrm{tr}(W_q^{\otimes l}(u_i^l)\mathcal{D}_i^l)\ d\mu_l((u_i^l, D_i^l)_{i=1}^{M_l}) \geq 1 - \lambda \tag{136}$$

holds. It follows the existence of a sequence $(u_i^l, D_i^l)_{l\in\mathbb{N}}$ of (l, M_l)-codes for the discrete memoryless cq-channel W_q satisfying

$$\liminf_{l\to\infty}\frac{1}{l}\log M_l = \overline{C}_{A,d}(\mathcal{A},\lambda) - \delta \qquad \text{and} \tag{137}$$

$$\liminf_{l\to\infty}\frac{1}{M_l}\sum_{i=1}^{M_l}\mathrm{tr}(W_q^{\otimes l}(u_i^l)D_i^l) \geq 1-\lambda. \tag{138}$$

By virtue of the strong converse theorem for single cq-DMCs given in [29] (also to be found and independently obtained in [26]), for any $\lambda \in [0,1)$, $\delta > 0$ it follows

$$\overline{C}_{A,r}(\mathcal{A},\lambda) - \delta = \liminf_{l\to\infty}\frac{1}{l}\log M_l \tag{139}$$

$$\leq \max_{p\in\mathfrak{P}(\mathbf{X})}(p, W_q) \tag{140}$$

and, since $W_q \in \mathcal{W}$ was arbitrary,

$$\overline{C}_{A,r}(\mathcal{A},\lambda) - \delta \leq \min_{W\in\mathcal{W}}\max_{p\in\mathfrak{P}(\mathbf{X})}\chi(p,W) \tag{141}$$

$$= \max_{p\in\mathfrak{P}(\mathbf{X})}\min_{W\in\mathcal{W}}\chi(p,W).. \tag{142}$$

The equality in (142) holds by (129). Since δ was an arbitrary positive number, we are done. □

The following lemma contains the essence of the derandomization procedure.

Lemma 10 (Random Code Reduction). *Let $\mathcal{A} = \{A_s\}_{s\in\mathbf{S}}$ be an AVcqC, $l \in \mathbb{N}$, μ_l an (l, M_l) random code for $\mathcal{A}$ and $1 > \varepsilon_l \geq 0$ with*

$$e(\mu_l,\mathcal{A}) := \inf_{s^l\in\mathbf{S}^l}\int\frac{1}{M_l}\sum_{i=1}^{M_l}\mathrm{tr}(A_{s^l}(x_i^l)D_i^l)d\mu_l((x_i^l,D_i^l)_{i=1}^{M_l}) \geq 1-\varepsilon_l. \tag{143}$$

Let $n, m \in \mathbb{R}$. Then if $4\varepsilon_l \leq l^{-m}$ and $2\log|\mathbf{S}| < l^{n-m-1}$ there exist l^n (l, M_l)-deterministic codes $(x_{1,j}^l,\ldots,x_{M_l,j}^l,D_{1,j}^l,\ldots,D_{M_l,j}^l)$ $(1 \leq j \leq l^n)$ for $\mathcal{A}$ such that

$$\frac{1}{l^n}\sum_{j=1}^{l^n}\frac{1}{M_l}\sum_{i=1}^{M_l}\mathrm{tr}(A_{s^l}(x_{i,j}^l)D_{i,j}^l) \geq 1-l^{-m} \qquad \forall s^l\in\mathbf{S}^l. \tag{144}$$

Proof. Set $\varepsilon := 2l^{-m}$. By the assumptions of the lemma we have

$$e(\mu_l,\mathcal{A}) := \min_{s^l\in\mathbf{S}^l}\int\frac{1}{M_l}\sum_{i=1}^{M_l}\mathrm{tr}(A_{s^l}(x_i^l)D_i^l)d\mu_l((x_i^l,D_i^l)_{i=1}^{M_l}) \geq 1-\varepsilon_l. \tag{145}$$

For a fixed $K \in \mathbb{N}$, consider K independent random variables Λ_i with values in $((\mathbf{X}^l)^{M_l}) \times \mathcal{M}_{M_l}(\mathcal{H}^{\otimes l}))$ which are distributed according to μ_l.

Define, for each $s^l \in \mathbf{S}^l$, the function $p_{s^l} : ((\mathbf{X}^l)^{M_l}) \times \mathcal{M}_{M_l}(\mathcal{H}^{\otimes l})) \to [0,1]$, $(x_1^l, \ldots, x_{M_l}^l, D_1^l, \ldots, D_{M_l}^l) \mapsto \frac{1}{M_l}\sum_{i=1}^{M_l} \mathrm{tr}(A_{s^l}(x_i^n)D_i^l)$.

We get, by application of Markovs inequality, for every $s^l \in \mathbf{S}^l$:

$$\mathbb{P}(1 - \frac{1}{K}\sum_{j=1}^{K} p_{s^l}(\Lambda_j) \geq \varepsilon/2) = \mathbb{P}(2^{K-\sum_{j=1}^{K} p_{s^l}(\Lambda_j)} \geq 2^{K\varepsilon/2}) \tag{146}$$

$$\leq 2^{-K\varepsilon/2}\mathbb{E}(2^{(K-\sum_{j=1}^{K} p_{s^l}(\Lambda_j))}). \tag{147}$$

The Λ_i are independent and it holds $2^t \leq 1 + t$ for every $t \in [0,1]$ as well as $\log(1+\varepsilon_l) \leq 2\varepsilon_l$ and so we get

$$\mathbb{P}(1 - \frac{1}{K}\sum_{j=1}^{K} p_{s^l}(\Lambda_j) \geq \varepsilon/2) \leq 2^{-K\varepsilon/2}\mathbb{E}(2^{K-\sum_{j=1}^{K} p_{s^l}(\Lambda_j)}) \tag{148}$$

$$= 2^{-K\varepsilon/2}\mathbb{E}(2^{\sum_{j=1}^{K}(1-p_{s^l}(\Lambda_j))}) \tag{149}$$

$$= 2^{-K\varepsilon/2}\mathbb{E}(2^{(1-p_{s^l}(\Lambda_1))})^K \tag{150}$$

$$\leq 2^{-K\varepsilon/2}\mathbb{E}(1 + (1 - p_{s^l}(\Lambda_1)))^K \tag{151}$$

$$\leq 2^{-K\varepsilon/2}(1+\varepsilon_l)^K \tag{152}$$

$$\leq 2^{-K\varepsilon/2}2^{K\varepsilon/4} \tag{153}$$

$$= 2^{-K\varepsilon/4}. \tag{154}$$

Therefore,

$$\mathbb{P}(\frac{1}{K}\sum_{j=1}^{K} p_{s^l}(\Lambda_j) \geq 1 - \varepsilon/2) \geq 1 - |\mathbf{S}|^l 2^{-K\varepsilon/4}. \tag{155}$$

By assumption, $2\log|\mathbf{S}| \leq l^{(n-m-1)}$ and thus the above probability is larger than zero, so there exists a realization $\Lambda_1, \ldots, \Lambda_{l^n}$ such that

$$\frac{1}{l^n}\sum_{i=1}^{l^n}\frac{1}{M_l}\mathrm{tr}(W_{s^l}(x_i^l)D_i^l) \geq 1 - \frac{1}{l^m}. \tag{156}$$

□

Now we pass to the proof of Theorem 2. If $C_{A,r}(\mathcal{A}) = 0$ or $C_{A,d}(\mathcal{A}) = 0$ there is nothing to prove. So, let $\overline{C}_{A,r}(\mathcal{A}) > 0$ and $\overline{C}_{A,d}(\mathcal{A}) > 0$. Then we know that, to every $l \in \mathbb{N}$, there exists a deterministic code for $\mathcal{A}$ that, for sake of simplicity, is denoted by $(x_1^l, \ldots, x_{l^2}^l, D_1, \ldots, D_{l^2})$, such that

$$\min_{s^l\in\mathbf{S}^l}\frac{1}{l^2}\sum_{i=1}^{l^2}\mathrm{tr}(A_{s^l}(x_i^l)D_i^l) \geq 1 - \varepsilon_l \tag{157}$$

and $\varepsilon_l \searrow 0$. Also, by Lemma 9, to every $\varepsilon > 0$ there is a sequence $(\mu_m)_{m\in\mathbb{N}}$ of random codes for transmission of messages over $\mathcal{A}$ using the average error probability criterion and an $m_0 \in \mathbb{N}$ such that

$$\liminf_{m\to\infty} \frac{1}{m} \log M_m \geq \overline{C}_{A,r}(\mathcal{A}) - \varepsilon \tag{158}$$

$$\int \frac{1}{M_m} \sum_{j=1}^{M_m} \operatorname{tr}(A_{s^m}(x_j^m) D_j^l) d\mu_m((x_1^m, \dots, x_{M_m}^m, D_1^l, \dots, D_{M_m}^l)) \geq 1 - 2^{-mc} \tag{159}$$

for all $m \geq m_0$ with a suitably chosen (and possibly very small) $c > 0$. This enables us to define the following sequence of codes: Out of the random code, by application of Lemma 10 and for a suitably chosen $m_1 \geq m_0$ such that the preliminaries of Lemma 10 are fulfilled, we get for every $m \geq m_1$ a discrete random code supported only on the set $\{(y_{1,j}^m, \dots, y_{M_m,j}^m, E_{1,j}, \dots, E_{M_m,j})\}_{j=1}^{m^2}$ such that

$$\liminf_{m\to\infty} \frac{1}{l} \log M_m \geq \overline{C}_{A,r}(\mathcal{A}) - \varepsilon \tag{160}$$

$$\frac{1}{m^2} \sum_{j=1}^{m^2} \frac{1}{M_m} \sum_{i=1}^{M_m} \operatorname{tr}(A_{s^m}(y_{i,j}^m) E_{i,j}) \geq 1 - \frac{1}{m} \qquad \forall m \geq m_1. \tag{161}$$

Now all we have to do is combine the two codes: For $l, m \in \mathbb{N}$, define an $(l + m, \frac{1}{l^2 M_m})$-deterministic code with the doubly-indexed message set $\{i, j\}_{i=1,j=1}^{l^2, M_m}$ by the following sequence:

$$((x_i^l, y_{ij}^m), D_i^l \otimes E_{ij})_{i=1,j=1}^{l^2, M_m}. \tag{162}$$

For the average success probability, by Lemma 8 it then holds

$$\min_{(s^l, s^m) \in \mathbf{S}^{l+m}} \frac{1}{l^2 M_m} \sum_{i=1}^{l^2} \sum_{j=1}^{M_m} \operatorname{tr}(A_{(s^l, s^m)}((x_i^l, y_{ij}^m)) D_i^l \otimes E_{ij}) \geq 1 - 2 \max\{\varepsilon_l, \frac{1}{m}\}. \tag{163}$$

Now let there be sequences $(l_t)_{t\in\mathbb{N}}$ and $(m_t)_{t\in\mathbb{N}}$ such that $l_t = o(l)$ and $l_t + m_t = t$ f.a. $t \in \mathbb{N}$. Define a sequence of $(t, \frac{1}{l_t^2 M_{m_t}})$-deterministic codes $(\hat{x}_1^t, \dots, \hat{x}_{l_t^2 M_{m_t}}^t, \hat{D}_1, \dots, \hat{D}_{l_t^2, M_{m_t}})$ for $\mathcal{A}$ by applying, for each $t \in \mathbb{N}$, the above described procedure with $m = m_t$ and $l = l_t$. Then

$$\liminf_{t\to\infty} \frac{1}{t} \log l_t^2 M_{m_t} \geq R \qquad \text{and} \tag{164}$$

$$\lim_{t\to\infty} \min_{s^t \in \mathbf{S}^t} \frac{1}{l_t^2 M_{m_t}} \sum_{k=1}^{l_t^2 M_{m_t}} \operatorname{tr}(A_{s^t}(\hat{x}_k^t) \hat{D}_k) = 1. \tag{165}$$

6.2 M-Symmetrizability

In this section, we prove Theorem 4.

Proof. We adapt the strategy of [22], that has already been successfully used in [6]. Assume $\mathcal{A}$ is m-symmetrizable. Let $l \in \mathbb{N}$. Take any $a^l, b^l \in \mathbf{X}^l$. Then there exist corresponding probability distributions $p(\cdot|a_1), \ldots, p(\cdot|a_l), p(\cdot|b_1), \ldots, p(\cdot|b_l) \in \mathfrak{P}(\mathbf{S})$ such that the probability distributions $p(\cdot|a^l), p(\cdot|b^l) \in \mathfrak{P}(\mathbf{S}^l)$ defined by $p(s^l|a^l) := \prod_{i=1}^l p(s_i|a_i)$, $p(s^l|b^l) := \prod_{i=1}^l p(s_i|b_i)$ satisfy

$$\sum_{s^l \in \mathbf{S}^l} p(s^l|a^l) A_{s^l}(a^l) = \sum_{s^l \in \mathbf{S}^l} p(s^l|b^l) A_{s^l}(b^l) \tag{166}$$

and thereby lead, for every two measurement operators $D_a, D_b \geq 0$ satisfying $D_a + D_b \leq \mathbb{1}_{\mathcal{H}^{\otimes l}}$, to the following inequality:

$$\sum_{s^l \in \mathbf{S}^l} p(s^l|a^l) \mathrm{tr}(A_{s^l}(a^l) D_a) = \sum_{s^l \in \mathbf{S}^l} p(s^l|b^l) \mathrm{tr}(A_{s^l}(b^l) D_a) \tag{167}$$

$$\leq \sum_{s^l \in \mathbf{S}^l} p(s^l|b^l) \mathrm{tr}(A_{s^l}(b^l)(\mathbf{1}_{\mathcal{H}^{\otimes l}} - D_b)) \tag{168}$$

$$= 1 - \sum_{s^l \in \mathbf{S}^l} p(s^l|b^l) \mathrm{tr}(A_{s^l}(b^l) D_b). \tag{169}$$

Let a sequence of (l, M_l) codes for message transmission over $\mathcal{A}$ using the maximal error probability criterion satisfying $M_l \geq 2$ and $\min_{i \in [M_l]} \min_{s^l \in \mathbf{S}^l} \mathrm{tr}(A_{s^l}(x_i^l) D_i^l) = 1 - \varepsilon_l$ be given, where $\varepsilon_l \searrow 0$. Then from the above inequality we get

$$1 - \varepsilon_l \leq 1 - (1 - \varepsilon_l) \qquad \Leftrightarrow \qquad \varepsilon_l \geq 1/2. \tag{170}$$

Therefore, $C_{A,d}(\mathcal{A}) = 0$ has to hold.

Now, assume that $\mathcal{A}$ is not m-symmetrizable. Then there are $x, y \in \mathbf{X}$ such that

$$\mathrm{conv}(\{\mathcal{A}_s(x)\}_{s \in \mathbf{S}}) \cap \mathrm{conv}(\{\mathcal{A}_s(y)\}_{s \in \mathbf{S}}) = \varnothing. \tag{171}$$

The rest of the proof is identical to that in [6] with $\hat{l}$ set to one. □

6.3 Relation to the Zero-Error Capacity

A remarkable feature of classical arbitrarily varying channels is their connection to the zero-error capacity of (classical) d.m.c.s, which was established by Ahlswede in [3, Theorem 3].

We shall first give a reformulation of Ahlswede's original result and then consider two straightforward generalizations of it result, one for cq-channels, the other for quantum channels. In both cases it is shown, that no such straightforward generalization is possible.

Ahlswede's Original Result. Ahlswede's result can be formulated using the following notation. For two finite sets $\mathbf{A}, \mathbf{B}$, $C(\mathbf{A}, \mathbf{B})$ stands for the set of channels from $\mathbf{A}$ to $\mathbf{B}$, i.e. each element of $W \in C(\mathbf{A}, \mathbf{B})$ defines a set of output probability distributions $\{W(\cdot|a)\}_{a\in\mathbf{A}}$. With slight abuse of notation, for each $D \subset \mathbf{B}$ and $a \in \mathbf{A}$, $W(D|a) := \sum_{b\in D} W(b|a)$. The (finite) set of extremal points of the (convex) set $C(\mathbf{A}, \mathbf{B})$ will be written $E(\mathbf{A}, \mathbf{B})$.

For two channels $W_1, W_2 \in C(\mathbf{A}, \mathbf{B})$, their product $W_1 \otimes W_2 \in C(\mathbf{A}^2, \mathbf{B}^2)$ is defined through $W_1 \otimes W_2(b^2|a^2) := W_1(b_1|a_1)W_2(b_2|a_2)$. An arbitrarily varying channel (AVC) is, in this setting, defined through a set $\mathbb{W} = \{W_s\}_{s\in\mathbf{S}} \subset C(\mathbf{A}, \mathbf{B})$ (we assume $\mathbf{S}$ and, hence, $|\mathbb{W}|$, to be finite). The different realizations of the channel are written

$$W_{s^l} := W_{s_1} \otimes \ldots \otimes W_{s_l} \qquad (s^l \in \mathbf{S}^l) \tag{172}$$

and, formally, the AVC $\mathbb{W}$ consists of the set $\{W_{s^l}\}_{s^l\in\mathbf{S}^l,\ l\in\mathbb{N}}$.

An (l, M_l)-code for the AVC $\mathbb{W}$ is given by a set $\{a_i^l\}_{i=1}^{M_l} \subset \mathbf{A}^l$ called the 'codewords' and a set $\{D_i^l\}_{i=1}^{M_l}$ of subsets of $\mathbf{B}^l$ called the 'decoding sets', that satisfies $D_i^l \cap D_j^l = \varnothing$, $i \neq j$.

A nonnegative number $R \in \mathbb{R}$ is called an achievable maximal-error rate for the AVC $\mathbb{W}$, if there exists a sequence of (l, M_l) codes for $\mathbb{W}$ such that both

$$\liminf_{l\to\infty} \frac{1}{l} \log M_l \geq R \qquad \text{and} \qquad \lim_{l\to\infty} \min_{s^l\in\mathbf{S}^l} \min_{1\leq i\leq M_l} W_{s^l}(D_i^l|x_i^l) = 1. \tag{173}$$

The (deterministic) maximal error capacity $C_{\max}(\mathbb{W})$ of the AVC $\mathbb{W}$ is, as usually, defined as the supremum over all achievable maximal-error rates for $\mathbb{W}$.

Much stronger requirements concerning the quality of codes can be made. An (l, M_l)-code is said to have zero error for the AVC $\mathbb{W}$, if for all $1 \leq i \leq M_l$ and $s^l \in \mathbf{S}^l$ the equality $W_{s^l}(D_i^l|x_i^l) = 1$ holds.

The zero error capacity $C_0(\mathbb{W})$ of the AVC $\mathbb{W}$ is defined as

$$C_0(\mathbb{W}) := \lim_{l\to\infty} \max\{\frac{1}{l} \log M_l : \exists\ (l, M_l)\text{−code with zero error for } \mathbb{W}\}. \tag{174}$$

The above definitions carry over to single channels $W \in C(\mathbf{A}, \mathbf{B})$ by identifying W with the set $\{W\}$.

In short form, the connection [3, Theorem 3] between the capacity of certain arbitrarily varying channels and the zero-error capacity of stationary memoryless channels can now be reformulated as follows:

Theorem 7. *Let $W \in C(\mathbf{A}, \mathbf{B})$ have a decomposition $W = \sum_{s\in\mathbf{S}} q(s)W_s$, where $\{W_s\}_{s\in\mathbf{S}} \subset E(\mathbf{A}, \mathbf{B})$ and $q(s) > 0\ \forall s \in \mathbf{S}$. Then for the AVC $\mathbb{W} := \{W_s\}_{s\in\mathbf{S}}$:*

$$C_0(W) = C_{\max}(\mathbb{W}). \tag{175}$$

Conversely, for every AVC $\mathbb{W} = \{W_s\}_{s\in\mathbf{S}} \subset E(\mathbf{A}, \mathbf{B})$ and every $q \in \mathfrak{P}(\mathbf{S})$ with $q(s) > 0\ \forall s \in \mathbf{S}$, equation (175) holds for the channel $W := \sum_{s\in\mathbf{S}} q(s)W_s$.

Remark 7. *Let us note at this point, that the original formulation of the theorem did not make reference to extremal points of the set of channels, but rather used the equivalent notion "channels of* $0-1$*-type".*

Remark 8. *By choosing* $W \in E(\mathbf{A},\mathbf{B})$*, one gets the equality* $C_0(W) = C_{max}(W)$*. The quantity* $C_{max}(W)$ *being well-known and easily computable, it may seem that Theorem 7 solves Shannons's zero-error problem. This is not the case, as one can verify by looking at the famous pentagon channel that was introduced in [27, Figure 2.]. The pentagon channel is far from being extremal. That its zero-error capacity is positive [27] is due to the fact that it is not a member of the relative interior* $riE(\mathbf{A},\mathbf{B})$.

Recently, in [6], this connection was investigated with a focus on entanglement and strong subspace transmission over arbitrarily varying quantum channels. The complete problem was left open, although partial results were obtained.

A No-Go Result for cq-Channels. We will show below that, even for message transmission over AVcqCs, there is (in general) no equality between the capacity $C_0(W)$ of a channel $W \in CQ(\mathbf{X},\mathcal{H})$ and any AVcqC $\mathcal{A} = \{A_s\}_{s\in\mathbf{S}}$ constructed by choosing the set $\{A_s\}_{s\in\mathbf{S}}$ to be a subset of the set of extremal points of $CQ(\mathbf{X},\mathcal{H})$ such that

$$W = \sum_{s\in\mathbf{S}} \lambda(s) A_s \tag{176}$$

holds for a $\lambda \in \mathfrak{P}(\mathbf{S})$. Observe that the requirement that each A_s ($s \in \mathbf{S}$) be extremal in $CQ(\mathbf{X},\mathcal{H})$ is a natural analog of the decomposition into channels of $0-1$-type that is used in the second part of [3].

A first hint why the above statement is true can be gained by looking at the method of proof used in [3], especially equation (22) there. The fact that the decoding sets of a code for an arbitrarily varying channel as described in [3] have to be mutually disjoint, together with the perfect distinguishability of different non-equal outputs of the special channels that are used in the second part of this paper, is at the heart of the argumentation.

The following lemma shows why, in our case, it is impossible to make a step that is comparable to that from [3, equation (21)] to [3, equation (22)].

Lemma 11. *Let* $\mathcal{A} = \{A_s\}_{s\in\mathbf{S}}$ *be an AVcqC with* $C_{A,d}(\mathcal{A}) > 0$ *and* $0 < R < C_{A,d}(\mathcal{A})$*. To every sequence of* (l, M_l) *codes satisfying* $\liminf_{l\to\infty} \frac{1}{l}\log M_l \geq R$ *and* $\lim_{l\to\infty} \min_{i\in[M_l]} \min_{s^l\in\mathbf{S}^l} \mathrm{tr}(A_{s^l}(x_i^l)D_i^l) = 1$ *there is another sequence of* (l, M_l) *codes with modified decoding operators* $\tilde{D}_i^l$ *such that*

$$1) \qquad \liminf_{l\to\infty} \frac{1}{l}\log M_l \geq R \tag{177}$$

$$2) \qquad \lim_{l\to\infty} \min_{i\in[M_l]} \min_{s^l\in\mathbf{S}^l} \mathrm{tr}(A_{s^l}(x_i^l)\tilde{D}_i^l) = 1 \tag{178}$$

$$3) \qquad \forall\, i \in [M_l],\ l \in \mathbb{N},\ \ \mathrm{tr}(A_{s^l}(x_i^l)\tilde{D}_i^l) < 1 \tag{179}$$

Proof. Just use, for some $c > 0$, the transformation $\tilde{D}_i^l := (1 - 2^{-lc})D_i^l + 2^{-lc}\frac{1}{M_l}(\mathbb{1}_{\mathcal{H}^{\otimes l}} - D_0^l)$. □

After this preliminary statement, we give an explicit example that shows where the construction in equation (176) must fail.

Lemma 12. *Let $\mathbf{X} = \{1, 2\}$ and $\mathcal{H} = \mathbb{C}^2$. Let $\{e_1, e_2\}$ be the standard basis of $\mathcal{H}$ and $\psi_+ := \sqrt{1/2}(e_1 + e_2)$. Define $W \in CQ(\mathbf{X}, \mathcal{H})$ by $W(1) = |e_1\rangle\langle e_1|$ and $W(2) = |\psi_+\rangle\langle\psi_+|$. Then the following hold.*

1. *W is extremal in $CQ(\mathbf{X}, \mathcal{H})$*
2. *For every set $\{A_s\}_{s\in\mathbf{S}} \subset CQ(\mathbf{X}, \mathcal{H})$ and every $\lambda \in \mathfrak{P}(\mathbf{S})$ such that (176) holds, $\{A_s\}_{s\in\mathbf{S}} = \{W\}$.*
3. *$C_0(W) = 0$, but $C_{A,d}(\{W\}) > 0$.*

Proof. 1) Let, for an $x \in (0, 1)$ and $W_1, W_2 \in CQ(\mathbf{X}, \mathcal{H})$,

$$W = xW_1 + (1 - x)W_2. \tag{180}$$

Then, clearly,

$$|e_1\rangle\langle e_1| = xW_1(1) + (1 - x)W_2(1) \quad \Longrightarrow \quad W_1(1) = W_2(1) = W(1) \tag{181}$$

and

$$|\psi_+\rangle\langle\psi_+| = xW_1(2) + (1 - x)W_2(2) \quad \Longrightarrow \quad W_1(2) = W_2(2) = W(2), \tag{182}$$

so $W = W_1 = W_2$.
2) is equivalent to 1).
3) It holds $\mathrm{tr}\{W(i)W(j)\} > 1/2$ $(i, j \in \mathbf{X})$. Let $l \in \mathbb{N}$. Assume there are two codewords $a^l, b^l \in \mathbf{X}^l$ and corresponding decoding operations $C, D \geq 0$, $C + D \leq \mathbb{1}_{\mathbb{C}^2}^{\otimes l}$, such that

$$\begin{aligned} &\mathrm{tr}\{W^{\otimes l}(a^l)C\} = \mathrm{tr}\{W^{\otimes l}(b^l)D\} = 1 \\ &(\Longrightarrow \mathrm{tr}\{W^{\otimes l}(a^l)D\} = \mathrm{tr}\{W^{\otimes l}(b^l)C\} = 0). \end{aligned} \tag{183}$$

Then we may add a third operator $E := \mathbb{1}_{\mathbb{C}^2}^{\otimes l} - C - D$ and it holds that

$$\mathrm{tr}\{W^{\otimes l}(a^l)E\} = \mathrm{tr}\{W^{\otimes l}(b^l)E\} = 0. \tag{184}$$

From equations (184) and (183) we deduce the following:

$$\begin{aligned} \sqrt{E}W^{\otimes l}(a^l)\sqrt{E} &= \sqrt{E}W^{\otimes l}(b^l)\sqrt{E} \\ = \sqrt{D}W^{\otimes l}(a^l)\sqrt{D} &= \sqrt{C}W^{\otimes l}(b^l)\sqrt{C} = 0. \end{aligned} \tag{185}$$

With these preparations at hand, we are led to the following chain of inequalities:

$$
\begin{aligned}
0 &< \mathrm{tr}\{W^{\otimes l}(a^l)W^{\otimes l}(b^l)\} && (186)\\
&= \mathrm{tr}\{W^{\otimes l}(C+D+E)(a^l)W^{\otimes l}(b^l)(C+D+E)\} && (187)\\
&= \langle CW^{\otimes l}(a^l), W^{\otimes l}(b^l)C\rangle_{HS} + \langle CW^{\otimes l}(a^l), W^{\otimes l}(b^l)D\rangle_{HS}\\
&\quad + \langle CW^{\otimes l}(a^l), W^{\otimes l}(b^l)E\rangle_{HS} + \langle DW^{\otimes l}(a^l), W^{\otimes l}(b^l)C\rangle_{HS}\\
&\quad + \langle DW^{\otimes l}(a^l), W^{\otimes l}(b^l)D\rangle_{HS} + \langle DW^{\otimes l}(a^l), W^{\otimes l}(b^l)E\rangle_{HS}\\
&\quad + \langle EW^{\otimes l}(a^l), W^{\otimes l}(b^l)C\rangle_{HS} + \langle EW^{\otimes l}(a^l), W^{\otimes l}(b^l)D\rangle_{HS}\\
&\quad + \langle EW^{\otimes l}(a^l), W^{\otimes l}(b^l)E\rangle_{HS} && (188)\\
&= 0, && (189)
\end{aligned}
$$

as can be seen from a repeated application of the Cauchy-Schwarz-inequality to every single one of the above terms and use of equation (185). Thus, by contradiction, $C_0(W) = 0$ has to hold.

Now assume that the AVcqC $\{W\}$ is m-symmetrizable. This is the case only if

$$W(1) = W(2) \tag{190}$$

holds, which is clearly not the case. Thus, $C_{A,d}(\{W\}) > 0$. □

A No-Go Result for Quantum Channels. We now formulate a straightforward analogue of Theorem 7 for quantum channels. To this end, let us introduce some notation. We heavily rely on [6]. The set of completely positive and trace-preserving maps from $\mathcal{B}(\mathcal{H})$ to $\mathcal{B}(\mathcal{K})$ (where both $\mathcal{H}$ and $\mathcal{K}$ are finite-dimensional) is denoted $\mathcal{C}(\mathcal{H},\mathcal{K})$. For a Hilbert space $\mathcal{H}$, $S(\mathcal{H})$ denotes the set of vectors of unit lenght in it.

An arbitrarily varying quantum channel (AVQC) is defined by any set $\mathfrak{I} = \{\mathcal{N}_s\}_{s\in\mathbf{S}} \subset \mathcal{C}(\mathcal{H},\mathcal{K})$ and formally given by $\{\mathcal{N}_{s^l}\}_{s^l\in\mathbf{S}^l, l\in\mathbb{N}}$, where

$$\mathcal{N}_{s^l} := \mathcal{N}_{s_1} \otimes \ldots \otimes \mathcal{N}_{s_l} \qquad (s^l \in \mathbf{S}^l). \tag{191}$$

Let $\mathfrak{I} = \{\mathcal{N}_s\}_{s\in\mathbf{S}}$ be an AVQC. An $(l,k_l)-$*strong subspace transmission code* for $\mathfrak{I}$ is a pair $(\mathcal{P}^l,\mathcal{R}^l) \in \mathcal{C}(\mathcal{F}_l,\mathcal{H}^{\otimes l}) \times \mathcal{C}(\mathcal{K}^{\otimes l},\mathcal{F}_l')$, where $\mathcal{F}_l$, $\mathcal{F}_l'$ are Hilbert spaces and $\dim\mathcal{F}_l = k_l$, $\mathcal{F}_l \subset \mathcal{F}_l'$.

Definition 17. *A non-negative number R is said to be an achievable strong subspace transmission rate for the AVQC $\mathfrak{I} = \{\mathcal{N}_s\}_{s\in\mathbf{S}}$ if there is a sequence of $(l,k_l)-$strong subspace transmission codes such that*

1. $\liminf_{l\to\infty} \frac{1}{l}\log k_l \geq R$ *and*
2. $\lim_{l\to\infty} \inf_{s^l\in\mathbf{S}^l} \min_{\psi\in S(\mathcal{F}_l)} \langle\psi, \mathcal{R}^l \circ \mathcal{N}_{s^l} \circ \mathcal{P}^l(|\psi\rangle\langle\psi|)\psi\rangle = 1.$

The random strong subspace transmission capacity $\mathcal{A}_{\mathrm{s,random}}(\mathfrak{I})$ of $\mathfrak{I}$ is defined by

$$\mathcal{A}_{\mathrm{s,det}}(\mathfrak{I}) := \sup\left\{R : \begin{array}{l}R \textit{ is an achievable strong subspace}\\ \textit{transmission rate for } \mathfrak{I}\end{array}\right\}. \tag{192}$$

Self-evidently, we will also need a notion of zero-error capacity:

Definition 18. *An (l,k) zero-error quantum code (QC for short) $(\mathcal{F},\mathcal{P},\mathcal{R})$ for $\mathcal{N}\in\mathcal{C}(\mathcal{H},\mathcal{K})$ consists of a Hilbert space $\mathcal{F}$, $\mathcal{P}\in\mathcal{C}(\mathcal{F},\mathcal{H}^{\otimes l})$, $\mathcal{R}\in\mathcal{C}(\mathcal{K}^{\otimes l},\mathcal{F})$ with $\dim\mathcal{F}=k$ such that*

$$\min_{x\in\mathcal{F},||x||=1}\langle x,\mathcal{R}\circ\mathcal{N}^{\otimes l}\circ\mathcal{P}(|x\rangle\langle x|)x\rangle=1. \tag{193}$$

The zero-error quantum capacity $Q_0(\mathcal{N})$ of $\mathcal{N}\in\mathcal{C}(\mathcal{H},\mathcal{K})$ is now defined by

$$Q_0(\mathcal{N}):=\lim_{l\to\infty}\frac{1}{l}\log\max\{\dim\mathcal{F}:\exists(l,k)\text{ zero-error QC for }\mathcal{N}\}. \tag{194}$$

Conjecture 1. Let $\mathcal{N}\in\mathcal{C}(\mathcal{H},\mathcal{K})$ have a decomposition $\mathcal{N}=\sum_{s\in\mathbf{S}}q(s)\mathcal{N}_s$, where each $\mathcal{N}_s$ is extremal in $\mathcal{C}(\mathcal{H},\mathcal{K})$ and $q(s)>0\ \forall s\in\mathbf{S}$. Then for the AVQC $\mathfrak{I}:=\{\mathcal{N}_s\}_{s\in\mathbf{S}}$:

$$Q_0(\mathcal{N})=\mathcal{A}_{\mathrm{s,det}}(\mathfrak{I}). \tag{195}$$

Conversely, for every AVQC $\mathfrak{I}=\{\mathcal{N}_s\}_{s\in\mathbf{S}}$ with $\mathcal{N}_s$ being extremal for every $s\in\mathbf{S}$ and every $q\in\mathfrak{P}(\mathbf{S})$ with $q(s)>0\ \forall s\in\mathbf{S}$, equation (195) holds for the channel $\mathcal{N}:=\sum_{s\in\mathbf{S}}q(s)\mathcal{N}_s$.

Remark 9. *One could formulate weaker conjectures than the one above. A crucial property of extremal classical channels that was used in the proof of Theorem 7 was that $W_{s^l}(\cdot|x_i^l)$ is a dirac-measure for every codeword x_i^l, if only $\{W_{s^l}\}_{s\in\mathbf{S}}\subset E(\mathbf{A},\mathbf{B})$.*

This property gets lost for the extremal points of $\mathcal{C}(\mathcal{H},\mathcal{K})$ (see the channels that are used in the proof of Theorem 8), but could be regained by restriction to channels consisting of only one single Kraus operator.

This conjecture leads us to the following theorem:

Theorem 8. *Conjecture 1 is wrong.*

Remark 10. *As indicated in Remark 9, there could still be interesting connections between (for example) the deterministic strong subspace transmission capacity of AVQCs and the zero-error entanglement transmission of stationary memoryless quantum channels.*

Proof. Let $\mathcal{H}=\mathcal{K}=\mathbb{C}^2$. Let $\{e_0,e_1\}$ be the standard basis of $\mathbb{C}^2$. Consider, for a fixed but arbitrary $x\in[0,1]$ the channel $\mathcal{N}_x\in\mathcal{C}(\mathcal{H},\mathcal{K})$ defined by Kraus operators $A_1:=\sqrt{1-x^2}|e_0\rangle\langle e_1|$ and $A_2:=|e_0\rangle\langle e_0|+x|e_1\rangle\langle e_1|$. As was shown in [30], this channel is extremal in $\mathcal{C}(\mathcal{H},\mathcal{K})$. It is also readily seen from the definition of Kraus operators, that it approximates the identity channel $id_{\mathbb{C}^2}\in\mathcal{C}(\mathcal{H},\mathcal{K})$:

$$\lim_{x\to1}||\mathcal{N}_x-id_{\mathbb{C}^2}||_\diamond=0. \tag{196}$$

Now, on the one hand, $\mathcal{N}_x$ being extremal implies $\mathrm{span}(\{A_i^*A_j\}_{i,j=1}^2)=M(\mathbb{C}^2)$ for all $x\in[0,1)$ (where $M(\mathbb{C}^2)$ denotes the set of complex 2×2 matrices) by

[12, Theorem 5]. This carries over to the channels $\mathcal{N}_x^{\otimes l}$ for every $l \in \mathbb{N}$: Let the Kraus operators of $\mathcal{N}_x^{\otimes l}$ be denoted $\{A_{i^l}\}_{i^l \in \{1,2\}^l}$, then

$$\text{span}(\{A_{i^l}^* A_{j^l}\}_{i^l, j^l \in \{1,2\}^l}) = \{M : M \text{ is complex } 2^l \times 2^l\text{–matrix}\}. \quad (197)$$

On the other hand, it was observed e.g. in [17], that for two pure states $|\phi\rangle\langle\phi|, |\psi\rangle\langle\psi| \in \mathcal{S}((\mathbb{C}^2)^{\otimes l})$, the subspace spanned by them can be transmitted with zero error only if

$$|\psi\rangle\langle\phi| \perp \text{span}(\{A_{i^l}^* A_{j^l}\}_{i^l, j^l \in \{1,2\}^l}). \quad (198)$$

This is in obvious contradiction to equation (197), therefore $Q_0(\mathcal{N}_x) = 0 \ \forall x \in [0, 1)$.

On the other hand, from equation (196) and continuity of $\mathcal{A}_{\text{s,det}}(\cdot)$ in the specifying channel set ([6], though indeed only the continuity results of [23] that were also crucial in the development of corresponding statements in [6] are really needed here) we see that there is an $X \in [0, 1)$ such that for all $x \geq X$ we have $\mathcal{A}_{\text{s,det}}(\{\mathcal{N}_x\}) > 0$. Letting $x = X$ we obtain $Q_0(\mathcal{N}_X) = 0$ and $\mathcal{A}_{\text{s,det}}(\{\mathcal{N}_X\}) > 0$, so $Q_0(\mathcal{N}_X) \neq \mathcal{A}_{\text{s,det}}(\mathcal{N}_X)$ in contradiction to the statement of the conjecture. □

Acknowledgments. This work was supported by the DFG via grant BO 1734/20-1 (I.B, H.B.) and by the BMBF via grant 01BQ1050 (I.B., H.B., J.N.).

References

1. Ahlswede, R.: Certain results in coding theory for compound channels I. In: Proceedings of Colloquium on Information Theory, Debrecen (1967); J. Bolyai Math. Soc. 1, 35–60, Budapest (1968)
2. Ahlswede, R., Wolfowitz, J.: The structure of capacity functions for compound channels. In: Proc. of the Internat. Symposium on Probability and Information Theory at McMaster University, pp. 12–54 (1969)
3. Ahlswede, R.: A note on the existence of the weak capacity for channels with arbitrarily varying channel probability functions and its relation to Shannon's zero error capacity. The Annals of Mathematical Statistics 41(3) (1970)
4. Ahlswede, R.: Elimination of correlation in random codes for arbitrarily varyingchannels. Z. Wahrscheinlichkeitstheorie Verw. Gebiete 44, 159–175 (1978)
5. Ahlswede, R.: Coloring Hypergraphs: a new approach to multi-user source coding-II. Journal of Combinatorics, Information & System Sciences 5(3), 220–268 (1980)
6. Ahlswede, R., Bjelaković, I., Boche, H., Nötzel, J.: Quantum capacity under adversarial noise: arbitrarily varying quantum channels. Commun. Math. Phys. (in print), `http://arxiv.org/abs/1010.0418`
7. Ahlswede, R., Blinovsky, V.: Classical capacity of classical-quantum arbitrarily varying channels. IEEE Trans. Inf. Theory 53(2), 526–533 (2007)
8. Bhatia, R.: Matrix Analysis. Springer (1997)
9. Bjelaković, I., Boche, H.: Classical capacities of averaged and compound quantum channels. IEEE Trans. Inf. Theory 55(7), 3360–3374 (2009)
10. Bjelaković, I., Boche, H., Nötzel, J.: Quantum capacity of a class of compound channels. Phys. Rev. A 78, 042331 (2008)

11. Bjelaković, I., Boche, H., Nötzel, J.: Entanglement transmission and generation under channel uncertainty: universal quantum channel coding. Commun. Math. Phys. 292, 55–97 (2009)
12. Choi, M.-D.: Completely positive linear maps on complex matrice. Linear Algebra and Its Applications 10, 285–290 (1975)
13. Christandl, M.: The structure or bipartite quantum states - insights from group theory and cryptography, Dissertation (2006), `http://arxiv.org/abs/quant-ph/0604183v1`
14. Csiszár, I., Körner, J.: Information Theory; Coding Theorems for Discrete Memoryless Systems. Akadémiai Kiadó, Budapest/Academic Press Inc., New York (1981)
15. Datta, N., Dorlas, T.: Coding theorem for a class of quantum channels with long-term memory. J. Phys. A: Math. Gen. 40, 8147–8164 (2007)
16. Datta, N., Hsieh, M.-H.: Universal coding for transmission of private information. J. Math. Phys. 51, 122202 (2010)
17. Duan, R., Severini, S., Winter, A.: Zero-error communication via quantum channels, non-commutative graphs and a quantum Lovász θ function, arXiv:1002.2514v2
18. Fannes, M.: A continuity property of the entropy density for spin lattice systems. Comm. Math. Phys. 31, 291–294 (1973)
19. Hayashi, M.: Optimal sequence of POVMs in the sense of Stein's lemma in quantum hypothesis testing, arXiv: quant-ph/0107004 (2001)
20. Hayashi, M., Nagaoka, H.: General formulas for capacity of classical-quantum channels. IEEE Trans. Inf. Th. 49, 1753 (2003)
21. Hayashi, M.: Universal coding for classical-quantum channel. Comm. Math. Phys. 289, 1087–1098 (2009)
22. Kiefer, J., Wolfowitz, J.: Channels with arbitrarily varying channel probability functions. Information and Control 5, 44–54 (1962)
23. Leung, D., Smith, G.: Continuity of quantum channel capacities. Commun. Math. Phys. 292, 201–215 (2009)
24. Milman, V.D., Schechtman, G.: Asymptotic theory of finite dimensional normed spaces. Lecture Notes in Mathematics, vol. 1200. Springer (1986)
25. Ogawa, T., Hayashi, M.: A new proof of the direct part of Stein's lemma in quantum hypothesis testing, arXiv:quant-ph/0110125 (2001)
26. Ogawa, T., Nagaoka, H.: Strong converse to the quantum channel coding theorem. IEE Trans. Inf. Th. 45, 2486–2489 (1999)
27. Shannon, C.E.: The zero error capacity of a noisy channel. IRE Trans. Information Theory IT-2, 8–19 (1956)
28. Webster, R.: Convexity. Oxford University Press (1994)
29. Winter, A.: Coding theorem and strong converse for quantum channels. IEEE Trans. Inf. Th. 45, 2481 (1999)
30. Wolf, M.M., Cirac, J.I.: Dividing quantum channels. Commun. Math. Phys. 279, 147–168 (2008)

On the Value of Multiple Read/Write Streams for Data Compression

Travis Gagie

Department of Computer Science and Engineering, Aalto University, Finland

travis.gagie@aalto.fi

Dedicated to the memory of Rudolf Ahlswede

Abstract. We study whether, when restricted to using polylogarithmic memory and polylogarithmic passes, we can achieve qualitatively better data compression with multiple read/write streams than we can with only one. We first show how we can achieve universal compression using only one pass over one stream. We then show that one stream is not sufficient for us to achieve good grammar-based compression. Finally, we show that two streams are necessary and sufficient for us to achieve entropy-only bounds.

Keywords: data compression, online algorithms, external-memory algorithms, read/write streams.

1 Introduction

Massive datasets seem to expand to fill the space available and, in situations where they no longer fit in memory and must be stored on disk, we may need new models and algorithms. Grohe and Schweikardt [21] introduced read/write streams to model situations in which we want to process data using mainly sequential accesses to one or more disks. As the name suggests, this model is like the streaming model (see, e.g., [28]) but, as is reasonable with datasets stored on disk, it allows us to make multiple passes over the data, change them and even use multiple streams (i.e., disks). As Grohe and Schweikardt pointed out, sequential disk accesses are much faster than random accesses — potentially bypassing the von Neumann bottleneck — and using several disks in parallel can greatly reduce the amount of memory and the number of accesses needed. For example, when sorting, we need the product of the memory and accesses to be at least linear when we use one disk [27,20] but only polylogarithmic when we use two [9,21]. Similar bounds have been proven for a number of other problems, such as checking set disjointness or equality; we refer readers to Schweikardt's survey [35] of upper and lower bounds with one or more read/write streams, Heinrich and Schweikardt's paper [23] relating read/write streams to classic complexity theory, and Beame and Huynh's paper [4] on the value of multiple read/write streams for approximating frequency moments.

H. Aydinian, F. Cicalese, and C. Deppe (Eds.): Ahlswede Festschrift, LNCS 7777, pp. 284–297, 2013.

Since sorting is an important operation in some of the most powerful data compression algorithms, and compression is an important operation for reducing massive datasets to a more manageable size, we wondered whether extra streams could also help us achieve better compression. In this paper we consider the problem of compressing a string s of n characters over an alphabet of size σ when we are restricted to using $\log^{\mathcal{O}(1)} n$ bits of memory and $\log^{\mathcal{O}(1)} n$ passes over the data. Throughout, we write log to mean $\log_2$ unless otherwise stated. In Section 2, we show how we can achieve universal compression using only one pass over one stream. Our approach is to break the string into blocks and compress each block separately, similar to what is done in practice to compress large files. Although this may not usually significantly worsen the compression itself, it may stop us from then building a fast compressed index (see [29] for a survey) unless we somehow combine the indexes for the blocks, and stop us clustering by compression [11] (since concatenating files should not help us compress them better if we then break them into pieces again). In Section 3 we use a vaguely automata-theoretic argument to show one stream is not sufficient for us to achieve good grammar-based compression. Of course, by 'good' we mean here something stronger than universal compression: we want to build a context-free grammar that generates s and only s and whose size is nearly minimum. In a paper with Gawrychowski [17] we showed that with constant memory and logarithmic passes over a constant number of streams, we can build a grammar whose size is at most quadratic in the minimum. Finally, in Section 4 we show that two streams are necessary and sufficient for us to achieve entropy-only bounds. Along the way, we show we need two streams to find strings' periods or compute the Burrows-Wheeler Transform. As far as we know, this is the first paper on compression with read/write streams, and among the first papers on compression in any streaming model; we hope the techniques we have used will prove to be of independent interest.

2 Universal Compression

An algorithm is called universal with respect to a class of sources if, when a string is drawn from any of those sources, the algorithm's redundancy per character approaches 0 with probability 1 as the length of the string grows. One class that is often considered, and which we consider in this section, is that of FSMX sources [31]. A kth-order FSMX source is a finite-state source in which the current state is determined by at most the preceding k characters emitted. Since the kth-order empirical entropy $H_k(s)$ of s is the minimum self-information per character of s with respect to a kth-order Markov source (see [34]), an algorithm is universal with respect to FSMX sources if it stores any string s in $nH_k(s)+o(n)$ bits for any fixed σ and k. The kth-order empirical entropy of s is also our expected uncertainty about a randomly-chosen character of s when given the k preceding characters. Specifically,

$$H_k(s) = \begin{cases} (1/n)\sum_a \mathsf{occ}(a,s)\log\frac{n}{\mathsf{occ}(a,s)} & \text{if } k=0,\\ (1/n)\sum_{|w|=k}|w_s|H_0(w_s) & \text{otherwise,}\end{cases}$$

where $\mathsf{occ}(a, s)$ is the number of times character a occurs in s, and w_s is the concatenation of those characters immediately following occurrences of k-tuple w in s.

In a previous paper [19] we showed how to modify the well-known LZ77 compression algorithm [36] to use sublinear memory while still storing s in $nH_k(s) + \mathcal{O}(n \log\log n / \log n)$ bits for any fixed σ and k. Our algorithm uses nearly linear memory and so does not fit into the model we consider in this paper, but we mention it here because it fits into some other streaming models (see, e.g., [28]) and, as far as we know, was the first compression algorithm to do so. In the same paper we proved several lower bounds using ideas that eventually led to our lower bounds in Sections 3 and 4 of this paper.

Theorem 1 (Gagie and Manzini, 2007). *We can achieve universal compression using one pass over one stream and $\mathcal{O}(n/\log^2 n)$ bits of memory.*

To achieve universal compression with only polylogarithmic memory, we use an algorithm due to Gupta, Grossi and Vitter [22]. Although they designed it for the RAM model, we can easily turn it into a streaming algorithm by processing s in small blocks and compressing each block separately.

Theorem 2 (Gupta, Grossi and Vitter, 2008). *In the RAM model, we can store any string s in $nH_k(s) + \mathcal{O}(\sigma^k \log n)$ bits, for all k simultaneously, using $\mathcal{O}(n)$ time.*

Corollary 1. *We can achieve universal compression using one pass over one stream and $\mathcal{O}(\log^{1+\epsilon} n)$ bits of memory.*

Proof. We process s in blocks of $\log^\epsilon n$ characters, as follows: we read each block into memory, apply Theorem 2 to it, output the result, empty the memory, and move on to the next block. (If n is not given in advance, we increase the block size as we read more characters.) Since Gupta, Grossi and Vitter's algorithm uses $\mathcal{O}(n)$ time in the RAM model, it uses $\mathcal{O}(n \log n)$ bits of memory and we use $\mathcal{O}(\log^{1+\epsilon} n)$ bits of memory. If the blocks are $s_1, \ldots, s_b$, then we store all of them in a total of

$$\sum_{i=1}^{b} \left(|s_i| H_k(s_i) + \mathcal{O}(\sigma^k \log\log n)\right) \leq nH_k(s) + \mathcal{O}(\sigma^k n \log\log n / \log^\epsilon n)$$

bits for all k simultaneously. Therefore, for any fixed σ and k, we store s in $nH_k(s) + o(n)$ bits. □

A bound of $nH_k(s) + \mathcal{O}(\sigma^k n \log\log n / \log^\epsilon n)$ bits is not very meaningful when k is not fixed and grows as fast as $\log\log n$, because the second term is $\omega(n)$. Notice, however, that Gupta et al.'s bound of $nH_k(s) + \mathcal{O}(\sigma^k \log n)$ bits is also not very meaningful when $k \geq \log n$, for the same reason. As we will see in Section 4, it is possible for s to be fairly incompressible but still to have $H_k(s) = 0$ for $k \geq \log n$. It follows that, although we can prove bounds that hold for all

$X_7 \to X_6 X_5$
$X_6 \to X_5 X_4$
$X_5 \to X_4 X_3$
$X_4 \to X_3 X_2$
$X_3 \to X_2 X_1$
$X_2 \to a$
$X_1 \to b$

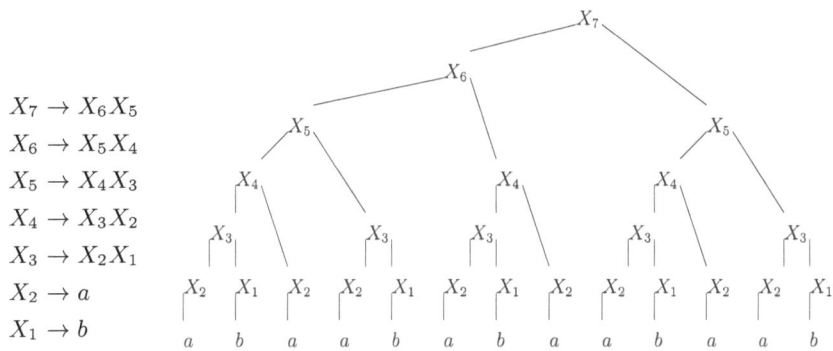

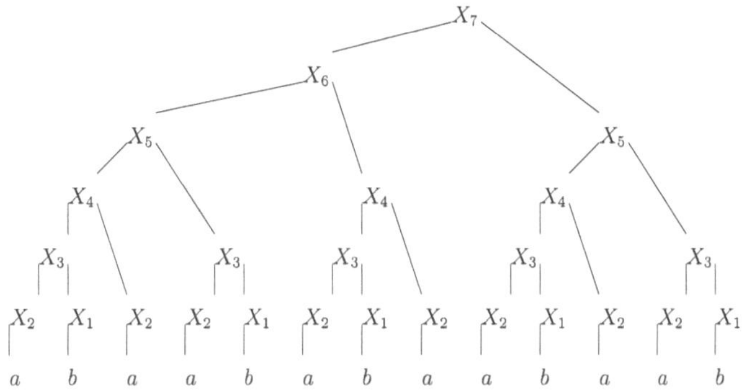

Fig. 1. An context-free grammar (left) generating Fibonacci word *abaababaabaab*, and the corresponding parse tree (right)

k simultaneously, those bounds cannot guarantee good compression in terms of $H_k(s)$ when $k \geq \log n$.

By using larger blocks — and, thus, more memory — we can reduce the $\mathcal{O}(\sigma^k n \log\log n / \log^\epsilon n)$ redundancy term in our analysis, allowing k to grow faster than $\log\log n$ while still having a meaningful bound. Specifically, if we process s in blocks of c characters, then we use $\mathcal{O}(c \log n)$ bits of memory and achieve a redundancy term of $\mathcal{O}(\sigma^k n \log c / c)$, allowing k to grow nearly as fast as $\log_\sigma c$ while still having a meaningful bound.

Corollary 2. *We can achieve universal compression with any redundancy term in $\sigma^k n / \log^{\mathcal{O}(1)} n$, using one pass over one stream and polylogarithmic memory.*

We will show later, in Theorem 15, that the tradeoff described above is nearly optimal: if we use m bits of memory and p passes over one stream and our redundancy term is $\mathcal{O}(\sigma^k r)$, then $mpr = \Omega(n/f(n))$ for any function f that increases without bound. It is not clear to us, however, whether we can modify Corollary 1 to take advantage of multiple passes. That is, with multiple passes over one stream, can we achieve better bounds on the memory and redundancy than we can with one pass?

3 Grammar-Based Compression

Charikar et al. [8] and Rytter [33] independently showed how to build a nearly minimal context-free grammar APPROX that generates s and only s. Specifically, their algorithms yield grammars that are an $\mathcal{O}(\log n)$ factor larger than the smallest such grammar OPT, which has size $\Omega(\log n)$ bits. Figure 1 shows a context-free grammar generating only the Fibonacci word *abaababaabaab*.

Theorem 3 (Charikar et al., 2005; Rytter, 2003). *In the RAM model, we can approximate the smallest grammar with* $|\mathsf{APPROX}| = \mathcal{O}(|\mathsf{OPT}|^2)$ *using* $\mathcal{O}(n)$ *time.*

In this section we prove that, if we use only one stream, then in general our approximation must be superpolynomially larger than the smallest grammar. Our idea is to show that periodic strings whose periods are asymptotically slightly larger than the product of the memory and passes, can be encoded as small grammars but, in general, cannot be compressed well by algorithms that use only one stream. The period of s is the length $\ell \leq n$ of the shortest string t such that $s = t^{\lfloor n/\ell \rfloor} t'$, where t' is a proper prefix of t and $t^{\lfloor n/\ell \rfloor} t'$ means t repeated $\lfloor n/\ell \rfloor$ times followed by t'. We call t the repeated substring of s. Our argument is based on the following two lemmas.

Lemma 1. *If s has period ℓ, then the size of the smallest grammar for that string is*

$$\mathcal{O}(\ell \log \sigma + \log n \log \log n)$$

bits.

Proof. Let t and t' be as described above. We can encode a unary string $X^{\lfloor n/\ell \rfloor}$ as a grammar G_1 with $\mathcal{O}(\log n)$ productions of total size $\mathcal{O}(\log n \log \log n)$ bits. We can also encode t and t' as grammars G_2 and G_3 with $\mathcal{O}(\ell)$ productions of total size $\mathcal{O}(\ell \log \sigma)$ bits. Suppose S_1, S_2 and S_3 are the start symbols of G_1, G_2 and G_3, respectively. By combining those grammars and adding the productions $S_0 \rightarrow S_1 S_3$ and $X \rightarrow S_2$, we obtain a grammar with $\mathcal{O}(\ell + \log n)$ productions of total size $\mathcal{O}(\ell \log \sigma + \log n \log \log n)$ bits that maps S_0 to s. □

Lemma 2. *Consider a lossless compression algorithm that uses only one stream, and a machine performing that algorithm. We can uniquely recover any substring from*

- *its length;*
- *for each pass, the machine's memory configurations when it reaches and leaves the part of the stream that initially holds that substring;*
- *all the output the machine produces while over that part.*

Proof. Let t be the substring and assume, for the sake of a contradiction, that there exists another substring t' with the same length that takes the machine between the same configurations while producing the same output. Then we can substitute t' for t in s without changing the machine's complete output, contrary to our specification that the compression be lossless. □

Lemma 2 implies that, for any substring, the size of the output the machine produces while over the part of the stream that initially holds that substring, plus twice the product of the memory and passes (i.e., the number of bits needed to store the memory configurations), must be at least that substring's Kolmogorov complexity (i.e., the length of the shortest program that generates it). Therefore, if a substring is not compressible by more than a constant factor (as is the case for most strings) and is asymptotically larger than the product of the memory and passes, then the size of the output for that substring must be at least proportional to the substring's length. In other words, the algorithm cannot take full advantage of similarities between substrings to achieve better compression. In particular, if s is periodic with a period that is asymptotically slightly larger than the product of the memory and passes, and s's repeated substring is not compressible by more than a constant factor, then the algorithm's complete output must be $\Omega(n)$ bits. By Lemma 1, however, the size of the smallest grammar that generates s and only s is bounded in terms of the period.

Theorem 4. *With one stream, we cannot approximate the smallest grammar with*

$$|\mathsf{APPROX}| \leq |\mathsf{OPT}|^{\mathcal{O}(1)}.$$

Proof. Suppose an algorithm uses only one stream, m bits of memory and p passes to compress s, with $mp = \log^{\mathcal{O}(1)} n$, and consider a machine performing that algorithm. Furthermore, suppose s is binary and periodic with period $mp \log n$ and its repeated substring t is not compressible by more than a constant factor. Lemma 2 implies that the machine's output while over a part of the stream that initially holds a copy of t, must be $\Omega(mp \log n - mp) = \Omega(mp \log n)$. Therefore, the machine's complete output must be $\Omega(n)$ bits. By Lemma 1, however, the size of the smallest grammar that generates s and only s is $\mathcal{O}(mp \log n + \log n \log \log n) \subset \log^{\mathcal{O}(1)} n$ bits. Since $n = \log^{\omega(1)} n$, the algorithm's complete output is superpolynomially larger than the smallest grammar. □

As an aside, we note that a symmetric argument shows that, with only one stream, in general we cannot decode a string encoded as a small grammar. To see why, instead of considering a part of the stream that initially holds a copy of the repeated substring t, consider a part that is initially blank and eventually holds a copy of t. (Since s is periodic and thus very compressible, its encoding takes up only a fraction of the space it eventually occupies when decompressed; without loss of generality, we can assume the rest is blank.) An argument similar to the proof of Lemma 2 shows we can compute t from the machine's memory configurations when it reaches and leaves that part, so the product of the memory and passes must again be greater than or equal to t's complexity.

Theorem 5. *With one stream, we cannot decompress strings encoded as small grammars.*

Theorem 4 also has the following corollary, which may be of independent interest.

Corollary 3. *With one stream, we cannot find strings' periods.*

Proof. Consider the proof of Theorem 4. Notice that, if we could find s's period, then we could store s in $\log^{\mathcal{O}(1)} n$ bits by writing n and one copy of its repeated substring t. It follows that we cannot find strings' periods. □

Corollary 3 may at first seem to contradict work by Ergün, Muthukrishnan and Sahinalp [12], who gave streaming algorithms for determining approximate periodicity. Whereas we are concerned with strings which are truly periodic, however, they were concerned with strings in which the copies of the repeated substring can differ to some extent. To see why this is an important difference, consider the simple case of checking whether s has period $n/2$ (i.e., whether or not it is a square). Suppose we know the two halves of s are either identical or differ in exactly one position, and we want to determine whether s truly has period $n/2$; then we must compare each corresponding pair of characters and, by a crossing-sequences argument (see, e.g., [27] for details of a similar argument), this takes $\Omega(n/m)$ passes. Now suppose we care only whether the two halves of s match only in nearly all positions; then we need compare only a few randomly-chosen pairs to decide correctly with high probability.

Theorem 6. *With one stream, we cannot even check strings' periods.*

In the conference version of this paper [16] we left as an open problem proving whether or not multiple streams are useful for grammar-based compression. As we noted in the introduction, in a subsequent paper with Gawrychowski [17] we showed that with constant memory and logarithmic passes over a constant number of streams, we can approximate the smallest grammar with $|\mathsf{APPROX}| = \mathcal{O}(|\mathsf{OPT}|^2)$, answering our question affirmatively.

4 Entropy-Only Bounds

Kosaraju and Manzini [25] pointed out that proving an algorithm universal does not necessarily tell us much about how it behaves on low-entropy strings. In other words, showing that an algorithm encodes s in $nH_k(s)+o(n)$ bits is not very informative when $nH_k(s) = o(n)$. For example, although the well-known LZ78 compression algorithm [37] is universal, $|\mathsf{LZ78}(1^n)| = \Omega(\sqrt{n})$ while $nH_0(1^n) = 0$. To analyze how algorithms perform on low-entropy strings, we would like to get rid of the $o(n)$ term and prove bounds that depend only on $nH_k(s)$. Unfortunately, this is impossible since, as the example above shows, even $nH_0(s)$ can be 0 for arbitrarily long strings.

It is not hard to show that only unary strings have $H_0(s) = 0$. For $k \geq 1$, recall that $H_k(s) = (1/n)\sum_{|w|=k} |w_s| H_0(w_s)$. Therefore, $H_k(s) = 0$ if and only if each distinct k-tuple w in s is always followed by the same distinct character. This is because, if a w is always followed by the same distinct character, then w_s is unary, $H_0(w_s) = 0$ and w contributes nothing to the sum in the formula. Manzini [26] defined the kth-order modified empirical entropy $H_k^*(s)$ such that each context w contributes at least $\lfloor \log |w_s| \rfloor + 1$ to the sum. Because modified empirical entropy is more complicated than empirical entropy — e.g., it allows for variable-length contexts — we refer readers to Manzini's paper for the full definition. In our proofs in this paper, we use only the fact that

$$nH_k(s) \leq nH_k^*(s) \leq nH_k(s) + \mathcal{O}(\sigma^k \log n) .$$

Manzini showed that, for some algorithms and all k simultaneously, it is possible to bound the encoding's length in terms of only $nH_k^*(s)$ and a constant g_k that depends only on σ and k; he called such bounds 'entropy-only'. In particular, he showed that an algorithm based on the Burrows-Wheeler Transform (BWT) [7] stores any string s in at most $(5+\epsilon)nH_k^*(s)+\log n+g_k$ bits for all k simultaneously (since $nH_k^*(s) \geq \log(n-k)$, we could remove the $\log n$ term by adding 1 to the coefficient $5+\epsilon$).

Theorem 7 (Manzini, 2001). *Using the BWT, move-to-front coding, run-length coding and arithmetic coding, we can achieve an entropy-only bound.*

The BWT sorts the characters in a string into the lexicographical order of the suffixes that immediately follow them. When using the BWT for compression, it is customary to append a special character \$ that is lexicographically less than any in the alphabet. For a more thorough description of the BWT, we again refer readers to Manzini's paper. In this section we first show how we can compute and invert the BWT with two streams and, thus, achieve entropy-only bounds. We then show that we cannot achieve entropy-only bounds with only one stream. In other words, two streams are necessary and sufficient for us to achieve entropy-only bounds.

One of the most common ways to compute the BWT is by building a suffix array. In his PhD thesis, Ruhl introduced the StreamSort model [32,2], which is similar to the read/write streams model with one stream, except that it has an extra primitive that sorts the stream in one pass. Among other things, he showed how to build a suffix array efficiently in this model.

Theorem 8 (Ruhl, 2003). *In the StreamSort model, we can build a suffix array using $\mathcal{O}(\log n)$ bits of memory and $\mathcal{O}(\log n)$ passes.*

Corollary 4. *With two streams, we can compute the BWT using $\mathcal{O}(\log n)$ bits of memory and $\mathcal{O}(\log^2 n)$ passes.*

Proof. We can compute the BWT in the StreamSort model by appending \$ to s, building a suffix array, and replacing each value i in the array by the $(i-1)$st character in s (replacing either 0 or 1 by \$, depending on where we start counting). This takes $\mathcal{O}(\log n)$ bits of memory and $\mathcal{O}(\log n)$ passes. Since we can sort with two streams using $\mathcal{O}(\log n)$ bits memory and $\mathcal{O}(\log n)$ passes (see, e.g., [35]), it follows that we can compute the BWT using $\mathcal{O}(\log n)$ bits of memory and $\mathcal{O}(\log^2 n)$ passes. □

We note as an aside that, once we have the suffix array for a periodic string, we can easily find its period. To see why, suppose s has period ℓ, and consider the suffix u of s that starts in position $\ell+1$. The longest common prefix of s and u has length $n-\ell$, which is maximum; if another suffix v shared a longer common prefix with s, then s would have period $n-|v|<\ell$. It follows that, if the first position in the suffix array contains i, then the $(\ell+1)$st position contains $i-1$ (assuming s terminates with \$, so u is lexicographically less than s). With two streams we can easily find the position $\ell+1$ that contains $i-1$ and then check that s is indeed periodic with period ℓ.

Corollary 5. *With two streams, we can compute a string's period using $\mathcal{O}(\log n)$ bits and $\mathcal{O}(\log^2 n)$ passes.*

Now suppose we are given a permutation π on $n+1$ elements as a list $\pi(1), \ldots, \pi(n+1)$, and asked to rank it, i.e., to compute the list $\pi^0(1), \ldots, \pi^n(1)$, where $\pi^0(1) = 1$ and $\pi^i(1) = \pi(\pi^{i-1}(1))$ for $i \geq 1$. This problem is a special case of list ranking (see, e.g., [3]) and has a surprisingly long history. For example, Knuth [24, Solution 24] described an algorithm, which he attributed to Hardy, for ranking a permutation with two tapes. More recently, Bird and Mu [5] showed how to invert the BWT by ranking a permutation. Therefore, reinterpreting Hardy's result in terms of the read/write streams model gives us the following bounds.

Theorem 9 (Hardy, c. 1967). *With two streams, we can rank a permutation using $\mathcal{O}(\log n)$ bits of memory and $\mathcal{O}(\log^2 n)$ passes.*

Corollary 6. *With two streams, we can invert the BWT using $\mathcal{O}(\log n)$ bits of memory and $\mathcal{O}(\log^2 n)$ passes.*

Proof. The BWT has the property that, if a character is the ith in $\mathsf{BWT}(s)$, then its successor in s is the lexicographically ith in $\mathsf{BWT}(s)$ (breaking ties by order of appearance). Therefore, we can invert the BWT by replacing each character by its lexicographic rank, ranking the resulting permutation, replacing each value i by the ith character of $\mathsf{BWT}(s)$, and rotating the string until \$ is at the end. This takes $\mathcal{O}(\log n)$ memory and $\mathcal{O}(\log^2 n)$ passes. □

Since we can compute and invert move-to-front, run-length and arithmetic coding using $\mathcal{O}(\log n)$ bits of memory and $\mathcal{O}(1)$ passes over one stream, by combining Theorem 7 and Corollaries 4 and 6 we obtain the following theorem.

Theorem 10. *With two streams, we can achieve an entropy-only bound using $\mathcal{O}(\log n)$ bits of memory and $\mathcal{O}(\log^2 n)$ passes.*

It follows from Theorem 10 and a result by Hernich and Schweikardt [23] that we can achieve an entropy-only bound using $\mathcal{O}(1)$ bits of memory, $\mathcal{O}(\log^3 n)$ passes and four streams. It follows from their theorem below that, with more streams, we can even reduce the number of passes to $\mathcal{O}(\log n)$.

Theorem 11 (Hernich and Schweikardt, 2008). *If we can solve a problem with logarithmic work space, then we can solve it using $\mathcal{O}(1)$ bits of memory and $\mathcal{O}(\log n)$ passes over $\mathcal{O}(1)$ streams.*

Corollary 7. *With $\mathcal{O}(1)$ streams, we can achieve an entropy-only bound using $\mathcal{O}(1)$ bits of memory and $\mathcal{O}(\log n)$ passes.*

Proof. To compute the ith character of $\mathsf{BWT}(s)$, we find the ith lexicographically largest suffix. To find this suffix, we loop through all the suffixes and, for each, count how many other suffixes are lexicographically less. Comparing two suffixes character by character takes $\mathcal{O}(n^2)$ time, so we use a total of $\mathcal{O}(n^4)$ time; it does not matter now how much time we use, however, just that we need only a constant number of $\mathcal{O}(\log n)$-bit counters. Since we can compute the BWT with logarithmic work space, it follows from Theorem 11 that we can compute it — and thereby achieve an entropy-only bound — with $\mathcal{O}(1)$ bits of memory and $\mathcal{O}(\log n)$ passes over $\mathcal{O}(1)$ streams. □

Although we have not been able to prove an $\Omega(\log n)$ lower bound on the number of passes needed to achieve an entropy-only bound with $\mathcal{O}(1)$ streams, we have been able to prove such a bound for computing the BWT. Our idea is to reduce sorting to the BWT, since Grohe and Schweikardt [21] showed we cannot sort n numbers with $o(\log n)$ passes over $\mathcal{O}(1)$ streams. It is trivial, of course, to reduce sorting to the BWT if the alphabet is large enough — e.g., linear in n — but our reduction is to the more reasonable problem of computing the BWT of a ternary string.

Theorem 12. *With $\mathcal{O}(1)$ streams, we cannot compute the BWT using $o(\log n)$ passes, even for ternary strings.*

Proof. Suppose we are given a sequence of n numbers $x_1, \ldots, x_n$, each of $2 \log n$ bits. Grohe and Schweikardt showed we cannot generally sort such a sequence using $o(\log n)$ passes over $\mathcal{O}(1)$ tapes. We now use $o(\log n)$ passes to turn $x_1, \ldots, x_n$ into a ternary string s such that, by calculating $\mathsf{BWT}(s)$, we sort $x_1, \ldots, x_n$. It follows from this reduction that we cannot compute the BWT using $o(\log n)$ passes, either.

With one pass, $O(\log n)$ bits of memory and two tapes, for $1 \leq i \leq n$ and $1 \leq j \leq 2 \log n$, we replace the jth bit $x_i[j]$ of x_i by $x_i[j]\ 2\ x_i\ i\ j$, writing 2 as a single character, x_i in $2 \log n$ bits, i in $\log n$ bits and j in $\log \log n + 1$ bits; the resulting string s is of length $2n \log n(3 \log n + \log \log n + 2)$. The only characters followed by 2s in s are the bits at the beginning of replacement phrases, so the last $2n \log n$ characters of $\mathsf{BWT}(s)$ are the bits of $x_1, \ldots, x_n$; moreover, since the lexicographic order of equal-length binary strings is the same as their numeric order, the $x_i[j]$ bits will be arranged by the x_i values, with ties broken by the i values (so if $x_i = x_{i'}$ with $i < i'$, then every $x_i[j]$ comes before every $x_{i'}[j']$) and further ties broken by the j values; therefore, the last $2n \log n$ bits of the transformed string are $x_1, \ldots, x_n$ in sorted order. □

To show we need at least two streams to achieve entropy-only bounds, we use De Bruijn cycles in a proof much like the one for Theorem 4. We used De Bruijn cycles in a similar way in a previous paper [15] to prove a lower bound on redundancy. A σ-ary De Bruijn cycle of order k is a cyclic sequence in which every possible k-tuple appears exactly once. For example, Figure 2 shows binary De Bruijn cycles of orders 3 and 4. Our argument this time is based on Lemma 2 and the results below about De Bruijn cycles. We note as a historical aside that Theorem 13 was first proven for the binary case in 1894 by Flye Sainte-Marie [14], but his result was later forgotten; De Bruijn [6] gave a similar proof for that case in 1946, then in 1951 he and Van Aardenne-Ehrenfest [1] proved the general version we state here.

0 0	1 0 0 0 0 1
1 0	1 0
0 1	1 0
1 1	1 0 1 0 1 1

Fig. 2. Examples of binary De Bruijn cycles of orders 3 and 4

Lemma 3. *If $s \in d^*$ for some binary σ-ary De Bruijn cycle d of order k, then $nH_k^*(s) = \mathcal{O}(\sigma^k \log n)$.*

Proof. By definition, each distinct k-tuple is always followed by the same distinct character; therefore, $nH_k(s) = 0$ and $nH_k^*(s) = \mathcal{O}(\sigma^k \log n)$. □

Theorem 13 (Van Aardenne-Ehrenfest and De Bruijn, 1951). *There are $\left(\sigma!^{\sigma^{k-1}} / \sigma^k\right)$ σ-ary De Bruijn cycles of order k.*

Corollary 8. *We cannot store most kth-order De Bruijn cycles in $o(\sigma^k \log \sigma)$ bits.*

Proof. By Stirling's Formula, $\log\left(\sigma!^{\sigma^{k-1}} / \sigma^k\right) = \Theta(\sigma^k \log \sigma)$. □

Since there are σ^k possible k-tuples, kth-order De Bruijn cycles have length σ^k, so Corollary 8 means that we cannot compress most De Bruijn cycles by more than a constant factor. Therefore, we can prove a lower bound similar to Theorem 4 by supposing that s's repeated substring is a De Bruijn cycle, then using Lemma 3 instead of Lemma 1.

Theorem 14. *With one stream, we cannot achieve an entropy-only bound.*

Proof. As in the proof of Theorem 4, suppose an algorithm uses only one stream, m bits of memory and p passes to compress s, with $mp = \log^{\mathcal{O}(1)} n$, and consider a machine performing that algorithm. This time, however, suppose s is binary and periodic with period $mp\, f(n)$, where $f(n) = \mathcal{O}(\log n)$ is a function that increases without bound; furthermore, suppose s's repeated substring t is a kth-order De Bruijn cycle, $k = \log(mp\, f(n))$, that is not compressible by more than a constant factor. Lemma 2 implies that the machine's output while over a part of the stream that initially holds a copy of t, must be $\Omega(mp\, f(n) - mp) = \Omega(mp\, f(n))$. Therefore, the machine's complete output must be $\Omega(n)$ bits. By Lemma 3, however, $nH_k^*(s) = \mathcal{O}(2^k \log n) = \mathcal{O}(mp\, f(n) \log n) \subset \log^{\mathcal{O}(1)} n$. □

Recall that in Section 2 we asserted the following claim, which we are now ready to prove.

Theorem 15. *If we use m bits of memory and p passes over one stream and achieve universal compression with an $\mathcal{O}(\sigma^k r)$ redundancy term, for all k simultaneously, then $mpr = \Omega(n / f(n))$ for any function f that increases without bound.*

Proof. Consider the proof of Theorem 14: $nH_k(s) = 0$ but we must output $\Omega(n)$ bits, so $r = \Omega(n/\sigma^k) = \Omega(n/(mp\, f(n)))$. □

Notice Theorem 14 also implies a lower bound for computing the BWT: if we could compute the BWT with one stream then, since we can compute move-to-front, run-length and arithmetic coding using $\mathcal{O}(\log n)$ bits of memory and $\mathcal{O}(1)$ passes over one stream, we could thus achieve an entropy-only bound with one stream, contradicting Theorem 14.

Corollary 9. *With one stream, we cannot compute the BWT.*

5 Recent and Future Work

In the conference version of this paper [16] we closed with a brief discussion of three entropy-only bounds that we proved with Manzini [18]. Our first bound was an improved analysis of the BWT followed by move-to-front, run-length and arithmetic coding (which lowered the coefficient from $5 + \epsilon$ to $4.4 + \epsilon$), but our other bounds (one of which had a coefficient of $2.69 + \epsilon$) were analyses of the BWT followed by algorithms which we were not sure could be implemented with $\mathcal{O}(1)$ streams. We now realize that, since both of these other algorithms can be computed with logarithmic work space, it follows from Theorem 11 that they can indeed be computed with $\mathcal{O}(1)$ streams.

After having proven that we cannot compute the BWT with one stream, we promptly started working with Ferragina and Manzini on a practical algorithm [13] that does exactly that. However, that algorithm does not fit into the streaming models we have considered in this paper; in particular, the product of the internal memory and passes there is $\mathcal{O}(n \log n)$ bits, but we use only n bits of workspace on the disk. The existence of a practical algorithm for computing the BWT in external memory raises the question of whether we can query BWT-based compressed indexes quickly in external memory. Chien et al. [10] proved lower bounds for indexed pattern matching in the external-memory model, but that model does not distinguish between sequential and random access to blocks. The read/write-streams model is also inappropriate for analyzing the complexity of this task, since we can trivially use only one pass over one stream if we leave the text uncompressed and scan it all with a classic sequential pattern-matching algorithm. Orlandi and Venturini [30] recently showed how we can store a sample of the BWT that lets us estimate what parts of the full BWT we need to read in order to answer a query. If we modify their data structure slightly, we can make it recursive; i.e., with a smaller sample we can estimate what parts of the sample we need to read in order to estimate what parts of the full BWT we need to read. Suppose we store on disk a set of samples whose sizes increase exponentially, finishing with the BWT itself. We use each sample in turn to estimate what parts of the next sample we need to read, then read them into internal memory using only one pass over the next sample. This increases the size of the whole index only slightly and lets us answer queries by reading few blocks and in the order they appear on disk. We are currently working to optimize and implement this idea.

Acknowledgments. Many thanks to Ferdinando Cicalese, Paolo Ferragina, Paweł Gawrychowski, Roberto Grossi, Ankur Gupta, Andre Hernich, Giovanni Manzini, Jens Stoye and Rossano Venturini, for helpful discussions, and to the anonymous referees. This research was done while the author was at the University of Eastern Piedmont in Alessandria, Italy, supported by the Italy-Israel FIRB Project "Pattern Discovery Algorithms in Discrete Structures, with Applications to Bioinformatics", and at Bielefeld University, Germany, supported by the Sofja Kovalevskaja Award from the Alexander von Humboldt Foundation and the German Federal Ministry of Education and Research.

References

1. van Aardenne-Ehrenfest, T., de Bruijn, N.G.: Circuits and trees in oriented linear graphs. Simon Stevin 28, 203–217 (1951)
2. Aggarwal, G., Datar, M., Rajagopalan, S., Ruhl, M.: On the streaming model augmented with a sorting primitive. In: Proceedings of the 45th Symposium on Foundations of Computer Science, pp. 540–549 (2004)
3. Arge, L., Bender, M.A., Demaine, E.D., Holland-Minkley, B., Munro, J.I.: An optimal cache-oblivious priority queue and its application to graph algorithms. SIAM Journal on Computing 36(6), 1672–1695 (2007)
4. Beame, P., Huynh, T.: On the value of multiple read/write streams for approximating frequency moments. In: Proceedings of the 49th Symposium on Foundations of Computer Science, pp. 499–508 (2008)
5. Bird, R.S., Mu, S.-C.: Inverting the Burrows-Wheeler transform. Journal of Functional Programming 14(6), 603–612 (2004)
6. de Bruijn, N.G.: A combinatorial problem. Koninklijke Nederlandse Akademie van Wetenschappen 49, 758–764 (1946)
7. Burrows, M., Wheeler, D.J.: A block-sorting lossless data compression algorithm, Technical Report 24, Digital Equipment Corporation (1994)
8. Charikar, M., Lehman, E., Liu, D., Panigrahy, R., Prabhakaran, M., Sahai, A., Shelat, A.: The smallest grammar problem. IEEE Transactions on Information Theory 51(7), 2554–2576 (2005)
9. Chen, J., Yap, C.-K.: Reversal complexity. SIAM Journal on Computing 20(4), 622–638 (1991)
10. Chien, Y.-F., Hon, W.-K., Shah, R., Vitter, J.S.: Geometric Burrows-Wheeler Transform: Linking range searching and text indexing. In: Proceedings of the Data Compression Conference, pp. 252–261 (2008)
11. Cilibrasi, R., Vitányi, P.: Clustering by compression. IEEE Transactions on Information Theory 51(4), 1523–1545 (2005)
12. Ergün, F., Muthukrishnan, S., Sahinalp, S.C.: Sublinear Methods for Detecting Periodic Trends in Data Streams. In: Farach-Colton, M. (ed.) LATIN 2004. LNCS, vol. 2976, pp. 16–28. Springer, Heidelberg (2004)
13. Ferragina, P., Gagie, T., Manzini, G.: Lightweight data indexing and compression in external memory. Algorithmica 63(3), 707–730 (2012)
14. Flye Sainte-Marie, C.: Solution to question nr. 48. L'Intermédiare de Mathématiciens 1, 107–110 (1894)
15. Gagie, T.: Large alphabets and incompressibility. Information Processing Letters 99(6), 246–251 (2006)

16. Gagie, T.: On the Value of Multiple Read/Write Streams for Data Compression. In: Kucherov, G., Ukkonen, E. (eds.) CPM 2009. LNCS, vol. 5577, pp. 68–77. Springer, Heidelberg (2009)
17. Gagie, T., Gawrychowski, P.: Grammar-Based Compression in a Streaming Model. In: Dediu, A.-H., Fernau, H., Martín-Vide, C. (eds.) LATA 2010. LNCS, vol. 6031, pp. 273–284. Springer, Heidelberg (2010)
18. Gagie, T., Manzini, G.: Move-to-Front, Distance Coding, and Inversion Frequencies Revisited. In: Ma, B., Zhang, K. (eds.) CPM 2007. LNCS, vol. 4580, pp. 71–82. Springer, Heidelberg (2007)
19. Gagie, T., Manzini, G.: Space-Conscious Compression. In: Kučera, L., Kučera, A. (eds.) MFCS 2007. LNCS, vol. 4708, pp. 206–217. Springer, Heidelberg (2007)
20. Grohe, M., Koch, C., Schweikardt, N.: Tight lower bounds for query processing on streaming and external memory data. Theoretical Computer Science 380(1-3), 199–217 (2007)
21. Grohe, M., Schweikardt, N.: Lower bounds for sorting with few random accesses to external memory. In: Proceedings of the 24th Symposium on Principles of Database Systems, pp. 238–249 (2005)
22. Gupta, A., Grossi, R., Vitter, J.S.: Nearly tight bounds on the encoding length of the Burrows-Wheeler Transform. In: Proceedings of the 4th Workshop on Analytic Algorithmics and Combinatorics, pp. 191–202 (2008)
23. Hernich, A., Schweikardt, N.: Reversal complexity revisited. Theoretical Computer Science 401(1-3), 191–205 (2008)
24. Knuth, D.E.: The Art of Computer Programming, 2nd edn., vol. 3. Addison-Wesley (1998)
25. Kosaraju, R., Manzini, G.: Compression of low entropy strings with Lempel-Ziv algorithms. SIAM Journal on Computing 29(3), 893–911 (1999)
26. Manzini, G.: An analysis of the Burrows-Wheeler Transform. Journal of the ACM 48(3), 407–430 (2001)
27. Munro, J.I., Paterson, M.S.: Selection and sorting with limited storage. Theoretical Computer Science 12, 315–323 (1980)
28. Muthukrishnan, S.: Data Streams: Algorithms and Applications. Foundations and Trends in Theoretical Computer Science. Now Publishers (2005)
29. Navarro, G., Mäkinen, V.: Compressed full-text indexes. ACM Computing Surveys 39(1) (2007)
30. Orlandi, A., Venturini, R.: Space-efficient substring occurrence estimation. In: Proceedings of the 30th Symposium on Principles of Database Systems, pp. 95–106 (2011)
31. Rissanen, J.: Complexity of strings in the class of Markov sources. IEEE Transactions on Information Theory 32(4), 526–532 (1986)
32. Ruhl, J.M.: Efficient algorithms for new computational models, PhD thesis, Massachusetts Institute of Technology (2003)
33. Rytter, W.: Application of Lempel-Ziv factorization to the approximation of grammar-based compression. Theoretical Computer Science 302(1-3), 211–222 (2003)
34. Savari, S.: Redundancy of the Lempel-Ziv incremental parsing rule. IEEE Transactions on Information Theory 43(1), 9–21 (1997)
35. Schweikardt, N.: Machine models and lower bounds for query processing. In: Proceedings of the 26th Symposium on Principles of Database Systems, pp. 41–52 (2007)
36. Ziv, J., Lempel, A.: A universal algorithm for sequential data compression. IEEE Transactions on Information Theory 23(3), 337–343 (1977)
37. Ziv, J., Lempel, A.: Compression of individual sequences via variable-rate coding. IEEE Transactions on Information Theory 24(5), 530–536 (1978)

How to Read a Randomly Mixed Up Message

Matthias Löwe

Westfälische Wilhelms-Universität Münster, Fachbereich Mathematik,
Einsteinstraße 62, 48149 Münster, Germany
maloewe@math.uni-muenster.de

Dedicated to the memory of Rudolf Ahlswede

Abstract. We review results on scenery reconstruction obtained over the past decade and place the problem into the context of information retrieval, when the message is damaged by a special form of a reading error.

Keywords: scenery reconstruction, random walk, random scenery, random environment.

1 Introduction

The following question has its roots in ergodic theory, in particular in the so called $T-T^{-1}$-problem solved by Kalikow [8]. It is nowadays, however considered a problem on the interface of probability theory, statistics, combinatorics, and information theory, that is interesting in its own rights.

To phrase it, let us put the problem into the context of information theory. Let us assume an infinitely long message over an m-ary alphabet is sent via an information channel. However, the receiver is not able to read the message directly, but only a corrupted version, where the reading head jumps to the left and to the right in some random fashion. This, of course, constitutes an extreme form of noise and the question for the general communication model for one sender in [1]: "How many messages can we transmit over a noisy channel?" must now be turned into: "Can we transmit information over such a channel at all?"

This seems to be even more problematic as we fix the error rate to be zero, i.e. we want to identify the original message correctly with probability one.

To put the model in a more mathematical framework, let $m \geq 2$. The alphabet is then given by the set $\{0, \ldots, m-1\}$ and the (doubly infinite) message will be denoted by

$$\xi : \mathbb{Z} \to \{0, \ldots, m-1\}.$$

Later we will also discuss the case of "higher dimensional" messages $\xi : \mathbb{Z}^d \to \{0, \ldots, m-1\}$, these we will most often call sceneries and their letters will be called colors. The second ingredient one needs is a discrete time-stochastic

H. Aydinian, F. Cicalese, and C. Deppe (Eds.): Ahlswede Festschrift, LNCS 7777, pp. 298–312, 2013.

process, which for the time being we choose to be symmetric nearest neighbor random walk $S = (S_n)$ on $\mathcal{Z}$, i.e. $S_0 = 0$ and

$$P(S_n = z \pm 1 | S_{n-1} = z) = \frac{1}{2} \qquad \text{for all } z \in \mathcal{Z}, \text{ and } n \in \mathbb{N}.$$

The difficulty of the present problem originates from the fact that neither the message (or scenery) ξ, nor the realization of the random walk path is known to the receiver. The only thing he does know is what the reader head produces, the observations, the so called record (or, since in the context of scenery reconstruction, the letter of the alphabet are referred to as colors, the color record)

$$\chi := (\chi_n)_{n \in \mathbb{N}} := (\xi(S_n))_{n \in \mathbb{N}}.$$

From here the problem splits into three different questions:

1. Can we distinguish two given messages ξ and η by their record, i.e. is there a test, working correctly with probability one, that tells whether we have red the corrupted version of ξ or of η?
2. Are there messages, where this does not work, i.e. messages that cannot be told apart form their record?
3. Can we even reconstruct the message ξ from the record χ with probability one?

In the next section of this little survey we will concentrate on the first two of these questions. The third problem will be addressed in its easiest form in the third section. Extensions of this problem in dimension one will be given in Section 4. Section 5 eventually contains some results on higher dimensional versions of the problem. As a matter of fact, only dimension $d = 2$ seems to be understood in some aspects, while results for $d \geq 3$ are very sparse and we can only formulate some guidelines for future research.

2 The Distinction Problem

To discuss the question, whether two messages can be distinguished by their record (almost surely), let us make precise what we mean by distinguishing two messages.

To this end for two messages ξ and η let us consider the measures Q^l_ξ and Q^l_η induced by the records $(\xi_n)_{n \geq l}$ and $(\eta_n)_{n \geq l}$. We will say that we are able to distinguish ξ from η, if, for each l, Q^l_ξ and Q^l_η are orthogonal, i.e., if for each l there are measurable sets $N = N_l$, with

$$Q^l_\xi(N) = 0 \quad \text{and } Q^l_\eta(N^c) = 0.$$

Note that we impose this condition for each l to avoid trivial cases, where e.g. the origin has a different letter for ξ and η. Moreover, this has the advantage that it turns the event "χ and η are indistinguishable" into a tail event, which by the 0-1-law has either probability 0 or 1.

We start with an almost trivial observation that has been the basis for a conjecture, that resisted a proof or disproof for more than a decade: Since already in the first step, we do not whether S_n jumps to the right or to the left, we are not able to distinguish two messages that differ only up to reflection symmetry. Moreover, by the above condition also sceneries that differ from each other by a shift of $\mathcal{Z}$ can not be told apart. This motivates

Definition 1. *Two messages ξ and η are called equivalent ($\xi \sim \eta$), if there is $a \in \mathcal{Z}$ and $s \in \{-1, +1\}$ such that*

$$\xi(a + sz) = \eta(z) \qquad \text{for all } z \in \mathcal{Z}.$$

In dimension $d \geq 2$ two messages or sceneries ξ and η are called equivalent, if there is $a \in \mathcal{Z}^d$ and a linear map $M : \mathcal{Z}^d \to \mathcal{Z}^d$ with $|\, det(M)| = 1$ such that

$$\xi(a + Mz) = \eta(z) \qquad \text{for all } z \in \mathcal{Z}^d.$$

The above discussion shows that equivalent messages or sceneries cannot be distinguished (which, however, is only a partial answer to Question 2 in the introduction). Den Hollander and Keane [9] conjectured that the converse is also true: If two messages are indistinguishable, then they are also equivalent.

This conjecture is supported by two very strong results that show that many non-equivalent messages can actually be distinguished. The term "many" in this statement is most often characterized probabilistically. To this end we will often choose a message at random. This means (if not stated otherwise), that we have an a priori measure $\hat{\rho}$ on the letters $\{0, \ldots, m-1\}$ charging every letter and we select a message ξ according to the product measure

$$\rho(\xi) = \prod_{z \in \mathcal{Z}} \hat{\rho}(\xi_z).$$

With these definitions Benjamini and Kesten were able to prove:

Theorem 1 (Benjamini and Kesten [3]). *For an alphabet of two letters, i.e. $m = 2$, every fixed message ξ can be distinguished from ρ-every message η.*

It should be remarked that the above theorem is true under more general assumptions. First of all, $m = 2$ is no restriction - as a matter of fact, the more letters one has, the easier is it to distinguish messages. On a heuristic level, this is easily explained: The larger m, the more characteristic is a letter for a vertex, and the more information do we gain, when we read it. In the extreme case $m = \infty$ every letter would be characteristic of its location in a huge part of the scenery.

Secondly, Benjamini and Kesten show the same result in dimension 2 (see [3]). Finally, as Kesten points out in [11], also the condition that (S_n) is a nearest neighbor random walk may be relaxed. In fact, jumps may be permitted, if the expectation of the increments of the random walk is zero (to ensure recurrence of the random process) and the jump size is bounded.

If the dimension is larger than $d = 2$, the task of distinguishing two sceneries, becomes increasingly difficult, since simple random walk is no longer recurrent for $d \geq 3$ and there is a positive fraction of points we do not even see. However, as Benjamini and Kesten show, this can be compensated by a larger number of colors.

Theorem 2 (Benjamini and Kesten [3]). *For ever fixed $d \geq 3$ there is a number of colors $m_0(d)$, such that for all $m \geq m_0(d)$ every fixed m-color scenery ξ can be distinguished from ρ-every m-color scenery η.*

It is conjectured, that this results is wrong for a "small" number of colors, but to prove this and to show, how small "small" really is, remains one of the major challenges in this area.

We refrain from giving proofs for the above theorems. They rely on a careful analysis of the statistics of the record read by the random walk (this is nowadays known in great detail, see e.g. [5]). With its help, an even stronger result can be shown: Even a single defect in a random message can be detected in $d = 1$, if there are enough letters:

Theorem 3 (Kesten [10]). *In dimension $d = 1$ assume that the alphabet has $m \geq 5$ letters. Then for ρ-almost every ξ and every $z_0 \in \mathcal{Z}$ the message ξ can be distinguished from ξ', where*

$$\xi'(z) \begin{cases} = \xi(z) & z \neq z_0 \\ \neq \xi(z_0) & z = z_0 \end{cases}$$

This gives strong evidence that the above mentioned conjecture of den Hollander and Keane could be correct. However, as a consequence of the following result by Lindenstrauss, it cannot be true.

Theorem 4 (Lindenstrauss [14]). *In dimension $d = 1$ and with $m = 2$ colors there is a uncountable set of pairwise indistinguishable sceneries.*

As for every fixed scenery ξ there is obviously only a countable number of equivalent sceneries, Lindenstrauss'es theorem refutes den Hollander's and Keane's conjecture.

3 Reconstruction of Random Messages

Lindenstrauss'es result shows that to even reconstruct a random scenery can be a difficult, if not impossible task, as sceneries that cannot be distinguished from a non-equivalent scenery, cannot be reconstructed (up to equivalence). On the other hand, after a moment of reflection, one realizes that the set of sceneries that cannot be reconstructed may be uncountable (as Lindenstrauss showed), yet be of measure 0.

The first observation when considering the reconstruction of a randomly mixed up message is, that, since distinction is only possible up to equivalence, we can, of course also only reconstruct messages up to equivalence. However, it is amazing that this is possible in many situations:

Theorem 5 (see Matzinger [21], [19], [20], and Löwe/Matzinger [18]). *Let $d = 1$ and $2 \leq m \leq \infty$. Then, either if the letters are taken to be i.i.d. and $m \geq 2$ (see [21], [19], [20]), or, if the letters are correlated, $m \geq 3$, and an additional technical assumption is fulfilled (see [18] for details), there is a measurable function*

$$\mathcal{A} : \{0, \ldots, m-1\}^{\mathbb{N}} \to \{0, \ldots, m-1\}^{\mathbb{Z}}$$

from the observations to the messages, such that for almost all ξ it holds

$$\mathcal{P}(\mathcal{A}(\chi) \sim \xi) = 1.$$

(Here $\mathcal{P}$ is a probability measure on the set of random walks).

Remark 1. *Note that Theorem 5 also improves the number of colors in Theorem 3 to $m = 2$.*

These theorems have a variety of proofs depending on the assumptions. However, the core ideas are similar and we present them in a situation that admits for a "proof from the book".

Sketch of the Proof: We will start by assuming that there are two special letters $A, B \notin \{0, \ldots, m-1\}$ that only exist once in the message. Moreover, assume we know that the locations of these special letter x_1 and x_2 are L steps apart from each other and that we only need to reconstruct the message between these locations.

Then we would wait, until we see an A in the observations and we would know that the reading head is at position x_1. When we read a B afterwards, we know that we are in x_2. Even more: If we read the letter B exactly L steps after having read the letter A, we know that the reading head must have moved from x_1 to x_2 *directly*. In such a direct crossing it has read exactly the desired message between A and B (see Figure 1).

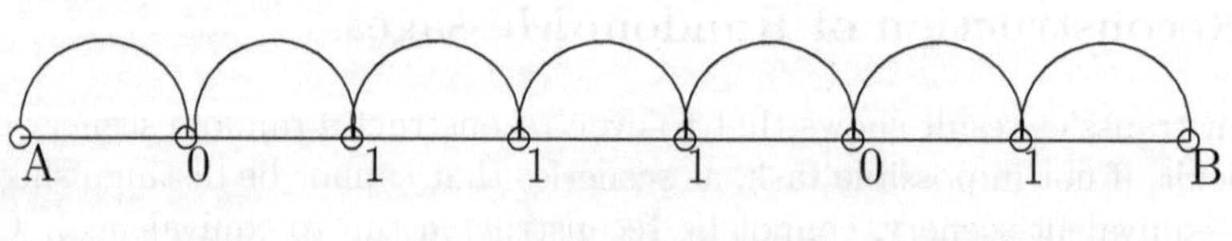

Fig. 1. A direct crossing between two special colors

The random walk is, of course not obliged to walk from x_1 to x_2 directly. However, this event has positive probability 2^{-L}. Moreover, symmetric, simple random walk is *recurrent* in $\mathcal{Z}$, i.e. we will see the point x_0 infinitely often and will have infinitely many chances to walk to x_1 in a straight fashion. By Borel-Cantelli then with probability one, we will sooner or later see such a straight walk and we are done.

Moreover, the assumption that we know the distance of the special letters is not essential. Indeed, by the very same argument as above, the shortest crossing between an A and a successive B stems form a shortest crossing of the reader head from position A to position B.

The major problem in the above arguments is, of course, the assumption that we have these special letters A and B, which we do not (and in fact we would need infinitely many of them, as eventually we want to reconstruct the entire message). In a nutshell: The art of message or scenery reconstruction is to find a substitute for these special letters.

We will here give an idea, that was used by Matzinger in [19] and by Matzinger and Löwe in [18]. It requires that $m \geq 3$. Take the m-regular unrooted tree T_m and take an arbitrary vertex in T_m and call it the origin o. Label the vertices V of T_m now in such a way with

$$\phi : V \to \{0, \ldots, m-1\},$$

that $\phi(o) = \xi(S_0)$ and for each letter each vertex has a neighbor carrying this letter (of course, there is more than one such labeling, which one we take is unessential). In such a labeled tree the message ξ corresponds to a nearest neighbor path $\tilde{R} : \mathcal{Z} \to V$, by $\tilde{R}(0) = o$ and $\phi(\tilde{R}(z)) = \xi(z)$ for all $z \in \mathcal{Z}$. If knew $\tilde{R}$, we would indeed know ξ up to equivalence: We could arbitrarily choose one vertex on $\tilde{R}$ put the letters in one direction successively on $\mathcal{Z}^+$ and in the other direction on $\mathcal{Z}^-$.

However, we do not know $\tilde{R}$ but just $R = \tilde{R} \circ S$, i.e. the nearest neighbor random walk on the path R induced by the observations. The key observation is now that R walks on the nearest neighbor path $\tilde{R}$ and this will help to reconstruct $\tilde{R}$ and thus ξ.

To this end we fix two vertices v and w in R. Since they also lie on $\tilde{R}$, there are $x_1, x_2 \in \mathcal{Z}$ such that $\phi(v) = \xi(x_1)$ and even $\tilde{R}(x_1) = v$ as well as $\phi(w) = \xi(x_2)$ and $\tilde{R}(x_2) = w$. In order to learn the message between x_1 and x_2 it would thus suffice to explore the labels of the tree between v and w by R. There are two major obstacles: It is conceivable that R moves back and forth between v and w and there are two possible reasons: The message may repeat itself between x_1 and x_2, e.g. by having a pattern like 0101, or the random walk S_n jumps back and forth. To distinguish between these two cases, we recall the trick we have met in the situation with the two special letters. Since random walk in dimension one is recurrent, we will see the vertices x_1 and x_2 infinitely often. Each time we see one of them, we have a positive probability to walk directly to the other point, inducing a straight walk of R from v to w or from w to v. Thus, with probability one, we will see such a shortest crossing and the shortest crossing of R between v and w corresponds to a part of the message.

There is, however, a second difficulty. The vertices v and w may correspond to more than one stretch in the message (e.g. to x_1 and x_2 and to y_1 and y_2). In this case there could be various shortest crossing between v and w corresponding to different parts of the original message. In the worst case, they would be e.g. of the same length, but with a different sequence of letters between v and w. Then the algorithm would return a non-unique piece of the message, i.e. it would not return anything. This may indeed happen for certain v and w. However, if we take v and w further and further apart from each other, the probability for this to happen decreases to zero. Indeed: if for two sequences (v_n) and (w_n) in the image of R with

$$dist(v_n, w_n) \to \infty,$$

for all n the letters between v_n and w_n always correspond to two different parts of the scenery, then there are points in $\tilde{R}$ that are visited infinitely often. This is however not possible with probability one, since the random path $\tilde{R}$ is transient. □

This argument does, of course, not work for $m = 2$ letters, since the resulting R is recurrent. A result for $m = 2$ can obtained, by using monochromatic blocks as markers or special characters (see [21], [20]). We will come back to a similar argument in the next Section.

4 Extensions in Dimension One

The argument in the proof in the previous section was in this way first given in [18]. It leads to

Theorem 6 (Löwe/Matzinger [18]). *Let $m \geq 3$. If the random path $\tilde{R}$ is transient, almost every message can be reconstructed up to equivalence.*

This extension is interesting, since also allows for messages with correlated letters (and seems to be the only result in this direction).

Apart from this generalization, there are two natural directions, in which one would extend the result of Theorem 5. The first follows Kesten's remark that distinguishing two messages is also possible, if the random walk is not symmetric nearest neighbor random walk. An obvious question is, whether also reconstruction of a message works under the weaker assumptions. The following answer was given in a joint paper with Matzinger and Merkl.

To formulate it, let μ be a probability measure over $\mathbb{Z}$ supported over a finite set $\mathcal{M} := \operatorname{supp} \mu \subseteq \mathbb{Z}$ and let $S = (S(k))_{k \in \mathbb{N}}$ be a random walk on $\mathcal{Z}$ starting in the origin and with independent increments having the distribution μ. We assume that $E[S(1)] = 0$; thus S is recurrent. Furthermore we assume that $\operatorname{supp} \mu$ has the greatest common divisor 1, thus S can reach every $z \in \mathbb{Z}$ with positive probability. Eventually we assume that the message $\xi = (\xi(j))_{j \in \mathbb{Z}}$ consists of i.i.d. random variables, independent of S, uniformly distributed over $\{0, \ldots, m-1\}$. Then the following theorem holds

Theorem 7 (Löwe/Matzinger/Merkl, see [16]). *If $m > |\mathcal{M}|$, then there exists a measurable map $\mathcal{A} : \mathcal{C}^{\mathbb{N}} \to \mathcal{C}^{\mathbb{Z}}$ such that*

$$P[\mathcal{A}(\chi) \sim \xi] = 1. \tag{1}$$

From a point of view of information theory the condition $m > |\mathcal{M}|$ is very satisfactory. It basically states the the information one obtains by reading a letter is larger than the loss of information from not knowing where the random walk jumps. However, it should be emphasized that a similar result can be obtained with an alphabet consisting of two letters, only (see [12]).

One idea of the proof of Theorem 7 is, to use blocks that consist of the same letter ("monochromatic islands") as markers, i.e. as a replacement for the missing special characters. Monochromatic islands of arbitrary size exist in the message. Moreover they can be detected from the observations quite faithfully, since on a monochromatic island of size n the random walk typically produces order n^2 observations of the same letter, while this exponentially unlikely, if the monochromatic island is much smaller. On the other hand, these islands are typical of the region, where they are. Indeed, if we have a monochromatic island of some color (letter), the next monochromatic island of the same size and color is in average exponentially (in the size of the island) far away, since producing such an island is exponentially unlikely. Hence for a long time, we can be pretty sure, that if, we see a long observation of one letter, we are on a certain monochromatic island.

This can be used to implement the second main idea of the proof, namely to again collect shortest crossings between monochromatic islands. However, they will in general not be unique. If the maximum step size of S is L, then we will collect L shortest crossings, and a major part of the work is to handle the combinatorics of these shortest crossings and to play some sort of jigsaw puzzle to get them in the right order. For more details, the reader is referred to [16].

A second natural generalization of the result in the previous section concerns the question of reading with errors. This inspired among others by the Rényi-Berlekamp-Ulam game which is a classical model for the problem of determining the minimum number of queries to find an unknown member in a finite set when answers may be erroneous (see e.g. [2] or [26]) and by the random coin tossing games considered in [6], which will be briefly discussed in the next section. Hence the question is: Can we still reconstruct a random message, if it is not only mixed up (again by a random walk, possibly with jumps, as in Theorem 7, but if there is also a percentage of letters which are not read correctly. Building on the techniques developed in [16], Matzinger and Rolles were able to show:

Theorem 8 (Matzinger/Rolles [23]). *Under the assumptions of Theorem 7 there exists a $\delta_0 > 0$ and a measurable map $\mathcal{A} : \mathcal{C}^{\mathbb{N}} \to \mathcal{C}^{\mathbb{Z}}$ such that for all $\delta < \delta_0$, if each letter is read correctly with probability $1 - \delta$ (independently of all other reading errors),*

$$P[\mathcal{A}(\chi) \sim \xi] = 1. \tag{2}$$

For a generalization of this theorem we refer the reader to [7].

5 Extension to Several Dimensions

So far we have omitted another natural generalization of the message reconstruction problem, i.e. the question whether reconstruction is also possible in higher dimensions. We will first address the question in $d = 2$ and then talk about possible extensions. In this section we will also call our messages "sceneries", since this naming seems more appropriate in view of the high dimensional character of the messages.

At first glance one might get the impression that there is hardly any difference between $d = 1$ and $d = 2$. In fact, simple random walk in one dimension is recurrent as is simple random walk in dimension $d = 2$. However, there are some important differences. The first is only of technical nature: In dimension one we have often used the idea, that a straight walk from a vertex x_1 to a vertex x_2 is one that reads a piece of scenery directly. However, in $d = 2$ there are many possible shortest crossings from a vertex x_1 to a vertex x_2, if these are not in a line and the corresponding jigsaw is much more complex than (the already complex) one for random walks with jumps.

The second obstacle is however, much more fundamental: In dimension $d = 1$ the local time of the origin, i.e. the number of visits by time n, is of order $\sqrt{n}$, while in $d = 2$ it is of logarithmic order, only. This is important, since the major problem in scenery reconstruction is to determine, where the random walk actually is, to be able to use the information we obtain from the color record. We have seen, that we can extract knowledge about the current location by markers (that serve as substitute for special letters/colors). These markers however, are only reliable in a finite time horizon, as we have seen in the sketch of the proof of Theorem 7. As the same piece of scenery will be repeated also elsewhere in the scenery with probability one, reading a marker is only a good indicator of a certain region, if the random walk (with high probability) has not reached the next area where the marker occurs. We have seen that we can construct markers that are typical of a region of exponential length. In dimension one we will therefore return to this marker exponentially often, before it becomes unreliable.

In dimension two, this is dramatically different. Here we will only see the marker polynomially often, and therefore will only be able to collect polynomially many observations about the neighborhood of the marker (and thus extend our knowledge of the scenery), before we cannot trust it anymore.

This may have also been the basis of a conjecture of Harry Kesten, who suspected that scenery reconstruction is possible in dimension $d = 1$, but may be impossible in dimension $d = 2$.

A similar phenomenon is observed in random coin tossing. Here we attach coins to the vertices of an infinite graph (in our case this graph will always be $\mathcal{Z}$ or $\mathcal{Z}^2$). All the coins are fair, but the coin in the origin may have a bias. So for each time $n \in \mathbb{N}$ and all $z \in \mathcal{Z}$ or $z \in \mathcal{Z}^2$, resp., we have independent random variables X_n^z, with $\mathcal{P}(X_n^z = \pm 1) = \frac{1}{2}$ for all n and $z \neq 0$, while

$$\mathcal{P}(X_n^0 = 1) = \theta.$$

Moreover, there is nearest neighbor random walk $S = (S_n)$ on $\mathcal{Z}$ or $\mathcal{Z}^2$, resp., which again is not directly observable. The observations we have are the coin tosses seen along a random walk path, i.e. $(\zeta_n) = (X_n^{S_n})$. The question now is: Can we detect, whether $\theta = \frac{1}{2}$? The surprising answer given by Harris and Keane [6] is:

Theorem 9 (Harris/Keane [6]). *If $d = 1$, there exists $\theta_0 < 1$, such that, if $\theta \geq \theta_0$, there is a test, that works with probability 1 correctly, showing that $\theta \neq \frac{1}{2}$.*

Such a test does not exist in dimension $d = 2$.

Furthermore, Levin, Pemantle, and Peres were able to show that not only dimension (or, more precisely the question, whether the renewal probabilities are square summable) plays an important role, but also the size of the bias has an influence of the result: Indeed, if θ is smaller than some value $\theta_c > \frac{1}{2}$, the distributions $\mathcal{P}_\theta^{(\zeta_n)}$ and $\mathcal{P}_{\frac{1}{2}}^{(\zeta_n)}$ are mutually absolutely continuous, implying that a bias of size θ cannot be detected almost surely (see [13]).

Hence it does not seem unlikely, that Kesten's conjecture is true. However, in [17] we were able to prove that reconstruction is possible, given we have enough colors. More precisely:

Theorem 10 (Löwe/Matzinger [17]). *Assume that $d = 2$ and the scenery ξ consists of i.i.d. random variables that are uniformly distributed over the set $\{0, \ldots, m-1\}$. Then there exists $m_0 \in \mathbb{N}$ such that if $m \geq m_0$, there exists a measurable function*

$$\mathcal{A} : \{0, \ldots, m-1\}^{\mathbb{N}} \to \{0, \ldots, m-1\}^{\mathcal{Z}^2}$$

such that

$$P(\mathcal{A}(\chi) \sim \xi) = 1. \tag{3}$$

Given the arguments above, this result is, of course, satisfactory. In view of the information theoretic argument given in the previous section one could conjecture that $m_0 = 5$ could be a good bound for m_0, since then in every step again, we would gain more information by the new color we read than we loose by not knowing where the random walk jumps. In the proof of Theorem 10 we did not give a numerical value for m_0. However, if one follows the arguments given there one arrives at a bound for m_0 of the order of 10^n, where n is a large two-digits number. Hence there is obviously space for improvement.

The proof of Theorem 10 follows some basic ideas that seem worth mentioning.

First of all, we show that we do not need to give a *perfect* algorithm, i.e. one that reconstructs the scenery correctly with probability 1, a good one, i.e. one reconstructing the unknown scenery correctly with probability larger than $\frac{1}{2}$, is sufficient. Indeed, if we are given such a good algorithm, we can apply it to $(\chi_n)_{n\geq 0}$, then to $(\chi_n)_{n\geq 1}$, to $(\chi_n)_{n\geq 2}$, and so on. Since the algorithm is good, by the ergodic theorem the majority of the reconstructions will be equivalent to the actual scenery and we just return as output one element of this equivalence class.

The actual reconstruction algorithm works inductively. More precisely, we start reconstructing the scenery in a discrete ball B_{r_0} of size r_0 and then show, how the knowledge of the scenery inside B_r helps to reconstruct on B_{r+1}.

For the beginning of the induction we need the large number of colors. To this fix a small $\varepsilon > 0$ and r_0 large enough that B_{r_0} contains sufficiently many words, i.e. sequences of letters that can be read horizontally in the scenery, of length $\log n$ (this is a technical condition, that can be found in [17], for the purpose of this illustration we should simply imagine r_0 to be large enough). Since simple random walk in $d = 2$ is recurrent, we will see every point in B_{r_0} infinitely often with probability one. Hence, there is a time T_0 after which simple random walk has crossed every path of length 2 inside B_{r_0} with probability larger than $1 - \varepsilon$. Now we take m_0 so large, that all different vertices we have seen by time T_0 have different colors with probability $1 - \varepsilon$. If this is the case, each vertex in B_{r_0} has its own unique color and we can indeed reconstruct the scenery inside B_{r_0}.

To understand this, see that we can reconstruct three basic schemes. The first is the origin with its nearest neighbors. Indeed, we know the color of the origin and as we have crossed all edges between the origin and a neighboring point, we also know the colors of the neighbors of the origin. Even more, we know which of these colors are opposite to each other, and which are not, since for the colors of opposite points there is only one way to reach one from the other in two steps (and this way passes the origin), while for points that are not opposite, there are two ways. But this this knowledge suffices to reconstruct the scenery in the origin and its four neighbors up to symmetries of $\mathcal{Z}^2$, i.e. up to equivalence.

Next, knowing the colors of three vertices of a square, we can also find the color of the fourth vertex. So, if in Figure 2 we know the color of the vertices 1,2, and 3, we also know the vertex of color 4, since it is the only color adjacent to the colors of 1 and 3 and different from the color of vertex 2.

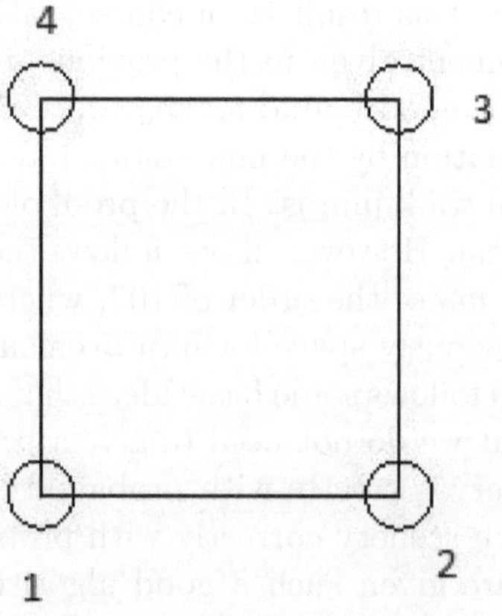

Fig. 2. The colors of a square

With this tool at our disposal we are able to extend the knowledge of the scenery in the origin and its four neighbors to the knowledge of the scenery on the square of side length 2, centered in the origin. Indeed, the four missing colors can be identified as the missing color in a square of side length one.

Eventually, there is a third scheme on which we can reconstruct the scenery and this is the 4-star (see Figure 3).

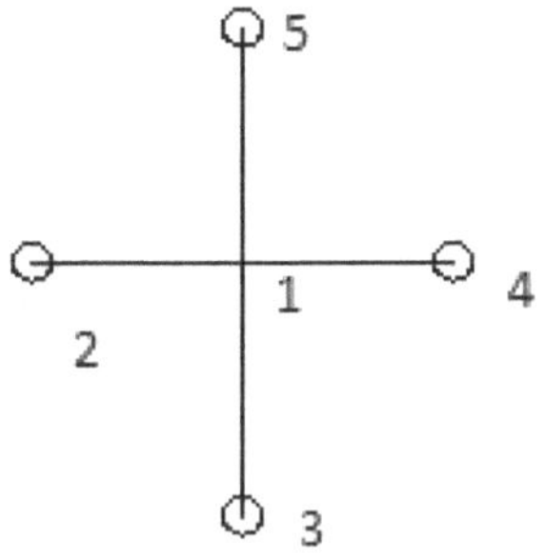

Fig. 3. The colors of 4-star

If there we know the color of the vertex 1 in the center as well as the colors of vertices 2,3, and 4, we will also be able to detect the color of vertex 5, as it is the only color adjacent to the color of 1, but different from the color of vertices 2,3 and 4. This allows to extend the knowledge of the scenery to four more vertices.

Alternating the reconstruction on a square and a 4-star, we are eventually able to reconstruct the scenery on the entire ball B_{r_0}. This completes the beginning of the induction.

For the induction step $r \mapsto r+1$, we use the knowledge of B_r to find markers that tell that we are close to the part of the scenery we already know. To this end we build the collection of all words, i.e. subsequent colors of length $c \log r$ (for an appropriate constant c) that can be read horizontally when we are in B_r. If we read enough of such words, we take this as an indicator, that we are inside B_r. This indicator is reliable roughly for a time interval of length e^{r^2}, since the next ball of radius r in which the colors resemble those of B_r is roughly distance e^{r^2} apart. It can be shown (using precise estimates on the local time of two-dimensional random walk as in [4] and the computation of the maximal disc covered by random walk in two dimensions up to time t as in [25]) that within this time window we return sufficiently often to the origin and whenever we are inside B_r, we have sufficiently many walks to the boundary. These walks can be followed closely, since we know the scenery inside B_r. We extend the walks by one step. This first new color then is the color of a boundary point. For a more technical explanation the reader is referred to [17] or [15].

It should be mentioned that Theorem 10 up to now is the only higher-dimensional result on scenery reconstruction in a strict sense.

In dimensions $d \geq 3$ of course, we cannot expect to obtain scenery reconstruction for symmetric, simple random walk, as random walk in these dimensions in no longer recurrent, and, in particular, we do not even see all the vertices. A possible setup in which the reconstruction problem makes sense also in higher dimensions was recently presented by Pachon and Popov in [24]. As underlying random process they take supercritical branching random walk. To be more precise, they consider random walkers, that in each step either branch into two walkers or do nothing. After that all the walkers perform a usual random walk step. This process is recurrent for all d, i.e. each point in $\mathcal{Z}^d$ is seen by the entire process infinitely often, even though every single random walk path is transient. The color record in this setup is given by all the observation of all the walkers together with the information which of the walkers has read them.

Under these assumptions Pachon and Popov prove in [24] a mild form of scenery reconstruction. Namely they show that with probability 1 a random two-color scenery can be reconstructed, if not only the current color can observed by each walker, but also an entire block (of a certain size) of scenery around each of the walkers. This result may not seem to be too surprising, yet the technique of the authors adds a new element to the know scenery reconstruction methods. They identify so-called "good" blocks, i.e. blocks where the scenery is sufficiently diverse that is can be reconstructed locally, and then they prove that these good blocks percolate.

In the same spirit Pachon, Popov, and Matzinger ([22]) are able to drop the assumption that each walker sees a block of the scenery and replace it by a lager number of colors, namely $2d$. This result shows that improvement over the results in [24] is possible.

On the other hand, one can hardly avoid the impression, that with an exponentially growing number of walkers and a block of scenery around each of the walkers' positions that can be observed, one has a certain amount of information overkill. It is certainly conceivable that there is space for improvements: First, one should be able to reconstruct with the direct observations, only, i.e. without entire blocks of scenery that are observed. Secondly, one should be able to limit the number of walkers in such a way, that the process is still recurrent. These are subjects for future research.

Another direction which is currently being investigated is a combination of random walk in random scenery with the very popular subject of random walk in random environment (see e.g. [27] for a short survey). Here challenging problems and interesting phenomena can be expected.

Acknowledgement. The authors thanks tow anonymous referees for a very careful reading of the text.

References

1. Ahlswede, R.: General theory of information transfer: updated. Discrete Appl. Math. 156(9), 1348–1388 (2008), `http://dx.doi.org/10.1016/j.dam.2007.07.007`, `doi:10.1016/j.dam.2007.07.007`
2. Ahlswede, R., Cicalese, F., Deppe, C.: Searching with lies under error cost constraints. Discrete Appl. Math. 156(9), 1444–1460 (2008), `http://dx.doi.org/10.1016/j.dam.2007.04.033`, doi:10.1016/j.dam.2007.04.033

3. Benjamini, I., Kesten, H.: Distinguishing sceneries by observing the scenery along a random walk path. J. Anal. Math. 69, 97–135 (1996), `http://dx.doi.org/10.1007/BF02787104`, doi:10.1007/BF02787104
4. Erdős, P., Taylor, S.J.: Some problems concerning the structure of random walk paths. Acta Math. Acad. Sci. Hungar. 11, 137–162 (unbound insert) (1960)
5. Fleischmann, K., Mörters, P., Wachtel, V.: Moderate deviations for a random walk in random scenery. Stochastic Process. Appl. 118(10), 1768–1802 (2008), `http://dx.doi.org/10.1016/j.spa.2007.11.001`, doi:10.1016/j.spa.2007.11.001
6. Harris, M., Keane, M.: Random coin tossing. Probab. Theory Related Fields 109(1), 27–37 (1997), `http://dx.doi.org/10.1007/s004400050123`, doi:10.1007/s004400050123
7. Hart, A., Matzinger, H.: Markers for error-corrupted observations. Stochastic Process. Appl. 116(5), 807–829 (2006), `http://dx.doi.org/10.1016/j.spa.2005.11.012`, doi:10.1016/j.spa.2005.11.012
8. Kalikow, S.A.: T, T^{-1} transformation is not loosely Bernoulli. Ann. of Math. (2) 115(2), 393–409 (1982), `http://dx.doi.org/10.2307/1971397`, doi:10.2307/1971397
9. Keane, M., den Hollander, W.T.F.: Ergodic properties of color records. Phys. A 138(1-2), 183–193 (1986), `http://dx.doi.org/10.1016/0378-4371(86)90179-2`, doi:0.1016/0378-4371(86)90179-2
10. Kesten, H.: Detecting a single defect in a scenery by observing the scenery along a random walk path. In: Itô's Stochastic Calculus and Probability Theory, pp. 171–183. Springer, Tokyo (1996)
11. Kesten, H.: Distinguishing and reconstructing sceneries from observations along random walk paths. In: Microsurveys in Discrete Probability, Princeton, NJ, 1997. DIMACS Ser. Discrete Math. Theoret. Comput. Sci, vol. 41, pp. 75–83. Amer. Math. Soc., Providence (1998)
12. Lember, J., Matzinger, H.: Information recovery from a randomly mixed up message-text. Electron. J. Probab. 13(15), 396–466 (2008), `http://dx.doi.org/10.1214/EJP.v13-491`, doi:10.1214/EJP.v13-491
13. Levin, D.A., Pemantle, R., Peres, Y.: A phase transition in random coin tossing. Ann. Probab. 29(4), 1637–1669 (2001), `http://dx.doi.org/10.1214/aop/1015345766`, doi:10.1214/aop/1015345766
14. Lindenstrauss, E.: Indistinguishable sceneries. Random Structures Algorithms 14(1), 71–86 (1999), `http://dx.doi.org/10.1002/(SICI)1098-2418(1999010)14:1<71::AID-RSA4>3.0.CO;2-9` , doi:10.1002/(SICI)1098-2418(1999010)14:1<71::AID-RSA4>3.0.CO;2-9
15. Löwe, M.: Rekonstruktion zufälliger Landschaften. Math. Semesterber. 48(1), 29–48 (2001), `http://dx.doi.org/10.1007/PL00009931`, doi:10.1007/PL00009931
16. Löwe, M., Matzinger, H., Merkl, F.: Reconstructing a multicolor random scenery seen along a random walk path with bounded jumps. Electron. J. Probab. 9(15), 436–507 (2004) (electronic), `http://www.math.washington.edu/~ejpecp/EjpVol9/paper15.abs.html`
17. Löwe, M., Matzinger III, H.: Scenery reconstruction in two dimensions with many colors. Ann. Appl. Probab. 12(4), 1322–1347 (2002), `http://dx.doi.org/10.1214/aoap/1037125865`, doi:10.1214/aoap/1037125865
18. Löwe, M., Matzinger III, H.: Reconstruction of sceneries with correlated colors. Stochastic Process. Appl. 105(2), 175–210 (2003), `http://dx.doi.org/10.1016/S0304-4149(03)00003-6`, doi:10.1016/S0304-4149(03)00003-6

19. Matzinger, H.: Reconstructing a three-color scenery by observing it along a simple random walk path. Random Structures Algorithms 15(2), 196–207 (1999), http://dx.doi.org/10.1002/(SICI)1098-2418(199909)15:2<196::AID-RSA5>3.3.CO;2-R, doi:10.1002/(SICI)1098-2418(199909)15:2<196::AID-RSA5>3.3.CO;2-R
20. Matzinger, H.: Reconstructing a two-color scenery by observing it along a simple random walk path. Ann. Appl. Probab. 15(1B), 778–819 (2005), http://dx.doi.org/10.1214/105051604000000972, doi:10.1214/105051604000000972
21. Matzinger, H.F.: Reconstruction of a one dimensional scenery seen along the path of a random walk with holding, ProQuest LLC, Ann Arbor, MI, Thesis (Ph.D.)–Cornell University (1999)
22. Matzinger, H., Popov, S., Pachon, A.: Reconstruction of a many-dimensional scenery with branching random walk (preprint) (submitted)
23. Matzinger, H., Rolles, S.W.W.: Reconstructing a random scenery observed with random errors along a random walk path. Probab. Theory Related Fields 125(4), 539–577 (2003), http://dx.doi.org/10.1007/s00440-003-0257-3, doi:10.1007/s00440-003-0257-3
24. Popov, S., Pachon, A.: Scenery reconstruction with branching random walk. Stochastics 83(2), 107–116 (2011), http://dx.doi.org/10.1080/17442508.2010.544973, doi:10.1080/17442508.2010.544973
25. Révész, P.: Estimates of the largest disc covered by a random walk. Ann. Probab. 18(4), 1784–1789 (1990)
26. Spencer, J., Winkler, P.: Three thresholds for a liar. Combin. Probab. Comput. 1(1), 81–93 (1992), http://dx.doi.org/10.1017/S0963548300000080, doi:10.1017/S0963548300000080
27. Zeitouni, O.: Random walks in random environments. J. Phys. A 39(40), R433–R464 (2006), http://dx.doi.org/10.1088/0305-4470/39/40/R01, doi:10.1088/0305-4470/39/40/R01

Multiple Objects: Error Exponents in Hypotheses Testing and Identification

Evgueni Haroutunian and Parandzem Hakobyan

Institute for Informatics and Automation Problems,
National Academy of Sciences of the Republic of Armenia
{evhar,par_h}@ipia.sci.am

Dedicated to the memory of Rudolf Ahlswede

Abstract. We survey a series of investigations of optimal testing of multiple hypotheses concerning various multiobject models.

These studies are a prominent instance of application of methods and techniques developed in Shannon information theory for solution of typical statistical problems.

Keywords: multiple hypotheses testing, LAO tests, many independent objects, dependent objects, multiobject model, identification of distribution, testing with rejection of decision, arbitrarily varying object.

1 Introduction

"One can conceive of Information Theory in the broad sense as covering the theory of Gaining, Transferring, and Storing Information, where the first is usually called Statistics." [2].

Shannon information theory and mathematical statistics interaction revealed to be effective. This interplay is mutually fruitful, in some works results of probability theory and statistics were obtained with application of information-theoretical methods and there are studies where statistical results provide ground for new findings in information theory [13], [15], [17]–[20], [36], [40], [51], [56], [59]–[61].

This paper can serve an illustration of application of information-theoretical methods in statistics: on one hand this is analogy in problem formulation and on the other hand this is employment of technical tools of proof, specifically of the method of types [16], [18].

It is often necessary in statistical research to make decisions regarding the nature and parameters of stochastic model, in particular, the probability distribution of the object. Decisions can be made on the basis of results of observations of the object. The vector of results is called a sample. The correspondence between samples and hypotheses can be designed based on some selected criterion. The procedure of statistical hypotheses detection is called test.

The classical problem of statistical hypothesis testing refers to two hypotheses. Based on data samples a statistician makes decision on which of the two proposed

H. Aydinian, F. Cicalese, and C. Deppe (Eds.): Ahlswede Festschrift, LNCS 7777, pp. 313–345, 2013.

hypotheses must be accepted. Many mathematical investigations, some of which have also applied significance, were implemented in this direction [52].

The need of testing of more than two hypotheses in many scientific and applied fields has essentially increased recently. As an instance microarray analysis could be mentioned [23].

The decisions can be erroneous due to randomness of the sample. The test is considered as good if the probabilities of the errors in given conditions are as small as possible.

Frequently the problem is solved for the case of a tests sequence, where the probabilities of error decrease exponentially as 2^{-NE}, the number of observations N tends to the infinity. We call the exponent of error probability E the *reliability*. In case of two hypotheses, when there is a trad off between the reliabilities corresponding to two possible error probabilities, it is an accepted way to fix the value of one of the reliabilities and try to make the tests sequence get the greatest value of the remaining reliability. Such a test is called *logarithmically asymptotically optimal* (LAO). Such optimal tests were considered first by Hoeffding [50], examined later by Csiszár and Longo [19], Tusnady [59], [60] (he called such test series *exponentially rate optimal* (ERO)), Longo and Sgarro [54]. The term LAO for testing of two hypotheses was proposed by Birge [11]. Amongst papers on testing, associated with information theory, we can also note works of Blahut [12], Natarajan [56], Gutman [26], Anantharam [8], Tuncel [58], Fu and Shen [24], Han [27], Westover [63] and of many others. Some objectives in this direction were first suggested in original introductory article by Dobrushin, Pinsker and Shirjaev [22].

The problem has common features with the issue studied in the information theory on interrelation between the rate R of the code and the exponent E of the error probability. In information theory the relation $E(R)$ according to Shannon is called the *reliability function*, also *rate-reliability function*, while $R(E)$ is named the *E-capacity*, or the *reliability-rate function*, as it was introduced by Haroutunian [29], [35], [45].

The concept of simultaneous investigation of some number of objects of the same type, evidently, was first formulated by Ahlswede and Haroutunian [6] for reliable testing of distributions of multiple items. Of course frequently a statistician can test independent objects separately, but for dependent objects the simultaneous testing is necessary. Therefor investigation must be started from the first simpler case. Combined examination of common properties of many similar objects may be attractive and effective in plenty of other statistical situations.

The organization of this paper is as follows. We start with the definitions and notations in the next section. In section 3 we introduce the problem of multihypotheses testing concerning one object. In section 4 we consider the reliability approach to multihypotheses testing for many independent and dependent objects. Section 5 is dedicated to the problem of statistical identification under condition of optimality. Section 6 is devoted to description of characteristics of

LAO hypotheses testing with permission of rejection of decision for the model consisting of one and of more independent objects.

2 Definitions and Notations

We denote finite sets by script capitals. The cardinality of a set $\mathcal{X}$ is denoted as $|\mathcal{X}|$. Random variables (RVs), which take values in finite sets $\mathcal{X}$, $\mathcal{S}$ are denoted by X, S. Probability distributions (PDs) are denoted by Q, P, G, V, W, $Q \circ V$ (V and W will be used for conditional PDs and notation $Q \circ V$ for joint PD).

Let PD of RV X, characterizing an object, be $Q \stackrel{\Delta}{=} \{Q(x),\ x \in \mathcal{X}\}$, and conditional PD of RV X be $V \stackrel{\Delta}{=} \{V(x|s),\ x \in \mathcal{X},\ s \in \mathcal{S}\}$ for given value of state s of the object.

The Shannon entropy $H_Q(X)$ of RV X with PD Q is:

$$H_Q(X) \stackrel{\Delta}{=} -\sum_{x \in \mathcal{X}} Q(x) \log Q(x).$$

The conditional entropy $H_{P,V}(X \mid S)$ of RV X for given RV S with corresponding PDs is:

$$H_{P,V}(X \mid S) \stackrel{\Delta}{=} -\sum_{x \in \mathcal{X}, s \in \mathcal{S}} P(s) V(x|s) \log V(x|s).$$

The divergence (Kullback-Leibler information, or "distance") of PDs Q and G on $\mathcal{X}$ is:

$$D(Q||G) \stackrel{\Delta}{=} \sum_{x \in \mathcal{X}} Q(x) \log \frac{Q(x)}{G(x)},$$

and conditional divergence of the PD $P \circ V = \{P(s)V(x|s),\ x \in \mathcal{X},\ s \in \mathcal{S}\}$ and PD $P \circ W = \{P(s)\, W(x|s), x \in \mathcal{X},\ s \in \mathcal{S}\}$ is:

$$D(P \circ V||P \circ W) = D(V||W|P) \stackrel{\Delta}{=} \sum_{x,s} P(s)V(x|s) \log \frac{V(x|s)}{W(x|s)}.$$

For our investigations we use the method of types, one of the important technical tools in Shannon theory [18,16]. The type $Q_{\mathbf{x}}$ of a vector $\mathbf{x} = (x_1, ..., x_N) \in \mathcal{X}^N$ is a PD (the empirical distribution)

$$Q_{\mathbf{x}} = \left\{ Q_{\mathbf{x}}(x) = \frac{N(x|\mathbf{x})}{N}, x \in \mathcal{X} \right\},$$

where $N(x|\mathbf{x})$ is the number of repetitions of symbol x in vector $\mathbf{x}$.

The joint type of vectors $\mathbf{x} \in \mathcal{X}^N$ and $\mathbf{s} = (s_1, s_2, ..., s_N) \in \mathcal{S}^N$ is the PD

$$P_{\mathbf{s},\mathbf{x}} = \left\{ \frac{N(s, x|\mathbf{s}, \mathbf{x})}{N},\ x \in \mathcal{X},\ s \in \mathcal{S} \right\},$$

where $N(s,x|\mathbf{s},\mathbf{x})$ is the number of occurrences of symbols pair (s,x) in the pair of vectors $(\mathbf{x},\mathbf{s})$. The conditional type of $\mathbf{x}$ for given $\mathbf{s}$ is a conditional PD

$$V_{\mathbf{x}|\mathbf{s}} = \{V_{\mathbf{x}|\mathbf{s}}(x|s), x \in \mathcal{X}, s \in \mathcal{S}\},$$

defined by the relation $N(s,x|\mathbf{s},\mathbf{x}) = N(s|\mathbf{s})V_{\mathbf{x}|\mathbf{s}}(x|s)$ for all $x \in \mathcal{X}$, $s \in \mathcal{S}$.

We denote by $\mathcal{Q}^N(\mathcal{X})$ the set of all types of vectors in $\mathcal{X}^N$ for given N, by $\mathcal{P}^N(\mathcal{S})$ – the set of all types of vectors $\mathbf{s}$ in $\mathcal{S}^N$ and by $\mathcal{V}^N(\mathcal{X}|\mathbf{s})$ – the set of all possible conditional types of vectors $\mathbf{x}$ in $\mathcal{X}^N$ for given $\mathbf{s} \in \mathcal{S}^N$. The set of vectors $\mathbf{x}$ of type Q is denoted by $\mathcal{T}_Q^N(X)$ and the family of vectors $\mathbf{x}$ of conditional type V for given $\mathbf{s} \in \mathcal{S}^N$ of type P by $\mathcal{T}_{P,V}^N(X \mid \mathbf{s})$. The set of all possible PDs Q on $\mathcal{X}$ and PDs P on $\mathcal{S}$ is denoted, correspondingly, by $\mathcal{Q}(X)$ and $\mathcal{P}(\mathcal{S})$.

We need the following frequently used inequalities [18]:

$$\mid \mathcal{Q}^N(\mathcal{X}) \mid \leq (N+1)^{|\mathcal{X}|}, \tag{1}$$

$$\mid \mathcal{V}^N(\mathcal{X}|\mathbf{s}) \mid \leq (N+1)^{|\mathcal{S}||\mathcal{X}|}, \tag{2}$$

for any type $Q \in \mathcal{Q}^N(\mathcal{X})$

$$(N+1)^{-|\mathcal{X}|} \exp\{NH_Q(X)\} \leq \mid \mathcal{T}_Q^N(X) \mid \leq \exp\{NH_Q(X)\}, \tag{3}$$

and for any type $P \in \mathcal{P}^N(\mathcal{S})$ and $V \in \mathcal{V}^N(\mathcal{X}|\mathbf{s})$

$$(N+1)^{-|\mathcal{S}||\mathcal{X}|} \exp\{NH_{P,V}(X|S)\} \leq \mid \mathcal{T}_{P,V}^N(X \mid \mathbf{s}) \mid \leq \exp\{NH_{P,V}(X|S)\}. \tag{4}$$

3 LAO Testing of Multiple Hypotheses for One Object

Generalization of results on two hypotheses noted in above. The problem of optimal testing of *multiple* hypotheses was proposed by Dobrushin [21], and was investigated in [30] – [34]. The problem of multiple hypotheses LAO testing for a discrete stationary Markov source of observations was solved by Haroutunian [31] – [33].

Here for clearness we expose the results on multiple hypotheses LAO testing for the case of the most simple invariant object.

Let $\mathcal{X}$ be a finite set of values RV X. M possible PDs $G_m = \{G_m(x),\ x \in \mathcal{X}\}$, $m = \overline{1,M}$, of RV X characterizing the object are known. The statistician must detect one among M alternative hypotheses G_m, using sample $\mathbf{x} = (x_1, ..., x_N)$ of results of N independent observations of the object.

The procedure of decision making is a non-randomized test $\varphi_N(\mathbf{x})$, it can be defined by division of the sample space $\mathcal{X}^N$ on M disjoint subsets $\mathcal{A}_m^N = \{\mathbf{x} : \varphi_N(\mathbf{x}) = m\}$, $m = \overline{1,M}$. The set $\mathcal{A}_m^N$ consists of all samples $\mathbf{x}$ for which the hypothesis G_m must be adopted. We study the probabilities $\alpha_{l|m}(\varphi_N)$ of the erroneous acceptance of hypothesis G_l provided that G_m is true

$$\alpha_{l|m}(\varphi_N) \triangleq G_m^N(\mathcal{A}_l^N),\ l,m = \overline{1,M},\ \ m \neq l. \tag{5}$$

The probability to reject the hypothesis G_m, when it is true, is also considered

$$\alpha_{m|m}(\varphi_N) \stackrel{\Delta}{=} \sum_{l\neq m} \alpha_{l|m}(\varphi_N) = G_m^N(\overline{\mathcal{A}_m^N}) = (1 - G_m^N(\mathcal{A}_m^N)). \tag{6}$$

A quadratic matrix of M^2 error probabilities $\{\alpha_{l|m}(\varphi_N),\ m = \overline{1,M},\ l = \overline{1,M}\}$ is the *power* of the tests.

Error probability exponents of the infinite sequence φ of tests, which we call *reliabilities*, are defined as follows:

$$E_{l|m}(\varphi) \stackrel{\Delta}{=} \lim_{N\to\infty} \inf \left\{ -\frac{1}{N} \log \alpha_{l|m}(\varphi_N) \right\},\quad m,l = \overline{1,M}. \tag{7}$$

We see from (6) and (7) that

$$E_{m|m}(\varphi) = \min_{l\neq m} E_{l|m}(\varphi),\quad m = \overline{1,M}. \tag{8}$$

The matrix

$$\mathbf{E}(\varphi) = \begin{pmatrix} E_{1|1} \ \dots\ E_{l|1} \ \dots\ E_{M|1} \\ \dots\dots\dots\dots\dots\dots \\ E_{1|m} \dots E_{l|m} \dots E_{M|m}, \\ \dots\dots\dots\dots\dots\dots \\ E_{1|M} \dots E_{l|M} \dots E_{M|M} \end{pmatrix}$$

called the *reliabilities matrix* of the tests sequence φ is the object of our investigation.

We name a sequence φ^* of tests is LAO if for given positive values of $M-1$ diagonal elements of matrix $\mathbf{E}(\varphi^*)$ $E_{1|1}$, $E_{2|2}$, ..., $E_{M-1|M-1}$, the procedure provides maximal values for all other elements of it.

Now we form the LAO test by constructing decision sets noted $\mathcal{R}_m^{(N)}$. Given strictly positive numbers $E_{m|m}$, $m = \overline{1, M-1}$, we define the following regions:

$$\mathcal{R}_m \stackrel{\Delta}{=} \{Q:\ D(Q||G_m) \leq E_{m|m}\},\quad m = \overline{1, M-1}, \tag{9}$$

$$\mathcal{R}_M \stackrel{\Delta}{=} \{Q:\ D(Q||G_m) > E_{m|m},\quad m = \overline{1, M-1}\}, \tag{10}$$

$$\mathcal{R}_m^{(N)} \stackrel{\Delta}{=} \mathcal{R}_m \bigcap \mathcal{Q}^N(\mathcal{X}),\quad m = \overline{1,M}, \tag{11}$$

and corresponding values:

$$E_{m|m}^* = E_{m|m}^*(E_{m|m}) \stackrel{\Delta}{=} E_{m|m},\quad m = \overline{1, M-1}, \tag{12}$$

$$E^*_{m|l} = E^*_{m|l}(E_{m|m}) \stackrel{\Delta}{=} \inf_{Q \in \mathcal{R}_m} D(Q||G_l), \quad l = \overline{1,M}, \;\; m \neq l, \;\; m = \overline{1, M-1}, \tag{13}$$

$$E^*_{M|m} = E^*_{M|m}(E_{1|1}, E_{2|2}, ..., E_{M-1|M-1}) \stackrel{\Delta}{=} \inf_{P \in \mathcal{R}_M} D(Q||G_m), \quad m = \overline{1, M-1}, \tag{14}$$

$$E^*_{M|M} = E^*_{M|M}(E_{1|1}, E_{2|2}, ..., E_{M-1|M-1}) \stackrel{\Delta}{=} \min_{m: m=\overline{1,M-1}} E^*_{M|m}. \tag{15}$$

Theorem 3.1 [34]. *If for the described model all conditional PDs G_m, $m = \overline{1,M}$, are different in the sense that, $D(G_l||G_m) > 0$, $l \neq m$, and the positive numbers $E_{1|1}, E_{2|2}, ..., E_{M-1|M-1}$ are such that the following $M-1$ inequalities, called compatibility conditions, hold*

$$E_{1|1} < \min_{m=\overline{2,M}} D(G_m||G_1),$$
$$E_{m|m} < \min \left[\min_{l=\overline{1,m-1}} E^*_{l|m}(E_{l|l}), \; \min_{l=\overline{m+1,L}} D(G_l||G_m) \right], \quad m = \overline{2, M-1}, \tag{16}$$

then there exists a LAO sequence φ^* of tests, the reliabilities matrix of which $\mathbf{E}(\varphi^*) = \{E^*_{m|l}\}$ is defined in (12)–(15) and all elements of it are positive.

When one of inequalities (16) is violated, then at least one element of matrix $\mathbf{E}(\varphi^*)$ is equal to 0.

The proof of Theorem 3.1 given in to the Appendix.

It is worth to formulate the following useful property of reliabilities matrix of the LAO test.

Remark 3.1 [40]. *The diagonal elements of the reliabilities matrix of the LAO test in each row are equal only to the element of the last column:*

$$E^*_{m|m} = E^*_{M|m}, \text{ and } E^*_{m|m} < E^*_{l|m}, \quad l = \overline{1, M-1}, \;\; l \neq m, \;\; m = \overline{1,M}. \tag{17}$$

That is the elements of the last column are equal to the diagonal elements of the same row and due to (8) are minimal in this row. Consequently the first $M-1$ elements of the last column also can be used as given parameters for construction of a LAO test.

4 The Reliability Approach to Multihypotheses Testing for Many Objects

In [6] Ahlswede and Haroutunian proposed a new aspect of the statistical theory – investigation of models with many objects. This work developed the ideas of papers on Information theory [1], [5], of papers on many hypotheses testing [30]-[34] and of book [9], devoted to research of sequential procedures solving decision problems such as ranking and identification. The problem of hypotheses testing for the model consisting of two independent and of two strictly dependent objects (when they cannot admit the same distribution) with two possible hypothetical distributions were solved in [6]. In [40] specific characteristics of the model consisting of $K(\geq 2)$ objects each independently of others following one of given $M(\geq 2)$ probability distributions were explored. In [49] the model composed of stochastically related objects was investigated. The result concerning two independent Markov chains is presented in [37]. In this section we expose these results.

4.1 Multihypotheses LAO Testing for Many Independent Objects

Let us now consider the model with three independent similar objects. For brevity we solve the problem for three objects, the generalization of the problem for K independent objects will be discussed hereafter along the text.

Let X_1, X_2 and X_3 be independent RVs taking values in the same finite set $\mathcal{X}$, each of them with one of M hypothetical PDs $G_m = \{G_m(x),\ \ x \in \mathcal{X}\}$. These RVs are the characteristics of the objects. The random vector (X_1, X_2, X_3) assumes values $(x^1, x^2, x^3) \in \mathcal{X}^3$.

Let $(\mathbf{x_1}, \mathbf{x_2}, \mathbf{x_3}) \stackrel{\Delta}{=} ((x_1^1, x_1^2, x_1^3), \ldots, (x_n^1, x_n^2, x_n^3), \ldots, (x_N^1, x_N^2, x_N^3))$, $x_n^k \in \mathcal{X}$, $k = \overline{1,3}$, $n = \overline{1,N}$, be a vector of results of N independent observations of the family (X_1, X_2, X_3). The test has to determine unknown PDs of the objects on the base of observed data. The detection for each object should be made from the same set of hypotheses: G_m, $m = \overline{1,M}$. We call this procedure the *compound test* for three objects and denote it by Φ_N, it can be composed of three individual tests φ_N^1, φ_N^2, φ_N^3 for each of the three objects. The test φ_N^i, $i = \overline{1,3}$, is a division of the space $\mathcal{X}^N$ into M disjoint subsets $\mathcal{A}_m^i$, $m = \overline{1,M}$. The set $\mathcal{A}_m^i$, $m = \overline{1,M}$, contains all vectors $\mathbf{x}_i$ for which the hypothesis G_m is adopted. Hence test Φ_N is realised by division of the space $\mathcal{X}^N \times \mathcal{X}^N \times \mathcal{X}^N$ into M^3 subsets $\mathcal{A}_{m_1,m_2,m_3} = \mathcal{A}_{m_1}^1 \times \mathcal{A}_{m_2}^2 \times \mathcal{A}_{m_3}^3$, $m_i = \overline{1,M}$, $i = \overline{1,3}$. We denote the infinite sequence of compound tests by Φ. When we have K independent objects the test Φ is composed of tests $\varphi^1, \varphi^2, \ldots, \varphi^K$.

The probability of the falsity of acceptance of hypotheses triple $(G_{l_1}, G_{l_2}, G_{l_3})$ by the test Φ_N provided that the triple of hypotheses $(G_{m_1}, G_{m_2}, G_{m_3})$ is true, where $(m_1, m_2, m_3) \neq (l_1, l_2, l_3)$, $m_i, l_i = \overline{1,M}$, $i = \overline{1,3}$, is:

$$\begin{aligned}
\alpha_{l_1,l_2,l_3|m_1,m_2,m_3}(\Phi_N) &\stackrel{\Delta}{=} G_{m_1}^N \circ G_{m_2}^N \circ G_{m_3}^N \left(\mathcal{A}_{l_1,l_2,l_3}^N\right) \\
&\stackrel{\Delta}{=} G_{m_1}^N \left(\mathcal{A}_{l_1}^N\right) \cdot G_{m_2}^N \left(\mathcal{A}_{l_2}^N\right) \cdot G_{m_3}^N \left(\mathcal{A}_{l_3}^N\right) \\
&= \sum_{\mathbf{x}_1 \in \mathcal{A}_{l_1}^N} G_{m_1}^N(\mathbf{x}_1) \sum_{\mathbf{x}_2 \in \mathcal{A}_{l_2}^N} G_{m_2}^N(\mathbf{x}_2) \sum_{\mathbf{x}_3 \in \mathcal{A}_{l_3}^N} G_{m_3}^N(\mathbf{x}_3).
\end{aligned}$$

The probability to reject a true triple of hypotheses $(G_{m_1}, G_{m_2}, G_{m_3})$ by analogy with (6) is defined as follows:

$$\alpha_{m_1,m_2,m_3|m_1,m_2,m_3}(\Phi_N) \stackrel{\Delta}{=} \sum_{(l_1,l_2,l_3) \neq (m_1,m_2,m_3)} \alpha_{l_1,l_2,l_3|m_1,m_2,m_3}(\Phi_N). \qquad (18)$$

We study corresponding reliabilities $E_{l_1,l_2,l_3|m_1,m_2,m_3}(\Phi)$ of the sequence of tests Φ,

$$E_{l_1,l_2,l_3|m_1,m_2,m_3}(\Phi) \triangleq \lim_{N\to\infty} \inf \left\{ -\frac{1}{N} \log \alpha_{l_1,l_2,l_3|m_1,m_2,m_3}(\Phi_N) \right\},$$

$$m_i, l_i = \overline{1,M}, \quad i = \overline{1,3}. \tag{19}$$

Definitions (18) and (19) imply (compare with (8)) that

$$E_{m_1,m_2,m_3|m_1,m_2,m_3}(\Phi) = \min_{(l_1,l_2,l_3)\neq(m_1,m_2,m_3)} E_{l_1,l_2,l_3|m_1,m_2,m_3}(\Phi). \tag{20}$$

Our aim is to analyze the reliabilities matrix $\mathbf{E}(\Phi^*) = \{E_{l_1,l_2,l_3|m_1,m_2,m_3}(\Phi^*)\}$ of LAO test sequence Φ^* for three objects. We call the test sequence LAO for the model with many objects if for given positive values of certain part of elements of reliabilities matrix the procedure provides maximal values for all other elements of it.

Let us denote by $\mathbf{E}(\varphi^i)$ the reliabilities matrices of the sequences of tests φ^i, $i = \overline{1,3}$. The following Lemma is a generalization of Lemma from [6].

Lemma 4.1. *If elements $E_{l|m}(\varphi^i)$, $m, l = \overline{1,M}$, $i = \overline{1,3}$, are strictly positive, then the following equalities hold for $\mathbf{E}(\Phi)$, $\Phi = (\varphi^1, \varphi^2, \varphi^3)$, $l_i, m_i = \overline{1,M}$:*

$$E_{l_1,l_2,l_3|m_1,m_2,m_3}(\Phi) = \sum_{i=\overline{1,3}:\ m_i\neq l_i} E_{l_i|m_i}(\varphi^i), \tag{21}$$

The proof of Lemma 4.1 is given in Appendix.

Now we shall show how we can construct the LAO test from the set of compound tests when $3(M-1)$ strictly positive elements of the reliabilities matrix $E_{M,M,M|m,M,M}$, $E_{M,M,M|M,m,M}$ and $E_{M,M,M|M,M,m}$, $m = \overline{1,M-1}$, are preliminarily given.

The following subset of tests:

$$\mathcal{D} = \{\Phi : E_{m|m}(\varphi^i) > 0, \quad m = \overline{1,M}, \ i = \overline{1,3}\}$$

is distinguished by the property that when $\Phi \in \mathcal{D}$ the elements $E_{M,M,M|m,M,M}(\Phi)$, $E_{M,M,M|M,m,M}(\Phi)$ and $E_{M,M,M|M,M,m}(\Phi)$, $m = \overline{1,M-1}$, of the reliabilities matrix are strictly positive.

Indeed, because $E_{m|m}(\varphi^i) > 0$, $m = \overline{1,M}$, $i = \overline{1,3}$, then in view of (8) $E_{M|m}(\varphi^i)$ are also strictly positive. From equality (21) we obtain that the noted elements are strictly positive for $\Phi \in \mathcal{D}$ and $m = \overline{1,M-1}$

$$E_{M,M,M|m,M,M}(\Phi) = E_{M|m}(\varphi^1), \tag{22}$$

$$E_{M,M,M|M,m,M}(\Phi) = E_{M|m}(\varphi^2), \tag{23}$$

$$E_{M,M,M|M,M,m}(\Phi) = E_{M|m}(\varphi^3). \tag{24}$$

For given positive elements

$$E_{M,M,M|m,M,M},\ E_{M,M,M|M,m,M},\ E_{M,M,M|M,M,m},\ m=\overline{1,M-1},$$

define the following family of decision sets of PDs:

$$\mathcal{R}_m^{(i)} \triangleq \{Q : D(Q||G_m) \le E_{M,M,M|m_1,m_2,m_3},\ m_i = m,\ m_j = M,\ \ i\neq j, j=\overline{1,3}\}$$

$$m=\overline{1,M-1},\ \ i=\overline{1,3}, \tag{25}$$

$$\mathcal{R}_M^{(i)} \triangleq \{Q : D(Q||G_m) > E_{M,M,M|m_1,m_2,m_3},\ m_i = m,\ m_j = M,\ i\neq j, j=\overline{1,3},$$

$$m=\overline{1,M-1}\},\ \ i=\overline{1,3}. \tag{26}$$

Define also the elements of the reliability matrix of the compound LAO test for three objects:

$$\begin{aligned}
E^*_{M,M,M|m,M,M} &\triangleq E_{M,M,M|m,M,M},\\
E^*_{M,M,M|M,m,M} &\triangleq E_{M,M,M|M,m,M},\\
E^*_{M,M,M|M,M,m} &\triangleq E_{M,M,M|M,M,m},
\end{aligned} \tag{27}$$

$$E^*_{l_1,l_2,l_3|m_1,m_2,m_3} \triangleq \inf_{Q\in R_{l_i}^{(i)}} D(Q||G_{m_i}),$$

$$i=\overline{1,3},\ m_k = l_k,\ m_i \neq l_i,\ i\neq k,\ k\in[[1,2,3]-i], \tag{28}$$

$$E^*_{m_1,m_2,m_3|l_1,l_2,m_3} \triangleq \sum_{i\neq k} \inf_{Q\in R_{l_i}^{(i)}} D(Q||G_{m_i}),$$

$$m_k = l_k,\ m_i\neq l_i,\ k=\overline{1,3},\ i\in[[1,2,3]-k], \tag{29}$$

$$E^*_{l_1,l_2,l_3|m_1,m_2,m_3} \triangleq \sum_{i=1}^{3} \inf_{Q\in R_{l_i}^{(i)}} D(Q||G_{m_i}),\ m_i\neq l_i, i=\overline{1,3}. \tag{30}$$

The following theorem is a generalization and improvement of the corresponding theorem proved in [6] for the case $K=2,\ M=2$.

Theorem 4.1 [40]. *For considered model with three objects, if all distributions $G_m,\ m=\overline{1,M}$, are different, (and equivalently $D(G_l||G_m)>0,\ l\neq m,\ l,m=\overline{1,M}$), then the following statements are valid:*

a) when given strictly positive elements $E_{M,M,M|m,M,M}$, $E_{M,M,M|M,m,M}$ and $E_{M,M,M|M,M,m}$, $m=\overline{1,M-1}$, meet the following conditions

$$\max(E_{M,M,M|1,M,M}, E_{M,M,M|M,1,M}, E_{M,M,M|M,M,1})$$

$$< \min_{l=\overline{2,M}} D(G_l||G_1), \quad (31)$$

and for $m = \overline{2, M-1}$,

$$E_{M,M,M|m,M,M} < \min\left[\min_{l=\overline{1,m-1}} E^*_{l,m,m|m,m,m}, \min_{l=\overline{m+1,M}} D(G_l||G_m)\right], \quad (32)$$

$$E_{M,M,M|M,m,M} < \min\left[\min_{l=\overline{1,m-1}} E^*_{m,l,m|m,m,m}, \min_{l=\overline{m+1,M}} D(G_l||G_m)\right], \quad (33)$$

$$E_{M,M,M|M,M,m} < \min\left[\min_{l=\overline{1,m-1}} E^*_{m,m,l|m,m,m}, \min_{l=\overline{m+1,M}} D(G_l||G_m)\right], \quad (34)$$

then there exists a LAO test sequence $\Phi^* \in \mathcal{D}$, the reliability matrix of which $\mathbf{E}(\Phi^*)$ is defined in (27)–(30) and all elements of it are positive,

b) if even one of the inequalities (31)–(34) is violated, then there exists at least one element of the matrix $\mathbf{E}(\Phi^*)$ equal to 0.

For the proof of Theorem 4.1 see Appendix.

When we consider the model with K independent objects the generalization of Lemma 4.1 will take the following form.

Lemma 4.2. *If elements $E_{l_i|m_i}(\varphi^i)$, $m_i, l_i = \overline{1,M}$, $i = \overline{1,K}$, are strictly positive, then the following equalities hold for $\Phi = (\varphi^1, \varphi^2, ..., \varphi^K)$:*

$$E_{l_1,l_2,...,l_K|m_1,m_2,...,m_K}(\Phi) = \sum_{i=\overline{1,K}:\ m_i \neq l_i} E_{l_i|m_i}(\varphi^i).$$

For given $K(M-1)$ strictly positive elements $E_{M,M,...,M|m,M,...,M}$, $E_{M,M,...,M|M,m,...,M}$, ... , $E_{M,...,M,M,|M,M...,m}$, $m = \overline{1, M-1}$, for K independent objects we can find the LAO test Φ^* in a way similar to the case of three independent objects.

Comment 4.1. The idea to renumber K-distributions from 1 to M^K and consider them as PDs of one complex object offers an alternative way of testing for models with K objects. We can give $M^K - 1$ diagonal elements of the corresponding large matrix $\mathbf{E}(\Phi)$ and apply Theorem 3.2 concerning one composite object. In this direct algorithm the number of the preliminarily given elements of the matrix $\mathbf{E}(\Phi)$ would be greater (because $M^K - 1 > K(M-1)$, $M \geq 2, K \geq 2$) but the procedure of calculations would be longer than in our algorithm presented above in this section. Advantages of our approach to the problem is optimality in calculations and the possibility to define the LAO tests for each of the separate objects. The approach with renumbering of K-vectors of hypotheses does not have the last opportunity. At the same time in the case of direct algorithm there is opportunity for the investigator to preliminarily given greater number of elements of the matrix $\mathbf{E}(\Phi)$. In applications one of two approaches may be used in conformity with preferences of the investigator.

Example. Some illustrations of outlined results given calculations for an example concerning two objects. The set $\mathcal{X} = \{0,1\}$ contains two elements and the following PDs are given on $\mathcal{X}$: $G_1 = \{0,10;0,90\}$, $G_2 = \{0,85;0,15\}$, $G_3 = \{0,23;0,77\}$. As it follows from relations (27)–(30), several elements of the reliability matrix are functions of one of given elements, there are also elements which are functions of two, or three given elements. For example, in Fig. 1 and Fig. 2 the results of calculations of functions $E_{1,2|2,1}(E_{3,3|1,3}, E_{3,3|3,2})$ and $E_{1,2|2,2}(E_{3,3|1,3})$ are presented. For these distributions we have $\min(D(G_2||G_1), D(G_3||G_1)) \approx 2,2$ and $\min(E_{2,2|2,1}, D(G_3||G_2)) \approx 1,4$. We see that, when the inequalities (31) or (34) are violated, $E_{1,2|2,1} = 0$ and $E_{1,2|2,2} = 0$.

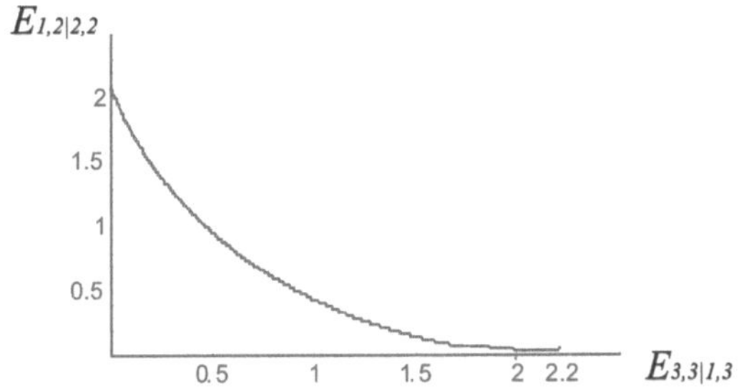

Fig. 1.

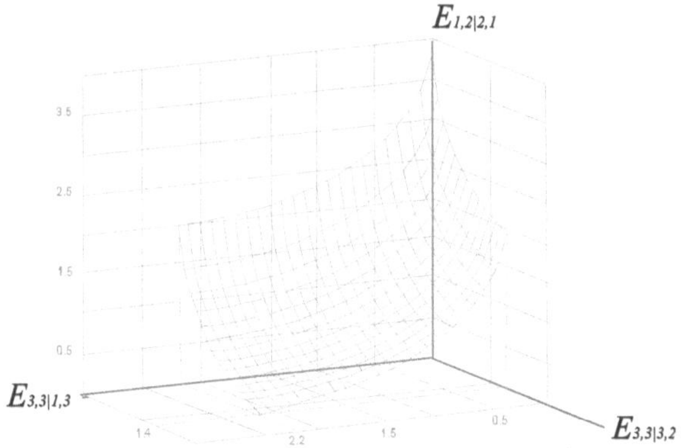

Fig. 2.

4.2 Multihypotheses LAO Testing for Two Dependent Objects

We consider characteristics of procedures of LAO testing of probability distributions of two related (*stochastically, statistically* and *strictly dependent*) objects. We use these terms for different kinds of dependence of two objects.

Let X_1 and X_2 be RVs taking values in the same finite set $\mathcal{X}$ and $\mathcal{Q}(\mathcal{X})$ be the space of all possible distributions on $\mathcal{X}$.

Let $(\mathbf{x_1}, \mathbf{x_2}) = ((x_1^1, x_1^2), (x_2^1, x_2^2), ...(x_N^1, x_N^2))$ be a sequence of results of N independent observations of the pair of objects.

First we consider the model, which consists of two *stochastically* related objects. We name so the following more general dependence. There are given M_1 PDs

$$G_{m_1} = \{G_{m_1}(x^1),\ x^1 \in \mathcal{X}\},\ \ m_1 = \overline{1, M_1}.$$

The first object is characterized by RV X_1 which has one of these M_1 possible PDs, the second object is dependent on the first and is characterized by RV X_2 which can have one of $M_1 \times M_2$ conditional PDs

$$G_{m_2|m_1} = \{G_{m_2|m_1}(x^2|x^1),\ x^1, x^2 \in \mathcal{X}\},\ m_1 = \overline{1, M_1},\ m_2 = \overline{1, M_2}.$$

The joint PD of the pair of objects is

$$G_{m_1,m_2} = G_{m_1} \circ G_{m_2|m_1} = \{G_{m_1,m_2}(x^1, x^2),\ x^1, x^2 \in \mathcal{X}\},$$

where

$$G_{m_1,m_2}(x^1, x^2) = G_{m_1}(x^1) G_{m_2|m_1}(x^2|x^1),\ \ m_1 = \overline{1, M_1},\ m_2 = \overline{1, M_2}.$$

The probability $G^N_{m_1,m_2}(\mathbf{x_1}, \mathbf{x_2})$ of N-vector $(\mathbf{x_1}, \mathbf{x_2})$ is the following product:

$$\begin{aligned} G^N_{m_1,m_2}(\mathbf{x_1}, \mathbf{x_2}) &\triangleq G^N_{m_1}(\mathbf{x_1}) G^N_{m_2|m_1}(\mathbf{x_2}|\mathbf{x_1}) \\ &\triangleq \prod_{n=1}^{N} G_{m_1}(x_n^1) G_{m_2|m_1}(x_n^2|x_n^1), \end{aligned}$$

with $G^N_{m_1}(\mathbf{x_1}) = \prod\limits_{n=1}^{N} G_{m_1}(x_n^1)$, $G^N_{m_2 m_1}(\mathbf{x_2}|\mathbf{x_1}) = \prod\limits_{n=1}^{N} G_{m_2|m_1}(x_n^2|x_n^1)$.

In somewhat particular case, when X_1 and X_2 are related *statistically* [46], [62], the second object depends on the index of PD of the first object but does not depend on value x^1 taken by the first object. The second object is characterized by RV X_2 which can have one of $M_1 \times M_2$ conditional PDs $G_{m_2|m_1} = \{G_{m_2|m_1}(x^2),\ x^2 \in \mathcal{X}\}$, $m_1 = \overline{1, M_1}$, $m_2 = \overline{1, M_2}$.

In the third case of the *strict* dependence, which is a special case of *statistical* dependence, the objects X_1 and X_2 can have only different distributions from the same given family of M PDs G_1, G_2, ..., G_M.

Discussed in Comment 4.1 the *direct approach* to LAO testing of PDs for two related objects, consisting in consideration of the pair of objects as one composite object and then in use of Theorem 3.1, is applicable for the third cases [47]. But now we consider another approach.

Let us remark that test Φ^N can be composed of a pair of tests φ_1^N and φ_2^N for the separate objects: $\Phi^N = (\varphi_1^N, \varphi_2^N)$. Denote by φ^1, φ^2 and Φ the infinite sequences of tests for the first, the second and the pair of objects, respectively.

Let X_1 and X_2 be related *stochastically*. For the object characterized by X_1 the non-randomized test $\varphi_N^1(\mathbf{x_1})$ can be determined by partition of the sample space $\mathcal{X}^N$ on M_1 disjoint subsets $\mathcal{A}_{l_1}^N = \{\mathbf{x_1} : \varphi_N^1(\mathbf{x_1}) = l_1\}$, $l_1 = \overline{1, M_1}$, i.e. the set $\mathcal{A}_{l_1}^N$ consists of vectors $\mathbf{x_1}$ for which the PD G_{l_1} is adopted. The probability $\alpha_{l_1|m_1}(\varphi_N^1)$ of the erroneous acceptance of PD G_{l_1} provided that G_{m_1} is true, $l_1, m_1 = \overline{1, M_1}$, $m_1 \neq l_1$, is defined by the set $\mathcal{A}_{l_1}^N$

$$\alpha_{l_1|m_1}(\varphi_N^1) \stackrel{\Delta}{=} G_{m_1}^N(\mathcal{A}_{l_1}^N).$$

We define the probability to reject G_{m_1}, when it is true, as follows

$$\alpha_{m_1|m_1}(\varphi_N^1) \stackrel{\Delta}{=} \sum_{l_1 : l_1 \neq m_1} \alpha_{l_1|m_1}(\varphi_N^1) = G_{m_1}^N(\overline{\mathcal{A}_{m_1}^N}). \tag{35}$$

The corresponding error probability exponents are:

$$E_{l_1|m_1}(\varphi^1) \stackrel{\Delta}{=} \lim_{N \to \infty} \inf \left\{ -\frac{1}{N} \log \alpha_{l_1|m_1}(\varphi_N^1) \right\}, \quad m_1, l_1 = \overline{1, M_1}. \tag{36}$$

It follows from (35) and (36) that

$$E_{m_1|m_1}(\varphi^1) = \min_{l_1 : l_1 \neq m_1} E_{l_1|m_1}(\varphi^1), \quad l_1, m_1 = \overline{1, M_1}.$$

For construction of the LAO test we assume given strictly positive numbers $E_{m_1|m_1}$, $m = \overline{1, M_1 - 1}$ and define regions $\mathcal{R}_{l_1}$, $l = \overline{1, M_1}$ as in (9)–(10).

For the second object characterized by RV X_2 depending on X_1 the non-randomized test $\varphi_N^2(\mathbf{x_2}, \mathbf{x_1}, l_1)$, based on vectors $(\mathbf{x_1}, \mathbf{x_2})$ and on the index of the hypothesis l_1 adopted for X_1, can be given for each l_1 and $\mathbf{x_1}$ by division of the sample space $\mathcal{X}^N$ on M_2 disjoint subsets

$$\mathcal{A}_{l_2|l_1}^N(\mathbf{x_1}) \stackrel{\Delta}{=} \{\mathbf{x_2} : \varphi_2^N(\mathbf{x_2}, \mathbf{x_1}, l_1) = l_2\}, \; l_1 = \overline{1, M_1}, \; l_2 = \overline{1, M_2}. \tag{37}$$

Let

$$\mathcal{A}_{l_1, l_2}^N \stackrel{\Delta}{=} \{(\mathbf{x_1}, \mathbf{x_2}) : \mathbf{x_1} \in A_{l_1}^N, \; \mathbf{x_2} \in \mathcal{A}_{l_2|l_1}^N(\mathbf{x_1})\}. \tag{38}$$

The probabilities of the erroneous acceptance for $(l_1, l_2) \neq (m_1, m_2)$ are

$$\alpha_{l_1, l_2|m_1, m_2} \stackrel{\Delta}{=} G_{m_1, m_2}^N(\mathcal{A}_{l_1, l_2}^N).$$

The corresponding reliabilities are denoted $E_{l_1, l_2|m_1, m_2}$ and are defined as in (19).

We can upper estimate the probabilities of the erroneous acceptance for $(l_1,l_2) \neq (m_1,m_2)$

$$G^N_{m_1,m_2}(\mathcal{A}^N_{l_1,l_2}) = \sum_{(\mathbf{x}_1,\mathbf{x}_2)\in\mathcal{A}^N_{l_1,l_2}} G^N_{m_1}(\mathbf{x}_1)G^N_{m_2|m_1}(\mathbf{x}_2|\mathbf{x}_1)$$
$$= \sum_{\mathbf{x}_1\in\mathcal{A}^N_{l_1}} G^N_{m_1}(\mathbf{x}_1)G^N_{m_2|m_1}(\mathcal{A}^N_{l_2|l_1}(\mathbf{x}_1)|\mathbf{x}_1)$$
$$\leq \max_{\mathbf{x}_1\in\mathcal{A}^N_{l_1}} G^N_{m_2|m_1}(\mathcal{A}^N_{l_2|l_1}(\mathbf{x}_1)|\mathbf{x}_1) \sum_{\mathbf{x}_1\in\mathcal{A}^N_{l_1}} G^N_{m_1}(\mathbf{x}_1)$$
$$= G^N_{m_1}(\mathcal{A}^N_{l_1}) \max_{\mathbf{x}_1\in\mathcal{A}^N_{l_1}} G^N_{m_2|m_1}(\mathcal{A}^N_{l_2|l_1}(\mathbf{x}_1)|\mathbf{x}_1).$$

These upper estimates of $\alpha_{l_1,l_2|m_1,m_2}(\Phi_N)$ for each $(l_1,l_2) \neq (m_1,m_2)$ we denote by

$$\beta_{l_1,l_2|m_1,m_2}(\Phi_N) \stackrel{\Delta}{=} G^N_{m_1}(\mathcal{A}^N_{l_1}) \max_{\mathbf{x}_1\in\mathcal{A}^N_{l_1}} G^N_{m_2|m_1}(\mathcal{A}^N_{l_2|l_1}(\mathbf{x}_1)|\mathbf{x}_1).$$

Consequently we can deduce that for $l_1,m_1 = \overline{1,M_1},\ l_2,m_2 = \overline{1,M_2}$, new parameters

$$F_{l_1,l_2|m_1,m_2}(\Phi) \stackrel{\Delta}{=} \lim_{N\to\infty} \inf\{-\frac{1}{N}\log\beta^N_{l_1,l_2|m_1,m_2}(\Phi_N)\},$$

are lower estimates for reliabilities $E_{l_1,l_2|m_1,m_2}(\Phi)$. We can introduce for $l_1,m_1 = \overline{1,M_1},\ l_2,m_2 = \overline{1,M_2},\ m_2 \neq l_2$,

$$\beta_{l_2|l_1,m_1,m_2}(\varphi^2_N) \stackrel{\Delta}{=} \max_{\mathbf{x}_1\in\mathcal{A}^N_{l_1}} G^N_{m_2|m_1}(\mathcal{A}^N_{l_2|l_1}(\mathbf{x}_1)|\mathbf{x}_1),$$

and also consider

$$\beta_{m_2|l_1,m_1,m_2}(\varphi^2_N) \stackrel{\Delta}{=} \max_{\mathbf{x}_1\in\mathcal{A}^N_{l_1}} G^N_{m_2|m_1}(\overline{\mathcal{A}^N_{m_2|l_1}(\mathbf{x}_1)}|\mathbf{x}_1)$$

$$= \sum_{l_2\neq m_2} \beta_{l_2|l_1,m_1,m_2}(\varphi^2_N). \tag{39}$$

The corresponding estimates of the reliabilities of test φ^2_N are the following

$$F_{l_2|l_1,m_1,m_2}(\varphi^2) \stackrel{\Delta}{=} \lim_{N\to\infty} \inf\left\{-\frac{1}{N}\log\beta_{l_2|l_1,m_1,m_2}(\varphi^2_N)\right\},$$

$$l_1,m_1 = \overline{1,M_1},\ l_2,m_2 = \overline{1,M_2},\ m_2 \neq l_2. \tag{40}$$

It is clear from (39) and (40) that for every $l_1,m_1 = \overline{1,M_1},\ l_2,m_2 = \overline{1,M_2}$

$$F_{m_2|l_1,m_1,m_2}(\varphi^2) = \min_{l_2:l_2\neq m_2} F_{l_2|l_1,m_1,m_2}(\varphi^2). \tag{41}$$

For given positive numbers $F_{l_2|l_1,m_1,l_2}$, $l_2=\overline{1,M_2-1}$, $Q\in\mathcal{R}_{l_1}$ and each pair $l_1,m_1=\overline{1,M_1}$ let us define the following regions and values:

$$\mathcal{R}_{l_2|l_1}(Q)\triangleq\{V:\ D(V||G_{l_2|l_1}|Q)\leq F_{l_2|l_1,m_1,l_2}\},\ l_2=\overline{1,M_2-1},\tag{42}$$

$$\mathcal{R}_{M_2|l_1}(Q)\triangleq\left\{V:\ D(V||G_{l_2|l_1}|Q)>F_{l_2|l_1,m_1,l_2},\ l_2=\overline{1,M_2-1}\right\},\tag{43}$$

$$F^*_{l_2|l_1,m_1,l_2}(F_{l_2|l_1,m_1,l_2})\triangleq F_{l_2|l_1,m_1,l_2},\ l_2=\overline{1,M_2-1},\tag{44}$$

$$F^*_{l_2|l_1,m_1,m_2}(F_{l_2|l_1,m_1,l_2})\triangleq\inf_{Q\in\mathcal{R}_{l_1}}\inf_{V\in\mathcal{R}_{l_2|l_1}(Q)}D(V||G_{m_2|m_1}|Q),$$
$$m_2=\overline{1,M_2},\ ,m_2\neq l_2,\ l_2=\overline{1,M_2-1},\tag{45}$$

$$F^*_{M_2|l_1,m_1,m_2}(F_{1|l_1,m_1,1},\ldots,F_{M_2-1|l_1,m_1,M_2-1})$$
$$\triangleq\inf_{Q\in\mathcal{R}_{l_1}}\inf_{V\in\mathcal{R}_{M_2|l_1}(Q)}D(V||G_{m_2|m_1}|Q),$$
$$m_2=\overline{1,M_2-1},\tag{46}$$

$$F^*_{M_2|l_1,m_1,M_2}(F_{1|l_1,m_1,1},\ldots,F_{M_2-1|l_1,m_1,M_2-1})$$
$$\triangleq\min_{l_2=\overline{1,M_2-1}}F^*_{l_2|l_1,m_1,M_2}.\tag{47}$$

We denote by $\mathbf{F}(\varphi_2)$ the matrix of lower estimates for elements of matrix $\mathbf{E}(\varphi_2)$.

Theorem 4.2 [49]. *If for given $m_1,l_1=\overline{1,M_1}$, all conditional PDs $G_{l_2|l_1}$, $l_2=\overline{1,M_2}$, are different in the sense that $D(G_{l_2|l_1}||G_{m_2|m_1}|Q)>0$, for all $Q\in\mathcal{R}_{l_1}$, $l_2\neq m_2$, $m_2=\overline{1,M_2}$, when the strictly positive numbers $F_{1|l_1,m_1,1}$, $F_{2|l_1,m_1,2}$,...,$F_{M_2-1|l_1,m_1,M_2-1}$ are such that the following compatibility conditions hold*

$$F_{1|l_1,m_1,1}<\min_{l_2=\overline{2,M_2}}\inf_{Q\in\mathcal{R}_{l_1}}D(G_{l_2|l_1}||G_{1|m_1}|Q),\tag{48}$$

$$F_{m_2|l_1,m_1,m_2}<\min\left(\min_{l_2=\overline{m_2+1,M_2}}\inf_{Q\in\mathcal{R}_{l_1}}D(G_{l_2|l_1}||G_{m_2|m_1}|Q),\right.$$
$$\left.\min_{l_2=\overline{1,m_2-1}}F^*_{l_2|l_1,m_1,m_2}(F_{l_2|l_1,m_1,l_2})\right),\ m_2=\overline{1,M_2-1},\tag{49}$$

then there exists a sequence of tests $\varphi^{2,*}$, such that the lower estimates are defined in (4.29)– (4.32) are strictly positive.

Inequalities (48), (49) are necessary for existence of the test sequence with matrix $\mathbf{F}(\varphi^{2,*})$ of positive lower estimates having given elements $F_{l_2|l_1,m_1,l_2}$, $l_2=\overline{1,M_2-1}$ in diagonal.

Let us define the following subsets of $\mathcal{Q}(\mathcal{X})$ for given strictly positive elements $E_{M_1,l_2|l_1,l_2}$, $l_1=\overline{1,M_1-1}$, $l_2=\overline{1,M_2-1}$:

$$\mathcal{R}_{l_1} \triangleq \{Q : D(Q||G_{l_1}) \leq E_{M_1,l_2|l_1,l_2}\},\ l_1 = \overline{1, M_1 - 1},\ l_2 = \overline{1, M_2 - 1},$$

$$\mathcal{R}_{M_1} \triangleq \{Q : D(Q||G_{l_1}) > E_{M_1,l_2|l_1,l_2},\ l_1 = \overline{1, M_1 - 1},\ l_2 = \overline{1, M_2 - 1}\},$$

and for given strictly positive elements $F_{l_1,M_2|l_1,l_2}$, $l_1 = \overline{1, M_1 - 1}$, $l_2 = \overline{1, M_2 - 1}$ and each $Q \in \mathcal{R}_{l_1}$, $l_1 = \overline{1, M_1}$, the following regions of condition PDs V:

$$\mathcal{R}_{l_2|l_1}(Q) \triangleq \{V : D(V||G_{l_2|l_1}|Q) \leq F_{l_1,M_2|l_1,l_2}\},\ l_1 = \overline{1, M_1 - 1},\ l_2 = \overline{1, M_2 - 1},$$

$$\mathcal{R}_{M_2|l_1}(Q) \triangleq \{V : D(V||G_{l_2|l_1}|Q) > F_{l_1,M_2|l_1,l_2},\ l_1 = \overline{1, M_1 - 1},\ l_2 = \overline{1, M_2 - 1}\}.$$

Assume also that

$$F^*_{l_1,M_2|l_1,l_2} \triangleq F_{l_1,M_2|l_1,l_2}, \tag{50}$$

$$E^*_{M_1,l_2|l_1,l_2} \triangleq E_{M_1,l_2|l_1,l_2},\ l_1 = \overline{1, M_1 - 1},\ l_2 = \overline{1, M_2 - 1}, \tag{51}$$

$$E^*_{l_1,l_2|m_1,l_2} \triangleq \inf_{Q:Q \in R_{l_1}} D(Q||G_{m_1}),\quad m_1 \neq l_1, \tag{52}$$

$$F^*_{l_1,l_2|l_1,m_2} \triangleq \inf_{Q \in \mathcal{R}_{l_1}} \inf_{V:V \in R_{l_2/l_1}(Q)} D(V||G_{m_2/m_1}|Q),\ m_2 \neq l_2, \tag{53}$$

$$F^*_{l_1,l_2|m_1,m_2} \triangleq F^*_{m_1,l_2|m_1,m_2} + E^*_{l_1,m_2|m_1,m_2},\quad m_i \neq l_i,\ i = 1, 2, \tag{54}$$

$$F^*_{m_1,m_2|m_1,m_2} \triangleq \min_{(l_1,l_2) \neq (m_1,m_2)} F^*_{l_1,l_2|m_1,m_2}. \tag{55}$$

Theorem 4.3 [46]. *If all PDs G_{m_1}, $m_1 = \overline{1, M_1}$, are different, that is $D(G_{l_1}||G_{m_1}) > 0$, $l_1 \neq m_1$, $l_1, m_1 = \overline{1, M_1}$, and all conditional PDs $G_{l_2|l_1}$, $l_2 = \overline{1, M_2}$, are also different for all $l_1 = \overline{1, M_1}$, in the sense that $D(G_{l_2|l_1}||G_{m_2|m_1}|Q) > 0$, $l_2 \neq m_2$, then the following statements are valid.*

When given strictly positive elements $E_{M_1,l_2|m_1,l_2}$ and $F_{l_1,M_2|l_1,m_2}$, $m_1 = \overline{1, M_1 - 1}$, $m_2 = \overline{1, M_2 - 1}$, meet the following conditions

$$E_{M_1,l_2|1,l_2} < \min_{l_1=\overline{2,M_1}} D(G_{l_1}||G_1), \tag{56}$$

$$F_{l_1,M_2|l_1,1} < \min_{l_2=\overline{2,M_2}} \inf_{Q \in \mathcal{R}_{l_1}} D(G_{l_2|l_1}||G_{1|m_1}|Q), \tag{57}$$

$$E_{M_1,l_2|m_1,l_2} < \min[\min_{l_1=\overline{1,m_1-1}} E^*_{l_1,l_2|m_1,l_2},\ \min_{l_1=\overline{m_1+1,M_1}} D(G_{l_1}||G_{m_1})],$$
$$m_1 = \overline{2, M_1 - 1}, \tag{58}$$

$$F_{l_1,M_2|l_1,m_2} < \min[\min_{l_2=\overline{1,m_2-1}} F^*_{l_1,l_2|l_1,m_2},\ \min_{l_2=\overline{m_2+1,M_2}} \inf_{Q \in \mathcal{R}_{l_1}} D(G_{l_2|l_1}||G_{m_2|m_1}|Q)],$$
$$m_2 = \overline{2, M_2 - 1}, \tag{59}$$

then there exists a LAO test sequence Φ^**, the matrix of lower estimates of which*

$$\mathbf{F}(\Phi^*) = \{F_{l_1,l_2|m_1,m_2}(\Phi^*)\}$$

is defined in (50)-(55) and all elements of it are positive.

When even one of the inequalities (56)-(59) is violated, then at least one element of the lower estimate matrix $\mathbf{F}(\Phi^*)$ *is equal to* 0.

When X_1 and X_2 are related statistically [46], [62] we will have instead of (39), (40) $\mathcal{A}^N_{l_2|l_1} = \{\mathbf{x_2} : \varphi^N_2(\mathbf{x_2}, l_1) = l_2\}$, $l_1 = \overline{1, M_1}$, $l_2 = \overline{1, M_2}$, and $\mathcal{A}^N_{l_1,l_2} \stackrel{\Delta}{=} \{(\mathbf{x}_1, \mathbf{x}_2) : \mathbf{x}_1 \in \mathcal{A}^N_{l_1}, \mathbf{x}_2 \in \mathcal{A}^N_{l_2|l_1}(\mathbf{x}_1)\}$. In that case we have error probabilities

$$\begin{aligned} G^N_{m_1,m_2}(\mathcal{A}^N_{l_1,l_2}) &\stackrel{\Delta}{=} \sum_{(\mathbf{x}_1,\mathbf{x}_2)\in\mathcal{A}^N_{l_1,l_2}} G^N_{m_1}(\mathbf{x}_1) G^N_{m_2|m_1}(\mathbf{x}_2) \\ &= \sum_{\mathbf{x}_1\in\mathcal{A}^N_{l_1}} G^N_{m_1}(\mathbf{x}_1) \sum_{\mathbf{x}_2\in\mathcal{A}^N_{l_2|l_1}} G^N_{m_2|m_1}(\mathbf{x}_2) \\ &= G^N_{m_2|m_1}(\mathcal{A}^N_{l_2|l_1}) G_{m_1}(\mathcal{A}^N_{l_1}), \ (l_1, l_2) \neq (m_1, m_2). \end{aligned}$$

For the second object the conditional probabilities of the erroneous acceptance of PD $G_{l_2|l_1}$ provided that $G_{m_2|m_1}$ is true, for $l_1, m_1 = \overline{1, M_1}$, $l_2, m_2 = \overline{1, M_2}$, are the following

$$\alpha^N_{l_2|l_1,m_1,m_2}(\varphi^2_N) \stackrel{\Delta}{=} G^N_{m_2|m_1}(\mathcal{A}^N_{l_2|l_1}), \ l_2 \neq m_2.$$

The probability to reject $G_{m_2|m_1}$, when it is true is denoted as follows

$$\alpha^N_{m_2|l_1,m_1,m_2}(\varphi^2_N) \stackrel{\Delta}{=} G^N_{m_2|m_1}(\overline{\mathcal{A}_{m_2|l_1}}) = \sum_{l_2\neq m_2} \alpha^N_{l_2|l_1,m_1,m_2}(\varphi^2_N).$$

Thus in the conditions and in the results of Theorems 4.2 and 4.3 we will have just divergences
$D(G_{l_2|l_1}||G_{m_2|m_1})$, instead of conditional divergences $\inf_{Q\in\mathcal{R}_{l_1}} D(G_{l_2|l_1}||G_{m_2|m_1}|Q)$, $\inf_{Q\in\mathcal{R}_{l_1}} D(V||G_{m_2|m_1}|Q)$ $D(V||G_{m_2|m_1})$ and $E_{l_2|l_1,m_1,m_2}(\Phi)$, $E_{l_1,l_2|m_1,m_2}(\Phi)$, $l_1, m_1 = \overline{1, M_1}$, $l_2, m_2 = \overline{1, M_2}$ will be in place of $F_{l_2|l_1,m_1,m_2}(\Phi)$, $F_{l_1,l_2|m_1,m_2}(\Phi)$, $l_1, m_1 = \overline{1, M_1}$, $l_2, m_2 = \overline{1, M_2}$. And in that case regions defined in (42), (43) will be changed as follows:

$$\begin{aligned} \mathcal{R}_{l_2|l_1} &\stackrel{\Delta}{=} \{V : \ D(V||G_{l_2|l_1}) \leq E_{l_2|l_1,m_1,l_2}\}, \ \ l_2 = \overline{1, M_2 - 1}, \\ \mathcal{R}_{M_2|l_1} &\stackrel{\Delta}{=} \{V : \ D(V||G_{l_2|l_1}) > E_{l_2|l_1,m_1,l_2}, \ \ l_2 = \overline{1, M_2 - 1}\}, \end{aligned}$$

In case of two statistically dependent objects the corresponding regions will be

$$\mathcal{R}_{l_1} \triangleq \{Q : D(Q||G_{l_1}) \leq E_{M_1,l_2|l_1,l_2}\},\ l_1 = \overline{1, M_1 - 1},\ l_2 = \overline{1, M_2 - 1},$$

$$\mathcal{R}_{l_2|l_1} \triangleq \{V : D(V||G_{l_2|l_1}) \leq E_{l_1,M_2|l_1,l_2}\},\ l_1 = \overline{1, M_1 - 1},\ l_2 = \overline{1, M_2 - 1},$$

$$\mathcal{R}_{M_1} \triangleq \{Q : D(Q||G_{l_1}) > E_{M_1,l_2|l_1,l_2},\ l_1 = \overline{1, M_1 - 1},\ l_2 = \overline{1, M_2 - 1}\},$$

$$\mathcal{R}_{M_2|l_1} \triangleq \{V : D(V||G_{l_2|l_1}) > E_{l_1,M_2|l_1,l_2},\ l_1 = \overline{1, M_1 - 1},\ l_2 = \overline{1, M_2 - 1}\}.$$

So in this case the matrix of reliabilities
$\mathbf{E}(\Phi*) = \{E*_{l_1,l_2|m_1,m_2},\ l_1, m_1 = \overline{1, M_1},\ l_2, m_2 = \overline{1, M_2}\}$, will have the following elements:

$$E^*_{l_1,M_2|l_1,l_2} \triangleq E_{l_1,M_2|l_1,l_2},$$

$$E^*_{M_1,l_2|l_1,l_2} \triangleq E_{M_1,l_2|l_1,l_2},$$

$$l_1 = \overline{1, M_1 - 1},\ l_2 = \overline{1, L_2 - 1},$$

$$E^*_{l_1,l_2|m_1,l_2} \triangleq \inf_{Q:Q\in R_{l_1}} D(Q||G_{m_1}),\ m_1 \neq l_1,$$

$$E^*_{l_1,l_2|l_1,m_2} \triangleq \inf_{V:V\in R_{l_2|l_1}} D(V||G_{m_2|m_1}),\quad m_2 \neq l_2,$$

$$E^*_{l_1,l_2|m_1,m_2} \triangleq E^*_{m_1,l_2|m_1,m_2} + E^*_{l_1,m_2|m_1,m_2},\ m_i \neq l_i,\ i = 1, 2,$$

$$E^*_{m_1,m_2|m_1,m_2} \triangleq \min_{(l_1,l_2)\neq(m_1,m_2)} E^*_{l_1,l_2|m_1,m_2}.$$

Theorem 4.4. [46] *If all PDs G_{m_1}, $m_1 = \overline{1, M_1}$, are different, that is $D(G_{l_1}||G_{m_1}) > 0$, $l_1 \neq m_1$, $l_1, m_1 = \overline{1, M_1}$, and all conditional PDs $G_{l_2|l_1}$, $l_2 = \overline{1, M_2}$, are also different for all $l_1 = \overline{1, M_1}$, in the sense that $D(G_{l_2|l_1}||G_{m_2|m_1}) > 0$, $l_2 \neq m_2$, then the following statements are valid.*

When given strictly positive elements $E_{M_1,l_2|l_1,l_2}$ and $E_{l_1,M_2|l_1,l_2}$, $l_1 = \overline{1, M_1 - 1}$, $l_2 = \overline{1, M_2 - 1}$, meet the following compatibility conditions

$$E_{M_1,l_2|1,l_2} < \min_{l_1=\overline{2,M_1}} D(G_{l_1}||G_1),$$

$$E_{l_1,M_2|l_1,1} < \min_{l_2=\overline{2,M_2}} D(G_{l_2|l_1}||G_{1|m_1}),$$

$$E_{M_1,l_2|m_1,l_2} < \min\left[\min_{l_1=\overline{1,m_1-1}} E^*_{l_1,l_2|m_1,l_2},\ \min_{l_1=\overline{m_1+1,M_1}} D(G_{l_1}||G_{m_1})\right],$$

$$m_1 = \overline{2, M_1 - 1},$$

$$E_{l_1,M_2|l_1,m_2} < \min\left[\min_{l_2=\overline{1,m_2-1}} E^*_{l_1,l_2|l_1,m_2},\ \min_{l_2=\overline{m_2+1,M_2}} D(G_{l_2|l_1}||G_{m_2|m_1})\right],$$

$$m_2 = \overline{2, M_2 - 1},$$

then there exists a LAO test sequence Φ^*, the matrix of which $\mathbf{E}(\Phi^*)$ is stated above and all elements of it are positive.

When even one of the compatibility conditions is violated, then at least one element of the matrix $\mathbf{E}(\Phi^*)$ *is equal to* 0.

5 Identification of Distribution for One and for Many Objects

In [9] Bechhofer, Kiefer, and Sobel presented investigations on sequential multiple-decision procedures. This book concerns principally with a particular class of problems referred to as ranking problems. Chapter 10 of the book by Ahlswede and Wegener [7] is devoted to statistical identification and ranking. Problems of distribution identification and distributions ranking for one object applying the concept of optimality developed in [11], [50], [30]–[33] were solved in [6]. In papers [41], [48], [49] and [55] identification problems for models composed with two independent, or strictly dependent objects were investigated.

In [6], [41], [49] and [55] models considered in [9] and [7] and variations of these models inspired by the pioneering paper by Ahlswede and Dueck [5], applying the concept of optimality developed in [11], [30]–[33], [50], were studied.

First we formulate the concept of the identification for one object, which was considered in [6]. There are known $M \geq 2$ possible PDs, related with the object in consideration. Identification gives the answer to the question whether r-th PD occured, or not. This answer can be given on the base of a sample $\mathbf{x}$ by a test $\varphi_N^*(\mathbf{x})$. More precisely, identification can be considered as an answer to the question: is the result l of testing algorithm equal to r (that is $l = r$), or not (that is $l \neq r$).

There are two types of error probabilities of identification for each $r = \overline{1, M}$: $\alpha_{l\neq r|m=r}(\varphi_N)$ the probability to accept l different from r, when r is in reality, and the probability $\alpha_{l=r|m\neq r}(\varphi_N)$ that r is accepted by test φ_N, when r is not correct.

The probability $\alpha_{l\neq r|m=r}(\varphi_N)$ coincides with the error probability of testing $\alpha_{r|r}(\varphi_N)$ (see (6)) which is equal to $\sum\limits_{l:l\neq r} \alpha_{l|r}(\varphi_N)$. The corresponding reliability $E_{l\neq r|m=r}(\varphi)$ is equal to $E_{r|r}(\varphi)$ which satisfies the equality (8).

And what is the reliability approach to identification? It is necessary to determine the dependence of optimal reliability $E^*_{l=r|m\neq r}$ upon given $E^*_{l\neq r|m=r} = E^*_{r|r}$, which can be assigned a value satisfying conditions analogical to (16).

The result from paper [6] is:

Theorem 5.1. *In the case of distinct hypothetical PDs* $G_1, G_2, ..., G_M$, *for a given sample* $\mathbf{x}$ *we define its type* Q, *and when* $Q \in \mathcal{R}_l^{(N)}$ *(see (9)–(11)) we accept the hypothesis* l. *Under condition that the a priori probabilities of all* M *hypotheses are positive the reliability of such identification* $E_{l=r|m\neq r}$ *for given* $E_{l\neq r|m=r} = E_{r|r}$ *is the following:*

$$E_{l=r|m\neq r}(E_{r|r}) = \min_{m:m\neq r} \inf_{Q:D(Q\|G_r)\leq E_{r|r}} D(Q\|G_m), \quad r = \overline{1, M}.$$

We can accept the supposition of positivity of a priori probabilities of all hypotheses with no loss of generality, because the hypothesis which is known to have probability 0, that is impossible, must not be included in the studied family.

Now let us consider the model consisting of two independent objects. Let hypothetical characteristics of objects X_1 and X_2 be independent RVs taking values in the same finite set $\mathcal{X}$ with one of M PDs. Identification means that the statistician has to answer the question whether the pair of distributions (r_1, r_2) occurred or not. Now the procedure of testing for two objects can be used. Let us study two types of error probabilities for each pair (r_1, r_2), $r_1, r_2 = \overline{1, M}$. We denote by $\alpha^{(N)}_{(l_1,l_2)\neq(r_1,r_2)|(m_1,m_2)=(r_1,r_2)}$ the probability, that pair (r_1, r_2) is true, but it is rejected. Note that this probability is equal to $\alpha_{r_1,r_2|r_1,r_2}(\Phi_N)$. Let $\alpha^{(N)}_{(l_1,l_2)=(r_1,r_2)|(m_1,m_2)\neq(r_1,r_2)}$ be the probability that (r_1, r_2) is identified, when it is not correct. The corresponding reliabilities are $E_{(l_1,l_2)\neq(r_1,r_2)|(m_1,m_2)=(r_1,r_2)} = E_{r_1,r_2|r_1,r_2}$ and $E_{(l_1,l_2)=(r_1,r_2)|(m_1,m_2)\neq(r_1,r_2)}$. Our aim is to determine the dependence of $E_{(l_1,l_2)=(r_1,r_2)|(m_1,m_2)\neq(r_1,r_2)}$ on given $E_{r_1,r_2|r_1,r_2}(\Phi_N)$.

Let us define for each r, $r = \overline{1, M}$, the following expression:

$$A(r) = \min\left[\min_{l=\overline{1,r-1}} D(G_l||G_r), \min_{l=\overline{r+1,M}} D(G_l||G_r)\right].$$

Theorem 5.2 [41]. *For the model consisting of two independent objects if the distributions G_m, $m = \overline{1, M}$, are different and the given strictly positive number $E_{r_1,r_2|r_1,r_2}$ satisfy condition*

$$E_{r_1,r_2|r_1,r_2} < \min\left[A(r_1),\ A(r_2)\right],$$

then the reliability $E_{(l_1,l_2)=(r_1,r_2)|(m_1,m_2)\neq(r_1,r_2)}$ is defined as follows:

$$E_{(l_1,l_2)=(r_1,r_2)|(m_1,m_2)\neq(r_1,r_2)}\left(E_{r_1,r_2|r_1,r_2}\right)$$

$$= \min_{m_1\neq r_1, m_2\neq r_2}\left[E_{m_1|r_1}(E_{r_1,r_2|r_1,r_2}), E_{m_2|r_2}(E_{r_1,r_2|r_1,r_2})\right],$$

where $E_{m_1|r_1}(E_{r_1,r_2|r_1,r_2})$ and $E_{m_2|r_2}(E_{r_1,r_2|r_1,r_2})$ are determined by (13).

Now we will present the lower estimates of the reliabilities for LAO identification for the *dependent* object which can be then applied for deducing the lower estimates of the reliabilities for LAO identification of two *related* objects. There exist two error probabilities for each $r_2 = \overline{1, M_2}$: the probability $\alpha_{l_2\neq r_2|l_1,m_1,m_2=r_2}(\varphi^2_N)$ to accept l_2 different from r_2, when r_2 is in reality, and the probability $\alpha_{l_2=r_2|l_1,m_1,m_2\neq r_2}(\varphi^2_N)$ that r_2 is accepted, when it is not correct.

The upper estimate $\beta_{l_2\neq r_2|l_1,m_1,m_2=r_2}(\varphi^2_N)$ for $\alpha_{l_2\neq r_2|l_1,m_1,m_2=r_2}(\varphi^2_N)$ is already known, it coincides with $\beta_{r_2|l_1,m_1,r_2}(\varphi^2_N)$ which is equal to $\sum\limits_{l_2:l_2\neq r_2} \beta_{l_2|l_1,m_1,r_2}(\varphi^2_N)$. The corresponding $F_{l_2\neq r_2|l_1,m_1,m_2=r_2}(\varphi^2)$ is equal to $F_{r_2|l_1,m_1,r_2}(\varphi^2)$, which satisfies the equality (20).

We determine the optimal dependence of $F^*_{l_2=r_2|l_1,m_1,m_2\neq r_2}$ upon given $F^*_{l_2\neq r_2|l_1,m_1,m_2=r_2}$.

Theorem 5.3 [49]. *In case of distinct PDs $G_{1|l_1}, G_{2|l_1}, ..., G_{M_2|l_1}$, under condition that a priori probabilities of all M_2 hypotheses are strictly positive, for each $r_2 = \overline{1, M_2}$ the estimate of $F_{l_2=r_2|l_1,m_1,m_2\neq r_2}$ for given $F_{l_2\neq r_2|l_1,m_1,m_2=r_2} = F_{r_2|l_1,m_1,r_2}$ is the following:*

$$F_{l_2=r_2|l_1,m_1,m_2\neq r_2}(F_{r_2|l_1,m_1,r_2}) =$$

$$\min_{m_2:m_2\neq r_2} \inf_{Q\in\mathcal{R}_{l_1}} \inf_{V:D(V\|G_{r_2|l_1}|Q)\leq F_{r_2|l_1,m_1,r_2}} D(V\|G_{m_2|m_1}|Q).$$

The result of the reliability approach to the problem of identification of the probability distributions for two *related* objects is the following.

Theorem 5.4. *If the distributions G_{m_1} and $G_{m_2|m_1}$, $m_1 = \overline{1, M_1}$, $m_2 = \overline{1, M_2}$, are different and the given strictly positive number $F_{r_1,r_2|r_1,r_2}$ satisfies the condition*

$$E_{r_1|r_1} < \min\left[\min_{l=\overline{1,r_1-1}} D(G_{r_1}||G_{l_1}), \min_{l_1=\overline{r_1+1,M_1}} D(G_{l_1}||G_{r_1})\right],$$

$$F_{r_2|l_1,m_1,r_2} < \min\left[\inf_{Q\in\mathcal{R}_{l_1}} \min_{l_2=\overline{1,r_2-1}} D(G_{r_2|m_1}||G_{l_2|l_1}|Q),\right.$$

$$\left.\inf_{Q\in\mathcal{R}_{l_1}} \min_{l_2=\overline{r_2+1,M_2}} D(G_{l_2|l_1}||G_{r_2|m_1}|Q)\right],$$

then the lower estimate $F_{(l_1,l_2)=(r_1,r_2)|(m_1,m_2)\neq(r_1,r_2)}$ of the reliability $E_{(l_1,l_2)=(r_1,r_2)|(m_1,m_2)\neq(r_1,r_2)}$ can be calculated as follows

$$F_{(l_1,l_2)=(r_1,r_2)|(m_1,m_2)\neq(r_1,r_2)}\left(F_{r_1,r_2|r_1,r_2}\right)$$
$$= \min_{m_1\neq r_1, m_2\neq r_2} \left[E_{r_1|m_1}(F_{r_1,r_2|r_1,r_2}), F_{r_2|l_1,m_1,m_2}(F_{r_1,r_2|r_1,r_2})\right],$$

where $E_{r_1|m_1}(F_{r_1,r_2|r_1,r_2})$ and $F_{r_2|l_1,m_1,m_2}(F_{r_1,r_2|r_1,r_2})$ are determined respectively by (13) and (45).

The particular case, when X_1 and X_2 are related *statistically*, was studied in [46], [62].

6 Multihypotheses Testing for Arbitrarily Varying Source and for Cases With Possibility of Decision Rejection

This section is devoted to description of characteristics of LAO hypotheses testing with permission of decision rejection for the model consisting of one or more objects. In subsection 6.1 we consider multiple statistical hypotheses testing with possibility of rejecting to make choice between hypotheses concerning distribution of a discrete arbitrarily varying source. The multiple hypotheses testing problem with possibility of rejection of decision for arbitrarily varying object with side information and for the model of two or more independent objects was examined by Haroutunian, Hakobyan and Yessayan [43], [44]. These works were induced by the paper of Nikulin [57] concerning two hypotheses testing with refusal to take decision. An asymptotically optimal classification with rejection of decision was considered by Gutman [26].

6.1 Many Hypothesis Testing with Rejection of Decision by Informed Statistician for Arbitrarily Varying Source

"A broad class of statistical problems arises in the framework of hypothesis testing in the spirit of identification for different kinds of sources, with complete or partial side information or without it" [2].

The problem concerning arbitrarily varying sources solved in [39] was induced by the ideas of the paper of Ahlswede [1]. Fu and Shen [24] explored the case of two hypotheses testing when side information is absent. The case of two hypotheses with side information about states was considered in [3]. In [38] Haroutunian and Grigoryan generalized results from [24], [31]–[33] for multihypotheses LAO testing by a non-informed statistician for an arbitrarily varying Markov source. In the same way as in [24] and [54] from result on LAO testing, the rate-reliability and the reliability-rate functions for arbitrarily varying source with side information were obtained in [39].

The arbitrarily varying source is a generalized model of the discrete memoryless source. In the first one, the source outputs distributions depends on the source state. The latter varies within a finite set from one time instant to the next in an arbitrary manner.

Let $\mathcal{X}$ be a finite set of values of RV X, and $\mathcal{S}$ be an alphabet of states of the object. M possible conditional PDs of the characteristic X of the object depending on values s of states, are given:

$$W_m \stackrel{\Delta}{=} \{W_m(x|s),\ x \in \mathcal{X},\ \ s \in \mathcal{S}\},\ \ m = \overline{1,M},\ \ |\mathcal{S}| \geq 1,$$

but it is not known which of these alternative hypotheses W_m, $m = \overline{1,M}$, is real PD of the object. The statistician must select one among M hypotheses, or he can withdraw any judgement. It is possible, for instance, when it is supposed that real PD is not in the family of M given PDs. An answer must be given using the sample $\mathbf{x} \stackrel{\Delta}{=} (x_1, x_2, ...x_N)$ and the vector of states of the object $\mathbf{s} \stackrel{\Delta}{=} (s_1, s_2, ..., s_N)$, $s_n \in \mathcal{S}$, $n = \overline{1,N}$.

The procedure of decision making is a non-randomized test $\varphi_N(\mathbf{x}, \mathbf{s})$, it can be defined by division of the sample space $\mathcal{X}^N$ for each $\mathbf{s}$ on $M+1$ disjoint subsets $\mathcal{A}_m^N(\mathbf{s}) = \{\mathbf{x} : \varphi_N(\mathbf{x}, \mathbf{s}) = m\}$, $m = \overline{1, M+1}$. The set $\mathcal{A}_l^N(\mathbf{s})$, $l = \overline{1,M}$, consists of vectors $\mathbf{x}$ for which the hypothesis W_l is adopted, and $\mathcal{A}_{M+1}^N(\mathbf{s})$ includes vectors for which the statistician refuses to give a certain answer.

We study the probabilities of the erroneous acceptance of hypothesis W_l provided that W_m is true

$$\alpha_{l|m}(\varphi_N) \stackrel{\Delta}{=} \max_{\mathbf{s} \in \mathcal{S}^N} W_m^N \left(\mathcal{A}_l^N(\mathbf{s})|\mathbf{s}\right),\ m, l = \overline{1,M},\ \ m \neq l. \qquad (60)$$

When decision is declined, but the hypothesis W_m is true, we consider the following probability of error:

$$\alpha_{M+1|m}(\varphi_N) \stackrel{\Delta}{=} \max_{\mathbf{s}\in\mathcal{S}^N} W_m^N\left(\mathcal{A}_{M+1}^N(\mathbf{s})|\mathbf{s}\right).$$

If the hypothesis W_m is true, but it is not accepted, or equivalently while the statistician accepted one of hypotheses W_l, $l=\overline{1,M}$, $l \neq m$, or refused to make decision, then the probability of error is the following:

$$\alpha_{m|m}(\varphi_N) \stackrel{\Delta}{=} \sum_{l:\, l\neq m} \alpha_{l|m}(\varphi_N) = \max_{\mathbf{s}\in\mathcal{S}^N} W_m^N\left(\overline{\mathcal{A}_m^N(\mathbf{s})}|\mathbf{s}\right), \quad m=\overline{1,M}. \tag{61}$$

Corresponding reliabilities are defined similarly to (6):

$$E_{l|m}(\varphi) \stackrel{\Delta}{=} \lim_{N\to\infty} \inf\left\{-\frac{1}{N}\log \alpha_{l|m}(\varphi_N)\right\}, \quad m=\overline{1,M}, \quad l=\overline{1,M+1}. \tag{62}$$

It also follows that for every test φ

$$E_{m|m}(\varphi) = \min_{l=\overline{1,M+1},\ l\neq m} E_{l|m}(\varphi), \quad m=\overline{1,M}. \tag{63}$$

The matrix

$$\mathbf{E}(\varphi) = \begin{pmatrix} E_{1|1} & \dots E_{l|1} & \dots & E_{M|1}, \ E_{M+1|1} \\ \dots\dots\dots\dots\dots\dots\dots\dots \\ E_{1|m} & \dots E_{l|m} & \dots & E_{M|m}, \ E_{M+1|m} \\ \dots\dots\dots\dots\dots\dots\dots\dots \\ E_{1|M} & \dots E_{l|M} & \dots & E_{M|M} \ E_{M+1|M} \end{pmatrix}$$

is the reliabilities matrix of the tests sequence φ for the described model.

We call the test LAO for this model if for given positive values of certain M elements of the matrix $\mathbf{E}(\varphi)$ the procedure provides maximal values for other elements of it.

For construction of the LAO test positive elements $E_{1|1}$, ..., $E_{M|M}$ are supposed to be given preliminarily. The optimal dependence of error exponents was determined in [43]. This result can be easily generalized for the case of an arbitrarily varying Markov source.

6.2 Multiple Hypotheses LAO Testing with Rejection of Decision for Many Independent Objects

For brevity we consider the problem for two objects, the generalization of the problem for K independent objects will be discussed along the text.

Let X_1 and X_2 be independent RVs taking values in the same finite set $\mathcal{X}$ with one of M PDs $G_m \in \mathcal{P}(\mathcal{X})$, $m=\overline{1,M}$. These RVs are the characteristics of corresponding independent objects. The random vector (X_1, X_2) assumes values $(x^1, x^2) \in \mathcal{X}\times\mathcal{X}$. Let $(\mathbf{x^1},\mathbf{x^2}) \stackrel{\Delta}{=} \left((x_1^1,x_1^2),...,(x_n^1,x_n^2),...,(x_N^1,x_N^2)\right)$, $x_n^k \in \mathcal{X}$,

$k = \overline{1,2}, n = \overline{1,N}$, be a vector of results of N independent observations of the pair of RVs (X_1, X_2). On the base of observed data the test has to determine unknown PDs of the objects or withdraw any judgement. The selection for each object should be made from the same set of hypotheses: G_m, $m = \overline{1,M}$. We call this procedure the compound test for two objects and denote it by Φ_N, it can be composed of two individual tests φ_N^1, φ_N^2 for corresponding objects. The test φ_N^i, $i = \overline{1,2}$, can be defined by division of the space $\mathcal{X}^N$ into $M+1$ disjoint subsets $\mathcal{A}_m^i$, $m = \overline{1,M+1}$. The set $\mathcal{A}_m^i$, $m = \overline{1,M}$ contains all vectors $\mathbf{x}^i$ for which the hypothesis G_m is adopted and $\mathcal{A}_{M+1}^i$ includes all vectors for which the test refuses to take a certain answer. Hence Φ_N is division of the space $\mathcal{X}^N \times \mathcal{X}^N$ into $(M+1)^2$ subsets $\mathcal{A}_{m_1,m_2} = \mathcal{A}_{m_1}^1 \times \mathcal{A}_{m_2}^2$, $m_i = \overline{1,M+1}$. We again denote the infinite sequences of tests by Φ, φ^1, φ^2.

Let $\alpha_{l_1,l_2|m_1,m_2}(\Phi_N)$ be the probability of the erroneous acceptance of the pair of hypotheses (G_{l_1}, G_{l_2}) by the test Φ_N provided that the pair of hypotheses (G_{m_1}, G_{m_2}) is true, where $(m_1, m_2) \neq (l_1, l_2)$, $m_i = \overline{1,M}$, $l_i = \overline{1,M}$, $i = \overline{1,2}$:

$$\begin{aligned}\alpha_{l_1,l_2|m_1,m_2}(\Phi_N) &= G_{m_1} \circ G_{m_2}(\mathcal{A}_{l_1,l_2}) \\ &= G_{m_1}^N(\mathcal{A}_{l_1}) \cdot G_{m_2}^N(\mathcal{A}_{l_2}).\end{aligned}$$

When the pair of hypotheses (G_{m_1}, G_{m_2}), $m_1, m_2 = \overline{1,M}$ is true, but we decline the decision the corresponding probabilities of errors are:

$$\begin{aligned}\alpha_{M+1,M+1|m_1,m_2}(\Phi_N) &= G_{m_1} \circ G_{m_2}(\mathcal{A}_{M+1,M+1}) \\ &= G_{m_1}^N(\mathcal{A}_{M+1}^1) \cdot G_{m_2}^N(\mathcal{A}_{M+1}^2).\end{aligned}$$

or

$$\alpha_{M+1,l_2|m_1,m_2}(\Phi_N) = G_{m_1}^N(\mathcal{A}_{M+1}^1) \cdot G_{m_2}^N(\mathcal{A}_{l_2}^2)$$

or

$$\alpha_{l_1,M+1|m_1,m_2}(\Phi_N) = G_{m_1}^N(\mathcal{A}_{l_1}^1) \cdot G_{m_2}^N(\mathcal{A}_{M+1}^2).$$

If the pair of hypotheses (G_{m_1}, G_{m_2}) is true, but it is not accepted, or equivalently while the statistician accepted one of hypotheses (G_{l_1}, G_{l_2}), or refused to make decision, then the probability of error is the following:

$$\alpha_{m_1,m_2|m_1,m_2}(\Phi_N) = \sum_{(l_1,l_2)\neq(m_1,m_2)} \alpha_{l_1,l_2|m_1,m_2}(\Phi_N), \tag{64}$$
$$l_i = \overline{1,M+1},\ m_i = \overline{1,M},\ i = \overline{1,2}.$$

We study reliabilities $E_{l_1,l_2|m_1,m_2}(\Phi)$ of the sequence of tests Φ,

$$E_{l_1,l_2|m_1,m_2}(\Phi) \triangleq \lim_{N\to\infty} \inf \left\{ -\frac{1}{N} \log \alpha_{l_1,l_2|m_1,m_2}(\Phi_N) \right\}, \tag{65}$$
$$m_i, = \overline{1,M},\ l_i = \overline{1,M+1},\quad i = \overline{1,2}.$$

Definitions (64) and (65) imply that

$$E_{m_1,m_2|m_1,m_2}(\Phi) = \min_{(l_1,l_2)\neq(m_1,m_2)} E_{l_1,l_2|m_1,m_2}(\Phi). \tag{66}$$

We can erect the LAO test from the set of compound tests when $2M$ strictly positive elements of the reliability matrix $E_{M+1,m|m,m}$ and $E_{m,M+1|m,m}$, $m = \overline{1,M}$, are preliminarily given (see [43]).

Remark 6.1. It is necessary to note that the problem of reliabilities investigation for LAO testing of many hypotheses with possibility of rejection of decision for the model consisting of two or more independent objects can not be solved by the direct method of renumbering.

7 Conclusion and Open Problems

In this paper, we described solutions of a part of possible problems concerning algorithms of distributions optimal testing for certain classes of one, or multiple objects. For the same models PD optimal identification is discussed again in the spirit of error probability exponents optimal dependence.

These investigations can be continued in plenty directions. "Paper [6] is a start" [2].

Some problems formulated in [6] and [45], particularly, concerning the remote statistical inference formulated by Berger [10], examined in part by Ahlwede and Csiszár [4] and Han and Amari [28] still remain open.

All our results concern with discrete distributions, but it is necessary to study many objects with general distributions as in [27]. For multiple objects multistage [42] and sequential testing [14] can be also considered. Problems for many objects are present in statistics with fuzzy data [25], Bayesian detection of multiple hypotheses [53] and geometric interpretations of tests [63].

Acknowledgments. The authors are grateful to the referees for their helpful comments.

References

1. Ahlswede, R.: Coloring hypergraphs: a new approach to multi-user source coding. Journal of Combinatorics, Information and System Sciences I 4(1), 76–115 (1979), II 5(3), 220–268 (1980)
2. Ahlswede, R.: Towards a general theory of information transfer. Shannon Lecture at ISIT in Seatle 2006, IEEE Information Theory Society Newsletter 57(3), 6–28 (2007)
3. Ahlswede, R., Aloyan, E., Haroutunian, E.: On Logarithmically Asymptotically Optimal Hypothesis Testing for Arbitrarily Varying Sources with Side Information. In: Ahlswede, R., Bäumer, L., Cai, N., Aydinian, H., Blinovsky, V., Deppe, C., Mashurian, H. (eds.) Information Transfer and Combinatorics. LNCS, vol. 4123, pp. 547–552. Springer, Heidelberg (2006)
4. Ahlswede, R., Csiszár, I.: Hypothesis testing with Communication Constraints. IEEE Trans. on Inf. Theory 32(4), 533–542 (1997)
5. Ahlswede, R., Dueck, G.: Identification via channels. IEEE Trans. on Inf. Theory 35(1), 15–29 (1989)

6. Ahlswede, R., Haroutunian, E.: On Logarithmically Asymptotically Optimal Testing of Hypotheses and Identification. In: Ahlswede, R., Bäumer, L., Cai, N., Aydinian, H., Blinovsky, V., Deppe, C., Mashurian, H. (eds.) Information Transfer and Combinatorics. LNCS, vol. 4123, pp. 553–571. Springer, Heidelberg (2006)
7. Ahlswede, R., Wegener, I.: Search Problems. Wiley, New York (1987)
8. Anantharam, V.: A large deviations approach to error exponent in source coding and hypotheses testing. IEEE Trans. on Inf. Theory 36(4), 938–943 (1990)
9. Bechhofer, R., Kiefer, J., Sobel, M.: Sequential identification and ranking procedures. The University of Chicago Press, Chicago (1968)
10. Berger, T.: Decentralized estimation and decision theory. Presented at IEEE Seven Springs Workshop on Information Theory, Mt. Kisco, NY (1979)
11. Birgé, L.: Vitesses maximales de décroissence des erreurs et tests optimaux associeés. Z. Wahrsch. Verw. Gebiete 55, 261–273 (1981)
12. Blahut, R.: Hypothesis testing and information theory. IEEE Trans. on Inf. Theory 20, 405–417 (1974)
13. Blahut, R.: Principles and practice of information theory. Addison-Wesley, Reading (1987)
14. Brodsky, B., Darkhovsky, B.: Nonparametric methods in change-point problems. Kluwer Academic Publishers, Dordrecht (1993)
15. Cover, T., Thomas, J.: Elements of Information Theory, 2nd edn. Wiley, New York (2006)
16. Csiszár, I.: The method of types. IEEE Trans. on Inf. Theory 44(6), 2505–2523 (1998)
17. Csiszár, I.: Information thoretic methods in probability and statistics. Shannon Lecture at ISIT in Ulm Germany 1997, IEEE Information Theory Society Newsletter 48(1), 1, 21–30 (1998)
18. Csiszár, I., Körner, J.: Information Theory: Coding Theorems for Discrete Memoryless Systems. Academic Press, New York (1981)
19. Csiszár, I., Longo, G.: On the error exponent for source coding and for testing simple statistical hypotheses. Studia Sc. Math. Hungarica 6, 181–191 (1971)
20. Csiszár, I., Shields, P.: Information theory and statistics: a tutorial. Foundations and Trends in Communications and Information Theory 1(4) (2004)
21. Dobrushin, R.: On optimal testing of multiple statistical hypotheses. Personal Communication (1987)
22. Dobrushin, R., Pinsker, M., Shiryaev, A.: Application of the notion of entropy in the problems of detecting a signal in noise. Lithuanian Mathematical Transactions 3(1), 107–122 (1963) (in Russian)
23. Dudoit, S., Shaffer, J., Boldrick, J.: Multiple hypothesis testing in microarray experiments. Statistical Science 18(1), 71–103 (2003)
24. Fu, F.-W., Shen, S.-Y.: Hypothesis testing for arbitrarily varying source with exponential-type constraint. IEEE Trans. on Inf. Theory 44(2), 892–895 (1998)
25. Grzegorzewski, P., Hryniewicz, O.: Testing statistical hypotheses in fuzzy environment. Mathware and Soft Computing 4(2), 203–217 (1997)
26. Gutman, M.: Asymptotically optimal classification for multiple tests with empirically observed statistics. IEEE Trans. on Inf. Theory 35(2), 401–408 (1989)
27. Han, T.: Hypothesis testing with the general source. IEEE Trans. on Inf. Theory 46(7), 2415–2427 (2000)
28. Han, T., Amari, S.: Statistical inference under multiterminal data compression. IEEE Trans. on Inf. Theory 44(6), 2300–2324 (1998)

29. Haroutunian, E.: Upper estimate of transmission rate for memoryless channel with countable number of output signals under given error probability exponent. In: 3rd All Union Conf. on Theory of Information Transmission and Coding, Uzhgorod, pp. 83–86. Publishing Hous of the Uzbek Ac. Sc, Tashkent (1967) (in Russian)
30. Haroutunian, E.: Many statistical hypotheses: interdependence of optimal test's error probabilities exponents. Abstract of the Report on the 3rd All-Union School-seminar, Program-algorithmical Software for Applied Multi-variate Statistical Analysis, Part 2, Tsakhkadzor, pp. 177–178, (1988) (in Russian)
31. Haroutunian, E.: On asymptotically optimal testing of hypotheses concerning Markov chain. Izvestiya Akademii Nauk Armenii, Mathemtika 23(1), 76–80 (1988) (in Russian)
32. Haroutunian, E.: On asymptotically optimal criteria for Markov chains. In: The First World Congress of Bernoulli Society vol. 2(3), pp. 153–156 (1989) (in Russian)
33. Haroutunian, E.: Asymptotically optimal testing of many statistical hypotheses concerning Markov chain. In: 5th Inter. Vilnius Conferance on Probability Theory and Mathem. Statistics, vol. 1, (A-L), pp. 202–203 (1989) (in Russian)
34. Haroutunian, E.: Logarithmically asymptotically optimal testing of multiple statistical hypotheses. Problems of Control and Information Theory 19(5-6), 413–421 (1990)
35. Haroutunian, E.: On bounds for E-capacity of DMC. IEEE Trans. on Inf. Theory 53(11), 4210–4220 (2007)
36. Haroutunian, E.: Information Theory and Statistics. In: International Encyclopedia of Statistical Science, pp. 643–645. Springer (2010)
37. Haroutunian, E., Grigoryan, N.: On reliability approach for testing of many distributions for pair of Markov chains. Transactions of IIAP of NAS of RA, Mathematical Problems of Computer Science 29, 89–96 (2007)
38. Haroutunian, E., Grigoryan, N.: On arbitrarily varying Markov source coding and hypothesis LAO testing by non-informed statistician. In: Proc. of IEEE Int. Symp. Inform. Theory, Seoul, pp. 981–985 (2009)
39. Haroutunian, E., Hakobyan, P.: On multiple hypotheses testing by informed statistician for arbitrarly varying object and application to source coding. Transactions of IIAP of NAS of RA and of YSU, Mathematical Problems of Computer Science 23, 36–46 (2004)
40. Haroutunian, E., Hakobyan, P.: Multiple hypotheses LAO testing for many independent object, Scholarly Research Exchange, Article ID 921574 (2009)
41. Haroutunian, E., Hakobyan, P.: Remarks about reliable identification of probability distributions of two independent objects. Transactions of IIAP of NAS of RA, Mathematical Problems of Computer Science 33, 91–94 (2010)
42. Haroutunian, E., Hakobyan, P., Hormozi Nejad, F.: On two-stage logarithmically asymptotically optimal testing of multiple hypotheses. Transactions of IIAP of NAS of RA, Mathematical Problems of Computer Science 37, 25–34 (2012)
43. Haroutunian, E., Hakobyan, P., Yessayan, A.: Many hypotheses LAO testing with rejection of decision for arbitrarily varying object. Transactions of IIAP of NAS of RA, Mathematical Problems of Computer Science 35, 77–85 (2011)
44. Haroutunian, E., Hakobyan, P., Yessayan, A.: On multiple hypotheses LAO testing with rejection of decision for many independent objects. In: Proceedings of International Conference CSIT 2011, pp. 141–144 (2011)
45. Haroutunian, E., Haroutunian, M., Harutyunyan, A.: Reliability criteria in information theory and in statistical hypothesis testing. Foundations and Trends in Communications and Information Theory 4(2-3) (2008)

46. Haroutunian, E., Yessayan, A.: On logarithmically asymptotically optimal hypothesis testing for pair of statistically dependent objects. Transactions of IIAP of NAS of RA, Mathematical Problems of Computer Science 29, 97–103 (2007)
47. Haroutunian, E., Yessayan, A.: On optimal hypothesis testing for pair of stochastically coupled objects. Transactions of IIAP of NAS of RA, Mathematical Problems of Computer Science 31, 49–59 (2008)
48. Haroutunian, E., Yessayan, A., Hakobyan, P.: On reliability approach to multiple hypothesis testing and identification of probability distributions of two stochastically coupled objects. International Journal Informations Theories and Applications 17(3), 259–288 (2010)
49. Haroutunian, E., Yessayan, A.: On reliability approach to multiple hypothesis testing and identification of probability distributions of two stochastically related objects. In: Proc. of IEEE Int. Symp. Inform. Theory, Seint-Peterburg, pp. 2671–2675 (2011)
50. Hoeffding, W.: Asymptotically optimal tests for multinomial distributions. The Annals of Mathematical Statistics 36, 369–401 (1965)
51. Kullback, S.: Information Theory and Statistics. Wiley, New York (1959)
52. Lehman, E., Romano, J.: Testing Statistical Hypotheses, 3rd edn. Springer (2005)
53. Leong, C., Johnson, D.: On the asymptotics of M-hypothesis Bayesian detection. IEEE Transactions on Information Theory 43(1), 280–282 (1997)
54. Longo, G., Sgarro, A.: The error exponent for the testing of simple statistical hypotheses: a combinatorial approach. Journal of Combinatorics, Information and System Sciences 5(1), 58–67 (1980)
55. Navaei, L.: On reliable identification of two independent Markov chain distributions. Transactions of IIAP of NAS of RA, Mathematical Problems of Computer Science 32, 74–77 (2009)
56. Natarajan, S.: Large derivations, hypothesis testing, and source coding for finite Markov chains. IEEE Trans. on Inf. Theory 31(3), 360–365 (1985)
57. Nikulin, M.: On one result of L. N. Bolshev from theory of statistical hypothesis testing. Studies on Mathematical Statistics, Notes of Scientific Seminars of Saint-Petersburg Branch of the Mathematical Institute 7, 129–137 (1986) (in Russian)
58. Tuncel, E.: On error exponents in hypothesis testing. IEEE Trans. on Inf. Theory 51(8), 2945–2950 (2005)
59. Tusnády, G.: On asymptotically optimal tests. Annals of Statatistics 5(2), 385–393 (1977)
60. Tusnády, G.: Testing statistical hypotheses (an information theoretic approach). Preprint of the Mathematical Institute of the Hungarian Academy of Sciences, Budapest (1979, 1982)
61. Verdú, S.: Multiuser Detection. Cambridge University Press (1998)
62. Yessayan, A.: On reliability approach to identification of probability distributions of two statistically dependent objects. Transactions of IIAP of NAS of RA, Mathematical Problems of Computer Science 32, 65–69 (2009)
63. Westover, B.: Asymptotic Geometry of Multiple Hypothesis Testing. IEEE Trans. on Inf. Theory 54(7), 3327–3329 (2008)

Appendix

Proof of Theorem 3.1. Probability $G_m^N(\mathbf{x})$ for $\mathbf{x} \in \mathcal{T}_Q^N(X)$ can be presented as follows:

$$\begin{aligned} G_m^N(\mathbf{x}) &= \prod_{n=1}^{N} G_m(x_n) \\ &= \prod_{x} G_m(x)^{N(x|\mathbf{x})} \\ &= \prod_{x} G_m(x)^{NQ(x)} \\ &= \exp\left\{ N \sum_{x} \left(-Q(x) \log \frac{Q(x)}{G_m(x)} + Q(x) \log Q(x) \right) \right\} \\ &= \exp\{-N[D(Q \parallel G_m) + H_Q(X)]\}. \end{aligned} \tag{67}$$

Let us consider the sequence of tests $\varphi_N^*(\mathbf{x})$ defined by the sets

$$\mathcal{B}_m^{(N)} \triangleq \bigcup_{Q \in \mathcal{R}_m^{(N)}} \mathcal{T}_Q^N(X), \quad m = \overline{1, M}. \tag{68}$$

Each $\mathbf{x}$ is in one and only one of $\mathcal{B}_m^{(N)}$, that is

$$\mathcal{B}_l^{(N)} \bigcap \mathcal{B}_m^{(N)} = \varnothing, \quad l \neq m, \qquad \text{and} \qquad \bigcup_{m=1}^{M} \mathcal{B}_m^{(N)} = \mathcal{X}^N.$$

Indeed, for $l = \overline{1, M-2}$, $m = \overline{2, M-1}$, for each $l < m$ let us consider arbitrary $\mathbf{x} \in \mathcal{B}_l^{(N)}$. It follows from (9) and (11) that there exists type $Q \in \mathcal{Q}^N(\mathcal{X})$ such that $D(Q||G_l) \leq E_{l|l}$ and $\mathbf{x} \in \mathcal{T}_Q^N(X)$. From (13) and (16) we have $E_{m|m} < E_{l|m}^*(E_{l|l}) < D(Q||G_m)$. From the definition of $\mathcal{B}_m^{(N)}$ we see that $\mathbf{x} \notin \mathcal{B}_m^{(N)}$. Definitions (11), (14) and (16) show also that

$$\mathcal{B}_M^{(N)} \bigcap \mathcal{B}_m^{(N)} = \varnothing, \quad m = \overline{1, M-1}.$$

Now, let us remark that for $m = \overline{1, M-1}$, using (1), (3), (5)–(7) and (67) we can estimate $\alpha_{m|m}^{(N)}(\varphi^*)$ as follows:

$$
\begin{aligned}
\alpha_{m|m}(\varphi_N^*) &= G_m^N\left(\overline{\mathcal{B}_m^{(N)}}\right) \\
&= G_m^N\left(\bigcup_{Q:D(Q||G_m)>E_{m|m}} \mathcal{T}_Q^N(X)\right) \\
&\leq (N+1)^{|\mathcal{X}|} \sup_{Q:D(Q||G_m)>E_{m|m}} G_m(\mathcal{T}_Q^N(X)) \\
&\leq (N+1)^{|\mathcal{X}|} \sup_{Q:D(Q||G_m)>E_{m|m}} \exp\{-ND(Q||G_m)\} \\
&\leq \exp\left\{-N[\inf_{Q:D(Q||G_m)>E_{m|m}} D(Q||G_m) - o_N(1)]\right\} \\
&\leq \exp\left\{-N[E_{m|m} - o_N(1)]\right\},
\end{aligned}
$$

where $o_N(1) \to 0$ with $N \to \infty$.

For $l = \overline{1, M-1}$, $m = \overline{1, M}$, $l \neq m$, using (1), (3), (5)–(7) and (67), we can obtain similar estimates:

$$
\begin{aligned}
\alpha_{l|m}(\varphi_N^*) &= G_m^N\left(\mathcal{B}_l^{(N)}\right) \\
&= G_m^N\left(\bigcup_{Q:D(Q||G_l)\leq E_{l|l}} \mathcal{T}_Q^N(X)\right) \\
&\leq (N+1)^{|\mathcal{X}|} \sup_{Q:D(Q||G_m)\leq E_{l|l}} G_m^N\left(\mathcal{T}_Q^N(X)\right) \\
&\leq (N+1)^{|\mathcal{X}|} \sup_{Q:D(Q||G_m)\leq E_{l|l}} \exp\{-ND(Q||G_m)\} \\
&= \exp\left\{-N\left(\inf_{Q:D(Q||G_m)\leq E_{l|l}} D(Q||G_m) - o_N(1)\right)\right\}.
\end{aligned} \tag{69}
$$

Now let us prove the inverse inequality:

$$
\begin{aligned}
\alpha_{l|m}(\varphi_N^*) &= G_m^N\left(\mathcal{B}_l^{(N)}\right) \\
&= G_m^N\left(\bigcup_{Q:D(Q||G_l)\leq E_{l|l}} \mathcal{T}_Q^N(X)\right) \\
&\geq \sup_{Q:D(Q||G_l)\leq E_{l|l}} G_m^N\left(\mathcal{T}_Q(X)\right) \\
&\geq (N+1)^{-|\mathcal{X}|} \sup_{Q:D(Q||G_l)\leq E_{l|l}} \exp\left\{-ND\left(Q||G_m\right)\right\} \\
&= \exp\left\{-N\left(\inf_{Q:D(Q||G_l)\leq E_{l|l}} D(Q||G_m) + o_N(1)\right)\right\}.
\end{aligned} \tag{70}
$$

Taking into account (69), (70) and the continuity of the functional $D(Q||G_l)$ we obtain that

$$\lim_{N\to\infty}\{-N^{-1}\log\alpha_{l|m}(\varphi_N^*)\}$$

exists and in correspondence with (13) equals to $E_{l|m}^*$. Thus $E_{l|m}(\varphi^*)=E_{l|m}^*$, $m=\overline{1,M}$, $l=\overline{1,M-1}$, $l\neq m$. Similarly we can obtain upper and lower bounds for $\alpha_{M|m}(\varphi_N^*)$, $m=\overline{1,M}$. Applying the same reasoning we get that the reliability $E_{M|m}(\varphi^*)=E_{M|m}^*$. By the definition (8) $E_{M|M}(\varphi^*)=E_{M|M}^*$. The proof of the first part of the theorem will be accomplished if we demonstrate that the sequence of tests φ^* is LAO, that is for given $E_{1|1},\ldots,E_{M-1|M-1}$ and every sequence of tests φ for all $l,m\in\overline{1,M}$, $E_{l|m}(\varphi)\leq E_{l|m}^*$.

Let us consider any other sequence φ^{**} of tests which is defined by the sets $\mathcal{D}_1^{(N)},\ldots,\mathcal{D}_M^{(N)}$ ($\mathcal{D}_i^{(N)}\bigcap\mathcal{D}_j^{(N)}=\varnothing$, $i,j=\overline{1,M}$, $i\neq j$), such that

$$E_{l|m}(\varphi^{**})\geq E_{l|m}^*,\quad m,l=\overline{1,M}. \tag{71}$$

a) Let us examine the sets $D_m^{(N)}\bigcap\mathcal{B}_m^{(N)}$, $m=\overline{1,M-1}$. This intersection cannot be empty, because in that case

$$\begin{aligned}\alpha_{m|m}(\varphi_N^{**}) &= G_m^N\left(\overline{\mathcal{D}}_m^{(N)}\right)\\ &\geq G_m^N\left(\mathcal{B}_m^{(N)}\right)\\ &\geq \exp\{-N(E_{m|m}+o_N(1))\}.\end{aligned}$$

Let us show that

b) $\mathcal{D}_l^{(N)}\bigcap\mathcal{B}_m^{(N)}=\varnothing$, $m,l=\overline{1,M-1}$, $m\neq l$.

Suppose the contrary. If there exists Q such that $D(Q||G_m)\leq E_{m|m}$ and

1. $\mathcal{T}_Q^N(X)\subseteq\mathcal{D}_l^{(N)}$, then

$$\begin{aligned}\alpha_{l|m}(\varphi_N^{**}) &= G_m^N\left(\mathcal{D}_l^{(N)}\right)\\ &> G_m^N\left(\mathcal{T}_Q^N(X)\right)\\ &\geq \exp\{-N(E_{m|m}+o_N(1))\}.\end{aligned}$$

2. $\varnothing\neq\mathcal{D}_l^{(N)}\bigcap\mathcal{T}_Q^N(X)\neq\mathcal{T}_Q^N(X)$, we also obtain that

$$\begin{aligned}\alpha_{l|m}(\varphi_N^{**}) &= G_m^N\left(\mathcal{D}_l^{(N)}\right)\\ &> G_m^N\left(\mathcal{D}_l^{(N)}\right)\bigcap\mathcal{T}_Q^N(X)\\ &\geq \exp\{-N(E_{m|m}+o_N(1))\}.\end{aligned}$$

Thus it follows that $E_{l|m}(\varphi^{**})\leq E_{m|m}$, which in turn according to (8) provides that $E_{l|m}(\varphi^{**})=E_{m|m}$. From condition (16) it follows that $E_{m|m}<E_{l|m}^*$, for all $l=\overline{1,m-1}$, hence $E_{l|m}(\varphi^{**})<E_{l|m}^*$ for all $l=\overline{1,m-1}$, which contradicts to (71).

c) $\mathcal{D}_M^{(N)} \bigcap \mathcal{B}_m^{(N)} = \varnothing$, $m = \overline{1, M-1}$.

If there exists Q such that $D(Q||G_m) \leq E_{m|m}$ and $\mathcal{T}_Q^N(X) \subseteq \mathcal{D}_M^{(N)}$ $\left(\text{or } \varnothing \neq \mathcal{D}_l^{(N)} \bigcap \mathcal{T}_Q^N(X) \neq \mathcal{T}_Q^N(X)\right)$, then

$$\begin{aligned}\alpha_{M|m}(\varphi_N^{**}) &= G_m^N\left(\mathcal{D}_M^{(N)}\right)\\ &> G_m^N(\mathcal{T}_Q^N(X))\\ &\geq \exp\{-N(E_{m|m} + o_N(1))\}.\end{aligned}$$

Thus it follows that $E_{M|m}(\varphi^{**}) \leq E_{m|m} \Longleftrightarrow E_{M|m}(\varphi^{**}) = E_{m|m}$. By the definition (8) and conditions (16) for the optimal exponent $E^*_{M|m}$ the following inequalities $E_{m|m} \leq E^*_{M|m}$ is true. Hence $E_{M|m}(\varphi^{**}) \leq E^*_{M|m}$, which contradicts to (71).

According to $D_i^{(N)} \bigcap \mathcal{D}_j^{(N)} = \varnothing$, $B_i^{(N)} \bigcap \mathcal{B}_j^{(N)} = \varnothing$, $i \neq j$, $i, j = \overline{1, M}$ and a), b), c) we obtain that $\mathcal{D}_m^{(N)} = B_m^{(N)}$, $m = \overline{1, M}$

The proof of the second part of the Theorem 3.1 is simple. If one of the conditions (16) is violated, then from (13)– (15) it follows that at least one of the elements $E_{m|l}$ is equal to 0. For example, let in (16) the m-th condition be violated. It means that $E_{m|m} \geq \min\limits_{l=\overline{m+1,M}} D(G_l||G_m)$, then there exists $l^* \in \overline{m+1, M}$ such that $E_{m|m} \geq D(G_l^*||G_m)$. From latter and (13) we obtain that $E^*_{m|l} = 0$.

The theorem is proved.

Proof of Lemma 4.1. It follows from the independence of the objects that

$$\alpha_{l_1,l_2,l_3|m_1,m_2,m_3}(\Phi_N) = \prod_{i=1}^{3} \alpha_{l_i|m_i}(\varphi_N^i), \text{ if } m_i \neq l_i, \tag{72}$$

$$\alpha_{l_1,l_2,l_3|m_1,m_2,m_3}(\Phi_N) = (1 - \alpha_{l_k|m_k}(\varphi_N^k)) \prod_{i \in [[1,2,3]-k]} \alpha_{l_i|m_i}(\varphi_N^i),$$

$$m_k = l_k,\ m_i \neq l_i,\ k = \overline{1,3},\ i \neq k, \tag{73}$$

$$\alpha_{l_1,l_2,l_3|m_1,m_2,m_3}(\Phi_N) = \alpha_{l_i|m_i}(\varphi_N^i) \prod_{k \in [[1,2,3]-i]} (1 - \alpha_{l_k|m_k}(\varphi_N^i)),$$

$$m_k = l_k,\ m_i \neq l_i,\ i = \overline{1,3}. \tag{74}$$

Remark that here we consider also the probabilities of right (not erroneous) decisions. Because $E_{l|m}(\varphi^i)$ are strictly positive then the error probability $\alpha_{l|m}(\varphi_N^i)$ tends to zero, when $N \longrightarrow \infty$. According to this fact we have

$$\begin{aligned}\lim_{N\to\infty} \inf \left\{-\frac{1}{N} \log\left(1 - \alpha_{l|m}(\varphi_N^i)\right)\right\} &= \lim_{N\to\infty} \inf \frac{\alpha_{l|m}(\varphi_N^i)}{N} \frac{\log\left(1 - \alpha_{l|m}(\varphi_N^i)\right)}{-\alpha_{l|m}(\varphi_N^i)}\\ &= 0.\end{aligned} \tag{75}$$

From definitions (19), equalities (72)–(74), applying (75) we obtain relations (21).

The Lemma is proved.

Proof of Theorem 4.1. The test $\Phi^* = (\varphi^{1,*}, \varphi^{2,*}, \varphi^{3,*})$, where $\varphi^{i,*}$, $i = \overline{1,3}$ are LAO tests of objects X_i, belongs to the set $\mathcal{D}$. Our aim is to prove that such Φ^* is a compound LAO test. Conditions (31)–(34) imply that inequalities analogous to (16) hold simultaneously for tests for three separate objects.

Let the test $\Phi \in \mathcal{D}$ be such that $E_{M,M,M|m,M,M}(\Phi) = E_{M,M,M|m,M,M}$ $E_{M,M,M|M,m,M}(\Phi) = E_{M,M,M|M,m,M}$, and $E_{M,M,M|m,M,M}(\Phi) = E_{M,M,M|M,M,m}$, $m = \overline{1, M-1}$.

Taking into account (22)–(24) we can see that conditions (31)–(34) for every $m = \overline{1, M-1}$ may be replaced by the following inequalities:

$$E_{M|m}(\varphi^i) < \min\left[\min_{l=\overline{1,m-1}} \inf_{Q:D(Q||G_m)\le E_{M|m}(\varphi^i)} D(Q||G_l), \min_{l=\overline{m+1,M}} D(G_l||G_m)\right]. \tag{76}$$

According to Remark 3.1 for LAO test $\varphi^{i,*}$, $i = \overline{1,3}$, we obtain that (76) meets conditions (16) of Theorem 3.1 for each test $\Phi \in \mathcal{D}$, $E_{m|m}(\varphi^i) > 0$, $i = \overline{1,3}$, hence it follows from (7) that $E_{m|l}(\varphi^i)$ are also strictly positive. Thus for a test $\Phi \in \mathcal{D}$ conditions of Lemma 4.1 are fulfilled and the elements of the reliability matrix $\mathbf{E}(\Phi)$ coincide with elements of matrix $\mathbf{E}(\varphi^i)$, $i = \overline{1,3}$, or sums of them. Then from definition of LAO test it follows that $E_{l|m}(\varphi^i) \le E_{l|m}(\varphi^{i,*})$, then $E_{l_1,l_2,l_3|m_1,m_2,m_3}(\Phi) \le E_{l_1,l_2,l_3|m_1,m_2,m_3}(\Phi^*)$. Consequently Φ^* is a LAO test and $E_{l_1,l_2,l_3|m_1,m_2,m_3}(\Phi^*)$ verifies (27)–(30).

b) When even one of the inequalities (31)–(34) is violated, then at least one of inequalities (76) is violated. Then from Theorem 3.2 one of elements $E_{m|l}(\varphi^{i,*})$ is equal to zero. Suppose $E_{3|2}(\varphi^{1,*}) = 0$, then the element $E_{3,m,l|2,m,l}(\Phi^*) = E_{3|2}(\varphi^{1,*}) = 0$.

The Theorem is proved.

Family Complexity and VC-Dimension*

Christian Mauduit[1,2] and András Sárközy[3]

[1] Institut de Mathématiques de Luminy, CNRS, UMR 6206, 163 avenue de Luminy, Case 907, F-13288 Marseille Cedex 9, France
[2] Instituto de Matemática Pura e Aplicada, IMPA–CNRS, UMI 2924 Estrada Dona Castorina 110, 22460-320 Rio de Janeiro, RJ, Brasil
mauduit@iml.univ-mrs.fr
[3] Eötvös Loránd University, Department of Algebra and Number Theory, H-1117 Budapest, Pázmány Péter sétány 1/C, Hungary
sarkozy@cs.elte.hu

Dedicated to the memory of Rudolf Ahlswede and Levon H. Khachatrian

Abstract. In 2003 Ahlswede, Khachatrian, Mauduit and Sárközy introduced the notion of family complexity of binary sequences, and in 2006 Ahlswede, Mauduit and Sárközy extended this definition to sequences of k symbols. Since that several further related papers have been published on this subject. In this paper our main goal is to present a survey of all these papers. We will also answer a question of Csiszár and Gách on the connection of family complexity and VC-dimension.

Keywords: pseudorandom, family complexity, VC-dimension.

1 Introduction

In 2003 Ahlswede, Khachatrian, Mauduit and Sárközy [2] introduced and studied the notion of *family complexity* (or briefly *f-complexity*) of families of binary sequences. In [3] Ahlswede, Mauduit and Sárközy extended the notion of family complexity to sequences of k symbols. Levon Khachatrian had died before [2] appeared and in 2010 Rudolf Ahlswede also died. The two authors of [2] still alive, Christian Mauduit and András Sárközy dedicate this paper to the memory of Rudy and Levon.

In Sections 1 and 2 we will recall the definitions and results presented in [2], resp. [3]. In Sections 3, 4, 5, 6, 7 we will survey the related papers written since that. Finally in Section 8 we will study a related problem which was raised by Péter Gách and Imre Csiszár during the July 2011 Ahlswede memorial conference. They asked the following questions: What is the connection between family complexity and VC-dimension? Are they not related or even equivalent? The last section will be devoted to the discussion of this problem.

* Research partially supported by Hungarian National Foundation for Scientific Research, grants K72731 and K100291, and the Agence Nationale de La Recherche, grant ANR-10-BLAN 0103 MUNUM.

2 The Notion and Basic Properties of the Family Complexity of Binary Sequences

Mauduit and Sárközy [18] introduced the following measures of pseudorandomness of binary sequences

$$E_N = (e_1, e_2, \dots, e_N) \in \{-1, +1\}^N :$$

The *well-distribution measure* of E_N is defined as

$$W(E_N) = \max_{a,b,t} \left| \sum_{j=0}^{t-1} e_{a+jb} \right| \tag{1}$$

where the maximum is taken over all a, b, t such that $a, b, t \in \mathbb{N}$ and $1 \le a \le a + (t-1)b \le N$, while *the correlation measure of order* k of E_N is defined as

$$C_k(E_N) = \max_{M,\mathbf{D}} \left| \sum_{n=1}^{M} e_{n+d_1}\, e_{n+d_2} \, \cdots \, e_{n+d_k} \right|$$

where the maximum is taken over all $\mathbf{D} = (d_1, \dots, d_k)$ and M such that $0 \le d_1 < \dots < d_k \le N - M$. The *combined* (well-distribution-correlation) *pseudorandom measure of order* k was also introduced:

$$Q_k(E_N) = \max_{b,t,\mathbf{D}} \left| \sum_{j=0}^{t} e_{jb+d_1} \dots e_{jb+d_k} \right| \tag{2}$$

where the maximum is taken over all b, t and $\mathbf{D} = (d_1, \dots, d_k)$ such that all the subscripts $jb + d_\ell$ belong to $\{1, 2, \dots, N\}$. (Note that $Q_1(E_N) = W(E_N)$ and clearly $C_k(E_N) \le Q_k(E_N)$.) Then the sequence E_N is considered as a "good" pseudorandom sequence if both measures $W(E_N)$ and $C_k(E_N)$ (at least for small k) are "small" in terms of N. (This terminology is justified by the fact that for a truly random sequence $E_N \in \{-1, +1\}^N$ both $W(E_N)$ and, for fixed k, $C_k(E_N)$ are around $N^{1/2}$ with near 1 probability.)

In [18] it was also shown that the Legendre symbol forms a "good" pseudorandom sequence. More precisely, if p is an odd prime, $N = p - 1$, $e_n = \left(\frac{n}{p}\right)$ for $n = 1, 2, \dots, N$ and $E_N = (e_1, e_2, \dots, e_N)$, then we have

$$W(E_N) \ll N^{1/2} \log N \quad \text{and} \quad C_k(E_N) \ll k N^{1/2} \log N .$$

Later many other "good" pseudorandom binary sequences have been constructed by different authors. However, in many applications, e.g., in cryptography it is not enough to construct a "few" good sequences; instead, one needs large families of them.

Goubin, Mauduit and Sárközy [10] constructed the first *large family of binary sequences* with strong pseudorandom properties. They proved:

Theorem 1. *If p is a prime number, $f(x) \in \mathbb{F}_p[x]$ ($\mathbb{F}_p$ being the field of the modulo p residue classes) has degree k (> 0) and no multiple zero in $\overline{\mathbb{F}}_p$ (= the algebraic closure of $\mathbb{F}_p$), and the binary sequence $E_p = (e_1, \dots, e_p)$ is defined by*

$$e_n = \begin{cases} \left(\frac{f(n)}{p}\right) & \text{for } (f(n), p) = 1 \\ +1 & \text{for } p \mid f(n), \end{cases} \tag{3}$$

then we have

$$W(E_p) < 10kp^{1/2} \log p.$$

Moreover, assume that for $\ell \in \mathbb{N}$ one of the following assumptions holds:
(i) $\ell = 2$;
(ii) $\ell < p$, *and 2 is a primitive root modulo p;*
(iii) $(4k)^\ell < p$.
Then we also have

$$C_\ell(E_p) < 10k\ell p^{1/2} \log p.$$

(The crucial tool in the proof was Weil's theorem [23].)

Since that many further constructions have been given for large families of binary sequences with strong pseudorandom properties. However, in many applications it is not enough to know that our family $\mathcal{F}$ of "good" binary sequences is large; it can be much more important to know that $\mathcal{F}$ has a "rich", "complex" structure. Thus in [2] Ahlswede, Khachatrian, Mauduit and Sárközy introduced a quantitative measure of a property of families of binary sequences which plays an especially important role in cryptography:

Definition 1. *The* family complexity *or briefly* f-complexity $\Gamma(\mathcal{F})$ *of a family $\mathcal{F}$ of binary sequences $E_N \in \{-1, +1\}^N$ is defined as the greatest integer j so that for any* specification

$$e_{i_1} = \varepsilon_1, \dots, e_{i_j} = \varepsilon_j \quad (0 < i_1 < \dots < i_j \le N)$$

there is at least one $E_N = (e_1, \dots, e_N) \in \mathcal{F}$ which satisfies it. The f-complexity of $\mathcal{F}$ is denoted by $\Gamma(\mathcal{F})$. (If there is no $j \in \mathbb{N}$ with the property above then we set $\Gamma(\mathcal{F}) = 0$.)

It is explained in [2] why is it important to know in cryptography that a family of binary sequences is of large complexity.

It follows easily from Definition 1 that

$$2^{\Gamma(\mathcal{F})} \le |\mathcal{F}|$$

whence

$$\Gamma(F) \le \frac{\log |\mathcal{F}|}{\log 2}. \tag{4}$$

In [2] we also showed that a variant of the family of the binary sequences studied in Theorem 1 also has *large f-complexity:*

Theorem 2. *Let p be a prime number, and $K \in \mathbb{N}$, $L \in \mathbb{N}$,*

$$(4K)^L < p.$$

Consider all the polynomials $f(x) \in \mathbb{F}_p[x]$ with the properties that

$$0 < \deg f(x) \le K$$

(where $\deg f(x)$ *denotes the degree of $f(x)$) and $f(x)$ has no multiple zero in $\overline{\mathbb{F}}_p$. For each of these polynomials $f(x)$, consider the binary sequence $E_p = E_p(f) = (e_1, \dots, e_p) \in \{-1, +1\}^p$ defined by (3), and let $\mathcal{F}$ denote the family of all the binary sequences obtained in this way. Then for all $E_p \in \mathcal{F}$ we have*

$$W(E_p) < 10Kp^{1/2} \log p$$

and

$$C_\ell(E_p) < 10KLp^{1/2} \log p \quad \text{for all} \quad \ell \in \mathbb{N},\ 1 \le \ell \le L.$$

Moreover, we have

$$\Gamma(\mathcal{F}) \ge K. \tag{5}$$

We derived this theorem from Theorem 1 by using Lagrange interpolation.

It is easy to see that the family $\mathcal{F}$ defined in Theorem 2 satisfies

$$|\mathcal{F}| \le (1 + o(1))p^{K+1}. \tag{6}$$

It follows from (4) and (6) that

$$\Gamma(\mathcal{F}) \le (1 + o(1)) \frac{K \log p}{\log 2}. \tag{7}$$

Observe that this upper bound for $\Gamma(\mathcal{F})$ is greater than the lower bound in (5) by only a factor $c \log p$, so that the lower bound (5) in Theorem 2 is nearly optimal.

In [2] we also studied the cardinality of a smallest family achieving a prescribed f-complexity. Among others we proved (by using Ahlswede's covering lemma [1]):

Theorem 3. *The cardinality $S(N, K)$ of a smallest family $\mathcal{F} \in \{-1, +1\}^N$ with f-complexity $\Gamma(\mathcal{F}) = K$ satisfies*

$$2^K \le S(N, K) \le 2^K \log\left(\binom{N}{K} 2^K\right) \le 2^K K \log N \quad (\textit{for } K \ge 4).$$

3 The Family Complexity of Families of Sequences of k Symbols

In [19] Mauduit and Sárközy extended the study of pseudorandomness from binary sequences to sequences of k symbols ("letters"). First they extended the definitions of the measures of pseudorandomness described in Section 2 to this more general situation. It is not at all trivial how to do this extension and, indeed, in [19] we introduced two different ways of extension which are nearly equivalent. Here we will present only one of them which is more suitable for our purpose.

Let $k \in \mathbb{N}$, $k \geq 2$, and let $\mathcal{A} = \{a_1, a_2, \dots, a_k\}$ be a finite set ("alphabet") of k symbols ("letters") and consider a sequence $E_N = (e_1, e_2, \dots, e_N) \in \mathcal{A}^N$ of these symbols. Write

$$x(E_N, a, M, u, v) = \left|\{j\colon\ 0 \leq j \leq M-1, e_{u+jv} = a\}\right|$$

and for $\mathbf{W} = (a_{i_1}, \dots, a_{i_\ell}) \in \mathcal{A}^\ell$ and $\mathbf{D} = (d_1, \dots, d_\ell)$ with non-negative integers $d_1 < \dots < d_\ell$,

$$g(E_N, \mathbf{W}, M, \mathbf{D}) = \left|\left\{n\colon\ 1 \leq n \leq M, \left(e_{n+d_1}, \dots, e_{n+d_\ell}\right) = \mathbf{W}\right\}\right|.$$

Then the *f-well-distribution* ("*f*" for "frequency") measure of E_N is defined as

$$\delta(E_N) = \max_{a,M,u,v} \left| x(E_N, a, M, u, v) - \frac{M}{k} \right|$$

where the maximum is taken over all $a \in \mathcal{A}$ and u, v, M with $u + (M-1)v \leq N$, while the *f-correlation measure* of order ℓ of E_N is defined by

$$\gamma_\ell(E_N) = \max_{\mathbf{W},M,\mathbf{D}} \left| g(E_N, \mathbf{W}, M, \mathbf{D}) - \frac{M}{k^\ell} \right|$$

where the maximum is taken over all $\mathbf{W} \in \mathcal{A}^\ell$, and $\mathbf{D} = (d_1, \dots, d_\ell)$ and M such that $M + d_\ell \leq N$.

We showed in [19] that in the special case $k = 2$, $\mathcal{A} = \{-1, +1\}$ the f-measures $\delta(E_N)$, $\gamma_\ell(E_N)$ are between two constant multiples of the binary measures $W(E_N)$, resp. $C_\ell(E_N)$, so that, indeed, the f-measures can be considered as extensions of the binary measures.

In [19] we also constructed a k symbol sequence with strong pseudorandom properties by using (multiplicative) characters. (This construction is the generalization of the Legendre symbol construction presented in [18] in the special case $k = 2$.)

In [3] Ahlswede, Khachatrian, Mauduit and Sárközy showed that if k is a prime number then this construction can be extended to a *large family* of k symbol sequences possessing strong pseudorandom properties, i.e., they proved the analog of Theorem 1 in this case. (The case of composite k is more difficult, it would need further work and ideas to cover this case.) They proved the following result:

Theorem 4. *Assume that k, p are prime numbers, $k \mid p-1$, χ is a (multiplicative) character modulo p of order k, $H \in \mathbb{N}$, $H < p$. Consider all the polynomials $f(x) \in \mathbb{F}_p[x]$ with the properties that*

$$0 < \deg f(x) \leq H$$

and in $\overline{\mathbb{F}}_p$ the multiplicity of each zero of $f(x)$ is less than k. For each of these polynomials $f(x)$ define the sequence $E_p = E_p(f) = (e_1, \ldots, e_p)$ on the k letter alphabet of the k-th (complex) roots of unity by

$$e_n = \begin{cases} \chi(f(n)) & \text{for } (f(n), p) = 1 \\ +1 & \text{for } p \mid f(n). \end{cases}$$

Then we have

$$\delta(E_p) < 11Hp^{1/2}\log p.$$

Moreover, if $\ell \in \mathbb{N}$ and
(i) *either*

$$(4H)^{\ell} < p$$

(ii) *or k is a primitive root modulo p and $\ell < p$,*
then also

$$\gamma_{\ell}(E_p) < 10\ell Hkp^{1/2}\log p$$

holds.

Next in [3] we extended the notion of family complexity to the k symbol case:

Definition 2. *If $\mathcal{A}$ is a set of k symbols, $N, t \in \mathbb{N}$, $(\varepsilon_1, \varepsilon_2, \ldots, \varepsilon_t) \in \mathcal{A}^t$, $i_1, i_2, \ldots, i_t$ are positive integers with $1 \le i_1 < \cdots < i_t \le N$, and we consider sequences $E_N = (e_1, \ldots, e_N) \in \mathcal{A}^N$ with*

$$e_{i_1} = \varepsilon_1, \ldots, e_{i_t} = \varepsilon_t \tag{8}$$

then $\left(e_{i_1}, \ldots, e_{i_t}; \varepsilon_1, \ldots, \varepsilon_t\right)$ is said to be a specification *of E_N of length t or a t-*specification *of E_N.*

Definition 3. *The f-*complexity of a family *$\mathcal{F}$ of sequences $E_N \in \mathcal{A}^N$ of k symbols is defined as the greatest integer t so that for any t-specification there is at least one $E_N \in \mathcal{F}$ which satisfies it. The f-complexity of $\mathcal{F}$ is denoted by $\Gamma_k(\mathcal{F})$. (If there is no $t \in \mathbb{N}$ with the property above then we set $\Gamma_k(\mathcal{F}) = 0$.)*

We proved in [3] that the family $\mathcal{F}$ of the polynomials f described in Theorem 4 is also of large f-complexity:

Theorem 5. *Assume that all the conditions in Theorem 4 hold, consider all the polynomials f satisfying the conditions in the theorem, and let $\mathcal{F}$ denote the family of the sequences $E_p(f)$ assigned to these polynomials. Then we also have*

$$\Gamma_k(\mathcal{F}) \ge H.$$

Finally, in [3] we also studied the cardinality of a smallest family achieving a prescribed f-complexity.

4 A Result of Gyarmati

In [12] Gyarmati improved on Theorem 2 considerably. By combining elementary algebra and Weil's theorem in a very ingenious way, she proved:

Theorem 6. *Let $p \geq 3$ be a prime. Consider all polynomials $f(x)$ such that*

$$0 \leq \deg f(x) \leq K$$

and $f(x)$ has no multiple zero in $\overline{\mathbb{F}}_p$. For each of these polynomials $f(x)$, consider the binary sequence $E_p = E_p(f) = (e_1, e_2, \dots, e_p) \in \{-1, +1\}^p$ defined by (3), and let $\mathcal{F}$ denote the family of all binary sequences obtained in this way. Then

$$\Gamma(\mathcal{F}) \geq \frac{K}{2\log 2} \log p - O\big(K \log(K \log p)\big). \tag{9}$$

This tightens the gap between the lower and upper bounds in (4) and (6): it follows from (6) and (9) that

$$\frac{K}{2\log 2} \log p - O\big(K \log(K \log p)\big) \leq \Gamma(\mathcal{F}) \leq (1 + o(1)) \frac{K \log p}{\log 2}$$

where the lower and upper bounds for $\Gamma(\mathcal{F})$ differ only by a factor $\frac{1}{2} + o(1)$.

5 Two Further Results on the Family Complexity of Families of Binary Sequences

In [11] Gyarmati studied the following construction: Let p be an odd prime, g be a fixed primitive root modulo p, and let ind n denote the base g index of n modulo p, i.e., define ind n by $1 \leq \text{ind } n \leq p-1$ and $n \equiv g^{\text{ind } n} \pmod p$. Let $f(x) \in \mathbb{F}_p[x]$, and define the binary sequence $E_{p-1} = E_{p-1}(f) = (e_1, e_2, \dots, e_{p-1})$ by

$$e_n = \begin{cases} +1 & \text{if } 1 \leq \text{ind } f(n) \leq (p-1)/2 \\ -1 & \text{if } (p+1)/2 \leq \text{ind } f(n) \leq p-1 \text{ or } p \mid f(n). \end{cases} \tag{10}$$

She showed that if the polynomial $f(x)$ satisfies certain assumptions then the binary sequence $E_{p-1} = E_{p-1}(f)$ associated with it possesses certain strong pseudorandom properties. Moreover, she showed:

Theorem 7. *Consider all polynomials $f(x) \in \mathbb{F}_p[x]$ with $0 < \deg f(x) \leq K$. For each of these polynomials $f(x)$ consider the binary sequence $E_p(f)$ defined by (10), and let $\mathcal{F}$ denote the family of all the binary sequences obtained in this way. Then we have*

$$\Gamma(\mathcal{F}) > K.$$

In [8] Folláth gave a construction by using finite fields of characteristic 2:

Theorem 8. *Let $\mathbb{F}_q$ be a finite field of characteristic 2 such that its multiplicative group is of prime order (so that q is of form $q = 2^k$ and $q - 1 = 2^k - 1$ is a Mersenne prime). Let χ be a non-principal additive character and α be a primitive element of $\mathbb{F}_q$, let $f(x) \in \mathbb{F}_q[x]$ be of odd degree $d \geq \log q$, and for $i < d$ let the coefficient of the term x^i in $f(x)$ be 0 if and only if i is even. Define the binary sequence $E_{q-1} = E_{q-1}(f) = (e_1, e_2, \ldots, e_{q-1}) \in \{-1, +1\}^{q-1}$ by*

$$e_n = \chi(f(d^n)) \quad (\text{for } n = 1, 2, \ldots, q-1). \tag{11}$$

Then we have

$$Q_k(E_{q-1}) \leq 9dq^{1/2} \log q$$

(where Q_k is defined by (2)).

Thus under the assumptions of the theorem the binary sequence $E_{q-1}(f)$ possesses strong pseudorandom properties. In [9] Folláth also studied *families* of binary sequences defined in this way. He introduced the following definition:

Definition 4. *A polynomial $f(x) \in \mathbb{F}_q[x]$ of the form $f(x) = \sum_{i=0}^{d} a_i x^{2i+1}$ is said to be a* ***comb polynomial****.*

He proved:

Theorem 9. *Consider all the comb polynomials $f(x) \in \mathbb{F}_q[x]$ of degree at most d, and for each of these polynomials define the binary sequence $E_{q-1}(f) = (e_1, e_2, \ldots, e_{q-1})$ by (11). Denote the family of all these binary sequences $E_{q-1}(f)$ by $\mathcal{F}$. Then we have*

$$\Gamma(\mathcal{F}) \geq \left[\frac{d+1}{2}\right].$$

6 Families of Subsets of the Integers Not Exceeding N

In [6] Dartyge and Sárközy introduced and studied the notion of pseudorandomness of subsets of integers not exceeding N. They defined the measures of pseudorandomness of subsets $\mathcal{R} \subset \{1, 2, \ldots, N\}$ in the following way:

Define the binary sequence

$$E_N = E_N(\mathcal{R}) = (e_1, e_2, \ldots, e_N) \in \left\{1 - \frac{|\mathcal{R}|}{N}, -\frac{|\mathcal{R}|}{N}\right\}^N$$

by

$$e_n = \begin{cases} 1 - \frac{|\mathcal{R}|}{N} & \text{for } n \in \mathcal{R}, \\ -\frac{|\mathcal{R}|}{N} & \text{for } n \notin \mathcal{R} \end{cases} \qquad (n = 1, 2, \ldots, N).$$

Then the well-distribution measure of $\mathcal{R}$ is defined as

$$W(\mathcal{R}, N) = \max_{a,b,t} \left|\sum_{j=0}^{t-1} e_{a+jb}\right|$$

where the maximum is taken over all $a, b, t \in \mathbb{N}$ such that $1 \le a \le a+(t-1)b \le N$, and for $k \in \mathbb{N}$, $k \ge 2$ the correlation measure of order k of $\mathcal{R}$ is defined as

$$C_k(\mathcal{R}, N) = \max_{M,D} \left| \sum_{n=1}^{M} e_{n+d_1} e_{n+d_2} \cdots e_{n+d_k} \right|$$

where the maximum is taken over all $M \in \mathbb{N}$ and $D = (d_1, d_2, \dots, d_k) \in \mathbb{Z}^k$ such that $0 \le d_1 < d_2 \cdots < d_k \le N - M$. The subset $\mathcal{R}$ is considered as a "good" pseudorandom subset of $\{1, 2, \dots, N\}$ if both $W(\mathcal{R}, N)$ and $C_k(\mathcal{R}, N)$ (at least for "small" k) are "small" in terms of N. In [6] constructions were given for subsets with strong pseudorandom properties in terms of these measures.

In [7] Dartyge, Mosaki and Sárközy constructed *large families* of subsets with strong pseudorandom properties, and they also introduced and studied the notion of family complexity in this situation:

Definition 5. *Let $\mathcal{F}$ be a family of subsets of $\{1, 2, \dots, N\}$. Then the family complexity $\Gamma(\mathcal{F})$ of $\mathcal{F}$ is defined the greatest $k \in \mathbb{N}$ such that for every $\mathcal{A} \subset \{1, 2, \dots, N\}$ with $|\mathcal{A}| = k$ and every subset $\mathcal{B}$ of $\mathcal{A}$ there is an $\mathcal{R} \in \mathcal{F}$ such that $\mathcal{R} \cap \mathcal{A} = \mathcal{B}$ (in other words, for every $\mathcal{A} \subset \{1, 2, \dots, N\}$ with $|\mathcal{A}| = k$ and every partition $\mathcal{A} = \mathcal{B} \cup \mathcal{C}$, $\mathcal{B} \cap \mathcal{C} = \varnothing$ of $\mathcal{A}$ there is an $\mathcal{R} \in \mathcal{F}$ such that $\mathcal{B} \subset \mathcal{R}$ and $\mathcal{C} \subset \overline{\mathcal{R}} \overset{def}{=} \{1, 2, \dots, N\} \setminus \mathcal{R}$).*

They proved that if p is an odd prime, $f(x) \in \mathbb{F}_p[x]$ is of degree $d \ge 2$, $r \in \mathbb{Z}$, $s \in \mathbb{N}$, $s < p/2$ and $\mathcal{R} = \mathcal{R}(f) \subset \{1, 2, \dots, p\}$ is defined by

$n \in \mathcal{R}$ if there is $h \in \{r, r+1, \dots, r+s-1\}$ with $f(n) \equiv h \pmod p$

and

$n \notin \mathcal{R}$ otherwise,

then the subset $\mathcal{R}(f)$ (of $\{1, 2, \dots, p\}$) possesses strong pseudorandom properties in terms of measures defined above. Moreover, for all the polynomials $f(x) \in \mathbb{F}_p[x]$ with $2 \le d = \deg f(x) \le D$, consider the subset $\mathcal{R}(f) \subset \{1, 2, \dots, p\}$ and let $\mathcal{F}_D$ denote the family of these subsets $\mathcal{R}(f)$. They estimated the family complexity of $\mathcal{F}_D$:

Theorem 10. *The family complexity of this family $\mathcal{F}_D$ of subsets of $\{1, 2, \dots, p\}$ satisfies the inequality*

$$\min(p, D+1) \le \Gamma(\mathcal{F}_D) \le \min\left(p, (D+1)\frac{\log p}{\log 2}\right).$$

Dartyge, Mosaki and Sárközy also studied the following construction: let p be an odd prime, $d \in \mathbb{N}$, $d < p$, $r \in \mathbb{Z}$, $s \in \mathbb{N}$, $s < p$, $\mathcal{A} \subset \mathbb{F}_p$, $|\mathcal{A}| = d$ and $f(x) = \prod_{a \in \mathcal{A}} (x - a)$. Define $\mathcal{R}' = \mathcal{R}'(f) \subset \{1, 2, \dots, p\}$ by

$n \in \mathcal{R}'$ if $(f(n), p) = 1$ and there is $h \in \{r, r+1, \dots, r+s-1\}$ with $hf(n) \equiv 1 \pmod p$

and

$n \notin \mathcal{R}'$ otherwise.

They proved that these subsets $\mathcal{R}'(f) \subset \{1, 2, \ldots, p\}$ possess strong pseudorandom properties in terms of the measures introduced above, and conjectured that the family of the subsets $\mathcal{R}'(f)$ assigned to the polynomials f of the above form is of large family complexity. They presented computational evidences supporting this conjecture but could not prove it. They also proved that a slightly modified form of the construction (which uses polynomials of less special form) is also of large complexity:

Theorem 11. *Assume that p is a prime, $D \in \mathbb{N}$, $D \geq 2$, $\mathcal{S} \subset \mathbb{F}_p$ and*

$$D \leq \frac{|\mathcal{S}|(p - |\mathcal{S}|)}{p}.$$

Write $\mathcal{R}(f, \mathcal{S}) = \{1 \leq n \leq p : \text{ there is } h \in \mathcal{S} \text{ with } f(n) \equiv h \pmod p\}$, and define the family $\mathcal{F}(D, \mathcal{S})$ by

$$\mathcal{F}(D, \mathcal{S}) = \{\mathcal{R}(f, \mathcal{S}) : f(x) \in \mathbb{F}_p[x], \deg f(x) \leq D, f(x) \text{ has no multiple zero}\}.$$

Then we have

$$\Gamma(\mathcal{F}(D, \mathcal{S})) \geq D + 1.$$

(It is also shown that the subsets belonging to this modified family still possess reasonably good pseudorandom properties, although not as strong ones as the subsets belonging to the original family.)

7 The Family Complexity of Families of Binary Lattices

In [17] Hubert, Mauduit and Sárközy extended the theory of pseudorandomness to n dimensions. They introduced the following definitions:

Denote by I_N^n the set of n-dimensional vectors whose coordinates are integers between 0 and $N - 1$:

$$I_N^n = \{\mathbf{x} = (x_1, \ldots, x_n) : x_i \in \{0, 1, \ldots, N - 1\}\}.$$

This set is called an *n-dimensional N-lattice* or briefly an *N-lattice*. In [15] the definition was extended to more general lattices in the following way: Let $\mathbf{u}_1, \mathbf{u}_2, \ldots, \mathbf{u}_n$ be n linearly independent n-dimensional vectors over the field of the real numbers such that the i-th coordinate of $\mathbf{u}_i$ is a positive integer and the other coordinates of $\mathbf{u}_i$ are 0, so that $\mathbf{u}_i$ is of the form $(0, \ldots, 0, z_i, 0, \ldots, 0)$ (with $z_i \in \mathbb{N}$). Let $t_1, t_2, \ldots, t_n$ be integers with $0 \leq t_1, t_2, \ldots, t_n < N$. Then we call the set

$$B_N^n = \{\mathbf{x} = x_1\mathbf{u}_1 + \cdots + x_n\mathbf{u}_n : x_i \in \mathbb{N} \cup \{0\}, 0 \leq x_i|\mathbf{u}_i| \leq t_i (< N) \text{ for } i = 1, \ldots, n\}$$

an *n-dimensional box N-lattice* or briefly a *box N-lattice*.

In [17] the definition of binary sequences was extended to more dimensions by considering functions of type

$$\eta(\mathbf{x}):\ I_N^n \to \{-1,+1\}.$$

If $\mathbf{x} = (x_1,\dots,x_n)$ so that $\eta(\mathbf{x}) = \eta((x_1,\dots,x_n))$ then we may simplify the notation slightly by writing $\eta(\mathbf{x}) = \eta(x_1,\dots,x_n)$. Such a function can be visualized as the lattice points of the N-lattice replaced by $+1$ and -1 thus they are called *binary N-lattices.*

In [17] Hubert, Mauduit and Sárközy introduced the following measures of pseudorandomness of binary lattices (here we present the definition in the same slightly modified form as in [15]). Consider the binary lattice η: $I_N^n \to \{-1,+1\}$. Then define the *pseudorandom measure of order k* of η by

$$Q_k(\eta) = \max_{B,\mathbf{d}_1,\dots,\mathbf{d}_k} \left| \sum_{\mathbf{x}\in B} \eta(\mathbf{x}+\mathbf{d}_1)\dots\eta(\mathbf{x}+\mathbf{d}_k) \right|$$

where the maximum is taken over all distinct $\mathbf{d}_1,\dots,\mathbf{d}_k \in I_N^n$ and all box N-lattices B such that $B+\mathbf{d}_1,\dots,B+\mathbf{d}_k \subseteq I_N^n$. Note that in the one-dimensional special case $Q_k(\eta)$ is the same as the combined pseudorandom measure (2) for every k and, in particular, $Q_1(\eta)$ is the well-distribution measure W in (1).

Then η is said to have strong pseudorandom properties if for fixed n and k and "large" N the measure $Q_k(\eta)$ is "small" (much smaller than the trivial upper bound N^n). Indeed, it was shown in [17] that for a truly random binary lattice defined on I_N^n and for fixed k the measure $Q_k(\eta)$ is "small": it is less than $N^{n/2}$ multiplied by a logarithmic factor.

Several constructions have been given for *large families* of binary lattices with strong pseudorandom properties. The notion of *family complexity* was also extended to the n-dimensional case [13]:

Let $\mathcal{F}$ be a family of binary lattices η: $I_N^n \to \{-1,+1\}$, let $j \le N^n$, let $\mathbf{x}_1,\mathbf{x}_2,\dots,\mathbf{x}_j$ be j distinct vectors from I_N^n, and let $(\varepsilon_1,\varepsilon_2,\dots,\varepsilon_j) \in \{-1,+1\}^j$. If we consider binary lattices η: $I_N^n \to \{-1,+1\}$ with

$$\eta(\mathbf{x}_1) = \varepsilon_1,\ \eta(\mathbf{x}_2) = \varepsilon_2,\ \dots,\ \eta(\mathbf{x}_j) = \varepsilon_j, \tag{12}$$

then

Definition 6. (12) *is said to be a* specification *of length j of η.*

Definition 7. *The* family complexity *or* f-complexity *of a family $\mathcal{F}$ of binary lattices η: $I_N^n \to \{-1,+1\}$, denoted by $\Gamma(\mathcal{F})$, is defined as the greatest integer j so that for any specification* (12) *of length j there is at least one $\eta \in \mathcal{F}$ which satisfies it.*

Then it is easy to see that again (4) holds.

So far two constructions have been given for large families of binary lattices with strong pseudorandom properties so that their family is also of large f-complexity. The first one is the n-dimensional analog of the construction in Theorem 2. More precisely, first Mauduit and Sárközy [20] proved the n-dimensional analog of Theorem 1:

Theorem 12. *Assume that $q = p^n$ is the power of an odd prime, $f(x) \in \mathbb{F}_q[x]$ has degree ℓ with $0 < \ell < p$, and $f(x)$ has no multiple zero in $\overline{\mathbb{F}}_q$. Denote the quadratic character of $\mathbb{F}_q$ by γ (setting also $\gamma(0) = 0$). Consider the linear vector space formed by the elements of $\mathbb{F}_q$ over $\mathbb{F}_p$, and let $v_1, \dots, v_n$ be a basis of this vector space (i.e., assume that $v_1, v_2, \dots, v_n$ are linearly independent over $\mathbb{F}_p$). Define the n-dimensional binary p-lattice η: $I_p^n \to \{-1, +1\}$ by*

$$\eta(\mathbf{x}) = \eta((x_1, \dots, x_n)) = \begin{cases} \gamma(f(x_1 v_1 + \dots + x_n v_n)) & \text{for } f(x_1 v_1 + \dots + x_n v_n) \neq 0 \\ +1 & \text{for } f(x_1 v_1 + \dots + x_n v_n) = 0. \end{cases} \tag{13}$$

Assume also $k \in \mathbb{N}$ and

$$4^{n(k+\ell)} < p. \tag{14}$$

Then we have

$$Q_k(\eta) < k\ell\left(q^{1/2}(1 + \log p)^n + 2\right). \tag{15}$$

(Indeed, this is a combination of Theorems 1 and 2 in [20].)

Now define p, q, n as above, and set

$$L = \frac{1}{2 \log 4} \frac{\log p}{n}. \tag{16}$$

Let $\mathcal{F}_L$ denote the family of the binary lattices η assigned to the *monic* polynomials f satisfying the conditions in Theorem 12 with

$$0 < \deg f = \ell < L. \tag{17}$$

Then for every k with

$$k < L \tag{18}$$

(14) holds, thus by Theorem 12 all these lattices η satisfy (15) for every k satisfying (18), so that all these lattices η possess strong pseudorandom properties in this sense.

We proved in [13] that this family $\mathcal{F}_L$ is also of *large complexity* and, indeed, this is so far any number K with $0 < K < p$ in place of the number L defined by (16):

Theorem 13. *Assume that $q = p^n$ is the power of an odd prime, let*

$$0 < K < p,$$

and consider all the polynomials $f(x) \in \mathbb{F}_q[x]$ such that

$$0 < \deg f < K$$

and $f(x)$ has no multiple zero in $\overline{\mathbb{F}}_q$. To each of these polynomials f assign the binary lattice η defined by (13) as described in Theorem 12, and let $\mathcal{F}_K$ denote the family of these binary lattices. Then we have

$$\Gamma(\mathcal{F}_K) > \frac{K-1}{2 \log 2} \log q - cK \log(K \log q). \tag{19}$$

We proved this theorem by adapting Gyarmati's method used in the one-dimensional case in [12].

Note that the number of polynomials $f \in \mathbb{F}_q[x]$ with $\deg f < K$ is clearly at most q^{K+1}, thus we have

$$|\mathcal{F}_K| \le \left|\{f\colon\ f \in \mathbb{F}_q[x],\ \deg f < K\}\right| \le q^{K+1}.$$

It follows from (2) that

$$\Gamma(\mathcal{F}_K) \le \frac{\log|\mathcal{F}_K|}{\log 2} \le \frac{(K+1)\log q}{\log 2}$$

so that the lower bound (19) is best possible apart from a constant factor at most.

In [14] Gyarmati, Mauduit and Sárközy studied *another large family* of binary lattices with strong pseudorandom properties. First we proved the analog of Theorem 13 for this construction, i.e., we showed that the construction generates binary lattices of strong pseudorandom properties.

Let $q = p^n$ be the power of an odd prime. We will consider the field $\mathbb{F}_q$ of order q, its prime field of order p will be denoted by $\mathbb{F}_p$ (and we will identify $\mathbb{F}_p$ with the field of the modulo p residue classes, and we write i for the residue class $\equiv i (\mathrm{mod}\ p)$). Fix a basis $v_1, v_2, \dots, v_n$ of the linear vector space formed by $\mathbb{F}_q$ over $\mathbb{F}_p$ (i.e., $v_1, v_2, \dots, v_n$ are linearly independent over $\mathbb{F}_p$). Let $\varphi\colon\ I_p^n \to \mathbb{F}_q$ be the mapping defined so that for $\mathbf{x} = (x_1, \dots, x_n) \in I_p^n$ we have

$$\varphi(\mathbf{x}) = \varphi\big((x_1, \dots, x_n)\big) = x_1 v_1 + \dots + x_n v_n \in \mathbb{F}_q;$$

clearly, this is a bijection. Define the boxes $B_1, B_2, \dots, B_n$ by

$$B_1 = \left\{\sum_{i=1}^n u_i v_i\colon\ 0 \le u_1 \le \frac{p-3}{2}, u_2, \dots, u_n \in \mathbb{F}_p\right\},$$

$$B_j = \left\{\sum_{i=1}^n u_i v_i\colon u_1 = \dots = u_{j=1} = \frac{p-1}{2},\ 0 \le u_j \le \frac{p-3}{2},\ u_{j+1}, \dots, u_n \in \mathbb{F}_p\right\}$$

and write

$$\mathcal{B} = \bigcup_{j=1}^n B_j.$$

Let $f(z) = \mathbb{F}_q[z]$ be a non-constant polynomial, and define the binary lattice $\eta:\ I_p^n \to \{-1, +1\}$ by

$$\eta(\mathbf{x}) = \eta_f(\mathbf{x}) = \begin{cases} +1 & \text{if } f(\varphi(\mathbf{x})) \in \mathcal{B} \\ -1 & \text{if } f(\varphi(\mathbf{x})) \notin \mathcal{B}. \end{cases} \tag{20}$$

Theorem 14. *Let $k, \ell \in \mathbb{N}$ with*

$$2 \le \ell < p \tag{21}$$

and

$$2 \leq k \leq \ell - 1, \tag{22}$$

let $f(z) \in \mathbb{F}_q[z]$ *be of degree* ℓ, *and define* η *by* (20). *Then we have*

$$Q_k(E_N) < 2^k \ell n^k q^{1/2} (\log p + 2)^{n+k}.$$

Unfortunately, it is known that a rather strong upper bound for k of type (22) is necessary, i.e., the correlation of large order can be large.

Our next goal in [14] was to construct, by using (20), a large family of n-dimensional binary p-lattices η each of them having strong pseudorandom properties, more precisely, we wanted $Q_k(\eta)$ to be "small" for every η belonging to the family and every $k \in \mathbb{N}$ less than a certain parameter $K \in \mathbb{N}$. By Theorem 14 the lattice $\eta = \eta_f$ in (20) satisfies this requirement if conditions (21) and (22) in Theorem 14 hold with $\ell = \deg f < p$ and K in place of $k+1$: $K \leq \ell < p$. On the other hand, if $\ell = \deg f$ increases, then the computational complexity of the construction also increases, thus we have to keep $\ell = \deg f$ possibly small. To balance these two requirements, we took polynomials of degree exactly K, i.e., we considered the family

$$\mathcal{G}_k = \{\eta : \ \eta = \eta_f \text{ is of form (19) with } f \in \mathbb{F}_q[x], \ \deg f = K\}.$$

Note that the coefficients of f can be chosen in $(q-1)q^K$ ways so that

$$|\mathcal{G}_k| = (q-1)q^K. \tag{23}$$

The next goal was to define a *subfamily* $\mathcal{H}_K$ of $\mathcal{G}_K$ which is just slightly smaller than $\mathcal{G}_K$, and which is of high family complexity (it follows from $\mathcal{H}_k \subseteq \mathcal{G}_k$ that $\Gamma(H_k) \leq \Gamma(\mathcal{G}_k)$ so that then $\mathcal{G}_k$ is also of high complexity). Then, indeed, both the *lattices* belonging to the family $\mathcal{H}_K$ and the *family* itself are of strong pseudorandom properties.

Define $\mathcal{S}^+$ and $\mathcal{S}^-$ as the set of the polynomials of the following form:

$$\mathcal{S}^+ = \{x^K + x^2 g(x) + x + 1 : \ g(x) \in \mathbb{F}_q[x], \ \deg g(x) \leq K-3 \text{ or } g(x) \equiv 0\},$$
$$\mathcal{S}^- = \{x^K + x^2 g(x) - x - 1 : \ g(x) \in \mathbb{F}_q[x], \ \deg g(x) \leq K-3 \text{ or } g(x) \equiv 0\},$$

and let

$$\mathcal{S} = \mathcal{S}^+ \cup \mathcal{S}^-$$

and

$$\mathcal{H}_K = \{\eta : \ \eta = \eta_f \text{ with some } f \in \mathcal{S}\}.$$

Then clearly

$$|\mathcal{S}| = |\mathcal{S}^+| + |\mathcal{S}^-| = 2q^{K-2},$$

and one can also show that

$$|\mathcal{H}_K| = |\mathcal{S}| = 2q^{K-2} \tag{24}$$

which is, indeed, just slightly smaller than $|\mathcal{G}_k|$ in (23).

By adapting the interpolation method used in [2] we proved that $\mathcal{H}_K$ is of high complexity:

Theorem 15. *Define $q = p^n$, $\mathcal{S}$ and $\mathcal{H}_K$ as above, and assume that $K \in \mathbb{N}$ is such that*

$$3 < K < p.$$

Then we have

$$\Gamma(\mathcal{H}_K) \geq K - 2. \tag{25}$$

Note that by (4) and (24)

$$\Gamma(\mathcal{H}_K) \leq \frac{\log |\mathcal{H}_K|}{\log 2} = \frac{\log 2 + (K-2)\log q}{\log 2} < \frac{2}{\log 2}(K-2)\log q$$

so that our lower bound (25) is worse than the best possible one by at most a factor $c \log q$.

8 The Connection between Family Complexity and VC-Dimension

In this section our goal is to answer the questions of Csiszár and Gách presented at the end of the introduction.

The notion of *VC-dimension* originates from a paper of Vapnik and Chervonenkis [22]. Alon and Spencer [4, p. 243] formulate its definition in the following way:

Definition 8. "*A* range space *S is a pair (X, R), where X is a (finite or infinite) set and R is a (finite or infinite) family of subsets of X. The members of X are called* points *and those of R are called* ranges. *If A is a subset of X then $P_R(A) = \{r \cap A : r \in R\}$ is the* projection *of R on A. In case this projection contains all subsets of A we say that A is* shattered. *The* Vapnik–Chervonenkis dimension *(or* VC-dimension*) of S, denoted by* $\mathrm{VC}(S)$, *is the maximum cardinality of a shattered subset of X. If there are arbitrarily large shattered subsets then* $\mathrm{VC}(S) = \infty$."

(See also [5, p. 86–96], [16] and [21].)

In order to compare the family complexity (of binary sequences) with the VC-dimension first we have to "translate" Definition 1 into the language used in Definition 8. Let $N \in \mathbb{N}$, let $X = \{x_1, x_2, \ldots, x_N\}$ be a set with $|X| = N$, let $\mathcal{F}$ be a family of binary sequences $E_N = (e_1, e_2, \ldots, e_N) \in \{-1, +1\}^N$, and define the mapping $\varphi : \mathcal{F} \to \{A : A \subset X\}$ by

$$\varphi(E_N) = \varphi\big((e_1, e_2, \ldots, e_n)\big) = \{x_i : 1 \leq i \leq n,\ e_i = +1\}.$$

Write $R = R(\mathcal{F}) = \{\varphi(E_N) : E_N \in \mathcal{F}\}$, and consider the range space $S = S(\mathcal{F}) = (X, R)$. Using the terminology and notation above we may redefine the notion of family complexity in the following way:

Definition 9. *The* family complexity $\Gamma(\mathcal{F})$ *of a family* $\mathcal{F}$ *of* binary sequences $E_N \in \{-1,+1\}^N$ *is defined as the greatest positive integer* k *such that every subset* A *of* X *with* $|A| \le k$ *is shattered in* $S(\mathcal{F})$. *(If there is no positive integer* k *with this property, i.e.,* X *has a one-element subset* $\{x_i\}$ *which is not shattered or, in other words, no* $r \in R(\mathcal{F})$ *contains it, then we set* $\Gamma(\mathcal{F}) = 0$.*)*

This definition can be extended to any range space $S = (X, R)$.

Definition 10. *The* family complexity Γ_S *of a* finite range space $S = (X, R)$ *is defined as the greatest positive integer* k *such that every subset* A *of* X *with* $|A| \le k$ *is shattered.*

This is the definition of family complexity to be compared with the definition of the VC-dimension in Definition 8.

Since *every* subset of X of cardinality Γ_S is shattered while $\mathrm{VC}(S)$ is the cardinality of the *maximal* shattered subset of X thus clearly:

Proposition 1. *For any finite range space* S *we have*

$$\Gamma_S \le \mathrm{VC}(S). \tag{26}$$

One may have equality in (26):

Example 1. Let $k, N \in \mathbb{N}$, $1 \le k \le N$, $|X| = N$, $R = \{A : \; A \subset X, \; |A| \le k\}$ *and* $S = (X, R)$. *Then clearly we have*

$$\Gamma_S = \mathrm{VC}(S) = k. \tag{27}$$

(Note that this is not the only case when equality holds in (26). If k, N, X, R are defined as in Example 1 and B is any subset of X with $|B| = k + 2$ then writing $R' = R \cup B$ and $S = (X, R')$, again (27) holds.)

On the other hand, $\mathrm{VC}(S)$ can be much greater than Γ_S:

Example 2. Let $N \in \mathbb{N}$, $|X| = N$, $X = \{x_1, x_2, \dots, x_N\}$, $R = \{B : B \subset \{x_1, x_2, \dots, x_{N-1}\}\}$ *and* $S = (X, R)$. *Then no* $B \subset R$ *contains* x_N *and* $\{x_1, x_2, \dots, x_{N-1}\}$ *is shattered, thus we have*

$$\Gamma_S = 0, \quad \mathrm{VC}(S) = N - 1.$$

(We note that the difference between $\mathrm{VC}(S)$ and Γ_S is maximal in this case. If $\mathrm{VC}(S) = N$, then (27) holds with $k = N$.)

Examples 1 and 2 can be generalized:

Proposition 2. *Let* $k \in \mathbb{N} \cup \{0\}$, $\ell, N \in \mathbb{N}$, $k \le \ell < N$, $|X| = N$, $X = \{x_1, x_2, \dots, x_N\}$, $U = \{A : \; A \subset X, \; |A| \le k\}$, $V = \{B : \; B \subset \{x_1, x_2, \dots, x_\ell\}\}$, $R = U \cup V$ *and* $S = (X, R)$. *Then we have*

$$\Gamma_S = k, \quad \mathrm{VC}(S) = \ell. \tag{28}$$

(We leave the proof to the reader.) (26) and (28) answer the question of Csiszár and Gách completely: writing $\Gamma_s = k$, $VC(S) = \ell$ we always have $k \leq \ell$ and, on the other hand, for every pair $k \leq \ell (< N)$ there is an S satisfying (28).

Finally, we conclude by asking a few related questions:

Problem 1. *Let $N \in \mathbb{N}$ be large, $|X| = N$, let R be a truly random family of subsets of X (i.e., the subsets of X are chosen independently and with probability $\frac{1}{2}$), and let $S = (X, R)$. What can one say about the expected value of Γ_S and $\mathrm{VC}(S)$? How much greater is $\mathrm{VC}(S)$ than Γ_S? Is it true that $\mathrm{VC}(S) - \Gamma_S \to +\infty$ as $N \to +\infty$?*

References

1. Ahlswede, R.: Coloring hypergraphs: a new approach to multi-user source coding, Part I. J. Combinatorics, Information and System Sciences 4, 76–115 (1979); Part II, J. Combinatorics, Information and System Sciences 5, 220–268 (1980)
2. Ahlswede, R., Khachatrian, L.H., Mauduit, C., Sárközy, A.: A complexity measure for families of binary sequences. Period. Math. Hungar. 46, 107–118 (2003)
3. Ahlswede, R., Mauduit, C., Sárközy, A.: Large Families of Pseudorandom Sequences of k Symbols and Their Complexity – Part II. In: Ahlswede, R., Bäumer, L., Cai, N., Aydinian, H., Blinovsky, V., Deppe, C., Mashurian, H. (eds.) Information Transfer and Combinatorics. LNCS, vol. 4123, pp. 308–325. Springer, Heidelberg (2006)
4. Alon, N., Spencer, J.H.: The Probabilistic Method, 3rd edn. Wiley, Hoboken (2008)
5. Csiszár, I., Körner, J.: Information Theory, Coding Theorems for Discrete Memoryless Systems. Academic Press, New York (1981)
6. Dartyge, C., Sárközy, A.: On pseudo-random subsets of the set of integers not exceeding N. Periodica Math. Hungar. 54, 183–200 (2007)
7. Dartyge, C., Mosaki, E., Sárközy, A.: On large families of subsets of the set of the integers not exceeding N. Ramanujan J. 18, 209–229 (2009)
8. Folláth, J.: Construction of pseudorandom binary sequences using additive characters over $GF(2^k)$ I. Periodica Math. Hungar. 57, 73–81 (2008)
9. Folláth, J.: Construction of pseudorandom binary sequences using additive characters over $GF(2^k)$ II. Periodica Math. Hungar. 60, 127–135 (2010)
10. Goubin, L., Mauduit, C., Sárközy, A.: Construction of large families of pseudorandom binary sequences. J. Number Theory 106, 56–69 (2004)
11. Gyarmati, K.: On a family of pseudorandom binary sequences. Periodica Math. Hungar. 49, 45–63 (2004)
12. Gyarmati, K.: On the complexity of a family related to the Legendre symbol. Periodica Math. Hungar. 58, 209–215 (2009)
13. Gyarmati, K., Mauduit, C., Sárközy, A.: Measures of pseudorandomness of families of binary lattices I (Definitions, a construction using quadratic characters). Publ. Math. Debrecen (to appear)
14. Gyarmati, K., Mauduit, C., Sárközy, A.: Measures of pseudorandomness of families of binary lattices II (A further construction). Publ. Math. Debrecen (to appear)
15. Gyarmati, K., Sárközy, A., Stewart, C.L.: On Legendre symbol lattices. Unif. Distrib. Theory 4, 81–95 (2009)
16. Haussler, H., Welzl, E.: ε-nets and simplex range queries. Discrete Comp. Geom. 2, 127–151 (1987)

17. Hubert, P., Mauduit, C., Sárközy, A.: On pseudorandom binary lattices. Acta Arith. 125, 51–62 (2006)
18. Mauduit, C., Sárközy, A.: On finite pseudorandom binary sequences I, Measure of pseudorandomness, the Legendre symbol. Acta Arith. 82, 365–377 (1997)
19. Mauduit, C., Sárközy, A.: On finite pseudorandom sequences of k symbols. Indag. Mathem. 13, 89–101 (2002)
20. Mauduit, C., Sárközy, A.: On large families of pseudorandom binary lattices. Unif. Distrib. Theory 2, 23–37 (2007)
21. Sauer, N.: On the density of families of sets. J. Combin. Theory, Ser. A 13, 145–147 (1972)
22. Vapnik, V.N., Chervonenkis, A.Y.: On the uniform convergence of relative frequencies of events to their probabilities. Theory Probab. Appl. 16, 264–280 (1971)
23. Weil, A.: Sur les courbes algébriques et les variétés qui s'en déduisent, Act. Sci. Ind., 1041, Hermann, Paris (1948)

The Restricted Word Shadow Problem

Rudolf Ahlswede* and Vladimir Lebedev**

IPPI (Institute for Information Transmission Problems of the Russian Academy of Sciences (Kharkevich Institute)), Bol'shoi Karetnyi per. 19, Moscow, Russia
lebedev37@mail.ru

Abstract. Recently we introduced and studied the shadow minimization problem under word-subword relation. In this paper we consider this problem for the restricted case and give optimal solution.

Keywords: minimal shadow.

1 Introduction

In [2], [3] the minimal shadow problem for the word-subword relation was introduced. The shadow problem for words has not been studied before, whereas its analogs for sets ([8], [5], [6], [7]), sequences ([1]), and vector spaces for finite fields ([4]) are well-known.

For an alphabet $\mathcal{X} = \{0, 1, \cdots, q-1\}$ we consider the set $\mathcal{X}^k$ of words $x^k = x_1 x_2 \cdots x_k$ of length k. A word x^n is an n-subword of y^k if there exist a^i and b^{k-n-i} such that $y^k = a^i x^n b^{k-n-i}$, where $i \in \{0, 1, \cdots, k-n\}$.

Definition 1. *[3] The shadow of y^k is the set of all its n-subword:*

$$shad_{k,n}(y^k) = \{x^n : x^n \text{ is an n-subword of } y^k\} \tag{1}$$

and for any subset $A \subset \mathcal{X}^k$ we define its shadow

$$shad_{k,n}(A) = \bigcup_{a^k \in A} shad_{k,n}(a^k). \tag{2}$$

In [3] we studied the problem of finding optimal or at least asymptotically optimal lower bounds on the cardinality of N-sets $A \subset \mathcal{X}^k$, that is the function

$$\underset{k,n}{\triangle}(q, N) = \min\{|\mathrm{shad}_{k,n}(A)| : A \subset \mathcal{X}^k, |A| = N\}. \tag{3}$$

* Rudolf Ahlswede, one of the world's top information theorists at the University of Bielefeld, died December 18, 2010.

** Supported in part by the Russian Foundation for Basic Research, project no 12-01-00905.

H. Aydinian, F. Cicalese, and C. Deppe (Eds.): Ahlswede Festschrift, LNCS 7777, pp. 364–370, 2013.

Theorem 1. *[3] For integers* $N = q^{l+v} + q^{l+v-1}(l-v)(q-1)$ *and* $k = l+m+v > 2l \geq 2v$*, where* $v = k-n$*, we have*

$$\frac{1}{q^v} N \leq \underset{k,n}{\Delta}(q,N) \leq \frac{1}{q^v}\left(1 + \frac{v}{l-v+1}\right) N. \tag{4}$$

In this paper we are interested in the restricted word shadow problem.

Let us denote by $\mathcal{X}_w^k$ the set of words $a^k \in \mathcal{X}^k$ of weight (the number of nonzero symbols) $wt(a^k) = w$.

We consider first the binary case, so $\mathcal{X} = \{0,1\}$.

Definition 2. *For integers k,n,w,N with* $1 \leq w \leq n < k$ *and* $1 \leq N \leq \binom{k}{w}$ *we define*

$$S(k,n,w,N) = \min\{|shad^w_{k,n}(A)| : A \subset \mathcal{X}_w^k,\ |A| = N\} \tag{5}$$

where

$$shad^w_{k,n}(A) = \bigcup_{a^k \in A} shad^w_{k,n}(a^k) \tag{6}$$

and

$$shad^w_{k,n}(y^k) = \{x^n : x^n \in \mathcal{X}_w^n \text{ is an } n\text{-subword of } y^k\} \tag{7}$$

When k, n and w are specified we also use sometimes $S(N)$ for $S(k,n,w,N)$.

In this paper we solve the restricted word shadow minimization problem, namely, we determine the function $S(k,n,w,N)$ for all parameters. We also observe that our result can be easily generalized to arbitrary alphabet size.

2 The Restricted Word Shadow Problem

Let $a = a_1a_2\cdots a_m \in \mathcal{X}^m$ and $b = b_1b_2\cdots b_n \in \mathcal{X}^n$ then we denote by $ab = c = c_1c_2\cdots c_{m+n} \in \mathcal{X}^{m+n}$ with

$$c_1 = a_1, \cdots, c_m = a_m, c_{m+1} = b_1, \cdots, c_{m+n} = b_n.$$

For a subset $A \subset \mathcal{X}^n$ we denote by $AB = \{ab : a \in A, b \in B\}$.

It turns out to be very convenient to introduce the sets

$$A(\epsilon,\delta) = \epsilon\mathcal{X}^{k-2}_{w-\epsilon-\delta}\delta \ \text{ for } \ \epsilon,\delta \in \mathcal{X} = \{0,1\}. \tag{8}$$

So words in $A(\epsilon,\delta)$ are from $\mathcal{X}_w^k$, start with ϵ, end with δ, and between these two letters have a word of length $k-2$ and weigth $w-\epsilon-\delta$.

Thus we have the partition

$$\mathcal{X}_w^k = \bigcup_{\epsilon,\delta \in \mathfrak{X}} A(\epsilon,\delta), \text{ where } |A(\epsilon,\delta)| = \binom{k-2}{w-\epsilon-\delta}, \tag{9}$$

or explicitly

$$|A(1,1)| = \binom{k-2}{w-2}, |A(1,0)| = |A(0,1)| = \binom{k-2}{w-1}, |A(0,0)| = \binom{k-2}{w}.$$

Consider the following partition of $\mathcal{X}_w^k$

$$\mathcal{X}_w^k = \bigcup_s J_s \tag{10}$$

where

$$J_1 = A(1,1), J_2 = A(1,10) \cup A(01,1), J_3 = A(1,100) \cup A(01,10) \cup A(001,1),$$

and so on.

Thus, for J_{s+1} we have

$$J_{s+1} = \bigcup_{i=0}^{s} A(0^i 1, 10^{s-i}).$$

Finally, for the last partition class $J_{k-w=1}$ we have

$$J_{k-w+1} = \bigcup_{i=0}^{k-w} A(0^i 1, 10^{k-w-i}).$$

In other words, we have the following partition classes:

$$1\,b\,1 \tag{11}$$

$$\begin{matrix} 1\,b\,1\,0 \\ 0\,1\,b\,1 \end{matrix} \tag{12}$$

$$\begin{matrix} 1\,b\,1\,0\,0 \\ 0\,1\,b\,1\,0 \\ 0\,0\,1\,b\,1 \end{matrix} \tag{13}$$

for any b from $\mathcal{X}_{w-2}^{k-2}$, $\mathcal{X}_{w-2}^{k-3}$ and $\mathcal{X}_{w-2}^{k-4}$ respectively.

We continue this procedure and for any b from $\mathcal{X}_{w-2}^{k-c-1}$ we take

$$\bigcup_{i=0}^{s} 0^i 1b10^{s-i}. \tag{14}$$

For example if $s = 4$ then we take

$$\begin{matrix} 1\,b\,1\,0\,0\,0\,0 \\ 0\,1\,b\,1\,0\,0\,0 \\ 0\,0\,1\,b\,1\,0\,0 \\ 0\,0\,0\,1\,b\,1\,0 \\ 0\,0\,0\,0\,1\,b\,1 \end{matrix} \tag{15}$$

The last partition class is given by

$$\bigcup_{i=0}^{k-w} 0^i 1^w 0^{k-w-i}. \tag{16}$$

2.1 Case $n = k - 1$

Note that for the case $n = k-1$ the shadow of $A \subset \mathcal{X}_w^k$ can be defined as $\mathrm{shad}^w(A) = \mathrm{shad}_L^w(A) \cup \mathrm{shad}_R^w(A)$ where $\mathrm{shad}_L^w(A) = \{a_2a_3\cdots a_k : wt(a_2\cdots a_k) = w\}$ and $\mathrm{shad}_R(A) = \{a_1a_3\cdots a_{k-1} : w(a_1a_3\cdots a_{k-1}) = w\}$.

In this case we have a nice graph illustration for our partition.

Consider a graph $G = (V, E)$ associated with this word-subword relation: the vertex set V is $\mathcal{X}_w^k$. Two vertices a^k and b^k form an edge $(a^k, b^k) \in E$ if and only if $\mathrm{shad}_{k,k-1}^w(a^k) \cap \mathrm{shad}_{k,k-1}^w(b^k) \neq \varnothing$. Note that there is one to one correspondence between edges in E and elements from $\mathcal{X}_w^{k-1}$.

It follows from the partition described above that the graph G consists of $\binom{k-2}{w-2}$ isolated vertices P_0, $\binom{k-3}{w-2}$ paths of length 1: P_1, $\binom{k-4}{w-2}$ paths of length 2: P_2 and so on.

Given integer $1 \le N \le \binom{n}{w}$ the restricted word shadow problem for the case $n = k-1$ is equivalent to the problem of finding N vertices of the graph G that are incident with minimal number of edges.

We order all vertices of the graph G in the following way. We start with vertices from P_0 in arbitrary order. Then we consider set P_1 from the first partition class in arbitrary order and order vertices from P_1 in compliance with (12): (first $1b10$ and then $01b1$). Then we do the same with sets $P_2, P_3, \ldots, P_{k-w+1}$. We take sets from $(s+1)$-th partition class in arbitrary order and for the set P_s order vertices from P_s in compliance with (14).

It is not hard to see now that the first N vertices, in the described ordering, have minimum number of edges incident with them. This clearly gives us an optimal solution to the problem.

Hence for $\mathrm{N} \le \binom{k-2}{w-2}$ we have $S(k, k-1, w, N) = 0$ and for $\binom{k-2}{w-2} < \mathrm{N} \le \binom{k-2}{w-2} + 2\binom{k-3}{w-2}$ we have for $z \in \mathbb{N}$

$$S(k, k-1, w, N) = \begin{cases} z & , \text{ if } N = 2z + \binom{k-2}{w-2} \\ z+1 & , \text{ if } N = 2z + 1 + \binom{k-2}{w-2}. \end{cases}$$

Now easy calculation gives us the following numerical formulation of our result.

Theorem 2. *For*

$$\binom{k-2}{w-2}+\ldots+(c-1)\binom{k-c}{w-2} \le N \le \binom{k-2}{w-2}+\ldots(c-1)\binom{k-c}{w-2}+c\binom{k-c-1}{w-2}$$

we have

$$S^c := S\left(\binom{k-2}{w-2}+\ldots+(c-1)\binom{k-c}{w-2}\right) = \binom{k-3}{w-2}+\ldots+(c-2)\binom{k-c}{w-2}$$

and

$$S(k,k-1,w,N) = \begin{cases} S^c+(c-1)z & , \text{ if } N = cz+\binom{k-2}{w-2}+\ldots+(c-1)\binom{k-c}{w-2} \\ S^c+(c-1)z+m & , \text{ if } N = cz+m+\binom{k-2}{w-2}+\ldots+(c-1)\binom{k-c}{w-2} \end{cases}$$

where $m = 1, 2, \ldots, c-1$ *and* $c = 2, 3, \ldots, k-w+1$.

2.2 General Case: $w \le n \le k-1$

For general case we have that

$$\text{shad}_{k,n}^{w} J_{s+1}(k) = J_{s-v+1}(n) \tag{17}$$

where $v = k - n$.

Thus the described above ordering of $\mathcal{X}_w^k$ also gives us an optimal solution to the problem in this general case. The set of first N vectors from $\mathcal{X}_w^k$ has the minimal possible restricted shadow and so we have

Theorem 3. *For* $N \le \binom{k-2}{w-2}+\ldots+v\binom{k-v-1}{w-2}$ *we have* $S(N) = 0$ *and for*

$$\binom{k-2}{w-2}+\ldots+(c-1)\binom{k-c}{w-2} \le N \le \binom{k-2}{w-2}+\ldots(c-1)\binom{k-c}{w-2}+c\binom{k-c-1}{w-2}$$

we have

$$S^c = S\left(\binom{k-2}{w-2}+\ldots+(c-1)\binom{k-c}{w-2}\right) = \binom{k-2-v}{w-2}+\ldots+(c-v-1)\binom{k-c}{w-2}$$

and

$$S(k,n,w,N) = \begin{cases} S^c+(c-1)z & , \text{ if } N = cz+\binom{k-2}{w-2}+\ldots+(c-1)\binom{k-c}{w-2} \\ S^c+(c-1)z+m & , \text{ if } N = cz+m+\binom{k-2}{w-2}+\ldots+(c-1)\binom{k-c}{w-2} \end{cases}$$

where $m = 1, 2, \ldots, c-1$, $v = k - m$ *and* $c = 2, 3, \ldots, k-w+1$.

Remark 1. *We note that our result can be easily extended to the q-ary case* $\mathcal{X} = \{0, 1, \ldots, q-1\}$. *Consider only the case* $n = k - 1$.

For integers $k, w, N \in \mathbb{N}$ with $1 \leq w \leq k$ and $1 \leq N \leq \binom{k}{w}$ we define

$$S_q(N) = \min\{|\text{shad}^w(A)| : A \subset \mathcal{X}_w^k, |A| = N\}.$$

The proof goes along the same lines as the proof of Theorem 2. We just replace a symbol 1 in (11)-(16) to any nonzero symbol from $\mathcal{X}$. So we have

Theorem 4. *For*

$$(q-1)^w \binom{k-2}{w-2} + \ldots + (c-1)(q-1)^w \binom{k-c}{w-2} < N \leq$$

$$\leq (q-1)^w \binom{k-2}{w-2} + \ldots (c-1)(q-1)^w \binom{k-c}{w-2} + c(q-1)^w \binom{k-c-1}{w-2}$$

we have

$$S_q(\gamma) = (q-1)^w \binom{k-3}{w-2} + \ldots + (c-2)(q-1)^w \binom{k-c}{w-2},$$

where

$$\gamma := (q-1)^w \binom{k-2}{w-2} + \ldots + (c-1)(q-1)^w \binom{k-c}{w-2}$$

and

$$S_q(N) = \begin{cases} S_q(\gamma) + (c-1)z & , \text{ if } N = cz + \gamma \\ S_q(\gamma) + (c-1)z + m , & \text{if } N = cz + m + \gamma, \end{cases}$$

where $m = 1, 2, \ldots, c-1$ *and* $c = 2, 3, \ldots, k - w + 1$.

References

1. Ahlswede, R., Cai, N.: Shadows and isoperimetry under the sequence-subsequence relation. Combinatorica 17(1), 11–29 (1997)
2. Ahlswede, R., Lebedev, V.: Shadows under the word-subword relation. In: Twelfth International Workshop on Algebraic and Combinatorial Coding Theory, Akademgorodok, Novosibirsk, pp. 16–19 (2010)
3. Ahlswede, R., Lebedev, V.: Shadows under the word-subword relation. Problems of Information Transmission 48(1), 30–45 (2012)
4. Chowdhury, A., Patkos, B.: Shadows and intersections in vector spaces. J. Combin. Theory Ser. A 117(8), 1095–1106 (2010)
5. Kruskal, J.B.: The Number of Simplices in a Complex, Mathematical Optimization Techniques, Berkeley, Los Angeles, pp. 251–278 (1963)

6. Katona, G.: A theorem on finite sets. In: Theory of Graphs, Proc. Colloq. Tihany 1966, pp. 187–207. Akadémiai Kiadó (1968)
7. Lindström, B.A., Zetterström, H.O.: A combinatorial problem in the k-adic number systems. Proc. Amer. Math. Soc. 18, 166–170 (1967)
8. Schützenberger, M.P.: A characteristic property of certain polynomials of E.F. Moore and C.E. Shannon, RLE Quarterly Progress Report, No. 55, Research Laboratory of Electronics, M.I.T., pp. 117–118 (1959)

Mixed Orthogonal Arrays, k-Dimensional M-Part Sperner Multifamilies, and Full Multitransversals

Harout Aydinian[1,⋆], Éva Czabarka[2,⋆⋆,⋆⋆⋆], and László A. Székely[2,†]

[1] University of Bielefeld, POB 100131, D-33501 Bielefeld, Germany
ayd@math.uni-bielefeld.de
[2] University of South Carolina, Columbia, SC 29208, USA
{czabarka,szekely}@math.sc.edu

Dedicated to the memory of Rudolf Ahlswede

Abstract. Aydinian *et al.* [J. Combinatorial Theory A **118**(2)(2011), 702–725] substituted the usual BLYM inequality for L-Sperner families with a set of M inequalities for $(m_1, m_2, \ldots, m_M; L_1, L_2, \ldots, L_M)$ type M-part Sperner families and showed that if all inequalities hold with equality, then the family is homogeneous. Aydinian *et al.* [Australasian J. Comb. **48**(2010), 133–141] observed that all inequalities hold with equality if and only if the transversal of the Sperner family corresponds to a simple mixed orthogonal array with constraint M, strength $M-1$, using m_i+1 symbols in the ith column. In this paper we define k-dimensional M-part Sperner multifamilies with parameters $L_P : P \in \binom{[M]}{k}$ and prove $\binom{M}{k}$ BLYM inequalities for them. We show that if $k < M$ and all inequalities hold with equality, then these multifamilies must be homogeneous with profile matrices that are strength $M-k$ mixed orthogonal arrays. For $k = M$, homogeneity is not always true, but some necessary conditions are given for certain simple families. These results extend to products of posets which have the strong normalized matching property. Following the methods of Aydinian *et al.* [Australasian J. Comb. **48**(2010), 133–141], we give new constructions to simple mixed orthogonal arrays with constraint M, strength $M-k$, using m_i+1 symbols in the ith column. We extend the convex hull method to k-dimensional M-part Sperner multifamilies, and allow additional conditions providing new results even for simple 1-part Sperner families.

⋆ This author was supported by DFG-project AH46/7-1 "General Theory of Information Transfer".

⋆⋆ This author was supported by a PIRA grant of the University of South Carolina and the hospitality of the Rheinische Friedrich-Wilhelms Universität, Bonn.

⋆⋆⋆ The last two authors acknowledge financial support from the grant #FA9550-12-1-0405 from the U.S. Air Force Office of Scientific Research (AFOSR) and the Defense Advanced Research Projects Agency (DARPA).

† This author was supported in part by the NSF DMS contracts No. 0701111, No. 1000475, and by the Alexander von Humboldt-Stiftung at the Rheinische Friedrich-Wilhelms Universität, Bonn.

H. Aydinian, F. Cicalese, and C. Deppe (Eds.): Ahlswede Festschrift, LNCS 7777, pp. 371–401, 2013.

Keywords: transversal, packing, extremal set theory, Sperner theory, mixed orthogonal array, packing array, BLYM inequality, multiset.

1 Notations

We will use $[n] = \{1, 2, \ldots, n\}$ and $[n]^\star = \{0, 1, \ldots, n-1\}$, and let $\binom{X}{\ell}$ denote the family of all ℓ-element subsets of the set X.

We will talk about multisets, where every element appears with some positive integer multiplicity. We will use the notation $[\![\cdot]\!]$ to emphasize that we talk about a multiset. If $\mathcal{A}$ is a multiset, then the support set $\text{supp}(\mathcal{A})$ of $\mathcal{A}$ is the simple set containing all elements of $\mathcal{A}$. We denote the multiplicity of an object x in a multiset $\mathcal{A}$ by $\#[x, \mathcal{A}]$. Clearly, $x \notin \mathcal{A}$ iff $\#[x, \mathcal{A}] = 0$.

If $P(\cdot)$ is a proposition and k_C are non-negative integers, then the notation $[\![C^{k_C} : P(C)]\!]$ stands for the multiset we obtain by taking all objects C with multiplicity k_C that satisfy $P(\cdot)$. Clearly, if $\mathcal{A}$ is a multiset, then $[\![C^{\#[C,\mathcal{A}]} : P(C)]\!]$ will only contain elements of $\mathcal{A}$. If $P(\cdot)$ is a Boolean polynomial on ℓ sets and $k_{C_1,\ldots,C_\ell}$ are non-negative integers, then $[\![C^{k_{C_1,\ldots,C_k}} : C = P(C_1, \ldots, C_\ell)]\!]$ denotes the multiset where every C appears with multiplicity $\sum_{(C_1,\ldots,C_\ell)} k_{C_1,\ldots,C_\ell}$ where the sum is taken over all different ℓ-tuples $(C_1, \ldots, C_\ell)$ for which $C = P(C_1, \ldots, C_\ell)$.

For a multiset $\mathcal{A}$, the *size* or *cardinality* of $\mathcal{A}$ is $|\mathcal{A}| = \sum_{x \in \mathcal{A}} \#[x, \mathcal{A}]$.

We use $\uplus$ to denote disjoint unions of multisets; if $\mathcal{A}$ and $\mathcal{B}$ are multisets, then $\mathcal{A} \uplus \mathcal{B}$ denotes the multiset obtained by $\#[x, \mathcal{A} \uplus \mathcal{B}] = \#[x, \mathcal{A}] + \#[x, \mathcal{B}]$. Clearly, if $\mathcal{A}$ and $\mathcal{B}$ are disjoint (simple) sets, then $\uplus$ is the usual (disjoint) union.

For multisets $\mathcal{A}$ and $\mathcal{B}$, $\mathcal{A} \cup \mathcal{B}$ denotes the multiset obtained by $\#[x, \mathcal{A} \cup \mathcal{B}] = \max(\#[x, \mathcal{A}], \#[x, \mathcal{B}])$.

For multisets $\mathcal{A}$ and $\mathcal{B}$, $\mathcal{A} \cap \mathcal{B}$ denotes the multiset obtained by $\#[x, \mathcal{A} \cap \mathcal{B}] = \min(\#[x, \mathcal{A}], \#[x, \mathcal{B}])$.

For multisets $\mathcal{A}$ and $\mathcal{B}$, $\mathcal{A} \setminus \mathcal{B}$ denotes the multiset obtained by $\#[x, \mathcal{A} \setminus \mathcal{B}] = \max(0, \#[x, \mathcal{A}] - \#[x, \mathcal{B}])$.

A multiset $\mathcal{B}$ of subsets of X is a *multichain* of length $|\mathcal{B}|$, if the elements of $\mathcal{B}$ are pairwise comparable (i.e. the different elements of $\mathcal{B}$ form a chain in the usual sense, and elements may occur with higher multiplicity then 1).

A multiset $\mathcal{B}$ is called an *antichain* if it is a simple set forming an antichain. Antichains are always simple sets.

Finally, if $\mathcal{F}$ is a multiset and $k(F)$ is a real-valued function on $\text{supp}(\mathcal{F})$, then we use the notation

$$\sum_{F \in \mathcal{F}} k(F) := \sum_{F \in \text{supp}(\mathcal{F})} k(F) \cdot \#[F, \mathcal{F}].$$

2 Definitions: k-Dimensional Multitransversals and Mixed Orthogonal Arrays

Let us be given $1 \le n_1, \dots, n_M$, an integer $k \in [M]$, and set for the rest of the paper $\pi_M = \prod_{i=1}^{M} [n_i]^\star$. For each $P \in \binom{[M]}{k}$ let us be given an integer L_P such that $1 \le L_P$. A multiset $\mathcal{T}$ with $\text{supp}(\mathcal{T}) \subseteq \pi_M$ is called a *k-dimensional multitransversal*[1] *on* π_M with these parameters if for every $P \in \binom{[M]}{k}$, fixing $b_j \in [n_j]^\star$ arbitrarily for every $j \in [M] \setminus P$, we have that

$$\left| \left[\!\left[(i_1, \dots, i_M)^{\#[(i_1,\dots,i_M),\mathcal{T}]} : i_j = b_j \text{ for all } j \in [M] \setminus P \right]\!\right] \right| \le L_P. \quad (1)$$

If we want to emphasize that $\mathcal{T}$ is a set and not a multiset (i.e. every element of $\mathcal{T}$ has multiplicity 1), then we call it a *k-dimensional transversal* or a *k-dimensional simple transversal.*

It is easy to see that if $\mathcal{T}$ is a k-dimensional multitransversal, then we have the inequalities

$$\forall P \in \binom{[M]}{k} \qquad |\mathcal{T}| \le L_P \prod_{j \notin P} n_j. \quad (2)$$

A k-dimensional multitransversal is called *full*, if equality holds for at least one inequality set by a $P \in \binom{[M]}{k}$. It is clear from the definitions that equality in one inequality (i.e. having a full transversal) implies equalities in all inequalities iff

$$\frac{1}{L_P} \prod_{j \in P} n_j \text{ does not depend on the choice of } P. \quad (3)$$

The k-dimensional multitransversals above have intimate connection to mixed orthogonal arrays. Consider sets S_i of n_i symbols ($i = 1, \dots, M$) and consider an $N \times M$ matrix T, whose i^{th} column draws its elements from the set S_i. This matrix is called a *mixed* (or *asymmetrical*) *orthogonal array* or *MOA* (the notion of *orthogonal array with variable numbers of symbols* is also used), of *strength* d, *constraint* M and *index set* $\mathbb{L}$, if for any choice of d different columns $j_1, \dots, j_d$ each sequence $(a_{j_1}, \dots, a_{j_d}) \in S_{j_1} \times \dots \times S_{j_d}$ appears exactly $\lambda(j_1, \dots, j_d) \in \mathbb{L}$ times after deleting the other $M - d$ columns. In the case of equal symbol set sizes (and therefore constant λ) we have the classical definition of orthogonal arrays. A (mixed) orthogonal array is *simple*, if the matrix T has no repeated rows. The following proposition easily follows from the definitions.

Proposition 1. *If the parameters $n_1, \dots, n_M, \{L_P : P \in \binom{[M]}{k}\}$ satisfy the condition* (3)*, then any full k-dimensional multitransversal is a MOA with symbol sets $S_i = [n_i]^\star$, of constraint M, strength $M - k$, and index set $\mathbb{L} = \{L_P : P \in \binom{[M]}{k}\}$, with $\lambda(j_1, \dots, j_{M-k}) = L_{[M]\setminus\{j_1,\dots,j_{M-k}\}}$. Furthermore, if the transversal is simple, then so is the MOA.*

[1] This concept is different from the *transversal design* in [14] even for the simple transversals.

Moreover, if a MOA T is given with symbol sets S_i, (where $n_i = |S_i|$), of constraint M, strength d, with an index set $\mathbb{L}$, then T corresponds to a full $(M-d)$-dimensional multitransversal with parameters n_i and $L_P = \lambda([M] \setminus P)$. Furthermore, if T is simple, then so is the corresponding multitransversal.

Orthogonal arrays were introduced by Rao [18,19], the terminology was introduced by Bush [5,6]. Cheng [7] seems to be the first author to consider MOAs. MOAs are widely used in planning experiments. Known mixed orthogonal arrays of practical size are available on the web [15]. The standard reference work for (mixed) orthogonal arrays is the monograph of Hedayat, Sloane and Stufken [14]. Constructions for MOAs usually use finite fields and few MOAs of strength > 2 are known.

An alternative formulation to k-dimensional (simple) transversals is the following: a set of length M codewords from π_M, such that for every $P \in \binom{[M]}{k}$ set of character positions, if the characters are prescribed in any way for the $i \notin P$ character positions, at most L_P of our codewords show all the prescribed values. In particular, if L_P is identically 1, then a k-dimensional transversal is a code of minimum Hamming distance $k+1$ (see [20]).

Also, k-dimensional transversals are *packing arrays* and their complements are *covering arrays* (for the definitions, see [14]).

3 k-Dimensional M-Part Sperner Multifamilies

Let us be given an underlying set X of cardinality n (often just $X = [n]$), and a fixed partition $X_1, \ldots, X_M$ of X with $|X_i| = m_i$. Set $n_i = m_i + 1$ (this convention will be used throughout the paper from now on).

Assume that $\mathbb{C}_i$ is a (simple) chain in the subset lattice of X_i, for $i \in P$, where $P \subseteq [M]$. We define the product of these chains as

$$\prod_{i \in P} \mathbb{C}_i = \left\{ \biguplus_{i \in P} A_i : A_i \in \mathbb{C}_i \right\}.$$

Let us be given for every $P \in \binom{[M]}{k}$ a positive integer L_P.

We call a multifamily of subsets of X, $\mathcal{F}$, a *k-dimensional M-part Sperner multifamily with parameters* $\{L_P : P \in \binom{[M]}{k}\}$, if for all $P \in \binom{[M]}{k}$, for all (simple) chains $\mathbb{C}_j$ in X_j $(j \in P)$ and for all fixed sets $D_i \subseteq X_i$ $(i \notin P)$ we have that

$$\left| \left[\!\!\left[F^{\#[F,\mathcal{F}]} : \Big(F \cap \biguplus_{j \in P} X_j\Big) \in \prod_{j \in P} \mathbb{C}_j, \forall i \in [M] \setminus P \ X_i \cap F = D_i \right]\!\!\right] \right| \leq L_P. \quad (4)$$

A *k-dimensional M-part Sperner family* or a *simple k-dimensional M-part Sperner family* $\mathcal{F}$ is a Sperner multifamily where $\#[F,\mathcal{F}] \in \{0,1\}$. For simple families, for dimension $k = 1$ we get back the concept of M-part

$(m_1,\dots,m_M;L_1,\dots,L_M)$-Sperner families from [1], and restricting further with $M=1$, we get back the concept of the classical L-Sperner families.

The *profile vector* of a subset F of X is the M-dimensional vector

$$(|F\cap X_1|,\dots,|F\cap X_j|,\dots,|F\cap X_M|)\in\pi_M.$$

The *profile array* $\mathbb{P}(\mathcal{F})=(p_{i_1,\dots,i_M})_{(i_1,\dots,i_M)\in\pi_M}$ of a multifamily $\mathcal{F}$ of subsets of X is an M-dimensional array, whose entries count with multiplicity the elements of $\mathcal{F}$ with a given profile vector:

$$p_{i_1,i_2,\dots,i_M}=\left|\left[\!\left[F^{\#[F,\mathcal{F}]}:\ \ \forall j\ \ |F\cap X_j|=i_j\right]\!\right]\right|. \tag{5}$$

A multifamily $\mathcal{F}$ of subsets of X is called *homogeneous*, if the profile vector of a set determines the multiplicity of the set in $\mathcal{F}$, i.e. for each $(i_1,\dots,i_M)\in\pi_M$ there is a nonnegative integer $r_{i_1,\dots,i_M}$ such that if the profile vector of F is $(i_1,\dots,i_M)$ then $\#[F,\mathcal{F}]=r_{i_1,\dots,i_M}$. In a homogeneous multifamily $\mathcal{F}$, we have that $p_{i_1,i_2,\dots,i_M}=r_{i_1,\dots,i_M}\prod\limits_{j=1}^{M}\binom{m_j}{i_j}$. For simple families, $r_{i_1,\dots,i_M}\in\{0,1\}$, and this concept of homogeneity simplifies to the usual concept.

Given a homogeneous k-dimensional M-part Sperner multifamily $\mathcal{F}$ with parameters $\{L_P : P\in\binom{[M]}{k}\}$, we observe that the multiset containing each $(i_1,\dots,i_M)$ with multiplicity $r_{i_1,\dots,i_M}$ is a k-dimensional multitransversal with these parameters, and every k-dimensional multitransversal comes from a homogeneous k-dimensional M-part Sperner multifamily. The multifamily is a (simple) family precisely when the corresponding multitransversal is in fact a simple transversal.

The following sections contain results on k-dimensional M-part Sperner families. Using the proper definitions, some of these results can be extended from k-dimensional M-part Sperner multifamilies on $\prod_{j=1}^{M}[m_i]$ to k-dimensional M-part Sperner families on product of ranked posets with the strong normalized matching property.

In the spirit of [3], the definitions above can easily be extended to products of ranked posets X_i with rank function R_i, where where n_i is the number of ranks in X_i (the elements of X_i have rank $0,1,\dots,n_i-1$), and we denote the number of elements of rank j in X_i by N_i^j. W A multifamily $\mathcal{F}$ with supp$(\mathcal{F})\subseteq\prod_{j=1}^{M}X_j$ is a k-dimensional M-part Sperner family on $\prod_{j=1}^{M}X_j$ with parameters $L_P : P\in\binom{[M]}{k}$, if for every $P\in\binom{[M]}{k}$, every chain $\mathbb{C}_j$ in X_j and every element $D_i\in X_i$ we have that

$$\left|\left[\!\left[F^{\#[F,\mathcal{F}]}:F=(F_1,\dots,F_M),\forall j\in P\ \ F_j\in\mathbb{C}_j,\forall i\notin P\ \ F_i=D_i\right]\!\right]\right|\le L_P.$$

The *profile vector* of $(F_1,\dots,F_M)\in\prod_{j=1}^{M}X_j$ is the M-dimensional vector

$$(R_1(F_1),R_2(F_2),\dots,R_M(F_M))\in\pi_M.$$

The *profile array* $\mathbb{P}(\mathcal{F})=(p_{i_1,\dots,i_M})_{(i_1,\dots,i_M)\in\pi_M}$ of a multifamily $\mathcal{F}$ of subsets of $\prod_{j=1}^{M}X_j$ is an M-dimensional array, whose entries count with multiplicity the elements of $\mathcal{F}$ with a given profile vector:

$$p_{i_1,i_2,\dots,i_M} = \left| \left[\!\left[F^{\#[F,\mathcal{F}]} : F = (F_1,\dots,F_M),\ \forall j\ R_j(F_j) = i_j \right]\!\right] \right|.$$

As before, a multifamily $\mathcal{F}$ of subsets of X is called *homogeneous*, if the profile vector of a set determines the multiplicity of the set in $\mathcal{F}$; and in a homogeneous family we have nonnegative integers $r_{i_1,\dots,i_M}$ such that $p_{i_1,\dots,i_M} = r_{i_1,\dots,i_M} \prod_{j=1}^{M} N_j^{i_j}$.

Note that using these definitions the results of Section 4 extend *mutatis mutandis* to multifamilies of products of ranked posets X_i, if we replace $\binom{m_j}{i_j}$ with $N_j^{i_j}$ in the equations, and assume that each X_i has the strong normalized matching property.

4 New Sperner Type Results

In Sections 4, 5 and 7 we do not break the narrative with lengthy proofs and leave those to Sections 8, 9 and 10. We start with the following:

Theorem 1. [BLYM inequalities] *Given a k-dimensional M-part Sperner multifamily $\mathcal{F}$ with parameters $\{L_P : P \in \binom{[M]}{k}\}$, the following inequalities hold:*

$$\forall P \in \binom{[M]}{k} \qquad \sum_{(i_1,\dots,i_M)\in\pi_M} \frac{p_{i_1,\dots,i_M}}{\prod_{j=1}^{M} \binom{m_j}{i_j}} \le \frac{L_P}{\prod_{j\in P} n_j} \prod_{j=1}^{M} n_j. \tag{6}$$

For simple families, the special case of this theorem for $k = 1$ was found by Aydinian, Czabarka, P. L. Erdős, and Székely in [1], Theorem 6.1. The special case for $M = 1$ was first in print in [11], and the special case $L = M = 1$ is the Bollobás–Lubell–Meshalkin–Yamamoto (BLYM) inequality [4,16,17,22]. Note that the single classical BLYM inequality has been substituted by a *family* of inequalities. Cases of equality can be characterized as follows:

Theorem 2. *Given integers $1 = k \le M$ or $2 \le k \le M - 1$, let $\mathcal{F}$ be a k-dimensional M-part Sperner multifamily with parameters $\{L_P : P \in \binom{[M]}{k}\}$ satisfying all inequalities in* (6) *with equality. Then the following are true:*

(i) *$\mathcal{F}$ is homogeneous;*
(ii) *$\frac{L_P}{\prod_{j\in P} n_j}$ does not depend on the choice of P;*
(iii) *the k-dimensional multitransversal corresponding to $\mathcal{F}$ is a MOA with symbol sets $S_i = [n_i]^\star$, of constraint M, strength $M - k$, and index set $\mathbb{L} = \{L_P : P \in \binom{[M]}{k}\}$, with $\lambda(j_1,\dots,j_{M-k}) = L_{[M]-\{j_1,\dots,j_{M-k}\}}$.*

Any MOA, as described in (iii) *is a k-dimensional multitransversal on π_M with parameters $\{L_P : P \in \binom{[M]}{k}\}$, and it corresponds to the profile array of a homogeneous k-dimensional M-part Sperner multifamily $\mathcal{F}$ with parameters $\{L_P : P \in \binom{[M]}{k}\}$ on a partitioned $(m_1 + \ldots + m_M)$-element underlying set, which satisfies all inequalities in* (6) *with equality.*

Under this correspondence, simple k-dimensonal M-part Sperner families correspond to simple MOAs.

Note that the last sentence is obvious and part (ii) follows directly from the conditions of the theorem.

The special case of this theorem for $k = 1$ and for simple families and simple transversals was found in [1], Theorem 6.2 but failed to mention (iii). Note also that (iii) turns trivial for $M = k = 1$, as the array in question has a single column. Conclusion (i) for the special case $L = k = M = 1$ restricted to simple families is known as the strict Sperner theorem, already known to Sperner [21]; for $M = 1$, $L > 1$, it was discovered by Paul Erdős [9]. However, Theorem 2 does not hold for $k = M \geq 2$, as the following example shows.

Example 1. Let $k = M \geq 2$ and $L_{[M]} = 1$ with $|X_i| = m_i$ for $i \in [M]$, and assume $m_M \geq 2$. For integers r, s with $1 \leq r \leq m_M - 1$ and $2 \leq s \leq \min\left(n_1, \ldots, n_{M-1}, \binom{m_M}{r}\right)$, consider a partition $\binom{X_M}{r} = \mathcal{B}_1 \uplus \ldots \uplus \mathcal{B}_s$; and for each $j \in [M-1]$, fix an s-element set $\{i_1^{(j)}, \ldots, i_s^{(j)}\} \subseteq [n_j]^\star$. Define a k-dimensional k-part Sperner family $\mathcal{F}$ as follows:

$$\mathcal{F} = \biguplus_{\ell=1}^{s} \left(\left(\prod_{j=1}^{M-1} \binom{X_j}{i_\ell^{(j)}} \right) \times \mathcal{B}_\ell \right).$$

This $\mathcal{F}$ is not homogeneous (e.g. the number of elements in $\mathcal{F}$ with profile vector $(i_1^{(1)}, i_1^{(2)}, \ldots, i_1^{(M-1)}, r)$ is $|\mathcal{B}_1| \cdot \prod_{j=1}^{M-1} \binom{n_j}{i_1^{(j)}}$, and, as $0 < |\mathcal{B}_1| < \binom{n_M}{r}$, this is not a multiple of $\binom{n_M}{r} \cdot \prod_{j=1}^{M-1} \binom{n_j}{i_1^{(j)}}$). However,

$$\sum_{(i_1,\ldots,i_M)\in\pi_M} \frac{p_{i_1,\ldots,i_M}}{\prod_{j=1}^{M} \binom{m_j}{i_j}} = \sum_{\ell=1}^{s} \frac{|\mathcal{B}_\ell| \prod_{j=1}^{M-1} \binom{m_j}{i_\ell^{(j)}}}{\binom{m_M}{r} \prod_{j=1}^{M-1} \binom{m_j}{i_\ell^{(j)}}} = \frac{\sum_{\ell=1}^{s} |\mathcal{B}_\ell|}{\binom{m_M}{r}} = 1 = L_{[M]},$$

therefore $\mathcal{F}$ still satisfies (6) with (a single) equality.

The above example can be easily extended to $L_{[M]} > 1$. Although we did not characterize cases of equality in (6) for $k = M$, in the case $L_{[M]} = 1$ we are able to give a necessary condition for an M–dimensional M-part Sperner family to satisfy equality in (6).

Theorem 3. *Let $\mathcal{F}'$ be a k-dimensional M-part Sperner family with $k = M$ and $L_{[M]} = 1$, satisfying the equality*

$$\sum_{E\in\mathcal{F}'} \frac{1}{\prod_{i=1}^{k} \binom{m_i}{|E\cap X_i|}} = 1. \tag{7}$$

Then for each $i \in [M]$, the trace $\mathcal{F}'_{X_i} := \{F \cap X_i : F \in \mathcal{F}'\}$ of $\mathcal{F}'$ on X_i is a union of full levels of 2^{X_i}.

For the proof of Theorem 2 we need to prove a special case that is also a straightforward generalization of the BLYM for 1-part L-Sperner families, as stated below.

Lemma 1. *Let $\mathcal{F}$ be a multifamily of subsets of $[n]$ containing no multichain of length $L+1$. Then we have*

$$\sum_{F\in\mathcal{F}} \frac{1}{\binom{n}{|F|}} \le L, \tag{8}$$

with equality if and only if $\mathcal{F}$ is homogeneous.

Proof. The inequality part follows from Theorem 1, $k = M = 1$. Suppose now we have equality in (8). We claim then that $\mathcal{F}$ can be partitioned into L or less antichains. (In fact, this is the multiset analogue of the well-known dual version of Dilworth's Theorem.) We now mimic the inductive proof that works for simple families. For $L = 1$, $\mathcal{F}$ has to be a simple family and the claim is exactly the strict Sperner Theorem. Let $L > 1$ and assume that the statement is true for all $1 \le L' < L$. Consider the (simple) set $\mathcal{F}_1$ of maximal elements in $\mathcal{F}$ (note that the multiplicity of each element in $\mathcal{F}_1$ is one by definition). Then $\mathcal{F}_2 := \mathcal{F} \setminus \mathcal{F}_1$ contains no multichain of length L. Thus we have

$$\sum_{F\in\mathcal{F}_1} \frac{1}{\binom{n}{|F|}} \le 1 \text{ and } \sum_{F\in\mathcal{F}_2} \frac{1}{\binom{n}{|F|}} \le L-1. \tag{9}$$

But we also have

$$L = \sum_{F\in\mathcal{F}} \frac{1}{\binom{n}{|F|}} = \sum_{F\in\mathcal{F}_1} \frac{1}{\binom{n}{|F|}} + \sum_{F\in\mathcal{F}_2} \frac{1}{\binom{n}{|F|}},$$

therefore equality holds in both inequalities at (9), and by the induction hypothesis both $\mathcal{F}_1$ and $\mathcal{F}_2$ are homogeneous. The lemma follows. □

5 Convex Hull of Profile Matrices of M-Part Multifamilies

The vertices of the convex hull of profile matrices of different kind of families were described by P. L. Erdős, Frankl, and Katona [10], facilitating the optimization of linear functions of the entries of profile matrices of members of the family in question. P. L. Erdős and Katona [12] adapted the method for M-part Sperner families, and recently Aydinian, Czabarka, P. L. Erdős, and Székely adapted it for 1-dimensional M-part $(m_1,\ldots,m_M;L_1,\ldots,L_M)$ Sperner families. The purpose of this section is to generalize these results for k-dimensional M-part Sperner multifamilies, and even further.

Let $X = X_1 \uplus X_2 \uplus \cdots \uplus X_M$ be a partition of the n-element underlying set X, where $|X_i| = m_i \ge 1$ and $m_1+\ldots+m_M = n$. Let $\mathcal{F}$ be a multifamily of subsets of X. The profile array $\mathbb{P}(\mathcal{F}) := (p_{i_1,\ldots,i_M})_{(i_1,\ldots,i_M)\in\pi_M}$ can be identified with a point or its location vector in the Euclidean space $\mathbb{R}^N$, where $N = \prod_{j=1}^M n_i$.

Let $\alpha \subseteq \mathbb{R}^N$ be a finite point set. Let $\langle\alpha\rangle$ denote the *convex hull* of the point set, and $\varepsilon(\alpha) = \varepsilon(\langle\alpha\rangle)$ its *extreme points*. It is well-known that $\langle\alpha\rangle$ is equal to the set of all convex linear combinations of its extreme points.

Let $\mathbb{A}$ be a family of multifamilies of subsets of X. Let $\mu(\mathbb{A})$ denote the set of all profile-matrices of the multifamilies in $\mathbb{A}$, i.e.

$$\mu(\mathbb{A}) = \{\mathbb{P}(\mathcal{F}) : \mathcal{F} \in \mathbb{A}\}.$$

Then the extreme points $\varepsilon(\mu(\mathbb{A}))$ are integer vectors and they are profile matrices of multifamilies from $\mathbb{A}$.

In [12], P. L. Erdős and G.O.H. Katona developed a general method to determine the extreme points $\varepsilon(\mu(\mathbb{A}))$ for families of simple families. We adapt their results to a more general setting. Let I be a multiset with $\mathrm{supp}(I) \subseteq \pi_M$. Let $T(I)$ denote the M-dimensional array, in which the entry $t_{i_1,\ldots,i_M}(I) = \#[(i_1,\ldots,i_M), I]$. Furthermore, let $S(I)$ be the M-dimensional array, in which $S_{i_1,\ldots,i_M}(I) = t_{i_1,\ldots,i_M}(I)\binom{m_1}{i_1}\cdots\binom{m_M}{i_M}$. Recall that a multifamily of subsets of an M-partitioned underlying set is called *homogeneous*, if for any set, the sizes of its intersections with the partition classes already determine the (possibly 0) multiplicity with which the set belongs to the multifamily. It is easy to see that a homogeneous multifamily $\mathcal{F}$ on X has $\mathbb{P}(\mathcal{F}) = S(I)$ for a certain multiset I with $\mathrm{supp}(I) \subseteq \pi_M$.

We say that $\mathfrak{L}$ is a *product-permutation* of X, if the ordered n-tuple $\mathfrak{L} = (x_1,\ldots,x_n)$ is a permutation of $X = X_1 \uplus X_2 \uplus \cdots \uplus X_M$ such that $X_j = \{x_i : i = m_1 + \cdots + m_{j-1} + 1, \ldots, m_1 + \cdots + m_j\}$ i.e. is $\mathfrak{L}$ is a juxtaposition of permutations of X_1, X_2,...,X_M, in this order. Furthermore, we say that a subset $H \subseteq X$ is *initial with respect* to $\mathfrak{L}$, if for all $j = 1, 2, \ldots, M$ we have

$$H \cap X_j = \left\{x_{m_1+\cdots+m_{j-1}+1}, \ldots, x_{m_1+\cdots+m_{j-1}+|H\cap X_j|}\right\},$$

i.e. $H \cap X_j$ is an initial segment in the permutation of X_j. For a multifamily $\mathcal{H}$ on X, define $\mathcal{H}(\mathfrak{L}) = [\![H^{\#[H,\mathcal{H}]} : H \text{ is initial with respect to } \mathfrak{L}]\!]$. Similarly, for an $\mathbb{A}$ family of multifamilies on X, let $\mathbb{A}(\mathfrak{L}) := \{\mathcal{H}(\mathfrak{L}) : \mathcal{H} \in \mathbb{A}\}$.

Lemma 2 (cf. [12] Lemma 3.1). *Suppose that for a finite family $\mathbb{A}$ of M-part multifamilies the set $\mu(\mathbb{A}(\mathfrak{L}))$ does not depend on the choice of $\mathfrak{L}$. Then*

$$\mu(\mathbb{A}) \subseteq \left\langle\left\{S(I) : \mathrm{supp}(I) \subseteq \pi_M \text{ and } T(I) \in \mu(\mathbb{A}(\mathfrak{L}))\right\}\right\rangle \tag{10}$$

holds.

The next theorem follows easily from this lemma:

Theorem 4 (cf. [12] Theorem 3.2). *Suppose that a finite family $\mathbb{A}$ of M-part multifamilies satisfies the following two conditions:*

$$\text{the set } \mu(\mathbb{A}(\mathfrak{L})) \text{ does not depend on } \mathfrak{L}, \text{ and} \tag{11}$$

for all I with $\operatorname{supp}(I) \subseteq \pi_M$, $T(I) \in \mu(\mathbb{A}(\mathfrak{L}))$ *implies* $S(I) \in \mu(\mathbb{A})$. (12)

Then

$$\varepsilon(\mu(\mathbb{A})) = \varepsilon\Big(\Big\{S(I) : \operatorname{supp}(I) \subseteq \pi_M, T(I) \in \mu(\mathbb{A}(\mathfrak{L}))\Big\}\Big). \tag{13}$$

Consequently, among the maximum size elements of $\mathbb{A}$, there are homogeneous ones, and the profile matrices of maximum size elements of $\mathbb{A}$ are convex linear combinations of the profile matrices of homogeneous maximum size elements.

Proof. The identity

$$\langle \mu(\mathbb{A}) \rangle = \Big\langle \Big\{S(I) : \operatorname{supp}(I) \subseteq \pi_M \text{ with } T(I) \in \mu(\mathbb{A}(\mathfrak{L}))\Big\} \Big\rangle$$

follows from (10) and (12). If two convex sets are equal, then so are their extreme points. □

For any finite set Γ, a *Γ-multiplicity constraint* $\mathbb{M}_\Gamma$ is

$$\mathbb{M}_\Gamma = \{(A_\gamma \geq 0, \{\alpha^\gamma_{(i_1,\ldots,i_M)} \geq 0 : (i_1,\ldots,i_M) \in \pi_M\}) : \gamma \in \Gamma\}.$$

We say that a multiset $\mathcal{F}$ with $\operatorname{supp}(\mathcal{F}) \subseteq X$ satisfies the Γ-multiplicity constraint $\mathbb{M}_\Gamma$, if

$$\forall \gamma \in \Gamma \qquad \sum_{(i_1,\ldots,i_M)\in\pi_M} \alpha^\gamma_{i_1,\ldots,i_M} \cdot \max\{\#[F,\mathcal{F}] : \forall j \in [M]\ |F \cap X_j| = i_j\} \leq A_\gamma.$$

Analogously, a multiset I with $\operatorname{supp}(I) \subseteq \pi_M$ satisfies the Γ-multiplicity constraint $\mathbb{M}_\Gamma$, if

$$\forall \gamma \in \Gamma \qquad \sum_{(i_1,\ldots,i_M)\in\pi_M} \alpha^\gamma_{i_1,\ldots,i_M} \cdot \#[(i_1,\ldots,i_M), I] \leq A_\gamma.$$

It is easy to see that simple families can be characterized by the following condition: For all $(i_1,\ldots,i_M) \in \pi_M$, $\max\{\#[F,\mathcal{F}] : \forall j\ |F \cap X_j| = i_j\} \leq 1$. This in turn can be written in the form of a Γ-multiplicity constraint by $\Gamma = \pi_M$, $A_\gamma = 1$, $\alpha^\gamma_\lambda = \delta_{\gamma,\lambda}$ using the Kronecker δ notation.

Theorem 5. *To the family $\mathbb{A}$ of k-dimensional M-part Sperner multifamilies with parameters L_P for $P \in \binom{[M]}{k}$ satisfying a fixed Γ-multiplicity constraint $\mathbb{M}_\Gamma$, Theorem 4 applies. In other words, all extreme points of $\mu(\mathbb{A})$ come from homogeneous multifamilies.*

This theorem implies the results of [12] and [1] on the convex hull with one exception: there not just all extreme points came from homogeneous families, but all homogeneous families provided extreme points. This is not the case, however, for multifamilies, but characterizing which homogeneous families are extreme is hopeless. For *simple families*, however, we can characterize these extreme points.

We say that an I k-dimensional M-part multitransversal on π_M with Γ-multiplicity constraint $\mathbb{M}_\Gamma$ is *lexicographically maximal (LEM)*, if the support set $\mathrm{supp}(I)$ has an ordering $\boldsymbol{j}_1, \boldsymbol{j}_2, \ldots, \boldsymbol{j}_s$, such that for every $I^\star$ k-dimensional M-part multitransversal on π_M with Γ-multiplicity constraint $\mathbb{M}_\Gamma$, the following holds:
(i) If $\boldsymbol{j}_1 \in \mathrm{supp}(I^\star)$, then $\#[\boldsymbol{j}_1, I] \geq \#[\boldsymbol{j}_1, I^\star]$, and
(ii) for every $1 \leq \ell \leq s-1$, if $\{\boldsymbol{j}_1, \boldsymbol{j}_2, \ldots, \boldsymbol{j}_\ell\} \subseteq \mathrm{supp}(I^\star)$ and $\#[\boldsymbol{j}_h, I] = \#[\boldsymbol{j}_h, I^\star]$ for $h = 1, 2, \ldots, \ell$, then $\#[\boldsymbol{j}_{\ell+1}, I] \geq \#[\boldsymbol{j}_{\ell+1}, I^\star]$.

Lemma 3. *For a family $\mathbb{A}$ of k-dimensional M-part Sperner multifamilies with parameters L_P for $P \in \binom{[M]}{k}$ satisfying a Γ-multiplicity constraint $\mathbb{M}_\Gamma$, the profile matrices $S(I)$, of LEM k-dimensional multitransversals I with* $\mathrm{supp}(I) \subseteq \pi_M$ *that satisfy the same Γ-multiplicity constraint, are extreme points of $\mu(\mathbb{A})$.*

For simple k-dimensional M-part Sperner families $\mathcal{F}$, i.e. when the Γ-multiplicity constraint $\mathbb{M}_\Gamma$ *includes* the conditions $\max\{\#[F, \mathcal{F}] : \forall j\ |F \cap X_j| = i_j\} \leq 1$ for all $(i_1, \ldots, i_M) \in \pi_M$, every I k-dimensional M-part Sperner multitransversal with parameters L_P for $P \in \binom{[M]}{k}$ satisfying a Γ-multiplicity constraint $\mathbb{M}_\Gamma$ has the LEM property. This finally derives the convex hull results of [1] and [12] from our results. Note, however, that the Γ-multiplicity constraint provides new results even for the classical $M = 1$ case. For completeness, we state explicitly our result for simple families.

Theorem 6. *The extreme points of the convex hull of profile matrices of all k-dimensional M-part simple Sperner families with a Γ-multiplicity constraint $\mathbb{M}_\Gamma$ are exactly the profile matrices of the homogeneous families corresponding to k-dimensional M-part simple transversals with the same Γ-multiplicity constraint $\mathbb{M}_\Gamma$. Therefore, among the maximum size k-dimensional M-part Sperner families with a Γ-multiplicity constraint, there are homogeneous ones.*

6 Applications of the Convex Hull Method

Although the previous section reduces the problem of finding the maximum size of such families to a "number" problem from a "set" problem, however, we assert that the problem is still "combinatorial" due to the complexity of transversals:

Problem 1. *For a $(t_1, \ldots, t_M) \in \pi_M$, set the weight $W(t_1, \ldots, t_M) = \prod_{i=1}^{M} \binom{m_i}{t_i}$. Find a set of codewords $C \subseteq \pi_M$ with the largest possible sum of weights, such that for every $P \in \binom{[M]}{k}$ set of character positions, if the characters are prescribed in any way for the $i \notin P$ character positions, at most L_P from C show all the prescribed values.*

In view of Theorem 6, Problem 1 is equivalent to finding maximum size k-dimensional M-part simple Sperner families. Recall that this problem is not solved even for the case $L = 1, k = 1, M \geq 3$ (see [1] for a survey of results). Note also that there are examples in [1] without a full 1-dimensional transversal defining a maximum size homogeneous family, unlike in the case $M = 2, L = 1$.

Our results allow us to prove that certain maximum size families must always be homogeneous.

Theorem 7. *Let $1 \leq k < M$ or $k = M = 1$. If every maximum size homogeneous k-dimensional M-part Sperner family (alternatively: Sperner multifamily) satisfies inequality* (6) *with equality, then every maximum size k-dimensional M-part Sperner family (Sperner multifamily) is homogeneous.*

Proof. Fix a $P \in \binom{[M]}{k}$ and let $C = L_P \prod_{j \notin P} n_j$. By the assumptions, the value of C is independent of P. Let $\mathcal{F}$ be a maximum size family/multifamily with profile array $\mathbb{P}(\mathcal{F}) = (p_{(i_1,\ldots,i_M)})$. Let $\mathcal{G}_1, \ldots, \mathcal{G}_s$ be an enumeration of all maximum size homogeneous families/multifamilies, and let $I_1, \ldots, I_s$ be the M-dimensional transversals/multitransversals on π_M for which $\mathbb{P}(\mathcal{G}_j) := (p^{(j)}_{(i_1,\ldots,i_M)}) = S(I_j)$. By the assumptions for each $j \in [s]$ we have

$$\sum_{(i_1,\ldots,i_M)\in\pi_M} \frac{p^{(j)}_{(i_1,\ldots,i_M)}}{\prod_{\ell=1}^{M}\binom{m_\ell}{i_\ell}} = C.$$

By Theorems 5 and 6 we have $\lambda_j \geq 0$ such that $\sum_j \lambda_j = 1$ and $\mathbb{P}(\mathcal{F}) = \sum_{j=1}^{s} \lambda_j \mathbb{P}(\mathcal{G}_j)$. Therefore

$$\begin{aligned}\sum_{(i_1,\ldots,i_M)\in\pi_M} \frac{p_{(i_1\ldots,i_M)}}{\prod_{\ell=1}^{M}\binom{m_\ell}{i_\ell}} &= \sum_{(i_1,\ldots,i_M)\in\pi_M} \frac{\sum_{j=1}^{s}\lambda_j p^{(j)}_{(i_1\ldots,i_M)}}{\prod_{\ell=1}^{M}\binom{m_\ell}{i_\ell}} \\ &= \sum_{j=1}^{s}\left(\lambda_j \sum_{(i_1,\ldots,i_M)\in\pi_M} \frac{p^{(j)}_{(i_1\ldots,i_M)}}{\prod_{\ell=1}^{M}\binom{m_\ell}{i_\ell}}\right) \\ &= C\sum_{j=1}^{s}\lambda_j = C,\end{aligned}$$

and $\mathcal{F}$ is homogeneous by Theorem 2. □

We state some simple results for the case when all parameters $L_P = 1$.

Theorem 8. *Consider the (simple) M-part families such that for all $E, F \in \mathcal{F}$, if $E \neq F$ then there is a $j \in [M]$ such that $E \cap X_j \neq F \cap X_j$. If $\mathcal{F}$ has maximum size among these families, then*

$$|\mathcal{F}| = \prod_{i=1}^{M}\binom{m_i}{\lfloor m_i/2\rfloor}. \tag{14}$$

Moreover, $\mathcal{F}$ is a maximum size homogeneous family precisely when $\mathbb{P}(\mathcal{F}) = S((\ell_1,\ldots,\ell_M))$ where for each $i \in [M]$, $\ell_i \in \{\lfloor m_i/2\rfloor, \lceil m_i/2\rceil\}$. In particular, when all m_i are even, the maximum size family is unique and homogeneous.

Proof. A family $\mathcal{G}$ satisfies the conditions precisely when the family is a M-dimensional M-part Sperner family with $L = 1$. By Theorem 6, there are homogeneous families among such maximum size families. So let $\mathcal{F}$ be a homogeneous maximum size family. Then $\mathbb{P}(\mathcal{F}) = S(I)$ for some $I \subseteq \pi_M$. It follows from the conditions that $|I| = 1$, so $I = \{(i_1, \ldots, i_M)\}$ and $|\mathcal{F}| = \prod_{j=1}^{M} \binom{m_j}{i_j}$. (14) follows, moreover the homogeneous maximum size families are precisely the ones listed in the theorem.

Since by Theorem 6 the profile array $\mathbb{P}(\mathcal{F})$ is the convex combination of the profile matrices of maximum size families, it follows that for m_i even the maximum size family is unique. □

Note that it is easy to create a nonhomogeneous maximum size family when at least one of the m_i is odd along the lines of Example 1.

For the next result we will use the following, which follows easily by induction on K.

Lemma 4. *Let K, M be positive integers and for each $i \in [K]$ and $j \in [M]$ let a_{ij} be nonnegative reals such that $a_{1j} \geq a_{2j} \geq \cdots \geq a_{K,j}$ and S_K denotes the set of permutations on $[K]$. Then*

$$\max\{\sum_{\ell=1}^{K} \prod_{j=1}^{M} a_{\pi_j(\ell),j} : \forall j \in [M]\ \pi_j \in S_K\} = \sum_{\ell=1}^{K} \prod_{j=1}^{M} a_{\ell j}.$$

□

Theorem 9. *Assume that $m_M = \min m_i$ and consider the $(M-1)$-dimensional M-part Sperner families with parameters $L_{[M]\setminus\{i\}} = 1 : i \in [M]$. If $\mathcal{F}$ is of maximum size amongst these families, then*

$$|\mathcal{F}| = \sum_{i=0}^{m_M} \prod_{j=1}^{M} \binom{m_j}{\lceil \frac{m_j}{2} \rceil + (-1)^i \lceil \frac{i}{2} \rceil}.$$

Moreover, if $\mathcal{F}$ is a maximum size homogeneous family, then $\mathbb{P}(\mathcal{F}) = S(I)$ for some $I = \{(b_{i1}, \ldots, b_{iM}) : i \in [n_M]^\star\}$ where for each fixed $j \in [M]$ the b_{ij} are n_M different integers from $[n_j]^\star$ such that $\binom{m_j}{b_{ij}} = \binom{m_j}{\lceil \frac{m_j}{2} \rceil + (-1)^i \lceil \frac{i}{2} \rceil}$.

If in addition $m_1 = \ldots = m_M$, then all maximum size families are homogeneous.

Proof. Theorem 6 implies that amongst the maximum size families there are homogeneous ones. Let $\mathcal{F}$ be a (not necessarily maximum size) homogeneous $(M-1)$-dimensional M-part Sperner family with all parameters 1, and let I be the transversal for which $\mathbb{P}(\mathcal{F}) = S(I)$. Then if $\boldsymbol{i} = (i_1, \ldots, i_M)$ and $\boldsymbol{i}' = (i'_1, \ldots, i'_M)$ are elements of I such that for some $\ell \in [M]$ $i_\ell = i'_\ell$, we must have that $\boldsymbol{i} = \boldsymbol{i}'$. Therefore there is a $K \leq n_M$ such that $I = \{(b_{i1}, \ldots, b_{iM}) : i \in [K]^\star\}$ where for each fixed $j \in [M]$ the b_{ij} are n_M different integers from $[n_j]^\star$

and $|\mathcal{F}| = \sum_{\ell=1}^{K} \prod_{j=1}^{M} \binom{m_j}{b_{\ell j}}$. The statement about maximum size homogeneous families follows from Lemma 4 and the fact that

$$\binom{m_j}{\lceil \frac{m_j}{2} \rceil + (-1)^0 \lceil \frac{0}{2} \rceil} \geq \binom{m_j}{\lceil \frac{m_j}{2} \rceil + (-1)^1 \lceil \frac{1}{2} \rceil} \geq \ldots \geq \binom{m_j}{\lceil \frac{m_j}{2} \rceil + (-1)^{m_j} \lceil \frac{m_j}{2} \rceil}.$$

The rest follows from Theorem 7. □

7 New k-Dimensional Transversals and Mixed Orthogonal Arrays

Aydinian, Czabarka, Engel, P. L. Erdős, and Székely [2] ran into MOAs as they faced the problem of constructing 1-dimensional full transversals for $M > 2$. Using the indicator function of the k-dimensional transversal in (1) instead of the transversal itself, it is easy to see that the existence of "fractional full k-dimensional transversal" is trivial. Therefore the construction problem of full k-dimensional transversals is a problem of integer programming. For $M = 2$, such construction was found [13] using matching theory, which does not apply for $M > 2$. [2] observed Proposition 1 for $k = 1$ (the property "simple" was assumed tacitly) and constructed 1-dimensional full transversals for any parameter set, and infinitely many MOAs with constraint M and strength $M - 1$. The key element of the construction was the elementary Lemma 5, which only uses properties of the fractional part $\langle x \rangle = x - \lfloor x \rfloor$ function of a real number x. This lemma will be heavily used again in this paper.

Lemma 5. [Engel's Lemma.] *Let n be a positive integer, μ, α, β be real numbers such that $0 < \mu$ and $0 \leq \beta \leq 1 - \mu$. Then*

$$\left| \left\{ i \in [n]^\star : \left\langle \alpha + \frac{i}{n} \right\rangle \in [\beta, \beta + \mu) \right\} \right| \in \{\lfloor \mu n \rfloor, \lceil \mu n \rceil\}.$$

All our constructions for full k-dimensional transversals and simple MOAs are based on the following construction.

Construction 1. *For $n_1, \ldots, n_M$ positive integers, $0 < \mu \leq 1$ real, and $0 \leq \beta \leq 1 - \mu$, define*

$$\mathbb{C}(n_1, \ldots, n_M; \beta, \mu) := \left\{ (i_1, \ldots, i_M) \in \pi_M : \left\langle \sum_{j=1}^{M} \frac{i_j}{n_j} \right\rangle \in [\beta, \beta + \mu) \right\}. \quad (15)$$

For the case $k = 1$, [2] showed that for any $i \in [M]$, any $L_i \in [n_i]$, any $0 < \mu \leq \min\{\frac{L_i}{n_i} : i \in [M]\}$, any $0 \leq \beta \leq 1 - \mu$, the construction in (15) is a 1-dimensional transversal for the given parameters, moreover, if $\mu = \min\{\frac{L_i}{n_i} : i \in [M]\}$, then this 1-dimensional transversal is full.

The following facts are almost immediate from the construction:

Proposition 2. *Let $n_1, \ldots, n_M$ be positive integers, $k \in [M]$, and $\{L_P : P \in \binom{[M]}{k}\}$ be given such that $1 \leq L_P \leq \prod_{i \in P} n_i$ are integers. If there is a $0 < \mu_0 \leq 1$ such that for each $0 \leq \beta \leq 1 - \mu_0$ the construction $\mathbb{C}(n_1, \ldots, n_M; \beta, \mu_0)$ is a full k-dimensional transversal with these L_P parameters, then*

(i) *$\mathbb{C}(n_1, \ldots, n_M; \beta, \mu)$ is a k-dimensional transversal with these parameters for every $0 < \mu < \mu_0$ and $0 \leq \beta \leq 1 - \mu$.*
(ii) *π_M can be partitioned into $\lceil \frac{1}{\mu_0} \rceil$ k-dimensional transversals with these parameters, and $\lfloor \frac{1}{\mu_0} \rfloor$ of these are full.*
(iii) *With $\alpha = \min\left\{\frac{L_P}{\prod_{i \in P} n_i} : P \in \binom{[M]}{k}\right\}$, we have $\lfloor \frac{1}{\alpha} \rfloor \leq \frac{1}{\mu_0} \leq \lceil \frac{1}{\alpha} \rceil$. In particular, if $\frac{1}{\alpha}$ is an integer, all k-dimensional transversals in the partition in (ii) are full.*

Proof. (i) follows from the fact that for every β, μ in (i), exists a $0 \leq \beta' \leq 1 - \mu_0$, such that $[\beta, \beta + \mu) \subseteq [\beta', \beta' + \mu_0)$. (Here we did not use the fullness in the hypothesis.) For (ii), we use the fact that $[0, 1)$ can be partitioned into $\lceil \frac{1}{\mu_0} \rceil$ half-open intervals, $\lfloor \frac{1}{\mu_0} \rfloor$ of which have length μ_0. Finally, (iii) follows from (2) and (ii). □

We arrived at the following generalization of Engel's lemma (Lemma 5):

Lemma 6. *Let $n_1, \ldots, n_k$ be positive integers, $N = \mathrm{lcm}(n_1, \ldots, n_k)$, $K = \prod_{i=1}^{k} n_k$ and $\ell = \frac{K}{N}$. If α, β, μ are real numbers with $0 < \mu < 1$ and $0 \leq \beta \leq 1 - \mu$, then*

$$\left| \left\{ (i_1, \ldots, i_k) \in \pi_M : \left\langle \alpha + \sum_{j=1}^{k} \frac{i_j}{n_j} \right\rangle \in [\beta, \beta + \mu) \right\} \right| \in \{\ell \lfloor \mu N \rfloor, \ell \lceil \mu N \rceil\}.$$

The proof of Lemma 6 is postponed to Section 10. Based on Lemma 6, the following theorem gives a sufficient criterion to use (15) to construct full k-dimensional transversals. For $k = 1$ it gives back the construction in [2].

We set a generic notation here for the rest of this section and Section 10. Let us be given $n_1, \ldots, n_M \geq 1$ integers, a $k \in [M]$, and for every $P \in \binom{[M]}{k}$ let the integer L_P be given such that $1 \leq L_P \leq \prod_{i \in P} n_i$. For every $P \in \binom{[M]}{k}$, set $K_P = \prod_{i \in P} n_i$, $N_P = \mathrm{lcm}\{n_i : i \in P\}$, and $\ell_P = \frac{K_P}{N_P}$.

Theorem 10. *Assume that a $\mu > 0$ is given such that*

$$\forall\, P \in \binom{[M]}{k} \quad \ell_P \lceil \mu N_P \rceil \leq L_P. \tag{16}$$

Then for any $0 \leq \beta \leq 1 - \mu$, $\mathbb{C}(n_1, \ldots, n_M; \beta, \mu)$ is a k-dimensional transversal with the given parameters L_P. Moreover, if $\mu = \min_{P \in \binom{[M]}{k}} \frac{L_P}{K_P}$, then it is a full transversal.

Note that condition (16) easily implies that $\mu \leq \min_{P\in\binom{[M]}{k}} \frac{L_P}{K_P}$. The proof of Theorem 10 is also postponed to Section 10.

Corollary 1. *If the n_i numbers are pairwise relatively prime, then for $\mu = \min_{P\in\binom{[M]}{k}} \frac{L_P}{K_P}$ and for any $0 \leq \beta \leq 1-\mu$, $\mathbb{C}(n_1,\dots,n_M;\beta,\mu)$ is a full k-dimensional transversal with these parameters.*

Proof. It is enough to check that (16) holds in Theorem 10 for this μ. Let $P \in \binom{[M]}{k}$. From the fact that the n_i numbers are relatively prime, it follows that $K_P = N_P$ and $\ell_P = 1$. Therefore $\ell_P\lceil \mu N_P\rceil \leq \lceil \frac{L_P}{N_P}\cdot N_P\rceil = L_P$. □

Corollary 1 ensures that we have a full k-dimensional transversal for all $k \in [M]$ and all allowed settings of $\{L_P : P \in \binom{[M]}{k}\}$ whenever $n_1,\dots,n_M$ are relatively prime. Unfortunately, this does not allow us to chose parameters that give MOAs, i.e. for values of L_P such that $\frac{K_P}{L_P}$ is constant. We can still use the construction in (15) to find such transversal, but we need to put more restrictions on the possible values of the L_P.

Corollary 2. *Assume that there is a constant $0 < \mu \leq 1$ such that for each $P \in \binom{[M]}{k}$, μN_P is an integer and $L_P = \mu K_P$. Then, for every $0 \leq \beta \leq 1-\mu$, $\mathbb{C}(n_1,\dots,n_M;\beta,\mu)$ is a full k-dimensional transversal, and provides a simple MOA of strength $M-k$.*

Proof. The condition on μ gives $\ell_P\lceil\mu N_P\rceil = \ell_P\mu N_P = \mu K_P = L_P$ and $\mu = \min_P \frac{L_P}{K_P}$; the statement follows from Theorem 10. □

While the conditions of the theorem may at first glance seem restrictive, we can easily satisfy them. For a given $k \in [M]$ we chose a sequence of integers $j_1, j_2, \dots, j_M$, and set $n_i = \prod_{v=1}^{i} j_v$. Set q as one of the divisors of n_k and $\mu = \frac{1}{q}$. It is clear that this choice of μ satisfies the conditions of Theorem 2, since for each $P \in \binom{[M]}{k}$ we have that $N_P = \operatorname{lcm}\{n_i : i \in P\} = n_{\max P}$. By the choice of the n_i's and the fact that $k \leq \max P$, n_k divides N_P. Since μn_k is an integer, so is μN_P. Thus, for each $P \in \binom{[M]}{k}$ if we chose $L_P = \mu K_P$, then the construction gives a simple MOA with the given parameters.

We also give two "generic" constructions to create new full k-dimensional multitransversals and MOAs from already known ones, under some numerical conditions: "linear combination", and "tensor product". The correctness of these constructions is straightforward from the definitions.

Proposition 3. [Linear Combination for Transversals.]

(i) *Let $j \in \mathbb{Z}^+$ and for each $\ell \in [j]$ let $\mathcal{T}_\ell$ be a k-dimensional multitransversal on π_M with parameters $L_P^{(\ell)} : P \in \binom{[M]}{k}$. Assume that for all $\ell \in [j]$ positive reals α_ℓ are given such that for all $(i_1,\dots,i_M) \in \pi_M$ the quantity $\sum_{\ell=1}^{j} \alpha_\ell \cdot \#[(i_1,\dots,i_M),\mathcal{T}_\ell]$ is an integer, and let*

$$\mathcal{T}^\star = [\![(i_1,\dots,i_M)^{\sum_{\ell=1}^{j}\alpha_\ell\cdot\#[(i_1,\dots,i_M),\mathcal{T}_\ell]} : (i_1,\dots,i_M) \in \pi_M]\!].$$

Then $\mathcal{T}^\star$ *is a* k*-dimensional multitransversal on* π_M *with parameters* $L_P^\star := \lfloor \sum_{\ell=1}^{j} \alpha_\ell L_P^{(\ell)} \rfloor : P \in \binom{[M]}{k}$.

(ii) *Assume further that each* $\mathcal{T}_\ell$ *above is a full multitransversal and there is a common* $A \in \binom{[M]}{k}$ *on which all* $\mathcal{T}^\ell$ *simultaneously meet the bound, i.e.*

$$\forall \ell \in [j] \;\; L_A^\ell \prod_{j \notin A} n_j = \min_{P \in \binom{[M]}{k}} \left(L_P^{(\ell)} \prod_{j \notin P} n_j \right).$$

Then $\mathcal{T}^\star$ *is a full multitransversal as well.* □

Since the condition is true when the α_ℓ are all integers, this means in particular that if $\mathcal{T}_1$ and $\mathcal{T}_2$ are both k-dimensional multitransversals, then so is $\mathcal{T}_1 \uplus \mathcal{T}_2$.

Proposition 4. [Linear Combination for MOAs.] *Let* $j \in \mathbb{Z}^+$ *and for each* $\ell \in [j]$ *let* $\mathcal{T}_\ell$ *be a full* k*-dimensional multitransversal on* π_M *with parameters* $L_P^{(\ell)} : P \in \binom{[M]}{k}$ *such that* $L_P^\ell \cdot \prod_{j \notin P} n_j$ *is independent of* P *(i.e.* $\mathcal{T}_\ell$ *is a MOA). Let nonzero reals* α_ℓ *be given for all* $\ell \in [j]$ *such that for all* $(i_1, \ldots, i_M) \in \pi_M$ *the quantity* $\sum_{\ell=1}^{j} \alpha_\ell \cdot \#[(i_1, \ldots, i_M), \mathcal{T}_\ell]$ *is a non-negative integer, and let* $\mathcal{T}^\star = \sum_{\ell=1}^{j} \alpha_\ell \mathcal{T}_\ell$ *be defined as*

$$\mathcal{T}^\star = [\![(i_1, \ldots, i_M)^{\sum_{\ell=1}^{j} \alpha_\ell \cdot \#[(i_1, \ldots, i_M), \mathcal{T}_\ell]} : (i_1, \ldots, i_M) \in \pi_M]\!].$$

Then $\mathcal{T}^\star$ *is a full* k*-dimensional multitransversal on* π_M *with parameters* $L_P^\star = \sum_{\ell=1}^{j} \alpha_\ell L_P^{(\ell)} : P \in \binom{[M]}{k}$*, moreover,* $L_P^\star \prod_{j \notin P} n_j$ *is independent of* P *(with other words,* $\mathcal{T}^\star$ *is a MOA).* □

In Proposition 4, chose $j = 2$, and MOAs $\mathcal{T}_1$ and $\mathcal{T}_2$ such that $\#[\boldsymbol{i}, \mathcal{T}_2] \geq \#[\boldsymbol{i}, \mathcal{T}_1]$ for all $\boldsymbol{i} \in \pi_M$. Then setting $\alpha_\ell = (-1)^\ell$ for $\ell \in [2]$ satisfies the conditions of Proposition 3 and $\mathcal{T}^\star = \sum_{\ell=1}^{2} \alpha_\ell \mathcal{T}_\ell = \mathcal{T}_2 \setminus \mathcal{T}_1$; this type of linear combination is exactly the relative complementation on MOAs. Accordingly, if a MOA contains another one with the same strength as a subarray, erasing the rows of the subarray results in a new MOA.

Proposition 3 allows us to use the construction in (15) to build simple MOAs different from the ones in (15).

Corollary 3. *Let* $n_1, \ldots, n_M$*, and* $0 < \mu < 1$ *be given such that they satisfy the conditions of Corollary 2. For a fixed positive integer* Q*, and for each* $i \in [2Q+1]$ *let* β_i *be given such that* $0 \leq \beta_1 < \beta_2 < \cdots < \beta_{2Q+1} < \beta_1 + \mu \leq 1$ *and* $\beta_{2Q+1} \leq 1 - \mu$*. Define* $\mathcal{I} \subseteq [0, 1)$ *by*

$$\mathcal{I} = \left(\bigcup_{\ell=1}^{Q} \Big[\beta_{2\ell-1}, \beta_{2\ell} \Big) \right) \cup \Big[\beta_{2\ell+1}, \beta_1 + \mu \Big) \cup \left(\bigcup_{\ell=1}^{Q} \Big[\beta_{2\ell} + \mu, \beta_{2\ell+1} + \mu \Big) \right).$$

Then the following is a k*-dimensional transversal on* π_M *with parameters* $L_P = \mu \prod_{q \in P} n_q$ *and provides a simple MOA of strength* $M - k$*:*

$$\mathcal{T} = \left\{ (i_1, \ldots, i_M) \in \pi_M : \left\langle \sum_{r=1}^{M} \frac{i_r}{m_r} \right\rangle \in \mathcal{I} \right\}.$$

Proof. For $\ell \in [2Q+1]$ let $\mathcal{T}_\ell = \mathbb{C}(n_1,\dots,n_M;\beta_\ell,\mu)$. By Corollary 2, each $\mathcal{T}_\ell$ is a full k-dimensional transversal on π_M with parameters $L_P = \mu\prod_{j\in P} n_j$ satisfying the conditions of Proposition 3. Also, using $\alpha_\ell = (-1)^{\ell+1}$ we obtain that $\mathcal{T} = \sum_{\ell=1}^{2Q+1}\alpha_\ell\mathcal{T}_\ell$. The statement follows from Proposition 3 and the fact that $\sum_{\ell=1}^{2Q+1}\alpha_\ell = 1$. □

Proposition 5. [Tensor product.]

(i) *Let $\mathcal{T}_1$ and $\mathcal{T}_2$ be k-dimensional multitransversals on $\prod_{j=1}^M [n_j^{(1)}]^\star$ and $\prod_{j=1}^M [n_j^{(2)}]^\star$ with parameters $L_P^{(1)}$ and $L_P^{(2)}$ ($P \in \binom{[M]}{k}$), respectively. Then*

$$\mathcal{T} = \left[\!\left[(a_1 n_1^{(2)} + b_1,\dots,a_M n_M^{(2)} + b_M)^{\#[(a_1,\dots,a_M),\mathcal{T}_1]\cdot\#[(b_1,\dots,b_M),\mathcal{T}_2]}\right]\!\right]$$

is a k-dimensional multitransversal on $\prod_{j=1}^M [n_j^{(1)} n_j^{(2)}]^\star$ with parameters $L_P = L_P^{(1)} L_P^{(2)}$.

(ii) *Assume that $\mathcal{T}_1$ and $\mathcal{T}_2$ above are full multitransversals, and assume that there exists an $A \in \binom{[M]}{k}$, in which both meet the bound set by A, i.e. for $i \in \{1,2\}$ we have*

$$L_A^{(i)} \prod_{j\notin A} n_j^{(i)} = \min_{P\in\binom{[M]}{k}} \left(L_P^{(i)} \prod_{j\notin P} n_j^{(i)} \right).$$

Then $\mathcal{T}$ is a full multitransversal as well. □

Condition (ii) holds, in particular, if (3) holds for both $\mathcal{T}_1$ and $\mathcal{T}_2$, therefore the tensor product of MOAs of the same constraint and the same strength is a MOA of the same constraint and the same strength, using in the ith column of $\mathcal{T}$ the Cartesian product of the symbol sets of the i-th columns of $\mathcal{T}_1$ and $\mathcal{T}_2$ with appropriate multiplicities.

8 Proofs of the Sperner Type Results

In the proofs of this section we will frequently make use of the following structure. Let $\mathcal{F}$ be a multifamily on $\biguplus_{i\in[M]} X_i$. Fix a $D \subseteq [M]$ and let $F \subseteq X \setminus \biguplus_{i\in D} X_i$. We define

$$\mathcal{F}(F;D) = \left[\!\left[(E\setminus F)^{\#[E,\mathcal{F}]} : E \cap \biguplus_{i\in[M]\setminus D} X_i = F\right]\!\right].$$

The following lemma is clear from the definitions.

Lemma 7. *Let $\mathcal{F}$ be a k-dimensional M-part Sperner multifamily with parameters $L_P : P \in \binom{M}{k}$. Fix $k \le N \le M$, a $D \in \binom{[M]}{N}$ and let $F \subseteq X \setminus \biguplus_{i\in D} X_i$.*

(i) *$\mathcal{F}(F;D)$ is a k-dimensional N-part Sperner multifamily on $\biguplus_{i\in D} X_i$ with parameters $L_P : P \in \binom{D}{k}$.*

(ii) *If $\mathcal{F}$ is a simple family, so is $\mathcal{F}(F;D)$.* □

Proof of Theorem 1: First assume $M = k$, and call our multifamily $\mathcal{F}'$ instead of $\mathcal{F}$. For each $i \in [M]$, there are $m_i!$ (simple) chains of maximum size (i.e. of length n_i) in X_i. We count the number of ordered $(k+1)$-tuples in the following multiset in two ways:

$$\left[\!\!\left[(E, \mathbb{C}_1, \ldots, \mathbb{C}_k)^{\#[E,\mathcal{F}']} : \; E \in \prod_{i\in[M]} \mathbb{C}_i, \text{ where } \; \mathbb{C}_i \text{ is a chain of size } n_i \text{ in } X_i \right]\!\!\right].$$

Since each chain product $\prod_{i=1}^{M} \mathbb{C}_i$ contains at most $L_{[M]}$ sets from $\mathcal{F}'$ by definition, the number of such $(k+1)$-tuples is at most $L_{[M]} \prod_{i\in[M]} m_i!$. Since each $E \in \mathcal{F}'$ can be extended to precisely $\prod_{i\in[M]} |E \cap X_i|!(m_i - |E \cap X_i|)!$ chain products with each chain being maximum size, we have that

$$\sum_{E\in\mathcal{F}'} \prod_{i\in[M]} |E \cap X_i|!(m_i - |E \cap X_i|)! \le L_{[M]} \prod_{i\in[M]} m_i!$$

from which the claimed inequality follows in the form

$$\sum_{E\in\mathcal{F}'} \frac{1}{\prod\limits_{i\in[M]} \binom{m_i}{|E\cap X_i|}} \le L_{[M]}.$$

Now assume $M > k$ and take an arbitrary $P \in \binom{[M]}{k}$ to prove the theorem for our multifamily $\mathcal{F}$. Take an $F \subseteq X \setminus \bigcup_{i\in P} X_i$, and assume $f_i = |F \cap X_i|$ for $i \notin P$. By Lemma 7, $\mathcal{F}(F;P)$ is a k-dimensional k-part Sperner multifamily with parameter L_P, and therefore, using Theorem 1 we get

$$\sum_{E\in\mathcal{F}(F;P)} \frac{1}{\prod\limits_{i\in P} \binom{m_i}{|E\cap X_i|}} \le L_P. \tag{17}$$

From this we can write for any fixed sequence f_i $(i \notin P)$:

$$\sum_{\substack{F:F\subseteq X\setminus\bigcup_{i\in P} X_i \\ |F\cap X_i|=f_i, i\notin P}} \sum_{E\in\mathcal{F}(F;P)} \frac{1}{\prod\limits_{i\in P} \binom{m_i}{|E\cap X_i|} \prod\limits_{i\notin P} \binom{m_i}{f_i}} \le L_P.$$

Finally, summing up the previous inequality for $f_i = 0, 1, \ldots, m_i$, for all $i \notin P$, we obtain the theorem. □

To prove Theorem 3, we first need the following definitions: Let $\mathcal{F}$ be an non-empty M-part multifamily on X, and let $j \in [M]$. We define $\mathrm{high}_j(\mathcal{F})$ and

$\text{low}_j(\mathcal{F})$ as the largest and smallest levels in X_j that the trace $\mathcal{F}_{X_j}$ in X_j intersects. With other words,

$$\text{high}_j(\mathcal{F}) = \max\left\{q \in [n_j]^\star : \mathcal{F}_{X_j} \cap \binom{X_j}{q} \neq \varnothing\right\},$$

$$\text{low}_j(\mathcal{F}) = \min\left\{q \in [n_j]^\star : \mathcal{F}_{X_j} \cap \binom{X_j}{q} \neq \varnothing\right\}.$$

First, we will need the following:

Lemma 8. *Let $M > 1$, $j \in [M-1]$ and let $\mathcal{F}'$ be an M-dimensional M-part Sperner family with $\text{high}_j(\mathcal{F}') > \text{low}_j(\mathcal{F}')$ that satisfies* (7), *and let $E_0 \in \mathcal{F}'_{X_M}$ be fixed. Then there is an M-dimensional M-part Sperner family $\mathcal{F}$ that also satisfies* (7) *such that for all $i \in [M] \setminus \{j\}$ we have $\mathcal{F}_{X_i} \subseteq \mathcal{F}'_{X_i}$, $E_0 \in \mathcal{F}_{X_M}$ and* $\text{high}_j(\mathcal{F}) - \text{low}_j(\mathcal{F}) = \text{high}_j(\mathcal{F}') - \text{low}_j(\mathcal{F}') - 1$.

Proof. Let $t = \text{high}_j(\mathcal{F}')$ and $\mathcal{B} = \{B_1, \ldots, B_s\} = \mathcal{F}'_{X_j} \cap \binom{X_j}{t}$. For $i \in [s]$ let $\mathcal{E}_i = \{E \in \mathcal{F}' : E \cap X_j = B_i\}$ and $\mathcal{E} = \cup_{i=1}^s \mathcal{E}_i$. Given $A \subseteq X_j$, we define

$$w(A) = \sum_{E \in \mathcal{F}' : E \cap X_j = A} \frac{1}{\prod_{i:i \neq j} \binom{m_i}{|E \cap X_j|}}.$$

We also assume (w.l.o.g.) that $w(B_1) \geq \ldots \geq w(B_s)$; we will use $w_i := w(B_i)$. Using this notation we can rewrite (7) as

$$\sum_{A \subseteq X_j} \frac{w(A)}{\binom{m_j}{|A|}} = 1,$$

or equivalently

$$\sum_{i=1}^{s} \frac{w_i}{\binom{m_j}{t}} + \delta = 1; \quad \delta := \sum_{A \in \mathcal{F}'_{X_j} \setminus \mathcal{B}} \frac{w(A)}{\binom{m_j}{|A|}}. \tag{18}$$

Recall the following well-known fact (see e.g. [8]) that for every $t \in [n]$ and a subset $\mathcal{A} \subseteq \binom{[n]}{t}$ we have

$$\frac{|\partial(\mathcal{A})|}{\binom{n}{t-1}} \geq \frac{|\mathcal{A}|}{\binom{n}{t}}, \tag{19}$$

where $\partial(\mathcal{A})$, called the *lower shadow* of $\mathcal{A}$, is defined as $\partial(\mathcal{A}) = \{E \in \binom{[n]}{t-1} : E \subsetneq F \text{ for some } F \in \mathcal{A}\}$. Moreover, equality in (19) holds if and only if $\mathcal{A} = \binom{[n]}{t}$.

A similar inequality holds for the *upper shadow* $\sigma(\mathcal{A})$ of $\mathcal{A}$ defined as $\sigma(\mathcal{A}) = \{E \in \binom{[n]}{t+1} : E \supsetneq F \text{ for some } F \in \mathcal{A}\}$, that is $|\sigma(\mathcal{A})|/\binom{n}{t+1} \geq |\mathcal{A}|/\binom{n}{t}$ (with equality if and only if $\mathcal{A} = \binom{[n]}{t}$).

Let us denote $\mathbb{B}_i = \{B_1, \ldots, B_i\}$; $i = 1, \ldots, s$ (thus $\mathbb{B}_i \subsetneq \mathbb{B}_{i+1}$ and $\mathbb{B}_s = \mathcal{B}$). We define then the following partition of $\partial(\mathcal{B}) = \mathbb{B}'_1 \cup \ldots \cup \mathbb{B}'_s$:

$$\mathbb{B}'_1 = \partial(\mathbb{B}_1), \ \mathbb{B}'_i = \partial(\mathbb{B}_i) \setminus \partial(\mathbb{B}_{i-1}); \ i = 2, \ldots, s.$$

Then, in view of (19), we have

$$\frac{\sum_{\ell=1}^{i}|\mathbb{B}'_\ell|}{\binom{m_j}{t-1}} = \frac{|\partial(\mathbb{B}_i)|}{\binom{m_j}{t-1}} \geq \frac{|\mathbb{B}_i|}{\binom{m_j}{t}} = \frac{i}{\binom{m_j}{t}};\ i=1,\ldots,s, \tag{20}$$

with strict inequality if $s < \binom{m_j}{t}$.

Recall that $\mathcal{A}$ is an M-dimensional M-part Sperner family with parameter $L_{[M]} = 1$ precisely when for all $A, B \in \mathcal{A}$ with $A \neq B$ there is an $\ell \in [M]$ such that $A \cap X_\ell$ and $B \cap X_\ell$ are incomparable by the subset relation.

For ease of description, let us represent each family $\mathcal{E}_i$, defined above, by the direct product $\mathcal{E}_i = \{B_i\} \times \mathcal{H}_i$, where $\mathcal{H}_i = \mathcal{F}'(B_i; [M] \setminus \{j\})$ is an $(M-1)$-dimensional $(M-1)$-part Sperner family in the partition set $\biguplus_{i\in[M]\setminus\{j\}} X_i$.

We now construct a new family $\mathcal{E}^\star$ from $\mathcal{E}$ as follows. We replace each $\mathcal{E}_i$ by $\mathcal{E}_i^\star := \mathbb{B}'_i \times \mathcal{H}_i$; $i = 1, \ldots, s$ and define $\mathcal{E}^\star = \cup_{i=1}^{s} \mathcal{E}_i^\star$. Observe now that for each $A^\star \in \mathcal{E}_i^\star$ there is an $A \in \mathcal{E}_i$ such that $A^\star \cap X_\ell = A \cap X_\ell$ for all $\ell \in [M] \setminus \{j\}$ and $A^\star \cap X_j \subsetneq A \cap X_j$. This implies that $\mathcal{E}^\star \cap \mathcal{F}' = \varnothing$, since $\mathcal{F}'$ is an M-dimensional M-part Sperner family with parameter $L_{[M]} = 1$.

Moreover, it is not hard to see that $\mathcal{F}^\star := (\mathcal{F}' \setminus \mathcal{E}) \cup \mathcal{E}^\star$ is an M-dimensional M-part Sperner family with parameter $L^\star_{[M]} = 1$. If we have that A, B are different elements of $\mathcal{F}' \setminus \mathcal{E}$, then the required property follows from the fact that A, B are both elements of $\mathcal{F}'$. If $A^\star, B^\star$ are different elements of $\mathcal{E}^\star$, then either $A^\star \cap X_j$ and $B^\star \cap X_j$ are both incomparable, or $A^\star, B^\star \in \mathcal{E}_i$ for some i, in which case the corresponding sets $A, B \in \mathcal{E}_i \subseteq \mathcal{F}'$ agree with $A^\star, B^\star$ on $X \setminus X_j$ and $A^\star \cap X_j = B^\star \cap X_j = B_i$, from which the required property follows. Finally, take $A^\star \in \mathcal{E}_i^\star$ for some i and $B \in \mathcal{F}' \setminus \mathcal{E}$, and let $A \in \mathcal{E}_i \subseteq \mathcal{F}'$ be the corresponding set. If $A \cap X_j$ and $B \cap X_j$ are comparable, then from the fact that t was the largest level of $\mathcal{F}'_{X_j}$ we get that $B \cap X_j \subseteq A^\star \cap X_j \subsetneq B_i = A \cap X_j$; and from the fact that A, B are both elements of $\mathcal{F}'$ and $A \setminus X_j = A^\star \setminus X_j$ the required property follows.

Therefore $\mathcal{F}^\star$ is an M-dimensional M-part Sperner family with parameter 1. Thus, for $\mathcal{F}^\star$ the following inequality must hold:

$$\sum_{E\in\mathcal{F}^\star} \frac{1}{\prod_{i=1}^{M}\binom{m_i}{|E\cap X_i|}} = \sum_{i=1}^{s} \frac{|\mathbb{B}'_i|\cdot w_i}{\binom{m_j}{t-1}} + \delta \leq 1. \tag{21}$$

On the other hand, (20) together with $w_1 \geq \ldots \geq w_s \geq 0 =: w_{s+1}$ implies that

$$\sum_{\ell=1}^{s} \frac{|\mathbb{B}'_\ell|\cdot w_\ell}{\binom{m_j}{t-1}} = \sum_{i=1}^{s}\sum_{\ell=1}^{i} \frac{|\mathbb{B}'_\ell|\cdot(w_i - w_{i+1})}{\binom{m_j}{t-1}} \geq \sum_{i=1}^{s} \frac{i\cdot(w_i - w_{i+1})}{\binom{m_j}{t}} = \sum_{i=1}^{s} \frac{w_i}{\binom{m_j}{t}}. \tag{22}$$

In fact, the latter means that $\mathcal{B} = \binom{X_j}{t}$, otherwise we have strict inequality in (22) a contradiction with (21), in view of (18). Thus, for the new family $\mathcal{F}^\star$ we have

$$\sum_{E\in\mathcal{F}^\star} \frac{1}{\prod_{i=1}^{M}\binom{m_i}{|E\cap X_i|}} = 1.$$

Moreover, $\text{high}_j(\mathcal{F}^\star) = \text{high}_j(\mathcal{F}') - 1$ and $\text{low}_j(\mathcal{F}^\star) = \text{low}_j(\mathcal{F}')$, so $\text{high}_j(\mathcal{F}^\star) - \text{low}_j(\mathcal{F}^\star) = \text{high}_j(\mathcal{F}') - \text{low}_j(\mathcal{F}') - 1$. In addition, for all $\ell \in [M] \setminus \{j\}$ we have $\mathcal{E}^\star_{X_\ell} \subseteq \mathcal{E}_{X_\ell}$, therefore $\mathcal{F}^\star_{X_\ell} \subseteq \mathcal{F}'_{X_\ell}$. Therefore, if $E_0 \in (\mathcal{F}' \setminus \mathcal{E})_{X_M}$, i.e. the trace of $\mathcal{F}' \setminus \mathcal{E}$ in X_M contains E_0, then setting $\mathcal{F} := \mathcal{F}^\star$ will give the required family.

If $E_0 \notin (\mathcal{F}' \setminus \mathcal{E})_{X_M}$, then, since $\mathcal{F}'_{X_M} \setminus \mathcal{E}_{X_M} \subseteq (\mathcal{F}' \setminus \mathcal{E})_{X_M}$ and $E_0 \in \mathcal{F}'_{X_M}$ we must have that $E_0 \in \mathcal{E}_{X_M}$. Similar to the described "pushing down" transformation in $\mathcal{F}'$ we can apply "pushing up" transformation with respect to the smallest level $\mathcal{D}$ in $\mathcal{F}'_{X_j}$, replacing it by its upper shadow $\mathfrak{S}(\mathcal{D})$ to obtain the new family $\mathcal{F}$. Since $\mathcal{D} \neq \mathcal{B}$, we now have $\mathcal{E} \subseteq \mathcal{F}$, therefore $E_0 \in \mathcal{F}_{X_M}$. All other required conditions follow as before. □

Proof of Theorem 3: Let $\mathcal{F}'$ be an M-dimensional M-part Sperner family with parameter 1 satisfying (7). Without loss of generality assume, contrary to the statement of the theorem, that the trace $\mathcal{F}'_{X_M}$ of $\mathcal{F}'$ in X_M contains an incomplete level, i.e. there is a $y_M \in [n_M]^\star$ such that for $\mathcal{G} = \mathcal{F}' \cap \binom{X_M}{y_M}$ we have that $\varnothing \subsetneq \mathcal{G} \subsetneq \binom{X_M}{y_M}$. Fix an $E_0 \in \mathcal{G}$.

Let $\mathcal{F}^{(0)} := \mathcal{F}'$. We will define a sequence $\mathcal{F}^{(1)}, \ldots, \mathcal{F}^{(M-1)}$ of M-dimensional M-part Sperner families such that for each $\ell \in [M-1]$ the following hold:

(i) Equality (7) holds for $\mathcal{F}^{(\ell)}$, with other words

$$\sum_{E \in \mathcal{F}^{(\ell)}} \frac{1}{\prod_{i=1}^{M} \binom{m_i}{|E \cap X_i|}} = 1. \tag{23}$$

(ii) There is a $y_\ell \in [n_\ell]^\star$ such that $\mathcal{F}^{(\ell)}_{X_\ell} \subseteq \binom{X_\ell}{y_\ell}$, with other words the trace of $\mathcal{F}^{(\ell)}$ in X_ℓ consist of a single (not necessarily full) level.

(iii) For each $i \in [M] \setminus \{\ell\}$, $\mathcal{F}^{(\ell)}_{X_i} \subseteq \mathcal{F}^{(\ell-1)}_{X_i}$.

(vi) $E_0 \in \mathcal{F}^{(\ell)}_{X_M}$.

Once this sequence is defined, it follows that for all $j \in [M-1]$ we have that $\mathcal{F}^{(M-1)}_{X_j} \subseteq \binom{X_j}{y_j}$, also $E_0 \in \left(\mathcal{F}^{(M-1)}_{X_M} \cap \binom{X_M}{y_M}\right) \subseteq \mathcal{G} \subsetneq \binom{X_M}{y_M}$, therefore the trace of $\mathcal{F}^{(M-1)}$ in X_M contains an incomplete level.

Also, for all $F \in X \setminus X_M$ we must have that $\mathcal{F}^{(M-1)}(F; \{M\})$ is a 1-dimensional 1-part Sperner family with parameter 1, therefore it satisfies (8) with the parameter set to 1. In view of these facts, using (23) for $\ell = M-1$ we get that

$$\begin{aligned} 1 &= \sum_{E \in \mathcal{F}^{(M-1)}} \frac{1}{\prod_{i=1}^{M} \binom{m_i}{|E \cap X_i|}} \\ &= \frac{1}{\prod_{i=1}^{M-1} \binom{m_i}{t_i}} \sum_{F \subseteq X \setminus X_M} \left(\sum_{E \in \mathcal{F}^{(M-1)}(F;\{M\})} \frac{1}{\binom{m_M}{|E \cap X_M|}} \right) \\ &\leq \frac{1}{\prod_{i=1}^{M-1} \binom{m_i}{t_i}} \sum_{F \subseteq X \setminus X_M} 1 = 1. \end{aligned}$$

This implies that for all $F \subseteq X \setminus X_M$, inequality (8) holds with equality for $\mathcal{F}^{(M-1)}(F,\{M\})$, so by Lemma 1 we get that $\mathcal{F}^{(M-1)}(F,\{M\})$ is a full level. Since

$$\mathcal{F}^{(M-1)}_{X_M} = \bigcup_{F \subseteq X \setminus X_M} \mathcal{F}^{(M-1)}(F,\{M\}),$$

this implies that $\mathcal{F}^{(M-1)}_{X_M}$ must consist of full levels only, a contradiction.

Note that $\mathcal{F}^{(0)}$ is defined, it satisfies (7), and it does not need to satisfy any other conditions. All that remains to show is that $\mathcal{F}^{(\ell)}$ can be defined for each $\ell \in [M-1]$ such that it satisfies the conditions (i)–(iv).

To this end, assume that $j \in [M-1]$ and $\mathcal{F}^{(j-1)}$ is already given satisfying all required conditions. Let $Q = \mathrm{high}_j(\mathcal{F}^{(j-1)}) - \mathrm{low}_j(\mathcal{F}^{(j-1)})$. If $Q = 0$, then $\mathcal{F}^{(j-1)}_{X_j}$ consists of a single, not necessarily full, level, and we set $\mathcal{F}^{(j)} = \mathcal{F}^{(j-1)}$; (i)–(iv) are clearly satisfied.

If $Q > 0$, then let $\mathcal{K}^{(0)} = \mathcal{F}^{(j-1)}$. By Lemma 8 we can define a sequence $\mathcal{K}^{(1)}, \ldots, \mathcal{K}^{(Q)}$ of M-dimensional M-part Sperner families with parameter 1 such that for all $\ell \in [Q]$ the following hold:

(a) $\mathcal{K}^{(\ell)}$ satisfies (7).
(b) For all $i \in [M] \setminus \{j\}$ we have $\mathcal{K}^{(\ell)}_{X_i} \subseteq \mathcal{K}^{(\ell-1)}_{X_i}$.
(c) $E_0 \in \mathcal{K}^{(\ell)}_{X_M}$.
(d) $\mathrm{high}_j(\mathcal{K}^{(\ell)}) - \mathrm{low}_j(\mathcal{K}^{(\ell)}) = \mathrm{high}_j(\mathcal{K}^{(\ell-1)}) - \mathrm{low}_j(\mathcal{K}^{(\ell-1)}) - 1$.

It follows that $\mathrm{high}_j(\mathcal{K}^{(Q)}) = \mathrm{low}_j(\mathcal{K}^{(Q)})$ and we set $\mathcal{F}^{(j)} = \mathcal{K}^{(Q)}$; (i)–(iv) are clearly satisfied. □

It only remains to prove Theorem 2. We will start with a series of lemmata. The first lemma states for multifamilies what Theorem 6.2 in [1] stated for simple families:

Lemma 9. *Let* $1 \leq M$ *and* $\mathcal{F}$ *be a* 1*-dimensional* M*-part Sperner multifamily with parameters* $L_{\{i\}}$ *for* $i \in [M]$ *satisfying* (6) *with equalities, i.e.*

$$\forall i \in [M] \qquad \sum_{(i_1,\ldots,i_M)\in\pi_M} \frac{p_{i_1,\ldots,i_M}}{\prod_{j=1}^{M}\binom{m_j}{i_j}} = \frac{L_{\{i\}}}{n_i}\prod_{j=1}^{M} n_j. \tag{24}$$

Then $\mathcal{F}$ *is homogeneous.*

Proof. For $M = 1$ the statement is proved in Lemma 1. Let $M \geq 2$ and take an arbitrary $F \in \mathcal{F}$. We set $F_i = F \cap X_i$ and $G_i = F \setminus F_i$. By Lemma 7 that for each $j \in [M]$, $\mathcal{F}(G_j;\{j\})$ is a (1-dimensional 1-part) Sperner multifamily with parameter $L_{\{j\}}$. From the proof of Theorem 1 and (24) we get that equality must hold in (17), i.e.

$$\sum_{E\in\mathcal{F}(G_j;\{j\})} \frac{1}{\binom{m_i}{|E\cap X_i|}} = L_{\{j\}},$$

which by Lemma 1 implies that $\mathcal{F}(G_j;\{j\})$ is homogeneous. In particular this means that for all $A \in \mathcal{F}$ and for all $j \in [M]$ if B is a set such that $|A \cap X_j| = |B \cap X_j|$ and for all $i \in [M] \setminus \{j\}$ we have $A \cap X_i = B \cap X_i$, then $\#[A,\mathcal{F}] = \#[B,\mathcal{F}]$. If A, B are sets with the same profile vector, we define the sequence $A = Y_0, Y_1, \dots, Y_M = B$ by $Y_i = (Y_{i-1} \setminus X_i) \uplus (B \cap X_i)$ for all $i \in [M]$. It follows that $\#[Y_{i-1},\mathcal{F}] = \#[Y_i,\mathcal{F}]$, and so $\#[A,\mathcal{F}] = \#[B,\mathcal{F}]$. Thus $\mathcal{F}$ is homogeneous. □

Lemma 10. *Let $1 \le k$ and let $\mathcal{F}$ be a k-dimensional $(k+1)$-part Sperner multifamily with parameters $L_{[k+1]\setminus\{i\}}$ for $i \in [k+1]$ satisfying* (6) *with equality, i.e.*

$$\forall i \in [k+1] \qquad \sum_{(i_1,\dots,i_{k+1})\in\pi_{k+1}} \frac{p_{i_1,\dots,i_{k+1}}}{\prod_{j=1}^{k+1}\binom{m_j}{i_j}} = L_{[k+1]\setminus\{i\}}n_i. \tag{25}$$

Then $\mathcal{F}$ is homogeneous.

Proof. The proof is by induction on k. For $k = 1$, it is proved in Lemma 9. By Lemma 7 we have that for each $j \in [M]$ and each $F \subseteq X_j$, $\mathcal{F}(F;[k+1]\setminus\{j\})$ is a k-dimensional k-part Sperner multifamily with parameter $L_{[k+1]\setminus\{j\}}$. From the proof of Theorem 1 and (25) we get that equality must hold in (17), i.e.

$$\sum_{E\in\mathcal{F}(F;[k+1]\setminus\{j\})} \frac{1}{\prod_{i:i\neq j}\binom{m_i}{|E\cap X_i|}} = L_{[k+1]\setminus\{j\}}. \tag{26}$$

Fixing a maximal chain $F_0 \subsetneq F_1 \subsetneq \cdots \subsetneq F_{m_j}$ in X_j, we get that $\mathcal{F}' = \biguplus_{q=0}^{m_j} \mathcal{F}(F_q;[k+1]\setminus\{j\})$ is a $(k-1)$-dimensional k-part Sperner multifamily with parameters $L'_{[k+1]\setminus\{j,\ell\}} := L_{[k+1]\setminus\{\ell\}} : \ell \in [k+1]\setminus\{j\}$, moreover, using (26) for each $F = F_q$ we get that

$$\sum_{E\in\mathcal{F}'} \frac{1}{\prod_{i:i\neq j}\binom{m_i}{|E\cap X_i|}} = \sum_{q=0}^{m_j}\left(\sum_{E\in\mathcal{F}(F_q;[k+1]\setminus\{j\})} \frac{1}{\prod_{i:i\neq j}\binom{m_i}{|E\cap X_i|}}\right) = n_j L_{[k+1]\setminus\{j\}}.$$

By (25) we have that $L_{[k+1]\setminus\{j\}}n_j = L_{[k+1]\setminus\{\ell\}}n_\ell = L'_{[k+1]\setminus\{j,\ell\}}n_\ell$, therefore $\mathcal{F}'$ is homogeneous by the induction hypothesis. In particular this means that for all $A \in \mathcal{F}$ for all $j \in [M]$ if B is a set such that $|A \cap X_j| = |B \cap X_j|$ and for all $i \in [M]\setminus\{j\}$ we have $A \cap X_i = B \cap X_i$, then the $\#[A,\mathcal{F}] = \#[B,\mathcal{F}]$. This implies, as in the proof of Lemma 9, that $\mathcal{F}$ is homogeneous. □

Lemma 11. *Let $2 \le k \le M-1$ and let $\mathcal{F}$ be a k-dimensional M-part Sperner multifamily with parameters L_P for $P \in \binom{[M]}{k}$ satisfying* (6) *with equalities. Then $\mathcal{F}$ is homogeneous.*

Proof. The proof is essentially the same as the proof of Lemma 10. If $M = k+1$, we are done by Lemma 10. If $M > k+1$, by Lemma 7 we get that for each

$D \in \binom{[M]}{k+1}$ and $F \subseteq X \setminus \biguplus_{i\in D} X_i$, $\mathcal{F}(F;D)$ is a k-dimensional $(k+1)$-part Sperner multifamily with parameters $L_P : P \in \binom{D}{k}$. Fix an $F \subseteq X \setminus \biguplus_{i\in D} X_i$, and set $\mathcal{F}' = \mathcal{F}(F;D)$. For any $j \in D$ and $G \subseteq X_j$ we have that $\mathcal{F}'(G; D\setminus\{j\}) = \mathcal{F}(F \uplus G; D\setminus\{j\})$ and $\mathcal{F}'(G;D\setminus\{j\})$ is a k-dimensional k-part Sperner family with parameter $L_{D\setminus\{j\}}$. From the proof of Theorem 1 and the fact that for $\mathcal{F}$ the inequality (6) holds with equality we get that equality must hold for all $j \in D$ and all $G \subseteq X_j$ for $\mathcal{F}'(G;D\setminus\{j\})$ in (17), i.e.

$$\sum_{E\in\mathcal{F}'(G;D\setminus\{j\})} \frac{1}{\prod\limits_{\ell\in[M]\setminus(D\cup\{j\})} \binom{m_\ell}{|E\cap X_\ell|}} = L_{D\setminus\{j\}}. \tag{27}$$

Fixing a maximal chain $G_0 \subsetneq G_1 \subsetneq \cdots \subsetneq G_{m_j}$ in X_j we get that $\mathcal{F}^\star = \biguplus_{i=0}^{m_j} \mathcal{F}'(G_i; D\setminus\{j\})$ is a $(k-1)$-dimensional k-part Sperner multifamily with parameters $L^\star_{P^\star} := L_{P^\star\cup\{j\}} : P^\star \in \binom{D\setminus\{j\}}{k-1}$, moreover, using (27) for each G_i we get that

$$\sum_{E\in\mathcal{F}^\star} \frac{1}{\prod\limits_{\ell\in D\setminus\{j\}} \binom{m_\ell}{|E\cap X_\ell|}} = \sum_{i=1}^{m_j}\left(\sum_{E\in\mathcal{F}(G_i;D\setminus\{j\})} \frac{1}{\prod\limits_{\ell\in D\setminus\{j\}} \binom{m_\ell}{|E\cap X_\ell|}}\right) = L_{D\setminus\{j\}} n_j.$$

Fix any $P^\star \in \binom{D\setminus\{j\}}{k-1}$. Then $P^\star = D\setminus\{i,j\}$ for some $i \in D\setminus\{j\}$, and from the conditions of the theorem we get that

$$L^\star_{P^\star} n_i = L_{P^\star\cup\{j\}} n_i = L_{D\setminus\{i\})} n_i = L_{D\setminus\{j\}} n_j,$$

therefore $\mathcal{F}^\star$ is homogeneous by the induction hypothesis. This means that if $A \in \mathcal{F}$ and B is a set with the same profile vector as A, and $A\cap X_i = B \cap X_i$ for at least $M-k-1 \geq 1$ values of i, then $\#[A,\mathcal{F}] = \#[B,\mathcal{F}]$. As before, we get that $\mathcal{F}$ is homogeneous. □

Proof of Theorem 2: Lemmata 9, 10 and 11 together prove part (i), and, as remarked earlier, part (ii) follows from the conditions.
(iii): By part (i), equality in (6) implies homogeneity, i.e. that for any $(i_1,\ldots,i_M) \in \pi_M$ there is a positive integer $r_{i_1,\ldots,i_M}$ such that every set in $\mathcal{F}$ that has profile vector $(i_1,\ldots,i_M)$ appears with multiplicity $r_{i_1,\ldots,i_M}$, and also equality in (17). Equality in (17) means that for any chain product $\mathbb{C} := \prod_{i=1}^M \mathbb{C}_i$ where $\mathbb{C}_i$ is a maximal chain in X_i, any given $P \in \binom{[M]}{k}$ and any subset $F \subseteq X \setminus \cup_{i\in P} X_i$, each subproduct $\prod_{j\in P} \mathbb{C}_j$ of maximal chains is covered exactly L_P times by the elements of $\mathcal{F}(F;P)$, that is

$$\left|\left[\!\!\left[E^{\#[E,\mathcal{F}(F;P)]} : E \in \prod_{j\in P} \mathbb{C}_j \right]\!\!\right]\right| = L_P. \tag{28}$$

For a given chain product $\mathbb{C} = \prod_{i\in[M]} \mathbb{C}_i$ of maximum-size chains $\mathbb{C}_i$ in X_i, we define

$$\mathcal{F}[\mathbb{C}] = \left[\!\left[F^{\#[F,\mathcal{F}]} : F \in \mathbb{C} \right]\!\right].$$

Each $F \in \mathcal{F}[\mathbb{C}]$ is uniquely determined from its profile vector $(f_1, \ldots, f_M)$. Let $\mathcal{T}_{\mathbb{C}}$ denote the multiset of all profile vectors of the sets in $\mathcal{F}[\mathbb{C}]$, where each profile vector appears with the multiplicity of its corresponding set in $\mathcal{F}[\mathbb{C}]$. Since $\mathcal{F}$ is homogeneous, $\mathcal{T}_{\mathbb{C}}$ does not depend on the choice of $\mathbb{C}$.

We can describe now property (28) of $\mathcal{F}[\mathbb{C}]$ in terms of its profile vectors as follows: for each subset $\{i_1 < \ldots < i_{M-k}\} \in \binom{[M]}{M-k}$, and each $(M-k)$-tuple of coordinate values $(f_{i_1}, \ldots, f_{i_{M-k}}) \in \prod_{j=1}^{M-k} [n_{i_j}]^\star$ the set of vectors in $\mathcal{T}_{\mathbb{C}}$ where the i_j-th coordinate is f_{i_j} for $j \in [M-k]$ has size $L_{M\setminus\{i_1,\ldots,i_{M-k}\}}$. Let $\mathcal{T}$ denote the transversal corresponding to the homogeneous multifamily $\mathcal{F}$. Then clearly $\mathcal{T} = \mathcal{T}_{\mathbb{C}}$ for every product of maximal chains $\mathbb{C} = \prod_{i=1}^M \mathbb{C}_i$.

We infer now that the k-dimensional multitransversal $\mathcal{T}$ is a simple MOA with symbol sets $S_i = \{0, 1, 2\ldots, m_i\}$, of constraint M, strength $M-k$, and index set $\mathbb{L} = \{L_P : P \in \binom{[M]}{k}\}$, with $\lambda(j_1, \ldots, j_{M-k}) = L_{[M]\setminus\{j_1,\ldots,j_{M-k}\}}$. This completes the proof of part (iii).

It is also clear that any MOA with the parameters described above is a k-dimensional multitransversal corresponding to a homogeneous k-dimensional M-part Sperner multifamily $\mathcal{F}$ with parameters $\{L_P : P \in \binom{[M]}{k}\}$ on a partitioned $(m_1 + \ldots + m_M)$-element underlying set, where the multiplicity of each element in $F \in \mathcal{F}$ is the same as the multiplicity of its profile vector $(f_1, \ldots, f_M)$ in the multitransversal, which satisfies equality in (6). □

9 Proofs of the Convex Hull Results

Proof of Lemma 2: It will suffice to show that for every multifamily $\mathcal{H} \in \mathbb{A}$, there are non-negative coefficients $\lambda(I)$ for every I with $\mathrm{supp}(I) \subseteq \pi_M$ and $T(I) \in \mu(\mathbb{A}(\mathfrak{L}))$, such that $\sum_I \lambda(I) = 1$ and

$$\sum_I \lambda(I) S(I) = \mathbb{P}(\mathcal{H}). \tag{29}$$

To this end, fix an $\mathcal{H} \in \mathbb{A}$ and for all $H \subseteq X$ let $\mathcal{H}_H = [\![H^{\#[H,\mathcal{H}]}]\!]$, with other words $\mathcal{H}_H$ has H with the same multiplicity as $\mathcal{H}$, and it has no other elements. Consider the sum

$$\sum_{(\mathfrak{L},H)} \frac{\mathbb{P}(\mathcal{H}_H)}{\prod_{j=1}^{M} (m_j!)} \tag{30}$$

for all ordered pairs $(\mathfrak{L}, H)$, where $\mathfrak{L}$ is a product-permutation, $H \subseteq X$, and H is initial with respect to the product-permutation $\mathfrak{L}$. We evaluate (30) in two ways. The first way is:

$$\sum_{(\mathfrak{L},H)} \frac{\mathbb{P}(\mathcal{H}_H)}{\prod_{j=1}^{M}(m_j!)} = \sum_{\mathfrak{L}} \frac{1}{\prod_{j=1}^{M}(m_j!)} \left(\sum_{\substack{H \subseteq X: \\ H \text{ is initial for } \mathfrak{L}}} \mathbb{P}(\mathcal{H}_H) \right)$$
$$= \sum_{\mathfrak{L}} \frac{\mathbb{P}(\mathcal{H}(\mathfrak{L}))}{\prod_{j=1}^{M}(m_j!)}. \tag{31}$$

Observe that $\mathbb{P}(\mathcal{H}(\mathfrak{L})) \in \mu(\mathbb{A}(\mathfrak{L}))$, and therefore for every $\mathfrak{L}$ there is a unique I such that $T(I) = \mathbb{P}(\mathcal{H}(\mathfrak{L}))$. Collecting the identical terms in the right side of (31),

$$\sum_{\mathfrak{L}} \frac{\mathbb{P}(\mathcal{H}(\mathfrak{L}))}{\prod_{j=1}^{M}(m_j!)} = \sum_{T(I)\in\mu(\mathbb{A}(\mathfrak{L}))} \lambda(I)T(I), \tag{32}$$

where $\lambda(I)$ is the proportion of the $\prod_{j=1}^{M}(m_j!)$ product-permutations such that $\mathbb{P}(\mathcal{H}(\mathfrak{L})) = T(I)$, thus $\sum_{T(I)\in\mu(\mathbb{A}(\mathfrak{L}))}\lambda(I) = 1$. Consider a fixed set H with profile vector $(i_1, i_2, \ldots, i_M)$. There are exactly $\prod_{j=1}^{M}(i_j! \cdot (m_j - i_j)!)$ product-chains to which H is initial. Using this, we also get:

$$\sum_{(\mathfrak{L},H)} \frac{\mathbb{P}(\mathcal{H}_H)}{\prod_{j=1}^{M}(m_j!)} = \sum_{H:H\subseteq X} \sum_{\substack{\mathfrak{L}: \\ H \text{ is initial for } \mathfrak{L}}} \frac{\mathbb{P}(\mathcal{H}_H)}{\prod_{j=1}^{M}(m_j!)}$$
$$= \sum_{H:H\subseteq X} \frac{\prod_{j=1}^{M}(i_j! \cdot (m_j - i_j)!)}{\prod_{j=1}^{M}(m_j!)} \cdot \mathbb{P}(\mathcal{H}_H) \tag{33}$$
$$= \sum_{H:H\subseteq X} \frac{\mathbb{P}(\mathcal{H}_H)}{\prod_{j=1}^{M}\binom{m_j}{i_j}} = \left(\frac{p_{i_1,\ldots,i_M}(\mathcal{H})}{\binom{m_1}{i_1}\cdots\binom{m_M}{i_M}} \right)_{(i_1,\ldots,i_M)\in\pi_M}. \tag{34}$$

Combining (30), (31), (32), (33), and (34), we obtain

$$\left(\frac{p_{i_1,\ldots,i_M}(\mathcal{H})}{\binom{m_1}{i_1}\cdots\binom{m_M}{i_M}} \right)_{(i_1,\ldots,i_M)\in\pi_M} = \sum_{T(I)\in\mu(\mathbb{A}(\mathfrak{L}))} \lambda(I)T(I),$$

which implies for all $(i_1, \ldots, i_M) \in \pi_M$ that

$$p_{i_1,\ldots,i_M}(\mathcal{H}) = \sum_{T(I)\in\mu(\mathbb{A}(\mathfrak{L}))} \left(\lambda(I) \left(\prod_{j=1}^{M} \binom{m_j}{i_j} \right) t_{i_1,\ldots,i_M}(I) \right).$$

This proves (29). □

Proof of Theorem 5: First observe that $\mu(\mathbb{A}(\mathfrak{L}))$ does not depend on $\mathfrak{L}$, so (11) holds. Next we have to show (12), i.e. we have to show that if $T(I) \in \mu(\mathbb{A}(\mathfrak{L}))$ for some I with $\text{supp}(I) \subseteq \pi_M$ and all product-permutation $\mathfrak{L}$, then $S(I) \in \mu(\mathbb{A})$.

Assume I satisfies $\text{supp}(I) \subseteq \pi_M$ and $T(I) \in \mu(\mathbb{A}(\mathfrak{L}))$ for all product-permutation $\mathfrak{L}$. Then for each product-permutation $\mathfrak{L}$ there is an $\mathcal{H}_{\mathfrak{L}} \in \mathbb{A}$ such that $T(I) = \mathbb{P}(\mathcal{H}_{\mathfrak{L}}(\mathfrak{L}))$. Since $\mathcal{H}_{\mathfrak{L}}$, and therefore $\mathcal{H}_{\mathfrak{L}}(\mathfrak{L})$ as well, satisfies $\mathbb{M}_\Gamma$, we must have that I satisfies $\mathbb{M}_\Gamma$. Let $\mathcal{F}_{S(I)}$ be the homogeneous multifamily that realizes the profile array $S(I)$, then for all $(i_1,\dots,i_M) \in \pi_M$ we have $\max\{\#[F,\mathcal{F}_{S(I)}] : \forall j\ |F \cap X_j| = i_j\} = \#[(i_1,\dots,i_M), I]$, consequently, $\mathcal{F}_{S(I)}$ satisfies $\mathbb{M}_\Gamma$. Thus $S(I) \notin \mu(\mathbb{A})$ implies that the homogeneous multifamily $\mathcal{F}_{S(I)}$ is not a k-dimensional M-part multifamily with parameters $L_P : P \in \binom{[M]}{k}$. This means that there is a $P_0 \in \binom{[M]}{k}$, sets D_i for all $i \notin P_0$ and chains $\mathbb{C}_j$ for all $j \in P_0$ such that

$$\left| \left[\!\left[F^{\#[F,\mathcal{F}_{S(I)}]} : \left(F \cap \biguplus_{j \in P_0} X_j\right) \in \prod_{j \in P_0} \mathbb{C}_j, \forall i \in [M] \setminus P_0\ X_i \cap F = D_i \right]\!\right] \right| > L_{P_0}. \tag{35}$$

Take now a product-permutation $\mathfrak{L}_0$ in which all sets D_i ($i \notin P_0$) and all elements of the chains $\mathbb{C}_j$ ($j \in P_0$) are initial with respect to $\mathfrak{L}_0$. Since $\mathbb{P}(\mathcal{F}_{S(I)}(\mathfrak{L}_0)) = T(I)$ we can rewrite (35) as

$$\sum_{\substack{(i_1,\dots,i_M) \in \pi_M: \\ \forall j \notin P_0\ i_j = |D_j|}} t_{i_1,\dots,i_M}(I) > L_{P_0}. \tag{36}$$

As $T(I) = \mathbb{P}(\mathcal{H}_{\mathfrak{L}_0}(\mathfrak{L}_0))$, (36) gives that the following is larger than L_{P_0}

$$\left| \left[\!\left[F^{\#[F,\mathcal{H}_{\mathfrak{L}_0}(\mathfrak{L}_0)]} : \left(F \cap \biguplus_{j \in P_0} X_j\right) \in \prod_{j \in P_0} \mathbb{C}_j, \forall i \in [M] \setminus P_0\ X_i \cap F = D_i \right]\!\right] \right|. \tag{37}$$

However, from $\mathcal{H}_{\mathfrak{L}_0} \in \mathbb{A}$ we get that $\mathcal{H}_{\mathfrak{L}_0}$, and consequently $\mathcal{H}_{\mathfrak{L}_0}(\mathfrak{L}_0)$ must be k-dimensional M-part Sperner multifamilies with parameters $L_P : P \in \binom{[M]}{k}$, contradicting that the quantity in (37) is larger than L_{P_0}. □

Proof of Lemma 3: Let $\mathbb{A}$ be family of k-dimensional M-part Sperner multifamilies that satisfy a Γ-multiplicity contraints $\mathbb{M}_\Gamma$, and let I be a k-dimensional multitransversal on π_M with the same parameters L_P satisfying the same Γ-multiplicity constraint $\mathbb{M}_\Gamma$. Let $\mathfrak{L}$ be a fixed product-permutation, for each $(i_1,\dots,i_M) \in I$ let $H_{(i_1,\dots,i_M)}$ be the (unique) initial set with respect to $\mathfrak{L}$ with profile vector $(i_1,\dots,i_M)$ and let

$$\mathcal{H}_{\mathfrak{L}} = [\![H_{(i_1,\dots,i_M)}^{t_{i_1,\dots,i_M}(I)}]\!].$$

It follows that $\mathcal{H}_{\mathfrak{L}}(\mathfrak{L}) = \mathcal{H}_{\mathfrak{L}}$, $\mathbb{P}(\mathcal{H}_{\mathfrak{L}}) = T(I)$, and from the properties of I we have that $\mathcal{H}_{\mathfrak{L}} \in \mathbb{A}$. Therefore we get that $T(I) \in \mu(\mathbb{A}(\mathfrak{L}))$. By Theorem 4, the vector $S(I)$ is present in the set on the right hand side of (13), whose extreme points agree with those of $\mu(\mathbb{A})$, and by Theorem 5, $S(I) \in \mu(\mathbb{A})$. All that remains to be shown is that if $S(I) = \sum_{T(I_u)\in\mu(\mathbb{A}(\mathfrak{L}))} \lambda(I_u)S(I_u)$ with $\lambda(I_u) \geq 0$ and $\sum_{T(I_u)\in\mu(\mathbb{A}(\mathfrak{L}))} \lambda(I_u) = 1$, then I is among the I_u's, and all others come with a zero coefficient. $S(I) = \sum_{T(I_u)\in\mu(\mathbb{A}(\mathfrak{L}))} \lambda(I_u)S(I_u)$ means that for all $(i_1,\ldots,i_M) \in \pi_M$ we have

$$t_{i_1,\ldots,i_M}(I)\prod_{j=1}^{M}\binom{m_j}{i_j} = \sum_{T(I_u)\in\mu(\mathbb{A}(\mathfrak{L}))} \lambda(I_u)t_{i_1,\ldots,i_M}(I_u)\prod_{j=1}^{M}\binom{m_j}{i_j},$$

which implies that

$$T(I) = \sum_{T(I_u)\in\mu(\mathbb{A}(\mathfrak{L}))} \lambda(I_u)T(I_u).$$

Let the ordering $\operatorname{supp}(I) = \{\boldsymbol{j}_1,\boldsymbol{j}_2,\ldots,\boldsymbol{j}_s\}$ show that I has the LEM property. Then for all u, $T_{\boldsymbol{j}_1}(I) \geq T_{\boldsymbol{j}_1}(I_u)$, and as the coefficients sum to 1, for all u, $T_{\boldsymbol{j}_1}(I) = T_{\boldsymbol{j}_1}(I_u)$. This argument repeats to $\boldsymbol{j}_2,\ldots,\boldsymbol{j}_s$. Hence for all u, $\operatorname{supp}(I) \subseteq \operatorname{supp}(I_u)$. If $\operatorname{supp}(I)$ is a proper subset of $\operatorname{supp}(I_u)$, then we must have $\lambda(I_u) = 0$. Therefore for all the I_u that have $\lambda(I_u) \neq 0$ we must have $\operatorname{supp}(I_u) = \operatorname{supp}(I)$, and consequently $I_u = I$. □

10 Proofs for the Results on Transversals

We start with two lemmata.

Lemma 12. *Let n_1, n_2 be positive integers, and set $\ell = \gcd(n_1,n_2)$, $m_i = \frac{n_i}{\ell}$ and $N = \operatorname{lcm}(n_1,n_2) = \frac{n_1n_2}{\ell}$. For every $j \in [N]^\star$, there are exactly ℓ vectors $(a_1,a_2) \in \pi_2$, such that $\left\langle \frac{a_1}{n_1} + \frac{a_2}{n_2}\right\rangle = \frac{j}{N}$.*

Proof. Since m_1, m_2 are relatively prime, for any integer $j \in [N]^\star$ we have integers z_1, z_2 such that $z_1m_2 + z_2m_1 = j$, therefore $\frac{z_1}{n_1} + \frac{z_2}{n_2} = \frac{j}{N}$. Taking $a_i \in [n_i]^\star$ such that $a_i \equiv z_i \mod n_i$ we obtain that the required vectors (a_1,a_2) exist for any j. It is also clear that for any $(a_1,a_2) \in \pi_2$ there is some $j \in [N]^\star$ such that $\left\langle \frac{a_1}{n_1} + \frac{a_2}{n_2}\right\rangle = \frac{j}{N}$.

So we define for any $j \in [N]^\star$

$$\mathcal{D}_j = \left\{(a_1,a_2) \in \pi_2 : \left\langle \frac{a_1}{n_1} + \frac{a_2}{n_2}\right\rangle = \frac{j}{N}\right\}.$$

Fix $j \in [N]^\star$ and $(x_1,x_2) \in \mathcal{D}_j$. For any $(y_1,y_2) \in \pi_2$ we have that $(y_1,y_2) \in \mathcal{D}_j$ iff $\frac{y_1-x_1}{n_1} + \frac{y_2-x_2}{n_2}$ is an integer.

The $\mathcal{D}_j$ are nonempty and partition π_2. If for each $j, j' \in [N]^\star$, there is an injection from $\mathcal{D}_j$ to $\mathcal{D}_{j'}$, then $|\mathcal{D}_j| = |\mathcal{D}_{j'}|$, and consequently $|\mathcal{D}_j| = \frac{n_1n_2}{N} = \ell$, which proves our statement. So we will construct such an injection.

Let $j, j' \in [N]^\star$. Fix an $(a_1, a_2) \in \mathcal{D}_j$ and a $(b_1, b_2) \in \mathcal{D}_{j'}$. We define the map $\phi : \mathcal{D}_j \to \pi_2$ by $\phi(c_1, c_2) = (d_1, d_2) \in \pi_2$ iff $d_i \equiv c_i + (b_i - a_i) \mod n_i$. Clearly, the map is a well-defined injection, moreover, $\phi(a_1, a_2) = (b_1, b_2)$.

Assume that $(d_1, d_2) \in \phi(\mathcal{D}_j)$. Then $(d_1, d_2) = \phi(c_1, c_2)$ for some $(c_1, c_2) \in \mathcal{D}_j$, and $d_i - b_i \equiv c_i - a_i \mod n_i$. Thus $\left(\frac{d_1-b_1}{n_i} + \frac{d_2-b_2}{n_2}\right) - \left(\frac{c_1-a_1}{n_1} + \frac{c_2-a_2}{n_2}\right)$ is an integer. Since $(c_1, c_2) \in \mathcal{D}_j$, this implies $\frac{d_1-b_1}{n_i} + \frac{d_2-b_2}{n_2}$ is also an integer, with other words $(d_1, d_2) \in \mathcal{D}_{j'}$. Therefore $\phi(\mathcal{D}_j) \subseteq \mathcal{D}_{j'}$. □

Lemma 13. *Let $n_1, n_2, \ldots, n_k$ be given, and set $K = \prod_{j=1}^k n_j$, $N =$ lcm $(n_1, \ldots, n_k)$ and $\ell = \frac{K}{N}$. For each $j \in [N]^\star$ we have that there are exactly ℓ vectors $(a_1, \ldots, a_k) \in \pi_k$ such that*

$$\left\langle \sum_{i=1}^{k} \frac{a_i}{n_i} \right\rangle = \frac{j}{N}.$$

Proof. We prove the statement by induction on k. The statement is clearly true for $k = 1$ (when $N = n_1$ and $\ell = 1$); and it was proved in Lemma 12 for $k = 2$. So assume that $k > 2$ and we know the statement already for all $1 \leq k' \leq k-1$.

It is clear that for any $(a_1, \ldots, a_k) \in \pi_k$ we have precisely one $j \in [N]^\star$ such that $\left\langle \sum_{i=1}^k \frac{a_i}{n_i} \right\rangle = \frac{j}{N}$. Let $K_1 = \prod_{j=1}^{k-1} n_j$, $N_1 = \mathrm{lcm}(n_1, \ldots, n_{k-1})$ and $\ell_1 = \frac{K_1}{N_1}$, and $\ell_2 = \gcd(N_1, n_k)$. Then $K = K_1 n_k$, $N = \mathrm{lcm}(N_1, n_k)$ and $\ell = \frac{K_1 n_k}{\mathrm{lcm}(N_1, n_k)} = \frac{K_1}{N_1} \cdot \frac{N_1 n_k}{\mathrm{lcm}(N_1, n_k)} = \ell_1 \ell_2$.

Fix a $j \in [N]^\star$. Note that for integers a_i, $\left\langle \sum_{i=1}^{k-1} \frac{a_i}{n_i} \right\rangle \in \{\frac{j'}{n} : j' \in [N_1]^\star\}$, and for real numbers c, d we have $\langle\langle c\rangle + \langle d\rangle\rangle = \langle c + d\rangle$. By Lemma 12, there are precisely ℓ_2 pairs $(b, a_k) \in [N_1]^\star \times [n_k]^\star$ such that $\left\langle \frac{b}{N_1} + \frac{a_k}{n_k} \right\rangle = \frac{j}{N}$. By the induction hypothesis for each $b \in [N_1]$ there are precisely ℓ_1 values $(a_1, \ldots, a_{k-1}) \in \pi_{k-1}$ such that $\left\langle \sum_{j=1}^{k-1} \frac{a_j}{n_j} \right\rangle = \left\langle \frac{b}{N_1} \right\rangle$. Since $\ell_1 \ell_2 = \ell$, the statement follows. □

Proof to Lemma 6: By Lemma 13 the statement is equivalent with

$$\left| \left\{ j \in [N]^\star : \left\langle \alpha + \frac{j}{N} \right\rangle \in [\beta, \beta + \mu) \right\} \right| \in \{\lceil \mu N \rceil, \lfloor \mu N \rfloor\}$$

which follows from Lemma 5. □

Proof of Theorem 10: Assume that μ satisfies condition (16) and $0 \leq \beta \leq 1 - \mu$. Fix $P \in \binom{[M]}{k}$ and for each $j \notin P$ fix a $b_j \in [n_j]$. Then Condition (1) follows from Lemma 6 using $\alpha = \sum_{j \notin P} \frac{b_j}{n_j}$; thus $\mathbb{C}(n_1, \ldots, n_M; \beta, \mu)$ is a k-dimensional transversal with the given parameters L_P.

Assume now further that for $P_0 \in \binom{[M]}{k}$ we have that $\mu = \frac{L_{P_0}}{K_{P_0}}$ (as this is equivalent with $\mu = \min_P \frac{L_P}{K_P}$). Then we have that

$$L_{P_0} \geq d_{P_0} \lceil \mu N_{P_0} \rceil = d_{P_0} \left\lceil \frac{L_{P_0}}{d_{P_0}} \right\rceil \geq L_{P_0},$$

which implies that μN_{P_0} is an integer, i.e. by Lemma 6 our transversal is full. □

Acknowledgements. This research started at the "Search Methodologies II" workshop at the Zentrum für interdisziplinäre Forschung of Universität Bielefeld, where the last two authors met Professor Ahlswede for the last time. Special thanks go to Professor Charles Colbourn for his encouragement to continue our investigation in this direction.

References

1. Aydinian, H., Czabarka, É., Erdős, P.L., Székely, L.A.: A tour of M-part L-Sperner families. J. Combinatorial Theory A 118(2), 702–725 (2011)
2. Aydinian, H., Czabarka, É., Engel, K., Erdős, P.L., Székely, L.A.: A note on full transversals and mixed orthogonal arrays. Australasian J. Combin. 48, 133–141 (2010)
3. Aydinian, H., Erdős, P.L.: AZ-identities and Strict 2-part Sperner Properties of Product Posets. To appear in Order
4. Bollobás, B.: On generalized graphs. Acta Mathematica Academiae Scientiarum Hungaricae 16(34), 447–452 (1965)
5. Bush, K.A.: Orthogonal arrays, Ph.D.Thesis, North Carolina State University (1950)
6. Bush, K.A.: A generalization of a theorem due to MacNeish. Ann. Math. Stat. 23(2), 293–295 (1952)
7. Cheng, C.-S.: Orthogonal arrays with variable numbers of symbols. Ann. Statistics 8(2), 447–453 (1980)
8. Engel, K.: Sperner Theory. Encyclopedia of Mathematics and its Applications, vol. 65, pp. x+417. Cambridge University Press, Cambridge (1997)
9. Erdős, P.: On a lemma of Littlewood and Offord. Bull. of the Amer. Math. Soc. 51, 898–902 (1945)
10. Erdős, P.L., Frankl, P., Katona, G.O.H.: Extremal hypergraph problems and convex hulls. Combinatorica 5, 11–26 (1985)
11. Erdős, P.L., Füredi, Z., Katona, G.O.H.: Two-part and k-Sperner families - new proofs using permutations. SIAM J. Discrete Math. 19, 489–500 (2005)
12. Erdős, P.L., Katona, G.O.H.: Convex hulls of more-part Sperner families. Graphs and Combinatorics 2, 123–134 (1986)
13. Füredi, Z., Griggs, J.R., Odlyzko, A.M., Shearer, J.M.: Ramsey-Sperner theory. Discrete Mathematics 63, 143–152 (1987)
14. Hedayat, A.S., Sloane, N.J.A., Stufken, J.: Orthogonal Arrays: Theory and Applications. Springer Series in Statistics, pp. xxiv+416. Springer, New York (1999)
15. Kuhfeld, W.F.: Orthogonal Arrays List, `www.support.sas.com/techsup/technote/ts723.pdf`
16. Lubell, D.: A short proof of Sperner's lemma. J. Comb. Theory 1(2), 299 (1966)
17. Meshalkin, L.D.: Generalization of Sperner's theorem on the number of subsets of a finite set. Theory of Probability and its Applications 8(2), 203–204 (1963)
18. Rao, C.R.: M.A. Thesis, Calcutta University (1943)
19. Rao, C.R.: Factorial experiments derivable from combinatorial arrangements of arrays. Suppl. J. Royal Stat. Soc. 9(1), 128–139 (1947)
20. MacWilliams, F.J., Sloane, N.J.A.: The Theory of Error Correcting Codes, Part I: xv+369, Part II: ix+391. North-Holland, Amsterdam (1977)
21. Sperner, E.: Ein Satz über Untermengen einer endlichen Menge. Math. Z. 27, 544–548 (1928)
22. Yamamoto, K.: Logarithmic order of free distributive lattice. J. Math. Soc. Japan 6, 343–353 (1954)

Generic Algorithms for Factoring Strings

David E. Daykin[1], Jacqueline W. Daykin[2,3,*],
Costas S. Iliopoulos[3,5], and W.F. Smyth[3,4,5,**]

[1] Department of Mathematics, University of Reading, UK
[2] Department of Computer Science, Royal Holloway, University of London, UK
J.Daykin@cs.rhul.ac.uk, jackie.daykin@kcl.ac.uk
[3] Department of Informatics, Kings College London, UK
c.iliopoulos@kcl.ac.uk
[4] Department of Computing & Software, McMaster University, Canada
smyth@mcmaster.ca
[5] Department of Mathematics & Statistics,
University of Western Australia, Perth, Australia

Dedicated to the memory of Rudolf Ahlswede

Abstract. In this paper we describe algorithms for factoring words over sets of strings known as circ-UMFFs, generalizations of the well-known Lyndon words based on lexorder, whose properties were first studied in 1958 by Chen, Fox and Lyndon. In 1983 Duval designed an elegant linear-time sequential (RAM) Lyndon factorization algorithm; a corresponding parallel (PRAM) algorithm was described in 1994 by Daykin, Iliopoulos and Smyth. In 2003 Daykin and Daykin introduced various circ-UMFFs, including one based on V-words and V-ordering; in 2011 linear string comparison and sequential factorization algorithms based on V-order were given by Daykin, Daykin and Smyth. Here we first describe generic RAM and PRAM algorithms for factoring a word over any circ-UMFF; then we show how to customize these generic algorithms to yield optimal parallel Lyndon-like V-word factorization.

Keywords: circ-UMFF, complexity, factor, generic, lexicographic order, Lyndon word, optimal, parallel algorithm, PRAM, RAM, sequential algorithm, V-order, V-word.

1 Introduction

In this paper we design and analyze sequential and parallel algorithms for factoring strings over particular sets of strings known as circ-UMFFs, a generalization of the Lyndon words $\mathcal{L}$. We introduce a generic algorithmic framework for computing factorizations applicable to any circ-UMFF. The input to an algorithm

[*] In loving memory of my father David.

[**] The work of the fourth author was supported in part by a grant from the Natural Sciences and Engineering Research Council of Canada.

H. Aydinian, F. Cicalese, and C. Deppe (Eds.): Ahlswede Festschrift, LNCS 7777, pp. 402–418, 2013.

is a string $\boldsymbol{x}$ and a "factorization family" $\mathcal{W}$, and the result is the factorization of $\boldsymbol{x}$ over $\mathcal{W}$.

A factorization family $\mathcal{W}$ is an infinite set of *strings* (equivalently *words*) into which any string on the same alphabet can be factored. If all such factorizations are unique, the set is called a Unique Maximal Factorization Family (UMFF) [DD03], and if moreover $\mathcal{W}$ contains exactly one rotation of every primitive string, then it is called a circ-UMFF [DD08]. In this paper we study factorizations over circ-UMFFs.

We characterize a circ-UMFF as an infinite set $\mathcal{W}$ of strings on a given alphabet Σ, $|\Sigma| \geq 2$, that is closed, according to specified rules, under the reciprocal operations of concatenation and factorization. In particular,

* $\lambda \in \Sigma \Longrightarrow \lambda \in \mathcal{W}$;
* (concatenation) $\boldsymbol{u}, \boldsymbol{v} \neq \boldsymbol{u} \in \mathcal{W} \Longrightarrow$ exactly one of $\boldsymbol{uv}, \boldsymbol{vu} \in \mathcal{W}$;
* (factorization) $\boldsymbol{w} \in \mathcal{W}$ and $|\boldsymbol{w}| > 1 \Longrightarrow$ there exist $\boldsymbol{u}, \boldsymbol{v} \neq \boldsymbol{u} \in \mathcal{W}$ such that $\boldsymbol{uv} = \boldsymbol{w}$.

The study of circ-UMFFs originated with Lyndon words [CFL58, L83], which became prominent in stringology due to their combinatorial richness and the many applications of Lyndon decomposition (factorization) [S03]. Lyndon words have a deep connection with de Bruijn sequences [dB46]; infinite Lyndon words have been considered in [SMDS94, BMP07]. Their versatility is expressed by the wide range of applications of the Lyndon factorization, including: musicology [C04], algorithms for digital geometry [BLPR09], cryptanalysis [P05], the Burrows-Wheeler transform related to bioinformatics and compression [CDP05, GS09, K09], and string matching [CP91, BGM11]; for applications of Lyndon words related to free Lie algebras, see [R93]. Both sequential [Du83, D11] and CRCW parallel RAM algorithms [DIS94] have been proposed for computing the Lyndon factorization of a string. Like the Lyndon words $\mathcal{L}$, the elements of a circ-UMFF $\mathcal{W}$ are totally ordered, and also share the property that the maximal factorization of any string over $\mathcal{W}$ yields factors that are monotone non-increasing (in $\mathcal{W}$-order) [DD08, DDS09]. Thus circ-UMFFs can be described in terms of dictionaries [DDS09], leading to possible applications to searching and sorting problems on words, as shown for Lyndon words in [Du83].

Various classes and properties of circ-UMFFs have recently been identified [DD03, DD08, DDS09]. This paper confirms the suggestion made in [DDS09] that the parallel Lyndon factorization algorithm of [DIS94] can be extended generically to string factorization over general circ-UMFFs. The extension turns out to be straightforward, since the theoretical results for Lyndon words given in [DIS94], on which the parallel algorithm is based, can be proved also in the more general context of circ-UMFFs, as we show in Section 2. Thus in Section 4, rather than repeat all the details of the algorithm, we confine ourselves to providing an overview that explains the main steps. At the same time we correct a conceptual error in [DIS94]: the use of a "cut-point" rather than a concatenation point ("concat-point") as introduced here to guide the merged factorization of $\boldsymbol{uv}$, given the individual factorizations of $\boldsymbol{u}$ and $\boldsymbol{v}$. Furthermore, a generic sequential factorization algorithm for circ-UMFFs is given in Section 3.

The existence of a generic algorithmic framework paves the way for developing efficient algorithms honed to specific factorization families as they arise. In particular, we consider the V-word circ-UMFF, studied combinatorially in [DD03] and for sequential algorithms in [DDS11, DDS12]. V-words are analogous structures to Lyndon words, and interestingly, similar applications are emerging: V-words have arisen naturally in music analysis [CT03]; they also provide a variant of the classic lexicographic Burrows-Wheeler transform [DS12]. In Section 5 we customize the generic parallel algorithm of Section 4 for this particular circ-UMFF, resulting in a new optimal V-word PRAM algorithm.

The PRAM [J92] has had a revival of interest since, due to advances in fabricating and programming these architectures, it is no longer considered an unrealistic model. The PRAM paradigm has featured as the underlying architecture in commercial products: see for example [N08]. And PRAM algorithmics are also being developed for chip technologies in order to support desktop supercomputing targeted at modern multicore processors [V08].

2 Unique Maximal Factorization Families

We begin with a summary of relevant UMFF theory from [DDS09], and then extend some results on Lyndon words [L83] to circ-UMFFs. We use standard terminology and notation from combinatorics on words: alphabet Σ, Σ^*, Σ^+, border, concatenate, period, repetition, and so on (see [S03]). All examples based on the alphabet Σ of lower-case letters use the standard English-language order $a < b < \cdots < z$.

A string $\boldsymbol{w}$ is a ***factor*** of a string $\boldsymbol{x}[1..n]$ if and only if $\boldsymbol{w} = \boldsymbol{x}[i..j]$ for $1 \le i \le j \le n$. Note that a factor is nonempty. If $\boldsymbol{x} = \boldsymbol{w_1 w_2 \cdots w_k}$, $1 \le k \le n$, then $\boldsymbol{w_1 w_2 \cdots w_k}$ is said to be a ***factorization*** of $\boldsymbol{x}$; if every factor $\boldsymbol{w_j}$, $1 \le j \le k$, belongs to a set $\mathcal{W}$, then the factorization is denoted $F_{\mathcal{W}}(\boldsymbol{x})$.

Definition 1. *A subset $\mathcal{W} \subseteq \Sigma^+$ is a* ***factorization family*** *(FF) if and only if for* every *nonempty string $\boldsymbol{x}$ on Σ there exists a factorization $F_{\mathcal{W}}(\boldsymbol{x})$.*

For some string $\boldsymbol{x}$ and some FF $\mathcal{W}$, suppose $\boldsymbol{x} = \boldsymbol{w_1 w_2 \cdots w_k}$, where $\boldsymbol{w_j} \in \mathcal{W}$ for every $1 \le j \le k$. For some $1 \le k' \le k$, write $\boldsymbol{x} = \boldsymbol{u w_{k'} v}$, where $\boldsymbol{u} = \boldsymbol{w_1 w_2 \cdots w_{k'-1}}$ (empty if $k' = 1$) and $\boldsymbol{v} = \boldsymbol{w_{k'+1} w_{k'+2} \cdots w_k}$ (empty if $k' = k$). Suppose that there does not exist a suffix $\boldsymbol{u'}$ of $\boldsymbol{u}$ nor a prefix $\boldsymbol{v'}$ of $\boldsymbol{v}$ such that $\boldsymbol{u' w_{k'} v'} \ne \boldsymbol{w_{k'}}$ and $\boldsymbol{u' w_{k'} v'} \in \mathcal{W}$; then $\boldsymbol{w_{k'}}$ is said to be a ***max factor*** of $\boldsymbol{x}$. If *every* factor $\boldsymbol{w_{k'}}$ is max, then the factorization $F_{\mathcal{W}}(\boldsymbol{x})$ is itself said to be ***max***. Observe that a max factorization must be unique: there exists no other max factorization of $\boldsymbol{x}$ that uses only elements of $\mathcal{W}$.

Definition 2. *Let $\mathcal{W}$ be an FF on an alphabet Σ. Then $\mathcal{W}$ is a* ***unique maximal factorization family*** *(UMFF) if and only if there exists a max factorization $F_{\mathcal{W}}(\boldsymbol{x})$ for* every *string $\boldsymbol{x} \in \Sigma^+$.*

We will assume throughout, that when factoring over an UMFF, the factorization is chosen to be the one which is maximal.

Lemma 1. *(The* **xyz** *Lemma [DD03]) An FF $\mathcal{W}$ is an UMFF if and only if whenever $\boldsymbol{xy}, \boldsymbol{yz} \in \mathcal{W}$ for some nonempty $\boldsymbol{y}$, then $\boldsymbol{xyz} \in \mathcal{W}$.*

Corollary 1. *([DDS09]) Suppose $\boldsymbol{x} = \boldsymbol{u_1 u_2 \cdots u_m}$ and $\mathcal{W}$ is an UMFF, where for every $1 \leq j \leq m$, $\boldsymbol{u_j} \in \mathcal{W}$. Then we can write the max factorization $F_{\mathcal{W}}(\boldsymbol{x}) = \boldsymbol{w_1 w_2 \cdots w_k}$, where*

$$\boldsymbol{w_1} = \boldsymbol{u_{j_0+1}} \cdots \boldsymbol{u_{j_1}}, \boldsymbol{w_2} = \boldsymbol{u_{j_1+1}} \cdots \boldsymbol{u_{j_2}}, \ \ldots \ , \boldsymbol{w_k} = \boldsymbol{u_{j_{k-1}+1}} \cdots \boldsymbol{u_{j_k}},$$

$0 = j_0 < j_1 < j_2 < \cdots < j_{k-1} < j_k = m.$

This corollary generalizes Theorem 2.2 in [DIS94] and will be applied in the parallel algorithms of Sections 4 and 5 to factor the concatenation of a pair of previously factored substrings. See also Lemmas 7 and 8.

Definition 3. *An UMFF $\mathcal{W}$ over Σ^+ is a* ***circ-UMFF*** *if and only if it contains exactly one rotation of every primitive string $\boldsymbol{x} \in \Sigma^+$.*

Definition 4. *If a circ-UMFF $\mathcal{W}$ contains strings $\boldsymbol{u}$, $\boldsymbol{v}$ and $\boldsymbol{uv}$, we write $\boldsymbol{u} <_{\mathcal{W}} \boldsymbol{v}$ (called $\mathcal{W}$-**order**).*

Theorem 1. *([DD08]) Let $\mathcal{W}$ be a circ-UMFF.*
(1) If $\boldsymbol{u} \in \mathcal{W}$ then $\boldsymbol{u}$ is border-free (no period less than $|\boldsymbol{u}|$).
(2) If $\boldsymbol{u}, \boldsymbol{v} \in \mathcal{W}$ and $\boldsymbol{u} \neq \boldsymbol{v}$ then $\boldsymbol{uv}$ is primitive (not a repetition).
(3) If $\boldsymbol{u}, \boldsymbol{v} \in \mathcal{W}$ and $\boldsymbol{u} \neq \boldsymbol{v}$ then $\boldsymbol{uv} \in \mathcal{W}$ or $\boldsymbol{vu} \in \mathcal{W}$ (but not both).
(4) If $\boldsymbol{u}, \boldsymbol{v}, \boldsymbol{uv} \in \mathcal{W}$ then $\boldsymbol{u} <_{\mathcal{W}} \boldsymbol{v}$ and $<_{\mathcal{W}}$ is a total order of $\mathcal{W}$.
(5) If $\boldsymbol{w} \in \mathcal{W}$ and $|\boldsymbol{w}| \geq 2$ then there exist $\boldsymbol{u}, \boldsymbol{v} \in \mathcal{W}$ with $\boldsymbol{w} = \boldsymbol{uv}$.

Taken together with Definition 4, Theorem 1(4) extends to all circ-UMFFs the classical result [Du83], given as Theorem 2.1 in [DIS94], that for $\boldsymbol{u}, \boldsymbol{v} \in \mathcal{L}$, $\boldsymbol{uv} \in \mathcal{L} \Longleftrightarrow \boldsymbol{u} <_{\mathcal{L}} \boldsymbol{v}$.

The algorithms developed here relate to the following classes of circ-UMFFs introduced in [DDS09]:

Definition 5. *A circ-UMFF $\mathcal{W}$ is said to be* ***Type Flight Deck*** *if and only if $\boldsymbol{w}[1..n] \in \mathcal{W}$ implies $\boldsymbol{w}[1] \leq_{\mathcal{W}} \boldsymbol{w}[i]$ for every $i \in 2..n$.*

For example, the Lyndon word circ-UMFF, based on lexicographic order (***lex-order***), is Type Flight Deck, as is also the V-order circ-UMFF discussed in Section 5.

Definition 6. *A circ-UMFF $\mathcal{W}$ is said to be* ***Type Acrobat*** *if and only if it contains elements $\boldsymbol{uv_1}$, $\boldsymbol{w}$ and $\boldsymbol{uv_2}$, nonempty $\boldsymbol{u}$ not a prefix of $\boldsymbol{w}$, such that*

$$\boldsymbol{uv_1} <_{\mathcal{W}} \boldsymbol{w} <_{\mathcal{W}} \boldsymbol{uv_2}.$$

Types Flight Deck and Acrobat are not necessarily incompatible. However, if the prefix $\boldsymbol{u}$ in Definition 6 is in fact a single letter λ, then since by Theorem 1(4) both $\lambda \boldsymbol{v_1 w}$ and $\boldsymbol{w} \lambda \boldsymbol{v_2}$ are elements of $\mathcal{W}$, and since either $\lambda <_{\mathcal{W}} \boldsymbol{w}[1]$ or $\boldsymbol{w}[1] <_{\mathcal{W}} \lambda$, it follows from Definition 5 that $\mathcal{W}$ cannot be Type Flight Deck.

The following lemma, trivially true for Lyndon words $\mathcal{L}$, will be applied in procedure Contest in Section 4.

Lemma 2. *Let $\mathcal{W}$ be a Flight Deck circ-UMFF and suppose $\boldsymbol{u} \in \mathcal{W}$, where $\boldsymbol{u}[1] = \lambda$. If for some $\boldsymbol{v} = \boldsymbol{v}[1..m]$, $\lambda <_{\mathcal{W}} \boldsymbol{v}[i]$ for every $i \in 1..m$, then $\boldsymbol{uv} \in \mathcal{W}$.*

Proof. Suppose that $\boldsymbol{uv}$ is a repetition. Since λ does not occur in $\boldsymbol{v}$, it follows that $\boldsymbol{u}$ must have period less than $|\boldsymbol{u}|$, contradicting Theorem 1(1).

Assume therefore that $\boldsymbol{uv}$ is primitive. Then, according to Definition 3, some rotation of $\boldsymbol{uv}$ is in $\mathcal{W}$. Since $\mathcal{W}$ is Type Flight Deck the chosen rotation must start with λ. If λ occurs only once in $\boldsymbol{uv}$, then the rotation is $\boldsymbol{uv}$. Otherwise, write $\boldsymbol{u} = \lambda \boldsymbol{y_1} \lambda \boldsymbol{y_2} \cdots \lambda \boldsymbol{y_t}$ for $t \geq 2$, where $\boldsymbol{y_i} \in \Sigma^*$ for $1 \leq i \leq t-1$, and $\boldsymbol{y_t} \in \Sigma^+$. Now let $\boldsymbol{r} = \lambda \boldsymbol{y_i} \cdots \lambda \boldsymbol{y_t} \boldsymbol{v} \lambda \boldsymbol{y_1} \cdots \lambda \boldsymbol{y_{i-1}}$ be a rotation of $\boldsymbol{uv}$ starting with λ. If $\boldsymbol{r} \in \mathcal{W}$, then applying Lemma 1 to $\boldsymbol{r}$ and $\boldsymbol{u}$ implies that the word $\lambda \boldsymbol{y_i} \cdots \lambda \boldsymbol{y_t} \boldsymbol{v} \lambda \boldsymbol{y_1} \cdots \lambda \boldsymbol{y_{i-1}} \lambda \boldsymbol{y_i} \cdots \lambda \boldsymbol{y_t}$ with border $\lambda \boldsymbol{y_i} \cdots \lambda \boldsymbol{y_t}$ belongs to $\mathcal{W}$, again contradicting Theorem 1(1). We conclude that $\boldsymbol{uv} \in \mathcal{W}$. □

Concatenation properties of circ-UMFFs will be used when extending existing factors in the algorithms (see Lemmas 7 and 8):

Lemma 3. *([DD08]) Suppose that $\boldsymbol{w}$ is an element of a circ-UMFF $\mathcal{W}$. If $\boldsymbol{u_1}, \boldsymbol{u_2}, \ldots, \boldsymbol{u_{k_1}}$ are all the proper prefixes of $\boldsymbol{w}$ in increasing order of length that belong to $\mathcal{W}$, and if $\boldsymbol{v_1}, \boldsymbol{v_2}, \ldots, \boldsymbol{v_{k_2}}$ are all the proper suffixes of $\boldsymbol{w}$ in decreasing order of length that belong to $\mathcal{W}$, then*

$$\boldsymbol{u_1} <_{\mathcal{W}} \boldsymbol{u_2} <_{\mathcal{W}} \cdots <_{\mathcal{W}} \boldsymbol{u_{k_1}} <_{\mathcal{W}} \boldsymbol{w} <_{\mathcal{W}} \boldsymbol{v_1} <_{\mathcal{W}} \boldsymbol{v_2} <_{\mathcal{W}} \cdots <_{\mathcal{W}} \boldsymbol{v_{k_2}}.$$

Lemma 4. *([DD08]) Let $\mathcal{W}$ be a circ-UMFF. If $\boldsymbol{u_1}, \boldsymbol{u_2}, ..., \boldsymbol{u_m} \in \mathcal{W}$, $\boldsymbol{u_1} <_{\mathcal{W}} \boldsymbol{u_2} <_{\mathcal{W}} \cdots <_{\mathcal{W}} \boldsymbol{u_m}$, $m \geq 2$, and if $k_1, k_2, \ldots, k_m$ are positive integers, then*

$$\boldsymbol{u_1}^{k_1} \boldsymbol{u_2}^{k_2} \cdots \boldsymbol{u_m}^{k_m} \in \mathcal{W}.$$

The following result generalizes the Lyndon factorization theorem [CFL58] to circ-UMFFs, and together with Corollary 1 is relevant to the algorithms of Sections 3, 4 & 5.

Theorem 2. *([DDS09]) Let $\mathcal{W}$ be a circ-UMFF and let $\boldsymbol{x} = \boldsymbol{u_1 u_2} \cdots \boldsymbol{u_m}$, with each $\boldsymbol{u_j} \in \mathcal{W}$. Then $F_{\mathcal{W}}(\boldsymbol{x}) = \boldsymbol{u_1 u_2} \cdots \boldsymbol{u_m}$ if and only if $\boldsymbol{u_1} \geq_{\mathcal{W}} \boldsymbol{u_2} \geq_{\mathcal{W}} \cdots \geq_{\mathcal{W}} \boldsymbol{u_m}$.*

The following two results, useful algorithmically, give conditions under which different sections of a string can be factored independently — as we shall see, the generic parallel algorithm starts by factoring "blocks" of the input string. The first lemma corresponds to Theorem 2.5 for Lyndon words in [DIS94].

Lemma 5. *Suppose $\boldsymbol{x} = \boldsymbol{uwv} = \boldsymbol{u_1 u_2} \cdots \boldsymbol{u_m w v_1 v_2} \cdots \boldsymbol{v_{m'}}$, with $\boldsymbol{w}$ an element of a circ-UMFF $\mathcal{W}$, and $F_{\mathcal{W}}(\boldsymbol{u}) = \boldsymbol{u_1 u_2} \cdots \boldsymbol{u_m}$, $F_{\mathcal{W}}(\boldsymbol{v}) = \boldsymbol{v_1 v_2} \cdots \boldsymbol{v_{m'}}$. Then $F_{\mathcal{W}}(\boldsymbol{x}) = F_{\mathcal{W}}(\boldsymbol{u}) \boldsymbol{w} F_{\mathcal{W}}(\boldsymbol{v})$ if and only if these conditions all hold:*
(1) no nonempty suffix $\boldsymbol{s}$ of $\boldsymbol{u}$ and nonempty prefix $\boldsymbol{w'}$ of $\boldsymbol{w}$ such that $\boldsymbol{sw'} \in \mathcal{W}$;
(2) no nonempty prefix $\boldsymbol{p}$ of $\boldsymbol{v}$ and nonempty suffix $\boldsymbol{w''}$ of $\boldsymbol{w}$ such that $\boldsymbol{w''p} \in \mathcal{W}$;
(3) no such $\boldsymbol{swp} \in \mathcal{W}$.

Proof. Repeated applications of Lemma 1 and Corollary 1. □

The next lemma shows that if $\boldsymbol{u} <_{\mathcal{W}} \boldsymbol{v}$, then $\boldsymbol{u}$ is less in $\mathcal{W}$-order than any right extension of $\boldsymbol{v}$ that is also in $\mathcal{W}$; this is a property related to the operation of procedure Right-Extension, described in Section 4.

Lemma 6. *([DDS09]) Suppose $\boldsymbol{u} \in \mathcal{W}$ and $\boldsymbol{v} \in \mathcal{W}$, where $\mathcal{W}$ is a circ-UMFF. If $\boldsymbol{u} <_{\mathcal{W}} \boldsymbol{v}$, then for every string $\boldsymbol{w}$ such that $\boldsymbol{vw} \in \mathcal{W}$, $\boldsymbol{u} <_{\mathcal{W}} \boldsymbol{vw}$.*

Like Lemmas 5 and 6, the following definitions pertain to the final factorization of $F_{\mathcal{W}}(\boldsymbol{uv})$ over $\mathcal{W}$, given factorizations $F_{\mathcal{W}}(\boldsymbol{u})$ and $F_{\mathcal{W}}(\boldsymbol{v})$:

Definition 7. *Suppose $F_{\mathcal{W}}(\boldsymbol{u}) = \boldsymbol{u_1 u_2} \cdots \boldsymbol{u_s}$, $F_{\mathcal{W}}(\boldsymbol{v}) = \boldsymbol{v_1 v_2} \cdots \boldsymbol{v_t}$.*
(1) j is the ***cut-point*** *of $F_{\mathcal{W}}(\boldsymbol{uv})$ if and only if it is the greatest integer in $1..t$ such that $\boldsymbol{v_j} >_{\mathcal{W}} \boldsymbol{u_s}$ ($j = 0$ if $\boldsymbol{u_s} \geq_{\mathcal{W}} \boldsymbol{v_1} \geq_{\mathcal{W}} \boldsymbol{v_2} \geq_{\mathcal{W}} \cdots \geq_{\mathcal{W}} \boldsymbol{v_t}$).*
(2) k is the ***concat-point*** *of $F_{\mathcal{W}}(\boldsymbol{uv})$ if and only if it is the greatest integer in $1..t$ such that $\boldsymbol{u_s v_1 v_2} \cdots \boldsymbol{v_k} \in \mathcal{W}$ ($k = 0$ if $\boldsymbol{u_s v_1} \notin \mathcal{W}$).*

Since $\boldsymbol{u_s v_1} \notin \mathcal{W} \iff \boldsymbol{u_s} \geq_{\mathcal{W}} \boldsymbol{v_1}$, we see that cut-point $j = 0$ and concat-point $k = 0$ describe the same condition and must be equal. On the other hand, $k > 0$ is not necessarily the same as $j > 0$, as shown by the following simple examples:

Example 1. *Let $\mathcal{W}$ be the Lyndon circ-UMFF, and consider $\boldsymbol{u} = a$, $\boldsymbol{v} = cab$. The Lyndon factorizations of $\boldsymbol{u}$ and $\boldsymbol{v}$, respectively, are $F_{\mathcal{W}}(\boldsymbol{u}) = \boldsymbol{u_1} = a$ (hence $s = 1$), $F_{\mathcal{W}}(\boldsymbol{v}) = \boldsymbol{v_1 v_2} = (c)(ab)$ (hence $t = 2$). We seek $F_{\mathcal{W}}(\boldsymbol{uv})$. Since $\boldsymbol{v_2} = ab > a = \boldsymbol{u_1}$, we see that the cut-point $j = 2$; on the other hand, since $ac \in \mathcal{W}$ while $acab \notin \mathcal{W}$, the concat-point $k = 1$. Using the concat-point k, we can conclude that the Lyndon factorization of $\boldsymbol{uv} = acab$ is $(ac)(ab)$.*

Example 2. *Similarly if $F_{\mathcal{W}}(\boldsymbol{u}) = \boldsymbol{u_1} = a$, $F_{\mathcal{W}}(\boldsymbol{v}) = \boldsymbol{v_1 v_2} = (b)(ab)$, we again find $j = 2$, $k = 1$, and the Lyndon factorization $F_{\mathcal{W}}(\boldsymbol{uv}) = (ab)(ab)$.*

In [DIS94] (procedure Contest, line 9) the cut-point was used to guide the computation of $F_{\mathcal{W}}(\boldsymbol{uv})$ for Lyndon decomposition, thus leading, as these two examples show, to incorrect results in some cases. In the remainder of this section we show that in fact the concat-point can be used to compute $F_{\mathcal{W}}(\boldsymbol{uv})$ correctly, not just in the Lyndon case, but for any circ-UMFF factorization.

The first result generalizes Theorem 2.2 in [DIS94] from Lyndon words to circ-UMFFs, at the same time replacing cut-point by concat-point; although this theorem is just a special case of Corollary 1, we include it because the proof illustrates the algorithmic use of right and left extensions of factors in order to form $F_{\mathcal{W}}(\boldsymbol{uv})$.

Theorem 3. *Suppose $\mathcal{W}$ is a circ-UMFF, $F_{\mathcal{W}}(\boldsymbol{u}) = \boldsymbol{u_1 u_2} \cdots \boldsymbol{u_s}$, $F_{\mathcal{W}}(\boldsymbol{v}) = \boldsymbol{v_1 v_2} \cdots \boldsymbol{v_t}$, $s \geq 1$, $t \geq 1$. Then for some $i \in 0..s$ and concat-point $k \in 0..t$,*

$$F_{\mathcal{W}}(\boldsymbol{uv}) = \boldsymbol{u_1 u_2} \cdots \boldsymbol{u_i w v_{k+1} v_{k+2}} \cdots \boldsymbol{v_t},$$

where $\boldsymbol{w} = \boldsymbol{u_{i+1}} \cdots \boldsymbol{u_s v_1 v_2} \cdots \boldsymbol{v_k}$.

Proof. If $\boldsymbol{u_s} \geq_{\mathcal{W}} \boldsymbol{v_1}$ $(i = s, k = 0)$, it follows immediately from Theorem 2 that $F_{\mathcal{W}}(\boldsymbol{uv}) = \boldsymbol{u_1u_2}\cdots\boldsymbol{u_sv_1v_2}\cdots\boldsymbol{v_t}$, with $\boldsymbol{w} = \varepsilon$, the empty string. Assume therefore that $\boldsymbol{u_s} <_{\mathcal{W}} \boldsymbol{v_1}$, and note from Corollary 1 that substrings of factors do not need to be considered.

(1) Right Extension. We determine the concat-point $k \in 1..t$ such that $\boldsymbol{w}' = \boldsymbol{u_sv_1}\cdots\boldsymbol{v_k} \in \mathcal{W}$ and $\boldsymbol{w}' \geq_{\mathcal{W}} \boldsymbol{v_{k+1}} \geq_{\mathcal{W}} \cdots \geq_{\mathcal{W}} \boldsymbol{v_t}$. For $k < t$, by the definition of concat-point we have $\boldsymbol{w}'\boldsymbol{v_{k+1}} \notin \mathcal{W}$, so that $F_{\mathcal{W}}(\boldsymbol{u_sv}) = \boldsymbol{w}'\boldsymbol{v_{k+1}}\cdots\boldsymbol{v_t}$. For $k = t$, $F_{\mathcal{W}}(\boldsymbol{u_sv}) = \boldsymbol{w}'$.

(2) Left Extension. We may suppose $F_{\mathcal{W}}(\boldsymbol{u}) = \boldsymbol{u_1u_2}\cdots\boldsymbol{u_{s-h}u_s^h}$ for some $h \in 1..s$. From Lemma 4, $\boldsymbol{u_s^h v_1}\cdots\boldsymbol{v_k} \in \mathcal{W}$. If $h = s$ (that is, $i = s - h = 0$), $\boldsymbol{w} = \boldsymbol{u_1^s v_1}\cdots\boldsymbol{v_k}$ and $F_{\mathcal{W}}(\boldsymbol{uv}) = \boldsymbol{wv_{k+1}}\cdots\boldsymbol{v_t}$. Otherwise, by Corollary 1, find the least $i \in 0..s-h$ such that $\boldsymbol{w} = \boldsymbol{u_{i+1}}\cdots\boldsymbol{u_{s-h}u_s^h v_1v_2}\cdots\boldsymbol{v_k} \in \mathcal{W}$. □

We give two examples to show that the extensions of Theorem 3 are nontrivial for circ-UMFFs of both types Flight Deck and Acrobat.

Example 3. *For the Lyndon circ-UMFF $\mathcal{L}$, consider $\boldsymbol{u}, \boldsymbol{v}$ such that $F_{\mathcal{L}}(\boldsymbol{u}) = \boldsymbol{u_1u_2u_3u_4} = (c)(bc)(bc)(b)$ and $F_{\mathcal{L}}(\boldsymbol{v}) = \boldsymbol{v_1v_2v_3} = (c)(c)(a)$. Extending right, we find that $\boldsymbol{v_4u_1u_2} = bcc \in \mathcal{L}$, but not $\boldsymbol{v_4u_1u_2u_3} = bcca$, so that the concat-point $k = 2$. Extending left, we see that $\boldsymbol{v_2v_3v_4u_1u_2} = bcbcbcc \in \mathcal{L}$, but not cbcbcbcc. Thus $\boldsymbol{w} = bcbcbcc$ and $F_{\mathcal{L}}(\boldsymbol{uv}) = (c)\boldsymbol{w}(a)$.*

For the second example, we introduce the ***co-Lyndon circ-UMFF*** $\hat{\mathcal{L}}$ [DDS09], where $\boldsymbol{u} \in \hat{\mathcal{L}}$ if and only if the reversed string $\overline{\boldsymbol{u}}$ is a Lyndon word, and $\boldsymbol{u} <_{\hat{\mathcal{L}}} \boldsymbol{v}$ if and only if $\overline{\boldsymbol{u}} > \overline{\boldsymbol{v}}$ in lexorder. $\hat{\mathcal{L}}$ is type Acrobat.

Example 4. *For the co-Lyndon circ-UMFF $\hat{\mathcal{L}}$, suppose*

$$F_{\hat{\mathcal{L}}}(\boldsymbol{u}) = pika >_{\hat{\mathcal{L}}} goose >_{\hat{\mathcal{L}}} owl \geq_{\hat{\mathcal{L}}} owl,$$

$F_{\hat{\mathcal{L}}}(\boldsymbol{v}) = gorilla$. Extending right, since $owl <_{\hat{\mathcal{L}}} gorilla$, $\boldsymbol{u_4v_1} = owlgorilla$ and the concat-point $k = 1$. Extending left, we find $(owl)^2 <_{\hat{\mathcal{L}}} gorilla$, then $goose <_{\hat{\mathcal{L}}} (owl)^2 gorilla$. But since $pika >_{\hat{\mathcal{L}}} goose(owl)^2 gorilla$, we complete the left extension with $\boldsymbol{w} = gooseowlowlgorilla$ and $F_{\hat{\mathcal{L}}}(\boldsymbol{uv}) = (pika)\boldsymbol{w}$.

In both of these examples, both right extension and left extension are somehow uniform over a range; that is, if some longest substring belongs to $\mathcal{W}$, then so do a range of its prefixes/suffixes. The following two lemmas make this idea more precise, and provide justification for the novel use of binary search in the parallel algorithm of Section 4 for factoring over circ-UMFFs.

Lemma 7. *Suppose $\mathcal{W}$ is a circ-UMFF with $\boldsymbol{u}, \boldsymbol{v_1}, \boldsymbol{v_2}, \ldots, \boldsymbol{v_t} \in \mathcal{W}$, $t \geq 1$. Let $\boldsymbol{v} = \boldsymbol{v_1v_2}\cdots\boldsymbol{v_t}$, $\boldsymbol{w} = \boldsymbol{uv}$ and suppose further that $F_{\mathcal{W}}(\boldsymbol{v}) = \boldsymbol{v_1v_2}\cdots\boldsymbol{v_t}$, $F_{\mathcal{W}}(\boldsymbol{w}) = \boldsymbol{w}$. Then $\boldsymbol{uv_1v_2}\cdots\boldsymbol{v_i} \in \mathcal{W}$ for every $i \in 1..t-1$.*

Proof. Recall from Theorem 2 that $F_{\mathcal{W}}(\boldsymbol{v}) = \boldsymbol{v_1v_2}\cdots\boldsymbol{v_t}$ if and only if $\boldsymbol{v_1} \geq_{\mathcal{W}} \boldsymbol{v_2} \geq_{\mathcal{W}} \cdots \geq_{\mathcal{W}} \boldsymbol{v_t}$. Since $F_{\mathcal{W}}(\boldsymbol{w}) = \boldsymbol{w}$ it follows from Lemma 3 that $\boldsymbol{u} <_{\mathcal{W}} \boldsymbol{w} <_{\mathcal{W}} \boldsymbol{v_t}$, and so, since $\boldsymbol{v_t} \leq_{\mathcal{W}} \boldsymbol{v_1}$, that $\boldsymbol{u_1} = \boldsymbol{uv_1} \in \mathcal{W}$ (Definition 4 and

Theorem 1(4)). But since $\boldsymbol{w} = \boldsymbol{u_1 v_2} \cdots \boldsymbol{v_t} \in \mathcal{W}$, it follows, again from Theorem 2 and Lemma 3, that $\boldsymbol{u_1} <_{\mathcal{W}} \boldsymbol{v_2}$, hence $\boldsymbol{u_2} = \boldsymbol{u_1 v_2} \in \mathcal{W}$. Continuing in this way, we conclude that $\boldsymbol{u_{t-1}} = \boldsymbol{u v_1 v_2} \cdots \boldsymbol{v_{t-1}} \in \mathcal{W}$. This completes the proof. □

Lemma 8. *Suppose $\mathcal{W}$ is a circ-UMFF with $\boldsymbol{u_1}, \boldsymbol{u_2}, \ldots, \boldsymbol{u_s}, \boldsymbol{v} \in \mathcal{W}$, $s \geq 1$. Let $\boldsymbol{u} = \boldsymbol{u_1 u_2} \cdots \boldsymbol{u_s}$, $\boldsymbol{w} = \boldsymbol{uv}$ and suppose further that $F_{\mathcal{W}}(\boldsymbol{u}) = \boldsymbol{u_1 u_2} \cdots \boldsymbol{u_s}$, $F_{\mathcal{W}}(\boldsymbol{w}) = \boldsymbol{w}$. Then $\boldsymbol{u_i u_{i+1}} \cdots \boldsymbol{u_s v} \in \mathcal{W}$ for every $i \in 2..s$.*

Proof. By Theorem 2, $\boldsymbol{u_1} \geq_{\mathcal{W}} \boldsymbol{u_2} \geq_{\mathcal{W}} \cdots \geq_{\mathcal{W}} \boldsymbol{u_s}$. Since $F_{\mathcal{W}}(\boldsymbol{uv}) = \boldsymbol{w}$, it follows from Lemma 3 that $\boldsymbol{u_1} <_{\mathcal{W}} \boldsymbol{w} <_{\mathcal{W}} \boldsymbol{v}$, and so, since $\boldsymbol{u_s} \leq_{\mathcal{W}} \boldsymbol{u_1}$, that $\boldsymbol{v_1} = \boldsymbol{u_s v} \in \mathcal{W}$ (Definition 4 and Theorem 1(4)). By Theorem 2 and Lemma 3 again, $\boldsymbol{u_1} <_{\mathcal{W}} \boldsymbol{w} <_{\mathcal{W}} \boldsymbol{v_1}$, and therefore, since $\boldsymbol{u_{s-1}} \leq_{\mathcal{W}} \boldsymbol{u_1}$, we conclude that $\boldsymbol{v_2} = \boldsymbol{u_{s-1} v_1} \in \mathcal{W}$. Continuing in this way, we find $\boldsymbol{v_{s-1}} = \boldsymbol{u_2 u_3} \cdots \boldsymbol{u_s v} \in \mathcal{W}$. □

3 Generic Sequential Algorithm

The generic parallel algorithm for factorization with respect to circ-UMFFs outlined in Section 4 depends for its execution on parallel invocations of a sequential factorization algorithm. Thus in Figure 1 we give pseudocode for Algorithm DDIS that in $O(n^3)$ time computes the sequential factorization $F_{\mathcal{W}}(\boldsymbol{x})$ of a given string $\boldsymbol{x} = \boldsymbol{x}[1..n]$ in terms of a known circ-UMFF $\mathcal{W}$. This timing depends

```
— Algorithm DDIS
i←1
while i ≤ n do
    j←i; k←i; β[j]←0; jmax←i

    while j ≤ n do
        if IN-W(x, β, i, j) then
            jmax←j; j←j+1; jlen←j−i
            while x[j] = x[k] do
                β[j]←β[j−1]+1; j←j+1; k←k+1
            jfreq← ⌊(k−i)/jlen⌋ ; k←i
        else
            j←j+1
         — Border computation.
        b←β[j−1]+1
        while b > 1 and x[j] ≠ x[i + b − 1] do
            b←β[b−1]+1
        if x[j] = x[i + b − 1] then β[j]←b else β[j]←0

    output jmax
    for j'←1 to jfreq do
        jmax←jmax+jlen; output jmax
    i←jmax+1
```

Fig. 1. Given a string $\boldsymbol{x}$ and a circ-UMFF $\mathcal{W}$, output $F_{\mathcal{W}}(\boldsymbol{x})$

upon being able to identify in $O(j-i)$ time whether or not the current substring $\boldsymbol{x}[i..j]$ of $\boldsymbol{x}$ is an element of $\mathcal{W}$ (the Boolean function IN-$\mathcal{W}$, described below in Figure 2).

Algorithm DDIS performs a left-to-right scan of $\boldsymbol{x}$, at each position i identifying the largest integer $j = jmax$ such that $\boldsymbol{x}[i..j] \in \mathcal{W}$. This is done by scanning all positions $j \in i..n$. For each candidate value of $jmax$ found, the corresponding length $jlen$ of the factor is computed, and in addition the maximum number $jfreq$ of repeating occurrences of the factor is counted. For each range $i..n$ in $\boldsymbol{x}$ the border array $\beta[i..n]$ is incrementally computed to facilitate determining whether or not the current substring $\boldsymbol{x}[i..j] \in \mathcal{W}$ (Theorem 1(1)).

```
function IN-W(x, β, i, j)
  — A single letter always belongs to the UMFF.
if j = i then
      INC←TRUE; return TRUE
  — Apply Lemma 3.
elsif INC and x[j-1] <_W x[j] then
      return TRUE
else
      INC←FALSE
        — Apply Theorem 1(1) and Lemma 3.
      if β[j] > 0 or x[j] <_W x[i] then
          return FALSE
        — Apply rules specific to the circ-UMFF W.
      else return BELONGS(W, x[i..j])
```

Fig. 2. Determine whether or not $\boldsymbol{x}[i..j] \in \mathcal{W}$

The Boolean function IN-$\mathcal{W}$ (Figure 2) tries first to apply rules that apply to all circ-UMFFs, each of them implemented in constant time per position j. These rules may involve letter comparisons that without loss of generality are decided using the order of the underlying alphabet. In case these rules are insufficient to determine whether or not $\boldsymbol{x}[i..j] \in \mathcal{W}$, rules specific to the current circ-UMFF must be applied, that we suppose require a total of $O(j-i)$ time.

For examples of customizing and optimizing this generic algorithm, we have Duval's well known linear Lyndon factorization algorithm [Du83] (an alternative exposition of this algorithm is given in [S03], pages 158–166), and the linear-time V-word factorization algorithm of Section 5.

4 Generic Parallel Algorithm

In this section we give an overview of the CRCW parallel RAM factorization algorithm over a circ-UMFF $\mathcal{W}$. As explained earlier, the structure of this algorithm mirrors closely that of the corresponding Lyndon algorithm described in [DIS94]. The reason for this is that the theoretical results on which the algorithm's correctness and complexity is based hold also in this more general case;

specifically, Corollary 1 (along with Theorem 3) and Theorems 1 and 2, Lemmas 3, 6 and 7–8, and in the case of a Flight Deck circ-UMFF Lemma 2. Since much of the pseudocode is virtually identical to that of the Lyndon algorithm, we explain here the main ideas and modifications and refer the reader to [DIS94] for fuller details.

The algorithm forms a binary comparison tree T. The input string $\boldsymbol{x}$ is partitioned into consecutive substrings (blocks) that form the leaves of T, each of which is factored. At each level of T, factors in pairwise adjacent siblings are concatenated using $\mathcal{W}$-order wherever possible so as to achieve maximal factors (as in Definition 2). The computation iterates from the leaves upwards towards the root, where finally we have the required unique factorization of $\boldsymbol{x}$.

The algorithm consists of the following five main procedures, the first two executed in parallel, the others sequential:

Initialization. Partition the input string into $\lceil n/\log n \rceil$ blocks, each of length $\lceil \log n \rceil$; factor each block based on $\mathcal{W}$-order using a sequential algorithm (see Section 3) executed in parallel.

Contest between Two Substrings. A parallel computation moving upwards in T that compares factors in adjacent siblings in T.

Right-Extension of a Factor. Search T recursively downwards to locate the blocks whose factorizations need to be recalculated. The method depends on finding the concat-point by binary search: deals primarily with the righthand of a pair of substrings.

Left-Extension of a Factor. Similarly search T to locate the block which needs to be refactored: deals primarily with the lefthand of a pair of substrings.

Final-Extension of a Factor. Recompute the factorization of a substring using binary search to achieve a left/right concatenation.

Embedded in the procedures Contest, Right-Extension, Left-Extension and Final-Extension is a procedure **String Comparison** (depending on implementation) that determines the $\mathcal{W}$-order (Definition 4) of two strings (for example, lexorder in the case of the Lyndon factorization).

4.1 Initialization

Given a string $\boldsymbol{x} = \boldsymbol{x}[1..n]$, we compute $F_{\mathcal{W}}(\boldsymbol{x})$ in parallel by partitioning $\boldsymbol{x}$ into $k = \lceil n/\log n \rceil$ adjacent ***blocks*** $\boldsymbol{X_i}$, $i = 1, 2, \ldots, k$, each of length $\lceil \log n \rceil$. For simplicity of presentation assume WLOG that k is a power of 2 (achievable by appropriate padding of the string).

Assign a processor p_i to each block $\boldsymbol{X_i}$, $1 \leq i \leq k$. Processor p_i computes $F_{\mathcal{W}}(\boldsymbol{X_i})$ using a sequential factorization algorithm for $\mathcal{W}$ — either an efficient algorithm for the given circ-UMFF is supplied (as in Duval's algorithm for Lyndon decomposition [Du83], or [D11]), or if no such algorithm for $\mathcal{W}$ is known, the generic sequential algorithm of Section 3 can be used. These sequential factorizations are computed in parallel for all blocks. The sequential algorithm applied to a block $\boldsymbol{X_i}$ yields the start (and end) positions in $\boldsymbol{X_i}$ of all of its maximal factors over $\mathcal{W}$. The initialization therefore requires $\lceil n/\log n \rceil$ processors and time $O(S(\log n))$, where $S(\log n)$ is the worst-case sequential time complexity for a single block ($S(\log n) = \log n$ for Lyndon decomposition).

Let a_i be the start position in $\boldsymbol{x}$ of the rightmost occurrence in block $\boldsymbol{X_i}$ of the least factor of $F_{\mathcal{W}}(\boldsymbol{X_i})$. Each a_i can be determined in constant time as a byproduct of the sequential factorization algorithm executed for block $\boldsymbol{X_i}$.

4.2 Data Structures

The data structures used are similar to those described in [DIS94] (though we replace the integer array FA of [DIS94] by a bit array):

* The blocks $\boldsymbol{X_i}$ are the leaves of the full binary tree T. T is the basic data structure of the procedure Contest described below and underlies the processor allocation technique.
* A bit array $FA = FA[1..n]$ specifies the ***start positions*** and ***end positions*** of factors of $\boldsymbol{x}$ that are output at any stage of the computation. Thus any factor can be described using only constant space. Suppose that a provisional factorization $F'_{\mathcal{W}}(\boldsymbol{x}) = \boldsymbol{w_1 w_2} \cdots \boldsymbol{w_m}$ has been computed, where

  $$\boldsymbol{w_1} = \boldsymbol{x}[1..k_1],\ \boldsymbol{w_2} = \boldsymbol{x}[k_1{+}1..k_2], \ldots, \boldsymbol{w_m} = \boldsymbol{x}[k_{m-1}{+}1..k_m].$$

  The start and end positions $(k_t{+}1, k_{t+1})$ of $\boldsymbol{w_t}$ are specified as follows:

  $$FA[j] = \begin{cases} 1\ if\ j = k_t + 1, \\ 0\ otherwise; \end{cases}$$

  where $0 \le t \le m-1$, $k_0 = 0$, $k_m = n$. A sequence of pointers *next* links adjacent factors: for $FA[j] = 1$, $next(j) = j'$, where j' is the least position greater than j such that $FA[j'] = 1$ ($j' = n{+}1$ if no such position exists); the *end* position of the factor beginning at j is $j' - 1$. The array is initialized by the parallel execution of Initialization, whereby processor p_i sets the bits in the ith "block" in FA corresponding to $F_{\mathcal{W}}(\boldsymbol{X_i})$.
* Various data structures may be used for String Comparison $\boldsymbol{u} <_{\mathcal{W}} \boldsymbol{v}$ required in the parallel procedure Contest and the sequential procedures Right-Extension, Left-Extension and Final-Extension. In [DIS94] the Four Russians technique [IS92] is proposed to compare strings of length n on a bounded alphabet in $O(1)$ time, while for an unbounded alphabet the parallel construction of a merged suffix tree using the CRCW PRAM model [IS92] is proposed that can be constructed in $O(\log n \log\log n)$ time using $O(n/\log n)$ processors; using this tree, sequential comparison requires $O(\log\log n)$ time. It is because the CRCW PRAM model may be required in particular cases such as [DIS94] that we have proposed its usage here; otherwise the EREW model would suffice. For V-order (see Section 5), the use of a doubly-linked list [DDS11] allows string comparison in time proportional to string length. Otherwise the methodology and data structures of function IN-$\mathcal{W}$ can be used.

4.3 Contest between Factorizations of Adjacent Blocks

This procedure depends on the monotonicity of factors in a factorization (Theorem 1(4), Lemma 3, Theorem 2), as well as on the fact that they cannot be subdivided (Corollary 1).

Processing starts at the leaves by comparing the adjacent blocks in parallel: for $i = 1, 3, \ldots, k-1$, each last word w_i starting at a_i in $\boldsymbol{X_i}$ stages a contest with the corresponding last word w_{i+1} starting at a_{i+1} in $\boldsymbol{X_{i+1}}$. The result of each contest is the starting position $a_{i'}$ of the last factor in the factorization of the doubled substring of size $2\lceil \log n \rceil$, along with the updated array FA. At each subsequent stage the adjacent doubled blocks are similarly refactored. At the root — that is, at stage $\lceil \log n \rceil$ — we have computed the required factorization $F_{\mathcal{W}}(\boldsymbol{x})$. Note that only at the leaves are substrings of size $\lceil \log n \rceil$ for all blocks $\boldsymbol{X_i}$ stored; at internal tree nodes, only the index a_i of the rightmost factor of the associated substring is stored.

Figure 3 gives a corrected version of the corresponding procedure in [DIS94], while generalizing to circ-UMFFs along with introducing a Flight Deck condition. Note that $w_r(s,t)$ denotes the substring of $\boldsymbol{x}$ which starts at index a_r and has length $a_t - a_s$ (the rightmost factor in a block starting at index a_i is simply w_i).

procedure Contest ($F_{\mathcal{W}}(\boldsymbol{X_i})$, a_i $\forall i$)

Input: $F_{\mathcal{W}}$(block $\boldsymbol{X_i}$) and a_i for $1 \le i \le \lceil n/log\ n \rceil$ specified by the array FA.
Output: the factorization $F_{\mathcal{W}}(\boldsymbol{x})$ given in the array FA.

```
begin
    repeat
        forall distinct pairs of siblings a_i, a_j with i < j pardo
            if w_i ≥_W w_j then
                a_j becomes the parent node; (Lemma 3)
                Right-Ext (w_i(i, j), a_j); (Lemma 7, Theorem 2)
                elsif W is a Flight Deck circ-UMFF and
                w_i[1] <_W w_j[1] then
                    a_i becomes the parent node; (Lemma 2)
                    next(FA[a_i]) := next(FA[a_j]);
                    end(FA[a_i]) := end(FA[a_j]);
                else Right-Ext (w_i(i, end(FA[a_j]) + 1), end(FA[a_j]) + 1);
            odpar
            when all computations are complete move up to the next level
            of the tree;
    until the root is reached
end
```

Fig. 3. The contest between factorizations of adjacent blocks

The following example shows the necessity of extending to the rightmost factor of the right substring in Contest. Note that even though the Lyndon words are Type Flight Deck, nevertheless the condition of Lemma 2 is not satisfied.

Example 5. *Let $\mathcal{L}$ be the Lyndon circ-UMFF. Suppose $F_{\mathcal{L}}(\boldsymbol{u}) = \boldsymbol{u_1} = ab$; and $F_{\mathcal{L}}(\boldsymbol{v}) = \boldsymbol{v_1} \geq_{\mathcal{L}} \boldsymbol{v_2}$, where $\boldsymbol{v_1} = b$ and $\boldsymbol{v_2} = ac$. Then $ab <_{\mathcal{L}} b$ and since $abb <_{\mathcal{L}} ac$ we must extend again to the right to achieve the final factorization, namely $F_{\mathcal{L}}(\boldsymbol{uv}) = abbac$.*

4.4 Right-Extension

Given $F_{\mathcal{W}}(\boldsymbol{u}) = \boldsymbol{u_1} \geq_{\mathcal{W}} \cdots \geq_{\mathcal{W}} \boldsymbol{u_s}$ and $F_{\mathcal{W}}(\boldsymbol{v}) = \boldsymbol{v_1} \geq_{\mathcal{W}} \cdots \geq_{\mathcal{W}} \boldsymbol{v_t}$ we perform a "right-extension" from $\boldsymbol{u_s}$ to the factors of $F_{\mathcal{W}}(\boldsymbol{v})$. Note that we extend right repeatedly across a factorization as compared to a factor as in Lemma 6. As in [DIS94], procedure Right-Extension (Right-Ext) is a recursive sequential algorithm based on the divide and conquer technique, as suggested by Theorem 3.

4.5 Left-Extension

Left-Extension is a recursive sequential procedure which possibly modifies the factorization associated with the lefthand of a pair of sibling tree nodes. Similar to Right-Extension it applies divide and conquer to travel down the binary comparison tree T to locate a leaf. Right-Extension may have concatenated the smallest factor (in $\mathcal{W}$-order) in the left substring with a sequence of the largest factors in the right substring. Hence Left-Extension must locate the leftmost factor in the left substring satisfying the conditions of Theorem 3.

4.6 Final-Extension

This corresponds to the procedure of the same name in [DIS94], but implemented here with binary search. It is invoked at leaf level where the size of a block is $\lceil \log n \rceil$. For right concatenation, the procedure begins at position a_i marking the rightmost factor $\boldsymbol{w_i}$ in the lefthand string L, then applies binary search up to position a_j in the righthand string R to determine the unique concat-point k such that $\boldsymbol{w_i v_1 v_2} \cdots \boldsymbol{v_k} \in \mathcal{W}$ and $\boldsymbol{w_i v_1 v_2} \cdots \boldsymbol{v_k v_{k+1}} \notin \mathcal{W}$, as in Theorem 3 and Lemma 7. The output is the updated factorization defined by FA over the range $a_i..a_j - 1$. Left concatenation analogously starts with the possibly extended $\boldsymbol{w_i}$ and applies binary search leftwards to determine the leftmost position in L that can be the start point of the factor, as in Theorem 3 and Lemma 8. Since binary search requires logarithmic time, the time complexity is $O\big(\log(\lceil \log n \rceil) IN_{\mathcal{W}}(\log n)\big)$, where $IN_{\mathcal{W}}$ denotes the time required by function IN-$\mathcal{W}$.

4.7 Correctness and Complexity

Correctness follows from Theorems 1, 2, 3, and Corollary 1, along with Lemmas 3, 7, 8 applied to the analogue of the algorithm given in [DIS94].

The worst-case time of the parallel algorithm depends on:

- the cost of initialization (parallel sequential algorithms applied to blocks): $O(S(\log n))$.
- the time required to refactor a pair of factorizations using $\lceil n/\log n\rceil$ processors:
$$O(\lceil \log n\rceil IN_{\mathcal{W}}(\log n)\lceil n/\log n\rceil) = O(nIN_{\mathcal{W}}(\log n)).$$

So, for example, applying linear sequential Lyndon decomposition, the complexity of the Lyndon CRCW PRAM algorithm for bounded alphabets in [DIS94] is $O(log\ n)$, and since the number of processors is $\lceil n/log\ n\rceil$, the cost is $O(n)$ and optimal.

5 Parallel Lyndon-Like V-Order Factorization

In this section we show how to customize the generic parallel algorithm of Section 4 so as to factor strings over the V-order circ-UMFF $\mathcal{V}$ into maximal V-words ([DaD96], [DD03], [DDS11], [DDS12]). First we define these terms, and then explain the algorithmics.

Let $\boldsymbol{u} = u_1u_2...u_n$ be a string over a totally ordered alphabet Σ. Define $h \in 1..n$ by $h = 1$ if $u_1 \leq u_2 \leq \cdots \leq u_n$; otherwise, by the unique value h such that $u_{h-1} > u_h \leq u_{h+1} \leq u_{h+2} \leq \ ... \leq u_n$. Let $\boldsymbol{u}^* = u_1u_2...u_{h-1}u_{h+1}...u_n$, where the star $*$ indicates deletion of the "V" letter u_h. Write $\boldsymbol{u}^{s*}$ for $(...(\boldsymbol{u}^*)^*...)^*$ with $s \geq 0$ stars [1]. Let $\mu\boldsymbol{u} = max\{u_1, u_2, ..., u_n\}$, say $\mu\boldsymbol{u} = g$, and let $k = \nu\boldsymbol{u}$ be the number of occurrences of g in $\boldsymbol{u}$. Then the sequence $\boldsymbol{u}, \boldsymbol{u}^*, \boldsymbol{u}^{2*}, ...$ ends $g^k, ..., g^2, g^1, g^0 = \boldsymbol{\varepsilon}$. In the *star tree* each string $\boldsymbol{u}$ over Σ labels a vertex, and there is a directed edge from $\boldsymbol{u}$ to $\boldsymbol{u}^*$, with $\boldsymbol{\varepsilon}$ as the root.

Definition 8. *We define* V-order $\prec$ *between distinct strings* $\boldsymbol{u}, \boldsymbol{v}$. *First* $\boldsymbol{v} \prec \boldsymbol{u}$ *if* $\boldsymbol{v}$ *is in the path* $\boldsymbol{u}, \boldsymbol{u}^*, \boldsymbol{u}^{2*}, ..., \boldsymbol{\varepsilon}$. *If* $\boldsymbol{u}, \boldsymbol{v}$ *are not in a path, there exist smallest* s, t *such that* $\boldsymbol{u}^{(s+1)*} = \boldsymbol{v}^{(t+1)*}$. *Put* $\mathbf{c} = \boldsymbol{u}^{s*}$ *and* $\mathbf{d} = \boldsymbol{v}^{t*}$; *then* $\mathbf{c} \neq \mathbf{d}$ *but* $|\mathbf{c}| = |\mathbf{d}| = m$ *say. Let* j *be the greatest* i *in* $1 \leq i \leq m$ *such that* $\mathbf{c}[i] \neq \mathbf{d}[i]$. *If* $\mathbf{c}[j] < \mathbf{d}[j]$ *in* Σ *then* $\boldsymbol{u} \prec \boldsymbol{v}$. *Clearly* $\prec$ *is a total order.*

We introduce a natural analogue of Lyndon words, derived from V-order, known as ***V-words***, which like Lyndon words are of Type Flight Deck (Definition 5); we denote the set of V-words as the circ-UMFF $\mathcal{V}$.

Definition 9. *([DD03]) A string* $\boldsymbol{w}$ *over* Σ *is a* ***V-word*** *if it is the unique minimum in* V*-order* $\prec$ *among the set of rotations of* $\boldsymbol{w}$.

A fundamental operation in the algorithm is procedure String Comparison for determining concatenation (Definition 4). In [DDS11] we apply the technique of "longest matching suffix" using a doubly linked list to compute this comparison in linear time.

[1] Note that this star operator, as defined in [DaD96], [DD03], [DDS11] and [DDS12], is distinct from the Kleene star operator.

Corresponding to Section 4.1, the initialization process of the parallel algorithm first partitions the input string into consecutive blocks; each block is then sequentially factored using a linear V-word factorization algorithm (procedure VF(x) in [DDS11]), and the maximum positions a_i are computed — all with cost $O(\log n)$.

— Parallel Lyndon-Like V-order Factorization Algorithm

Initialization (Apply procedure VF(x) to each block in parallel)
begin
  repeat
    forall distinct pairs of sibling strings
      $((u_1 \geq_V u_2 \geq_V ... \geq_V u_i), (v_1 \geq_V v_2 \geq_V ... \geq_V v_j))$
    pardo
    $u_{i'}$:= Right-Extension($u_i, v_1 \geq_V v_2 \geq_V ... \geq_V v_j$); (Corollary 1)
    Left-Extension($u_1 \geq_V u_2 \geq_V ... \geq_V u_{i-1}, u_{i'}$) (Theorem 2)
  until the root is reached
end

Fig. 4. Compute the factorization of $x[1..n]$ into V-words in parallel

As the parallel algorithm progresses up the comparison tree T (Figure 4), a processor is assigned to a distinct pair of sibling nodes each containing pointers to the least factors in factored substrings, $(u_1 \geq_V u_2 \geq_V \cdots \geq_V u_i)$ and $(v_1 \geq_V v_2 \geq_V \cdots \geq_V v_j)$, a_i and a_j, say. Procedure Right-Extension applies binary search (Lemma 7) to find the concat-point k, $1 \leq k \leq j$, where $u_i v_1 v_2 \cdots v_k \in \mathcal{V}$ and $u_i v_1 v_2 \cdots v_k \geq_{\mathcal{V}} v_{k+1} \geq_{\mathcal{V}} \cdots \geq_{\mathcal{V}} v_j$. Procedure Left-Extension (Lemma 8) then checks the left sibling node for any further concatenation resulting in $u_1 \geq_{\mathcal{V}} u_2 \geq_{\mathcal{V}} \cdots \geq_{\mathcal{W}} u_{i''-1} \geq u_{i''} \cdots u_i v_1 v_2 \cdots v_k \geq_{\mathcal{V}} \cdots \geq_{\mathcal{V}} v_j$.

This progression up the tree is achieved by applying Corollary 1, which we will demonstrate for the circ-UMFF $\mathcal{V}$ and $m = 2$. Suppose the strings $u, v \in \mathcal{V}$, so that they have V-form: $u = fx_1fx_2...fx_j$ and $v = gy_1gy_2...gy_k$ (every letter in x_i, y_j is less than f, g respectively) and they satisfy *lex-extension* $\prec_{LEX}$ (lexorder of substrings, which are compared using V-order) [DD03], [DDS11]. If $u = v$ then to factor uv maximally, with the least number of factors, yields $u \geq_{\mathcal{V}} v$, for otherwise we get a bordered word. If $u <_{\mathcal{V}} v$ then, applying Definition 4 and Theorem 1(3), the factorization is the string (uv), and note here that the letters f, g satisfy $f \succeq g$ and, if they are distinct, $f <_{\mathcal{V}} g$. Consider the case $u >_{\mathcal{V}} v$. Since $uv \notin \mathcal{V}$, therefore by Theorem 1(3), $vu \in \mathcal{V}$ and hence $f \preceq g$, $f \geq_{\mathcal{V}} g$. If $f \prec g$, $f >_{\mathcal{V}} g$, then from their V-form, the factorization of uv is $u >_{\mathcal{V}} v$. So assume $f = g$, and $w = x_1x_2...x_jy_1y_2...y_k$ is not Lyndon under $\prec_{LEX}$. Hence some distinct rotation r of w satisfies $r \prec_{LEX} w$. Clearly r cannot start with x_1. Further, r cannot start with $x_s \neq x_1$ for otherwise u is not a V-word satisfying the Lyndon condition. Similarly r cannot start with $y_t \neq y_1$ for otherwise v is not a V-word. Finally, r cannot start with a proper suffix of any substring $\boldsymbol{x_p}$ or $\boldsymbol{y_q}$, for otherwise we get a bordered word from Lemma 1. Hence $\boldsymbol{r}$ starts with $\boldsymbol{y_1}$ and so the factorization of $\boldsymbol{uv}$ is $(\boldsymbol{u})(\boldsymbol{v})$.

Theorem 2 together with Lemmas 3 and 7 can be applied to binary search for the concat-point at the final level of recursion as follows. Suppose that $\boldsymbol{u_i} \geq_{\mathcal{V}} \boldsymbol{v_j}$, then by Lemma 3, $\boldsymbol{u_i v_1 v_2 ... v_j} \notin \mathcal{V}$. We then subdivide the $\boldsymbol{v}$ factors giving a potential concat-point k, $\boldsymbol{u_i} <_{\mathcal{V}} \boldsymbol{v_k}$, and then test whether $\boldsymbol{u_i v_1 v_2 ... v_k} \in \mathcal{V}$; if not, continue subdividing. Similarly Lemma 8 can then be applied with binary search to extend leftwards and complete the refactoring.

The time required is the cost of initialization, again $O(\log n)$, plus the worst-case cost of refactoring a pair of strings using $O(n/\log n)$ processors: binary search for concat-points ($O(\log n)$) and V-order factorization of leaves (also $O(\log n)$). Thus, like Lyndon factorization, the cost is an optimal $O(n)$.

6 Future Research

In this paper we have extended parallel factorization from Lyndon words to the more general class of circ-UMFFs. In particular, we have used this algorithmic technique to implement PRAM factorization based on V-order. We believe it would be of interest to customize our sequential and parallel generic algorithms to other Binary, Flight Deck and Acrobat classes of circ-UMFFs (see [DD03, DDS09]). We conclude by conjecturing that the set "$\mathcal{W}(acaab)$" is a new ternary Flight Deck circ-UMFF, called "$\mathcal{W}(acaab)$" since it contains $acaab$, and is defined by starting with the words: $\mathcal{W} = \{a <_{\mathcal{W}} b <_{\mathcal{W}} c,\ a <_{\mathcal{W}} ab,\ a <_{\mathcal{W}} ac, ...,\ ac <_{\mathcal{W}} ab,\ ac <_{\mathcal{W}} aab, ...\}$ (see ([DD08], Section 5 for the construction of a circ-UMFF).

References

[BGM11] Breslauer, D., Grossi, R., Mignosi, F.: Simple Real-Time Constant-Space String Matching. In: Giancarlo, R., Manzini, G. (eds.) CPM 2011. LNCS, vol. 6661, pp. 173–183. Springer, Heidelberg (2011)

[BLPR09] Brlek, S., Lachaud, J.-O., Provençal, X., Reutenauer, C.: Lyndon + Christoffel = digitally convex. Pattern Recognition 42(10), 2239–2246 (2009)

[BMP07] Brlek, S., Melançon, G., Paquin, G.: Properties of the extremal infinite smooth words. Discrete Math. Theor. Comput. Sci. 9(2), 33–50 (2007)

[C04] Chemillier, M.: Periodic musical sequences and Lyndon words. Soft Computing - A Fusion of Foundations, Methodologies and Applications 8(9), 611–616 (2004) ISSN 1432-7643 (Print) 1433-7479 (Online)

[CDP05] Crochemore, M., Désarménien, J., Perrin, D.: A note on the Burrows-Wheeler transformation. Theoret. Comput. Sci. 332(1-3), 567–572 (2005)

[CFL58] Chen, K.T., Fox, R.H., Lyndon, R.C.: Free differential calculus, IV — the quotient groups of the lower central series. Ann. Math. 68, 81–95 (1958)

[CP91] Crochemore, M., Perrin, D.: Two-way string-matching. J. Assoc. Comput. Mach. 38(3), 651–675 (1991)

[CT03] Chemillier, M., Truchet, C.: Computation of words satisfying the "rhythmic oddity property" (after Simha Arom's works). Inf. Proc. Lett. 86, 255–261 (2003)

[D11] Daykin, D.E.: Algorithms for the Lyndon unique maximal factorization. J. Combin. Math. Combin. Comput. 77, 65–74 (2011)

[DaD96] Danh, T.-N., Daykin, D.E.: The structure of V-order for integer vectors. In: Hilton, A.J.W. (ed.) Congressus Numerantium, vol. 113, pp. 43–53. Utilitas Mat. Pub. Inc., Winnipeg (1996)

[dB46] de Bruijn, N.G.: A combinatorial problem. Koninklijke Nederlandse Akademie v. Wetenschappen 49, 758–764 (1946)

[DD03] Daykin, D.E., Daykin, J.W.: Lyndon-like and V-order factorizations of strings. J. Discrete Algorithms 1, 357–365 (2003)

[DD08] Daykin, D.E., Daykin, J.W.: Properties and construction of unique maximal factorization families for strings. Internat. J. Found. Comput. Sci. 19(4), 1073–1084 (2008)

[DDS09] Daykin, D.E., Daykin, J.W., (Bill) Smyth, W.F.: Combinatorics of Unique Maximal Factorization Families (UMFFs). Fund. Inform. 97(3), 295–309 (2009); Special Issue on Stringology, Janicki, R., Puglisi, S.J., Rahman, M.S. (eds.)

[DDS11] Daykin, D.E., Daykin, J.W., Smyth, W.F.: String Comparison and Lyndon-Like Factorization Using V-Order in Linear Time. In: Giancarlo, R., Manzini, G. (eds.) CPM 2011. LNCS, vol. 6661, pp. 65–76. Springer, Heidelberg (2011)

[DDS12] Daykin, D.E., Daykin, J.W., Smyth, W.F.: A linear partitioning algorithm for Hybrid Lyndons using V-order. Theoret. Comput. Sci. (in press, 2012), doi:10.1016/j.tcs.2012.02.001

[DIS94] Daykin, J.W., Iliopoulos, C.S., Smyth, W.F.: Parallel RAM algorithms for factorizing words. Theoret. Comput. Sci. 127(1), 53–67 (1994)

[DS12] Daykin, J.W., Smyth, W.F.: A bijective variant of the Burrows-Wheeler transform using V-Order (submitted)

[Du83] Duval, J.P.: Factorizing words over an ordered alphabet. J. Algorithms 4, 363–381 (1983)

[GS09] Gil, J., Scott, D.A.: A bijective string sorting transform (submitted)

[IS92] Iliopoulos, C.S., Smyth, W.F.: Optimal algorithms for computing the canonical form of a circular string. Theoret. Comput. Sci. 92(1), 87–105 (1992)

[J92] JaJa, J.: Introduction to Parallel Algorithms. Addison-Wesley (1992) ISBN 0-201-54856-9

[K09] Kufleitner, M.: On bijective variants of the Burrows-Wheeler transform. In: Proc. Stringology, pp. 65–79 (2009)

[L83] Lothaire, M.: Combinatorics on Words. Addison-Wesley, Reading (1983), 2nd edn. Cambridge University Press, Cambridge (1997)

[N08] `http://developer.nvidia.com/cuda`

[P05] Perret, L.: A chosen ciphertext attack on a public key cryptosystem based on Lyndon words. In: Proceedings of International Workshop on Coding and Cryptography (WCC 2005), pp. 235–244 (2005)

[R93] Reutenauer, C.: Free Lie Algebras, London Math. Soc. Monographs, New Ser., vol. 7. Oxford University Press (1993)

[S03] Smyth, B.: Computing Patterns in Strings. Pearson (2003)

[SMDS94] Siromoney, R., Matthew, L., Dare, V.R., Subramanian, K.G.: Infinite Lyndon words. Inform. Process. Lett. 50, 101–104 (1994)

[V08] `http://www.umiacs.umd.edu/~vishkin/index.shtml`

On Data Recovery in Distributed Databases

Sergei L. Bezrukov[1], Uwe Leck[1], and Victor P. Piotrowski[2]

[1] Dept. of Mathematics and Computer Science, University of Wisconsin-Superior
{sbezruko,uleck}@uwsuper.edu
[2] National Science Foundation, Arlington, VA, USA
vpiotrow@nsf.gov

Dedicated to the memory of Rudolf Ahlswede

Abstract. We present an approach for data encoding and recovery of lost information in a distributed database system. The dependencies between the informational redundancy of the database and its recovery rate are investigated, and fast recovery algorithms are developed.

Keywords: database, iterative code, RAID.

1 Introduction

We consider the graph-theoretical model of a distributed database system by which we mean a graph G whose vertices correspond to the nodes of the database where the information is stored (e.g. computer disk drives or computing centers). These nodes are interconnected by a network modeled by the edges of G. Each node of the database can be in one of two states: healthy or faulty. If a node is in the healthy state, one has full access to the data stored in it. Otherwise, no data stored in that node is accessible. We investigate the possibility to recover this data by using the data stored in some other nodes.

One of the simplest ways to improve the database reliability is frequently used in practice: duplication of the data stored in each node in other nodes. This provides the fastest recovery rate in terms of the number of other node accesses, as one just needs to read the data from some node mirroring the faulty one. However, this approach noticeably increases the database's informational redundancy, since if s nodes could fail simultaneously, then it is necessary to duplicate s times the data stored in each node of the network.

The main goal of our research is to decrease the redundancy of the database system by decreasing the recovery rate while keeping it within acceptable limits. We distinguish two recovery modes: online and offline. The online recovery starts after every data request to a faulty node. The main goal is to complete the request as soon as possible by recovering just the queried node data and not taking care of recovering the other faulty nodes (if any). In contrast to that, the offline maintenance mode is scheduled at regular intervals (e.g., when the database access rate is low), and its goal is the fastest data recovery for all faulty nodes (if any).

H. Aydinian, F. Cicalese, and C. Deppe (Eds.): Ahlswede Festschrift, LNCS 7777, pp. 419–431, 2013.

To be more specific, we assume that the database network is represented by an n-dimensional $k \times k \times \cdots \times k$ grid R. A similar architecture is used, for example, in the RAID technology. Our approach is based on a coding theory technique and assigning symbols of some large alphabet to the nodes. Thus, we consider the data stored in the system as a word of length k^n over this alphabet. Furthermore, we assume that the information is encoded using n-dimensional iterative codes [2], whose every inner code is capable to recover up to $t \leq k$ erasures. We only deal with combinatorial aspects of the data recovery problem specified below and do not address the encoding/decoding of data and its technical implementation. Without loss of generality, we assume that in the online mode the goal is to recover the node corresponding to the origin of R.

Definition 1. *A collection T of faulty points of R is called a* ***configuration****.*

Each configuration corresponds to the set of erasures in the code word.

Definition 2. *The process of recovery of the erasures in a code word of any inner code is called a* ***decoding step****. After every decoding step the initial configuration T is transformed into a smaller one $T' \subset T$ by deleting from T all points belonging to some line ℓ (i.e., the collection of all points of R that agree in $n-1$ coordinates) provided that $|\ell \cap T| \leq t$.*

We assume that every decoding step takes the same time.

Definition 3. *The configuration T is called* ***t-decodable*** *(or simply* ***decodable*** *if it is clear from the context what t is) if for any point $f \in T$ there exists a sequence of decoding steps $s_1, \ldots, s_r$, (which is called the* ***decoding scheme****) resulting in a configuration T' with $f \notin T'$. The minimum number r of steps required to remove f from T is called the* ***complexity of T with respect to f****. If f=(0,...,0), then r is simply called the* ***complexity of*** *T and denoted by $L(T)$.*

We call a subset $D \subseteq T$ a **deadlock** if for any line ℓ of R the condition $D \cap \ell \neq \varnothing$ implies $|D \cap \ell| > t$. Clearly, no point of of a deadlock D can be eliminated in a decoding step. Obviously, every non-decodable configuration T contains a deadlock. Let $D(T)$ denote the (inclusion-)maximal deadlock of T for non-decodable T. Note that $D(T)$ unique. Indeed, if there are two maximal deadlocks D_1 and D_2, then the set $D' = D_1 \cup D_2$ is a deadlock, too. This contradicts the maximality of D_1 and D_2 unless $D_1 = D_2$.

For the rest of the paper we assume that $n = 2$. We will refer to the horizontal and vertical lines of R also as rows and columns, respectively.

Let $F(m, k, t) = \max(L(T))$, where the maximum is taken over all t-decodable configurations $T \subseteq R$ of size m. In other words, F equals the minimum number of decoding steps needed in the worst case to recover the origin in a decodable configuration of size m. In practice, this is proportional to the recovery time, provided that the access time is the same for each node of R.

It may appear more natural to define $F(m, k, t)$ as the maximum of $L(T)$ over all configurations that allow the recovery of the origin (instead of being completely decodable). Note that this is equivalent to the given definition which follows by the observation that if T is non-decodable but allows the recovery of the origin, then $L(T) = L(T \setminus D(T))$.

2 Preliminary Results

Theorem 1. *Every configuration of size m is decodable if and only if $m \leq (t+1)^2 - 1$.*

Proof. The bound is necessary, since if T is a $(t+1) \times (t+1)$ subgrid of R, then no decoding step can be applied.

The sufficiency of the condition follows from the fact that for any configuration T of size $m < (t+1)^2$, there exists a line ℓ such that $0 < |\ell \cap T| \leq t$. Indeed, by changing the order of rows and columns of T, without loss of generality we can assume that $T \subseteq Q$, where $Q = [0, a_1) \times [0, a_2)$ for some a_1, a_2 with $0 < a_i \leq k$ for $i = 1, 2$, and every line of Q contains some point of T. If every line in Q contains at least $t+1$ points of T then $a_i \geq t+1$ for $i = 1, 2$, so $|T| \geq (t+1)^2$.

Hence, we can perform a decoding step along the line ℓ and get a decodable subconfiguration of T of smaller size. The rest of the proof can done by induction on m. ∎

Note that Theorem 1 can easily be extended to $n \geq 2$, where the inequality becomes $m \leq (t+1)^n - 1$.

Lemma 1. *If T is a t-decodable configuration, then it is possible to remove all points from T in at most $2k - t$ decoding steps.*

Proof. Let T be t-decodable. Consider in some decoding sequence for T the situation when for the $(k-t)$-th time a row has been chosen. (If that situation never occurs, then we are done because T is completely decoded in less than $k-t$ horizontal and at most k vertical steps.) After this, on any column at most t points of T remain which can be decoded by choosing columns only. That is, T can be decoded by choosing $k-t$ rows and at most k columns. ∎

As an immediate consequence we obtain a sharp upper bound for the size of a t-decodable configuration.

Theorem 2. *If T is a t-decodable configuration, then $|T| \leq k^2 - (k-t)^2$.*

Proof. Let T be a t-decodable configuration. As in every decoding step at most t points can be removed from T, Lemma 1 implies

$$|T| \leq (2k-t)t = k^2 - (k-t)^2.$$ ∎

The bound in Theorem 2 is tight, and it is attained for the configuration obtained from R by deleting a $(k-t) \times (k-t)$ subgrid.

By Theorem 1, every configuration of size $m \leq (t+1)^2 - 1$ is decodable. For $(t+1)^2 \leq m \leq k^2 - (k-t)^2$ some configurations of that size m are decodable, and some are not. For m exceeding the threshold stated in Theorem 2, no configuration is decodable.

Theorem 3. *If* $2t \le m \le k^2 - (k-t)^2$, *then*

$$F(m,k,t) \ge \min\left\{\left\lceil \frac{m-1}{2} \right\rceil - t + 2,\ k - t + 1\right\}.$$

Proof. It is straightforward to verify that for every initial segment of length $m \le k^2 - (k-t)^2$ of the following numbering of points its complexity matches the given lower bound.

First we number the points on the axes in alternating order, that is,

$$(0,0),\ (0,1),\ (1,0),\ (0,2),\ (2,0), \ldots, (0,k-1),\ (k-1,0).$$

Then we proceed in similar fashion with the remaining points in the first row and in the first column and number them alternating between the row and the column. In general, after numbering all points of the l-th row and l-th column we proceed with numbering the remaining points of the $(l+1)$-st row and the $(l+1)$-st column in alternating order. ■

Let T be any configuration of m faulty nodes. Denote by $L(T,t)$ the minimum number of decoding steps to delete the origin from T by using the inner codes capable of correcting up to t errors. Let $L(m,t) = \max_{|T|=m} L(T,t)$. In other terms, $L(m,t) = F(m,\infty,t)$.

Theorem 4. $L(m,1) = \lfloor \frac{m-1}{2} \rfloor + 1.$

Proof. Let T be a 1-decodable configuration of maximum complexity for its size. We convert T into a configuration of the same size and with no smaller complexity, all whose points are on the axes.

Assume that the last step in some optimal decoding scheme for T was applied along the y-axis, and denote

$$\begin{aligned} A &= \{(x,0) \in T\}, \\ B &= \{(0,y) \in T\}, \\ C &= \{(x,y) \mid y > 0 \text{ and } (x,0) \in A\}. \end{aligned}$$

We first show that $C \cap T = \varnothing$. Indeed, assume $z \in C \cap T$. If z was not deleted from T in our decoding scheme, we can obviously put it on the y-axis without decreasing the complexity. Otherwise, since the column through z contains two points of T, it can only be removed from T by applying a decoding scheme along the row through z. For $t = 1$ the only reason to apply a row decoding step through z is to be able to apply a column decoding step through z afterwards in order to remove some point of A. But since the last decoding step was applied along the y-axis, one or both decoding steps involving z could be skipped and we would get a decoding scheme for T requiring fewer steps.

Therefore, no decoding step was performed along the columns passing through the points of A. In other words, no point from the x-axis is removed. Further note that the decoding steps could be rearranged so that the removal of the points of B is done during the very last decoding steps. Denote by D the set of points in T which were removed in decoding steps before the first point of B is removed. One has

$$L(T) = L(m, 1) = |B| + |D|.$$

Since T could be also decoded by applying the decoding steps along the columns through the points of A, this scheme should not require fewer steps than the optimal one. In other terms,

$$|B| + |D| \leq |A|.$$

Hence, if we place the points of D on the y-axis, the resulting configuration would have complexity $|B| + |D|$ and could be decoded by applying the decoding steps along the rows only. If there are some other points of T outside the axes, we put them all on the x-axis. The resulting configuration has all its points on axes and is of complexity $L(m, 1)$. It is easily seen that among such configurations the one that has $\lfloor \frac{m-1}{2} \rfloor$ points on the x-axis (excluding the origin) and $\lceil \frac{m-1}{2} \rceil$ points on the y-axis has maximum complexity. ∎

3 The Absolute Complexity of a Configuration

Definition 4. *The* ***absolute complexity*** $A(T)$ *of a configuration* T *is the minimum number of decoding steps necessary to delete all points of* T.

Let $G(m, k, t) = \max A(T)$, where the maximum is taken over all t-decodable configurations T of size m in R. The next statement explores the function $G(m, k, t)$.

Lemma 2. *If* $m < k^2 - (k - t)^2$ *then*

$$G(m, k, t) \leq G(m + 1, k, t) \leq G(m, k, t) + 1.$$

Proof. To prove the first inequality, consider a configuration T of size m and absolute complexity $G(m, k, t)$, where $m < k^2 - (k - t)^2$. If at every decoding step of some decoding scheme exactly t points are removed from T, the inequality is obvious. Assume that at some decoding step less than t points are removed from T and consider a decoding scheme in which such a step occurs as early as possible. Let q be that decoding step number, and assume it is performed along some column C of R. We show that there exists a point $x \in R \setminus T$ such that the configuration $T' = T \cup \{x\}$ is decodable.

Let B be the set of points removed at step q, and let $y \in C \setminus B$. Note that $|B| < t$. If a decoding step along the row through y was not performed before step q, then $y \notin T$ and adding it to T leads to a decodable configuration of the same absolute complexity. Otherwise, $k - t + 1$ decoding steps were performed prior to step q along the rows through the points in $C \setminus B$. Consider the decoding step number $q - 1$. We can assume that it is performed along a row through some point $y \in C \setminus B$. Then swapping steps q and $q - 1$, we get a decoding scheme where the first decoding step that removes less than t points from T is performed along the row through y. By a similar argument, among the steps preceding the q-th there must be $k-t+1$ along columns (including C). In other words, during the first $q-2$ steps the decoding was performed along some $k-t$ rows and some $k-t$ columns, and $2t(k-t)$ points are removed from T (so $q = 2(k-t)+2$). Hence, after the first $q-2$ decoding steps the remaining part T'' of T is a subset of the intersection of some t rows and t columns of R, implying $|T''| \leq t^2$. None of these rows and columns was used in a previous decoding step. Thus, if $|T''| < t^2$ we can add a point to T'' leaving it decodable. Finally, if $|T''| = t^2$, then $m = |T| = |T''| + 2t(k-t) = k^2 - (k-t)^2$, a contradiction.

Therefore, the absolute complexity z of $T' = T \cup \{x\}$ satisfies $z \leq G(m+1, k, t)$. Obviously, deleting a point from T' leads to a configuration of no larger absolute complexity. Let us delete x from T' and obtain T. Hence,

$$G(m,k,t) \leq z \leq G(m+1,k,t).$$

To show the second inequality from the statement, let T be a decodable configuration of size $m+1$ and absolute complexity $G(m+1,k,t)$. After one decoding step one obtains a decodable configuration of size $m' \leq m$. Using the first inequality from the statement, one has

$$G(m+1,k,t) \leq 1 + G(m',k,t) \leq 1 + G(m,k,t),$$

which completes the proof. ■

Let $m(l,k,t)$ denote the minimum cardinality of a decodable configuration of complexity l in R. Note that $m(l,k,t)$ is well-defined by Lemma 2 for $1 \leq l \leq \max A(T)$, where the maximum is taken over all t-decodable configurations T in R. For fixed k,t we abbreviate $m(l,k,t)$ by $m(l)$.

According to the next lemma, the largest l for which $m(l)$ is defined is $2k-t$.

Lemma 3. $\max A(T) = 2k-t$, *where the maximum is taken over all t-decodable configurations T in R.*

Proof. Note that $A(T) \leq 2k-t$ follows immediately from Lemma 1.

Finally, $T = \{(i,j) : 0 \leq i,j \leq k-1, \ \min\{i,j\} \leq t-1\}$ is a decodable configuration of size $k^2-(k-t)^2 = (2k-t)t$. As any decoding step removes at most t points, T can not be completely decoded in less than $2k-t$ steps. ■

In what follows, $m(l)$ will be determined completely.

Obviously, $m(l) = l$ for $l \leq k$, and an extremal configuration consists of l points placed on a diagonal of R. The next step is to evaluate $m(k+1)$.

Theorem 5

$$m(k+1) = \begin{cases} 2t+k-1, & \textit{if } \ 1 \leq t \leq k/2, \\ 2t+k, & \textit{if } \ k/2 < t < k. \end{cases}$$

Proof. Let T be a decodable configuration with $A(T) = k+1$ and $|T| = m(k+1)$. Then at least one column of R intersects T in at least $t+1$ points, since otherwise a trivial column-by-column decoding leads to $A(T) \leq k$. Similarly, at least one row of R intersects T in at least $t+1$ points. Without loss of generality, we can assume that each of the coordinate axes contains at least $t+1$ points from T. If there are $x+t$ points from T on the horizontal axis for some $x \geq 1$, and $y+t$ points from T on the vertical axis for some $y \geq 1$, then at least $x+y+1$ decoding steps are needed just to delete these points from T. In each of the remaining $k+1-(x+y+1) = k-x-y$ steps at least one point is removed from T, so

$$|T| \geq (x+t)+(y+t)-1+k-x-y = 2t+k-1.$$

This bound it tight for $t \leq k/2$, since the configuration

$$\{(i,0) \mid i = 0,1,\ldots,2t-1\} \cup \{(0,j) \mid j = 0,1,\ldots,k-1\}$$

is of size $2t+k-1$ and has absolute complexity $k+1$.

Assume that $t > k/2$ and that T is a configuration of size $2t+k-1$ and absolute complexity $k+1$. Without loss of generality, we can again assume that both axes intersect T in at least $t+1$ points. There exists an algorithm for decoding T which removes the points from each axis as soon as only t points are left on the axis. Since the number of decoding steps is at least $k+1$ and in two of those steps t points are removed, exactly one point must be removed in every other step, and the total number of steps must be exactly $k+1$. This implies that there is no line L with $2 \leq |T \cap L| \leq t$ because otherwise we could remove the points from $L \cap T$ in the very first decoding step.

Assume that there is a column V different from the axis and such that $|T \cap V| \geq t+1$. Since $t > k/2$ there must be a row H different from the axis and such that $|T \cap H| \geq 2$. Hence, $|T \cap H| \geq t+1$, so for the part $T' \subseteq T$ consisting of the intersections of T with H, V and the coordinate axis, one has $2t+k-1 = |T| \geq |T'| \geq 4t$. This contradicts $t > k/2$.

Consequently, every column different from the axis contains at most one point of T. This implies $2k-1 \geq |T| = 2t+k-1$, contradicting $t > k/2$.

Finally, $m(k+1) = 2t+k$ for $t > k/2$ follows by the observation that the configuration

$$\{(i,0) \mid i = 0,1,\ldots,t\} \cup \{(j,0) \mid j = 0,1,\ldots,t\} \cup \{(h,h) \mid h = 0,1,...,k-1\}$$

has size $2t+k$ and absolute complexity $k+1$. ∎

Lemma 4. *If $l < 2k-t$, then $m(l) < m(l+1)$.*

Proof. Assume the contrary, $m(l+1) \leq m(l)$. Let T be a decodable configuration of complexity $l+1$ such that $|T| \leq m(l)$. Our assumption implies that such configuration exists. Let T' be obtained from T by making the first step in a decoding of T that uses $A(T)$ steps. Then $|T'| < |T| \leq m(l)$ and $A(T') = A(T)-1 = l$. This contradicts the definition of $m(l)$. ∎

Corollary 1. *If $k < l \leq 2k - 2t + 1$ and $2 \leq t \leq k/2$, then $m(l) = l + 2t - 2$.*

Proof. Theorem 5 and Lemma 4 imply that for $l = k + 1 + s$ one has $m(l) \geq m(k+1) + s = 2t + k - 1 + s = l + 2t - 2$. The matching upper bound is attained for the following configuration:

$$\{(0,0), (1,0), \ldots, (2t+s-1, 0)\} \cup \{(0,0), (0,1), \ldots, (0, 2t+s-1)\} \cup$$
$$\{(2t+s, 2t+s), (2t+s+1, 2t+s+1), \ldots, (k-1, k-1)\}.$$

This configuration is of cardinality $l + 2t - 2$ and complexity l. The construction works as long as $2t + s \leq k$, which is equivalent to $l \leq 2k - 2t + 1$. ∎

In the next theorem we establish general lower and upper bounds for the minimum size of a configuration of the maximum absolute complexity $2k - t$.

Theorem 6. $2k + t - 1 \leq m(2k - t) \leq 2k + (t+1)^2 - 4$ *holds for any* $1 \leq t < k$.

Proof. For the upper bound, the following configuration can be used:

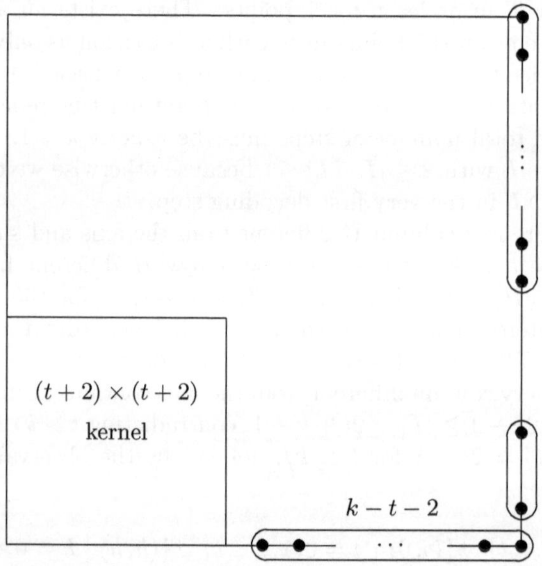

Fig. 1. A construction for the upper bound of $m(2k - t, k, t)$

The kernel consists of all points of the $(t+2) \times (t+2)$ corner subgrid excluding the diagonal points $\{(i, t+1-i) \mid i = 0, 1, \ldots, t-1\}$ and the points $\{(j, 0) \mid j = t, t+1\}$ of the bottom line. Hence, every column of the kernel has $t+1$ points, the bottom row has t points, the next row has $t+2$ points, and all other rows have $t+1$ points. The size of this configuration is $(t+2)(t+1) + (t-1) + 2(k-t-2) = 2k + (t+1)^2 - 4$.

The only way to decode the kernel is to start with its bottom line, for which $k-t-2$ rightmost points on the horizontal axis must be deleted first. For this, in turn, we have to delete all the points of the last column, for which $(k-t-2)+1$ steps are needed. After this the kernel can be decoded in $t+4$ steps. This implies that the absolute complexity of the configuration is $2k-t$.

To prove the lower bound, we proceed by induction on k. If $k=t+1$, then $m(2k-t)=m(k+1)$ which is equal to $2t+k=2k+t-1$ by Theorem 5.

Assume that $k \geq t+2$ and that $m(2(k-1)-t, k-1, t) \geq 2k+t-3$. Let T be a configuration of size $m(2k-t)$ and absolute complexity $2k-t$ in the $k \times k$ grid R. If every row of R intersects T in at least $t+1$ points, then $|T|=m(2k-t) \geq k(t+1) \geq 2k+t-1$, and we are done. That is, we can assume that there is a row r that intersects T in at most t points. For symmetry, we can also assume that there is a column c with $|c \cap T| \leq t$. Let T' be the configuration (in a $(k-1) \times (k-1)$ grid) obtained from T by removing the points on r and c. As $A(T)=2k-t$ and $A(T') \leq 2(k-1)-t$, it follows that $A(T')=2(k-1)-t$ and that there are at least two points of T in $r \cup c$. Using the induction hypothesis, we obtain

$$m(2k-t)=|T| \geq |T'|+2 \geq m(2(k-1)-t, k-1, t)+2 \geq 2k-t-1. \quad \blacksquare$$

In the case when $t=2$ the maximum complexity of a decodable configuration is $2k-2$. According to the previous results, we know the function $m(l)$ for $l \leq 2k-2t+1=2k-3$. Consequently, the only remaining case for $t=2$ is the value of $m(2k-2)$. By Theorem 6, we know that if $t=2$, then $2k+1 \leq m(2k-2) \leq 2k+5$. The exact values for all k are given in Theorem 7.

Theorem 7

$$m(2k-2,k,2)=\begin{cases} 7, & \textit{if } k=3 \\ 10, & \textit{if } k=4 \\ 14, & \textit{if } k=5 \\ 16, & \textit{if } k=6 \\ 2k+5, & \textit{if } k \geq 7. \end{cases}$$

Proof. It is an easy exercise to show that the absolute complexities of the configurations shown in Figure 2 are the same as in the statement, thus establishing an upper bound for $m(2k-2,k,2)$ when $3 \leq k \leq 7$. Furthermore, the upper bound of $2k+5$ for $k \geq 7$ follows from Theorem 6. So, we concentrate on lower bounds.

The value for $k=3$ follows from Theorem 5 or Theorem 6.

To prove the theorem for larger k, we first make a general observation. Let T be a minimum-size configuration of absolute complexity $2k-t$ in a $k \times k$ grid. The pigeonhole principle, together with the already established upper bound for $|T|$, implies that there is a row r that contains at most 2 points from T. If there is an empty column, then making the first decoding step along r results in a decodable configuration in a $(k-1) \times (k-1)$-grid. Then the complexity of T does not exceed $2(k-1)-t+1=2k-t-1$, contradicting the choice of T. Hence, without loss of generality we can assume that every column of the grid has a non-empty intersection with T.

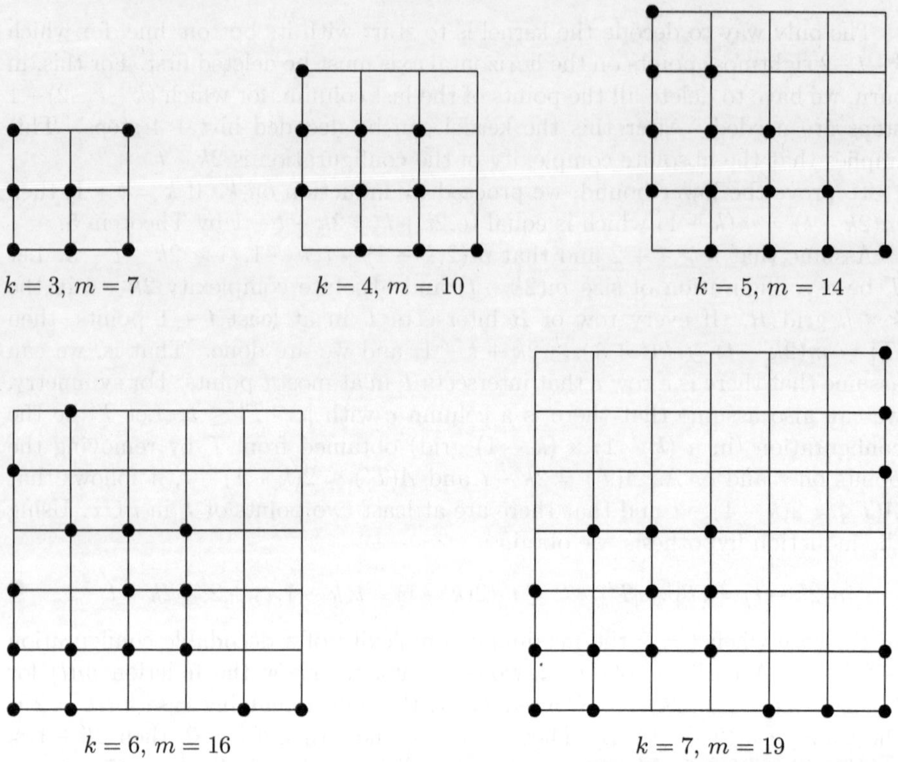

Fig. 2. Minimum-size configurations of complexity $2k - 2$ for $k = 3, 4, 5, 6, 7$

Next, we consider the case $k = 4$. It is sufficient to show that every decodable configuration of size 9 can be decoded in 5 steps. If the first two decoding steps can be applied to some row and some column and result in the removal of at at least 3 points from T, then we end up with a configuration of size at most 6 in a 3×3 grid which can be decoded in 3 steps. Hence, T can be decoded in 5 steps. By this, without loss of generality we can assume that no row or column of the grid contains exactly 2 points of T. Let c_i be the size of the intersection of the i-th column of the grid with T. Clearly, $c_1 + c_2 + c_3 + c_4 = 9$, and without loss of generality, we can assume $c_1 \geq c_2 \geq c_3 \geq c_4$. Taking into account that $c_i \in \{1, 3, 4\}$ for $i = 1, 2, 3, 4$, it turns out that the only choice for (c_1, c_2, c_3, c_4) is $(4, 3, 1, 1)$. It follows some row of the grid contains exactly 2 points from T, and we are done.

The cases $k = 5$ and $k = 7$ can be handled similarly, although there are more subcases to consider.

In the proof of the lower bound in Theorem 6 we presented an argument that shows that $m(2k-2, k, 2) \geq m(2(k-1)-2, k-1, 2)$ for $k \geq 4$. By this, the claim for $k = 6$ is implied by the result for $k = 5$, and the claim for $k \geq 7$ follows by straight forward induction on k, for which $k = 7$ serves as the base case. ■

4 Complexity of Data Recovery

Here we deal with the complexity aspects of the data recovery problem. The following lemma plays a central role in our analysis.

Lemma 5. *Let $T \subseteq R$ be a configuration. Assume there exists a line ℓ of R such that $0 < |T \cap \ell| \leq t$, and let $T' = T \setminus (T \cap \ell)$. Then T is decodable if and only if T' is decodable.*

Proof. Indeed, if T is decodable, then so it is T' since $T' \subset T$. In fact, the decoding scheme for T can be applied to decode T'. On the other hand, if T' is decodable, then T can be decoded by decoding along ℓ in the first step and then applying the decoding scheme for T'. ■

By Lemma 5, if T is decodable, then we can convert T into the empty set using feasible decoding steps in any order.

Recall from introduction that $D(T)$ denotes the (inclusion-)maximal deadlock of a non-decodable configuration T.

Lemma 6. *If T, ℓ, and T' are as in Lemma 5, then $D(T) = D(T')$.*

Proof. If T is decodable, then Lemma 5 implies $D(T) = D(T') = \varnothing$. Assume that T is not decodable. Since $T' \subseteq T$ we have $D(T') \subseteq D(T)$. Assume there is some $p \in D(T) \setminus D(T')$. This means that some line ℓ' intersecting $D(T)$ after deleting all points on ℓ intersects $D(T)$ in less points than before. But this is impossible as no point on ℓ belongs to $D(T)$. ■

By Lemma 6, the deadlock $D(T)$ consists precisely of those points that cannot be deleted from T in decoding steps. Thus, we get a polynomial time algorithm for constructing $D(T)$ by applying decoding steps in any order, and this way shrinking T down to $D(T)$. As a corollary, we get solutions to the following Problems 1 and 2 in polynomial time.

PROBLEM 1.
Instance: A list of faulty nodes $T \subseteq R$, k, t, and a point $p \in T$.
Question: Is it possible to recover the contents of the node p?
For this, we just need to check whether $p \in D(T)$.

PROBLEM 2.
Instance: A set $T \subseteq R$ of faulty nodes, k, and t.
Question: Is it possible to recover the data from all faulty nodes?
For this, we just need to check whether $D = \varnothing$.

PROBLEM 3.
Instance: A list $T \subseteq R$ of faulty nodes and an integer s.
Question: Is there a decoding scheme of complexity at most s?

Proposition 1. *Problem 3 is NP-complete.*

Proof. The statement follows directly from the NP-completeness of the problem considered in [1]. ∎

PROBLEM 4.
Instance: A list $T \subseteq R$ of faulty nodes, a node $p \in T$, and integer an s.
Question: Is there a decoding scheme of complexity at most s to restore the content of node p?

Theorem 8. *Problem 4 is NP-complete.*

Proof. Obviously, the problem is in NP. We reduce the EXACT-3-COVER Problem, which is known to be NP-complete [3], to our problem. Given an instance $\{X_1, \ldots, X_r\}$ of 3-subsets of $U = \{1, \ldots, 3m\}$, for $i \in U$ denote $d_i = |\{j \mid i \in X_j\}|$. We construct an instance of our problem for $t = \max\{\max_i d_i, 3\}$. Consider the set T of faulty nodes defined as follows

$$T = A \cup B \cup C \cup D \cup E,$$

where

$$A = \{(i, 0) \mid i = 0, \ldots 3m + t - 1\},$$
$$B = \bigcup_{i=1}^{r} (X_i \times i),$$
$$C = \bigcup_{i=1}^{3m} \{(i, r + j) \mid j = 1, \ldots, t - d_i\},$$
$$D = \{(0, r + j) \mid j = 1, \ldots, t + 1\},$$
$$E = \{(i, j) \mid 3m + 1 \le i \le 3m + t + 1,\ r + 1 \le j \le r + t + 1\}.$$

Obviously, construction can be accomplished in polynomial time. We show that for the selected t the node $(0, 0)$ can be restored in $4m + 1$ decoding steps iff there exists an exact cover $X = \{X_{i_1}, \ldots, X_{i_m}\}$ of U.

Indeed, assuming the existence of the exact cover, let us apply m decoding steps through the rows $y = i_j$, $j = 1, \ldots, m$. Now every column $x = i$ for $i = 1, \ldots, 3m$ contains exactly t faulty nodes, and they can be restored in the next $3m$ steps. After those decoding steps the number of faulty nodes on the x-axis becomes t and the origin can be restored right away.

On the other hand, assume the node $(0, 0)$ can be restored in $4m + 1$ steps. Note that no decoding step can be applied for a row or column through the set $D \cup E$, since every such row/column contains at least $t + 1$ faulty nodes. In particular, we cannot restore the nodes of C by using the row decodings. Therefore, the only way to restore the origin is to apply a decoding step along the x-axis. This can be only done after restoring the nodes $(i, 0)$ for $i = 1, \ldots, 3m$ by using the decodings via the columns through the set C, which requires $3m$

steps. But for this every such column must have at most t faulty nodes, which could be only established in the remaining m steps by applying the decodings via the rows through the set B, namely via the rows corresponding to the exact cover of U. ∎

Acknowledgements. We are grateful to the referee for suggesting the reduction from EXACT-3-COVER in the proof of Theorem 8.

References

1. Fernau, H.: Complexity of a $\{0, 1\}$-matrix problem. Austral. J. Combin. 29, 273–310 (2004)
2. MacWilliams, F.J., Sloane, N.J.A.: The Theory of Error-Correcting Codes, Parts I, II (3rd repr.), vol. 16, XX, 762. North-Holland Mathematical Library, Amsterdam etc., North-Holland, Elsevier (1985)
3. Papadimitriou, C.H., Steiglitz, K.: Combinatorial Optimization: Algorithms and Complexity. Dover (1998)

An Unstable Hypergraph Problem with a Unique Optimal Solution*

Carlos Hoppen[1,**], Yoshiharu Kohayakawa[2,***], and Hanno Lefmann[3]

[1] Instituto de Matemática, Universidade Federal do Rio Grande do Sul, Avenida Bento Gonçalves, 9500, 91501–970 Porto Alegre, RS, Brazil
choppen@ufrgs.br

[2] Instituto de Matemática e Estatística, Universidade de São Paulo, Rua do Matão, 1010, 05508-090 São Paulo, Brazil
yoshi@ime.usp.br

[3] Fakultät für Informatik, Technische Universität Chemnitz, Straße der Nationen 62, D-09107 Chemnitz, Germany
Lefmann@Informatik.TU-Chemnitz.de

Dedicated to the memory of Rudolf Ahlswede

Abstract. There is a variety of problems in extremal combinatorics for which there is a unique configuration achieving the optimum value. Moreover, as the size of the problem grows, configurations that "almost achieve" the optimal value can be shown to be "almost equal" to the extremal configuration. This phenomenon, known as *stability*, has been formalized by Simonovits [A Method for Solving Extremal Problems in Graph Theory, Stability Problems, *Theory of Graphs* (Proc. Colloq., Tihany, 1966), 279–319] in the context of graphs, but has since been considered for several combinatorial structures. In this work, we describe a hypergraph extremal problem with an unusual combinatorial feature, namely, while the problem is unstable, it has a unique optimal solution up to isomorphism. To the best of our knowledge, this is the first such example in the context of (simple) hypergraphs.

More precisely, for fixed positive integers r and ℓ with $1 \leq \ell < r$, and given an r-uniform hypergraph H, let $\kappa(H, 4, \ell)$ denote the number of 4-colorings of the set of hyperedges of H for which any two hyperedges in the same color class intersect in at least ℓ elements. Consider the function $\mathrm{KC}(n, r, 4, \ell) = \max_{H \in \mathcal{H}_{n,r}} \kappa(H, 4, \ell)$, where the maximum runs over the family $\mathcal{H}_{n,r}$ of all r-uniform hypergraphs on n vertices. We show that, for n large, there is a unique n-vertex hypergraph H for which $\kappa(H, 4, \ell) = \mathrm{KC}(n, r, 4, \ell)$, despite the fact that the problem of determining $\mathrm{KC}(n, r, 4, \ell)$ is unstable for every fixed $r > \ell \geq 2$.

Keywords: colorings, hypergraphs, stability theorem.

* This work was partially supported by the University of São Paulo, through the MaCLinC/NUMEC project.

** Partially supported by FAPERGS (Proc. 11/1436-1) and CNPq (Proc. 484154/2010-9).

*** Partially supported by CNPq (Proc. 308509/2007-2, 484154/2010-9).

H. Aydinian, F. Cicalese, and C. Deppe (Eds.): Ahlswede Festschrift, LNCS 7777, pp. 432–450, 2013.

1 Introduction

In this work, we study combinatorial problems of the following type. Given a class $\mathcal{C}$ of combinatorial objects and a real function f on $\mathcal{C}$, we wish to maximize f over the elements of $\mathcal{C}$. The following is a common feature of a large number of problems of this type. The function being optimized admits a unique optimal solution; moreover, every solution that "almost" achieves the optimal value must also be structurally "close" to the optimal solution. This phenomenon is called *stability* and, in a combinatorial perspective, has been first formally introduced for graphs by Simonovits [18].

Definition 1. *Let f be a function on finite graphs that is invariant under isomorphism and let $\mathcal{P}_f(n)$ be the problem of maximizing f over the class of all n-vertex graphs. The problem $\mathcal{P}_f(n)$ is said to be* stable *if, for every $\varepsilon > 0$, there exist a positive integer n_0 and a constant $\delta > 0$ for which the following holds. Let G be a graph on $n \geq n_0$ vertices such that $f(H) - f(G) < \delta f(H)$, where H is an n-vertex graph that maximizes f. Then there is a graph H' that is isomorphic to H with the property that H' and G differ in at most εn^2 edges.*

One of the first problems studied under the lens of stability is the *Turán problem*, which, for a fixed graph F, asks for the maximum number $\text{ex}(n, F)$ of edges in an F-free n-vertex graph, which is also known as the *Turán number* of F. In his original work [18], Simonovits proved, among other more general results, that, for complete graphs K_ℓ, the problem of determining $\text{ex}(n, K_\ell)$ is stable. More precisely, he has shown the following.

Theorem 1. *For any $\varepsilon > 0$ and $\ell \geq 3$ there exist $\delta > 0$ and n_0 such that the following holds for any $n \geq n_0$. If $G = (V, E)$ is an n-vertex graph with no copy of the complete graph K_ℓ and with at least $(1 - \delta)\,\text{ex}(n, K_\ell)$ edges, then there exists a partition $V = V_1 \cup \ldots \cup V_{\ell-1}$ of the vertex set V such that the number of edges with both endpoints in the same partition class is at most εn^2.*

Theorem 1 has been obtained independently by Erdős [5] (for a recent, streamlined proof, see Füredi [9]). Stability has also been observed for Turán numbers of several other forbidden graphs and hypergraphs, such as color critical graphs and the 3-uniform Fano plane (see Füredi and Simonovits [11], and Keevash and Sudakov [16]). Moreover, it has been considered for other graph and hypergraph problems (see, for instance, Füredi [10], Simonovits [19] and Keevash [14], and the references therein), as well as for combinatorial structures such as linear spaces (see Blokhuis, Brouwer, Szőnyi, and Weiner [2]), finite geometries (see Szőnyi and Weiner [20]) and set systems (see Keevash and Mubayi [15]).

On the other hand, it is known that, if multiple edges are allowed, then there exist graphs F for which there are classes of F-free graphs on n vertices that are almost extremal, but which are structurally far apart. For example, Brown and Simonovits [4] have studied the case where the forbidden graph F is a triangle with exactly one double edge, and where maximality was considered with respect to all n-vertex graphs for which every edge appears at most twice. They showed in [4] that there is a single extremal n-vertex graph, namely the graph $T_2^{(2)}(n)$ obtained by doubling the edges of the *n-vertex bipartite Turán graph*, the balanced complete bipartite graph on n vertices. The graph $T_2^{(2)}(n)$ has $2\lceil n/2\rceil\lfloor n/2\rfloor$ edges. However, the (simple) complete graph K_n, with $(n^2-n)/2$ edges, is within $o(n^2)$ of the extremal value, even though $\Omega(n^2)$ edges must be modified to turn it into a copy of $T_2^{(2)}(n)$.

We consider r-uniform hypergraphs, where a *hypergraph* $H=(V,E)$ is given by its *vertex set* V, typically $V=[n]=\{1,\dots,n\}$, and its set E of *hyperedges*, whose elements are subsets of the vertex set. A hypergraph $H=(V,E)$ is said to be *r-uniform* if each hyperedge $e\in E$ has cardinality r. As before, for a fixed r-uniform hypergraph F and a positive integer n, the *Turán number* $\mathrm{ex}(n,F)$ is defined as the maximum number of hyperedges in an n-vertex, r-uniform hypergraph not containing any copy of F. Hypergraphs $H=(V,E)$ on n vertices with $\mathrm{ex}(n,F)$ hyperedges that do not contain a copy of F are called *F-extremal.*

Concerning forbidden hypergraphs, a well-known problem raised by Turán [21] is to determine $\mathrm{ex}\left(n,K_4^{(3)}\right)$, where $K_4^{(3)}$ is the complete 3-uniform hypergraph on four vertices. The best known lower bound is $\mathrm{ex}\left(n,K_4^{(3)}\right)\geq\frac{5}{9}\binom{n}{3}$ and there are exponentially many constructions achieving this [3,7,8,17]. Thus, if one could show that $\mathrm{ex}\left(n,K_4^{(3)}\right)=\left(\frac{5}{9}+o(1)\right)\binom{n}{3}$ holds, instability would occur here and there might even be several extremal configurations.

The main contribution of this paper is the statement of a natural problem that is provably unstable, but which has a unique optimal solution up to isomorphism. To the best of our knowledge, the problem that we consider here is the first such example in the context of simple hypergraphs, that is, of hypergraphs for which multiple hyperedges are not allowed. To be precise, for a fixed family $\mathcal{F}$ of r-uniform hypergraphs, an r-uniform host hypergraph H and an integer k, we investigate the number $c_{k,\mathcal{F}}(H)$ of k-colorings of the set of hyperedges of H with no monochromatic copy of any $F\in\mathcal{F}$. Then, for every positive integer n, we may consider the function

$$c_{k,\mathcal{F}}(n)=\max_{H\in\mathcal{H}_{n,r}} c_{k,\mathcal{F}}(H),$$

the largest number of k-colorings of H with no monochromatic copy of any $F\in\mathcal{F}$ among all hypergraphs H in the class $\mathcal{H}_{n,r}$ of r-uniform n-vertex hypergraphs. For instance, if H is a graph and $\mathcal{F}$ only contains the single path of length two, then each color class has to be a matching and $c_{k,\mathcal{F}}(H)$ is the number of proper k-edge colorings of H. The number $c_{k,\mathcal{F}}(n)$ would be the largest number of distinct proper k-edge colorings of an n-vertex graph.

Here, we follow our work in [13] and we forbid pairs of hyperedges of the same color that share fewer than ℓ vertices, *thus forcing every color class to be ℓ-intersecting.* Formally, for fixed integers ℓ and r with $1\leq\ell<r$ and $i\in\{0,\dots,\ell-1\}$, let $F_{r,i}$ be the r-uniform hypergraph on $2r-i$ vertices with two hyperedges sharing exactly i vertices, and let $\mathcal{F}_{r,\ell}=\{F_{r,i}\colon i=0,\dots,\ell-1\}$.

Let $\kappa(H,k,\ell)$ denote the number of k-colorings of the set of hyperedges of a hypergraph H with no monochromatic copy of any $F \in \mathcal{F}_{r,\ell}$, thus $\kappa(H,k,\ell) = c_{k,\mathcal{F}_{r,\ell}}(H)$. These $\mathcal{F}_{r,\ell}$-avoiding colorings with k colors are called *(k,ℓ)-Kneser colorings*. Let $\mathrm{KC}(n,r,k,\ell) = \max\{\kappa(H,k,\ell)\colon H \in \mathcal{H}_{n,r}\}$, where $\mathcal{H}_{n,r}$ is the family of all r-uniform hyperpergraphs on n vertices; and hence $\mathrm{KC}(n,r,k,\ell) = c_{k,\mathcal{F}_{r,\ell}}(n)$. In other words, the quantity $\kappa(H,k,\ell)$ is the number of (k,ℓ)-Kneser colorings of a hypergraph H, while $\mathrm{KC}(n,r,k,\ell)$ is the largest number of (k,ℓ)-Kneser colorings amongst all r-uniform n-vertex hypergraphs. Our main concern is to determine which n-vertex r-uniform hypergraphs H maximize $\kappa(H,k,\ell)$. The hypergraphs H that satisfy $\kappa(H,k,\ell) = \mathrm{KC}(n,r,k,\ell)$ are called *(r,k,ℓ)-extremal hypergraphs.*

Observe that, as every color class in a (k,ℓ)-Kneser coloring does not contain two sets that share fewer than ℓ vertices, this problem is related to the classical Erdős-Ko-Rado Theorem [6] and its generalizations (see Ahlswede and Khachatrian [1]). Recall that, for n large, up to isomorphism, the unique extremal hypergraph for $\mathrm{ex}\,(n,\mathcal{F}_{r,\ell})$ is the hypergraph on $[n]$ whose hyperedges are all the r-element subsets of $[n]$ containing a fixed ℓ-element set. Since this hypergraph has exactly $\binom{n-\ell}{r-\ell}$ hyperedges and every k-coloring of its hyperedges trivially satisfies the required property, we have

$$\mathrm{KC}(n,r,k,\ell) \geq k^{\binom{n-\ell}{r-\ell}}. \tag{1}$$

We have shown in [13] that equality holds in (1) if and only if $k \in \{2,3\}$, and, for $k \geq 4$, we have determined structural properties of (r,k,ℓ)-extremal hypergraphs. Central in our arguments is the notion of an *ℓ-cover.*

Definition 2. *For a positive integer ℓ, an* ℓ-cover *of a hypergraph H is a set C of ℓ-subsets of vertices of H such that every hyperedge of H contains an element of C. A* minimum ℓ-cover *of a hypergraph H is an* ℓ-cover *of minimum cardinality.*

Note that this definition coincides with the definition of a vertex cover of a graph or hypergraph when $\ell = 1$.

Definition 3. *Fix integers $n, r \geq 2$ and $1 \leq \ell < r$. Let C be a collection of ℓ-subsets of an n-element set V. The* (C,r)-complete hypergraph *$H_{C,r}(n)$ is the r-uniform hypergraph on V that contains as hyperedges all the r-sets of V containing some member of C.*

In the theorem below, the family $\mathcal{H}_{r,4,\ell}(n)$ consists of all n-vertex r-uniform hypergraphs $H = H_{C,r}(n)$ for $C = \{t_1, t_2\}$, where the distinct sets t_1 and t_2 are ℓ-subsets of the vertex set.

Theorem 2. *[13] Let $1 \leq \ell < r$ be fixed integers. For every $\varepsilon > 0$, there is $n_0 > 0$ such that, for any $n > n_0$ and every hypergraph $H \in \mathcal{H}_{r,4,\ell}(n)$,*

$$\kappa(H,4,\ell) \geq (1-\varepsilon)\,\mathrm{KC}(n,r,4,\ell).$$

Conversely, there exist $n_1 > 0$ and $\varepsilon_0 > 0$ such that, if $n > n_1$ and H is an n-vertex r-uniform hypergraph satisfying $\kappa(H,4,\ell) \geq (1-\varepsilon_0)\cdot \mathrm{KC}(n,r,4,\ell)$, then $H \in \mathcal{H}_{r,4,\ell}(n)$.

Theorem 2 may be naturally interpreted in terms of stability, which in our framework may be formalized as follows. Here, for two sets A and B, we write $A \triangle B$ for their symmetric difference $(A \setminus B) \cup (B \setminus A)$.

Definition 4. *Let r, k and ℓ be fixed. The problem $\mathcal{P}_{n,r,k,\ell}$ of determining* $\mathrm{KC}(n,r,k,\ell)$ *is* stable *if, for every $\varepsilon > 0$, there exist $\delta > 0$ and $n_0 > 0$ such that the following is satisfied. Let H^* be an (r,k,ℓ)-extremal hypergraph on $[n]$, where $n \geq n_0$, and let H be an r-uniform hypergraph on $[n]$ satisfying $\kappa(H,k,\ell) > (1-\delta)\,\mathrm{KC}(n,r,k,\ell)$. Then $|E(H) \triangle E(H')| < \varepsilon |E(H')|$ for some hypergraph H' that is isomorphic to H^*.*

With this definition, Theorem 2 implies that $\mathcal{P}_{n,r,4,\ell}$ is unstable for every $r > \ell \geq 2$, as the values of $\kappa(H,4,\ell)$ are asymptotically equal for hypergraphs H in $\mathcal{H}_{r,4,\ell}(n)$, but two hypergraphs in this class for which the cover elements have different intersection sizes are far from being isomorphic. We remark that, in the case $\ell = 1$, the problem $\mathcal{P}_{n,r,4,\ell}$ has been proven to be stable in [13]. An $(r,4,1)$-extremal hypergraph on n vertices is given by a (C,r)-complete hypergraph on n vertices with vertex cover $C = \{\{v\},\{w\}\}$ for distinct vertices v and w; moreover, if n is large enough, this extremal hypergraph is unique up to isomorphism.

We may now state the main result of this paper. Despite the instability of $\mathcal{P}_{n,r,4,\ell}$, we show that, for every $\ell \geq 2$ and $k = 4$ colors, the problem $\mathcal{P}_{n,r,4,\ell}$ has a unique solution up to isomorphism, namely the set of hypergraphs in $\mathcal{H}_{r,4,\ell}(n)$ for which the two elements in the ℓ-cover have intersection of size $\ell - 1$.

Theorem 3. *Let $r > \ell \geq 2$ be integers. Then, there exists n_0 such that, for all $n \geq n_0$, the extremal hypergraph for the problem $\mathcal{P}_{n,r,4,\ell}$ is the hypergraph $H_{C,r}(n)$ with $C = \{t_1, t_2\}$, where t_1 and t_2 are ℓ-sets with $|t_1 \cap t_2| = \ell - 1$.*

The remainder of this paper is structured as follows. In Section 2, we introduce the main tools for proving Theorem 3. The proof of Theorem 3 is the subject of Section 3, while Section 4 is devoted to final remarks and open problems. Along the way, we give a new self-contained proof of Theorem 2, which, despite not being significantly different from the work in [13], takes advantage of simplified calculations for the case of $k = 4$ colors.

2 Preliminaries

The objective of this section is to introduce most of the ingredients that will be used to prove Theorem 3. Although it is not crucial for our result, we shall provide a proof of Theorem 2 for completeness. The arguments here are reminiscent of the ones in the proof of Theorem 1.7 in [13], which deals with a more general case, but calculations are less involved owing to the restriction to the case of $k = 4$ colors. In this simplification, we incorporate some of the ideas used for proving Theorem 1.2 in [12], which addresses a similar problem in the context of graphs.

We concentrate on the proof of the second statement of Theorem 2, namely that every n-vertex r-uniform hypergraph H satisfying $\kappa(H,4,\ell) \geq (1-\varepsilon_0)\,\cdot$

$\mathrm{KC}(n,r,4,\ell)$ lies in $\mathcal{H}_{r,4,\ell}(n)$. To this end, we show that the largest number of $(4,\ell)$-Kneser colorings of H is achieved when the size of a minimum ℓ-cover $C=\{t_1,\ldots,t_c\}$ is equal to two, i.e., $c=2$, and that the number of $(4,\ell)$-Kneser colorings is substantially smaller when this is not the case. To conclude this, we shall prove that the bulk of the $(4,\ell)$-Kneser colorings of a $(4,\ell)$-extremal hypergraph consists of colorings such that every color appears 'many' times and that, when this happens, the coloring must be 'star-like', in the sense that, for every given color σ, there must be a cover element t_i contained in all the hyperedges colored σ. As a consequence of this fact, we deduce that extremal hypergraphs are (C,r)-*complete*, that is, they contain all possible hyperedges covered by ℓ-sets in C. In particular, a hypergraph of this type must lie in $\mathcal{H}_{r,4,\ell}(n)$.

In the remainder of this section, we follow this plan to reach the desired conclusion. First, we show that, for r and ℓ fixed, a $(4,\ell)$-Kneser colorable hypergraph has a small ℓ-cover.

Lemma 1. *Let r and ℓ be positive integers with $\ell<r$, and let $H=(V,E)$ be an r-uniform $(4,\ell)$-colorable hypergraph. Then H has an ℓ-cover $C=\{t_1,\ldots,t_c\}$ with cardinality $c\le 4r^\ell$ and with $|\bigcup_{i=1}^c t_i|\le 4r$.*

Proof. The $(4,\ell)$-Kneser colorability of H ensures that there cannot be more than four hyperedges that pairwise intersect in fewer than ℓ vertices. Hence there is a set $S\subseteq E$ of at most four hyperedges such that every hyperedge of H is ℓ-intersecting with some element of S. In particular, the set $C=\{t\colon t\subset e\in S, |t|=\ell\}$ is an ℓ-cover of H with cardinality $|C|\le 4\binom{r}{\ell}\le 4r^\ell$ and $\left|\bigcup_{s\in S} s\right|\le 4r$. □

Proof of Theorem 2. Let $C=\{t_1,\ldots,t_c\}$ be a minimum ℓ-cover in the hypergraph H. Let $V_C=\bigcup_{i=1}^c t_i$ be the set of vertices of H that appear in C, where $|V_C|\le 4r$ by Lemma 1. The set E of hyperedges of H will be split into $E=E'\cup F$, where $e\in E$ is assigned to E' if $|e\cap V_C|=\ell$ and it is assigned to F if $|e\cap V_C|>\ell$. Since each element of F has intersection at least $(\ell+1)$ with V_C, we have that the size of F is bounded above by

$$|F|\le\binom{|V_C|}{\ell+1}\binom{n-\ell-1}{r-\ell-1}\le\binom{4r}{\ell+1}\binom{n-\ell-1}{r-\ell-1}, \tag{2}$$

which, for n large, is smaller than the largest possible size of E', namely $\binom{|V_C|}{\ell}\binom{n-|V_C|}{r-\ell}$. As a consequence, the contribution of the $(4,\ell)$-Kneser colorings of F is negligible, and we will focus on the structure of the colorings of $H'=H\setminus F$.

Let $E'_i\subseteq E'$ be the set of hyperedges of H' containing the cover element $t_i\in C$. Clearly, any $(4,\ell)$-Kneser coloring of H is the combination of a $(4,\ell)$-Kneser coloring of $H'=H\setminus F$ with a coloring of the hyperedges in F with at most four colors. We know by (2) that there are at most

$$4^{|F|}\le 4^{\binom{4r}{\ell+1}\binom{n-\ell-1}{r-\ell-1}}, \tag{3}$$

colorings of the latter type, thus we now concentrate on $(4,\ell)$-Kneser colorings of H'.

Fix a $(4,\ell)$-Kneser coloring Δ of H'. For each ℓ-set $t_i \in C$ and each color $\sigma \in \{1,\ldots,4\}$, we say that σ is *substantial* for t_i with respect to Δ if the number of hyperedges in E_i' that have color σ is larger than

$$L = (r-\ell)\binom{n-\ell-1}{r-\ell-1}. \tag{4}$$

We call a color σ *influential* if it is substantial for some cover element.

Our first auxiliary result shows that, if σ is substantial for $t_i \in C$, then all hyperedges with color σ must contain t_i. Hence, given a color σ, there is at most one cover element for which it is substantial, so that the subgraph of H' induced by σ is a "star" centered at the cover element t_i.

Lemma 2. *Let Δ be a $(4,\ell)$-Kneser coloring of H. If the color σ is substantial for a cover element t_i and e is a hyperedge of H' with color σ, then $t_i \subseteq e$.*

Proof. Let $E_{i,\sigma}'$ be the set of all hyperedges in $E_i' \subseteq E'$, which contain the cover element t_i. Suppose for a contradiction that a hyperedge $e \in E'$ has color σ, but $t_i \not\subseteq e$, and let $t_{i'}$ be an element in the ℓ-cover C contained in e. By definition, the number of hyperedges h in $E_{i,\sigma}'$, whose intersection with e has size at least ℓ, is at most

$$U = \binom{r-\ell}{\ell - |t_i \cap t_{i'}|}\binom{n-2\ell+|t_i\cap t_{i'}|}{r-(\ell+|t_i\setminus t_{i'}|)},$$

since any such hyperedge h must contain at least $\ell - |t_i \cap t_{i'}|$ elements of $e \setminus t_{i'}$. Taking the maximum over all possible sizes $|t_i \cap t_{i'}|$, we have, for n sufficiently large,

$$U \leq \max_{0\leq m\leq \ell-1} \binom{r-\ell}{\ell-m}\binom{n-2\ell+m}{r-2\ell+m} = L.$$

This contradicts the hypothesis that σ is substantial for t_i. □

Lemma 3. *There exists $0<\gamma<1$, and there exists n_0 such that the following holds for $n \geq n_0$. Let H be an n-vertex and r-uniform hypergraph and let $\mathcal{C}$ be the class of $(4,\ell)$-Kneser colorings of H for which at least one of the colors is not influential. Then*

$$|\mathcal{C}| \leq 4^{(1-\gamma)\binom{n-\ell}{r-\ell}}.$$

Proof. Consider the set $\mathcal{C}$ of all $(4,\ell)$-Kneser colorings of H for which a color σ is not influential. Note that, in this case, σ may appear at most L times in each set E_i', so that the number of ways of coloring the hyperedges of H' with color σ is bounded above by

$$\sum_{(a_1,\ldots,a_c)}\left(\prod_{i=1}^{c}\binom{\binom{n-|V_C|}{r-\ell}}{a_i}\right),$$

where the sum is such that each a_i ranges from 0 to L. With (4) and $c \leq 4r^\ell$ by Lemma 1, for n sufficiently large, an upper bound on this value is given by

$$\sum_{(a_1,\ldots,a_c)} \binom{n}{r-\ell}^{\sum_{i=1}^{c} a_i} = \left(\sum_{p=0}^{L} \binom{n}{r-\ell}^p\right)^c \leq 2^c \cdot \binom{n}{r-\ell}^{Lc}$$
$$\leq 2^{c+Lcr\log_2 n} \leq 2^{4r^\ell + 4r^\ell \frac{r^2}{n-\ell}\binom{n-\ell}{r-\ell}\log_2 n}$$
$$\leq 2^{5\frac{r^{\ell+2}}{n-\ell}\binom{n-\ell}{r-\ell}\log_2 n}.$$

In particular, if we have $j \leq 4$ colors that are not influential, the number of ways of assigning these colors to the hyperedges of H' is bounded above by

$$2^{5j\frac{r^{\ell+2}}{n-\ell}\binom{n-\ell}{r-\ell}\log_2 n} \leq 2^{\frac{20r^{\ell+2}}{n-\ell}\binom{n-\ell}{r-\ell}\log_2 n}. \tag{5}$$

Now, suppose that we have $j \leq 3$ colors to assign to the hyperedges in H' that contain some ℓ-set in C with no restriction on the number of colors used. As these hyperedges are ℓ-intersecting, they may be colored in at most $j^{\binom{n-\ell}{r-\ell}}$ ways. Analogously, if we assign j_1 colors to the hyperedges that contain cover element t_1, and j_2 to the hyperedges that contain cover element t_2, where $|t_1 \cup t_2| = \ell + m$, $m \geq 1$, and $(j_1 + j_2)$ colors to the hyperedges that contain $(t_1 \cup t_2)$, the number of possible colorings of the hyperedges is at most

$$j_1^{\binom{n-\ell}{r-\ell}} \cdot j_2^{\binom{n-\ell}{r-\ell}} \cdot (j_1 + j_2)^{\binom{n-\ell-m}{r-\ell-m}}. \tag{6}$$

Note that, by summing over all possible distributions of the colors, for n large this is smaller than $3^{\binom{n-\ell}{r-\ell}}$ whenever $j_1 + j_2 \leq 3$. This includes with Lemma 2 the case, where we assign one color to each cover element t_1, t_2, t_3, giving at most six colorings. Now, we may easily derive an upper bound on the cardinality of $\mathcal{C}$ based on the following facts: (i) there is at least one color that is not influential; (ii) no color can be substantial for two or more cover elements; (iii) the number of ways of assigning noninfluential colors is small.

Combining equations (3), (5), and (6) we have

$$|\mathcal{C}| \leq 4^{\binom{4r}{\ell+1}\binom{n-\ell-1}{r-\ell-1}} \cdot 2^{\frac{20r^{\ell+2}}{n-\ell}\binom{n-\ell}{r-\ell}} \cdot 3^{\binom{n-\ell}{r-\ell}} \leq 4^{(1-\gamma)\binom{n-\ell}{r-\ell}},$$

where the last inequality holds for every fixed $\gamma < 1 - 1/\log_3 4$, if $n \geq n_0(\gamma, r, \ell)$ is sufficiently large. □

As a consequence of this result, we may show that, for a hypergraph to have many $(4,\ell)$-Kneser colorings, the number of hyperedges covered by each cover element cannot be too small. As it turns out, we shall see later that $(r,4,\ell)$-extremal hypergraphs are (C,r)-complete with respect to their minimum covers C (see Lemma 6).

Lemma 4. *Let H be an r-uniform n-vertex hypergraph with minimum ℓ-cover C, where n is sufficiently large, and let H' and E_i be defined as above. If $\kappa(H,4,\ell) = \mathrm{KC}(n,r,4,\ell)$, then, for every cover element t_i, we have $|E_i'| > 4L$.*

Proof. Recall from inequality (1) the easy lower bound on $\mathrm{KC}(n,r,4,\ell)$ based on a maximum family of mutually ℓ-intersecting r-subsets of $[n]$, namely

$$\mathrm{KC}(n,r,4,\ell) \geq 4^{\binom{n-\ell}{r-\ell}}.$$

Let H be an r-uniform n-vertex hypergraph with minimum ℓ-cover $C = \{t_1,\ldots,t_c\}$ and, for a contradiction, assume without loss of generality that $|E'_c| \leq 4L$. By equation (3), we know that

$$\kappa(H,4,\ell) \leq 4^{\binom{4r}{\ell+1}\binom{n-\ell-1}{r-\ell-1}} \cdot \kappa(H',4,\ell),$$

so that we concentrate on $\kappa(H',4,\ell)$.

Let $\mathcal{C}_1$ be the family of $(4,\ell)$-Kneser colorings of H' for which there is a color that is substantial for t_c, and let $\mathcal{C}_2$ contain the remaining $(4,\ell)$-Kneser colorings of H', so that $\kappa(H',4,\ell) = |\mathcal{C}_1| + |\mathcal{C}_2|$. We shall bound the number of $(4,\ell)$-Kneser colorings in each of these two families separately.

We start with colorings in $\mathcal{C}_1$. Consider the hypergraph $\hat{H}$ obtained by deleting all the hyperedges of H' that contain t_c. Given a coloring Δ in $\mathcal{C}_1$, consider the coloring $\hat{\Delta}$ of $\hat{H}$ obtained by ignoring the colors of the deleted hyperedges. Let $\hat{\mathcal{C}}_1$ be the set of all $(4,\ell)$-Kneser colorings of $\hat{H}$ obtained in this way. If σ is substantial for t_c with respect to Δ, then it cannot be substantial for any other cover elements in C, and hence σ cannot be influential with respect to $\hat{\Delta}$. In particular, if we apply Lemma 3 to $\hat{H}$ with some $0<\gamma<1$, we deduce that

$$|\hat{\mathcal{C}}_1| \leq 4^{(1-\gamma)\binom{n-\ell}{r-\ell}}$$

for $n \geq n_1$. Moreover, each coloring of $\hat{\mathcal{C}}_1$ corresponds to at most 4^{4L} colorings of $\mathcal{C}_1$. In particular, for n sufficiently large, we have

$$|\mathcal{C}_1| \leq 4^{4L} \cdot |\hat{\mathcal{C}}_1| \leq 4^{\left(\frac{4r^2}{n-\ell}+(1-\gamma)\right)\binom{n-\ell}{r-\ell}} \leq 4^{(1-\gamma/2)\binom{n-\ell}{r-\ell}}. \tag{7}$$

Now, let Δ be a coloring in $\mathcal{C}_2$ and let σ be the color of an edge whose single ℓ-subset in C is t_c. Since σ is not substantial for cover element t_c, Lemma 2 implies that σ cannot be influential with respect to Δ, so that, by applying Lemma 3 to H' with the same value of γ, we have

$$|\mathcal{C}_2| \leq 4^{(1-\gamma)\binom{n-\ell}{r-\ell}}$$

for $n \geq n_2$. As a consequence, with (7) for n sufficiently large, we have

$$\kappa(H',4,\ell) = |\mathcal{C}_1| + |\mathcal{C}_2| \leq 4^{(1-\gamma/2)\binom{n-\ell}{r-\ell}} + 4^{(1-\gamma)\binom{n-\ell}{r-\ell}} \leq 4^{(1-\gamma/3)\binom{n-\ell}{r-\ell}},$$

from which we deduce that

$$\kappa(H,4,\ell) \leq 4^{\binom{4r}{\ell+1}\binom{n-\ell-1}{r-\ell-1}} \cdot \kappa(H',4,\ell) \leq 4^{(1-\gamma/4)\binom{n-\ell}{r-\ell}} < 4^{\binom{n-\ell}{r-\ell}},$$

hence $|E'_i| > 4L$ for $i = 1,\ldots,c$. □

We shall now prove that, for $(r,4,\ell)$-extremal hypergraphs, the size of a minimum ℓ-cover is at most two. This will be improved in Lemma 7, where we show that it must be equal to two.

Lemma 5. *Let H be an r-uniform n-vertex hypergraph with minimum ℓ-cover C, where n is sufficiently large. If $\kappa(H,4,\ell)=\mathrm{KC}(n,r,4,\ell)$, then $|C|\in\{1,2\}$.*

Proof. Let H be a r-uniform n-vertex hypergraph with minimum ℓ-cover $C=\{t_1,\ldots,t_c\}$ where $c\geq 3$. By Lemma 4, we may suppose that the sets E_i' associated with the cover elements t_i satisfy $|E_i'|>4L$. In particular, by the pigeonhole principle, for every $(4,\ell)$-Kneser coloring of H and every cover element t_i, there must be a color that is substantial for t_i. As every color may be substantial for at most one cover element, we have $c\leq 4$.

Note that, in the case $c=4$, this implies that every $(4,\ell)$-Kneser coloring is such that E_i' is monochromatic, so that, with (3), we have

$$\kappa(H,4,\ell)\leq 4^{\binom{4r}{\ell+1}\binom{n-\ell-1}{r-\ell-1}}\cdot 4! < 4^{\binom{n-\ell}{r-\ell}}$$

for large n. Analogously, if $c=3$, three of the colors are needed for each of the three sets E_i', while the fourth color is either not influential or is used to color the edges of some E_i'. For the first class we may use Lemma 3 with some $0<\gamma<1$ while the second can be bounded directly. We have

$$\kappa(H,4,\ell)\leq 4^{(1-\gamma)\binom{n-\ell}{r-\ell}}+4^{\binom{4r}{\ell+1}\binom{n-\ell-1}{r-\ell-1}}\cdot\frac{3\cdot 4!}{2!}\cdot 2^{\binom{n-\ell}{r-\ell}}<4^{\binom{n-\ell}{r-\ell}}$$

for n sufficiently large, thus $c\leq 2$, which concludes the proof. □

Remark 1. *The proof of Lemma 5 actually shows more. Indeed, it implies that, for some $0<\gamma<1$, there is n_0 with the following property. If H is an r-uniform n-vertex hypergraph with minimum ℓ-cover C of size $|C|\geq 3$, where $n\geq n_0$, then*

$$\kappa(H,k,\ell)<4^{(1-\gamma)\binom{n-\ell}{r-\ell}}. \tag{8}$$

Lemma 6. *Let $H=(V,E)$ be an r-uniform n-vertex hypergraph with minimum ℓ-cover C satisfying $\kappa(H,4,\ell)=\mathrm{KC}(n,r,4,\ell)$. Then there exists n_0, such that for every integer $n\geq n_0$ the hypergraph H is (C,r)-complete, i.e., every r-subset of V containing some cover element $t\in C$ is a hyperedge of H.*

Proof. By Lemma 5 we may suppose that $H=(V,E)$ is an r-uniform hypergraph with minimum ℓ-cover C, where $|C|\in\{1,2\}$. Note that the result is immediate if $|C|=1$, as in this case every 4-coloring of the hyperedges of H is ℓ-intersecting, so that

$$\kappa(H,4,\ell)=4^{|E|}\leq 4^{\binom{n-\ell}{r-\ell}},$$

with equality only if $|E|=\binom{n-\ell}{r-\ell}$, that is, only if H is (C,r)-complete.

Now, let $C=\{t_1,t_2\}$, and for a contradiction, assume that H is not (C,r)-complete. By Lemma 4, we may assume that every element in C covers more than $4L$ hyperedges not covered by any other element of C. With the assumption that the hypergraph H is not (C,r)-complete, let e be an r-subset of V containing $t_i \in C$ that is not a hyperedge of H, and define E_i' as before. Let Δ be a $(4,\ell)$-Kneser coloring of H. By the pigeonhole principle, at least one of the colors, say σ, appears more than L times in E_i'. By Lemma 2, we know that all the hyperedges assigned color σ by Δ must contain cover element t_i, so that Δ may be extended to a $(4,\ell)$-Kneser coloring of $H \cup \{e\}$ by assigning color σ to e.

Furthermore, there is at least one $(4,\ell)$-Kneser coloring of H using exactly two colors, namely the one that assigns color 1 to all hyperedges containing t_1 and color 2 to all remaining hyperedges. In this case, we have at least three options to color e, one using a color already used, and two using a new color. As a consequence, the hypergraph $H \cup \{e\}$ has more $(4,\ell)$-Kneser colorings than H, establishing that such an r-uniform hypergraph H cannot be $(r,4,\ell)$-extremal. □

Remark 2. *The proof of Lemma 6 may be easily adapted to show that* $\kappa(H,4,\ell) \leq \frac{2}{3}\,\mathrm{KC}(n,r,4,\ell)$ *if H is not complete. Indeed, we can show this directly with multiplicative constant $1/4$ in the case when the minimum ℓ-cover has size one. In the case when the minimum ℓ-cover C has size two, we may easily show that the bulk of the colorings of the (C,r)-complete hypergraph are the colorings for which each cover element has two substantial colors. In other words, when the graph H in the proof of Lemma 6 is almost complete, most colorings can be extended in two different ways when the new edge e is included. Actually, since the remaining colorings are "rare", the multiplicative constant $1/2$ could be replaced by $(1/2+\varepsilon)$ for every $\varepsilon > 0$.*

By the work done so far, we already know that, for n sufficiently large, an $(r,4,\ell)$-extremal hypergraph H^* on n vertices is of the form $H_{C,r}(n)$, for some set C of ℓ-subsets of $[n]$, where $|C| \in \{1,2\}$. We want to show that two is the correct size of C.

Lemma 7. *Let C_1, C_2 be sets of ℓ-subsets of $[n]$ such that $|C_1| = 1$ and $|C_2| = 2$. Then, for n sufficiently large, we have* $\kappa(H_{C_2,r}(n),4,\ell) \geq 5 \cdot \kappa(H_{C_1,r}(n),4,\ell)$.

Proof. Let $C_1 = \{t_1\}$ be the minimum ℓ-cover of $H_1 = H_{C_1,r}(n)$. We may view $H_2 = H_{C_2,r}(n)$ as the hypergraph obtained from H_1 by the addition of all hyperedges containing t_2 for some ℓ-set of $[n]$ that is distinct from t_1.

Let $\mathcal{C}_1$ and $\mathcal{C}_2$ denote the sets of $(4,\ell)$-Kneser colorings of H_1 and H_2 for which every color is influential. By Lemma 3, we have

$$|\mathcal{C}_1| \leq \kappa(H_1,4,\ell) \leq |\mathcal{C}_1| + 4^{(1-\gamma)\binom{n-\ell}{r-\ell}}$$

$$|\mathcal{C}_2| \leq \kappa(H_1,4,\ell) \leq |\mathcal{C}_2| + 4^{(1-\gamma)\binom{n-\ell}{r-\ell}}.$$

Since we know that $|\mathcal{C}_1| \gg 4^{(1-\gamma)\binom{n-\ell}{r-\ell}}$ for n sufficiently large, we shall obtain our result if we show that $|\mathcal{C}_2| \geq 6|\mathcal{C}_1|$. To this end, we consider the following mapping $\phi\colon \mathcal{C}_1 \to \mathcal{C}_2$, for which the image of Δ_1 under ϕ is a coloring Δ_2 with colors 1 and 2 substantial for t_1 and colors 3 and 4 substantial for t_2. The hyperedges that contain $t_1 \cup t_2$ have the same color with respect to Δ_1 and Δ_2. A hyperedge e containing t_1, but not containing t_2, receives color 1 if it has color 1 or 3 with respect to Δ_1, otherwise it is assigned color 2. On the other hand, a hyperedge e containing t_2, but not containing t_1, receives color 3 if $(e \setminus t_2) \cup t_1$ has color 2 or 3 with respect to Δ_1, otherwise it is assigned color 4. It is immediate from this definition that edges with the same color with respect to Δ_2 are ℓ-intersecting.

The crucial fact about this construction is that the color of e that was originally assigned by Δ_1 is uniquely determined by the colors of e and $(e \setminus t_1) \cup t_2$ in Δ_2. The injectivity of ϕ is an easy consequence of this fact, and hence $|\mathcal{C}_1| \leq |\mathcal{C}_2|$. Now, note that we may interchange the roles of the colors in the above mapping, that is, instead of 1 and 2, any two colors could be assigned to the hyperedges containing t_1 and the remaining two colors could be used for hyperedges containing t_2. Since all the colors are substantial for t_1 with respect to the colorings in $\mathcal{C}_1$, the colorings created with these new mappings are all distinct from the colorings created through the original mapping. By the argument above we get $|\mathcal{C}_2| \geq 6|\mathcal{C}_1|$, which is the desired inequality and concludes the proof. □

If we combine the last result with Remarks 1 and 2, we see that there exist $\varepsilon_0 > 0$ and an integer n_1 such that $\kappa(H, 4, \ell) \leq (1 - \varepsilon_0) \cdot \mathrm{KC}(n, r, 4, \ell)$, whenever $H \notin \mathcal{H}_{r,4,\ell}(n)$ and $n \geq n_1$. To conclude the proof of Theorem 2, we still need to prove that, given $\varepsilon > 0$, every hypergraph $\mathcal{H}_{r,4,\ell}(n)$ is within $\varepsilon \cdot \mathrm{KC}(n, r, 4, \ell)$ of being maximum. This will be an easy consequence of the precise calculations in the following section.

3 Main Result

In this section, we shall prove the main result in our work, namely Theorem 3, which is restated below. To this end, we need to count the number of $(4, \ell)$-Kneser colorings of a hypergraph $H \in \mathcal{H}_{r,4,\ell}(n)$ with high accuracy. The following class of colorings turns out to be instrumental in accomplishing this task.

Definition 5. *Let $H = (V, E)$ be a hypergraph with ℓ-cover $C = \{t_1, \ldots, t_c\}$. A* star coloring *of H is a Kneser coloring such that, for every color σ, all the hyperedges of H with color σ contain some fixed element $t_i = t_i(\sigma)$ of the ℓ-cover C. A Kneser coloring of H that is not a star coloring is called a* non-star coloring.

Theorem 4. Let $r > \ell \geq 2$ be integers. Then, there exists n_0, such that for all $n \geq n_0$, the extremal hypergraph for the problem $\mathcal{P}_{n,r,4,\ell}$ is the (C, r)-complete hypergraph $H_{C,r}(n)$ where $C = \{t_1, t_2\}$ is an ℓ-cover such that $|t_1 \cap t_2| = \ell - 1$.

Proof. Let $H^* = (V, E)$ be a complete, r-uniform hypergraph on n vertices with ℓ-cover $C = \{t_1, t_2\}$ where $|t_1 \cap t_2| = y \leq \ell - 1$. Our proof consists of two main steps. We first show that the number of star colorings of H^* is maximum when $y = \ell - 1$. In the second step, we find an upper bound on the number of non-star colorings of H^*, which allows us to show that the gap on the number of star colorings cannot be bridged through non-star colorings, so that $\kappa(H^*, 4, \ell)$ is indeed maximum when $y = \ell - 1$.

To compute the number of star colorings, observe that we may either assign two colors to each cover element, or three colors to one cover element and one color to the other. Let $E_1^* = \{e \in E : t_1 \subset e, t_2 \not\subset e\}$ and $E_2^* = \{e \in E : t_1 \not\subset e, t_2 \subset e\}$ be the sets of hyperedges containing exactly one of the cover elements.

We begin with the case where two colors are assigned to each cover element, which can be done in $\binom{4}{2}$ ways. We first assume that $y \geq 2\ell - r$, which implies that there exist hyperedges in H^* containing $t_1 \cup t_2$, namely $\binom{n-2\ell+y}{r-2\ell+y}$ of them. Once the colors have been fixed, the number of distinct colorings of the hyperedges that contain $t_1 \cup t_2$ is $4^{\binom{n-2\ell+y}{r-2\ell+y}}$. Also, we may color the hyperedges in E_1^* in $2^{\binom{n-\ell}{r-\ell}-\binom{n-2\ell+y}{r-2\ell+y}}$ ways (the same holds for hyperedges in E_2^*). Note that, in this argument, the colorings for which both E_1^* and E_2^* are monochromatic are counted exactly twice, so that we need to substract the term $\binom{4}{2} \cdot 2 \cdot 4^{\binom{n-2\ell+y}{r-2\ell+y}}$. This amounts to

$$\binom{4}{2} \cdot 2^{2\left(\binom{n-\ell}{r-\ell}-\binom{n-2\ell+y}{r-2\ell+y}\right)} 4^{\binom{n-2\ell+y}{r-2\ell+y}} - \binom{4}{2} \cdot 2 \cdot 4^{\binom{n-2\ell+y}{r-2\ell+y}} \tag{9}$$

distinct colorings. In the case when $y < 2\ell - r$, that is, in the case when $|t_1 \cup t_2| > r$ and no hyperedge contains both cover elements, the same formula holds observing that $\binom{n-2\ell+y}{r-2\ell+y} = 0$.

We now consider the colorings for which three colors are assigned either to E_1^* or to E_2^*. The number of ways of assigning the colors is $2\binom{4}{3}$, and we need to count the number of 3-colorings of the r-sets in E_1^* (or in E_2^*) for which all three colors are used, as the other colorings have already been counted in the case where two colors were assigned to each cover element. By inclusion-exclusion, we see that there are

$$2 \cdot \binom{4}{3} \cdot \left(3^{\binom{n-\ell}{r-\ell}-\binom{n-2\ell+y}{r-2\ell+y}} - 3 \cdot 2^{\binom{n-\ell}{r-\ell}-\binom{n-2\ell+y}{r-2\ell+y}} + 3\right) \cdot 4^{\binom{n-2\ell+y}{r-2\ell+y}} \tag{10}$$

such colorings. Again, this formula holds for all values of y, but it does not depend on y if $y < 2\ell - r$ (as $\binom{n-2\ell+y}{r-2\ell+y} = 0$).

Combining equations (9) and (10), we deduce that the number $S(y)$ of star colorings of H^* is given by

$$S(y) = \binom{4}{2} \cdot 2^{2(\binom{n-\ell}{r-\ell} - \binom{n-2\ell+y}{r-2\ell+y})} 4^{\binom{n-2\ell+y}{r-2\ell+y}} - \binom{4}{2} \cdot 2 \cdot 4^{\binom{n-2\ell+y}{r-2\ell+y}} + \tag{11}$$

$$+2 \cdot \binom{4}{3} \cdot \left(3^{\binom{n-\ell}{r-\ell} - \binom{n-2\ell+y}{r-2\ell+y}} - 3 \cdot 2^{\binom{n-\ell}{r-\ell} - \binom{n-2\ell+y}{r-2\ell+y}} + 3\right) \cdot 4^{\binom{n-2\ell+y}{r-2\ell+y}}$$

$$= 6 \cdot 4^{\binom{n-\ell}{r-\ell}} + 8 \cdot 3^{\binom{n-\ell}{r-\ell}} \left(\frac{4}{3}\right)^{\binom{n-2\ell+y}{r-2\ell+y}} \tag{12}$$

$$\times \left(1 - 3\left(\frac{2}{3}\right)^{\binom{n-\ell}{r-\ell} - \binom{n-2\ell+y}{r-2\ell+y}} + 3\left(\frac{1}{3}\right)^{\binom{n-\ell}{r-\ell} - \binom{n-2\ell+y}{r-2\ell+y}} - \frac{3 \cdot 3^{\binom{n-2\ell+y}{r-2\ell+y}}}{2 \cdot 3^{\binom{n-\ell}{r-\ell}}}\right).$$

We know that $y \leq \ell - 1$ and that the last three terms within the brackets in (12) have a nonpositive sum and each tends to 0 as n increases. We conclude that, for n sufficiently large,

$$6 \cdot 4^{\binom{n-\ell}{r-\ell}} + 7 \cdot 3^{\binom{n-\ell}{r-\ell}} \left(\frac{4}{3}\right)^{\binom{n-2\ell+y}{r-2\ell+y}}$$

$$\leq S(y) \leq 6 \cdot 4^{\binom{n-\ell}{r-\ell}} + 8 \cdot 3^{\binom{n-\ell}{r-\ell}} \left(\frac{4}{3}\right)^{\binom{n-2\ell+y}{r-2\ell+y}}. \tag{13}$$

Hence, for large n, we have $S(y-1) \leq S(y)$ for every y, with equality only if $\binom{n-2\ell+y}{r-2\ell+y} = 0$, that is, only if there are no hyperedges containg $t_1 \cup t_2$ when $|t_1 \cap t_2| = y$. Observe that, for $y = \ell - 1$, we have $r - 2\ell + y = r - \ell - 1 \geq 0$, so that $\binom{n-2\ell+y}{r-2\ell+y} \geq 1$ in this case. Therefore $S(y)$ is maximum for $y = \ell - 1$.

We now show that, if $|t_1 \cap t_2| < \ell - 1$, then H^* is not $(r, 4, \ell)$-extremal. To this end, we find an upper bound on the number of non-star colorings of H^*. For such a coloring Δ, there exists at least one pair (a, b) of r-sets $a, b \in E$ of the same color such that $|a \cap b| \geq \ell$ with $t_1 \subset a$ and $t_2 \subset b$, but $t_1 \cup t_2 \not\subseteq a$ and $t_1 \cup t_2 \not\subseteq b$. Let $|a \cap (t_2 \setminus t_1)| = q$ and $|b \cap (t_1 \setminus t_2)| = p$, where we may assume $p \leq q$ by symmetry. Thus

$$p + y \leq \ell - 1 \quad \text{and} \quad q + y \leq \ell - 1. \tag{14}$$

Let

$$\mathcal{F}_1(b) = \{e \in E \,|\, t_1 \subset e, t_2 \not\subset e, \Delta(e) = \Delta(b)\} \tag{15}$$

$$\mathcal{F}_2(a) = \{e \in E \,|\, t_2 \subset e, t_1 \not\subset e, \Delta(e) = \Delta(a)\}. \tag{16}$$

Lemma 7 *There exists a constant $C > 0$ and a positive integer n_0 such that, for all $n \geq n_0$, we have*

$$|\mathcal{F}_1(b)| \leq C \cdot n^{r-2\ell+p+y} \qquad \text{and} \qquad |\mathcal{F}_2(a)| \leq C \cdot n^{r-2\ell+q+y}. \tag{17}$$

Notice that $r \geq 2\ell - p - y$ and $r \geq 2\ell - q - y$. Indeed, since a and b intersect in at least ℓ elements, we must have $|(a \setminus t_1) \cap (b \setminus t_1)| \geq \ell - (p + y)$, hence a must contain at least $\ell + (\ell - (p + y)) = 2\ell - p - y$ elements, which implies that $r \geq 2\ell - p - y$. Similarly we obtain $r \geq 2\ell - q - y$.

Proof. Given r-sets a and b, there is a constant $C > 0$ such that

$$|\mathcal{F}_1(b)| \le \binom{r-p-y}{\ell-(p+y)}\binom{n-(2\ell-p-y)}{r-(2\ell-p-y)} \le C \cdot n^{r-2\ell+p+y},$$

as, ignoring overcounting, we can choose $\ell-(p+y)$ elements from the set $b \setminus t_1$ in $\binom{r-p-y}{\ell-(p+y)}$ ways, and the remaining $r-(2\ell-p-y)$ elements in at most $\binom{n-(2\ell-p-y)}{r-(2\ell-p-y)}$ ways. Similarly, the second inequality follows, namely

$$|\mathcal{F}_2(a)| \le \binom{r-q-y}{\ell-(q+y)}\binom{n-(2\ell-q-y)}{r-(2\ell-q-y)} \le C \cdot n^{r-2\ell+q+y}.$$

□

With the r-sets a and b fixed, subfamilies $\mathcal{G}_1 \subseteq \mathcal{F}_1(b)$ and $\mathcal{G}_2 \subseteq \mathcal{F}_2(a)$ may be assigned the same color as a and b, provided that $|a' \cap b'| \ge \ell$ for all r-sets $a' \in \mathcal{G}_1$ and $b' \in \mathcal{G}_2$.

The r-sets a and b may be chosen in at most $\binom{n-\ell}{r-\ell}^2 \le n^{2r-2\ell}$ ways, the subfamilies $\mathcal{G}_1$ and $\mathcal{G}_2$ may be fixed in at most $2^{|\mathcal{F}_1(b)|+|\mathcal{F}_2(a)|}$ ways and their color may be chosen in four ways. As $p \le q$, for n large, with Lemma 7 we deduce that the total number of choices is at most

$$\begin{aligned} 4n^{2r-2\ell} \cdot 2^{|\mathcal{F}_1(b)|+|\mathcal{F}_2(a)|} &\le 4n^{2r-2\ell} \cdot 2^{Cn^{r-2\ell+p+y}+Cn^{r-2\ell+q+y}} \\ &\le 4n^{2r-2\ell} \cdot 2^{2Cn^{r-2\ell+q+y}}. \end{aligned} \tag{18}$$

Having fixed the color of a and b and all r-sets in $\mathcal{G}_1 \cup \mathcal{G}_2$, we may use this color only for r-sets covering $t_1 \cup t_2$, which appear if $r \ge 2\ell - y$. We may finish the coloring in two ways.

On the one hand, we may use the remaining three colors for a star coloring of the set of uncolored hyperedges. This can be done in at most

$$2\binom{3}{2} \cdot 2^{\binom{n-\ell}{r-\ell}-\binom{n-2\ell+y}{r-2\ell+y}} \cdot 4^{\binom{n-2\ell+y}{r-2\ell+y}} = 6 \cdot 2^{\binom{n-\ell}{r-\ell}+\binom{n-2\ell+y}{r-2\ell+y}} \tag{19}$$

ways; hence, with (18) and (19), for n sufficiently large, the number of such non-star colorings is at most

$$24n^{2r-2\ell} \cdot 2^{2Cn^{r-2\ell+q+y}} \cdot 2^{\binom{n-\ell}{r-\ell}+\binom{n-2\ell+y}{r-2\ell+y}}. \tag{20}$$

On the other hand, assume that there exists another pair $(a_1, b_1) \neq (a, b)$ of r-sets $a_1, b_1 \in E$ with $|a_1 \cap b_1| \ge \ell$ and $t_1 \subset a_1$, and $t_2 \subset b_1$, but $t_1 \cup t_2 \not\subseteq a_1$ and $t_1 \cup t_2 \not\subseteq b_1$, and with $|a_1 \cap (t_2 \setminus t_1)| = q'$ and $|b_1 \cap (t_1 \setminus t_2)| = p'$, say $p' \le q'$, where $\Delta(a_1) = \Delta(b_1) \neq \Delta(a)$. Let the families $\mathcal{F}_1(b_1)$ and $\mathcal{F}_2(a_1)$ be defined as in (15) and (16). As above, r-sets in subfamilies $\mathcal{G}_1' \subseteq \mathcal{F}_1(b_1)$ and $\mathcal{G}_2' \subseteq \mathcal{F}_2(a_1)$ may be assigned the same color as a_1 and b_1, provided that $|a' \cap b'| \ge \ell$ for all r-sets $a' \in \mathcal{G}_1'$ and $b' \in \mathcal{G}_2'$. The pair (a_1, b_1) may be chosen in at most $n^{2r-2\ell}$

ways. Again by Lemma 7 with $p' \leq q'$, for n sufficiently large, we have that, for some constant $C' > 0$,

$$2^{|\mathcal{F}_1(b_1)|+|\mathcal{F}_2(a_1)|} \leq 2^{C'n^{r-2\ell+p'+y}+C'n^{r-2\ell+q'+y}} \leq 2^{2C'n^{r-2\ell+q'+y}}. \tag{21}$$

Combining this with (18) and (21) gives us, for n sufficiently large, at most

$$\begin{aligned} &4n^{2r-2\ell} \cdot 2^{2Cn^{r-2\ell+q+y}} \cdot 3 \cdot 2 \cdot n^{2r-2\ell} \cdot 2^{2C'n^{r-2\ell+q'+y}} \cdot 4^{\binom{r-2\ell+y}{r-2\ell+y}} \\ &= 24n^{4r-4\ell} \cdot 2^{2Cn^{r-2\ell+q+y}+2C'n^{r-2\ell+q'+y}} \cdot 4^{\binom{r-2\ell+y}{r-2\ell+y}} \end{aligned} \tag{22}$$

such non-star colorings, since the two colors not used so far can be taken for those remaining hyperedges not covering $t_1 \cup t_2$ in at most two ways. Indeed, if we use one of the two remaining colors for another pair (a_2, b_2) of r-sets that is distinct from the pairs (a, b) and (a_1, b_1), where $t_1 \cup t_2 \not\subseteq a_2$ and $t_1 \cup t_2 \not\subseteq b_2$, then by Lemma 7 with (14), for some constant $C^* > 0$ this color can be used for at most $C^* n^{r-\ell-1}$ many r-sets containing t_1 but not t_2, or t_2 but not t_1, respectively. However, this leaves at least $\binom{n-\ell}{r-\ell} - C^* n^{r-\ell-1}$ uncolored r-sets containing t_1 or t_2, but not $t_1 \cup t_2$, which cannot be colored properly with a single color.

Hence, combining (20) and (22) with the inequalities $q + y \leq \ell - 1$ and $q' + y \leq \ell - 1$ given in (14), we have that, for some constant $C'' > 0$, the total number of non-star colorings of H^* is bounded above by

$$\begin{aligned} &24n^{2r-2\ell} \cdot 2^{\binom{n-\ell}{r-\ell}+\binom{n-2\ell+y}{r-2\ell+y}} \cdot 2^{2Cn^{r-2\ell+q+y}} \\ &\quad +24n^{4r-4\ell} \cdot 2^{2Cn^{r-2\ell+q+y}+2C'n^{r-2\ell+q'+y}} \cdot 4^{\binom{r-2\ell+y}{r-2\ell+y}} \\ &\quad \leq 24 \cdot 2^{C''n^{r-\ell-1}} \left(n^{2r-2\ell} \cdot 2^{\binom{n-\ell}{r-\ell}+\binom{n-\ell-1}{r-\ell-1}} + n^{4r-4\ell} \cdot 4^{\binom{r-\ell-1}{r-\ell-1}} \right). \end{aligned} \tag{23}$$

For fixed p and q (respectively p' and q') the number of possibilities for choosing the intersections $|a \cap (t_2 \setminus t_1)| = q$ and $|b \cap (t_1 \setminus t_2)| = p$ as well as $|a_1 \cap (t_2 \setminus t_1)| = q'$ and $|b_1 \cap (t_1 \setminus t_2)| = p'$ is bounded from above by a constant, which affects the upper bound (23) by at most a constant factor.

The non-star colorings are not enough to bridge the gap between $S(\ell - 1)$ and $S(\ell - 2)$, as by (13) we have

$$\begin{aligned} S(\ell-1) - S(\ell-2) &\geq 7 \cdot 3^{\binom{n-\ell-1}{r-\ell}} \cdot 4^{\binom{n-\ell-1}{r-\ell-1}} - 8 \cdot 3^{\binom{n-\ell}{r-\ell}} \cdot \left(\frac{4}{3}\right)^{\binom{n-\ell-2}{r-\ell-2}} \\ &= 3^{\binom{n-\ell-1}{r-\ell}} \cdot 4^{\binom{n-\ell-2}{r-\ell-2}} \left(7 \cdot 4^{\binom{n-\ell-2}{r-\ell-1}} - 8 \cdot 3^{\binom{n-\ell-2}{r-\ell-1}}\right). \end{aligned}$$

Since $7 \cdot 4^{\binom{n-\ell-2}{r-\ell-1}} - 8 \cdot 3^{\binom{n-\ell-2}{r-\ell-1}} \geq 1$ for $r > \ell$ and n sufficiently large, we have

$$S(\ell-1) - S(\ell-2) \geq 3^{\binom{n-\ell-1}{r-\ell}} \cdot 4^{\binom{n-\ell-2}{r-\ell-2}},$$

which is much larger than the upper bound in (23).

Therefore, the number of $(4, \ell)$-Kneser colorings of H^* is maximized for $y = \ell - 1$, which finishes the proof of Theorem 3. □

To conclude this section, we argue that the proof of Theorem 3 implies the missing part of the proof of Theorem 2. Recall that we have shown in Section 2 that there exist $\varepsilon_0 > 0$ and an integer n_1 such that $\kappa(H, 4, \ell) \leq (1 - \varepsilon_0) \cdot \mathrm{KC}(n, r, 4, \ell)$, whenever $H \notin \mathcal{H}_{r,4,\ell}(n)$. However, we did not show that, given $\varepsilon > 0$, every hypergraph $\mathcal{H}_{r,4,\ell}(n)$ is within $\varepsilon \cdot \mathrm{KC}(n, r, 4, \ell)$ of being maximum. Now, let $H \in \mathcal{H}_{r,4,\ell}(n)$. The $(4, \ell)$-Kneser colorings of H may be split into star colorings and non-star colorings, and we know from the proof of Theorem 3 that the number of non-star colorings is very small in comparison with the number of star colorings (this would also be a consequence of Lemma 3, as all $(4, \ell)$-Kneser colorings for which all colors are influential must be star colorings). Moreover, the formula (13) for $S(y)$ shows that, regardless of the value of y, the number of star colorings of H is given by $6 \cdot 4^{\binom{n-\ell}{r-\ell}} + o\left(4^{\binom{n-\ell}{r-\ell}}\right)$, which leads to the desired conclusion.

4 Final Remarks and Open Problems

In this paper, we have described an extremal problem for hypergraphs that, in spite of being provably unstable, has a unique optimal solution. Even though we have concentrated on the problem $\mathcal{P}_{n,r,k,\ell}$ in the case when $k = 4$ and $r > \ell \geq 2$, there is a full characterization of the stability of this problem for general r, k and ℓ.

Theorem 5. *[13] Let $k \geq 2$ and $r > \ell$ be positive integers.*

(i) If $k \in \{2, 3\}$, then $\mathcal{P}_{n,r,k,\ell}$ is stable.
(ii) If $k = 4$, then $\mathcal{P}_{n,r,k,\ell}$ is stable if and only if $\ell = 1$.
(iii) If $k \geq 5$, then $\mathcal{P}_{n,r,k,\ell}$ is stable if and only if $r \geq 2\ell - 1$.

In the current paper, we were able to carry out the calculations for the case $k = 4$ and $\ell > 1$, and we have determined the unique optimal configuration. For other values of k, r and ℓ, and n large, the unique (r, k, ℓ)-extremal hypergraphs on n vertices are known for all instances for which the problem is stable, while the family of hypergraphs that are asymptotically optimal has been found for every instance such that the problem is unstable (see [13]). For a better description of the situation, let $H_{n,r,k,\ell}$ be the (C, r)-complete r-uniform hypergraph on n vertices with minimum ℓ-cover C of size $\lceil k/3 \rceil$, where distinct ℓ-sets in C have empty intersection. This hypergraph plays an important role, as it is the unique (r, k, ℓ)-extremal hypergraph for the problem $\mathcal{P}_{n,r,k,\ell}$ whenever n is sufficiently large and the problem is stable. Furthermore, even when the problem is unstable, it has been proved in [13] that the hypergraph $H_{n,r,k,\ell}$ is asymptotically optimal. On the other hand, Theorem 3 implies that $H_{n,r,k,\ell}$ is not optimal in the case $k = 4$. This behavior is not accidental, and it is possible to show that, for n sufficiently large, the hypergraph $H_{n,r,k,\ell}$ is never (r, k, ℓ)-extremal when the problem $\mathcal{P}_{n,r,k,\ell}$ is unstable. To prove this, one may show that a (C, r)-complete hypergraph with minimum ℓ-cover C of size $\lceil k/3 \rceil$ for which every two cover elements have intersection of size $2\ell - r - 1$ has more (k, ℓ)-colorings than $H_{n,r,k,\ell}$. As a matter of fact, we conjecture that the pairwise intersections of the cover elements should be as large as possible to achieve maximality.

Conjecture 1. If $k \geq 5$, r and ℓ are positive integers with $\ell < r < 2\ell$, then a hypergraph $H = H_{C,r}(n)$ such that

$$\kappa(H, k, \ell) = \text{KC}(n, r, k, \ell)$$

must satisfy $|C| = c(k) = \lceil k/3 \rceil$ and $|t_i \cap t_j| = 2\ell - r - 1$ for every distinct $t_i, t_j \in C$.

As it turns out, even if Conjecture 1 is true in general, there may be several configurations of ℓ-sets in C whose pairwise intersections have size $2\ell - r - 1$. In particular, this conjecture would be a first step in establishing that, for sufficiently large n, there is always a unique optimal configuration for the problem $\mathcal{P}_{n,r,k,\ell}$.

Acknowledgements. We thank Miklós Simonovits for valuable comments and for directing us to related literature.

References

1. Ahlswede, R., Khachatrian, L.H.: The complete intersection theorem for systems of finite sets. European J. Combin. 18(2), 125–136 (1997)
2. Blokhuis, A., Brouwer, A., Szőnyi, T., Weiner, Z.: On q-analogues and stability theorems. Journal of Geometry 101, 31–50 (2011)
3. Brown, W.G.: On an open problem of Paul Turán concerning 3-graphs. Studies in pure mathematics, pp. 91–93. Birkhäuser, Basel (1983)
4. Brown, W.G., Simonovits, M.: Extremal multigraph and digraph problems. Paul Erdős and his Mathematics, II (Budapest, 1999), Bolyai Soc. Math. Stud., János Bolyai Math. Soc. 11, 157–203 (2002)
5. Erdős, P.: Some recent results on extremal problems in graph theory. In: Results, Theory of Graphs, Internat. Sympos., Rome, pp. 117–123 (English) (1966), pp. 124–130 (French). Gordon and Breach, New York (1967)
6. Erdős, P., Ko, C., Rado, R.: Intersection theorems for systems of finite sets. Quart. J. Math. Oxford Ser. 2(12), 313–320 (1961)
7. Fon-Der-Flaass, D.G.: A method for constructing (3, 4)-graphs. Mat. Zametki 44(4), 546–550, 559 (1988)
8. Frohmader, A.: More constructions for Turán's (3, 4)-conjecture. Electron. J. Combin. 15(1), Research Paper 137, 23 (2008)
9. Füredi, Z.: A new proof of the stability of extremal graphs: Simonovits's stability from Szemerédi's regularity. Talk at the Conference in Honor of the 70th Birthday of Endre Szemerédi, Budapest, August 2-7 (2010)
10. Füredi, Z.: Turán type problems, Surveys in combinatorics, Guildford. London Math. Soc. Lecture Note Ser., vol. 166, pp. 253–300. Cambridge Univ. Press, Cambridge (1991)
11. Füredi, Z., Simonovits, M.: Triple systems not containing a Fano configuration. Combin. Probab. Comput. 14(4), 467–484 (2005)
12. Hoppen, C., Kohayakawa, Y., Lefmann, H.: Edge colourings of graphs avoiding monochromatic matchings of a given size. Combin. Probab. Comput. 21(1-2), 203–218 (2012)

13. Hoppen, C., Kohayakawa, Y., Lefmann, H.: Hypergraphs with many Kneser colorings. European Journal of Combinatorics 33(5), 816–843 (2012)
14. Keevash, P.: Hypergraph Turán problems, Surveys in combinatorics 2011. London Math. Soc. Lecture Note Ser., vol. 392, pp. 83–140. Cambridge Univ. Press, Cambridge (2011)
15. Keevash, P., Mubayi, D.: Set systems without a simplex or a cluster. Combinatorica 30(2), 175–200 (2010)
16. Keevash, P., Sudakov, B.: The Turán number of the Fano plane. Combinatorica 25(5), 561–574 (2005)
17. Kostochka, A.V.: A class of constructions for Turán's (3,4)-problem. Combinatorica 2(2), 187–192 (1982)
18. Simonovits, M.: A method for solving extremal problems in graph theory, stability problems. In: Theory of Graphs, Proc. Colloq., Tihany, 1966, pp. 279–319. Academic Press, New York (1968)
19. Simonovits, M.: Some of my favorite Erdős theorems and related results, theories. In: Paul Erdős and his Mathematics, II, Bolyai Soc. Math. Stud. (Budapest, 1999) János Bolyai Math. Soc. 11, 565–635 (2002)
20. Szőnyi, T., Weiner, Z.: On stability theorems in finite geometry, 38 pp. (2008) (preprint)
21. Turán, P.: Research problems. Magyar Tud. Akad. Mat. Kutató Int. Közl 6, 417–423 (1961)

Multiparty Communication Complexity of Vector–Valued and Sum–Type Functions

Ulrich Tamm

German Language Department of Business Informatics,
Marmara University, Istanbul, Turkey
tamm@ieee.org

Dedicated to the memory of Rudolf Ahlswede

Abstract. Rudolf Ahlswede's work on communication complexity dealt with functions defined on direct sums: vector–valued functions and sum–type functions. He was interested in single–letter characterizations and provided several lower bound techniques to this aim. In this paper we shall review these lower bounds and extend them to the "number in hand" multiparty model of communication complexity.

Keywords: communication complexity, direct sum functions, tensor product.

1 Introduction

Sum-type functions f_n and vector-valued functions f^n are defined on the powers $\mathcal{X}^n, \mathcal{Y}^n$ of the sets from the domain of some basic function $f : \mathcal{X} \times \mathcal{Y} \to \mathcal{Z}$. Elements of $\mathcal{X}^n$ and $\mathcal{Y}^n$ are denoted as x^n and y^n, respectively. Hence, e. g., $x^n = (x_1, \dots, x_n)$ for some $x_1, \dots, x_n \in \mathcal{X}$. With this notation

$$f^n(x^n, y^n) = \big(f(x_1, y_1), \dots, f(x_n, y_n)\big), \qquad f_n(x^n, y^n) = \sum_{i=1}^{n} f(x_i, y_i),$$

where it is required that the range $\mathcal{Z}$ is a subset of an additive group G.

Motivated by the communication complexity of the Hamming distance [8], in a series of papers Rudolf Ahlswede ([1] - [7]) and his group in Bielefeld ([17] - [20]) studied the communication complexity of sum–type and vector–valued functions. The results are summarized in [21]. Rudolf Ahlswede was mainly interested in a single–letter characterization basing the communication complexity of f_n and f^n on the communication complexity of the function f. To this aim he and his coauthors demonstrated that several lower bounds behave multiplicatively. These results and also their applications yielding the exact communication complexity for special functions as Hamming distance and set intersection are presented in Section 2.

H. Aydinian, F. Cicalese, and C. Deppe (Eds.): Ahlswede Festschrift, LNCS 7777, pp. 451–462, 2013.

A further line of research leading to direct sum methods in communication complexity goes back to the question if it is easier to solve communication problems simultaneously than separately, cf. [15], pp. 42 - 48. Recall the definition of a vector-valued function $f^n((x_1, \dots, x_n), (y_1, \dots, y_n)) = (f(x_1, y_1), \dots, f(x_n, y_n))$. An obvious upper bound on the communication complexity $C(f^n)$ is obtained by evaluating each component $f(x_i, y_i)$ separately and communicating the result for component i using the optimal protocol for f. Can we do better by considering all components simultaneously? Ahlswede et al. [3], [5] using data compression could show that for set intersection it is $C(f) = 2$ but $C(f^n) = \lceil n \cdot \log_2 3 \rceil$.

The measure $\limsup_{n\to\infty} \frac{1}{n} C(f^n)$ is also called amortized communication complexity (see [13]). One of the main open problems in communication complexity is the question if there can exist a significant gap between the communication complexity and the amortized communication complexity of a function. Direct sum methods in communication complexity are also useful in the comparison of lower bound techniques and the study of their power. The famous log-rank conjecture states that the gap between the rank lower bound and the communication complexity cannot be too big.

The last problem was recently extended to the "number in hand" model of multiparty communication complexity [12]. Yao's model of communication complexity can be generalized to several multiparty models depending on the information accessible to each person. Most well studied is the "number on the forehead" model in which each person knows all inputs but her own, for instance [10]. The "number in hand" model, in which each person knows just her own input, was not so popular in the beginning but later found an important application in streaming [9]. The problem with "number in hand" is that a generalization of the lower bound techniques is rather difficult. The most powerful lower bound in two-party communication complexity is the rank lower bound. But the rank of a matrix is generalized by a tensor rank (3 and higher dimensional matrices), which is not so easy to determine. Besides, the matrix rank is multiplicative under the tensor product (very important for functions on direct sums). This is no longer the case for higher dimensional tensors, cf. [11].

Vector-valued and sum-type functions can straightforwardly be extended to functions in 3 and more arguments. In Section 3 we shall study the multiparty communication complexity of several generalizations of the Hamming distance and set intersection, the two functions mainly discussed by Ahlswede and his coauthors. Fortunately, a bound introduced by Ahlswede and Cai [3] via the independence number can replace the rank lower bound in this case, such that sharp lower bounds are still possible. This will be demonstrated with four boolean functions in more than two arguments.

For sum–type functions the independence number is not an appropriate lower bound. Since also the rank lower bound is not easily applicable, it hence remains to study largest monochromatic rectangles. In Section 4 a generalization of Ahlswede's 4-word property is presented. This yields tight bounds on the size of the largest monochromatic rectangles for some functions only if there is an even number of persons involved in the multiparty communication. For odd numbers one dimension can not be included in the formula.

As an example that it may occur that three-party communication behaves much like two-party communication the pairwise comparison of the inputs is analyzed in Section 5.

2 Bounds on Communication Complexity

The notion of communication complexity was introduced by Yao in 1979 [22]. Since then it found many applications in Computer Science, for which we refer to the books by Kushilevitz and Nisan [15] or by Hromkovic [14]. The communication complexity of a function $f : \mathcal{X} \times \mathcal{Y} \to \mathcal{Z}$ (where $\mathcal{X}$, $\mathcal{Y}$, and $\mathcal{Z}$ are finite sets), denoted as $C(f)$, is the number of bits that two persons, P_1 and P_2, have to exchange in order to compute the function value $f(x, y)$, when initially P_1 only knows $x \in \mathcal{X}$ and P_2 only knows $y \in \mathcal{Y}$. To this aim they follow a predetermined interactive protocol in which the set of messages a person is allowed to send at each instance of time form a prefix code.

Upper bounds are usually obtained by special protocols. Often, the trivial protocol, in which one person sends all the bits of his input and the other person returns the result, is at least asymptotically optimal.

Lower bounds are expressed via the function matrix $M(f) = \big(f(x,y)\big)_{x\in\mathcal{X},y\in\mathcal{Y}}$. and the function value matrices $M_z(f) = (a_{xy})_{x\in\mathcal{X},y\in\mathcal{Y}}$ for all $z \in \mathcal{Z}$ defined by $a_{xy} = \begin{cases} 1 & \text{if } f(x,y) = z \\ 0 & \text{if } f(x,y) \neq z. \end{cases}$

Yao [22] already showed that $C(f) \geq \log D(f)$, where the decomposition number $D(f)$ denotes the minimum size of a partition of $\mathcal{X} \times \mathcal{Y}$ into monochromatic rectangles, i. e., products $A \times B$ of pairs $A \subset \mathcal{X}, B \subset \mathcal{Y}$ on which the function is constant. The decomposition number usually is hard to determine, however, further lower bounds can be derived from it. Immediately, we have

$$C(f) \geq \left\lceil \log \frac{|\mathcal{X}| \cdot |\mathcal{Y}|}{Lmr\big(M(f)\big)} \right\rceil, \tag{1}$$

where $Lmr\big(M(f)\big)$ denotes the size of the largest monochromatic rectangle in the function matrix $M(f)$.

In order to make induction proofs possible Ahlswede weakened the conditions on the rectangles. He no longer required that the function is constant on the rectangle $A \times B$ but that the so called 4-word- property has to be fulfilled, i. e., for all $a, a' \in A$, $b, b' \in B$

$$f(a,b) - f(a',b) - f(a,b') + f(a',b') = 0$$

Denoting by $Lfw(f)$ the size of the largest rectangle, on which the 4-word-property holds, we obtain

$$C(f) \geq \left\lceil \log \frac{|\mathcal{X}| \cdot |\mathcal{Y}|}{Lfw(f)} \right\rceil. \tag{2}$$

A z–independent set $\{(x^{(1)}, y^{(1)}), \ldots, (x^{(N)}, y^{(N)})\}$ for the function value z in $M(f)$ is a set of pairs with $f(x^{(i)}, y^{(i)}) = z$ for all $i = 1, \ldots, N$ such that no two members of the set are in the same monochromatic rectangle. Denoting the size of a z–independent set by $ind(M_z(f))$ and $Ind(f) = \sum_{z \in Z} ind(M_z(f))$ we obtain [3]

$$C(f) \geq \lceil \log Ind(f) \rceil. \tag{3}$$

$C(f)$ can also be lower bounded by the rank of the corresponding function matrices

$$C(f) \geq \lceil \log r(f) \rceil, \text{ where } r(f) = \sum_{z \in Z} \text{rank} M_z(f) \tag{4}$$

It can be shown that the function f has the same communication complexity as the function g defined by $g(x, y) = c^{f(x,y)}$ for all x, y, when the number c is chosen appropriately ($c \neq 0, |c| \neq 1$). So it is also possible to lower bound $C(f)$ by the rank of $M(g) = \exp(M(f), c) = (c^{f(x,y)})_{x \in \mathcal{X}, y \in \mathcal{Y}}$, the exponential transform of the matrix $M(f)$. This yields

$$C(f) \geq \lceil \log \text{rank} \exp(M(f), c) \rceil. \tag{5}$$

Central in the following arguments is the observation that the function matrices of the vector-valued and sum-type functions can be expressed in terms of the Kronecker product or tensor product, defined for two matrices $A = (a_{ij})_{i,j}$ and $B = (b_{kl})_{k,l}$ as $A \otimes B = (a_{ij} \cdot b_{kl})_{i,j,k,l}$. The n-fold Kronecker product of a matrix is denoted as $A^{\otimes n}$. We have (cf. [3], [4], [17], [19])

$$M_{(z_1,\ldots,z_n)}(f^n) = M_{z_1}(f) \otimes M_{z_2}(f) \otimes \cdots \otimes M_{z_n}(f) \tag{6}$$

$$M_z(f_n) = \sum_{\substack{(z_1,\ldots,z_n) \\ z_1 + \cdots + z_n = z}} M_{z_1}(f) \otimes \cdots \otimes M_{z_n}(f) \tag{7}$$

$$\exp\big(M(f_n), c\big) = \big[\exp(M(f), c)\big]^{\otimes n} \tag{8}$$

It can be shown that the parameters in the bounds (2) - (5) behave multiplicatively, since the rank and hence also $r(f) = \sum_{z \in Z} \text{rank} M_z(f)$ are multiplicative under the Kronecker product.

Theorem 1. *([3], [4]):*

$$Lfw(f_n) = n \cdot Lfw(f) \tag{9}$$

$$rank \exp\big(M(f_n), c\big) = \big(rank\big[\exp(M(f), c)\big]\big)^n \tag{10}$$

$$r(f^n) = r(f)^n \tag{11}$$

$$Ind(f^n) \geq Ind(f)^n \tag{12}$$

Using these bounds Ahlswede et al. analyzed several sum – type functions especially the Hamming distance and set intersection defined by the basic function matrices $M(h) = \begin{pmatrix} 0\,1 \\ 1\,0 \end{pmatrix}$ and $M(si) = \begin{pmatrix} 0\,0 \\ 0\,1 \end{pmatrix}$

For sum-type and vector–valued functions defined on more than two arguments the corresponding function tensors (higher dimensional function matrices) can again be described in terms of the tensor product, where $A \otimes B$ is the tensor obtained by multiplying each entry of A with each entry of B. This way, the descriptions (6), (7), and (8) generalize. However, the tensor product now is no longer multicplicative (cf. [11]), that means, $\text{rank}(A \otimes B)$ can be smaller than $\text{rank}(A) \cdot \text{rank}(B)$. So, the rank lower bounds (10) and (11) can no longer be applied. However, for vector–valued functions still (12) and for sum–type functions (9) can be used.

3 Multiparty Communication Complexity of Vector-Valued Functions

In this section we try to extend Rudolf Ahlswede's methods to determine the multi–party communication complexity of some functions defined on direct sums. First is repeated the result for symmetric difference and the set intersection, since we need it for the multiparty protocols below.

Theorem 2. *([3], [5]):*

$$C(h^n) = 2n, \qquad C(si^n) = \lceil \log 3 \rceil \tag{13}$$

Proof: For the symmetric difference h^n, the trivial protocol requires $\lceil \log 2^n \rceil + \lceil \log 2^n \rceil = 2n$ bits of communication. With the rank lower bound (4) and (11) this can be shown to be optimal, since

$$C(h^n) \geq \log r(h^n) = n \cdot \log r(h) = n \cdot \log(\text{rank} M_0(h) + \text{rank} M_1(h))$$

$$= n \cdot \log(\text{rank} \begin{pmatrix} 1\,0 \\ 0\,1 \end{pmatrix} + \text{rank} \begin{pmatrix} 0\,1 \\ 1\,0 \end{pmatrix}) = n \cdot \log 4 = 2n$$

For set-intersection the rank lower bound yields

$$C(si^n) \geq \lceil \log r(si^n) \rceil = \lceil n \cdot \log r(si) \rceil = \lceil n \cdot \log(\text{rank} M_0(si) + \text{rank} M_1(si)) \rceil$$

$$= \lceil n \cdot \log(\text{rank} \begin{pmatrix} 1\,1 \\ 1\,0 \end{pmatrix} + \text{rank} \begin{pmatrix} 0\,0 \\ 0\,1 \end{pmatrix}) \rceil = \lceil n \cdot \log 3 \rceil$$

In order to obtain the same upper bound, we shall modify the trivial protocol, which would require $2n$ bits of transmission. Again, in the first round person P_1 encodes his input $x^n \in \{0, 1\}^n$. P_2 then knows both values and hence is able to compute the result $si^n(x^n, y^n)$, which is returned to P_1. However, in knowledge of x^n the set of possible function values is reduced to the set $S(x^n) = \{y^n : y^n \subset x^n\}$. Hence, only $\lceil \log S(x^n) \rceil$ bits have to be reserved for the transmission of $si^n(x^n, y^n)$ such that P_1 can assign longer messages to elements with few subsets. So, in contrast to the trivial protocol, the messages $\{\phi_1(x^n) : x^n \in \{0,1\}^n\}$ are now of variable length. Since the prefix property has to be guaranteed, Kraft's inequality for prefix codes yields a condition, from which the upper bound can be derived. Specifically, we require that to each x^n there corresponds a message $\phi_1(x^n)$ of (variable) length $l(x^n)$ such that for all $x^n \in \{0,1\}^n$ the sum $l(x^n) + \lceil \log S(x^n) \rceil$ takes a fixed value, L say. Kraft's inequality states that a prefix code exists, if $\sum_{x^n} 2^{-l(x^n)} \leq 1$. This is equivalent to $\sum_{x^n} 2^{\lceil \log S(x^n) \rceil} \leq 2^L$. With the choice $L = \lceil \log 3^n \rceil$ Kraft's inequality holds.

The functions in more than 2 arguments below are canonical extensions of the symmetric difference (r and s below) and the set intersection function (basic functions t and u). Namely, the basic functions defined on the product $\{0,1\} \times \{0,1\} \times \cdots \times \{0,1\}$ are the following boolean functions:

1) $r(x_1, x_2, \ldots, x_k) = x_1 + x_2 + \cdots + x_k \bmod 2$

2) $s(x_1, x_2, \ldots, x_k) = \begin{cases} 1 \, , x_1 = x_2 = \cdots = x_k \\ 0 \, , \text{ else} \end{cases}$

3) $t(x_1, x_2, \ldots, x_k) = \begin{cases} 1 \, , x_1 = x_2 = \cdots = x_k = 1 \\ 0 \, , \text{ else} \end{cases}$

4) $u(x_1, x_2 \ldots, x_k) = \begin{cases} 1 & \text{if at least half of the inputs } x_i = 1 \\ 0 & \text{else} \end{cases}$

The big problem is that for more than 2 parties communicating the rank lower bound loses much of its power. The function matrices are now replaced by tensors (i.e. , higher dimensional matrices). The rank of a matrix can be extended to tensors, but it is not so easy to determine any more. The most efficient methods to determine the rank of a matrix - eigenvalues and diagonalization of matrices - cannot be applied any more. As for matrices, the rank of a tensor can be combinatorially expressed as the minimal number of rank 1 tensors whose sum is the tensor, the rank of which has to be determined. Unfortunately, this tensor rank does not behave multiplicatively under the tensor product. This means that the rank lower bound for sum–type and vector–valued functions can not be easily applied any more.

As an alternative the independence number may be considered for vector–valued functions. Following the argumentation by Ahlswede and Cai in [3], as for functions in two arguments $Ind(f^n) \geq Ind(f)^n$ also holds for vector–valued functions in $k > 2$ arguments, such that we have the lower bound $C(f^n) \geq n \cdot \log Ind(f)$. In general, the independence number is, of course, very difficult to determine, but for basic functions over small alphabets it may yield sharp bounds, as in the following theorem.

Theorem 3.

$$C(r^n) = k \cdot n$$

$$\lceil n \cdot \log(k+2) \rceil \leq C(s^n) \leq \lceil n \cdot (2 \log k) + k - 3 \rceil$$

$$\lceil n \cdot \log(k+1) \rceil \leq C(t^n) \leq \lceil n \cdot \log(k+1) \rceil + k - 2$$

$$C(u^n) = \lceil n \cdot \log 6 \rceil \textit{ for } k = 3.$$

Proof: Obviously, no two of the 2^k entries of the function tensor of r can be contained in a monochromatic rectangle such that the independence number $Ind(r) = 2^k$. Hence $C(r^n) \geq \log Ind(r)^n = \log 2^{kn} = k \cdot n$, which is also the complexity of the trivial protocol, in which all k persons transmit all their inputs.

Next, let us consider the vector–valued function $t^n(x_1^n, x_2^n, \ldots x_k^n)$, which gives the intersection of the k sets represented by the binary strings $x_1^n, \ldots, x_k^n$. The function tensor of the basic function t contains exactly one entry 1 namely for $x_1 = x_2 = \cdots = x_k = 1$, i.e., the all–1 vector of length k . All other entries are 0. The k neighbours of the all–1 vector, i.e. all $(x_1, \ldots, x_k)$ with exactly one $x_i = 0$ and all other $x_j = 1$ obviously must be contained in different monochromatic rectangles. Since also the all-1 vector must be contained in a separate monochromatic rectangle, the independence number $Ind(t) = k + 1$ and hence $C(t^n) \geq \lceil n \log Ind(t) \rceil = \lceil n \log(k+1) \rceil$.

A protocol that almost achieves this lower bound is again obtained by assigning an appropriate prefix code to the messages in the trivial protocol. As for the set intersection function si^n in two arguments, again Person 1 can assign longer messages to inputs with few 1s. The other persons then can determine the exact value following an optimal protocol for set intersection of $k-1$ sets. For $k = 2$ we already know that $\lceil n \cdot \log 3 \rceil$ bits are optimal. So, for $k = 3$, Person 1 transmits $l(x)$ bits, say for an input x. Since the total number of bits transmitted should be a fixed value L, say, $L = l(x) + f(x)$, where $f(x)$ is the number of bits the other persons should still transmit to agree on the result. In order to guarantee the existence of a prefix code, Kraft's inequality $\sum_x 2^{-l(x)} \leq 1$ must hold. This is equivalent to $\sum_x 2^{-(L-f(x))} \leq 1$ or $\sum_x 2^{f(x)} \leq 2^L$. Now if Person 1 has an input $x = x_1$ with exactly i many 1's then by the protocol for si we know already that $f(x) = \lceil i \cdot \log 3 \rceil$ bits are enough to determine the set intersection of the remaining two sets by persons 2 and 3. So, Kraft's inequality reduces to $\sum_i \binom{n}{i} 2^{\lceil i \log 3 \rceil} \leq 2^L$. This can be assured by the choice $L = \lceil n \log 4 \rceil + 1$. Analogously, for $k > 3$ we inductively obtain from Kraft's inequality $\sum_i \binom{n}{i} 2^{\lceil i \cdot \log(k) \rceil + k - 3} \leq 2^L$, which is fulfilled for $L = \lceil n \cdot \log(k+1) \rceil + k - 2$.

For the function s^n the first person can send all the n bits of her input x_1. In knowledge of this the other $k-1$ persons have to determine for each component either the function t or $1-t$. Hence, their task is to evaluate a function equivalent to set intersection t^n on $k-1$ arguments, which can be done with $\lceil n \cdot \log k \rceil + k - 3$ bits of communication by the previous considerations. However, there is a gap to the lower bound, since a maximal independent set only has size $k+2$ - the two 1's (for $x_1 = x_2 = \cdots = x_k = 0$ or 1, respectively) plus the k 0's adjacent (at Hamming distance 1 to the all-one or all-zero vector) to one of these 1's.

The same protocol as for s^n can be used for u^n in the case of $k = 3$ inputs. Again after Person 1 has transmitted all the bits of its input in each component the function t or $1 - t$ must be computed, which can be done with $\lceil n \cdot \log 3 \rceil$ bits of communication. Here, the situation is better than for the function s, since we can find an independent set of size 6 in the function tensor of the basic function u: the three 1's $u(1,1,0) = u(1,0,1) = u(0,1,1) = 1$ and the three 0's $u(0,0,1) = u(0,1,0) = u(1,0,0) = 0$ must be contained in different monochromatic rectangles, such that $C(u^n) \geq \lceil n \log Ind(u) \rceil = \lceil n \cdot \log 6 \rceil$, which is exactly the complexity of the protocol described above.

Remarks

1) Unfortunately, for $k > 3$, the function u^n is not so nicely analyzable.

2) The lower bound for the function s^n is not so easy to improve as already the case $k = 3$ demonstrates. Here $\lceil n \log 5 \rceil \leq C(f) \leq \lceil n \log 6 \rceil$. However, there is a decomposition of the function tensor of s into just 5 monochromatic rectangles: $\{0\} \times \{0\} \times \{0\}$ and $\{1\} \times \{1\} \times \{1\}$ for the two 1s and $\{0\} \times \{1\} \times \{0,1\}$, $\{1\} \times \{0,1\} \times \{0\}$ as well as $\{0,1\} \times \{0\} \times \{1\}$ for the 0s.

4 Largest Monochromatic Rectangles for Multiparty Sum–Type Functions and a Generalization of Ahlswede's 4-Word Property

In order to be able to inductively determine the size of monochromatic rectangles Ahlswede and his coauthors[2], [6] introduced the weaker 4-word property It is no longer required that the function is constant on the rectangle $A \times B$ but that for all $a, a' \in A$, $b, b' \in B$

$$f(a,b) - f(a',b) - f(a,b') + f(a',b') = 0$$

This 4-word property behaves multiplicatively for sum-type functions in the sense that if the 4-word property holds on a rectangle $A \times B$ for the basic function f then it also holds on the rectangle $A^n \times B^n$ for the sum–type function f_n. Indeed, if $M(f,R,n)$ is the size of the largest rectangle with the 4-word property in f_n it can be shown that $M(f,R,n) = M(f,R,1)^n$. This often allows to determine exactly the size of the largest monochromatic rectangle for sum–type functions and hence bound the communication complexity from below.

In my PhD thesis [17] following a question posed by Rudolf Ahlswede an extension of the 4-word property to functions in more than 2 arguments was derived. Actually, this is just the 4-word property applied to the two-dimensional projections of higher dimensional rectangles. For instance, for a basic function $f : X_1 \times X_2 \times X_3 \times X_4 \rightarrow R$ in 4 arguments this yields an 8-word property, namely:

A rectangle (A,B,C,D) with $A \subset X_1$, $B \subset X_2$, $C \subset X_3$, $D \subset X_4$ fulfills the 8-word property if for all $a, a' \in A$, $b, b' \in B$, $c, c' \in C$ and $d, d' \in D$ it holds

$$f(a,b,c_*,d_*) - f(a',b,c_*,d_*) - f(a,b',c_*,d_*) + f(a',b',c_*,d_*) = 0$$

for $c_* \in \{c, c'\}$ and $d_* \in \{d, d'\}$ and

$$f(a_*,b_*,c,d) - f(a_*,b_*,c',d) - f(a_*,b_*,c,d') + f(a_*,b_*,c',d') = 0$$

for $a_* \in \{a, a'\}$ and $b_* \in \{b, b'\}$.

This can be straightforwardly generalized to a 2t-word property for functions in an even number t of arguments. The proof follows the lines of the one in [6]. Again, the 2t-word property is multiplicative in the above sense that $M(f,R,n) = M(f,R,1)^n$ for the sum-type function f_n.

Unfortunately, these rectangles usually become too large in order to prove asymptotically tight lower bounds for the communication complexity of sum–type functions in more than two arguments. However, for some very natural functions the size of the largest monochromatic rectangles can be determined. Let discuss sum-type functions with the basic boolean functions $f : \{0,1\}^n \times \{0,1\}^n \times \{0,1\}^n \times \{0,1\}^n \to \{0,1\}$ from the previous section.

1) $r(x,y,z,w) = x+y+z+w \bmod 2$. Then for $n = 2m$ r_n takes the constant value m on the rectangle $A \times B \times C \times D$ with $A = \{00, 11\}^m$, $B = C = D = \{01, 10\}^m$. Hence, the size of the largest monochromatic rectangle of the sum-type function r_n is at least 2^{2n}. On the other hand, obviously $\{01,\} \times \{0\} \times \{0\} \times \{0\}$ is a maximal 8-word set (of size 2) for the basic function r, such that the maximal 8-word set for r_n can have size at most 2^{2n}. Hence the above configuration yields the largest monochromatic rectangle.

2) $s(x,y,z,w) = \begin{cases} 1 & ,x=y=z=w \\ 0 & ,\text{else} \end{cases}$. Here s_n takes the constant value 0 on the rectangle $\{0\}^n \times \{1\}^n \times \{0,1\}^n \times \{0,1\}^n$. This yields a monochromatic rectangle of size 4^n. On the other hand, with the largest 8-word set $\{01,\} \times \{0\} \times \{0,1\} \times \{0\}$ for the basic function s it can be shown that there is no larger monochromatic rectangle.

It would be interesting to find an analogue for the 4-word property also for sum–type functions with an odd number of arguments. For instance we conjecture the monochromatic rectangles in the function matrices of the sum–type function f_n in the following three examples for basic functions $f : \{0,1\} \times \{0,1\} \times \{0,1\} \to \{0,1\}$ in three arguments to be optimal, but there is no suitable lower bound, so far.

1') $r(x,y,z) = x+y+z \bmod 2$. For $n = 2m$ the sum–type function r_n takes the constant value m on the rectangle $\{01, 10\}^m \times \{01, 10\}^m \times \{01, 10\}^m$, hence the largest monochromatic rectangle has size at least $2^{\frac{3}{2}n}$.

2') $s(x,y,z) = \begin{cases} 1 & ,x=y=z \\ 0 & ,\text{else} \end{cases}$.

On $\{0\}^n \times \{1\}^n \times \{0,1\}^n$ s_n takes the constant value 0. Hence, the size of the largest monochromatic rectangle in the function matrix of s_n is at least 2^n. Another configuration achieving this bound with constant value m for $n = 4m$ is $\{0000, 1111\}^m \times \{0011, 1100\}^m \times \{0101, 0110, 1001, 1010\}^m$.

3') $u(x, y, z) = 1$ iff at least two of the arguments are 1 (and 0 else). Again for $n = 2m$ u_n takes the constant value m on the rectangle $\{01, 10\}^m \times \{01, 10\}^m \times \{01, 10\}^m$, which means that the size of the largest monochromatic rectangle is at least $2^{\frac{3}{2}n}$.

5 Communication Complexity of Pairwise Comparison

Let there be k persons $P_1, P_2, \ldots, P_k$ each holding a binary string $x_i \in \{0,1\}^n$ $(i = 1, \ldots, k)$. Their task is to pairwisely compare their strings in the "number in hand" model with a minimum amount of data exchange. So we have to determine the communication complexity $C(f)$ of the function $f : \{0,1\}^n \times \ldots \{0,1\}^n \to \{0,1\}^{\binom{k}{2}}$ where $f(x_1, \ldots x_k) = (f_{(i,j)}(x_i, x_j))_{i<j\in\{1,\ldots,k\}}$ with $f_{(i,j)}(x_i, x_j) = 1$ iff $x_i = x_j$ (and 0 iff $x_i \neq x_j$).

A lower bound is obviously $C(f) \geq \lfloor \frac{k}{2} \rfloor n + 1$, since f automatically compares the two strings obtained by concatenating the first $\lfloor \frac{k}{2} \rfloor$ and the next $\lfloor \frac{k}{2} \rfloor$ inputs in the two - party comunication model. Here the trivial protocol is optimal for the equality function.

Theorem 4. $\lim_{n\to\infty} \frac{1}{n} C(f) = \lfloor \frac{k}{2} \rfloor$

Proof: With the following "divide and conquer" protocol it can be shown that this lower bound is asympotically optimal. This is somehow surprising, since for odd k one might expect some additional communication. This is, however, negligible - only a $\sqrt{n}$ term:

For $k = 2$ Person 1 transmits her complete string and Person 2 returns the result.

For $k \geq 3$ Person 1 transmits the first $\lceil \sqrt{n} \rceil$ bits of her input. The other persons then send a 0 if their inputs coincide on these $\lceil \sqrt{n} \rceil$ bits or a 1, respectively, if this is not the case. If all other persons have sent a 0, then Person 1 transmits the next $\lceil \sqrt{n} \rceil$ bits of her input and the other persons respond in the same way. After Person 1 has transmitted, say, $t\lceil \sqrt{n} \rceil$ bits for the first time some of the other persons, say $k - i$ of them, will answer with a 1. Their $k - i$ inputs then have to be compared on $n - (t-1)\lceil \sqrt{n} \rceil$ bits, the other i inputs have to be compared on $n - t\lceil \sqrt{n} \rceil$ bits.

Let $M(k, n)$ be the number of bits transmitted during this protocol in the worst case. Then

$$M(k,n) \leq \begin{cases} \frac{k}{2}n + a_k & , k \text{ even} \\ \frac{k-1}{2}n + b_k\sqrt{n} + c_k & , k \text{ odd} \end{cases}$$

for certain numbers a_k, b_k, c_k only depending on the number of partys k and not on n. With this, the asymptotic statement of the theorem is immediate.

The above formula for $M(k,n)$ can be proven by induction. Obviously $M(2,n) = n+1$. Further, $M(3,n) \leq n+3\lceil\sqrt{n}\rceil+1$. To see this, assume that after Person 1 has sent $t\lceil\sqrt{n}\rceil$ bits for the first time the other Persons do not reply with 0 both. So at least one of their inputs does not coincide with x_1 on the last $\lceil\sqrt{n}\rceil$ bits transmitted. If only one person, Person 3 say, sent a 1, it is clear that x_3 is different from x_1 and x_2, which then have to be compared on the remaining $t - \lceil\sqrt{n}\rceil$ bits.

If both persons replied with 1, then x_1 is different from x_2 and x_3, which then still have to be compared on $n-(t-1)\lceil\sqrt{n}\rceil$ bits. This is obviously the worst case and here still $M(2, n-(t-1)\lceil\sqrt{n}\rceil+1)$ further bits must be exchanged to obtain the result. Hence

$$M(3,n) \leq t\lceil\sqrt{n}\rceil+2t+n-(t-1)\lceil\sqrt{n}\rceil+1 = n+\lceil\sqrt{n}\rceil+2t+1 \leq n+3\lceil\sqrt{n}\rceil+1.$$

For $k \geq 4$ the above protocol yields the recursion

$$M(k,n) \leq \max_t \max_{i=1,\dots,k} t\lceil\sqrt{n}\rceil + (k-1)t + M(i, n-t\lceil\sqrt{n}\rceil) + M(k-i, n-(t-1)\lceil\sqrt{n}\rceil).$$

from which the numbers a_k, b_k, and c_k can be recursively calculated with several case investigations (k, i even or odd).

References

1. Ahlswede, R.: On code pairs with specified Hamming distances. Colloquia Math. Soc. J. Bolyai 52, 9–47 (1988)
2. Ahlswede, R., Mörs, M.: Inequalities for code pairs. European J. Combinatorics 9, 175–188 (1988)
3. Ahlswede, R., Cai, N.: On communication complexity of vector-valued functions. IEEE Trans. Inf. Theory 40(6), 2062–2067 (1994)
4. Ahlswede, R., Cai, N.: 2-Way communication complexity of sum-type functions for one processor to be informed. Probl. Inf. Transmission 30(1), 1–10 (1994)
5. Ahlswede, R., Cai, N., Tamm, U.: Communication complexity in lattices. Appl. Math. Letters 6(6), 53–58 (1993)
6. Ahlswede, R., Cai, N., Zhang, Z.: A general 4–word–inequality with consequences for 2–Way communication complexity. Advances in Applied Mathematics 10, 75–94 (1989)
7. Ahlswede, R., Zhang, Z.: Code pairs with specified parity of the Hamming distances. Discr. Math. 188, 1–11 (1998)
8. Ahlswede, R., El Gamal, A., Pang, K.F.: A two–family extremal problem in Hamming space. Discr. Math. 49, 1–5 (1984)
9. Alon, N., Matias, M., Szegedy, M.: The space complexity of approximating the frequency moments. J. Comput. Syst. Sci. 58(1), 137–147 (1999)
10. Babai, L., Frankl, P., Simon, J.: Complexity classes in communication complexity theory. In: Proc. IEEE FOCS, pp. 337–347 (1986)
11. Chen, L., Chitambar, E., Duan, R., Ji, Z., Winter, A.: Tensor rank and stochastic entaglement catalysis for multipartite pure states. Physical Review Letters 105 (2010)
12. Draisma, J., Kushilevitz, E., Weinreb, E.: Partition arguments in multiparty communication complexity. Theoretical Computer Science 412, 2611–2622 (2011)
13. Feder, T., Kushilevitz, E., Naor, M., Nisan, N.: Amortized communication complexity. SIAM J. Comp. 24(4), 736–750 (1995)

14. Hromkovic, J.: Communication Complexity and Parallel Computing. Springer (1997)
15. Kushilevitz, E., Nisan, N.: Communication Complexity. Cambridge University Press (1997)
16. Nisan, N., Wigderson, A.: On rank vs communication complexity. Combinatorica 15(4), 557–566 (1995)
17. Tamm, U.: Communication complexity of sum-Type functions. PhD thesis, Bielefeld (1991)
18. Tamm, U.: Still another rank determination of set intersection matrices with an application in communication complexity. Appl. Math. Letters 7, 39–44 (1994)
19. Tamm, U.: Communication complexity of sum - type functions invariant under translation. Inform. and Computation 116(2), 162–173 (1995)
20. Tamm, U.: Deterministic communication complexity of set intersection. Discr. Appl. Math. 61, 271–283 (1995)
21. Tamm, U.: Communication complexity of functions on direct sums. In: Althöfer, I., Cai, N., Dueck, G., Khachatrian, L., Pinsker, M., Sárközy, A., Wegener, I., Zhang, Z. (eds.) Numbers, Information and Complexity, pp. 589–602. Kluwer (2000)
22. Yao, A.C.: Some complexity questions related to distributive computing. In: Proc. ACM STOC, pp. 209–213 (1979)

Threshold Functions for Distinct Parts: Revisiting Erdős–Lehner

Éva Czabarka[1], Matteo Marsili[2], and László A. Székely[1,*]

[1] Department of Mathematics, University of South Carolina, Columbia, SC 29208, USA
{czabarka,szekely}@math.sc.edu

[2] The Abdus Salam International Centre for Theoretical Physics, Strada Costiera 11, 34014, Trieste, Italy
marsili@ittp.it

Dedicated to the memory of Rudolf Ahlswede

Abstract. We study four problems: put n distinguishable/non-distinguishable balls into k non-empty distinguishable/non-distinguishable boxes randomly. What is the threshold function $k = k(n)$ to make almost sure that no two boxes contain the same number of balls? The non-distinguishable ball problems are very close to the Erdős–Lehner asymptotic formula for the number of partitions of the integer n into k parts with $k = o(n^{1/3})$. The problem is motivated by the statistics of an experiment, where we only can tell whether outcomes are identical or different.

Keywords: integer partition with distinct parts, integer composition with distinct parts, set partition with distinct class sizes, random set partition, random integer partition, random integer composition, random function, threshold function.

1 Motivation

Consider a generic experiment where the state of a complex system is probed with repeated experiments providing outcome sequence $x_1, x_2, \ldots, x_n$. The experimenter can tell only whether two outcomes are identical or different. So each outcome can be thought of as a sample of i.i.d. draws from an unknown probability distribution over a discrete space. The order of outcomes carry no valuable information for us. We want to understand this process under the condition that

* The first and the third authors acknowledge financial support from the grant #FA9550-12-1-0405 from the U.S. Air Force Office of Scientific Research (AFOSR) and the Defense Advanced Research Projects Agency (DARPA). The third author was also supported in part by the NSF DMS contract 1000475. This research started at the MAPCON 12 conference, where the authors enjoyed the hospitality of the Max Planck Institute for the Physics of Complex Systems, Dresden, Germany. We thank Danny Rorabaugh for his comments.

H. Aydinian, F. Cicalese, and C. Deppe (Eds.): Ahlswede Festschrift, LNCS 7777, pp. 463–471, 2013.

k different outcomes are observed out of n experiments. As an example, think of a botanist collecting specimen of flowers in a yet unexplored forest. In order to classify observations into species, all that is needed is an objective criterion to decide whether two specimens belong to the same species or not. More generally, think of unsupervised data clustering of a series of observations.

In the extreme situation where $k = n$ and all outcomes are observed only once, the classification is not very informative. At the other extreme, when all the outcomes belong to the same class, the experimenter will think he/she has discovered some interesting regularity. In general, the information that a set of n repeated experiments yields, is the size m_i of each class (or cluster), i.e. the numbers $m_i := |\{\ell : x_\ell = i\}|$ (with $\sum_{i=1}^{k} m_i = n$), as the relative frequency of the observations is what allows to make comparative statements. We expect that when the number k of classes is large there will be several classes of the same size, i.e. that will not be discriminated by the experiment, whereas when k is small each class will have a different size[1].

This is clearly a problem that can be rephrased in terms of distributions of balls (outcomes) into boxes (classes). We consider, in particular, the null hypothesis of random placement of balls into boxes. In this framework, the question we ask is what is the critical number of boxes $k_c(n)$ such that for $k \ll k_c$ we expect to find that all boxes i contain a different number m_i of balls whereas for $k \gg k_c$ boxes with the same number of balls will exist with high probability.

2 Introduction

Recall the surjective version (no boxes are empty) of the twelvefold way of counting [10] p.41: putting n distinguishable/non-distinguishable balls into k non-empty distinguishable/non-distinguishable boxes correspond to four basic problems in combinatorial enumeration according to Table 1. Our concern in all four type of problems is the threshold function $n = n(k)$ that makes almost sure that no two boxes contain the same number of balls for a randomly and uniformly selected ball placement. Although studying all distinct parts is a topical issue for integer partitions [2] and compositions [6], it is hard to find any corresponding results for surjections and set partitions except [8]. Our results regarding the threshold functions are summed up in Table 1 in parentheses. We have to investigate only three problems, since every k-partition of an n-element set corresponds to exactly $k!$ surjections from $[n]$ to $[k]$—namely those surjections, whose inverse image partition is the k-partition in question. (Also, the threshold function is the same for compositions and partitions, although the number of compositions corresponding to a partition may vary from 1 to $k!$.)

[1] This observation can be made precise in information theoretic terms. The label X of one outcome, taken at random from the sample, is a random variable whose entropy $H[X]$ quantifies its information content. The size m_X of the class containing X, clearly has a smaller entropy $H[m] \leq H[X]$, by the data processing inequality [3]. When k is small, we expect that $H[X] = H[m]$ whereas when k is large, $H[X] > H[m]$.

Therefore the threshold function for set partitions is the same as the threshold function for surjections.

Our proofs use the first and the second moment method, the second moment method in the form (due to Chung and Erdős, see [11] p.76) below. For events $A_1, A_2, ..., A_N$, the following inequality holds:

$$P\left(\bigcup_{i=1}^{N} A_i\right) \geq \frac{\left(\sum_{i=1}^{N} P(A_i)\right)^2}{\sum_{i=1}^{N} P(A_i) + 2\sum_{1\leq i<j\leq N} P(A_i \cap A_j)}. \tag{1}$$

We always will assume

$$n \geq \binom{k+1}{2} > \frac{k^2}{2}, \tag{2}$$

otherwise the parts clearly cannot have distinct sizes. We did not even attempt to obtain a limiting distribution $f(c)$ when n is c times the threshold function—though such an estimate should be possible to obtain. Although there are deep asymptotic results on random functions and set partitions based on generating functions (e.g. see Sachkov [9]), we do not see how to apply generating functions for our threshold problems. Our results corroborate some formulae of Knessl and Kessler [7], who used techniques from applied mathematics to obtain heuristics for partition asymptotics from basic partition recursions.

Table 1. Threshold functions for distinct parts for the four surjective cases in the twelvefold way of counting

		k non-empty boxes	
		distinguishable	non-distinguishable
n balls	distinguishable	surjections ($n = k^5$)	set partitions ($n = k^5$)
	non-distinguishable	integer compositions ($n = k^3$)	integer partitions ($n = k^3$)

3 Threshold Function for Integer Compositions

Recall that the number of compositions of the integer n into k positive parts is $\mathcal{C}(n,k) = \binom{n-1}{k-1}$. Let $A_{ij}(t)$ denote the event that the ith and j^{th} parts are equal t in a random composition of n into k positive parts. Using the first moment method, it is easy to see that

$$P(\exists \text{ equal parts}) = P\left(\bigcup_{i<j}\bigcup_{t} A_{ij}(t)\right) \leq \sum_{i<j}\sum_{t} P(A_{ij}(t)) = \binom{k}{2}\sum_{t\geq 1}\frac{\mathcal{C}(n-2t,k-2)}{\mathcal{C}(n,k)}$$
$$= \binom{k}{2}\sum_{t\geq 1}\frac{\binom{n-2t-1}{k-3}}{\binom{n-1}{k-1}} \leq \binom{k}{2}\frac{\binom{n-2}{k-2}}{\binom{n-1}{k-1}} = o(1),$$

as $n/k^3 \to \infty$. We make an elementary claim here that we use several times and leave its proof to the Reader.

Claim 1. *Assume that we have an infinite list of finite sequences of non-negative numbers,* $a_1(n)$, $a_2(n)$, ..., $a_{N(n)}(n)$ *for* $n = 1, 2, ...$, *such that none of the sequences is identically zero. Assume that the number of increasing and decreasing intervals of these finite sequences is bounded, and that* $\max_i a_i(n) = o\Big(\sum_i a_i(n)\Big)$ *as* $n \to \infty$. *Then, for any fixed* k, *as* $n \to \infty$. *we have*

$$\sum_{i:k|i} a_i(n) = \Big(\frac{1}{k} + o(1)\Big) \sum_i a_i(n).$$

Using the claim we can get a more precise estimate

$$\sum_{t\geq 1} \binom{n-2t-1}{k-3} = \frac{1+o(1)}{2} \binom{n-2}{k-2}. \tag{3}$$

Next we use (1) and (3) to show that $P(\exists \text{ equal parts}) \to 1$ as $n/k^3 \to 0$. The numerator of (1) is the square of

$$\sum_{i<j} \sum_t P(A_{ij}(t)) = \binom{k}{2} \sum_{t\geq 1} \frac{\binom{n-2t-1}{k-3}}{\binom{n-1}{k-1}} = (1+o(1)) \frac{k^3}{4n} \tag{4}$$

that grows to infinity. Therefore we can neglect the same term without square in the denominator of (1). A second negligible term in the denominator arises if $i < j$ and $u < v$ make only 3 distinct indices (note that they must occur with the same t). The corresponding sum of the probabilities is estimated by

$$k \binom{k-1}{2} \sum_{t\geq 1} \frac{\mathcal{C}(n-3t, k-3)}{\mathcal{C}(n,k)} \leq \frac{k^3}{2} \sum_{t\geq 1} \frac{\binom{n-3t-1}{k-4}}{\binom{n-1}{k-1}} < \frac{k^3}{2} \frac{\binom{n-3}{k-3}}{\binom{n-1}{k-1}} < \frac{k^5}{n^2} = o\left(\Big(\frac{k^3}{n}\Big)^2\right). \tag{5}$$

A third negligible term in the denominator arises if all four indices are distinct, but the four corresponding parts are all the same:

$$\binom{k}{4} \sum_{t\geq 1} \frac{\mathcal{C}(n-4t, k-4)}{\mathcal{C}(n,k)}.$$

This can be estimated by $O(k^7/n^3)$ like the estimate in (5), and is similarly negligible. The significant term in the denominator is

$$\binom{k}{2} \binom{k-2}{2} \sum_{\ell\geq 1} \sum_{t\geq 1, t\neq \ell} \frac{\mathcal{C}(n-2\ell-2t, k-4)}{\mathcal{C}(n,k)} \tag{6}$$

corresponding to the cases analysis when 2-2 parts are the same. This term will not change asymptotically when we add the $t = \ell$ cases to the summation. So (6) is asymptotically equal to (using Claim 1 again)

$$\frac{k^4}{4} \sum_{\ell\geq 1} \sum_{t\geq 1} \frac{\binom{n-2\ell-2t-1}{k-5}}{\binom{n-1}{k-1}} \sim \frac{k^4}{8} \sum_{\ell\geq 1} \frac{\binom{n-2\ell-2}{k-4}}{\binom{n-1}{k-1}} \sim \frac{k^4}{16} \frac{\binom{n-3}{k-3}}{\binom{n-1}{k-1}} \sim \frac{k^6}{16n^2}. \tag{7}$$

We conclude that the numerator in (1) is asymptotically equal to its denominator (7), proving that $P(\exists \text{ equal parts}) \to 1$ as $n/k^3 \to 0$.

4 Erdős–Lehner and the Threshold Function for Integer Partitions

Let $\mathcal{D}(n,k)$ denote the number of compositions of n into k distinct positive terms. In the previous section we proved

$$\lim_{n/k^3 \to 0} \frac{\mathcal{D}(n,k)}{\mathcal{C}(n,k)} = 1 \text{ and} \tag{8}$$

$$\lim_{n/k^3 \to \infty} \frac{\mathcal{D}(n,k)}{\mathcal{C}(n,k)} = 0. \tag{9}$$

Let $p(n,k)$ denote the number of partitions of n into k positive terms and $q(n,k)$ denote the number of partitions of n into k distinct positive terms. For $1 \le x_1 \le x_2 \le \ldots \le x_k$, the well-known bijection $x_1 + x_2 + \ldots + x_k \to (x_1) + (x_2 + 1) + (x_3 + 2) + \ldots + (x_k + k - 1)$ shows that $q(n,k) = p\big(n - \binom{k}{2}, k\big)$.

A theorem of Erdős and Lehner ([4], see also in [2]) asserts that for $k = o(n^{1/3})$, the following asymptotic formula holds:

$$p(n,k) \sim \frac{1}{k!}\mathcal{C}(n,k) = \frac{1}{k!}\binom{n-1}{k-1}. \tag{10}$$

Gupta's proof to Erdős–Lehner ([5], see also in [2]) obtains

$$\frac{1}{k!}\mathcal{C}(n,k) \le p(n,k) = q\left(n + \binom{k}{2}, k\right) \le \frac{1}{k!}\mathcal{C}\left(n + \binom{k}{2}, k\right) \tag{11}$$

from the asymptotic equality of

$$\binom{n-1}{k-1} \sim \binom{n + \binom{k}{2} - 1}{k-1}, \tag{12}$$

the leftmost and rightmost terms in (11), under the assumption $k = o(n^{1/3})$.

To get the $n = k^3$ threshold function for integer partitions, we first show that for $n/k^3 \to 0$,

$$\begin{aligned} k!q(n,k) &= \mathcal{D}(n,k), \\ \mathcal{D}(n,k) &= o(\mathcal{C}(n,k)), \text{ from (9)}, \\ \text{and } \mathcal{C}(n,k) &\le k!p(n,k). \end{aligned}$$

To do the case $n/k^3 \to \infty$, i.e. $k = o(n^{1/3})$, we use Erdős–Lehner twice and also (12):

$$q(n,k) = p\left(n - \binom{k}{2}, k\right) \sim \frac{1}{k!}\binom{n - \binom{k}{2} - 1}{k-1} \sim \frac{1}{k!}\binom{n-1}{k-1} \sim p(n,k).$$

5 Threshold Function for Surjections

Let $F(n,k)$ denote the number of $[n] \to [k]$ surjections. It is well-known from the Bonferroni inequalities that $k^n - n(k-1)^n \le F(n,k) \le k^n$ and hence for $n >> k\log k$, $F(n,k) = (1+o(1))k^n$ uniformly. Take an $f : [n] \to [k]$ random surjection. Let $A_{ij}(t)$ denote the event that for $1 \le i < j \le k$ we have $|f^{-1}(i)| = |f^{-1}(j)| = t$. Observe that $P(A_{ij}(t)) = \sum_{t\ge 1}\binom{n}{2t}\binom{2t}{t}\frac{F(n-2t,k-2)}{F(n,k)}$ and recall $\binom{2t}{t} \sim \frac{2^{2t}}{\sqrt{\pi t}}$. Observe[2] that

$$P(\exists i<j\colon |f^{-1}(i)| = |f^{-1}(j)|) = P\left(\bigcup_{i<j}\bigcup_{t} A_{ij}(t)\right) \le \sum_{i<j}\sum_{t} P(A_{ij}(t))$$
$$\sim \binom{k}{2}\sum_{t\ge 1}\binom{n}{2t}\binom{2t}{t}\frac{(k-2)^{n-2t}}{k^n} \tag{13}$$
$$\sim \binom{k}{2}\left(1-\frac{2}{k}\right)^n\sum_{t\ge 1}\binom{n}{2t}\left(\frac{2}{k-2}\right)^{2t}\frac{1}{\sqrt{\pi t}}.$$

Let $b(n,i)$ denote the term $\binom{n}{i}p^i(1-p)^{n-i}$ from the binomial distribution with $p = 1-\frac{2}{k}$. It is easy to see that the core summation in (13) is

$$\sum_{t\ge 1}\binom{n}{2t}\left(\frac{2}{k-2}\right)^{2t}\frac{1}{\sqrt{\pi t}} = \left(1-\frac{2}{k}\right)^{-n}\sum_{t\ge 1}\frac{b(n,2t)}{\sqrt{\pi t}}.$$

Observe that for this binomial distribution $\mu = np = \frac{2n}{k}$ and $\sigma < \sqrt{np} = \sqrt{\frac{2n}{k}}$. Recall from [1] the large deviation inequality for sums of independent Bernoulli random variables:

$$P\left(|Y-\mu| > \epsilon\mu\right) < 2e^{c_\epsilon \mu},$$

where $c_\epsilon = \min\{\ln(\epsilon^\epsilon(1+\epsilon)^{(1+\epsilon)}), \epsilon^2/2\}$. We select $\epsilon = \frac{1}{\ln n}$, with which for sufficiently large n, $c_\epsilon = \epsilon^2/2 = \frac{1}{2\ln^2 n}$. Set $A = (1-\epsilon)\mu$ and $B = (1+\epsilon)\mu$. As $[A,B]$ includes the range where the normal convergence takes place, $\sum_{A\le t\le B} b(n,t) \sim 1$. By Claim 1, $\sum_{A\le 2t\le B} b(n,2t) \sim 1/2$. Also, if $A \le t \le B$, then $t \sim \frac{2n}{k}$. By the large deviation inequality above and $k < \sqrt{n}$ from (2), we obtain

$$\sum_{\substack{t\ge 1\\ t\notin[A,B]}} b(n,t) = o\left(\sqrt{\frac{k}{2n}}\sum_{A\le t\le B} b(n,t)\right),$$

[2] Note that for large t (i.e. $n-2t = O(k\ln k)$) the approximation for $P(A_{ij}(t))$ is not accurate. The same problem occurs for small t ($t = O(1)$), because of the estimate of $\binom{2t}{t}$. The corresponding terms, however, are negligible both in the sum of probabilities and in (13).

and combining with Claim 1 we obtain

$$\sum_{\substack{t\geq 1\\ t\notin[A,B]}} b(n,t) = o\left(\sqrt{\frac{k}{2n}} \sum_{A\leq 2t\leq B} b(n,2t)\right).$$

Putting together these arguments:

$$\sum_{t\geq 1} \frac{b(n,2t)}{\sqrt{\pi t}} \sim \sqrt{\frac{k}{2\pi n}} \sum_{A\leq 2t\leq B} b(n,2t) \sim \sqrt{\frac{k}{2\pi n}} \sum_{t\geq 1} b(n,2t) \sim \frac{1}{2}\sqrt{\frac{k}{2\pi n}} \sum_{t\geq 1} b(n,t) \sim \frac{1}{2}\sqrt{\frac{k}{2\pi n}}.$$

We obtain asymptotic formula for the upper bound with

$$\sum_{i<j}\sum_{t} P(A_{ij}(t)) = (1+o(1))\frac{k^2}{4}\sqrt{\frac{k}{2n\pi}} \tag{14}$$

which goes to zero as $n/k^5 \to \infty$.

Next we use (1) and (14) to show that $P(\exists i<j\colon |f^{-1}(i)| = |f^{-1}(j)|) \to 1$ as $n/k^5 \to 0$. The numerator of (1) is the square of (14) that grows to infinity. Therefore we can neglect the same term without square in the denominator of (1).

A second negligible term in the denominator arises if $i<j$ and $u<v$ make only 3 distinct indices (note that they must occur with the same t). The corresponding sum of the probabilities is estimated by

$$k\binom{k-1}{2}\sum_{t\geq 1}\binom{n}{3t}\frac{(3t)!}{(t!)^3}\frac{F(n-3t,k-3)}{F(n,k)} \leq k^3\sum_{t\geq 1}\binom{n}{3t}\frac{(3t/e)^{3t}\sqrt{6\pi t}}{(t/e)^{3t}(\sqrt{2\pi t})^3}\frac{(k-3)^{n-3t}}{k^n}$$

$$\leq k^3\sum_{t\geq 1}\binom{n}{3t}\left(\frac{3}{k-3}\right)^{3t}\frac{\sqrt{3}}{2\pi t}\frac{(k-3)^n}{k^n}. \tag{15}$$

We have from the Binomial Theorem $\sum_t \binom{n}{t}\left(\frac{3}{k-3}\right)^t = \left(1+\frac{3}{k-3}\right)^n$. Working with every third term in a binomial distribution with $p=\frac{3}{k}$ like we worked above with every second, one obtains the upper bound for (15)

$$O\left(\frac{k^4}{n}\right) = o\left(\left(\sum_{i<j}\sum_{t} P(A_{ij}(t))\right)^2\right),$$

using (14) as $n/k^5 \to 0$.

A third negligible term in the denominator arises if $i<j$ and $u<v$ are 4 distinct indices, but the corresponding parts (set sizes) are all equal. The corresponding term is

$$\binom{k}{4}\sum_{t\geq 1}\binom{n}{4t}\frac{(4t)!}{(t!)^4}\frac{F(n-4t,k-4)}{F(n,k)}.$$

This is easily estimated by $O(k^{11/2}/n^{3/2})$ like the estimate in (15), and is similarly negligible compared to (14) as $n/k^5 \to 0$. The significant term in the denominator is

$$\binom{k}{2}\binom{k-2}{2}\sum_{\ell\geq 1}\sum_{t\geq 1, t\neq \ell}\binom{n}{2\ell}\binom{2\ell}{\ell}\binom{n-2\ell}{2t}\binom{2t}{t}\frac{F(n-2\ell-2t,k-4)}{F(n,k)} \quad (16)$$

corresponding to the cases when for 4 distinct indices 2-2 parts (set sizes) are the same. The siginificant term will not change asymptotically when we add the $t=\ell$ cases to the summation.

We do not repeat below arguments about the binomial distribution which we went through before. So (16) is asymptotically equal to

$$\frac{k^4}{4}\sum_{\ell\geq 1}\binom{n}{2\ell}\frac{4^\ell}{\sqrt{\ell\pi}}\sum_{t\geq 1}\binom{n-2\ell}{2t}\frac{4^t}{\sqrt{t\pi}}\frac{(k-4)^{n-2\ell-2t}}{k^n} \quad (17)$$

$$\sim \frac{k^4}{4}\sum_{\ell\geq 1}\binom{n}{2\ell}\frac{4^\ell}{\sqrt{\ell\pi}}\frac{(k-4)^{n-2\ell}}{k^n}\sum_{t\geq 1}\binom{n-2\ell}{2t}\left(\frac{2}{k-4}\right)^{2t}\frac{1}{\sqrt{t\pi}} \quad (18)$$

$$\sim \frac{k^4}{4}\sum_{\ell\geq 1}\binom{n}{2\ell}\frac{4^\ell}{\sqrt{\ell\pi}}\frac{(k-4)^{n-2\ell}}{k^n}\cdot\frac{1}{2}\left(1+\frac{2}{k-4}\right)^{n-2\ell}\sqrt{\frac{k-2}{2\pi(n-2\ell)}} \quad (19)$$

$$\sim \frac{k^4}{8\pi}\left(1-\frac{2}{k}\right)^n\sum_{\ell\geq 1}\binom{n}{2\ell}\left(\frac{2}{k-2}\right)^{2\ell}\sqrt{\frac{k-2}{2\ell(n-2\ell)}} \quad (20)$$

$$\sim \frac{k^4}{8\pi}\left(1-\frac{2}{k}\right)^n\cdot\frac{1}{2}\left(1+\frac{2}{k-2}\right)^n\sqrt{\frac{k-2}{\frac{4n}{k}(n-4\frac{n}{k})}}\sim\frac{k^5}{32\pi n} \quad (21)$$

We conclude that the numerator in (1), which is (14) squared in our setting, is asymptotically equal to its denominator (17), proving that $P(\exists i<j\colon |f^{-1}(i)|=|f^{-1}(j)|)\to 1$ as $n/k^5\to 0$.

References

1. Alon, N., Spencer, J.H.: The Probabilistic Method, 2nd edn. John Wiley and Sons, New York (2000)
2. Andrews, G.E.: The Theory of Partitions. Cambridge University Press (1998), see also Encyclopedia of Mathematics and Its Applications, vol. 2. Addison-Wesley (1976)
3. Cover, T.M., Thomas, J.A.: Elements of Information Theory. John Wiley & Sons (2006)
4. Erdős, P., Lehner, J.: The distribution of the number of summands in the partitions of a positive integer. Duke Math. J. 8, 335–345 (1941)
5. Gupta, H.: On an asymptotic formula in partitions. Proc. Indian Acad. Sci. A16, 101–102 (1942)
6. Hitczenko, P., Stengle, G.: Expected number of distinct parts sizes in random integer composition. Combinatorics, Probability and Computing 9, 519–527 (2000)

7. Knessl, C., Keller, J.B.: Partition asymptotics from recursion equations. SIAM J. Appl. Math. 50(2), 323–338 (1990)
8. Knopfmacher, A., Odlyzko, A.M., Pittel, B., Richmond, L.B., Stark, D., Szekeres, G., Wormald, N.C.: The asymptotic number of set partitions with unequal block sizes. Electr. J. Combinatorics 6, #R2, 36 (1999)
9. Sachkov, V.N.: Probabilistic Methods in Combinatorial Analysis. Encyclopedia of Mathematics and its Applications, vol. 56. Cambridge University Press (1997)
10. Stanley, R.P.: Enumerative Combinatorics, vol. I. Cambridge University Press (1997)
11. Szpankowski, W.: Average Case Analysis of Algorithms on Sequences. Wiley (2011)

On Some Structural Properties of Star and Pancake Graphs

Elena Konstantinova

Sobolev Institute of Mathematics, Novosibirsk-90, 630090 Russia
e_konsta@math.nsc.ru

Dedicated to the memory of Rudolf Ahlswede

Abstract. In this paper we give a report on some results for the Star and Pancake graphs obtained after the conference "Search Methodologies" which was held in October, 2010. The graphs are defined as Cayley graphs on the symmetric group with the generating sets of all prefix–transpositions and prefix–reversals, correspondingly. They are also known as the Star and Pancake networks in computer science. In this paper we give the full characterization of perfect codes for these graphs. We also investigate a cycle structure of the Pancake graph and present an explicit description of small cycles as well as their number in the graph.

Keywords: Cayley graph, Pancake graph, Star graph, Star network, Pancake network, cycle embedding, efficient dominating set, perfect code.

1 Introduction

I was one of participants of the conference "Search Methodologies" which was held in the frame of the ZiF Project "Search Methodologies" in October, 2010. It was the last time when I met with Rudi Ahlswede. He wrote in a short description of the project:

> *In the three decades which passed since 1979 there has been an explosion of developments in search for instance in Computer Science, Image Reconstruction, Machine Learning, Information Theory, and Operations Research.*

And one more quote:

> *A search structure is defined by a space of objects searched for and a space of tests (questions). In specifying a search problem performance criteria have to be chosen. Furthermore we distinguish combinatorial and probabilistic models.*

As my main research interests are Cayley graphs and investigations of their structural properties, I gave a talk concerning search combinatorial problems on

H. Aydinian, F. Cicalese, and C. Deppe (Eds.): Ahlswede Festschrift, LNCS 7777, pp. 472–487, 2013.

Cayley graphs. So, the space of objects is given by Cayley graphs and the space of questions is given by combinatorial problems. In this paper I would like to give a report on some problems which were announced at that time as open but now they are solved. This is my report to Rudi.

2 Main Definitions, Main Problems

We consider Cayley graphs defined as follows. Let G be a group, and $S \subset G$ be a set of generators (a generating set) such that $e \notin S$ and $S = S^{-1}$. Then in the *Cayley graph* $Cay(G, S) = (V, E)$ vertices correspond to the elements of the group, i.e. $V = G$, and edges correspond to the action of the generators, i.e. $E = \{\{g, gs\} : g \in G, s \in S\}$. Cayley graphs are connected $|S|$–regular vertex–transitive graphs. In a Cayley graph $Cay(G, S)$ the diameter is the maximum, over $g \in G$, of the length of a shortest expression for g as a product of generators.

There are many well–known open combinatorial problems on Cayley graphs such as the diameter problem (the problem of searching the diameter); the hamiltonian problem (the problem of searching the hamiltonian cycle); the problem of searching perfect codes, etc. There are no answers in a general case, and moreover sometimes it is difficult to solve a problem even for a specific class of graphs. For example, the problem of determining the diameter of the so–called Pancake graphs, posed in [DW75] is still open.[1]

The Pancake graph $P_n = Cay(Sym_n, PR), n \geqslant 2$, is the Cayley graph on the symmetric group Sym_n of permutations $\pi = [\pi_1 \pi_2 \ldots \pi_n]$, where $\pi_i = \pi(i)$, $1 \leqslant i \leqslant n$, with the generating set $PR = \{r_i \in Sym_n : 2 \leqslant i \leqslant n\}$ of all prefix–reversals r_i reversing the order of any substring $[1, i], 2 \leqslant i \leqslant n$, of a permutation π when multiplied on the right, i.e. $[\pi_1 \ldots \pi_i \pi_{i+1} \ldots \pi_n] r_i = [\pi_i \ldots \pi_1 \pi_{i+1} \ldots \pi_n]$. It is a connected vertex–transitive $(n-1)$-regular graph of order $n!$. The *distance* $d = d(\pi, \tau)$ between two vertices π and τ in P_n is defined as the least number of prefix–reversals transforming π into τ, i.e. $\pi r_{i_1} r_{i_2} \ldots r_{i_d} = \tau$, and the diameter is $diam(P_n) = \max\{d(\pi, \tau) : \pi, \tau \in Sym_n\}$.

Some upper and lower bounds on the diameter of the Pancake graph as well as its exact values for $2 \leqslant n \leqslant 19$ are known [GP79, HS97, AKSK06, C11]. One of the main difficulties in solving this problem is a complicated cycle structure of the graph, which dramatically affects to the vertex distributions in the metric spheres $S_i = \{\pi \in Sym : d(I, \pi) = i\}$ of radius $1 \leqslant i \leqslant diam(P_n)$ centered at the identity permutation $I = [1 \ldots n]$. It is known [KF95, STC06], that all cycles of length l, where $6 \leqslant l \leqslant n!$, can be embedded in P_n, $n \geqslant 3$, but there are no cycles of length 3, 4, and 5. Moreover, as it will be shown below, each of vertices

[1] During my talk I have also mentioned about Rubik's cube for which the diameter of its Cayley graph was unknown. However, at the end of my talk Prof. Martin Milanič has found the information in Google that this problem was solved recently! While it had been known since 1995, that 20 was a lower bound on the diameter, it was proved in 2010 by Tomas Rokicki, Herbert Kociemba, Morley Davidson, and John Dethridge through extensive computer calculations that this bound is a sharp upper bound. (see http://tomas.rokicki.com/)

of P_n belongs to exactly one 6–cycle. From this, one can immediately obtain that $|S_1| = n-1$, $|S_2| = (n-1)(n-2)$, $|S_3| = (n-1)(n-2)^2 - 1$. However, to get $|S_i|$ for $4 \leqslant i \leqslant diam(P_n)$ we need to know an explicit description of cycles. This problem was open some time ago. Now some results for small cycles are obtained and they will be presented in Section 4.

Another problem which was open for the Pancake graph as well as for the Star graph came from coding theory. First let us give the definition of Star graphs and then to describe the problem.

The Star graph $S_n = Cay(Sym_n, PT)$, $n \geqslant 2$, is the Cayley graph on the symmetric group Sym_n of permutations $\pi = [\pi_1\pi_2 \ldots \pi_n]$, where $\pi_i = \pi(i)$, $1 \leqslant i \leqslant n$, with the generating set $PT = \{t_i \in Sym_n : 2 \leqslant i \leqslant n\}$ of all prefix–transpositions t_i transposing the 1st and ith elements, $2 \leqslant i \leqslant n$, of a permutation π when multiplied on the right, i.e. $[\pi_1\pi_2 \ldots \pi_{i-1}\pi_i\pi_{i+1} \ldots \pi_n]t_i = [\pi_i\pi_2 \ldots \pi_{i-1}\pi_1\pi_{i+1} \ldots \pi_n]$. It is a connected vertex–transitive $(n-1)$–regular graph of order $n!$ and diameter $diam(S_n) = \lfloor \frac{3(n-1)}{2} \rfloor$ [AK89]. This graph is hamiltonian; the set of all its even cycles was obtained in [JLD91].

The Star and Pancake graphs have a hierarchical structure such that for any $n \geqslant 3$ a graph $\Gamma_n \in \{S_n, P_n\}$ consists of n copies $\Gamma_{n-1}(i) = (V^i, E^i)$, $1 \leqslant i \leqslant n$, where the vertex set is presented by permutations with the fixed last element:

$$V^i = \{[\pi_1 \ldots \pi_{n-1} i], \text{ where } \pi_k \in \{1, \ldots, n\} \backslash \{i\} : 1 \leqslant k \leqslant n-1\}, \quad (1)$$

with $|V^i| = (n-1)!$, and the edge set is presented by the set:

$$E^i = \{\{[\pi_1 \ldots \pi_{n-1} i], [\pi_1 \ldots \pi_{n-1} i] g_j\}, \text{ where } g_j \in \{t_j, r_j\} : 2 \leqslant j \leqslant n-1\},$$

with $|E^i| = \frac{(n-1)!(n-2)}{2}$. There are $(n-2)!$ *external* edges between any two copies $\Gamma_{n-1}(i)$, $\Gamma_{n-1}(j)$, $i \neq j$. These edges are defined by the generating element g_n. The generating elements g_j, $2 \leqslant j \leqslant n-1$, define *internal* edges within all copies $\Gamma_{n-1}(i)$, $1 \leqslant i \leqslant n$. The copies $\Gamma_{n-1}(i)$ are also called $(n-1)$*–copies*.

In [DS03] it was shown the existence of perfect codes, or efficient dominating sets, in Cayley graphs on symmetric groups having a hierarchical structure. In particular, the Star and Pancake graphs should have perfect codes. The structure of these perfect codes was described but the full characterization of all such codes was not obtained. At the first time the characterization of all perfect codes in the Pancake graph was presented in 2010 [K10]. Recently it was shown that the same result could be applied to the Star graph [KS12]. In this paper we prove that there are exactly n efficient dominating sets of cardinality $(n-1)!$ in Γ_n, $n \geqslant 3$. Moreover, we describe all these sets in the next Section and investigate their distance properties.

3 Efficient Dominating Sets in the Star and Pancake Graphs

Let us give additional definitions. An *independent set* is a set of vertices in a graph, no two of which are adjacent. An independent set D of vertices in a graph Γ is an *efficient dominating set* (or *perfect code* [DS03]) if each vertex not in D is adjacent to exactly one vertex in D. The *domination number* $\gamma(\Gamma)$ in a graph Γ is the number of vertices in the smallest dominating set for Γ where a *dominating set* D is a set of vertices such that every vertex not in D is adjacent to at least one vertex in D. It is obvious, that efficient dominating sets are the smallest sets among all dominating sets.

It was shown in [AK96] that the cardinality of the minimal dominating sets in the Star graph is equal to $(n-1)!$. In [Q06] minimal efficient dominating sets were used in broadcasting algorithms for multiple messages on the n-dimensional Star and Pancake networks (graphs). These networks have been considered in computer science as models for interconnection networks [AK89, LJD93] such that processors are labeled by permutations of length n, and two processors are connected when the label of one is obtained from the other by a prefix–transposition or a prefix–reversal.

As above, we put $\Gamma_n \in \{S_n, P_n\}$ and $g_i \in \{t_i, r_i\}$, $2 \leqslant i \leqslant n$, and define sets:

$$D_k = \{[k\,\pi_2 \, \ldots \, \pi_n],\, \pi_j \in \{1,\ldots,n\}\backslash\{k\} : 2 \leqslant j \leqslant n\}, \qquad 1 \leqslant k \leqslant n, \tag{2}$$

consisting of all permutations with the fixed first element. These sets have the following evident properties.

Property 1. $|D_k| = (n-1)!$ for any $1 \leqslant k \leqslant n$.
Property 2. $|D_{k_1} \bigcap D_{k_2}| = 0$, $k_1 \neq k_2$, moreover any vertex from D_{k_1} is adjacent to exactly one vertex from D_{k_2}.
Property 3. $\bigcup_{k=1}^{n} D_k = V(\Gamma_n)$.
Property 4. D_k is independent for any $1 \leqslant k \leqslant n$.

The last property follows from the definitions of the graphs since any generating element g_i, $2 \leqslant i \leqslant n$, transposes the 1st and ith elements of a permutation when multiplied on the right. Hence, any two permutations from D_k, $1 \leqslant k \leqslant n$, are not adjacent which means that the set is independent.

The *open neighborhood* $N(A)$ of a set $A \subseteq V(\Gamma)$ in a graph $\Gamma = (V,E)$ is defined as the subset of vertices in $V(\Gamma) \setminus A$ adjacent to some vertex in A, namely:

$$N(A) = \{v \in V(\Gamma) \setminus A \mid \{v,u\} \in E(\Gamma) \text{ for all } u \in A\}.$$

In particular, the sets (1) and (2) have the following properties.

Property 5. $N(V^k) = D_k$ for any $1 \leqslant k \leqslant n$.

The distance $d(v,u)$ between vertices $v,u \in V(\Gamma_n)$ is defined as the minimum number of generating elements transforming v in u. If X is an efficient dominating set in Γ_n then the following property holds.

Property 6. $d(u,v) \geqslant 3$ for any $u,v \in X$.

This is true because $d(u,v) \neq 1$ since X is independent and $d(u,v) \neq 2$ since X is efficient dominating. Let us also define the following sets:

$$D_k^i = \{[k\,\pi_2 \, \ldots \, \pi_{n-1}\, i],\, \pi_j \in \{1,\ldots,n\}\backslash\{k,i\} : 2 \leqslant j \leqslant n\}, \tag{3}$$

where $1 \leqslant k \neq i \leqslant n$, consisting of all permutations with the first and the last fixed elements. There are $n\,(n-1)$ such sets in $\Gamma_n, n \geqslant 3$, of cardinality $(n-2)!$. There is the following evident connection between sets (2) and (3).

Property 7. $D_k = \bigcup_{i=1, i\neq k}^{n} D_k^i$ for any $1 \leqslant k \leqslant n$.

The following property is also held for vertices of the set (3).

Property 8. Any vertex from D_k^i is adjacent to exactly one vertex from D_i^k and exactly to one vertex from D_j^i for any $j \neq i, k$.

The main theorem of this Section is formulated as follows.

Theorem 1. *There are only n efficient dominating sets in Γ_n given by (2).*

Proof. First of all let us show that the sets given by (2) are efficient dominating sets. It was proved in more generality in [DS03]. Since $\Gamma_n, n \geqslant 3$, is $(n-1)$–regular, hence any vertex from some dominating set X dominates itself and $n-1$ adjacent vertices, so $|X| \geqslant \frac{n!}{n} = (n-1)!$. By Property 1, sets (2) have the smallest cardinality $(n-1)!$, and by Property 4 they are independent. To prove that D_k are the efficient dominating sets, we have to show that for any $k = 1, \ldots, n$ any vertex $\pi = [\pi_1 \ldots \pi_n] \notin D_k$ is adjacent to exactly one vertex from D_k, i.e. there is the only vertex $\pi^* \in D_k$ such that $\{\pi, \pi^*\} \in E(\Gamma_n)$. Indeed, since $\pi \notin D_k$, hence $\pi_1 \neq k$, and there exists the only $\pi_i = k$ for some $2 \leqslant i \leqslant n$. Then by multiplying π on a generating element g_i on the right we have π^* with the first element k, i.e. $\pi^* \in D_k$. So $\{\pi, \pi^*\} \in E(\Gamma_n)$ and such a vertex $\pi^* \in D_k$ is the only one for a vertex $\pi \notin D_k$, $1 \leqslant k \leqslant n$. Thus, the sets (2) are efficient dominating sets in $\Gamma_n, n \geqslant 3$.

Now we prove that there are no other efficient dominating sets in $\Gamma_n, n \geqslant 3$. Let X be an efficient dominating set. Since the sets D_k, $1 \leqslant k \leqslant n$, partition the whole symmetric group, hence $X \bigcap D_k \neq \varnothing$. We choose such a k and conclude that $X = D_k$. Since $D_k = \bigcup_{i=1, i\neq k}^{n} D_k^i$ by Property 7 for any $1 \leqslant k \leqslant n$, then it is sufficient to prove that $X_k^i = D_k^i$ for any $1 \leqslant i \neq k \leqslant n$, where $X_k^i = X \bigcap D_k^i$. There are two cases: (a) $X_k^i = D_k^i$; (b) $X_k^i \subset D_k^i$.

(a) By Property 8, $D_i^k = N(D_k^i) \bigcap V^k$ and $V^k \setminus D_i^k = N(N(D_k^i)) \bigcap V^k$, where V^k is defined by (1). Since $D_k^i \subseteq X$, and by Property 6 we have $d(u, v) \geqslant 3$ for any two vertices $u, v \in X$, then $D_i^k \bigcap X = \varnothing$ and $(V^k \setminus D_i^k) \bigcap X = \varnothing$, i.e. $X \bigcap V^k = \varnothing$. Moreover, by Property 5 we have $N(V^k) = D_k$ for any $1 \leqslant k \leqslant n$ and any vertex in V^k is adjacent to exactly one vertex in D_k, hence $D_k \subseteq X$. Since $|D_k| = |X| = (n-1)!$, then finally we have $X = D_k$.

(b) In this case our goal is to show that if we assume that there is an efficient dominating set $X_k^i \subset D_k^i$, $1 \leqslant i \neq k \leqslant n$, then it should be empty, i.e. we have to show that $|X_k^i| = 0$ in this case. To prove this we consider the following sets:

$$A^k = N(X_k^i) \bigcap V^k, \quad A^k \subset D_i^k; \tag{4}$$

$$A^i = N(X_i^k)\bigcap V^i, \quad A^i \subset D_k^i; \tag{5}$$

$$B^i = N(C^i)\bigcap V^i \bigcap X, \quad B^i \subset V^i \setminus D_k^i; \tag{6}$$

$$B^k = N(C^k)\bigcap V^k \bigcap X, \quad B^k \subset V^k \setminus D_i^k, \tag{7}$$

where $C^i = D_k^i \setminus (X_k^i \bigcup A^i)$, and $C^k = D_i^k \setminus (X_i^k \bigcup A^k)$. A schematic representation of the sets (4)–(7) is given on Figure 1 where one–to–one correspondences between sets are shown by arrows. The correspondence means that any vertex from one set is adjacent to exactly one vertex from another set. From the definition of the open neighborhood and by Property 8 it follows that the following pairs have this correspondence: A^k and X_k^i, A^i and X_i^k, B^i and C^i, B^k and C^k, which means that $|A^k| = |X_k^i|$, $|A^i| = |X_i^k|$, $|B^i| = |C^i|$, $|B^k| = |C^k|$. Moreover, since $|C^i| = |D_k^i| - (|X_k^i| + |A^i|)$ and $|C^k| = |D_i^k| - (|X_i^k| + |A^k|)$ then $|C^i| = |C^k|$, and hence $|B^i| = |B^k|$.

Now we show that sets $V^k \bigcap X$ and $B^k \bigcup X_i^k$ are the same. Indeed, $B^k \bigcup X_i^k \subseteq V^k \bigcap X$ by (7). On the other hand, since X is an efficient dominating set, then from Property 6 we have $|(V^k \bigcap X) \bigcap N(A^k)| = 0$ and $|(V^k \bigcap X) \bigcap N(X_i^k)| = 0$, so

$$((V^k \bigcap X) \setminus X_i^k) \subset N(C^k)\bigcap V^k,$$

that means $V^k \bigcap X \subseteq (N(C^k) \bigcap V^k \bigcap X) \bigcup X_i^k = B^k \bigcup X_i^k$. Thus, the sets are the same, and hence $|V^k \bigcap X| = |B^k \bigcup X_i^k|$.

We also show that $V^i \bigcap X = B^i \bigcup X_k^i$, since we have $B^i \bigcup X_k^i \subseteq V^i \bigcap X$ by (6), and we have $|(V^i \bigcap X) \bigcap N(A^i)| = 0$ and $|(V^i \bigcap X) \bigcap N(X_k^i)| = 0$ by Property 6, so

$$((V^i \bigcap X) \setminus X_k^i) \subset N(C^i)\bigcap V^i,$$

that means $V^i \bigcap X \subseteq (N(C^i) \bigcap V^i \bigcap X) \bigcup X_k^i = B^i \bigcup X_k^i$. Thus, the sets are the same, and hence $|V^i \bigcap X| = |B^i \bigcup X_k^i|$.

In a general case, for any $b \neq i,k$ any vertex from $\bigcup_{a \neq i,k} D_b^a$ either belongs to $\bigcup_{a \neq i,k}(X \bigcap V^a)$, or it is adjacent to exactly one vertex of this set. Indeed, for any $a \neq i,k$, there are two cases: 1) $X_b^a = D_b^a$; 2) $X_b^a \subset D_b^a$. In the first case, D_b^a belongs to $\bigcup_{a \neq i,k}(X \bigcap V^a)$. In the second case, $D_b^a = X_b^a \bigcup A^a \bigcup C^a$, i.e. for any vertex $x \in D_b^a$ there are three possibilities:

1) $x \in X_b^a \subset \bigcup_{a \neq i,k}(X \bigcap V^a)$;
2) $x \in A^a$, hence x is adjacent to exactly one vertex from $X_a^b \subset \bigcup_{a \neq i,k}(X \bigcap V^a)$;
3) $x \in C^a$, hence x is adjacent to exactly one vertex from $B^a \subset \bigcup_{a \neq i,k}(X \bigcap V^a)$.

On the other hand, the same arguments could be applied to any vertex from $\bigcup_{a \neq i,k}(X \bigcap V^a)$: either it belongs to the set $\bigcup_{a \neq i,k} D_b^a$, or it is adjacent to the only vertex from this set. This means that the cardinalities of these sets should be the same:

$$|\bigcup_{a \neq i,k} D_b^a| = |\bigcup_{a \neq i,k} (X \bigcap V^a)| = |X| - |X \bigcap V^k| - |X \bigcap V^i| =$$

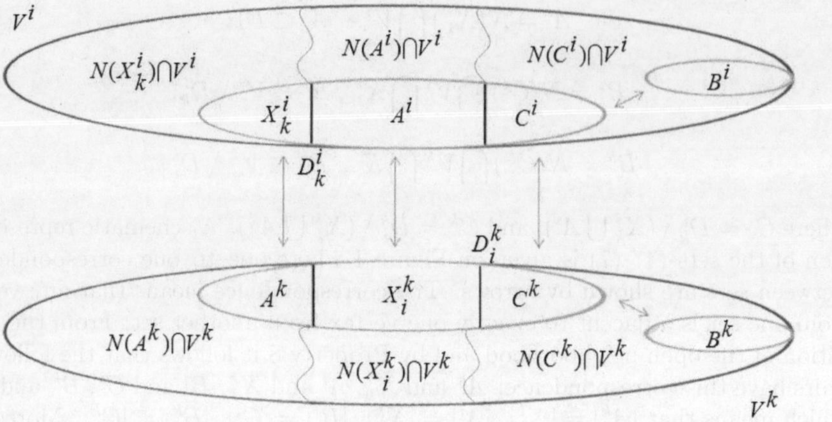

Fig. 1. A schematic representation of the sets

$$= (n-1)! - |X \bigcap V^k| - |X \bigcap V^i| = (n-1)! - (|X_i^k| + |B^k|) - (|X_k^i| + |B^i|) =$$
$$= (n-1)! - ((n-2)! - |X_i^k|) - ((n-2)! - |X_k^i|) = (n-1)! - 2(n-2)! + |X_i^k| + |X_k^i|.$$

On the other hand, we have:

$$|\bigcup_{a \neq i,k} D_b^a| = |D_b| - |D_b^k| - |D_b^i| = (n-1)! - 2(n-2)!.$$

Thus, $|X_k^i| = 0$ that contradicts to our assumption that $|X_k^i|$ is a non–empty set and this completes the proof. □

One of the referees of the manuscript suggested that "the proof might be quite shorter ". I do believe that this is true but no one could find it yet.

Corollary 1. $\gamma(\Gamma_n) = (n-1)!,\ n \geqslant 3$.

We finish this Section by giving the distance characterization for vertices of a given efficient dominating set in the Star and Pancake graphs. Let π^d be a permutation being at the distance d from a permutation π in Γ_n.

3.1 Distances in the Efficient Dominating Sets of the Star Graph

In the Star graph S_n the distance between vertices of a given efficient dominating set is obtained as follows. Let $\pi \in S_n$ and let

$$\pi^{d-1} = \pi\, t_j\, t_{i_1}\, t_{i_2} \ldots t_{i_{d-2}} \notin D_k, \tag{8}$$

where $2 \leqslant j, i_s \leqslant n$, and $j \neq i_1$, $i_s \neq i_{s+1}$ for any $1 \leqslant s < d-2$. We define p_s, $0 \leqslant s \leqslant d-1$, such that $p_0 = p_1 = j$, and for any $s \geqslant 2$ the following holds:

$$p_s = \begin{cases} i_s, & i_{s-1} = j, \\ j, & i_{s-1} \neq j. \end{cases} \tag{9}$$

Lemma 1. *In the Star graph $S_n, n \geqslant 3$, for any $\pi \in D_k, 1 \leqslant k \leqslant n$, and for any $3 \leqslant d \leqslant diam(S_n)$ there exists $\pi^d \in D_k$ presented as follows:*

$$\pi^d = \pi^{d-1} t_{p_{d-1}}, \tag{10}$$

where π^{d-1} corresponds (8) and p_{d-1} is defined by (9).

Proof. Let $\pi^1 = \pi t_j$. There are two cases. If $i_s \neq j$ for any s, $2 \leqslant s \leqslant d-1$, then $\pi^* = \pi^1 t_{i_1} t_{i_2} \ldots t_{i_s}$ with $\pi^*_j = k$. Hence, $\pi^* t_{p_s} \in D_k$, where $p_s = j$. If $i_s = j$ for some s, then $\pi^* = \pi^1 t_{i_1} t_{i_2} \ldots t_{i_s} \in D_k$, $\pi^{**} = \pi^* t_{i_{s+1}}$ with $\pi^{**}_{i_{s+1}} = k$. Thus, $\pi^{**} t_{p_s} \in D_k$, where $p_s = i_{s+1}$. □

3.2 Distances in the Efficient Dominating Sets of the Pancake Graph

Now we consider distances between vertices of a given efficient dominating set in the Pancake graph P_n. Let $\pi \in D_k$, $1 \leqslant k \leqslant n$, and let

$$\pi^{d-1} = \pi r_j r_{i_1} r_{i_2} \ldots r_{i_{d-2}} \notin D_k, \tag{11}$$

where $2 \leqslant j, i_s \leqslant n$, $j \neq i_1$ and $i_s \neq i_{s+1}$ for any $1 \leqslant s < d-2$. We define q_s, $0 \leqslant s \leqslant d-2$, by the following recurrent way. We put

$$q_0 = j, \qquad q_1 = \begin{cases} i_1 - j + 1, \text{ if } i_1 > j, \\ j, \qquad\qquad\quad \text{if } i_1 < j. \end{cases} \tag{12}$$

and for any $2 \leqslant s < d-2$ define

$$q_s = \begin{cases} i_s - q_{s-1} + 1, \text{ if } i_s > q_{s-1}, \\ q_{s-1}, \qquad\qquad\quad \text{if } i_s < q_{s-1}. \end{cases} \tag{13}$$

Then we have the following result.

Theorem 2. *In the Pancake graph $P_n, n \geqslant 3$, for any $\pi \in D_k, 1 \leqslant k \leqslant n$, and any $3 \leqslant d \leqslant diam(P_n)$ there exists $\pi^d \in D_k$ such that*

$$\pi^d = \pi^{d-1} r_{q_{d-2}}, \tag{14}$$

where π^{d-1} is defined by (11) and q_{d-2} is defined by (13),(14), moreover

$$q_{d-2} = \begin{cases} j, \text{ if } i_s < j \text{ for any } 1 \leqslant s \leqslant d-2 \text{ in } (12),(13), \\ q, \text{ if } j < i_1 < i_2 \ldots < i_{d-2} \text{ in } (12),(13). \end{cases} \tag{15}$$

where

$$q = \begin{cases} \sum_{p=0}^{\lfloor (d-2)/2 \rfloor} i_{2p+1} - \sum_{p=1}^{\lfloor (d-2)/2 \rfloor} i_{2p} - j + 1, \text{ if } d \text{ is odd}, \\ \sum_{p=1}^{(d-2)/2} i_{2p} - \sum_{p=0}^{(d-2)/2-1} i_{2p+1} + j, \qquad \text{if } d \text{ is even}. \end{cases} \tag{16}$$

Proof. By Property 6 any two vertices from D_k, $1 \leqslant k \leqslant n$, are at the distance at least three from each other. Indeed, let $\pi = [k\,\pi_2 \ldots \pi_n] \in D_k$. Vertices at the distance one from π don't belong to D_k since they are presented as follows:

$$\pi^1 = \pi\, r_j = [\pi_j\, \pi_{j-1} \ldots \pi_2\, k\, \pi_{j+1} \ldots \pi_n], \pi^1_j = k,\ 2 \leqslant j \leqslant n.$$

Vertices at the distance two from π also don't belong to D_k since they have one of the following representations:

$$\pi^2 = \pi^1\, r_{i_1} = [\pi_{j-i_1+1} \ldots \pi_j\, \pi_{j-i_1} \ldots \pi_2\, k\, \pi_{j+1} \ldots \pi_n], \tag{17}$$

where $\pi^2_j = k$, $2 \leqslant i_1 < j \leqslant n$, or

$$\pi^2 = \pi^1\, r_{i_1} = [\pi_{i_1} \ldots \pi_{j+1}\, k\, \pi_2 \ldots \pi_j\, \pi_{i_1+1} \ldots \pi_n], \tag{18}$$

where $\pi^2_{i_1-j+1} = k$, $2 \leqslant j < i_1 \leqslant n$. Permutations (17),(18) are at the distance one from permutations belonging D_k since multiplying them on r_j and r_{i_1-j+1}, correspondingly, on the right we obtain permutations with the first element k. Thus, for any $\pi, \tau \in D_k$, $1 \leqslant k \leqslant n$, we have $d(\pi, \tau) \geqslant 3$.

On the other hand, multiplying these permutations on r_{i_2}, where $i_2 \neq j$ and $i_2 \neq i_1 - j + 1$, on the right we obtain permutations $\pi^3 = \pi^2\, r_{i_2} \notin D_k$ such that

1) $\pi^3_j = k$, if $i_2 < j$ in (17);
2) $\pi^3_{i_2-j+1} = k$, if $i_2 > j$ in (17);
3) $\pi^3_{i_1-j+1} = k$, if $i_2 < i_1 - j + 1$ in (18);
4) $\pi^3_{i_2-i_1+j} = k$, if $i_2 > i_1 - j + 1$ in (18),

where $i_2 - i_1 + j = i_2 - (i_1 - j + 1) + 1$. Since positions of element k in π^3 are known, so all permutations $\pi^4 \in D_k$ are obtained by multiplying on the corresponding prefix–reversals r_j, r_{i_2-j+1}, r_{i_1-j+1}, $r_{i_2-i_1+j}$ on the right. Let us note that $\pi^3 = \pi^2\, r_{i_2} \notin D_k$ are presented by (11), and $\pi^4 \in D_k$ are obtained by multiplying on prefix–reversals defined by (12),(13) on the right. Taking into account the same arguments, we obtain a permutation $\pi^d = \pi^{d-1}\, r_{q_{d-2}} \in D_k$, where $3 \leqslant d \leqslant diam(P_n)$ for any given $\pi \in D_k$. Moreover, if $i_s < j$, $1 \leqslant s \leqslant d-2$, then the position of element k in π^s doesn't change when π^s is multiplied on r_{i_s} on the right, hence $q_{d-2} = j$ in (13). In the case when $j < i_1 < i_2 \ldots < i_{d-2}$, the position of element k is changed every time when π^s is multiplied on r_{i_s} on the right, hence by (12),(13) the value q_{d-2} will be defined as $i_{d-2} - (i_{d-3} - \ldots - (i_1 - j + 1) + \ldots + 1) + 1)$ for some given d. Moreover, if d is even then all even values i_s in the expression above will be positive while all odd values will be negative, and vice versa, if d is odd then all even values i_s will be negative while all odd will be positive. Thus, we have $q_{d-2} = q$, where q is defined by (16). □

4 Small Cycles of the Pancake Graph

In this Section we present results on the full characterization of cycles of length 6, 7, 8 and 9 in the Pancake graph. To describe cycles the following cycle representation via a product of generating elements was used in [KM10]. A cycle of length l is also called an *l–cycle*. A sequence of prefix–reversals $C_l = r_{i_0} \dots r_{i_{l-1}}$, where $2 \leqslant i_j \leqslant n$ and $i_j \neq i_{j+1}$ $((j+1) \bmod l)$ for any $0 \leqslant j \leqslant l-1$, such that $\pi r_{i_0} \dots r_{i_{l-1}} = \pi$, where $\pi \in Sym_n$, is called a *form* of l–cycle. Any l–cycle can be represented by $2\,l$ its forms (not necessarily distinct) with respect to a vertex and a direction. *The canonical form C_l of an l–cycle* is called a form with a lexicographically maximal sequence of indices $i_0 \dots i_{l-1}$. Two cycles in a graph are *independent* if they do not have any vertex in common.

The main results of this Section are presented by the following theorems.

Theorem 3. *[KM10] The Pancake graph $P_n, n \geqslant 3$, has $\frac{n!}{6}$ independent 6–cycles of the canonical form*

$$C_6 = r_3 r_2 r_3 r_2 r_3 r_2. \tag{19}$$

Moreover, each of vertices of P_n belongs to exactly one 6–cycle.

Theorem 4. *[KM10] The Pancake graph $P_n, n \geqslant 4$, has $n!(n-3)$ distinct 7–cycles of the canonical form*

$$C_7 = r_k r_{k-1} r_k r_{k-1} r_{k-2} r_k r_2, \tag{20}$$

where $4 \leqslant k \leqslant n$. Moreover, each of vertices of P_n belongs to $7(n-3)$ distinct 7–cycles and there are $\frac{n!}{8} \leqslant N_7 \leqslant \frac{n!}{7}$ independent 7–cycles.

As one can see, the descriptions of 6–cycles and 7–cycles are not so complicated. However, the situation is changed dramatically for 8–cycles and 9–cycles.

Theorem 5. *[KM12] Each of vertices of the Pancake graph P_n, $n \geqslant 4$, belongs to $N_8 = \frac{n^3+12n^2-103n+176}{2}$ distinct 8–cycles of the following canonical forms:*

$$\begin{array}{ll}
C_8^1 = r_k r_j r_i r_j r_k r_{k-j+i} r_i r_{k-j+i}, & 2 \leqslant i < j \leqslant k-1,\ 4 \leqslant k \leqslant n; \\
C_8^2 = r_k r_{k-1} r_2 r_{k-1} r_k r_2 r_3 r_2, & 4 \leqslant k \leqslant n; \\
C_8^3 = r_k r_{k-i} r_{k-1} r_i r_k r_{k-i} r_{k-1} r_i, & 2 \leqslant i \leqslant k-2,\ 4 \leqslant k \leqslant n; \\
C_8^4 = r_k r_{k-i+1} r_k r_i r_k r_{k-i} r_{k-1} r_{i-1}, & 3 \leqslant i \leqslant k-2,\ 5 \leqslant k \leqslant n; \\
C_8^5 = r_k r_{k-1} r_{i-1} r_k r_{k-i+1} r_{k-i} r_k r_i, & 3 \leqslant i \leqslant k-2,\ 5 \leqslant k \leqslant n; \\
C_8^6 = r_k r_{k-1} r_k r_{k-i} r_{k-i-1} r_k r_i r_{i+1}, & 2 \leqslant i \leqslant k-3,\ 5 \leqslant k \leqslant n; \\
C_8^7 = r_k r_{k-j+1} r_k r_i r_k r_{k-j+1} r_k r_i, & 2 \leqslant i < j \leqslant k-1,\ 4 \leqslant k \leqslant n; \\
C_8^8 = r_4 r_3 r_4 r_3 r_4 r_3 r_4 r_3. &
\end{array}$$

Moreover, there are $\frac{n!(n^3+12n^2-103n+176)}{16}$ distinct 8–cycles and $\frac{n!}{8}$ independent 8–cycles in the Pancake graph.

Theorem 6. *[KM11] Each of vertices of the Pancake graph $P_n, n \geqslant 4$, belongs to $N_9 = \frac{8n^3-45n^2+61n-12}{2}$ distinct 9–cycles of the following canonical forms:*

$C_9^1 = r_k r_{k-1} r_i r_{k-1} r_k r_i r_{i-1} r_{i+1} r_2,$ $3 \leqslant i \leqslant k-2, 5 \leqslant k \leqslant n;$
$C_9^2 = r_2 r_{k-i+2} r_k r_{i-2} r_{i-1} r_i r_{i-1} r_k r_{k-i+2},$ $4 \leqslant i \leqslant k-1, 5 \leqslant k \leqslant n;$
$C_9^3 = r_k r_{k-i} r_{k-1} r_{k-j+i-1} r_{k-j} r_k r_{j-i+1} r_j r_i,$ $2 \leqslant i < j \leqslant k-2, 5 \leqslant k \leqslant n;$
$C_9^4 = r_k r_{k-1} r_i r_{i-1} r_{k-1} r_k r_i r_{i+1} r_2,$ $3 \leqslant i \leqslant k-2, 5 \leqslant k \leqslant n;$
$C_9^5 = r_k r_{k-1} r_{k-2} r_{k-1} r_{k-2} r_k r_3 r_k r_{k-2},$ $4 \leqslant k \leqslant n;$
$C_9^6 = r_k r_{k-1} r_{k-2} r_i r_k r_2 r_k r_i r_{k-1},$ $2 \leqslant i \leqslant k-3, 5 \leqslant k \leqslant n;$
$C_9^7 = r_k r_{k-j+i} r_k r_j r_i r_k r_{k-j} r_{k-i} r_{j-i},$ $2 \leqslant i \leqslant j-2, i+2 \leqslant j \leqslant k-2, 6 \leqslant k \leqslant n;$
$C_9^8 = r_k r_{k-j+i} r_{k-j} r_k r_j r_i r_k r_{k-i} r_{j-i},$ $2 \leqslant i \leqslant j-2, i+2 \leqslant j \leqslant k-2, 6 \leqslant k \leqslant n;$
$C_9^9 = r_k r_{k-j+i} r_{k-j+1} r_k r_j r_i r_k r_{k-i+1} r_{j-i+1},$ $2 \leqslant i < j \leqslant k-1, 4 \leqslant k \leqslant n;$
$C_9^{10} = r_k r_{k-1} r_k r_{k-1} r_k r_{k-1} r_{k-3} r_k r_3,$ $5 \leqslant k \leqslant n.$

Moreover, there are $\frac{n!(8n^3-45n^2+61n-12)}{18}$ *distinct 9–cycles in the Pancake graph.*

Proofs of these results are based on the hierarchical structure of the Pancake graph. Below we give proofs for Theorem 3 and Theorem 4. To present them we need some new definitions and notations.

A *segment* $[\pi_i \dots \pi_j]$ of a permutation $\pi = [\pi_1 \dots \pi_i \dots \pi_j \dots \pi_n]$ consists of all elements entered into between π_i and π_j inclusive. Any permutation can be written as a sequence of singleton and multiple segments which are presented by $\{i, j, k\}$ and $\{\alpha, \beta, \gamma\}$, respectively. For example, $\pi = [i\pi_2\pi_3\pi_4 j\pi_6\pi_7\pi_8 k]$ can be presented as $\pi = [i\alpha j\beta k]$ where $\alpha = [\pi_2\pi_3\pi_4]$, $\beta = [\pi_6\pi_7\pi_8]$. If $\overline{\alpha}$ is the inversion of a segment α then $\overline{\overline{\alpha}} = \alpha$. Let us denote the number of elements in a segment α as $|\alpha|$. We also put $\overline{\pi} = \pi r_n$ and $\overline{\tau} = \tau r_n$.

Lemma 2. *[KM10] Let two distinct permutations π and τ belong to the same $(n-1)$–copy of $P_n, n \geqslant 3$, and let $d(\pi, \tau) \leqslant 2$, then $\overline{\pi}, \overline{\tau}$ belong to distinct $(n-1)$–copies of the graph.*

Proof. Let $\pi, \tau \in P_{n-1}(i)$, $1 \leqslant i \leqslant n$. If $d(\pi, \tau) = 1$ and if we put $\pi = [j\alpha k\beta i]$ then $\tau = [k\overline{\alpha}j\beta i]$ where $j \neq k \neq i$. So, $\overline{\pi} = [i\overline{\beta}k\overline{\alpha}j]$ and $\overline{\tau} = [i\overline{\beta}j\alpha k]$, which means that $\overline{\pi}, \overline{\tau}$ belong to distinct copies $P_{n-1}(j)$ and $P_{n-1}(k)$. If $d(\pi, \tau) = 2$ then there is a permutation ω in $P_{n-1}(i)$ adjacent to π and τ. Permutations π and τ are obtained from ω by multiplying on different (not equal to r_n) prefix–reversals on the right. Thereby, the first elements of π and τ should be different hence $\overline{\pi} = \pi\, r_n$ and $\overline{\tau} = \tau\, r_n$ should be different, i.e. they belong to the distinct $(n-1)$–copies of P_n. □

4.1 6–Cycles of the Pancake Graph

In this section we present the proof of Theorem 3.

Proof. If $n = 3$ then $P_3 \cong C_6$ and there is the only 6–cycle presented as $[123] \overset{r_2}{\to} [213] \overset{r_3}{\to} [312] \overset{r_2}{\to} [132] \overset{r_3}{\to} [231] \overset{r_2}{\to} [321] \overset{r_3}{\to} [123]$ for which the canonical form is $C_6 = r_3 r_2 r_3 r_2 r_3 r_2$.

Let us show that there are no other forms of 6–cycles in $P_n, n \geqslant 4$. First of all, we prove that a 6–cycle doesn't appear on vertices of two distinct $(n-1)$–copies. Indeed, if $\pi, \tau \in P_{n-1}(i)$ and $\overline{\pi}, \overline{\tau} \in P_{n-1}(j)$ then $d(\pi, \tau) \neq 1$ and $d(\pi, \tau) \neq 2$

by Lemma 2, and hence $d(\pi,\tau) \geqslant 3$. Suppose that there is a 6–cycle containing vertices $\pi,\tau,\overline{\pi},\overline{\tau}$. So if $d(\pi,\tau) = 3$ then $\overline{\pi},\overline{\tau}$ are adjacent in $P_{n-1}(j)$ and by Lemma 2 vertices $\pi = \overline{\pi}r_n, \tau = \overline{\tau}r_n$ belong to distinct $(n-1)$–copies but this is not true since $\pi,\tau \in P_{n-1}(i)$. If $d(\pi,\tau) = 4$ then $\overline{\pi} = \overline{\tau}$ but this is not possible since $\pi \neq \tau$. Thus, a 6–cycle doesn't appear on vertices of two distinct $(n-1)$–copies.

Now let us prove that a 6–cycle doesn't appear on vertices of three distinct $(n-1)$–copies. Let $\pi,\tau \in P_{n-1}(i), \pi \neq \tau$ such that $d(\pi,\tau) \leqslant 2$ then by Lemma 2 vertices $\overline{\pi},\overline{\tau}$ belong to distinct $(n-1)$–copies. We consider two cases.

If $d(\pi,\tau) = 1$ then vertices $\pi,\tau,\overline{\pi},\overline{\tau}$ might be belong to a 6–cycle if and only if $d(\overline{\pi},\overline{\tau}) = 3$. Show that this is not true. Let $\pi = [j\alpha k\beta i]$ then $\tau = [k\overline{\alpha}j\beta i]$ and $\overline{\pi} = [i\overline{\beta}k\overline{\alpha}j] \in P_{n-1}(j)$, $\overline{\tau} = [i\overline{\beta}j\alpha k] \in P_{n-1}(k)$. The shortest path starting at $\overline{\pi}$ and belonging to $P_{n-1}(k)$ should contain vertices $\omega = [k\beta i\overline{\alpha}j]$ and $\overline{\omega} = [j\alpha i\overline{\beta}k] \in P_{n-1}(k)$, i.e. $d(\overline{\pi},\overline{\omega}) = 2$. It is evident that there is no a prefix–reversal transforming $\overline{\omega}$ into $\overline{\tau}$, i.e. $d(\overline{\omega},\overline{\tau}) \neq 1$, and hence $d(\overline{\pi},\overline{\tau}) \neq 3$.

If $d(\pi,\tau) = 2$ then vertices $\pi,\tau,\overline{\pi},\overline{\tau}$ might be belong to a 6–cycle if and only if $d(\overline{\pi},\overline{\tau}) = 2$. However this is not possible since by Lemma 2 vertices $\pi = \overline{\pi}r_n$ and $\tau = \overline{\tau}r_n$ belong to distinct $(n-1)$–copies. Thus, a 6–cycle doesn't appear on vertices of three distinct $(n-1)$–copies.

It is also evident that a 6–cycle doesn't appear on vertices of four and more distinct $(n-1)$–copies since there should be at least four external edges as well as at least one edge in each of $(n-1)$–copies so we have a 8–cycle.

Thus, there is the only canonical form, namely $r_3r_2r_3r_2r_3r_2$, to describe 6–cycles in $P_n, n \geqslant 3$. These cycles are independent for $n \geqslant 4$ since prefix–reversals $r_i, 4 \leqslant i \leqslant n$, define external edges for 6–cycles which means that each of vertices of P_n belongs to exactly one 6–cycle. □

4.2 7–Cycles of the Pancake Graph

In this section we give the proof of Theorem 4.

Proof. We prove Theorem 4 by the induction on the dimension k of the Pancake graph P_k when $k \geqslant 4$. If $k = 3$ then there are no 7–cycles in $P_3 \cong C_6$.

If $k = 4$ then Theorem says that each of vertices of P_4 belongs to 7 distinct 7–cycles. Since P_n is a vertex–transitive graph then it is enough to check this fact for any its vertex. In particular, all 7–cycles containing the identity permutation [1234] are presented in the Table 1. They could be found easily by considering vertex distributions of P_4 in metric spheres centered at the identity permutation $I = [1234]$. The canonical form for all cycles presented in Table 1 is $C_6 = r_4r_3r_4r_3r_2r_4r_2$ that corresponds to (20) when $k = 4$.

Now we assume that Theorem is hold for $k = n-1$ and prove that it is hold also for $k = n$ using the hierarchical structure of P_n.

By the induction assumption, any vertex of any $(n-1)$–copy belongs to $7((n-1)-3) = 7(n-4)$ distinct 7–cycles of this copy. However, besides 7–cycles belonging to the same $(n-1)$–copy there may also be 7–cycles belonging to distinct $(n-1)$–copies of the graph. The following three cases are possible.

Table 1. 7–cycles in P_4 containing the identity permutation [1234]

?	vertex description	prefix–reversal description
1	[1234]-[4321]-[2341]-[1432]-[3412]-[4312]-[2134]	$r_4r_3r_4r_3r_2r_4r_2$
2	[1234]-[3214]-[4123]-[2143]-[1243]-[3421]-[4321]	$r_3r_4r_3r_2r_4r_2r_4$
3	[1234]-[4321]-[2341]-[3241]-[1423]-[4123]-[3214]	$r_4r_3r_2r_4r_2r_4r_3$
4	[1234]-[3214]-[2314]-[4132]-[1432]-[2341]-[4321]	$r_3r_2r_4r_2r_4r_3r_4$
5	[1234]-[2134]-[4312]-[3412]-[2143]-[4123]-[3214]	$r_2r_4r_2r_4r_3r_4r_3$
6	[1234]-[4321]-[3421]-[1243]-[4213]-[3124]-[2134]	$r_4r_2r_4r_3r_4r_3r_2$
7	[1234]-[2134]-[4312]-[1342]-[2431]-[3421]-[4321]	$r_2r_4r_3r_4r_3r_2r_4$

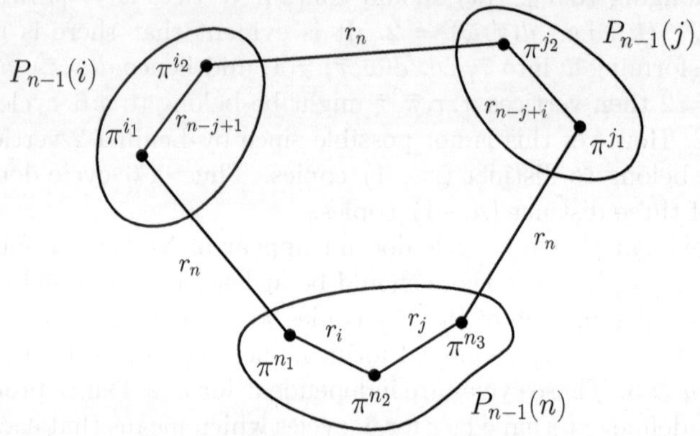

Fig. 2. Case 2 of the proof of Theorem 4

Case 1. Suppose that a sought 7–cycle C_7^* is formed on vertices from two copies $P_{n-1}(i)$ and $P_{n-1}(j)$, $1 \leqslant i \neq j \leqslant n$, such that either two vertices of C_7^* belong to $P_{n-1}(i)$ and other five vertices belong to $P_{n-1}(j)$, or three vertices of C_7^* belong to $P_{n-1}(i)$ and other four vertices belong to a copy $P_{n-1}(j)$. In the both cases we have $d(\pi, \tau) \leqslant 2$ for any vertices $\pi, \tau \in P_{n-1}(i)$ belonging to C_7^*. Then by Lemma 2 vertices $\overline{\pi}, \overline{\tau}$ belong to distinct $(n-1)$–copies that contradicts to our assumption. Therefore, a 7–cycle does not occur in this case.

Case 2. Suppose that a sought 7–cycle C_7^* is formed on vertices from three distinct $(n-1)$–copies such that two vertices π^{i_1}, π^{i_2} belong to $P_{n-1}(i)$, two vertices π^{j_1}, π^{j_2} belong to $P_{n-1}(j)$, the other three vertices $\pi^{n_1}, \pi^{n_2}, \pi^{n_3}$ belong to $P_{n-1}(n)$, where $1 \leqslant i < j \leqslant n$ (see Figure 2).

Let us describe a sought cycle. Since P_n is a vertex–transitive graph then there is no loss of generality in taking $\pi^{n_2} = I_n = [\alpha i \beta j \gamma n]$, where $|\alpha| = i-1$, $|\beta| = j-i-1$, $|\gamma| = n-j-1$. By Lemma 2 vertices π^{n_1} and π^{n_3} are adjacent to vertices from distinct $(n-1)$–copies $P_{n-1}(i)$ and $P_{n-1}(j)$, hence these vertices are presented as follows:

$$\pi^{n_1} = \pi^{n_2} r_i = [i\overline{\alpha}\beta j\gamma n], \quad \text{where} \quad \pi^{n_1}_j = j,$$

$$\pi^{n_3} = \pi^{n_2} r_j = [j\overline{\beta}i\overline{\alpha}\gamma n], \quad \text{where} \quad \pi^{n_3}_{j-i+1} = i.$$

Their adjacent vertices in copies $P_{n-1}(i)$ and $P_{n-1}(j)$ are presented as follows:

$$\pi^{i_1} = \pi^{n_1} r_n = [n\overline{\gamma}j\overline{\beta}\alpha i], \quad \text{where} \quad \pi^{i_1}_{n-j+1} = j,$$

$$\pi^{j_1} = \pi^{n_3} r_n = [n\overline{\gamma}\alpha i\beta j], \quad \text{where} \quad \pi^{j_1}_{n-j+i} = i.$$

A vertex π^{i_2} should be adjacent to the vertex π^{i_1} and to one of vertices, say π^{j_2}, from the copy $P_{n-1}(j)$:

$$\pi^{i_2} = \pi^{i_1} r_{n-j+1} = [j\gamma n\overline{\beta}\alpha i], \quad \text{where} \quad \pi^{i_2}_1 = j.$$

On the other hand, a vertex π^{j_2} should be adjacent to the vertex π^{j_1}. Moreover, since it is also adjacent to π^{i_2} hence π^{j_2} has the following view:

$$\pi^{j_2} = \pi^{j_1} r_{n-j+i} = [i\overline{\alpha}\gamma n\beta j], \quad \text{where} \quad \pi^{j_2}_1 = i.$$

By our assumption, the vertices π^{i_2} and π^{j_2} are incident to the same external edge which means that a permutation $\pi^* = \pi^{i_2} r_n = [i\overline{\alpha}\beta n\overline{\gamma}j]$ should coincide with the permutation π^{j_2}. This is possible only in the case when segments β and γ are empty, i.e. $|\beta| = j - i - 1 = 0$ and $|\gamma| = n - j - 1 = 0$. From this we have $j = n - 1$ and $i = j - 1 = n - 2$, and a 7–cycle is presented as follows: $\pi^{i_1} \xrightarrow{r_2} \pi^{i_2} \xrightarrow{r_n} \pi^{j_2} \xrightarrow{r_{n-1}} \pi^{j_1} \xrightarrow{r_n} \pi^{n_3} \xrightarrow{r_{n-1}} \pi^{n_2} \xrightarrow{r_{n-2}} \pi^{n_1} \xrightarrow{r_n} \pi^{i_1}$. Its canonical form $C_7 = r_n r_{n-1} r_n r_{n-1} r_{n-2} r_n r_2$ coincide with (20) when $k = n$.

Case 3. Suppose that a sought 7–cycle is formed on vertices from four or more $(n-1)$–copies. It follows from the hierarchical structure of the graph that any its vertex is incident to the only external edge. So any 7–cycle in this graph should contain at least two vertices of the same $(n-1)$–copy and hence a 7–cycle does not occur in this assumption.

Thus, the only canonical form $r_n r_{n-1} r_n r_{n-1} r_{n-2} r_n r_2$ representing seven cycles of the length 7 and containing vertices from three distinct $(n-1)$–copies of the graph P_n is found. It is evident that any vertex of P_n belongs to all these cycles. By the induction assumption, any vertex of any $(n-1)$–copy belongs to $7(n-4)$ distinct 7–cycles from this copy. Therefore, any vertex of P_n belongs to $7(n-4)+7 = 7(n-3)$ distinct 7–cycles of the canonical form (20) that completes the proof on the main fact of Theorem 4.

Since each of vertices belongs to $7(n-3)$ distinct 7–cycles and there are $n!$ vertices in P_n, hence there are $n!7(n-3)$ cycles of length 7. However, each cycle was enumerated seven times, so totally there are $7(n-3)$ distinct 7–cycles.

It is also easy to show that there are three independent 7–cycles in P_4. For example, the following three 7–cycles are independent in P_4:

$$\begin{aligned} C_7^1 &= [1234] - [2134] - [4312] - [1342] - [2431] - [3421] - [4321], \\ C_7^2 &= [3241] - [2341] - [1432] - [3412] - [2143] - [4123] - [1423], \\ C_7^3 &= [4213] - [2413] - [3142] - [4132] - [2314] - [1324] - [3124]. \end{aligned}$$

It follows from the hierarchical structure of $P_n, n \geqslant 4$, that there are $\frac{n!}{24}$ copies of P_4 and each of them has exactly three independent 7–cycles. So, totally there are at least $\frac{n!}{8}$ independent 7–cycles that gives the lower bound. The upper bound is obtained in assumption that each of vertices of $P_n, n \geqslant 7$, belongs to exactly one 7–cycle. □

We omit the proof of Theorems 4 and 5 since it takes almost 30 pages to present them together. The following idea was used in proofs.

Since $P_3 \cong C_6$ and due to the hierarchical structure, P_4 has four copies of P_3, each of which obviously cannot contain 8– or 9–cycles. However, P_4 has 8– and 9–cycles consisting of paths within copies of P_3 as well as external edges between these copies. In general, any 8– or 9–cycle of $P_n, n \geqslant 4$, must consist of paths within subgraphs that are isomorphic to P_{k-1} for some $4 \leqslant k \leqslant n$, joined by external edges between these subgraphs. Hence, all 8– and 9–cycles of $P_n, n \geqslant 4$, could be found recursively by considering 8– or 9–cycles within each $P_k, 4 \leqslant k \leqslant n$, consisting of vertices from some copies of P_{k-1}. Among all paths belonging to 8– and 9–cycles within subgraphs that are isomorphic to P_{k-1} for some $4 \leqslant k \leqslant n$, paths of length three between vertices of a given form are the most important cases. So, some results concerning paths of length three between vertices of a given form are also used in proofs.

Acknowledgments. The author thanks Harout Aydinian, Ferdinando Cicalese, and Christian Deppe for the chance to write a paper in memory of Rudi. The author also thanks the referees for their helpful comments and suggestions.

This paper was supported by grant 12-01-00448 of the RFBR and by Interdisciplinary Integration Project of Fundamental Investigations of SB RAS No.21.

References

[AK89] Akers, S.B., Krishnamurthy, B.: A group–theoretic model for symmetric interconnection networks. IEEE Trans. Comput. 38(4), 555–566 (1989)

[AK96] Arumugam, S., Kala, R.: Domination parameters of Star graph. Ars Combinatoria 44(1), 93–96 (1996)

[AKSK06] Asai, S., Kounoike, Y., Shinano, Y., Kaneko, K.: Computing the Diameter of 17-Pancake Graph Using a PC Cluster. In: Nagel, W.E., Walter, W.V., Lehner, W. (eds.) Euro-Par 2006. LNCS, vol. 4128, pp. 1114–1124. Springer, Heidelberg (2006)

[C11] Cibulka, J.: On average and highest number of flips in pancake sorting. Theoretical Computer Science 412, 822–834 (2011)

[DS03] Dejter, I.J., Serra, O.: Efficient dominating sets in Cayley graphs. Discrete Appl. Math. 129, 319–328 (2003)

[DW75] Dweighter, H.: E 2569 in: Elementary problems and solutions. Amer. Math. Monthly 82(1), 1010 (1975)

[GP79] Gates, W.H., Papadimitriou, C.H.: Bounds for sorting by prefix–reversal. Discrete Math. 27, 47–57 (1979)

[HS97] Hyedari, M.H., Sudborough, I.H.: On the diameter of the pancake network. Journal of Algorithms 25(1), 67–94 (1997)

[JLD91] Jwo, J.S., Lakshmivarahan, S., Dhall, S.K.: Embedding of cycles and grids in star graphs. J. Circuits, Syst., Comput. 1(1), 43–74 (1991)

[KF95] Kanevsky, A., Feng, C.: On the embedding of cycles in pancake graphs. Parallel Computing 21, 923–936 (1995)

[K10] Konstantinova, E.: Perfect codes in the pancake networks, `http://www.math.uniri.hr/NATO-ASI/abstracts/Konstantinova_abstract.pdf`

[KM10] Konstantinova, E.V., Medvedev, A.N.: Cycles of length seven in the Pancake graph. Diskretn. Anal. Issled. Oper. 17(5), 46–55 (2010) (in Russian)

[KM11] Konstantinova, E.V., Medvedev, A.N.: Cycles of length nine in the Pancake graph. Diskretn. Anal. Issled. Oper. 8(6), 33–60 (2011) (in Russian)

[KM12] Konstantinova, E.V., Medvedev, A.N.: Small cycles in the Pancake graph. Submitted to Ars Mathematica Contemporanea

[KS12] Konstantinova, E.V., Savin, M.Y.: Efficient dominating sets of the Star and Pancake graphs. Submitted to Diskretn. Anal. Issled. Oper. (in Russian)

[LJD93] Lakshmivarahan, S., Jwo, J.S., Dhall, S.K.: Symmetry in interconnection networks based on Cayley graphs of permutation groups: a survey. Parallel Comput. 19(4), 361–407 (1993)

[Q06] Qiu, K.: Optimal broadcasting algorithms for multiple messages on the star and pancake graphs using minimum dominating sets. Congress Numerantium 181, 33–39 (2006)

[STC06] Sheu, J.J., Tan, J.J.M., Chu, K.T.: Cycle embedding in pancake interconnection networks. In: Proc. 23rd Workshop on Combinatorial Mathematics and Computation Theory, Taiwan, pp. 85–92 (2006)

Threshold and Majority Group Testing

Rudolf Ahlswede*, Christian Deppe[1],**, and Vladimir Lebedev[2],***

[1] Department of Mathematics, University of Bielefeld
cdeppe@math.uni-bielefeld.de
[2] IPPI (Institute for Information Transmission Problems
of the Russian Academy of Sciences (Kharkevich Institute))
lebedev37@mail.ru

Abstract. We consider two generalizations of group testing: threshold group testing (introduced by Damaschke [11]) and majority group testing (a further generalization, including threshold group testing and a model introduced by Lebedev [20]).

We show that each separating code gives a nonadaptive strategy for threshold group testing for some parameters. This is a generalization of a result in [4] on "guessing secrets", introduced in [9].

We introduce threshold codes and show that each threshold code gives a nonadaptive strategy for threshold group testing. Threshold codes include also the construction of [6]. In contrast to [8], where the number of defectives is bounded, we consider the case when the number of defectives are known. We show that we can improve the rate in this case.

We consider majority group testing if the number of defective elements is unknown but bounded (otherwise it reduces to threshold group testing). We show that cover-free codes and separating codes give strategies for majority group testing. We give a lower bound for the rate of majority group testing.

Keywords: group testing, pooling, threshold group testing, separating codes, cover-free codes.

1 Introduction

Group testing is of interest for many applications like in molecular biology. For an overview of results and applications we refer to the books [13] and [14].

The classical group testing problem is to find the unknown subset $\mathcal{D}$ of all defective elements in the set $[N] = \{1, 2, \ldots, N\}$. We consider the case if the cardinality of $|\mathcal{D}| = D$ is known.

For a *test set* $\mathcal{S} \subset [N]$ a test $t_{\mathcal{S}}$ is the function $t_{\mathcal{S}} : 2^{[N]} \to \{0, 1\}$ defined by

$$t_{\mathcal{S}}(\mathcal{D}) = \begin{cases} 0\,, & \text{if } |\mathcal{S} \cap \mathcal{D}| = 0 \\ 1\,, & \text{otherwise.} \end{cases}$$

* Rudolf Ahlswede, one of the world's top information theorists at the University of Bielefeld, died December 18, 2010.

** Supported by the DFG in the project "General Theory of Information Transfer".
*** Supported in part by the Russian Foundation for Basic Research, project no 12-01-00905.

H. Aydinian, F. Cicalese, and C. Deppe (Eds.): Ahlswede Festschrift, LNCS 7777, pp. 488–508, 2013.

We call a binary $n \times N$ matrix $X = (m_{ij})_{1 \le i \le n, 1 \le j \le N}$ a nonadaptive group testing strategy, where the row $m_i = (m_{i1}, m_{i2}, \dots, m_{iN})$ of the matrix represent the ith test set $S_i = \{j : m_{ij} = 1\}$. The testresult is a map $t : \{(m_{ij})_{1 \le i \le n, 1 \le j \le N} : m_{ij} \in \{0,1\}\} \times [N] \to \{v^n \in \{0,1\}^n\}$ with $t(X, D) = (t_1, \dots, t_n)$, where $t_i = t_{S_i}(\mathcal{D})$. A strategy is called successful, if we can uniquely determine $\mathcal{D}$. This means for all $\mathcal{D}_1, \mathcal{D}_2 \subset [N]$ with $|\mathcal{D}_1| = |\mathcal{D}_2| = D$

$$t(X, \mathcal{D}_1) \neq t(X, \mathcal{D}_2).$$

We remind the reader of the concepts of adaptive and nonadaptive strategies.

Strategies are called adaptive if the results of the first $k-1$ tests determine the kth test. Strategies in which we choose all tests independently are called nonadaptive. We consider here only nonadaptive strategies.

In the present paper we study two generalizations of group testing which are quite natural.

In **threshold group testing** the integers $0 \le l < u$ are given and a test $t_{\mathcal{S}}$ is the function $t_{\mathcal{S}} : 2^{[N]} \to \{0, 1, \{0,1\}\}$, defined by

$$t_{\mathcal{S}}(\mathcal{D}) = \begin{cases} 0 & , \text{ if } |\mathcal{S} \cap \mathcal{D}| \le l \\ 1 & , \text{ if } |\mathcal{S} \cap \mathcal{D}| \ge u \\ \{0,1\} & , \text{ otherwise} \\ & \quad \text{(meaning that the result can be arbitrarily 0 or 1).} \end{cases}$$

We call $g = u - l - 1$ the gap. In threshold group testing it is not possible to find the set $\mathcal{D}$ of all defective elements if $g = u - l - 1 > 0$ (see [11]). It is only possible to find $\mathcal{P} \subset [N]$ with $|\mathcal{P} \backslash \mathcal{D}| \le g$ and $|\mathcal{D} \backslash \mathcal{P}| \le g$. Therefore a nonadaptive strategy $X = (m_{ij})_{1 \le i \le n, 1 \le j \le N}$ is called a successful nonadaptive threshold strategy, if for all $\mathcal{D}_1, \mathcal{D}_2 \subset [N]$ with $|\mathcal{D}_1 \backslash \mathcal{D}_2| > g$, $|\mathcal{D}_2 \backslash \mathcal{D}_1| > g$, and $|\mathcal{D}_1| = |\mathcal{D}_2| = D$

$$t(X, \mathcal{D}_1) \neq t(X, \mathcal{D}_2).$$

In [8] a stronger definition of successful strategies is used and do not include all successful strategies in our sense, but with this idea the author got relations to known designs. A good survey of all known results in threshold group testing is given in [12].

In this paper all logarithms have the basis 2. In [11] a strategy for threshold group testing without gap with $O(u^2 + D \log N)$ tests was given. In [5] a strategy with $O(D \log N)$ tests was given. In [2] a very easy strategy with $D \log(N-D+1)$ tests and a strategy with $(u-1)\lceil \log(N-D+1) \rceil + \lceil \log \binom{N-u+1}{D-u+1} \rceil + D - u + 1$ were given.

In **majority group testing** there are two functions $f_1, f_2 : \{0, 1, \dots, N\} \to \mathbb{R}^+$ which put weights on the number $D = |\mathcal{D}| \in \{0, 1, \dots, N\}$ of defective elements and $f_1(D) < f_2(D)\ \forall D \in [0, 1, \dots, N]$.

They describe the structure of tests $t_{\mathcal{S}} : 2^{[N]} \to \{0, 1, \{0,1\}\}$ as follows

$$t_{\mathcal{S}}(\mathcal{D}) = \begin{cases} 0 & , \text{ if } |\mathcal{S} \cap \mathcal{D}| \le f_1(D) \\ 1 & , \text{ if } |\mathcal{S} \cap \mathcal{D}| \ge f_2(D) \\ \{0,1\} & , \text{ otherwise} \\ & \quad \text{(meaning that the result can be arbitrarily 0 or 1).} \end{cases}$$

We use the name majority group testing because first the case $f_1 = \frac{D}{2} - 1$ and $f_2 = \frac{D}{2} + 1$ was considered in [20] (see below). Clearly majority group testing is a generalization of threshold group testing. We get threshold group testing as a special case by setting $f_1(D) = l$ and $f_2(D) = u$. Furthermore the models are equivalent if the number $\mathcal{D}$ of defectives is known. In majority group testing, in particular also for threshold group testing if $g \neq 0$, it is not possible to find the set $\mathcal{D}$ of all defective elements. We can find a family of subsets $\mathbb{F} \subset 2^{[N]}$, which contains $\mathcal{D}$. This set depends on f_1 and f_2, on $\mathcal{D}$, and on the strategy used. In this case we call a strategy successful, if we can find an $\mathbb{F}$ with the smallest possible size in the worst case.

Another model is when f_1 and f_2 depend on $\mathcal{S}$ instead of D, where the gapless model is regarded in the paper [17].

A special case of majority group testing was introduced by Lebedev [20] as follows

$$t_{\mathcal{S}}(\mathcal{D}) = \begin{cases} 0 & , \text{ if } |\mathcal{S} \cap \mathcal{D}| < \frac{D}{2} \\ 1 & , \text{ if } |\mathcal{S} \cap \mathcal{D}| > \frac{D}{2} \\ \{0,1\} & , \text{ if } |\mathcal{S} \cap \mathcal{D}| = \frac{D}{2}. \end{cases}$$

It was shown in [20] that a (w, w) separating code gives a successful nonadaptive strategy if it is assumed that D is odd and that $D < 2w$. (See Section 5 for other special cases studied in [20]).

In [4] it was shown that for **guessing secrets** (that means $l = 0$ and $u = D$ for threshold group testing) a (D, D) separating code gives a successful nonadaptive strategy. We generalize the guessing secret case in **Section 2** and prove that for threshold group testing a $(u, D-l)$ separating code gives a successful nonadaptive strategy if $D = u+l$. This improves the result of [6] for this special case, because the authors use a $(u, D-l)$ cover-free code for the strategy, which has a smaller rate than a separating code. We improve this by a new concept in the next section.

In **Section 3** we **introduce threshold codes** and show that these codes give nonadaptive strategies for threshold group testing, if the number of defectives are known. Furthermore we give an upper bound for the rate of this construction.

In **Section 4** we consider **majority group** testing for $f_1(D) = \lceil \frac{D}{k} \rceil - 1$ and $f_2(D) = \lfloor \frac{D}{k} \rfloor + 1$ where $2 \leq k \in \mathbb{N}$. We first give conditions for a successful nonadaptive strategy. Then we give a lower bound for its rate. Again we find relations to separating codes and cover-free codes.

We assume that $D \geq u$, because otherwise all answers are arbirary and it is not possible to identify a defective element.

Finally, in the appendix we compare threshold codes with the construction introduced in [6] and give bounds for special cases.

2 Nonadaptive Threshold Group Testing Using Separating Codes

It is obvious that it is not possible to identify a defective element if $D \leq u-1$. In [11] it is shown that if $D \geq u$ we can find a set $\mathcal{P}$ such that

$$|\mathcal{D} \backslash \mathcal{P}| \leq g \text{ and } |\mathcal{P} \backslash \mathcal{D}| \leq g \tag{1}$$

or more general we can find a family $\mathbb{F}$ of subsets of $[N]$ with

$$\mathcal{D} \in \mathbb{F} \quad \text{and} \quad \forall \mathcal{P}, \mathcal{P}' \in \mathbb{F}: \ |\mathcal{P}' \backslash \mathcal{P}| \leq g \text{ and } |\mathcal{P} \backslash \mathcal{P}'| \leq g. \tag{2}$$

It is shown in [11] that all answers given for a strategy can be the same for all sets in the family $\mathbb{F}$ as for the set $\mathcal{D}$ of defective elements. Thus we cannot distinguish these sets.

We consider here the case where the number D of defectives is known.

Definition 1. *$n_{Th}(N, l, u, D)$ is the minimal number of tests of a nonadaptive strategy for threshold group testing with lower bound l and upper bound u (see the definition in the introduction) to find a family $\mathbb{F}$ which fulfills (2), if there are D defective elements. $R_{Th} = R_{Th}(l, u, D) = \sup_N \frac{\log N}{n_{Th}(N,l,u,D)}$ denotes the maximal achievable rate of a nonadaptive strategy for threshold group testing for given D, u, l.*

Definition 2. *An $n \times N$ matrix $X = (m_{ij})_{1 \leq i \leq n, 1 \leq j \leq N}$ is called a (w, r) separating code of size $n \times N$, if for any pair of subsets $I, J \subset [N]$ such that $|I| = w$, $|J| = r$, and $I \cap J = \varnothing$, there exists a row index $k \in [n]$ such that $m_{ki} = 1$ $\forall i \in I$ and $m_{kj} = 0$ $\forall j \in J$ or vice versa.*

By $n_S(N, w, r)$ we denote the minimal number of rows of a (w, r) separating code with N columns and by R_S the corresponding maximal achievable rate. The following theorem is not an immediate following of the result of [6], but it should be possible to prove this with their methods (using hypergraphs). We will give another proof.

Theorem 1. *Let $D = u + l$, then $n_{Th}(N, l, u, D) \leq n_S(N, u, D - l)$.*

Proof. The unknown set $\mathcal{D}$ of defective elements are chosen by an adversary.

Let $(m_{ij})_{\substack{i=1,\ldots,n \\ j=1,\ldots,N}}$ be a $(u, D-l)$ separating code of size $n \times N$.

We use the n rows as test sets (written in binary representation) for our strategy and show that we can find a family $\mathbb{F}$ of sets such that (2) is fulfilled.

Let $\mathbb{F}_0 = \{\mathcal{A}_1, \mathcal{A}_2, \ldots, \mathcal{A}_{\binom{N}{D}}\}$ be the family of all D-element subsets of $\{1, 2, \ldots, N\}$.

First consider $\mathcal{A}_1$ and search for the set $\mathcal{A}_i$ with the smallest index $i > 1$, such that $|\mathcal{A}_1 \backslash \mathcal{A}_i| > g = u - l - 1$.

Now we compare these two sets.

Case 1: $u - l \leq |\mathcal{A}_1 \backslash \mathcal{A}_i| < u$.
Set $\mathcal{I} = (\mathcal{A}_1 \backslash \mathcal{A}_i) \cup \mathcal{B}$, where $\mathcal{B} \subset \mathcal{A}_i \cap A_1$, such that $|\mathcal{I}| = u$ and set $\mathcal{J} \subset \mathcal{A}_i \backslash \mathcal{B}$, such that $|\mathcal{J}| = D - l$. This is possible because $|\mathcal{B}| \leq l$.

Case 2: $|\mathcal{A}_1 \backslash \mathcal{A}_i| \geq u$.
Set $\mathcal{I} \subset \mathcal{A}_1 \backslash \mathcal{A}_i$ such that $|\mathcal{I}| = u$ and set $\mathcal{J} \subset \mathcal{A}_i$, such that $|\mathcal{J}| = D - l$.

There exists a row (a test set $\mathcal{S}$), because of the properties of separating codes, such that $\mathcal{I} \subset \mathcal{S}$ ($m_{ki} = 1$) and $\mathcal{J} \not\subset \mathcal{S}$ ($m_{kj} = 0$) or vice versa, where $\mathcal{S}$ is the subset which corresponds to the row.

Case ($\mathcal{I} \subset \mathcal{S}$ and $\mathcal{J} \not\subset \mathcal{S}$): If the result is 1 then we continue our strategy with the family $\mathbb{F}_1 = \mathbb{F}_0 \backslash \{\mathcal{A}_i\}$. Otherwise we continue with the family $\mathbb{F}_1 = \mathbb{F}_0 \backslash \{\mathcal{A}_1\}$.

Case ($\mathcal{I} \not\subset \mathcal{S}$ and $\mathcal{J} \subset \mathcal{S}$): If the result is 1 then we continue our strategy with the family $\mathbb{F}_1 = \mathbb{F}_0 \backslash \{\mathcal{A}_1\}$. Otherwise we continue with the family $\mathbb{F}_1 = \mathbb{F}_0 \backslash \{\mathcal{A}_i\}$.

If $\mathcal{A}_1 \in \mathbb{F}_1$ we search again for the set with the smallest index $i > 1$, such that $|\mathcal{A}_1 \backslash \mathcal{A}_i| > g = u - l - 1$, if such a set exists.

Otherwise we continue with $\mathcal{A}_2$. We stop at step s if there are no sets $\mathcal{A}$ and $\mathcal{B}$ in the family $\mathbb{F}$ such that $|\mathcal{A} \backslash \mathcal{B}| > g$ and $|\mathcal{B} \backslash \mathcal{A}| > g$. The remaining family $\mathbb{F}_s$ has the claimed properties:

We did not exclude the set $\mathcal{D}$ which contains all defective elements from $\mathbb{F}_0$ for the following reason. If we compare $\mathcal{D}$ and $\mathcal{A}$ and the result of our test is 1, we remove $\mathcal{A}$, because more than u elements of $\mathcal{D}$ are in the test set. If the result is 0 we also remove $\mathcal{A}$, because then less than l elements are in the test set $\mathcal{S}$. Therefore our remaining family $\mathbb{F}_s$ contains the set with all defective elements and for $\mathbb{F}_s$ (2) holds. □

Note that in the proof the exact order in which we compare the sets is irrelevant.

The following is an upper bound for n_S (the authors use the terminology (N, u)-universal sets, which are (u, u)-separating codes).

Theorem [22]. $n_S(N, u, u) \leq u 2^{2u} \log N$.

In conjunction with our Theorem 1 this implies (this was shown in [4] for $l = 0$ only)

Corollary 1. *If $D = u + l$, then $n_{Th}(N, l, u, D) \leq u 2^u \log N$ and $R_{Th} \geq \frac{1}{u 2^{2u}}$.*

By random choice of a separating code (see [10]) we get a lower bound for R_S and thus we get from Theorem 1

Corollary 2. *If $D = u + l$, then*

$$R_{Th} \geq R_S \geq \frac{-\log(1 - 2^{-(2u-1)})}{2u - 1}. \tag{3}$$

Sketch of the Proof of [10]. Let n, N be positive integers and $X = (m_{ij})_{1\leq i\leq n, 1\leq j\leq N}$ an $n\times N$ matrix. We say that a column c_j is (w,r)-bad if there exists a pair of sets $\mathcal{S}, \mathcal{T} \in [N]$ with $|\mathcal{S}| = w$, $|\mathcal{T}| = r$, $\mathcal{S}\cap\mathcal{T} = \varnothing$, and $j \in |\mathcal{S}|$ for which there is no row r_k such that that $m_{ki} = 1$ $\forall i \in I$ and $m_{kj} = 0$ $\forall j \in J$ or vice versa. Otherwise the column is called (w,r)-good.

Consider now a random $(m_{ij})_{1\leq i\leq n, 1\leq j\leq N}$ an $n\times N$ matrix, where the elements are independently identically distributed random variables ($P(m_{ik} = 1) = p$ and $P(m_{ik} = 0) = 1-p$). Therefore

$$P(c_j \text{ is bad }) \leq \binom{N-1}{w+r-1}\binom{w+r-1}{w}(1-(p^r(1-p)^w + p^w(1-p)^w))^n \quad (4)$$

We set $n = \lceil \frac{1+\log(\binom{N-1}{w+r-1}\binom{w+r-1}{w})}{-\log(1-(p^r(1-p)^w+p^w(1-p)^w))} \rceil$, then the right hand-side of (4) does not exceed $\frac{1}{2}$. Thus, there exists a matrix with $\frac{N}{2}$ (w,r)-good columns. Taking the rate we will get the result. □

It is also possible to consider this model with errors. Therefore we need the following definition.

Definition 3. *An $n\times N$ matrix $(m_{ij})_{1\leq i\leq n, 1\leq j\leq N}$ is called a $(w,r;z)$ separating code of size $n\times N$, if for any pair of subsets $I, J \subset [N]$ such that $|I| = w$, $|J| = r$, and $I\cap J = \varnothing$, there exist z row indices $k \in [n]$ such that $m_{ki} = 1$ $\forall i \in I$ and $m_{kj} = 0$ $\forall j \in J$ or vice versa.*

By $n_S(N,w,r;z)$ we denote the minimal number of rows of a $(w,r;z)$ separating code with N columns and by R_S the corresponding maximal achievable rate. Obviously, $n_S(N,w,r;1) = n_S(N,w,r)$.

$n_{Th}(N,l,u,D,e)$ denotes the minimal number of tests of a nonadaptive strategy for threshold group testing with lower bound l and upper bound u to find a family $\mathbb{F}$ which fulfills (2), if there are D defective elements and there are at most e falsy answers.

Theorem 2. *Let $D = u + l$, then $n_{Th}(N,l,u,D,e) \leq n_S(N,u,D-l;2e+1)$.*

Sketch of the Proof: We modify the algorithm of the proof of Theorem 1 as follows. At the step, where we know that there exists a row (a test set $\mathcal{S}$), we will find now $2e+1$ rows. Instead of having one result, we have now $2e+1$ results. We continue now our algorithm in dependence of the result of the majority. In this way the algorithm can correct e errors. □

3 A General Lower Bound for the Rate for Nonadaptive Threshold Group Testing

In the previous section we got a lower bound for the rate for threshold group testing if $D = u + l$.

Definition 4. *An $n \times N$ matrix $(m_{ij})_{1\leq i\leq n, 1\leq j\leq N}$ is called a (w,r) cover-free code of size $n \times N$, if for any pair of subsets $\mathcal{I}, \mathcal{J} \subset [N]$ such that $|\mathcal{I}| = w$, $|\mathcal{J}| = r$, and $\mathcal{I} \cap \mathcal{J} = \varnothing$, there exists a row index $k \in [n]$ such that $m_{ki} = 1$ $\forall i \in \mathcal{I}$ and $m_{kj} = 0$ $\forall j \in \mathcal{J}$.*

Please notice that the only difference in the definition of a (w,r) cover-free code of size $n \times N$ and a (w,r) separating code of size $n \times N$ is the vice versa at the end. Therefore every cover-free code is also a separating code.

$n_c(N,w,r)$ denotes the minimal number of rows among all (w,r) cover-free codes with N columns.

Threshold group testing without gap is a special case of the complex group testing model, which was introduced in [26]. In complex group testing we have a set of N elements and a family $\mathcal{P}$ of defective subsets of this set. The test gives a positive result, if it includes all elements of a defecive subset. The goal is to find all defective subsets.

Let $\mathcal{D}$ be the set of defective elements in threshold group testing with the upper bound u and the lower bound l. If we choose $\mathcal{P} = \binom{\mathcal{D}}{u}$ then threshold group testing without gap and complex group testing are the same. Therefore the bounds for complex group testing in [16] can be used for threshold group testing without gap. For $u = 3$ it is the same bound as in [6].

In [6] it is shown that every $(u, D'-l)$ cover-free code is a nonadaptive strategy for threshold group testing, if D is unknown but bounded by D'. This implies

Theorem [6]. $n'_{Th}(N,l,u,D') \leq n_c(N,u,D'-l)$, where n'_{Th} denotes the minimal number of tests of a nonadaptive strategy for threshold group testing with lower bound l, upper bound u, and D bounded by D'.

Applying a bound for cover-free codes it is shown in [6] that

$$n'_{Th}(N,l,u,D') \leq \left(\frac{u+D'-l}{D'-l}\right)^{D'-l} \left(\frac{u+D'-l}{u}\right)^{u} \cdot \left(1 + (u+D'-l)\log\left(\frac{N}{u+D'-l}+1\right)\right).$$

For the rate this gives

$$R_{Th} \geq \frac{\left(\left(\frac{D'-l}{D'-l+u}\right)^{D'-l}\left(\frac{u}{D'-l+u}\right)^{u}\right)}{D'-l+u}. \tag{5}$$

The best known lower bound for the rate of cover-free codes is given in [15] (see also [24], [25] for constructions of cover-free codes) by

$$R_c \geq \frac{-\log\left(1 - \left(\frac{D'-l}{D'-l+u}\right)^{D'-l}\left(\frac{u}{D'-l+u}\right)^{u}\right)}{D'-l+u-1}. \tag{6}$$

One gets this lower bound for the rate by random choice of a $(u, D'-l)$ cover-free code (see Sketch of proof of Corollary 2).

We consider the case when D is known and derive another lower bound for the rate for threshold group testing. Let $(m_{ij})_{1\le i\le n, 1\le j\le N}$ be an $n\times N$ matrix. We denote by $r_i = (m_{i1}, \dots, m_{iN})$ the ith row and by $c_j = (m_{1j}, \dots m_{nj})$ the jth column.

Definition 5. *We call an $n \times N$ matrix a (D, u, l)-threshold code, if for all $\mathcal{A}, \mathcal{B} \subset \{1, 2, \dots, N\}$, $|\mathcal{A}| = |\mathcal{B}| = D$, and $|\mathcal{A}\backslash\mathcal{B}| \ge u - l$ there exists an $i \in \{1, 2, \dots, n\}$ such that*

$$\begin{array}{c}(\sum_{a\in\mathcal{A}} m_{ia} \ge u \text{ and } \sum_{b\in\mathcal{B}} m_{ib} \le l)\\ \text{or}\\ (\sum_{a\in\mathcal{A}} m_{ia} \le l \text{ and } \sum_{b\in\mathcal{B}} m_{ib} \ge u).\end{array} \tag{7}$$

We call the rows of the matrix tests and the columns codewords.

In the previous section we have shown how to get a nonadaptive group testing strategy in case $D = u + l$ by an $(u, D-l)$ separating code. A (D, u, l) threshold code is defined in such a way that it gives a nonadaptive strategy for threshold group testing for every u, l, D. Therefore we get the following

Lemma 1. *Every (D, u, l) threshold code gives a nonadaptive strategy for threshold group testing if the number D of defectives is known.*

Sketch of the Proof. We take an (D, u, l) threshold code of size $n \times N$ and give a strategy for a threshold group testing with N elements, D defectives, and thresholds u, l. The n rows are test sets. Like in the proof of Theorem 1 we start with $\mathcal{A}_1$ of the family $\mathbb{F}_0 = \{\mathcal{A}_1, \mathcal{A}_2, \dots, \mathcal{A}_{\binom{N}{D}}\}$. We do the same procedure like in the proof of Theorem 1 to end with a family $\mathbb{F}_s$ which fulfill the properties of a solution of threshold group testing.□

Now we want to find a lower bound for the rate $R = \frac{\log N}{n}$ of a (D, u, l) threshold code. First we calculate the rate for codes with a weaker condition (7'), that is if (7) holds only for all $\mathcal{A}$ and $\mathcal{B}$ with $|\mathcal{A} \cap \mathcal{B}| = z$ for some z fixed.

Given an integer N, what is the minimal number (of rows) n such that a threshold code of size $n \times N$ fulfills this weaker condition?

We say that the jth column c_j of the threshold code is bad if there exists a pair of sets $\mathcal{A}, \mathcal{B} \subset \{1, 2, \dots, N\}$ with $|\mathcal{A}| = |\mathcal{B}| = D$ and for which (7') is not true for any row. Otherwise we call c_j good. Consider a random matrix $(X_{ij})_{1\le i\le n, 1\le j\le N}$ where the X_{ij}'s are independent identically distributed random variables. We choose $P(X_{ij} = 1) = p$ and $P(X_{ij} = 0) = q$.

Let $\mathcal{A}, \mathcal{B} \subset [N]$ with $|\mathcal{A} \cap \mathcal{B}| = D - u + l$. Then every test (row) of a (D, u, l)-threshold code contains exactly l 1s inside of the positions corresponding to $|\mathcal{A} \cap \mathcal{B}|$. If there are less, then in the first set we cannot have more than u

1s, and if there are more then in the second set we will have more than l 1s. Therefore in this case

$$P(c_j \text{ is bad} \wedge |\mathcal{A} \cap \mathcal{B}| = D - u + l) = \tag{8}$$
$$\binom{D}{D-u+l}\binom{N-1}{D+u-l-1} \cdot \binom{D+u-l-1}{D} \cdot \left(1 - 2\binom{D-u+l}{l} p^l q^{D-u} p^{u-l} q^{u-l}\right)^n.$$

If we assign

$$n = n_* = -\frac{\log\left(\binom{N-1}{D+u-l-1}\binom{D+u-l-1}{D}\right)}{\log\left(1 - 2\binom{D-u+l}{l} p^u q^{D-l}\right)} + 1$$

then the right-hand side of (8) does not exceed $\frac{1}{2}$ and the average number of bad columns does not exceed $\frac{N}{2}$. Thus there exists a matrix which has at least $\frac{N}{2}$ good columns. By using $R \geq \lim_{N\to\infty} \frac{\log \frac{N}{2}}{n_*}$ we get

$$R \geq \frac{-\log\left(1 - 2\binom{D-u+l}{l} p^u q^{D-l}\right)}{D + u - l - 1}. \tag{9}$$

We want to consider the general case. We say that c_j is bad if there exists a pair of sets $\mathcal{A}, \mathcal{B} \subset \{1, 2, \ldots, N\}$ with $|\mathcal{A}| = |\mathcal{B}| = D$ and for which (7) is not true for any row. Clearly

$$P(c_j \text{ is bad}) = \sum_{k=0}^{D-u+l} P(c_j \text{ is bad} \wedge |\mathcal{A} \cap \mathcal{B}| = k). \tag{10}$$

If $|\mathcal{A} \cap \mathcal{B}| = k$ we get

$$P(c_j \text{ is bad} \wedge |\mathcal{A} \cap \mathcal{B}| = k) = \binom{D}{k} \cdot \binom{N-1}{2D-k-1} \cdot \binom{2D-k-1}{D}.$$

$$\left(1 - 2\left(\sum_{j=0}^{\min\{k,l\}} \binom{k}{j} p^j q^{k-j} \left(\sum_{i=u-j}^{D-k} \binom{D-k}{i} p^i q^{D-k-i}\right)\left(\sum_{t=0}^{l-j} \binom{D-k}{t} p^t q^{D-k-t}\right)\right)\right)^n. \tag{11}$$

Now we need an upper bound

$$P(c_j \text{ is bad}) \leq (D - u + l + 1) \max_{k\in\{0,1,\ldots,D-u+l\}} P(c_j \text{ is bad} \wedge |\mathcal{A} \cap \mathcal{B}| = k). \tag{12}$$

We calculate for each k the rate like for $k = D-u+l$. The factor $(D-u+l+1)$ in (12) does not change the rate. Therefore the minimal of these rates gives a bound for the rate in the general case.

Hence we get

Theorem 3. *Let $0 \leq l < u \leq D$ be given, then*

$$R_{Th} \geq R_T = \max_{0\leq p\leq 1} \min_{0\leq k\leq D-u+l} \tag{13}$$
$$\frac{-\log(1 - 2(\sum_{j=0}^{\min\{k,l\}} \binom{k}{j} p^j q^{k-j} (\sum_{i=u-j}^{D-k} \binom{D-k}{i} p^k q^{D-k-i})(\sum_{t=0}^{l-j} \binom{D-k}{t} p^t q^{D-k-t})))}{2D - k - 1}.$$

Like for separating codes we can generalize threshold codes to threshold codes with errors. We will now compare our new construction with the one already known.

4 Majority Group Testing

We remind the reader of the definition of majority group testing in Section 1.

This is a generalization of the model considered in [20].

We consider $f_1(D) = \lceil \frac{D}{k} \rceil - 1$ and $f_2(D) = \lfloor \frac{D}{k} \rfloor + 1$ and write $f(D) = \frac{D}{k}$.

$$t_{\mathcal{S}}(\mathcal{D}) = \begin{cases} 0 & \text{, if } |\mathcal{S} \cap \mathcal{D}| < \frac{D}{k} = f(D) \\ 1 & \text{, if } |\mathcal{S} \cap \mathcal{D}| > \frac{D}{k} = f(D) \\ \{0,1\} & \text{, otherwise.} \end{cases}$$

If D is known this problem can be reduced to threshold group testing:

1. For $D \mod k \equiv 0$ we set $l = \frac{D}{k} - 1$ and $u = \frac{D}{k} + 1$. Therefore we get a strategy by a $(\frac{D}{k} + 1, \frac{k-1}{k}D + 1)$ cover-free code, by a $(\frac{D}{2} + 1, \frac{D}{2} + 1)$ separating code for $k = 2$, or by a $(D, \frac{D}{k} + 1, \frac{k-1}{k}D + 1)$ threshold code.
2. For $D \mod k \equiv s$ with $0 < s < k$ we set $l = \frac{D-s}{k}$ and $u = \frac{D+k-s}{k}$. Therefore we get a strategy by a $(\frac{D+k-s}{k}, \frac{(k-1)D+s}{k})$ cover-free code, by a $(\frac{D+1}{2}, \frac{D+1}{2})$ separating code for $k = 2$, or by a $(D, \frac{D+k-s}{k}, \frac{(k-1)D+s}{k})$ threshold code.

Now we will consider the case when D is bounded by some $D' < N$. The number of tests depends on D'.

First we consider the case $k = 2$.

It is clear that as in threshold group testing it is not always possible to determine the set of defectives.

Definition 6. *We say that two sets $\mathcal{A}, \mathcal{B}$ are* **indistinguishable** *if for any strategy (or equivalently if for asking all sets) it may happen that the answers are all the same (i.e. the adversary can answer this way). By a* **solution** *we mean a family $\mathbb{F}$ of sets that contains $\mathcal{D}$ and any pair of set F_1 and F_2 are indistinguishable. We call such an $\mathbb{F}$ also* **completely indistinguishable**.

A **successful strategy** *is a strategy which finds a solution.*

The next theorem gives conditions for a solution.

Theorem 4. *Let $\mathcal{D} \subset [N]$ be the set of defectives and $f(D) = \frac{D}{2}$. We can determine a solution $\mathbb{F}$ such that for all sets $\mathcal{P}_1, \mathcal{P} \in \mathbb{F}$ with $|\mathcal{P}_1| \geq |\mathcal{P}|$ the following holds*

1. If $\mathcal{P} \subset \mathcal{P}_1$ and $\mathcal{P}$ is even then

$$|\mathcal{P}_1 \backslash \mathcal{P}| \leq 2. \tag{14}$$

2. If $\mathcal{P} \subset \mathcal{P}_1$ and $\mathcal{P}$ is odd then

$$|\mathcal{P}_1 \backslash \mathcal{P}| \leq 1. \tag{15}$$

3. If $|\mathcal{P} \backslash \mathcal{P}_1| = 1$ and $\mathcal{P}$ is even then

$$|\mathcal{P}_1 \backslash \mathcal{P}| \leq 1. \tag{16}$$

Proof. First we show that there exists a strategy, such that we get a family $\mathbb{F}$ which satisfies (14), (15), and (16). We can immediately throw out all sets of size bigger then $\mathcal{D}'$. So we consider the family $\mathbb{F}_0 = \cup_{j=0}^{\mathcal{D}'} \binom{[N]}{j}$ of all possible sets of defectives. If there are two sets in $\mathbb{F}_0$ which do not fulfill (14), (15), and (16) we show that there exists a test such that one will be removed and we show that the set of defectives will not be removed. Let $w = |\mathcal{P}_1 \backslash \mathcal{P}|$.

1. Let $\mathcal{P} \subset \mathcal{P}_1$ and $|\mathcal{P}| = 2a$.
   It is enough to have a test set $\mathcal{S}$ which includes exactly $a-1$ elements of $\mathcal{P}$ (thus does not include other $a+1$ elements of $\mathcal{P}$) and exactly $3 + \lceil (w-3)/2 \rceil$ elements of $\mathcal{P}_1 \backslash \mathcal{P}$. If the result of this test is "1" we continue with $\mathbb{F}_1 = \mathbb{F}_0 \backslash \mathcal{P}$, otherwise we continue with $\mathbb{F}_1 = \mathbb{F}_0 \backslash \{\mathcal{P}_1\}$. Now we have to show that we will not remove $\mathcal{D}$.
   If $\mathcal{P} = \mathcal{D}$ then $|\mathcal{S} \cap \mathcal{D}| = a - 1 < \frac{D}{2} = a$ and the result is "0".
   If $\mathcal{P}_1 = \mathcal{D}$ then $|\mathcal{S} \cap \mathcal{D}| = a + 2 + \lceil (w-3)/2 \rceil > \frac{D}{2} = a + \frac{w}{2}$, because $|\mathcal{P}_1 \backslash \mathcal{P}| = w \geq 3$ by assumption and the result is "1".

   Now for the case $|\mathcal{P}_1 \backslash \mathcal{P}| = 2$ (thus $|P_1| = 2a+2$): If $|\mathcal{S} \cap \mathcal{P}_1| > |\mathcal{P}_1|/2$ then $|\mathcal{S} \cap \mathcal{P}_1| \geq a+2$ and so $|\mathcal{S} \cap \mathcal{P}| \geq a$.
   If $|\mathcal{S} \cap \mathcal{P}_1| < |\mathcal{P}_1|/2$ then $|\mathcal{S} \cap \mathcal{P}_1| \leq a$ and so $|\mathcal{S} \cap \mathcal{P}| \leq a$.
   Thus sets P and P_1 are indistinguishable.
2. Let $\mathcal{P} \subset \mathcal{P}_1$ and $|\mathcal{P}| = 2a+1$.
   It is enough to have a test set $\mathcal{S}$ which includes exactly a elements of $\mathcal{P}$ and exactly $2 + \lceil (w-2)/2 \rceil$ elements of $\mathcal{P}_1 \backslash \mathcal{P}$. If the result of this test is "1" we continue with $\mathbb{F}_1 = \mathbb{F}_0 \backslash \{\mathcal{P}\}$, otherwise we continue with $\mathbb{F}_1 = \mathbb{F}_0 \backslash \{\mathcal{P}_1\}$. Now we have to show that we will not remove $\mathcal{D}$.
   If $\mathcal{P} = \mathcal{D}$ then $|\mathcal{S} \cap \mathcal{D}| = a < \frac{D}{2} = a + \frac{1}{2}$ and the result is "0".
   If $\mathcal{P}_1 = \mathcal{D}$ then $|\mathcal{S} \cap \mathcal{D}| = a + 2 + \lceil \frac{w-2}{2} \rceil > \frac{D}{2} = a + \frac{w+1}{2}$, because $|\mathcal{P}_1 \backslash \mathcal{P}| = w \geq 2$ by assumption and the result is "1".
   For the case $|\mathcal{P}_1 \backslash \mathcal{P}| = 1$ (thus $|\mathcal{P}_1| = 2a+2$): If $|\mathcal{S} \cap \mathcal{P}_1| > |\mathcal{P}_1|/2$ then $|\mathcal{S} \cap \mathcal{P}_1| \geq a+2$ and so $|\mathcal{S} \cap \mathcal{P}| \geq a+1$.
   If $|\mathcal{S} \cap \mathcal{P}_1| < |\mathcal{P}_1|/2$ then $|\mathcal{S} \cap \mathcal{P}_1| \leq a$ and so $|\mathcal{S} \cap \mathcal{P}| \leq a$.
   Thus sets P and P_1 are indistinguishable.
3. Let $\mathcal{P} \not\subset \mathcal{P}_1$, $|\mathcal{P} \backslash \mathcal{P}_1| = 1$, and $|\mathcal{P}| = 2a$.
   It is enough to have a test set $\mathcal{S}$ which includes exactly $a-1$ elements of $\mathcal{P} \cap \mathcal{P}_1$, includes exactly $2 + \lceil (w-2)/2 \rceil$ elements of $\mathcal{P}_1 \backslash \mathcal{P}$ and does not include the element from $\mathcal{P} \backslash \mathcal{P}_1$. If the result of this test is "1" we continue with $\mathbb{F}_1 = \mathbb{F}_0 \backslash \{\mathcal{P}\}$, otherwise we continue with $\mathbb{F}_1 = \mathbb{F}_0 \backslash \{\mathcal{P}_1\}$. Now we have to show that we will not remove $\mathcal{D}$.

If $\mathcal{P} = \mathcal{D}$ then $|\mathcal{S} \cap \mathcal{D}| = a - 1 < \frac{D}{2} = a$ and the result is "0".
If $\mathcal{P}_1 = \mathcal{D}$ then $|\mathcal{S} \cap \mathcal{D}| = a - 1 + 2 + \lceil (w-2)/2 \rceil > \frac{D}{2} = a + \frac{w-1}{2}$, because $|\mathcal{P}_1 \backslash \mathcal{P}| = w \geq 2$ by assumption and the result is "1".
For the case $|\mathcal{P}_1 \backslash \mathcal{P}| = 1$ (thus $|P_1| = 2a$): If $|S \cap \mathcal{P}_1| > |\mathcal{P}_1|/2$ then $|S \cap \mathcal{P}_1| \geq a + 1$ and so $|S \cap \mathcal{P}| \geq a$.
If $|S \cap \mathcal{P}_1| < |\mathcal{P}_1|/2$ then $|S \cap \mathcal{P}_1| \leq a - 1$ and so $|S \cap \mathcal{P}| \leq a$.
Thus sets P and P_1 are indistinguishable.

4. Let $\mathcal{P} \not\subset \mathcal{P}_1$, $|\mathcal{P} \backslash \mathcal{P}_1| = 1$, and $|\mathcal{P}| = 2a + 1$.
It is enough to have a test set $\mathcal{S}$ which includes exactly a elements of $\mathcal{P} \cap \mathcal{P}_1$, includes exactly $1 + \lceil (w-1)/2 \rceil$ elements of $\mathcal{P}_1 \backslash \mathcal{P}$ and not includes the element from $\mathcal{P} \backslash \mathcal{P}_1$. If the result of this test is "1" we continue with $\mathbb{F}_1 = \mathbb{F}_0 \backslash \{\mathcal{P}\}$, otherwise we continue with $\mathbb{F}_1 = \mathbb{F}_0 \backslash \{\mathcal{P}_1\}$. Now we have to show that we will not remove $\mathcal{D}$.
If $\mathcal{P} = \mathcal{D}$ then $|\mathcal{S} \cap \mathcal{D}| = a < \frac{D}{2} = a + \frac{1}{2}$ and the result is "0".
If $\mathcal{P}_1 = \mathcal{D}$ then $|\mathcal{S} \cap \mathcal{D}| = a + 1 + \lceil (w-1)/2 \rceil > \frac{D}{2} = a + \frac{w}{2}$, because $|\mathcal{P}_1 \backslash \mathcal{P}| = w \geq 1$ by assumption and the result is "1".
5. Let $\mathcal{P} \not\subset \mathcal{P}_1$, $|\mathcal{P} \backslash \mathcal{P}_1| > 1$, and $|\mathcal{P}| = 2a$.
   (a) $|\mathcal{P} \cap \mathcal{P}_1| = l \geq a - 1$.
   It is enough to have a test set $\mathcal{S}$ which includes exactly $a - 1$ elements of $\mathcal{P} \cap \mathcal{P}_1$, no other elements of $\mathcal{P}$ and includes exactly $\lfloor |\mathcal{P}_1|/2 \rfloor - a + 2$ elements of $\mathcal{P}_1 \backslash \mathcal{P}$. If the result of this test is "1" we continue with $\mathbb{F}_1 = \mathbb{F}_0 \backslash \{\mathcal{P}\}$, otherwise we continue with $\mathbb{F}_1 = \mathbb{F}_0 \backslash \{\mathcal{P}_1\}$. Now we have to show that we will not remove $\mathcal{D}$.
   If $\mathcal{P} = \mathcal{D}$ then $|\mathcal{S} \cap \mathcal{D}| = a - 1 < \frac{D}{2} = a$ and the result is "0".
   If $\mathcal{P}_1 = \mathcal{D}$ then $|\mathcal{S} \cap \mathcal{D}| = \lfloor |\mathcal{P}_1|/2 \rfloor + 1 > \frac{D}{2}$ and the result is "1".
   (b) $|\mathcal{P} \cap \mathcal{P}_1| = l < a - 1$.
   It is enough to have a test set $\mathcal{S}$ which includes all elements of $\mathcal{P} \cap \mathcal{P}_1$, $a - 1 - |\mathcal{P} \cap \mathcal{P}_1|$ elements of $\mathcal{P}$, and includes exactly $\lfloor |\mathcal{P}_1|/2 \rfloor - a + 2$ elements of $\mathcal{P}_1 \backslash \mathcal{P}$. If the result of this test is "1" we continue with $\mathbb{F}_1 = \mathbb{F}_0 \backslash \{\mathcal{P}\}$, otherwise we continue with $\mathbb{F}_1 = \mathbb{F}_0 \backslash \{\mathcal{P}_1\}$. Now we have to show that we will not remove $\mathcal{D}$.
   If $\mathcal{P} = \mathcal{D}$ then $|\mathcal{S} \cap \mathcal{D}| = a - 1 < \frac{D}{2} = a$ and the result is "0".
   If $\mathcal{P}_1 = \mathcal{D}$ then $|\mathcal{S} \cap \mathcal{D}| = \lfloor |\mathcal{P}_1|/2 \rfloor + 1 > \frac{D}{2}$ and the result is "1".
6. Let $\mathcal{P} \not\subset \mathcal{P}_1$, $|\mathcal{P} \backslash \mathcal{P}_1| > 1$, and $|\mathcal{P}| = 2a + 1$.
   (a) $|\mathcal{P} \cap \mathcal{P}_1| = l \geq a$.
   It is enough to have a test set $\mathcal{S}$ which includes exactly a elements of $\mathcal{P} \cap \mathcal{P}_1$, no other elements of $\mathcal{P}$ and includes exactly $\lfloor |\mathcal{P}_1|/2 \rfloor - a + 1$ elements of $\mathcal{P} \backslash \mathcal{P}_1$. If the result of this test is "1" we continue with $\mathbb{F}_1 = \mathbb{F}_0 \backslash \{\mathcal{P}\}$, otherwise we continue with $\mathbb{F}_1 = \mathbb{F}_0 \backslash \{\mathcal{P}_1\}$. Now we have to show that we will not remove $\mathcal{D}$.
   If $\mathcal{P} = \mathcal{D}$ then $|\mathcal{S} \cap \mathcal{D}| = a < \frac{D}{2} = a + \frac{1}{2}$ and the result is "0".
   If $\mathcal{P}_1 = \mathcal{D}$ then $|\mathcal{S} \cap \mathcal{D}| = \lfloor |\mathcal{P}_1|/2 \rfloor + 1 > \frac{D}{2}$ and the result is "1".
   (b) $|\mathcal{P} \cap \mathcal{P}_1| = l < a$.
   It is enough to have a test set $\mathcal{S}$ which includes all elements of $\mathcal{P} \cap \mathcal{P}_1$, $a - |\mathcal{P} \cap \mathcal{P}_1|$ elements of $\mathcal{P}$, and includes exactly $\lfloor |\mathcal{P}_1|/2 \rfloor - a + 1$ elements

of $\mathcal{P}_1 \backslash \mathcal{P}$. If the result of this test is "1" we continue with $\mathbb{F}_1 = \mathbb{F}_0 \backslash \{\mathcal{P}\}$, otherwise we continue with $\mathbb{F}_1 = \mathbb{F}_0 \backslash \{\mathcal{P}_1\}$. Now we have to show that we will not remove $\mathcal{D}$.
If $\mathcal{P} = \mathcal{D}$ then $|\mathcal{S} \cap \mathcal{D}| = a < \frac{D}{2} = a + \frac{1}{2}$ and the result is "0".
If $\mathcal{P}_1 = \mathcal{D}$ then $|\mathcal{S} \cap \mathcal{D}| = \lfloor |\mathcal{P}_1|/2 \rfloor + 1 > \frac{D}{2}$ and the result is "1".

In all cases it is not always possible to distinguish the sets of possible solutions. □

Remarks

1. In the corresponding matrix of a strategy for majority group testing with $f(D) = \frac{D}{2}$ it is possible to exchange the zeros and the ones.
2. The proof of Theorem 4 shows that for a successful strategy we have to have for all disjoint pairs $\mathcal{J}, \mathcal{I} \subset [N]$ with the size $\left\lfloor \frac{D'}{2} \right\rfloor + 1$ two test sets such that the elements of $\mathcal{J}$ are contained in one test set and no element of $\mathcal{I}$ is contained in the other test set or vice versa. This is exactly a separating code and therefore we have the following

Theorem 5. *Let $f(D) = \frac{D}{2}$ and D be bounded by D', which is known, then a $(\left\lfloor \frac{D'}{2} \right\rfloor + 1, \left\lfloor \frac{D'}{2} \right\rfloor + 1)$ separating code gives a nonadaptive strategy for majority group testing.*

Now let us consider the case $k > 2$.

As before we have conditions for a solution. They are given by the following

Theorem 6. *Let $\mathcal{D} \subset [N]$ be the set of defectives and $f(D) = \frac{D}{k}$. We can determine a solution $\mathbb{F}$ such that for all sets $\mathcal{P}_1, \mathcal{P} \in \mathbb{F}$ with $|\mathcal{P}_1| \geq |\mathcal{P}| = ak + s$ ($s = 0$ or $s = k - 1$) the following holds*

1. *If $\mathcal{P} \subset \mathcal{P}_1$ and $s = 0$ or $s = k - 1$ then*

$$|\mathcal{P}_1 \backslash \mathcal{P}| \leq 1. \tag{17}$$

2. *If $|\mathcal{P} \backslash \mathcal{P}_1| = 1$ and $s = 0$ then*

$$|\mathcal{P}_1 \backslash \mathcal{P}| \leq 1. \tag{18}$$

Proof.

1. Let $\mathcal{P} \subset \mathcal{P}_1$ and $|\mathcal{P}| = ak + s$.
   (a) $s = 0$:
   For the case $|\mathcal{P}_1 \backslash \mathcal{P}| = 1$ (thus $|P_1| = ak + 1$): If $|\mathcal{S} \cap \mathcal{P}_1| > |\mathcal{P}_1|/k$ then $|\mathcal{S} \cap \mathcal{P}_1| \geq a + 1$ and so $|\mathcal{S} \cap \mathcal{P}| \geq a$.
   If $|\mathcal{S} \cap \mathcal{P}_1| < |\mathcal{P}_1|/k$ then $|\mathcal{S} \cap \mathcal{P}_1| \leq a$ and so $|\mathcal{S} \cap \mathcal{P}| \leq a$.
   Thus sets P and P_1 are indistinguishable.

(b) $s = k - 1$:
For the case $|\mathcal{P}_1 \backslash \mathcal{P}| = 1$ (thus $|P_1| = (a+1)k$): If $|S \cap \mathcal{P}_1| > |\mathcal{P}_1|/k$ then $|S \cap \mathcal{P}_1| \geq a + 2$ and so $|S \cap \mathcal{P}| \geq a + 1$.
If $|S \cap \mathcal{P}_1| < |\mathcal{P}_1|/k$ then $|S \cap \mathcal{P}_1| \leq a$ and so $|S \cap \mathcal{P}| \leq a$.
Thus sets P and P_1 are indistinguishable.

2. Let $\mathcal{P} \not\subset \mathcal{P}_1$, $|\mathcal{P} \backslash \mathcal{P}_1| = 1$, and $s = 0$.
For the case $|\mathcal{P}_1 \backslash \mathcal{P}| = 1$ we have $|P| = ak$ and $|P_1| = ak$:
If $|S \cap \mathcal{P}_1| > |\mathcal{P}_1|/k$ then $|S \cap \mathcal{P}_1| \geq a + 1$ and so $|S \cap \mathcal{P}| \geq a$.
If $|S \cap \mathcal{P}_1| < |\mathcal{P}_1|/k$ then $|S \cap \mathcal{P}_1| \leq a - 1$ and so $|S \cap \mathcal{P}| \leq a$.
Thus sets P and P_1 are indistinguishable.

The proof that there exists a strategy, such that we get a family $\mathbb{F}$ which fulfills (17) and (18) follows the same ideas as the proof of Theorem 4. We immediately throw out all sets of size bigger then $\mathcal{D}'$. Then we consider the family $\mathbb{F}_0 = \cup_{j=0}^{D'} \binom{[N]}{j}$ of all possible sets of defectives. If there are two sets in $\mathbb{F}_0$ which do not fulfill (17) and (18) we show that there exists a test such that one will be removed and we show that the set of defectives will not be removed.

It is enough to have a test set $\mathcal{S}$ which includes $\lfloor |\mathcal{P}_1|/k \rfloor + 1$ elements of $\mathcal{P}_1$ and less then $|\mathcal{P}_1|/k$ elements of $\mathcal{P}$. We first include to S elements from $\mathcal{P}_1 \backslash \mathcal{P}$ and then from $\mathcal{P} \cap \mathcal{P}_1$. Then we do not include to S elements from $\mathcal{P} \backslash \mathcal{P}_1$ and then from $\mathcal{P} \cap \mathcal{P}_1$. For others elements from $[N]$ it is not important the set S includes them or not. We always have that if $\mathcal{P} = \mathcal{D}$ then the result is "0" and if $\mathcal{P}_1 = \mathcal{D}$ then the result is "1".

The analysis of different cases like in Theorem 4 shows that it is possible to constract this set S. □

The proof of Theorem 6 shows that every $(\left\lfloor \frac{D'}{k} \right\rfloor + 1, D' - \left\lceil \frac{D'}{k} \right\rceil + 1)$ cover-free code gives a strategy and thus the following.

Theorem 7. *Let $f(D) = \frac{D}{k}$ and D be bounded by D', which is known, then a $(\left\lfloor \frac{D'}{k} \right\rfloor + 1, D' - \left\lceil \frac{D'}{k} \right\rceil + 1)$ cover-free code is a nonadaptive strategy for majority group testing.*

References

1. Ahlswede, R., Deppe, C., Lebedev, V.: Majority Group Testing with Density Tests. In: Proceedings of the IEEE ISIT 2011, pp. 291–295 (2011)
2. Ahlswede, R., Deppe, C., Lebedev, V.: Finding one of defective elements in some group testing models. Problems of Information Transmission 48(2), 173–181 (2012)
3. Ahlswede, R., Wegener, I.: Suchprobleme, Teubner Verlag, Stuttgart (1979), Russian Edition: Zadatsi Poiska, MIR (1982), English Edition: Search Problems. Wiley-Interscience Series in Discrete Mathematics and Optimization (1987)
4. Alon, N., Guruswami, V., Kaufman, T., Sudan, M.: Guessing secrets efficiently via list decoding. In: 13th SODA, pp. 254–262 (2002)
5. Chang, H., Chen, H.-B., Fu, H.-L., Shi, C.-H.: Reconstruction of hidden graphs and threshold group testing. J. on Comb. Opt. 22(2), 270–281 (2010)

6. Chen, H.-B., Fu, H.-L.: Nonadaptive algorithms for threshold group testing. Discrete Appl. Math. 157(7), 1581–1585 (2009)
7. Chen, H.-B., Fu, H.-L., Hwang, F.K.: An upper bound of the number of tests in pooling designs for the error-tolerant complex model. Opt. Lett. 2, 425–431 (2008)
8. Cheraghchi, M.: Improved constructions for non-adaptive threshold group testing. In: ICALP, vol. 1, pp. 552–564 (2010)
9. Chung, F., Graham, R., Leighton, F.T.: Guessing secrets. Electronic J. on Combinatorics 8, 1–25 (2001)
10. Cohen, G., Schaathun, H.G.: Asymptotic overview on separating codes, Technical report, No. 248 from Department of Informatics, University of Bergen (2003)
11. Damaschke, P.: Threshold Group Testing. In: Ahlswede, R., Bäumer, L., Cai, N., Aydinian, H., Blinovsky, V., Deppe, C., Mashurian, H. (eds.) General Theory of Information Transfer and Combinatorics. LNCS, vol. 4123, pp. 707–718. Springer, Heidelberg (2006)
12. Deppe, C., D'yachkov, A., Lebedev, V., Rykov, V.: Superimposed codes and threshold group testing. In: Aydinian, H., Cicalese, F., Deppe, C. (eds.) Ahlswede Festschrift. LNCS, vol. 7777, pp. 509–533. Springer, Heidelberg (2013)
13. Du, D.Z., Hwang, F.K.: Combinatorial Group Testing and its Applications, 2nd edn. Series on Applied Mathematics, vol. 12. World Scientific Publishing Co. Pte. Ltd, Hackensack (2000)
14. Du, D.Z., Hwang, F.K.: Pooling Designs and Nonadaptive Group Testing. Important Tools for DNA Sequencing. Series on Applied Mathematics, vol. 18. World Scientific Publishing Co. Pte. Ltd, Hackensack (2006)
15. D'yachkov, A., Macula, A., Vilenkin, P., Torney, D.: Families of finite sets in which no intersection of l sets is covered by the union of s others. J. Combin. Theory Ser. A. 99(2), 195–218 (2002)
16. D'yachkov, A., Macula, A., Vilenkin, P., Torney, D.: Two models of nonadaptive group testing for designing screening experiments. In: Proc. 6th Int.Workshop on Model-Orented Designs and Analysis, pp. 63–75 (2001)
17. Gerbner, D., Keszegh, B., Pálvölgyi, D., Wiener, G.: Density-Based Group Testing. In: Aydinian, H., Cicalese, F., Deppe, C. (eds.) Ahlswede Festschrift. LNCS, vol. 7777, pp. 543–556. Springer, Heidelberg (2013)
18. Kautz, W., Singleton, R.: Nonrandom binary superimposed codes. IEEE Trans. Information Theory 10(4), 363–377 (1964)
19. Lebedev, V.S.: An asymptotic upper bound for the rate of (w, r)-cover-free codes. Probl. Inf. Transm. 39(4), 317–323 (2003)
20. Lebedev, V.S.: Separating codes and a new combinatorial search model. Probl. Inf. Transm. 46(1), 1–6 (2010)
21. Mitchell, C.J., Piper, F.C.: Key storage in secure networks. Discrete Appl. Math. 21(3), 215–228 (1988)
22. Naor, M., Schulman, L.J., Srinivasan, A.: Splitters and near-optimal derandomization. In: Proceedings. of th 36th Annual Symposium on Foundations of Computer Science, pp. 182–191 (1995)
23. Stinson, D.R.: Generalized cover-free families, In honour of Zhu Lie. Discrete Math. 279(1-3), 463–477 (2004)
24. Stinson, D.R., van Trung, T., Wei, R.: Secure frameproof codes, key distribution patterns, group testing algorithms and related structures. Journal of Statistical Planning and Inference 86, 595–617 (2000)
25. Stinson, D.R., Wei, R., Zhu, L.: Some new bounds for cover-free families. J. Combin. Theory A. 90, 224–234 (2000)
26. Torney, D.C.: Sets pooling designs. Ann. Combin. 3, 95–101 (1999)

Appendix

In this appendix we will compare threshold codes with the construction of [6] and give bounds for the rate of threshold codes in dependence of R_c. First we consider a threshold group testing with the thresholds u and l without error and D defectives elements. In [6] it is shown that a cover-free code is a successful strategy.

The bound for the rate of cover-free codes is derived in the same way as we did for threshold codes. Recall that the best known bound for the rate, using cover-free codes is

$$R_{Th} \geq R_c = \frac{-\log\left(1-\left(\frac{D-l}{D-l+u}\right)^{D-l}\left(\frac{u}{D-l+u}\right)^{u}\right)}{D-l+u-1}. \tag{19}$$

It holds $-\log(1-x) > x$ if $x > 0$. Thus we want to compare

$$R_{Th} \geq B_{Th} = \max_{0\leq p\leq 1}\min_{0\leq k\leq D-u+l} \tag{20}$$
$$\frac{2(\sum_{j=0}^{\min\{k,l\}}\binom{k}{j}p^j q^{k-j}(\sum_{i=u-j}^{D-k}\binom{D-k}{i}p^i q^{D-k-i})(\sum_{t=0}^{l-j}\binom{D-k}{t}p^t q^{D-k-t})))}{2D-k-1}$$

and

$$R_c \geq B_c = \frac{\left(\frac{D-l}{D-l+u}\right)^{D-l}\left(\frac{u}{D-l+u}\right)^{u}}{D-l+u-1}. \tag{21}$$

Note first that, since $u \geq l+1$, for all $0 \leq k \leq D-u+l$

$$F_T = \frac{2}{2D-k-1} \geq \frac{2}{2D-1} \geq F_c = \frac{1}{D-l+u-1} \geq \frac{1}{D}. \tag{22}$$

We have $\frac{F_T}{F_c} > 1$, for instance for $u = l+1$ and $k = D-1$ $\frac{F_T}{F_c} = 2$, or $\lambda = 0$, $\mu = 1$, and therefore $\kappa = 0$, as it occurs in the result of [4].

We start with our analysis for $u = l$, that is in relative quantities

$$\lambda = \frac{l}{D},\ \mu = \frac{u}{D},\ \kappa = \frac{k}{D}, \tag{23}$$

and for the probabilities in (21)

$$\frac{u}{D-l+u} = \frac{\mu}{1-\lambda+\mu} = \mu = p,\quad 1-\mu = q. \tag{24}$$

We begin first with the **entropy description** of the lower bounds for $u = l+1$

$$B_c = F_c\mu^{\mu D}(1-\mu)^{1-\mu D} = F_c 2^{-h(\mu)D}, \tag{25}$$

$$B_{Th} \geq F_T \min_{0 \leq k \leq D-u+l} \sum_{j=0}^{\min\{k,l\}} \binom{k}{j} p^j q^{k-j} \cdot \qquad (26)$$

$$(\sum_{i=u-j}^{D-k} \binom{D-k}{i} p^i q^{D-k-i}) \cdot$$

$$(\sum_{t=0}^{l-j} \binom{D-k}{t} p^t q^{D-k-t}).$$

For $u = l+1$ we have

$$B_{Th}(l) \geq F_T \min_{0 \leq k \leq l} \sum_{j=0}^{k} \binom{k}{j} \mu^j (1-\mu)^{k-j}$$

$$(1 - E(D,u,l,n))E(D,u,l,n),$$

where $E(D,u,l,n) = \sum_{t=0}^{l-j} \binom{D-k}{t} \mu^t (1-\mu)^{k-j}$ and $\mu = \frac{u}{n}$.

Argument. We choose the optimal j for all sums (max. entropy principle). If $j^* = \mu k = \lambda \kappa D$, then

$$\binom{k}{j^*} \mu^{j^*} (1-\mu)^{k-j^*} \geq \frac{1}{k+1} \geq \frac{1}{D} \qquad (27)$$

$$\binom{D-k}{u-j^*} \mu^{u-j^*} (1-\mu)^{D-k-u+j^*} \geq \frac{1}{D-k+1} \geq \frac{1}{D}$$

$$\binom{D-k}{l-j^*} \mu^{l-j^*} (1-\mu)^{D-k-l+j^*} \geq \frac{1}{D-k+1} \geq \frac{1}{D}.$$

Consequently

$$\frac{B_{Th}(l)}{B_c} \geq \frac{F_T}{F_c} 2^{h(\mu)D - o(D)}. \qquad (28)$$

Therefore we have

Theorem 8. *In the case without gap, that is $u = l+1$, it holds*

$$B_{Th} \geq B_c 2^{h(\frac{u}{n})D - o(D)}. \qquad (29)$$

We write now

$$B_{Th} \leq F_T \min_{0 \leq k \leq D-u+l} E_1 E_2 E_3$$

with $E_1 = \sum_{j=0}^{\min\{k,l\}} \binom{k}{j} p^j q^{k-j}$, $E_2(j) = \sum_{i=u-j}^{D-k} \binom{D-k}{i} p^i q^{D-k-i}$,
and $E_3(j) = \sum_{t=0}^{l-j} \binom{D-k}{t} p^t q^{D-k-t}$.

To approximate E_1 for all k observe first that we need

$$j(\kappa) \leq \kappa p D \text{ and } j(\kappa) \leq \lambda D. \tag{30}$$

Let $\kappa \leq 1-\mu+\lambda$ and

$$p = p(\mu,\lambda) = \frac{\lambda}{1-\mu+\lambda} \tag{31}$$

then for $j = \frac{\lambda}{1-\mu+\lambda}\kappa D$ the inequalities in (30) hold.

Lemma 2. *1. $\lambda \leq p(\mu,\lambda) \leq \mu$.*
2. $p(\mu,\lambda) \leq \frac{\mu+\lambda}{2}$ for $\mu \leq \frac{1}{2}$.
3. The inequality $\sqrt{\lambda\mu} \leq p(\mu,\lambda)$ does not always hold.

Proof.

1. Since $\mu \geq \lambda$, $1-\mu+\lambda \leq 1$ and $\lambda \leq p(\mu,\lambda)$ then

   $$\lambda = (1-\mu+\lambda)p(\mu,\lambda) \leq (1+(\lambda-\mu))\mu$$

   or equivalently

   $$\lambda - \mu \leq (\lambda-\mu)\mu,$$

   which holds, because $0 \leq \mu \leq 1$ and $\lambda - \mu$ is negative.
2. The inequality is equivalent to

   $$\lambda \leq \frac{\mu+\lambda}{2} - \frac{\mu^2-\lambda^2}{2} \tag{32}$$

   or to

   $$\lambda - \lambda^2 \leq \mu - \mu^2,$$

   which holds, because $\lambda \leq \mu \leq \frac{1}{2}$ and thus

   $$\lambda(1-\lambda) \leq \mu(1-\mu).$$

3. Counterexample: $\mu = \frac{1}{2}$ and $\lambda = \frac{1}{4}$.

It remains to estimate E_2 and E_3 from below

$$E_2 = \sum_{i=u-j(\kappa)}^{D-k} \binom{D-k}{i} p^i q^{D-k-i},$$

where $u - j(\kappa) = (\mu - \frac{\lambda}{1-\mu+\lambda}\kappa)D$.

$$E_2 \geq \binom{(1-\kappa)D}{(\mu-p\kappa)D} p^{(\mu-p\kappa)D}(1-p)^{(1-\mu-(1-p)\kappa)D}$$

$$E_2 \geq 2^{D(1-\kappa)[h(\frac{\mu-p\kappa}{1-\kappa})+\frac{\mu-p\kappa}{1-\kappa}\log p+\frac{1-\mu-(1-p)\kappa}{1-\kappa}\log(1-p)]}.$$

We set

$$f(\kappa,\mu,\lambda,p) = h(\frac{\mu - p\kappa}{1-\kappa}) + \frac{\mu - p\kappa}{1-\kappa}\log p + \frac{1-\mu-(1-p)\kappa}{1-\kappa}\log(1-p)]$$

and want to find $\min_\kappa f(\kappa,\mu,\lambda,p)$. Let $r = \frac{\mu - p\kappa}{1-\kappa}$, then we have

$$\begin{aligned}&-r\log r - (1-r)\log(1-r) + r\log p + (1-r)\log 1-p\\ &= r\log\frac{p}{r} + (1-r)\log\frac{1-p}{1-r}\\ &= -D((r,1-r)||(p,1-p)),\end{aligned}$$

where $D((r,1-r)||(p,1-p))$ is called information divergence. For E_3 we have

$$E_3 = \sum_{t=0}^{l-j(\kappa)} \binom{D-k}{t} p^t q^{D-k-t},$$

where $l - j(\kappa) = (\lambda - p\kappa)D$. Therefore

$$E_3 \geq 2^{D(1-\kappa)[h(\frac{\lambda-p\kappa}{1-\kappa}) + \frac{\lambda-p\kappa}{1-\kappa}\log p + \frac{1-\kappa-\lambda+p\kappa)}{1-\kappa}\log(1-p)]}.$$

We set

$$b(\kappa,\mu,\lambda,p) = h(\frac{\lambda - p\kappa}{1-\kappa}) + \frac{\lambda - p\kappa}{1-\kappa}\log p + \frac{1-\kappa-\lambda+p\kappa}{1-\kappa}\log(1-p)]$$

and want to find $\min_\kappa b(\kappa,\mu,\lambda,p)$. Let $s = \frac{\lambda - p\kappa}{1-\kappa}$, then we have

$$\begin{aligned}&-s\log s - (1-s)\log(1-s) + s\log p + (1-s)\log\frac{1-p}{1-s}\\ &= s\log\frac{p}{s} + (1-s)\log\frac{1-p}{1-s}\\ &= -D((s,1-s)||(p,1-p)).\end{aligned}$$

It follows

Proposition 1.

$$B_{Th} \to F_T 2^{(-D((r,1-r)||(p,1-p)) - D((s,1-s)||(p,(1-p)))(1-\kappa)},$$

if $n \to \infty$.

Therefore we compare

$-D((r,1-r)||(p,1-p)) - D((s,1-s)||(p,(1-p)))(1-\kappa)$ with $-\max(h(\lambda),h(\mu))$.

We need some basic calculation for the bound on B_c.

Lemma 3. *Denote* $P = P(\lambda, \mu) = \frac{\mu}{1-\lambda+\mu}$, *then*

$$B_c/F_c = \left(\frac{u}{D-l+u}\right)^u \left(\frac{D-l}{D-l+u}\right)^{D-l} = 2^{-h(P)(1-\lambda+\mu)D}.$$

Proof. We have

$$\begin{aligned}
&\left(\frac{\mu}{1-\lambda+\mu}\right)^{\mu D} \left(\frac{1-\lambda}{1-\lambda+\mu}\right)^{(1-\lambda)D} \\
&= P^{PD(1-\lambda+\mu)}(1-P)^{(1-P)D(1-\lambda+\mu)} \\
&= 2^{-h(P)(1-\lambda+\mu)D}.
\end{aligned}$$

We note that $1-\lambda+\mu > 1$.
So far $P = \frac{\mu}{1-\lambda+\mu}$ and we have

$$h(P)(1-\lambda+\mu) > h(\mu) > h(\lambda).$$

Now we want to show that

$$h(P)(1-\lambda+\mu) > \max_{\kappa}(1-\kappa)(D((r,1-r)||(p,1-p)+D((s,1-s)||(p,1-p))$$

with $P = \frac{\mu}{1-\lambda+\mu}$, $r = \frac{\mu-p\kappa}{1-\kappa}$, $s = \frac{\lambda-p\kappa}{1-\kappa}$, and $p = \frac{\lambda}{1-\mu+\lambda}$.
We consider the special case $D = u+l$. Then we have $1 = \mu+\lambda$, $0 \leq \kappa \leq 2\lambda$, $P = \frac{1}{2}$, and $p = \frac{1}{2}$.

For $\mu = \lambda$ we have

$$h(P)(1-\lambda+\mu) = 2\mu > 0 = (1-\kappa)2D((\frac{1}{2},\frac{1}{2})||(\frac{1}{2},\frac{1}{2})).$$

For $\lambda < \mu$ we have $r = \frac{2\mu-\kappa}{2-2\kappa}$ and $s = \frac{2\lambda-\kappa}{2-2\kappa}$.

We have to show that for $0 \leq \kappa \leq \lambda$

$$2\mu > \max_{\kappa} -(1-\kappa)(D((\frac{2\mu-\kappa}{2-2\kappa},1-\frac{2\mu-\kappa}{2-2\kappa})||(\frac{1}{2},\frac{1}{2})+D((\frac{2\lambda-\kappa}{2-2\kappa},1-\frac{2\lambda-\kappa}{2-2\kappa})||(\frac{1}{2},\frac{1}{2})).$$

This is equivalent to (because $\lambda = 1-\mu$)

$$2 \leq h(\frac{2\mu-\kappa}{2-2\kappa}) + h(\frac{2(1-\mu)-\kappa}{2-2\kappa}) + \frac{2\mu}{1-\kappa} \tag{33}$$

$$2 \leq h(\frac{2\mu-\kappa}{2-2\kappa}) + h(1-\frac{2\mu-\kappa}{2-2\kappa}) + \frac{2\mu}{1-\kappa}$$

$$2 \leq 2h(1-\frac{2\mu-\kappa}{2-2\kappa}) + \frac{2\mu}{1-\kappa}$$

$$1-\frac{2\mu}{2-2\kappa} \leq h(1-\frac{2\mu-\kappa}{2-2\kappa}). \tag{34}$$

(33) is true for $\mu > 1 - \kappa$. For $\mu \le 1 - \kappa$ it holds $\frac{1}{2} \le \frac{2\mu-\kappa}{2-2\kappa} \le 1$. Therefore

$$1 - \frac{2\mu - \kappa}{2 - 2\kappa} \le h(1 - \frac{2\mu - \kappa}{2 - 2\kappa})$$

and (34) holds.

For $\kappa = 0$ and $\mu = 1$, that means $\lambda = 0$, the two terms are the same. In general we get

Theorem 9. *Let* $1 = \mu + \lambda$ *and* $\lambda \le \mu$, *then*

$$B_{Th} \ge 2^{2\lambda} B_c.$$

The goal of this appendix was to compare B_{Th} and B_c. It is clear that $B_{Th} \ge B_c$ because every (D, u, l) threshold code of size $n \times N$ is also a $(u, D - l)$ cover-free code of size $n \times N$. The quotient of the rates depends on u and l. We were only able to give bounds of the quotient in special cases.

Superimposed Codes and Threshold Group Testing

Arkadii D'yachkov[1], Vyacheslav Rykov[2], Christian Deppe[4,⋆], and Vladimir Lebedev[3,⋆⋆]

[1] Moscow State University, Faculty of Mechanics and Mathematics, Department of Probability Theory, Moscow, 119992, Russia
agd-msu@yandex.ru
[2] Department of Mathematics, University of Nebraska at Omaha, 6001 Dodge St., Omaha, NE 68182-0243, USA
vrykov@mail.unomaha.edu
[3] IPPI Institute for Information Transmission Problems of the Russian Academy of Sciences (Kharkevich Institute), Bol'shoi Karetnyi per. 19, Moscow 101447, Russia
lebedev37@yandex.ru
[4] Universität Bielefeld, Fakultät für Mathematik, Universitätsstraße 25, 33615 Bielefeld, Germany
cdeppe@math.uni-bielefeld.de

Dedicated to the memory of Rudolf Ahlswede

Abstract. We will discuss superimposed codes and non-adaptive group testing designs arising from the potentialities of compressed genotyping models in molecular biology. The given paper was motivated by the 30th anniversary of D'yachkov-Rykov recurrent upper bound on the rate of superimposed codes published in 1982. We were also inspired by recent results obtained for non-adaptive threshold group testing which develop the theory of superimposed codes.

Keywords: group testing, compressed genotyping, screening experiments, search designs, superimposed codes, rate of codes, rate of designs, bounds on the rate, shortened RC-code, threshold search designs.

1 Introduction

We consider superimposed codes and non-probabilistic, non-adaptive group testing. This model is also termed as combinatorial or deterministic design of experiment and pooling designs. The theory of designing screening experiments (DSE) ([35], [36], [37]) can be located in applied mathematics in the border region of search [2] and information theory [41]. Problems are equivalent to certain coding problems from information theory. In particular coding theory for

⋆ Supported by the DFG in the project "General Theory of Information Transfer".
⋆⋆ Supported in part by the Russian Foundation for Basic Research, project no 12-01-00905.

H. Aydinian, F. Cicalese, and C. Deppe (Eds.): Ahlswede Festschrift, LNCS 7777, pp. 509–533, 2013.

so-called multiple-access channels (MAC) [6] finds application in practical problems here. There are also applications of the theory for instance in engineering, electronics, medicine, and biology. Many examples of practical search problems serve to illustrate the mathematical models. We will elaborate on some models especially important for the practice. Furthermore we turn our attention to the information theoretical aspects of the theory; in particular we distinctly emphasize the connection of this theory to MAC. This connection is very useful for the obtaining of important results, which are of coding theoretical interest as well.

In many "processes" which are dependent on a large number of factors, it is natural, that one assumes a small number of "significant" factors, which really control the process, and considers the influence of the other factors as mere "experiment errors". Experiments to identify the significant factors are called screening experiments. A typical problem from DSE theory is the following. Among t factors there are p "significant", which need to be identified. By tests which examine arbitrary subsets of the factors, it can be determined whether significant factors appear in it or not. Now one tries to perform these experiments as economical as possible; a basal criterion at this is: how many tests are at least necessary to identify all significant factors in the most unfavorable case? Often the test outcomes are not determined uniquely; outer influences or human error lead to distortions that can be modeled mathematically by random noise. It can also be, that the significant factors cannot be determined with certainty. For this, "strategies" that identify the desired objects correctly with "high probability" are examined. Obviously this model corresponds to the general group testing model. The t factors correspond to the elements and the p significant factors correspond to defects. Hence the paper by Dorfman [9] can be seen as a pioneering work of the DSE theory.

A great directing influence on the theory had the cycle of examinations of applications of information theory in statistics by A. Rényi in the sixties. Of these papers the titles about mathematical search problems have an immediate reference to our topic (see [40]). Even though a very simplified model (search for *one* significant factor) was treated, many of Rényi's methods and notations could be adopted to DSE theory. Rényi also introduced so-called random strategies, where the tests are chosen independently by a probability distribution. Such strategies that, as we will see, are analogous to so-called random codes in information theory [41], make existence statements for nearly optimal "regular" strategies possible. They are used almost exceptionless in DSE theory, since a construction of optimal strategies in most cases is very difficult and yet unsolved. In the middle of the seventies Maljutov published a series of papers about special models that cleared up the connection to information theory ([35], [36], [37]).

The aim of our paper is to present the principal combinatorial results for the symmetric search model[1]. In Section 2, we give a brief survey of necessary

[1] We don't discuss here the general noisy symmetric model of non-adaptive search designs which can be described using the terminology of MAC [6]. An interested reader is referred to [11]. The information theory problems for non-symmetric search model are considered in [37].

definitions and bounds on the rate of superimposed codes which are the *base for studying* of non-adaptive group testing models.

In Section 3, we introduce the concept of non-adaptive group testing designs arising from the potentialities of compressed genotyping models in molecular biology and establish a universal upper bound on their rate. The universal bound is prescribed by D'yachkov-Rykov [17] recurrent upper bound on the rate of classical superimposed codes.

In Section 4, we remind our constructions of superimposed codes based on shortened RS-codes. These constructions are presented in papers [15]-[13], where we essentially extended optimal and suboptimal constructions of classical superimposed codes suggested in [28]. Note that we included in [15]-[14] the detailed tables with parameters of the best known superimposed codes. We don't mention other authors because, unfortunately, we don't know any papers containing relevant results, i.e., the similar or improved tables of parameters. Any extension of our tables is the important open problem.

In Section 5, the threshold group testing model is discussed. We apply the conventional terminology of superimposed code theory to refine the description of a new lower bound on the rate of threshold designs recently obtained in [4].

1.1 Notations, Definitions and Relevant Issues

Let $[n]$ be the set of integers from 1 to n and the symbol $\triangleq$ denote definitional equalities. For integers $N \geq 2$ and $t \geq 2$, symbols $\Omega_j \subset [N]$, $j = 1, 2, \ldots, t$, denote subsets of $[N]$. Subsets Ω_j, $j \in [t]$, are identified with binary columns $\mathbf{x}(j) \triangleq (x_1(j), x_2(j), \ldots, x_N(j))$ in which

$$x_i(j) \triangleq \begin{cases} 1 \text{ if } i \in \Omega_j, \\ 0 \text{ if } i \notin \Omega_j, \quad i \in [N]. \end{cases}$$

An incidence matrix $X \triangleq \|x_i(j)\|$, $i \in [N]$, $j \in [t]$, is called a *code* with t codewords (columns) $\mathbf{x}(1), \mathbf{x}(2), \ldots, \mathbf{x}(t)$ of length N corresponding to a *family of subsets* $\Omega_1, \Omega_2, \ldots, \Omega_t$.

Let $P \subset [t]$ be an arbitrary fixed subset of $[t]$ and $|P|$ be its size, i.e.,

$$P \triangleq \{p_1, p_2, \ldots, p_{|P|}\} \subset [t], \quad 1 \leq p_1 < p_2 < \cdots < p_{|P|} \leq t.$$

Denote by $\mathcal{P}(t, \leq s)$ ($\mathcal{P}(t, = s)$) the collection of all $\sum_{i=0}^{s} \binom{t}{i}$ ($\binom{t}{s}$) subsets P of size $|P| \leq s$ ($|P| = s$). Let $N \geq 2$ be an integer and $\mathcal{A} = \{A_1, A_1, \ldots, A_N\}$, $A_i \subset [t]$, $i \in [N]$, be a fixed family of subsets of $[t]$. Subsets A_i are identified with binary rows $\mathbf{x}_i \triangleq (x_i(1), x_i(2), \ldots, x_i(t))$ in which

$$x_i(j) \triangleq \begin{cases} 1 \text{ if } j \in A_i, \\ 0 \text{ if } j \notin A_i, \quad i \in [N], \ j \in [t]. \end{cases}$$

We will identify the family $\mathcal{A}$ with its incidence matrix (code) $X = \|x_i(j)\|$, $i \in [N]$, $j \in [t]$.

In the theory of *group testing* [25] (*designing screening experiments* [11]) the given, in advance, family $\mathcal{A} = \{A_1, A_1, \ldots, A_N\}$ is interpreted as a *non-adaptive search design* consisting of N group tests (experiments) A_i, $i \in [N]$. An experimenter wants to construct group tests A_i, $i \in [N]$, to carry out the corresponding experiments and then to identify an *unknown subset* $P \subset [t]$ with the help of test outcomes provided that $P \subset \mathcal{P}(t, \leq s)$ or $P \subset \mathcal{P}(t, = s)$, where $s \ll t$. If for each test A_i, $i \in [N]$, its outcome *depends only on the size of intersection*

$$|P \cap A_i| = \sum_{m=1}^{|P|} x_i(p_m), \quad i \in [N],$$

then we will say that a *symmetric model* [11] of non-adaptive search design is considered.

2 Superimposed (z, u)-Codes

In this section we give a brief survey of necessary definitions and bounds on the rate of superimposed codes which are the *base for studying* of non-adaptive group testing models.

Let z and u be positive integers such that $z + u \leq t$.

Definition 1. *[13] A family of subsets $\Omega_1, \Omega_2, \ldots, \Omega_t$, where $\Omega_j \subseteq [N]$, $j \in [t]$, is called an (z, u)*–cover-free family *if for any two non-intersecting subsets $Z, U \subset [t]$, $Z \cap U = \varnothing$, such that $|Z| = z$, $|U| = u$, the following condition holds:*

$$\bigcap_{j \in U} \Omega_j \not\subseteq \bigcup_{j \in Z} \Omega_j.$$

An incidence matrix $X = \|x_i(j)\|$, $i \in [N]$, $j \in [t]$, corresponding to (z, u)–cover-free family is called a superimposed (z, u)-code.

The following evident necessary and sufficient condition for Definition 1 takes place.

Proposition 1. *[13] Any binary $(N \times t)$-matrix X is a superimposed (z, u)-code if and only if for any two subsets Z, $U \subset [t]$, such that $|Z| = z$, $|U| = u$ and $Z \cap U = \varnothing$ the matrix X contains a row $\mathbf{x}_i = (x_i(1), x_i(2) \ldots, x_i(t)$, for which*

$$x_i(j) = 1 \quad \text{for all} \quad j \in U, \qquad x_i(j) = 0 \quad \text{for all} \quad j \in Z.$$

Let $t(N, z, u)$ be the maximal possible size of superimposed (z, u)-codes. For fixed $1 \leq u < z$, define a *rate* of (z, u)-codes:

$$R(z, u) \triangleq \varlimsup_{N \to \infty} \frac{\log_2 t(N, z, u)}{N}.$$

For the classical case $u = 1$, superimposed $(z, 1)$–codes and their applications were introduced by W.H Kautz, R.C. Singleton in [28]. Further, these codes along with new applications were investigated in [17]-[11]. The best known upper and lower bounds on the rate $R(z, 1)$ can be found in [17],[21] and [13] .

2.1 Recurrent Upper Bounds on $R(z,1)$ and $R(z,u)$

Let $h(\alpha) \triangleq -\alpha\log_2\alpha - (1-\alpha)\log_2(1-\alpha)$, $0<\alpha<1$, be the binary entropy. To formulate an *upper bound* on the rate $R(z,1)$, $z \ge 1$, we introduce the function [17]

$$\mathsf{f}_z(\alpha) \triangleq h(\alpha/z) - \alpha\, h(1/z), \quad z = 1,2,\ldots,$$

of argument α, $0<\alpha<1$.

Theorem 1. *[17]-[18] (Recurrent upper bound on $R(z,1)$). If $z = 1,2,\ldots,$ then the rate $R(z,1) \le \overline{R}(z,1)$, where*

$$\overline{R}(1,1) = R(1,1) = 1, \quad \overline{R}(2,1) \triangleq \max_{0<\alpha<1} \mathsf{f}_2(\alpha) = 0.321928 \qquad (1)$$

and sequence $\overline{R}(z,1)$, $z = 3,4,\ldots,$ is defined as the unique solution of recurrent equation

$$\overline{R}(z,1) = \mathsf{f}_z\left(1 - \frac{\overline{R}(z,1)}{\overline{R}(z-1,1)}\right). \qquad (2)$$

Up to now the recurrent sequence $\overline{R}(z,1)$, $z = 1,2,\ldots,$ defined by (1)-(2) and called a *recurrent upper bound* has been the best known upper bound on the rate $R(z,1)$. The reciprocal values of $\overline{R}(z,1)$ taken from [18], are given in **Table 1.**

Table 1.

z	$1/\overline{R}(z,1)$	z	$1/\overline{R}(z,1)$	z	$1/\overline{R}(z,1)$	z	$1/\overline{R}(z,1)$
2	3.1063	6	12.0482	10	24.5837	14	40.3950
3	5.0180	7	14.8578	11	28.2402	15	44.8306
4	7.1196	8	17.8876	12	32.0966	16	49.4536
5	9.4660	9	21.1313	13	36.1493	17	54.2612

Applying Theorem 1 and the corresponding calculus arguments, we proved

Theorem 2. *[17]-[18] (Non-recurrent upper bound on $R(z,1)$) For any $z \ge 2$, the rate $R(z,1)$ satisfies inequality*

$$R(z,1) \le \frac{2\log_2[e(z+1)/2]}{z^2}, \qquad z = 2,3,\ldots,$$

which leads to the asymptotic inequality

$$R(z,1) \le \frac{2\log_2 z}{z^2}\,(1+o(1)), \quad z\to\infty.$$

Theorem 3. *[31] (Recurrent inequality for $R(z,u)$)*
If $z \ge u \ge 2$, then for any $i \in [z-1]$ and $j \in [u-1]$, the rate

$$R(z,u) \le \frac{R(z-i,u-j)}{R(z-i,u-j) + \frac{(i+j)^{i+j}}{i^i \cdot j^j}}. \qquad (3)$$

Recurrent inequality (3) and the known numerical values of recurrent upper bound $\overline{R}(z,1)$, $z=1,2,\ldots$, defined by (1)-(2), give numerical values of the best known upper bound $\overline{R}(z,u)$ on the rate $R(z,u)$, $z \ge u \ge 2$. An asymptotic consequence from the given upper bound is presented by

Theorem 4. *[24] If $z \to \infty$ and $u \ge 2$ is fixed, then*

$$R(z,u) \le \overline{R}(z,u) \le \frac{(u+1)^{u+1}}{2\,e^{u-1}} \cdot \frac{\log_2 z}{z^{u+1}} \cdot (1+o(1)).$$

2.2 Random Coding Lower Bounds on $R(z,u)$ and $R(z,1)$

Theorem 5. *[13] A random coding lower bound on the rate $R(z,u)$ has the form:*

$$R(z,u) \ge \underline{R}(z,u) \triangleq -(z+u-1)^{-1}\log_2\left(1-\frac{z^z\,u^u}{(z+u)^{z+u}}\right), \quad 2 \le u < z.$$

If $u \ge 2$ is fixed and $z \to \infty$, then the asymptotic form of the given lower bound is

$$R(z,u) \ge \underline{R}(z,u) = \frac{e^{-u}\cdot u^u \cdot \log_2 e}{z^{u+1}} \cdot (1+o(1)).$$

If $u=1$, then the best known random coding lower bound on the rate $R(z,1)$ is given by

Theorem 6. *[16] For any $z=1,2,\ldots$, the rate $R(z,1) \ge \underline{R}(z,1) \triangleq \frac{A(z)}{z}$, where*

$$A(z) \triangleq \max_{0<\alpha<1,\ 0<Q<1}\left\{-(1-Q)\log(1-\alpha^z) + z\left(Q\log\frac{\alpha}{Q} + (1-Q)\log\frac{1-\alpha}{1-Q}\right)\right\}.$$

If $z \to \infty$, then the rate

$$R(z,1) \ge \underline{R}(z,1) = \frac{1}{z^2\log e}(1+o(1)) = \frac{0.693}{z^2}(1+o(1)).$$

In the first and second rows of Table 2, we give values of $\underline{R}(s,1) < 1/s$, $s = 2,3\ldots,8$, along with the corresponding values of $\overline{R}(s,1) < 1/s$, $s = 2,3\ldots,8$, taken from Table 1.

Table 2.

s	2	3	4	5	6	7	8
$\underset{\sim}{R}_1(\le s) = \underline{R}(s,1)$	.182	.079	.044	.028	.019	.014	.011
$\widetilde{R}_1(\le s) = \overline{R}(s,1)$	.3219	.1993	.1405	.1056	.0830	.0673	.0559
$\underset{\sim}{R}_2(\le s) = \underline{R}(s-1,2)$	-	.0321	.0127	.0068	.0037	.0024	.0015
$\widetilde{R}_2(\le s) = \overline{R}(s-1,2)$	-	.1610	.0745	.0455	.0287	.0204	.0146
$\underset{\sim}{R}_3(\le s) = \underline{R}(s-2,3)$	-	-	.0127	.0046	.0020	.0010	.0001
$\widetilde{R}_3(\le s) = \overline{R}(s-2,3)$	-	-	.0745	.0387	.0183	.0109	.0067
$\underline{R}(F_0^1,=s)$	.302	.142	.082	.053	.037	.027	.021

3 $(F^l, \leq s)$–Designs, $(F^l, = s)$–Designs and $\mathcal{D}^l_s$–Codes

In this section we introduce the concept of non-adaptive group testing designs arising from the potentialities of compressed genotyping models in molecular biology and establish a universal upper bound on their rate. The universal bound is prescribed by our recurrent upper bound on the rate of classical superimposed codes. Using notations of Section 1, we give

Definition 2. *Let l, $1 \leq l < s < t$ be integers and $F^l = F^l(n)$, $n = 0, 1, \ldots, l$, be an arbitrary fixed function of integer argument $n = 0, 1, \ldots, l$ such that for any $n = 0, 1, \ldots, l-1$, its value $F^l(n) \neq F^l(l)$. Define the vector*

$$\mathbf{y}^\ell(P, \mathcal{A}) \triangleq \left(y^\ell_1, y^\ell_2, \ldots, y^\ell_N\right), \quad y^\ell_i \triangleq \begin{cases} F^\ell(n) \text{ if } |P \cap A_i| = n, & n = 0, 1, \ldots, \ell-1, \\ F^\ell(\ell) \text{ if } |P \cap A_i| \geq \ell, & i \in [N]. \end{cases}$$

or

$$\mathbf{y}^\ell(P, X) \triangleq \left(y^\ell_1, y^\ell_2, \ldots, y^\ell_N\right), \quad y^\ell_i \triangleq \begin{cases} F^\ell(n) \text{ if } \sum\limits_{m=1}^{|P|} x_i(p_m) = n, & n = 0, 1, \ldots, \ell-1, \\ F^\ell(\ell) \text{ if } \sum\limits_{m=1}^{|P|} x_i(p_m) \geq \ell, & i \in [N]. \end{cases}$$

*A code X of length N and size t is called an $(F^l, \leq s)$–*design, *($(F^l, = s)$–*design*), $1 \leq l < s < t$,* for group testing model *if $\mathbf{y}^l(P, X) \neq \mathbf{y}^l(P', X)$ for any*

$$P \neq P', \quad P \in \mathcal{P}(t, \leq s), \; P' \in \mathcal{P}(t, \leq s) \quad \left(P \in \mathcal{P}(t, = s), \; P' \in \mathcal{P}(t, = s)\right).$$

Remark 1. *$(F^l, \leq s)$–design and $(F^l, = s)$–design are examples, which can be interpreted as compressed genotyping [27] models in molecular biology.*

Remark 2. *In [26], a special $(F^l, \leq s)$–design is considered. The authors introduce the ranges $(0 \triangleq r_0 < r_1 < r_2 < \cdots < r_k \triangleq p)$ and set*

$$\begin{aligned} F^\ell(r_0 + 1) = \ldots &= F^\ell(r_1) = 1 \\ F^\ell(r_1 + 1) = \ldots &= F^\ell(r_2) = 2 \\ \vdots = \ldots &= \vdots \\ F^\ell(r_{k-1} + 1) = \ldots &= F^\ell(r_k) = k \end{aligned}$$

This model can be viewed as an adder model followed by a quantizer.

Let $1 \leq l < s < t$ be integers. For any set $\mathcal{S} \subset [t]$ of size $|\mathcal{S}| = s$, we denote by $\binom{\mathcal{S}}{l}$ the collection of all $\binom{s}{l}$ l–subsets of the set $\mathcal{S}$.

Definition 3. *[19] A family of subsets $\Omega_1, \Omega_2, \ldots, \Omega_t$ is called an $\mathcal{D}^l_s$–*family *if for any $\mathcal{S} \subset [t]$, $|\mathcal{S}| = s$, and any $j \notin \mathcal{S}$*

$$\Omega_j \not\subseteq \bigcup_{\binom{\mathcal{S}}{l}} \left\{ \bigcap_{k=1}^{l} \Omega_{j_k} \right\},$$

where

$$\binom{\mathcal{S}}{l} \triangleq \{(j_1, j_2, \ldots, j_l) \,:\, j_i \in \mathcal{S}, \quad j_1 < j_2 < \cdots < j_l\}.$$

An incidence matrix $X = \|x_i(j)\|$, $i \in [N]$, $j \in [t]$, *corresponding to* $\mathcal{D}_s^l$*– family is called a* superimposed $\mathcal{D}_s^l$-code *(briefly,* $\mathcal{D}_s^l$*-code).*

One can easily check the following

Proposition 2. *Any binary* $(N \times t)$*-matrix* X *is a* $\mathcal{D}_s^l$*–code,* $1 \le l < s < t$*, if and only if for any collection of* $s+1$ *integers* $j_1, j_2, \ldots, j_s, j_{s+1}$*,* $j_k \ne j_m$*,* $j_k \in [t]$*, there exists* $i \in [N]$ *such that*

$$x_i(j_{s+1}) = 1, \qquad \sum_{k=1}^{s} x_i(j_k) \le l-1.$$

For $l = 1$ *and* $s = 2, 3 \ldots$*, the definition of* $\mathcal{D}_s^1$*-code coincides with the definition of superimposed* $(s,1)$*-code. In addition, if* $1 \le l < s-1$*, then any* $\mathcal{D}_s^l$*-code is a* $\mathcal{D}_s^{l+1}$*-code.*

Remark 3. *For* $s > l \ge 2$*,* $\mathcal{D}_s^l$*-codes were suggested in [19] for the study of some communication systems with random multiple access.*

3.1 Universal Upper Bound for $(F^l, \le s)$–Designs

Let $t(N, \mathcal{D}_s^l)$, $t(N, F^l, \le s)$ and $t(N, F^l, = s)$ be the maximal size of superimposed $\mathcal{D}_s^l$–codes, $(F^l, \le s)$–designs and $(F^l, = s)$–designs. For fixed $1 \le l < s$, define the corresponding *rates:*

$$R\left(\mathcal{D}_s^l\right) \triangleq \overline{\lim_{N\to\infty}} \frac{\log_2 t\left(N, \mathcal{D}_s^l\right)}{N}, \quad 1 \le l < s,$$

$$R\left(F^l, \le s\right) \triangleq \overline{\lim_{N\to\infty}} \frac{\log_2 t\left(N, F^l, \le s\right)}{N}, \qquad R\left(F^l, = s\right) \triangleq \overline{\lim_{N\to\infty}} \frac{\log_2 t\left(N, F^l, = s\right)}{N}.$$

Obviously, for any $1 \le \ell < s$, the following inequalities hold:

$$t\left(N, F^\ell, \le s\right) \le t\left(N, F^\ell, = s\right), \quad R\left(F^\ell, \le s\right) \le R\left(F^\ell, = s\right) \le \frac{\log_2(\ell+1)}{s}. \tag{4}$$

Proposition 3. *[19]*

If $1 \le l < s-1$*, then any* $(F^l, \le s)$*–design is a superimposed* $\mathcal{D}_{s-1}^l$*–code, i.e.,*

$$t\left(N, F^l, \le s\right) \le t\left(N, \mathcal{D}_{s-1}^l\right), \qquad R\left(F^l, \le s\right) \le R\left(\mathcal{D}_{s-1}^l\right), \quad 1 \le l < s-1.$$

Proof. By contradiction. If a code $X = \|x_i(j)\|$, $i \in [N]$, $j \in [t]$ doesn't satisfy the definition of $\mathcal{D}_{s-1}^l$–code, then in virtue of Proposition 1, there exists a collection of s integers $j_1, j_2, \ldots, j_{s-1}, j_s$, $j_k \neq j_m$, $j_k \in [t]$, such that for any $i \in [N]$,

$$x_i(j_s) = 1 \quad \Longrightarrow \quad \sum_{k=1}^{s-1} x_i(j_k) \geq l.$$

Hence, for $(s-1)$-subset $P \triangleq \{j_1, j_2, \ldots, j_{s-1}\} \subset [t]$ and s-subset $P' \triangleq \{j_1, j_2, \ldots, j_{s-1}, j_s\} \subset [t]$, the vector $\mathbf{y}^l(P, X) = \mathbf{y}^l(P', X)$. This contradicts to the definition of $(F^l, \leq s)$–design.

Theorem 7. *(De Bonis, Vaccaro [8]) For any $1 \leq l < s$, the rate $R(\mathcal{D}_s^l)$ of superimposed $\mathcal{D}_s^l$–codes satisfies inequality*

$$R(\mathcal{D}_s^l) \leq R\left(\left\lfloor \frac{s}{l} \right\rfloor, 1\right),$$

where $R(z,1)$, $z \geq 1$, is the rate of classical superimposed $(z,1)$-codes.

Proposition 3 and Theorem 7 lead to inequalities:

$$R(F^\ell, \leq s) \leq R(\mathcal{D}_{s-1}^\ell) \leq R\left(\left\lfloor \frac{s-1}{\ell} \right\rfloor, 1\right) \leq \overline{R}\left(\left\lfloor \frac{s-1}{\ell} \right\rfloor, 1\right), \quad 1 \leq \ell \leq s, \tag{5}$$

where $\overline{R}(z,1)$ is the recurrent upper bound on the rate $R(z,1)$ presented by Theorem 1. For instance, if $(l=3, s=10)$ or $(l=3, s=13)$, then Table 2 shows that

$$\overline{R}(3,1) = .199 < .200 = 2/10 \quad \text{or} \quad \overline{R}(4,1) = .140 < .154 = 2/13,$$

i.e., for $l=3$ and $s=3k+1$, $k=3,4,\ldots$, bound (5) improves the trivial bound (4).

From inequalities (4)-(5), it follows

Proposition 4. *(Universal upper bound). For any $(F^l, \leq s)$–design, the rate*

$$R(F^l, \leq s) \leq \min\left\{\frac{\log_2(l+1)}{s}, \overline{R}\left(\left\lfloor \frac{s-1}{l} \right\rfloor, 1\right)\right\}, \quad 1 \leq l < s,$$

and the asymptotic inequality

$$R(F^l, \leq s) \leq \frac{2l^2 \log_2 s}{s^2}(1+o(1)), \quad l = 1, 2, \ldots, \qquad s \to \infty,$$

holds.

4 Constructions of Superimposed (z,u)-Codes and $\mathcal{D}_s^l$–Codes

4.1 Superimposed $(s,1)$-Codes and $\mathcal{D}_s^l$–Codes Based on Shortened Reed-Solomon Codes

Let $\mathcal{Q}$ be the set of all primes or prime powers ≥ 2, i.e.,

$$\mathcal{Q} \triangleq \{2,3,4,5,7,8,9,11,13,16,17,19,23,25,27,29,31,32,37,\ldots\}.$$

Let $q \in \mathcal{Q}$ and $2 \leq k \leq q+1$ be fixed integers for which there exists the q-ary Reed-Solomon code (RS-code) B of size q^k, length $(q+1)$ and the Hamming distance $d = q-k+2 = (q+1)-(k-1)$ [34]. We will identify the code B with an $((q+1) \times q^k)$–matrix whose columns, (i.e., $(q+1)$-sequences from the alphabet $\{0, 1, 2, \ldots, q-1\}$) are the codewords of B. Therefore, the maximal possible number of positions (rows) where its two codewords (columns) can coincide, called a *coincidence* of code B, is equal to $k-1$.

Fix an arbitrary integer $r = 0, 1, 2, \ldots, k-1$ and introduce the *shortened* RS-code $\tilde{B}$ of size $t = q^{k-r}$, length $n = q+1-r$ that has the same Hamming distance $d = q-k+2$. Code $\tilde{B}$ is obtained by the *shortening* of the *subcode* of B which contains $0's$ in the first r positions (rows) of B. Obviously, the coincidence of $\tilde{B}$ is equal to

$$\lambda \triangleq n - d = (q+1-r) - d = q+1-r-(q-k+2) = k-r-1. \qquad (6)$$

Consider the following standard transformation of the q-ary code $\tilde{B}$. Each symbol of the q-ary alphabet $\{0, 1, 2, \ldots, q-1\}$ is substituted for the corresponding binary column of the length q and weight 1, namely:

$$0 \Leftrightarrow \underbrace{(1, 0, 0, \ldots, 0)}_{q}, \quad 1 \Leftrightarrow \underbrace{(0, 1, 0, \ldots, 0)}_{q}, \quad \ldots \quad q-1 \Leftrightarrow \underbrace{(0, 0, 0, \ldots, 1)}_{q}.$$

As a result we have a binary constant-weight code X of size t, length N and weight w, where

$$t = q^{k-r} = q^{\lambda+1}, \quad N = n \cdot q = (q+1-r)q, \quad w = n = q+1-r. \qquad (7)$$

From Propositions 1-2 and (6), it follows

Proposition 5. *Let integers* $1 \leq \ell < s$ *satisfy inequalities*

$$s[(k-1)-r] \leq \ell\,(q+1-r) - 1, \quad 2 \leq k \leq q+1, \quad 0 \leq r \leq k-1. \qquad (8)$$

Then the binary constant-weight code X *with parameters* (7) *is a* $\mathcal{D}_s^\ell$*–code if* $2 \leq \ell < s$*, or* X *is a superimposed* $(s, 1)$*-code if* $\ell = 1$.

For $l = 1$, the detailed tables with parameters of the best known superimposed $(s, 1)$-codes (or $\mathcal{D}_s^1$–codes) based on Proposition 5 are presented in [15]-[14]. Table 3 gives an example of such table. In Table 3, we mark by the **boldface type** two *triples* of superimposed code parameters known from [28]. The rest triples of superimposed code parameters from Table 3 can be found in [15]-[14].

For the general case of superimposed (z, u)-codes, $2 \leq u < z$, the construction similar to Proposition 5 was developed in [13]. Another significant constructions of superimposed (z, u)-codes, $2 \leq u < z$, were suggested in [29]-[30]. Table 4 gives parameters of the best known superimposed (z, u)-codes if $u = 2, 3$ and $z = 2, 3, \ldots 9$.

4.2 Examples of $\mathcal{D}_s^l$–Codes

Example 1. *If* $q = 5$, *then for the pair* $(l = 2,\ s = 3)$, *inequalities (8) are fulfilled with* $k = 5$ *and* $r = 2$. *Therefore, the construction of Proposition 4 yields a binary constant-weight* $\mathcal{D}_3^2$*–code* X *with parameters*

$$t = q^{k-r} = 5^3 = 125, \quad N = n{\cdot}q = (q{+}1{-}r)q = 4{\cdot}5 = 20, \quad w = n = q{+}1{-}r = 4. \tag{9}$$

Parameters (9) give the following lower bound on the maximal size: $t\left(20, \mathcal{D}_3^2\right) \geq 125$.

Example 2. *If* $q = 7$, *then for the pair* $(l = 2,\ s = 4)$, *inequalities (8) are fulfilled with* $k = 6$ *and* $r = 3$. *Therefore, the construction of Proposition 4 yields a binary constant-weight* $\mathcal{D}_4^2$*–code* X *with parameters*

$$t = q^{k-r} = 7^3 = 343, \quad N = n{\cdot}q = (q{+}1{-}r)q = 5{\cdot}7 = 35, \quad w = n = q{+}1{-}r = 5. \tag{10}$$

Parameters (10) give the following lower bound on the maximal size: $t\left(35, \mathcal{D}_4^2\right) \geq 343$.

Example 3. *If* $q = 8$, *then for two pairs of integers* $(l = 2,\ s = 6)$ *and* $(l = 3,\ s = 10)$, *inequalities (8) are fulfilled with* $k = 5$ *and* $r = 2$. *Therefore, the construction of Proposition 4 yields a binary constant-weight* $\mathcal{D}_6^2$*–code* X *and a binary constant-weight* $\mathcal{D}_{10}^3$*–code* X *with parameters*

$$t = q^{k-r} = 8^3 = 512, \quad N = n{\cdot}q = (q{+}1{-}r)q = 7{\cdot}8 = 56, \quad w = n = q{+}1{-}r = 7. \tag{11}$$

Parameters (11) give the following lower bounds on the maximal size $t\left(N, \mathcal{D}_s^l\right)$ of $\mathcal{D}_s^l$–codes:

$$t\left(56, \mathcal{D}_6^2\right) \geq 512, \qquad t\left(56, \mathcal{D}_{10}^3\right) \geq 512.$$

For comparison, if $(u = 1,\ z = 6)$ and $N = 56$, then the best known lower bound on the size of optimal superimposed $(6,1)$-codes, calculated in [15], is $t(56, 6, 1) \geq 64$. In addition, this example shows that for $l = 3$, the parameter $s = 10$ of $\mathcal{D}_{10}^3$–code X can exceed the corresponding code weight $w = 7$.

4.3 Parameters of Constant-Weight superimposed $(s,1)$-Codes $2 \leq s \leq 8$, of Weight w, Length N, Size $t = q^{\lambda+1}$, $2^m \leq t < 2^{m+1}$, $5 \leq m \leq 30$, Based on the q-ary Shortened Reed-Solomon Codes

Table 3 also contains numerical values of the rate for several obtained codes, namely: the values of fraction $\frac{m}{N}$, $m = 12, 20, 25, 29$. The comparison with lower $\underline{R}(s,1)$ and upper $\overline{R}(s,1)$ bounds from Table 2 (their values are included in Table 3 as well) yields the following conclusions:

Table 3.

s	2	3	4	5	6	7	8
$\underline{R}(s,1)$	.182	.079	.044	.028	.019	.014	.011
$\overline{R}(s,1)$	.322	.199	.140	.106	.083	.067	.056
m	q, λ, N	q, λ, N	q, λ, N	q, λ, N	q, λ, N	q, λ, N	q, λ, N
5	–	7, 1, 28	7, 1, 35	7, 1, 42	7, 1, 49	–	–
6	4, 2, 20	8, 1, 32	8, 1, 40	8, 1, 48	8, 1, 56	9, 1, 72	11, 1, 99
7	–	–	13, 1, 65	13, 1, 78	13, 1, 91	13, 1, 104	13, 1, 117
8	7, 2, 35	7, 2, 49	–	16, 1, 96	16, 1, 112	16, 1, 128	16, 1, 144
9	8, 2, 40	8, 2, 56	8, 2, 72	–	23, 1, 161	23, 1, 184	23, 1, 207
10	–	11, 2, 77	11, 2, 99	11, 2, 121	–	–	–
11	7, 3, 49	–	13, 2, 117	13, 2, 143	13, 2, 169	–	–
12	8, 3, 56	9, 3, 90	16, 2, 144	16, 2, 176	16, 2, 208	16, 2, 240	**16, 2, 272**
$\frac{12}{N}$	.214	.133	.083	.068	.058	.050	.044
13	–	11, 3, 110	–	23, 2, 253	23, 2, 299	23, 2, 345	23, 2, 391
14	–	13, 3, 130	13, 3, 169	–	27, 2, 351	27, 2, 405	27, 2, 459
15	8, 4, 72	–	–	–	–	32, 2, 480	32, 2, 544
16	–	16, 3, 160	16, 3, 208	16, 3, 256	19, 3, 361		
17	11, 4, 99	–	–	–	–	–	–
18	13, 4, 117	13, 4, 169	–	23, 3, 368	23, 3, 437	23, 3, 506	25, 3, 625
19	–	–	–	27, 3, 432	27, 3, 513	27, 3, 594	27, 3, 675
20	11, 5, 121	16, 4, 208	**16, 4, 272**	–	32, 3, 608	32, 3, 704	32, 3, 800
$\frac{20}{N}$	.165	.096	.074	-	.034	.028	.025
21	–	–	19, 4, 323				41, 3, 1025
22	13, 5, 143	–	23, 4, 391	23, 4, 483	–	–	–
23	–	–	25, 4, 425	25, 4, 525	25, 4, 625	–	–
24	–	16, 5, 256	–	27, 4, 609	29, 4, 725	29, 4, 841	–
25	13, 6, 169	19, 5, 304	–	–	32, 4, 800	32, 4, 928	32, 4, 1056
$\frac{25}{N}$	.148	.082	-	-	.031	.027	.024
26	–	–	–	–	37, 4, 925	37, 4, 1073	37, 4, 1221
27	–	–	23, 5, 483	–	–	43, 4, 1247	43, 4, 1419
28	16, 6, 208	–	27, 5, 702	25, 5, 650	–	–	49, 4, 1617
29	–	19, 6, 361	29, 5, 609	29, 5, 754	31, 5, 961	–	–
$\frac{29}{N}$	–	.080	.048	.038	.030	–	–
30	–	–	–	32, 5, 832	32, 5, 992	–	–

– if $s = 2$ and $m \le 15$, then the values $\frac{m}{N}$ exceed the random coding rate $\underline{R}(2,1) = .182$;

– if $s \ge 3$ and $m \le 30$, then the values $\frac{m}{N}$ exceed the random coding rate $\underline{R}(s,1)$.

4.4 Size t and Length N of Superimposed (z, u)-Codes, $u = 2, 3$ and $z = 2, 3, \ldots 9$

Table 4.

(2,2)	(3,2)	(4,2)	(5,2)	(6,2)	(7,2)	(8,2)	(9,2)
t, N	t, N	t, N	t, N	t, N	t, N	t, N	t, N
8, 14	7, 21	11, 55	11, 55	20, 190	26, 260	16, 120	38, 703
9, 18	8, 28	13, 65	16, 120	25, 210	50, 350	32, 496	82, 738
10, 20	10, 30	17, 68	26, 130	49, 294	64, 448	65, 520	120, 1090
12, 22	16, 42	22, 77	48, 246	63, 385	80, 568	81, 648	166, 1562
16, 26	21, 56	25, 100	62, 330	79, 497	118, 882	119, 981	250, 2531
18, 30	24, 76	47, 205	78, 434	117, 792	164, 1308	165, 1430	282, 2933
22, 34	49, 147	64, 252	121, 605	169, 1014	256, 1800	256, 2040	361, 3249
24, 37	–	–	–	–	–	–	–
32, 43	(3,3)	(4,3)	(5,3)	(6,3)	(7,3)	(8,3)	(9,3)
40, 50	t, N	t, N	t, N	t, N	t, N	t, N	t, N
48, 59	7, 35	12, 220	16, 560	17, 680	19, 969	20, 1140	22, 1540
56, 65	8, 54	13, 253	19, 612	20, 816	21, 1071	21, 1330	23, 1771
64, 68	11, 66	23, 253	25, 700	26, 910	27, 1170	22, 1386	45, 14190
80, 76	16, 112	24, 532	31, 3951	32, 4683	52, 11313	53, 12757	54, 14352
112, 96	22, 176	169, 3289	50, 8830	51, 10008	529, 25740	729, 73125	729, 81900
128, 100	23, 399	–	256, 8960	361, 15504	–	–	–
144, 109	121, 660	–	–	–	–	–	–
512, 126	–	–	–	–	–	–	–

5 Threshold Group Testing Model

5.1 Superimposed (z, u)-Codes and $(F_0^w, \leq s)$-Designs

Let w be an integer with $1 \leq w < s$. Let the function $F^w \triangleq F_0^w = F_0^w(n)$ takes binary values, namely:

$$F_0^w(n) \triangleq \begin{cases} 0 \text{ if} & n = 0, 1, \ldots, w-1, \\ 1 \text{ if} & n = w. \end{cases}$$

If $w \geq 2$, then the given particular case is called a *threshold group testing model* [7]. For the non-adaptive threshold group testing model which is the principal model for applications [27], a *refined form* of Definition 2 can be written as follows.

Definition 4. *Let w, $1 \leq w < s < t$ be integers. For code $X = \|x_i(k)\|$, $k \in [t]$, $i \in [N]$, and a subset $P \in \mathcal{P}(t, \leq s)$, define the i-th* outcome of non-adaptive threshold group testing

$$y_i^w(P, X) \triangleq \begin{cases} 0 \text{ if } \sum\limits_{k \in P} x_i(k) \leq w-1, \\ 1 \text{ if } \sum\limits_{k \in P} x_i(k) \geq w, \quad i \in [N]. \end{cases}$$

A code X is called a $(F_0^w, \leq s)$-design, *($(F_0^w, = s)$-design) if for any* $P \neq P'$, $P, P' \in \mathcal{P}(t, \leq s) \setminus \mathcal{P}(t, \leq w-1)$ ($P \in \mathcal{P}(t, = s)$, $P' \in \mathcal{P}(t, = s)$), *there exists an index* $i \in [N]$, *where* $y_i^w(P, X) \neq y_i^w(P', X)$.

An important connection between $(F_0^w, \leq s)$-designs and superimposed $(s-w+1, w)$–codes is described by

Proposition 6. *([3], [33]) If* $1 \leq w < s$, *then any superimposed* $(s-w+1, w)$–*code is a* $(F_0^w, \leq s)$-*design, i.e.*

$$t(N, s-w+1, w) \leq t(N, F_0^w, \leq s), \qquad R(s-w+1, w) \leq R(F_0^w, \leq s).$$

The lower bound of Theorem 5 and Propositions 6 lead to the following lower bound on the rate of $(F_0^w, \leq s)$–designs.

Proposition 7. *(Random coding bound) For any* $1 \leq w < s$, *the rate*

$$R(F_0^w, \leq s) \geq R(s-w+1, w) \geq -\frac{1}{s}\log_2\left[1 - \frac{(s-w+1)^{s-w+1} \cdot w^w}{(s+1)^{s+1}}\right], \quad 1 \leq w < s. \tag{12}$$

If $w \geq 1$ *is fixed and* $s \to \infty$, *then the asymptotic form of the given lower bound is*

$$R(F_0^w, \leq s) \geq \frac{e^{-w} \cdot w^w \cdot \log_2 e}{s^{w+1}} \cdot (1 + o(1)). \tag{13}$$

5.2 Bounds on the Rate of $(F_0^1, \leq s)$ and $(F_0^1, = s)$-Designs

If $w = 1$ and $s \geq 2$, then the universal upper bound of Proposition 4 lead to the inequalities :

$$R(F_0^1, \leq s) \leq \min\{1/s\,;\, \overline{R}(s-1, 1)\}, \qquad s = 2, 3, \ldots,$$

where $\overline{R}(z, 1)$, $z = 1, 2, \ldots$, is the recurrent upper bound from Theorem 1. Hence, the asymptotic upper bound

$$R(F_0^1, \leq s) \leq \overline{R}(s-1, 1) = \frac{2 \cdot \log_2 s}{s^2} \cdot (1 + o(1)), \quad s \to \infty,$$

holds.

The best known asymptotic random coding lower bounds on $R(F_0^1, \leq s)$ and $R(F_0^1, = s)$ along with the best known upper bound on $R(F_0^1, = s)$ can be found in [21]-[16] (see, also [11]). These bounds have the form: These bounds have the form:

$$R(F_0^1, \leq s) \geq \underline{R}(s, 1) = \frac{1}{s^2 \cdot \log_2 e} \cdot (1 + o(1)) = \frac{0.693}{s^2} \cdot (1 + o(1)), \quad s \to \infty, \tag{14}$$

$$R\left(F_0^1, = s\right) \geq \underline{R}\left(F_0^1, = s\right) = \frac{2}{s^2 \cdot \log_2 e} \cdot (1+o(1)) = \frac{1.386}{s^2} \cdot (1+o(1)), \quad s \to \infty, \tag{15}$$

$$R\left(F_0^1, = s\right) \leq \overline{R}\left(F_0^1, = s\right) = \frac{4 \cdot \log_2 s}{s^2} \cdot (1 + o(1)), \quad s \to \infty. \tag{16}$$

Notice that $\underline{R}(s, 1)$ is defined in Theorem 6 and numerical values of $\underline{R}\left(F_0^1, = s\right)$, $s = 2, 3, \ldots, 8$, are given in Table 2. For the particular case $w = 1$, bound (14) is better than the lower bound (13) of Proposition 7.

In addition, applying the corresponding non-asymptotic results [11], one can calculate numerical values of upper bound (16), i.e., $\overline{R}\left(F_0^1, = s\right)$, $s \geq 1$, which lead to the inequality: $R\left(F_0^1, = s\right) < 1/s$ if $s \geq 11$. For $s = 2$, the nontrivial inequality $R\left(F_0^1, = 2\right) < 0.4998 < 1/2$ was proved in [5]. For $3 \leq s \leq 10$, the inequality $R\left(F_0^1, = s\right) < 1/s$ can be considered as our conjecture.

5.3 Lower Bound on the Rate of $\left(F_0^w, \leq s\right)$–Designs

For $(F_0^w, \leq s)$–designs, $w \geq 2$, the lower bound (12) of Proposition 7 can be improved [4]. An improvement is obtained with the help of the following auxiliary concepts.

Definition 5. *[4] Let w, $1 \leq w < s < t/2$ be integers. For code $X = \|x_i(k)\|$, $k \in [t]$, $i \in [N]$, and a subset $P \in \mathcal{P}(t, \leq s)$, define the i-th* outcome of non-adaptive threshold group testing

$$y_i^w\left(P, X\right) \triangleq \begin{cases} 0 \text{ if } \sum\limits_{k \in P} x_i(k) \leq w - 1, \\ 1 \text{ if } \sum\limits_{k \in P} x_i(k) \geq w, \quad i \in [N]. \end{cases}$$

A code X is called a threshold $(w, \leq s)$–design *of length N and size t if for any $P \neq P'$ with*

$$|P| \geq |P'| \quad P, P' \in \mathcal{P}(t, \leq s) \setminus \mathcal{P}(t, \leq w - 1)$$

there exists an index $i \in [N]$, where the i-th outcome of non-adaptive threshold group testing is

$$y_i^w\left(P, X\right) = 1 \quad \text{and} \quad y_i^w\left(P', X\right) = 0.$$

Let $t_w(N, \leq s)$ denote the maximal possible size of threshold $(w, \leq s)$–designs. For fixed $1 \leq w < s$, define the corresponding *rate*:

$$R_w\left(\leq s\right) \triangleq \overline{\lim_{N \to \infty}} \frac{\log_2 t_w(N, \leq s)}{N}.$$

Obviously, any threshold $(w, \leq s)$–designs is a $(F_0^w, \leq s)$–design and the rate

$$R\left(F_0^w, \leq s\right) \geq R_w\left(\leq s\right), \quad 1 \leq w < s. \tag{17}$$

Definition 6. *[4] Let w, $1 \le w < s < t/2$ be integers. A binary $(N \times t)$-matrix X is called a* superimposed $\mathcal{M}_s^w$-code *(briefly,* $\mathcal{M}_s^w$-code*) if for any two non-intersecting subsets $Z, U \in \mathcal{P}(t, \le s)$, $Z \cap U = \varnothing$, such that $w \le |U| \le s$, $|Z| \le |U|$ and for any element $j \in U$, the matrix X contains a row $\mathbf{x}_i = (x_i(1), x_i(2)\ldots, x_i(t))$, $i \in [N]$, for which*

$$x_i(j) = 1, \quad \sum_{k \in U} x_i(k) = w \quad \text{and} \quad x_i(k) = 0 \quad \text{for all} \quad k \in Z.$$

Let $t(N, \mathcal{M}_s^w)$ denote the maximal size of $\mathcal{M}_s^w$–codes. For fixed $1 \le w < s$, introduce

$$R(\mathcal{M}_s^w) \triangleq \overline{\lim_{N \to \infty}} \frac{\log_2 t(N, \mathcal{M}_s^w)}{N}, \qquad 1 \le w < s.$$

called a *rate* of $\mathcal{M}_s^w$–codes. The evident connection between $\mathcal{M}_s^w$–codes and superimposed $(2s - w, 1)$-codes is given by

Proposition 8. *[4]* **1.** *Let $2 \le s < t/2$. If $w = 1$, then any $\mathcal{M}_s^1$-code X of size t is a superimposed $(2s-1, 1)$-code and, vice versa, any superimposed $(2s-1, 1)$-code X of size t is a $\mathcal{M}_s^1$-code, i.e., the rate $R(\mathcal{M}_s^1) = R(2s-1, 1)$.* **2.** *If $2 \le w < s < t/2$, then any $\mathcal{M}_s^w$-code X of size t is a superimposed $(2s - w, 1)$-code, i.e., the rate $R(\mathcal{M}_s^w) \le R(2s - w, 1)$.*

Proposition 9. *[4] If $1 \le w < s < t/2$, then any $\mathcal{M}_s^w$-code X of size t is a threshold $(w, \le s)$-design, i.e. the rate $R(\mathcal{M}_s^w) \le R_w(\le s)$.*

Proof of Proposition 9. Let $X = \|x_i(k)\|$, $k \in [t]$, $i \in [N]$, be an arbitrary $\mathcal{M}_s^w$-code. Consider arbitrary subsets: $P, P' \in \mathcal{P}(t, \le s)$, $P \ne P'$, and such that

$$|P| \ge |P'|, \quad P, P' \in \mathcal{P}(t, \le s) \setminus \mathcal{P}(t, \le w-1), \quad w \le |P| \le s, \quad w \le |P'| \le |P|.$$

Fix an arbitrary $j \in P \setminus P'$, $j \notin P'$ and define non-intersecting subsets $U \triangleq P$ and $Z \triangleq P' \setminus P$. We have

$$w \le |U| \le s, \quad j \in U, \quad U \cap Z = \varnothing, \quad Z \subset P', \quad P' \setminus Z \subset U, \quad |Z| \le |P'| \le |P| = |U|.$$

Definition 6 of $\mathcal{M}_s^w$-code implies that there exists an index $i \in [N]$ such that

$$\left(\sum_{k \in U} x_i(k) = w, \sum_{k \in Z} x_i(k) = 0, \; x_i(j) = 1, \; \sum_{k \in P' \setminus Z} x_i(k) \le w - 1 \right) \Rightarrow$$

$$\Rightarrow \left(\sum_{k \in P} x_i(k) = w, \; \sum_{k \in P'} x_i(k) \le w - 1 \right) \Rightarrow \left(y_i(P, X) = 1, \; y_i(P', X) = 0 \right),$$

i.e., code X is a threshold $(w, \le s)$–design.

Proposition 9 is proved.

We say that a column is $\mathcal{M}_s^w$ – bad if the above property not hold. If $\beta \triangleq \Pr\{x_i(k) = 1\}$ and $1 - \beta \triangleq \Pr\{x_i(k) = 0\}$, then one can check that for any $j \in [t]$, the probability

$$\Pr\{\mathbf{x}(j) \text{ is } \mathcal{M}_s^w - \text{bad}\} \le \sum_{u=w}^{s} \sum_{z=0}^{u} \binom{t-1}{u+z-1} \binom{u+z-1}{u-1} \times$$

$$\times \left[1 - \binom{u-1}{w-1} \beta^w (1-\beta)^{u+z-w}\right]^N.$$

The given inequality leads to the following *random coding lower bound* on the rate of $\mathcal{M}_s^w$-codes:

Proposition 10. *For any β, $0 < \beta < 1$, the rate $R(\mathcal{M}_s^w)$ satisfies inequality*

$$R(\mathcal{M}_s^w) \ge \min_{w \le u \le s;\ 0 \le z \le u} \left\{ \frac{-\log_2 \left[1 - \binom{u-1}{w-1} \beta^w (1-\beta)^{u+z-w}\right]}{u+z-1} \right\} \ge \min_{w \le u \le s} L_w(\beta, u),$$

where

$$L_w(\beta, u) \triangleq \left\{ \frac{-\log_2 \left[1 - \binom{u-1}{w-1} \beta^w (1-\beta)^{2u-w}\right]}{2u-1} \right\}, \quad w \le u \le s, \quad 0 < \beta < 1. \tag{18}$$

From (17) and Propositions 9-10 it follows a lower bound on the rate of $(F_0^w, \le s)$-designs :

$$R(F_0^w, \le s) \ge \underline{R}(F_0^w, \le s) \triangleq \max_{0<\beta<1} \min_{w \le u \le s} L_w(\beta, u) =$$

$$= \max_{0<\beta<1} \min_{w \le u \le s} \left\{ \frac{-\log_2 \left[1 - \binom{u-1}{w-1} \beta^w (1-\beta)^{2u-w}\right]}{2u-1} \right\}, \quad 1 \le w < s. \tag{19}$$

The calculation of numerical values for lower bound (19) is an open problem.

Theorem 8. *For fixed $w = 2, \ldots$ and $s \to \infty$, the lower bound $\underline{R}(F_0^w, \le s)$ defined by (18) – (19) has the form :*

$$\underline{R}(F_0^w, \le s) = \frac{1}{s^2} \cdot \frac{3(w-2)^w (\ln 2)^w \log_2 e}{(w-1)! 2^{2w-1}} \cdot (1 + o(1)). \tag{20}$$

Proof of Theorem 8
Consider the following partition of the interval $w \le u \le s$:

$$\frac{s}{2^m} < u \le \frac{s}{2^{m-1}},$$

where $m = 1, 2, \ldots, \log_2 s/w$.

Define a random code as follows: for each m we consider N_m by t submatrix with probabilities

$\Pr\{x_i(k) = 1\} \triangleq \beta_m$ and $\Pr\{x_i(k) = 0\} \triangleq 1 - \beta_m$.

For fixed m, define the corresponding rate:

$$R_m \triangleq \overline{\lim_{N\to\infty}} \frac{\log_2 t_w(N_m, \leq s)}{N_m}.$$

We have $N = \sum N_m$ so $R^{-1} = \sum R_m^{-1}$.

Let us now $w \geq 2$ is fixed and $s \to \infty$.

Consider $u = xz$ where $1/2 \leq x \leq 1$ and take $\beta = \frac{w}{\gamma z}$.

We want to find the minimum of

$$\frac{x^{w-1} z^{w-1} (w)^w \log_2 e}{2xz(w-1)!\gamma^w z^w e^{2xw/\gamma}}.$$

The function $y(x) = \frac{x^{w-2}}{e^{2wx/\gamma}}$ has maximum at $x = \frac{(w-2)\gamma}{2w}$ and $y(1) = e^{-\frac{2w}{\gamma}}$, $y(1/2) = e^{-\frac{w}{\gamma}} 2^{2-w}$. So for $\gamma = \frac{w}{(w-2)\ln 2}$ it gives minimum and $min_{1/2 \leq x \leq 1} y(x) = y(1) = y(1/2) = \frac{1}{2^{2w-4}}$.

From this follows that for the parameters $\beta_m = \frac{(w-2)\cdot 2^{m-1} \ln 2}{s}$ for $m = 1, 2, \ldots \log(s/w)$

we have

$$R_m = \frac{2^{2m-2}}{s^2} \cdot \frac{(w-2)^w (\ln 2)^w \log_2 e}{(w-1)! 2^{2w-3}} \cdot (1 + o(1));$$

So the sum $\sum R_m^{-1}$ is equal to

$$\frac{(w-1)! 2^{2w-3}}{(w-2)^w (\ln 2)^w \log_2 e} \sum_{m=1}^{M} \frac{s^2}{2^{2m-2}} = s^2 \frac{4(w-1)! 2^{2w-3}}{3(w-2)^w (\ln 2)^w \log_2 e} \cdot (1 + o(1)).$$

Thus we have

$$R = \frac{1}{s^2} \cdot \frac{3(w-2)^w (\ln 2)^w \log_2 e}{(w-1)! 2^{2w-1}} \cdot (1 + o(1)).$$

Theorem 8 is proved.

5.4 Another Threshold Design Model

Theorem 8 improves the asymptotic bound from Proposition 7. In this section we slightly change the definition of a threshold design and show that the results are different.

Let w, $1 \leq w < s < t/2$, be integers. For a comparison of Definitions 4 and 5 , introduce

Definition $\widetilde{5}$. A code X is called a *threshold* $\overline{(w, \le s)}$*-design*, of length N and size t if for any $P \ne P'$ with

$$P \setminus P' \ne \varnothing, \quad P, P' \in \mathcal{P}(t, \le s) \setminus \mathcal{P}(t, \le w-1),$$

there exists an index $i \in [N]$, where the i-th outcome of non-adaptive threshold group testing is

$$y_i^w(P, X) = 1 \quad \text{and} \quad y_i^w(P', X) = 0.$$

Let $\widetilde{t}_w(N, \le s)$, be the maximal size of threshold $\overline{(w, \le s)}$–designs. For fixed $1 \le w < s$, define the corresponding *rate*

$$\widetilde{R}_w(\le s) \triangleq \overline{\lim_{N \to \infty}} \frac{\log_2 \widetilde{t}_w(N, \le s)}{N}.$$

The following important property is given by

Proposition 11. *If* $1 \le w < s < t/2$, *then* **(1)** *any superimposed* $(s-w+1, w)$*–code* X *of size* t *is a threshold* $\overline{(w, \le s)}$*–design and, vice versa,* **(2)** *any threshold* $\overline{(w, \le s)}$*–design* X *of size* t *is a superimposed* $(s-w+1, w)$*–code, i.e., the rate* $\widetilde{R}_w(\le s) = R(s-w+1, w)$.

Evidently, any threshold $\overline{(w, \le s)}$–design is a threshold $(w, \le s)$–design. Therefore, in virtue of Proposition 11, the rate

$$\widetilde{R}_w(\le s) = R(s-w+1, w) \le R_w(\le s) \le R(F_0^w, \le s).$$

Denote by $\underline{R}(z, u)$, $1 \le u \le z$, the lower bound on $R(z, u)$ formulated in Theorems 5 and 6. Let $\overline{R}(z, u)$ be the upper bound on $R(z, u)$ given by Theorem 3. For parameters $w = 1, 2, 3$ and $s = w+1, w+2, \ldots, 8$, numerical values of lower bound $\underline{\widetilde{R}}_w(\le s) \triangleq \underline{R}(s-w+1, w)$ and upper bound $\overline{\widetilde{R}}_w(\le s) \triangleq \overline{R}(s-w+1, w)$ on the rate $\widetilde{R}_w(\le s) = R(s-w+1, w)$ are presented in Table 2.

Proof of Proposition 11. **(1)** Let $X = \|x_i(k)\|$, $k \in [t]$, $i \in [N]$, be a superimposed $(s-w+1, w)$–code. Consider arbitrary subsets: $P, P' \in \mathcal{P}(t, \le s)$, $P \ne P'$, and such that

$$P \setminus P' \ne \varnothing, \quad P, P' \in \mathcal{P}(t, \le s) \setminus \mathcal{P}(t, \le w-1), \quad w \le |P| \le s, \quad w \le |P'| \le s.$$

Fix an arbitrary subset $U \subset P$ such that $|U| = w$, and $U \setminus P' \ne \varnothing$. Note that the size of intersection $|P' \cap U| \le w-1$.

Consider the set $P' \setminus (P' \cap U)$. Introduce a set Z, $Z \subset [t]$, of size $|Z| = s-(w-1)$, where the intersection $Z \cap U = \varnothing$, as follows.

1. If $|P' \setminus (P' \cap U)| \ge s-(w-1)$, then we choose the set Z, $Z \subseteq P' \setminus (P' \cap U)$, $Z \cap U = \varnothing$, as an arbitrary fixed subset of size $|Z| = s-(w-1)$. Let a

row i, $i \in [N]$ corresponds to the pair (U, Z) in Definition 1 of superimposed $(s-w+1, w)$-code X. One can easily see that

$$\sum_{k\in P} x_i(k) \ge \sum_{k\in U} x_i(k) = w, \; \sum_{k\in P'} x_i(k) \le |P'|-|Z| \le s-[s-(w-1)] = w-1.$$

Hence, $(\, y_i(P, X) = 1, \; y_i(P', X) = 0\,)$.

2. If $|P' \setminus (P' \cap U)| < s-(w-1)$, then we choose the set Z, $Z \supset P' \setminus (P' \cap U)$, as an arbitrary fixed superset of size $|Z| = s-(w-1)$. Let a row i, $i \in [N]$ corresponds to the pair (U, Z) in Definition 1 of superimposed $(s-w+1, w)$-code X. One can easily see that

$$\sum_{k\in P} x_i(k) \ge \sum_{k\in U} x_i(k) = w, \; \sum_{k\in P'} x_i(k) \; = |P' \cap U| \le w-1.$$

Hence, $(\, y_i(P, X) = 1, \; y_i(P', X) = 0\,)$.

Arguments 1. and 2. imply that code X is a threshold $\overline{(w, \le s)}$–design. Therefore, the statement **(1)** of Proposition 11 is proved.

(2) Let $X = \|x_i(k)\|$, $k \in [t]$, $i \in [N]$, be a threshold $\overline{(w, \le s)}$–design. Consider two arbitrary non-intersecting sets U and Z, where

$$U \subset [t], \quad |U| = w, \quad Z \subset [t], \quad |Z| = s-(w-1), \quad U \cap Z = \varnothing,$$

and fix an element $j \in U$. Introduce subsets $P, P' \in \mathcal{P}(t, \le s) \setminus \mathcal{P}(t, \le w-1)$ as follows:

$$P \triangleq U, \quad P' \triangleq (U \backslash j) \cup Z, \quad P \backslash P' \ne \varnothing, \quad |P| = w, \quad |P'| = (w-1)+s-(w-1) = s.$$

Definition $\tilde{\mathbf{5}}$ of threshold $\overline{(w, \le s)}$–design means that there exists an index $i \in [N]$ such that

$$(\, y_i(P, X) = 1, \; y_i(P', X) = 0\,) \Rightarrow \left(\sum_{k\in P} x_i(k) \ge w, \; \sum_{k\in P'} x_i(k) \le w-1 \right) \Rightarrow$$

$$\Rightarrow \left(\sum_{k\in U} x_i(k) \ge w, \; \sum_{k\in U\backslash j} x_i(k) + \sum_{k\in Z} x_i(k) \le w-1 \right) \Rightarrow$$

$$\Rightarrow \quad x_i(k) = 1, \; k \in U, \; |U| = w; \qquad x_i(k) = 0, \; k \in Z, \; |Z| = s-(w-1).$$

Hence, code X is a superimposed $(s-w+1, w)$-code, i.e., statement **(2)** is established.

Proposition 11 is proved.

5.5 Improved Bounds for $(F^w, = s)$–Designs with Gap

In threshold group testing [7], numbers $0 \leq l < s$ are given and a test t_A is the function $t_A : 2^{[t]} \to \{0, 1, \{0, 1\}\}$, defined by

$$t_A(P) = \begin{cases} 0 & , \text{ if } |P \cap A| \leq l \\ 1 & , \text{ if } |P \cap A| \geq w \\ \{0,1\} & , \text{ otherwise} \\ & \text{(meaning that the result can be arbitrary 0 or1).} \end{cases}$$

It is not possible to find the set P exactly if the gap $g \triangleq w - l - 1 > 0$. We can find a set P' such that $|P \backslash P'| \leq g$.

In this Section we present the results of [1] where the authors considered the case $g > 0$ provided that the number of defectives is known, i.e., $|P| = s$.

Definition 7. *Let l, w, $0 \leq l < w < s < t$ be integers. For code $X = \|x_i(k)\|$, $k \in [t]$, $i \in [N]$, and a subset $P \in \mathcal{P}(t, \leq s)$, define the i-th* outcome of non-adaptive threshold group testing

$$y_i^w(P, X) \triangleq \begin{cases} 0 \text{ if } \sum\limits_{k \in P} x_i(k) \leq l, \\ 1 \text{ if } \sum\limits_{k \in P} x_i(k) \geq w, \quad i \in [N]. \end{cases}$$

Let $g = w - l + 1$, a code X is called a $(F_g^w, \leq s)$-design, *($(F_g^w, = s)$*-design*) of length N and size t if for any $P, P' \in \mathcal{P}(t, = s)$ with $|P \backslash P'| > g$, there exists an index $i \in [N]$, where $y_i^w(P, X) \neq y_i^w(P', X)$.*

Definition 8. *Let z and u be positive integers such that $z + u \leq t$. A family of subsets $\Omega_1, \Omega_2, \ldots, \Omega_t$, where $\Omega_j \subseteq [N]$, $j \in [t]$, is called an (z, u)*–separating family *if for any two non-intersecting subsets Z, $U \subset [t]$, $Z \cap U = \varnothing$, such that $|Z| = z$, $|U| = u$, the following condition holds:*

$$\bigcap_{j \in U} \Omega_j \not\subseteq \bigcup_{j \in Z} \Omega_j \quad \text{or} \quad \bigcap_{j \in Z} \Omega_j \not\subseteq \bigcup_{j \in U} \Omega_j$$

An incidence matrix $X = \|x_i(j)\|$, $i \in [N]$, $j \in [t]$, corresponding to (z, u)-separating family is called a separating (z, u)-code.

The following evident necessary and sufficient condition for Definition 8 takes place.

Remark 4. *Any binary $(N \times t)$-matrix X is a separating (z, u)-code if and only if for any two subsets Z, $U \subset [t]$, such that $|Z| = z$, $|U| = u$ and $Z \cap U = \varnothing$ the matrix X contains a row $\mathbf{x}_i = (x_i(1), x_i(2) \ldots, x_i(t)$, for which*

$$(x_i(j) = 1 \quad \forall j \in U, \; x_i(j) = 0 \quad \forall j \in Z) \;\text{ or }\; (x_i(j) = 0 \quad \forall j \in U, \; x_i(j) = 1 \quad \forall j \in Z)$$

Let $\hat{t}(N, z, u)$ be the maximal possible size of separating (z, u)-codes. For fixed $1 \leq u < z$, define a *rate* of separating (z, u)-codes:

$$\hat{R}(z, u) \triangleq \overline{\lim_{N \to \infty}} \frac{\log_2 \hat{t}(N, z, u)}{N}.$$

Proposition 12. *([1]) If $s = w + l$, then any separating $(s - l, w)$-code is a $(F_g^w, = s)$-design, i.e.*

$$t(N, s-l, w) \le t(N, F_0^w, = s), \qquad R(s-w+1, w) \le R(F_0^w, = s).$$

In [1] the authors also consider the case $s \neq w + l$. They introduce threshold codes.

Definition 9. *We call an $N \times t$ matrix $M = \|m_{ij}\|$ a (p, w, l)-threshold code, if for all $A, B \subset [t]$, $|A| = |B| = p$, and $|A \backslash B| \ge w - l$ there exists an $i \in [N]$ such that*

$$\left(\sum_{a \in A} m_{ia} \ge w \text{ and } \sum_{b \in B} m_{ib} \le l\right)$$
$$\text{or}$$
$$\left(\sum_{a \in A} m_{ia} \le l \text{ and } \sum_{b \in B} m_{ib} \ge w\right).$$

Proposition 13. *([1]) If $1 \le w < s$, then any (p, w, l)–threshold code is a $(F_g^w, = s)$-design*

6 Concluding Remarks

In this Section, we would like to distinguish the principal achievements for the theory of non-adaptive group testing models and superimposed codes obtained in the last decade.

1. In 2003, Vladimir Lebedev [31] proved Theorem 3 which established a recurrent inequality for the rate $R(z, u)$ of superimposed (z, u)-codes. This inequality and the best known numerical values [17,13] of upper bound on the rate $R(z, 1)$ gave the best known numerical values of upper bound on the rate $R(z, u)$, $z \ge u \ge 2$.
2. In 2004, Vladimir Lebedev and Hyun Kim [30] presented the best known and optimal constructions (see, Table 4) of superimposed (z, u)-codes, $z \ge u \ge 2$.
3. In 2004, Annalisa De Bonis and Ugo Vaccaro [8] proved Theorem 7 which established an upper bound on the rate of superimposed $\mathcal{D}_s^w$-codes via the rate $R(z, 1)$ of superimposed $(z, 1)$-codes. The result leads to the universal upper bound (Proposition 4) on the rate of group testing designs motivated by compressed genotyping models in molecular biology.
4. In 2010, Mahdi Cheraghchi [4] introduced the concepts of threshold $(w, \le s)$–designs and superimposed $\mathcal{M}_s^w$-codes and proved Proposition 9 which actually established an improved lower bound (19) on the rate of non-adaptive threshold group testing model.

Acknowledgements. All authors are grateful to Professor Rudolf Ahlswede for his lifetime friendship and encouragement. He wrote in 1979 the first book on search theory in German ([2]), describing the connection between several areas. The extensive literature is presented in such a way that the reader can quickly understand the range of questions and obtain a survey of them which is as comprehensive as possible. In 1982 the Russian edition was published by MIR. It includes also a supplement, *Information-theory Methods in Search Problems*, which was written by Maljutov. The English edition, published in 1987, includes also a section "Further reading", where articles and books are mentioned which inform the researcher about new developments and results which seem to carry the seed for further discoveries.

References

1. Ahlswede, R., Deppe, C., Lebedev, V.: Bounds for threshold and majority group testing. In: 2011 IEEE International Symposium on Information Theory, Sankt-Peterburg, pp. 69–73 (2011)
2. Ahlswede, R., Wegener, I.: Suchprobleme, Teubner (1979), MIR russ. edition (1981), Wiley engl. edition (1987)
3. Chen, H.B., Fu, H.L.: Nonadaptive algorithms for threshold group testing. Discrete Applied Mathematics 157, 1581–1585 (2009)
4. Cheraghchi, M.: Improved constructions for non-adaptive threshold group testing. In: Proceedings of the 37th International Colloquium on Automata, Languages and Programming (ICALP), arXiv:1002.2244 (2010)
5. Coppperersmith, D., Shearer, J.: New bounds for union-free families of sets. The Electronic Journal of Combinatorics 5(1), R39 (1998)
6. Csizár, I., Körner, J.: Information Theory: Coding Theorems for Discrete Memoryless Systems. Academiai Kiado, Budapest (1981)
7. Damaschke, P.: Threshold group testing. In: General Theory of Information Transfer and Combinatorics, pp. 707–718. Kluwer Academic Publishers (2006)
8. De Bonis, A., Vaccaro, U.: Optimal algorithms for group testing problems, and new bounds on generalized superimposed codes. IEEE Trans. Inf. Theory 52(10), 4673–4680 (2006)
9. Dorfman, R.: The detection of defective members of large populations. The Annals of Mathematical Statistics 14(4), 436–440 (1943)
10. D'yachkov, A.G.: Superimposed designs and codes for non-adaptive search of mutually obscuring defectives. In: 2003 IEEE International Symposium on Information Theory, Yokohama, p. 134 (2003)
11. D'yachkov, A.G.: Lectures on Designing Screening Experiments. Lecture Note Series 10, Monograph, p. 112, Combinatorial and Computational Mathematics Center, Pohang University of Science and Technology (POSTECH) (February 2004)
12. D'yachkov, A.G., Macula, A.J., Torney, D.C., Vilenkin, P.A.: Two models of non-adaptive group testing for designing screening experiments. In: Advances in Model Oriented Design and Analysis: Proceedings of the 6th International Workshop on Model Oriented Design and Analysis, Puchberg/Schneeberg, Austria, June 25-29, pp. 63–75. Physica-Verlag, Heidelberg (2001)
13. D'yachkov, A., Macula, A., Torney, D., Vilenkin, P.: Families of finite sets in which no intersection of l sets is covered by the union of s others. Journal of Combinatorial Theory, Series A 99, 195–218 (2002)
14. D'yachkov, A.G., Macula, A.J., Rykov, V.V.: New applications and results of superimposed code theory arising from the potentialities of molecular biology. In: Numbers, Information and Complexity, pp. 265–282. Kluwer Academic Publishers (2000)

15. D'yachkov, A.G., Macula, A.J., Rykov, V.V.: New constructions of superimposed codes. IEEE Trans. Inf. Theory 46(1), 284–290 (2000)
16. D'yachkov, A.G., Rashad, A.M.: Universal decoding for random design of screening experiments. Microelectronics and Reliability 29(6), 965–971 (1989)
17. D'yachkov, A.G., Rykov, V.V.: Bounds on the length of disjunct codes. Problemy Peredachi Informatsii 18(3), 7–13 (1982) (in Russian)
18. D'yachkov, A.G., Rykov, V.V.: A survey of superimposed code theory. Problems of Control and Inf. Theory 12(4), 229–242 (1983)
19. D'yachkov, A.G., Rykov, V.V.: Generalized superimposed codes and their application to random multiple access. In: Proc. of the 6th International Symposium on Information Theory, Part 1, Taschkent (1984)
20. D'yachkov, A.G., Rykov, V.V.: On a model of associative memory. Problemy Peredachi Inform 24(3), 107–110 (1988) (in Russian)
21. D'yachkov, A.G., Rykov, V.V., Rashad, A.M.: Superimposed distance codes. Problems of Control and Inform. Theory 18(4), 237–250 (1989)
22. D'yachkov, A.G., Rykov, V.V.: The capacity of the boolean associative memory. In: Proc. of the 5th International Conference on Artificial Neural Networks, Churchill Colledge, Cambridge, UK, pp. 158–160 (1997)
23. D'yachkov, A.G., Rykov, V.V.: Optimal superimposed codes and designs for Rényi's search model. Journal of Statistical Planning and Inference 100, 281–302 (2002)
24. D'yachkov, A.G., Vilenkin, P.A., Yekhanin, S.M.: Upper bounds on the rate of superimposed (s, ℓ)-codes, based on Engel's inequality. In: Proceedings of the 8th International Workshop Algebraic and Combinatorial Coding Theory, Tsarskoe Selo, Russia, pp. 95–99 (2002)
25. Du, D.-Z., Hwang, F.K.: Combinatorial group testing and its applications. World Scientific, Singapore (1993)
26. Emad, A., Milenkovic, O.: Semi-quantitative group testing. Arxiv-1202.2887 (2011)
27. Erlich, Y., Gordon, A., Brand, M., Hannon, G., Mitra, P.: Compressed genotyping. IEEE Trans. Inf. Theory 56(2), 706–723 (2010)
28. Kautz, W.H., Singleton, R.C.: Nonrandom Binary Superimposed Codes. IEEE Trans. Inf. Theory 10(4), 363–377 (1964)
29. Kim, H., Lebedev, V.S.: On the optimality of trivial (w, r) cover-free codes. Probl. Inf. Transm. 40(3), 195–201 (2004)
30. Kim, H., Lebedev, V.S.: On optimal superimposed codes. Journal of Combinatorial Designs 12(2), 79–91 (2004)
31. Lebedev, V.S.: An asymptotic upper bound on the rate of (w, r)-cover-free codes. Probl. Inf. Transm. 39(4), 317–323 (2003)
32. Lebedev, V.S.: Some tables for (w, r) superimposed codes. In: Proceedings of the 8th International Workshop, Algebraic and Combinatorial Coding Theory, Tsarskoe Selo, Russia, pp. 185–189 (2002)
33. Lebedev, V.S.: Separating codes and a new combinatorial search model. Probl. Inf. Transm. 46(1), 1–6 (2010)
34. MacWilliams, F.J., Sloane, N.J.A.: The theory of error-correcting codes. North Holland (1977)
35. Maljutov, M.B.: On planning screening experiments. In: Proceedings of the IEEE-USSR Joint Workshop on Information Theory, pp. 144–147. Inst. Electr. Electron. Engrs., New York (1976)
36. Maljutov, M.B., Mateev, P.S.: The design of screening experiments with a nonsymmetric response function. Dokl. Akad. Nauk SSSR 244(1), 42–46 (1979)

37. Maljutov, M.B., Mateev, P.S.: Design of screening experiments with a nonsymmetric response function. Mat. Zametki 27(1), 109–127 (1980)
38. Nguyen Quang, A., Zeisel, T.: Bounds on constant weight binary syperimposed codes. Probl. of Control and Inform. Theory 17(4), 223–230 (1988)
39. Rashad, A.M.: Random coding bounds on the rate for list-decoding superimposed codes. Problems of Control and Inform. Theory 19(2), 141–149 (1990)
40. Rényi, A.: On a problem of information theory. MTA Mat. Kut. Int. Kozl., 6B, 505–516 (1961)
41. Shannon, C.E.: A mathematical theory of communication. Bell System Technical Journal 27, 379–423 & 623–656 (1948)

New Construction of Error-Tolerant Pooling Designs

Rudolf Ahlswede* and Harout Aydinian**

Department of Mathematics, University of Bielefeld,
POB 100131, D-33501 Bielefeld, Germany
ayd@math.uni-bielefeld.de

Abstract. In this paper a new class of error-tolerant pooling designs associated with finite vector spaces is presented. We construct d^z-disjunct inclusion matrices using packings in finite projective spaces. For certain parameters our construction gives better performance than previously known ones. In particular, the construction gives a family of disjunct matrices with near optimal parameters.

Keywords: group testing, nonadaptive algorithm, pooling designs, d^z–disjunct matrix.

1 Introduction

Combinatorial group testing has various practical applications [4], [5]. In the classical group testing model we have a set $[n] = \{1, \ldots, n\}$ of n items containing at most d defective items. The basic problem of group testing is to identify the set of all defective items with a small number of group tests. Each *group test*, also called a *pool*, is a subset of items. It is assumed that there is a testing mechanism that for each subset $A \subset [n]$ gives one of two possible outcomes : *negative* or *positive*. The outcome is positive if A contains at least one defective and is negative otherwise.

A group testing algorithm is called *nonadaptive* if all tests are specified without knowledge of the outcomes of other tests. Traditionally, a nonadaptive group testing algorithm is called a *pooling design*. Pooling designs have many applications in molecular biology, such as DNA screening, nonunique probe selection, gene detection, etc. (see [1], [5], [7], [8]).

A pooling design is associated with a $(0,1)-$ inclusion matrix $M = (m_{ij})$, where the rows are indexed by tests $A_1, \ldots, A_t \subset [n]$, the columns are indexed by items $1, \ldots, n$, and $m_{ij} = 1$ if and only if $j \in A_i$. The major tool used for construction of pooling designs are $d-$*disjunct* matrices. Let M be a binary $t \times n$ matrix where the columns $C_1, \ldots, C_n$ are viewed as subsets of $[t] = \{1, \ldots, t\}$ represented by their characteristic vectors. Then M is called d–disjunct if no

* Rudolf Ahlswede, one of the world's top information theorists at the University of Bielefeld, died December 18, 2010.

** This author was supported in part by the DFG Project AH46/8-1.

H. Aydinian, F. Cicalese, and C. Deppe (Eds.): Ahlswede Festschrift, LNCS 7777, pp. 534–542, 2013.

column is contained in the union of d others. The notion of d–disjunctness was introduced by Kautz and Singleton [15]. They proved that a d-disjunct matrix M can identify up to d defective items. d–disjunct matrices are also known as d–cover free families studied in extremal set theory [3]. The maximal d for which M is d-disjunct is called the degree of disjunctness and is denoted by d_{max}. Note that d-disjunctness of a pooling design is a sufficient, but not a necessary condition for identification of d defectives. However a d–disjunct pooling design has an advantage of a very simple decoding. Removing from the set of items all items in negative pools we get all defectives (see [5] for details).

A pooling design is called *error-tolerant* if it can detect/correct some errors in test outcomes. Biological experiments are known to be unreliable (see [5]), which, in fact, is a practical motivation for constructing efficient error-tolerant pooling designs. For error correction in tests the notion of a *d^z–disjunct matrix* was introduced in [1], [18]. A d–disjunct matrix is called *d^z–disjunct* if for any $d+1$ of its columns $C_{i_1}, \ldots, C_{i_{d+1}}$ we have $|C_{i_1} \setminus (C_{i_2} \cup \ldots \cup C_{i_{d+1}})| \geq z$. In fact, the d^1–disjunctness is simply the d–disjunctness. A d^z–disjunct matrix can detect $z-1$ errors and correct $\lfloor \frac{z-1}{2} \rfloor$ errors (see e.g. [8] or [5]). Constructions of d^z–disjunct matrices are given by many authors (see [1], [7], [8], [10], [18], [19]).

Most known constructions of d^z–disjunct matrices are matrices with a constant column weight. Let M be a binary $t \times n$ matrix with a constant column weight k and let s be the maximum size of intersection (number of common ones) between two different columns. Kautz and Singleton [15] observed that then M is d-disjunct with $d = \lfloor \frac{k-1}{s} \rfloor$. Moreover, for integers $0 \leq s < k < t$ the maximum number $n(d,t,k)$ for which there exists such a disjunct matrix is upper bounded by

$$n(d,t,k) \leq \binom{t}{s+1} / \binom{k}{s+1}. \tag{1}$$

Note that the columns of M considered as the family $\mathcal{F}$ of k–subsets of $[t]$ (called blocks) form an $(s+1,k,t)$–packing, that is each $(s+1)$–subset of $[t]$ is contained in at most one block of $\mathcal{F}$. Note also that equality in (1.1) is attained if and only if $\mathcal{F}$ is an $(s+1,k,t)$–Steiner system (each $(s+1)$–subset is contained in precisely one block).

Thus, packing designs can be used for construction of d–disjunct matrices. However, not much is known about explicit constructions of $(s+1,k,t)$–packings in general (even for $s \geq 3$). Several other constructions (see [9, Ch.3]) of disjunct matrices are also based on combinatorial structures or error correcting codes. We note that $(s+1,k,t)$–packings can also be described in terms of error-correcting codes in the Johnson graph $J(n,k)$ (also called constant weight codes) with minimum distance $d_J = k - s$. It seems natural to try other distance regular graphs, for construction of d–disjunct matrices, using the idea of packings.

In this paper we construct new error-tolerant pooling designs associated with finite vector spaces. In Section 2 we briefly review some known constructions of disjunct matrices based on partial orders and determine the degree of disjunctness for the construction proposed by Ngo and Du [19]. Our main results are stated and proved in Section 3. We present a construction of d^z–disjunct matrices based on packings in finite projective spaces. For certain parameters the construction gives better performance than previously known ones.

2 d^z–Disjunct Matrices from Partial Orders

In [17] Macula proposed a simple direct construction of d–disjunct matrices based on containment relation in finite sets. Given integers $1 \le d < k < m$, let $M = (m_{ij})$ be an $\binom{m}{d} \times \binom{m}{k}$ matrix where the rows are indexed by elements of $\binom{[m]}{d}$, the columns are indexed by the elements of $\binom{[m]}{k}$, and $m_{ij} = 1$ if we have containment relation between the subsets corresponding to the ith row and the jth column, otherwise $m_{ij} = 0$. Note that each column has weight $\binom{k}{d}$ and each row has weight $\binom{m-d}{k-d}$. Macula showed that M is a d-disjunct matrix and $d_{max} = d$.

Similar constructions, using different posets, were given by several authors (see [5] Ch.4). Ngo and Du [19] extended Macula's construction to some geometric structures. In particular they considered the following construction of a d–disjunct matrix $M_q(m, d, k)$ associated with finite vector spaces. Let $GF(q)^m$ be the m–dimensional vector space over $GF(q)$. The set of all subspaces of $GF(q)^m$, called projective space, is denoted by $\mathcal{P}_q(m)$. Recall that $\mathcal{P}_q(m)$ ordered by containment is known as the poset of linear spaces (or linear lattice). Given an integer $0 \le k \le m$, the set of all k-dimensional subspaces (k–spaces for short) of $GF(q)^m$ is called a *Grassmannian* and denoted by $\mathrm{G}_q(m, k)$. Thus, we have $\bigcup_{0 \le k \le m} \mathrm{G}_q(m, k) = \mathcal{P}_q(m)$. A graph associated with $\mathrm{G}_q(m, k)$ is called the *Grassmann graph*, when two vertices (elements of $\mathrm{G}_q(m, k)$) V and U are adjacent iff $\dim(V \cap U) = k - 1$. It is known that the size of the Grassmannian $|\mathrm{G}_q(m, k)|$ is determined by the q-ary Gaussian coefficient $\begin{bmatrix} m \\ k \end{bmatrix}_q$; $k = 0, 1, \dots, m$ ($\begin{bmatrix} m \\ 0 \end{bmatrix}_q := 1$),

$$|\mathrm{G}_q(m, k)| = \begin{bmatrix} m \\ k \end{bmatrix}_q = \frac{(q^m - 1)(q^{m-1} - 1) \cdots (q^{m-k+1} - 1)}{(q^k - 1)(q^{k-1} - 1) \cdots (q - 1)}. \tag{2}$$

For integers $1 \le r < k < m$, the $\begin{bmatrix} m \\ r \end{bmatrix}_q \times \begin{bmatrix} m \\ k \end{bmatrix}_q$ incidence matrix $M_q(m, r, k) = (m_{ij})$ is defined as follows. The rows and the columns are indexed by the elements of $\mathrm{G}_q(m, r)$ and $\mathrm{G}_q(m, k)$ (given in a fixed ordering), respectively, and $m_{ij} = 1$ if we have containment relation, otherwise $m_{ij} = 0$. Note that each column of $M_q(m, r, k)$ has weight $\begin{bmatrix} k \\ r \end{bmatrix}_q$ and each row has weight $\begin{bmatrix} m-r \\ k-r \end{bmatrix}_q$. Ngo and Du [19] showed that $M_q(m, r, k)$ is an r–disjunct matrix. However D'yachkov et al. [8] observed that the degree of disjunctness of $M_q(m, r, k)$ can be much bigger than r. Moreover, the construction can in general tolerate many errors.

Theorem DHMVW [8]
For $k - r \ge 2$ and $d < \frac{q(q^{k-1}-1)}{q^{k-r}-1}$, the matrix $M_q(m, r, k)$ is d^z–disjunct with

$$z \ge \begin{bmatrix} k \\ r \end{bmatrix}_q - d \begin{bmatrix} k-1 \\ r \end{bmatrix}_q + (d-1) \begin{bmatrix} k-2 \\ r \end{bmatrix}_q. \tag{3}$$

The bound is tight for $d \le q + 1$.

Note that the maximum number d in (3) for which $z > 0$ is $d = \frac{q(q^{k-1}-1)}{q^{k-r}-1}$. Thus, the theorem tells us that $d_{max} \geq \frac{q(q^{k-1}-1)}{q^{k-r}-1}$. In fact, we determine d_{max} for every $M_q(m,r,k)$.

Theorem 1. *For integers $1 \leq r < k < m$, the degree of disjunctness of $M_q(m,r,k)$ equals*

$$d_{max} = \frac{q(q^r-1)}{q-1}. \tag{4}$$

Proof. Let $V \in \mathrm{G}_q(m,k)$. We wish to determine the minimum size of a set of k–spaces which cover (contain) all r–spaces of V. Suppose $U_1, \ldots, U_p \in \mathrm{G}_q(m,k)$ is a minimal covering of the r–spaces of V. Without loss of generality, we may assume that $\dim(U_i \cap V) = k-1$ for $i = 1, \ldots, p$. Therefore, $W_1 = U_1 \cap V, \ldots, W_p = U_p \cap V$ can be viewed as a set of hyperplanes of $\mathcal{P}_q(k)$ that cover all r–spaces of $\mathcal{P}_q(k)$. Let now $A_i \in \mathcal{P}_q(k)$ be the orthogonal space of W_i; $i = 1, \ldots, p$. Thus, $\mathcal{A} = \{A_1, \ldots, A_p\}$ is a set of one dimensional subspaces, that is points, in $\mathcal{P}_q(k)$. By the principle of duality, every $(k-r)$–space of $\mathcal{P}_q(k)$ contains an element of $\mathcal{A}$. To complete the proof we use the following result.

Theorem BB [2]. *Let $\mathcal{A} \subset GF(q)^m \setminus \{0\}$ have a non-empty intersection with every $(k-r)$–space of $\mathcal{P}_q(k)$. Then $|\mathcal{A}| \geq (q^{r+1}-1)/(q-1)$, with equality if and only if $\mathcal{A}$ consists of $(q^{r+1}-1)/(q-1)$ points of an $(r+1)$–space of $\mathcal{P}_q(k)$.*

It is clear now that $d_{max} = (q^{r+1}-1)/(q-1) - 1$. □

3 The Construction

Our construction of a disjunct matrix M is based on packings in $\mathcal{P}_q(m)$. For integers $0 \leq s < k < m$, a subset $\mathcal{C} \subset \mathrm{G}(m,k)$ (with the elements called blocks) is called an $[s+1,k,m]_q$–packing if each $(s+1)$–space of $\mathcal{P}_q(m)$ is contained in at most one block of $\mathcal{C}$. This clearly means that $\dim(V \cap U) \leq s$ for every distinct pair $V, U \in \mathcal{C}$. $\mathcal{C}$ is called an $[s+1,k,m]_q$–*Steiner structure* if each $(s+1)$–space of $\mathcal{P}_q(m)$ is contained in precisely one block of $\mathcal{C}$. Let $N(m,k,s)$ denote the maximum size of an $[s+1,k,m]_q$–packing. Note that in case $k = s+1$ we have a trivial $[k,k,m]_q$–packing and $N(m,k,k) = \left[{m \atop k}\right]_q$. An equivalent definition of an $[s+1,k,m]_q$–packing can be given in terms of the subspace distance $d_S(V,U)$ defined (in general for any $V, U \in \mathcal{P}_q(m)$) by $d_S(V,U) = \dim V + \dim U - 2\dim(V \cap U)$. Then clearly $d_S(V,U) \geq 2(k-s)$ for every pair of elements $V, U \in \mathcal{C}$. The following simple observation is an analogue of (1.1) for projective spaces. Let M be the incidence matrix of an $[s+1,k,m]_q$–packing $\mathcal{C}$ with $s \geq 1$, that is the $t \times n$ matrix where the rows (resp. columns) are indexed by the points of $\mathcal{P}_q(m)$ (resp. by the blocks of $\mathcal{C}$) given in a fixed ordering. Thus, $t = (q^m-1)/(q-1)$, $n = |C|$ and each column of M has weight $(q^k-1)/(q-1)$.

Lemma 1. *(i) For* $d \leq \lceil \frac{q^k-1}{q^s-1} \rceil - 1$, *the matrix* M *is* d^z*–disjunct with* $z = \frac{q^k-1-d(q^s-1)}{q-1}$.
(ii) The number of columns

$$n \leq N(m,k,s) \leq \begin{bmatrix} m \\ s+1 \end{bmatrix}_q \Big/ \begin{bmatrix} k \\ s+1 \end{bmatrix}_q \tag{5}$$

with both equalities if and only if $\mathcal{C}$ *is an* $[s+1,k,m]_q$*–Steiner structure.*

Proof. (i) By the definition of an $[s+1,k,m]_q$–packing, each $(s+1)$–space is contained in at most one k–space of $\mathcal{C}$. Therefore, any two columns in M have at most $(q^s-1)/(q-1)$ common ones. Hence, a column in M can be covered by at most $\lceil \frac{q^k-1}{q^s-1} \rceil$ other columns. This clearly means that $d_{max} = \lceil \frac{q^k-1}{q^s-1} \rceil - 1 \geq q^{k-s}$. It is also clear that for $d \leq d_{max}$ we have a d^z-disjunct matrix with $z \geq \frac{q^k-1-d(q^s-1)}{q-1}$.
(ii) Since the number of $(s+1)$–spaces contained in a k–space is $\begin{bmatrix} k \\ s+1 \end{bmatrix}_q$, we have the following simple packing bound $N(m,k,s) \leq \begin{bmatrix} m \\ s+1 \end{bmatrix}_q / \begin{bmatrix} k \\ s+1 \end{bmatrix}_q$. The equality in (5) is attained iff we have a partition of all $(s+1)$–spaces by the blocks of $\mathcal{C}$. □

A challenging problem is to find Steiner structures in $\mathcal{P}_q(n)$. Note that no nontrivial Steiner structures, except for the case $s=0$ (partition of $GF(q)^m$ by k–spaces) are known.

Theorem KK [16]. *Given integers* $0 \leq s < k \leq \frac{1}{2}m$ *and prime power* q, *there exists an explicit construction of an* $[s+1,k,m]_q$*–packing* $\mathcal{C}(m,k,s)$ *with*

$$|\mathcal{C}(m,k,s)| = q^{(s+1)(m-k)}. \tag{6}$$

The construction of such packings is based on Gabidulin codes [13]. The explicit description (in terms of subspace codes) is given in [16]. For completeness we describe here this construction, in terms of $[s+1,k,m]_q$–packings. Let $\mathbb{F}_q^{k\times r}$ denote the set of all $k\times r$ matrices over $GF(q)$. For $X, Y \in \mathbb{F}_q^{k\times r}$ the rank distance between X and Y is defined as $d_R(X,Y) = \text{rank}(X-Y)$. The rank-distance is a metric and codes in metric space $(\mathbb{F}_q^{k\times r}, d_R)$ are called rank-metric codes. It is known [13] that for a rank-metric code $\mathcal{C} \subseteq \mathbb{F}_q^{k\times r}$ with minimum distance $d_R(\mathcal{C})$ one has the Singleton bound $\log_q |\mathcal{C}| \leq \min\{k(r-d_R(\mathcal{C})+1), r(k-d_R(\mathcal{C})+1)\}$. Codes attaining this bound are called maximum-rank-distance codes (MRD). An important class of MDR codes are Gabidulin codes [13], which exist for all parameters k, r and $d_R \leq \min\{k,r\}$. The construction of an $[s+1,k,m]_q$–packing from an MRD code is as follows. Consider the space $\mathbb{F}_q^{k\times(m-k)}$ and let $m \geq 2k$. Then for any integer $0 \leq s \leq k-1$ there exists a Gabidulin code $\mathcal{C}_G \subset \mathbb{F}_q^{k\times(m-k)}$ of minimum distance $d_R = k-s$ and size $q^{(s+1)(m-k)}$. To each matrix $A \in \mathcal{C}_G$ we put into correspondence the matrix $[I_k|A] \in \mathbb{F}_q^{k\times m}$ (I_k is the $k\times k$ identity matrix). We define now the set of k–spaces $\mathcal{C}(m,k,s)_q = \{\text{rowspace}([I_k|A]) : A \in \mathcal{C}_G\}$. It can easily be observed now that $\dim(V \cap U) \leq s$ for all pairs $V, U \in \mathcal{C}(m,k,s)_q$. This means that $\mathcal{C}(m,k,s)_q$ is an $[s+1,k,m]_q$–packing with $|\mathcal{C}(m,k,s)_q| = |\mathcal{C}_G| = q^{(s+1)(m-k)}$. Similarly is described the $[s+1,k,m]_q$–packing $\mathcal{C}(m,k,s)_q$ for $k < m < 2k$. Note however, that for our purposes it is sufficient to consider the case $m \geq 2k$.

The following is a useful estimate for the Gaussian coefficients (see e.g. [16]).

Lemma 2. *For integers* $1 \leq k < m$ *we have*

$$q^{(m-k)k} < \begin{bmatrix} m \\ k \end{bmatrix}_q < \alpha(q) \cdot q^{(m-k)k}, \tag{7}$$

where $\alpha(2) = 4$ *and* $\alpha(q) = \frac{q}{q-2}$ *for* $q \geq 3$.

Lemma 2, applied to our upper bound (5), gives: $N(n,k,s) < \alpha(q) \cdot q^{(s+1)(m-k)}$. The latter implies that the packing $\mathcal{C}(m,k,s)_q$ is nearly optimal, that is

$$|\mathcal{C}(m,k,s)_q| > \frac{1}{\alpha(q)} N(n,k,s)_q.$$

Here actually $\lim \alpha(q) = 1$, as $q \to \infty$, yields asymptotic optimality.

Let $P(m,k,s)_q$ denote the incidence matrix of $\mathcal{C}(m,k,s)_q$. We summarize our findings in

Theorem 2. *Given integers* $0 < s+1 < k \leq \frac{1}{2}m$ *and a prime power* q, *we have*
(i) $P(m,k,s)_q$ *is a* d*–disjunct* $t \times n$ *matrix with* $d = q^{k-s}$, $t = \frac{q^m-1}{q-1}$, $n = q^{(s+1)(m-k)}$.
(ii) In case $k = s+1$ *(trivial packing) we have* $d = q$, $t = \frac{q^m-1}{q-1}$, $n = \begin{bmatrix} m \\ k \end{bmatrix}_q$.
(ii) For any $d \leq q^{k-s}$, *the matrix* $P(m,k,s)_q$ *is* d^z*–disjunct with* $z = \frac{q^k-1-d(q^s-1)}{q-1}$.

Next we wish to know how good our construction is. Let $t(d,n)$ denote the minimum number of rows for a d–disjunct matrix with n columns. In the literature known are the bounds asymptotic in n:

$$\Omega\left(\frac{d^2 \log n}{\log d}\right) \leq t(d,n) \leq O(d^2 \log n) \tag{8}$$

(log is always of base 2). The lower bound is proved in [9], [20], [12]. For the upper bound see [10], [14]. Constructions of good disjunct matrices (also referred to as superimposed codes) are given in D'yachkov et al [6], [7], where also the detailed tables with parameters of the best known superimposed codes are presented.

In fact, currently the best upper and lower bounds are due to D'yachkov and Rykov [9] and D'yachkov et al [10]. In particular, when $d \to \infty$ these bounds are:

$$\frac{d^2}{2\log d}(1+o(1))\log n \leq t(d,n) \leq d^2 \log e(1+o(1))\log n.$$

Let us take now in our construction $q = 2,\ m = 2k$. Then we have $d = 2^{k-s}$, $t = 2^{2k} - 1$, $n = 2^{(s+1)k}$ and hence

$$t < \frac{2^{2k}\log n}{(s+1)k} < \frac{2^{2s}}{s+1} \cdot \frac{d^2 \log n}{\log d}.$$

Corollary 1. *Given integer $s \geq 1$, our construction gives a class of d–disjunct $t \times n$ matrices with parameters $d = 2^{k-s}$, $t = 2^{2k}$, $n = 2^{(s+1)k}$ attaining the lower bound in (8).*

As an example, let us compare our construction with the construction in Ngo and Du [19], described in Section 2. The construction in [19] has parameters $d = \frac{q(q^r-1)}{q-1}$ (Theorem 2), $t = \left[{m \atop r}\right]_q > q^{(m-r)r}$, $n = \left[{m \atop k}\right]_q < \alpha(q)q^{(m-k)k}$ (Lemma 2). Without loss of generality we may assume that $1 \leq r < k \leq m/2$. We now fix some d and t (thus r and m are also fixed).

For the parameters in our construction we use the notation n_0, k_0, t_0, d_0. Both constructions are considered over $GF(q)$ and also we take $k_0 = k$. Thus, for $k > s+1$ we have $d_0 = q^{k-s}$, $t_0 = (q^{m_0} - 1)/(q-1)$, $n_0 = q^{(s+1)(m_0-k)}$. Note that the case $k = s+1$ (trivial packing) and $m_0 = m$ corresponds to the case $r = 1$. Consider now the case $r \geq 2$

Case $k - r \geq 2$: We put $s = k - r - 1$ and $m_0 = (m-r)r$. Then we have
$d_0 = q^{r+1}$, $t_0 = (q^{(m-r)r} - 1)/(q-1)$, and $n_0 = q^{(k-r)(mr-r^2-k)}$.

It is easy to check that $d_0 > d$, $t_0 < t$. We claim now that except for the case $r = 2, k = 4$ we have $n_0 > n$, or equivalently, $(m-r)(mr - r^2 - k) > (m-k)k + \log_q \alpha(q)$. Easy calculation shows that the latter is equivalent to $(m-r)(kr - r^2 - k) > \log_q \alpha(q)$ (where $m - r > 2$ and $\log_q \alpha(q) < 2$). Hence it is enough to show that $kr - r^2 - k \geq 1$. The latter inequality holds if $k \geq \lceil\frac{r^2+1}{r-1}\rceil = r+2$ for $r \geq 3$ (resp. if $k \geq 5$ for $r = 2$). Thus, it remains to consider the

Subcase $r = 2, k = 4$: We take $m_0 = 2m - 3$ and thus $t_0 = (q^{2m-3} - 1)/(q-1)$. It is easy to check now that $t_0 = (q^{2m-3} - 1)/(q-1) < \left[{m \atop 2}\right]_q = (q^m - 1)(q^{m-1} - 1)/(q^2-1)(q-1)$. Observe also that $n_0 = q^{2(2m-7)} > \alpha(q) \cdot q^{(m-4)4} > n$.

Case $k - r = 1$: In this case we take $s = k - r = 1$ and $m_0 = (m-r)r$. Then we note that the actual value of d_0 in our construction (see Lemma 1)
$d_0 \geq \dfrac{q^k - 1}{q-1} - 1 = \dfrac{q^{r+1} - 1}{q-1} - 1 = d$. Note now that $n = q^{2(mr-r^2-k)} = q^{2(mk-m-k^2+k-1)} > n$. Thus, for each triple (d, t, n) we have a better triple (d_0, t_0, n_0) (more defectives, less tests, and more items), moreover, in case $r \geq 2$ we have $n_0(d,t) \gg n(d,t)$ as $m \to \infty$.

Note on the other hand that the construction of Ngo and Du [19] has better performance than the construction of Macula [17] (this can easily be shown). As a measure of performance of a d-disjunct $t \times n$ matrix, in general one can take the ratio $(\log n)/t$.

Let us also compare our construction with another construction in [11], where the authors explore a novel use of ℓ–packings to construct d-disjunct matrices.

The construction is based on the known Steiner system $S(3, q+1, q^r+1)$ (for any prime power q and integer $r \geq 2$) and has the following parameters:

$$d = \binom{q+1}{2} - 1,\ t = \binom{q^r+1}{2} - 1,\ n = \frac{\binom{q^r+1}{3}}{\binom{q+1}{3}} = \frac{(q^{2r}-1)q^{r-1}}{q^2-1}.$$

Note that in order to get a nontrivial construction one has to take $r \geq 4$. We fix now some q and t, that is d and r are also fixed.

In our construction we have: $d_0 = q^{k-s}$, $t_0 = (q^m - 1)/(q-1)$, and $n_0 = q^{(s+1)(m-k)}$. We put $k = s+2 = r-1$ and $m = 2r-1$. Then we have $d_0 = q^2$, $t_0 = (q^{2r-1}-1)/(q-1)$, $n_0 = q^{(r-2)r}$. Observe now that $d_0 > d$, $t_0 < t$, $n_0 \geq n$ for $r > 4$.

In case $r = 4$ we take $k = s+2 = 4, m_0 = 8$. Thus, $d_0 = q^2, t_0 = (q^8 - 1)/(q-1), n_0 = q^{12}$. Correspondingly we have $d = \binom{q+1}{2} - 1, t = (q^4+1)q^4/2$, and $n = (q^8-1)q^3/(q^2-1)$. Note now that again we have $d_0 > d$, $t_0 < t$, and $n_0 > n$. Thus, for every triple (d, t, n) (in Construction [11]) we have a better triple (d_0, t_0, n_0), moreover $n_0(d,t) \gg n(d,t)$ as $r \to \infty$.

Acknowledgment. The second author would like to thank Prof. Arkadii D'yachkov for useful discussions.

References

1. Balding, D.J., Torney, D.C.: Optimal pooling designs with error detection. J. Combin. Theory Ser. A 74(1), 131–140 (1996)
2. Bose, R.C., Burton, R.C.: A characterization of flat spaces in a finite geometry and the uniqueness of the Hamming and the MacDonald codes. J. Combin. Theory 1, 96–104 (1996)
3. Erdős, P., Frankl, P., Füredi, Z.: Families of finite sets in which no set is covered by the union of r others. Isr. J. Math. 51(1-2), 79–89 (1985)
4. Du, D.-Z., Hwang, F.K.: Combinatorial Group Testing and its Applications, 2nd edn. World Scientific, Singapore (2000)
5. Du, D.-Z., Hwang, F.K.: Pooling Designs and Nonadaptive Group Testing-Important Tools for DNA sequencing. World Scientific (2006)
6. D'yachkov, A., Macula, A., Rykov, V.V.: New constructions of superimposed codes. IEEE Trans. Inf. Theory 46(1), 284–290 (2000)
7. D'yachkov, A., Macula, A., Rykov, V.V.: New applications and results of superimposed code theory arising from the potentialities of molecular biology. In: Numbers, Information, and Complexity, pp. 265–282. Kluwer Academic Publishers (2000)
8. D'yachkov, A., Hwang, F.K., Macula, A., Vilenkin, P., Weng, C.: A construction of pooling designs with some happy surprises. J. Comput. Biology 12, 1129–1136 (2005)
9. D'yachkov, A., Rykov, V.V.: Bounds on the length of disjunctive codes. Problems Inform. Transmission 18(3), 7–13 (1982)
10. D'yachkov, A., Rykov, V.V., Rashad, A.M.: Superimposed distance codes. Problems Control Inform. 18(4), 237–250 (1989)
11. Fu, H.-L., Hwang, F.K.: A novel use of t-packings to construct d-disjunct matrices. Discrete Applied Mathematics 154, 1759–1762 (2006)
12. Füredi, F.: On r-cover-free families. J. Comb. Theory A 73, 172–173 (1996)

13. Gabidulin, E.M.: Theory of codes with maximum rank distance. Problems Inform. Transmission 21(1), 1–12 (1985)
14. Hwang, F.K., Sós, V.T.: Non-adaptive hypergeometric group testing. Studia Scient. Math. Hungarica 22, 257–263 (1987)
15. Kautz, W.H., Singleton, R.C.: Nonrandom binary superimposed codes. IEEE Trans. Info. Theory 10, 363–377 (1964)
16. Koetter, R., Kschischang, F.R.: Coding for errors and erasures in random network coding. IEEE Trans. Inf. Theory 54(8), 3579–3591 (2008)
17. Macula, A.J.: A simple construction of d-disjunct matrices with certain constant weights. Discrete Math 162, 311–312 (1996)
18. Macula, A.J.: Error-correcting nonadaptive group testing with d^z-disjunct matrices. Discrete Appl. Math. 80, 217–222 (1996)
19. Ngo, H.Q., Du, D.-Z.: New constructions of non-adaptive and error-tolerance pooling designs. Discrete Math 243, 161–170 (2002)
20. Ruszinkó, M.: On the upper bound on the size of r-cover-free families. J. Comb. Theory A 66, 302–310 (1994)

Density-Based Group Testing

Dániel Gerbner[1,*], Balázs Keszegh[1,**],
Dömötör Pálvölgyi[2,***], and Gábor Wiener[3,†]

[1] Rényi Institute of Mathematics, Hungarian Academy of Sciences,
{gerbner.daniel,keszegh.balazs}@renyi.mta.hu
[2] Department of Computer Science, Eötvös University
dom@cs.elte.hu
[3] Department of Computer Science and Information Theory,
Budapest University of Technology and Economics
wiener@cs.bme.hu

Dedicated to the memory of Rudolf Ahlswede

Abstract. In this paper we study a new, generalized version of the well-known group testing problem. In the classical model of group testing we are given n objects, some of which are considered to be defective. We can test certain subsets of the objects whether they contain at least one defective element. The goal is usually to find all defectives using as few tests as possible. In our model the presence of defective elements in a test set Q can be recognized if and only if their number is large enough compared to the size of Q. More precisely for a test Q the answer is YES if and only if there are at least $\alpha|Q|$ defective elements in Q for some fixed α.

Keywords: group testing, search, query.

1 Introduction

The concept of group testing was developed in the middle of the previous century. Dorfman, a Swiss physician intended to test blood samples of millions of soldiers during World War II in order to find those who were infected by syphilis. His key idea was to test more blood samples at the same time and learn whether at least one of them are infected [4]. Some fifteen years later Rényi developed a theory of search in order to find which electrical part of his car went wrong. In

* Research supported by Hungarian Science Foundation EuroGIGA Grant OTKA NN 102029.

** Research supported by OTKA, grant NK 78439 and NN 102029 (EUROGIGA project GraDR 10-EuroGIGA-OP-003).

*** Research supported by Hungarian National Science Fund (OTKA), grant PD 104386 and NN 102029 (EUROGIGA project GraDR 10-EuroGIGA-OP-003), and the János Bolyai Research Scholarship of the Hungarian Academy of Sciences.

† Supported in part by the Hungarian National Research Fund and by the National Office for Research and Technology (Grant Number OTKA 67651).

H. Aydinian, F. Cicalese, and C. Deppe (Eds.): Ahlswede Festschrift, LNCS 7777, pp. 543–556, 2013.

his model – contrary to Dorfman's one – not all of the subsets of the possible defectives (electric parts) could be tested [7].

Group testing has now a wide variety of applications in areas like DNA screening, mobile networks, software and hardware testing.

In the classical model we have an underlying set $[n] = \{1, \dots, n\}$ and we suppose that there may be some defective elements in this set. We can test all subsets of $[n]$ whether they contain at least one defective element. The goal is to find all defectives using as few tests as possible. One can easily see that in this generality the best solution is to test every set of size 1. Usually we have some additional information like the exact number of defectives (or some bounds on this number) and it is also frequent that we do not have to find all defectives just some of them or even just to tell something about them.

In the case when we have to find a single defective it is well-known that the information theoretic lower bound is sharp: the number of questions needed in the worst case is $\lceil \log n \rceil$, which can be achieved by binary search. All logarithms appearing in the paper are binary.

Another well-known version of the problem is when the maximum size of a test is bounded. (Motivated by the idea that too large tests are not supposed to be reliable, because a small number of defectives may not be recognized there). This version can be solved easily in the adaptive case, but is much more difficult in the non-adaptive case. This latter version was first posed by Rényi. Katona [6] gave an algorithm to find the exact solution to Rényi's problem and he also proved the best known lower bound on the number of queries needed. The best known upper bound is due to Wegener [8].

In this paper we assume that the presence of defective elements in a test set Q can be recognized if and only if their number is large enough compared to the size of Q. More precisely for a test $Q \subseteq [n]$ the answer is YES if and only if there are at least $\alpha|Q|$ defective elements in Q. Our goal is to find at least m defective elements using tests of this kind.

Definition 1. *Let $g(n, k, \alpha, m)$ be the least number of questions needed in this setting, i.e. to find m defective elements in an underlying set of size n which contains at least k defective elements, where the answer is* YES *for a question $Q \subseteq [n]$ if there are at least $\alpha|Q|$ defective elements in Q, and* NO *otherwise.*

We suppose throughout the whole paper that $1 \leq m \leq k$ and $0 < \alpha < 1$. Let $a = \lfloor \frac{1}{\alpha} \rfloor$, that is, a is the largest size of a set where the answer NO has the usual meaning, namely that there are no defective elements in the set. It is obvious that if a set of size greater than k/α is asked then the answer is automatically NO, so we will suppose that question sets has size at most k/α.

It is worth mentioning that a similar idea appears in a paper by Damaschke [2] and a follow-up paper by De Bonis, Gargano, and Vaccaro [3]. Since their motivation is to study the concentration of liquids, their model deals with many specific properties arising in this special case and they are interested in the number of merging operations or the number of tubes needed in addition to the number of tests.

If $k = m = 1$, then the problem is basically the same as the usual setting with the additional property that the question sets can have size at most a: this is the above mentioned problem of Rényi. As we have mentioned, finding the optimal non-adaptive algorithm, or even just good bounds is really hard even in this simplest case of our model, thus in this paper we deal only with adaptive algorithms.

In the next section we give some upper and lower bounds as well as some conjectures depending on the choices of n, k, α, and m. In the third section we prove our main theorem, which gives a general lower and a general upper bound, differing only by a constant depending only on k. In the fourth section we consider some related questions and open problems.

2 Upper and Lower Bounds

First of all it is worth examining how binary search, the most basic algorithm of search theory works in our setting. It is easy to see that it does not work in general, not even for $m = 1$. If (say) $k = 2$ and $\alpha = 0.1$, then question sets have at most 20 elements (recall that we supposed that there are no queries containing more than k/α elements, since they give no information at all, because the answer for them is always NO), thus if n is big, we cannot perform a binary search.

However, if $k \geq n\alpha$, then binary search can be used.

Theorem 1. *If $\alpha \leq k/n$, then $g(n, k, \alpha, m) \leq \lceil \log n \rceil + c$, where c depends only on α and m, moreover if $m = 1$, then $c = 0$.*

Proof. We show that binary search can be used to find m defectives. That is, first we ask a set F of size $\lfloor n/2 \rfloor$ and then the underlying set is substituted by F if the answer is YES and by $\overline{F}$ if the answer is NO. We iterate this process until the size of the underlying set is at most $2m/\alpha$. Now we check that the condition $\alpha \leq k/n$ remains true after each step. Let $n' = \lfloor n/2 \rfloor$ be the size of the new underlying set and k' be the number of defectives there. If the answer was YES, then $k' \geq \alpha n'$, thus $\alpha \leq k'/n'$. If the answer was NO, then there are at least $k - \lceil \alpha n' \rceil + 1$ defectives in the new underlying set, that is $k' \geq k - \lceil \alpha n' \rceil + 1 \geq \alpha n - \lceil \alpha n' \rceil + 1 \geq \alpha n'$, thus $\alpha \leq k'/n'$ again.

Now if $m = 1$ we simply continue the binary search until we find a defective element, altogether using at most $\lceil \log n \rceil$ questions.

If $m > 1$, then we can find m defectives in the last underlying set using at most

$$c := \max_{n' \leq 2m/\alpha} g(n', m, \alpha, m)$$

further queries.

(Notice that since the size of the last underlying set is greater than m/α, it contains at least m defectives.) This number c does not depend on k, just on α and m and it is obvious that we used at most $\lceil \log n \rceil + c$ queries altogether. □

This theorem has an easy, yet very important corollary. If the answer for a question A is YES, then there are at least $\alpha|A|$ defective elements in A. If $\alpha|A| \geq m$, then we can find m of these defectives using $g(|A|, \alpha|A|, \alpha, m) \leq \log|A| + c$ questions, where c depends only on α and m. Basically it means that whenever we obtain a YES answer, we can finish the algorithm quickly.

The proof of Theorem 1 is based on the fact that if the ratio of the defective elements k/n is at least α, then this condition always remains true during binary search. If $k/n < \alpha$, then this trick does not work, however if the difference between k/n and α is small, a similar result can be proved for $m = 1$. Recall that $a = \lfloor 1/\alpha \rfloor$.

Theorem 2. *If* $k \geq \frac{n}{a} - \lfloor \log \frac{n}{a} \rfloor - 1$ *and* $k \geq 1$, *then* $g(n, k, \alpha, 1) \leq \lceil \log n \rceil + 1$.

The proof of the theorem is based on the following lemmas.

Lemma 1. *Let* $t \geq 0$ *be an integer. Then* $g(2^t a, 2^t - t, \alpha, 1) \leq t + \lceil \log a \rceil$.

Proof. We use induction on t. For $t = 0$ and $t = 1$ the proposition is true, since we can perform a binary search on a or $2a$ elements (by asking sets of size at most a we learn whether they contain a defective element). Suppose now that the proposition holds for t, we have to prove it for $t + 1$. That is, we have an underlying set of size $2^{t+1}a$ containing at least $2^{t+1} - t - 1$ defectives. Our first query is a set A of size $2^t a$. If the answer is YES , then we can continue with binary search. If the answer is NO, then there are less than $\alpha 2^t a \leq 2^t$ defectives in A, therefore there are at least $2^{t+1} - t - 1 - 2^t + 1 = 2^t - t$ defectives in $\overline{A}$. By the induction hypothesis $g(2^t a, 2^t - t, \alpha, 1) \leq t + \lceil \log a \rceil$, thus $g(2^{t+1}a, 2^{t+1} - t - 1, \alpha, 1) \leq t + 1 + \lceil \log a \rceil$ follows, finishing the proof of the lemma. □

Lemma 2. *Let* $t \geq 2$ *be an integer. Then* $g(2^t a, 2^t - t - 1, \alpha, 1) \leq t + \lceil \log a \rceil + 1$.

Proof. Let us start with asking three disjoint sets, each of cardinality $2^{t-2}a$. If the answer to any of these is YES, then we can continue with binary search, using $t - 2 + \lceil \log a \rceil$ additional questions. If all three answers are NO, then there are at least $2^t - t - 1 - 3(2^{t-2} - 1) = 2^{t-2} - (t - 2)$ defectives among the remaining $2^{t-2}a$ elements, hence we can apply Lemma 1. □

Proof (of Theorem 2). Let us suppose $n > 2a$ (otherwise binary search works) and let $t = \lfloor \log \frac{n}{a} \rfloor$, $r = n - 2^t a$. We have an underlying set of size $n = 2^t a + r$ containing at least $\frac{n}{a} - \lfloor \log \frac{n}{a} \rfloor - 1$ defectives. If $r = 0$, then by Lemma 2 we are done. Otherwise let the first query A contain r elements. A positive answer allows us to find a defective element by binary search on A using altogether at most $\lceil \log n \rceil + 1$ questions (actually, at most $\lceil \log n \rceil$ questions, because $r \leq n/2$). If the answer is negative then the new underlying set contains $2^t a$ elements, of which more than $\frac{n}{a} - \lfloor \log \frac{n}{a} \rfloor - \alpha r - 1 = 2^t + r/a - \alpha r - \lfloor \log \frac{n}{a} \rfloor - 1 \geq 2^t - \lfloor \log \frac{n}{a} \rfloor - 1$ are defective. Since $\lfloor \log \frac{n}{a} \rfloor = t$, the number of defectives is at least $2^t - t$, thus by Lemma 1 we need at most $t + \lceil \log a \rceil$ more queries to find a defective element, thus altogether we used at most $t + 1 + \lceil \log a \rceil \leq \lceil \log n \rceil + 1$ queries, from which the theorem follows. □

One might think that binary search is the best algorithm to find one defective if it can be used (i.e. for $k \geq n\alpha$). A counterexample for k really big is easy to give: if $k = n$ then we do not need any queries and for $m = 1, k = n - 1$ we need just one query. It is somewhat more surprising that $g(n, \alpha n, \alpha, 1) \geq \lceil \log n \rceil$ is not necessarily true.

For example, the case $n = 10, k = 4, \alpha = 0.4, m = 1$ can be solved using 3 queries: first we ask a set A of size 4. If the answer is YES, we can perform a binary search on A, if the answer is NO then there are at least 3 defectives among the remaining 6 elements and now we ask a set B of size 2. If the answer is YES then we perform a binary search on B, otherwise there are at least 3 defectives among the remaining 4 elements, so one query (of size 1) is sufficient to find a defective. However, a somewhat weaker lower bound can be proved:

Theorem 3. $g(n, k, \alpha, m) \geq \lceil \log(n - k + 1) \rceil$.

We prove the stronger statement that even if one can use any kind of yes-no questions, still at least $\lceil \log(n - k + 1) \rceil$ questions are needed. This is a slight generalization of the information theoretic lower bound.

Theorem 4. *To find one of k defective elements from a set of size n, one needs* $\lceil \log(n - k + 1) \rceil$ *yes-no questions in the worst case and this is sharp.*

Proof. Suppose there is an algorithm that uses at most q questions. The number of sequences of answers obtained is at most 2^q, thus the number of different elements selected by the algorithm as the output is also at most 2^q. This means that $n - 2^q \leq k - 1$, otherwise it would be possible that all k defective elements are among those ones that were not selected. Thus $q \geq \lceil \log(n - k + 1) \rceil$ indeed.

Sharpness follows easily from the simple algorithm that puts $k - 1$ elements aside and runs a binary search on the rest. □

This theorem has been independently proved in [1]. Theorem 3 is an immediate consequence of Theorem 4, but this is not true for the sharpness of the result. However, Theorem 3 is also sharp: if $\alpha \leq \frac{2}{n-k+1}$, then we can run a binary search on any $n - k + 1$ of the elements to find a defective.

We have seen in Theorem 1 that if $n \leq k/\alpha$, then binary search works (with some additional constant number of questions if $m > 1$). On the other hand, if n goes to infinity (with k and α fixed), then the best algorithm is linear.

Theorem 5. *For any* k, α, m

$$\frac{n}{a} + c_1 \leq g(n, k, \alpha, m) \leq \frac{n}{a} + c_2,$$

where c_1 *and* c_2 *depend only on* k, α, *and* m.

Proof. Upper bound: first we partition the underlying set into $\lfloor \frac{n}{a} \rfloor$ a-element sets and possibly one additional set of less than a elements. We ask each of these sets (at most $\lfloor \frac{n}{a} \rfloor + 1$ questions). Then we choose m sets for which we obtained a YES answer (or if there are less than m such sets, then we choose all of them).

We ask every element one by one in these sets (at most ma questions). One can easily see that we find at least m defective elements, using at most $\lfloor \frac{n}{a} \rfloor + ma + 1$ questions.

Lower bound: We use a simple adversary's strategy: suppose all the answers are NO and there are m elements identified as defectives. Let us denote the family of sets that were asked by $\mathcal{F}$. It is obvious that those sets of $\mathcal{F}$ that have size at most a contain no defective elements. Suppose there are i such sets. We use induction on i. There are $n' \geq n - ia$ elements not contained in these sets and we should prove that at least $\frac{n}{a} + c_1 - i \leq \frac{n'}{a} + c_1$ other questions are needed. Hence by the induction it is enough to prove the case $i = 0$.

Suppose $i = 0$. If there is a set A of size $k + 1$, such that $|A \cap F| \leq 1$ for all $F \in \mathcal{F}$, then any k-element subset of $|A|$ can be the set of the defective elements. In this case any element can be non-defective, a contradiction. Thus for every set A of size $k + 1$ there exists a set $F \in \mathcal{F}$, such that $|A \cap F| \geq 2$.

Let $b = \lfloor \frac{k}{\alpha} \rfloor$. We know that every set of $\mathcal{F}$ has size at most b. Then a given $F \in \mathcal{F}$ intersects at most $\sum_{j=2}^{k+1} \binom{b}{j} \binom{n-b}{k+1-j}$ $(k + 1)$-element sets in at least two points. This number is $O(n^{k-1})$, and there are $\Omega(n^{k+1})$ sets of size $k + 1$, hence $|\mathcal{F}| = \Omega(n^2)$ is needed.

It follows easily that there is an n_0, such that if $n > n_0$, then $|\mathcal{F}| \geq \frac{n}{a}$. Now let $c_1 = -n_0/a$. If $n > n_0$ then $|\mathcal{F}| \geq \frac{n}{a} \geq \frac{n}{a} + c_1$, while if $n \leq n_0$ then $|\mathcal{F}| \geq 0 \geq \frac{n}{a} + c_1$, thus the number of queries is at least $\frac{n}{a} + c_1$, finishing the proof. □

Remark. The theorem easily follows from Theorem 7, it is included here because of the much simpler proof.

It is easy to give a better upper bound for $m = 1$.

Theorem 6. *Suppose* $k + \log k + 1 \leq \lceil \frac{n}{a} \rceil$. *Then*

$$g(n, k, \alpha, 1) \leq \left\lceil \frac{n}{a} \right\rceil - k + \lceil \log a \rceil.$$

Proof. First we ask a set X of size ka. If the answer is YES, then we can find a defective element in $\lceil \log ka \rceil$ steps by Theorem 1. In this case the number of questions used is at most $1 + \lceil \log ka \rceil = 1 + \lceil \log k + \log a \rceil \leq 1 + \lceil \log k \rceil + \lceil \log a \rceil \leq \lceil \frac{n}{a} \rceil - k + \lceil \log a \rceil$, where the last inequality follows from the condition of the theorem.

If the answer is NO, then we know that there are at most $k - 1$ defectives in X, so we have at least one defective in $\overline{X}$. Continue the algorithm by asking disjoint subsets of X of size a, until the answer is YES or we have at most $2a$ elements not yet asked. In these cases using at most $\lceil \log 2a \rceil$ questions we can easily find a defective element, thus the total number of questions used is at most $1 + \lceil \frac{n-ka-2a}{a} \rceil + \lceil \log 2a \rceil = 1 + \lceil \frac{n}{a} \rceil - k - 2 + \lceil \log a \rceil + 1 = \lceil \frac{n}{a} \rceil - k + \lceil \log a \rceil$, finishing the proof. □

Note that if the condition of Theorem 6 does not hold (that is, $k + \log k + 1 > \lceil \frac{n}{a} \rceil$), then $k \geq \frac{n}{a} - \lfloor \log \frac{n}{a} \rfloor - 1$, hence $\lceil \log n \rceil + 1$ questions are enough by Theorem 2.

The exact values of $g(n, k, \alpha, m)$ are hard to find, even for $m = 1$. The algorithm used in the proof of Theorem 6 seems to be optimal for $m = 1$ if $k + \log k + 1 \leq \lceil \frac{n}{a} \rceil$. However, counterexamples with $1/\alpha$ not an integer are easy to find (consider i.e. $n = 24$, $k = 2$, $\alpha = \frac{2}{11}$).

Conjecture 1. *If $\frac{1}{\alpha}$ is an integer and $k + \log k + 1 \leq \lceil \frac{n}{a} \rceil$, then the algorithm used in the proof of Theorem 6 is optimal for $m = 1$.*

It is easy to see that Conjecture 1 is true for $k = 1$. For other values of k it would follow from the next, more general conjecture.

Conjecture 2. *If $\frac{1}{\alpha}$ is an integer, then $g(n, k, \alpha, 1) \leq g(n, k + 1, \alpha, 1) + 1$.*

Obviously, Conjecture 2 also fails if $1/\alpha$ is not an integer. One can see for example that $g(24, 1, 2/11, 1) = 7$ and $g(24, 2, 2/11, 1) = 5$.

3 The Main Theorem

In this section we prove a lower and an upper bound differing only by a constant depending only on k. For the lower bound we need the following simple generalization of the information theoretic lower bound.

Proposition 1. *Suppose we are given p sets $A_1, \ldots, A_p$ of size at least n, each one containing at least one defective and an additional set A_0 of arbitrary size containing no defectives. Let $m \leq p$. Then the number of questions needed to find at least m defectives is at least $\lceil m \log n \rceil$.*

Proof. Suppose that we are given the additional information that every set A_i $(i \geq 1)$ contains exactly one defective element. Now we use the information theoretic lower bound: there are $\prod_{i=1}^{p} |A_i|$ possibilities for the distribution of the defective elements at the beginning, and at most $\prod_{i=1}^{p-m} |A_{j_i}|$ at the end (suppose we have found defective elements in every set A_i except in $A_{j_1}, \ldots, A_{j_{p-m}}$), thus if we used l queries, then $2^l \geq n^m$, from which the proposition follows. □

Now we formulate the main theorem of the paper.

Theorem 7. *For any k, α, m*

$$\frac{n}{a} + m \log a - c_1(k) \leq g(n, k, \alpha, m) \leq \frac{n}{a} + m \log a + c_2(k),$$

where $c_1(k)$ and $c_2(k)$ depend only on k.

Proof. First we give an algorithm that uses at most $\frac{n}{a} + m \log a + c_2(k)$ queries, proving the upper bound. In the first part of the procedure we ask disjoint sets $A_1, A_2, \ldots, A_r$ of size a until either there were m YES answers or there are no more elements left. In this way we ask at most $\lceil \frac{n}{a} \rceil$ questions.

Suppose we obtained YES answers for the sets $A_1, A_2, \ldots A_{m_1}$ and NO answers for the sets $A_{m_1+1}, \ldots, A_r$. If $m_1 \geq m$, then in the second part of the procedure we use binary search in the sets $A_1, A_2, \ldots, A_m$ in order to find one defective element in each of them. For this we need $m\lceil \log a \rceil$ more questions.

If $m_1 < m$, then first we use binary search in the sets $A_1, A_2, \ldots, A_{m_1}$ in order to find defective elements $a_1 \in A_1, a_2 \in A_2, \ldots, a_{m_1} \in A_{m_1}$. Then we iterate the whole process using $S_1 = \cup_{i=1}^{m_1} A_i \setminus \{a_i\}$ as an underlying set, that is we ask disjoint sets $B_1, B_2, \ldots, B_t$ of size a until either we obtain $m - m_1$ YES answers or there are no more elements left. Suppose we obtained YES answers for the sets $B_1, B_2, \ldots B_{m_2}$ and NO answers for the sets $A_{m_2+1}, \ldots, A_t$. If $m_2 \geq m - m_1$, then in the second part of the procedure we use binary search in the sets $B_1, B_2, \ldots, B_{m-m_1}$ in order to find one defective element in each of them, while if $m_2 < m - m_1$, then first we use binary search in the sets $B_1, B_2, \ldots, B_{m_2}$ in order to find defective elements $b_1 \in B_1, b_2 \in B_2, \ldots, b_{m_2} \in A_{m_2}$ and continue the process using $S_2 = \cup_{i=1}^{m_2} B_i \setminus \{b_i\}$ as an underlying set, and so on, until we find $m = m_1 + m_2 + \ldots + m_j$ defective elements. Note that $m_i \geq 1$, $\forall i \leq j$, since $k \geq m$. We have two types of queries: queries of size a and queries of size less than a (used in the binary searches). The number of questions of size a is at most $\lceil \frac{n}{a} \rceil$ in the first part and at most $m_1 + m_2 + \ldots + m_{j-1} < m \leq k$ in the second part. The total number of queries of size less than a is at most $m\lceil \log a \rceil$, thus the total number of queries is at most $\lceil \frac{n}{a} \rceil + m\lceil \log a \rceil + k$, proving the upper bound.

To prove the lower bound we need the following purely set-theoretic lemma.

Lemma 3. *Let k, l, a be arbitrary positive integers and $\beta > 1$. Let now $\mathcal{H}$ be a set system on an underlying set S of size $c(k, l, \beta) \cdot a = k\beta(2^{kl} - 1)a$, such that every set of $\mathcal{H}$ has size at most βa and every element of S is contained in at most l sets of $\mathcal{H}$. Then we can select k disjoint subsets of S (called heaps) $K_1, K_2, \ldots, K_k$ of size βa, such that every set of $\mathcal{H}$ intersects at most one heap.*

Proof. Let us partition the underlying set into k heaps of size $\beta a(2^{kl} - 1)$ in an arbitrary way. Now we execute the following procedure at most $kl - 1$ times, eventually obtaining k heaps satisfying the required conditions. In each iteration we make sure that the members of a subfamily $\mathcal{H}'$ of $\mathcal{H}$ will intersect at most one heap at the end.

In each iteration we do the following. We build the subfamily $\mathcal{H}' \subseteq \mathcal{H}$ by starting from the empty subfamily and adding an arbitrary set of $\mathcal{H}$ to our subfamily until there exists a heap K_i such that $|K_i \cap \cup_{H \in \mathcal{H}'} H| \geq |K_i|/2$, that is K_i is at least half covered by $\mathcal{H}'$. We call K_i the selected heap. If the half of several heaps gets covered in the same step, then we select one where the difference of the number of covered elements and the half of the size of the heap is maximum.

Now we keep the covered part of the selected heap and keep the uncovered part of the other heaps and throw away the other elements. We also throw away the sets of the subfamily $\mathcal{H}'$ from our family $\mathcal{H}$, as we already made sure that the members of $\mathcal{H}'$ will not intersect more than one heap at the end. In this way we obtain smaller heaps but we only have to deal with the family $\mathcal{H} \setminus \mathcal{H}'$.

We prove by induction that after s iterations all heaps have size at least $\beta a(2^{kl-s} - 1)$. This trivially holds for $s = 0$. By the induction hypothesis, the heaps had size at least $\beta a(2^{kl-s+1}-1)$ before the sth iteration step. After the sth step the new size of the selected heap K is at least $|K|/2 \geq \beta a(2^{kl-s+1}-1)/2 \geq \beta a(2^{kl-s}-1)$. Now we turn our attention to the unselected heaps. Suppose the set we added last to $\mathcal{H}'$ is the set I. Clearly, $|K_j \cap \cup_{H\in\mathcal{H}'\setminus\{I\}} H| \leq |K_j|/2$ for all j. Let K be the selected heap and K_i be an arbitrary unselected heap. Now by the choice of K we have $|K_i \cap \cup_{H\in\mathcal{H}'} H| \leq |K_i|/2 + |I|/2$, otherwise $|K_i \cap \cup_{H\in\mathcal{H}'} H| + |K \cap \cup_{H\in\mathcal{H}'} H| > |K_i|/2 + |K|/2 + |I|$, which is impossible, since $|K_i \cap \cup_{H\in\mathcal{H}'} H| + |K \cap \cup_{H\in\mathcal{H}'} H| = |((K_i \cup K) \cap \cup_{H\in\mathcal{H}'\setminus\{I\}} H) \cup ((K_i \cup K) \cap I)| \leq |K_i|/2 + |K|/2 + |I|$.

Now since $|I| \leq \beta a$, the new size of the unselected heap K_i is $|K_i'| = |K_i \setminus \cup_{H\in\mathcal{H}'} H| \geq |K_i|/2 - \beta a/2 \geq \beta a(2^{kl-s+1}-1)/2 - \beta a/2 \geq \beta a(2^{kl-s}-1)$, finishing the proof by induction.

Now in each iteration we delete a family that covers the selected heap, thus any heap can be selected at most l times, since every element is contained in at most l sets. After $kl-1$ iterations the size of an arbitrary heap will be still at least βa. Furthermore, all but one heaps were selected exactly l times, thus any remaining set of $\mathcal{H}$ can only intersect the last heap. That is, heaps at this point satisfy the required condition for all sets of $\mathcal{H}$.

If we can iterate the process at most $kl-2$ times, then after the last possible iteration more than half of any heap is not covered by the union of the remaining sets. Deleting the covered elements from each heap we obtain heaps of size at least βa that satisfy the condition. □

Now we are in a position to prove the lower bound of Theorem 7. We use the adversary method, i.e. we give a strategy to the adversary that forces the questioner to ask at least $\frac{n}{a} + m\log a - c_1(k)$ questions to find m defective elements.

Recall that all questions have size at most $\lfloor k/\alpha \rfloor$ and now the adversary gives the additional information that there are exactly k defective elements.

During the procedure, the adversary maintains weights on the elements. At the beginning all elements have weight 0. Let us denote the set of the possible defective elements by S'. At the beginning $S' = S$. At each question A the strategy determines the answer and also adds appropriate weights to the elements of A. If a question A is of size at most $a = \lfloor 1/\alpha \rfloor$, then the answer is NO and weight 1 is given to all elements of A. If $|A| > a$, the answer is still NO and weight $a/\lfloor k/\alpha \rfloor$ is given to the elements of A. Thus after some r questions the sum of the weights is at most ra. If an element reaches weight 1, then the adversary says that it is not defective, and the element is deleted from S'. The adversary does that until there are still ca elements in S' but in the next step S' would become smaller than this threshold (the exact value of c will be determined later). Up to this point the number of elements thrown away is at least $n - ca - \lfloor k/\alpha \rfloor$, thus the number of queries is at least $\frac{n}{a} - c - \lfloor k/\alpha \rfloor/a \geq \frac{n}{a} - c - k$.

Let the set system $\mathcal{F}$ consist of the sets that were asked up to this point and let $\mathcal{F}' = \{F \cap S' \mid F \in \mathcal{F}, |F| > a\}$.

The following observations are easy to check.

Lemma 4

- $|S'| \geq ca$.
- *Every set $F \in \mathcal{F}'$ has size at most $\lfloor k/\alpha \rfloor \leq k(a+1) \leq 2ka$*
- *Every element of S' is contained in at most $\lfloor k/\alpha \rfloor / a \leq k(1+1/a) \leq 2k$ sets of $\mathcal{F}'$.*
- *Every k-set that intersects each $F \in \mathcal{F}'$ in at most one element is a possible set of defective elements.*

Now let $l := 2k$, $\beta := 2k$, and $c := c(k, l, \beta) = k\beta(2^{kl} - 1) = 2k^2(2^{2k^2} - 1)$. By the observations above, we can apply Lemma 3 with $\mathcal{H} = \mathcal{F}'$. The lemma guarantees the existence of heaps $K_1, K_2, \ldots, K_k$ of size $\beta a \geq a$, such that every transversal of the K_i's is a possible k-set of defective elements. Now by applying Proposition 1 with $A_i = K_i$ and $A_0 = S \setminus S'$, we obtain that the questioner needs to ask at least $\lceil m \log a \rceil$ more queries to find m defective elements.

Altogether the questioner had to use at least $\frac{n}{a} - c - k + m \log a$ queries, which proves the lower bound, since the number c depends only on k (the constant in the theorem is $c_1(k) = c + k$). □

The constant in the lower bound is quite large, by a more careful analysis one might obtain a better one. For example, we could redefine the weights, such that we give weight $a/|A|$ to the elements of A, thus still distributing weight at most a per asked set.

It is also worth observing that if $1/\alpha$ is an integer, then we can use Lemma 3 with $l = \beta = k$, instead of $l = \beta = 2k$. This way one can prove stronger results for small values of k and m if $1/\alpha$ is an integer. We demonstrate it for $k = 2$ in the next section. The following claim is easy to check.

Claim 1. *Let $\mathcal{H}$ be a set system on an underlying set S of size $3a$, consisting of disjoint sets of size at most $2a$. Then we can select 2 disjoint subsets of S (called heaps) K_1, K_2 of size at least a, such that every set of $\mathcal{H}$ intersects at most one heap.*

4 The Case $k = 2$, $m = 1$

In this section we determine the exact value $g(n, 2, \alpha, 1)$. Let $\delta = \lfloor 2\{\frac{1}{\alpha}\} \rfloor$, where $\{x\}$ denotes the fractional part of x.

Consider the following algorithm W, where n denotes the number of remaining elements:

If $n \leq 2^{\lceil \log a \rceil} + 1$, we ask a question of size $\lfloor n/2 \rfloor \leq a$, then depending on the answer we continue in the part that contains at least one defective element, and find that with binary search.

If $2^{\lceil \log a \rceil} + 2 \leq n \leq 2^{\lceil \log a \rceil + 1} + 1$, then we ask a question of size $2^{\lceil \log a \rceil} + 1$ (this falls between a and $2a + 1$). If the answer is YES, we put an element aside and continue with the remaining elements of the set we asked, otherwise we continue with the elements not in the set we asked. This way independent of whether we got a YES or NO answer, we have at most $2^{\lceil \log a \rceil}$ elements with at least one defective, hence we can apply binary search.

If $2^{\lceil \log a \rceil + 1} + 2 \leq n \leq 3a + \delta + 2^{\lceil \log a \rceil}$, then first we ask a question of size $2a + \delta$. If the answer is YES, we put an element aside and continue with the remaining elements of the set we asked, otherwise we continue with the elements not in the set we asked. This way independent of whether we got a YES or NO answer, we have at most $2^{\lceil \log a \rceil} + a$ elements with at least one defective. We continue with a set of size a, and after that we can finish with binary search.

If $n \geq 3a + \delta + 2^{\lceil \log a \rceil} + 1$, then we ask a question of size a. If the answer is NO, we proceed as above. If the answer is YES, we can find a defective element with at most $\lceil \log a \rceil$ further questions.

Counting the number of questions used in each case, we can conclude.

Claim 2. *If $n \leq 3a + \delta + 2^{\lceil \log a \rceil}$, then algorithm W takes only $\lceil \log(n-1) \rceil$ questions, thus according to Theorem 4 it is optimal.*

In fact a stronger statement is true. Note that the following theorem does not contradict to Conjecture 1, as the algorithm mentioned there uses the same number of steps as algorithm W in case $k = 2$, $1/\alpha$ is an integer and $\lceil n/a \rceil \geq 4$.

Theorem 8. *Algorithm W is optimal for any n.*

Proof. We prove a slightly stronger statement, that algorithm W is optimal even among those algorithms that have access to an unlimited number of extra non-defective elements. This is crucial as we use induction on the number of elements, n.

It is easy to check that the answer for a set that is greater than $2a + \delta$ is always NO, while if both defective elements are in a set of size $2a + \delta$, then the answer is YES. We say that a question is *small* if its size is at most a, and *big* if its size is between $a + 1$ and $2a + \delta$. Note that small questions test if there is at least one defective element in the set, while big questions test if both defective elements are in the set. Suppose by contradiction that there exists an algorithm Z that is better than W, i.e. there is a set of elements for which Z is faster than W. Denote by n the size of the smallest such set and by $z(n)$ the number of steps in algorithm Z. We will establish through a series of claims that such an n cannot exist. It already follows from Claim 2 that n has to be at least $3a + \delta + 2^{\lceil \log a \rceil} + 1$.

Note that for $n = 3a + \delta + 2^{\lceil \log a \rceil}$ algorithm W uses $\lceil \log(n-1) \rceil = \lceil \log(2a + \delta - 1) \rceil + 1$ questions. An important tool is the following lemma.

Lemma 5. *If $n \geq 3a + \delta + 2^{\lceil \log a \rceil} + 1$, then algorithm Z has to start with a big question. Moreover, it can ask a small question among the first $z(n) - \lceil \log(2a + \delta - 1) \rceil$ questions only if one of the previous answers was* YES.

Proof. First we prove that algorithm Z has to start with a big question. Suppose it starts with a small question. We show that in case the answer is NO, it cannot be faster than algorithm W. In this case after the first answer there are at least $n - a$ (and at most $n - 1$) elements which can be defective, and an unlimited number of non-defective elements, including those which are elements of the first question. By induction algorithm W is optimal in this case, and one can easily see that it cannot be faster if there are more elements, hence algorithm Z cannot be faster than algorithm W on $n - a$ elements plus one more question. On the other hand algorithm W clearly uses this many questions (as it starts with a question of size a), hence it cannot be slower than algorithm Z.

Similarly, to prove the moreover part, suppose that the first $z(n) - \lceil \log(2a + \delta - 1) \rceil$ answers are NO and one of these questions, A is small. Let us delete every element of A. By induction algorithm W is optimal on the remaining at least $n - a$ elements, hence similarly to the previous case, algorithm Z uses more questions than algorithm W on $n - a$ elements, hence cannot be faster than algorithm W. More precisely, we can define algorithm W', which starts with asking A, and after that proceeds as algorithm W. One can easily see that algorithm W' cannot be slower than algorithm Z or faster than algorithm W. □

Note that a YES answer would mean that $\lceil \log(2a + \delta - 1) \rceil$ further questions would be enough to find a defective with binary search, hence in the worst case, (when the most steps are needed) no such answer occurs among the first $z(n) - \lceil \log(2a + \delta - 1) \rceil$ questions anyway. Now we can finish the proof of the theorem with the following claim.

Claim 3. *If $n > 3a + \delta + 2^{\lceil \log a \rceil}$, then algorithm W is optimal.*

Proof. If not, then the smallest n for which W is not optimal must be of the form $2a + \delta + 2^{\lceil \log a \rceil} + za + 1$, where $z \geq 1$ integer. (This follows from the fact that the number of required questions is monotone in n if we allow the algorithm to have access to an unlimited number of extra non-defective elements.) By contradiction, suppose that algorithm Z uses only $\lceil \log(2a + \delta - 1) \rceil + z$ questions. Suppose the answer to the first z questions are NO. Then by to Lemma 5, these questions are big. Suppose that the $z + 1$st answer is also NO. We distinguish two cases depending on the size of the $z + 1$st question A. In both cases we will use reasoning similar to the one in Theorem 5.

Case 1. The $z + 1$st question is small. After the answer there are $\lceil \log(2a + \delta - 1) \rceil - 1$ questions left, so depending on the answers given to them, any deterministic algorithm can choose at most $2^{\lceil \log(2a+\delta-1) \rceil - 1}$ elements. Hence algorithm Z gives us after the $z + 1$st answer a set B of at most $2^{\lceil \log(2a+\delta-1) \rceil - 1}$ elements, which contains a defective.

Before starting the algorithm, all the $\binom{n}{2}$ pairs are possible candidates to be the set of defective elements. However, after the $z + 1$st question (knowing the algorithm) the only candidates are those which intersect B. The $z + 1$st question shows at most a non-defective elements, but all the pairs which intersect neither A nor B have to be excluded by the first z questions. Thus $\binom{n-|A|-|B|}{2} \geq \binom{2a+\delta+2^{\lceil \log a \rceil}+za+1-a-2^{\lceil \log(2a+\delta-1) \rceil - 1}}{2} \geq \binom{(z+1)a+\delta+1}{2}$ pairs should be excluded, but z questions can exclude at most $z\binom{2a+\delta}{2}$ pairs, which is less if $z \geq 1$.

Case 2. The $z+1$st question is big. After it we have $\lceil \log(2a+\delta-1)\rceil - 1$ questions left, so depending on the answers given to them, any deterministic algorithm can choose at most $2^{\lceil \log(2a+\delta-1)\rceil-1}$ elements. This means that we have to exclude with the first $z+1$ questions at least

$$\binom{2a+\delta+2^{\lceil \log a\rceil}+za+1-2^{\lceil \log(2a+\delta-1)\rceil-1}}{2} \geq \binom{(z+2)a+\delta+1}{2}$$

pairs. But they can exclude at most $(z+1)\binom{2a+\delta}{2}$ pairs, which is less if $z \geq 1$. □

This finishes the proof of the theorem. □

5 Open Problems

It is quite natural to think that $g(n,k,\alpha,m)$ is increasing in n but we did not manage to prove that. The monotonicity in k and m is obvious from the definition. On the other hand, we could have defined $g(n,k,\alpha,m)$ as the smallest number of questions needed to find m defectives assuming there are *exactly* k defectives (instead of *at least* k defectives) among the n elements, in which case the monotonicity in k is far from trivial. We conjecture that this definition gives the same function as the original one.

It might seem strange to look for monotonicity in α, but we have seen that for $m=1$ we can reach the information theoretic lower bound (which is $\lceil \log(n-k+1)\rceil$ in this setting) for $\alpha \leq 2/(n-k+1)$. All the theorems from Section 2 also suggest that the smaller α is, the faster the best algorithm is even for general m. Basically in case of a NO answer it is better if α is small, and in case of a YES answer the size of α does not matter very much, since the process can be finished fast. However, we could only prove Theorem 7 concerning this matter.

Another interesting question is if we can choose α. If $m=1$ then we should choose $\alpha \leq 1/(n-k+1)$, and as we have mentioned in the previous paragraph, we believe that a small enough α is the best choice.

Another possibility would be if we were allowed to choose a new α for every question. Again, we believe that the best solution is to choose the same, small enough α every time. This would obviously imply the previous conjecture.

Finally, a more general model to study is the following. We are given two parameters, $\alpha \geq \beta$. If at least an α fraction of the set is defective, then the answer is YES, if at most a β fraction, then it is NO, while in between the answer is arbitrary. With these parameters, this paper studied the case $\alpha = \beta$. This model is somewhat similar to the threshold testing model of [2], where instead of ratios α and β they have fixed values a and b as thresholds.

References

1. Ahlswede, R., Deppe, C., Lebedev, V.: Finding one of D defective elements in some group testing models. Probl. Inf. Transm. 48(2), 173–181 (2012)

2. Damaschke, P.: The Algorithmic Complexity of Chemical Treshold Testing. In: Bongiovanni, G., Bovet, D.P., Di Battista, G. (eds.) CIAC 1997. LNCS, vol. 1203, pp. 205–216. Springer, Heidelberg (1997)
3. De Bonis, A., Gargano, L., Vaccaro, U.: Efficient algorithms for chemical threshold testing problems. Theoret. Comput. Sci. 259(1-2), 493–511 (2001)
4. Dorfman, R.: The Detection of defective members of large populations. Ann. Math. Statistics 14, 436–440 (1943)
5. Du, D.Z., Hwang, F.K.: Combinatorial Group Testing and Its Applications, 1st edn. World Scientific (1994)
6. Katona, G.O.H.: On separating systems of a finite set. J. Combinatorial Theory 1, 174–194 (1966)
7. Katona, G.O.H.: Rènyi and the Combinatorial Search Problems. Studia Sci. Math. Hungar. 26, 363–378 (1991)
8. Wegener, I.: On separating systems whose elements are sets of at most k elements. Discrete Math 28, 219–222 (1979)

Group Testing with Multiple Mutually-Obscuring Positives

Hong-Bin Chen[1] and Hung-Lin Fu[2]

[1] Institute of Mathematics, Academia Sinica, Taipei 10617, Taiwan
hbchen@math.sinica.edu.tw
[2] Department of Applied Mathematics, National Chiao Tung University, Hsinchu 30050, Taiwan
hlfu@math.nctu.edu.tw

In memory of a great mathematician and information scientist Rudolf Ahlswede

Abstract. Group testing is a frequently used tool to identify an unknown set of defective (positive) elements out of a large collection of elements by testing subsets (pools) for the presence of defectives. Various models have been studied in the literature. The most studied case concerns only two types (defective and non-defective) of elements in the given collection. This paper studies a novel and natural generalization of group testing, where more than one type of defectives are allowed with an additional assumption that certain obscuring phenomena occur among different types of defectives. This paper proposes some algorithms for this problem, trying to optimize different measures of performance: the total number of tests required, the number of stages needed to perform all tests and the decoding complexity.

Keywords: pooling design, group testing, selectors.

1 Introduction

The classical group testing problem is described as follows: Given a set $\mathcal{N}$ of n items consisting of two types of items, a set $\mathcal{P}$ of positive items with $|\mathcal{P}| \leq d$ and the others being negative items, the goal is to identify $\mathcal{P}$ in an efficient manner by using group tests. A test (pool) can be applied to any subset of items in $\mathcal{N}$ with two possible outcomes: a negative outcome indicates that there is no positive in the test while a positive outcome indicates that at least one positive is in the test. The concept of group testing originated from the application of blood testing during World War II. Afterwards, it has been also found applications in molecular biology, including screening clone libraries [3], sequencing by hybridization [31], yeast one-hybrid screens [36], and recently, the mapping of protein-protein interactions [37]. Additionally, group testing has proved relevant in other fields such as multiple access communication [4], image compression [27] and more recently data gathering in sensor networks [28]. For general references, readers may refer to the books [18,19].

H. Aydinian, F. Cicalese, and C. Deppe (Eds.): Ahlswede Festschrift, LNCS 7777, pp. 557–568, 2013.

Due to a diversity of its applications, there has been many models that were proposed and studied in the literature. For example, the inhibitor model [5,6,8,26] where the presence of an inhibitor can somehow cancel the effect of positive elements, the complex model [1,2,10,12,35] where positive reactions are caused by certain sets of elements rather than a single one of elements, the threshold model [9,15,11,17] where two thresholds are given for the conditions of positive reactions and negative reactions to occur, the interference model [7,16,20] where two or more positive elements appearing in a pool can interfere with each other so that the positive reaction cannot be detected, and more others.

In this paper we study a generalization of group testing as follows. Consider a set $\mathcal{N}$ of n items which is known to contain s types of positive items $P_1, P_2, \cdots, P_s$, where $|P_i| \leq p_i$, and the others being negative items. The task is to classify all items in $\mathcal{N}$ with as few tests as possible. For a test $Q \subseteq \mathcal{N}$, define

$$I_Q \equiv \{i : Q \cap P_i \neq \varnothing\}.$$

Then the outcome of a test Q will be given according to the following rules:

- If $I_Q = \{i\}$, then the response will be "i-positive".
- If $I_Q = \varnothing$, then the response will be "negative".
- If $|I_Q| \geq 2$, then the response can be either "negative" or be "i-positive" for some $i \in I_Q$ (not knowing which i).

For example, given a test Q with $I_Q = \{1, 3, 4\}$, the outcome of the test Q can be any and exactly (but not knowing which) one of the four cases: negative, 1-positive, 3-positive and 4-positive. We refer to this problem as the *Multiple Mutually-Obscuring Positives* (MMOP) problem. Obviously, for the case $s = 1$, the MMOP problem is exactly the same as the classical group testing. For the case $s = 2$, it is coincident to the coin-weighing problem with test-type device (such as a spring balancer or electronic scale) where, given a set of coins and some of them are counterfeit (too heavy or too light), the task is not only to identify all counterfeit coins but also to make them classified as heavy or light. Notably, the MMOP problem is not a generalization of the "mutually-obscuring problem" discussed in [7,16,20].

In this paper, we provide a unified technique to deal with the MMOP problem for general s. In the next section, we propose an efficient *nonadaptive algorithm*, i.e., all tests are set up in advance and thus can be performed simultaneously without any information of outcomes of other tests. In particular, the proposed algorithm can be decoded in polynomial time. Section 3 provides a 2-stage algorithm for this problem. Instrumental to the result is based on a combinatorial structure, (k, m, n)*-selector*, first introduced by De Bonis, Gąsieniec and Vaccaro [6] in the context of designing efficient trivial 2-stage pooling strategies on the classic pooling design problem. We propose a new point of view for the mentioned selectors. This enables us to construct the combinatorial tool easily. Probabilistic constructions are provided and numerical results show that our constructions are slightly better than the currently best known result by De Bonis et al. [6].

2 Nonadaptive Algorithms

In order to present our results, we now introduce some notations and definitions. Note that n and p_i's are given in advance and we assume $\sum_{i=1}^{s} p_i \ll n$. Throughout this paper, a pooling design is represented by a 0-1 matrix M where columns are the set of objects, rows are the set of tests, and cell $(i,j) = 1$ signifies that the j-th object is in the i-th test and $(i,j) = 0$ for otherwise. For convenience, a column (row) can be treated as the set of row (column) indices where the column (row) has a 1, respectively. For any two columns C and C', we denote $C \cup C'$ as the boolean sum of C and C'. We say that a set X of columns appears (or is contained) in a row if all columns in X have a 1-entry in the row. A pool is called an i-positive pool if its outcome is i-positive, and a non-i-positive pool if it is not i-positive. For a column C, denote by $t_i(C)$ the number of i-positive pools in which column C appears. Likewise, denote by $t_{\bar{i}}(C)$ the number of non-i-positive pools in which column C appears.

Consider a fixed family $T = \{T_1, T_2, \cdots, T_t\}$ with $T_i \subseteq \mathcal{N}, 1 \le i \le t$. Let R_i, $1 \le i \le s$, be arbitrary disjoint subsets of $\mathcal{N}$ and let $J = \cup_{i=1}^{s} R_i$. We define the *syndrome vector* of J in T by $\phi_T(J) = (\phi_1(J), \phi_2(J), \cdots, \phi_t(J))$, where

$$\phi_j(J) = \{i : T_j \cap R_i \neq \varnothing\}.$$

For any two distinct J_0 and J_1, we say their syndromes $\phi_T(J_0)$ and $\phi_T(J_1)$ are *different*, denoted by $\phi_T(J_0) \not\sim \phi_T(J_1)$, if and only if there exists some $j \in [t]$ and $i \in \{0,1\}$ such that $\phi_j(J_{1-i}) = \{k\}$ for some $k \in [s]$ and $k \notin \phi_j(J_i)$. Denote by $\phi_T(J_0) \sim \phi_T(J_1)$ if they are *coincident* (not different).

Definition 1. *Let $T = \{T_1, T_2, \cdots, T_t\}$ with $T_i \subseteq \mathcal{N}, 1 \le i \le t$. We say the family T is MMOP-separable if*

$$\phi_T(J_0) \not\sim \phi_T(J_1)$$

for any two distinct $J_0, J_1 \subseteq \mathcal{N}$ with $J_0 = \cup_{i=1}^{s} R_i^0$ and $J_1 = \cup_{i=1}^{s} R_i^1$, where R_i^ℓ's, $\ell \in \{0,1\}$, are disjoint subsets of $\mathcal{N}$ and $|R_i^\ell| \le p_i$ for $1 \le i \le s$.

Lemma 1. *MMOP-separability is a sufficient and necessary condition for the MMOP problem.*

Proof. The lemma follows by definition immediately. □

We first present a lower bound on the number of tests required for any nonadaptive algorithms for the MMOP model. This lower bound is obtained by establishing a connection to disjunct matrices (equivalently, superimposed codes or cover-free families).

Definition 2. *[30] A binary matrix is called d-disjunct if for any $d+1$ columns $C_0, C_1, \cdots, C_d$,*

$$\left| C_0 \setminus \bigcup_{i=1}^{d} C_i \right| \ge 1.$$

It is well-known [21] that disjunct matrices of size $t \times n$ have a lower bound $t = \Omega(d^2 \log n / \log d)$ and an upper bound $t = O(d^2 \log n)$. The literature contains many studies (see [19] and references therein) on the construction of disjunct matrices (sometimes called superimposed codes and cover-free families). One of the most common approaches is to control the number of intersections of any two columns to guarantee the disjunctness property. For recognition, we refer disjunct matrices with this particular structure as "w/λ-disjunct".

Definition 3. *[30] A binary matrix is w/λ-disjunct if the following properties both hold: (1) every column has more than w 1-entries; (2) any two distinct columns intersect at no more than λ rows.*

It is easy to see that an w/λ-disjunct matrix is $\lfloor \frac{w}{\lambda} \rfloor$-disjunct and so has the lower bound $\Omega(d^2 \log n / \log d)$ on the number of rows where $d = \lfloor \frac{w}{\lambda} \rfloor$. Previous results [21,22,29,13,14,32] have also shown that this particular structure can achieve the best known upper bound $O(d^2 \log n)$ on the number of rows for d-disjunct matrices.

For short, let $d = \sum_{i=1}^{s} p_i$ in the rest of the paper. Note that asymptotic results presented in the paper are under the assumption that d is constant and n approaches to infinity.

Theorem 1. *Let $\mathcal{N}$ be a set of n items which is known to contain s types of positive items $P_1, P_2, \cdots, P_s$, where $|P_i| \leq p_i$, and the others being negative items. Then any nonadaptive algorithm for the MMOP problem requires $\Omega(d^2 \log n / \log d)$ tests to classify all positive items.*

Proof. To prove this theorem, it suffices to show that MMOP-separable implies $(d-1)$-disjunct. Then, by Lemma 1 and the well-known lower bound for disjunct matrices, we get the desired bound. Suppose to the contrary that there exists a family $T = \{T_1, T_2, \cdots, T_t\}$ of subset of $\mathcal{N}$ and its corresponding matrix M is not a $(d-1)$-disjunct matrix of size $t \times n$. Then there exist d columns $C_0, C_1, \cdots, C_{d-1}$ in M such that $|C_0 \setminus \bigcup_{j=1}^{d-1} C_i| = 0$. That means for every row T_i where C_0 appears there exists some $j \in \{1, 2, \cdots, d-1\}$ such that C_j also appears in the row T_i. Consider the two subsets $J_0 = \{C_1, \cdots, C_{d-1}\}$ and $J_1 = \{C_0, C_1, \cdots, C_{d-1}\}$. Then, obviously, J_0 and J_1 are distinct but their syndromes $\phi_T(J_0)$ and $\phi_T(J_1)$ are not different, i.e., there does not exist $j \in [t]$ and $i \in \{0, 1\}$ such that $\phi_j(J_{1-i}) = \{k\}$ for some $k \in [s]$ and $k \notin \phi_j(J_i)$. Thus, by Lemma 1, the $(d-1)$-disjunctness is a necessary condition for any nonadaptive strategy and the bound follows immediately. □

Next, we propose a nonadaptive algorithm for the considered problem. In particular, our algorithm can be decoded in polynomial time to recover all positives from outcomes.

Theorem 2. *Let $\mathcal{N}$ be a set of n items which is known to contain s types of positive items $P_1, P_2, \cdots, P_s$, where $|P_i| \leq p_i$, and the others being negative items. A (w/λ)-disjunct matrix of n columns with $w/\lambda > d$ can solve the MMOP problem using $O(d^2 \log n)$ tests and $O(d^2 n \log n)$ time.*

Proof. Let M be a (w/λ)-disjunct matrix of n columns with $w/\lambda > d$ and use it as the pooling design. Consider an arbitrary P_i for some i and let R_i be an element in P_i. Observe that R_i appears in a non-i-positive pool only when the pool contains another item $R_j \in P_j$ for some $j \neq i$. Since M is a (w/λ)-disjunct matrix, we have that $t_{\bar{i}}(R_i) \leq \lambda \sum_{k \neq i} p_k$. For an item $C \notin P_i$, C appears in an i-positive pool only when the pool contains some items of P_i. Hence, we conclude that $t_{\bar{i}}(C) \geq w - \lambda p_i$ since M is (w/λ)-disjunct.

By the above discussion along with the condition $w/\lambda > d$, we have that

$$t_{\bar{i}}(R_i) \leq \lambda \sum_{k \neq i} p_k < w - \lambda p_i \leq t_{\bar{i}}(C)$$

for any $R_i \in P_i$ and $C \notin P_i$. Thus, we can separate all items in P_i from those not in P_i through counting $t_{\bar{i}}(C)$ for each $C \in \mathcal{N}$. Since P_i is chosen arbitrarily, all items in $\mathcal{N}$ can be classified in a similar way.

For the decoding issue, our algorithm only needs to compute $t_{\bar{i}}(C)$ for each item C in $\mathcal{N}$. This can be done easily by going through each entry in the column C once. Hence, the decoding complexity is at most $O(d^2 n \log n)$ time. □

The following is the pseudo code of our decoding algorithm.

Algorithm 1 CLASSIFICATION

Use a (w/λ)-disjunct matrix with $w/\lambda > \sum_{i=1}^{s} p_i$ as a pooling design.
$P_i \leftarrow \varnothing$, for $1 \leq i \leq s$.
for each item $C \in \mathcal{N}$ **do**
  if $t_{\bar{i}}(C) \leq \lambda \sum_{j \neq i} p_j$ for some i **then**
    $P_i \leftarrow P_i \cup \{C\}$
Return P_i for $1 \leq i \leq s$.

3 A 2-Stage Algorithm

Theorem 1 shows that any nonadaptive algorithm for the MMOP problem requires $\Omega(d^2 \log n / \log d)$ tests. However, the information-theoretic lower bound for algorithms without any constraint on the number of stages reduces down to $\log_{s+1} \binom{n}{d} = \Theta(d \log(n/d) / \log(s+1))$. This section shall provide a 2-stage algorithm for the MMOP problem that uses $O(d \log(n/d))$ tests.

Definition 4. *[6] Given integers k, m and n with $1 \leq m \leq k \leq n$, we say that a binary $t \times n$ matrix M is a (k, m, n)-selector if any submatrix of M subject to k out of n arbitrary columns contains at least m distinct rows of the identity matrix I_k.*

The integer t is called the size of the (k, m, n)-selector. Denote by $t_s(k, m, n)$ the minimum size of a (k, m, n)-selector. De Bonis, Gąsieniec and Vaccaro [6] suggested a way to construct a (k, m, n)-selector by searching for a vertex cover on a properly defined hypergraph and derived the following result, which is the best known asymptotical bound.

Theorem 3. *[6] For any integers k, m, n with $1 \leq m \leq k < n$,*

$$t_s(k, m, n) < \frac{ek^2}{k - m + 1} \ln(n/k) + \frac{ek(2k - 1)}{k - m + 1},$$

where e is the base of natural logarithm.

By definition, we have the following result immediately.

Lemma 2. *A (k, m, n)-selector is a $(k - i, m - i, n)$-selector for any integer i with $1 \leq i \leq m - 1$.*

Theorem 4. *For all integers n and k with $n \geq k \geq 2d$, there exists a two-stage group testing algorithm for the MMOP problem that classifies all types of positives, and uses at most $t_s(k, 2d - 1, n) + k - d$ tests.*

Proof. Our algorithm works as follows. In the first stage, we test the pools associated with rows of a $(k, 2d - 1, n)$-selector M satisfying $n \geq k \geq 2d$ to find a set D' of at most $k - d$ suspicious candidates. In the second stage, all suspicious candidates are tested individually and simultaneously so as to classify all types of positives. To complete the proof, it suffices to show that after the first stage we can determine such a set D' that contains all positives with $|D'| \leq k - d$.

Let D be the set of actual positive items and let J be the set of the column indices associated with the set D. Suppose that there are exactly y pairwise distinct sets $J_1, J_2, \cdots, J_y$ of suspicious candidates and each of whose syndrome vectors agrees with the set D of all positives. Notice that $|J_i| \leq d$ for all i, and obviously $J \in \{J_1, J_2, \cdots, J_y\}$. If $|\bigcup_{i=1}^{y} J_i| \leq k - d$ as desired, then we are done. If $|\bigcup_{i=1}^{y} J_i| \geq k - d + 1$, then there must exist an integer ℓ with $2 \leq \ell \leq y$ satisfying

$$k - d + 1 \leq |\bigcup_{i=1}^{\ell} J_i| \leq k$$

because of $|J_i| \leq d$ and $k \geq 2d$. Let $|\bigcup_{i=1}^{\ell} J_i| = q$. Since $2d - 1 - (k - q) \geq d$, by Lemma 2, M is also a (q, d, n)-selector. Subject to the q columns associated with the set $\bigcup_{i=1}^{\ell} J_i$, the submatrix of M contains at least d distinct rows of the identity matrix I_q. That means at least d suspicious candidates that appear separately and individually in some rows without any other suspicious candidates of $\bigcup_{i=1}^{\ell} J_i$. For the coincidence of the syndromes of $J_1, J_2, \cdots, J_\ell$, each of these d or more suspicious candidates must belong to every J_i for $1 \leq i \leq \ell$, a contradiction to the assumption that J_i's are pairwise distinct sets with $|J_i| \leq d$ for each i. Hence, we can determine a set $D' = \bigcup_{i=1}^{y} J_i$ of cardinality at most $k - d$, as desired. □

By Theorem 4 with $k = 4d - 2$ and Theorem 3, we have the following result.

Corollary 1. *There exists a two-stage group testing algorithm for the MMOP problem that classifies all types of positives, and uses at most $O(d\log(n/d))$ tests.*

Note that for the particular case $s = 1$ (indeed the classic group testing problem) this result can be found in [22,23,24,25].

3.1 Improved Upper Bounds on the Size of (k, m, n)-Selectors

In this subsection, we exploit three different probabilistic methods to derive upper bounds on the size of (k, m, n)-selectors and present numerical results of these bounds and that in [6] for comparison. The three probabilistic methods can also be found in several studies [14,12,21,22,33,34] on different combinatorial structures.

Definition 5. *A matrix $[m_{ij}]$ is strongly (d, r)-disjunct if for any two disjoint sets $\mathcal{C}_1$ and $\mathcal{C}_2$ of columns with $|\mathcal{C}_1| = d$ and $|\mathcal{C}_2| = r$, there exists a row k and a column $c \in \mathcal{C}_2$ such that $m_{kc} = 1$ and $m_{kj} = 0$ for all $j \in \mathcal{C}_1 \cup \mathcal{C}_2 \setminus \{c\}$.*

Theorem 5. *A (k, m, n)-selector is the same as a strongly $(m-1, k-m+1)$-disjunct matrix of n columns.*

Proof. Suppose that M is a (k, m, n)-selector. For any two disjoint sets $\mathcal{C}_1$ and $\mathcal{C}_2$ of columns with $|\mathcal{C}_1| = m-1$ and $|\mathcal{C}_2| = k-m+1$, the submatrix $M_{\mathcal{C}_1\cup\mathcal{C}_2}$ subject to $\mathcal{C}_1 \cup \mathcal{C}_2$ must contain m distinct rows of an identity matrix I_k, by the definition of (k, m, n)-selectors. Therefore, there exists at least one row i and a column $c \in \mathcal{C}_2$ (by pigeonhole principle) such that $m_{ic} = 1$ and $m_{ij} = 0$ for all $j \in \mathcal{C}_1 \cup \mathcal{C}_2 \setminus \{c\}$, as desired.

Suppose that M is a strongly $(m-1, k-m+1)$-disjunct matrix of n columns. For any fixed set $\mathcal{K}$ of k columns, we want to find m distinct rows of the identity matrix I_k in $M_{\mathcal{K}}$. Partition arbitrarily $\mathcal{K}$ into two disjoint sets $\mathcal{C}_1 = \{c_{1,1}, c_{1,2}, \cdots, c_{1,m-1}\}$ and $\mathcal{C}_2 = \mathcal{K} \setminus \mathcal{C}_1$ with $|\mathcal{C}_1| = m-1$ and $|\mathcal{C}_2| = k-m+1$, then we can find a row i_1 and a column $c_{2,1} \in \mathcal{C}_2$, such that $m_{i_1 c_{2,1}} = 1$ and $m_{i_1 j} = 0$ for all $j \in \mathcal{C}_1 \cup \mathcal{C}_2 \setminus \{c_{2,1}\}$ since M is strongly $(m-1, k-m+1)$-disjunct. Exchanging the column $c_{1,1}$ with $c_{2,1}$ such that $\mathcal{C}_1 = \{c_{2,1}, c_{1,2}, \cdots, c_{1,m-1}\}$ and $\mathcal{C}_2 = \mathcal{K} \setminus \mathcal{C}_1$, similarly we can find another row i_2 and a column $c_{2,2} \in \mathcal{C}_2$ satisfying the desired property, as in row i_1. Notice that the column $c_{2,2}$ chosen from $\mathcal{C}_2$ must be different from the column $c_{2,1}$ which is in the updated $\mathcal{C}_1$. Keep doing this process until $c_{1,j}$'s are all removed to $\mathcal{C}_2$, we then obtain a set $\{i_1, i_2, \cdots, i_m\}$ of m distinct rows of the identity matrix I_k in $M_{\mathcal{K}}$. Hence M is a (k, m, n)-selector. □

Next, we derive upper bounds by probabilistic methods on the minimum number of rows of strongly (d, r)-disjunct matrices, and consequently on the minimum size of selectors. Construct a $t_1 \times 2n$ $0-1$ matrix M where each entry is defined to be 1 with probability p and 0 with probability $1-p$. We say that a column c_j is unsatisfied if there exists two disjoint sets $\mathcal{C}_1$ and $\mathcal{C}_2$ of columns with $c_j \in \mathcal{C}_1 \cup \mathcal{C}_2$, $|\mathcal{C}_1| = d$ and $|\mathcal{C}_2| = r$ such that Definition 5 is not true for any row. Then, given a fixed column c_j, the probability of the event that c_j is unsatisfied is

$$P(c_j \text{ is unsatisfied}) = \binom{2n-1}{d+r-1}\binom{d+r}{d}\left[1-rp(1-p)^{d+r-1}\right]^{t_1}. \quad (1)$$

By simple technique of calculus, we take $p = \frac{1}{d+r}$ to minimize Equation (1) and obtain

$$P(c_j \text{ is unsatisfied}) = \binom{2n-1}{d+r-1}\binom{d+r}{d}\left[1-\tfrac{r}{d+r}\cdot(\tfrac{d+r-1}{d+r})^{d+r-1}\right]^{t_1}. \quad (2)$$

Setting

$$t_1 = \frac{\ln(\binom{2n-1}{d+r-1}\binom{d+r}{d})}{-\ln(1-\frac{r}{d+r}\cdot(\frac{d+r-1}{d+r})^{d+r-1})} + \ln 2, \quad (3)$$

the right-hand side of (2) is less than 1/2, which implies that the expected value of the total number of unsatisfied columns in the matrix M does not exceed n. Hence, there must exist a matrix which is strongly (d,r)-disjunct and of size $t_1 \times n$.

Secondly, we construct a random $t_2 \times n$ $0-1$ matrix M where each entry is defined to be 1 with probability $\frac{1}{d+r}$ and 0 with probability $1-\frac{1}{d+r}$. Let $\mathcal{C}_1$ and $\mathcal{C}_2$ be two disjoint sets of columns with $|\mathcal{C}_1| = d$ and $|\mathcal{C}_2| = r$. Similarly, we say that the pair $(\mathcal{C}_1,\mathcal{C}_2)$ is unsatisfied if this pair does not satisfy the requirement of Definition 5. Then the probability of the event that $(\mathcal{C}_1,\mathcal{C}_2)$ is unsatisfied is

$$P((\mathcal{C}_1,\mathcal{C}_2) \text{ is unsatisfied}) = \left[1-\tfrac{r}{d+r}\cdot(\tfrac{d+r-1}{d+r})^{d+r-1}\right]^{t_2}. \quad (4)$$

Hence the expected value of the total number of unsatisfied pairs in the matrix M is

$$E[\text{unsatisfied pairs}] = \binom{n}{d+r}\binom{d+r}{d}\left[1-\frac{r}{d+r}\cdot(\frac{d+r-1}{d+r})^{d+r-1}\right]^{t_2}. \quad (5)$$

Setting

$$t_2 = \frac{\ln(\binom{n}{d+r}\binom{d+r}{d})}{-\ln(1-\frac{r}{d+r}\cdot(\frac{d+r-1}{d+r})^{d+r-1})}, \quad (6)$$

the right-hand side of Equation (5) is less than 1, which implies that the probability of the existence of a strongly (d,r)-disjunct matrix of size $t_2 \times n$ is greater than 0. Thus, there exists a strongly (d,r)-disjunct matrix of size $t_2 \times n$.
Remark: The above argument can be further extended to a high probability version. Given $0 < \epsilon < 1$, if we require $\mathrm{t}_2 = \frac{\ln(\binom{n}{d+r}\binom{d+r}{d})-\ln\epsilon}{-\ln(1-\frac{r}{d+r}\cdot(\frac{d+r-1}{d+r})^{d+r-1})}$, then Equation (5) is less than ϵ, which implies the desired probability is greater than $1-\epsilon$. This means that a strongly (d,r)-disjunct matrix can be efficiently constructed with probability as high as desired.

Thirdly, we will apply the Lovász Local Lemma to derive an upper bound on the number of rows of a strongly (d,r)-disjunct matrix. The Lovász Local

Lemma first proved by Erdős and Lovász is a powerful tool to prove the existence of combinatorial structures satisfying a prescribed collection of criteria. Here is the lemma in a symmetric form.

Lemma 3. *Let $A_1, A_2, \cdots, A_N$ be events in an arbitrary probability space with $P(A_i) \le p$ for all $1 \le i \le N$. Suppose that each event is mutually independent of all the other events except for at most μ of them. If $ep(\mu+1) \le 1$, then*

$$P\left(\bigcap_{i=1}^{N} \overline{A_i}\right) > 0,$$

where e denotes the base of natural logarithms.

Let $M = (m_{ij})$ be a random $t_3 \times n$ $0-1$ matrix with $P(m_{ij} = 1) = \frac{1}{d+r}$, $P(m_{ij} = 0) = 1 - \frac{1}{d+r}$, and the entries m_{ij} are mutually pairwise independent. For any two disjoint sets $\mathcal{C}_1$ and $\mathcal{C}_2$ of columns with $|\mathcal{C}_1| = d$ and $|\mathcal{C}_2| = r$, let $A_{\mathcal{C}_1,\mathcal{C}_2}$ be the event that ($\mathcal{C}_1$,$\mathcal{C}_2$) is unsatisfied. Obviously,

$$P(A_{\mathcal{C}_1,\mathcal{C}_2}) = \left[1 - \frac{r}{d+r} \cdot (\frac{d+r-1}{d+r})^{d+r-1}\right]^{t_3}$$

and

$$\mu + 1 = \left[\binom{n}{d+r}\binom{d+r}{d} - \binom{n-d-r}{d+r}\binom{d+r}{d}\right].$$

Setting

$$t_3 = \frac{\ln\left[\binom{n}{d+r}\binom{d+r}{d} - \binom{n-d-r}{d+r}\binom{d+r}{d}\right] + 1}{-\ln(1 - \frac{r}{d+r} \cdot (\frac{d+r-1}{d+r})^{d+r-1})}, \tag{7}$$

we have $ep(\mu + 1) \le 1$, and thus by Lemma 3 the desired probability is greater than 0. Consequently, there exists a strongly (d, r)-disjunct matrix of size $t_3 \times n$.

Remark: the purpose of the subsection is to exploit three known methods to derive upper bounds on the length of selectors and to compare the obtained bounds with the one by De Bonis et al. which is the best known result in general cases. Although the obtained results are not as strong as we like (they are asymptotically same), from another point of view, following a conventional analysis in the survey [23], the rate $\lim_{t\to\infty} \frac{\log n}{t}$ derived from the t_1 bound can be shown the best one among all the other ones.

The comparison in the following is based on their original forms, not simplified forms (asymptotic forms). The motivation is in calling awareness to the existence of better results and the alternative constructions (randomness approach).

We now present numerical results of the bounds proposed in this section and that of De Bonis et al. [6] with some parameters. For the sake of fairness, the bound of De Bonis et al. we use for comparison is the original form

$$t_s = \frac{\binom{n}{n/k}}{(k-m+1)\binom{n-k}{n/k-1}} \left[\ln\left(\binom{k-1}{k-m}\binom{n/k}{1}\binom{n-n/k}{k-1}\right) + 1\right],$$

where $k = d+r$ and $m = d+1$ in our setting of strongly (d, r)-disjunct matrices, in the proof in [6, Theorem 1]. The following tables present some numerical results to compare the bounds for some specific parameters. The numerical results show that our constructions are slightly better than the currently best known result by De Bonis et al. [6], at least for the considered parameters.

Table 1 lists the number t of rows obtained by the proposed methods for the case of $n = 300$ and $d = 3$. As shown, setting r an integer close to d seems to be the best choice to get a small number of required tests.

Table 1. The number of rows needed for fixed parameters $n = 300$ and $d = 3$

n	d	r	t_1	t_2	t_3	t_s
300	3	1	175	188	171	186
300	3	2	142	145	137	153
300	3	3	138	136	131	148
300	3	4	140	136	132	150
300	3	5	145	138	136	155
300	3	6	152	142	141	161

Table 2 lists the number of rows required for some small parameters by setting $r = d$. Given a fixed n, the t_2 bound is the best bound for large d, while the t_3 bound is the best for small d except for the case $d = 1$. Notice that the bounds t_2 and t_3 always yield better results than the t_s bound by De Bonis et al.

Table 2. The number of rows required for the $r = d$ case

n	d	r	t_1	t_2	t_3	t_s	min.
300	1	1	27	40	28	44	t_1
300	2	2	84	90	82	98	t_3
300	6	6	283	258	259	278	t_2
300	7	7	327	295	297	316	t_2
1000	1	1	31	48	32	54	t_1
1000	2	2	99	111	98	122	t_3
1000	5	5	287	276	270	302	t_3
1000	12	12	666	603	604	657	t_2
1000	13	13	716	646	647	702	t_2
1000	20	20	1047	926	930	1001	t_2

Acknowledgement. The authors would like to thank the anonymous reviewers for their careful reading and valuable comments.

References

1. Alon, N., Asodi, V.: Learning a hidden subgraph. SIAM J. Discrete Math. 18, 697–712 (2005)
2. Alon, N., Beigel, R., Kasif, S., Rudich, S., Sudakov, B.: Learning a hidden matching. SIAM J. Comput. 33, 487–501 (2004)
3. Barillot, E., Lacroix, B., Cohen, D.: Theoretical analysis of library screening using an n-dimensional pooling strategy. Nucleic Acids Research, 6241–6247 (1991)
4. Berger, T., Mehravari, N., Towsley, D., Wolf, J.: Random multipleaccess communication and group testing. IEEE Trans. Commun. 32, 769–779 (1984)
5. De Bonis, A.: New combinatorial structures with applications to efficient group testing with inhibitors. J. Combin. Optim. 15, 77–94 (2008)
6. De Bonis, A., Gąsieniec, L., Vaccaro, U.: Optimal two-stage algorithms for group testing problems. SIAM J. Comput. 34, 1253–1270 (2005)
7. De Bonis, A., Vaccaro, U.: Optimal algorithms for two group testing problems, and new bounds on generalized superimposed codes. IEEE Trans. Inform. Theory 52, 4673–4680 (2006)
8. Chang, H.L., Chen, H.B., Fu, H.L.: Identification and classification problems on pooling designs for inhibitor models. J. Comput. Biol. 17(7), 927–941 (2010)
9. Chang, H.L., Chen, H.B., Fu, H.L., Shi, C.H.: Reconstruction of hidden graphs and threshold group testing. J. Combin. Optim. 22, 270–281 (2011)
10. Chen, H.B., Du, D.Z., Hwang, F.K.: An unexpected meeting of four seemingly unrelated problems: graph testing, DNA complex screening, superimposed codes and secure key distribution. J. Combin. Optim. 14, 121–129 (2007)
11. Chen, H.B., Fu, H.L.: Nonadaptive algorithms for threshold group testing. Discrete Appl. Math. 157(7), 1581–1585 (2009)
12. Chen, H.B., Fu, H.L., Hwang, F.K.: An upper bound of the number of tests in pooling designs for the error-tolerant complex model. Optim. Lett. 2, 425–431 (2008)
13. Cheng, Y.X., Du, D.Z.: Efficient constructions of disjunct matrices with applications to DNA library screening. J. Comput. Biol. 14, 1208–1216 (2007)
14. Cheng, Y.X., Du, D.Z.: New constructions of one- and two-stage pooling designs. J. Comput. Biol. 15(2), 195–205 (2008)
15. Cheraghchi, M.: Improved Constructions for Non-adaptive Threshold Group Testing. In: Abramsky, S., Gavoille, C., Kirchner, C., Meyer auf der Heide, F., Spirakis, P.G. (eds.) ICALP 2010. LNCS, vol. 6198, pp. 552–564. Springer, Heidelberg (2010)
16. Damaschke, P.: Randomized group testing for mutually obscuring defectives. Inf. Process. Lett. 67, 131–135 (1998)
17. Damaschke, P.: Threshold Group Testing. In: Ahlswede, R., Bäumer, L., Cai, N., Aydinian, H., Blinovsky, V., Deppe, C., Mashurian, H. (eds.) Information Transfer and Combinatorics. LNCS, vol. 4123, pp. 707–718. Springer, Heidelberg (2006)
18. Du, D.Z., Hwang, F.K.: Combinatorial Group Testing and Its Applications, 2nd edn. World Scientific (2000)
19. Du, D.Z., Hwang, F.K.: Pooling Designs and Nonadaptive Group Testing - Important Tools for DNA Sequencing. World Scientific (2006)
20. D'yachkov, A.G.: Superimposed designs and codes for nonadaptive search of mutually obscuring defectives. In: Proc. 2003 IEEE Int. Symp. Inf. Theory, p. 134 (2003)
21. D'yachkov, A.G., Rykov, V.V.: Bounds on the length of disjunct codes. Problemy Peredachi Inform. 18(3), 7–13 (1982)

22. D'yachkov, A.G., Rykov, V.V.: A survey of superimposed code theory. Problems Control Inform. Theory 12, 229–242 (1983)
23. D'yachkov, A.G.: Lectures on designing screening experiments, Lecture Note Series 10, pages (monograph, pp. 112), Combinatorial and Computational Mathematics Center, Pohang University of Science and Technology (POSTECH), Korea Republic (2004)
24. D'yachkov, A.G., Rykov, V.V.: On superimposed codes. In: Fourth International Workshop Algebraic and Combinatorial Coding Theory, Novgorod, Russia, pp. 83–85 (1994)
25. D'yachkov, A.G., Rykov, V.V., Antonov, M.G.: New bounds on rate of the superimposed codes. In: The 10th All-Union Symposium for the Redundancy Problem in Information Systems, Papers, Part 1, St-Petersburg (1989)
26. Farach, M., Kannan, S., Knill, E., Muthukrishnan, S.: Group testing problem with sequences in experimental molecular biology. In: Proc. Compression and Complexity of Sequences, pp. 357–367 (1997)
27. Hong, E.H., Ladner, R.E.: Group testing for image compression. IEEE Trans. Image Process. 11, 901–911 (2002)
28. Hong, Y.W., Scaglione, A.: On multiple access for distributed dependent sensors: a content-based group testing approach. In: IEEE Information Theory Workshop, pp. 298–303 (2004)
29. Hwang, F.K., Sós, V.T.: Nonadaptive hypergeometric group testing. Studia Scient. Math. Hungarica 22 (1987)
30. Kautz, W.H., Singleton, R.R.: Nonrandom binary superimposed codes. IEEE Trans. Inform. Theory 10, 363–377 (1964)
31. Pevzner, P.A., Lipshutz, R.: Towards DNA Sequencing Chips. In: Privara, I., Ružička, P., Rovan, B. (eds.) MFCS 1994. LNCS, vol. 841, pp. 143–158. Springer, Heidelberg (1994)
32. Porat, E., Rothschild, A.: Explicit non-adaptive combinatorial group testing schemes. In: Proceedings of the 35th International Colloquium on Automata, Languages and Programming (ICALP), pp. 748–759 (2008)
33. Stinson, D.R., Wei, R.: Generalized cover-free families. Discrete Math., 463–477 (2004)
34. Stinson, D.R., Wei, R., Zhu, L.: Some new bounds for cover-free families. J. Combin. Theory Ser. A, 224–234 (2000)
35. Torney, D.C.: Sets pooling designs. Ann. Combin., 95–101 (1999)
36. Vermeirssen, V., Deplancke, B., Barrasa, M.I., Reece-Hoyes, J.S., et al.: Matrix and steiner-triple-system smart pooling assays for high-performance transcription regulatory network mapping. Nature Methods 4, 659–664 (2007)
37. Yu, H., Braun, P., Yildirim, M.A., Lemmens, I., Venkatesan, K., Sahalie, J., Hirozane-Kishikawa, T., Gebreab, F., Li, N., Simonis, N., et al.: High-quality binary protein interaction map of the yeast interactome network. Science 322, 104–110 (2008)

An Efficient Algorithm for Combinatorial Group Testing

Andreas Allemann

Lehrstuhl II für Mathematik, RWTH Aachen, D-52056 Aachen
allemann@zeuxis.de

Dedicated to the memory of Rudolf Ahlswede

Abstract. In the (d, n) group testing problem n items have to be identified as either good or defective and the number of defective items is known to be d. A test on an arbitrary group (subset) of items reveals either that all items in the group are good or that at least one of the items is defective, but not how many or which items are defective. We present a new algorithm which in the worst case needs less than $0.255d + \frac{1}{2}\log d + 5.5$ tests more than the information lower bound $\lceil \log \binom{n}{d} \rceil$ for $\frac{n}{d} \geq 2$. For $\frac{n}{d} \geq 38$, the difference decreases to less than $0.187d + \frac{1}{2}\log d + 5.5$ tests. For $d \geq 10$, this is a considerable improvement over the $d - 1$ additional tests given for the best previously known algorithm by Hwang. We conjecture that the behaviour for large n and d of the difference is optimal for $\frac{n}{d} \leq 4$. This implies that the $\frac{1}{2} - \log \frac{32}{27} = 0.255$ tests per defective given in the bound above are the best possible.

Keywords: group testing, nested algorithms, competitive analysis.

1 Introduction

Group testing is a class of search problems, in which we typically have a set of n items, each of which is either good or defective. A test on an arbitrary group (subset) of items reveals either that all items in the group are good or that at least one of the items is defective, but not how many or which items are defective. The aim is to identify all items as either good or defective using as few tests as possible. We focus on the *sequential* case, in which the results of preceding tests may be used to determine the next test group, and the *combinatorial* (worst case) group testing problem, in which the aim is to minimize the maximum number of tests required.

Let $\log x$ always denote $\log_2 x$.

In combinatorial group testing, it is typically assumed that there are exactly d defectives among the n items, and we try to minimize the worst case number of tests. This is called the (d, n) *group testing* problem. As there are $\binom{n}{d}$ possible sets of defectives and in t tests at most 2^t cases can be differentiated, at least $\lceil \log \binom{n}{d} \rceil$ tests are needed. This is called the *information lower bound.* The best

H. Aydinian, F. Cicalese, and C. Deppe (Eds.): Ahlswede Festschrift, LNCS 7777, pp. 569–596, 2013.

upper bound known in the literature was proven by Hwang [7] for his *generalized binary splitting algorithm*, which tests groups of size 2^a where the choice of the integer a depends on the proportion of defectives among the not yet identified items; whenever a test result is positive, a defective is identified by repeatedly halving the contaminated group. The number of tests needed by the algorithm exceeds the information lower bound at most by $d-1$. The generalized binary splitting algorithm belongs to the *nested class*, in which if a group of at least two items is known to contain a defective, a subset of this group is tested next.

In practical applications the exact number of defectives is hardly ever known. This is addressed by the *strict group testing* problem. The number of defectives among n items is unknown and we try to minimize the worst case number of tests needed if there happen to be at most d defectives. This is similar to the *generalized (d, n) group testing* problem, in which it is known that there are at most d defectives, with the crucial difference that the case of more than d defectives is not excluded but has to be detected as well. This avoids the unfortunate property of the (d, n) and generalized (d, n) group testing problems that algorithms designed for these problems typically miss defectives or may even report good items as defective if more than d items happen to be defective in a practical application. I proved in [1, section 2.2] that the optimal algorithm for the strict group testing problem needs exactly one additional test compared to the optimal algorithm for the (d, n) problem. As the proof is constructive, it provides explicit instructions to transform algorithms between these two problems.

For the case that nothing is known about the number of defectives, Du and Hwang [4] formulated the *competitive group testing* problem. A competitive algorithm requires for every number of items n and number of defectives d at most a fixed multiple, called *competitive ratio*, of the number of tests used by the optimal algorithm for the (d, n) group testing problem plus a constant. The lowest competitive ratio achieved so far is $1.5 + \varepsilon$, for which Schlaghoff and Triesch [10] gave competitive algorithms for all positive ε.

Many variations of these group testing problems and numerous applications in a wide range of fields have been examined in the literature. Du and Hwang [3] give a comprehensive overview of the combinatorial side of group testing in their book as of the year 2000.

In this paper we introduce a new algorithm for combinatorial group testing from my doctoral thesis [1]. We define the *split and overlap algorithm* for the strict group testing problem, but it can easily be adapted to the (d, n) or generalized (d, n) group testing problems by simply omitting all tests that become predictable due to the additional information about the number of defectives. In the beginning, the initial test group size m is chosen depending on the ratio $\frac{n}{d}$. The algorithm then repeatedly tests groups of size m and whenever a test result is positive at least one defective is identified in the contaminated group of size m by splitting it repeatedly, similarly to nested algorithms. However, unlike in nested algorithms, complex subalgorithms involving overlapping test groups are used on contaminated groups of certain sizes. These subalgorithms make it possible to deal efficiently with arbitrary test group sizes m, not just with powers of two, as in Hwang's generalized binary splitting algorithm.

The following estimates for the number of tests required by the split and overlap algorithm all refer to its application to the (d, n) group testing problem. We demonstrate that for the initial test group size m, the algorithm needs at most $\frac{1}{m}$ tests for each good item and a constant number of tests per defective identified plus 4 tests. This is less than $0.255d + \frac{1}{2}\log d + 5.5$ tests above the information lower bound $\lceil \log \binom{n}{d} \rceil$ for $\frac{n}{d} \geq 2$. For $\frac{n}{d} \geq 38$, the difference decreases to less than $0.187d + \frac{1}{2}\log d + 5.5$ tests. For $d \geq 10$, this is a considerable improvement over the $d-1$ tests given by Hwang [7] for his generalized binary splitting algorithm. We conjecture that the behaviour for large n and d of the difference between the number of tests required by the split and overlap algorithm and the information lower bound is optimal for $\frac{n}{d} \leq 4$. This implies that the $\frac{1}{2} - \log \frac{32}{27} = 0.255$ tests per defective given in the bound above are the best possible. Interestingly, leaving the nested class and using overlapping test groups seems to yield much bigger gains in combinatorial than in probabilistic group testing.

We build up the split and overlap algorithm modularly. After basic definitions, a nested subalgorithm, and a brief overview over the split and overlap algorithm, we introduce new efficient subalgorithms and a general method of scaling up arbitrary subalgorithms by powers of two. Using these as building blocks, we construct algorithms that repeatedly test groups of the same size and always follow the same procedure if a group is found to be contaminated. The split and overlap algorithm then works by choosing the best of these algorithms depending on $\frac{n}{d}$. At each stage, estimates of the maximum number of tests used are provided, and finally an estimate for the whole algorithm is given.

2 The Group Testing Model

Consider a set I of n items each being either good or defective. The state of the items can be determined only by testing subsets of I. A test can yield a negative result, indicating that all tested items are good, or a positive result, indicating that the group is contaminated, that is, at least one of the tested items is defective. We call an item *free* if there is no information about the item, i.e., it is not known to be either good or defective and does not belong to a contaminated group. All tests are considered to be error-free.

We consider the *strict group testing problem*: The number of defectives is unknown and an algorithm must identify all items as good or defective. Denote by $L_A(d, n)$ the maximum number of tests needed by the algorithm A in those cases in which there are at most d defectives. This implies that A detects in at most $L_A(d, n)$ tests if there are more than d defectives. Let $L(d, n) = \min_A L_A(d, n)$ denote the worst case number of tests of a minimax algorithm.

Denote by $M_A(d, n)$ the number of tests needed by A for the classical (d, n) group testing problem in which there are known to be at most d defectives. Let $M(d, n) = \min_A M_A(d, n)$ denote the worst case number of tests of a minimax algorithm for the (d, n) group testing problem.

3 Nested Algorithms

In the nested class introduced by Sobel and Groll [11], if a group with at least two items is known to be contaminated, then the next test has to be performed on a subset of this group.

Chang, Hwang, and Weng [2] gave the following binary splitting algorithm to identify one defective in a contaminated group L of $l \geq 2$ items.

Subalgorithm B_l

Let $k = \max\left(l - 2^{\lceil \log l \rceil - 1}, 2^{\lceil \log l \rceil - 2}\right)$.
Test a group $K \subset L$ of size k.
If the result is positive,
  apply B_k to K,
else
  apply B_{l-k} to $L \setminus K$.

B_1 identifies a defective without any further tests. If l is a power of 2, the algorithm B_l repeatedly halves the size of the contaminated group.

Lemma 1. *The subalgorithm B_l identifies a defective in at most $\lceil \log l \rceil$ tests. If $\lceil \log l \rceil$ tests are actually used, then at least $2^{\lceil \log l \rceil} - l$ good items are identified as well.*

The proof of Lemma 1 is by induction and is detailed in [2].

Hwang [7] gave the following generalized binary splitting algorithm for the (d, n) group testing problem.

Algorithm G

Denote by I the set of n items containing d defectives. Always remove items from I when they are identified as good or defective.

While $d \geq 1$:
  If $|I| \leq 2d - 2$, test all items in I individually and then stop.
  Let $k = \left\lfloor \log \frac{|I|-d+1}{d} \right\rfloor$.
  Test a group of size 2^k.
  If the result is positive,
    apply B_{2^k} to the contaminated group and set $d := d - 1$.

Du and Hwang [3, Section 2.4] proved that the generalized binary splitting algorithm G exceeds the information lower bound by at most $d - 1$ tests for $d \geq 2$. Interestingly, the application of a much more general result for hypergraphs by Triesch [12] to complete hypergraphs of rank d yields the same upper bound. The corresponding algorithm also belongs to the nested class, but differs considerably from Hwang's generalized binary splitting algorithm.

Du and Hwang [3, Section 2.5] gave a set of recursive equations that describe the number of tests required by a minimax nested algorithm and a procedure to find such an algorithm based on computing the solution to these equations. This algorithm repeatedly tests a group of some size l depending on the actual values of n and d and applies B_l if the group is found to be contaminated. Unfortunately, no estimate of the number of tests needed by this minimax nested algorithm is provided.

4 Overview of the Split and Overlap Algorithm

At the start of the split and overlap algorithm, we choose the *initial test group size* m depending only on the ratio $\frac{n}{d}$ of all items to the defectives.

If we do not know any contaminated group, we always test a group of m items as long as there are enough free items left. If the test result is negative, we test the next group of m items. Otherwise, if the result is positive, we proceed to identify at least one defective in the group of m items. As the nested algorithms in the last section, we always test subsets of the current contaminated group. However, contaminated groups of certain sizes are dealt with by special subalgorithms, which are more efficient due to the use of overlapping test groups. Some of these subalgorithms identify at least two or three defectives starting with as many disjoint contaminated groups of the same size l. For these subalgorithms we collect the required number of contaminated groups of size l by setting them aside as they occur. After identifying at least one defective or setting aside a contaminated group of size l we continue by testing the next group of the same size m as before.

When there are not enough free items left to repeat this procedure, we use binary splitting till the end in a way specified in Section 8.

We show in Section 9 that the algorithm with initial test group size m requires at most $\frac{1}{m}$ tests for the identification of each good item and a constant number of tests to identify each defective plus 5 tests.

5 Overlapping Subalgorithms

We can represent a subalgorithm as a binary tree in the following way. Each test is represented by a node that has two subtrees for the cases of a negative and positive test result, respectively. By convention, we always depict the negative test result on the left and the positive test result on the right side. The root corresponds to the start of the subalgorithm, when there are one or more contaminated groups of size l. Each leaf marks the end of the subalgorithm, when at least one defective has been identified and only contaminated groups of size l may be left.

At each node we display the part of the information test hypergraph that is relevant to the subalgorithm and specifically omit edges containing just a single item. Additional edges that are disjoint to the displayed edges can be omitted safely, as they do not interfere with the others as long as no items from them are included in the test group. The next test group is always indicated by a dashed line.

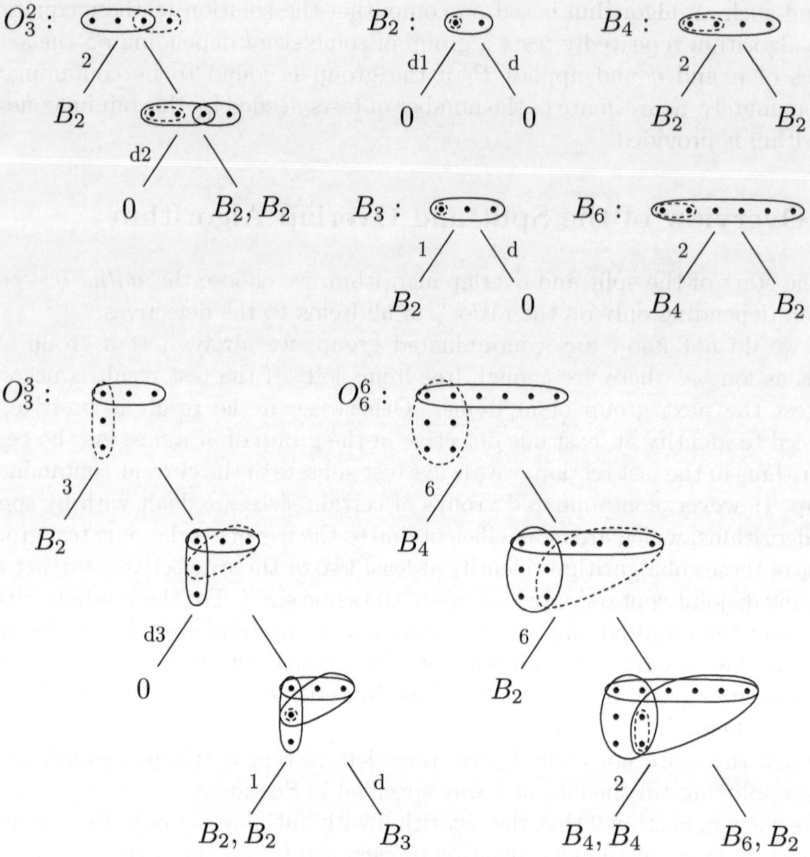

Fig. 1. The subalgorithms O_3^2,O_3^3, and O_6^6

The number of good items identified in the case of a negative test result, which always coincides with the size of the test group, is displayed above the connection to the left child. An item identified as defective is indicated by the letter 'd' above the connection to the corresponding child.

A zero on a leaf of the binary tree indicates that no contaminated group containing at least two items is left. Subtrees identical to subalgorithms already shown are represented by the name of the corresponding subalgorithm instead of the contaminated group; for example, B_2 indicates that an edge containing two items is left of which one is tested next. Several subalgorithms separated by commas indicate that the corresponding edges are dealt with independently one after the other. A contaminated group that is left over is represented by a number indicating its size.

Items from a contaminated group that become free again are reused for further tests whenever possible to reduce the number of free items required by the subalgorithms.

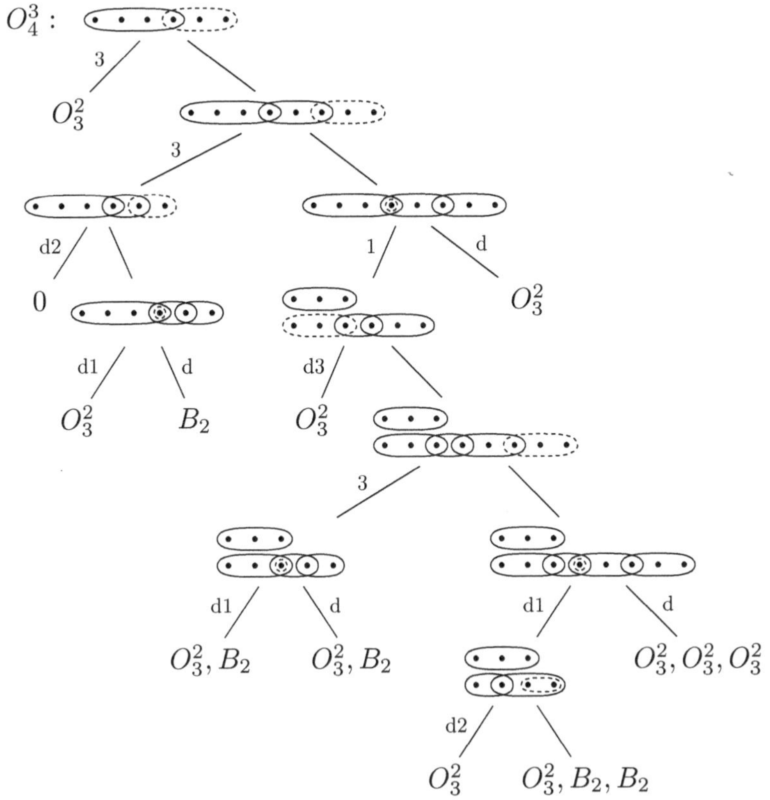

Fig. 2. The subalgorithm O_4^3

The nested algorithms in Section 3 use subalgorithm B_3 on contaminated groups of three items, that is, they first test a single item from the group. In the worst case, B_3 identifies one defective and one good item in two tests. The subalgorithms O_3^2 and O_3^3 depicted in Figure 1 improve on this by first testing a group of size two respectively three that contains exactly one of the three items in the original contaminated group. If one of the first two test results is negative, O_3^2 and O_3^3 identify one defective in two tests, like B_3, but at least two respectively three good items instead of one. Otherwise, in the case of two consecutive positive test results, they need four respectively five tests, but identify two defectives instead of one. In addition to the contaminated group of size three, O_3^2 requires one and O_3^3 requires two free items. The technique to derive O_6^6 in Figure 1 from O_3^3 is explained in Section 7.

On a contaminated group of four items, nested algorithms always use subalgorithm B_4, that is, they halve the group two times. In the worst case, B_4 identifies one defective and no good items in two tests. The algorithm that always tests a group of four items and then uses B_4 if the test result is positive needs three tests for the identification of each defective. Surprisingly, we can do better than

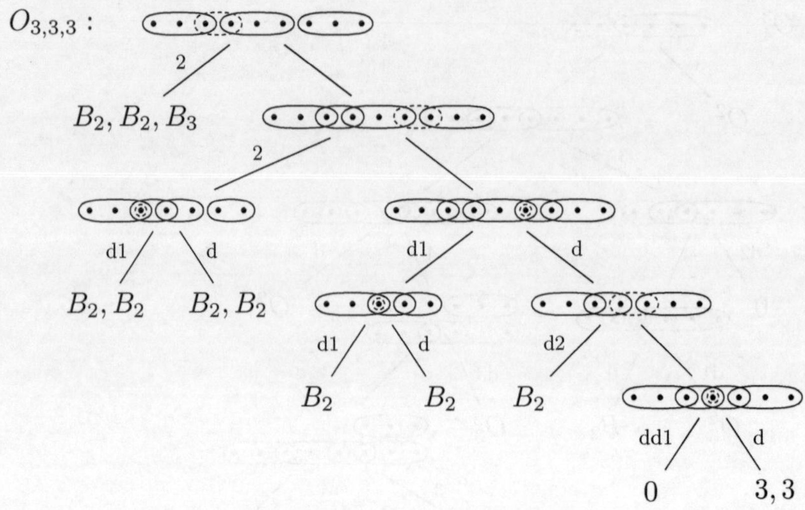

Fig. 3. The subalgorithm $O_{3,3,3}$

this. The subalgorithm O_4^3 depicted in Figure 2 starts by testing a group of size three that contains exactly one of the four items in the original contaminated group. It typically needs more tests than B_4 to identify a defective, but more than compensates for this by identifying good items as well. In addition to the contaminated group of size four, O_4^3 requires up to ten free items.

If a subalgorithm is used in an algorithm with the initial test group size m, the identification of each good item needs $\frac{1}{m}$ tests. The preceding subalgorithms improve on B_3 and B_4 only by identifying more good items. They are therefore best for small m. To construct subalgorithms on a contaminated group of size l that are good for large m, we have to reduce the number of tests needed to identify a defective. As we have to distinguish at least the l cases in which exactly one of the items in the group of size l is defective, we need at least $\lceil \log l \rceil$ tests, which is already achieved by subalgorithm B_l. To do better, we simultaneously identify at least one defective in each of k contaminated groups of size l. This requires at least $\lceil \log l^k \rceil$ tests in the worst case, that is, $\frac{1}{k} \lceil k \log l \rceil$ tests per defective.

The subalgorithm $O_{3,3,3}$ depicted in Figure 3 needs five tests to identify three defectives in three contaminated groups of size three. The $1\frac{2}{3}$ tests per defective compare favourably with the two tests needed by B_3, O_3^2, and O_3^3. Additionally, $O_{3,3,3}$ identifies at least one good item. If all five test results are positive, then no good items can be identified, and the subalgorithm identifies two defectives in one of the groups of size three and leaves the other two groups unchanged.

It would be even better to use eight tests to identify five defectives in five contaminated groups of size three, yielding $1\frac{3}{5}$ tests per defective. This would require the $3^5 = 243$ cases in which each of the five groups contains exactly one defective to be split by the first test in such a way that neither after a positive

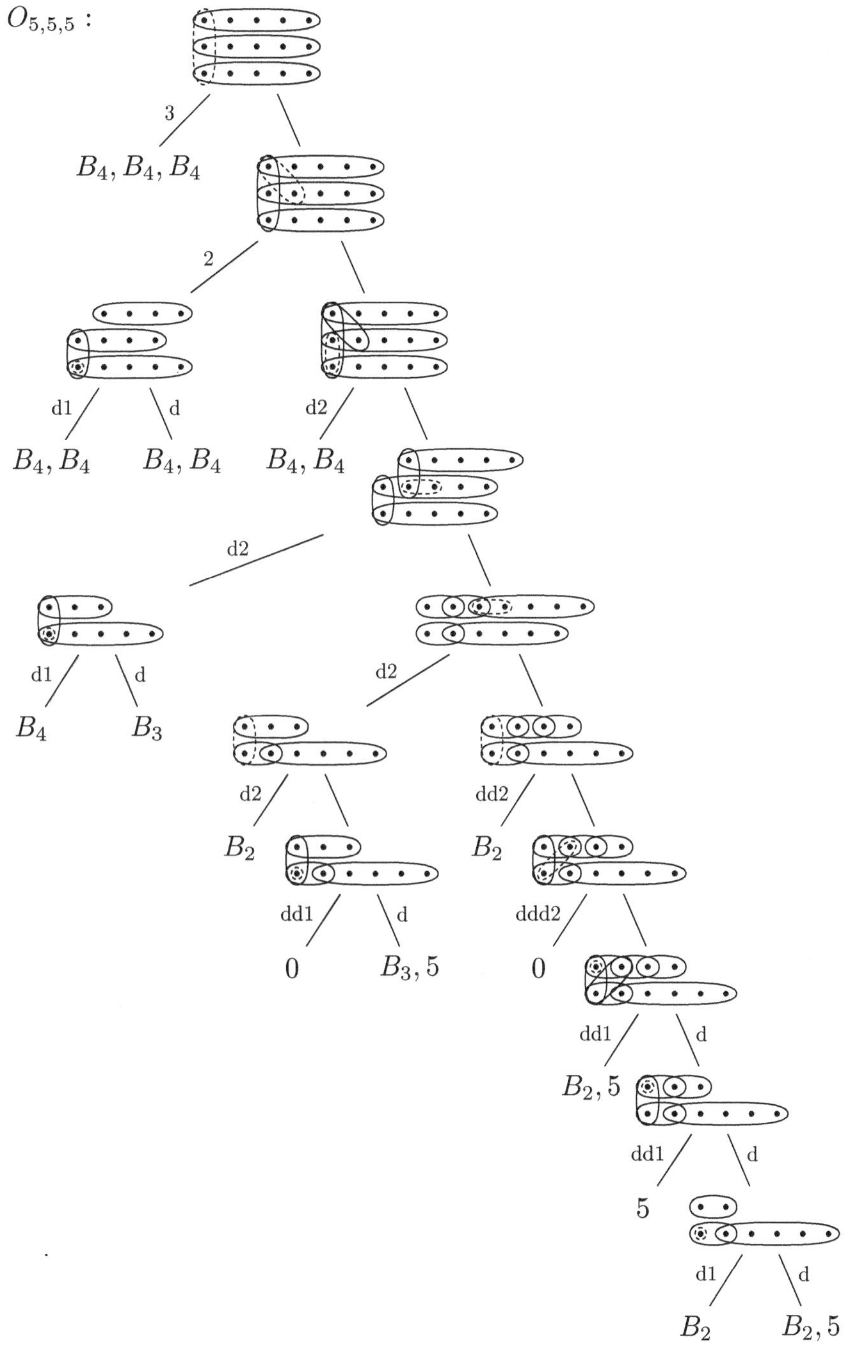

Fig. 4. The subalgorithm $O_{5,5,5}$

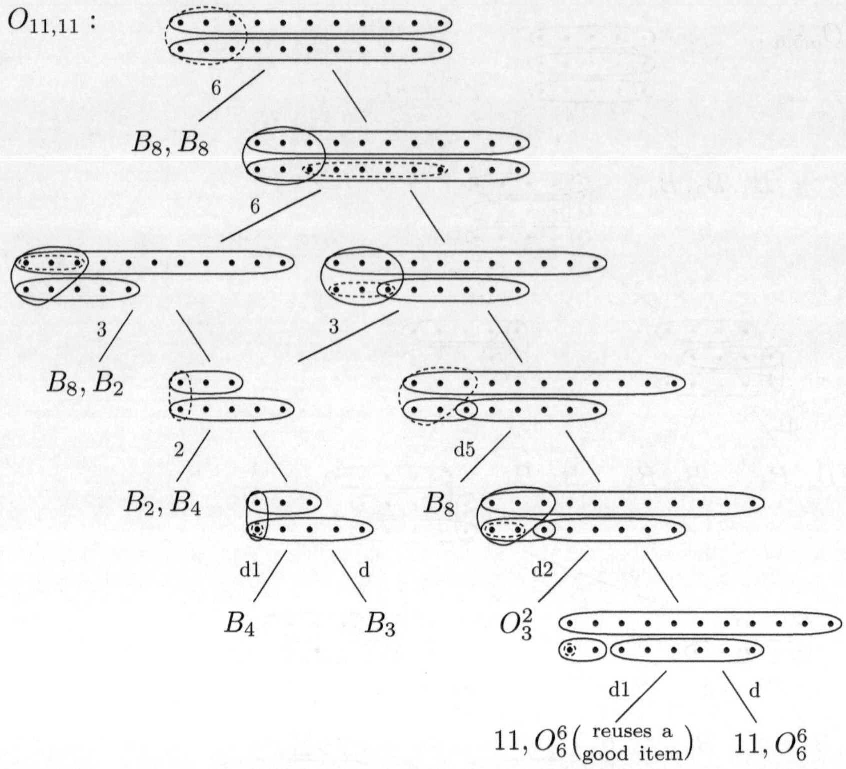

Fig. 5. The subalgorithm $O_{11,11}$

nor a negative test result more than $2^7 = 128$ cases remain, which cannot be accomplished by a group test.

Similarly to $O_{3,3,3}$, the subalgorithm $O_{5,5,5}$ depicted in Figure 4 starts with three contaminated groups of size five and typically identifies three defectives and at least two good items in seven tests. The $2\frac{1}{3}$ tests per defective are much better than the three tests required by B_5. Alternatively, $O_{5,5,5}$ may identify three defectives and at least one good item in nine tests while leaving one of the original groups of size five unchanged. All other possibilities are not relevant for the worst case, as we show in the next section.

The subalgorithm $O_{11,11}$ depicted in Figure 5 needs seven tests to identify two defectives and at least four good items in two contaminated groups of eleven items. Again, the $3\frac{1}{2}$ tests per defective are less than the four tests required by B_{11}. Alternatively, the subalgorithm may identify two defectives and at least six good items in nine tests while leaving one of the original groups of size eleven unchanged, whereas all other possibilities turn out not to be relevant for the worst case. If the results of the first five tests are positive and of the sixth test negative, only three of the items in the two groups of size eleven are not contained in any contaminated group, but subalgorithm O_6^6 needs four free items. To avoid requiring the existence of another free item, O_6^6 reuses an item already identified as good in the next test, which may reduce the number of good items identified by one but does not affect the worst case behaviour of $O_{11,11}$.

6 Cost Estimate of Subalgorithms

In the binary tree representation, each sequence of test results in the subalgorithm corresponds to a path from the root to a leaf. We denote the set of all these paths by T. From now on, we refer to paths from the root to a leaf simply as paths.

For the path $P \in T$ denote by d_P the number of defectives and by g_P the number of good items identified along the path; denote by t_P the number of tests needed; denote by s_P the number of contaminated sets of size l present in the root of the tree but not in the leaf at the end of the path, that is, the number of groups of size l used up to identify defectives. Let $\bar{g}_P = \frac{g_P}{d_P}$, $\bar{t}_P = \frac{t_P}{d_P}$, and $\bar{s}_P = \frac{s_P}{d_P}$ denote the corresponding values per identified defective.

In the following, we assume that the main algorithm always tests a group of size m if no contaminated groups are known and that all contaminated groups of size l required by the subalgorithm are obtained along the same path in the algorithm. Denote by $t_{m,l}$ the number of tests needed on this path until a contaminated group of size l is known including the first test on m items with positive result and denote by $g_{m,l}$ the number of good items identified in the process. If we assign a cost of $\frac{1}{m}$ tests to the identification of each good item, then $c_{m,l} = t_{m,l} - \frac{g_{m,l}}{m}$ describes the cost to get a contaminated group of size l.

If the tests follow the path P inside the subalgorithm, then d_P defectives are identified using $s_P t_{m,l} + t_P$ tests while identifying $s_P g_{m,l} + g_P$ good items. Then we get $f\left(c_{m,l}, \frac{1}{m}\right)$, the worst case cost to identify a defective, by taking the maximum over all paths in the subalgorithm.

$$\begin{aligned} f\left(c_{m,l}, \tfrac{1}{m}\right) &= \max_{P \in T} \frac{1}{d_P}\left(s_P t_{m,l} + t_P - \frac{s_P g_{m,l}}{m} - \frac{g_P}{m}\right) \\ &= \max_{P \in T} \bar{s}_P c_{m,l} + \bar{t}_P - \frac{\bar{g}_P}{m} \\ &= \max_{P \in T} f_P\left(c_{m,l}, \tfrac{1}{m}\right). \end{aligned}$$

Here, let

$$f_P(x, y) = \bar{t}_P + \bar{s}_P x - \bar{g}_P y$$

be the cost to identify a defective if path P is followed in the subalgorithm where $x = c_{m,l}$ and $y = \frac{1}{m}$ are determined by the main algorithm. This can be depicted as a plane in R^3. Then $f(x, y)$ is the maximum over the planes of every path in the subalgorithm. As $m \geq l$ and $c_{m,l} \geq 1$ always hold, we are interested only in the domain $D_l = \left\{(x, y) | 1 \leq x, 0 \leq y \leq \frac{1}{l}\right\}$.

The following lemma gives the conditions under which the plane belonging to the path $Q \in T$ is below the plane of the path $P \in T$ on the whole domain D_l.

Lemma 2. *$f_P(x, y) \geq f_Q(x, y)$ on D_l if and only if $\bar{s}_P \geq \bar{s}_Q$ and $\bar{s}_P + \bar{t}_P \geq \bar{s}_Q + \bar{t}_Q$ and $\bar{s}_P + \bar{t}_P - \frac{\bar{g}_P}{l} \geq \bar{s}_Q + \bar{t}_Q - \frac{\bar{g}_Q}{l}$.*

Table 1. Worst case costs of subalgorithms

	path P	$\bar{s}_P$	$\bar{t}_P$	$\bar{g}_P$	path Q	$\bar{s}_Q$	$\bar{t}_Q$	$\bar{g}_Q$
B_{2^k}	1…1	1	k	0				
O_3^2	01	1	2	2	1111	$\frac{1}{2}$	2	0
O_3^3	01	1	2	3	11011	$\frac{1}{2}$	$2\frac{1}{2}$	$\frac{1}{2}$
$O_{3,3,3}$	11110	1	$1\frac{2}{3}$	$\frac{1}{3}$	11111	$\frac{1}{2}$	$2\frac{1}{2}$	0
O_4^3	001	1	3	5	110110001	$\frac{1}{3}$	3	2
$O_{5,5,5}$	1011111	1	$2\frac{1}{3}$	$\frac{2}{3}$	111111110	$\frac{2}{3}$	3	$\frac{1}{3}$
$O_{11,11}$	1101101	1	$3\frac{1}{2}$	2	111111011	$\frac{1}{2}$	$4\frac{1}{2}$	3

Proof. $\bar{s}_P + \bar{t}_P \geq \bar{s}_Q + \bar{t}_Q$ is equivalent to $f_P(1,0) \geq f_Q(1,0)$, and $\bar{s}_P + \bar{t}_P - \frac{\bar{g}_P}{l} \geq \bar{s}_Q + \bar{t}_Q - \frac{\bar{g}_Q}{l}$ is equivalent to $f_P(1, \frac{1}{l}) \geq f_Q(1, \frac{1}{l})$, whereas $\bar{s}_P$ and $\bar{s}_Q$ are the gradients of $f_P(x,y)$ and $f_Q(x,y)$ in x-direction. Therefore the conditions on the right side are clearly sufficient. Conversely, $\bar{s}_P \geq \bar{s}_Q$ is necessary, as else $f_P(x,y) \geq f_Q(x,y)$ would be violated for x sufficiently large, whereas the other conditions are obviously necessary.

We call $W \subset T$ a *worst case set* of paths if on the whole domain D_l

$$\max_{P \in W} f_P(x,y) = \max_{P \in T} f_P(x,y),$$

that is, W shows the same worst case behaviour as T. We represent a path by the sequence of the test results with 0 standing for a negative and 1 for a positive test result. Table 1 lists paths constituting a worst case set for all subalgorithms presented in the last section and for B_{2^k} with $k \geq 1$ together with the parameters of their planes. This is shown in the following Lemma.

Lemma 3. *For each subalgorithm the paths given in Table 1 constitute a worst case set.*

Proof. For each subalgorithm, the application of Lemma 2 shows that the planes of all other paths are below one of the listed paths on the whole domain. There is only one exception in subalgorithm $O_{5,5,5}$. Denote by R the path 11111111101. Then $\bar{s}_R = \frac{3}{4}$, $\bar{t}_R = 2\frac{3}{4}$, and $\bar{g}_R = \frac{1}{4}$. With $y \leq \frac{1}{5}$ follows $f_P(x,y) \geq f_R(x,y)$ for $x \geq 2$ and $f_Q(x,y) \geq f_R(x,y)$ for $x \leq 2$. Hence, the plane of R is below the maximum of the planes of P and Q from Table 1 on the whole domain D_5.

7 Scaling Up Subalgorithms

We can get subalgorithms for further test sizes by scaling up known subalgorithms. To do this, we choose a subalgorithm and $k \geq 1$. In all group tests we substitute 2^k items for each item. Thus, the size of all test groups and the overlap with existing contaminated groups is multiplied by 2^k. Whenever the original subalgorithm identifies a defective, the scaled up version uses B_{2^k} to identify a defective in the corresponding group of 2^k items using k tests. If we scale up a subalgorithm on contaminated groups of size l, say X_l, we get a subalgorithm on contaminated groups of size $2^k l$, which we denote by $X_l \times B_{2^k}$. For example, the binary tree representation of $O_6^6 = O_3^3 \times B_2$ is depicted in Figure 1. If P is a path in X_l, we denote by $P \times B_{2^k}$ the path in $X_l \times B_{2^k}$ represented by the sequence of test results from P with k positive results inserted for each occurrence of B_{2^k} in $X_l \times B_{2^k}$. Note that $(X_l \times B_{2^k}) \times B_{2^j} = X_l \times (B_{2^k} \times B_{2^j}) = X_l \times B_{2^{k+j}}$.

The following lemma describes the worst case behaviour of $X_l \times B_{2^k}$.

Lemma 4. *Let W be a worst case set of paths in the subalgorithm X_l. Then*

$$W \times B_{2^k} = \{P \times B_{2^k} | P \in W\}$$

is a worst case set of paths in $X_l \times B_{2^k}$. For each path P in X_l the plane of the path $P' = P \times B_{2^k}$ is determined by $\bar{s}_{P'} = \bar{s}_P$, $\bar{t}_{P'} = \bar{t}_P + k$, and $\bar{g}_{P'} = 2^k \bar{g}_P$.

Proof. Along the path $P' = P \times B_{2^k}$ one defective is identified by the last test of each B_{2^k} that occurs in $X_l \times B_{2^k}$ exactly where X_l identifies a defective. Hence $d_{P'} = d_P$. The transition from P to P' does not change the number of contaminated groups of size l and $2^k l$ respectively that are present at the start and remain at the end of the path. Thus $\bar{s}_{P'} = \frac{s_{P'}}{d_{P'}} = \frac{s_P}{d_P} = \bar{s}_P$. The path P' contains k additional tests per identified defective compared to P. Therefore $\bar{t}_{P'} = \frac{t_{P'}}{d_{P'}} = \frac{t_P + k d_P}{d_P} = \bar{t}_P + k$. Good items are identified by tests with negative results only, whose test group sizes are all multiplied by 2^k in the transition from P to P'. The additional tests in P' all yield positive test results. Hence $\bar{g}_{P'} = \frac{g_{P'}}{d_{P'}} = \frac{2^k g_P}{d_P} = 2^k \bar{g}_P$.

Together, this leads to

$$\begin{aligned} f_{P'}(x,y) &= \bar{t}_{P'} + \bar{s}_{P'} x - \bar{g}_{P'} y \\ &= \bar{t}_P + k + \bar{s}_P x - 2^k \bar{g}_P y \\ &= f_P(x, 2^k y) + k. \end{aligned}$$

In this, $(x,y) \in D_{2^k l}$ if and only if $(x, 2^k y) \in D_l$. Denote by T and T' the set of all paths in X_l and $X_l \times B_{2^k}$, respectively. Then for all $(x,y) \in D_{2^k l}$

$$\begin{aligned} \max_{P' \in W \times B_{2^k}} f_{P'}(x,y) &= \max_{P \in W} f_P(x, 2^k y) + k \\ &= \max_{P \in T} f_P(x, 2^k y) + k \\ &= \max_{P' \in T \times B_{2^k}} f_{P'}(x,y) \\ &= \max_{P' \in T'} f_{P'}(x,y). \end{aligned}$$

The last equality is due to the fact that negative instead of positive test results in a B_{2^k} yield additional good items without changing anything else. Therefore, the cost of a path that contains such negative test results is always lower than the cost of the path from $T \times B_{2^k}$ whose test results in the B_{2^k} are always positive. Hence $W \times B_{2^k}$ is a worst case set.

This method of scaling up X_l is superior to repeatedly halving groups of size $2^k l$ and applying X_l to the resulting groups of size l, because it requires the same number of tests but additionally scales up the number of good items identified.

It is possible to scale up a subalgorithm by multiples other than powers of two by using some other subalgorithm instead of B_{2^k} in the construction above. However, this seems to result in comparatively less efficient algorithms.

We indicate the scaling up of one of the subalgorithms introduced in the last section by multiplying all its indices by the scaling factor, for instance, $O_{10,10,10} = O_{5,5,5} \times B_2$ and $O_{12}^{12} = O_3^3 \times B_4$.

8 Fixed Size Algorithms

We can combine subalgorithms in the following way. Assume that a contaminated group M of size m is known. Then we test a subset $L \subset M$ of size $l < m$. If the test result is positive, we continue with subalgorithm X_l on L. If the test result is negative, we have identified l good items and proceed with subalgorithm Y_{m-l} on $M \setminus L$. We denote this splitting of a contaminated group of size m by $S_m(X_l, Y_{m-l})$. By nesting these splittings we can build procedures for any start size m. If a subalgorithm, for instance, $O_{3,3,3}$, requires more than one group of size l, then the contaminated groups of size l are put aside until enough groups have been collected to execute the subalgorithm.

We denote by A_m an algorithm based on the best procedure with initial test group size m constructed in this way. Table 2 lists the procedures of A_m for selected initial test group sizes m with $m \leq 80$. The choice of these particular values for m is explained in Section 10. Procedures for $m \geq 80$ can be obtained by scaling up the procedures from Table 2 with $40 \leq m < 80$ in the same way as the subalgorithms in the last section. Then the procedure of $A_{2^k m} = A_m \times B_{2^k}$ with $k \geq 1$ has the same structure as the procedure of A_m with all test sizes multiplied by 2^k and each subalgorithm X_l substituted by $X_l \times B_{2^k}$.

A procedure can be represented as a binary tree. The root of the tree corresponds to the start, when no contaminated groups are known. At each node the size of the only contaminated group is given. Each leaf, except the one after a negative result in the first test, corresponds to the situation in which one contaminated group of some size l is known and is marked by the name of a subalgorithm that works on contaminated groups of size l. The binary tree representations of the procedures of A_{40}, A_{77}, and A_{80} are shown in Figures 6 and 7.

We denote the set of all paths from the root to a leaf by T, excluding the path that consists of one negative test result. For the path $Q \in T$ we denote by t_Q the length of the path, that is, the number of tests performed, and by g_Q the number of good items identified in the process. Furthermore we use the notation defined in Section 6. Let $W(Q)$ denote the worst case set of the subalgorithm

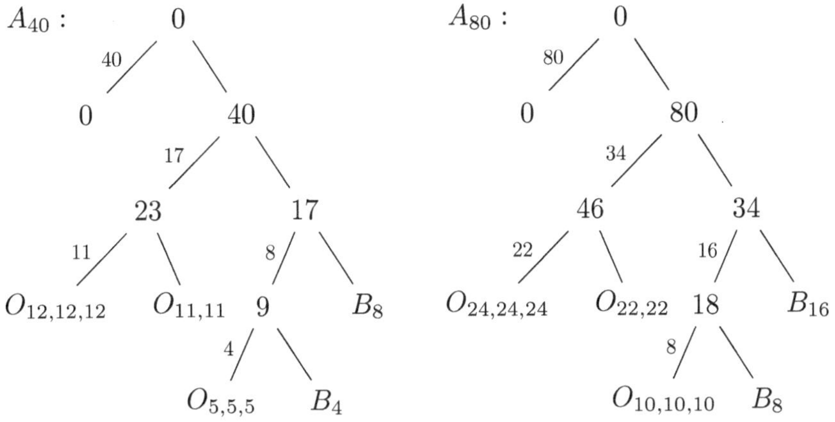

Fig. 6. The procedures of A_{40} and A_{80}

at the leaf of path Q that is based on Table 1 and Lemmas 3 and 4. We define the worst case cost c_m to identify a defective by

$$\begin{aligned} c_m &= \max_{Q\in T} \max_{P\in W(Q)} f_P\left(t_Q - \tfrac{g_Q}{m}, \tfrac{1}{m}\right) \\ &= \max_{Q\in T} \max_{P\in W(Q)} \bar{s}_P\left(t_Q - \tfrac{g_Q}{m}\right) + \bar{t}_P - \tfrac{\bar{g}_P}{m}. \end{aligned}$$

The values of c_m are listed in Table 2 for $m \le 80$. The values of c_m for $m \ge 80$ can be obtained by the following Lemma.

Lemma 5. *For $m \ge 40$ and $k \ge 1$,*

$$c_{2^k m} = c_m + k.$$

Proof. Denote by Q and Q' paths in the binary tree representations of the procedures of A_m and $A_{2^k m} = A_m \times B_{2^k}$ respectively that describe the same sequence of test results. Then $t_{Q'} = t_Q$ and $g_{Q'} = 2^k g_Q$. For $P \in W(Q)$ and $P' = P \times B_{2^k}$ Lemma 4 provides $\bar{s}_{P'} = \bar{s}_P$, $\bar{t}_{P'} = \bar{t}_P + k$, and $\bar{g}_{P'} = 2^k \bar{g}_P$. Together, this results in $f_{P'}\left(t_{Q'} - \frac{g_{Q'}}{2^k m}, \frac{1}{2^k m}\right) = f_P\left(t_Q - \frac{g_Q}{m}, \frac{1}{m}\right) + k$. Substitution in the definition of $c_{2^k m}$ proves the lemma.

We denote the maximum number of free items required by the procedure of A_m by n_m. In most procedures in Table 2, the free items required by a subalgorithm like O_3^3 can be drawn from items that belong to the m items of the initial test group but have become free again. This is not affected by scaling up a procedure. Therefore, in general $n_m = m$. The only exceptions are A_3 and A_4, for which $n_3 = 4$ and $n_4 = 14$ according to the description of O_3^2 and O_4^3 in Section 5.

When less than n_m free items are left, the procedure of A_m cannot be executed any more. Therefore, we continue by testing groups of size l_m and using B_{l_m} to identify a defective in the case of a positive test result where l_m is the largest

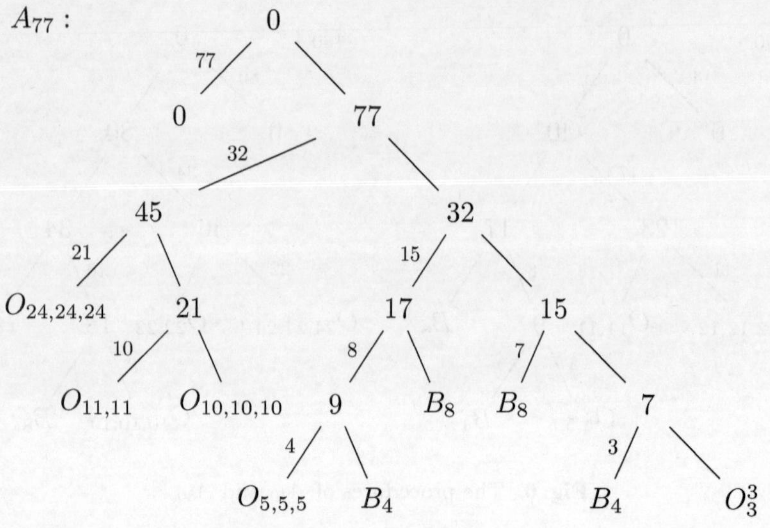

Fig. 7. The procedure of A_{77}

integer such that the cost of identifying a defective is at most c_m, that is, not greater than for the procedure of A_m. This leads to the definition

$$l_m = \max(2^{\lfloor c_m \rfloor - 1}, 2^{\lceil c_m \rceil - 1} - \lceil (\lceil c_m \rceil - c_m)m \rceil).$$

The values of l_m for $m \leq 80$ are listed in Table 2. As $l_m < m$ for $m > 2$, the cost of identifying a good item is greater with this method than by using the procedure of A_m.

In the following, we give a description of the *fixed size split and overlap algorithm* A_m where m can be any value for which a procedure is given in Table 2 or one of the values greater than or equal 40 from Table 2 multiplied by a power of two.

Algorithm A_m

Denote by I the set of n items. Always remove items from I when they are identified as good or defective. Denote by $\mathcal{C}$ a collection of contaminated groups that is initially empty.

Part 1

While I contains at least n_m free items,
  test a group M of m free items.
  If the result is positive,
    continue the procedure on M.
    If the final subalgorithm requires $k \geq 2$ contaminated groups of size l,
      add the contaminated group of size l to $\mathcal{C}$.
      If $\mathcal{C}$ contains k groups of size l,
        apply the subalgorithm and remove the k groups from $\mathcal{C}$.

Table 2. The fixed size algorithms A_m for $m \leq 80$

m	c_m	n_m	l_m	procedure of A_m
1	1	1	1	B_1
2	2	2	2	B_2
3	$2\frac{1}{2}$	4	2	O_3^2
4	$2\frac{5}{6}$	14	3	O_4^3
7	$3\frac{4}{7}$	7	5	$S_7(O_3^3, B_4)$
12	$4\frac{1}{3}$	12	8	$S_{12}(O_{5,5,5}, S_7(O_3^3, B_4))$
15	$4\frac{28}{45}$	15	10	$S_{15}(O_6^6, S_9(B_4, O_{5,5,5}))$
19	5	19	16	$S_{19}(B_8, S_{11}(O_{5,5,5}, O_{6,6,6}))$
24	$5\frac{1}{3}$	24	16	$S_{24}(O_{10,10,10}, S_{14}(O_6^6, B_8))$
30	$5\frac{28}{45}$	30	20	$S_{30}(O_{12}^{12}, S_{18}(B_8, O_{10,10,10}))$
40	$6\frac{1}{40}$	40	32	$S_{40}(S_{17}(B_8, S_9(B_4, O_{5,5,5})), S_{23}(O_{11,11}, O_{12,12,12}))$
47	$6\frac{25}{94}$	47	32	$S_{47}(S_{20}(S_9(B_4, O_{5,5,5}), O_{11,11}),$ $S_{27}(O_{12,12,12}, S_{15}(S_7(O_3^3, B_4), B_8)))$
49	$6\frac{16}{49}$	49	32	$S_{49}(S_{21}(O_{10,10,10}, O_{11,11}), S_{28}(O_{12}^{12}, B_{16}))$
54	$6\frac{25}{54}$	54	35	$S_{54}(S_{23}(O_{11,11}, O_{12,12,12}),$ $S_{31}(S_{14}(O_6^6, B_8), S_{17}(B_8, S_9(B_4, O_{5,5,5}))))$
62	$6\frac{41}{62}$	62	43	$S_{62}(S_{25}(O_{12,12,12}, S_{13}(O_{6,6,6}, S_7(O_3^3, B_4))),$ $S_{37}(B_{16}, S_{21}(O_{10,10,10}, O_{11,11})))$
77	$6\frac{75}{77}$	77	62	$S_{77}(S_{32}(S_{15}(S_7(O_3^3, B_4), B_8), S_{17}(B_8, S_9(B_4, O_{5,5,5}))),$ $S_{45}(S_{21}(O_{10,10,10}, O_{11,11}), O_{24,24,24}))$
80	$7\frac{1}{40}$	80	64	$S_{80}(S_{34}(B_{16}, S_{18}(B_8, O_{10,10,10})), S_{46}(O_{22,22}, O_{24,24,24}))$

Part 2

For each group L from $\mathcal{C}$,
apply $B_{|L|}$ to L and remove L from $\mathcal{C}$.

Part 3

While $|I| \geq 1$,
test a group of size $l = \min(l_m, |I|)$.
If the result is positive,
apply B_l to the contaminated group.

Part 1 is the main part of the algorithm, in which the procedure is applied repeatedly as long as enough free items are left. In Part 2, the contaminated groups left over from Part 1 are used up by identifying a defective in each group. Finally, Part 3 is just a simple nested algorithm to identify the small number of remaining items as good or defective.

The fixed size algorithm A_m does not use any knowledge about the number of defectives d.

9 Cost Estimate of Fixed Size Algorithms

The following theorem gives a bound for the number of tests needed by the fixed size algorithm A_m to solve the strict group testing problem.

Theorem 1. $L_{A_m}(d,n) < c_m d + \frac{1}{m}(n-d) + 5.$

This means that A_m needs a constant number of tests for the identification of each defective item, $\frac{1}{m}$ tests for each good item, and at most 5 tests due to restrictions near the end, when the number of unidentified items becomes small.

We derive an algorithm A'_m for the (d,n) group testing problem from A_m by simply omitting all tests whose result can be deduced in advance from the knowledge that there are exactly d defectives. [1, Lemma 2.4] states that A'_m always needs at least one test less than A_m, leading to the following corollary of Theorem 1.

Corollary 1. $M_{A'_m}(d,n) < c_m d + \frac{1}{m}(n-d) + 4.$

For the proof of Theorem 1, we fix an arbitrary path S from the root to a leaf in the binary tree representation of the fixed size algorithm A_m. Denote by S_1, S_2, and S_3 the subpaths of this path that fall into Part 1, 2, and 3 of A_m, respectively. Let t_{S_i} be the length of S_i, and d_{S_i} and g_{S_i} the number of defective and good items identified along S_i. Denote by $\mathcal{C}_1$ the content of $\mathcal{C}$ at the end of Part 1 of A_m, that is, the collection of contaminated groups left over from Part 1.

An inspection of the procedures listed in Table 2 reveals that each subalgorithm used in the procedure of A_m that requires at least two groups of size l occurs only once in the procedure. Therefore there is exactly one path Q leading to this subalgorithm in the procedure. We denote by $c_{m,l} = t_Q - \frac{g_Q}{m}$ the cost to obtain a contaminated group of size l for this subalgorithm.

Lemmas 6, 7, and 8 estimate the number of tests needed in S_1, S_2, and S_3. Their proofs are all based on the partition of the path S at the points at which no contaminated groups except those in $\mathcal{C}$ are known. In Part 1 of A_m, each execution of the procedure of A_m forms a subpath in this partition except that each subalgorithm requiring at least two contaminated groups forms a separate subpath. In Part 2, each execution of a B_l is a subpath of the partition. In Part 3, each subpath is either a single test with negative result or a test with positive result together with the following execution of B_l. On each subpath in this partition the number of tests is estimated using the number of identified defective and good items and the cost of contaminated groups added to or removed from $\mathcal{C}$.

Lemma 6. $t_{S_1} \leq c_m d_{S_1} + \frac{1}{m} g_{S_1} + \sum_{L \in \mathcal{C}_1} c_{m,|L|}.$

Proof. It suffices to show for all subpaths R in the partition of S_1 that

$$c_m d_R + \tfrac{1}{m} g_R + k_R c_{m,l} \geq t_R$$

where k_R denotes the number of contaminated groups of size l that are added to $\mathcal{C}$ minus the number that are removed from $\mathcal{C}$ along the subpath R. The estimations of the subpaths follow the description of the fixed size algorithm A_m.

If the first test result in the procedure is negative, m good items and no defectives are identified in one test. Since $0 + \frac{m}{m} = 1$, the above inequality is satisfied.

The rest of the proof treats the case that the first test result in the procedure is positive. Denote by Q the path in the binary tree representation of the procedure that corresponds to the test results.

If the subalgorithm at the end of path Q requires only one contaminated group, it is executed immediately. Denote by P the path followed in the subalgorithm, along which the contaminated group is always used up. Then $s_P = 1$, and d_P defectives and $g_Q + g_P$ good items are identified using $t_Q + t_P$ tests. This leads to the estimation

$$\begin{aligned} c_m d_P + \frac{g_Q + g_P}{m} &\geq \left(\bar{s}_P \left(t_Q - \frac{g_Q}{m}\right) + \bar{t}_P - \frac{\bar{g}_P}{m}\right) d_P + \frac{g_Q + g_P}{m} \\ &= s_P \left(t_Q - \frac{g_Q}{m}\right) + t_P - \frac{g_P}{m} + \frac{g_Q + g_P}{m} \\ &= t_Q + t_P. \end{aligned}$$

In this, the inequality $c_m \geq \bar{s}_P \left(t_Q - \frac{g_Q}{m}\right) + \bar{t}_P - \frac{\bar{g}_P}{m}$ follows from the definition of c_m in the last section.

On the other hand, if the subalgorithm at the end of path Q requires $k \geq 2$ contaminated groups of size l, the contaminated group of size l obtained along Q is added to the collection $\mathcal{C}$ of contaminated groups. Then g_Q good items and no defectives are identified using t_Q tests. Inserting the definition $c_{m,l} = t_Q - \frac{g_Q}{m}$ yields $0 + \frac{g_Q}{m} + c_{m,l} = t_Q$.

If k contaminated groups of size l are present after this, the subalgorithm is executed. Denote by P the path followed in the subalgorithm and by Q the path by which all k contaminated groups have been obtained. Then d_P defectives and g_P good items are identified using t_P tests, while s_P contaminated groups of size l are used up. This leads to the estimation

$$\begin{aligned} c_m d_P + \frac{g_P}{m} - s_P c_{m,l} &\geq \left(\bar{s}_P \left(t_Q - \frac{g_Q}{m}\right) + \bar{t}_P - \frac{\bar{g}_P}{m}\right) d_P \\ &\quad + \frac{g_P}{m} - s_P \left(t_Q - \frac{g_Q}{m}\right) \\ &= t_P. \end{aligned}$$

This process is repeated as long as enough free items are available. The addition of the inequalities for all subpaths in the partition of S_1 yields the statement of the lemma.

Lemma 7. $t_{S_2} < c_m d_{S_2} + \frac{1}{m} g_{S_2} - \sum_{L \in \mathcal{C}_1} c_{m,|L|} + 3.25.$

Proof. For $L \in \mathcal{C}_1$ with $l = |L|$ denote by R the subpath of S_2 that contains exactly the tests in B_l on L. By Lemma 1 B_l needs at most $\lceil \log l \rceil$ tests, in which case at least $2^{\lceil \log l \rceil} - l$ good items are identified as well. All other cases lead to a lower overall cost, hence $t_R - \frac{g_R}{m} \leq \lceil \log l \rceil - \frac{2^{\lceil \log l \rceil} - l}{m}$. In all cases, exactly one defective is identified and the contaminated group L of size l is used up.

Let

$$a_{m,l} = \min\left(0, \lceil \log l \rceil - \frac{2^{\lceil \log l \rceil} - l}{m} + c_{m,l} - c_m\right)$$

denote the maximum additional cost incurred by identifying one defective using B_l on a contaminated group of size l instead of following the procedure of A_m. Summation over all $L \in \mathcal{C}_1$ yields

$$t_{S_2} - c_m d_{S_2} - \tfrac{1}{m} g_{S_2} + \sum_{L \in \mathcal{C}_1} c_{m,|L|} \leq \sum_{L \in \mathcal{C}_1} a_{m,|L|}.$$

It remains to show $\sum_{L \in \mathcal{C}_1} a_{m,|L|} < 3.25$.

Contaminated groups of size l can appear in $\mathcal{C}_1$ if and only if the procedure of A_m contains a path leading to a subalgorithm that processes $k \geq 2$ groups of size l. Then $\mathcal{C}_1$ can contain up to $k - 1$ groups of size l, as k groups would have been eliminated in Part 1 of A_m by the application of the subalgorithm. For instance, the procedure of A_{77} depicted in Figure 7 uses subalgorithms $O_{5,5,5}$, $O_{10,10,10}$, $O_{11,11}$, and $O_{24,24,24}$, resulting in the estimation

$$\begin{aligned}\sum_{L \in \mathcal{C}_1} a_{77,|L|} &\leq 2a_{77,5} + 2a_{77,10} + a_{77,11} + 2a_{77,24} \\ &= 2 \cdot \tfrac{7}{11} + 2 \cdot \tfrac{41}{77} + \tfrac{32}{77} + 2 \cdot \tfrac{18}{77} \\ &= 3\tfrac{17}{77} \\ &< 3.25.\end{aligned}$$

A similar estimation shows $\sum_{L \in \mathcal{C}_1} a_{m,|L|} < 3.25$ for each m listed in Table 2. These estimations extend to all scaled up algorithms: For $m \geq 40$ and $k \geq 1$, Lemma 5 states $c_{2^k m} = c_m + k$, whereas the beginning of its proof shows $c_{2^k m, 2^k l} = c_{m,l}$. Inserting in the definition of $a_{m,l}$ leads to $a_{2^k m, 2^k l} = a_{m,l}$.

Lemma 8. $t_{S_3} \leq c_m d_{S_3} + \frac{1}{m} g_{S_3} + 1.75$.

Proof. If the test on a group of size $l = \min(l_m, |I|)$ yields a positive result, denote by R the subpath of S_3 that contains this test and the tests belonging to the following execution of B_l. The estimation of the overall cost of R is the same as at the beginning of the proof of Lemma 7 plus one for the first test. Together with $l \leq l_m$ this results in

$$
\begin{aligned}
t_R - \frac{g_R}{m} &\leq \lceil \log l \rceil + 1 - \frac{2^{\lceil \log l \rceil} - l}{m} \\
&\leq \lceil \log l_m \rceil + 1 - \frac{2^{\lceil \log l_m \rceil} - l_m}{m} \\
&= \max\left(\lfloor c_m \rfloor - 1 + 1 - \frac{2^{\lfloor c_m \rfloor - 1} - 2^{\lfloor c_m \rfloor - 1}}{m}, \right. \\
&\qquad \left. \lceil c_m \rceil - 1 + 1 - \frac{2^{\lceil c_m \rceil - 1} - 2^{\lceil c_m \rceil - 1} + \lceil (\lceil c_m \rceil - c_m) m \rceil}{m} \right) \\
&\leq \max\left(\lfloor c_m \rfloor, \lceil c_m \rceil - \frac{(\lceil c_m \rceil - c_m) m}{m} \right) \\
&\leq c_m.
\end{aligned}
$$

On the other hand, denote by $\mathcal{R}$ the set of all subpaths of S_3 that consist of a single test of a group of size $l = \min(l_m, |I|)$ with negative result. All but the last of the tests in $\mathcal{R}$ are on groups of size l_m, as for $|I| < l_m$ all remaining items are tested and identified as good. Denote by I_1 and I_2 the items remaining in I at the end of Part 1 and 2, respectively. Then $|\mathcal{R}| \leq \frac{|I_2|}{l_m} \leq \frac{|I_1|}{l_m}$. The set I_1 contains less than n_m free items and the items in the groups in $\mathcal{C}_1$. Therefore

$$
|I_1| < n_m + \sum_{L \in \mathcal{C}_1} |L| \leq \begin{cases} 3l_m \text{ for } m \neq 4 \\ 5l_m \text{ for } m = 4. \end{cases}
$$

It can easily be checked that the last inequality holds for all m in Table 2, which extends to $m \geq 80$ because for $k \geq 1$

$$
\begin{aligned}
l_{2^k m} &= \max\left(2^{\lfloor c_m + k \rfloor - 1}, 2^{\lceil c_m + k \rceil - 1} - \lceil (\lceil c_m + k \rceil - (c_m + k))\, 2^k m \rceil \right) \\
&= \max\left(2^k 2^{\lfloor c_m \rfloor - 1}, 2^k 2^{\lceil c_m \rceil - 1} - \lceil 2^k (\lceil c_m \rceil - c_m)\, m \rceil \right) \\
&\geq 2^k l_m.
\end{aligned}
$$

The inequality $l_m > \frac{5}{8} m$ can be checked in the same way.

Together this yields $|\mathcal{R}| \leq 3$ for $m \neq 4$ and $|\mathcal{R}| \leq 5$ for $m = 4$, leading to

$$
\sum_{R \in \mathcal{R}} t_R - \frac{g_R}{m} \leq \begin{cases} 3 - \frac{2l_m + 1}{m} < 1.75 \text{ for } m \neq 4 \\ 5 - \frac{4 \cdot 3 + 1}{4} = 1.75 \text{ for } m = 4. \end{cases}
$$

Adding the estimation from above for all subpaths of S_3 beginning with a positive test result yields the statement of the lemma.

Proof. [Proof of Theorem 1]For each path S of A_m that identifies d defectives in n items the addition of the inequalities of Lemmas 6, 7, and 8 yields

$$
t_S < c_m d + \tfrac{1}{m}(n - d) + 5.
$$

Maximizing over all paths yields the theorem.

An analysis of the arguments above for a path that achieves equality in Lemma 6 shows that $L_{A_m}(d,n) > c_m d + \frac{1}{m}(n-d) - 1$ for $\frac{n}{d} \geq m$. For sufficiently small $\frac{n}{d}$, some algorithms A_m need fewer tests than this in the worst case, as there are not enough good items to follow a path achieving equality in Lemma 6. However, in the main algorithm A in the next section A_m is used only for $\frac{n}{d} \geq m$.

A possible modification of Part 2 of the fixed size algorithm A_m is to use subalgorithms that require multiple contaminated groups not necessarily of the same size to identify several defectives together. This leads to a lower constant in Lemma 7 and therefore in Theorem 1. However, this does not improve c_m.

10 Main Algorithm

We suppose $d > 0$ and denote by $r = \frac{n}{d}$ the initial ratio of all items to defective items. To find the best fixed size algorithm for a given ratio r, we compare the number of tests needed by different algorithms in the worst case. The difference between the upper bounds from Theorem 1 for $L_{A_{m'}}(d,n)$ and $L_{A_m}(d,n)$ with $m \neq m'$ is

$$(c_{m'} - c_m)\, d + \left(\tfrac{1}{m'} - \tfrac{1}{m}\right)(n-d) = d\left(c_{m'} - c_m - \frac{m'-m}{mm'}(r-1)\right).$$

Let

$$r_{m,m'} = \frac{mm'}{m'-m}\,(c_{m'} - c_m) + 1$$

denote the value of r for which the above expression is zero, that is, the upper bounds for $L_{A_{m'}}(d,n)$ and $L_{A_m}(d,n)$ are equal.

For $r \geq m$ the difference between $L_{A_m}(d,n)$ and the upper bound from Theorem 1 is less than 6. Thus, if we suppose $m' > m$, then for each $r \geq m$ there exists $d_0(r)$ such that for all $d \geq d_0(r)$ and $n = rd$

$$L_{A_m}(d,n) < L_{A_{m'}}(d,n) \text{ for } r < r_{m,m'} \text{ and}$$
$$L_{A_m}(d,n) > L_{A_{m'}}(d,n) \text{ for } r > r_{m,m'}.$$

Therefore we can describe the range of ratios r for which A_m is best by

$$r_m^{\min} = \max_{m'<m} r_{m',m} \text{ and}$$
$$r_m^{\max} = \min_{m'>m} r_{m,m'},$$

except $r_1^{\min} = 0$. There is always some $m' > m$ with $r_m^{\max} = r_{m'}^{\min}$. For $m \leq 80$, the values $r_m^{\min}$ and $r_m^{\max}$ can be found in Table 3 together with α and $\bar{\alpha}$, which are defined in the next section. For $m \geq 40$ and $k \geq 1$, substituting $c_{2^k m} = c_m + k$ from Lemma 5 leads to $r_{2^k m, 2^k m'} - 1 = 2^k\left(r_{m,m'} - 1\right)$. Thus

$$r_{2^k m}^{\min} - 1 = 2^k\left(r_m^{\min} - 1\right) \text{ for } m \geq 47, \text{ and}$$
$$r_{2^k m}^{\max} - 1 = 2^k\left(r_m^{\max} - 1\right) \text{ for } m \geq 40.$$

Table 3. Ranges of A_m in the main algorithm A for $m \le 80$

m	c_m	$r_m^{\min}$ / $r_m^{\max}$	$\alpha(r)$	$\bar{\alpha}(r)$	m	c_m	$r_m^{\min}$ / $r_m^{\max}$	$\alpha(r)$	$\bar{\alpha}(r)$
1	1				24	$5\frac{1}{3}$			
		3	0.246	0.558			$35\frac{2}{3}$	0.199	0.220
2	2				30	$5\frac{28}{45}$			
		4	0.255	0.473			$49\frac{1}{3}$	0.181	0.196
3	$2\frac{1}{2}$				40	$6\frac{1}{40}$			
		5	0.224	0.391			$65\frac{5}{7}$	0.174	0.185
4	$2\frac{5}{6}$				47	$6\frac{25}{94}$			
		$7\frac{8}{9}$	0.229	0.329			$70\frac{3}{4}$	0.173	0.184
7	$3\frac{4}{7}$				49	$6\frac{16}{49}$			
		$13\frac{4}{5}$	0.225	0.280			$73\frac{1}{5}$	0.174	0.184
12	$4\frac{1}{3}$				54	$6\frac{25}{54}$			
		$18\frac{1}{3}$	0.179	0.220			84	0.174	0.183
15	$4\frac{28}{45}$				62	$6\frac{41}{62}$			
		$27\frac{11}{12}$	0.198	0.224			$100\frac{8}{15}$	0.180	0.187
19	5				77	$6\frac{75}{77}$			
		$31\frac{2}{5}$	0.208	0.232			$105\frac{2}{3}$	0.175	0.181
24	$5\frac{1}{3}$				80	$7\frac{1}{40}$			

This leads to the following *split and overlap algorithm* A for the strict group testing problem that always uses the best fixed size algorithm A_m for the actual ratio r. We refer to A in the following as the *main algorithm* to highlight the difference to the fixed size algorithm A_m.

Algorithm A

If $d = 0$,
  test the group of all items.
  If the result is negative, then stop, else set $d := 1$.
Let $r = \frac{n}{d}$.
Choose m satisfying $r_m^{\min} < r \le r_m^{\max}$.
Execute A_m.

In the main algorithm A the initial test group size m, which is used if no contaminated group is known, remains constant even if the ratio of good to defective unidentified items changes during the execution. In contrast, the generalized binary splitting algorithm G from Hwang presented in Section 3 adapts the test group size to the ratio of remaining good to defective items. Surprisingly, this does not seem to be necessary to obtain a good algorithm for the worst case scenario.

The list of fixed size algorithms A_m in Tables 2 and 3 is the result of a computation in the advanced functional programming language Haskell that searches for the best algorithms for each $m \ge 1$ using all subalgorithms and techniques introduced in this chapter, that is, those with the lowest cost c_m for identifying a defective. Then the above definitions can be extended to all $m \ge 1$, and the algorithms listed in Tables 2 and 3 are just those with $r_m^{\min} < r_m^{\max}$. For example, A_5 with the procedure $S_5(B_2, O_3^3)$ and A_6 with the procedure O_6^4 are made redundant by A_4 and A_7. It is possible to continue this process for $m > 80$ instead of scaling up algorithms for smaller m, but this yields only minor further improvements.

11 Cost Estimate of Main Algorithm

The minimum number of tests needed for the (d,n) group testing problem is known only for the trivial cases $M(0,n) = M(n,n) = 0$ and for a large proportion of defectives, that is, a low ratio $r = \frac{n}{d}$ of all items to defectives. Presume $0 < d < n$. For $r \le \frac{21}{8} = 2.625$, Du and Hwang [5] showed $M(d,n) = n-1$, which can be reached by testing all items individually. Leu, Lin, and Weng [8] extended this to $r \le \frac{43}{16} = 2.687\ldots$ for $d \ge 193$, Riccio and Colbourn [9] to $r < \log_{\frac{3}{2}} 3 = 2.709\ldots$ for sufficiently large d depending on r. All these results are based on a Lemma by Hu, Hwang, and Wang [6], who also conjectured $M(d,n) = n-1$ for $r \le 3$ and proved $M(d,n) < n-1$ for $r > 3$. The latter is achieved by a variant of the fixed size algorithm A_2 omitting all predictable tests, as shown by Du and Hwang [3, Section 3.5]. However, the proof of the conjecture remains elusive and calls for a different approach.

These results transfer to the strict group testing problem by [1, Theorem 2.1], which states $L(d,n) = L(d,n) + 1$ for $0 \le d < n$. For $d = n$, all items have to be tested individually, thus $L(n,n) = n$.

The main algorithm A is optimal in all these cases: Applied to the strict group testing problem, it performs only one test on the group of all defectives for $d = 0$ and tests all items individually for $r \le 3$.

The only general lower bound that is known for the (d,n) group testing problem is the information lower bound $\left\lceil \log \binom{n}{d} \right\rceil$. Therefore we compare the number of tests needed by A with $\log \binom{n}{d}$. First we prove some lemmas.

Lemma 9. *For $x > 0$,*

$$\left(\frac{x+1}{x}\right)^{x} \le e \le \left(\frac{x+1}{x}\right)^{x+1}.$$

This can be shown easily using Taylor's Formula.

Lemma 10. *For $0 < d \le \frac{n}{2}$,*

$$\log \binom{n}{d} > d \log \left(r \left(\frac{r}{r-1} \right)^{r-1} \right) - \frac{1}{2} \log d - 1.5.$$

The Lemma can be shown using Stirling's Formula, for a detailed proof see [1, section 3.9].

It can be shown by similar estimates that the difference between the two sides of the inequality in Lemma 10 is less than one.

We denote by

$$\alpha_m(r) = c_m + \frac{r-1}{m} - \log \left(r \left(\frac{r}{r-1} \right)^{r-1} \right)$$

the average number of tests per defective by which the fixed size algorithm A_m exceeds the information lower bound for large d, which we call the *loss per defective*. Similarly,

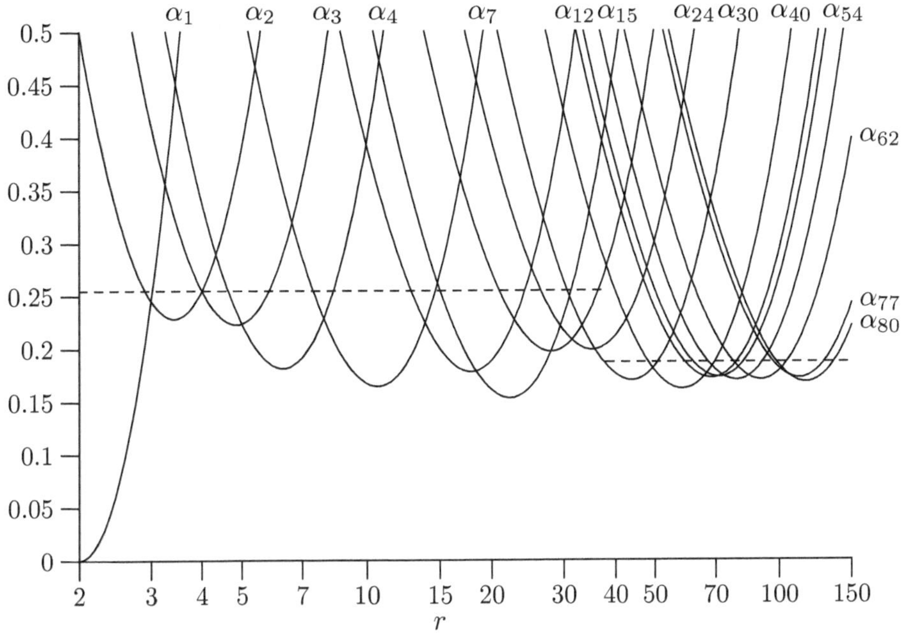

Fig. 8. The loss per defective $\alpha(r)$

$$\alpha(r) = \min_m \alpha_m(r)$$

denotes the loss per defective of the main algorithm A. Figure 8 shows a graph of $\alpha(r)$ and $\alpha_m(r)$.

In addition, let

$$\bar{\alpha}(r) = \min_m \left(c_m + \frac{r-1}{m} \right) - \log\left((r-1)e\right)$$

denote the upper bound of the loss per defective of all scaled up versions of the fixed size algorithm used for the ratio r, as shown by the following Lemma.

Lemma 11. *For $r \geq r_{47}^{\min}$ and $k \geq 1$,*

$$\alpha(2^k(r-1)+1) \leq \bar{\alpha}(r).$$

Proof. Choose $m \geq 47$ satisfying $r_m^{\min} < r \leq r_m^{\max}$. From the last section follows $r_{2^k m}^{\min} < 2^k(r-1)+1 \leq r_{2^k m}^{\max}$. Inserting in the definition of $\alpha_m(r)$ using

$r\left(\frac{r}{r-1}\right)^{r-1} = (r-1)\left(\frac{r}{r-1}\right)^{r}$ then yields

$$\begin{aligned}\alpha(2^k(r-1)+1) &= \alpha_{2^k m}(2^k(r-1)+1)\\ &= c_{2^k m} + \frac{2^k(r-1)}{2^k m}\\ &\quad - \log\left(2^k(r-1)\left(\frac{2^k(r-1)+1}{2^k(r-1)}\right)^{2^k(r-1)+1}\right).\end{aligned}$$

The application of Lemma 5, which states $c_{2^k m} = c_m + k$, and Lemma 9 with $x = 2^k(r-1)$ results in

$$\begin{aligned}\alpha(2^k(r-1)+1) &\leq c_m + k + \frac{r-1}{m} - \log\left(2^k(r-1)e\right)\\ &= \bar{\alpha}(r).\end{aligned}$$

The following theorem gives an upper bound for the loss per defective $\alpha(r)$.

Theorem 2. $\alpha(r) \leq \frac{1}{2} - \log\frac{32}{27} < 0.255$ *for* $r \geq 2$ *and*
$\alpha(r) < 0.187$ *for* $r \geq 38$.

Proof. $\alpha_m(r)$ is convex, as $\alpha_m''(r) = \frac{\log e}{r(r-1)} > 0$ for $r > 1$. Therefore $\alpha(r)$ is convex between $r_m^{\min}$ and $r_m^{\max}$ for all m, implying

$$\alpha(r) \leq \max\left(\alpha(r_m^{\min}), \alpha(r_m^{\max})\right) \text{ for } r_m^{\min} \leq r \leq r_m^{\max}.$$

For $m \leq 80$ Table 3 lists $\alpha(r_m^{\min})$, whereas for $47 \leq m \leq 80$ and $k \geq 1$ by Lemma 11 $\alpha(r_{2^k m}^{\min}) \leq \bar{\alpha}(r_m^{\min})$, which is also listed in Table 3. Since $r_m^{\max} = r_{m'}^{\min}$ for some $m' > m$, this extends to $\alpha(r_m^{\max})$.

This leads to $\alpha(r) \leq \alpha(4) = \frac{1}{2} - \log\frac{32}{27}$. Together with $\alpha(38) < 0.185$ follows $\alpha(r) < 0.187$ for $r \geq 38$.

The bounds for $\alpha(r)$ given in Theorem 2 are shown in Figure 8 as a dashed line.

A simple calculation shows that $\alpha_m(r)$ assumes its minimum at $\frac{1}{2^{\frac{1}{m}}-1} + 1$. The lowest minimum value of $\alpha_m(r)$ with $m \geq 2$ is the minimum of $\alpha(r)$ for $r \geq 2$:

$$\min_{r\geq 2} \alpha(r) = \alpha_{15}(22.1) < 0.154.$$

Now we can estimate the difference between the number of tests required by the split and overlap algorithm A for the strict group testing problem and $\log\binom{n}{d}$.

Theorem 3. *For* $0 < d \leq \frac{n}{2}$,

$$L_A(d,n) - \log\binom{n}{d} < \alpha(r)\, d + \frac{1}{2}\log d + 6.5.$$

Proof. Assume that the main algorithm A chooses the fixed size algorithm A_m with the initial test group size m. Applying Theorem 1 and Lemma 10 yields

$$\begin{aligned} L_A(d,n) - \log \binom{n}{d} &= L_{A_m}(d,n) - \log \binom{n}{d} \\ &< c_m d + \tfrac{1}{m}(n-d) + 5 \\ &\quad - d \log \left(r \left(\frac{r}{r-1} \right)^{r-1} \right) + \frac{1}{2} \log d + 1.5 \\ &= \alpha_m(r) d + \frac{1}{2} \log d + 6.5 \\ &= \alpha(r)\, d + \frac{1}{2} \log d + 6.5. \end{aligned}$$

As with the fixed size algorithm A_m in Section 9, we can derive from A an algorithm A' for the (d, n) group testing problem by simply omitting all tests that become predictable due to the knowledge that there are exactly d defectives. From [1, Lemma 2.4] then follows $M_{A'}(d,n) \leq L_A(d,n) - 1$, leading to the following corollary of Theorem 3.

Corollary 2. *For* $0 < d \leq \frac{n}{2}$,

$$M_{A'}(d,n) - \log \binom{n}{d} < \alpha(r)\, d + \frac{1}{2} \log d + 5.5.$$

For $d \geq 10$, this is considerably better than the $d-1$ additional tests for Hwang's generalized binary splitting algorithm G presented in Section 3.

The difference between the two sides of the inequalities in Theorem 3 and Corollary 2 is less than 7, as the corresponding differences in Theorem 1 and Lemma 10, which are used in the proof of Theorem 3, are less than 6 and 1, respectively. This leads to the following corollary.

Corollary 3. *For* $r \geq 2$,

$$\lim_{\substack{n,d \to \infty \\ \frac{n}{d} \to r}} \frac{1}{d} \left(M_{A'}(d,n) - \log \binom{n}{d} \right) = \alpha(r).$$

For small r the necessity of integral test sizes significantly restricts the choice of good algorithms, motivating the following conjecture.

Conjecture 1. For $2 \leq r \leq 4$,

$$\lim_{\substack{n,d \to \infty \\ \frac{n}{d} \to r}} \frac{1}{d} \left(M(d,n) - \log \binom{n}{d} \right) = \alpha(r).$$

This implies that the number of tests per defective needed by A_2 and A_3 is optimal and that there exists no fixed size algorithm with initial test group size 4 requiring less than $2\frac{3}{4}$ tests per defective. Furthermore, the conjecture implies that the general upper bound $\frac{1}{2} - \log \frac{32}{27}$ for $\alpha(r)$ given in Theorem 2 is the best possible for any group testing algorithm.

Acknowledgements. I wish to thank Prof. Dr. Eberhard Triesch for many fruitful discussions and helpful hints as well as for having introduced me to the fascinating area of group testing.

References

1. Allemann, A.: Improved upper bounds for several variants of group testing, Ph.D. thesis, RWTH Aachen (2003)
2. Chang, X.M., Hwang, F.K., Weng, J.F.: Group testing with two and three defectives. Ann. N. Y. Acad. Sci. 576, 86–96 (1989)
3. Du, D.Z., Hwang, F.K.: Combinatorial group testing and its applications, 2nd edn. World Scientific, Singapore (2000)
4. Du, D.Z., Hwang, F.K.: Competitive group testing. Disc. Appl. Math. 45, 221–232 (1993)
5. Du, D.Z., Hwang, F.K.: Minimizing a combinatorial function. SIAM J. Alg. Disc. Methods 3, 523–528 (1982)
6. Hu, M.C., Hwang, F.K., Wang, J.K.: A boundary problem for group testing. SIAM J. Alg. Disc. Methods 2, 81–87 (1981)
7. Hwang, F.K.: A method for detecting all defective members in a population by group testing. J. Amer. Statist. Assoc. 67, 605–608 (1972)
8. Leu, M.G., Lin, C.Y., Weng, S.Y.: Note on a conjecture for group testing. Ars Combin. 64, 29–32 (2002)
9. Riccio, L., Colbourn, C.J.: Sharper bounds in adaptive group testing. Taiwanese J. Math. 4, 669–673 (2000)
10. Schlaghoff, J., Triesch, E.: Improved results for competitive group testing. Combin. Probab. Comput. 14, 191–202 (2005)
11. Sobel, M., Groll, P.A.: Group testing to eliminate efficiently all defectives in a binomial sample. Bell System Tech. J. 38, 1179–1252 (1959)
12. Triesch, E.: A group testing problem for hypergraphs of bounded rank. Disc. Appl. Math. 66, 185–188 (1996)

Randomized Post-optimization for t-Restrictions

Charles J. Colbourn and Peyman Nayeri

School of Computing, Informatics, and Decision Systems Engineering Arizona State University P.O. Box 878809, Tempe, Arizona 85287-8809
{colbourn,nayeri}@asu.edu

Dedicated to the memory of Rudolf Ahlswede

Abstract. Search, test, and measurement problems in sparse domains often require the construction of arrays in which every t or fewer columns satisfy a simply stated combinatorial condition. Such *t-restriction problems* often ask for the construction of an array satisfying the t-restriction while having as few rows as possible. Combinatorial, algebraic, and probabilistic methods have been brought to bear for specific t-restriction problems; yet in most cases they do not succeed in constructing arrays with a number of rows near the minimum, at least when the number of columns is small. To address this, an algorithmic method is proposed that, given an array satisfying a t-restriction, attempts to improve the array by removing rows. The key idea is to determine the necessity of the entry in each cell of the array in meeting the t-restriction, and repeatedly replacing unnecessary entries, with the goal of producing an entire row of unnecessary entries. Such a row can then be deleted, improving the array, and the process can be iterated. For certain t-restrictions, it is shown that by determining conflict graphs, entries that are necessary can nonetheless be changed without violating the t-restriction. This permits a richer set of ways to improve the arrays. The efficacy of these methods is demonstrated via computational results.

Keywords: covering array, hash family, frameproof code, disjunct matrix.

1 Introduction

In combinatorial search, testing, and measurement problems, numerous problems of the following type arise. An $N \times k$ array is defined. Let Δ be a finite alphabet not containing $\star$. For $1 \leq i \leq N$, there is a finite alphabet $\Sigma_i \subseteq \Delta$ for which the ith row contains only symbols in $\Sigma_i \cup \{\star\}$. (When $\Sigma_1 = \cdots = \Sigma_N = \Sigma$, the array is *homogeneous*, otherwise it is *heterogeneous*.) For $1 \leq j \leq k$, there is a finite alphabet Δ_j not containing $\star$ for which the jth column contains only symbols in $\Delta_j \cup \{\star\}$. (When $\Delta_1 = \cdots = \Delta_k = \Delta$, the array is *uniform*, otherwise it is *nonuniform*.) Without loss of generality, $\Sigma_i \subseteq \cup_{j=1}^{k} \Delta_j$ and $\Delta_j \subseteq \cup_{i=1}^{N} \Sigma_i$.

H. Aydinian, F. Cicalese, and C. Deppe (Eds.): Ahlswede Festschrift, LNCS 7777, pp. 597–608, 2013.

If for some i, j with $1 \le i \le N$ and $1 \le j \le k$, we have $\Sigma_i \cap \Delta_j = \varnothing$, the (i, j) cell is permitted only to contain $\star$.

Within this framework, one considers restrictions on what must appear in some row within every subset of t columns. Such 'restriction' problems are considered in [3], but we use a somewhat more general definition here.

Let t be an integer, called the *strength.* A *t-restriction* is a list $(\mathcal{P}_1, \dots, \mathcal{P}_\tau)$ of subsets of Δ^t, called *demands* [3]. For every selection $S = (i_1, \dots, i_t)$ of t distinct column indices, the set of possible t-tuples that could arise is $\Delta_{i_1} \times \cdots \times \Delta_{i_t}$. Then the $N \times k$ array $A = (a_{ij})$ *satisfies the t-restriction* $(\mathcal{P}_1, \dots, \mathcal{P}_\tau)$ if and only if for all t-tuples $(x_1, \dots, x_t)$ of distinct column indices, and for $1 \le \ell \le \tau$, for each $\mathcal{P}_\ell$ with $\mathcal{P}_\ell \cap (\Delta_{x_1} \times \cdots \times \Delta_{x_t}) \ne \varnothing$, there exists an r with $1 \le r \le N$ for which $(a_{r,x_1}, \dots, a_{r,x_t}) \in \mathcal{P}_\ell$. The generality of the definition arises from the flexibility in specifying t-restrictions.

We enumerate a few well-studied examples.

Disjunct Matrix [10]: The demand is $\{(\delta_1, \dots, \delta_t) \in \{0,1\}^t : \delta_1 = \cdots = \delta_{t-1} = 0, \delta_t = 1\}$;
Frameproof Code [15]: The demand is $\{(\delta_1, \dots, \delta_t) \in \{0,1\}^t : \delta_1 = \cdots = \delta_{t-1},\ \delta_t \ne \delta_1\}$;
Covering Array [6]: Demands are all members of Δ^t;
Perfect Hash Family, PHF **[18]:** The demand is $\{(\delta_1, \dots, \delta_t) \in \Delta^t : \delta_i \ne \delta_j \text{ for } i \ne j\}$;

For covering arrays, requiring only a subset $\mathbb{S} \subseteq \Delta^t$ to be covered yields $\mathbb{S}$-*quilting arrays* [8]. For disjunct matrices (equivalently, superimposed codes or cover-free families), numerous t-restriction problems arise in search theory [1,2]. For hash families when the t columns to be separated are partitioned into ℓ classes $C_1, \dots, C_\ell$ of sizes $w_1, \dots, w_\ell$ (with $t = \sum_{i=1}^{\ell} w_i$) and we only require $\delta_i \ne \delta_j$ when i and j are in different classes, we obtain a $\{w_1, \dots, w_\ell\}$*-separating hash family,* $\{w_1, \dots, w_\ell\}$-SHF [12,16]. When on the t columns, the number of distinct symbols that arise is at most m, we have m*-strengthening hash families* [7]. A hash family that is $\{w_1, \dots, w_s\}$-separating *for all* $\{w_1, \dots, w_s\}$ with $\sum_{i=1}^{s} w_i = t$ is a (t, s)*-distributing hash family,* (t, s)-DHF [5]. An s-strengthening (t, s)-DHF is a (t, s)*-partitioning hash family,* (t, s)-PaHF [5].

These examples only scratch the surface. Numerous problems in combinatorial search and group testing [2,10] and in combinatorial cryptography [3,15] fall into this framework. Evidently, treating each such problem individually is problematic, and one wants general techniques to address the construction of arrays for t-restrictions. One general technique, explored in many of these contexts, is a recursive method using column replacement via hash families (e.g., [6]). But these techniques rely on knowing solutions for few columns to produce solutions for many.

Simple greedy or random algorithms produce solutions, but they cannot be expected to minimize the number of rows. We propose a general technique here to "post-optimize" an array, reducing its number of rows. We demonstrate that the reduction obtained is worthwhile, and sometimes dramatic.

2 Post-optimization

In [14], an heuristic method for reducing the number of rows in a covering array is developed. It relies on the fact that certain entries of the array may not be needed to ensure coverage. Such entries can be changed arbitrarily, with the result that other entries that were previously required, are no longer needed. The method exploits this to produce entire rows that are not needed. These can be deleted, improving the size of the array, and the process can be repeated. For covering arrays, the method has surprising success, and therefore we wish to apply the technique more generally. Here we develop it for general t-restriction problems.

2.1 Necessity Analysis

Consider an $N \times k$ array A on symbol set $\Delta \cup \{\star\}$ that meets the t-restriction $(\mathcal{P}_1, \ldots, \mathcal{P}_\tau)$. Evidently if A contains a row that consists entirely of $\star$ symbols, this row is not used to meet any of the requirements, and can be removed. The primary objective of our method is repeatedly to produce such an all-$\star$ row for removal. To do this, we consider the necessity of each entry.

When one of the demands is met for columns $(x_1, \ldots, x_t)$ for a single row of the array, the t entries in these columns in this row are *strictly necessary* to meet the demand. One might hope that all entries of the array not determined to be necessary in this way can be changed to $\star$, since they are not "needed". However once one is changed to $\star$, further entries may now become necessary. Indeed determining the maximum number of entries that can be simultaneously changed to $\star$ is NP-hard [14].

We therefore adopt a more useful notion of necessity. Let ρ be a permutation of $\{1, \ldots, N\}$, the row indices. For each demand and each tuple of t columns, there is a *first* row (under ρ) in which this demand is met; the entries in the t columns of this row are *necessary*. A single scan of the array now suffices to determine all necessary entries. All others are *unnecessary*, and all can be changed to $\star$ while ensuring that all demands are still met.

In determining necessity in this way, every demand must be checked in each t-tuple of columns. This can often be accomplished by considering only a subset of the t-tuples, as follows: Let π be a permutation of $\{1, \ldots, t\}$. If there are two demands $\mathcal{P}_a$ and $\mathcal{P}_b$ so that $\mathcal{P}_b = \{(\nu_{\pi_1}, \ldots, \nu_{\pi_t}) : (\nu_1, \ldots, \nu_t) \in \mathcal{P}_a\}$, then we can either (1) not check demand $\mathcal{P}_b$ if demand $\mathcal{P}_a$ is checked, or (2) not check the t-tuple $(x_{\pi_1}, \ldots, x_{\pi_t})$ of columns if $(x_1, \ldots, x_t)$ is checked. In practice this reduces the effort to determine necessity for most cases of interest.

2.2 Generic Post-optimization

We may be very lucky, and find that after marking unnecessary entries, we have an entire row of $\star$ entries. But this should not be expected. Following ideas from the special case of covering arrays [14], we employ two observations. Let A be an $N \times k$ array that satisfies the t-restriction $(\mathcal{P}_1, \ldots, \mathcal{P}_\tau)$. First, reordering the rows of A results in an array that still satisfies the t-restriction (provided, of course, that the specifications of the row alphabets are permuted in the same manner as are the rows). And secondly, an entry of $\star$ in row r and column c can be replaced by any symbol in $\Sigma_r \cap \Delta_c$, and the resulting array still satisfies the t-restriction.

```
input : t-restriction with demands P_1, ..., P_τ,
        N × k array A satisfying the t-restriction,
        ITERATION_LIMIT – number of iterations to be performed,
        LOCAL_LIMIT – number of iterations allowed with no row removal
output: M × k array C satisfying the t-restriction with M ≤ N

ρ ⟵ identity C ⟵ A noImprovementCounter ⟵ 0
maxUnnecessaryElements ⟵ 0 for i ← 1 ITERATION_LIMIT do
    Locate necessary and unnecessary entries in C using row order ρ Change all
    unnecessary entries to ⋆ currentMax ⟵ maximum number of ⋆s in a row of
    C if currentMax > maxUnnecessaryElements then
        maxUnnecessaryElements ⟵ currentMax
        noImprovementCounter ⟵ 0
    else
        add 1 to noImprovementCounter
    endif
    if C contains any rows consisting entirely of ⋆s then
        Remove all such rows from C, adjusting N and
        ρ maxUnnecessaryElements ⟵ 0 Nominate a row of array C and adjust
        ρ to make this row the last
    endif
    for every ⋆ at position (r, c) in C with r ≠ ρ(N) do
        if C(ρ(N), c) ∈ Δ_c ∩ Σ_r then
            C(r, c) ⟵ C(ρ(N), c)
        else
            C(r, c) ⟵ random value in Δ_c ∩ Σ_r
        endif
    endfor
    if noImprovementCounter ≥ LOCAL_LIMIT then
        Choose permutation ρ of {1, ..., N} at random
        noImprovementCounter ⟵ 0
    else
        Choose ρ at random, without changing ρ(N)
    endif
endfor
```

Algorithm 1. A generic post-optimization algorithm for k-restriction problems

These form the basis of a remarkably simple algorithm for post-optimization. We repeatedly change $\star$ entries to entries in Δ, implicitly reorder the rows of the array, mark unnecessary entries, and delete any rows that now contain only $\star$. A more precise version is shown in Figure 1.

Because progress occurs when a row is eliminated, a worthwhile intermediate goal is to attempt to make a row with as many $\star$ entries as we can. However, row reordering could result in a row with many $\star$ entries having none in the next iteration. Therefore the algorithm nominates one row, retained at the end of the row order, in which it repeatedly attempts to increase the number of $\star$s. By so doing, the method may become trapped in a local optimum, where no further $\star$ entries are formed in the nominated row. For this reason, a means to escape such local optima by requiring progress is included; when progress has apparently stalled, a complete row reordering is done, resulting in a new row becoming nominated.

We report computational results for Algorithm 1 in §4. Prior to doing so, we examine an interesting variant of the method.

3 Post-optimization with Conflict Graphs

In Algorithm 1, every entry is deemed to be either necessary or unnecessary. Consider the first 3×5 array on $\Delta = \{0, 1, 2\}$ in Figure 1; this array is a $\{1,2\}$-SHF . Every entry is strictly necessary. But the entry in the (3,4) position can nonetheless be changed, from 1 to 0, forming a second array that also satisfies the demand.

2 1 1 0 1	2 1 1 0 1
0 2 1 1 0	0 2 1 1 0
1 1 2 1 0	1 1 2 **0** 0

Fig. 1. $\{1,2\}$-SHFs

Once changed, there is a possibility that an unnecessary entry appears, and progress can be made. Next we explore transformations that permit the replacement of entries while still satisfying the t-restriction, for a certain type of t-restriction.

3.1 Conflict Graphs

A demand $\mathcal{P}_\ell$ is *totally symmetric* if, for every permutation π of the symbols in Δ, $(\pi(\delta_1), \ldots, \pi(\delta_t)) \in \mathcal{P}_\ell$ if and only if $(\delta_1, \ldots, \delta_t) \in \mathcal{P}_\ell$. We consider now only those t-restrictions with totally symmetric demands $\mathcal{P}_1, \ldots, \mathcal{P}_\tau$. When the demands are all totally symmetric, the symbols within any row can be permuted arbitrarily while still satisfying the t-restriction. A new row produced in this manner handles neither more nor fewer of the demands. We are interested in modifying the row to handle all of the demands that it currently does, but possibly to handle more. To do this, we develop conflict graphs, focussing on SHFs. Roughly speaking, edges indicate a requirement for columns to contain different symbols; we make this precise now.

Let A be an $N \times k$ array, a $\{w_1, \ldots, w_s\}$-SHF with $t = \sum_{i=1}^{s} w_i$. Let $z_j = \sum_{i=1}^{j} w_i$. Then the demand to be satisfied is $\{(\delta_1, \ldots, \delta_t) \in \Delta^t : \delta_a \neq \delta_b \text{ or } z_j < a, b \leq z_{j+1} \text{ for } j \in \{0, \ldots, s-1\}\}$. We construct a collection of graphs $G_1, \ldots, G_N$, one for each row, as follows. Each G_i contains k vertices, representing the column indices $\{1, \ldots, k\}$. As before, for every t-tuple of columns, we determine the *first* row in which the demand is met. Suppose that the demand is first met for columns $(x_1, \ldots, x_t)$ in row r. For row r to continue to meet this demand, it must be the case that the symbols $(\sigma_1, \ldots, \sigma_t)$ satisfy $\sigma_a \neq \sigma_b$ except possibly when $z_j < a, b \leq z_{j+1}$ for $j \in \{0, \ldots, s-1\}$. To represent this, we place an edge in G_r between vertices x_a and x_b for all $1 \leq a, b \leq t$, *except* when $z_j < a, b \leq z_{j+1}$ for $j \in \{0, \ldots, s-1\}$. Once all t-tuples of columns are processed in this way, the graphs $G_1, \ldots, G_N$ are the *conflict graphs* of array A for this demand. When the t-restriction consists of multiple (separating) demands, each can be processed in the same way, possibly adding further edges to the conflict graphs; this results in conflict graphs for the entire t-restriction. To connect with our earlier discussion, when vertex c is isolated (is incident on no edges) in G_r, this is precisely the same as saying that the (r, c) entry of A is unnecessary.

Interpret row r of A as a vertex colouring of G_r in $|\Sigma_r|$ colours, where the entry a_{rc} is treated as a colour of vertex c in G_r. This is a proper colouring. More importantly, suppose that we form *any* proper colouring of G_r; this can be interpreted as a row – and this row must meet all of the demands met for the first by the original row. When the new colouring is not simply a permutation of the original one, the new row may meet more demands than does the original! Each conflict graph can be assigned a new proper colouring independently, producing a new array of the same size satisfying the t-restriction. Thus, even when no unnecessary entries arise, we can transform the array – and perhaps form unnecessary entries. Extending the post-optimization process to incorporate these recolouring transformations, while still nominating a row in which to maximize the number of $\star$ entries, opens a further avenue to seek improvements. Before pursuing this further, we consider a small extension.

As developed thus far, conflict graphs are suitable for SHFs with multiple separation requirements, and therefore for PHFs and DHFs as well. For strengthening and partitioning hash families, however, we encounter a difficulty. It is not the case that any recolouring of the conflict graphs will serve. Indeed in these situations, the demand requires not only that a certain separation be accomplished, but also that not too many symbols (colours) are used in the separation; just recolouring the conflict graph properly does not ensure the latter. PaHFs admit an easy modification to the conflict graphs. When a demand is met in columns $(x_1, \ldots, x_t)$ in row r, this demand can only be met if vertices x_i and x_j receive different colours when $a_{rx_i} \neq a_{rx_j}$, *and receive the same colour when* $a_{rx_i} = a_{rx_j}$. Then in forming the conflict graphs, whenever we find that x_i and x_j must receive the same colour in G_r, we identify (coalesce) x_i and x_j into a single vertex, ensuring that they receive the same colour. (This can be effectively implemented using the disjoint set forest method [9].) Any recolouring of the (coalesced) conflict graphs continues to meet all demands.

One could also accommodate more general strengthening requirements in this approach, by adding vertices to the conflict graph to ensure that when a demand is met, not too many different colours are assigned to the corresponding columns. However, this appears to increase the size of the conflict graphs exponentially, so we do not pursue it here. Instead we focus on SHFs and PaHFs.

3.2 Recolouring Conflict Graphs

Vertex colouring is NP-complete [11]. However, our interest is in finding a colouring of a graph G_r in $|\Sigma_r|$ colours, when G_r is known to be $|\Sigma_r|$-colourable. Simply permuting the colours changes nothing. We want a *non-trivial* recolouring, a proper colouring that is not simply a permutation of the original. Unfortunately, deciding the existence of a non-trivial recolouring is also NP-complete. To see this, one can use the fact that deciding whether a 3-SAT formula has a second satisfying assignment, given one satisfying assignment, is NP-complete [19]. Then using the well-known reduction from 3-SAT to vertex colouring [11], one finds that deciding the existence of a non-trivial vertex recolouring is also NP-complete.

With this complexity in mind, we do not make a concerted effort to find non-trivial recolourings for the conflict graphs. Rather we use a simple greedy approach. For each G_r, collapse multiple edges (if present) and sort the vertices in nonincreasing order by degree, breaking ties at random. Now process the vertices in this order, assigning each in turn a colour chosen at random from those not already assigned to one of its neighbours. If none is available for some vertex, no colouring is produced. We repeat this process until either a colouring is produced, or a limit on the number of attempts is reached. When a colouring is found, its colours are interpreted as symbols to replace row r. When no colouring is found, the row is left unchanged.

Adding this recolouring method to the generic post-optimization strategy produces a variant, *recolouring post-optimization.*

4 Computational Results

A specialization of generic post-optimization has been surprisingly successful at improving covering arrays [13,14]. Here we focus on applications to hash families, but remind the reader that there is a wide variety of t-restriction problems in which the methods could be employed. We always treat homogeneous, uniform hash families with N rows, k columns, v symbols, and a restriction of strength t. C++ implementations of both generic and recolouring post-optimization were tested using an 8-core Intel Xeon processor clocked at 2.66GHz with 4MB of cache, bus speed 1.33GHz, and 16GB of memory. Testing proceeds by first generating an array one row at a time, choosing rows at random, until all demands are met. Post-optimization is then applied to improve the solution, if possible. Except when noted, post-optimization was executed for one minute on a single core.

Perfect hash families have been very extensively studied (for example, [18] and references therein). For strengths $t \in \{5, 6\}$, $v = t$, and $k \leq 25$, generic post-optimization of random arrays rarely produces arrays that are competitive with the best known sizes; this should be expected, given the computational effort invested [4,18]. What is surprising is that recolouring post-optimization not only recreates many of the best known results, but constructs a 10×10 PHF with $t = v = 5$, shown in Figure 2. This improves on the previous best known 13×10 and 11×9 PHFs [17].

We expect the most useful applications to arise for t-restriction problems that have not been extensively researched before. In producing a $\{w_1, \ldots, w_s\}$-SHF, one could naturally use a PHF of strength $t = \sum_{i=1}^{s} w_i$, provided that the number of symbols is at least t. We therefore compare cases for $\{2, 2, 1\}$-, $\{4, 1\}$-, and $\{3, 2\}$-SHFs, with the best known results for PHFs [17]. Table 1 gives a selection of results from generic post-optimization. Columns headed by 'G' are from generic post-optimization, those headed 'I' are sizes of the initial random array, and the one headed 'B' is the best known result from [17].

Naturally as v is decreased, the number of rows generally increases, as one would expect. Because of the randomness of post-optimization, it can happen that a solution with fewer rows is found even when k is increased of v is decreased;

Table 1. Generic post-optimization for SHFs

	$\{1,1,1,1,1\}$			$\{2,2,1\}$						$\{4,1\}$				$\{3,2\}$			
$v \rightarrow$	5	5	5	5	5	4	4	3	3	5	4	3	2	5	4	3	2
$k \downarrow$	B	G	I	G	I	G	I	G	I	G	G	G	G	G	G	G	G
5	1	1	9	1	7	6	18	15	80	3	3	3	5	4	5	6	10
6	3	3	28	3	24	7	40	15	99	3	3	5	6	5	5	10	15
7	6	6	49	6	24	7	58	28	212	3	4	5	7	7	7	11	15
8	8	8	73	8	42	12	75	28	236	4	6	7	8	7	9	13	22
9	11	12	107	13	45	22	98	28	371	4	6	9	9	7	11	16	27
10	13	15	109	13	41	22	92	28	417	5	7	9	10	8	11	18	31
11	16	20	125	17	55	33	118	75	338	7	7	12	20	12	14	21	35
12	21	25	141	19	59	36	112	84	385	7	9	13	21	11	16	23	39
13	26	30	138	22	71	43	127	98	502	8	9	14	13	12	16	25	43
14	32	36	166	25	71	47	136	108	454	8	10	17	22	12	17	26	45
15	35	40	165	27	71	50	135	126	453	10	12	18	23	14	18	29	49
16	39	45	181	28	74	54	147	138	542	9	12	18	24	15	20	30	52
17	44	50	185	32	79	57	146	150	513	10	13	19	25	16	20	33	55
18	49	56	200	33	77	62	156	164	469	11	14	22	24	16	22	33	59
19	53	61	217	35	78	67	158	177	526	11	14	22	26	17	23	34	62
20	57	64	263	37	78	71	163	188	442	11	15	24	27	18	23	35	64
21	61	70	244	39	95	75	151	204	471	12	15	25	27	18	25	38	66
22	64	75	254	42	77	82	168	220	484	13	16	25	29	19	25	39	69
23	68	81	244	44	84	83	179	233	618	12	17	27	28	20	27	40	73
24	71	84	294	44	98	86	213	233	606	13	17	27	28	21	27	41	77
25	74	90	248	50	88	92	187	247	670	14	18	28	31	22	29	43	78

2	3	4	3	0	4	1	0	2	1
0	1	1	3	4	2	2	4	3	0
1	4	2	0	3	0	3	4	2	1
1	1	2	3	3	4	0	4	2	0
2	4	4	1	1	2	0	3	0	3
1	1	4	3	4	0	0	2	2	3
2	1	0	3	2	0	4	1	4	3
4	3	0	2	4	1	3	1	2	0
4	2	0	3	0	3	2	1	4	1
2	4	1	4	3	2	3	0	0	1

4	4	1	4	0	3	2	4	5	1	0
5	0	5	1	3	2	0	2	2	3	4
5	4	3	1	2	3	3	0	2	3	0
3	1	4	0	4	0	2	5	3	5	1
4	1	0	0	4	3	5	2	5	1	3
3	0	0	2	2	5	2	2	4	1	1
0	1	5	4	0	1	0	5	3	3	2
5	2	1	4	0	3	3	5	2	0	4
5	2	3	5	0	0	1	4	3	5	1

Fig. 2. A 10×10 PHF with $t = v = 5$, and a 9×11 $(6, 2)$-DHF with $v = 6$

for example, a 13×13 $\{4, 1\}$-SHF with $v = 3$ was found, having fewer rows than the 21×12 solution with $v = 3$ and the 14×13 solution with $v = 4$. Here one could treat the 13×13 solution as having $v = 4$, or delete a column to obtain a 13×12 solution with $v = 3$. We have not recorded such implications in the tabulation, so as to focus on the results of post-optimization. Recolouring post-optimization can often improve these results, sometimes substantially: The 247×25 $\{2, 2, 1\}$-SHF with $v = 3$ improves to a 213×25 solution, a 13% reduction in the number of rows.

Restrictions with more than one demand can also be treated. Suppose, for example, that we want an array that is $\{4, 1\}$- and $\{3, 2\}$-separating with $k = 25$ and $v = 5$. Rather than using the 74×25 PHF, we could combine the 14×25 $\{4, 1\}$-SHF and the 22×25 $\{3, 2\}$-SHF to produce a 36×25 solution. Better yet, generic post-optimization using the two demands simultaneously yields a 22×25 solution in one minute. Similarly, a 49×25 array that is $\{2, 2, 1\}$-, $\{4, 1\}$-, and $\{3, 2\}$-SHF with $v = 5$ was found, which unexpectedly has fewer rows than the 50×25 $\{2, 2, 1\}$-SHF in Table 1.

Distributing hash families impose a number of separation demands simultaneously. Table 2 show results for $(6, s)$-DHFs with $s \in \{2, 3, 6\}$. The case when $s = 6$ is the PHF, and the known result from [17] is reported. The remaining values are from post-optimization. When $v = 6$, both generic and recolouring post-optimization typically produce solutions with fewer rows than the PHF. Notably, recolouring post-optimization often yields a much smaller result than does generic post-optimization, supporting the belief that conflict graph recolouring can avoid many of the local optima encountered in the generic method. Because these cases have strength $t = 6$, when $k = 20$ we are examining 27,907,200 6-tuples of columns to check demands. Hence one might expect that our standard one minute limit on computation time does not permit many iterations! Indeed permitting five minutes rather than one improves the 116×20 $(6, 3)$-DHF with $v = 6$ to an 88×20 solution.

Recolouring post-optimization also improves a 15×11 $(6, 2)$-DHF with $v = 6$ to a 9×11 solution, shown in Figure 2, and it improves a 36×20 $(6, 2)$-DHF with $v = 6$ to a 24×20 solution. In the interests of conserving space, we do not provide a complete list.

Table 2. Generic and recolouring post-optimization on (t, s)-DHFs

	Known	Generic									Recolouring			
	(6, 6)	(6, 2)					(6, 3)				(6, 3)			
$k \downarrow v \rightarrow$	6	6	5	4	3	2	6	5	4	3	6	5	4	3
6	1	1	7	12	15	31	10	17	31	90	1	16	28	89
7	4	8	8	11	20	42	11	17	39	113	4	16	37	101
8	8	7	15	18	27	50	17	25	54	177	10	22	48	175
9	13	9	13	20	32	57	16	31	64	224	16	29	63	223
10	18	11	16	24	39	72	21	38	83	291	20	37	81	278
11	24	15	19	26	43	84	26	46	99	352	25	44	95	340
12	27	15	21	30	49	96	34	94	150	563	30	53	113	403
13	39	19	24	32	55	110	37	66	140	516	36	61	133	473
14	53	20	26	39	62	123	44	81	162	590	40	72	150	539
15	64	25	29	43	69	136	55	93	185	676	47	83	173	615
16	77	26	32	45	72	151	63	113	211	765	52	90	190	685
17	86	29	35	47	81	164	71	112	236	853	64	101	214	769
18	94	30	39	53	89	182	87	130	273	1021	70	112	233	832
19	106	32	46	58	94	196	101	148	300	1021	74	127	256	914
20	120	36	53	63	104	213	116	158	321	1121	75	136	282	984

Table 3. Post-optimization on (t, s)-PaHFs. Columns labelled R are recolouring, the rest generic.

	(4, 2)-				(5, 2)-				(5, 4)-			(6, 5)-		(7, 6)-	
$k \downarrow v \rightarrow$	4	3	2	2R	5	4	3	2	5	5R	4	6	5	7	6
5	10	10	10	10	15	15	15	15	10	10	10				
6	12	12	10	10	15	24	15	16	16	15	18	15	15		
7	12	12	11	11	21	20	21	20	24	23	25	23	31	21	21
8	14	11	12	11	29	29	29	28	29	28	21	41	55	38	45
9	14	15	15	11	36	35	35	35	33	33	46	53	80	66	98
10	17	11	16	11	43	40	39	39	43	42	56	74	100	112	162
11	17	16	17	17	47	45	44	44	49	48	70	93	196	220	271
12	19	18	18	17	50	49	49	46	61	55	81	122	187	313	388
13	22	21	19	19	54	54	53	52	66	64	91	152	242	573	580
14	22	21	21	20	56	57	56	54	73	70	106	177	283	909	1047
15	23	22	21	20	61	62	61	58	80	76	118	209	345		
16	24	24	23	21	68	63	62	61							
17	26	25	23	22	69	70	68	64							
18	26	25	24	23	75	73	70	69							
19	27	26	25	24	88	80	75	72							
20	30	28	25	25	92	81	80	74							
21	29	29	26	26	129	103	85	80							
22	29	30	27	27	157	127	99	82							
23	32	30	28	28	119	126	96	83							
24	34	33	29	28	208	150	105	86							
25	35	33	30	29	258	159	123	89							

Table 3 reports on the results of post-optimization on randomly generated PaHFs; there is no known result with which to compare. Examining the $(4,2)$- and $(5,2)$-PaHFs, it is striking that allowing more symbols often yields a larger number of rows; yet it is clear that one can simply not use the extra symbols, and so an array on fewer symbols remains a solution on more. The behaviour is an artifact of the random selection process for the initial array. Indeed when more symbols are provided, the chance increases that, while columns are separated, too many symbols are used to do so. This could be overcome by using a better greedy method to make the initial array, for example the methods in [7]. This does not occur in every case examined. For $(6,5)$- and $(7,6)$-PaHFs, better results are obtained with more symbols permitted. In these cases, in a separation $\binom{t}{2} - 1$ pairs must have different values, but only one pair requires the same. Hence selecting rows uniformly at random yields a better initial solution in these cases.

Recolouring post-optimization (in the two columns marked R in Table 3) yields improvements beyond those obtained by generic post-optimization. Coalescing vertices in the conflict graphs appears to have lessened the benefit of recolouring; nonetheless it is striking that improvements remain possible.

5 Conclusion

Arrays for t-restrictions permeate many different applications. General tools to construct them include greedy methods and random methods, but both appear to yield arrays with an unnecessarily large number of rows. Naturally more sophisticated methods can typically be devised for a specific t-restriction, but requires careful analysis of the specifics of the restriction. Therefore we have developed a general technique, focussing first on unnecessary entries and then on changeable entries, to eliminate rows repeatedly. Even with modest investments of computation time, and even starting with poor input arrays, these post-optimization methods yield useful arrays. We anticipate that their main value is in improving solutions found by methods other than simple random techniques, as has been the case with covering arrays. However, the real strength of the methods is their ability to deal with arbitrary t-restriction problems. Applications beyond the realms of hash families and covering arrays appear well worth further research.

Acknowledgements. Thanks to Daniel Horsley and Violet Syrotiuk for helpful discussions about this research.

References

1. Ahlswede, R., Deppe, C., Lebedev, V.: Threshold and Majority Group Testing. In: Aydinian, H., Cicalese, F., Deppe, C. (eds.) Ahlswede Festschrift. LNCS, vol. 7777, pp. 488–508. Springer, Heidelberg (2013)
2. Ahlswede, R., Wegener, I.: Search Problems. Wiley Interscience (1987)
3. Alon, N., Moshkovitz, D., Safra, S.: Algorithmic construction of sets for k-restrictions. ACM Transactions on Algorithms 2, 153–177 (2006)

4. Colbourn, C.J.: Constructing perfect hash families using a greedy algorithm. In: Li, Y., Zhang, S., Ling, S., Wang, H., Xing, C., Niederreiter, H. (eds.) Coding and Cryptology, pp. 109–118. World Scientific, Singapore (2008)
5. Colbourn, C.J.: Distributing hash families and covering arrays. J. Combin. Inf. Syst. Sci. 34, 113–126 (2009)
6. Colbourn, C.J.: Covering arrays and hash families, Information Security and Related Combinatorics. In: NATO Peace and Information Security, pp. 99–136. IOS Press (2011)
7. Colbourn, C.J., Horsley, D., Syrotiuk, V.R.: Strengthening hash families and compressive sensing. Journal of Discrete Algorithms 16, 170–186 (2012)
8. Colbourn, C.J., Zhou, J.: Improving two recursive constructions for covering arrays. Journal of Statistical Theory and Practice 6, 30–47 (2012)
9. Cormen, T.H., Leiserson, C.E., Rivest, R.L., Stein, C.: Introduction to Algorithms, 3rd edn. The MIT Press (2009)
10. Du, D.-Z., Hwang, F.K.: Combinatorial group testing and its applications, 2nd edn. World Scientific Publishing Co. Inc., River Edge (2000)
11. Karp, R.M., Miller, R.E., Thatcher, J.W.: Reducibility among combinatorial problems. Journal of Symbolic Logic 40(4), 618–619 (1975)
12. Liu, L., Shen, H.: Explicit constructions of separating hash families from algebraic curves over finite fields. Designs, Codes and Cryptography 41, 221–233 (2006)
13. Nayeri, P., Colbourn, C.J., Konjevod, G.: Randomized Postoptimization of Covering Arrays. In: Fiala, J., Kratochvíl, J., Miller, M. (eds.) IWOCA 2009. LNCS, vol. 5874, pp. 408–419. Springer, Heidelberg (2009)
14. Nayeri, P., Colbourn, C.J., Konjevod, G.: Randomized postoptimization of covering arrays. European Journal of Combinatorics 34, 91–103 (2013)
15. Stinson, D.R., Van Trung, T., Wei, R.: Secure frameproof codes, key distribution patterns, group testing algorithms and related structures. J. Statist. Plann. Infer. 86, 595–617 (2000)
16. Stinson, D.R., Zaverucha, G.M.: Some improved bounds for secure frameproof codes and related separating hash families. IEEE Transactions on Information Theory, 2508–2514 (2008)
17. Walker II, R.A.: Phftables, `http://www.phftables.com` (accessed March 10, 2012)
18. Walker II, R.A., Colbourn, C.J.: Perfect hash families: Constructions and existence. Journal of Mathematical Cryptology 1, 125–150 (2007)
19. Yato, T., Seta, T.: Complexity and completeness of finding another solution and its application to puzzles. IEICE - Transactions on Fundamentals of Electronics, Communications and Computer Sciences E86-A(5), 1052–1060 (2003)

Search for Sparse Active Inputs: A Review

Mikhail Malyutov

Math. Dept., Northeastern University, 360 Huntington Ave., Boston, MA 02115
m.malioutov@neu.edu

Dedicated to the memory of Rudolf Ahlswede

Abstract. The theory of Compressed Sensing (highly popular in recent years) has a close relative that was developed around thirty years earlier and has been almost forgotten since – the design of screening experiments. For both problems, the main assumption is sparsity of active inputs, and the fundamental feature in both theories is the threshold phenomenon: reliable recovery of sparse active inputs is possible when the rate of design is less than the so-called capacity threshold, and impossible with higher rates.

Another close relative of both theories is *multi-access information transmission.* We survey a collection of tight and almost tight screening capacity bounds for both *adaptive* and *non-adaptive* strategies which correspond to either having or not having feedback in information transmission. These bounds are inspired by results from multi-access capacity theory. We also compare these bounds with the simulated performance of two analysis methods: (i) linear programming relaxation methods akin to basis pursuit used in compressed sensing, and (ii) greedy methods of low complexity for both non-adaptive and adaptive strategies.

Keywords: search for sparse active inputs, multi - access communication, compressive sensing, group testing, capacity.

1 Introduction and History Sketch

The idea of using 'sparsity' of factors actively influencing various phenomena appears repeatedly throughout a diverse range of applications in fields from computational biology to machine learning and engineering. Notably, [12] used the assumption of sparsity of contaminated blood donors to dramatically reduce the number of experiments needed to **adaptively** screen them out. Troubleshooting complex electronic circuits using a non-adaptive identification scheme was considered in [48] under the assumption that only a few elements (a sparse subset) become defective. A recent application of these ideas enabled affordable genetic screening to successfully eliminate lethal genetic diseases prevalent in an orthodox Jewish community in New York city, as described in [13].

Successful optimization of industrial output for **dozens of real world applications** was reported in [4]. The authors of that paper used the sparsity

H. Aydinian, F. Cicalese, and C. Deppe (Eds.): Ahlswede Festschrift, LNCS 7777, pp. 609–647, 2013.

assumption of non-negligible (active) coefficients in the multivariate quadratic regression model, randomized design, and proposed the **Random Balance Method** (RBM) of analysis of outputs. The method was essentially a visual greedy inspection of scatterplots to identify the most active inputs and allowed consequent optimization of the output. RBM, inspired by the Fisher's celebrated idea of randomization in estimation, was officially buried by the leading Western statisticians (G.E.P. Box, D.R. Cox, O. Kempthorne, W. Tukey et al) in their unjustified disparaging discussion following the publication of [4] in the first volume of Technometrics. As a result, F.E. Satterthwaite, the author of RBM, suffered a breakdown and was confined for the rest of his life to a psychiatric clinic.

In the pioneering annotated overview of Western literature on regression and design [39][1], V.V. Nalimov characterized RBM as a psycho-physiological triumph of the experimenter's intuition over mathematical arguments and predicted that mathematicians would never understand RBM's effectiveness, which Nalimov's team confirmed on many simulated and applied examples.

A partial combinatorial justification of RBM was soon obtained in [38] (see also our section 1.2) and immediately continued in [32]. A.N. Kolmogorov made me responsible to report to him on mathematical aspects of what was going on in Nalimov's Department of Kolmogorov's huge Lab. This task made me interested in the subject and resulted in my many publications on optimal design of experiment. [32] first introduced information-theoretic grounds for the effectiveness of RBM. [32], together with organizing the donors' blood group testing for hepatitis in the Moscow Blood Transfusion Center and my joint (later patented) work on built-in troubleshooting systems for the construction of complex redundant circuits in the first Soviet aircraft carriers [48], this became a stimulus to almost 40 years of my and my pupils' continuing research in this exciting area.

The funding for the latter project motivated A.G. D'yachkov and V.V. Rykov to join my **seminar at the Kolmogorov Lab** in Moscow University. Many fundamental results were obtained by its participants. While I restricted myself to finding probabilistic performance bounds in various screening adaptive and nonadaptive models with **positive error probability**, D'yachkov and Rykov's main results were directed to combinatorial and algebraic methods of **errorless** design construction for noiseless models. The Russian giants of Information Theory (Kolmogorov, Pinsker and Dobrushin) supported our activity substantially. G. Katona introduced us to previous research on Search Theory of the Erdős-Rényi school (implementable mostly for finding **one active input** under their *homogeneity condition*). In his last published paper, C. Shannon announced that he had found the Multi-Access capacity region, without disclosing it. This problem was solved in the famous paper [1]. I. Csiszár pointed to the relevance of this discovery to our problem.

R. Ahlswede was the first outstanding Western researcher to recognize the results of our Moscow team. He endorsed its description in my Addendum to the Russian translation of [2] and recommended its inclusion into the English

[1] This book was later translated back into English.

translation of their book (this was sabotaged by the Soviet copyright selling agency). R. Ahlswede's diploma student J. Viemeister at Bielefeld University prepared a detailed survey [45] of our results spanning around 200 pages in 1982. My stay in the instrumental depot of R. Ahlswede's home in Bielefeld during the period: Fall 1993 - Spring 1994, eliminated my depression caused by the collapse of Soviet science.

Ironically enough, W. Tukey, one of RBM's harshest critics, soon after the fatal outcome of the aforementioned discussion paper became a supervisor of an undergraduate student named David Donoho in Princeton. More than forty years after that sad story, Donoho initiated and successfully marketed an enthusiastically accepted revival of related ideas under the name 'Compressed Sensing' for estimation in over-parameterized linear models with L_1-penalty or Linear Programming (LP) analysis (basis pursuit). These ideas were also proposed around the same time in [43] under the name of LASSO for related statistical problems. Neither of these authors were apparently aware of the connection to the RBM method. Thanks to D. Donoho's popularization, sparsity is now a well established assumption in statistical applications! LP with proven moderate degree polynomial complexity under the so-called RIP condition of [5] on the design matrix allow some **upper bounds** for the sample size of the design matrix.

The threshold phenomenon was observed in [11] as a result of intensive simulation performed for randomized designs. Connection to the Shannon's celebrated justification for a closely related phenomenon arising in information transmission using his notion of **channel capacity** started to be noticed only later. Some attempts to find **lower** performance bounds for compressive sensing using information-theoretic tools were later made in [46]. I am unaware of the situations, where these lower bounds are asymptotically equivalent to the upper bounds under the RIP condition in contrast to our lower bounds for the STI analysis. Discussion of the relation between our and the most popular compressive model is in Remark 2 after the Theorem 2.1.1. In section 4 we briefly outline our doubts on validity of s/t asymptotics discussed in compressive sensing unless the unknown parameters of the linear model are **incommensurable**. Numerous publications on compressive sensing during the last decade are listed on www.dsp.ece.rice.edu/cs.

Asymptotically sharp capacity bounds inspired by the *Multi-Access Capacity Region construction* were obtained in [26,30], for brute-force (BF) analysis, and in [31,33], for separate testing of inputs methods (STI). These bounds can greatly enhance current understanding of the threshold phenomenon in sparse recovery. Some of these results for a particular case of $\cup$-model and BF analysis were rediscovered in [3].

These bounds are obtained for *asymptotically optimal designs* (which turn out to include *random designs*) and thus imply *upper bounds* for the performance of recovery under *arbitrary designs*. Random designs are a simple natural choice for pre-specified experimental design and moreover they provide the ability to effectively apply STI analysis via asymptotically optimal empirical mutual information maximization between inputs and the output. STI replaces the visual inspection in RBM [4]. STI and greedy STI capacities *outperform* Linear Programming relaxation for randomized designs in a wide range of models (see our section 3-4). They also admit a straightforward generalization to *nonparametric noisy models* including *'colored noise'*.

The outline of the present paper is as follows. Examples will always precede the general theory and tables will be made transparent by corresponding plots. We start by setting up elementary noiseless nonadaptive models and discuss their relation to the compressive sensing in sections 2.1-2. Noisy models are introduced in section 3 under known noise distribution. Capacity comparison simulation under several methods of analysis and IID unknown noise will be covered in sections 4-5 followed by the study of colored noise and finite $t, N \to \infty$ in sections 6-7. Section 8 on adaptive designs starts with an elementary asymptotically optimal adaptive search description for active inputs of an unknown boolean function, followed by discussion of the example when capacity of adaptive search exceeds that for any nonadaptive ones. Finally, we outline adaptive search for active inputs of a nonparametric noisy additive function, which is similar to the adaptive strategy for general unknown function studied in [35]. The outlines of more involved proofs will be found in Appendices 9.1-5.

2 Popular Elementary Models

The first popular models of screening include the Boolean sum (or simply $\cup$-model), and the forged coin model (FC-model). The $\cup$-model, was used [12] to model the pooled *adaptive* testing of patients' blood for the presence of a certain antigen. Dorfman's innovative adaptive design reduced experimental costs by an order of magnitude! (the $\cup$-model is also known as "group testing" in search theory or as "superimposed codes" in information theory).

Another combinatorial example is *nonadaptive* search for a subset of forged coins (with their weights exceeding the weight of genuine coins by one unit) using weighings of the minimal number of subsets of coins, as popularized in [14].

Let us formalize these two examples to show their interrelation. Introduce an $(N \times t)$ nonadaptive design matrix X with entries $x_i(a)$ as indicators of participation of a-th coin (patient) in the i-th trial, $i = 1, \ldots, N, a = 1, \ldots, t$. Let $A \in \Lambda(s,t)$ be the unordered subset of *active* inputs (AIs), i.e. forged coins (or, respectively, of sick patients) and denote the corresponding part of the i -th X's row as $x_i(A)$. The output y_i of the i-th trial can then be described by the formula:

$$y_i = g(x_i(A)), \tag{1}$$

where in the first example, g is a boolean sum $\cup$:

$$g(x(A)) = \cup_{a \in A} x(a), \tag{2}$$

and in the second example, g represents ordinary summation:

$$g(x(A)) = \sum_{a \in A} x_i(a). \tag{3}$$

Both functions are symmetric.

We first survey nonadaptive designs (as opposed to adaptive Dorfman's pooled blood testing strategies) for the $\cup$-model that were successfully applied in various settings, e.g. for quality improvement of complex circuits, for trouble-shooting of large circuits with redundancy [48], etc. The FC-model was used to model announcing existence of information packets ready for transmission through multi-access phase-modulated communication channels in ALOHA-type systems.

Introduce *'T-weakly separating'* $(N \times t)$ design matrices allowing identification of the true s-subset of active inputs with probability $\geq 1-\gamma$ under uniform distributions of active s-subsets and analysis T of outputs, and minimum $N^T(s,t,\gamma)$ of N for them. Of special interest is finding the maximal rate or (**capacity**) $C^T(s) = \lim_{t\to\infty} \log t/N^T(s,t,\gamma)$. $C^T(s)$ exists, does not depend on $\gamma > 0$ and is evaluated in the most general model with a nonparametric response function and arbitrary unknown measurement noise for at least two methods of analysis : Brute Force (BF) and Separate Testing of Inputs (STI). The capacity has not yet been found for the L_1 minimization analysis, although results of [17] reduce it to a more feasible problem.

The theory of *'strongly separating designs' (SSD) for sure identification ($\gamma = 0$) of every significant s-subset* is far from being complete: existing lower and upper bounds for the minimal number $N(s,t)$ in SSD differ several times apparently in *all models but one* (sum modulo 2) including the two elementary ones described before with $s > 1$. Apparently, randomly generated strongly separating designs require asymptotically larger N than the best combinatorial ones.

The early upper bounds $\bar{N}^{BF}(s,t,\gamma)$ for $N^{BF}(s,t,\gamma)$ and BF-analysis in many elementary models including the two mentioned above can be found in a survey [24] with lower bounds obtained thanks to the subadditivity of the entropy. The entropy of the prior uniform A distribution is $\log\binom{t}{s}$ while the entropy of each output cannot exceed 1 for the $\cup$-model and entropy of the hypergeometric distribution for the FC-model. An early upper bound for the $\cup$-model $NBF(s,t,\gamma) \leq s\log t + 2s\log s - s\log\gamma$ was obtained in [23] and strengthened in [25]. For the FC-model, the asymptotic capacity expression is $\bar{N}^{BF}(s,t,\gamma) = s\log t/H(B_s^{(1/2)})(1+o(1))$ as $t \to \infty$, where $B_s^{(1/2)}(\cdot)$ is the Binomial distribution with parameter 1/2, and $H(B_s^{(1/2)})$ is its binary entropy.

2.1 Noiseless Search in General Linear Model Under *U*

Consider the Linear (in inputs and parameters) model

$$y_i = \sum_{a\in A} b_a x_i(a), \tag{4}$$

with binary carriers $x_i(a) = \pm 1, i = 1,\ldots,N, a = 1,\ldots,t$.

Assume first the incommensurability of the active coefficients: **U**. The only solution to $\sum_{a\in A} b_a\theta(a) = 0$ with rational coefficients $\theta(a)$ is $\theta(a) \equiv 0$.

Under U, the cardinality of $\{\sum_{a\in A} b_a\theta(a), \theta(a) = \pm 1\}$, is 2^s. This maximal cardinality of the range of the outputs holds almost surely if active coefficients $b_a, a \in A$, are chosen randomly with whatever non-degenerate continuous distribution is taken in R^s. Their *values* can be determined in a few experiments but their *assignment* to inputs is harder to determine.

Remarks. For the FC case $b_a \equiv 1$, the cardinality of the same distinct linear combinations is $s+1$. These two models with maximal and minimal cardinality of the range of the function $g(\cdot, b_1, \ldots, b_s)$ are the ones that have been studied the best. Two main differences between the linear model under U and the previous two models are the *lack of symmetry* leading to necessity of determining ordered s-subsets of AIs and the *presence of nuisance active non-null coefficients.*

Theorem 2.1.1. [38] Under condition U and entries $x_i(a) = \pm 1$ of the design matrix chosen independently and equally likely, the system of equations (4) determines the set A_s unambiguously with probability not less than $1-\gamma$, if

$$N \geq \bar{N}^{BF}(s,t,\gamma) = s + \log([t-s+1]/\gamma). \tag{5}$$

Sketch of Proof

We use the following lemma in proving *all our upper bounds for N with $\gamma > 0$.*

Lemma 2.1.1. The mean of MEP over the ensemble of random designs equals the same mean of conditional error probabilities $\mathcal{P}_{[s]}$, when $[s] := \{1, \ldots, s\}$ is the true set of SI's.

The proof is straightforward because of the symmetry of the ensemble of random designs.

Thus we can assume without loss of generality (WLOG) that $A_s = [s]$. The next step is the following combinatorial lemma easily proved by induction:

Lemma 2.1.2. Any s-dimensional hyperplane can intersect the unit cube in R^N in no more than 2^s vertices.

Consider the following events:

$B_s = \{\mathbf{X} \in R^{N \times t}$: the first $s \times s$-minor is non-degenerate $\}$,

$C_{j,k} : \{\mathbf{X} : N$- column $\mathbf{x}(j)$ is linearly independent of the first s columns,

$D = B_s \Pi_{j>s} C_{j,s}$.

Meshalkin argues: D holds under U with probability exceeding $1-\gamma$ and identities in equations (4) hold, if nonzero values are assigned only for the first s coefficients under U and D.

Remarks. 1. The same lemma 2.1.2 and a straightforward criterion of [42] of a noiseless linear model to be strongly separating allowed the proof in [24] of an upper bound for N in randomly generated binary design to be strongly separating without any condition on cardinality of all possible outputs:

$N \geq 2s[\log(te/2s) + 1]$. I am unaware of any lower bound for this model.

2. Meshalkin generalized his result for random q-ary γ-separating designs obtaining for them the upper bound $N(s,t,\gamma) \leq s + \log_q[(t-s+1)/\gamma]$ and discussed the practical meaning of this bound in his two-page note of 1977.

Continuous design levels for the linear model can increase the capacity of screening. Moreover, the BF analysis can find the subset of AIs in a bounded number of experiments as $t \to \infty$ with prior probability 1 over significant coefficients, although this fact cannot be generalized for the linear model with noise, in contrast to the Meshalkin's theorem for discrete designs.

Adaptive designs for the linear model with the same capacity as nonadaptive ones were constructed in [32] (under a weaker assumption than U on active coefficients). The first information-theoretic interpretation of AIs recovery in sparse models was outlined there for the General Linear model.

The subadditivity of the entropy implies an asymptotic lower bound of about $\log t$, since the entropy of each trial cannot exceed s.

Random designs provide us with about s bits of information per each of the first $\log t$ independent trials on an unknown subset A due to the fact that under condition U, all 2^s outputs corresponding to distinct combinations of values taken by the AIs are different. In other words, the outputs signal about the values taken by *each AI* in the first independent experiments. The additivity of entropy for independent trials suggests that since initial uncertainty is $\log(t)_s = s \log t(1 + o(1))$, we can resolve the initial uncertainty in about $\log t$ steps. Thus lower and upper bounds asymptotically coincide as $t \to \infty$.

2.2 STI

'Identifying' in the definition of γ-separating designs means unique restoration of $A \in \Lambda(s,t)$ by T- analysis with probability $\geq 1 - \gamma$, which may involve searching through all $\binom{t}{s}$ possible subsets of all variables therefore becoming extremely computationally intensive for even moderate values of s as $t \to \infty$. Thus studying capacity under simplified methods such as historically the first Separate Testing of Inputs (STI) or greedy STI with analysis complexity of $O(t \log t)$ discussed further in I. 3 (or more recent Linear Programming (LP) with proven moderate degree polynomial complexity under the so-called RIP condition of [5] on the design matrix) is an important problem. The analysis problem becomes even more critical in more general models with noise and nuisance parameters. Finding the capacity for linear models under LP analysis is still a challenge. Hope for progress in this direction is based on applying the recent criterion [17] instead of the RIP condition [5] which seems not suitable for capacity evaluation.

3 I.I.D. Noisy Sparse Recovery Setting

Suppose that we can assign an arbitrary t-tuple of binary inputs $\mathbf{x} := (x(j), j \in [t]), [t] := \{1, \ldots, t\}$ and measure the **noisy output** Z in a measurable space $\mathcal{Z}$ such that $P(z|y)$ is its conditional distribution given *'intermediate output'* $y = g(x(A))$ as in (1), and $P(z|x(A))$ is their superposition (Multi Access Channel (MAC)), where $\mathbf{x}(\mathbf{A})$ is an s-tuple of $\mathbf{x}$ of AIs. Conversely, every MAC can be decomposed uniquely into such a superposition. We omit the obvious generalization to q-ary inputs and we mark distributions related to a particular MAC by the superscript $\mathbf{m} \in \mathcal{M}$, where $\mathcal{M}$ is a convex set of MACs.

Successive measurements $\zeta_i, i = 1, \ldots, N$, are independent given an N-sequence of input t-tuples ($(N \times t)$-design).

To avoid rather mild algebraic technicalities, we review first the case of a symmetric $g(x(A))$ and finite output alphabet, referring to [31], [36] for a general case.

For a sequence $\mathcal{D}$ of N-designs, $N = 1, \ldots$, introduce the asymptotic rate

$$C^T(\mathcal{D}) = \limsup_{t \to \infty} \frac{\log t}{N_s^T(\gamma)} \tag{6}$$

of a test T. Here $N_s^T(\gamma)$ is the minimal sample size such that the probability of T to misidentify s-tuple A over the measure product of uniform A prior and IID noise distributions (mean error probability (MEP)) is less than γ , $0 < \gamma < 1$. For a general class of tests and designs, AR does not depend on γ, $0 < \gamma < 1$, and remains the same for *slowly decreasing* $\gamma(t) : |\log \gamma(t)| = o(\log t)$ *as* $t \to \infty$.

It is well-known that for **known MAC** $P(z|x(A))$, the Maximum Likelihood (ML)-decision minimizes the MEP for any design. It replaces the BF analysis for the known MAC distribution. Thus C^{ML} is maximal among all tests, if applicable. If MAC is unknown, the **universal nonparametric test** $\mathcal{T}_s$ inspired by a similar development in [7] provides the maximal $C^{\mathcal{T}_s}$ under arbitrary finite inputs and assumptions.

i. Strict positivity of the cross-entropy between the output distributions corresponding to different $g(\cdot)$ and the so-called
ii. compactifiability of continuous output distributions [36].

Test $\mathcal{T}_s$ chooses as AIs the s-subset A maximizing the Empirical Shannon Information (ESI)

$$\mathcal{I}_s(\tau_N^N(A)) = \sum_{x(A) \in \mathcal{B}^{|A|}} \sum_{z \in \mathcal{Z}_N} \tau(x(A), z) \log(\tau(z|x(A))/\tau(z)), \tag{7}$$

where $\tau_N(\cdot)$ is the marginal quantized empirical distribution of the output.

The universal STI decision chooses maximal values of the $\mathcal{I}$ for an a-th input and output, $a = 1, \ldots, t$. $\mathcal{I}$ is defined as $\mathcal{I}_s$ with summation now extended over all values of $x(a)$ instead of s-tuples.

One of the intuitive ideas behind the choice of the statistics $\mathcal{T}$ and $\mathcal{T}_s$ is that for s- subset of AIs (and its subsets), $\mathcal{T}$ and $\mathcal{T}_s$ are strictly positive while for an s-subsets of inactive inputs, $\mathcal{T}$ and $\mathcal{T}_s$ for inactive inputs asymptotically vanish for large samples. The less transparent idea is that the large deviation probabilities for this test are unexpectedly easy to bound from above using Sanov's Large Deviations theorem and its conditional version (see 9.1-4).

3.1 Capacity of STI, Known Symmetric MAC

In Separate Testing Inputs for each input k we test the null 'randomness' hypothesis: product distribution $Q_0(x,z) = P(\xi(k) = x)P(\zeta = z)$ versus the 'activeness' alternative

$$Q_k(x,z) = P(x)P(\zeta = z|\xi(k) = x). \tag{8}$$

Due to the independence of trials, these distributions generate the product marginal distributions

$$Q_j(\mathbf{x}(k,\mathbf{z})), j = 0, k. \tag{9}$$

We denote the mean and variance over Q_j as $\mathbf{E}_j$ and σ_j^2 respectively, the likelihood ratios $l_k(\mathbf{x}(k,\mathbf{z}))$ and critical regions

$$\Delta_k = \{\mathbf{x},\mathbf{z} : l_k(\mathbf{x},\mathbf{z}) > \max\{\kappa_u(k), \max_{m \neq k} l_m(\mathbf{x},\mathbf{z})\}, k = 1,\ldots,s, \tag{10}$$

and their complement Δ_0.

The critical value is

$$\kappa_u(k) = \mathbf{E}l_k - u\sigma(l_k). \tag{11}$$

Of course, $\mathbf{E}l_k = \mathbf{I}(\zeta) \wedge \xi(k) = N\mathbf{I}(\zeta \wedge \xi(k))$. The variance is also additive.

Thus $\kappa_u(k) = NL_k - u\sigma_k\sqrt{N}$.

Let us define the STI decision as $(f(1),\ldots,f(t))$, where $f(k) = \sum k\mathbf{1}_k$, and $\mathbf{1}_k$ is the indicator of set Δ_k.

The decision is correct if $f(\lambda_i) = i, i = 1,\ldots,s$, where λ_i is the i-th AI and $f(j) = 0$ for all other j.

Definition. AI k is hidden in noise(HiN) if $Q_k = Q_0$ for all values p of the randomization parameter.

[31] gives elementary examples of HiN AI, a necessary and sufficient condition for AI to be HiN and, as a corollary, sufficient conditions for non-existence of HiN AIs.

It is proved in [31] that **there are no HiN AIs in symmetric models**.

Theorem 3.1.1.[31] For a symmetric model $N^{STI}(s,t,\gamma)/\log t \to 1/L$ as $t \to \infty$, where $L = \mathbf{E}l_k$ does not depend on k.

Proof. (sketch) Of the three types of errors, only

$e_{0k} = P((\xi(j),\zeta) \in \Delta_k, j > s), k \in [s]$, depends on the design rate as $t \to 0$.

$e_{0k} \leq PQ_0(l_k(\mathbf{x},\mathbf{z}) > \kappa_u(k)) \leq$

$\leq \sum_{\Delta_k} Q_0(\mathbf{x},\mathbf{z}) = \sum_{\Delta_k} \exp\left(l_k(\mathbf{x},\mathbf{z})\right) \leq (1 - e_{k0}) \exp -\kappa_u(k) \leq \exp -\kappa_u(k)$.

For $u \sim N^{\alpha/2}, 0 < \alpha < 1$, it follows that $(t-s)e_{0k} < \exp -\epsilon(logt)^{\alpha}$ for some $\epsilon > 0$ as $t \to \infty$ implying the necessary bound for MEP.

Remark. The asymptotic expression of theorem 1 can be applied to finite t with caution since the residual is a power of $\log t$ decaying very slowly as t grows. In the following tables and plots we simply add 5 to the asymptotic expression for $N(s,t,\gamma)$ believing that 5 is more than the residual.

Examples. 1. For the ∪ - model

$$L(s) = \mathbf{H}(\zeta) - \mathbf{H}(\zeta|\xi) = (s+1)/(2s) + (2^{-1/s} - 1/2)\log(1 - 2^{-(s-1)/s}). \quad (12)$$

$L(2) = .38$ is around .76 of the corresponding value for the BF-analysis. $L(s)s \to \ln 2$. Thus L(s) is around .7 of the corresponding value for the BF-analysis.

2. For the FC-model, the conditional entropy does not depend on the input value. Thus

$$L(s) = \mathbf{H}(\zeta) - \mathbf{H}(\zeta|\xi) = -\sum_{k=0}^{s} B_s^{(1/2)}(k)\log[B_s^{(1/2)}(k)] +$$

$$1/2\{\sum_{k=1}^{s} B_{s-1}^{(1/2)}(k-1)\log[B_{s-1}^{(1/2)}(k-1)] + \sum_{k=0}^{s-1} B_{s-1}^{(1/2)}(k)\log[B_{s-1}^{(1/2)}(k)]\}$$

$$= -\sum_{k=0}^{s} B_s^{(1/2)}(k)\log[B_s^{(1/2)}(k)] + \sum_{k=1}^{s} B_{s-1}^{(1/2)}(k)\log[B_{s-1}^{(1/2)}(k)]$$

$$= H(B_s^{(1/2)}) - H(B_{(s-1)}^{1/2}). \quad (13)$$

$s : H(B_s)$	2:1.81	3:2.03	4:2.20	5:2.33	6:2.45	7:2.54

The comparative performance of STI vs. BF for this model is worse: for large s its AR is around const× $\log s$ times less. This follows from the logarithmic growth of $H(B_s^{(1/2)})$ implying the almost constancy of $sL(s)$ as $s \to \infty$.

The values $\log(t)/L + 5$ for the cases 1. and 2. and s=2,...,7, are tabulated at the end of this section.

3. **What is most amazing is the situation for the general linear model with significant coefficients having a continuous non-degenerate distribution. This model is discussed in a series of recent papers on sparse representations and our section 2.1, where the general case parameters can be restored by STI with the same asymptotic rate as by the BF analysis.**

1. Principal term of $N^{STI}(s,t)$ for the ∪- model.

s/t	100	200	400	800	1600	3200	6400	12800
2	22.3	25.0	27.6	30.2	32.8	35.4	38.0	40.6
3	32.1	36.1	40.2	44.3	48.4	52.4	56.5	60.6
4	41.7	47.2	52.8	58.3	63.8	69.3	74.9	80.4
5	51.3	58.3	65.3	72.3	79.2	86.2	93.2	100.2
6	60.9	69.4	77.8	86.2	94.6	103.0	111.5	119.9
7	70.5	80.4	90.3	100.1	110.0	119.9	129.7	139.6

2. Principal term of $N^{STI}(s,t)$ for the FC-model.

s/t	100	200	400	800	1600	3200	6400	12800
2	13.2	14.4	15.7	16.9	18.1	19.4	20.6	21.8
3	35.2	39.7	44.3	48.8	53.4	57.9	62.5	67.0
4	44.1	50.0	55.8	61.7	67.6	73.5	79.4	85.3
5	56.1	63.8	71.5	79.2	86.9	94.6	102.3	110.0
6	60.4	68.7	77.0	85.4	93.7	102.0	110.4	118.7
7	78.8	89.9	101.0	112.2	123.3	134.4	145.5	156.6

Greedy STI for the $\cup$ Model. We give only an illustrative example suggesting that a more elaborate greedy procedure has polynomial complexity of small degree with larger values of s.

For s=2 in the $\cup$ model, the Greedy STI is a two-step procedure with MEP $< \gamma$, analysis complexity $O(t \log t)$ and the same randomization parameter and number of experiments as for the BF analysis. The STI is applied on the first stage yielding (according to our sketch of the proof of theorem 3.1.1) not more than $t^{-L(2)+\epsilon}$ suspicious inputs satisfying all equations for arbitrary $\epsilon > 0$, with probability approaching 1, while the probability of losing at least one active input is negligible as $t \to \infty$. Applying (12) for s=2, we see that the number of pairs of these suspicious inputs is less than t and BF analysis applied to all of them does not yield a single pair of active inputs with small probability. A modification of this idea works for s=3, and so on.

3.2 Linear Programming (LP) Relaxation, Hybrid LP-BF

Suppose instead of BF analysis, we use Linear Programming (LP) relaxations for both problems. For the linear version of the problem, we use the popular ℓ_1-norm relaxation of sparsity. The problem in (3) can be represented as

$$\min |A|, \text{ such that } y_i = \sum_{a \in A} x_i(a), \ \forall i. \tag{14}$$

Here $|A|$ represents the number of elements in A. Instead, we define the indicator vector I_A such that $I_A(a) = 1$ iff $a \in A$, and focus on the ℓ_1-norm of I_A, i.e. on $\sum_a I_A(a)$. Note that the range of I_A is $\{0, 1\}$, so it is always nonnegative, and instead of $\sum_a |I_A(a)|$ we can use $\sum_a I_A(a)$. We solve the relaxed problem

$$\min \sum_a I_A(a), \text{ such that } y_i = \sum_{a \in A} x_i(a), \ \forall i. \tag{15}$$

This type of relaxation has received a considerable amount of attention in many fields including statistics (Lasso regression, [43]) and signal processing (basis pursuit, [9,10], [21]). While the much simpler linear programming relaxation is not guaranteed to solve the original combinatorial problem, theoretical condition RIP was developed showing that if the unknown sparse signal is sparse enough, then the linear programming relaxation exactly recovers the unknown signal ([11] et al). This problem has also been studied in the random design simulation setting, and bounds were developed on recovery of the unknown signal with high probability summarized in the following sentence from [11]: "for many (design) matrices there is a threshold phenomenon: if the sparsest solution is sufficiently sparse, it can be found by linear programming". Our numerical check of this statement suggests that there are at least two thresholds: one for the brute force analysis and another for the linear programming. For the non-linear problem we are forced to relax not only the sparsity of the indicator vector I_A but also the measurement model.

Since $y_i = \cup_{a\in A} x_i(a)$, it then must hold that $y_i \le \sum_{a\in A} x_i(a)$. Hence, our first relaxation is

$$\min \sum_a I_A(a) \text{ such that } y_i \le \sum_{a\in A} x_i(a), \quad 0 \le I_A(a) \le 1. \tag{16}$$

We also note that if $y_i = 0$, then it must hold that all $x_i(a) = 0$, and the inequality $y_i \le \sum_{a\in A} x_i(a)$ holds with equality. Hence, a stronger relaxation is obtained by enforcing this equality constraint.

$$\min \sum_a I_A(a) \text{ such that } 0 \le I_A(a) \le 1, \text{ and } y_i \le \sum_{a\in A} x_i(a), \quad \text{if } y_i \ne 0 \text{ and} \tag{17}$$

$$y_i = \sum_{a\in A} x_i(a) \quad \text{if } y_i = 0. \tag{18}$$

Thus, taking into account particular features of this nonlinear model **before applying linear programming** is essential. To our knowledge, bounds for the performance of this linear programming relaxation of the nonlinear screening problem have not been studied in the literature.

Interesting relations between combinatorial properties of design matrices and correctness of the LP solution are found in [22].

3.3 N^{LP} and N^{LP-BF} Simulations for ∪- Model

This and the next section describe results of [34]. Simulation of N^* for several analysis methods was organized roughly speaking as follows. For every s (from 2 to the upper limit depending on resources), a random IID repeated sample of 100 ($N \times t$) binary design matrices was generated with optimal randomization : $P(0) = 2^{-1/s}$ for the ∪-model and $P(0) = 1/2$ for FC and Linear models, starting with N around 15 or 20. If, for a given t, the frequency of cases with *unique correct solution to the AIs recovery problem* exceeded 95 percent, then N for the next sample of 100 random matrices was chosen one less; otherwise N was chosen one more until the procedure converges. This procedure was then repeated for twice larger t, and so on.

The entries in parenthesis inside the following (and two later similar tables) are the empirical standard deviations for N^* obtained in 10 independent simulations.

s/t	100	200	400	800	1600	3200	6400	12800
2	49(1.32)	64(1.58)	70(1.58)	82(1.65)	94(2.71)	109(1.56)	123(1.57)	138(1.64)
3	75(2.01)	94(1.78)	114(1.97)	142(1.51)	168(2.22)	197(2.17)	231(1.45)	256(2.17)
4	96(2.20)	126(1.94)	157(1.65)	192(2.10)	237(2.59)	281(2.80)	325(2.10)	372(2.30)
5		145(3.78)	190(3.01)	242(2.41)	296(2.51)	365(1.81)	428(3.37)	496(3.26)
6		175(1.55)	220(2.18)	283(3.00)	358(2.22)	431(1.69)	518(2.32)	602(1.85)
7		190(3.34)	248(3.29)	321(3.41)	405(2.62)	501(1.65)	607(1.26)	779(1.45)

Computations took several weeks to perform on a Windows XP PC.

A more intelligent algorithm, described in the previous section, treats zero outputs as described in section 3.4. Both $N^{LP}s$ and computing times were reduced considerably. We emphasize that this method of analysis is 'half-way' to the BF taking into account neat details of the $\cup$- model, which seems always necessary for non-linear models. The table is as follows.

s/t	100	200	400	800	1600	3200	6400	12800
2	23(1.03)	26(1.35)	28(0.63)	30(0.88)	32(0.88)	34(0.74)	35(0.85)	38(1.32)
3	36(1.10)	39(1.35)	43(1.35)	45(1.43)	48(1.08)	51(1.07)	54(1.45)	56(1.18)
4	49(1.15)	52(1.55)	55(1.89)	58(1.49)	63(1.58)	66(2.06)	68(1.78)	74(1.81)
5	61(2.32)	67(1.81)	72(1.90)	76(0.97)	80(1.32)	85(1.32)	87(2.25)	91(2.13)
6	74(2.64)	80(2.50)	86(1.97)	89(2.87)	93(1.70)	100(1.58)	105(1.62)	111(2.15)
7	91(2.18)	96(2.75)	101(2.25)	108(2.79)	113(1.64)	118(2.04)	124(2.38)	129(2.21)

The table below represents the early upper bound [23], p. 145, for $N(s,t,\gamma) \leq s\log_2 t + 2s\log_2 s - s\log_2(0.05)$.

s/t	100	200	400	800	1600	3200	6400	12800
2	25.9	27.9	29.9	31.9	33.9	35.9	37.9	39.9
3	42.4	45.4	48.4	51.4	54.4	57.4	60.4	63.4
4	59.9	63.9	67.9	71.9	75.9	79.9	83.9	87.9
5	78.05	83.05	88.05	93.05	98.05	103.05	108.05	113.05
6	96.8	102.8	108.8	114.8	120.8	126.8	132.8	138.8
7	116.1	123.1	130.1	137.1	144.1	151.1	158.1	165.1

In [25], Remark 3, p. 166, a more accurate upper bound is given: $N(s,t,\gamma) \leq \log\binom{t-s}{s} - \log(\gamma - t^{-c_{17}})$, where constant c_{17} depends only on s and is obtained as a result of a transformation chain consisting of 17 steps. For $\gamma = 0.05$, replacing $-\log(\gamma - t^{-c_{17}})$ for our big $t's$ with a larger value 5, we get the following table:

s/t	100	200	400	800	1600	3200	6400	12800
2	17.21	19.25	21.27	23.28	25.28	27.29	29.29	31.29
3	22.17	25.26	28.30	31.32	34.34	37.34	40.34	43.35
4	26.66	30.83	34.91	38.95	42.97	46.98	50.99	54.99
5	30.79	36.06	41.19	46.25	51.28	56.30	61.30	66.31
6	34.60	41.00	47.19	53.28	59.3252	65.35	71.36	77.37
7	38.14	45.69	52.95	60.08	67.14	74.18	81.19	88.20

Comparing this last table and the second table we conclude that the BF capacity of ∪-screening is **smaller** than the simulated N^* under the intelligent improvement of LP analysis.

3.4 N^{LP} Simulation for the FC Model

We know only an *asymptotic* upper bound for the FC-model BF- capacity obtained in [24], Theorem 5.4.1, (this follows also from our general result [31] on *'ordinarity'* of symmetrical models). That result used a non-trivial entropy $H(B_s(p))$ of the binomial distribution maximization result by P. Mateev: $\max_{0<p<1} H(B_s(p)) = H(B_s(1/2)) := a_s$ for all $s \geq 1$: If for some $0 < \beta < 1, 0 < \epsilon$

$$N \geq s \log t/a_s + \kappa(\epsilon)(\log t/a_s)^{(1+\beta)/2}$$

s/t	100	200	400	800	1600	3200	6400	12800
2	16(0.67)	17(0.67)	18(0.57)	20(0.70)	22(0.52)	24(0.74)	27(0.42)	28(0.78)
3	22(1.06)	24(0.85)	25(0.95)	29(0.70)	31(0.88)	33(0.82)	35(0.85)	38(0.57)
4	28(1.29)	32(0.70)	33(0.92)	35(0.82)	39(0.63)	41(0.79)	44(0.71)	49(0.85)
5	35(1.76)	36(0.92)	40(0.82)	44(0.95)	47(0.88)	52(0.52)	57(0.52)	62(0.70)
6	44(1.81)	44(0.95)	50(0.88)	51(1.49)	53(1.05)	58(0.88)	64(0.42)	69(0.92)
7	51(1.99)	53(1.51)	55(1.17)	59(1.16)	62(1.07)	65(0.97)	73(0.85)	79(0.99)

then $\gamma \leq \epsilon(s \log t/a_s)^{\beta}$. Thus, $logt/N(s,t,\gamma) \to a_s/s$ as $t \to \infty$ and for β close to 1, the second additive term grows (or declines, if $\kappa < 0$) with a rate of about $(\log t)^{1/2}$.

$s : H(B_s)$	2:1.81	3:2.03	4:2.20	5:2.33	6:2.45	7:2.54

Using the preceding $H(B_s)$ table, we prepare the $\log \binom{t}{s}/a_s$ table which is asymptotically equivalent to $N(s,t,\gamma)$ (as $t \to \infty$) for any $0 < \gamma < 1$.

s/t	100	200	400	800	1600	3200	6400	12800
2	8.18	9.52	10.86	12.19	13.52	14.86	16.19	17.53
3	9.55	11.22	12.88	14.54	16.20	17.86	19.51	21.17
4	10.79	12.78	14.76	16.73	18.71	20.68	22.65	24.62
5	11.90	14.21	16.50	18.79	21.06	23.34	25.62	27.89
6	12.92	15.54	18.14	20.72	23.30	25.87	28.44	31.02
7	13.85	16.78	19.67	22.55	25.42	28.28	31.15	34.01

As in the ∪-model, $N^*(s,t)$ exceeds $N^{BF}(s,t,\gamma)$ significantly (by more than two times). The asymptotic nature of this last table could not influence the comparison. We observe that for large t (when the accuracy of our approximation is better), the ratio becomes even larger.

Figures 1,2 summarize our tables for respectively s=2, 7. The simulated straightforward N^{LP} values are very large and are beyond the frames of our figures. The lower *straight line* corresponds to BF analysis, the upper one – to STI.

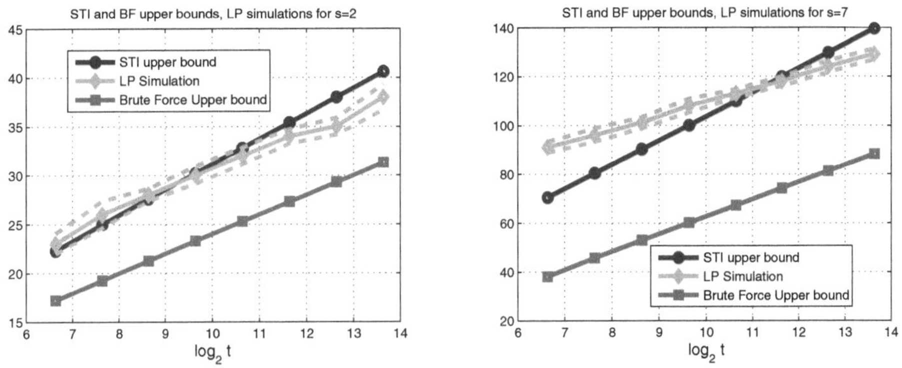

Fig. 1. Simulation for the ∪ model

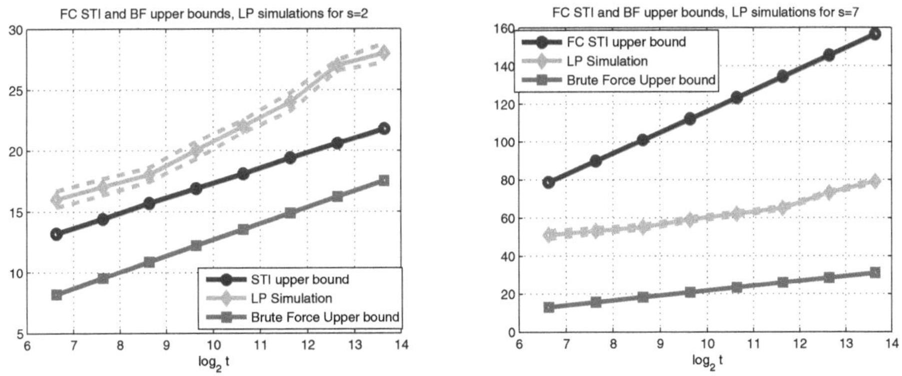

Fig. 2. Simulation for the FC model

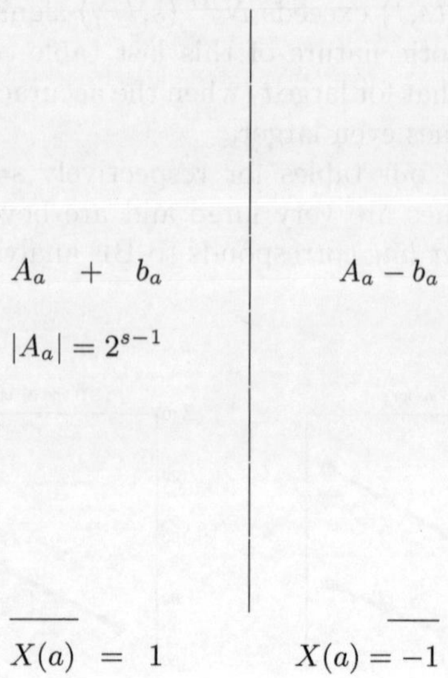

Fig. 3. *Outline of limiting scatter diagrams: a significant variable $X(a)$, no noise*

4 STI in Noiseless Linear Model under Condition U

Due to the presence of unknown active coefficients, we do not know the noise distribution and **must apply the universal decision $\mathcal{T}$.** We first consider the STI-detection of AIs under condition U on significant coefficients. We examine all pairs $(x^N(a), z^N)$ where $x^N(a)$ is the binary input column and z^N is the output column with components taking 2^s values.

Let us assume for definiteness WLOG that variables $x(1), \ldots, x(s)$ are AI. Fixing the value of one of them (say, the first) we still have 2^{s-1} equally likely combinations of the rest with coefficients ± 1.

In the left - (respectively right - hand) side of the scatter diagrams corresponding to $x(a) = \pm 1$, we have non-overlapping sets of outputs $y - b_a$ $(y + b_a)$, with coefficients b_a, $a = 1, \ldots, s$, y, where $y \in A_a$ and A_a is the set of linear combinations of significant variables different from $X(a)$. The cardinality of $|A_a|$ is 2^{s-1}. Hence, for each significant variable $X(a)$, we have a separate partition of the outputs $\mathcal{Z}$ into two subsets $\{\pm b_a + A_a\}$ displayed on the scatter diagrams.

It is clear that $\mathbf{I}(x_1 \wedge Y) = \mathbf{H}(y) - \mathbf{H}(Y|x_1) = 1$. Thus $C^{STI} = C^{BF} > C^{LP}$. The last inequality follows from the results of the simulation in Fig 3.

The first equality follows also from an elementary combinatorial argument: a particular inactive input can be regarded as active by STI if 2^s output values of its **both** partial scatterplots shrink to 2^{s-1} for random design with $N > (1+\epsilon)\log t$ rows, $\epsilon > 0$. These events are independent and have probability 1/2 for every row. Thus, it happens for at least one of the inactive inputs with probability $\leq (t-s)2^{-N} \leq t^{-\epsilon}$. Since $\epsilon > 0$ is arbitrary, the STI capacity for the random design ≥ 1.

Falsity of s/t =const as $t \to \infty$ asymptotics for Linear model and **binary design**. Meshalkin's upper bound shows that under U, any fraction $ct, c < 1$, can be restored with $N(s,t,\gamma)$ less than t for sufficiently large t. However, if U is not valid, then the information per measurement I is strictly less than s implying asymptotically that $N^{BF}(s,t,\gamma) \geq s\log t/I = (s/I)\log t$. Putting $s = kt$ for whatever $k < 1$ implies $N^{BF}(s,t,\gamma) \geq (kt/I)\log t > t$ for sufficiently large t, thereby showing that the ct-sparse designs and BF analysis are more economical, than the trivial ones **only if** U **holds**, which can only doubtfully be taken for granted in practice (although valid with Probability 1 under non-degenerate prior distribution of significant coefficients).

The left - hand part of figure 4 shows simulated performance for the linear model with incommensurable significant coefficients $1, e, \sqrt{2}, \pi$. In the right - hand side of figure 3, Meshalkins upper bound $\bar{N}^{BF}(s,t,0.05)$ (6) is also plotted.

4.1 Greedy STI, Its Simulation for Linear Model

STI involves $O(t\log t)$ operations as opposed to $O(t^s \log t)$ for BF. Models in [4] were usually multivariate polynomials of the second order involving a lot of parameters to model industrial production in sufficient detail. Certain steps can be taken after STI (greedy STI) to further improve its performance without significantly increasing the computational burden (see, e.g. [24,18]). The greedy STI consists of testing for significance the inputs which did not show activeness in the original STI one by one when applied to the revised data with the effects of the inputs *shown to be active by STI before* removed from the outputs. The idea of greedy STI is that the noise created by randomly varying other AIs may prevent the finding of less active inputs. After removal of this noise, AI detection becomes feasible. Especially fruitful is this update for incommensurable active parameters of the linear model, as shown in our Figure 4. LP N-values shown in yellow are significantly greater.

4.2 Simulation Design for Greedy STI

1. Create a random $(N \times\ t)$-matrix of $1's$ and $-1's$.
2. Set $s = 4,\ b_s = [1\ e\ \pi\ \sqrt{2}]$
3. Compute $y_i = \sum b_s x_{s_i},\ i = 1, \ldots, N$.
4. Count the number of $1's$ and $-1's$ for each $X^N(k),\ k = 1, \ldots, t$, separately, and compute $\tau(X^N(k))$,
   Count the number of various pairs $(X^N(k), Y^N)$ and compute $\tau(X^N(k), Y^N)$, $i = 1, \ldots, N$.
   Count the number of similar outputs and compute their empirical distribution.
5. Compute the ESI, $\mathcal{I}(X^N(k), Y^N),\ k = 1, \ldots, t$. Then,
   sort $\mathcal{I}(X^N(k), Y^N),\ k \in [s]$ in ascending order and sort $\mathcal{I}(X^N(k), Y^N),\ j > s$, in descending order.

If $\min_{k\in[s]} \mathcal{I}(X^N(k), Y^N) > \max_{j>s} \mathcal{I}(X^N(k), Y^N)$, there is no error, otherwise delete the $\min_{k\in[s]} \mathcal{I}(X^N(k), Y^N)$, and its corresponding coefficient b_s.

6. Set $z_i = y_i - \sum b_s x_{s_i}$, $i = 1, \ldots, N-1$, with the new b_s. Then go to step 4. If there are errors, calculate the error frequency for 100 runs,
7. If $P(\text{error}) \geq 5$ percent, increase N and start all over, for otherwise we get N^*.

4.3 N^* Simulation for STI in Linear Model

The left plot shows that LP curve is higher than STI under U.

N^* **table: Upgraded (greedy) STI**, incommensurable parameters

s/t	100	200	400	800	1600	3200	6400	12800
4	28	30	32	33	34	38	39	41

A similar N^* simulation for the STI analysis in the noiseless linear model with **commensurable** active coefficients 1, 2, 3 is described in [33].

In particular, $N^* = 18$ for t=100 and $N^* = 35$ for t=1600. Thus the suitable relation for these two (N, t)- points is approximately $\log t/N = 0.4$. The slope is roughly the Shannon information between input and output for this set of active coefficients and random equally likely binary design.

5 Noisy BF- and STI Recovery Capacity, Unknown MAC

Notation for the Main Results of This Section

We usually reserve Greek letters for random variables, corresponding Latin letters for their sample values, with bold capitals reserved for matrices and bold lower case (non-bold) letters for columns or rows. $P^{\mathbf{m}}_\beta$ is the joint distribution of Ξ, ζ under the random design with $P_\beta(x_i(\alpha) = 1) = \beta$ for all i, α. Lemmas are enumerated anew in each section.

$$I^{\mathbf{m}}_\beta(\xi(A) \wedge \zeta) = \mathbf{E}_{P_\beta} \log \frac{P^{\mathbf{m}}_\beta(\zeta|\xi([s]))}{P_\beta(\zeta)}, \tag{19}$$

where $P_\beta(\cdot)$ is the marginal distribution of ζ, expected value is over $P^{\mathbf{m}}_\beta$, and

$$\mathcal{C}(s) = \max_{\beta\in B} \inf_{\mathbf{m}\in\mathcal{M}} I^{\mathbf{m}}_\beta(X(a) \wedge \zeta). \tag{20}$$

Define $R(\mathbf{X}_t) = \log t/N$.

Theorem 5.0.1. *For a symmetric MAC and $\varepsilon > 0$, if $R(\mathbf{X}_t) < \mathcal{C}(s) - \varepsilon$ then MEP($\mathcal{T}_s$) decreases exponentially in N as $t \to \infty$. If $R(\mathbf{X}_t) > \mathcal{C}_s(s) + \varepsilon$, then MEP($\mathcal{T}$) $\geq p > 0$.*

I. Csiszár (in 1978) pointed to the close relation between our set up and that of MAC capacity region construction in [1]. One of the subtle *differences* is in

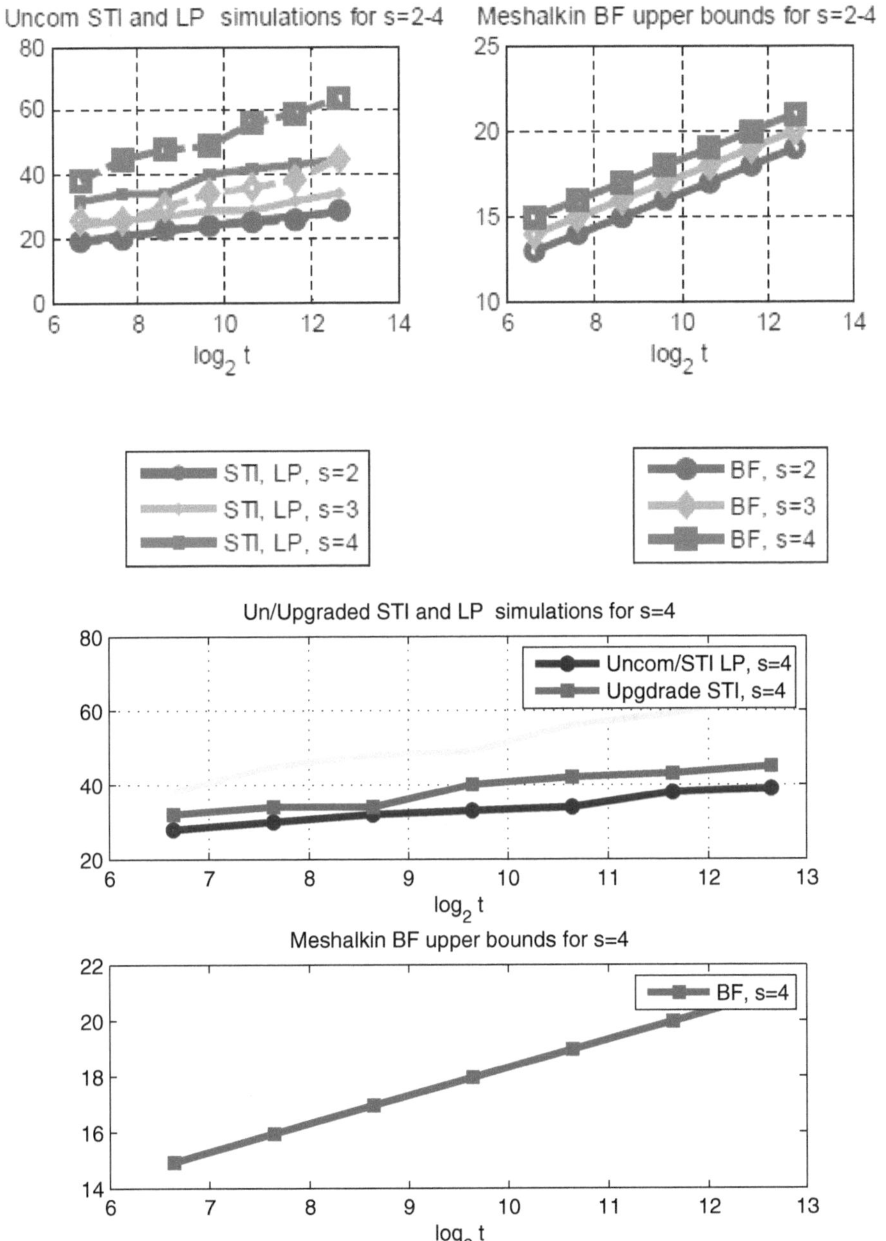

Fig. 4. Greedy STI vs. STI for Linear model under $\mathcal{U}$

the condition of *identity of rates and code distributions* used by all senders in our set up. The capacity found in theorem 1 *is the intersection of the corresponding MAC capacity region with the main diagonal*, but establishing this involves the nontrivial proof of the so-called **ordinarity** of symmetric MAC.

5.1 Ordinary Models

A model is called ordinary if the max in the game value is attained on the empty set, implying that the capacity is $\mathbf{I}(Z \wedge \mathbf{x}[s])$.

[31] starts with a consequence of the Chain Rule.

Lemma 5.2.1. $\mathbf{I}(\zeta \wedge \xi([s])|\xi([v])) = \sum_{u=v+1}^{s} \mathbf{I}(\zeta \wedge \xi(u))|\xi([u-1]))$.
Three equivalent necessary and sufficient conditions of a model being ordinary are then formulated, of which we show only one:

Theorem 5.1.2. A MAC is ordinary, iff for every randomization parameter p and every permutation $\sigma = (i_1, \ldots, i_s) \in \Lambda(s,s)$ of $[s]$ the following holds:

$$\mathbf{H}(\xi(\sigma(u+1))|z, x(\sigma[u]) \leq \mathbf{H}(\xi(\sigma u))|z, \mathbf{x}(\sigma[u-1])), u = 1, \ldots, s-1, \quad (21)$$

where $\sigma u = i_u$.

Corollary. All symmetric models are ordinary.

Proof. A well-known inequality

$$\mathbf{I}(\xi(u+1) \wedge \xi(u)|z, x([u-1])) \geq 0$$

and equality

$$\mathbf{H}(x(u)|z, x([u-1]) = \mathbf{H}(x(u+1)|z, x([u-1])),$$

are valid because of symmetry, and imply (21).

Elementary examples of non-ordinary MAC are given in [31].

5.2 Capacity in Possibly Asymmetric Models

Theorem 5.1 below is a generalization of theorem 3.1.1 to possibly asymmetric models using decision $\mathcal{T}_s$. Its proof is in section 9.2. As a result of Lemma 2.1.1, we can assume WLOG that $A_s = [s]$.

Introduce $L := \max_{0<\beta<1} \min_{1\leq k \leq s} E_\beta L_k$.

Theorem 5.1. *For any* $\varepsilon > 0$*, if* $R(\mathbf{X}_t) < L - \varepsilon$*, then MEP(*$\mathcal{T}$*) decreases exponentially in* N *as* $t \to \infty$*.*

For generally *asymmetric MACs*, identical *AR* of $\mathcal{T}_s$ and ML, (if applicable), are described in terms of conditional Shannon information (CSI) as $t \to \infty$. Under the fixed set $[s] = \{1, \ldots, s\}$ of AIs, let $\mathcal{V}(v,s)$ denote the set of ordered subsets $V \subset [s]$ of the cardinality $|V| = v$, $\mathcal{V}(s) = \cup_{r=0}^{s-1} \mathcal{V}(r,s), andletx(V)$ be a function $x(i), i \in V$.

$$I_{\mu}^{\mathbf{m}}(V) = \mathbf{E}_{\mu} I_{P_{\beta}}^{\mathbf{m}}(Z \wedge X(V)|X(V^c)),$$

where $V^c = [s] \setminus V, \mu$ is the class B^* of probability distribution μ on $B = \{0,1\}$. CSI is

$$I_{\beta}^{\mathbf{m}}(V) = I_{P_{\beta}}^{\mathbf{m}}(Z \wedge X(V)|X(V^c)),$$

where

$$I_P^{\mathbf{m}}(X \wedge Y|Z) = \mathbf{E}_P^{\mathbf{m}} \log \frac{P(X|YZ)}{P(X|Z)},$$

where $P(\cdot)$ is the joint density of X, Y, Z. Let $\mathcal{C}_1(s)$ be the value of the game in which the first player chooses $\beta \in B$ and the second chooses $V \in \mathcal{V}(s)$ with pay-off function

$$J_{\beta}^{\mathbf{m}}(V) = \frac{I_{\beta}^{\mathbf{m}}(V)}{|V|},$$

i.e.,

$$\mathcal{C}_s(s) = \sup_{\mu \in B^*} \inf_{\mathbf{m} \in \mathcal{M}} \min J_{\mu}^{\mathbf{m}}(V),$$

where min is taken over $V \subset \mathcal{V}(s)$.

Let $\mu^*(s)$ be the maximin measure on B in the preceding formula.

The following result generalizes Theorem 5.0.1 to not - necessarily symmetric MACs.

Theorem 5.2. *For any $\varepsilon > 0$, if $R(\mathbf{X_t}) < \mathcal{C}_s(s) - \varepsilon$, then MEP($\mathcal{T}_s$) decreases exponentially in N as $t \to \infty$. If $R(\mathbf{X_t}) > \mathcal{C}_1(s) + \varepsilon$, then MEP($\mathcal{T}_s$) $\geq p > 0$.*

For symmetric (and more generally ordinary) models, $\mathcal{C}(s) = \mathcal{C}_1(s)$ and the capacity is attained by the simpler ensemble P_{β} of random designs. The non-parametric ESI-decision was proved to attain the capacity in [33] for discrete output distributions, and was generalized to continuous distributions in [36].

5.3 Lower Bounds for $N^{BF}(s, t.\gamma)$

To make the idea in a general case more transparent, let us start with the lower bound for the $\cup$- model. Any γ-separating $N \times t$ design and uniform distribution on the set of unordered s-tuples $\Lambda_0(s,t)$ induce measures $\pi(\cdot)$ on the output columns η. The entropy satisfies

$$\mathbf{H}(\eta) \geq -(1-\gamma)\log\gamma + \min\{-\gamma\log\gamma, 1\}. \tag{22}$$

Proof. γ-separability implies that at least $(1-\gamma) \times |\Lambda_0|$ columns $\eta(\lambda)$ (corresponding to separated s-tuples) are distinct and equally likely. Their entropy is not less than the first summand. Entropy is minimal, if all other columns coincide which proves the bound.

The entropy of any **binary** component of the output columns does not exceed 1. The subadditivity of entropy yields the lower bound $N \geq \frac{\mathbf{H}(\eta)}{\max_{1\leq i\leq N}\mathbf{H}(\eta_i)} \sim \log\binom{t}{s} \sim s\log t$.

The only change in the scheme given above for the FC-model is in finding $\max \mathbf{H}(\eta_i)$, which is shown to be $\mathbf{H}(B_s(1/2))(1+o(1))$ by Mateev.

Given a row of the $(N \times t)$ - design matrix X for a general symmetric model, the uniform distribution of A implies the distribution $\pi_{\mathbf{x}}(\cdot)$ on $Y = \{y(\mathbf{x}(A))\}$ which is defined by

$$\pi_{\mathbf{x}}(c) = b^{-1} \sum_{A \in \Lambda(s,t)} \mathbf{1}_A(c)$$

where $b = \|Y\|$, $\mathbf{1}_A(c)$ is the indicator of the event $\{A : y(\mathbf{x}(A)) = c$.

Denoting $p = \sum_{j=1}^{t} |x|_j/t$ and $(B)_m = \Pi_{i=1}^{m}(B-i+1)$ for natural B, we get

$$\pi_{\mathbf{x}}(c) = \sum (|\mathbf{x}|)_{|A|}(t-|\mathbf{x}|_{s-|A|})/(t)_s := \pi_p(c) \tag{23}$$

where the sum is extended over $A : y(\mathbf{x}(A)) = c$

This Hypergeometric-like distribution $\pi_p(c)$ for sampling without replacement converges to that of sampling with replacement

$$\tilde{\pi}_p(c) := \sum p^{|A|}(1-p)^{s-|A|}. \tag{24}$$

As $t \to \infty, s =$ const with rate $O(1/t)$ of convergence. The sum extends over $A : y(\mathbf{x}(A)) = c$.

If X is γ - separating, then taking $\mathbf{x} = \mathbf{x}_i, i = 1, \ldots, N$ and applying well-known rules of elementary information theory [16], (theorems 4.3.1, 4.3.3 and 4.2.1), we get the following chain of inequalities [25] (called the 'Folk theorem' in [3]):

$$(1-\gamma)\log N - h(\gamma) \leq \mathbf{H}(\hat{A} \wedge \delta(X, \mathbf{z})) \leq (\mathbf{y}(X(A)) \wedge \mathbf{Z}) \leq \sum \mathbf{I}(y(\mathbf{y}_i(A)) \wedge Z_i) \leq \tag{25}$$

$N \max_{0<p<1} I(\pi_p(\cdot)) \sim N \max_{0<p<1} I(\tilde{\pi}_p(\cdot))$,
where $h(p) := -p\log p - (1-p)\log(1-p)$ and $\max I(\tilde{\pi}_p())$ can be easily evaluated. The first (Fano) inequality bounds from below the MEP given a complexity of a decision rule. The second 'Markov Chain' inequality takes into account that our decision is based solely on the design matrix and outputs of the model. The third step in the chain is due to the subadditivity of the Shannon Information for independent trials. Hence, we proved:

Theorem 5.4.1. $\lim N^{BF}(s,t,\gamma)/s\log t \geq 1/\max I(\tilde{\pi}_p(\cdot))$ as $t \to \infty, s =$ const.

For asymmetric models, this lower bound is generally not tight.

To strengthen it, we complement our previous chain of inequalities by a set of new ones for conditional decisions, which we only sketch. Two detailed proofs are in [30], pp. 93–97, and the third one is in [45], pp. 80-87.

Consider the random s-tuple $w, w \cap [s] = v$ and the random ensemble P_μ of design matrices. First, we prove that the ordered subset v and its complement v^c in w are asymptotically independent. Let $X(v)$ be a function $X(i), i \in v$.

We have a Markov sequence for an arbitrary ordered s-tuple $\mathbf{w}$ and its subset $\mathbf{v}$:

$$\mathbf{v}^c \to X(\mathbf{v}^c) \to (z, X\{\mathbf{v}\}) \to \delta(z, X(v) := \arg \max_{w \cap [s] = v} P(z|X(w)).$$

MEP $\leq \gamma$ for the latter conditional decision, if X is γ - separating. Repeating the previous arguments based on [16], theorems 4.3.1, 4.3.3 and 4.2.1, we prove that the *set of asymptotic inequalities* $N \geq \log t / J_\mu(v)(1 + o(1)), v \in [s]$, must **hold simultaneously** for the number N of rows in X.

Thus, $N \geq \log t / (sup_\mu \inf_v J_\mu(v)(1 + o(1))$ showing the tightness of our upper bounds proved in section 9.4.

6 The STI Capacity for Colored Noise

Given an arbitrary intermediate vector–output $\mathbf{y}$ (see section 3), the sequence of $\mathbf{z}$'s of the conditional distribution $\mathcal{P}_{\zeta - \mathbf{y}}$ is that of a stationary ergodic random string (SES) taking values from a finite alphabet $\mathcal{Z}$.

Our lower bounds hold with the 'entropy rate' $\lim_{N \to \infty}(I^{\mathbf{m}}_\beta(X_1^N(\lambda) \wedge \zeta_1^N)/N)$ instead of constant $I^{\mathbf{m}}_\beta(X(\lambda) \wedge \zeta)$. The limit exists as a result of the stationarity of the pair (IID X_i, ζ_i).

The asymptotically optimal STI- tests are as follows:

We choose a weakly universal compressor (UC) [47], denote

$$\mathcal{U} = \mathcal{B} \times \mathcal{Z},$$

and consider for a given $j = 1, \ldots, N$ two N-sequences with letters from $\mathcal{U}$:

$$u_j{}^N := (x_j(i), z(i)), i = 1 \ldots, N;$$

and

$$v_j^N := (x_j(i)(\times) z(i)), i = 1 \ldots, N,$$

taken from the original joint distribution and the generated product-distribution. We digitize them into binary sequences $\mathbf{U}_j^M, \mathbf{V}_j^M$ of appropriate length and evaluate the CCC homogeneity statistic (see further) of the product P_0 and original distributions $P_1 = P_1^j$.

Arbitrary UC maps source SES-strings $\mathbf{V}^{\mathbf{M}}$ into compressed binary strings $\mathbf{V}_{\mathbf{c}}^{\mathbf{M}}$ of approximate length $|\mathbf{V}_c^M| = -\log P(\mathbf{V}^M)) = L^M$ thereby *generating the approximate Log likelihood of source* $\mathbf{V}^M$ – the main inference tool about P_0.

Consider a **query** binary SES $\mathbf{U}^M$ distributed as P_1 and test whether the homogeneity hypothesis $P_0 = P_1$ contradicts the data or not. Let us partition $\mathbf{y}^M$ into several **slices** $\mathbf{U}_i, i = 1, \ldots, S$, of identical length n divided by 'brakes' - strings of relatively small-length δ to provide approximate independence of slices (brakes of length 2k are sufficient for k-MC). Introduce concatenated strings $\mathbf{C}_i = (\mathbf{V}^M, \mathbf{U}_i)$. Define $CCC_i = |C_i| - |\mathbf{V}^M|$ CCC-statistic and $\overline{CCC}$ = average of all CCC_i. Similarly, $\overline{CCC}^0$ = average of all CCC_i^0 with $\mathbf{U}_i$ replaced with independent P_0- distributed slices of the same length. Finally, homogeneity $\bar{R} = \overline{CCC} - \overline{CCC}^0$. The $\bar{R}$ test is shown in [29] and in section 9.5 to have the same exponential tail under P_0 as the Likelihood Ratio test, provided the error probability under alternative is positive and arbitrarily small. Thus, the $\bar{R}$ test is asymptotically efficient in terms of capacity.

6.1 Applications

We propose applying the methods analyzed in the last section to several problems of strong practical relevance for many industrial applications:

i. 'Tagging' the change-point in users' profiles in a large computer network possibly caused by unauthorized intrusion into the system for a more detailed follow up study.

ii. Monitoring large corpora of texts, e.g. on-line forums or the phone call traffic in some areas for 'tagging' matches to specific profiles of interest for a more detailed follow up study.

7 Static Search for $N \to \infty$ and Finite t

If $t = const$, the parametric set (including the set of AIs, the set of active parameters $b_a, a \in A$, and the unknown error distribution) are fixed as $N \to \infty$. The MEP here involves averaging the conditional error probabilities under the *fixed set* of different parameters (s-tuples of variables). Its Bayesian nature remains essential as $N \to \infty$. In particular, for studying the MEP of test $\mathcal{T}_s$, we need to average the conditional error probabilities under various AIs with the ones under inactive variables that have certain non-degenerate weights.

The asymptotic upper and lower bounds for the error *exponent* of the Bayes test with arbitrary strictly positive priors coincide and are expressed in terms of the Chernoff Information, see the proof in [6], pp. 308-309. The upper bound is based on Sanov's theorem ([6], pp. 292-294). The lower bound follows from finding the worst empirical distribution in Sanov's theorem.

Let us formulate this general result. Consider two distributions $P_i, i = 1, 2,$ on the finite sample space S and a sample x^N of N IID RV's obeying one of the above laws. Let $d(x^N)$ be an arbitrary decision on which of the distributions is correct. The Bayes Error Probability $aP_1(d = P_2) + (1-a)P_2(d = P_1)$ for any fixed prior weights $a, 1-a, a > 0$, is denoted by Q_N. It holds for the optimal decision:

$$\lim_{N\to\infty} \log Q_N/N = -\mathcal{C}(P_1, P_2) \equiv -\mathcal{C}(P_2, P_1),$$

where

$$\mathcal{C}(\cdot) = \min_{\rho\in\mathcal{E}} D(\rho||P_1), \tag{26}$$

and $\mathcal{E}$ is the set of distributions on S

$$\mathcal{E} = \{\rho : D(\rho||P_1) \geq D(\rho||P_2)\}. \tag{27}$$

$D(\rho||P) = E_\rho \log(\rho/P)$ is Divergence (cross-entropy) between ρ and P.

Now we formulate the main results of this section.

Let us make a specific choice concerning the distribution of $X^N(\lambda)$ in $\{-1,1\}^N$; namely, we assume that

(a) the vectors $\{X^N(\lambda)\}$ are i.i.d. for $\lambda = 1, \ldots, t$;

(b) $\sum_{i=1}^N X_i(\lambda) = (2\beta - 1)N$ for some fixed $\beta \in (0,1)$; and

(c) $X^N(\lambda)$ is chosen uniformly from all vectors satisfying (b).

Theorem 7.1. [28] *The ML-test based on s maximal values of the likelihood of* $(X(\lambda), Z)$ *and test* $\mathcal{T}$ *introduced in section 3 have the same error exponent*

$$EEP = \min_{a \in A} \mathcal{C}^\beta{}_a,$$

where

$$\mathcal{C}_a^\beta = \mathcal{C}(P(x(a), z); P(x(a)\bar{P}(z)), a \in A, a \in A.$$

Here $\mathcal{C}^\beta$ *denotes the* β*-weighted Chernoff Information of the conditional joint and product-input-output distributions given the input.*

The elementary case $t = 2, s = 1$ with no noise shows that the random designs are no longer optimal. Hence the problem of finding the best exponent of the search error probability remains a challenge.

8 Adaptive Search for AIs

The adaptive counterpart of the theory outlined earlier was developed in a sequence of our papers as an extension of elementary search strategy for AIs in an **unknown binary output noiseless model** suggested by L. Bassalygo and described in [20] . Its publication was delayed for several years until the former Soviet KGB gave permission (apparently due to KGB applications unknown to me). Suppose that the function $g(\cdot) : B^t \to B$ is binary and unknown. (This corresponds to a 'compound channel' in Information Theory). The asymptotically optimal adaptive restoration algorithm of all active inputs of $g(\cdot)$ and of the function $g(\cdot)$ itself with the Mean Length $s \log_2 t(1 + o(1))$ is as follows. A random search finds a t-tuple $\mathbf{x} \in B^t$ such that $g(x)$ differs from $g(0, \ldots, 0)$. Then, each non-zero entry of x is tested for activeness in the obvious way to yield at least one AI. This procedure is repeated s times until all AIs are found.

The noisy generalization of the problem is straightforward provided we admit $MEP < \gamma$ and $\gamma > 0$. The generalizations to a multi-valued noisy linear functions ([37] and [27]), extended to arbitrary unknown functions in [35] use the two-stage-loop generally suboptimal sequential algorithm which we describe further.

The first random search phase separates B^t into s subsets containing one AI each. Then, the elementary nonrandom algorithm finds each AI. If the error occurs, or the length of the algorithm is excessive, everything is repeated anew. We describe these algorithms more accurately below.

Suppose that fixing an input sequence

$$\mathbf{x}^N = (\mathbf{x}(1), \dots, \mathbf{x}(t))$$

we observe the output sequence of random variables

$$Z^N = (Z_1, \dots, Z_N)$$

from a measurable space $\mathcal{Z}$ distributed according to the transition probability density

$$\mathbf{F}^N(z^N|\mathbf{x}^N) = \prod_{i=1}^{N} \mathbf{F}(z_i|f_S(\mathbf{x}_i)),$$

where $\mathbf{F}(\cdot|\cdot)$ is a transition density function with respect to a σ–finite measure μ on $\mathcal{Z}$, satisfying the conditions formulated later. We use an adaptive design of experiments π consisting of a Markov stopping time N with respect to the filtration $\mathcal{F}_n$ generated by the measurements and randomizations of the design points until the time n, and $\mathcal{F}_{i-1}$- measurable randomized rules of choice of $\mathbf{x}_i, i = 1, \dots, N$.

A general suboptimal *adaptive* search strategy is outlined below. Its advantage over static ones is that codings of different AIs need not be the same. It follows from [27] (adapting the MAC strategy [15] for the FC model) that the capacity of adaptive search generally exceeds the capacity of the corresponding static one.

The best cooperation between codings of different variables maximizing capacity has been found neither in MAC theory nor in our parallel settings, even for s=2, see references in [44].

The algorithm consists of two phases. The aim of Phase 1 is to partition $[t] = \{1, ..., t\}$ into s disjoint sets T_α, $\alpha = 1, \dots, s$, containing *exactly one index* of an AI each. Such a partition is called *correct*; all others are called *false*.

We bound the MEP from above. The mean length (MEAL) of our procedure is $O((\log t)^{-1})$ as $t \to \infty$, while the MEAL of Phase 1 is $O(\log \log t)$. We do not optimize the performance in the Phase 1 because its MEAL has a smaller rate than that of the whole procedure as $t \to \infty, s = \text{const}$.

Phase 1 is composed of a random number of similar loops, each of them consisting of two stages. The first stage of a loop is a partition of $[t]$ into s subsets while stage 2 is testing for the presence of an AI in each subset. If the test rejects the presence of an AI in *at least one subset* or the test involves too many experiments, the next loop begins anew. Note that the presence of an AI in each subset implies that each subset contains a single AI because the number s is regarded as known.

We present our strategy for AIs separation in a general nonparametric *linear* model outlined in [37] and [27]. For the generalization to nearly maximally general model see [35,36]. Consider real variables $x(j), x(j) \in X = \{x : |x| \leq 1\}$, $j = 1, \dots, t$. Choosing a sequence $\mathbf{x}_i \in X^t$, we perform measurements of a nonparameric additive function

$$g(\mathbf{x}_i) = \sum_{\alpha \in A} g_\alpha(x_i(\alpha)), |A| = s,$$

corrupted by additive noise, i.e. we observe $z_i = g(\mathbf{x}_i) + \varepsilon_i; \varepsilon_i, i = 1, \ldots, N$, are independent random variables.

If $\alpha \in A$, then we call $x(\alpha)$ an *active variable* and $g_\alpha(\cdot)$ an *active function.* Assume that all the active functions on X belong to the class $G(\Delta, \mu)$, described by the following conditions

G1. $g(\cdot)$ is continuous on X and its modulus of continuity does not exceed $\mu(\cdot)$.
G2. $r(g(\cdot), X) = \max_x g(x) - \min_x g(x) \geq \Delta > 0$.

The algorithm consists of two phases. The aim of phase 1 is now to separate indices of AIs and also find two points $\overline{x}_\alpha$ and $\underline{x}_\alpha$ in each subset T_α of variables such that

$$g_\alpha(\overline{x}_\alpha) - g_\alpha(\underline{x}_\alpha) \geq r(g_\alpha, X) - \delta,$$

where $\delta > 0$ is chosen to be sufficiently small depending on the parameters in the problem. In the second phase of AIs identification, all the variables in T_α are chosen randomly to be $\overline{x}_\alpha$ or $\underline{x}_\alpha$ independently for different variables or measurements with certain randomization probabilities p_α as a part of a general strategy of simultaneous detection of AIs in each T_α with MEAL=$O(\log t)$ and MEP=$(\log t)^{-1}$. This is a suboptimal strategy for the second phase. The optimal strategy has not yet been found.

Let us outline phase 1 of our strategy in the case of an unknown discrete system. Assuming for simplicity that $t = s\ell, \ell \in \mathbf{N}$, we use the random partition of the set of indices into s subsets T_α each containing ℓ indices. After making the partition, we repeat independently several series of measurements for each subset T_α with the following design. We choose independent random equiprobable allocations of levels 0, 1 for any variable in T_α^c. For each fixed given series allocation $\mathcal{A}$ in T_α^c, we repeat measurements $y_j(\mathcal{A})$ at a random independent sequence of levels of variables $x_i, i \in T_\alpha$ putting $\mathbf{P}(X_i = 0) = \beta$. The mixed choice of $\beta = \beta_0$ provides the maximal AR of the search.

We then test adaptively the hypothesis of existence of an AI in each T_α. If for at least one T_α this hypothesis is rejected, or the forced termination of experiments is used, we start the whole procedure over again.

Let Us Outline First the Idea Behind the Procedure. If the allocation $\mathcal{A}$ of variables' levels in a particular T_α^c is chosen properly (that is, in such a way that the underlying function $f(\cdot)$ depends essentially on the unique AI in T_α), then the distribution of independent outputs in a given series is a *mixture of those levels, corresponding to distinct values of the AI in* T_α *with probabilities* β.

On the other hand, if T_α is devoid of AIs, the distribution of outputs in the given series is the same for a fixed allocation of levels in T_α^c, that is attached to *the same sequence of levels of AIs and hence to a unique value* y *of the function* $f(\cdot)$. If the parameter $\beta = \beta_0$ is chosen in such a way that all of these distributions $P(\cdot|y)$ are distinct from the mixtures described above, we can discriminate reliably, e.g., by the maximum likelihood test (provided the noise distribution is known). Note that the hypotheses $H_{i\alpha}$, stating that $i = 0, 1$ AIs are present in T_α, are *generally composite.* The first of them consists of $n, n = |\mathcal{Y}_f|$ distinct distributions $P(\cdot|y)$, while the number of distributions contained in $H_{1\alpha}$ does not exceed $n(n-1)/2$ under a fixed parameter β.

We need not estimate the error probability of the test (for presence of AIs) under the actual number of AIs in T_α exceeding 1. Undesirable influence of this event on the order of magnitude of the MEAL of the procedure is eliminated by a suitable truncation.

Repetitions of random allocations in T_α^c are needed to avoid situations (with a large probability when M is large) , such as, for example, when $f(a_1, a_2) = a_1 \cap a_2$, where the value $a_2 = 0$ involves $f(a_1, a_2) = 0$, irrespective of the value of a_1.

For discrimination between the presence and absence of AIs in T_α, the vector statistic $\mathbf{S}$ is used with the following components:

$$S_n^{ijy} := \sum_{k=1}^{n} \log F_{ij}^{\beta_0}(Z_k)/F(Z_k|y); i, j, y \in \mathcal{Y}.$$

For a fixed $\gamma > 0$, define the minimal value τ_α of $n \in \mathbf{N}$ such that either

$$\max_{i \neq j \in \mathcal{Y}} \max_{y \in \mathcal{Y}} \min S_n^{ijy} \geq -\log \gamma, \tag{28}$$

or

$$-\max_{y \in \mathcal{Y}} \min_{i \neq j \in \mathcal{Y}} -S_n^{ijy} \geq -\log \gamma, \tag{29}$$

and its truncated version

$$N_\alpha := \min\{\tau_\alpha, M_1\}, M_1 = (1+\epsilon)K_1(\beta_0)|\log \lambda|, \epsilon > 0,$$

which we choose as the stopping time in exploring subset $T_\alpha, \alpha \in [s]$.

Our technical tools for exploring means of RVs τ_α (and similar RVs for the second phase) and the deviations from their means include elementary large deviation bounds for martingales and the following bound for the overshoot at the first passage time to the remote quadrant.

Let a finite family of random sequences

$$\mathbf{X}_i := (X_i(l), l = 1, \ldots, l), i = 0, 1, \ldots, X_0(l) := 0,$$

be the sum of a strictly increasing linear non-random vector process and a square-integrable martingale family $\{Y(l)\}$ under the filtration $\mathcal{F} = (\mathcal{F}_n, n \in \mathbf{N})$, i.e.

$$X_i(l) = i \cdot A_l + Y_i(l),$$

where $A_l > 0$, $l = 1, \ldots, l$, $\mathbf{E}(Y_i(l)|\mathcal{F}_{i-1}) = Y_{i-1}(l)$ a.e. satisfying for a fixed number D, all $i \in \mathbf{N}$ and $l = 1 \ldots, L$, the following inequality a.e.:

$$\mathbf{E}[(Y_{i+1}(l) - Y_i(l))^2|\mathcal{F}_i] \leq D^2.$$

We study here the mean of the first passage time $\tau = \tau_x$ over a high level x by the process

$$Z_i := \min_{l=1,\ldots,L} X_i(l).$$

Theorem 8.4

$$\mathbf{E}(\tau) \le x \cdot A^{-1} + K_1 D L x^{1/2} + B$$

for any positive x, where K_1 is a universal constant, and

$$A := \min_{l=1,\ldots,L} A_l, \qquad B := A^{-1} \max_{l=1,\ldots,L} A_l.$$

Acknowledgements. The author is greatly indebted to Prof. E. Gover for correcting my English and to Dr. D. Malioutov for his help in writing several MATLAB codes and making us familiar with the literature on L_1 analysis. Intensive statistical simulation work of Dr. Hanai Sadaka was crucial in preparing this paper.

References

1. Ahlswede, R.: Multi-way communication channels. In: Proceedings of 2nd International Symposium on Information Theory, Tsahkadzor, pp. 23–52. Akademiai Kiado, Budapest (1971, 1973)
2. Ahlswede, R., Wegener, I.: Suchprobleme. Teubner, Stuttgart (1979)
3. Atia, G., Saligrama, V.: Boolean compressed sensing and noisy group testing, arXiv:0907.1061v2 (2009)
4. Budne, T.A.: Application of random balance designs. Technometrics 1(2), 139–155 (1959)
5. Candes, E.J.: Compressive sampling. In: International Congress of Mathematicians, Madrid, Spain, August 22-30, vol. 3, pp. 1433–1452. Eur. Math. Soc., Zürich (2006)
6. Cover, T., Thomas, J.: Elements of Information Theory. Wiley, N.Y. (1991)
7. Csiszár, I., Körner, J.: Information Theory: Coding Theorems for Discrete Memoryless Systems. Academic Press and Akadémiai Kiadó, Budapest (1981)
8. Csiszár, I., Shields, P.: Information Theory and Statistics, a Tutorial. In: Foundations and Trend in Communications and Information Theory, vol. 1(4). Now Publishers Inc., Hanover (2004)
9. Chen, S.S., Donoho, D.L., Saunders, M.A.: Atomic de-composition by basis pursuit. SIAM J. Scientific Computing 20, 33–61 (1998)
10. Donoho, D.L., Elad, M.: Maximal sparsity Representation via L_1 Minimization. Proc. Nat. Acad. Sci. 100, 2197–2202 (2003)
11. Donoho, D.L., Tanner, J.: Sparse nonnegative solution of underdetermined linear equations by linear programming. Proc. Nat. Acad. Sci. 102(27) (2005)
12. Dorfman, R.: The detection of defective members of large populations. Ann. Math. Statist. 14(4), 436–440 (1943)
13. Erlich, Y., Gordon, A., Brand, M., Hannon, G.J., Mitra, P.P.: Compressed genotyping. IEEE Trans. on Inf. Theory 56(2), 706–723 (2010)
14. Erdős, P., Rényi, A.: On two problems of information theory. Publ. Math. Inst. of Hung. Acad. of Sci. 8, 229–243 (1963)
15. Gaarder, N., Wolf, J.: The capacity region of a multiple-access discrete memoryless channel can increase with feedback. IEEE Trans. on Inf. Theory 21(1), 100–102 (1975)

16. Gallager, R.G.: Information Theory and Reliable Communication. Wiley, New York (1968)
17. Kashin, B.S., Temlyakov, V.N.: A remark on compressed sensing. Mat. Zametki 82(6), 829–937 (2007)
18. Kerkyacharian, G., Mougeot, M., Picard, D., Tribouley, K.: Learning out of leaders. In: Multiscale, Nonlinear and Adaptive Approximation. Springer (2010)
19. Kolmogorov, A.N.: Three approaches to the quantitative definition of information. Probl. of Inf. Transmiss. 1, 3–11 (1965)
20. Leontovich, A.M., Victorova, I.A.: On the number of tests to detect significant variables of a boolean function. Probl. of Inform. Transmiss. 14, 85–97 (1978)
21. Malioutov, D.M., Cetin, M., Willsky, A.S.: Optimal sparse representations in general overcomplete bases. In: IEEE International Conference on Acoustics, Speech and Signal Processing, Montreal, Canada, vol. 2, pp. 793–796 (May 2004)
22. Malioutov, D.M., Malyutov, M.B.: Boolean compressed sensing: LP relaxation for group testing. In: IEEE International Conference on Acoustics, Speech and Signal Processing, Kyoto, Japan (March 2012)
23. Malyutov, M.B.: On planning of screening experiments. In: Proceedings of 1975 IEEE-USSR Workshop on Inf. Theory, pp. 144–147. IEEE Inc., N.Y. (1976)
24. Malyutov, M.B.: Mathematical models and results in theory of screening experiments. In: Malyutov, M.B. (ed.) Theoretical Problems of Experimental Design, Soviet Radio, Moscow, pp. 5–69 (1977)
25. Malyutov, M.B.: Separating property of random Matrices. Mat. Zametki 23, 155–167 (1978)
26. Malyutov, M.B.: On the maximal rate of screening designs. Theory Probab. and Appl. 14, 655–657 (1979)
27. Malyutov, M.B.: On sequential search for significant variables of the additive function. In: Probability Theory and Mathematical Statistics, pp. 298–316. World Scientific, Singapore (1996)
28. Malyutov, M.B.: Log-efficient search for active variables of linear model. J. of Statist. Plan. and Infer. 100(2), 269–280 (2002)
29. Malyutov, M.B.: Compression based homogeneity testing. Doklady of RAS 443(4), 1–4 (2012)
30. Malyutov, M.B., D'yachkov, A.G.: On weakly separating designs. In: Methods of Transmission and Processing Information, Nauka, Moscow, pp. 87–104 (1980) (in Russian)
31. Malyutov, M.B., Mateev, P.S.: Screening designs for non-symmetric response function. Mat. Zametki 27, 109–127 (1980)
32. Malyutov, M.B., Pinsker, M.S.: Note on the simplest model of the random balance method. In: Probabilistic Methods of Research. Moscow University Press (1972) (in Russian)
33. Malyutov, M.B., Sadaka, H.: Jaynes principle in testing active variables of linear model. Random Operators and Stochastic Equations 6, 311–330 (1998)
34. Malyutov, M.B., Sadaka, H.: Capacity of screening under linear programming analysis. In: Proceedings of the 6th Simulation International Workshop on Simulation, St. Petersburg, Russia, pp. 1041–1045 (2009)
35. Malyutov, M.B., Tsitovich, I.I.: Sequential search for significant variables of unknown function. Problems of Information Transmission 36(3), 88–107 (1997)
36. Malyutov, M.B., Tsitovich, I.I.: Non-parametric search for active inputs of unknown system. In: Multiconference on Systemics, Cybernetics and Informatics, Orlando, FL, July 23-26, vol. 11, pp. 75–83 (2000)

37. Malyutov, M.B., Wynn, H.P.: Adaptive search for active variables of additive model. In: Markov Processes and Related Fields, pp. 253–265. Birkhauser, Boston (1994)
38. Meshalkin, L.D.: Justification of the random balance method, pp. 316–318. Industrial Lab. (1970)
39. Nalimov, V.V., Chernova, N.A.: Statistical Methods of Planning Extremal Experiments, M., Nauka (1965) (in Russian)
40. Rissanen, J.: Universal coding, information, prediction and estimation. IEEE Trans. Inf. Theory 30(4), 629–636 (1984)
41. Ryabko, B., Astola, J., Malyutov, M.B.: Compression-Based Methods of Prediction and Statistical Analysis of Time Series: Theory and Applications, Tampere, TICSP series, No. 56 (2010)
42. Srivastava, J.N.: Search for non-negligible variables. Survey of Statistical Design and Linear Models, 507–519 (1975)
43. Tibshirani, R.: Regression shrinkage and selection via the LASSO. J. Royal. Statist. Soc. B 58, 267–288 (1996)
44. Venkataramanan, R., Pradhan, S.S.: A new achievable rate region for the multiple-access channel with noiseless feedback. IEEE Trans. on Inf. Theory 57(12), 8038–8054 (2011)
45. Viemeister, J.: Diplomarbeit, Fakulatät für Mathematik, Universität Bielefeld (1982)
46. Wainwright, M.: Information-theoretic bounds on sparsity recovery in the high-dimensional and noisy setting. In: Proc. 2007 IEEE International Symposium on Information Theory, Piscataway, pp. 961–965. IEEE Press, NJ (2007)
47. Ziv, J.: On classification and universal data compression. IEEE Trans. on Inf. Theory 34(2), 278–286 (1988)
48. Zubashich, V.F., Lysyansky, A.V., Malyutov, M.B.: Block–randomized distributed trouble–shooting construction in large circuits with redundancy. Izvestia of the USSR Acad. of Sci., Technical Cybernetics (6) (1976)

9 Appendices

9.1 Large Deviations via Types

In proving the theorem in section 9.2, we use half of the celebrated Sanov's theorem in the theory of large deviations, which (for a finite output alphabet) is easier to introduce via the theory of types. The *type* of a sequence $x_1^n \in A^n$ is its empirical distribution $\hat{P} = \hat{P}_{x_1^n}$; that is, the distribution defined by

$$\hat{P}(a) = \frac{|\{i\colon x_i = a\}|}{n}, \quad a \in A.$$

A distribution P on A is called an *n-type*, if it is the type of some $x_1^n \in A^n$. The set of all $x_1^n \in A^n$ of type P is called the *type class of the n-type* P and is denoted by $\mathcal{T}_P^n$.

We use the notation: $D(Q\|P) := E_Q(\log(Q/P))$,
$D(\Pi\|P) = \inf_{Q\in\Pi} D(Q\|P)$.

Sanov's Theorem. Let Π be a set of distributions on A whose closure is equal to the closure of its interior. Then for the empirical distribution of a sample from a strictly positive distribution P on A,

$$-\frac{1}{n}\log P\left(\hat{P}_n \in \Pi\right) \to D(\Pi\|P).$$

See proof in [6] or in [8].

9.2 Proof of Theorem 5.1

By Lemma 2.1.1 we can assume WLOG that $A_s = [s]$ and under the maxmin value β^* we have $EL_1 = I_1 \le I_2 \le \cdots \le I_s$.

The error of the test $\mathcal{T}$ occurs, if

$$\max_{k>s} \mathcal{I}(\tau(X^N(k), Z^N)) \ge \min_{j\in[s]} \mathcal{I}(\tau(X^N(j), Z^N)) \quad (30)$$

The event in (30) is included in the event

$$\bigcap_{T\ge 0}\left[\bigcup_{k>s}\{\mathcal{I}(\tau(X^N(k), Z^N)) \ge T\}\cup_{j\in[s]}\{\mathcal{I}(\tau(X^N(j), Z^N)) \le T\}\right]. \quad (31)$$

We bound the probability of (31) from above by

$$\sum_{k>s} P(\mathcal{I}(\tau(X^N(k), Z^N)) \ge I_1 - \varepsilon) + \sum_{j\in[s]} P(\mathcal{I}(\tau(X^N(j), Z^N)) \le I_1 - \varepsilon), \quad (32)$$

where $\varepsilon > 0$ is arbitrary. Let us estimate from above the first sum in (32). It is easy to check that

$$\begin{aligned} D((\tau(x(k), z)||\pi(x(k), \cdot)\pi(\cdot, z)) &= D((\tau(x(k), z)||\tau(x(k))\tau(z)) + \\ &+ D((\tau(z)||\pi(\cdot, z)) + D((\tau(x(k))||\pi(x, \cdot)) \\ &= \mathcal{I}(\tau(x(k), z)) + D((\tau(z)||\pi(\cdot, z)) + \\ &+ D((\tau(x(k))||\pi(x, \cdot)). \end{aligned} \quad (33)$$

By Sanov's theorem, the first sum in (32) does not exceed

$$(t - s)\exp\{-N \min_{\tau\in D_\varepsilon} D((\tau(x(k), z)||\pi(x(k), \cdot)\pi(\cdot, z))\}$$

$$\le (t - s)\exp\{-N(I_1 - \varepsilon)\}, \quad (34)$$

where $D_\varepsilon = \{\tau : \mathcal{I}(\tau) \le L - \varepsilon\}$ in view of (33), (32) and nonnegativity of $D(\cdot||\cdot)$. When $R = \frac{\ln t}{N} \le L - 2\varepsilon$, we have

$$(t - s)\exp\{-N(L - \varepsilon)\} \le \exp\{N((L - 2\varepsilon) - L - \varepsilon)\} = \exp\{-N\varepsilon\}.$$

By Sanov's theorem, the second sum in (32) is bounded from above by

$$s\ \exp\{-N \min_{ClD_\varepsilon^c} \mathcal{D}(\tau(x(\xi), z)||\pi(x, z))\}. \quad (35)$$

Now, we note by definition that $\pi(\cdot,\cdot) \notin ClD_\varepsilon^c$ for any $\varepsilon > 0$.
Since $D(\tau||\pi)$ is a convex function of the pair (τ,π) [6], it follows that D is jointly continuous in τ and π, and
$\min_{\tau \in ClD_\varepsilon^c} D(\tau||\pi) = D((\tau^*||\pi)$ is attained at some points $\tau^* \in D_\varepsilon$, which does not coincide with π. Hence $D(\tau^*(x(\xi),z)||\pi(x,z)) = \delta(\varepsilon) > 0$. We see that (35) does not exceed $s\ \exp\{-N\delta(\varepsilon)\}$. The total upper bound for (32) is

$$\exp\{-N\varepsilon\} + s\ \exp\{-N\delta(\varepsilon)\} \to 0\ as\ N \to \infty$$

for any rate $R \leq L - 2\varepsilon$. Therefore, $AR > L - 2\varepsilon$ for any $\varepsilon > 0$, and consequently, $AR \geq L$. The proof is now complete.

Before proving theorem 5.2 we sketch the appropriate half of the conditional form of Sanov's theorem.

9.3 Conditional Types

In theorem 5.2, we use the following straightforward generalization of the method of types (cf. [7]) that was given above. If $\mathcal{X}$ and $\mathcal{Y}$ are two finite sets, the joint type of a pair of sequences $\underline{X}^N \in \mathcal{X}^N$ and $\underline{Y}^N \in \mathcal{Y}^N$ is defined as the type of the sequence $\{(x_i,y_i)\}_{i=1}^N \in (\mathcal{X} \times \mathcal{Y})^N$. In other words, this is the distribution $\tau_{\underline{X}^N,\underline{Y}^N}$ on $\mathcal{X} \times \mathcal{Y}$ denoted by

$$\tau_{\underline{X}^N,\underline{Y}^N}(\lambda,\alpha) = n(\lambda,\alpha|\underline{X}^N,\underline{Y}^N) \quad for\ every\ \lambda \in \mathcal{X},\ \alpha \in \mathcal{Y}. \tag{36}$$

Joint types will often be given in terms of the type of $\underline{X}^N$ and a stochastic matrix $M : \mathcal{X} \to \mathcal{Y}$ such that

$$\tau_{\underline{X}^N,\underline{Y}^N}(\lambda,\alpha) = \tau_{\underline{X}^N}(\lambda)M(\alpha|\lambda) \quad for\ every\ \lambda \in \mathcal{X},\ \alpha \in \mathcal{Y}. \tag{37}$$

Notice that the joint type $\tau_{\underline{X}^N,\underline{Y}^N}$ uniquely determines $M(\alpha,\lambda)$ for those $\lambda \in \mathcal{X}$ which do occur in the sequence $\underline{X}^N$. For conditional probabilities of sequences $\underline{Y}^N \in \mathcal{Y}$ given a sequence $\underline{X}^N \in \mathcal{X}$, the matrix M in the equation given above will play the same rule as the type of $\underline{Y}^N$ does for unconditional probabilities. We say that $\underline{Y}^N \in \mathcal{Y}^N$ has conditional type M given $\underline{X}^N \in \mathcal{X}^N$ if

$$n(\lambda,\alpha|\underline{X}^N,\underline{Y}^N) = n(\lambda|\underline{X}^N)M(\alpha|\lambda) \quad for\ every\ \lambda \in \mathcal{X},\ \lambda \in \mathcal{Y}. \tag{38}$$

For any given $\underline{X}^N \in \mathcal{X}^N$ and stochastic matrix $M : \mathcal{X} \to \mathcal{Y}$, the set of sequences $\underline{Y}^N \in \mathcal{Y}^Y$ having conditional type M given $\underline{X}^N$ will be called the M-shell of $\underline{X}^N, \mathcal{T}_M(\underline{X}^N)$.

Let $\mathcal{M}^N$ denote the family of such matrices M. Then

$$|\mathcal{M}^N| \leq (N+1)^{|\mathcal{X}||\mathcal{Y}|}. \tag{39}$$

Remark. The conditional type of $\underline{Y}^N$ given $\underline{X}^N$ is not uniquely determined, if some $\lambda \in \mathcal{X}$ do not occur in $\underline{X}^N$. Nevertheless, the set $\mathcal{T}_M(\underline{X}^N)$ containing $\underline{Y}^N$ is unique. Notice that the conditional type is a generalization of types. In fact, if all the components of the sequence $\underline{X}^N$ equal x, the conditional type coincides with the set of sequences of type $M(\cdot|x) \in \mathcal{Y}^N$.

In order to formulate the basic size and probability estimates for M-shells, it will be convenient to introduce some notation. The average of the entropies of the rows of a stochastic matrix $M : \mathcal{X} \to \mathcal{Y}$ with respect to a distribution τ on $\mathcal{X}$ will be denoted by

$$H(M|\tau) := \sum_{x \in \mathcal{X}} \tau(x) H(M(\cdot|x)). \tag{40}$$

The similar average of the informational divergences of the corresponding rows of stochastic matrix $M_1 : \mathcal{X} \to \mathcal{Y}$ from $M_2 : \mathcal{X} \to \mathcal{Y}$ will be denoted by

$$D((M_1||M_2|\tau) := \sum_{x \in \mathcal{X}} \tau(x) D(M_1(\cdot|x)||M_2(\cdot|x)). \tag{41}$$

Notice that $H(M_1|\tau)$ is the conditional entropy $H(X|Y)$ of RV's X and Y such that X has distribution τ and Y has conditional distribution M on X. The quantity $D(M_1||M_2|\tau)$ is called the conditional informational divergence (conditional relative or cross- entropy).

The detailed proofs of the following statements are in [33]. The counterpart of the theorem on type class size for M-shell is

Statement 1. For every $\underline{X}^N \in \mathcal{X}^N$ and stochastic matrix $M : \mathcal{X} \to \mathcal{Y}$ such that $\mathcal{T}_M(\underline{X}^N)$ is non-void, we have

$$\frac{1}{(N+1)^{|\mathcal{X}||\mathcal{Y}|}} \exp\{NH(M|\tau_{\underline{X}^N})\} \le |\mathcal{T}_M(\underline{X}^N)| \le \exp\{NH(M|\tau_{\underline{X}^N})\}.$$

Statement 2. For every $\underline{X}^N \in \mathcal{X}^N$ and stochastic matrix $M_1 : \mathcal{X} \to \mathcal{Y}$, $M_2 : \mathcal{X} \to \mathcal{Y}$ such that $|\mathcal{T}_{M_1}(\underline{X}^N)|$ is non-void,

$$M_2(\underline{Y}^N|\underline{X}^N) = \exp\{-N(D((M_1||M_2|\tau_{\underline{X}^N}) + H(M_1|\tau_{\underline{X}^N}))\}, \tag{42}$$

if $\underline{Y}^N \in \mathcal{T}_{M_1}(\underline{X}^N)$,

$$\frac{1}{(N+1)^{|\mathcal{X}|^s|\mathcal{Z}|}} \exp\{-ND((M_1||M_2|\tau_{\underline{X}^N})\} \le$$

$$\le M_2(\mathcal{T}_M(\underline{X}^N)|\underline{X}^N) \le \exp\{-ND((M_1||M_2|\tau_{\underline{X}^N})\}. \tag{43}$$

Statement 3. (Conditional Sanov's Theorem (part)): Let $\underline{X}^N \in \mathcal{X}^N$ be i.i.d. with distribution matrix $M_2 : \mathcal{X} \to \mathcal{Y}$. Let E be a set of distributions closed in the Euclidean topology of $\mathcal{Y}^*$. Then

$$M_2(E|\underline{X}^N) \le (N+1)^{|\mathcal{X}||\mathcal{Y}|} \exp\{-ND(M_1^*||M_2|\tau_{\underline{X}^N})\}, \tag{44}$$

where

$$M_1^* = \arg\min_{M_1 \in E} D(M_1||M_2|\tau_{\underline{X}^N}) \tag{45}$$

is the distribution in E that is closest to the matrix distribution M_2 in terms of conditional relative entropy.

9.4 Proof of Theorem 5.2

We denote the set of unordered v-sets $\nu = (i_1, \dots, i_\nu)$, $1 \le i_1 < i_2 < \cdots < i_\nu < s$, $0 \le \nu < s$ by V. For an arbitrary $\nu \in V$ we define $I(\nu)$ as follows:

$$I(\nu) = I(Z \wedge X([s] \setminus \nu)|X(\nu)).$$

Let us consider the subset $\Lambda(\nu)$, $\nu = (i_1, \dots, i_\nu) \in V$ of the set $\Lambda_0 = \Lambda(s,t) \setminus [s]$ that consists of those $\lambda \in \Lambda_0$ for which $\{\lambda_1, \dots, \lambda_s\} \cap [s] = \{i_1, \dots, i_\nu\}$.
It is also obvious that $\Lambda(\nu) \cap \Lambda(\nu^*) = \varnothing$ if $\nu \ne \nu^*$ and $\Lambda_0 = \cup_V \Lambda(\nu)$.
We define

$$A(\nu) = \{(X^N, Z^N) : T_s(\mathcal{X}^N, Z^N) \in \Lambda(\nu)\}.$$

It is obvious that $A(\nu)$ is expressed in the form

$$A(\nu) = \{(\mathcal{X}^N, Z^N) : \cup_{\lambda \in \Lambda(\nu)} \mathcal{I}(\tau(X^N(\lambda), Z^N)) \ge \mathcal{I}(\tau(X^N([s]), Z^N))\}.$$

By conditioning, we get

$$P(A(\nu)) = \sum_{x^N(\nu), z} P(A(\nu)|x^N(\nu)) \; P(x^N(\nu). \tag{46}$$

The following inclusion is valid for arbitrary T:

$$A(\nu) \subseteq \{\mathcal{I}(\tau(x^N([s]), z^N)) \le T\} \cup \{\max_{\Lambda(\nu)} \mathcal{I}(\tau(x^N(\lambda), z^N) \ge T\}.$$

Consequently, for fixed $x^N(\nu)$ the conditional probability of the event $A(\nu)$ can be estimated as

$$\begin{aligned} P(A(\nu)|x^N(\nu)) &\le P(\mathcal{I}(\tau((x^N([s]), z^N)) \le T|*) + \\ &+ P(\cup_{\Lambda(\nu)} \mathcal{I}(\tau((x^N(\lambda), z^N)) \ge T|*) \\ &\le P(\mathcal{I}(\tau((x^N([s]), z^N)) \le T|*) + \\ &+ \sum_{\Lambda(\nu)} P(\mathcal{I}(\tau((x^N(\lambda), z^N)) \ge T|*), \end{aligned}$$

where $T = T(*), * := x^N(\nu)$.

$$\begin{aligned} P(A(\nu)) = \sum_{x^N(\nu), z} \Bigg[& P(\mathcal{I}(\tau((x^N([s]), z^N)) \le T|*) + \\ & + \sum_{\lambda \in \Lambda(\nu)} P(\mathcal{I}(\tau((x^N(\lambda), z^N)) \ge T|*) \Bigg] P(x^N(\nu)). \end{aligned} \tag{47}$$

Let us estimate from above the second sum in square brackets in (47), $T = I(\nu) - \varepsilon$. We have the corresponding conditional identities

$$
\begin{aligned}
&D(\tau(x^N(\lambda), z^N)||\pi(x^N(\lambda), \cdot)\pi(\cdot, z^N)|\tau_{x^N}(\nu)) = \\
&= \sum_{x\in\mathcal{X}} \tau_{x^N}(\nu)\{D(\tau(x^N(\lambda), z^N|\tau_{x^N}(\nu))||\pi(x^N(\lambda), \cdot)\pi(\cdot, z^N)|\tau_{x^N}(\nu))\} = \\
&= \sum_{x\in\mathcal{X}} \tau_{x^N}(\nu)\{D(\tau(x^N(\lambda), z^N)||\tau(x^N(\lambda))\tau(z^N)|\tau_{x^N}(\nu)) + \\
&+ D(\tau(z^N)||\pi(\cdot, z^N)|\tau_{x^N}(\nu)) + D((\tau(x^N(\lambda))|\tau_{x^N}(\nu))||\pi(x^N, \cdot)|\tau_{x^N}(\nu))\} = \\
&= D(\tau(x^N(\lambda), z^N)||\tau(x^N(\lambda)\tau(z^N)|\tau_{x^N}(\nu)) + \\
&+ D(\tau(z^N)||\pi(\cdot, z^N)|\tau_{x^N}(\nu)) + D(\tau(x^N(\lambda))||\pi(x^N, \cdot)|\tau_{x^N}(\nu)) = \\
&= \mathcal{I}(\tau(x^N(\lambda), z^N)|\tau_{x^N}(\nu)) + \\
&+ D(\tau(z^N)||\pi(\cdot, z^N)|\tau_{x^N}(\nu)) + D(\tau(x^N(\lambda))||\pi(x^N, \cdot)|\tau_{x^N}(\nu)). \qquad (48)
\end{aligned}
$$

By the conditional form of Sanov's theorem, as well as in view of (47), (48), nonnegativity of $D(\cdot|| \cdot |\cdot)$ and the inequality $|\Lambda(\nu)| \le t^{s-\nu}$, the second sum in square brackets in (47) does not exceed

$$
\begin{aligned}
&t^{s-\nu}(N+1)^{|\mathcal{X}|^s|\mathcal{Z}|} \times \\
&\times \exp\{-N \min_{\tau\in D_\varepsilon} D(\tau(x^N(\lambda), z^N)||\pi(x^N(\lambda))\pi(z^N)|\tau_{x^N}(\nu))\}\}.
\end{aligned}
$$

Here, $D_\varepsilon = \{\tau : \mathcal{I}(\tau) \le I(\nu) - \varepsilon\}$. Since the last bound does not depend on $\tau(x^N(\nu))$, we bound (using Jensen's inequality) the sum in (47) over types of $x^N(\nu)$ from above by

$$
t^{s-\nu}(N+1)^{|\mathcal{X}|^s|\mathcal{Z}|}\exp\{-N(I_\pi(\nu) - \varepsilon)\}. \qquad (49)
$$

If $R = \frac{\ln t}{N} \le I(\nu) - 2\varepsilon$, (49) is bounded from above by

$$
\exp\{-N[\varepsilon(s-\nu) - \frac{|\mathcal{X}|^s|\mathcal{Z}|\log(N+1)}{N}]\},
$$

which is exponentially small for sufficiently large N.
Also, by the conditional form of Sanov's theorem, the first term in (47) is bounded from above by

$$
\sum_{\tau_{x^N(\nu)}} (N+1)^{|\mathcal{X}|^s|\mathcal{Z}|}\times
$$

$$
\times \exp\{-N \min_{ClD_\varepsilon^c} D((\tau(x^N([s]), z^N)||\pi(x^N, z^N)|\tau_{x^N(\nu)})\}p(\tau_{x^N}(\nu)). \qquad (50)
$$

We note that, by definition, $\pi(\cdot, \cdot) \notin ClD_\varepsilon^c$ for any $\varepsilon > 0$.
Hence, $\min_{ClD_\varepsilon} D((\tau(x^N(\lambda), z^N)||\pi(x^N, z^N)|\tau_{x^N(\nu)}) = \delta(\varepsilon) > 0$ because of conditional D() continuity in all its variables. We see from Jensen's inequality that (50) does not exceed

$$
\exp\{-N(\delta(\varepsilon) - \frac{|\mathcal{X}|^s|\mathcal{Z}|\log(N+1)}{N})\}.
$$

The total upper bound for (50) is

$$\exp\{-N(\delta(\varepsilon) - \frac{|\mathcal{X}|^s|\mathcal{Z}|\log(N+1)}{N}) +$$
$$+\exp\{-N[\varepsilon(s-\nu) - \frac{|\mathcal{X}|^s|\mathcal{Z}|\log(N+1)}{N}]\} \to 0 \;\; as\ N \to \infty.$$

for any rate $R \leq I - 2\varepsilon$. Therefore, $AR > I - \varepsilon$ for any $\varepsilon > 0$, and consequently, $AR \geq I$. The proof is thus complete.

9.5 Compression Based Homogeneity Testing

We develop here the theory of CCC test used in section 7. Let $\mathbf{B} = \{0,1\}$, $\mathbf{x^N} \in \mathbf{B}^N = (x_1, ..., x_N)$ be a stationary ergodic random binary (**training**) string (SES) distributed as $P_0 = P$.

An arbitrary Universal Compressor (UC) satisfying the conditions in [47] maps source strings $\mathbf{x^N} \in \mathbf{B^N}$ into compressed binary strings $\mathbf{x_c^N}$ of approximate length $|\mathbf{x}_c^N| = -\log P(\mathbf{x}^N) = L^N$ thus *generating the approximate Loglikelihood of source* $\mathbf{x}^N$ – the main inference tool about P.

Consider a **query** binary SES $\mathbf{y}^M$ distributed as P_1 and test whether the homogeneity hypothesis $P_0 = P_1$ contradicts the data or not. Let us partition $\mathbf{y}^M$ into several **slices** $\mathbf{y}_i, i = 1, \ldots, S$, of identical length n divided by 'brakes' - strings of relatively small-length δ to provide approximate independence of slices (brakes of length 2k are sufficient for k-MC). Introduce concatenated strings $\mathbf{C}_i = (\mathbf{x}^N, \mathbf{y}_i)$. Define $CCC_i = |C_i| - |\mathbf{x}^N|$. CCC-statistic is the $\overline{CCC}$ = average of all CCC_i. Homogeneity of two texts can be tested by the test statistic $\mathcal{T} = \overline{CCC}/s$, where s is the standard deviation of $CCC_i, i = 1, \ldots, S$. Extensive experimentation with real and simulated data displayed in the second part of [41] demonstrates excellent $\mathcal{T}$-discrimination between homogeneity and its absence in spite of the lack of knowledge about P_0, P_1. Here, we sketch the proof of its consistency and exponential tail equivalence to that of the Likelihood Ratio Test (LRT) in full generality under certain natural assumption about the sizes of the training string and query slices.

Validity of our assumption in applications depends both on the compressor used and the source distribution. The main advantages of CCC-test are: (i.) its applicability for arbitrary UC and long memory sources, where the likelihood is hard to evaluate, and (ii.) its computational simplicity (as compared to statistics from [47]) enabling processing of multi-channel data simultaneously on line.

Preliminaries. Conditions on regular stationary ergodic distributed strings are as in [47]. SES are well-approximated by n-Markov Chains as $n \to \infty$.

[19] outlined a compressor construction adapting to an unknown IID distribution and gave a sketch of a version of Theorem 9.5.1 below for IID sources, connecting for the first time the notions of *complexity and randomness.* First practically implementable UC LZ-77/78 were invented during the 12 years after, and became everyday tools of computer work after a further ten years had elapsed.

Introduce $L^n = -\log P(x^n)$ and the entropy rate $h^n = H(P) = -\sum P(x^n)\log P(x^n)$.
Warning Without notice, we actually consider conditional probabilities and expectations of functions of the query slice with regard to the training text.

A compressor is called UC (in a weak sense), if it adapts to an **unknown** SES distribution, namely if for any $P \in \mathbf{P}$ and any $\epsilon > 0$, it holds that:

$$\lim_{n\to\infty} P(x \in \mathbf{B}^n : |x_c^n| - L^n \le n\epsilon) = 1.$$

Equivalently, UC attain for $P \in \mathbf{P}$ asymptotically the Shannon entropy lower bound:

$$\lim_{n\to\infty} P(|\mathbf{x^n}_c|/|\mathbf{x^n}| \to h) = 1 \quad as \quad |x| \to \infty.$$

This was established in the seminal works of C. Shannon in 1948-1949, where SES were first singled out as appropriate models of natural language.

J. Rissanen's pioneering publication on the **Minimum Description Length** principle [40] initiated applications of UC to statistical problems for SES sources. This has been continued in a series of recent papers of B. Ryabko and coauthors.

Kraft Inequality Lengths of any uniquely decodable compressed strings satisfy the following inequality : $\Sigma_{\mathbf{B}^n} 2^{-|\mathbf{x}_c^n|)} \le 1$.

As before, cross entropy $D(P_1||P_0 := \mathbf{E}_1 \log(P_1/P_0)$. Consider goodness of fit tests of P_0 vs. P_1.

'Stein's lemma' for SED [47](Proved first for the IID case by H. Cramer in 1938). *If $D(P_1||P_0) \ge \lambda$ and any $0 < \varepsilon < 1$, then the error probabilities of LRT satisfy simultaneously*

$$P_0(L_0 - L_1 > n\lambda) \le 2^{-n\lambda}$$

and

$$\lim P_1(L_0 - L_1 > n\lambda) \ge 1 - \varepsilon > 0.$$

No other test has lower values for both error probabilities (in terms of order of magnitude) than the two given here.

Theorem 9.5.1. [47]. Consider the test statistic $T = L_0^n - |\mathbf{x}_c^n| - n\lambda$. Then nonparametric goodness of fit test $T > 0$ has the same asymptotics of error probabilities as in the Stein lemma.

Main Results. 'Quasiclassical Approximation' assumption (QAA). The sizes of the training string N and query slices n grow in such a way that the joint distribution of $CCC_i, i = 1, \ldots, S$, converges in Probability to $P_1(L_0^n(\mathbf{y}))$: $N \to \infty$ and $n \to \infty$ is sufficiently smaller than N.

The intuitive meaning of this assumption is: given a very long training set, continuing it with a comparatively small query slice with alternative distribution P_1 does not actively affect the *encoding rule.* The typical theoretical relation between lengths is $n \le const \log N \to \infty$.

In practice, an appropriate slice size is determined by empirical optimization.

Define entropy rates $h_i = \lim h_i^n, h_i^n = \mathbf{E}L_i^n, i = 0, 1$.

Under QAA and $h_1 \geq h_0$, the following two statements are true for SES strings:

Theorem 9.5.2: Consistency *The mean CCC is strictly minimal as $n \to \infty$ provided that $P_0 = P_1$.*

Proof. $\mathbf{E}_1(CCC) - \mathbf{E}_0(CCC) = \sum(P_0(x) - P_1(x)) \log P_0(x) =$ $= -h_0^n + \sum P_1(x) \log P_1(x)(P_0/P_1) = h_1^n - h_0^n + D(P_1||P_0)$ which finishes the proof thanks to positivity of divergence, unless $P_0 = P_1$.

If $h_1 < h_0$, then inhomogeneity can be proved by estimating both conditional and unconditional complexities of strings.

Generate an artificial (n)-sequence $\mathbf{z^n}$ independent of $\mathbf{y^n}$. This $\mathbf{z^n}$ is distributed as P_0 and we denote by CCC^0 its CCC.

Also, assume that the brakes' negligible sizes are such that the joint distribution of S slices of size n converge to their product distribution in Probability.

Theorem 9.5.3. *Suppose $D(P_1||P_0) > \lambda$ and we reject homogeneity, if the 'compression version of the Likelihood Ratio' test $\bar{R} = \overline{CCC} - \overline{CCC}^0 > n\lambda$. Then the same error probability asymptotics as valid for LRT and for this test.*

Proof (sketch). Under negligible brakes and independent slices, probabilities multiply. To transparently outline our ideas (with some abuse of notation) replace the condition under the summation sign with a similar one for the whole query string: instead of $P_0(\bar{R} > 0) = \sum_{\mathbf{y},\mathbf{z}:\overline{CCC}-\overline{CCC}^0>\lambda} P_0(\mathbf{y})P_0(\mathbf{z})$, we write the condition under the summation as $CCC - CCC^0 > n\lambda$, which is approximated in Probability under P_0 by $L_n(\mathbf{y}) - |\mathbf{z}| > n\lambda$.

Thus, $P_0(\bar{R} > 0) \leq \sum_{\mathbf{z}} \sum_{P_0 \leq 2^{-n\lambda - |\mathbf{z}|}} P_0(\mathbf{y})P_0(\mathbf{z}) \leq 2^{-n\lambda} \sum_{\mathbf{z}} 2^{-|\mathbf{z}|} = 2^{-n\lambda}$ by the Kraft inequality. We refer to [47] for an accurate completion of our proof in a similar situation.

Informally again, $\lim P_1(\bar{R} > 0) = \lim P_1(n^{-1}(|\mathbf{y}| - |\mathbf{z}|) > \lambda = D(P_1||P_0) + \varepsilon, \varepsilon > 0$. $|\mathbf{y}|/n$ is in Probability P_1 around $-\log P_0(\mathbf{y}) = \mathbf{E_1}(-\log(\mathbf{P_0}(\mathbf{y}))) + \mathbf{r}$, $|\mathbf{z}|/n$ is in Probability P_0 around $-\log P_0(\mathbf{z}) = h_0^n + r'$. As in the Consistency proof, all the principal deterministic terms drop out, and we are left with the condition $r < \varepsilon + r'$. Its probability converges to 1 since both r, r' shrink to zero in the product $(\mathbf{y}, \mathbf{z})$-Probability as $n \to \infty$.

Search When the Lie Depends on the Target

Gyula O.H. Katona[1,*] and Krisztián Tichler[2,**]

[1] Rényi Institute, Budapest, Hungary
ohkatona@renyi.hu
[2] Eötvös University, Budapest, Hungary
ktichler@inf.elte.hu

Dedicated to the memory of Rudolf Ahlswede

Abstract. The following model is considered. There is exactly one unknown element in the n-element set. A question is a partition of S into three classes: (A, L, B). If $x \in A$ then the answer is "yes" (or 1), if $x \in B$ then the answer is "no" (or 0), finally if $x \in L$ then the answer can be either "yes" or "no". In other words, if the answer "yes" is obtained then we know that $x \in A \cup L$ while in the case of "no" answer the conclusion is $x \in B \cup L$. The mathematical problem is to minimize the minimum number of questions under certain assumptions on the sizes of A, B and L. This problem has been solved under the condition $|L| \geq k$ by the author and Krisztián Tichler in previous papers for both the adaptive and non-adaptive cases. In this paper we suggest to solve the problem under the conditions $|A| \leq a, |B| \leq b$. We exhibit some partial results for both the adaptive and non-adaptive cases. We also show that the problem is closely related to some known combinatorial problems. Let us mention that the case $b = n - a$ has been more or less solved in earlier papers.

Keywords: combinatorial search, search with lies.

1 Introduction

Let us start with the basic model of Search Theory. An n-element set S is given, one of its elements, say x, is distinguished, the goal is to find x. Questions of type "$x \in A$?" can be asked, where A is a subset of S. The unknown x should be determined on the base of the answers to these questions. In general one cannot use every subset A. A family $\mathcal{A} \subset 2^S$ is given, the question sets A can be chosen only from $\mathcal{A}$.

We show some "practical examples".

* The first author was supported by the Hungarian National Foundation for Scientific Research, grant number NK78439.

** The work of the second author was supported by the European Union and co-financed by the European Social Fund (ELTE TÁMOP-4.2.2/B-10/1-2010-0030).

H. Aydinian, F. Cicalese, and C. Deppe (Eds.): Ahlswede Festschrift, LNCS 7777, pp. 648–657, 2013.

1.1. Twenty question. Alfred chooses a person x, Paul has to find out who he/she is and he can ask questions like "is x a man?", "is x alive?", and so on... Alfred answers honestly, and Paul has to determine the person based on the answers to his questions. Obviously, S is the set of persons, A is the set of men, or the set of living persons, etc.

1.2. Chemical analysis. It is known that the given solution contains exactly one metal. This should be determined by chemical tests. Then S is the set of all metals, $x \in S$ is the one contained in the solution. A chemical test is e.g. when a certain other specific chemical is added to the solution. If $x \in A$ where A is a subset of S then the solution turns red, otherwise it does not.

A recent important variant of this example when the solution contains an unknown genetic sequence.

1.3. Criminal investigation. Given a crime, we have a set S of possible perpetrators. The real perpetrator, $x \in S$ should be found. Each evidence restricts x to be in a set $A \subset S$. For instance if a witness says that the perpetrator is bold then we know $x \in A$, where A is the set of bold ones among the possible perpetrators.

There are two basic ways to use the questions. In the *adaptive model* the choice of the next question may depend on the answers to the previous questions. The *search algorithm* starts with a question set A. If the answer is that $x \notin A$ then the next question is a certain A_0, otherwise A_1. If the answer to the question "$x \in A_0$?" is *no* then the question set A_{00} comes, and so on. That is the search algorithm consists of a binary tree structure of subsets of S where A is the root. The information obtained along the path from the root to a leave uniquely determines x. The complexity of such a search algorithm is the length of the longest path from the root to a leaf. The mathematical problem is to find the search algorithm with the least complexity using sets from $\mathcal{A}$. (Shortest algorithm in the worst case.)

In the *non-adaptive model* the question sets are given in advance: $A_1, A_2, \ldots, A_m$. Of course the knowledge if $x \in A_i (1 \leq i \leq m)$ must uniquely determine x. One can easily see that this holds iff $A_1, A_2, \ldots, A_m$ is a *separating family* that is, for any $x, y \in S, x \neq y$ there is an i such that exactly one of $x \in A_i$ and $y \in A_i$ holds. The mathematical problem is to find the minimum of m that is the size of the smallest separating subfamily of $\mathcal{A}$.

There is a very large number of variants of this basic model. The interested reader can find them in the survey paper [11] and in the monographs [10], [3], [2].

An important direction is when the answers to the questions "$x \in A$?" can be wrong. The first such problem was independently posed by Rényi and Ulam. Every subset can be chosen as a question set that is $\mathcal{A} = 2^S$, the search is adaptive, at most one of the answers can be wrong along a path leading from the root to a leave. The unknown x has to be found surely, with probability one. What is the minimum number of questions in the worst case? The problem is called the Rényi-Ulam game ([17], [16]). Berlekamp ([5], [6]) has basically solved the problem (gave good estimates). This problem has many variants, as well.

A general term for these types of problems, when the answer to the question can be erroneous *Search with Lies.* [9] is a good survey paper. The case when $\mathcal{A}$ consists of all sets of size at most k is treated in [12].

2 Our Model

In the models Search with Lies briefly introduced in the first section every question has the same chance to be incorrectly answered. In other words, the occurrence of a lie does not depend on the relationship of the question set A and the unknown element x. In our present model this is not true. For a given question there are certain unknowns x triggering the possibility of a false answer. If the unknown x is different from these then the answer must be correct. Let us show some examples continuing our examples in the previous section.

2.1. Twenty question. Suppose that Alfred has a famous transsexual in his mind as the unknown x. Paul asks "is he a man?". Alfred has to answer "yes" or "no". He will unintentionally lie, misleading Paul. (When Twenty Question was played on the Hungarian TV in the nineteen seventies, they had to introduce the the third possible answer "not characteristic" because of the protests concerning incorrect answers of this type.)

2.2. Chemical analysis. The outcome of the chemical test might sensitively depend on a parameter we cannot well control or sense. But only in the case of certain metals. For "good" metals the result of the test is correct, for the "bad" metals however it might be wrong.

2.3. Criminal investigation. The officer asks the witness if the perpetrator is bold. The witness might lie only if it is in his/her interest: the perpetrator is his/her relative or friend.

In the first section a question A divided S into two parts: into A and $\overline{A}$. If the answer was "yes" we learned that $x \in A$, if it was "no" then the conclusion was $x \in \overline{A}$. Here a question is a partition of S into three classes: (A, L, B). If $x \in A$ then the answer is "yes" (or 1), if $x \in B$ then the answer is "no" (or 0), finally if $x \in L$ then the answer can be either "yes" or "no". In other words, if the answer "yes" is obtained then we know that $x \in A \cup L$ while in the case of "no" answer the conclusion is $x \in B \cup L$.

The obvious problem is what the fastest algorithm using such questions is. If there is no limitation on the choice of these 3-partitions, then the easy answer is that only partitions with $L = \varnothing$ should be used and we are back to the old, trivial model. Therefore a natural assumption is that every L is large, that is, $|L| \geq k$ holds for every partition we can use. On the other hand it will be supposed that all the possible partitions (A, L, B) satisfying $|L| \geq k$ can be used as questions.

The adaptive case will be solved in Section 3 by exhibiting the best algorithm and proving that there is no better one in the worst case. The non-adaptive case is more difficult. In Section 4 we reduce the problem to a graph theoretical problem: a nearly perfect matching in the graph of the n-dimensional cube should be found which satisfies the additional condition that the number of edges in the matching is the same in all directions.

The results of this paper were first presented at the "Workshop on Combinatorial search" in Budapest in April 26th, 2005. Professor Rudolf Ahlswede liked them very much. Each time when we met he urged us to write them up but we kept postponing it. In the mean time he even solved some closely related problems in [1]. I hope he will like that this paper is published at least in his memorial volume.

3 The Adaptive Search

Suppose that $k \leq n-2$ holds. We start with the description of an algorithm. The starting question is an arbitrary partition (A, L, B) satisfying $|L| = k, |A| = \lceil \frac{n-k}{2} \rceil, |B| = \lfloor \frac{n-k}{2} \rfloor$. After obtaining the answer the unknown x will be restricted either to $A \cup L$ or to $B \cup L$ where $|A \cup L| = \lceil \frac{n+k}{2} \rceil$, $|B \cup L| = \lfloor \frac{n+k}{2} \rfloor$, both sizes are $< n$.

Suppose that x is already limited to a set $Z \subset S$ at a certain stage of the search. The next step of the algorithm will be determined distinguishing two cases depending on the size of Z. However in both cases the new L is chosen to minimize $|Z \cap L|$ since the incorrect answer in L is not interesting outside of Z.

1. $|Z| > n-k$. Choose L of size k in the following way: $S - Z \subset L$. Divide $Z - L$ into two parts A and B of sizes $\left\lceil \frac{|Z-L|}{2} \right\rceil$ and $\left\lfloor \frac{|Z-L|}{2} \right\rfloor$, respectively. This defines the next question (A, L, B).

2. $|Z| \leq n-k$. Choose an L of size k to be disjoint to Z. Divide Z into two parts U and V of sizes $\left\lceil \frac{|Z|}{2} \right\rceil$ and $\left\lfloor \frac{|Z|}{2} \right\rfloor$, respectively. Let the next question (A, L, B) in the algorithm be defined by $A = U, B = V \cup (S - Z - L)$.

After receiving the answer to this last question the unknown element x is restricted to a set Z' of size either $\lceil \frac{n-k}{2} \rceil$ or $\lfloor \frac{n-k}{2} \rfloor$ in the first case and of size either $\lceil \frac{|Z|}{2} \rceil$ or $\lfloor \frac{|Z|}{2} \rfloor$ in the second case. (Observe that all these four values are less than $|Z|$.)

The algorithm stops when $|Z|$ becomes 1.

Theorem 1. *Let $k \leq n-2$. The algorithm described above is the fastest adaptive search.*

Proof. A stronger statement will be proved, namely that this algorithm is the fastest if it is started from a position when the unknown element is restricted to a z-element subset Z. Let $f(n, k, z)$ denote the minimum number of questions in this situation in the worst case. Induction on z will be used.

Suppose that the unknown element is restricted to to a set Z where $|Z| = z$. We will prove that our algorithm is the shortest one, using the assumption that it is the shortest for smaller values of z. Let (A, Y, B) with $|Y| \geq k$ be the first question of an arbitrary algorithm. If the answer is "yes" then the unknown element is restricted to the set $Z \cap (A \cup Y)$, otherwise to $Z \cap (B \cup Y)$. By the inductional hypothesis at least

$$\max\{f(n, k, |Z \cap (A \cup Y)|), f(n, k, |Z \cap (B \cup Y)|)\} \tag{3.1}$$

more questions are needed.

Since $A \cup Y$ and $B \cup Y$ cover Z,

$$\max\{|Z \cap (A \cup Y)|, |Z \cap (B \cup Y)|\} \geq \left\lceil \frac{|Z|}{2} \right\rceil. \tag{3.2}$$

On the other hand either $|A|$ or $|B|$ is at most $\left\lceil \frac{n-|Y|}{2} \right\rceil \leq \lfloor \frac{n-k}{2} \rfloor$. Hence the smaller one of $|Z \cap A|$ and $|Z \cap B|$ is also at most $\lfloor \frac{n-k}{2} \rfloor$. This implies

$$\max\{|Z \cap (A \cup Y)|, |Z \cap (B \cup Y)|\} \geq |Z| - \left\lfloor \frac{n-k}{2} \right\rfloor. \tag{3.3}$$

Using the obvious fact that $f(n,k,z)$ is a monotone function of z, (3.1)-(3.3) imply

$$f(n,k,z) \geq 1 + \max\{f(n,k,|Z \cap (A \cup Y)|), f(n,k,|Z \cap (B \cup Y)|)\} \geq$$

$$1 + \max\left\{ f\left(n,k,\left\lceil \frac{z}{2} \right\rceil\right), f\left(n,k,z - \left\lfloor \frac{n-k}{2} \right\rfloor\right)\right\}.$$

Here $\lceil \frac{z}{2} \rceil \geq z - \lfloor \frac{n-k}{2} \rfloor$ holds if and only if $z \leq n-k$, following the separation in the definition of the algorithm proving that we cannot do anything better than our algorithm. □

One can conclude that the best algorithm decreases the size of Z by $\lfloor \frac{n-k}{2} \rfloor$ in each step until its size becomes at most $n-k$. Then the usual "halving" finishes the algorithm. Using the trivial fact $f(n,k,1) = 0$, this gives us a formula for the length of the algorithm.

Consequence. *Suppose $k \leq n-2$. The length of the fastest adaptive algorithm is*

$$f(n,k,n) = f(n,k) = \left\lceil \frac{n}{\lfloor \frac{n-k}{2} \rfloor} \right\rceil - 2 + \left\lceil \log_2 \left(n - \left\lfloor \frac{n-k}{2} \right\rfloor \left(\left\lceil \frac{n}{\lfloor \frac{n-k}{2} \rfloor} \right\rceil - 2 \right) \right) \right\rceil.$$

It is worth mentioning that this formula is basically identical with that of Theorem 3.8 in [11].

$k \leq n-2$ was supposed in Consequence 1. If $k = n$, the tests give no information, the unknown element cannot be found. The case $k = n-1$ is not really better. Let the question contain L as an arbitrary $n-1$-element set , the remaining one-element set is A. If the answer is "yes" then we obtained no information. On the other hand, if the one-element set is B then the answer "no" leaves us without information. That is in the worst case no information is gained from these questions.

4 The Non-Adaptive Search

In this case the "algorithm" consists of a series of questions

$$(A_1, L_1, B_1), (A_2, L_2, B_2), \ldots, (A_m, L_m, B_m) \tag{4.1}$$

such that the answers to these questions uniquely determine x in all cases. Take two distinct elements $x, y \in S$. If

$$\text{either} \quad x \in A_i, y \in B_i \quad \text{or} \quad x \in B_i, y \in A_i \tag{4.2}$$

holds for the question (A_i, L_i, B_i) we say that this question *really separates* x and y. If (4.2) holds then the answer to this question will be different when x is the unknown element and when it is y. In other words this question distinguishes x and y. On the other hand, if both x and y are in A_i (B_i) then the answer to the question is the same in the two cases (when x is the unknown or it is y). Finally, if one or both x and y are in L_i then we might obtain the same answer in the two cases, this question does not necessarily distinguishes x and y.

One can see from this that the answers to the set of questions (4.1) uniquely determine the unknown x iff (4.2) holds for every pair $x, y \in S$. We say in this case that (4.1) is a *really separating* set of questions. Our goal is to minimize m under the conditions that (4.1) is really separating and $|L_i| \geq k$, for given n, k. Let this minimum be denoted by $N(n, k)$.

It is useful to consider the "characteristic matrix" of the set of questions. The characteristic vector associated with the question (A, L, B) is a vector containing 1, $*$, and 0 in the jth coordinate if the jth element of S is in A, L, B, respectively. Let the $m \times n$ question-matrix Q have the characteristic vector associated with (A_i, L_i, B_i) in its ith row. Condition (4.2) is equivalent to the condition that for any pair of distinct columns of Q there is a row where the entries are 0, 1 or 1, 0 in the crossing points of this row and the two given columns. We say that that such a matrix is *$*$-less separating*. In these terms $N(n, k)$ is the minimum number of rows in an $m \times n$, $*$-less separating 0,$*$,1-matrix containing at least k stars in each row.

The following trivial lemma will be used later.

Lemma 1. $2^x \geq 2x$ *holds for every non-negative integer* x.

Proof. The statement is true for $x = 0, 1, 2$. For $x \geq 3$ one can use induction: $2^x = 2^{x-1} + 2^{x-1} \geq 2(x-1) + 2 = 2x$. □

Lemma 2. *If Q is an $m \times n$, $*$-less separating 0,$*$,1-matrix containing at least k stars in each row then*

$$2km \leq 2^m \tag{4.3}$$

holds.

Proof. Let m_j denote the number of $*$s in the jth column of Q. Replacing all $*$s in the jth column by either 0 or 1, 2^{m_j} different columns are obtained. Consider another, say the ℓth column. Since Q is $*$-less separating, the columns obtained from the ℓth column by replacing the $*$s by 0 or 1 must be different from the columns obtained from the jth column. Hence we have

$$\sum_{j=1}^{n} 2^{m_j} \leq 2^m. \tag{4.4}$$

Lemma 4.1 gives a lower estimate on the left hand side:

$$\sum_{j=1}^{n} 2^{m_j} \geq \sum_{j=1}^{n} 2m_j = 2\sum_{j=1}^{n} m_j. \tag{4.5}$$

The last sum in (4.5) is just the total number of $*$s in Q therefore it must be at least km (at least k in each of the m rows).

$$\sum_{j=1}^{n} m_j \geq km. \tag{4.6}$$

Inequalities (4.4)-(4.6) give (4.3). □

Lemma 3. *If Q is an $m \times n$, $*$-less separating 0,$*$,1-matrix containing at least k stars in each row then*

$$n + km \leq 2^m \tag{4.7}$$

holds.

Proof. It will be very similar to the proof of the previous lemma. We use here a tiny bit improved version of Lemma 4.1. When $x = 0$ then $2^0 = 1$ is used rather than $2^0 \geq 2 \cdot 0$. (4.5) becomes

$$\sum_{j=1}^{n} 2^{m_j} \geq \sum_{j=1}^{n} m_j + \sum_{j=1}^{n} m_j + (\text{the number of } j\text{s with } m_j = 0). \tag{4.8}$$

Here

$$\sum_{j=1}^{n} m_j + (\text{the number of } j\text{s with } m_j = 0) \geq n \tag{4.9}$$

since the non-zero m_js are decreased by replacing them by 1. Use (4.6) for the first term of the right hand side of (4.8) then (4.8) for the two other terms:

$$\sum_{j=1}^{n} 2^{m_j} \geq km + n.$$

(4.4) finishes the proof. □

It is somewhat surprising that these two easy conditions (Lemmas 4.2 and 4.3) are sufficient for the existence of a good Q.

Theorem 2. *Suppose $3 \leq m$. A Q $m \times n$, $*$-less separating 0,$*$,1-matrix containing at least k stars in each row exists if and only if both (4.3) and (4.7) hold.*

Proof. *Sketching why we need here a graph construction.* We only have to construct a matrix satisfying the conditions if the inequalities (4.3) and (4.7) hold. The matrix will contain one or zero $*$s in every column, and exactly k

$*$s in every row. The 0,1 columns of the matrix will be considered as points of the m-dimensional cube B_m. (Here $B_m = (V, E)$ is a graph where V consists of all 0,1 sequences of length m and two such vertices are adjacent if the sequence differ in exactly one position.) A column containing one $*$ can be considered as a pair of points, namely the points corresponding to the two columns obtained by replacing the $*$ by a 0 and a 1. These points are adjacent in B_m therefore the column containing exactly one $*$ can be considered as an edge of B_m. This edge has a *direction*, namely the index of the position of the $*$. It is obvious that two such edges cannot have a common point, otherwise the two columns would not be different by all substitutions. This shows that our matrix generates a matching in B_m. Since we want to have exactly k $*$s in every row, the number of edges in the desired matching should be the same in every direction.

A subgraph (in our case a matching) of B_m is called *balanced* if the number of edges in every direction is the same. We showed how these concepts came into the picture. Let us now formulate our main tool what was developed for the present purpose but its proof can be found in [14].

Theorem 3. *$B_m (m \geq 3)$ contains a balanced matching with*

$$\left\lfloor \frac{2^{m-1}}{m} \right\rfloor \tag{4.10}$$

edges in every direction.

The construction. Suppose that (4.3) and (4.7) hold. Start with the balanced matching in Theorem 4.2. By (4.3) k cannot exceed (4.10). Keep only k edges of the matching in each direction. If e is an edge of the matching in direction i then take a a corresponding column in Q having a $*$ in the ith row, its other 0,1 entries are the joint coordinates of the two endpoints of e. In this way we obtained an $m \times km$ $*$-less separating matrix. We need to add $n - km$ 0,1 columns (without a $*$) keeping the property. The existing km columns exclude $2km$ columns, what are obtained by replacing the $*$s by 0 or 1. There are $2^m - 2km$ other 0,1 columns for our disposal. However $n - km \leq 2^m - 2km$ follows from (4.7), the construction of Q can be completed. □

Consequence. *If $k \geq n - 2 \geq 1$ then the minimum length of the non-adaptive algorithm is*

$$N(n, k) = \min\{m: \ 2km \leq 2^m, n + km \leq 2^m\}.$$

The conditions on n and k ensure $m \geq 3$ by (4.7). Theorem 4.2 can be applied. □

5 Remarks

1. Pálvölgyi (unpublished) [15] gave an asymptotically good construction for the non-adaptive case. Bassalygo and Kabatianski (unpublished) [4] also solved a problem related to the non-adaptive case.

2. There were earlier attempts to model the situation described in the paper. Katona and Szemerédi [13] considered the non-adaptive case (formulated in terms of graphs), when the partitions $(A_1, L_1, B_1), (A_2, L_2, B_2), \ldots, (A_m, L_m, B_m)$ really separate every pair of elements x and y. It was proved that

$$\sum_{i=1}^{m} |A_i| + \sum_{i=1}^{m} |B_i| \geq n \log_2 n$$

that is if the cost of a test is the number of "real elements" then one cannot do better than taking $L_i = \varnothing$ for every i and "halve" the underlying set $\log_2 n$ times. For what powers of $|A_i|$ and $|B_i|$ is it still true? For recent improvements see [7] and [8].

3. We have to admit that the condition that all L's have size at least k is not realistic from a practical point of view. In a typical case many L_i's can be empty. However if the partitions with large L_i are numerous and situated adversely then it can reduce the ideal minimum length $\log_2 n$. Find conditions for that in both the adaptive and non-adaptive cases.

An interesting generalization of our model in the present paper is the following. Let a test be a family $\{A_1, A_2, \ldots, A_t\}$ where their union covers the underlying set (set of possible unknown elements). If the only unknown element is in $A_{i_1} \cap \ldots \cap A_{i_u}$ then the result of the test is any one of the indices $i_1, \ldots, i_u$. One can ask mathematical questions similar to the ones in our paper.

Acknowledgements. The first author is indebted to Professor Rudolf Ahlswede for his lifetime friendship and encouragement.

The second author is grateful to Professor Ahlswede for offering him a 6 month pre-doctoral fellowship within Marie-Curie program in 2004/05.

We are also indebted to the anonymous referees for helping to improve the presentation of the results.

References

1. Ahlswede, R.: General theory of information transfer: updated, General Theory of Information Transfer and Combinatorics. Special Issue of Discrete Applied Mathematics 156(92), 1348–1388 (2008), `http://www.math.uni-bielefeld.de/ahlswede/homepage/public/220.pdf`
2. Ahlswede, R., Wegener, I.: Search Problems. Wiley Interscience Series in Discrete Mathematics. John Wiley & Sons Inc. (1980)
3. Aigner, M.: Combinatorial Search. John Wiley & Sons, Inc., New York (1988)
4. Bassalygo, L., Kabatianski, G.: Personal communication
5. Berlekamp, E.R.: Block coding for the binary symmetric channel with noiseless, delayless feedback. In: Mann, H.B. (ed.) Error Correcting Codes. Wiley (1968)
6. Berlekamp, E.R., Hill, R., Karim, J.: The solution of a problem of Ulam on searching with lies. In: IEEE Int. Symp. on Inf. Theory, vol. 244. MIT, Cambridge (1998)
7. Bollobás, B., Scott, A.: On separating systems. European J. Combin. 28(4), 1068–1071 (2007)

8. Bollobás, B., Scott, A.: Separating systems and oriented graphs of diameter two. J. Combin. Theory Ser. B 97(2), 193–203 (2007)
9. Deppe, C.: Coding with feedback and searching with lies. In: Csiszár, I., Katona, G.O.H., Tardos, G. (eds.) Entropy, Search, Complexity. Bolyai Society Mathematical Studies, vol. 16, pp. 27–70 (2007)
10. Du, D.-Z., Hwang, F.K.: Combinatorial Group Testing. World Scientific (1993)
11. Katona, G.: Combinatorial Search Problems. In: Srivastava, J.N. (ed.) A Survey of Combinatorial Theory, pp. 285–308. North Holland/American Elsevier, Amsterdam/New York (1973)
12. Katona, G.O.H.: Search with small sets in presence of a liar. J. Statictical Planning and Inference 100, 319–336 (2002)
13. Katona, G., Szemerédi, E.: On a problem of graph theory. Studia Sci. Math. Hungar. 2, 23–28 (1967)
14. Katona, G.O.H., Tichler, K.: Existence of a balanced matching in the hypercube (submitted)
15. Pálvölgyi, D.: Personal communication
16. Rényi, A.: On a problem of information theory. MTA Mat. Int. Közl., 6B, 505–516 (1961)
17. Ulam, S.: Adventures of a Mathematician. Scribner, New York (1976)

A Heuristic Solution of a Cutting Problem Using Hypergraphs

Christian Deppe[1] and Christian Wischmann[2]

[1] Universität Bielefeld, Fakultät für Mathematik, Universitätsstraße 25, 33615 Bielefeld, Germany
cdeppe@math.uni-bielefeld.de
[2] Monroe Community College, 1000 East Henrietta Road, Rochester, New York 14623, USA
cwischmann@ymail.com

Dedicated to the memory of Rudolf Ahlswede

Abstract. We consider a cutting problem, which have a practical application. In this problem, items are being cut from larger items, for example textile patterns from a panel of cloth. We deal steel tubes of given length, which have to be sawed from longer steel tubes. These longer steel tubes come with given costs and the sawing has to be planned such that the total costs are minimized.

Keywords: cutting problem, one-dimensional, Gilmore and Gomory approach.

1 Introduction

The minimization of costs while cutting various materials is a common problem in practice. Additionally, it is analogous to the packing problem where, for example, a container has to be loaded with boxes as efficiently as possible. Packing and cutting problems are divided into three groups: 1-dimensional, 2-dimensional and 3-dimensional, depending on whether 1-dimensional bars, 2-dimensional surfaces or 3-dimensional objects are packed or cut. Problems of higher dimensions are thinkable, but do not occur in practice. 3-dimensional cutting problems arise, as mentioned, when containers are loaded with boxes, 2-dimensional ones when cutting patterns from various materials. 1-dimensional cutting problems arise, like in our case, when tubes are sawed from larger tubes. A survey of approaches for all these problems was given in [6].

In our case we examine a 1-dimensional cutting problem. More precisely, under side conditions detailed later on, a number of ordered customer lengths $l_1, .., l_n$ with given quantities $q_1, .., q_n$ will be cut from a set of choosable initial lengths $L_1, .., L_m$ with certain costs a piece. The total costs, which are calculated from the costs and amounts of the initial lengths, has to be minimized. The 1-dimensional cutting problem was first solved by Eisemann [4] using linear programming. Here,

H. Aydinian, F. Cicalese, and C. Deppe (Eds.): Ahlswede Festschrift, LNCS 7777, pp. 658–676, 2013.

we assign a variable to any possibility to cut customer lengths from initial lengths. In Eisemann's case, these initial lengths were given. Using the simplex method, the variables which produce a minimal solution are chosen from the set of all variables. The number of variables is very large, however.

1961 Gilmore and Gomory published an approach [7] which builds on Eisemann's solution. It shortens the time-consuming calculation and choice process by column generation. Eisemann's method checks every assumed solution as to whether there exists an unused variable whose substitution would result in an improvement. All existing variables have to checked for this. The algorithm of Gilmore and Gomory does not choose such a variable from all possibilities, but instead constructs it directly. This additional subproblem is solved by a knapsack problem. 1963 Gilmore and Gomory published a second part of their paper [8]. The authors responded in particular to the needs of the paper industry (limiting the number of saws and even machine usage) and present an alternate algorithm for the solution of the knapsack problem. As will be seen later, even with the help of the second paper not all side conditions can be introduced into the algorithm, whereas the limit on the number of saws can easily be inserted into the first algorithm.

Another difference from Gilmore and Gomory to Eisemann is the waiving of the restriction to whole number solutions. The production numbers can be arbitrary non-negative numbers. The solution has to be turned into integers afterwards, which can result in a loss of optimality. A small survey of methods which solve this second subproblem has been published by Wäscher and Gau [11].

The algorithm of Gilmore and Gomory is the standard approach of 1-dimensional cutting problems. Another, newer approach by Dyckhoff [5] shall be presented nevertheless. The idea of Gilmore, Gomory and Eisemann was, that all production processes are carried out at once. Dyckhoff's algorithm starts by producing one customer length after the other, and only calculates the next step after the previous has been finished. After one has been produced, we have a new set of customer lengths. The result is not a simplex algorithm with column generation by a knapsack problem, but one single large knapsack problem. The solutions are integers, however, in the case of unfavorable numbers, a knapsack problem with a large number of variables and equations occurs which results in a long calculation.

There is no polynomial bound to the runtime of the algorithm of Gilmore and Gomory [1]. Using the ellipsoid method [9] would result in a polynomial bound, in practice, however, the method of Gilmore and Gomory calculates the solution faster.

We will use the algorithm of Gilmore and Gomory as a starting point. The main problem of this paper will be to adjust the algorithm to the various side conditions. The free choice of initial lengths and the limitation of the number of saws have already been mentioned. Unfortunately, we will see that the adjustment is not readily possible. The key point is that the customer lengths cannot be distributed among the initial lengths arbitrarily. For logistic reasons, the entire order of one length has to be cut from the same initial length. To solve this problem, an additional, graph theoretical algorithm is introduced, which can model the problem precisely. Every possibility to cut a set of customer lengths from a certain initial length (this is called a batch), constitutes an edge of a hypergraph. The cost of the batch takes the role of an edge weight. The goal is to choose edges which are a partition of the hypergraph. The sum of the weights has to be minimal. No algorithm for this problem exists so far, nor one for a similar problem that could be transformed. We will therefore suggest a heuristic solution.

In [12] an extended overview with examples of the results we present here is given.

This paper is structured as follows: first the production process, as far as relevant, is described, so that the side conditions can be understood. What follows is the construction of a cost function, which will calculate the cost of an initial length. In the next step the algorithm of Gilmore and Gomory is described, including the adjustments for the side conditions. Then we will reformulate the problem into graph theory. Lastly, the procedure to turn the solution into whole numbers is described, which will then accommodate the remaining side conditions.

2 The Production of Precision Tubes

We will first describe how precision tubes are produced and then what conditions have to be accounted for in the optimization of the production. So far, the optimization is carried out without the aid of a computer algorithm. The orders of a week are combined into batches in advance.

2.1 The Production

The precision tubes are made from steel cylinders, called blocks. These blocks are processed into hollows in the warm shop. These hollows are then processed on into precision tubes in the cold drawing shop.

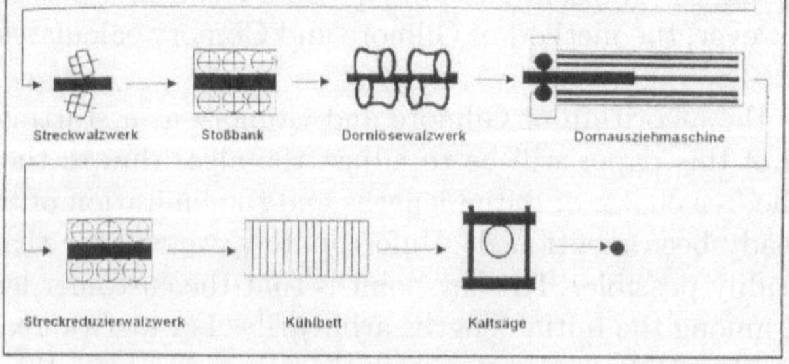

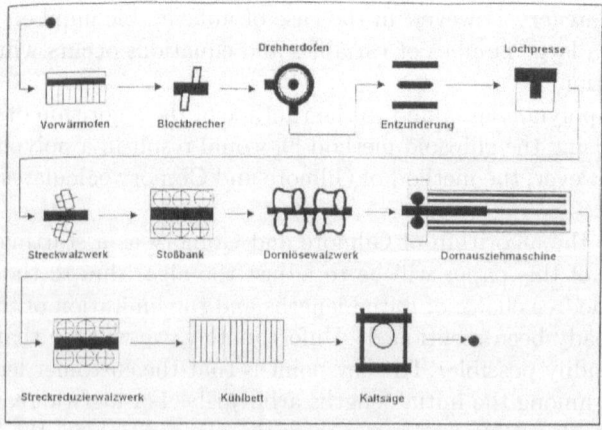

The Warm Shop

In the warm shop the block, which can weigh up to 180kg, is first heated and then brought into tube shape on the pushing bench. The pushing bench consists of rolls on which the heated block is put. A bar is then pushed into the block from one side, while the rolls stretch the block. The bar is then removed and the ends of the blocks cut off. Next the block is stretched again by moving rolls and cut into equal parts. These equal tubes are the hollows. Their measurement in terms of thickness and diameter are very imprecise, however, since they were not worked by an inner tool. A part of the production is sold directly as hollows. The majority is processed further in the drawing shop.

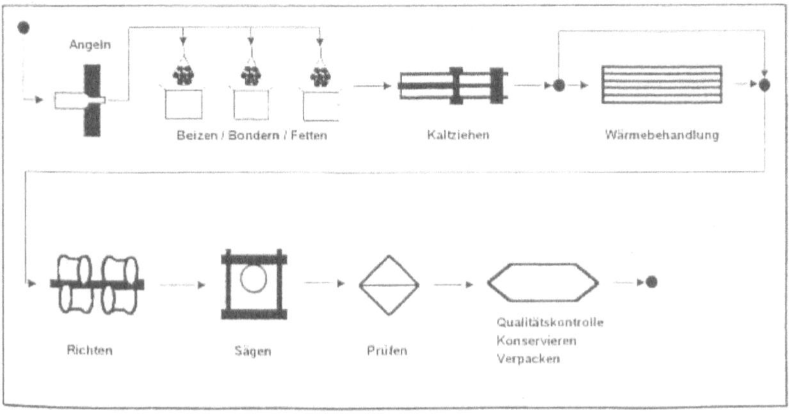

The Drawing Shop

In the drawing shops, the cooled hollows first have a tang, a kind of grip, worked into them and are provided with lubricants. Then the real production step takes place on one of several drawing benches. The longest bench can produce tubes of lengths up to 33 meters. The hollow is drawn through an outer and an inner tool, which stretches the tubes and gives it a precise diameter and thickness. The processed hollow is now a precision tube. After the draw, the ends of the precision tube are removed and the rest cut into the desired lengths. Since there are only 6 saws, one tube can be sawed into 5 smaller tubes (due to the cut off ends). These 5 tubes may, due to logistic reasons, only consist of 3 different lengths. After a negligible finishing process, the production of the precision tube is completed.

2.2 Side Conditions of the Optimization

There are four special conditions in which the given problem differs from the requirements of the algorithm of Gilmore and Gomory.

The first and most important difference is the combination to batches. A batch is simply a set of customer lengths with a quantity, which are processed from an initial length. The quantity of a customer length belonging to a batch has to be produced entirely from this batch, which means from this particular initial length. To assure optimality in general, Gilmore and Gomory will usually obtain each customer length from a number of initial lengths. The question how the customer lengths can be combined to batches is central. This adjustment happens within the graph theoretical algorithm.

The second difference is that, as mentioned, the initial lengths are not given, but freely selectable. Before the algorithm can be used, we therefore need a process that chooses adequate initial lengths. The necessary work happens within the construction of the cost function. This function will then be applied in the graph theoretical part.

Thirdly, it has to be paid attention to the fact that at most 3 different lengths and at most 5 tubes may be sawed. In the following, this condition is referred to shortly as 5-3-condition. The adjustment only requires a relatively small change in the knapsack problem.

The fourth condition is the integration of so-called variable customer lengths. The customers are offered two types of orders. Next to the normal type (length and quantity) variable lengths can be ordered as well. Here, the customer gives a tolerance of length per tube and a total length of the order. The actually produced length can vary within the tolerance and even vary within the batch. The quantity is then dependent on the choice. Both types of orders can be combined in a batch. The last section mostly deals with this problem.

3 The Cost Function

3.1 Notation

We come now to the construction of the cost function. The following parameters are relevant here.

First there are the measurements. Here, tk shall denote the thickness and od the outer diameter of the tube resp. the hollow. The restrictions of the machines are denoted by $\max_{\mathrm{DL}}$ for the maximum drawing length of a drawing bench, and $\max_{\mathrm{BW}}$ for the maximum block weight in the warm shop. T stands for the length of the hollow which is needed for the tang, which pulls the hollow. L stands for the loss at the beginning and end of the tube after the draw. The warm shop loss is always an almost constant 20kg, therefore no notation is introduced for it. Finally, mw denotes the weight of a hollow resp. tube per meter.

3.2 The Warm Shop

The linear approximation of the data gives $W^{\mathrm{II}}(b) = -1.778b+828$. This function calculates the cost of the warm shop per ton of steel depending on the utilized block weight b in kg.

The cost of just one block is then

$$W^{\mathrm{I}}(b) = \frac{-1.778b^2 + 828b}{1000}.$$

To calculate how heavy the block has to be for a particular drawing length, we first have to calculate how heavy the hollow, from which a particular precision tube is created, is. Let a be the length of the tube, then the weight of the hollow is

$$H(a) = \left(\frac{(a+L)\cdot \mathrm{tk}_t \cdot (\mathrm{od}_t - \mathrm{tk}_t)}{\mathrm{od}_h \cdot (\mathrm{od}_h - \mathrm{tk}_h)} + T\right) \cdot \mathrm{mw}_h.$$

The number of hollows per block is

$$N(a) = \left\lfloor \frac{\max_{\mathrm{BW}} - 20}{H(a)} \right\rfloor.$$

The utilized block weight is therefore

$$B(a) = N(a) \cdot H(a) + 20.$$

The warm shop costs for one precision tube is thus

$$W(a) = \frac{W^{\mathrm{I}}(B(a))}{N(a)}.$$

3.3 The Drawing Shop

The cost function for the drawing shop is

$$D^{\mathrm{I}}(d) = 35.83z^2 - 2.711z + 35.836.$$

This function gives the cost per ton of precision tubes depending on the deviation $0 \leq d \leq 1$ from the maximum drawing length. This cost function is only valid for one drawing bench ($\max_{\mathrm{DL}}$ here 33 meters). The difference from one bench to another is small, however.

If we replace d by $1 - \frac{a}{\max_{\mathrm{DL}}}$ we obtain the same function, but now in dependence of of the absolute drawing length.

$$D^{\mathrm{I}}(a) = \frac{35.83}{\max_{\mathrm{DL}}{}^2} a^2 - \frac{68.949}{\max_{\mathrm{DL}}} a + 68.955.$$

$D(a)$ gives the costs of one tube of length a instead of the costs of a ton of tubes

$$D(a) = \frac{a \cdot \mathrm{mw}_t}{1000} \cdot D^{\mathrm{I}}(a)$$

The total costs of a precision tube are therefore

$$C(a) = W(a) + D(a).$$

4 The Algorithm of Gilmore and Gomory

The algorithm of Gilmore and Gomory (1961) delivers, using then revised simplex method, production patterns with quantities, which produce the ordered customer lengths and whose total costs are minimal. The number of selected patterns is always equal to the number of different customer lengths. Their quantity can be 0, however.

4.1 Notations and Variables

The customer and initial lengths receive the following notations, which we have already used before partly. Since in the algorithm of Gilmore and Gomory the initial lengths are given, we will assign a variable to them, too.

m Number of used initial lengths
j Index of the used initial lengths
L_j The length of initial length j (w.l.o.g. we require $L_j > L_{j+1}$)
C_j The costs of initial length j

n Number of customer lengths and utilized patterns
i Index of customer lengths
l_i Length of customer length i
q_i Ordered quantity of customer length i

We are looking for cutting patterns after which one or more customer lengths are sawed from an initial length. A pattern is every natural solution of the inequality

$$\sum_{i=1}^{n} l_i y_{i,k} \leq L_j \text{ for one } j.$$

The variables here stand for:

k Index of the selected patterns
$y_{i,k}$ Multiplicity of customer length i in pattern k
c_k Costs of the pattern k (= costs of the respective initial length)
x_k Number of applications of pattern k

4.2 Course of the Algorithm

The revised simplex algorithm starts with a trivial solution:

The selected n patterns are the ones for which $y_{i,k} = \left\lfloor \frac{L_1}{l_i} \right\rfloor$ for $i = k$, and $y_{i,k} = 0$ otherwise.

The solution will now be improved step by step, by using column generation to construct a new pattern and replace a previously selected one. This happens until no new pattern can be found which would improve the solution.

Let A be the $(n \times n)$-matrix which consists of the currently selected pattern as column vectors $(y_{1,k}, ..., y_{n,k})^\top$.

Let $c = (c_1, ..., c_n)$ be the vector, which consists of the costs of the n patterns.

Column Generation. A new pattern p is a natural vector $(p_1, ..., p_n)$ with costs c_p. The pattern is a strict improvement of the solution if and only if:

$$c^\top A^{-1} p > c_p.$$

This is the case since $A^{-1}p$ is p represented as linear combination of the vectors in A. If this vector is multiplied by $c^\top$, we obtain the costs of this combination. If these costs are strictly greater than c_p, then adding p would be an improvement.

Now we will construct such a pattern directly. This happens by solving the following knapsack problem for some j:

$$L_j \geq l_1 p_1 + \cdots + l_n p_n$$

$$C_j < (c^\top A^{-1})_1 p_1 + \cdots + (c^\top A^{-1})_n p_n$$

The l_i correspond to the size resp. the costs of the items, which may not exceed a certain bound, the $(c^\top A^{-1})_i$ to the benefit, which may not be below a certain bound. The solution of the knapsack problem is done by the method of Dantzig [3], which will be presented in Section 4.4.

The initial lengths are processed one by one. If no pattern can be found for $j = 1$, we proceed with $j = 2$ and so on.

Choice of the Replaced Pattern. Let B be the matrix A expanded by a 0th row and column:

$$B_{0,0} := 1$$

$$B_{0,i} := -c_i$$

$$B_{i,0} := 0.$$

To find out which previously selected pattern shall be replaced by the new one, calculate for all $1 \leq i \leq n$ for which holds $(B^{-1}p)_i > 0$ and $x_i \geq 0$ the ratio $\frac{x_i}{(B^{-1}p)_i}$. The index of the minimal ratio is the index of the pattern which is to be removed. Let k be this index.

In case the choice of k is not unique, the lexicographic rule is applied [2]. For this, let $k_1, .., k_s$ be the candidates for k and β_{ij} the coefficients of B^{-1}. The index for which $\frac{\beta_{k_i 1}}{(B^{-1}p)_{k_i}}$ is minimal is chosen. If this is not unique either, restrict the candidates accordingly again and repeat the procedure with the next column of B^{-1}. Since the columns of B^{-1} are linear independent, a unique k will be determined.

End of the Algorithm. The solution is finally a vector $x = (x_0, x_1, .., x_n) := B^{-1} \cdot q$

Here, q is the vector $(0, q_1, .., q_n)$. In the solution x, x_0 gives the total costs and $x_1, .., x_n$ the multiplicity of the respective pattern.

The calculation of the new B^{-1} and x in each step can be accelerated. Define the $(n + 1) \times (n + 3)$-matrix

$$G := \left[B^{-1} \; x \; B^{-1}p' \right].$$

p' ist here the usual p with the additional 0th element. Now execute a Gauss elimination for the kth element of the last column. The new B^{-1} and x are the desired ones.

The $x_1, .., x_n$ now have to be rounded up. How the generated overproduction can be lowered will be explained later on.

4.3 Remark about Slack Variables

The common simplex algorithm uses slack variables as additional variables. As vectors they contain only zeros and a -1 at one position. They are necessary, since under certain circumstances, an optimal solution can be one with overproduction. Interestingly, slack variables can be omitted for the algorithm of Gilmore and Gomory. For every optimal solution which contains slack variables, there exists one with equal costs, but no slack variables. In [7], Gilmore and Gomory attribute this to the omission of the restriction to whole number solutions. This is incorrect, however, as the following remark will show.

Remark. From the property, that for every pattern $(y_1, ..., y_n)$ with at least one $y_i \geq 1$ we also have a pattern $(y_1, .., y_{i-1}, y_i - 1, y_{i+1}, .., y_n)$, and both have the same initial length and costs, follows that slack variables can be omitted.

Proof. As pattern with index j let the ith slack variables (-1 at the ith position) be part of the solution. This means the ith customer length is produced more often than necessary.

If there is a pattern j' in the solution with $y_{i,j'} \geq 1$ and $x_{j'} \geq x_j$, then the solution can be altered. The pattern j' gets produced x_j times less and a new pattern j is introduced, which, apart of one occurrence of l_i less, is identical with j' and is produced x_j times. The slack variable is removed. Both pattern use initial length j' and together are produced $x_{j'}$ times. The costs are the same.

If all patterns producing l_i have multiplicity $< x_j$, choose an arbitrary pattern as j'. Replace this pattern by the pattern which produces l_i one time less than j'. x_j is reduced by $x_{j'}$. Since the initial length of j' is not changed, the costs remain the same.

Repeat this step until the first case occurs. Since

$$\sum_{k \neq j} y_{i,k} \cdot x_k > x_j,$$

this will happen after finitely many steps. □

This proof differs only in one step from the one Gilmore and Gomory use. The difference is, that in the second case $y_{i,j'}$ is not reduced by 1, but 0 in entered immediately. This is why the removal of whole number solutions is necessary. The proof shows, that not non-whole number solutions are crucial, in fact they alone do not suffice. The above property holds for Gilmore and Gomory, too, of course, it is merely not used. What is used implicitly is the existence of a pattern $(y_1, .., y_{i-1}, 0, y_{i+1}, .., y_n)$.

Gilmore and Gomory continue to use slack variables, however. The reason is that when more than one optimal solution exists, the algorithm finds the one with the least utilized patterns. In every step of the algorithm, before the generation of a new pattern by the solution of the knapsack problem, it is first checked whether a slack variable would improve the current solution. If this check is removed, calculation time is saved. We will therefore not use slack variables, since the numbers in practice will only rarely allow more than one solution with equal costs. In Section 5.5 we will see that in their example, Gilmore and Gomory only used natural numbers ≤ 10. Should in practice such a suitable constellation occur, the solution which uses the least patterns is not automatically chosen anymore, however, it is also not always one with n patterns. In this case, the time needed to adjust the saws for a new pattern, which is not considered in the algorithm, has to be tolerated.

4.4 Solution of the Knapsack Problem by Dantzig and Adjustment

The knapsack problem is solved recursively, as in the work of Gilmore and Gomory [7]. They in turn refer to Dantzig [3].

Let $1 < t \leq n$ and let l be an arbitrary length. Then:

$$R_t(l) = \max_{0 \leq p_t \leq \lfloor \frac{l}{l_t} \rfloor} \{p_t(cA^{-1})_t + R_{t-1}(l - p_t l_t)\}.$$

Knapsack Heuristic. To increase the speed of the algorithm, it is not necessary to use the knapsack algorithm in each step. Patterns can be constructed using an arbitrary knapsack heuristic. If the heuristic does not find a new pattern at a certain point, it does not mean that there is none, but that the exact knapsack algorithm has to be used instead. This will usually only affect the last few steps of the algorithm of Gilmore and Gomory. The calculation time of the other steps is lowered in turn.

In this case we will use the knapsack heuristic of Dantzig [3], which Gilmore and Gomory used as well. Any other one is applicable as well. The heuristic works as follow:

We will switch the indices $1, .., n$ of the general knapsack problem so that:

$$\frac{cA_1^{-1}}{l_1} \geq \frac{cA_2^{-1}}{l_2} \geq \cdots \geq \frac{cA_n^{-1}}{l_n}.$$

Set the new pattern p as follows:

$$p_1 = \lfloor \frac{a}{l_1} \rfloor$$

$$p_2 = \lfloor \frac{a - l_1 q_{1,p}}{l_2} \rfloor$$

$$p_3 = \lfloor \frac{a - l_1 q_{1,p} - l_2 q_{2,p}}{l_3} \rfloor$$

and so on. Before the pattern can be used, we have to bring the indices back into the original order, of course.

Adjustment for the 5-3-Condition. The calculation of the solution of the knapsack problem will now be adjusted, so that the sum of the p_t cannot be larger than 5, and so that no more than 3 positive p_t exist.

To generalize, let g be the maximal number of produced customer lengths per precision tube (5 here) and h the maximal number of produced customer lengths of differing length per precision tube (3 here).

The new recursion step is now

$$R_{t,g,h}(l) = \max_{0 \leq p_t \leq r} \{p_t(cA^{-1})_t + R_{t-1,g-p_t,h'}(l - p_t l_t)\},$$

where

$$h' := \begin{cases} h & \text{if } p_t = 0 \\ h-1 & \text{if } p_t \geq 1 \end{cases}$$

and

$$r := \begin{cases} 0 & \text{if } h = 0 \\ \min\{\lfloor \frac{l}{l_t} \rfloor, g\} & \text{if } h \geq 1 \end{cases}.$$

What has to be solved then, is the problem $R_{n,5,3}(L_j)$ for one j.

The key in the adjustment is the restriction of p_t by r. If $h = 0$, then all following p_t are 0 as well. Otherwise r is additionally restricted by g. Both values are decreased appropriately in each recursion step.

In the heuristic it has to be watched that for every p_i, $p_i \leq 5$ and that the heuristic is stopped after the third positive p_i.

5 Graph Theoretical Approach

Since one customer length cannot be distributed among differing initial lengths, the goal is to partition the set of customer lengths into subsets, which will then equal the batches. For these subsets we then have to determine a suitable initial length onto which we will optimize the customer lengths via Gilmore and Gomory's algorithm.

5.1 Formulation

The problem can be described as a complete, weighted hypergraph $H = (V, E)$. In a hypergraph, V is the finite set of vertices. In our case the nodes will represent the various customer lengths l_i. The set E consists of a subset of the power set of V. The elements e of E are called hyper edges. While in a regular graph only two vertices can be connected by an edge, in a hypergraph a hyper edge can connect a number of vertices. Even one vertex alone constitutes an edge. The edges in our case stand for every possibility to combine customer lengths into a batch. Our hypergraph thus contains the entire power set (except the empty set).

Weighting means that a certain number is assigned to every hyper edge. Here the weights w_e of the edges e correspond to the minimal costs which would arise if the customer lengths in e were to be combined into one initial length.

The problem can now be solved by choosing a subset $F \subset E$, which constitutes a partition of the hypergraph and whose total weight is minimal.

No algorithm exists for this problem, nor for a similar problem which could be reformulated. The only possibility would be to check all possible F. For small n this might be feasible, but not for larger ones. Where the border lies for this is determined by the computer capacity.

One possibility to move this border would be to drop the completeness of the graph. In a practical case, it is improbable that batches will contain a large number of customer lengths. It would be possible to only allow edges up to a certain maximum grade Δ. Aside from having less possible partitions, this also reduces the required time to calculate the weights for all edges. This happens at the expense of the guarantee of optimality.

Calculation of the Edge Weights. Let an edge $e = \{k_1, .., k_t\}$ be given.

There is a property which can be utilized here. After the optimization, a chosen initial length has to contain at least one pattern which does not cause any waste. If all patterns would cause waste, the initial length could be shortened by the minimum waste, which would lower the costs. All initial lengths for which no pattern of the same length exists can be dropped therefore. Due to this, the choice of the initial lengths can be limited to the following candidates.

Calculate the length of all pattern $p = (p_1, ..., p_t)$, which consist of $k_1, ..., k_t$ and for which holds:

$$\max\left\{\max(k_i), \frac{\max(L)}{2}\right\} \leq \sum_{i=1}^{t} p_i k_i \leq \max(L).$$

Now, using Gilmore and Gomory, the customer lengths are optimized on each of these candidates and the one with minimal costs will be used. The calculated minimal costs give the weight w_e.

5.2 The Heuristic

For large n we will now introduce a heuristic, which will yield an at least acceptable solution quickly. First the edge weights are normalized:

$$\overline{w}_e := \frac{w_e}{\sum\limits_{\substack{i=1 \\ l_i \in e}}^{n} l_i \cdot q_i}.$$

$\overline{w}_e$ gives the costs per produced meter. The adjustment is necessary, since the absolute costs depend on the strongly varying total length of the customer orders and may differ greatly.

Course of the Heuristic. The heuristic will first construct a tree from the hypergraph, from which we will then be able to take the solution.

1. Consider every possibility to partition the vertices of the hypergraph into two sets M_1 and M_2. It shall hold $M_1 \dot{\cup} M_2 = V$ and $|M_1| = |M_2|$ for n even and $|M_1| = |M_2| + 1$ for n odd (w.l.o.g. for even n, l_1 is always in M_1, so that no bisection is calculated twice).
2. Choose a partition for which
$$\sum_{e \subset M_1} \overline{w}_e + \sum_{e \subset M_2} \overline{w}_e$$
is minimal.
   This is exactly the partition with minimal average costs per meter. (Dividing by $|\{e \subset M_1\}|$ and $|\{e \subset M_2\}|$ is unnecessary, since $|\{e \subset M_1\}|$ and $|\{e \subset M_2\}|$ are always the same.)
3. Continue with these two sets until n one-element sets are reached. This way we arrived at a tree with V as root and the leaves $l \in V$. Every vertex v of the tree stands for a bisection constructed before. A vertex is a neighbor of another if one of them originated from the other by a bisection.
   In the following, every vertex in the tree shall stand for the corresponding hyper edge in the graph. The weight w_e of the edge is transfered to the vertex of the tree.
4. The leaves of the tree are the starting solution. Choose the superior vertex v_1 to two currently chosen v_2 and v_3 with distance 2, if $w_{v_1} \leq w_{v_2} + w_{v_3}$.
5. If no more superior vertices can be chosen, the solution has been found.

5.3 An Alternative

Whenever the drawing length of a bench is changed, the change takes a certain time. The cost of this time could not be accounted for in the cost function.

There exists, however, a rule of thumb which is used in practice. Two batches are combined whenever the additional waste that occurs does not exceed a certain limit, even if the costs rise. This limit is given in kilograms and depends on the quality of the steel.

With the goal in mind to minimize the number of batches first and then the costs, one can proceed as follows.

During the calculation of the edge weights the waste in kilograms is calculated as well. If the value lies above the limit, the edge is deleted. Only edges with weight less than this limit appear now in the hypergraph. The number of edges will be reduced greatly this way. However, since an edge of degree 1 always has a waste of 0, there are no isolated vertices and there will still always exist a partition.

For small n all possible partitions are checked and the one which has the least edges is chosen. If this is not unique, choose the one with the lowest cost among these.

For large n the heuristic is used, which is shorter now. The difference is, that in step 2 we do have to divide this time. After this, the tree does not have to be constructed completely. As soon as in the process of bisections an M is found for which $M \in E(H)$, we can stop constructing this branch. At the end, again, the leaves of the tree yield the solution.

6 Variable Customer Length and Lowering of the Overproduction

6.1 Treatment of Variable Customer Lengths

The problem of variable customer lengths arises in two places. First, how do variable customer length take influence on the choice of the initial length? Second, how will they be treated in the algorithm of Gilmore and Gomory? Since the portion of variable length orders of the total order volume is relatively low, compatibility is more important than optimality. Let us write in the following $\min l_i$ for the lower bound of a variable length and $\max l_i$ for the upper one.

Variable Customer Length in the Choice of the Initial Length. In order to limit the length of the calculation it is important that the number of candidates does not increase too much.

Let an arbitrary hyper edge $l_1, .., l_k$ be given. W.l.o.g. let the first $1 \leq m \leq k$ lengths be variable. To take advantage of the entire range, first calculate all candidates for $l_i := \min l_i$ for $1 \leq i \leq m$, then with $l_i := \max l_i$ for the same i.

An additional opportunity presents itself to us here. Recalling the graph of costs per meter of tube, we can also admit the distinctive minima at which production is most cost effective.

Variable Customer Lengths in the Algorithm of Gilmore and Gomory. At this point we have to fix which lengths are actually used within the given range. On the one hand, the lengths should be long, to save costs, on the other hand longer lengths mean less possible pattern which can mean higher costs.

The choice of the length which is entered as parameter into the algorithm of Gilmore and Gomory, will be made as follows:

1. Let L be the given initial length. Calculate for all $1 \leq i \leq m$ all non-extendable pattern from the initial length $L - \min l_i$ which only use the customer lengths $\min l_i, l_{m+1}, .., l_k$.
2. Divide the waste of every pattern by 1+ the multiplicity of $\min l_i$. Let $\min\left\{\min l_i + \frac{W}{|\{\min l_i\}|+1}\right\}$ be the utilized length l_i.

6.2 Lowering of the Overproduction

If there is waste in a pattern which contains variable customer lengths, increase the chosen length by this waste, as long as it stays under $\max l_i$. If the pattern contains more than one, distribute the waste evenly.

Due to this lengthening and the rounding up of the x_i, overproduction occurs. This overproduction can be decreased by lowering the multiplicity of appropriate patterns. Since overproduction and waste mean the same loss, the lengthening of the variable customer lengths means that we have more possibilities to decrease multiplicities.

Sort the indices of the customer lengths within every batch downward by the total overproduction. Every pattern whose multiplicity can be lowered receives the indices of its element as a word, which means, that the ordered set of the indices is treated as letters. The multiplicity of the pattern with the lexicographically smallest word is now reduced as much as possible, then the next smallest word is treated the same until no multiplicities can be lowered anymore.

7 An Example

To illustrate the graph theoretical algorithm, we will give a short example. Under realistic conditions and 7 different customer lengths we will create the tree and the alternative tree. The procedure can be taken from these.

7.1 Example Parameters

The condition of the maximum usable block weight depends on the maximum block weight and this depends on the type of steel and its quality. The same holds for the allowable waste. The maximum drawing length depends on the utilized drawing bench.

Max. usable block length:	143.8 kg
Max. drawing length:	24.4 m
Max. all. waste per batch:	518 kg

The conditions on the measurements of the precision tubes and the hollow are the following:

	tube	hollow
Outer diam.:	25.4 mm	31.8 mm
Thickness:	2.27 mm	2.9 mm
Weight per m:	1.294 kg/m	2.065 kg/m

The previous conditions are only important to the calculation of the cost function. For the actual algorithm we primarily need the 7 different customer lengths and their quantities.

	Cust. L.	Quantity
1.	4.3 m	1778
2.	4.9 m	686
3.	9.5 m	525
4.	10.2 m	330
5.	11.0 m	330
6.	14.5 m	583
7.	15.5 m	750

To create the tree, the actual algorithm has been run manually. The maximal degree has been limited to 3. The calculations of the original algorithm of Gilmore and Gomory have been done using the computer program AMPL (which in turn uses CPLEX).

7.2 The Tree

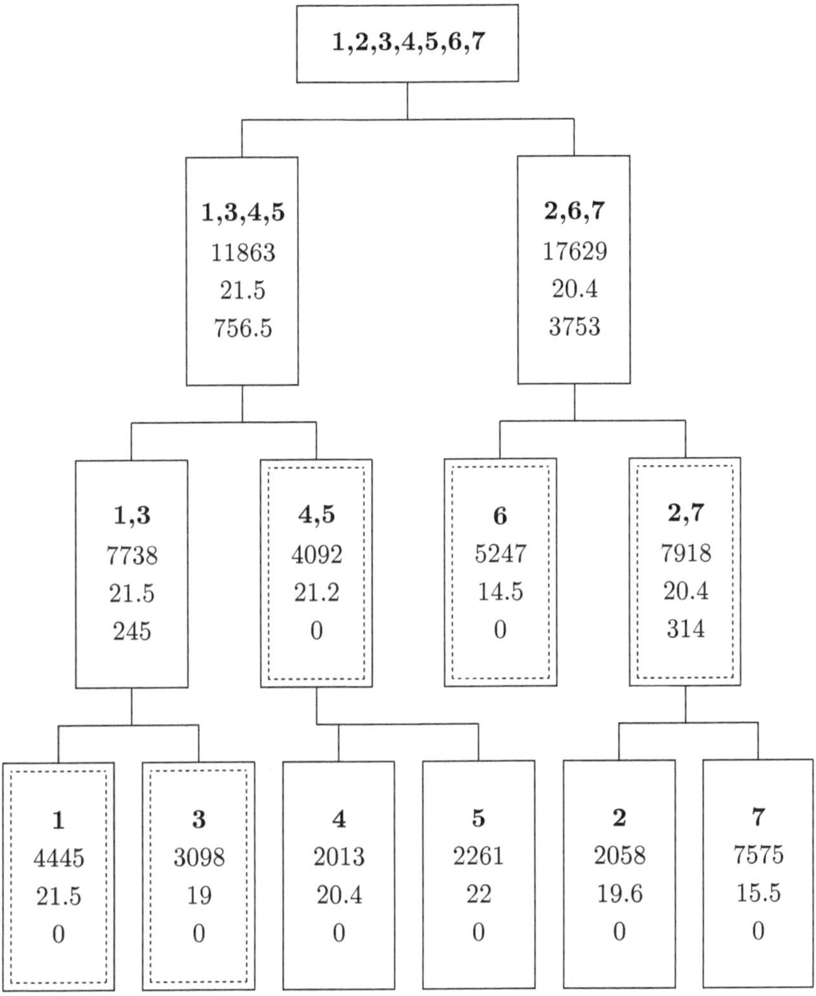

The numbers in the tree have the these meanings:

The highlighted numbers of the respective first line stand for the respective customer lengths. The second number shows how expensive the production of this batch would be and the third the used initial length. The last number gives the waste in kg.

The highlighted nodes are the optimal choice. As can be seen, the sum of the nodes **2** and **7** (resp. 4 and 5) is greater than the costs of the node **2,7**. This is not the case for **1** and **3**, however.

The node **1,2,3,4,5,6,7** has no calculated costs due to the previously stated reasons. The costs of the node **1,3,4,5** were calculated separately, but do not go into the creation of the tree, only into the choice of the batches.

7.3 The Tree (Alternative)

The tree of the alternative method of choice is only complete for presentation, the lowest row of nodes would not actually be created at all. The result differs from the first method in one point. The total waste belonging to node **1,3** of 245 kg is chosen to reduce the number of batches.

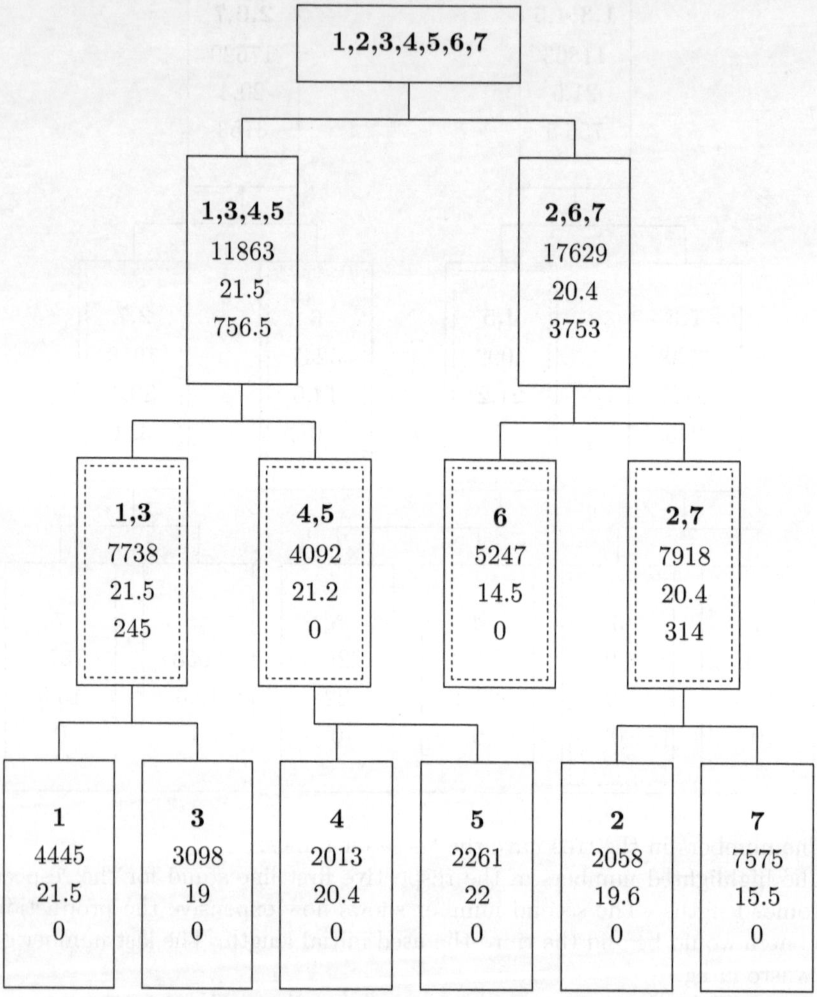

8 Continuing Thoughts

Two items can be listed here. The first concerns possible improvements of the algorithm, the second a production process that has been disregarded in this work.

The most obvious improvement would be that of an exact algorithm, not a heuristic one, which would solve the graph theoretical problem in acceptable run time. One approach would be an additional application of the simplex algorithm, where every edge of the graph would receive a variable. Via column generation the involved calculation of the edge weights could be lowered additionally.

The algorithm could also be improved if the decision over the respective initial length and the lengths of variable orders could be made directly during the optimization and not beforehand. This would require and entirely new algorithm.

Not considered in this work was the so-called multiple draw. In this situation, a hollow is first drawn on the bench and sawed as usual. After this, however, the tubes are drawn and stretched again. The new problems which would have to be solved in the algorithm due to this production process were avoided by omission. Even though the percental part of the total production volume is relatively low, the given algorithm is regrettably not complete.

9 Summary

We now possess an algorithm which delivers a solution for the given case of a 1-dimensional cutting problem, although not necessarily the optimal one. In this algorithm the given customer lengths represent the vertices of a hypergraph, from whose weighted edges a partition with minimal total weight is chosen. The elements of the partition correspond to the batches. The weighting of the edges are calculated using the adjusted algorithm of Gilmore and Gomory. The customer lengths of the respective edge (a heuristically chosen length in the case of variable customer lengths) and a set of candidates for the lengths of the drawn precision tube is given to the algorithm. The algorithm will yield a solution which does not necessarily consist of whole numbers for the multiplicities of the patterns, so this property is reinstated afterwards by a short process. Finally, from all the candidates the one is chosen whose total costs are minimal. This way we will receive weights for all edges. From these a partition is chosen by splitting the set of vertices in half iteratively such that the sum of the costs per meter of the edges, which only lie in one of the two subsets, is minimal. This stepwise bisection of the edges yields a tree, whose nodes give a small set of possible partitions into batches, from which the ideal solution is chosen.

References

1. Applegate, D.L., Buriol, L.S., Dillard, B.L., Johnson, D.S., Shor, P.W.: The cutting-stock approach to bin packing: theory and experiments. In: Proc. 5th Workshop on Algorithm Engineering and Experiments (ALENEX 2003), pp. 1–15 (2002)
2. Dantzig, G.B., Jaeger, A.: Lineare Programmierung und Erweiterungen. Springer (1966)
3. Dantzig, G.B., Wolfe, P.: Discrete variable extremum problems. Operations Research 5, 161–310 (1957)

4. Eisemann, K.: The trim problem. Management Science 3, 279–284 (1957)
5. Dyckhoff, H.: A new linear programming approach to the cutting stock problem. Operations Research 29, 1092–1104 (1981)
6. Dyckhoff, H., Finke, U.: Cutting and Packing in Production and Distribution: a Typology and Bibliography. Physica (1992)
7. Gilmore, P.C., Gomory, R.E.: A linear programming approach to the cutting-stock problem. Operations Research 9, 849–859 (1961)
8. Gilmore, P.C., Gomory, R.E.: A linear programming approach to the cutting-stock problem - part II. Operations Research 11, 863–887 (1963)
9. Grötschel, M., Lovász, L., Schrijver, A.: Geometric Algorithms and Combinatorial Optimization. Springer (1993)
10. Schimöller, T.: Kostenoptimierte Auslegung der Luppeneinsatzgebiete in der Präzisrohrfertigung. Diploma Thesis at Universität-Gesamthochschule Paderborn (1998)
11. Wäscher, G., Gau, T.: Heuristics for the integer one-dimensional cutting stock problem: a computational study. OR Spektrum 18, 131–144 (1996)
12. Wischmann, C.: Ein Verschnittproblem: heuristische Lösung unter Verwendung von Hypergraphen und des Verfahrens von Gilmore & Gomory. Diploma thesis (2005)

Remarks on History and Presence of Game Tree Search and Research

Ingo Althöfer

Fakultät für Mathematik und Informatik, Friedrich-Schiller-Universität Jena, 07737 Jena, Germany
ingo.althoefer@uni-jena.de

Dedicated to the memory of Rudolf Ahlswede

Abstract. One hundred years ago, in 1912 game tree search was introduced as a scientific field by Ernst Zermelo in a concise 4-page paper. Almost four decades later the first computers were there, and three more or less concrete proposals for a Chess computer program were made by Norbert Wiener, Claude Shannon, and Alan Turing. After a long march of craftsmanship, in 1997 computer Deep Blue beat the best human Chess player in a match with six games.

The other big classic in the world of games is Go from Asia. The approach from computer Chess does not work in Go. But in 2006 a Monte Carlo tree search procedure became the starting point of a triumph march. Within the following six years computer Go programs have reached a level near to that of the best western amateur players. Also in other games like Havannah, Monte Carlo search led to tremendous progress in computer playing strength.

We describe the origins of game tree search in the early 20th century and discuss some of the waves of progress. With the help of C. Donninger we also meditate about the twilight role of science and scientific research for progress in game programming.

Keywords: game tree search, computer Chess, computer Go, Monte Carlo game search.

1 Introduction

This paper concentrates on games with the following properties:

* There are 2 players.
* Both players have complete information all the time.
* The players move in turn.
* There are no elements of chance in the game.
* The game is zero sum, with only finitely many possible scores. The two most natural cases are games with the score spectra (win, loss) and (win, draw, loss).

H. Aydinian, F. Cicalese, and C. Deppe (Eds.): Ahlswede Festschrift, LNCS 7777, pp. 677–693, 2013.

* The game is finite in a two-fold sense. In each position there are only finitely many feasible moves. And each game can have only finite length.

Wellknown examples are Chess, Checkers, Nine Men's Morris, Go, other games of territory like Amazons, connection games like Hex and Havannah. Progress in game tree search and research happened in big waves. We present the main waves in the subsequent sections. Throughout the paper, we discuss mainly the developments in the classics Chess (Section 2) and Go (Section 3).

Rudolf Ahlswede never investigated combinatorial games, at least not academically. But he was an enthusiastic Chess player, for instance with Blitz sessions in the nights of Oberwolfach workshops. In the late 1980's he gave part of his research budget at Bielefeld University for my experiments in computer Chess and 3-Hirn Chess [3], [6]. I was his assistant, when in 1990 Ralph Gasser from the ETH Zurich visited our group and presented preliminary results from his retrograde analysis of the game "9 Men's Morris" (= Muehle, in German). In particular, Gasser reported that endgame positions with 6 stones vs 4 stones could be very complicated, including positions where the stronger side needed 157 moves for a win when both sides played perfectly. Such positions would be very difficult to win for human players.

Prof. Ahlswede did not believe this. Gasser had an Atari computer with him, and I proposed a bet: Ahlswede got the side with 6 stones and should try to win one of the 157-move positions against the machine. Our arrangement for the bet: Should Ahlswede win, he got one hundred Deutsche Mark. On the other hand he had to pay me ten D-Mark, when he was not able. So, we had a quota of 10:1. The machine was started and one of the positions with a win in 157 set up. Ahlswede made his first move, and immediately the program claimed on the monitor: Now it is a draw! Ahlswede tried for 80 more moves until he conceded the inevitable. After the exhibition we looked into the database and found that there were 15 feasible moves in the original position. Only one of them was winning, the other 14 gave a draw. Ahlswede did not hesitate to pay the ten D-Mark.

Two days later another professor from pure mathematics asked me in the lift: "There was this seminar talk by the ETH guy. At the strange end you bet 10:1 that your Professor would not win against the bot. Is it really true: you the assistant would give him 100 DM in one case, and get only 10 DM in the other case? In my younger days such bets were always the other way round." Rudolf Ahlswede did not have such prejudices about behaving (here betting) according to ranks.

2 Minimaxing in Game Trees

2.1 The Wave Inspired by Lasker-Tarrasch

In the early 20th century Chess played a major role in cultural life. To quote R. Leonard [42]: "The central thesis can be put simply: game theory saw the light of day as part of the rich Central European discussions of the psychology and mathematics of Chess and other games at the beginning of the 20th century." By the way, the focus of Leonard's book is mainly on "economic game theory", not on combinatorial games.

In 1908, one of the most popular matches for a Chess World Championship took place: between title holder E. Lasker and challenger S. Tarrasch. It was a clash of two different views on the game.

Tarrasch was a born teacher. He wrote several bestselling Chess books and formulated many rules of thumb which were adopted both by Chess masters and amateurs. One of Tarrasch's famous claims was: "In (almost) every Chess position there is a uniqe best move." He formulated it again and again, most pronounced in his last book [63, p. 306]. A crucial point: Tarrasch **never** formally defined what a **best move** is, and it seems he would not have been able when asked. On the other hand Lasker, who held a doctoral title in pure mathematics, was a very pragmatic fighter at the chess board. He was said to look not necessarily for some abstract best move but for a move that was most inconvenient for the current opponent. A year before the 1908 match Lasker had published his philosophical view on Chess and life in a small booklet, entitled "Kampf" (translated to "Struggle" in English) [40]. According to Leonard "Kampf" may be seen as a predecessor to game theory.

In those days, E. Zermelo was already famous within mathematics for his groundbreaking work on axiomatics and set theory [67]. Later for instance Goedel based his work on that of Zermelo. In 1912, Zermelo was invited to give one of the main talks at the fifth International Congress of Mathematics in Cambridge, UK. The audience was surprised when his did not speak on foundations of mathematics, but about best moves in Chess [68]. Here is some speculation why Zermelo found Chess a proper topic for a mathematician. In Goettingen, where Zermelo had got habilitation and his first professorship, he was active in Felix Klein's seminar. In winter semester 1909/1910, amongst others the psychology of Chess masters (including Tarrasch) was discussed in the seminar [36], [37]. It is likely that Zermelo did know about Tarraschs unformal claim on "the best move" - and wanted to formalize the term.

Zermelo describes Chess as a game in a tree. Each node is a position. There is an arrow from position X to position Y when there is a Chess move that transforms X to Y. Having a suitable stopping rule (for instance: a game is declared a draw when a position occurs a second time in that game) the tree is finite. For each terminal position the value is known: either a win for White or a win for Black or a draw. These values can be recursively backed up by a procedure called minimaxing. At the end, the values for all direct successors of the root node are known, and a best move is to go to one of the positions that give the best result for the root player.

For me it is not clear why Zermelo called his talk in Cambridge "Anwendung der Mengenlehre ..." ("application of set theory..."). Perhaps it was the most natural choice for him as a set theorist, or perhaps the title was a compromise to avoid scaring the audience already with a less mathematical title.

There is one contemporary of Tarrasch whose theoretical contribution had so far almost no influence on the development of computer Chess. In 1913, Oskar Cordel's "Three Moves Law" was published in his book [20]. Cordel claimed that during his analysis of opening lines he had found evidence that in most Chess positions there is either a unique best move or three almost equally best moves. So, having a position with exactly two best moves seemed to be an exception for him. As a Chess master Cordel was classes below Tarrasch, but as journalists, the two were serious rivals. Only recently people started to remember Cordel's law. In a study by E. Bleicher and L. Schreiber on Chess endgame databases (documented in [55]) it turned out that Cordel is in tendency right, although situations with exactly two best moves are not so seldom.

In 1911, the Spanish mathematician and engineer Torres y Quevedo built an automaton that was able to play the Chess endgame with king plus rook vs. king correctly for the stronger side. Unfortunately this machine remained an isolated achievement with no substantial follow-up. It is an open question how Quevedo was motivated to build the automaton. Was he influenced by the publicity of the Lasker-Tarrasch match? B. Randell, the author of [47] and [48], could not give an explanation when asked directly in February 2012.

2.2 The Early Computer Wave and Chess

During the last years of World War II the first freely programmable calculation machines had been built and used, for instance in code breaking. As soon as these computers existed creative minds began seriously to think about "interesting" applications. Computer Chess was one of the most attractive challenges.

In 1948, N. Wiener published his seminal book on cybernetics [66]. On the last two pages (p. 193 and 194) he describes how a Chess program could be developed using a depth-limited minimax search with an evaluation function. Interestingly, he did not use "Computer Chess" in the title of this section. Perhaps he had the fear that "Chess" might not look scientific enough. C. Shannon had less scruple. His famous article on computer Chess which appeared in the Philosophical Magazine [58] is likely the most-cited historical computer Chess paper. In a much more precise and technical way Shannon described what Wiener had proposed in handwaving manner. Interestingly, Shannon also made a comment on the very weak strength of a randomly moving Chess program. Perhaps he had tried some Monte Carlo approach, only to see that it would likely not work in computer Chess. In 1953 A. Turing wrote about his Chess machine and especially about his paper-and-pencil simulation of a whole Chess game [64]. Recently, M. Feist from the ChessBase company realized the programs of Shannon [25] and Turing [24] (with help by K. Thompson) so that they can be run on normal PC. Both engines are (of course) chanceless against state of the art programs on the same hardware. A 10-game match between the engines of Shannon and Turing on modern PC hardware ended in a 5-5 tie [9].

The principal approach of all three authors is easily understood with the Zermelo construction (from Subsection 2.1) in mind. In contrast to Zermelo they do not generate the whole Chess tree but only the top d levels, starting from the current board position. They "find" artificial evaluations for the positions at distance d from the root (for instance by weighted counts of the pieces on the board) and back them up by minimaxing as if they were the true values. At the root they make a move to that successor with the best backed-up value.

With this approach it took "only" less than fifty years of engineering and craftsmanship (with a few really clever ideas on the road) until in 1997 the best human player (Kasparov) lost a six game match against a Chess computer on super hardware named Deep Blue [35]. In some sense it might have gone faster; see the description by C. Donninger in Paragraph 4.1 on the complications, advantages, and disadvantages of secret research [23]. Five years after Deep Blue, commercial Chess programs had surpassed the best human players, when running on normal PC machines. I am proud that Stefan Meyer-Kahlen, one of my doctoral students, has won (so far) altogether 16 world championship titles in computer Chess (between 1996 and 2010) with his program "Shredder".

In the early game programming years Shannon also tried a completely different approach. Together with E.F. Moore he built analog computers for Hex and another connection game. The board is modeled by a two-dimensional electrical charge distribution, and the brightest lamp indicates the place where the next move should be played. Only sixty years later a theoretical analysis showed that this heuristic does not always find a best move, see Chapter 5 in [26].

2.3 The Wave of Retrograde Analysis

Retrograde analysis is similar to Zermelo's minimaxing in the way that it starts from end positions and works back by application of minimax. However, it takes into account that Chess is indeed not a game on a tree but on a directed graph where circuits are possible and frequent. This helps to keep the search space much smaller than the corresponding tree. In 1965 R. Bellman was the first to propose the retrograde approach for Chess endgames [14]. Five years later T. Ströhlein used retrograde analysis for a complete analysis of Chess endgames with only three or four pieces on the board [62].

In the meantime all Chess endgames with up to six pieces have been completely analysed. Even for some endgames with seven stones on the board the data bases have been generated. And there exists software, for instance the "Freezer" ([15], [16]) for analysing endgames with even more pieces when some of them have artificially reduced mobility.

Other games have been completely solved by retrograde analysis. R. Gasser found out that Nine Men's Morris ends in a draw when played perfectly [28], [29]. His results were independently verified by P. Stahlhacke [60]. Stahlhacke in turn solved the deeper Morris variant "Lasker Morris" [61], which had been proposed by E. Lasker in [41, p. 154]. J. Schaeffer [53] and his team solved the game Checkers in an endeavour of almost twenty years [52]. Checkers is a draw when both sides play perfectly. A verification of this analysis by an independent group is still due.

In the late 1970's human Chess masters were surprised to learn how difficult it is to win with king and queen against king and rook, when the rook side is played by a perfect data base. Looking at the analogous endgame on other board sizes than 8x8 gives a hint why this is the case. Whereas 3-piece endgames like king and rook versus king or king and queen versus king take at most linearly many moves (in n) on a board of size n x n, king plus queen versus king plus rook seems to be different. For quadratic boards with size up to n=15 retrograde analysis gives the following maximum lengths to mate (observe: in several other lists not the distance to mate is given, but the distance to conversion in a won sub-endgame.): 6x6 23; 7x7 29; 8x8 35; 9x9 44; 10x10 54; 11x11 69; 12x12 85; 13x13 108; 14x14 132; 15x15 205. It is an open question whether length_max(nxn) grows quadratically in n or even faster. More data points, especially for rectangular boards with m x n cells can be found in [12].

3 Evaluation by Random Games

3.1 The Monte Carlo Wave

In 1990, B. Abramson proposed a new way to evaluate a game position: Play many random games from this position to the very end and take record of all the final results. Abramson named the average over these results "Expected-Outcome" and used it as evaluation of the position [1]. He showed that Expected Outcome was a reasonable heuristic for the games Chess, Connect4, and Othello. His paper(s) did not have a big resonance because for the games from his investigation already strong traditional evaluation functions existed. Perhaps, also his correct but technical choice for the name ("Expected Outcome" instead of something more sketchy) was non-optimal from the viewpoint of marketing. Three years later and independently, the theoretical physicist B. Bruegmann looked for some interesting interlude after the completion of his Ph.D. Thesis. Without knowing the work of Abramson he invented "Monte-Carlo Go", both the procedure and the name. Bruegmann generated random games for each direct successor position of the root and played to the successor move with the best score. However, his random games were not purely random: more natural moves (like ko, other captures, shape moves) got higher probabilities to be played. On 9x9 Go board, Bruegmann's bot did not play completely bad, but also not really well. The report [18] was never regularly published but is **the** seminal paper of Monte-Carlo Go.

A real breakthrough came in 2006 with contributions by two independent groups: in [38] and [22] versions of Monte-Carlo were proposed which generated the game tree step by step. First, several random games from the successors of the root were played. Then, for the successors with the best scores the successors were generated and more random games from them were played. In the limit, the whole game tree would be generated, and the backed-up scores would converge to the game-theoretic values. R. Coulom did not only analyse his procedure theoretically but also applied it in a bot for the Computer Olympiad in 2006. For 9x9 Go his "CrazyStone" achieved the gold medal which was the starting signal for an unprecedented Monte-Carlo race on all board sizes that is still underway today (in 2012).

After the initial successes of Monte-Carlo Go, people quickly started to apply Monte-Carlo tree search to other games where normal evaluation functions were in trouble. The triumphal march of Monte-Carlo can best be seen from the winner lists in the Computer Olympiads. In Hex [34] and Amazons [10] it started in 2009, in computer Havannah Monte-Carlo was in the driver's seat from the very beginning in 2009 [32].

A side remark: In some 2-player games with elements of chance pure Monte Carlo without tree search was a helpful tool already for some decades in the last century. In Backgammon it started in the 1970's when some endgames were analysed by manual rollouts. In the late 1980's Scrabble programs used Monte-Carlo [59]. For card counting approaches in the casino game BlackJack Monte-Carlo has a history dating back to the 1950's.

3.2 Hot Spot Computer Go

The world of Go has a traditional rating system which is also used for finding proper handicaps. The ranks of weak players are counted in kyu degrees. Best kyu-level is 1-kyu, then comes 2-kyu, and so on. When two kyu players with different degrees (say k and m) play each other than the stronger player gives —k-m— handicap stones on the normal board of size 19x19. Next level above 1-kyu is 1-(amateur)-dan, then comes 2-dan, 3-dan, 7-dan. Here also the difference in degree tells how many handicap stones should be given. Professional players have their own professional dan levels, ranging from 1-p to 9-p. (Very rarely an "honor 10-p" degree is awarded.) Observe: in the professional ranking a difference of d levels does NOT mean that the stronger should give d handicap stones; the levels are much closer to each other.

In contrast to the best Chess programs 25 years ago, Go programmers have an easier life in finding appropriate human opponents for test play. There exist several internet servers where Go is played. One of the largest and most convenient is KGS: there computer players are welcome and find human opponents 24 hours per day. It is not uncommon that a bot plays 50 games per day on KGS - for a whole month this adds up to some 1,500 games. There is also one special bot tournament on KGS each month where the best programs compete. Some human players on KGS have specialized in beating bots. Being in combat with them helps the programmers a lot to weed out bot weaknesses.

Currently the strongest computer program is Zen. Appearing in public for the first time in March 2009, Zen has so far become the bot with the best ranks on KGS. In 2009 it was 1-dan and 2-dan; in 2010 it climbed from 2-dan to 4-dan. Early in 2012 Zen became 5-dan and only slightly later 6-dan in a "deep" version with strong hardware (26 cores).

In 2008, exhibition games of Go programs against professional players became popular. The humans are still clearly ahead, so the bots get handicap stones. A race started to get wins against pro players with as few handicap stones as possible. A list of all such games is maintained at [65]. The first sensation was when in August 2008 Myungwan Kim (8p) lost to MoGo (a program running on massively parallel hardware) at handicap 9. In September 2008 CrazyStone beat a 4p-player with handicap 8. In December 2008 CrazyStone won against the same player at handicap 7. In February 2009, MoGo was successful against Chun-Hsun Chou (9p) at handicap 7, and against a 1p at handicap 6. The following three steps forward were achieved by Zen: In July 2010 it won against a 4p at handicap 6, in August 2011 against a 6p at handicap 5, and in March 2012 against Masaki Takemiya (9p) on one day at handicaps 5 and 4.

Later in 2012, the leading bots Zen and CrazyStone were successful against pro players in several more games with handicap 4, during conferences and other events, see [65] for details. Remarkable is what happened during the European Go Congress in July and August 2012. Top European professional Catalin Taranu (5p) got the offer to decide about the number of handicap stones (5 or 4 or 3) in his games against bot CrazyStone. In case of handicap 5 Taranu would have got 300 Euro for a win, at handicap 4 200 Euro, and at handicap 3 only 100 Euro for a win. After two sparring games at handicap 4 (score 1:1) Taranu opted for handicap 4. He lost the first exhibition game and won a second one.

3.3 Havannah: Race for a Prize

In the late 1970's, game designer Christian Freeling (NL) invented an abstract board game for two players and called it Havannah. Havannah is an exceptional connection game. It is played on a six-sided board with hexagonal cells. The task for a player is to either form a ring or to connect two corners or to connect three sides by his pieces. The commercial version of Havannah was published by Ravensburger company in four editions during the 1980's and 1990's. This version was in the shortlist for the prestiguous German prize "Spiel des Jahres" ("Game of the Year") in 1981 and 1982 and has a board with side length 8 and 169 cells.

For many years, no computer program for Havannah had shown up, and Freeling came to the conclusion that his game was very difficult for machines to play. This led him to offer a prize of 1,000 Euro in 2002: He claimed that within ten years no bot would be able to beat him in at least one game in a series of ten games, played on a Havannah board with side length 10 and 291 cells. Until October 2008 nothing happened. Then, during the computer Olympiad in Beijing with another big step forward by Monte Carlo algorithms in Go, this author got the feeling that time was ripe to try with computer Havannah. Within less than a year he was able to motivate several programming teams from Germany, France, the Netherlands, Poland, Canada, and the U.S. to design Havannah bots. The internet game server LittleGolem.net became a good platform for sparring games and exchange of ideas.

In October 2012, finally one ten-game match was played between Freeling and three different bots: four games by "Lajkonik", programmed mainly by Marcin Ciura (Poland); four games by "Castro", programmed by Timo Ewalds from Edmonton; two games by "Wanderer", programmed by Richard Lorentz from California. Before the match, Freeling had had no problems to beat the bots in row. But to the surprise of most spectators, he showed some nerves and the bots a surprisingly strong performance. The machines were able to win three of the ten games (Lajkonik two, Castro one). Important detail: two of the three bot wins were at the last day of the five-day event, when Freeling was exhausted already. All three bots were based on Monte Carlo algorithms; without the "Monte Carlo working horse" the machines would have had no chance against Freeling. Three reports highlight different aspects of the Havannah match: [8] gives the historical development, [21] explains the programmer's view, and [39] is the protocol by the match organizers.

Humans are still ahead of the bots on large Havannah boards, but it may take only a few more years until they will be surpassed by machines. For the Monte Carlo game programming community it was a big piece of luck that Freeling had offered his prize. It turned out to be a challenge just of the right calibre.

4 Interplay between Science and Game Programming

A portion of the progress in computer Chess happened in the development of ever stronger commercial programs. Looking back on the last three decades, it even seems that the commercial branch of computer chess contributed much more to progress than the "establishment" of the scientific community.

4.1 Communication by Competition (by Guest Writer C. Donninger)

Dr. Christian Donninger has almost 25 years of experience in computer chess; he started in the late 1980's with his first chess program "Nimzo". Before that he had studied Mathematics and received a doctoral degree in Statistics from the "Technische Universitaet" in Vienna. In some phases during his computer Chess career he has simultaneously been working for Siemens, developing software for apparatus medicine. Donninger is an outspoken senior, knowing both the worlds of traditional science and commercial game programming. When Donninger wrote his referee report for this paper, he left anonymity and commented on the original version (text in [...] are explanations by I.A.):

"I miss somewhat a discussion of the harmful influence of the Artificial Intelligence community on computer Chess and computer Go. Progress started when programmers did not care anymore about the drosophila paradigm ["Drosophia paradigm" claimed that computer Chess is a role model for the research field of Artificial Intelligence (= AI), like the small fly Drosophila has been a role model in genetics. The hope was that a computer could be made to think like a human and that all techniques developed for computer Chess should be applicable to many other open problems in AI.]. From the perspective of a commercial Chess programmer the drosophila paradigm is completely ridiculous. It is clear that the mechanism of a program has nothing in common with human thinking. Only someone who had never tried to write a reasonably playing Chess program could come up with such an idea. So I [C.D.] ever wanted to know if the academic researchers really believed in this paradigm or if it was just a means to get research honors and funds... It was argued [in the original version of this article] that commercial computer Chess slowed down progress. The argument is in this form not valid. First of all, a commercial Chess programmer can dedicate hundred percent of his time to solve the real problem: playing as strong as possible. He does not have to pretend that he solves in fact something much more important than a mere game. The Chess programmers do not have to give lectures, write articles, ... Second: craftsmanship is not a high value in the academic world. Academics are usually poor programmers, with a few notable exceptions like D. Knuth. But solving the computer Chess problem is mainly a question of craftsmanship. The article [in its original form] calls this aspect "Engineering". The term craftsmanship is much more to the point. Personally I even find the term "Engineer" offending for my work.

In the computer Chess community another wellknown quote of mine [C.D.] is: "Those who publish something know nothing. And those who know something, do not publish." This is probably true not only in computer Chess. But, there is in fact a very intense form of communication between commercial programmers. The programs are improved by autoplay: one watches day after day, month after month the own program playing against the others. Over time, one knows the behaviour of opponent programs quite well and gets a very good feeling what they are doing and what they are not doing. If the new version of opponent X improves considerably, one analyses in detail the possible reasons and works as long till the own program has improved at least the same. This way of improving the programs resulted even in inbreeding effects.

Furthermore, one can always reverse-engineer a program by disassembling. For a skilled craftsman, in the computer Chess case disassembling is relatively straightforward, because one knows the general structure of the program. The recursive alpha-beta search is relatively easy to spot. It is then obvious at which point the move generator and the evaluation function are called ... E.g., back in early 2006 I disassembled Rybka 1.0 [the leading commercial program in that year] and informed the Rybka team: There is a bug in the mating routine, please fix this line accordingly. I have therefore known from the beginning that Rybka was a clone of "Fruit". [Fruit was a very strong open source program in 2005.] In the input-output unit the Rybka team had divided Fruit's node count by 16, reduced the search depth shown in the Fruit display by constant 3 and added some minor adjustments in the evaluation function. The mate bug mentioned above was also a new Rybka feature [compared with Fruit].

Also in other fields of technology there is in some way "Communication by Competition". This form of communication is rather effective and has a high signal to noise ratio. In "Communication by Competition" [CbC] one does not have to filter away the white noise of the AI paradigm which plays a significant role in academic papers. It was obvious that these AI-ideas were useless and nobody used them. In CbC every idea that is realized is interesting, because it had been tested at length before by a highly skilled craftsman or a highly competent company like e.g. Apple, Daimler-Benz, Siemens-Medical, ... One can be sure that this craftsman does not use a "galactic algorithm". In contrast, at least 75 to 95 percent of published algorithms (rate highly positively correlated with the academic prestige of the journal, see [57]) are galactic. [An algorithm is called "galactic" when it is asymptotically fast but slow in real-world instances. An example might be given by a linear runtime with very large coefficient, for instance $1,000,000,000,000,000n$ for input length n.] In computer Chess the galactic rate was even higher than 90 percent."Here the text by C. Donninger ends.

4.2 Deep Dead Roads in Game Tree Science

Mathematicians, also people from Information Theory, are familiar with the phenomenon described by Donninger above: Often deep and "clean" results are not important for applications; instead a few simple "tricks" are responsible for a large portion of the success. It may be particularly disappointing for researchers of high caliber - who put a lot of energy in understanding principle structures of some problem - to see when progress in real life relies mainly on craftsmanship and engineering. In game tree research the following theoretical approaches are examples for dead ends.

* **Combinatorial Game Theory.** Starting with analysis of the classical Nim game (with three heaps of matches), Combinatorial Game Theory developed over the whole 20th century, culminating in work by E. Berlekamp, J.H. Conway, and R. Guy [13]. There is even a thick book [19] trying to show how CGT can be applied to Go endgames. Only future can show if this deep and rich theory will ever play a role in mastering games of territory, like Go and Amazons.
* **Complexity Status of Games.** The 1970's and 1980's were golden decades for computational complexity theory, see the complexity bible of those years [30]. During that phase also most well-known games were proven to be computationally difficult [51], for instance: Chess on nxn board needs exponential time [27], Go with Japanese rules on nxn board requires exponential time [50], Hex on nxn board is PSpace-complete, see [49], based on a diploma thesis written in Bielefeld in 1978. None of these really nice results had any importance for game tree programming in practice.
* **Pathology in Game Tree Search.** Computer Chess practice shows that searching deeper in the game tree (before pruning and evaluating artificially) increases the playing strength [33]. This was evident already in the mid 1970's. Mathematicians tried to prove this phenomenon theoretically and stumbled across a paradox which they called "pathology in game tree search" [44], [45]: searching deeper led to worse results.

  Later it turned out that not the game tree search itself had this pathology but that instead a certain "simple" stochastic model was responsible: People assumed that the true game-theoretic values at the leaves of the tree were realizations of independent random variables, and also that the heuristic evaluations contained random noise, independently for each leave. In reality things are not this way: neighboring nodes and leaves of the the game tree tend to have positively correlated values, and also errors in the heuristic values are typically positively correlated for neighboring nodes.

G. Schrüfer [56] gave a very simple and convincing model which explained what happened in game trees from reality and what was responsible for increase or decrease of evaluation errors (he coined the terms "deepening positive" and "deepening negative" for the two cases). In [2] it was proved that any nontrivial Boolean function with independent random arguments has the property that independent evaluation errors are increased. So, the phenomenon is not restricted to game trees and even not restricted to trees with arbitrary back-up rules.

The pathology research, especially the results from the early years, had a terrible influence on the relationship between programmers and theorists: programmers knew from their bots that searching deeper improved playing strength. When people from theory now came and claimed the opposite this could only mean that theory was not useful in game tree search. The author of this paper was one of the victims of this development - with his theoretical research in the late 1980's.

* **Humanlike Chess Style.** Based on early studies by psychologist A. de Groot [31] it was clear that strong human Chess players have large sets of positional patterns in their brains, typically some ten-thousands. There were several attempts to build Chess programs with such data bases of "chunks". In the 1960' and 1970's H. Simon (co-winner of the Nobel prize for Economics in 1978) was one of the most prominent supporters of this approach, besides his attempts to design computer programs with "general problem solving" capabilities.

4.3 Game Design with Computer Help

The most natural goal in game programming is to concentrate on one game (for instance Chess or Go) and to make the program as strong as possible. However, programs can also be used to test prototypes of newly invented games. Computers can perform long series of selfplay, giving information about average game length, drawing quota, first player advantage and other natural parameters [4], [17]. For such tests it is not important that the bot plays the game almost optimally; a reasonable medium playing strength is completely sufficient. After a selfplay run the rules may be modified, a new selfplay run executed, rules modified again and so on. Within a few hours tremendous progress is possible - much faster than in the traditional design process with human-only test play. This author designed two commercially successful games with computer help: "EinStein würfelt nicht" (Edition Perlhuhn and 3-Hirn company, since 2004) and "Finale" (Noris company, 2005-2008) alias "Torjäger" (Kosmos company, since 2009). In C. Browne's Ph.D. thesis [17] the board game "Yavalath" is exhibited which was fully automatic generated by a computer program.

5 Conclusion and the Future

Starting with the seminal paper by Zermelo, a century with several waves of progress has passed by. The game programming scene is again in the middle of fast progress. Likely, already in five years people will read this essay and can tell: this and that happened indeed; here was a surprising breakthrough; and there game programmers are still stuck...

5.1 The Advantage of Being Late

The development in computer Go is about 25 years behind that in computer Chess. The internet with its advantages (quick communication, mailing lists, Go servers for realtime play) and also the advantage to see how progress happened in computer Chess make things easier for the Go programmers. Nevertheless, sometimes people seem to be blind for experiences from history and try to reinvent wheels and procedures. In particular, the Go world is still missing commercial Go bots with nice and pragmatic features for analysing games, despite everlasting efforts by this author [5].

5.2 Basic Research on Monte-Carlo Game Tree Search

Currently the success of Monte-Carlo game tree search is theoretically not really well understood. In many games it works surprisingly well, but programmers typically implement lots of special tricks to achieve a good performance. Even the non-tree version "pure Monte-Carlo" which plays random games only from the direct successors of the root is not well understood (see for instance the analysis of pure Monte-Carlo selfplay in Chapter 2 of the dissertation [26]). More abstract investigations are necessary here. For instance, [46] and [7] give surprising results for 2-player games where the turn order is not alternating but random: in this setting pure Monte-Carlo is asymptotically perfect for many games.

5.3 Waves to Come and Open Problems

* **Tools for Analysing a Human-Human Game.** In Chess, programs on normal computers are nowadays much stronger than most or even all human players. Nevertheless Chess between humans is still played, and computers have mainly become important tools for analysis (before and after a game) and opening preparation. For such tasks it is helpful to let the programs run in a mode where not only the best move is computed but instead the k best ones (for some appropriate value of k, for instance k=3), together with principal variations and evaluations. It is expected that a similar development will come for the game of Go, too. See [5] for several nice visualizations of multiple candidate moves in different games.
* **Opponent Modelling.** When computers are very strong in a game, where they have to play against less strong (human) opponents, it becomes an

important topic to exploit the special weaknesses of the opponents. Such opponent modelling includes "pressing for a win" in drawn positions and refined measures to avoid Monte Carlo "laziness" in handicap games [11].

* **Go at Super-Human Level.** When will Go programs play on one level with the best human players? Once that point is passed, how much better than humans can Go programs become? Will the bots once be able to give two or three handicap stones to the best human players? Will in that period "Combinatorial Game Theory" [19] play an important role again?
* **Quantum Computers.** Will they ever take a leading role in game tree search?

5.4 Yes, We Can!

In 1992, the discovery of America by Christopher Columbus 500 years ago was celebrated and critically discussed. Rudolf Ahlswede contributed with a very clear statement: "People like to criticize Columbus for many reasons. But the key point is: He had the courage and the energy to start sailing westwards." A similar statement is true in many other fields of life, also in mathematics and in game tree (re)search: Do not hesitate until doomsday, but make your first step! The world is mainly shaped by makers.

Acknowledgements. Two anonymous referees with their criticism, comments, and questions on the original version of this paper were very helpful. In particular, one of them even left anonymity: Dr. Christian "Chrilly" Donninger allowed me to include his view on the topic "progress by game competition" in Subsection 4.1.

Professor Rudolf Ahlswede was my teacher at Bielefeld University from 1983 to 1993. In his tolerant and unaffected research style he was the best possible advisor I could have had. In 1985, I had just become his assistant, I asked him for a topic for my doctoral research. His reply - after a short moment of thinking - was: "Of course, I could give you some special task or open problem. But I prefer when you search for your own subject." I was happy to accept.

In October 1989, it turned out that some money was left in the annual research budget of Rudolf Ahlswede's group at Bielefeld University. He asked me to think about useful ways to invest it. When I came up with a 3-Hirn experiment his first reaction was "You. Always with your Chess..". But then he accepted to invite International Chess Master Dr. Helmut Reefschläger for an experimental match which became a mile stone in the 3-Hirn development. Rudolf Ahlswede was always a man willing to look over and jump over fences, and he also encouraged others to jump over their fences. Thanks, thanks, thanks.

References

1. Abramson, B.: Expected-outcome: a general model of static evaluation. IEEE Transactions on Pattern Analysis and Machine Intelligence 12, 182–193 (1990)
2. Althöfer, I., Leader, I.: Correlation of Boolean functions and pathology in recursion trees. SIAM Journal of Discrete Mathematics 8, 526–535 (1995)

3. Althöfer, I.: Das Dreihirn - Entscheidungsteilung im Schach. Magazine "ComputerSchach und Spiele", 20–22 (December 1985) (text in German); translation of the title is "The Triple-Brain - decision sharing in Chess"
4. Althöfer, I.: Computer-aided game inventing. Technical report, FSU Jena, Fakultät Mathematik und Informatik (2003), http://www.minet.uni-jena.de/preprints/althoefer_03/CAGI.pdf
5. Althöfer, I., de Koning, J., Lieberum, J., Meyer-Kahlen, S., Rolle, T., Sameith, J.: Five visualisations of the k-best mode. ICGA Journal 12, 182–189 (2003), An extended version with examples from Go is available online at http://www.althofer.de/k-best-visualisations.html
6. Althöfer, I.: Improved game play by multiple computer hints. Theoretical Computer Science 313, 315–324 (2004)
7. Althöfer, I.: On games with random-turn order and Monte-Carlo perfectness. ICGA Journal 34, 179–190 (2011)
8. Althöfer, I.: On the histories of board game Havannah and computer Havannah. ICGA Journal 35 (2012)
9. Althöfer, I.: Shannon engine and Turing engine in an exhibition match. Submitted to ICGA Journal (2012), Preliminary version online available at http://www.althofer.de/shannon-turing-exhibition-match.pdf
10. Amazons at the Computer Olympiads, list of results online at http://www.grappa.univ-lille3.fr/icga/game.php?id=15
11. Baudis, P.: Balancing MCTS by dynamically adjusting komi value. ICGA Journal 34, 131–139 (2011)
12. Bleicher, E., Althöfer, I.: Retrograde analysis of the chess endgame with king plus queen versus king plus rook on mxn boards, results available online at http://www.althofer.de/chess-kq-kr.pdf
13. Berlekamp, E., Conway, J.H., Guy, R.: Winning ways for your mathematical play. Academic Press, New York (1982)
14. Bellman, R.: On the application of dynamic programming to the determination of optimal play in Chess and checkers. Proc. Nat. Academy of Sciences of the USA 53, 244–246 (1965)
15. Bleicher, E.: Analysis tool freezer for chess endgames, Commercially available since (2003), http://www.shredderchess.com/chess-program/freezer.html
16. Bleicher, E.: Building chess endgame databases for positions with many pieces using a-priori information. Technical Report, FSU Jena, Fakultaet Mathematik und Informatik (2004), http://www.minet.uni-jena.de/preprints/bleicher_04/FREEZER_.PDF
17. Browne, C.: Automated generation and evaluation of recombination games. Ph.D. Thesis (2008), http://www.cameronius.com/
18. Bruegmann, B.: Monte Carlo Go, Report, not offiially published (1993), http://www.althofer.de/Bruegmann-MonteCarloGo.pdf
19. Berlekamp, E., Wolfe, D.: Mathematical Go - Chilling Gets the Last Point. A.K. Peters/CRC Press (1997)
20. O. Cordel, Theorie und Praxis des Schachspiels, II. Band, A. Stein's Verlagsbuchhandlung, Potsdam, scan of page 302 (1913), http://www.althofer.de/cordel-p302.jpg
21. Ciura, M., Ewalds, T.: The Havannah prize match 2012 from programmer's perspective. ICGA Journal 35 (2012)
22. Coulom, R.: Efficient Selectivity and Backup Operators in Monte-Carlo Tree Search. In: van den Herik, H.J., Ciancarini, P., Donkers, H.H.L.M(J.) (eds.) CG 2006. LNCS, vol. 4630, pp. 72–83. Springer, Heidelberg (2007)

23. Donninger, C.: Null move and deep search: Selective-search heuristics for obtuse chess programs. ICCA Journal 16, 137–143 (1993)
24. Feist, M., Thompson, K.: The Turing engine (2007), description, engine and Turing Text from 1953 available online at http://www.chessbase.de/spotlight/spotlight2.asp?id=15
25. Feist, M.: The Shannon engine (2009), description and engine for download online at http://www.chessbase.de/nachrichten.asp?newsid=9711
26. Fischer, T.: Exakte Analyse von Heuristiken fuer kombinatorische Spiele. Doctoral dissertation, Fakultaet Mathematik und Informatik, FSU Jena (2011), http://www.althofer.de/dissertation_thomas-fischer.pdf
27. Fraenkel, A.S., Lichtenstein, D.: Computing a Perfect Strategy for nxn Chess Requires Time Exponential in N. In: Even, S., Kariv, O. (eds.) ICALP 1981. LNCS, vol. 115, pp. 278–293. Springer, Heidelberg (1981)
28. Gasser, R.: Harnessing computational resources for efficient exhaustive search. Doctoral dissertation, ETH Zurich (1994)
29. Gasser, R.: Solving Nine Men's Morris. In: Nowakowski, R.J. (ed.) Games of No Chance. Cambridge University Press (1998), http://library.msri.org/books/Book29/files/gasser.pdf
30. Garey, M.R., Johnson, D.S.: Computers and intractability - a guide to the theory of NP- completeness. Freeman (1979)
31. de Groot, A.: Het denken van den Schaker, een experimenteel-psychologische studie. Ph.D. thesis, University of Amsterdam (1946)
32. Game Havannah at the Computer Olympiads, List of results online at http://www.grappa.univ-lille3.fr/icga/game.php?id=37
33. Heinz, E.A.: Scalable Search in Computer Chess: Algorithmic Enhancements and Experiments at High Search Depths. Vieweg, Braunschweig (2000)
34. Game Hex at the Computer Olympiads, List of results online at http://www.grappa.univ-lille3.fr/icga/game.php?id=7
35. Hsu, F.-H.: Behind Deep Blue - Building the Computer that Defeated the World Chess Champion. Princeton University Press (2002)
36. Klein, F.: Handwritten seminar protocols for the period 1872-1912, scans available online at http://www.uni-math.gwdg.de/aufzeichnungen/klein-scans/
37. Klein, F.: Transcript of [36] for the winter semester 1909-1910, http://www.uni-bielefeld.de/idm/arge/klein29_cst.pdf
38. Kocsis, L., Szepesvári, C.: Bandit Based Monte-Carlo Planning. In: Fürnkranz, J., Scheffer, T., Spiliopoulou, M. (eds.) ECML 2006. LNCS (LNAI), vol. 4212, pp. 282–293. Springer, Heidelberg (2006)
39. Krabbenbos, J., van der Valk, T.: Report on the Havannah prize match 2012. ICGA Journal 35 (2012)
40. Lasker, E.: Kampf (Struggle) (2001) (reprint), available at Lasker Society http://www.lasker-gesellschaft.de/publikationen/emanuel-lasker-kampf/kampf.html (1907)
41. Lasker, E.: Brettspiele der Völker, Berlin (1931)
42. Leonard, R.: Von Neumann, Morgenstern and the creation of game theory: from chess to social science, 1900-1960. Cambridge University Press (2010)
43. Lichtenstein, D., Sipser, M.: Go is polynomial-space hard. Journal of the ACM 27, 393–401 (1980)
44. Nau, D.S.: An investigation of the causes of pathology in games. Artifical Intelligence 19, 257–278 (1982)
45. Pearl, J.: Heuristics - Intelligent Search Strategies for Computer Problem Solving. Adison-Wesley (1984)

46. Peres, Y., Schramm, O., Sheffield, S., Wilson, D.B.: Random-turn Hex and other selection games. American Mathematical Monthly 114, 373–387 (2007)
47. Randell, B.: From analytical engine to electronic digital computer: the contributions of Ludgate, Torres, and Bush. Annals of the History of Computing 4, 327–341 (1982)
48. Randell, B.: The Origins of Digital Computers, 3rd edn. Springer (1982)
49. Reisch, S.: Hex ist PSPACE-vollständig (Hex is PSPACE-complete). Acta Informatica 15, 167–191 (1981)
50. Robson, J.M.: The complexity of Go. In: Proceedings of the IFIP 9th World Computer Congress on Information Processing, pp. 413–417 (1983)
51. Robson, J.M.: Combinatorial games with exponential space complete decision problems. In: Proceedings of the Mathematical Foundations of Computer Science 1984, pp. 498–506. Springer, London (1984)
52. Schaeffer, J., Burch, N., Björnsson, Y., Kishimoto, A., Müller, M., Lake, R., Lu, P., Sutphen, S.: Checkers is solved. Science 317, 1518–1522 (2007)
53. Schaeffer, J.: One Jump Ahead - Challenging Human Supremacy in Checkers. Springer, New York (1997)
54. Schäfer, A.: Rock'n'Roll - a cross-platform engine for the board game 'EinStein würfelt nicht', `http://www.minet.uni-jena.de/preprints/althoefer_06/rockNroll.pdf`
55. Schreiber, L.: The generalized cordel property in discrete optimization. Doctoral dissertation, Fakultaet Mathematik und Informatik, FSU Jena (2012), `http://www.althofer.de/thesis-schreiber.pdf`
56. Schrüfer, G.: Presence and absence of pathology on game trees. In: Beal, D.F. (ed.) Advances in Computer Chess, vol. 4, pp. 101–112. Pergamon (1986)
57. Sedgewick, R.: Algorithms for the masses, `http://www.cs.princeton.edu/~rs/talks/AlgsMasses.pdf`
58. Shannon, C.E.: Programming a computer for playing chess. Philosophical Magazine, 7th series 41(314), 256–275 (1950)
59. Sheppard, B.: Towards perfect play of Scrabble. Doctoral dissertation, University of Maastricht (2002)
60. Stahlhacke, P.: Verification of Gasser's analysis of Nine Men's Morris. Personal Communication (1999)
61. Stahlhacke, P.: The game of Lasker Morris. Technical Report, FSU Jena (2003), `http://www.althofer.de/stahlhacke-lasker-morris-2003.pdf`
62. Ströhlein, T.: Untersuchungen kombinatorischer Spiele. Doctoral dissertation, TU Munich (1970)
63. Tarrasch, S.: Das Schachspiel, Deutsche Buchgemeinschaft (1931)
64. Turing, A.M.: Chess. In: Bowden, B.V. (ed.) Faster Than Thought, pp. 286–295. Pitman, London (1953)
65. Wedd, N.: List of human-computer Go exhibition games (since 1986), `http://www.computer-go.info/h-c/index.html`
66. Wiener, N.: Cybernetics, or control and communication in the animal and the machine. Wiley (1948)
67. Zermelo, E.: Beweis, da jede Menge wohlgeordnet werden kann. Mathematische Annalen 59, 514–516 (1904)
68. Zermelo, E.: Ueber eine Anwendung der Mengenlehre auf die Theorie des Schachspiels. In: Proceedings of the Fifth International Congress of Mathematicians, pp. 501–504. Cambridge University Press (1913)

Multiplied Complete Fix-Free Codes and Shiftings Regarding the 3/4-Conjecture

Michael Bodewig

Rheinisch-Westfälische Technische Hochschule Aachen,
Templergraben 55, 52062 Aachen, Germany
bodewig@math2.rwth-aachen.de

Dedicated to the memory of Rudolf Ahlswede

Abstract. Given a nonnegative sequence α of integers with Kraftsum at most 3/4, Ahlswede, Balkenhol and Khachatrian proposed the existence of a fix-free code with exactly α_n words for any length n. In this article complete thin fix-free codes are constructed and both so-called n-closed systems and multiplication are used to enlarge this class. In addition, a sufficient criterion is given in terms of elementary sequence-shifting preserving the fix-freedom of the associated code.

Keywords: 3/4-conjecture, fix-free code, complete code, Kraft sum.

1 Introduction

Being uniquely decipherable in spite of variable length, codes with words that are no prefixes of each other are an important subject in coding theory. In order to ensure a small average length by coding with these *prefix-free* codes, it was studied, for which sequences α there exists a prefix-free code C with exactly α_n words of length n for all n, shortly $C \sim \alpha$. Kraft's inequality solved this problem, proving that there exists a prefix-free $C \sim \alpha$ if and only if its *Kraftsum* $K(\alpha) := \sum_{i\in\mathbb{N}} \alpha_i \cdot 2^{-i}$ does not exceed 1, considering the binary case as in the whole article.

Sometimes it is desirable to decode from both sides simultaneously, a property fulfilled by *fix-free* codes which in addition to their prefix-freedom have no words that are suffixes of each other. Ahlswede, Balkenhol and Khachatrian proved in [1] that for each $\gamma > 3/4$ there exists a sequence α with $K(\alpha) < \gamma$ such that there is no fix-free code with length sequence α. Supporting their conjecture, they showed the existence of fix-free codes for special sequences with Kraftsum not exceeding 3/4. Following this, criteria for the existence of fix-free codes were given in [1–10], among these

$$K(\alpha) \leq 3/4 \quad , \quad 2\alpha_{l_{\min}(\alpha)} + \alpha_{l_{\min}(\alpha)+1} \geq 2^{l_{\min}(\alpha)} \quad \text{implying a fix-free } C \sim \alpha \quad (1)$$

and the sufficiency of the Kraftsum not exceeding 5/8, both proved by Yekhanin in [9] and [10] respectively.

H. Aydinian, F. Cicalese, and C. Deppe (Eds.): Ahlswede Festschrift, LNCS 7777, pp. 694–710, 2013.

A common strategy for proving the existence of a fix-free $C \sim \alpha$ is the following. Going successively through all lengths n, add α_n words of length n to the code C_{n-1} which has exactly α_i words of length i for all $i \leq n-1$, preserving the fix-freedom of the code. Adding the words of length n, one has to omit the words in $\Delta_{\mathrm{P}}^n(C_{n-1})$ and $\Delta_{\mathrm{S}}^n(C_{n-1})$, denoting the sets of n-words having a pre- and suffix in C_{n-1} respectively. Using Lemma 8 in [7] for the so-called *n-th bifix shadow* $\Delta_{\mathrm{B}}^n(C_{n-1}) := \Delta_{\mathrm{P}}^n(C_{n-1}) \cup \Delta_{\mathrm{S}}^n(C_{n-1})$ of a fix-free code C_{n-1}, we obtain

$$|\Delta_{\mathrm{B}}^n(C_{n-1})| = 2^{n+1} \cdot K(C_{n-1}) - \sum_{x,y \in C_{n-1}} |I_n(x,y)| \tag{2}$$

with $K(C_{n-1}) := K(\alpha')$ whenever $C_{n-1} \sim \alpha'$ and $I_n(x,y)$ denoting the set of n-words having x as a prefix and y as a suffix. The problem with the described strategy is that for an optimal C_{n-1} in the sense that $|\Delta_{\mathrm{B}}^n(C_{n-1})|$ is minimal, the code C_n, where we have added α_n words of length n to C_{n-1}, does not necessarily minimize $|\Delta_{\mathrm{B}}^{n+1}(\cdot)|$.

The beginning remarks in Section 2 reveal additional problems of the strategy discussed above, motivating a new approach using elementary shifting in the sequences preserving fix-freedom. A special case of this sufficient condition is proved and a new proof for Kraft's inequality deduced. Thin fix-free codes with Kraftsum 1, called *complete*, are constructed in Section 3. Using so-called n-closed systems, a method is presented for the construction of fix-free codes, not necessarily thin, with Kraftsum 1 from a given set of such codes. The results of multiplication with the constructed codes are presented in Section 4 and a discussion of prospective strategies is given in the last section.

2 Observations and Shifting

The two following remarks show that in general there is no satisfactory reduction from the $(n+1)$-th bifix shadow to the n-th bifix shadow and that the $(n+1)$-th bifix-shadow may contain all words of length $n+1$ even for small Kraftsum-values.

Remark 1. *For C fix-free and $n \geq l_{\max}(C)$ we have*

$$|\Delta_{\mathrm{B}}^n(C)| \leq |\Delta_{\mathrm{B}}^{n+1}(C)| \leq 4 \cdot |\Delta_{\mathrm{B}}^n(C)|$$

and the right inequality cannot be improved.

Proof. We have $v \in \Delta_{\mathrm{P}}^n(C)$ if and only if $v0, v1 \in \Delta_{\mathrm{P}}^{n+1}(C)$ and $w \in \Delta_{\mathrm{S}}^n(C)$ if and only if $0w, 1w \in \Delta_{\mathrm{S}}^{n+1}(C)$. This implies

$$|\Delta_{\mathrm{B}}^n(C)| \leq |\Delta_{\mathrm{P}}^n(C)| + |\Delta_{\mathrm{S}}^n(C)| = |\Delta_{\mathrm{P}}^{n+1}(C)| \leq |\Delta_{\mathrm{B}}^{n+1}(C)| \;,$$
$$|\Delta_{\mathrm{B}}^{n+1}(C)| \leq |\Delta_{\mathrm{P}}^{n+1}(C)| + |\Delta_{\mathrm{S}}^{n+1}(C)| = 2 \cdot (|\Delta_{\mathrm{P}}^n(C)| + |\Delta_{\mathrm{S}}^n(C)|) \leq 4 \cdot |\Delta_{\mathrm{B}}^n(C)| \;.$$

The example $|\Delta_{\mathrm{B}}^3(\{01\})| = 4 = 4 \cdot |\Delta_{\mathrm{B}}^2(\{01\})|$ shows that the right inequality cannot be improved. □

Regarding the left-hand side of Remark 1, the factor 1 cannot be raised to 18/13 or greater due to the example $C = \{001, 011, 1000, 1010, 1110\}$ for $n = 5$.

Let the *de Bruijn digraph* $B_2(n)$ be the digraph with all binary n-words as vertices and arcs (v, w) whenever $I_{n+1}(v, w) \neq \varnothing$. This adjacency condition will be used in the proof of the following remark.

Remark 2. *For any $\gamma > 1/2$ there exists a fix-free $C' \subseteq \{0,1\}^{l_{\max}(C')}$ with*

$$K(C') \leq \gamma \text{ and } \Delta_{\mathrm{B}}^{l_{\max}(C')+1}(C') = \{0,1\}^{l_{\max}(C')+1} ,$$

whereas $\Delta_{\mathrm{B}}^{l_{\max}(C)+1}(C) \neq \{0,1\}^{l_{\max}(C)+1}$ for any fix-free C with $K(C) \leq 1/2$.

Proof. For $n \in \mathbb{N}$ let W_n be a maximal loop-free independent set in $B_2(n)$ and let $\alpha^*(2, n) := |W_n|$ denote the loop-free independence number. Being a one-level code, $\{0,1\}^n \setminus W_n$ is fix-free such that (2) implies

$$\begin{aligned}
&|\Delta_{\mathrm{B}}^{n+1}(\{0,1\}^n \setminus W_n)| \\
&= 2^{n+2} \cdot K(\{0,1\}^n \setminus W_n) - \sum_{x,y \in \{0,1\}^n \setminus W_n} |I_{n+1}(x,y)| \\
&= 2^{n+2} \cdot (2^n - \alpha^*(2,n)) \cdot 2^{-n} - \sum_{x,y \in \{0,1\}^n} |I_{n+1}(x,y)| \\
&\quad + \sum_{x \in W_n, y \in \{0,1\}^n} (|I_{n+1}(x,y)| + |I_{n+1}(y,x)|) - \sum_{x,y \in W_n} |I_{n+1}(x,y)| \\
&= 2^{n+2} - 4\alpha^*(2,n) - 2^{n+1} + 4 \cdot |W_n| - 0 \\
&= 2^{n+1} .
\end{aligned}$$

Within the second to last step we used (2) with $|\Delta_{\mathrm{B}}^{n+1}(\{0,1\}^n)| = 2^{n+1}$ for the first sum,

$$\bigcup_{y \in \{0,1\}^n} I_{n+1}(x,y) = \{x0, x1\}$$

and $\bigcup_{y \in \{0,1\}^n} I_{n+1}(y,x) = \{0x, 1x\}$ for any $x \in W_n$ for the second sum and for the third sum, that W_n is a loop-free independent set in $B_2(n)$. For the calculation of the Kraftsum note that the independence number $\alpha(2, n)$ of $B_2(n)$ exceeds $\alpha^*(2, n)$ at most by two, since 0^n and 1^n have the only loops in $B_2(n)$:

$$\begin{aligned}
\lim_{n \to \infty} K(\{0,1\}^n \setminus W_n) &= 1 - \lim_{n \to \infty} \alpha^*(2,n) \cdot 2^{-n} \\
&\leq 1 - \lim_{n \to \infty} \alpha(2,n) \cdot 2^{-n} + \lim_{n \to \infty} 2^{-n+1} \\
&= \frac{1}{2} ,
\end{aligned}$$

using Theorem 3 in [1] to calculate the limit involving $\alpha(2, n)$. Hence, the definition $C' := \{0,1\}^n \setminus W_n$ with n large enough is sufficient.

Considering C fix-free with $K(C) < 1/2$ for the second statement, we obtain

$$|\Delta_{\mathrm{B}}^{l_{\max}(C)+1}(C)| \leq 2^{l_{\max}(C)+2} \cdot K(C) < 2^{l_{\max}(C)+2} \cdot 2^{-1} = 2^{l_{\max}(C)+1} \ .$$

Assuming $|\Delta_{\mathrm{B}}^{l+1}(C)| = 2^{l+1}$ in the case $K(C) = 1/2$ with $l := l_{\max}(C)$ leads to

$$2^{l+1} = |\Delta_{\mathrm{B}}^{l+1}(C)| = 2^{l+2} \cdot 2^{-1} - |\Delta_{\mathrm{P}}^{l+1}(C) \cap \Delta_{\mathrm{S}}^{l+1}(C)|$$

and hence $\{0,1\}^{l+1}$ would be a disjoint union of $\Delta_{\mathrm{P}}^{l+1}(C)$ and $\Delta_{\mathrm{S}}^{l+1}(C)$. Noticing that $0^{l+1} \in \Delta_{\mathrm{P}}^{l+1}(C)$ if and only if $0^{l+1} \in \Delta_{\mathrm{S}}^{l+1}(C)$, we derive a contradiction. □

Having faced some difficulties regarding the successive adding of words to a fix-free code, different approaches for proving the existence of fix-free codes will be shown. While constructive methods are presented in Section 3 and Section 4, the next remark provides a theoretical condition in terms of shifting.

Remark 3. *The 3/4-Conjecture is true, if the existence of a finite fix-free $C \sim \alpha$ with $K(\alpha) \leq 3/4$ implies the existence of fix-free $C^{(k)} \sim \alpha^{(k)}$ for all $k \in \mathbb{N}$, with*

$$\alpha^{(k)}_{l_{\max}(\alpha)} = \alpha_{l_{\max}(\alpha)} - k \geq 0 \ , \ \alpha^{(k)}_{l_{\max}(\alpha)+1} = 2k \ \text{ and } \ \alpha^{(k)}_i = \alpha_i \ \text{ otherwise.}$$

The statement remains true for prefix-free codes with the Kraftsum raised to 1.

Proof. Due to the 2-adic representation of the real numbers we can put without loss of generality $K(\alpha') = 3/4$ and $K(\alpha') = 1$ respectively. Considering the starting sequence $(1, 1, 0, \ldots) \sim \{0, 11\}$ and accordingly $(2, 0, \ldots) \sim \{0, 1\}$, any sequence α' with Kraftsum 3/4 and 1 respectively can be reached from this starting sequence by successive replacements in the following way. Going through any length of the starting sequence, test whether the entry compared to α' is too big or not, noting that it cannot be too small due to its Kraftsum. If the entry of the starting sequence is too big, decrease it by the necessary value k to equalize the entries and increase the next entry by $2k$, otherwise change nothing and move on to the next length. □

The advantage of the shifting-strategy is that the manipulation does not involve a raise of the Kraftsum. Note that the replacement of k words of length l_{max} by $2k$ words of length $l_{max} + 1$ is sufficient but not necessary for the shifting. In the following let $|w|$ denote the number of letters of w.

Theorem 1. *1. For any fix-free C with $K(C) \leq 2/3$ and $n > l_{\max}(C)$ there exists $w \in C$ replaceable by $2^{n-|w|}$ n-words in C so that C remains fix-free.*
2. For any prefix-free code C and $U \subseteq C$ the code

$$(C \setminus U) \cup \{uz \,|\, u \in U, \ z \in \{0,1\}\}$$

is prefix-free. Using Remark 3, this provides a new proof of Kraft's inequality.

Proof. 1. Defining l as the maximal number with

$$|\Delta_{\mathrm{B}}^n(C \setminus \{w\})| \geq |\Delta_{\mathrm{B}}^n(C)| - 2^{n-|w|} + l$$

for all $w \in C$, we consider $l \geq 1$ without loss of generality, since the claim otherwise would be proved already. For all $w \in C$ we have

$$\begin{aligned}
&2^{n-|w|} + l \\
&\leq |\Delta_{\mathrm{B}}^n(C \setminus \{w\})| - |\Delta_{\mathrm{B}}^n(C)| + 2^{n-|w|+1} \\
&\overset{(2)}{=} 2^{n-|w|+1} - 2^{n+1} \cdot K(\{w\}) - \sum_{x,y \in C \setminus \{w\}} |I_n(x,y)| + \sum_{x,y \in C} |I_n(x,y)| \\
&= -|I_n(w,w)| + \sum_{x \in C} (|I_n(x,w)| + |I_n(w,x)|) \\
&=: \psi(w) .
\end{aligned}$$

Choosing $\widetilde{w} \in C$ with $\psi(\widetilde{w}) = 2^{n-|\widetilde{w}|} + l$, we obtain

$$\begin{aligned}
&|\Delta_{\mathrm{B}}^n(C \setminus \{\widetilde{w}\})| \\
\overset{(2)}{=}\ & 2^{n+1} \cdot K(C) - 2^{n-|\widetilde{w}|+1} - \sum_{x,y \in C \setminus \{\widetilde{w}\}} |I_n(x,y)| \\
=\ & 2^{n+1} \cdot K(C) - 2^{n-|\widetilde{w}|+1} + \psi(\widetilde{w}) - \sum_{x,y \in C} |I_n(x,y)| \\
=\ & 2^{n+1} \cdot K(C) - 2^{n-|\widetilde{w}|+1} + 2^{n-|\widetilde{w}|} + l - 2^{-1} \cdot \sum_{y \in C} (\psi(y) + |I_n(y,y)|) \\
\leq\ & 2^{n+1} \cdot K(C) - 2^{n-|\widetilde{w}|} + l - 2^{-1} \cdot \sum_{y \in C} (2^{n-|y|} + l) \\
=\ & 2^{n+1} \cdot K(C) - 2^{n-|\widetilde{w}|} + l - 2^{n-1} \cdot K(C) - 2^{-1} \cdot |C| \cdot l \\
=\ & 3 \cdot 2^{n-1} \cdot K(C) - 2^{n-|\widetilde{w}|} - l \cdot (2^{-1} \cdot |C| - 1) \\
\overset{|C| \geq 2}{\leq}\ & 2^n - 2^{n-|\widetilde{w}|}
\end{aligned}$$

since $K(C) \leq 2/3$ and $l \geq 1$. Notice that the statement is trivial for $|C| = 1$.

2. Let $u \in U \subseteq C$. The words $u0$ and $u1$ have no prefixes of length smaller than $|u0|$ in $C \backslash U$, since these would have been prefixes of u in C. Moreover, $u0$ and $u1$ cannot be prefixes of another word in $C \backslash U$, since this word would have had u as a prefix in the original code. Moreover, the set $\{uz \mid u \in U,\ z \in \{0,1\}\}$ is prefix-free due to the prefix-freedom of U.

□

Comparing Theorem 1 to the preliminaries in Remark 3, we notice that the Kraftsum is 2/3 instead of 3/4, the shift is made from an unknown level $i \leq l_{\max}(\alpha)$ with $\alpha_i > 0$ instead of the determined level $l_{\max}(\alpha)$ and that we have $k = 1$. However, we have the advantage that the shift can be made not only onto the level $l_{\max}(\alpha) + 1$, but onto an arbitrary level larger than $l_{\max}(\alpha)$.

3 Complete Codes and n-Closed Systems

Since finding any fix-free code can be reduced to finding all maximal fix-free codes, it is important to consider these codes in view of the 3/4-Conjecture.

In [2] the class of maximal *thin* fix-free codes is studied extensively. Denoting the binary words including the empty word with $\{0,1\}^*$, a code $C \subseteq \{0,1\}^*$ is called thin, if there exists a word $w \in \{0,1\}^*$ with $\{xwy \mid x, y \in \{0,1\}^*\} \cap C = \varnothing$. The two basic parameters of a maximal thin fix-free code C are its *degree* and *kernel*. The degree is defined by

$$d(C) := \max\{L_C(w) \mid w \in \{0,1\}^*\}$$

where $L_C(w)$ is the number of prefixes of w, including the empty word, having no suffixes in C. The kernel is the set of all codewords that are also *internal factors* of the code, these are words where at least one beginning and one ending letter of some codeword are left out. Berstel and Perrin note that any maximal thin fix-free code is uniquely determined by the combination of its kernel and degree and that for a given degree there can be only finitely many maximal thin fix-free codes, all of them constructible by so-called *internal transformation*.

Since this recursive method does not deliver an overview over the constructed codes and sequences fitting them, a large class of complete thin fix-free codes and their sequences will be presented in Theorem 2. According to Kraft's inequality on the one hand and Proposition 2.1 in Chapter III and Theorem 5.10 in Chapter I of [2], among the thin fix-free codes the complete ones are exactly the maximal ones. The results within this and the next section can be viewed as a contribution to the 3/4-Conjecture in the sense that leaving out words in a complete thin fix-free code delivers many fix-free codes with Kraftsum at most 3/4.

Definition 1. *Let $m, k \in \mathbb{N}_0$ and $i \in \mathbb{N}$ with $2^{i-1} \geq k$. We define the sequences $\alpha^{(m,k,i)}$ and $\beta^{(m,k,i)}$ with $p := i + m - \lfloor \lim_{n \to k} \log_2(n) \rfloor \in \mathbb{N} \cup \{\infty\}$ by*

$$\alpha_l^{(m,k,i)} := \begin{cases} 0, & l < i+m \text{ or } l > i+p \ , \\ 2^{i-1} - k, & l = i+m \ , \\ 2^{i+m+1} - 3 \cdot 2^{i-1} + 2k, & l = i+m+1 < i+p-1 \ , \\ 2^{i+m+1} - 2^{i+1} + 4k, & l = i+m+1 \geq i+p-1 \ , \\ 2^{i-1}, & i+m+2 \leq l \leq i+p-2 \ , \\ 2^{p-m-1}k, & l = i+p-1 > i+m+1 \ , \\ 2^{i+1} - 2^{p-m}k, & l = i+p > i+m+1 \end{cases}$$

and

$$\beta_{i+m+1}^{(m,k,i)} = 2 \cdot \alpha_{i+m+1}^{(m,k,i)} - 2^{i+m+1} \ , \quad \beta_l^{(m,k,i)} = 2 \cdot \alpha_l^{(m,k,i)} \text{ for } l \neq i+m+1 \ .$$

Moreover, define the code

$$C_k^m(i) := \bigcup_{1 \leq r \leq 5, r \neq 3} K_r \cup \bigcup_{m+1 \leq j < p-1} K_{3,j} \ ,$$

where for fixed m, k, i, *for* $m+1 \leq j < p-1$ *and* $bi(w) := \sum_{i=0}^{|w|-1} 2^i \cdot w_{|w|-i}$, *denoting the binary value of* $w \in \{0,1\}^*$, *we put*

$$\begin{array}{lll}
K_1 & := \{w \in \{0,1\}^{i+m} & \mid k \leq bi(w) < 2^{i-1}\} \ , \\
K_2 & := \{w \in \{0,1\}^{i+m+1} & \mid bi(w) < k \text{ or } 2^i \leq bi(w) < 2^{i+m} + k \\
 & & \quad \text{or } 2^{i+m} + 2^{i-1} \leq bi(w)\} \ , \\
K_{3,j} & := \{w \in \{0,1\}^{i+j+1} & \mid 2^{i+j} + 2^{i-1} \leq bi(w) < 2^{i+j} + 2^i\} \ , \\
K_4 & := \{w \in \{0,1\}^{i+p-1} & \mid 2^{i-1} \leq bi(w) < 2^{p-m-1}k\} \ , \\
K_5 & := \{w \in \{0,1\}^{i+p} & \mid 2^{p-m-1}k \leq bi(w) < 2^i \\
 & & \quad \text{or } 2^{p-m-1}k + 2^{i+p-1} \leq bi(w) < 2^i + 2^{i+p-1}\} \ .
\end{array}$$

The connection between the codes and sequences just defined is revealed by the following theorem. Some special cases which will be mentioned in Corollary 1 and exemplary illustrations will be given both in Figure 1 and Figure 2.

Theorem 2. *Let* $m, k \in \mathbb{N}_0$ *and* $i \in \mathbb{N}$ *with* $2^{i-1} \geq k$. *Then* $C_k^m(i)$ *is a complete thin fix-free code which satisfies*

$$C_k^m(i) \sim \alpha^{(m,k,i)} \quad \text{and} \quad d\left(C_k^m(i)\right) = i + m + 1 \ .$$

Proof. Throughout this proof let

$$\begin{array}{rl}
w^{(1)} \in K_1 & \text{and} \quad w^{(4)} \in K_4 \ , \\
w^{(3,j)} \in K_{3,j} & \text{for} \quad m+1 \leq j < p-1 \ , \\
w^{(2,1)} \in K_2 & \text{with} \quad bi(w) < k \ , \\
w^{(2,2)} \in K_2 & \text{with} \quad 2^i \leq bi(w) < 2^{i+m} + k \ , \\
w^{(2,3)} \in K_2 & \text{with} \quad 2^{i+m} + 2^{i-1} \leq bi(w) \ , \\
w^{(5,1)} \in K_5 & \text{with} \quad 2^{p-m-1}k \leq bi(w) < 2^i \ , \\
w^{(5,2)} \in K_5 & \text{with} \quad 2^{p-m-1}k + 2^{i+p-1} \leq bi(w) < 2^i + 2^{i+p-1} \ .
\end{array}$$

In order to prove prefix-freedom, for a binary word w of length at most a let ${}^{P(a)}w$ denote an arbitrary child of w with length a. Using

$$2^{a-|w|} \cdot bi(w) \leq bi({}^{P(a)}w) < 2^{a-|w|} \cdot (bi(w) + 1)$$

for all $w \in \{0,1\}^*$, for $k > 0$ we get the following chain of inequalities for all $m+1 \leq j < p-1$, implying prefix-freedom of different $K_{3,j}$ and $C_k^m(i)$ overall:

$$\begin{array}{llll}
bi({}^{P(i+p)}w^{(2,1)}) & < 2^{p-m-1}k & \leq bi(w^{(5,1)}) & < 2^i \\
\leq bi({}^{P(i+p)}w^{(4)}) & < 2^{p-m}k & \leq bi({}^{P(i+p)}w^{(1)}) & < 2^{i+p-m-1} \\
\leq bi({}^{P(i+p)}w^{(2,2)}) & < 2^{i+p-1} + 2^{p-m-1}k & \leq bi(w^{(5,2)}) & < 2^{i+p-j-2} + 2^{i+p-1} \\
\leq bi({}^{P(i+p)}w^{(3,j)}) & < 2^{i+p-j-1} + 2^{i+p-1} & \leq bi({}^{P(i+p)}w^{(2,3)}) & .
\end{array}$$

For $k = 0$ and arbitrary $a \geq i+m+2$ instead of $i+p$, we obtain the subsequent correlation for all $m+1 \leq j \leq a-i-1$, implying prefix-freedom in this case:

$$\begin{array}{lll}
bi({}^{P(a)}w^{(1)}) & < 2^{a-m-1} & \leq bi({}^{P(a)}w^{(2,2)}) < 2^{a-1} + 2^{a-j-2} \\
\leq bi({}^{P(a)}w^{(3,j)}) & < 2^{a-1} + 2^{a-j-1} & \leq bi({}^{P(a)}w^{(2,3)}) \ .
\end{array}$$

For the proof of suffix-freedom let ${}_{S(a)}v$ denote the a-suffix of a word v with length at least a. For $k > 0$ and $w^{(5)} \in K_5$ we have $bi(w^{(4)}), bi({}_{S(i+p-1)}w^{(5)}) < 2^i$ implying the first $p-1$ letters of $w^{(4)}$ and ${}_{S(i+p-1)}w^{(5)}$ to be zero, whereas $K_4 = K_5 = \varnothing$ for $k = 0$. In order to obtain the $(i+m+1)$-suffixes of $w^{(4)}$ and ${}_{S(i+p-1)}w^{(5)}$ respectively, their first $(i+p-1)-(i+m+1) = p-m-2 < p-1$ letters are canceled and since these are zeros only, we obtain

$$bi({}_{S(i+m+1)}w^{(4)}) = bi(w^{(4)}) \quad \text{and} \quad bi({}_{S(i+m+1)}w^{(5)}) = bi({}_{S(i+p-1)}w^{(5)}) \ .$$

With these calculations we get the following chain of inequalities proving the suffix-freedom of $K_1 \cup K_2 \cup K_4 \cup K_5$, where $K_4 = K_5 = \varnothing$ for $k = 0$:

$$\begin{array}{llll} bi(w^{(2,1)}) & < k & \leq bi(0w^{(1)}) & < 2^{i-1} \\ \leq bi({}_{S(i+m+1)}w^{(4)}) & < 2^{p-m-1}k & \leq bi({}_{S(i+m+1)}w^{(5)}) & < 2^i \\ \leq bi(w^{(2,2)}) & < 2^{i+m}+k & \leq bi(1w^{(1)}) & < 2^{i+m}+2^{i-1} \\ \leq bi(w^{(2,3)}) \ . & & & \end{array}$$

Due to the binary value of the $K_{3,j}$-words, their first and $(j+2)$-th letter are ones where only zeros lie in between. On the one hand this implies that there cannot be any suffix-forbiddances within $\bigcup_{m+1\leq j<p-1} K_{3,j}$ due to the leading one and on the other hand we have

$$2^{i-1} \leq bi({}_{S(i+m+1)}w^{(3,j)}) < 2^i \quad \text{for all} \quad m+1 \leq j < p-1 \ .$$

Therefore, $w^{(3,j)}$ has no suffix in $K_1 \cup K_2$ and it is not a suffix in $K_4 \cup K_5$ since

$$bi(w^{(4)}), bi({}_{S(i+p-1)}w^{(5)}) < 2^i < bi(w^{(3,j)}) \ .$$

For the proof of $C_k^m(i) \sim \alpha^{(m,k,i)}$ note that $K_{3,p-2}$ and K_4 are disjoint due to different first letters and that for all $m+1 \leq j < p-2$ we have

$$\begin{aligned} |K_1| &= 2^{i-1} - k \ , \\ |K_2| &= k + 2^{i+m} + k - 2^i + 2^{i+m+1} - (2^{i+m} + 2^{i-1}) \\ &= 2^{i+m+1} - 3 \cdot 2^{i-1} + 2k \ , \\ |K_{3,j}| &= 2^{i+j} + 2^i - (2^{i+j} + 2^{i-1}) = 2^{i-1} \ , \\ |K_{3,p-2} \cup K_4| &\overset{k\neq 0}{=} 2^{i-1} + 2^{p-m-1}k - 2^{i-1} = 2^{p-m-1}k \ , \\ |K_5| &\overset{k\neq 0}{=} 2 \cdot (2^i - 2^{p-m-1}k) = 2^{i+1} - 2^{p-m}k \ . \end{aligned}$$

This implies the claim for $k < 2^{i-2}$ which is equivalent with $i+m+1 < i+p-1$. For the case $2^{i-2} \leq k < 2^{i-1}$, implying $i+m+1 = i+p-1$, note that K_2 and K_4 are disjoint and that

$$\begin{aligned} |C_k^m(i) \cap \{0,1\}^{i+m+1}| &= |K_2| + |K_4| \\ &= 2^{i+m+1} - 3 \cdot 2^{i-1} + 2k + 2^{p-m-1}k - 2^{i-1} \\ &= 2^{i+m+1} - 2^{i+1} + 4k \ . \end{aligned}$$

Within the case $k = 2^{i-1}$, which is equivalent with $i + m + 1 = i + p$, we have $K_2 \cup K_5 = \{0,1\}^{i+m+1}$ and therefore this case is also proved.

For $k < 2^{i-2}$ we show $K(C_k^m(i)) = 1$ by using the following chain of equality:

$$\begin{aligned} K(C_k^m(i)) &= \frac{\alpha_{i+m}^{(m,k,i)}}{2^{i+m}} + \frac{\alpha_{i+m+1}^{(m,k,i)}}{2^{i+m+1}} + \frac{\alpha_{i+p-1}^{(m,k,i)}}{2^{i+p-1}} + \frac{\alpha_{i+p}^{(m,k,i)}}{2^{i+p}} + \sum_{l=i+m+2}^{i+p-2} \frac{\alpha_l^{(m,k,i)}}{2^l} \\ &= (2^{-m-1} - 2^{-i-m}k) + (1 + 2^{-i-m}k - 3 \cdot 2^{-m-2}) + 2^{-i-m}k \\ &\quad +(2^{-p+1} - 2^{-i-m}k) + 2^i \cdot (1 - (1/2)^{i+p-1} - (1 - (1/2)^{i+m+2})) \\ &= 1 \,. \end{aligned}$$

Moreover, we have $K(C_{2^{i-1}}^m(i)) = K(\{0,1\}^{i+m+1}) = 1$ and within the case $2^{i-2} \leq k < 2^{i-1}$ it holds

$$\begin{aligned} K(C_k^m(i)) &= K(K_1) + K(K_2 \cup K_4) + K(K_5) \\ &= (2^{-m-1} - 2^{-i-m}k) + (1 - 2^{-m} + 2^{-i-m+1}k) + (2^{-p+1} - 2^{-i-m}k) \\ &= 1 \,. \end{aligned}$$

For $k > 0$ the finite code $C_k^m(i)$ is thin. Moreover, we have $v1^{i+m+2}w \notin C_0^m(i)$ for arbitrary binary words v and w, since the $K_{3,j}$-words are the only ones in $C_0^m(i)$ with length possibly larger than $i + m + 1$, but these have j zeros from position 2 to $j + 1$ and therefore contain maximally $(i + j + 1) - j = i + 1 < i + m + 2$ ones.

Since we have $0^{i+m+1} \in K_2 \subseteq C_k^m(i)$ for $k > 0$, the statement regarding the degree follows from Proposition 5.1 in Chapter III of [2] in this case. In the case $k = 0$, according to Theorem 3.1 in Chapter III of [2] we only have to calculate the number of prefixes without suffix in $C_0^m(i)$ for an arbitrary word which is not an internal factor of $C_0^m(i)$. As we have seen in the proof of thinness, 1^{i+m+2} is not an internal factor and the prefixes 1^j for $0 \leq j \leq i + m$ have no suffix in $C_0^m(i)$, in contrast to 1^{i+m+1} and 1^{i+m+2}, hence we have proved $d(C_k^m(i)) = i + m + 1$. □

Corollary 1. *The following choices of m, k and i are covered by Theorem 2:*

1. *The codes* $C_{2^{i-1}}^m(i) = \{0,1\}^{i+m+1}$ *for* $m \geq 0$ *and* $i \in \mathbb{N}$.
2. *Setting* $K\left(\alpha_j^{(0,2^{i-2},i)}\right) := K\left((0,\ldots,0,\alpha_j^{(0,2^{i-2},i)},0,\ldots)\right)$, $j \in \mathbb{N}$, $i \geq 2$:
$$K\left(\alpha_i^{(0,2^{i-2},i)}\right) = K\left(\alpha_{i+2}^{(0,2^{i-2},i)}\right) = \frac{1}{4} = \frac{1}{2} \cdot K\left(\alpha_{i+1}^{(0,2^{i-2},i)}\right) .$$
3. *The codes* $C_0^0(i) \sim \alpha^{(0,0,i)}$ *with* $\alpha_l^{(0,0,i)} = 2^{i-1}$ *for all* $l \geq i = l_{\min}(\alpha^{(0,0,i)})$.

Note that $C_k^m(i) \sim \alpha^{(m,k,i)}$ implies existence of a fix-free $C \sim \alpha$ for all α not exceeding the values of $\alpha^{(m,k,i)}$. The upper bound $2^{l_{\min}(\alpha)-1}$ estimated in Corollary 1 is the best possible uniform upper bound for a given minimal length $l_{\min}(\alpha)$ of an arbitrary sequence α. Moreover, this upper bound is an

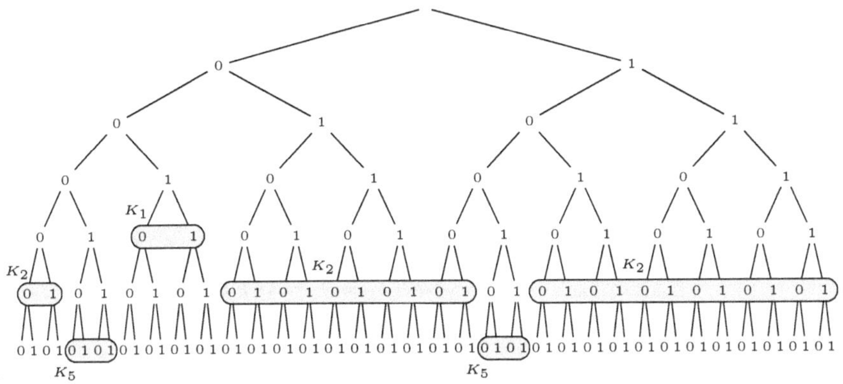

Fig. 1. The code $C_2^1(3)$

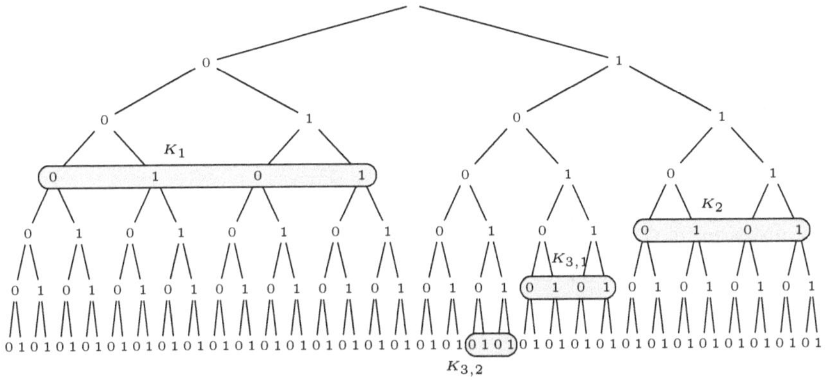

Fig. 2. The code $C_0^0(3)$ without $K_{3,j}$ for $j \geq 3$

improvement of the one given in Theorem II.4 of [6], apart from the number of words of maximal length.

The following definition will later be used as a useful instrument for the construction of new fix-free codes from a given class.

Definition 2. *For $T \subseteq \{0,1\}^*$ and $n \in \mathbb{N}$ define*

$$P_T^n := \left\{ w \in \{0,1\}^n \,\middle|\, w \in \Delta_{\mathrm{P}}^n(T) \text{ or } \exists t \in T : t \in \Delta_{\mathrm{P}}^{|t|}(\{w\}) \right\} ,$$

$$S_T^n := \left\{ w \in \{0,1\}^n \,\middle|\, w \in \Delta_{\mathrm{S}}^n(T) \text{ or } \exists t \in T : t \in \Delta_{\mathrm{S}}^{|t|}(\{w\}) \right\}$$

and call T n-closed system, if T is fix-free, $S_T^n \subseteq P_T^n$ and $K(P_T^n) \leq K(T)$.

Lemma 1. *1. If T is a n-closed system, then $T \cup (\{0,1\}^n \setminus P_T^n)$ is a fix-free code with Kraftsum 1. Vice versa there arises a n-closed system from a fix-free code with Kraftsum 1 by leaving out all n-words.*

2. *Let T be a n-closed system and $\overline{T}$ the set of all words in T where all ones and zeros are interchanged. If $T \cup \overline{T}$ is fix-free, then it is a n-closed system.*

Proof. 1. If T is a n-closed system, then $T \cup (\{0,1\}^n \setminus P_T^n) \subseteq T \cup (\{0,1\}^n \setminus S_T^n)$ is fix-free by the definition of P_T^n and S_T^n. Using that T and $\{0,1\}^n \setminus P_T^n$ are disjoint due to $T \cap \{0,1\}^n \subseteq P_T^n$, we get

$$\begin{aligned} K(T \cup (\{0,1\}^n \setminus P_T^n)) &= K(T) + K(\{0,1\}^n) - K(P_T^n) \\ &\geq K(T) + 1 - K(T) \\ &= 1 . \end{aligned}$$

We show that leaving out all words of an arbitrary length n from a fix-free code C with $K(C) = 1$ delivers a n-closed system T_n. As a subset of a fix-free code, T_n is fix-free. Since C is fix-free, $P_{T_n}^n \cup S_{T_n}^n = P_{C\setminus\{0,1\}^n}^n \cup S_{C\setminus\{0,1\}^n}^n$ and $C \setminus T_n = C \cap \{0,1\}^n$ are disjoint subsets of $\{0,1\}^n$. This implies the inequality $K(C \setminus T_n) + K(P_{T_n}^n \cup S_{T_n}^n) \leq 1$ and using $K(C) = 1$, we obtain $K(T_n) \geq K(P_{T_n}^n \cup S_{T_n}^n)$. The last inequality together with $K(P_{T_n}^n) \geq K(T_n)$ by the definition of $P_{T_n}^n$, implies $K(P_{T_n}^n) \leq K(T_n)$ and $S_{T_n}^n \subseteq P_{T_n}^n$.

2. We have $w \in S_{T\cup\overline{T}}^n$ if and only if w is or has a suffix in T or in $\overline{T}$ and hence

$$S_{T\cup\overline{T}}^n = S_T^n \cup S_{\overline{T}}^n .$$

In addition, $w \in S_{\overline{T}}^n$ means that w is or has a suffix in $\overline{T}$. This is equivalent to $\overline{w}$ being or having a suffix in T, which means $\overline{w} \in S_T^n$ and therefore $w \in \overline{S_T^n}$. Since the argumentation regarding the prefixes is the same, we get

$$S_{T\cup\overline{T}}^n = S_T^n \cup S_{\overline{T}}^n = S_T^n \cup \overline{S_T^n} \subseteq P_T^n \cup \overline{P_T^n} = P_T^n \cup P_{\overline{T}}^n = P_{T\cup\overline{T}}^n .$$

For $R := T \cup \overline{T}$ we get

$$\begin{aligned} K(R) &= K(\{t \in R \,|\, |t| < n\}) + K(\{t \in R \,|\, |t| = n\}) + K(\{t \in R \,|\, |t| > n\}) \\ &\overset{(*)}{\geq} K(\{p \in P_R^n \,|\, \exists x \in R \setminus P_R^n : p \in \Delta_{\mathrm{P}}^n(\{x\})\}) \\ &\quad + K(R \cap \{0,1\}^n) + K\left(\left\{p \in P_R^n \,|\, \exists x \in R \setminus P_R^n : x \in \Delta_{\mathrm{P}}^{|x|}(\{p\})\right\}\right) \\ &= K(P_R^n) , \end{aligned}$$

the last step following from the sets being disjoint due to the prefix-freedom of R. In the following we prove $(*)$. It is obvious that

$$K(\{t \in R \,|\, |t| < n\}) = K(\{p \in P_R^n \,|\, \exists x \in R \setminus P_R^n : p \in \Delta_{\mathrm{P}}^n(\{x\})\}) ,$$

since any x with the above property has $2^{n-|x|}$ children in P_R^n. Moreover, $K(\{t \in R \,|\, |t| = n\}) = K(R \cap \{0,1\}^n)$ is obvious. For the remaining inequality consider $p_0 \in P_R^n$ and $x_0 \in R \setminus P_R^n$ such that p_0 is a prefix of x_0. For $x_0 \in T$ we have $p_0 \in P_T^n$. Since T is a n-closed system, we have $K(P_T^n) \leq K(T)$ and therefore all the children of p_0 with length $|x_0|$ must be in T. The other case $x_0 \in \overline{T}$ is treated the same way so that all children of the elements in $p \in P_R^n$, p having at least one child in R, lie in R. □

Corollary 2. *For $k \in \mathbb{N}_0$ and $m, i \in \mathbb{N}$ with $2^{i-1} \geq k$ there exists a complete thin fix-free code B which satisfies*

$$B \sim \beta^{(m,k,i)} \quad \text{and} \quad d(B) = i + m + 1 \ .$$

Proof. We use the notation from the proof of Theorem 2. For $k = 2^{i-1}$ we have $\alpha^{(m,k,i)} = \beta^{(m,k,i)}$, so without loss of generality we put $k < 2^{i-1}$ from now on. According to Theorem 2 the code $C_k^m(i)$ is complete and fix-free, therefore

$$D_k^m(i) := C_k^m(i) \setminus \{0,1\}^{i+m+1} = K_1 \cup \bigcup_{m+1 \leq j < p-1} K_{3,j} \cup M \cup K_5$$

is an $(i+m+1)$-closed system due to Lemma 1, where $M = \varnothing$ if $k \geq 2^{i-2}$ and $M = K_4$ otherwise. In order to prove that $D := D_k^m(i) \cup \overline{D_k^m(i)}$ is an $(i+m+1)$-closed system, according to Lemma 1 it is only left to prove that D is fix-free.

Theorem 2 implies that $D_k^m(i)$ and $\overline{D_k^m(i)}$ are fix-free. Since $w \in D_k^m(i)$ is a prefix in $\overline{D_k^m(i)}$ if and only if $\overline{w} \in \overline{D_k^m(i)}$ is a prefix in $D_k^m(i)$, it suffices to consider the left half of the binary tree. Using $bi(\overline{w}) = 2^{|w|} - bi(w) - 1$ and the binary values from the proof of Theorem 2, we obtain the following chain of inequalities for $k > 0$, proving prefix-freedom with respect to $m > 0$:

$$\begin{aligned} & bi(w^{(5,1)}) < 2^i \leq bi({}^{P(i+p)}w^{(4)}) < 2^{p-m}k \\ \leq\ & bi({}^{P(i+p)}w^{(1)}) < 2^{i+p-1} - 2^{i+p-j-1} \leq bi({}^{P(i+p)}\overline{w^{(3,j)}}) < 2^{i+p-1} - 2^{i+p-j-2} \\ \leq\ & bi(\overline{w^{(5,2)}}) \ . \end{aligned}$$

Replacing the length $i+p$ by an arbitrary length $a \geq i+m+2$ yields the same argument for $k = 0$.

In order to prove suffix-freedom, we again use $bi(\overline{w}) = 2^{|w|} - bi(w) - 1$ and the binary values from the proof of Theorem 2 in order to obtain

$$\begin{aligned} & bi({}_{S(i+m+1)}w^{(4)}), bi({}_{S(i+m+1)}w^{(5)}), bi({}_{S(i+m+1)}w^{(3,j)}) \\ < \ & 2^i \\ \overset{m>0}{<} \ & 2^{i+m} - 2^{i-1} \\ \leq \ & bi(0\overline{w^{(1)}}), bi(1\overline{w^{(1)}}), bi({}_{S(i+m+1)}\overline{w^{(3,j)}}), bi({}_{S(i+m+1)}\overline{w^{(4)}}) \ . \end{aligned}$$

This proves suffix-freedom, since $\bigcup_{m+1 \leq j < p-1} K_{3,j} \cup K_4 \cup K_5$ cannot have $\overline{w^{(1)}}$ as suffix and $\overline{w^{(3,j)}}$, $\overline{w^{(4)}}$ cannot be a suffixes in $K_4 \cup K_5$ and K_5 respectively.

Therefore, we have proved D to be an $(i+m+1)$-closed system and we can complete it according to Lemma 1 to a complete fix-free code having $2 \cdot \alpha_l^{(m,k,i)}$ words of length $l \neq i+m+1$ and

$$2^{i+m+1} - \left[\left(2^{i+m+1} - \alpha_{i+m+1}^{(m,k,i)}\right) \cdot 2\right] = 2 \cdot \alpha_{i+m+1}^{(m,k,i)} - 2^{i+m+1}$$

words of length $i+m+1$, regarding that $D_k^m(l) \cap \overline{D_k^m(l)} = \varnothing$.

Let $B_k^m(i)$ denote the code constructed above with $B_k^m(i) \sim \beta^{(m,k,i)}$. For $k > 0$ the finite code $B_k^m(i)$ is thin. Moreover, we have $v1^{i+m+2}0^{i+m+2}w \notin B_0^m(i)$ for arbitrary v and w, since the $K_{3,j}$- and $\overline{K_{3,j}}$-words are the only ones in $B_0^m(i)$ with length possibly larger than $i+m+1$. But these $K_{3,j}$-words have j zeros from position 2 to $j+1$ and therefore contain maximally $(i+j+1)-j = i+1 < i+m+2$ ones and $\overline{K_{3,j}}$-words have j ones from position 2 to $j+1$ and therefore contain maximally $i+1$ zeros.

Since for $k > 0$ there are no words in D containing only zeros or only ones, we must have $0^{i+m+1}, 1^{i+m+1} \in B_k^m(i)$ and the statement regarding the degree follows from Proposition 5.1 in Chapter III of [2]. Using Theorem 3.1 in Chapter III of [2], the calculation of $d\left(B_0^m(i)\right)$ is reduced to the determination of the number of prefixes of $1^{i+m+2}0^{i+m+2}$ without suffix in $B_0^m(i)$, since $1^{i+m+2}0^{i+m+2}$ is not an internal factor of $B_0^m(i)$. The $(i+m)$-suffixes of the prefixes 1^j, $i+m \leq j \leq i+m+2$, and $1^{i+m+2}0^j$, $0 \leq j \leq i-1$, lie in $\overline{K_1}$ since $bi(1^{i+m-r}0^r) = 2^{i+m} - 2^r$, $0 \leq r \leq i+m$, and $\overline{K_1}$ covers exactly the $(i+m)$-words with binary value at least $2^{i+m} - 2^{i-1}$. Due to the binary values of the words in $K_{3,j}$, $j \geq m+1$, and in $\overline{K_1}$, the binary values of P_D^{i+m+1} on the right half of the binary tree are lower than $2^{i+m} + 2^{i-1}$ or are at least $2^{i+m+1} - 2^i$ respectively. Therefore, the $(i+m+1)$-suffixes of the prefixes $1^{i+m+2}0^j$, $i+1 \leq j \leq i+m-1$, are contained in the completion $B_0^m(i)$ of the $(i+m+1)$-closed system D, since their binary value is $bi(1^{i+m+1-j}0^j) = 2^{i+m+1} - 2^j$. The prefixes $1^{i+m+2}0^j$, $i+m \leq j \leq i+m+2$, have the suffix 0^{i+m} in $B_0^m(i)$. There are $i+m+1$ remaining prefixes, on the one hand 1^j, $0 \leq j \leq i+m-1$, which have no suffix in $B_0^m(i)$ due to their length, and on the other hand $1^{i+m+2}0^i$ which does not have a suffix of length at most $i+m+1$, as seen in the discussions of $(i+m)$- and $(i+m+1)$-suffixes above. In addition, for $r \geq i+m+2$, the r-suffix of $1^{i+m+2}0^i$ has binary value $bi(1^{r-i}0^i) = 2^r - 2^i > 2^{r-1} + 2^i$ for $m > 0$, whereas the binary value $2^{r-1} + 2^i$ is not exceeded by the words in $K_{3,j}$ for $j = r-i-1$. □

4 Multiplication of Fix-Free Codes

For $X, Y \subseteq \{0,1\}^*$ and two sequences α, β define the products

$$X \cdot Y := \{xy \in \{0,1\}^* \mid x \in X,\, y \in Y\}$$

and $\alpha \cdot \beta$ given by $(\alpha \cdot \beta)_i := \sum_{l,m \in \mathbb{N}:\, l+m=i} \alpha_l \cdot \beta_m$ for all $i \in \mathbb{N}$. According to Example 3.5 in Chapter III of [2], $X \cdot Y$ is a complete thin fix-free code, if the same holds for X, Y and within this case the degree is $d(X \cdot Y) = d(X) + d(Y)$.

Lemma 2. *The fix-free codes* $X \sim \alpha$ *and* $Y \sim \beta$ *satisfy*

$$X \cdot Y \sim \alpha \cdot \beta \text{ and } K(X \cdot Y) = K(X) \cdot K(Y) \ .$$

Proof. If X and Y are fix-free, any word $w \in X \cdot Y$ can be decomposed uniquely as $w = xy$ with $x \in X$ and $y \in Y$, implying $X \cdot Y \sim \alpha \cdot \beta$. Let α and β be sequences with $X \sim \alpha$ and $Y \sim \beta$. Then we get

$$\begin{aligned} K(X \cdot Y) &= K(\alpha \cdot \beta) \\ &= \sum_{i \in \mathbb{N}} \Bigg(\sum_{l,m \in \mathbb{N}:\, l+m=i} \alpha_l \cdot \beta_m \Bigg) \cdot 2^{-i} \\ &= \sum_{i \in \mathbb{N}} \sum_{l,m \in \mathbb{N}:\, l+m=i} (\alpha_l \cdot \beta_m \cdot 2^{-l} \cdot 2^{-m}) \\ &= \Bigg(\sum_{l \in \mathbb{N}} \alpha_l \cdot 2^{-l} \Bigg) \cdot \Bigg(\sum_{m \in \mathbb{N}} \beta_m \cdot 2^{-m} \Bigg) \\ &= K(X) \cdot K(Y) \ . \end{aligned}$$

□

This tells us that multiplication yields new classes of fix-free codes. The unsolved question arises, which sub-class of fix-free codes is sufficient for obtaining all fix-free codes with Kraftsum at most 3/4 by multiplication within this sub-class. For the next theorem we make use of the codes mentioned in Corollary 1 and write $\underline{n}$ for the natural numbers not exceeding n.

Theorem 3. *Let* $s \in (\mathbb{N}_0)^{\underline{n}} \setminus \{0\}$ *for* $n \in \mathbb{N}$. *Then* $C^s := \prod_{i \in \underline{n}} (C_0^0(i))^{s_i}$ *is a complete thin fix-free code which satisfies* $C^s \sim \alpha$ *and* $d(C^s) = l_{\min}(\alpha) + l$, *where* $l := \sum_{j \in \underline{n}} s_j$ *and*

$$\alpha_i = 2^{l_{\min}(\alpha) - l} \cdot \binom{i - l_{\min}(\alpha) + l - 1}{l - 1} \quad \textit{for all} \quad i \geq l_{\min}(\alpha) = \sum_{j \in \underline{n}} s_j \cdot j \ .$$

Proof. Let $s \neq 0$ be a nonnegative sequence. Due to Theorem 2, Lemma 2 and Example 3.5 in Chapter III of [2], the code C^s is thin, complete and fix-free with minimal length $l_{\min}(C^s) = \sum_{j \in \underline{n}} s_j \cdot j$ and degree $\sum_{j \in \underline{n}} s_j \cdot (j+1) = l_{\min}(\alpha) + l$. Estimating α by the according formal power series, we write $t := l_{\min}(\alpha) - l$ and for the number $a_k(l)$ of ordered partitions of k with l summands we have $a_k(l) = \binom{k-1}{l-1}$ for all $k \geq l$. We obtain

$$\begin{aligned}
\sum_{i\in\mathbb{N}} \alpha_i \cdot x^i &\overset{\text{Cor. 1}}{=} \prod_{i\in\underline{n}} \left(2^{i-1} \cdot \sum_{j\geq i} x^j\right)^{s_i} \\
&= 2^t \cdot x^t \cdot \left(\sum_{j\in\mathbb{N}} x^j\right)^l \\
&= 2^t \cdot x^t \cdot \sum_{k\in\mathbb{N}} (a_k(l) \cdot x^k) \\
&= \sum_{k\geq l} 2^t \cdot \binom{k-1}{l-1} \cdot x^{t+k} \\
&= \sum_{i\geq t+l} 2^t \cdot \binom{i-t-1}{l-1} \cdot x^i .
\end{aligned}$$

□

Corollary 3. *The subsequent choices for s in Theorem 3 yield the following entries of α for which there exists a fix-free $C^s \sim \alpha$:*

1. *The entries $\alpha_i = (i-k)\cdot 2^{k-1}$ for all $i \geq k+1 = l_{\min}(\alpha)$, choosing $s_1 = s_k = 1$ and $s_j = 0$ otherwise.*
2. *The entries $\alpha_i = i-1$ for all $i \in \mathbb{N}$, choosing $s_1 = 2$ and $s_j = 0$ otherwise.*
3. *The sequence α non-increasing for $l \geq l_{\min}(\alpha)$ with $K(\alpha) \leq 3/4$.*

Considering the non-increasing case, note that for $\alpha_{l_{\min}(\alpha)} > 2^{l_{\min}(\alpha)-1}$ the statement follows from Equation (1) and the opposite case is a consequence of the existence of the codes $C_0^0(i)$, $i \in \mathbb{N}$. Again notice that $C^s \sim \alpha$ implies the existence of a fix-free $C \sim \alpha'$ for all α' not exceeding the values of α.

So far, we have multiplied within a sub-class of our new codes, next we will use the multiplication of $C_0^0(1)$ with a fix-free code $E \sim \epsilon$. The result is the fix free code $C := C_0^0(1) \cdot E$ which fits the sequence $\alpha = (1, 1, \dots) \cdot \epsilon$ such that $\alpha_i = \sum_{j=1}^{i-1} \epsilon_j$ for $i \in \mathbb{N}$ and $K(\alpha) = K(\epsilon)$. Note that in none of the cases in the following remark it is necessary to demand $l_{\min}(\alpha) > 1$ due to Equation (1).

Corollary 4. *For the following choices of α with $K(\alpha) \leq 3/4$ there exists a fix-free $C \sim \alpha$:*

1. *The sequence α given by $\alpha_{l_{\min}(\alpha)} = \alpha_i < \alpha_k = \alpha_j$ for all $l_{\min}(\alpha) \leq i < k \leq j$ with $k \in \mathbb{N}$ fix.*
2. *The sequence α given by $\alpha_{k_i} = \alpha_j < \alpha_{k_{i+1}}$ for all $k_i \leq j < k_{i+1}$ and $i \in \mathbb{N}$, where $k_1 = l_{\min}(\alpha)$ and $k_{i+1} \geq 2k_i - 1$ for all $i \in \mathbb{N}$.*
3. *The sequence α non-decreasing with $\alpha_{l_{\min}(\alpha)} + \alpha_{l_{\min}(\alpha)+1} \geq 2^{l_{\min}(\alpha)-1}$, improving (1) for non-decreasing sequences.*
4. *The sequence α non-decreasing with $K((0,\dots,0,\alpha_{l_{\min}(\alpha)},0,\dots)) \geq 1/4$.*
5. *The sequence α non-decreasing with $\alpha_1 \neq 0$ or $\alpha_2 \neq 0$.*

Proof. 1. If we choose ϵ given by $K(\epsilon) \leq 3/4$ and $\epsilon_i > 0$ if and only if $i \in \{p, q\}$ with $p < q \in \mathbb{N}$, Theorem 12 in [4] implies the existence of a fix-free $E \sim \epsilon$. We obtain $\alpha_{i+1} - \alpha_i \neq 0$ if and only if $i \in \{p, q\}$. With $l_{\min}(\alpha) := p + 1$ and $k := q + 1$, the statement follows.

2. We choose ϵ given by $K(\epsilon) \leq 3/4$ and $\epsilon_m > 0$ if and only if we have that $m \in M = \{m_i \,|\, i \in \mathbb{N}\}$ where $m_{i+1} \geq 2m_i$ for all $i \in \mathbb{N}$. Then Lemma 4 in [1] delivers the existence of a fix-free $E \sim \epsilon$ and this implies that $\alpha_{m+1} - \alpha_m \neq 0$ if and only if $m \in M$. With $k_i := m_i + 1$ for all $i \in \mathbb{N}$ and $k_1 := l_{\min}(\alpha)$ the statement follows.
3. If we choose ϵ given by $K(\epsilon) \leq 3/4$ and $2\epsilon_k + \epsilon_{k+1} \geq 2^k$ with $k := l_{\min}(\epsilon)$, (1) implies the existence of a fix-free $E \sim \epsilon$. Using the relation between ϵ and α, we obtain the following condition on α:

$$2^{l_{\min}(\alpha)-1} = 2^k \leq 2\epsilon_k + \epsilon_{k+1} = 2\alpha_{k+1} + (\alpha_{k+2} - \alpha_{k+1}) = \alpha_{l_{\min}(\alpha)} + \alpha_{l_{\min}(\alpha)+1}$$

   and since ϵ is nonnegative, α is non-decreasing.
4. Since α is non-decreasing, the condition $\alpha_{l_{\min}(\alpha)} \geq 2^{l_{\min}(\alpha)-2}$ satisfies the condition of the preceding item.
5. For $l_{\min}(\alpha) = 1$ use (1) and $l_{\min}(\alpha) = 2$ implies $\alpha_{l_{\min}(\alpha)} \geq 2^{l_{\min}(\alpha)-2}$ again. □

5 Conclusion

We have noticed that proving a certain shifting to preserve fix-freedom, is a sufficient criterion regarding the 3/4-Conjecture. Since the proof seems to be difficult, it is feasible to start by finding special cases where the shifting is achieved by a simple replacement of words.

Moreover, the construction of complete thin codes yields large classes of fix-free codes, allowing words with Kraftsum at least 1/4 to be left out. The uniqueness verified in [2] motivates the construction of a complete thin fix-free code for given degree and kernel. In doing this, n-closed systems are a helpful instrument that could be generalized for adding more than just one level. Also binary trees can be used for giving an instant overview on the prefix-forbiddances, allowing to concentrate on the suffix-structure.

As Section 4 has demonstrated, the multiplication of fix-free codes enlarges this class and therefore it would be desirable to find all fix-free codes with an irreducible associated formal power series. Although we have restricted to the binary case in this article, most of the results are expected to be generalizable on the q-ary case.

Acknowledgements. For the introduction to this intriguing problem, the author would like to thank Eberhard Triesch as well as for his generous support of the research.

References

1. Ahlswede, R., Balkenhol, B., Khachatrian, L.: Some properties of fix-free codes. In: Proc. Int. Seminar Coding Theory and Combinator, Thahkadzor, Armenia, pp. 20–33 (1996)
2. Berstel, J., Perrin, D.: Theory of Codes. Academic Press, Orlando (1985)
3. Deppe, C., Schnettler, H.: On q-ary fix-free codes and directed deBruijn graphs. In: Proc. Int. Symp. Inf. Theory, Seattle, WA, pp. 1482–1485 (2006)
4. Harada, K., Kobayashi, K.: A note on the fix-free code property. IEICE Trans. Fundamentals E82-A(10), 2121–2128 (1999)
5. Khosravifard, M., Halabian, H., Gulliver, T.A.: A Kraft-type sufficient condition for the existence of D-ary fix-free codes. IEEE Trans. Inf. Theory 56(6), 2920–2927 (2010)
6. Kukorelly, Z., Zeger, K.: Sufficient conditions for existence of binary fix-free codes. IEEE Trans. Inf. Theory 51(10), 3433–3444 (2005)
7. Schnettler, H.: On the $\frac{3}{4}$-conjecture for fix-free codes: a survey. arXiv:0709.2598v1 [cs.IT]
8. Ye, C., Yeung, R.W.: Some basic properties of fix-free codes. IEEE Trans. Inf. Theory 47(1), 72–87 (2001)
9. Yekhanin, S.: Sufficient conditions of existence of fix-free codes. In: Proc. Int. Symp. Inf. Theory, Washington D.C, p. 284 (2001)
10. Yekhanin, S.: Improved upper bound for the redundancy of fix-free codes. IEEE Trans. Inf. Theory 50(11), 2815–2818 (2004)

Creating Order and Ballot Sequences

Ulrich Tamm

German Language Department of Business Informatics, Marmara University, Istanbul, Turkey
tamm@ieee.org

Dedicated to the memory of Rudolf Ahlswede

Abstract. Rudolf Ahlswede introduced the theory of creating order roughly at the same time as his theory of identification. He was always surprised that it did not achieve the same popularity as identification. We shall here present a multi-user model in which, contrasting to Ahlswede's original model, the size of the memory may vary in time. The influence of the maximum size of the memory device on the expected occurrence of the first 0 in the sequence produced by the organizer is studied. In the case that there is one outgoing bit in each time unit two steps of a simple random walk on the lattice can be combined to one step in a random walk for the exhaustion of the memory.

Keywords: creating order, permuting channel, Chebyshev polynomials, ballot sequences.

1 Introduction

Ahlswede, Ye, and Zhang [2] introduced the following model for creating order in sequence spaces. We are given a box containing (a fixed number) β balls labelled by letters from an alphabet of size α. In each time unit a person $\mathcal{O}$ – denoted as organizer – takes out one ball of the box which is replaced by a new ball thrown into the box by a second person $\mathcal{I}$. The aim of the organizer is to reduce the space of possible output sequences. As a measure for the efficiency of the ordering process the number of possible output sequences and the entropy of the output space have been studied (cf. also [3], [7], [15]). A related model in which the output sequence is regarded as a message from person $\mathcal{I}$ to a decoder $\mathcal{D}$ was considered by several authors studying the permuting channel, e. g. [1], [9], and [10].

In this paper a multi–user version of the original model for creating order is discussed. Now there are $s \geq 2$ persons $\mathcal{I}_1, \dots, \mathcal{I}_s$, say, throwing balls labelled either 0 or 1 into the box. We shall present the model using a slightly different terminology. In each time unit s sources $\mathcal{I}_1, \dots, \mathcal{I}_s$ produce one bit each. These s bits arrive at an organizer who in the same time unit has to choose one bit for output. This bit may be one of the s arriving bits or a bit stored in some memory device (the box), in which the bits not used so far may be stored. The

H. Aydinian, F. Cicalese, and C. Deppe (Eds.): Ahlswede Festschrift, LNCS 7777, pp. 711–724, 2013.

organizer follows a simple strategy: if it is possible the output must be a 1. So if one of the arriving bits is a 1, the organizer will put out a 1 for sure. The bits not used for output he may store in the memory device. If all the s sources produce a 0, then the organizer will take a look at the memory. If there is still a 1 contained he will put out a 1, otherwise he must put out a 0.

We assume that at the beginning the memory device is empty. At some point it may occur that no further 1 can be stored, since the device is full of 1's (a 0 may be replaced by a 1). In this case there is a maximum size or capacity of M bits which cannot be superceded. Of theoretical interest is also the not very realistic model, in which the memory device can store infinitely many bits.

Observe that contrasting to the original model for creating order the size of the memory device (or box) now may vary in time. A natural question is: how much influence does the maximum size of the memory have on the behaviour of the sequence of bits arranged by the organizer? Of course, in the strategy considered the organizer's aim is to produce the all–one sequence and we shall study how well he can manage to achieve this goal. As a new measure for the influence of the memory we consider the expected value of the first occurence of a 0 in this sequence. We shall denote this expectation as

$$E_0 = \sum_{t=1}^{\infty} t \cdot \text{Prob(first 0 at time t)}. \tag{1}$$

Further, we shall denote by

$$\underline{x}^i = (x_1^i, x_2^i, \ldots), \tag{2}$$

the sequences of bits produced by the sources $\mathcal{I}_i$, respectively and by

$$\underline{z} = (z_1, z_2, \ldots) \tag{3}$$

the sequence arranged by the organizer according to the above strategy. The results below are mainly derived from properties of the sequence

$$\underline{u} = (u_1, \ldots u_s, u_{s+1}, \ldots, u_{2s}, u_{2s+1}, \ldots) = (x_1^1, \ldots, x_1^s, x_2^1, \ldots, x_2^s, x_3^1, \ldots) \tag{4}$$

which is obtained by merging the sequences $\underline{x}^i, i = 1, \ldots, s$ into one sequence $\underline{u}$, where the bits in the positions $\equiv i \bmod s$ are those produced by $\mathcal{I}_i$. (Observe that the order of the incoming bits at each time unit does not matter, since the organizer waits until all the s bits have arrived. So $\underline{u}$ may also be defined by any other merging procedure which assigns the postions $(s-1)t+1, \ldots, st$ to the bits arriving in time unit t).

Our first result is obtained for two memoryless, independent and equiprobable sources, i. e., $\mathcal{I}_1$ and $\mathcal{I}_2$ produce 1 and 0 with equal probability $\frac{1}{2}$.In this case it is

$$\text{Prob(first 0 at time } t-1) = \frac{1}{4^t} \cdot a(0, t-1), \tag{5}$$

where $a(0, t-1)$ denotes the number of sequences $\underline{u}$ produced by the two sources leading to the all-one sequence as output with size of stock of ones 0 at time t.

Clearly an empty stock at time $t-1$ is a necessary condition for the occurrence of the first 0 at time unit t, which then happens with probability $\frac{1}{4}$ (4^{t-1} is the number of all possible sequences $\underline{u}$ until time $t-1$).

More exactly, $a(0,t-1) = a_M(0,t-1)$ also depends on the size of the memory M - here also the case of an infinite memory $M = \infty$ is possible. However, we shall mostly skip the index M, since in the respective chapters the parameter M will be indicated.

Analogously, the numbers $a(m,t)$ ($= a_M(m,t)$ are defined for the size of the stock of ones $m = 1,\dots,M, t = 1,2,\dots$. Let us denote by $\boldsymbol{u}_t = (u_{2t-1}, u_{2t})$ the two bits arriving at time t. Now observe that in each time unit the source inputs $\boldsymbol{u}_t = 10$ and $\boldsymbol{u}_t = 01$ do not change the size of the stock of ones, $\boldsymbol{u}_t = 00$ decreases the size by one bit (and is forbidden for $m = 0$, since in this case the organizer must put out a 0) and $\boldsymbol{u}_t = 11$ increases the size by one bit if $m < M$ (and does not change the size of the stock of ones if the capacity $m = M$ is reached. So we have a random walk in the memory size expressed by the recursion formulae for the numbers $a(m,t)$

$$\begin{pmatrix} a(0,t) \\ \vdots \\ a(M,t) \end{pmatrix} = A_{M+1} \cdot \begin{pmatrix} a(0,t-1) \\ \vdots \\ a(M,t-1) \end{pmatrix} \tag{6}$$

where

$$A_n = \begin{pmatrix} 2\,1\,0 \dots 0\,0\,0 \\ 1\,2\,1 \dots 0\,0\,0 \\ \vdots\;\vdots\;\vdots\;\ddots\;\vdots\;\vdots\;\vdots \\ 0\,0\,0 \dots 1\,2\,1 \\ 0\,0\,0 \dots 0\,1\,3 \end{pmatrix}, \tag{7}$$

with initial matrices $A_1 = (3)$, $A_2 = \begin{pmatrix} 2\,1 \\ 1\,3 \end{pmatrix}$.

Theorem 1. *If $\mathcal{I}_1$ and $\mathcal{I}_2$ are two memoryless, independent and symmetrical sources, then*

i)

$$E_0 = \frac{1}{4} \sum_{i=1}^{M+1} c_i \left(1 - \frac{\lambda_i^{(M+1)}}{4}\right)^{-2} \tag{8}$$

where $\lambda_i^{(M+1)}, i = 1,\dots,M+1$ are the eigenvalues

$$\lambda_i^{(M+1)} = 4 \cdot \sin \frac{i\pi}{2M+1} \tag{9}$$

of the matrix A_{M+1} and $c_i, i = 1,\dots,M+1$ are appropriate constants.

ii) E_0 does not exist if the size of the memory $M(t)$ at time t is bounded by a function $f(t)$ which exceeds every positve integer M from some time $t_0(M)$ on.

We shall prove Theorem 1 in the next section where we shall also take a closer look at the transition matrices defined in (7) and further matrices of a similar structure. The matrices A_n have been studied before in geometry and they can also be obtained from the squares of the transition matrices of the random walk which occurs in the study of the Brownian motion. Their characteristic polynomials are known to be combinations of Chebyshev polynomials. Further in Section 2 for small capacities $M = 0, 1$ the numbers $a(0, t)$ will be determined exactly for small capacities $M = 0, 1$ and also the case in which two identical nonsymmetric sources give bits to the organizer will be considered.

Part (ii) of Theorem 1 states that E_0 does not exist if the number of bits $M(t)$ that the memory device is able to store at time t is an arbitrarily slowly growing function (in fact, the condition is even weaker). If every incoming bit can be stored, i. e., $M(t) = t$ is linear in time, then there is a second proof based on some property of the merged sequence $\underline{u}$, which also extends to the situation of s incoming bits in each time unit.

Theorem 2. *Let there be s identical sources $\mathcal{I}_1, \ldots, \mathcal{I}_s$ producing one bit each per time unit with Prob$(X = 1) = p$, Prob$(X = 0) = 1-p$. If the memory device can store every incoming bit, then the expected value for the occurence of the first 0 in the sequence arranged by an organizer (if he puts out a 1 if possible) is*

$$E_0 \begin{cases} = \infty, & p = \frac{1}{s} \\ < \infty, & p < \frac{1}{s} \end{cases} \tag{10}$$

Theorem 2 will be proved in Section 3. It turns out that the numbers $a(0, t-1) = a_\infty(0, t-1)$ (defined as before for the finite memory case) are generalized Catalan numbers

$$C_t^{(s)} = \frac{1}{(s-1) \cdot t + 1} \cdot \binom{st}{t} \tag{11}$$

In Section 4 a model with more than one outgoing bits will be discussed Finally, in Section 5 some generalizations and open problems will be stated.

2 The Transition Matrices A_n

The transition matrices A_n and further matrices of a similar structure we shall consider in this section are closely related to Chebyshev polynomials $(t_n(x))_{n=1,2,\ldots}$ and Chebychev polynomials of the second kind $(u_n(x))_{n=1,2,\ldots}$. The reason is that these polynomials occur as determinants of special matrices, namely (cf. [5], p. 228)

$$t_n(x) = \frac{n}{2} \cdot \sum_{i=0}^{\lfloor \frac{n}{2} \rfloor} \frac{(-1)^i}{n-i} \binom{n-i}{i} (2x)^{n-2i} = \det T_n(x) \tag{12}$$

with

$$T_n(x) = \begin{pmatrix} x & 1 & 0 & \dots & 0 & 0 & 0 \\ 1 & 2x & 1 & \dots & 0 & 0 & 0 \\ \vdots & \vdots & \vdots & \ddots & \vdots & \vdots & \vdots \\ 0 & 0 & 0 & \dots & 1 & 2x & 1 \\ 0 & 0 & 0 & \dots & 0 & 1 & 2x \end{pmatrix}, \tag{13}$$

$$u_n(x) = \sum_{i=0}^{\lfloor \frac{n}{2} \rfloor} (-1)^i \binom{n-i}{i} (2x)^{n-2i} = \det U_n(x) \tag{14}$$

with

$$U_n(x) = \begin{pmatrix} 2x & 1 & 0 & \dots & 0 & 0 & 0 \\ 1 & 2x & 1 & \dots & 0 & 0 & 0 \\ \vdots & \vdots & \vdots & \ddots & \vdots & \vdots & \vdots \\ 0 & 0 & 0 & \dots & 1 & 2x & 1 \\ 0 & 0 & 0 & \dots & 0 & 1 & 2x \end{pmatrix}, \tag{15}$$

Further, in [12] it is shown that

$$u_n(x) + u_{n-1}(x) = \det V_n(x), \tag{16}$$

where

$$V_n(x) = \begin{pmatrix} 2x & 1 & 0 & \dots & 0 & 0 & 0 \\ 1 & 2x & 1 & \dots & 0 & 0 & 0 \\ \vdots & \vdots & \vdots & \ddots & \vdots & \vdots & \vdots \\ 0 & 0 & 0 & \dots & 1 & 2x & 1 \\ 0 & 0 & 0 & \dots & 0 & 1 & 2x+1 \end{pmatrix}, \tag{17}$$

Observe that the transition matrix $A_M = V_M(1)$ occurs for the special value $x = 1$.

The following well–known properties of the Chebyshev polynomials (cf. [12] and [11]) will be used later on.

$$t_n(x) = 2x \cdot t_{n-1}(x) - t_{n-2}(x), u_n(x) = 2x \cdot u_{n-1}(x) - u_{n-2}(x) \tag{18}$$

$$t_n(x) = u_n(x) - u_{n-2}(x) \tag{19}$$

$$4 \cdot \sin \frac{i\pi}{2n+1}, i = 1, \dots, , n \text{ are the eigenvalues of } V_n(x) \tag{20}$$

Proof of Theorem 1

i) With (6)

$$\begin{pmatrix} a(0,t) \\ \vdots \\ a(M,t) \end{pmatrix} = A_{M+1} \begin{pmatrix} a(0,t-1) \\ \vdots \\ a(m,t) \end{pmatrix} = A_{M+1}^t \begin{pmatrix} a(0,0) \\ \vdots \\ a(M,0) \end{pmatrix} = A_{M+1}^t \begin{pmatrix} 1 \\ 0 \\ \vdots \\ 0 \end{pmatrix}, \tag{21}$$

since in the beginning the memory device does not contain any 1. A_{M+1} is symmetric, hence there exists an orthogonal matrix $P = (p_{ij})_{i,j=1,\dots,M+1}$ with $P^T = P^{-1}$ such that

$$P \cdot A_{M+1} \cdot P^{-1} = \text{diag}(\lambda_1, \dots \lambda_{M+1}), \tag{22}$$

where $\text{diag}(\lambda_1, \dots, \lambda_{M+1})$ is the diagonal matrix consisting of the eigenvalues of A_{M+1}. Now $A^t_{M+1} = P^{-1} \cdot (\text{diag}(\lambda_1^t, \dots, \lambda_{M+1})^t) \cdot P$. With (21) and $c_i = p_{1i}^2$ then

$$a(0,t) = \sum_{i=1}^{M+1} c_i \cdot \lambda_i^t \tag{23}$$

Now using (23) and $\sum_{t=1}^{\infty} t \cdot x^{t-1} = \frac{1}{(1-x)^2}$

$$E_0 = \sum_{t=1}^{\infty} t \frac{a(0,t-1)}{4^t} = \frac{1}{4} \sum_{i=1}^{M+1} c_i \sum_{t=1}^{\infty} t \cdot (\frac{\lambda_i}{4})^{t-1} = \frac{1}{4} \sum_{i=1}^{M+1} c_i \cdot (1 - \frac{\lambda_i}{4})^{-2} \tag{24}$$

The eigenvalues of A_{M+1} have been listed under (20) and thus (i) is proved.

ii) Observe that the largest eigenvalues $\lambda_{max}^{(M+1)}$ of the matrices A_{M+1} form a sequence converging towards 4, hence the series of expected values for the occurrence of the first 0 $E_0^{(M)}$ is divergent. If the maximum size $M(t)$ of the memory device varies in time such that for every positive integer M it is $M(t) > M$ for some $t_0(M)$ on, then $E_0 > E_0^{(M)}$ for every fixed capacity M (with appropriate rescaling, set $t_0(M) = 0$).

Remark 1. *It is clear from the proof that the results of Theorem 1 also hold if we start with any number $m > 0$ of 1's in the memory device, only the constants c_i will change then. Implicitly this fact has already been used in the proof of part (ii).*

The matrices A_M can further be obtained as a submatrix of the squares of matrices

$$B_n = \begin{pmatrix} 0\,1\,0 \dots 0\,0\,0 \\ 1\,0\,1 \dots 0\,0\,0 \\ \vdots\;\vdots\;\vdots\;\ddots\;\vdots\;\vdots\;\vdots \\ 0\,0\,0 \dots 1\,0\,2 \\ 0\,0\,0 \dots 0\,1\,0 \end{pmatrix}, \tag{25}$$

with $B_1 = (0)$ and $B_2 = \begin{pmatrix} 0\,2 \\ 1\,0 \end{pmatrix}$.

To see this, observe that the recursion formulae in (6) are obtained by regarding the single time units t of the ordering process, which are represented by the two bits u_{2t-1}, u_{2t} of the sequence $\underline{u}$. It is also possible to obtain a recurrence from the single components of $\underline{u}$. In order to do so, consider the transitions (for $i = 1, 2, \dots; k = 0, 1, \dots, n$)

$$b(i+1,k)=\begin{cases} b(i,k+1) & k=0,\\ b(i,k-1)+b(i,k+1) & k=1,\dots n-2\\ b(i,k-1)+2\cdot b(i,k+1) & k=n-1\\ b(i,k-1), & k=n \end{cases} \tag{26}$$

which just define the matrices B_n. This is just a random walk on the line $\{0,\dots,n\}$ with absorption at the point 0 and reflection at point n, which occurs in the study of the Brownian motion (cf. [8]). Now observe that the recurrence formulae (6) are obtained from (26) by combining two consecutive instances $i, i+1$, i. e., $b(i+2,k)$ is obtained from the $b(i,.)$ via the same recursion as $a(t,m)$ is obtained from the $a(t-1,.)$. Since for even i $b(i,k)=0$ if k is odd, we have

Proposition 1. *Let* $B_{2n+1}^2 = (b_{ij})_{i,j=1,\dots 2n+1}$, *where* B_{2n+1} *and* A_n *are the matrices defined in (25) and (7). Then*

$$A_n = (b_{ij})_{i,j\ even}. \tag{27}$$

Moreover, by an appropriate permutation L *of lines and columns*

$$LB_{2n+1}^2L^T = \begin{pmatrix} A_n & 0\\ 0 & C_{n+1}\end{pmatrix}. \tag{28}$$

where

$$C_{n+1} = (b_{ij})_{i,j\ odd} = \begin{pmatrix} 1\,1\,0\dots0\,0\,0\\ 1\,2\,1\dots0\,0\,0\\ \vdots\,\vdots\,\vdots\,\ddots\,\vdots\,\vdots\,\vdots\\ 0\,0\,0\dots1\,2\,2\\ 0\,0\,0\dots0\,1\,2\end{pmatrix}, \tag{29}$$

with $C_2 = \begin{pmatrix}1\,2\\1\,2\end{pmatrix}$ *and* $C_3 = \begin{pmatrix}1\,1\,0\\1\,2\,2\\0\,1\,2\end{pmatrix}$.

Analogously, we can consider the matrices

$$\overline{A}_n = \begin{pmatrix} 2\,1\,0\dots0\,0\,0\\ 1\,2\,1\dots0\,0\,0\\ \vdots\,\vdots\,\vdots\,\ddots\,\vdots\,\vdots\,\vdots\\ 0\,0\,0\dots1\,2\,1\\ 0\,0\,0\dots0\,1\,2\end{pmatrix}, \tag{30}$$

$$\overline{B}_n = \begin{pmatrix} 0\,1\,0\dots0\,0\,0\\ 1\,0\,1\dots0\,0\,0\\ \vdots\,\vdots\,\vdots\,\ddots\,\vdots\,\vdots\,\vdots\\ 0\,0\,0\dots1\,0\,1\\ 0\,0\,0\dots0\,1\,0\end{pmatrix}, \tag{31}$$

$$\overline{C}_n = \begin{pmatrix} 1\,1\,0\ldots0\,0\,0 \\ 1\,2\,1\ldots0\,0\,0 \\ \vdots\;\vdots\;\vdots\;\ddots\;\vdots\;\vdots\;\vdots \\ 0\,0\,0\ldots1\,2\,1 \\ 0\,0\,0\ldots0\,1\,1 \end{pmatrix}, \tag{32}$$

where following the same lines it can be derived (with $\overline{B}_n^2 = (\bar{b}_{ij})_{i,j=1,\ldots,n}$)

$$L\overline{B}_{2n+1}^2 L^T = \begin{pmatrix} \overline{A}_n & 0 \\ 0 & \overline{C}_{n+1} \end{pmatrix}. \tag{33}$$

where

$$\overline{A}_n = (\bar{b}_{ij})_{i,j} \text{ even} \tag{34}$$

and

$$\overline{C}_{n+1} = (\bar{b}_{ij})_{i,j} \text{ odd} \tag{35}$$

The matrices B_n and $\overline{B}_n$ are well known as transition matrices for certain random walks (cf. [8] and [13], pp. 238–240) and the matrices A_n and $\overline{A}_n$ have an application in geometry (cf. [12]). Think of the vertices $P_1, \ldots, P_N$ of a regular N–gon drawn on a unit circle. Then the eigenvalues of $A_{\frac{N-1}{2}}$ (if N is odd) or $\overline{A}_{\frac{N-2}{2}}$ (if N is even) give the squares of the different distances $\overline{P_i, P_j}, i, j = 1, \ldots, N$.

From this property it is immediate that the largest eigenvalues $\lambda_{max}^{(n)}$ of the matrices A_n form a strictly increasing sequence with $\lim_{n\to\infty}\lambda_{max}^{(n)} = 4$. The reason is that the largest distance $\overline{P_1, P_{\frac{N-1}{2}}}$ tends to the diameter 2 of the unit circle. This property was the central argument in the proof of Theorem 1, where it was derived analytically using an explicit formula for the eigenvalues of A_n.

Now let us take a look at the characteristic polynomials.

Proposition 2. *For the characteristic polynomials of the matrices we consider holds*

$$\chi_{\overline{B}_n}(\lambda) = u_n(-\frac{\lambda}{2}) = \sum_{i=0}^{\lfloor \frac{n}{2} \rfloor} (-1)^{n-i} \binom{n-i}{i} \lambda^{n-2i} \tag{36}$$

$$\chi_{B_n}(\lambda) = t_n(-\frac{\lambda}{2}) = \sum_{i=0}^{\lfloor \frac{n}{2} \rfloor} (-1)^{n-i} (\binom{n-i}{i} + \binom{n-i-1}{i-1}) \lambda^{n-2i} \tag{37}$$

$$\chi_{A_n}(\lambda^2) = \frac{(-1)^{n+1}}{\lambda} \chi_{B_{2n+1}}(\lambda),\ \chi_{\overline{A}_n}(\lambda^2) = \frac{(-1)^{n+1}}{\lambda} \chi_{\overline{B}_{2n+1}}(\lambda), \tag{38}$$

$$\chi_{C_{n+1}}(\lambda) = -\lambda \chi_{A_n}(\lambda),\ \chi_{\overline{C}_{n+1}}(\lambda) = -\lambda \chi_{\overline{A}_n}(\lambda), \tag{39}$$

$$\mid \frac{1}{\lambda}(\chi_{B_{2n+1}}(\lambda))^2 \mid = \mid \chi_{B_{2n+1}^2}(\lambda^2) \mid,\ \mid \frac{1}{\lambda}(\chi_{\overline{B}_{2n+1}}(\lambda))^2 \mid = \mid \chi_{\overline{B}_{2n+1}^2}(\lambda^2) \mid, \tag{40}$$

Proof: By definition, (36) follows from the fact that $B_n = U_n(0)$.

Using the properties (18) and (19) of the Chebyshev polynomials, (37) can be derived via

$$\chi_{B_n}(\lambda) = -\lambda \cdot \chi_{\overline{B}_{n-1}}(\lambda) - 2\chi_{\overline{B}_{n-2}}(\lambda) = -\lambda \cdot u_{n-1}(\frac{\lambda}{2}) - 2u_{n-2}(\frac{\lambda}{2}) = u_n(\frac{\lambda}{2}) - 2u_{n-2}(\frac{\lambda}{2})$$

The characteristic polynomials in (38) can be derived using the recurrence $\chi_{A_n}(\lambda) = (2-\lambda) \cdot \chi_{A_{n-1}}(\lambda) - \chi_{A_{n-2}}(\lambda)$ which is the same as for $\chi_{\overline{A}_n}$, only the initial values differ: $\chi_{A_1}(\lambda) = 3 - \lambda$, $\chi_{A_2}(\lambda) = \lambda^2 - 5\lambda + 5$, whereas $\chi_{\overline{A}_1}(\lambda) = 2 - \lambda$, $\chi_{\overline{A}_2}(\lambda) = \lambda^2 - 4\lambda + 3$.

(39) can be derived using $\chi_{C_n}(\lambda) = (2-\lambda) \cdot \chi_{T_{n-1}(1)}(\lambda) - 2 \cdot \chi_{T_{n-2}(1)}(\lambda)$ and $\chi_{\overline{C}_n}(\lambda) = (1-\lambda) \cdot \chi_{T_{n-1}(1)}(\lambda) - \chi_{T_{n-2}(1)}(\lambda)$.

Finally (40), follows from (28), (33), (38), and (39).

Observe that for the coefficients r_i and $\overline{r}_i$ in the polynomials $\chi_{B_n}(\lambda) = \sum_{i=0}^{n} r_i \lambda^i$ and $\chi_{\overline{B}_n}(\lambda) = \sum_{i=0}^{n} \overline{r}_i \lambda^i$ it is

$$\sum_{i=0}^{n} | r_i | = L_n, \sum_{i=0}^{n} | \overline{r}_i | = F_n \tag{41}$$

where F_n and L_n denote the n–th Fibonacci and Lucas number, respectively defined by $F_n = F_{n-1} + F_{n-2}$ and $L_n = L_{n-1} + L_{n-2}$ with initial values $F_1 = F_2 = 1$ and $L_1 = 1, L_2 = 3$.

The Fibonacci numbers also occur in the analysis of the expected value E_0 for maximum memory size $M = 1$. Obviously, if there is no memory ($M = 0$), then the number $a(0,t) = 3^t$, where 3 is also the largest eigenvalue of $A_1 = (3)$. In this case $E_0 = \sum_{t=1}^{\infty} t \cdot (\frac{3}{4})^{t-1} \cdot \frac{1}{4} = 4$ is just the expected value of the geometric distribution with parameter $\frac{1}{4}$, since with probability $\frac{1}{4}$ two 0's arrive at time t.

Proposition 3. *For maximum memory size $M = 1$ the numbers $a(0,t)$ and $a(1,t)$ are $a(0,2t) = 5^t \cdot F_t$, $a(1,t) = 5^t \cdot F_{t+1}$,*

$$a(0,2t+1) = 5^t \cdot (2F_t + F_{t+1}),\ a(1,2t+1) = 5^t \cdot (F_t + 3F_{t+1}) \tag{42}$$

Proof

We have the transition matrices $A_M = \begin{pmatrix} 2 & 1 \\ 1 & 3 \end{pmatrix}$ Now $A_M^2 = \begin{pmatrix} 5 & 5 \\ 5 & 10 \end{pmatrix} = 5 \cdot \begin{pmatrix} 1 & 1 \\ 1 & 2 \end{pmatrix}$, from which follows that $A_M^{2t} = 5^t \cdot \begin{pmatrix} 1 & 1 \\ 1 & 2 \end{pmatrix}^t = 5^t \cdot \begin{pmatrix} F_t & F_{t+1} \\ F_{t+1} & F_{t+2} \end{pmatrix}^t$ by a well–known property of the Fibonacci numbers.

Finally let us take a look at the transition matrices in the case where two identical nonsymmetric sources give bits to the organizer. If p is the probality for the sources to produce a 1, the transition matrices can be obtained as in (28) from the square of the matrices

$$B'_n = \begin{pmatrix} 0 & 1-p & 0 & \dots & 0 & 0 & 0 \\ p & 0 & 1-p & \dots & 0 & 0 & 0 \\ \vdots & \vdots & \vdots & \ddots & \vdots & \vdots & \vdots \\ 0 & 0 & 0 & \dots & p & 0 & 1 \\ 0 & 0 & 0 & \dots & 0 & p & 0 \end{pmatrix}, \tag{43}$$

The matrices B'_n have been studied intensively in [8]. An exact determination of their eigenvalues seems to be quite difficult in the nonsymmetric case, however it is clear from Theorem 2 that for $M \longrightarrow \infty$ the sequence of expected values $E_0(M)$ is not divergent for $p < \frac{1}{2}$. The characteristic polynomial again is a combination of Chebyshev polynomials.

Proposition 4. *The characteristic polynomial of B'_n is*

$$\chi_{B'_n}(\lambda) = -\lambda(p(1-p))^{\frac{n-1}{2}} \cdot u_{n-1}(\frac{\lambda}{\sqrt{p(1-p)}}) - p(p(1-p))^{\frac{n-2}{2}} u_{n-2}(\frac{\lambda}{\sqrt{p(1-p)}}) \tag{44}$$

Proof: First, consider the matrix $\overline{B}'_n$ obtained from B'_n by changing the element $b'_{n-1,n}$ from 1 to $1-p$ The characteristic polynomial of the matrix $\overline{B}'_n$ obeys the three–term recurrence

$$\chi_{\overline{B}'_n}(\lambda) = -\lambda \cdot \chi_{\overline{B}'_{n-1}}(\lambda) - p(1-p) \cdot \chi_{\overline{B}'_{n-2}}(\lambda) \tag{45}$$

which as solution yields the weighted Chebyshev polynomial of the second kind

$$\chi_{\overline{B}'_n} = (p(1-p))^{\frac{n}{2}} \cdot u_n(\frac{\lambda}{\sqrt{p(1-p)}}) \tag{46}$$

Now observe that the characteristic polynomial χ'_{B_n} is

$$\chi_{B'_n}(\lambda) = -\lambda \cdot \chi_{\overline{B}'_{n-1}}(\lambda) - p \cdot \chi_{\overline{B}'_{n-2}}(\lambda) \tag{47}$$

3 Every Incoming Bit Can Be Stored

When the memory device can store every incoming bit, there is a simple criterion on the sequence $\underline{u}$, which guarantees that the organizer can put out the first 0 at time $t+1$. To see this, recall that there are two necessary conditions for the occurence of the first 0 at time unit t: 1) there are no further 1's in the memory device after the t–th bit has been put out by the organizer, 2) up to time t the all-one sequence has been arranged by the organizer. These conditions can be translated into conditions required from the sequence $\underline{u}$, namely

i)

$$\text{wt}(u_1, \dots, u_{st}) = t, \tag{48}$$

ii)

$$\mathrm{wt}(u_1, \ldots, u_{si}) \geq i \text{ for } i = 1, \ldots, t-1. \tag{49}$$

As usual, here the weight $\mathrm{wt}(\boldsymbol{x})$ of a $\{0,1\}$–vector $\boldsymbol{x}$ denotes the number of 1's in $\boldsymbol{x}$. By condition (i) no 1's can be left in the device, since t 1's have arrived at the organizer and t 1's have been used for output. The second condition (ii) assures that at all time units before it was possible to put out a 1 (by the same argumentation).

The number $a(0,t)$ of all sequences of length s fulfilling the conditions (i) and (ii) are well known in Combinatorial Theory, these are just the generalized Catalan numbers defined in (11) (cf. [6], pp. 343–350).

Proof of Theorem 2: Recall that the sources $\mathcal{I}_1, \ldots, \mathcal{I}_s$ each produce a 1 with probability p. Hence with probability $(1-p)^s$ only 0's arrive at the organizer at time unit $t+1$. In this case he has to put out a 1 if the memory device is empty, which happens by the preceding discussion with probability $a(0,t) \cdot p^t (1-p)^{(s-1)t}$. Hence the expected value for the occurence of the first 0 now is

$$E_0 = \sum_{t=1}^{\infty} (t+1) \cdot a(0,t) \cdot p^t (1-p)^{st} \tag{50}$$

We shall now use Stirling's formula

$$n! \approx \sqrt{2\pi n} \cdot (\frac{n}{e})^n \tag{51}$$

to find that the binomial coefficient $\binom{st}{t}$ in the Catalan numbers $a(0,t)$ is approximately

$$\binom{st}{t} \approx (\frac{s^s}{(s-1)^{s-1}})^t \cdot \frac{1}{\sqrt{t}} \tag{52}$$

Now let $p = \frac{1}{s'}$. Then with (52)

$$E_0 = \sum_{t=1}^{\infty} (t+1) a(0,t) (\frac{1}{s'})^t (1-\frac{1}{s'})^{st} = \sum_{t=1}^{\infty} \frac{t+1}{(s-1)t+1} \binom{st}{t} (\frac{(s'-1)^s}{(s')^{s+1}})^t \approx \sum_{t=1}^{\infty} \frac{S^t}{\sqrt{t}} \tag{53}$$

with

$$S = \frac{s^s \cdot (s'-1)^{s-1}}{(s')^s \cdot (s-1)^{s-1}} \tag{54}$$

Now if $p < \frac{1}{s}$ then $S < 1$ and the series (50) converges and hence the expected value E_0 exists. If $p = \frac{1}{s}$ then $S = 1$ and E_0 does not exist, since in this case $\frac{1}{\sqrt{t}}$ is the decisive term in the single summands which again yields a divergent series for (50).

4 More Than One Outgoing Bit

Finally, let us discuss a model for creating order in which there are $s > 2$ bits arriving in each time unit at the organizer who then has to put out a number l

$(s > l > 1)$ of bits. With the argumentation above, the number $a(0,t)$ in this case is the number of sequences of length st fulfilling the two conditions

$$\mathrm{wt}(u_1, \dots, u_{st}) = l \cdot t, \tag{55}$$

$$\mathrm{wt}(u_1, \dots, u_{si}) \geq l \cdot i \text{ for } i = 1, \dots, t-1. \tag{56}$$

For the analysis of the numbers $a(0,t)$ we need the concept of domination (or majorization). A sequence $a = (a_1, a_2, \dots, a_n)$ is dominated by a sequence $b = (b_1, b_2, \dots, b_n)$ if for all $m = 1, \dots, n$ it is $\sum_{i=1}^m a_i \leq \sum_{i=1}^m b_i$. We shall write $a \preceq b$ in this case.

Proposition 5. *When in each time unit s bits arrive at the organizer who in the same time unit has to put out a fixed number $l < s$ bits (using the strategy which prefers a 1 towards a 0), then the number $a(0,t)$ of sequences fulfilling (55) and (56) is*

$$a(0,t) = \sum_{(l,\dots,l) \preceq (i_1,\dots,i_t)} \binom{s}{i_1} \cdot \binom{s}{i_2} \cdots \binom{s}{i_t}$$

Proof: For $j = 1, \dots, t$ the binomial coefficient $\binom{s}{i_j}$ is the number of possible ways in which exactly i_j bits can arrive at the organizer in time unit j. In order to assure that the memory is exhausted at time t, i. e. (i') holds, it must be $i_1 + \dots + i_t = l \cdot t$ and in order to guarantee (ii') for all $j < t$, it must hold $i_1 + \dots + i_j \geq l \cdot i$. This just means that the sequence $(i_1, \dots, i_t)$ dominates the sequence $\underbrace{(l, \dots, l)}_{t}$

For the special case $l = 1$, we can derive from Proposition 5 the following identity for the generalized Catalan numbers

$$C_{t+1}^{(s)} = \sum_{(1,\dots,1) \preceq (i_1,\dots,i_t)} \binom{s}{i_1} \cdot \binom{s}{i_2} \cdots \binom{s}{i_t} \tag{57}$$

For the analysis of the parameter E_0 it would be nice to have such a closed expression also for the case $l > 1$, $\gcd(l,s) = 1$.

The random walk for the exhaustion of the memory is closely related to another random walk defined by a path in a lattice (where from point (x,y) one can either move to $(x+1, y+s-l)$ or to $(x+1, y-l)$. Essentially, an analysis for $l = 1$ outgoing bit in our model for creating order was possible since these two processes are closely related. The combination of s steps of the random walk defined by the lattice paths yields the one for the exhaustion of the memory and also the nonnegative paths in the lattice by Proposition 5 correspond to the sequences with the properties (48) and (49), such that finally the same counting function arises.

If there are $l > 2$ ($\gcd(l, s) = 1$) outgoing bits per time unit, the counting functions for the nonnegative paths in the corresponding lattice and the exhaustion of the memory, respectively, are different. Let, e. g., $s = 5$ and $l = 2$, then the numbers $a(0, t)$ of sequences fulfilling (55) and (56) for $s = 5$ and $l = 2$ start with $a(0, 1) = 10$, $a(0, 2) = 155$, $a(0, 3) = 2335$.

On the other hand, in order to enumerate the nonnegative paths of the corresponding elementary random walk in the lattice observe that such a walk yields a sequence $(a_1, \ldots, a_m)$ with $a_i \in \{s - l, -l\}$ and the nonnegative lattice paths correspond to such sequences with the additional property

$$\sum_{i=1}^{m} a_i = 0, \sum_{i=1}^{m'} a_i \geq 0 \text{ for } m' = 1, \ldots, m - 1$$

Observe that such a sum can only be 0 if $m = s \cdot n$ is divisible by s.

Letting again $s = 5$ and $l = 2$, there are two such sequences for $m = 5$ (namely $(3, 3 - 2, -2, -2)$ and $(3, -2, 3, -2, -2)$), 23 sequences for $m = 10$ and 377 sequences for $m = 15$, which fulfill These sequence had been first observed by Berlekamp [4] in the analysis of a special class of convolutional codes and thoroughly studied in [14].

5 Concluding Remarks

We analyzed a model for creating order in sequence spaces in which in each time unit two or more bits arrive at an organizer who then has to choose one bit for output. The case, where two bits are produced by symmetric sources is quite well understood.

Several generalizations are possible. In the s–user model ($s \geq 3$) also the case of a memory device with maximum size M may be considered. The analysis here is much more difficult, since now the transition matrices become more complicated, e. g., a reflection argument in order to obtain the transition matrix from the power of the matrix of an elementary random walk is no longer possible.

For the infinite–memory case an open problem connected to the model in which the organizer has to put out more than one bit per time unit has already been stated at the end of Section 3.

One might also consider nonidentical sources. Even for $s = 2$ sources producing a 1 with probalities p and q, respectively, this seems to be quite hard. For the case that every incoming bit can be stored one might ask for a condition on the probabilities p and q to assure the existence of E_0.

References

1. Ahlswede, R., Kaspi, A.: Optimal coding strategies for certain permuting channels. IEEE Trans. Inf. Theory 33, 310–314 (1987)
2. Ahlswede, R., Ye, J.P., Zhang, Z.: Creating order in sequence spaces with simple machines. Information and Computation 89(1), 47–94 (1990)

3. Ahlswede, R., Zhang, Z.: Contributions to a theory of ordering for sequence spaces. Problems of Control and Information Theory 18(4), 197–221 (1989)
4. Berlekamp, E.R.: A class of convolutional codes. Information and Control 6, 1–13 (1963)
5. Coxeter, H.S.M.: Regular Polytopes, Dover (1973)
6. Graham, R.L., Knuth, D.E., Patashnik, O.: Concrete Mathematics. Addison-Wesley (1988)
7. Hollmann, H.D.L., Vanroose, P.: Entropy reduction, ordering in sequence spaces, and semigroups of non–negative matrices, preprint 95-092, SFB 343, University of Bielefeld (1995)
8. Kac, M.: Random walk and the theory of Brownian motion. American Mathematical Monthly 54, 369–391 (1947)
9. Kobayashi, K.: Combinatorial structure and capacity of the permuting relay channel. IEEE Trans. Inf. Theory 33, 813–826 (1987)
10. Piret, P.: Two results on the permuting mailbox channel. IEEE Trans. Inf. Theory 35, 888–892 (1989)
11. Rivlin, T.J.: The Chebyshev polynomials. Wiley (1974)
12. Savio, D.Y., Suryanarayan, E.R.: Chebyshev polynomials and regular polygons. American Mathematical Monthly 100, 657–661 (1993)
13. Spitzer, F.: Principles of Random Walks. Springer (1976)
14. Tamm, U.: Lattice paths not touching a given boundary. J. Statistical Planning and Inference 105(2), 433–448 (2002)
15. Vanroose, P.: Ordering in sequence spaces: an overview. In: Althöfer, I., Cai, N., Dueck, G., Khachatrian, L., Pinsker, M., Sárközy, A., Wegener, I., Zhang, Z. (eds.) Numbers, Information and Complexity. Kluwer (2000)

Abschied

Alexander Ahlswede

Es füllten mich die Träume
wie immer mit altem Schwank,
im Schatten hoher Bäume
nahm ich den Zaubertrank.

Und nun trat plötzlich jenes
Los in mein stöberndes Licht:
zuviel erbittert Geschehenes,
und Hoffnung gibt es nicht.

Doch tragen die Schatten noch Bilder
vom geliebten Vater dahin,
und meine Seele wird milder,
gab er meiner Liebe doch Sinn.

Rudi

Beatrix Ahlswede Loghin

My name is Beatrix Ahlswede Loghin. I was married to Rudi Ahlswede from 1970 until 1984. Rudi and I are the parents of a son, Alexander Ahlswede.

Rudis death was sudden. There was no warning, no time to consider, to right wrongs, to express love and thanks. He left us quickly and undramatically. We have come together here today to reclaim a moment of closeness with him which death snatched away. Through the power of our remembrance, we evoke Rudi back into our world for this brief moment. Or, to quote T.S. Eliot : “History is now and England, with the drawing of this love, and the voice of this calling”.

Preparing this obituary I found myself pondering the question, again and again: how to go about this? A human being is so complex. Of all the myriad possibilities, moments, experiences, selves, of which we consist, which ones do we choose to share? What does one say? Isn't anything that we say a reduction, a limiting of this particular human beings complexity? Is not our life a great work of algebra, in which we ponder the great X, the mystery of our lives? And so I realized that I cannot write about Rudi, because I dont know “Rudi”. Even after all these years of experience with him, living with him, being in a family with him, I dont really know Rudi. All I know is my Rudi, my experience of him.

The Canadian writer, Margaret Atwood, gave this advice to young writers: “Say what is yours to tell”. That is all we can do, but also all we need to do: Say what is ours to tell.

“I come to bury Caesar, not to praise him”. No sooner are these words spoken, than Marc Antony of course begins to do just that praise Caesar, in Shakespeares historical drama. Nevertheless, I pondered the distinction. How does one speak of the dead? If we praise, we end up speaking only of the “nice”, “pleasant” attributes. A kind of “Rudi Ahlswede lite” version. Those of us who spent time with Rudi know that this was not his way. Rudis interaction with life was passionate. He loved “not wisely, but too well”. He was not given to strategic behavior, even though it would perhaps have been wiser at times. On the other hand, the dead are defenceless, they relinquish to us the power of definition, for we are still alive to tell the tale. Looking into my heart, I asked myself, “What is it really that you want to tell?” The answer that I found was this: I want to honor Rudis life here, I want to honor the complexity of his being. I want to acknowledge the difference Rudi made in my life.

But what does it mean to acknowledge someone? The Oxford dictionary states that to acknowledge means to take something which has been previously known to us and which we now feel bound to lay open or make public. It means to recognize a particular quality or relationship which we forgot or did not consciously see. And it means to own with gratitude.

What did I know then, and wish to lay open now? Which qualities did I forget or not consciously see? What can I own with gratitude? Of the rich tapestry of Rudis life, where do I begin to acknowledge? We cannot remember the entire sequence of life. We remember moments, special moments which - for some reason - stayed in our memory. So this is what I really want - to share with you some of these moments.

Thinking of Rudi, an image of a great mountain range comes to my mind, with invincible summits, terrifying plunges and depths, and a smattering of meadows in between. This image has been the defining core of my relationship with Rudi, beginning with our first meeting in the summer of 1967 in Columbus, Ohio. I was 18 years old and had just begun my freshman year at Ohio State University. Rudi was 29 years old and starting his first job in the US as an assistant professor in the Department of Mathematics.

At this time explosions were rocking the social and political fabric of American society. Afro-Americans, Latinos, Asian Americans and other groups were claiming their rightful place in American society, and protest against the Vietnam War was flaming up everywhere, even in politically conservative Ohio. I frequented a bar known as Larrys in Columbus, on High Street, refuge to those who considered themselves left-wing, or at least to the political left of the mainstream. In this bar, classical music, jazz and soul music was played, people of different races and nationalities congregated in cheerful bawdiness, and of course chess was played.

A mutual friend at Larrys Bar introduced us, and between long silences, in which he scrutinized his chess partners moves, Rudi told me a little about himself, his fascination with his research, information theory, and the discoveries he was making about life in the United States. The more I became embroiled in the political demonstrations against the Vietnam War, the more Rudi became interesting for me. My fellow demonstrators and I quoted Ho Chi Minh, Mao Tse Tung and Marx, but Rudi had actually read some of Karl Marxs writings, and he was able to put these writings into a philosophical context, showing the evolution of Hegels and Feuerbachs ideas. The great breadth of his knowledge left me stunned. I began to pay closer attention to Rudi. Not only had he read philosophy, but also literature, finding his own favourite writers and poets. In a conversation, Rudi would suddenly, just at the right moment, quote Schiller or Gottfried Benn, Goethe, Shakespeare, Thomas Wolfe or Nietzsche.

I was amazed, for he refuted all my conceptions of "typical" mathematicians. He told me more about himself. His parents owned a large farm in northern Germany. Born as the second son, he realized early in life that, much as he love the land with its wide open spaces, hills and cliffs and lush forests, he would have to leave it, as the farm would not be able to support two families. This realization was painful, tinged with bitterness. It forced him, at a very early age, to learn to create his own future. "God bless the child thats got his own", is a line from a Billie Holliday song. Rudi was such a blessed child - he had his own. He found his new world at school - his home became the world of books, the world of learning. And his aptitude in mathematics became apparent. At the age of ten he left his parents home and lived with another family in the nearby larger town, where he could attend the Gymnasium, the secondary school which would prepare him for a university education. Later, at Gymnasium, he often felt excluded because of his background as a farmers child. Some of his fellow students let him feel, very clearly, that he was lacking in social graces, that he came from an inferior social background. I think he never quite got over the pain of this discrimination. Learning became his passion. And this path led him from his humble elementary school in Dielmissen to the greatest universities in the world, to membership in the Russian Academy of Science. He had a fire in his mind, and this made conversations with him scintillating. This was the terrain where our minds met, and where I fell in love.

Many evenings, watching him sit in the turmoil of Larrys bar, he exuded a quality of tranquillity. He was above the fray, either focused on his chess game, or "in communion" with his own thoughts, which he would occasionally add to the paper lying before him. He clearly had something which very few others in the room had: a world of his own. He seemed incredibly strong and rooted in himself. Occasionally he would sit up, take notice of the life teeming around him, and then return again to this other, inner space.

This fascination with the world of mathematics became particularly evident one evening in the Spring of 1970. Richard Nixon had just announced the invasion of Cambodia. At universities around the country, massive strikes as a form of resistance took place. Soon the campus at Ohio State became a small battleground. Tanks rolled through the streets, students erected barricades and threw bricks and Molotov cocktails. Helicopters flew overhead, spraying the demonstrators with tear gas. Rudi and I sought refuge in the McDonalds on High Street, where we found Rudis colleague, Bogdan Baishanski, also seeking shelter. Demonstrators ran into the McDonalds, followed by night-stick brandishing police. We fled back onto the streets. In front of me, I saw Rudi and Bogdan running from the police, jumping over barricades, clearly illuminated by the searchlights of the helicopters flying over our heads, throwing more tear gas in our direction. Stumbling blindedly behind them, I noticed that, as they ran, they were deep in conversation - about the (at that time still unsolved) four color conjecture!

A short time later, Rudi had been stopped in the middle of the night while driving home, for making a right turn without a full stop. Because of an outstanding traffic violation, he was arrested and led off in handcuffs. I scrambled to find two hundred dollars with which to bail him out. When I arrived at the jail the next morning, Rudi emerged smiling. He told me about the "interesting" evening he had spent, stuffed in a holding cell with his fellow inmates. And, he told me proudly, he had gotten a new idea in jail which led to a significant break-through in the paper he was currently writing!

Years later I read in book written by someone who was researching happiness, that the happiest people are those who have something in their lives which so absorbs them that it permits them to completely forget themselves and the world around them. This process of forgetting oneself is called flow. I think

Rudi spent much of his life in this state. But of course this obliviousness to his surroundings left him vulnerable. Many times a date began with long searches in the parking lots around the Mathematics Department - Rudi simply could not remember where he had left the car that morning. Between us this was of course often a cause of exasperation on my part. One day, in a store, I noticed two young salesgirls giggling about Rudi, who was lost in space, smoking, and running his hands through his hair. A fierce determination to protect him in this vulnerability was born in me at that moment.

In this way, Rudi was like no one I had ever met. Years later, after we had moved to Germany, listening to my son and his friends recount funny anecdotes about Rudi, I realized that they were fascinated by precisely his way of being different from others, his eccentricity, to use another word. The word eccentric comes from the Greek words ek kentros, meaning not having the same center. Years later, after we had married, I stood in a market square with Rudi in Sicily, in Syracusa, the town where the great Archimedes had lived. He was killed when a Roman soldier accosted him in the market place, where he sat, drawing designs in the sand. Awed by Archimedes fame, the soldier asked if there was anything he could do for him. Archimedes is said to have answered: "Dont disturb my circles". This story impressed me greatly, for I was sure that Rudi would have given the same answer, and I recognized that he was a kindred spirit.

Shortly after we met, Rudi returned to Germany for a few weeks. He wrote to me that he was reading a book by Giordano Bruno, entitled "Heroic Passions". It seemed so fitting. Years later, when we lived in Rome, we spent many an hour at the Campo dei Fiori, where Bruno was burned at the stake for refusing to renounce his scientific ideas. I had no doubt that Rudi would have ended there too had he lived in this time. Rudi was never politically correct. He said what he thought and accepted the consequences. Rudi was incapable of inauthenticity. There was a wild, almost savage need in him to stay true to himself, a need which caused him much conflict and grief. But suppressing his beliefs in order to attain some goal was beyond him. He paid a huge price in his life for that and, at the same time, this is what made him so strong.

Rudi was the freest person I have ever met.

I saw Rudi for the last time on his last birthday, September 15, 2010. We spent the evening together, drinking a bottle of wine and talking of our son, of mutual old friends. The years passed by before our inner eyes. He was, as always, excited about life, looking forward to the new research he had embarked upon, and which he told me about, as always, with sparkling eyes. But something was different about this evening. After he finished talking, he asked me about myself. Amazed, I found myself telling Rudi about my life, my plans. He listened with a care and an attention that was new. We sat, side by side, companions of a shared life. I went home elated, feeling blessed and rich from this evening with Rudi.

Standing at his coffin in the cemetery, looking at his dead body, I realized there was only one word left to say to him: Thank you.

Gedenkworte für Rudolf Ahlswede

Konrad Jacobs

Universität Erlangen-Nürnberg, Lehrstuhl für Stochastik,
Cauerstr. 11, 91058 Erlangen
jackon@web.de

Rudi Ahlswede bin ich zum letzten Mal begegnet, als er am 30. Januar 2009 in Erlangen einen Festvortrag zur Nachfeier meines 80. Geburtstages hielt. Mathematiker wie Nichtmathematiker erlebten da einen Fürsten seines Fachs, der sein gewaltiges Souveränitätsgebiet begeistert und begeisternd durchstürmte und Ideen zu dessen fernerer Durchdringung und Ausweitung in großen Horizonten entwarf.

Ich möchte noch kurz ein wenig über die "Anfangsbedingungen" berichten, die Rudi bei seinem 1966 mit der Promotion endenden Stochastik-Studium in Göttingen vorfand. Das Fach Stochastik, damals Mathematische Statistik genannt, war nach Kriegsende in West-Deutschland m.W. nur durch die Göttinger Dozentur von Hans Münzner (1906 - 1997) vertreten und mußte somit praktisch neu aufgebaut werden. Das begann mit der Übernahme neugeschaffener Lehrstühle durch Leopold Schmetterer (1919 - 2004) in Hamburg und Hans Richter (1912 - 1978) in München, die beide ursprünglich Zahlentheoretiker waren und sich in ihr neues Fach einarbeiteten. Dieser "1. Welle" folgte eine zweite, in der Jungmathematiker, wie Klaus Krickeberg (* 1929) und ich (* 1928), die in ihrem ursprüng-lichen Arbeitsgebiet bereits eine gewisse Nachbarschaft zur Stochastik vorweisen konnten. Bei mir war das durch Arbeiten zur Ergoden- und Markov-Theorie gegeben. Als ich 1958 in Göttingen das Münznersche Klein-Institut im Keller des großen Mathematischen Instituts an der Bunsenstraße übernahm, war ich für meine neue Aufgabe eigentlich zu jung und unerfahren. Ein Student, der damals zu meiner kleinen Gruppe stieß, konnte nicht erwarten, von einem souveränen, erfahrenen Ordinarius umfassenden Rat zu erhalten: ich hatte ihm damals nur einen Schritt der Einarbeitung in neue Themengebiete voraus. Meinen Zugang zur Shannon'schen Informationstheorie, auf die ich Rudi und andere "anzusetzen" versuchte, hatte ich über die Ergodentheorie gefunden, die mit der Einführung der Entropie-Invarianten (1959) durch A.N. Kolmogorov (1903 - 1987) und Y. Sinai (* 1937) einen mich unmittelbar betreffenden Bezug zur Informationstheorie erhalten hatte, der in einem Uspehi-Artikel (1956) von A.Y. Chintchine (1894 - 1995) schon vorher systematisch ausgebreitet worden war; da diese Arbeit in Ostdeutschland sogleich ins Deutsche übersetzt worden war, hatten wir hier sprachlich sofort Zugang. Wesentlichere Impulse für uns ergaben sich allerdings aus dem Ergebnisbericht Coding Theorems of Information Theory (1961) von Jacob Wolfowitz (1910 - 1981). Nach seiner Promotion kam es zu intensiven Kontakten mit J. Wolfowitz, mit dem Rudi später mehere Arbeiten gemeinsam verfaßte, und dem er schließlich einen groß-artigen Nachruf widmete.

Da ich Studenten wie R. Ahlswede und V. Strassen nur geringfügig "voraus" war, hatte ich später das beglückendste Erlebnis, das einem akademischen Lehrer zuteil werden kann: von seinen "Schülern" überholt zu werden und von ihnen lernen zu können.

Auch nach der Erlanger Begegnung Anfang 2009 kam es immer wieder zu Telefonkontakten zwischen Rudi und mir. Bei einem der letzten (wohl 2010) schilderte ich ihm meine Erwägungen über die Frage, wie man sich als Mathematiker zu dem unvermeidlichen fachlichen Leistungsabfall - wie allmählich auch immer - nach der Emeritierung stellen solle. Ich hatte mich dafür entschieden, dann (bei mir nach 1993) nicht mehr forschungsaktiv zu sein, sondern mich anderen Interessengebieten zuzuwenden, wenn auch naturgemäß auf nunmehr amateurhaftem Niveau. Als ich ihn um seine Meinung hierzu fragte, kam die Antwort sogleich und in aller Entschiedenheit: seine Devise sei

Stirb in den Stiefeln!
(Die in your boots!).

Bei seinem Naturell kam nur in Frage, weiterzuarbeiten, so intensiv und so lange es nur angehen mochte. Rudi hatte noch eine Überfülle von Ideen und Problemen. In den Stiefeln, die ihm angewachsen waren, wäre er noch sehr lange weitermarschiert. So einen wie ihn vergißt man nie.

In Memoriam Rudolf Ahlswede 1938 - 2010*

Imre Csiszár[1], Ning Cai[2], Kingo Kobayashi[3], and Ulrich Tamm[4]

[1] Rényi Institute of Mathematics, Budapest, Hungary
csiszar@renyi.hu
[2] State Key Lab. of ISN, Xidian University Xian 710071, China
caining@mail.xidian.edu.cn
[3] The University of Electro-Communications Department of Information and Communication Engineering Chofugaoka 1-5-1 Chofu, Tokyo 182 Japan
kingo@ice.uec.ac.jp
[4] German Language Department of Business Informatics, Marmara University, Istanbul, Turkey
ulrich.tamm@yahoo.com

Rudolf Ahlswede, a mathematician, one of the truly great personalities of Information Theory, passed away on December 18, 2010 in his house in Polle, Germany, due to a heart attack. He is survived by his son Alexander. His untimely death, when he was still very actively engaged in research and was full with new ideas, is an irrecoverable loss for the IT community.

Ahlswede was born on September 15, 1938 in Dielmissen, Germany. He studied Mathematics, Philosophy and Physics in Göttingen, Germany, taking courses, among others, of the great mathematicians Carl Ludwig Siegel and Kurt Reidemeister. His interest in Information Theory was aroused by his advisor Konrad Jacobs, of whom many students became leading scientists in Probability Theory and related fields.

In 1967 Ahlswede moved to the US and became Assistant Professor, later Full Professor at Ohio State University, Columbus. His cooperation during 1967 - 1971 with J. Wolfowitz, the renowned statistician and information theorist, contributed to his scientific development. Their joint works included two papers on arbitrarily varying channels (AVCs), a subject to which Ahlswede repeatedly returned later.

His first seminal result was, however, the coding theorem for the (discrete memoryless) multiple-access channel (MAC). Following the lead of Shannons Two-Way Channel paper, this was one of the key results originating Multiuser Information Theory (others were those of T. Cover on broadcast channels and of D. Slepian and J. Wolf on separate coding of correlated sources), and it was soon followed by an extension to two-output MACs, requiring new ideas. Also afterwards, Ahlswede continued to be a major contributor to this research direction, in collaboration with J. Körner (visiting in Columbus in 1974) and later also with other members of the Information Theory group in Budapest, Hungary. In addition to producing joint papers enriching the field with new results

* This obituary first appeared in IEEE Information Theory Society Newsletter, Vol. 61, No. 1, 7-8 2011.

and techniques, this collaboration also contributed to the Csiszár- Körner book where several ideas are acknowledged to be due to Ahlswede or have emerged in discussions with him.

In 1975 Ahlswede returned to Germany, accepting an offer from Universität Bielefeld, a newly established "research university" with low teaching obligations. He was Professor of Mathematics there until 2003, and Professor Emeritus from 2003 to 2010. For several years he devoted much effort to building up the Applied Mathematics Division, which at his initiative included Theoretical Computer Science, Combinatorics, Information Theory, and Statistical Physics. These administrative duties did not affect his research activity. He was able to develop a strong research group working with him, including visitors he attracted as a leading scientist, and good students he attracted as an excellent teacher. In the subsequent years Ahlswede was heading many highly fruitful research projects, several of them regularly extended even after his retirement which is quite exceptional in Germany. The large-scale interdisciplinary project "General Theory of Information Transfer" (Center of Interdisciplinary Research, 2001 - 2004) deserves special mentioning. It enabled him to pursue very productive joint research with many guests and to organize several conferences. An impressive collection of new scientific results obtained within this project was published in the book "General Theory of Information Transfer and Combinatorics" (Lecture Notes in Computer Science, Springer, 2006).

During his research career Ahlswede received numerous awards and honours. He was recipient of the Shannon Award of the IEEE IT Society in 2006, and previously twice of the Paper Award of the IT Society (see below). He was member of the European Academy of Sciences, recipient of the 1998/99 Humboldt-Japan Society Senior Scientist Award, and he received honorary doctorate of the Russian Academy of Sciences in 2001. He was also honored by a volume of 50 articles on the occasion of his 60th birthday (Numbers, Information and Complexity, Kluwer, 2000.)

Ahlswedes research interests included also other fields of Applied and Pure Mathematics, such as Complexity Theory, Search Theory (his book "Search Problems" with I. Wegener is a classic), Combinatorics, and Number Theory. Many problems in these disciplines that aroused Ahlswedes interest had connections with Information Theory, and shedding light on the interplay of IT with other fields was an important goal for him. He was likely the first to deeply understand the combinatorial nature of many IT problems, and to use tools of Combinatorics to solve them.

In the tradition of giants as Shannon and Kolmogorov, Ahlswede was fascinated with Information Theory for its mathematical beauty rather than its practical value (of course, not underestimating the latter). In the same spirit, he was not less interested in problems of other fields which he found mathematically fascinating. This is not the right place to discuss his (substantial) results not related to IT. We just mention the celebrated Ahlswede-Daykin "Four Functions Theorem" having many applications in Statistical Physics and in Graph Theory, and the famous Ahlswede-Khachatrian "Complete Intersection Theorem". The latter provided the final solution of a problem of Paul Erdős, which had been very long-standing even though Erdős offered \$500 – for the solution (Ahlswede and Khachatrian collected). For more on this, and also on combinatorial results of information theoretic interest, see his book "Lectures on Advances in Combinatorics" with V. Blinovsky (Springer, 2008).

Even within strict sense Information Theory, Ahlswedes contributions are too wide-ranging for individual mentioning, they extend as far as the formerly exotic but now highly popular field of Quantum Information Theory. Still, many of his main results are one of the following two kinds.

On the one hand, Ahlswede found great satisfaction in solving hard mathematical problems. Apparently, this is why he returned again and again to AVCs, proving hard results on a variety of models. By his most famous AVC theorem, the (average error) capacity of an AVC either equals its random code capacity or zero. Remarkably, this needed no hard math at all, "only" a bright idea, the so-called elimination technique (a kind of derandomization). He was particularly proud of his solution of the AVC version of the Gelfand-Pinsker problem about channels with non-causal channel state information at the sender. To this, the elimination technique had to be combined with really hard math. Another famous hard problem he solved was the "zero excess rate" case of the Multiple Descriptions Problem (the general case is still unsolved).

On the other hand, Ahlswede was eager to look for brand new or at least little studied models, and was also pleased to join forces with coauthors suggesting work on such models. His most frequently cited result (with Cai, Li and Yeung), the Min- Cut-Max-Flow Theorem for communication networks with one source and any number of sinks, belongs to this category. So do also his joint results with Csiszár on hypothesis testing with communication constraints, and with Dueck on identification capacity, receiving the Best Paper Award of the IT Society in 1988 and 1990. Later on, Ahlswede has significantly broadened the scope of the theory of identification, for example to quantum channels (with Winter). Further, a two-part joint paper with Csiszár provides the first systematic study of the concept of common randomness, both secret and non-secret, relevant, among others, for secrecy problems and for identification capacity. The new kind of problems studied in these papers support Ahlswedes philosophical view that the real subject of information theory should be the broad field of "information transfer", which is currently unchartered and only some of its distinct areas (such as Shannons theory of information transmission and the Ahlswede-Dueck theory of identification) are in view. Alas, Rudi is no longer with us, and extending information theory to cover such a wide scope of yet unknown dimensions will be the task of the new generation.

Rudolf Ahlswede 1938-2010

Christian Deppe

Universität Bielefeld, Fakultät für Mathematik, Postfach 10 01 31, 33615 Bielefeld
cdeppe@math.uni-bielefeld.de

We, his friends and colleagues at the Department of Mathematics at the University of Bielefeld are terribly saddened to share the news that Professor Rudolf Ahlswede passed away in the early hours of Saturday morning 18th December, 2010.

Rudolf Ahlswede had after an excellent education in Mathematics, Physics, and Philosophy almost entirely at the University of Göttingen and a few years as an Assistant in Göttingen and Erlangen received a strong push towards research, when he moved to the US, taught there at the Ohio State University in Columbus and greatly profited from joint work in ***Information Theory*** with the distinguished statistician Jacob Wolfowitz at Cornell and the University of Illinois during the years 1967-1971 (see the obituary [A82]).

The promotion to full professor in Mathematics followed in 1972, but only after Rudolf Ahlswede convinced his faculty by his work in Classical Mathematics. Information Theory was not yet considered to be a part of it. A problem in p-adic analysis by K. Mahler found its solution in [AB75] and makes now a paragraph in his book [M81].[1]

For a short time concentrating on Pure Mathematics and quitting Information Theory was considered. But then came strong responses to multi-way channels [A71] and it became clear that Information Theory would always remain a favorite subject – it looked more interesting to Rudolf Ahlswede than many areas of Classical Mathematics. An account of this period is given in the books [W78], [CK81], and [CT06]. However, several hard problems in Multi-user Information Theory led Rudolf Ahlswede to ***Combinatorics***, which became the main subject in his second research stage starting in 1977. Writing joint papers, highly emphasized in the US, helped Rudolf Ahlswede to establish a worldwide network of collaborators. Finally, an additional fortunate development was an offer from the Universität Bielefeld in 1975, which for many years was the only research university in Germany with low teaching obligations, implying the possibility

[1] Ingo Althöfer heard this story from Rudolf Ahlswede in a version with more personal flavour: Rudolf Ahlswede was in the Math department of Ohio State, but his Mathematics (= Information Theory) was not fully accepted by some traditionalists in the department. Rudi decided to ask them: "Who is the strongest mathematician in the department?" Answer: "Kurt Mahler." So, he stepped in Mahler's office and asked him: "Please, give me an interesting open problem of yours." So did Mahler (1903-1988), and Ahlswede solved it within a few weeks. After that demonstration there were no problems for him to get full Professorship.

to teach only every second year. In a tour de force within half a year Rudolf Ahlswede shaped a main part of the Applied Mathematics Division with Professorships in Combinatorics, Complexity Theory (first position in Computer Science at the university), and Statistical Mechanics.

Among his students in those years were Ingo Althöfer (Habilitationspreis der Westfälisch-Lippischen Universitätsgesellschaft 1992), Ning Cai (IEEE Best Paper Award 2005), Gunter Dueck (IEEE Best Paper Award 1990; Wirtschaftsbuchpreis der Financial Times Deutschland 2006), Ingo Wegener (Konrad-Zuse-Medaille 2006), Andreas Winter (Philip Leverhulme Prize 2008) and Zhen Zhang. In the second stage 1977-87 the AD-inequality was discovered, made it into many text books like [B86], [A87], [AS92], [E97], and found many generalizations and number theoretical implications [AB08].

We cite from the book [B86] $ 19 The Four Function Theorem:

"At the first glance the FFT looks too general to be true and, if true, it seems too vague to be of much use. In fact, exactly the opposite is true: the Four Function Theorem (FFT) of Ahlswede and Daykin is a theorem from "the book". It is beautifully simple and goes to the heart of the matter. Having proved it, we can sit back and enjoy its power enabling us to deduce a wealth of interesting results. "

Combinatorics became central in the whole faculty, when the DFG-Sonderforschungsbereich 343 "Diskrete Strukturen in der Mathematik" was established in 1989 and lasted till 2000. The highlight of that third stage is among solutions of several number theoretical and combinatorial problems of P. Erdős [A01]. The most famous is the solution of the $4m$-Conjecture from 1938 of Erdős/Ko/Rado (see [E97], [CG98]), one of the oldest problems in combinatorial extremal theory and an answer to a question of Erdős (1962) in combinatorial number theory "What is the maximal cardinality of a set of numbers smaller than n with no $k+1$ of its members pairwise relatively prime?".

As a model most innovative seems to be in that stage Creating Order [AYZ90], which together with the Complete Intersection Theorem demonstrates two essential abilities, namely to shape new models relevant in science and/or technology and solving difficult problems in Mathematics.

In 1988 (with Imre Csiszár) and in 1990 (with Gunter Dueck) Rudolf Ahlswede received the Best Paper Award of the IEEE Information Theory Society. He received the Claude Elwood Shannon Award 2006 of the IEEE information Theory Society for outstanding achievements in the area of the information theory (see his Shannon Lecture [A06]).

A certain fertility caused by the tension between these two activities goes like a thread through Rudolf Ahlswede's work, documented in 235 published papers in roughly 4 stages from 1967-2010.

The last stage 1997-2010 was outshined by Network Information Flow [ACLY00] (see also [FS07a], [FS07b], [K]) and GTIT-updated [A08], which together with Creating Order [AYZ90] was linked with the goal to go from Search Problems to a Theory of Search.

The seminal paper [ACLY00] founded a new research direction in the year 2000, with many applications especially for the internet. It has been identified by Essential Science Indicators^{SM} as one of the most cited papers in the research area of "NETWORK INFORMATION FLOW". Research into network coding is growing fast, and Microsoft, IBM and other companies have research teams who are researching this new field. The most known application is the Avalanche program of Microsoft.

Rudolf Ahlswede had just started a new research project about quantum repeaters to bring his knowledge about physics and information theory together. Unfortunately he cannot work for the project anymore.

We lost a great scientist and a good friend. He will be missed by his colleagues and friends.

References

[A71] Ahlswede, R.: Multi-way communication channels. In: Proceedings of 2nd International Symposium on Inf. Theory, Thakadsor, Armenian SSR, Akademiai Kiado, Budapest, pp. 23–52 (1971, 1973)

[A82] Ahlswede, R., Wolfowitz, J.: IEEE Trans. Inf. Theory 28(5), 687–690 (1910-1981, 1982)

[A01] Ahlswede, R.: Advances on extremal problems in Number Theory and Combinatorics. In: European Congress of Mathematics, Barcelona, vol. I, pp. 147–175 (2000); Casacuberta, C., Miró-Roig, R.M., Verdera, J., Xambó-Descamps, S. (eds.): Progress in Mathematics 201. Birkhäuser, Basel (2001)

[A06] Ahlswede, R.: Towards a General Theory of Information Transfer, Shannon Lecture at ISIT in Seattle 13th July 2006. IEEE Inform. Theory Society Newsletter 57(3), 6–28 (2007)

[A08] Ahlswede, R.: General Theory of Information Transfer: updated, General Theory of Information Transfer and Combinatorics. Special Issue of Discrete Applied Mathematics 156(9), 1348–1388 (2008)

[AB08] Ahlswede, R., Blinovsky, V.: Lectures on Advances in Combinatorics, Universitext. Springer (2008)

[AB75] Ahlswede, R., Bojanic, R.: Approximation of continuous functions in p-adic analysis. J. Approximation Theory 15(3), 190–205 (1975)

[ACLY00] Ahlswede, R., Cai, N., Robert Li, S.Y., Yeung, R.W.: Network information flow. IEEE Trans. Inf. Theory 46(4), 1204–1216 (2000)

[AYZ90] Ahlswede, R., Ye, J.P., Zhang, Z.: Creating order in sequence spaces with simple machines. Information and Computation 89(1), 47–94 (1990)

[AZ89] Ahlswede, R., Zhang, Z.: Contributions to a theory of ordering for sequence spaces. Problems of Control and Information Theory 18(4), 197–221 (1989)

[AS92] Alon, N., Spencer, J.: The Probabilistic Method. Wiley-Interscience Series in Discrete Mathematics and Optimization. Wiley & Sons, Inc, New York (1992)

[A87] Anderson, I.: Combinatorics of Finite Sets, Oxford Science Publications. The Clarendon Press, Oxford University Press, New York (1987)

[B86] Bollobás, B.: Combinatorics, Set Systems, Hypergraphs, Families of Vectors and Combinatorial Probability. Cambridge University Press, Cambridge (1986)

[CG98] Chung, F., Graham, R.: Erdős on Graphs: His Legacy of Unsolved Problems. AK Peters (1998)

[CT06] Cover, T., Thomas, J.: Elements of Information Theory, 2nd edn. Wiley, New York (2006)

[CK81] Csiszár, I., Körner, J.: Information Theory. Coding Theorems for Discrete Memoryless Systems. In: Probability and Mathematical Statistics, Academic Press, New York (1981)

[E97] Engel, K.: Sperner Theory, Encyclopedia of Mathematics and its Applications, vol. 65. Cambridge University Press, Cambridge (1997)

[FS07a] Fragouli, C., Soljanin, E.: Network Coding Fundamentals, Foundations and Trends in Networking. Publishers Inc (2007)

[FS07b] Fragouli, C., Soljanin, E.: Network Coding Applications, Foundations and Trends in Networking. Publisher Inc. (2007)

[K] Kötter, R.: The Network Coding Home Page, http://www.ifp.illinois.edu/ koetter/NWC/

[M81] Mahler, K.: P-adic Numbers and their Functions, 2nd edn. Cambridge University Press (1981)

[W78] Wolfowitz, J.: Coding Theorems of Information Theory, 3rd edn. Springer, New York (1978)

Remembering Rudolf Ahlswede

Vladimir Blinovsky

Institute of Information Transmission Problems, Russian Academy of Sciences,
B. Karetnii Per 19, Moscow 127 994, Russia
vblinovs@yandex.ru

I might not sound original if I say that in life there are people whose demise leaves behind an irreplaceable emptiness. Trying to explain it in more details, the most painful part of this feeling is not to be able to imagine that now, when you dial that phone number, there is not anymore the familiar gruffish voice. And if you think, at first, that he is probably just away, mechanically you start considering the conferences occurring right now, and remember the last conversation - "Was he going to go somewhere last time?" You remember that he told: "I have been traveling a lot lately, but now I am at home, in Bielefeld, and I enjoyed the coming back home, to the German autumn, where the rain continuously drizzles, paints fade day by day, and with pleasure I drink the beer and look out of the window at the familiar street and think that Christmas is coming soon."

We have many photos at home with different scientific actions or simply friendly meetings, on the photos he is frequently with my father. The matter is that Rudi's scientific interests were closely connected to the researchs which were conducted at our institute at that time. He found here the right partners. He invited many researchers to Bielefeld for collaboration. Rudi always tried to make these trips not only interesting and productive from the scientific point of view. He organized meetings and conferences where it was possible to get acquainted with people, to exchange experiences. I remember that when I was a visitor and lived at the ZiF, at that time in the same place there were many scientists from Russia and from other countries, Hungary, Denmark, Italy. Ahlswede organized to meet once a week in an informal atmosphere in a guest lounge of the ZiF. But he also tried to make people acquainted with Germany, with the special charm of other places around Bielefeld which, can go unnoticed by the star-oriented Michelin's guide fans, but whose warmth foreign scientists could discover thanks to the trips with Rudi. After all he was born near Bielefeld. So we got acquainted on these trips with the German Renaissance, with cozy small towns, with history. For example, I remember the trip which began with visiting the possession of the visionary baron Münchhausen very popular in Russia because the baron served in Russia and the book of adventures was translated and published in Russia and adapted for children.

The Influence of Ahlswede to science and life of many mathematicians has deep and long-live consequences. He was the type of person who can change a life and deliver harmony in mathematical theories. His works is the genesis of Information Theory and Combinatorics which was very close to the main stream of research in the Institute of Information Transmission Problems at that time. Close cooperation with the Institute was one of the important activities of Rudi. He was recognized as Honorable Professor of the Institute, it was his first such award as he mentioned later. He appreciated the quality of scientific work in the Institute and many researchers from the Institute were frequent guests in Bielefeld. It was also the time when Russia became an open country and many people had their first experience to travel abroad. He also visited Russia many times. When Rudi was in Moscow I would invite him to go to the Bol'shoy theatre. Unfortunately during his stay there where no ballet in Bol'shoy, only operas. We went to see "Eugenii Onegin", by Chai'kovsky, it was in Russian. He listened the opera from the beginning to the end very enthusiastically and at the end Rudi said "Pushkin (the author of the same name poem) become my dream."

In a very popular German series, called "Crime scene" (in German "Tatort"), one of the episodes is "Bielefelder plot" ("Bielefelder Verschwörung"). The main point is that in the course of an investigation detectives face a hypothesis that the town of Bielefeld, indicated in all Germany maps, actually doesn't exist - possibly it is the thought-up strategic object and on this place anything isn't present, except for a field or the wood.

The inquisitive investigator as usual says "I don't trust, there can't be it." Intuition of the skilled police officer can't deceive, the city exist. And after looking for proofs, the investigator naturally finds them. And so if mathematicians from different places in the world are asked whether they know a city in Germany under the name of Bielefeld, each of them will tell: "Oh yes, it is that place where Ahlswede worked."

Rudi Ahlswede

Jacqueline W. Daykin

Department of Mathematics, University of Reading, UK Department of Computer Science, Royal Holloway, University of London, UK
J.Daykin@cs.rhul.ac.uk, jackie.daykin@kcl.ac.uk

Professor Dr. Rudolf Ahlswede. Rudi Ahlswede: this name will always remind us of a man of great intellectual prowess, totally confident in his own strengths, but ... also one who can be pretty mischievous!

My family met up with Rudis family around 1976 when Rudi was working with my father David Daykin. They clearly had good synergy, or chemistry as Rudi called it, both mathematically and in their personalities, along with much mutual respect. My father learnt a great deal from Rudi, as did most people who entered the circle of his aura and knowledge. They produced the celebrated 1978 "Four Functions" theorem, and I feel so proud that it carries the family name alongside that of Ahlswede. Funnily enough, these kindred spirits also had brief spells in jail: religious reasons for dad but irreligious for Rudi! Yes, they both liked to speak their mind, their convictions. And how lucky we are that Rudi did like to express himself and share his novel and ground-breaking ideas spanning: information theory, combinatorics, complexity and classical mathematics. He was guided not only by scientific truth, but also by aesthetic beauty. Even more, Rudi could talk for hours very knowledgeably on such topics as history, philosophy, literature and world affairs images of Renaissance man!

Life plays strange movies Rudi said to me. Well we were in one of those movies even when he said those words. It was 2010, and my father had died in the summer in true friendship Rudi came over to England to attend the memorial event at Reading University in the November. It was amazing for me to have contact with him again after so many years naturally he was very entertaining company, but also tremendously comforting regarding my recent loss. During those days I was also really fortunate to have Rudi advise me on my own research. How could we know at that time that we were actors in a movie that was about to end just 6 weeks later with the loss of Rudi? I feel truly honoured that I was present with my sons Paris and Alexander at the last talks given by Rudi, hosted by Reading University, Kings College and Queen Mary University of London; generous as ever with his acquired life experience, he also shared his company and wisdom with my sons, as he had done in previous years with my mother Valerie and sister Babs. Great memories - including us picking giant wild mushrooms in the forests of Germany with the Ahlswedes.

Coming to the Zif, University of Bielefeld, for Rudis memorial occasion in July 2011 was really enjoyable, especially seeing the ever-kind Trixie again, but it was also very nostalgic of course. What a delight to see how the young boy Sascha had grown into such a likeness of his dad no wonder Rudi had talked so fondly of him during our days together in that English autumn. Thank you Rudi for how you touched the lives of our family, for what you gave to the scientific community world-wide, and most importantly your own family too. You have left a rich legacy.

Yes, as we go through life we meet many people on the way and some leave a deeper impression than others: Rudis imprint on us all was permanent, so although he has gone ... he is still here. As is dad too of course. Indeed, it is quite poetic that my last joint paper with my father appears in the memorial volume in honour of the renowned Rudi Ahlswede.

The Happy Connection between Rudi and Japanese Researchers

Kingo Kobayashi[1] and Te Sun Han[2]

[1] The University of Electro-Communications Department of Information and Communication Engineering Chofugaoka 1-5-1 Chofu, Tokyo 182 Japan
`kingo@ice.uec.ac.jp`

[2] National Institute of Information and Communications Technology (NICT), Japan
`han@aoni.waseda.jp`

We would like here to summarize, in chronological order, the long-standing happy connection of Professor Rudolf Ahlswede to a Japanese research group, especially Japanese information theorists.

The first contact with Japanese researchers took place at Grignano, in 1979; Te Sun and Kingo first met there Rudi on the occasion of IEEE ISIT. This is our first experience to visit outside Japan and to join an international symposium. Subsequently, on response to Rudi's kind invitation, Te Sun visited Bielefeld for three months in 1980.

In 1984, Rudi visited Japan, and delivered several invited talks at the 7th SITA (=Symposium on Information Theory and its Applications), Kinugawa, and at University of Tokyo, a plenary talk : "Strategies for trap-door channels" by R. Ahlswede and A. Kaspi, and "The rate-distorsion region for multiple-descriptions without excess rate" by R. Ahlswede, respectively. In particular, the talk on the trapdoor channel has greatly interested Kingo and motivated him to enthusiastically study the permuting relay channel. As a result: Kingo published a paper related to the trapdoor channel, "Combinatorial structure and capacity of the permuting relay channel," *IEEE Trans. on Inform. Theory*, IT-33, No.6, pp.813-826, 1987. On the other hand, Te Sun and Shun-ichi Amari were greatly inspired by the talk on multiterminal hypothesis testing and multiterminal parameter estimation, which has occasioned for them to publish a series of papers on this topics in *IEEE Trans. on Inform. Theory*, 1987, 1989, 1995, 1998, etc.

In 1986, Te Sun and Kingo visited Germany to attend Oberwolfach Workshop on Information Theory that Rudi organized and invited us to. On the occasion of IEEE ISIT, Kobe in 1988, Rudi visited Japan, and joined a workshop at Hakone that we organized.

After IEEE IT Cornell Workshop in 1989, Te Sun, Kingo and Suguru Arimoto visited Germany to attend Oberwolfach Workshop on Information Theory which Rudi organized. Meanwhile, Te Sun visited Bielefeld University as a visiting fellow for joint research with Rudi between 1994 and 1995, and Kingo was staying at Bielefeld as a visiting fellow for joint research with him between 1993 to 1995. These collaborations led to publication of a joint paper, "Universal coding of integers and unbounded search trees," *IEEE Trans. on Inform. Theory*, R.Ahlswede, T. S. Han and K.Kobayashi, Vol.43, No.2, March, pp.669-682,1997.

It should be mentioned also that we made a remarkable success of a joint program between Japan and Germany, from 1995.10.1 to 1997.9.30, under the Japanese-German Cooperative Science Promotion Program "A mathematical study on fundamental structures of information" (Japanese side: S.Arimoto, T.S.Han, K.Kobayashi, etc.; German side: A.J.H.Vinck, R.Ahlswede, U.Tamm, etc.). During that period, several researchers from Bielefeld University, Ulrich Tamm, Ning Cai, Levon Khachatrian, etc., visited Japan, and exchanged research ideas with Japanese researchers by having small workshops several times.

In this connection, we would like to remind you of the paragraph "A visit of Te Sun Han for 6 months in Bielefeld in 1980 and of Kingo Kobayashi for two years in the 90's caused spreading of ideas and added to a flourishing school in Information Theory in Japan," which appeared in the preface of the book "Numbers, Information and Complexity," I. Althöfer, N. Cai, G. Dueck, L. Khachatrian, M. S. Pinsker, A. Sárközy, I. Wegener and Z. Zhang ed., Kluwer, 2000.

From Nov. 1, 1998 to March 5, 1999, Rudi visited University of Electro-Communications as a visiting scientist under the exchange program between JSPS (= Japan Society for the Promotion of Science) and the Humboldt Foundation, endowed with Japan Society Senior Scientist Award. In this opportunity, we held a Memorial Workshop for the 50th Anniversary of the Shannon Theory, SITA, Kofu, Japan, Jan. 22- 24, 1999. At this workshop, Rudi gave the special talk: "Some of my ideas in information theory."

On the other hand, Kingo, Hiroshi Nagaoka and Norihide Tokushige from Japan visited Bielefeld University to attend the Opening Conference on "General Theory of Information Transfer," in November, 2002. Moreover, Kingo and Hiroshi stayed in Bielefeld, August, 2003, to attend the workshop on "Information Theory and Some Friendly Neighbors - ein Wunschkonzert" on General Theory of Information Transfer and Combinatorics. Just before this workshop, we organized the Asia-Europe Workshop at Kamogawa Japan, followed by IEEE ISIT Kobe, to welcome many European researchers to Japan.

Furthermore, Kingo, Hiroyoshi Morita and Hirosuke Yamamoto visited Bielefeld University in April, 2004 to attend the workshop on General Theory of Information Transfer and Combinatorics, and Kingo attended at a workshop held on the occasion of Rudi's 66th birthday, Sep. 2004. At IEEE ISIT Seatle, June, 2006, Rudi won Shannon Award. In the same year, the seminal book, *General Theory of Information Transfer and Combinatorics,* R. Ahlswede et al. (Eds.), LNSC 4123, Springer, was published with the picture of the "trapdoor" on the cover of the book.

Kingo visited Germany to attend Dagstuhl workshop on Search Methodologies which Rudi organized in 2009, and also visited Bielefeld University in October to attend the 2nd workshop on Search Methodologies, in 2010.

Rudi put great significance on the research of "Search". He shed light on the logical structures of searching problems by revealing the essential, common key points underlying various kinds of search problems, and published the first book on search from a mathematical viewpoint in 1979, translated into Russian in 1982, and into English in 1987.

He chose "Search Problems" but not "The Theory of Searching" as the title of the book with the intention to focus on the interdisciplinary character of this study and to emphasize the recognition of the importance of this problem. After his great trail of multidimensional studies on (multiterminal, network and quantum) information theory, discrete structures of mathematics, number theory, etc., he came back to this theme of "searching" again, to organize two workshops on "Search Methodologies" held at Dagstuhl and Bielefeld in his last two years. It seems that he started his academic career with "search" and ended with "search."

At the end of the workshop, when Kingo asked Rudi if he is willing to organize another workshop, he replied with a childlike smile, "No, Enough!" In retrospect, was Rudi already feeling at that time a presentiment for the last sad accident?

From Information Theory to Extremal Combinatorics: My Joint Works with Rudi Ahlswede

Zhen Zhang

Communication Sciences Institute, Ming Hsieh Department of Electrical Engineering, Viterbi School of Engineering, University of Southern California
zhzhang@usc.edu

Rudolf Ahlswede, a giant in the fields of combinatorics and information theory, has left us for more than two years. I met Rudi for the first time in 1984 when he visited Cornell University. I came to the University of Bielefeld in 1986 and worked with Rudi for two years. After that I visited Rudi in Bielefeld several times in the 1990s. Rudi and I have common interest in both extremal combinatorics and information theory(actually Rudi introduced me to the theory of extremal combinatorics). During our long time collaborations, we published many joint papers. Rudi's unexpected death was an irrecoverable loss of the fields of both combinatorics and information theory. This article is written in memory of Rudi, in which I will recall some of our most important joint works.

1 Ahlswede-Zhang Identity

The Ahlswede-Zhang Identity was discovered in 1987 and published in 1990 ("An identity in combinatorial extremal theory", Adv. Math. 80 (1990), no. 2, 137151) . This identity turns unexpectedly the famous LYM inequality into an identity.

Let

$$[n] = \{1, 2, ..., n\},$$

and Ω_n be the family of all subsets of $[n]$, and Φ be the empty set. Let $\Phi \neq \mathcal{F} \subset \Omega_n$.

If $A \not\subseteq B, \forall A, B \in \mathcal{F}$ with $A \neq B$, then $\mathcal{F}$ is called an antichain. For any antichain $\mathcal{F}$, the following inequality holds:

$$\sum_{X \in \mathcal{F}} \frac{1}{\binom{n}{|X|}} \leq 1.$$

This is called the LYM-inequality (Lubell, Yamamoto, Meshalkin) Many generalizations of the LYM-inequality have been obtained. In particular, the LYM-inequality is a direct consequence of the AZ identity.

Theorem 1. *For every family $\mathcal{F}$ of non-empty subsets of $\Omega = \{1, 2, \ldots, n\}$*

$$\sum_{X \in 2^{\Omega}} \frac{W_{\mathcal{F}}(X)}{|X| \binom{n}{|X|}} = 1,$$

where $W_{\mathcal{F}}(X) = |\bigcup_{X \supset F \in \mathcal{F}} F|$.

This equation is called the AZ-identity. Note that when $\mathcal{F}$ is an antichain this equation becomes

$$\sum_{F \in \mathcal{F}} \frac{1}{\binom{n}{|F|}} + \sum_{X \notin \mathcal{F}} \frac{W_{\mathcal{F}}(X)}{|X|\binom{n}{|X|}} = 1,$$

and as the second term on the left is non-negative, we obtain the LYM inequality.

The AZ identity has been generalized in several ways and was found to be a powerful tool in extremal combinatorics.

2 Creating Order in Sequence Spaces

This series of works was done in 1987, too. At that time, I was helping Rudi to guide one of his Ph.D. students. I proposed a problem as the Ph.D. research topic for his student. The problem is the following: Let $\{X_n\}_{n=0}^{\infty}$ be a stationary random sequence(the simplest case is the i.i.d. sequence) with finite alphabet $\mathcal{X}$. Consider a finite state machine that operates as follows: An organizer who picks one element at time n as the output Y_n from a buffer of size m that stores m elements of $\mathcal{X}$ before X_n is added to the buffer as the input to replace the output removed. The output selection depends on the buffer content, the knowledge of some of the previous output(memory of the past) and may also depend on the future inputs(look ahead). This operation changes the order of the random variables of the random input sequence to create a new random output sequence. The problem is to find the output selection rules in such a way that the entropy rate of the output sequence is minimized.

The problem seems very difficult and we were able to solve only one of the simplest cases. When Rudi looked at the problem, he proposed to drop the randomness of the input sequence and consider the cardinality of the set of all possible output sequences instead of the entropy rate of the output. Then the problem is to minimize the cardinality of the output space. This made the problem much easier and tackleable. In our joint paper titled "Creating Order in Sequence Space with Simple Machines", (Information and Computation, Vol. 89, No. 1, 47-94, November, 1990) by Rudi Ahlswede, Ye and myself, the deterministic version of the problem was formulated and several cases were resolved.

Although there are some follow-up works, but just like Rudi mentioned in his Shannon Lecture that this research direction did not receive as much attention as it deserves.

3 Inherently Typical Subset Lemma

In the summer of 1995, Enhui Yang and I visited Rudi in Bielefeld, through our discussion with Rudi, we proposed a problem called identification with compressed data. During our research, we found that the problem is closely related to a very general isoperimetric problem originally proposed in the book "Information Theory: Coding Theorems for Discrete Memoryless Systems" by Imre Csiszár and Janos Körner which had been open for over 20 years. After a whole month of intensive work, the problem was resolved and a paper titled "Identification via Compressed Data" was published in Jan. 1997 in IEEE Trans. on Information Theory. The key for the solution of the problem is a lemma called inherently typical subset lemma. As stated in the paper, this new method goes considerably beyond the entropy characterization the image size characterization, and its extensions. It is conceivable that this new method has a strong impact on multiuser information theory.

In a survey paper by Imre Csiszár titled "The Method of Types", this lemma is cited as follows: A recent combinatorial result of Ahlswede, Yang, and Zhang is also easiest to state in terms of F-types. Their inherently typical subset lemma says, effectively, that given $\mathcal{X}$ and ϵ, there is a finite set $\mathcal{S}$ such that for sufficiently large n, to any $A \subset \mathcal{X}^n$ there exists a mapping $F : \mathcal{X}^* \to \mathcal{S}$ and an F-type such that

$$|A| \simeq |A \cap \mathcal{T}_{SX}^{n,F}| \geq exp\{nH(X|S) - \epsilon\}.$$

While this lemma is used to prove (the converse part of) a probabilistic result, it is claimed to also yield the asymptotic solution of the general isoperimetric problem for arbitrary finite alphabets and arbitrary distortion measures.

A conference paper by Rudi Ahlswede and myself titled "Asymptotical isoperimetric problem" was published in Proceedings of the 1999 IEEE Information Theory and Communications Workshop, in 1999. In this paper, the inherently typical subset lemma was used to resolve the general isoperimetric problem for arbitrary finite alphabets and arbitrary distortion measures. In the paper, only a sketch of the proof was given.

4 Other Works

During our 25 years collaboration since 1996, Rudi and I had written more than 20 joint papers. Besides the works mentioned above, we also worked on other extremal combinatorial problems such as the diameter problem (Rudolf Ahlswede, Ning Cai, Zhen Zhang: Diametric theorems in sequence spaces. Combinatorica 12(1):1-17. 1992), and many information theoretical problems such as the write efficient memories (Rudolf Ahlswede, Zhen Zhang: Coding for Write-Efficient Memory Inf. Comput. (IANDC) 83(1):80-97 (1989) , On multiuser write-efficient memories. IEEE Transactions on Information Theory (TIT) 40(3):674-686 (1994)), identification via channels (Rudolf Ahlswede, Zhen Zhang: New directions in the theory of identification via channels. IEEE Transactions on Information Theory (TIT) 41(4):1040-1050 (1995)) and so on.

Our last joint work was published in 2005 and I hope very much that there will be one more to come. After 13 years of silence on the general isoperimetric problem, recently Zhen Zhang and Kang Wei wrote a rigorous version of the proof and this will be submitted for publication with Rudi as one of the coauthors.

Mr. Schimanski and the Pragmatic Dean

Ingo Althöfer

Fakultät für Mathematik und Informatik, Friedrich-Schiller-Universität Jena, 07737 Jena Germanyingo.althoefer @ uni-jena.de

1 Mr. Schimanski, Sometimes Shy

When Prof. Ahlswede was Dean, he had regular contact with Mr. Schimanski who directed the administration of the department. They found out that they both liked card games a lot. But, of course the lunch break was typically too short for full joy. When at one day Schimmi again remembered: "Boss, we have to go back to office, our job is waiting", Ahlswede found an elegant alternative solution: "I will give you official order that you have to go to Cafeteria for card play with your dean from 1 p.m. to 2 p.m. every working day." The secretary Frau Mehler (who saw many strange deans come and go ...) typed the order, and Schimanski framed it, hanging the letter at the wall behind his office desk.

2 Teaching Elementary Math to Mr. Schimanski

One day in the early 1990's Ahlswede came to Mr. Schimanski's office: "Listen, now I can explain you what a geometric series is. Four weeks ago I was at Essen University, attending a conference in Numerical Mathematics. On the way back, police radar caught me on the Autobahn three times because of my speed. I got three fines: the first one was 100 D-Mark, the second one 200 D-Mark, and the third one 400 D-Mark. You see: 100 - 200 - 400, that is a geometric series. Each time the value is multiplied by the same scalar (= 2 in this example). By the way; I also learned from this series: I will never again drive to a conference for Numerical Mathematics!"

3 The Pragmatic Dean

Professor Ahlswede was Dean of the Bielefeld Math department in 1978 and again in 1992. During this second period he once had to direct a Habilitation talk, given by a young man from Algebra. The presentation took place in Lecture Hall 8, next to the paper shop in the big hall of Bielefeld's University building. The candidate was full of energy, and under his writing the long blackboard started to swing back and forth in some eigenfrequence.

Dean Ahlswede wanted to bring the blackboard back under control and asked the speaker to stop for a moment: "Here I have a doctoral dissertation, submitted just this afternoon. I hope we can fix the board by squeezing the thesis between board and wall." Said it, stood up, and immediately executed the plan. And it helped, indeed! What Ahlswede did not realize: Matthias Löwe, the author of the dissertation, was sitting in the audience and followed with disbelief what happened to his brainchild. But at the end everything turned well: The habilitation was successful, Löwe got his Dr. title and later also a professorship (at Münster University).

Broken Pipes*

Thomas M. Cover**

I met Rudi first in Tshahkadsor, Armenia, in the late 60s where we had several hours of intense conversation about the emerging field of multiuser information theory. I thought of him then and have thought of him since as one of the fastest and deepest thinkers in the field.

Perhaps I should tell a story about him in connection with his two month visit to Stanford University about 15 years ago. He was working on a number of problems in multiple user theory and information inequalities. I received a call at my home at 4 in the morning from Germany. I was informed that his house in Bielefeld, which was unoccupied at the time, had ice flowing out of the windows. Apparently, the pipes had broken, flooding the house, bursting through the windows and causing cascades of ice down the stairs, over the carpets and out the windows, making your house look like something out of Dr. Zhivago. Since this was an emergency, I called Rudi immediately. I described the damage in some detail. His first response was: "That's interesting, but wait till I tell you the result I got last night."

As I recall, it took him two days before he realized that the damage to his house would require that he has to go back and attend to it before he continued on with his research here.

So I would like to salute your devotion to information theory and your prolific contributions over the last 50 years.

---

* This memorial talk was given at the ITA Workshop, February 7, 2011.

** Thomas Cover, one of the world's top information theorists and a professor of electrical engineering and of statistics at Stanford University, died March 26 at Stanford Hospital at the age of 73.

Rudolf Ahlswede's Funny Character*

Ilya Dumer

Electrical Engineering Department, College of Engineering,
University of California Riverside

Most of us knew Rudi as a great information theorist. I wish to mention some other things that made him so special, such as his insatiable curiosity, funny character, and ability to reach to different scientific fields and many people.

I never worked with Rudi but visited him once in Bielefeld in 1993 when I was with Humboldt Foundation. We also met at many conferences. One of these encounters stands out for me. It was in Sweden in 1993 when Rudi offered me a ride back to Bielefeld. He, Ning Cai, and I soon arrived in Copenhagen, where we learned that Rudi wants to show us the city. It was an amazing intellectual tour de force. For two good hours Rudi guided the two of us through the narrow streets talking about history with such a detail as if he were present there at the times of Hans Christian Andersen.

In fact, Rudi was always much interested in many different subjects. He got a very rigorous classic education in languages, philosophy, and mathematics and it took him a few years to choose mathematics instead of philosophy. He kept reading and refreshing Encyclopedie Britannica through all his life. This curiosity and versatility have also motivated his research. For many of us, he is foremost an information theorist. Some mathematicians will consider him an equal authority in number theory or combinatorics. Yet, he also made some seminal contributions to coding theory. Jointly with Mark Pinsker and Leonid Bassalygo, he opened the whole new area of codes correcting localized errors. Within a few years, the three of them essentially closed the field by constructing exact bounds, almost optimal codes, and decoding algorithms with polynomial complexity.

This brings me to his outstanding ability to work with a huge variety of people. Throughout his life, he had many dozens of long-term visitors in Bielefeld. From the former Soviet Union alone, there were tens of visitors coming from all scientific centers (Moscow, Novosibirsk, Yerevan, and S.-Petersburg).

Finally, Rudi had a great sense of humor and an ability to make things look funnier. At the international conference in Tashkent in Uzbekistan in 1984, many participants got foodpoisoned. Unaware of this, all went for a long 4-hour bus trip to the ancient city of Samarkand. Outside of Tashkent, the life was quite simple at those times, to say the least, and normal public amenities were sometimes almost non-existent. Many participants were in desperate need of public restrooms but most laughed when Rudi sick himself declared that there was no other conference with such an urgent call for papers.

* This memorial talk was given at the ITA Workshop, February 7, 2011.

Last time, I talked to Rudi in September 2010 in Dublin. There was the same twinkle in his eye and the same excitement about his new grant and the new problems to be solved. Rudi carried this love for research and curiosity through all his life. To me, he always was a Renaissance man of many gifts and talents, and I will remember this feeling of fun and excitement that he could bring to his own life and the life of others.

Two Anecdotes of Rudolf Ahlswede

Ulrich Tamm

German Language Department of Business Informatics
Marmara University, Istanbul, Turkey
tamm@ieee.org

1 Rudolf Ahlswede Playing Cards

Rudolf Ahlswede was a very strong card player. With this passion he was not alone at the University of Bielefeld and even not in the Department of Mathematics. During the lunch break usually about 10 - 20 people met in the cafeteria including Rudolf Ahlswede and the head of our department's administration, Mr. Schimanski, who also was chairman of the local bridge club - hence, a very strong card player himself.

Both of them did not like too much losing a game. An advantage of partner games as bridge or the german game "Doppelkopf", which was actually played in the cafeteria, is that in this case one can shift the responsibility for the loss to the partner.

One day, as usual, there was a big crowd at the card players' table and Rudolf Ahlswede and Mr. Schimanski both were among the players. Suddenly, it was getting loud at their table and two persons stood up blaming each other for having played erroneously. First they addressed each other as "Rudi" and "Schimmi" (their nicknames among the card players) but soon they switched to the formal "Professor Ahlswede" and "Mr. Schimanski" (the way they addressed each other in the department). Since Rudolf Ahlswede was dean those days they had to spend the rest of the day in the same office.

2 Rudolf Ahlswede and the Computer

Rudolf Ahlswede was pushing computer science already in the 1970's. Among the students in his group were Ingo Wegener, Gunter Dueck, Rüdiger Reischuk, Friedhelm Meyer auf der Heide and other leading computer experts. Later, the department's computer laboratory was administered by his students. However, he himself did not use (and did not want to use) a computer himself. For a long time he had no email address - all his correspondence was done by his chair's secretary Mrs. Hollmann. He seemingly first wrote an email himself during his three-months visit to Japan in 1999 in order to keep the contact to his research group. All other computer work had to be done by his assistants. They even prepared the slides for his lectures.

In 2006 Rudolf Ahlswede received the most prestigious award in Information Theory - he was selected as Shannon Lecturer of the IEEE Information Theory

Society. I did not trust my eyes when I suddenly saw an email in my mailbox from the organizers of the ISIT conference in Seattle, where Rudolf Ahlswede had to present his Shannon Lecture. They asked me to tell him that he should not use printed slides for an overhead projector but provide an electronic file preferably in PowerPoint, which would be projected by a beamer. My first reaction was that I should keep out and direct the request to my colleagues from his group in Bielefed (I had already left Bielefeld five years before). However, since he always had used an overhead projector for his lectures, I was afraid that he might react very sensible. So, I told the organizers that he surely would not want to learn how to use PowerPoint and suggested them to inform him that they could not provide an overhead projector and instead would ask him to provide them the pdf-file of his slides which then would be projected by a beamer. They followed my suggestion and it worked well.

Later, I learnt the background: In a previous lecture Rudolf Ahlswede was so concentrated on his results that during his talk he had written some changes not on the slides but on the overhead projector. The organizers wanted to avoid such a situation and informed a German professor whom they thought to be rather close to Rudolf Ahlswede. He rejected (they always kill the people who bring the bad news - that was also the reason why I first wanted to forward the request to Bielefeld) and instead suggested to contact me in this matter.

Bibliography of Publications by Rudolf Ahlswede

1967

[1] Certain results in coding theory for compound channels, Proc. Colloquium Inf. Th. Debrecen (Hungary), 35–60.

1968

[2] Beiträge zur Shannonschen Informationstheorie im Fall nichtstationärer Kanäle, Z. Wahrscheinlichkeitstheorie und verw. Geb. 10, 1–42.

[3] The weak capacity of averaged channels, Z. Wahrscheinlichkeitstheorie und verw. Geb. 11, 61–73.

1969

[4] Correlated decoding for channels with arbitrarily varying channel probability functions, (with J. Wolfowitz), Information and Control 14, 457–473.

[5] The structure of capacity functions for compound channels, (with J. Wolfowitz), Proc. of the Internat. Symposium on Probability and Information Theory at McMaster University, Canada, April 1968, 12–54.

1970

[6] The capacity of a channel with arbitrarily varying channel probability functions and binary output alphabet, (with J. Wolfowitz), Z. Wahrscheinlichkeitstheorie und verw. Geb. 15, 186–194.

[7] A note on the existence of the weak capacity for channels with arbitrarily varying channel probability functions and its relation to Shannon's zero error capacity, Ann. Math. Stat., Vol. 41, No. 3, 1027–1033.

1971

[8] Channels without synchronization, (with J. Wolfowitz), Advances in Applied Probability, Vol. 3, 383–403.

[9] Group codes do not achieve Shannon's channel capacity for general discrete channels, Ann. Math. Stat., Vol. 42, No. 1, 224–240.

[10] Bounds on algebraic code capacities for noisy channels I, (with J. Gemma), Information and Control, Vol. 19, No. 2, 124–145.

[11] Bounds on algebraic code capacities for noisy channels II, (with J. Gemma), Information and Control, Vol. 19, No. 2, 146–158.

1973

[12] Multi–way communication channels, Proceedings of 2nd International Symposium on Information Theory, Thakadsor, Armenian SSR, Sept. 1971, Akademiai Kiado, Budapest, 23–52.

[13] On two–way communication channels and a problem by Zarankiewicz, Sixth Prague Conf. on Inf. Th., Stat. Dec. Fct's and Rand. Proc., Sept. 1971, Publ. House Chechosl. Academy of Sc., 23–37.

[14] A constructive proof of the coding theorem for discrete memoryless channels in case of complete feedback, Sixth Prague Conf. on Inf. Th., Stat. Dec. Fct's and Rand. Proc., Sept. 1971, Publ. House Czechosl. Academy of Sc., 1–22.

[15] The capacity of a channel with arbitrarily varying additive Gaussian channel probability functions, Sixth Prague Conf. on Inf. Th., Stat. Dec. Fct's and Rand. Proc., Sept. 1971, Publ. House Czechosl. Academy of Sc., 39–50.

[16] Channels with arbitrarily varying channel probability functions in the presence of noiseless feedback, Z. Wahrscheinlichkeitstheorie und verw. Geb. 25, 239–252.

[17] Channel capacities for list codes, J. Appl. Probability, lo, 824–836.

1974

[18] The capacity region of a channel with two senders and two receivers, Ann. Probability, Vol. 2, No. 5, 805–814.

[19] On common information and related characteristics of correlated information sources, (with J. Körner), presented at the 7th Prague Conf. on Inf. Th., Stat. Dec. Fct's and Rand. Proc., included in "Information Theory" by I. Csiszár and J. Körner, Acad. Press, 1981, General Theory of Information Transfer and Combinatorics, Lecture Notes in Computer Science, Vol. 4123, Springer Verlag, 2006, 664-677.

1975

[20] Approximation of continuous functions in p–adic analysis, (with R. Bojanic), J. Approximation Theory, Vol. 15, No. 3, 190–205.

[21] Source coding with side information and a converse for degraded broadcast channels, (with J. Körner), IEEE Trans. Inf. Theory, Vol. IT–21, 629–637.

[22] Two contributions to information theory, (with P. Gács), Colloquia Mathematica Societatis János Bolyai, 16. Topics in Information Theory, I. Csiszár and P. Elias Edit., Keszthely, Hungaria, 1975, 17–40.

1976

[23] Bounds on conditional probabilities with applications in multiuser communication, (with P. Gács and J. Körner), Z. Wahrscheinlichkeitstheorie und verw. Geb. 34, 157–177.

[24] Every bad code has a good subcode: a local converse to the coding theorem, (with G. Dueck), Z. Wahrscheinlichkeitstheorie und verw. Geb. 34, 179–182.

[25] Spreading of sets in product spaces and hypercontraction of the Markov operator, (with P. Gács), Ann. Prob., Vol. 4, No. 6, 925–939.

1977

[26] On the connection between the entropies of input and output distributions of discrete memoryless channels, (with J. Körner), Proceedings of the 5th Conference on Probability Theory, Brasov 1974, Editura Academeiei Rep. Soc. Romania, Bucaresti 1977, 13–23.

[27] Contributions to the geometry of Hamming spaces, (with G. Katona), Discrete Mathematics 17, 1–22.

[28] The number of values of combinatorial functions, (with D.E. Daykin), Bull. London Math. Soc., 11, 49–51.

1978

[29] Elimination of correlation in random codes for arbitrarily varying channels, Z. Wahrscheinlichkeitstheorie und verw. Geb. 44, 159–175.

[30] An inequality for the weights of two families of sets, their unions and intersections, (with D.E. Daykin), Z. Wahrscheinlichkeitstheorie und verw. Geb. 43, 183–185.

[31] Graphs with maximal number of adjacent pairs of edges, (with G. Katona), Acta Math. Acad. Sc. Hung. 32, 97–120.

1979

[32] Suchprobleme, (with I. Wegener), Teubner Verlag, Stuttgart, Russian Edition with Appendix by Maljutov 1981 (Book).

[33] Inequalities for a pair of maps $S \times S \to S$ with S a finite set, (with D.E. Daykin), Math. Zeitschrift 165, 267–289.

[34] Integral inequalities for increasing functions, (with D.E. Daykin), Math. Proc. Comb. Phil. Soc., 86, 391–394.

[35] Coloring hypergraphs: A new approach to multi–user source coding I, Journ. of Combinatorics, Information and System Sciences, Vol. 4, No. 1, 76–115.

1980

[36] Coloring hypergraphs: A new approach to multi–user source coding II, Journ. of Combinatorics, Information and System Sciences, Vol. 5, No. 3, 220–268.

[37] Simple hypergraphs with maximal number of adjacent pairs of edges, J. Comb. Theory, Ser. B, Vol. 28, No. 2, 164–167.

[38] A method of coding and its application to arbitrarily varying channels, J. Combinatorics, Information and System Sciences, Vol. 5, No. 1, 10–35.

1981

[39] To get a bit of information may be as hard as to get full information, (with I. Csiszár), IEEE Trans. Inf. Theory, IT–27, 398–408.

[40] Solution of Burnashev's problem and a sharpening of Erdős–Ko–Rado, Siam Review, to appear in a book by G. Katona. Recently included in General Theory of Information Transfer and Combinatorics, Lecture Notes in Computer Science, Vol. 4123, Springer Verlag, 2006, 1006-1009.

1982

[41] Remarks on Shannon's secrecy systems, Probl. of Control and Inf. Theory, Vol. 11, No. 4, 301–318.

[42] Bad Codes are good ciphers, (with G. Dueck), Probl. of Control and Inf. Theory, Vol. 11, No. 5, 337–351.

[43] Good codes can be produced by a few permutations, (with G. Dueck), IEEE Trans. Inf. Theory, IT–28, No. 3, 430–443.

[44] An elementary proof of the strong converse theorem for the multiple–access channel, J. Combinatorics, Information and System Sciences, Vol. 7, No. 3, 216–230.

[45] Jacob Wolfowitz (1910–1981), IEEE Trans. Inf. Theory, Vol. IT–28, No. 5, 687–690.

1983

[46] Note on an extremal problem arising for unreliable networks in parallel computing, (with K.U. Koschnick), Discrete Mathematics 47, 137–152.

[47] On source coding with side information via a multiple–access channel and related problems in multi–user information theory, (with T.S. Han), IEEE Trans. Inf. Theory, Vol. IT–29, No. 3, 396–412.

1984

[48] A two family extremal problem in Hamming space, (with A. El Gamal and K.F. Pang), Discrete mathematics 49, 1–5.

[49] Improvements of Winograd's Result on Computation in the Presence of Noise, IEEE Trans. Inf. Theory, Vol. IT–30, No. 6, 872–877.

1985

[50] The rate–distortion region for multiple descriptions without excess rate, IEEE Trans. Inf. Theory, Vol. IT–31, No. 6, 721–726.

1986

[51] Hypothesis testing under communication constraints, (with I. Csiszár), IEEE Trans. Inf. Theory, Vol. IT–32, No. 4, 533–543.

[52] On multiple description and team guessing, IEEE Trans. Inf. Theory, Vol. IT–32, No. 4, 543–549.

[53] Arbitrarily varying channels with states sequence known to the sender, invited paper at a Statistical Research Conference dedicated to the memory of Jack Kiefer and Jacob Wolfowitz, held at Cornell University, July 1983, IEEE Trans. Inf. Theory, Vol. IT–32, No. 5, 621–629.

1987

[54] Optimal coding strategies for certain permuting channels, (with A. Kaspi), IEEE Trans. Inf. Theory, Vol. IT–33, No. 3, 310–314.

[55] Search Problems, (with I. Wegener), English Edition of [32] with Supplement of recent Literature, R.L. Graham, J.K. Leenstra, R.E. Tarjan (Ed.), Wiley–Interscience Series in Discrete Mathematics and Optimization.

[56] Inequalities for code pairs, (with M. Moers), European J. of Combinatorics 9, 175–181.

[57] Eight problems in information theory
— a complexity problem
— codes as orbits
Contributions to "Open Problems in Communication and Computation", T.M. Cover and B. Gopinath (Ed.), Springer–Verlag.

[58] On code pairs with specified Hamming distances, Colloquia Mathematica Societatis János Bolyai 52, Combinatorics, Eger (Hungary), 9–47.

1989

[59] Identification via channels, (with G. Dueck), IEEE Trans. Inf. Theory, Vol. 35, No. 1, 15–29.

[60] Identification in the presence of feedback — a discovery of new capacity formulas, (with G. Dueck), IEEE Trans. Inf. Theory, Vol. 35, No. 1, 30–39.

[61] Contributions to a theory of ordering for sequence spaces, (with Z. Zhang), Problems of Control and Information Theory, Vol. 18, No. 4, 197–221.

[62] A general 4–words inequality with consequences for 2–way communication complexity, (with N. Cai and Z. Zhang), Advances in Applied Mathematics, Vol. 10, 75–94.

1990

[63] Coding for write–efficient memory, (with Z. Zhang), Information and Computation, Vol. 83, No. 1, 80–97.

[64] Creating order in sequence spaces with simple machines, (with Jian–ping Ye and Z. Zhang), Information and Computation, Vol. 89, No. 1, 47–94.

[65] An identity in combinatorial extremal theory, (with Z. Zhang), Adv. in Math., Vol. 80, No. 2, 137–151.

[66] On minimax estimation in the presence of side information about remote data, (with M.V. Burnashev), Ann. of Stat., Vol. 18, No. 1, 141–171.

[67] Extremal properties of rate–distortion functions, IEEE Trans. Inf. Theory, Vol. 36, No. 1, 166–171.

[68] A recursive bound for the number of complete K–subgraphs of a graph, (with N. Cai and Z. Zhang), "Topics in graph theory and combinatorics" in honour of G. Ringel on the occasion of his 70th birthday, R. Bodendiek, R. Henn (Eds), 37–39.

[69] On cloud–antichains and related configurations, (with Z. Zhang), Discrete Mathematics 85, 225–245.

1991

[70] Reusable memories in the light of the old AV– and new OV–channel theory, (with G. Simonyi), IEEE Trans. Inf. Theory, Vol. 37, No. 4, 1143–1150.

[71] On identification via multi–way channels with feedback, (with B. Verboven), IEEE Trans. Inf. Theory, Vol. 37, No. 5, 1519–1526.

[72] Two proofs of Pinsker's conjecture concerning AV channels, (with N. Cai), IEEE Trans. Inf. Theory, Vol. 37, No. 6, 1647–1649.

1992

[73] Diametric theorems in sequence spaces, (with N. Cai and Z. Zhang), Combinatorica, Vol. 12, No. 1, 1–17.

[74] On set coverings in Cartesian product spaces, General Theory of Information Transfer and Combinatorics, Lecture Notes in Computer Science, Vol. 4123, Springer Verlag, 2006, 926-937.

[75] Rich colorings with local constraints, (with N. Cai and Z. Zhang), J. Combinatorics, Information & System Sciences, Vol. 17, Nos. 3-4, 203–216.

1993

[76] Asymptotically dense nonbinary codes correcting a constant number of localized errors, (with L.A. Bassalygo and M.S. Pinsker), Proc. III International workshop "Algebraic and Combinatorial Coding Theory", June 22–28, 1992, Tyrnovo, Bulgaria, Comptes rendus de l' Académie bulgare des Sciences, Tome 46, No. 1, 35–37.

[77] The maximal error capacity of AV channels for constant list sizes, IEEE Trans. Inf. Theory, Vol. 39, No. 4, 1416–1417.

[78] Nonbinary codes correcting localized errors, (with L.A. Bassalygo and M.S. Pinsker), IEEE Trans. Inf. Theory, Vol. 39, No. 4, 1413–1416.

[79] Common randomness in information theory and cryptography, Part I: Secret sharing, (with I. Csiszár), IEEE Trans. Inf. Theory, Vol. 39, No. 4, 1121–1132.

[80] A generalization of the AZ identity, (with N. Cai), Combinatorica 13 (3), 241–247.

[81] On partitioning the n–cube into sets with mutual distance 1, (with S.L. Bezrukov, A. Blokhuis, K. Metsch, and G.E. Moorhouse), Applied Math. Lett., Vol. 6, No. 4, 17–19.

[82] Communication complexity in lattices, (with N. Cai and U. Tamm), Applied Math. Lett., Vol. 6, No. 6, 53–58.

[83] Rank formulas for certain products of matrices, (with N. Cai), Applicable Algebra in Engineering, Communication and Computing, 2, 1–9.

[84] On extremal set partitions in Cartesian product spaces, (with N. Cai), Combinatorics, Probability & Computing 2, 211–220.

1994

[85] Note on the optimal structure of recovering set pairs in lattices: the sandglass conjecture, (with G. Simonyi), Discrete Math., 128, 389–394.

[86] On extremal sets without coprimes, (with L.H. Khachatrian), Acta Arithmetica, LXVI 1, 89–99.

[87] The maximal length of cloud–antichains, (with L.H. Khachatrian), Discrete Mathematics, Vol. 131, 9–15.

[88] The asymptotic behaviour of diameters in the average, (with I. Althöfer), J. Combinatorial Theory, Series B, Vol. 61, No. 2, 167–177.

[89] 2–way communication complexity of sum–type functions for one processor to be informed, (with N. Cai), Problemy Peredachi Informatsii, Vol. 30, No. 1, 3–12.

[90] Messy broadcasting in networks, (with H.S. Haroutunian and L.H. Khachatrian), Special volume in honour of J.L. Massey on occasion of his 60th birthday. Communications and Cryptography (Two sides of one tapestry), R.E. Blahut, D.J. Costello, U. Maurer, T. Mittelholzer (Ed.), Kluwer Acad. Publ., 13–24.

[91] Binary constant weight codes correcting localized errors and defects, (with L.A. Bassalygo and M.S. Pinsker), Probl. Peredachi Informatsii, Vol. 30, No. 2, 10–13 (In Russian); Probl. of Inf. Transmission, 102–104.

[92] On sets of words with pairwise common letter in different positions, (with N. Cai), Proc. Colloquium on Extremal Problems for Finite Sets, Visograd, Bolyai Siciety Math. Studies, 3, Hungary, 25–38.

[93] On multi–user write–efficient memories, (with Z. Zhang), IEEE Trans. Inf. Theory, Vol. 40, No. 3, 674–686.

[94] On communication complexity of vector–valued functions, (with N. Cai), IEEE Trans. Inf. Theory, Vol. 40, No. 6, 2062–2067.

[95] On partitioning and packing products with rectangles, (with N. Cai), Combinatorics, Probability & Computing 3, 429–434.

[96] A new direction in extremal theory for graphs, (with N. Cai and Z. Zhang), J. Combinatorics, Information & System Sciences, Vol. 19, No. 3–4, 269–280.

[97] Asymptotically optimal binary codes of polynomial complexity correcting localized errors, (with L.A. Bassalygo and M.S. Pinsker), Proc. IV International workshop on Algebraic and Combinatorial Coding Theory, Novgorod, Russia, 1–3.

1995

[98] Localized random and arbitrary errors in the light of AV channel theory, (with L.A. Bassalygo and M.S. Pinsker), IEEE Trans. Inf. Theory, Vol. 41, No. 1, 14–25.

[99] Edge isoperimetric theorems for integer point arrays, (with S.L. Bezrukov), Applied Math. Letters, Vol. 8, No. 2, 75–80.

[100] New directions in the theory of identification via channels, (with Z. Zhang), IEEE Trans. Inf. Theory, Vol. 41, No. 4, 1040–1050.

[101] Towards characterising equality in correlation inequalities, (with L.H. Khachatrian), European J. of Combinatorics 16, 315–328.

[102] Maximal sets of numbers not containing $k+1$ pairwise coprime integers, (with L.H. Khachatrian), Acta Arithmetica LXX II, 1, 77–100.

[103] Density inequalities for sets of multiples, (with L.H. Khachatrian), J. of Number Theory, Vol. 55, No. 2., 170–180.

[104] A splitting property of maximal antichains, (with P.L. Erdős and N. Graham), Combinatorica 15 (4), 475–480.

1996

[105] Sets of integers and quasi–integers with pairwise common divisor, (with L.H. Khachatrian), Acta Arithmetica, LXXIV.2, 141–153.

[106] A counterexample to Aharoni's "Strongly maximal matching" conjecture, (with L.H. Khachatrian), Discrete Mathematics 149, 289.

[107] Erasure, list, and detection zero–error capacities for low noise and a relation to identification, (with N. Cai and Z. Zhang), IEEE Trans. Inf. Theory, Vol. 42, No. 1, 55–62.

[108] Optimal pairs of incomparable clouds in multisets, (with L.H. Khachatrian), Graphs and Combinatorics 12, 97–137.

[109] Sets of integers with pairwise common divisor and a factor from a specified set of primes, (with L.H. Khachatrian), Acta Arithmetica LXX V 3, 259–276.

[110] Cross–disjoint pairs of clouds in the interval lattice, (with N. Cai), The Mathematics of Paul Erdős, Vol. I; R.L. Graham and J. Nesetril, ed., Algorithms and Combinatorics B, Springer Verlag, Berlin/Heidelberg/New York, 155–164.

[111] Identification under random processes, (with V. Balakirsky), Problemy peredachii informatsii (special issue devoted to M.S. Pinsker), vol. 32, no. 1, 144–160, Jan.–March 1996; Problems of Information Transmission, Vol. 32, No. 1, 123–138.

[112] Report on work in progress in combinatorial extremal theory: Shadows, AZ–identity, matching. *Ergänzungsreihe* des SFB 343 "Diskrete Strukturen in der Mathematik", Universität Bielefeld, Nr. 95–004.

[113] Fault–tolerant minimum broadcast networks, (with L. Gargano, H.S. Haroutunian, and L.H. Khachatrian), Networks, Vol. 27, No. 4, 1293–1307.

[114] The complete nontrivial–intersection theorem for systems of finite sets, (with L.H. Khachatrian), J. Combinatorial Theory, Series A, 121–138.

[115] Incomparability and intersection properties of Boolean interval lattices and chain posets, (with N. Cai), European J. of Combinatorics 17, 677–687.

[116] Classical results on primitive and recent results on cross–primitive sequences, (with L.H. Khachatrian), The Mathematics of P. Erdős, Vol. I; R.L. Graham and J. Nesetril, ed., Algorithms and Combinatorics B, Springer Verlag, Berlin/Heidelberg/ New York, 104–116.

[117] Intersecting Systems, (with N. Alon, P.L. Erdős, M. Ruszinko, L.A. Székely), Combinatorics, Probability and Computing 6, 127–137.

[118] Some properties of fix–free codes, (with B. Balkenhol and L.H. Khachatrian), Proceedings First INTAS International Seminar on Coding Theory and Combinatorics 1996, Thahkadzor, Armenia, 20–33, 6–11 October 1996.

[119] Higher level extremal problems, (with N. Cai and Z. Zhang), Comb. Inf. & Syst. Sc., Vol. 21, No. 3–4, 185–210.

1997

[120] On interactive communication, (with N. Cai and Z. Zhang), IEEE Trans. Inf. Theory, Vol. 43, No. 1, 22–37.

[121] Identification via compressed data, (with E. Yang and Z. Zhang), IEEE Trans. Inf. Theory, Vol. 43, No. 1, 48–70.

[122] The complete intersection theorem for systems of finite sets, (with L.H. Khachatrian), European J. Combinatorics, 18, 125–136.

[123] Universal coding of integers and unbounded search trees, (with T.S. Han and K. Kobayashi), IEEE Trans. Inf. Theory, Vol. 43, No. 2, 669–682.

[124] Number theoretic correlation inequalities for Dirichlet densities, (with L.H. Khachatrian), J. Number Theory, Vol. 63, No. 1, 34–46.

[125] General edge–isoperimetric inequalities, Part 1: Information theoretical methods, (with Ning Cai), European J. of Combinatorics 18, 355–372.

[126] General edge–isoperimetric inequalities, Part 2: A local–global principle for lexicographical solutions, (with Ning Cai), European J. of Combinatorics 18, 479–489.

[127] Models of multi–user write–efficient memories and general diametric theorems, (with N. Cai), Information and Computation, Vol. 135, No. 1, 37–67.

[128] Shadows and isoperimetry under the sequence–subsequence relation, (with N. Cai), Combinatorica 17 (1), 11–29.

[129] Counterexample to the Frankl/Pach conjecture for uniform, dense families, (with L.H. Khachatrian), Combinatorica 17 (2), 299–301.

[130] Correlated sources help the transmission over AVC, (with N. Cai), IEEE Trans. Inf. Theory, Vol. 135, No. 1, 37–67.

1998

[131] Common randomness in Information Theory and Cryptography, Part II: CR capacity, (with I. Csiszár), IEEE Trans. Inf. Theory, Vol. 44, No. 1, 55–62.

[132] The diametric theorem in Hamming spaces — optimal anticodes, (with L.H. Khachatrian) Proceedings First INTAS International Seminar on Coding Theory and Combinatorics 1996, Thahkadzor, Armenia, 1–19, 6–11 October 1996; Advances in Applied Mathematics 20, 429–449.

[133] Information and Control: Matching channels, (with N. Cai), IEEE Trans. Inf. Theory, Vol. 44, No. 2, 542–563.

[134] Zero–error capacity for models with memory and the enlightened dictator channel, (with N. Cai and Z. Zhang), IEEE Trans. Inf. Theory, Vol. 44, No. 3, 1250–1252.

[135] Code pairs with specified parity of the Hamming distances, (with Z. Zhang), Discrete Mathematics 188, 1–11.

[136] Isoperimetric theorems in the binary sequences of finite lengths, (with Ning Cai), Applied Math. Letters, Vol. 11, No. 5, 121–126.

[137] The intersection theorem for direct products, (with H. Aydinian and L.H. Khachatrian), European Journal of Combinatorics 19, 649–661.

1999

[138] Construction of uniquely decodable codes for the two–user binary adder channel, (with V.B. Balakirsky), IEEE Trans. Inf. Theory, Vol 45, No. 1, 326–330.

[139] Arbitrarily varying multiple–access channels, Part I. Ericson's symmetrizability is adequate, Gubner's conjecture is true, (with N. Cai), IEEE Trans. Inf. Theory, Vol. 45, No. 2, 742–749.

[140] Arbitrarily varying multiple–access channels, Part II. Correlated sender's side information, correlated messages, and ambiguous transmission, (with N. Cai), IEEE Trans. Inf. Theory, Vol. 45, No. 2, 749–756.

[141] A pushing–pulling method: new proofs of intersection theorems, (with L.H. Khachatrian), Combinatorica 19(1), 1–15.

[142] A counterexample in rate–distortion theory for correlated sources, (with Ning Cai), Applied Math. Letters, Vol. 12, No. 7, 1–3.

[143] On maximal shadows of members in left–compressed sets, (with Zhen Zhang), Proceedings of the Rostock Conference, Discrete Applied Math. 95, 3–9.

[144] A counterexample to Kleitman's conjecture concerning an edge–isoperimetric problem, (with Ning Cai), Combinatorics, Probability and Computing 8, 301–305.

[145] Identification without randomization, (with Ning Cai), IEEE Trans. Inf. Theory, Vol. 45, No. 7, 2636–2642.

[146] On the quotient sequence of sequences of integers, (with L.H. Khachatrian and A. Sárközy), Acta Arithmetica, XCI.2, 117–132.

[147] On the counting function for primitive sets of integers, (with L.H. Khachatrian and A. Sárközy), J. Number Theory 79, 330–344.

[148] On the Hamming bound for nonbinary localized–error–correcting codes, (with L.A. Bassalygo and M.S. Pinsker), Problemy Per. Informatsii, Vol. 35, No. 2, 29–37, Probl. of Inf. Transmission, Vol. 35, No. 2, 117–124.

[149] Asymptotical isoperimetric problem, (with Z. Zhang), Proceedings 1999 IEEE ITW, Krüger National Park, South Africa, June 20–25, 85–87.

[150] Nonstandard coding method for nonbinary codes correcting localized errors, (with L. Bassalygo and M. Pinsker), Proceedings 1999 IEEE ITW, Krüger National Park, South Africa, June 20–25, 78–79.

2000

[151] On prefix–free and suffix–free sequences of integers, (with L.H. Khachatrian and A. Sárközy), Numbers, Information and Complexity, Special volume in honour of R. Ahlswede on occasion of his 60th birthday, editors I. Althöfer, N. Cai, G. Dueck, L.H. Khachatrian, M. Pinsker, A. Sárközy, I. Wegener, and Z. Zhang, Kluwer Acad. Publ., Boston, Dordrecht, London, 1–16.

[152] Splitting properties in partially ordered sets and set systems, (with L.H. Khachatrian), Numbers, Information and Complexity, Special volume in honour of R. Ahlswede on occasion of his 60th birthday, editors I. Althöfer, N. Cai, G. Dueck, L.H. Khachatrian, M. Pinsker, A. Sárközy, I. Wegener, and Z. Zhang, Kluwer Acad. Publ., Boston, Dordrecht, London, 29–44.

[153] The AVC with noiseless feedback and maximal error probability: A capacity formula with a trichotomy, (with N. Cai), Numbers, Information and Complexity, Special volume in honour of R. Ahlswede on occasion of his 60th birthday, editors I. Althöfer, N. Cai, G. Dueck, L.H. Khachatrian, M. Pinsker, A. Sárközy, I. Wegener, and Z. Zhang, Kluwer Acad. Publ., Boston, Dordrecht, London, 151–176.

[154] A diametric theorem for edges, (with L.H. Khachatrian), J. Comb. Theory, Series A 92, 1–16.

[155] Network information flow, (with Ning Cai, S.Y. Robert Li, and Raymond W. Yeung), Preprint 98–033, SFB 343 "Diskrete Strukturen in der Mathematik", Universität Bielefeld, IEEE Trans. Inf. Theory, Vol. 46, No. 4, 1204–1216.

2001

[156] On perfect codes and related concepts, (with H. Aydinian and L.H. Khachatrian),
Designs, Codes and Cryptography, 22, 221–237.

[157] Quantum data processing, (with Peter Löber), IEEE Trans. Inf. Theory, Vol. 47, No. 1, 474–478.

[158] On primitive sets of squarefree integers, (with L.H. Khachatrian and A. Sárközy), Periodica Mathematica Hungarica Vol. 42 (1–2), 99–115.

[159] Advances on extremal problems in number theory and combinatorics, European Congress of Mathematics, Barcelona 2000, Vol. I, 147–175, Carles Casacuberta, Rosa Maria Miró–Roig, Joan Verdera, Sebastiá Xambó–Descamps (Eds.), Progress in Mathematics, Vol. 201, Birkhäuser Verlag, Basel–Boston–Berlin.

[160] An isoperimetric theorem for sequences generated by feedback and feedback–codes for unequal error protection, (with N. Cai and C. Deppe), Problemy Peredachi Informatsii, No. 4, 63–70, 2001, Transmission Problems of Information Transmission, Vol. 37, No. 4, 332–338.

2002

[161] Strong converse for identification via quantum channels, (with A. Winter), IEEE Trans. Inf. Theory, Vol. 48, No. 3, 569–579.

[162] Parallel error correcting codes, (with B. Balkenhol and N. Cai), IEEE Trans. Inf. Theory, Vol. 48, No. 4, 959–962.

[163] Semi–noisy deterministic multiple–access channels: coding theorems for list codes and codes with feedback, (with N. Cai), IEEE Trans. Inf. Theory, Vol. 48, No. 8, 2953–2962.

[164] The t–intersection problem in the truncated Boolean lattice, (with C. Bey, K. Engel, and L.H. Khachatrian), European Journal of Combinatorics 23, 471–487.

[165] Unidirectional error control codes and related combinatorial problems, (with H. Aydinian and L.H. Khachatrian), in Proceedings of Eight International workshop on Algebraic and Combinatorial Coding Theory, 8–14 September, Tsarskoe Selo, Russia, 6–9.

2003

[166] Forbidden (0,1)–vectors in hyperplanes of $\mathbb{R}^n$: The restricted case, (with H. Aydinian and L.H. Khachatrian), Designs, Codes and Cryptography, 29, 17–28.

[167] Cone dependence — a basic combinatorial concept, (with L.H. Khachatrian), Designs, Codes and Cryptography, 29, 29–40.

[168] More about shifting techniques, (with H. Aydinian and L.H. Khachatrian), European Journal of Combinatorics 24, 551–556.

[169] On lossless quantum data compression and quantum variable–length codes, (with Ning Cai), Chapter 6 in "Quantum Information Processing", Gerd Leuchs, Thomas Beth (Eds.), Wiley–VCH Verlag, Weinheim, Germany, 66–78.

[170] Maximum number of constant weight vertices of the unit n–cube contained in a k–dimensional subspace, (with H. Aydinian and L.H. Khachatrian), Combinatorica, Vol. 23 (1), 5–22.

[171] A complexity measure for families of binary sequences (with L.H. Khachatrian, C. Mauduit, and A. Sárközy), Periodica Mathematica Hungarica, Vol. 46 (2), 107–118.

[172] Extremal problems under dimension constraints, (with H. Aydinian and L.H. Khachatrian), Discrete Mathematics, Special issue: EuroComb'01 – Edited by J. Nesetril, M. Noy and O. Serra, Vol. 273, No. 1–3, 9–21.

[173] Maximal antichains under dimension constraints, (with H. Aydinian and L.H. Khachatrian), Discrete Mathematics, Special issue: EuroComb'01 – Edited by J. Nesetril, M. Noy and O. Serra, Vol. 273, No. 1–3, 23–29.

[174] Large deviations in quantum information theory, (with V. Blinovsky), Probl. of Inf. Transmission, Vol. 39, Issue 4, 373–379.

2004

[175] On Bohman's conjecture related to a sum packing problem of Erdős, (with H. Aydinian and L.H. Khachatrian), Proceedings of the American Mathematical Society, Vol. 132, No. 5, 1257–1265.

[176] On shadows of intersecting families, (with H. Aydinian and L.H. Khachatrian), Combinatorica 24 (4), 555–566.

[177] On lossless quantum data compression with a classical helper, (with Ning Cai), IEEE Trans. Inf. Theory, Vol. 50, No. 6,

[178] On the density of primitive sets, (with L.H. Khachatrian and A. Sárközy), J. Number Theory 109, 319-361.

2005

[179] Katona's Intersection Theorem: Four Proofs, (with L.H. Khachatrian), Combinatorica 25 (1), 105-110.

[180] Forbidden (0,1)–vectors in Hyperplanes of $\mathbb{R}^n$: The unrestricted case, (with L.H. Khachatrian and H. Aydinian) Designs, Codes and Cryptography 37, 151-167.

[181] Nonbinary error correcting codes with noiseless feedback, localized errors or both, (with C. Deppe and V. Lebedev), Annals of European Academy of Sciences, No. 1, 285-309.

2006

[182] Search with noisy and delayed responses, (with N. Cai), General Theory of Information Transfer and Combinatorics, Lecture Notes in Computer Science, Vol. 4123, Springer Verlag, 695-703.

[183] Watermarking identification codes with related topics in common randomness, (with N. Cai), General Theory of Information Transfer and Combinatorics, Lecture Notes in Computer Science, Vol. 4123, Springer Verlag, 107-153.

[184] Large families of pseudorandom sequences of k symbols and their complexity, Part I, (with C. Mauduit and A. Sárközy), General Theory of Information Transfer and Combinatorics, Lecture Notes in Computer Science, Vol. 4123, Springer Verlag, 293-307.

[185] Large families of pseudorandom sequences of k symbols and their complexity, Part II, (with C. Mauduit and A. Sárközy), General Theory of Information Transfer and Combinatorics, Lecture Notes in Computer Science, Vol. 4123, Springer Verlag, 308-325.

[186] A Kraft–type inequality for d–delay binary search codes, (with N. Cai), General Theory of Information Transfer and Combinatorics, Lecture Notes in Computer Science, Vol. 4123, Springer Verlag, 704-706.

[187] Sparse asymmetric connectors in communication networks, (with H. Aydinian) General Theory of Information Transfer and Combinatorics, Lecture Notes in Computer Science, Vol. 4123, Springer Verlag, 1056-1062.

[188] A strong converse theorem for quantum multiple access channels, (with N. Cai), General Theory of Information Transfer and Combinatorics, Lecture Notes in Computer Science, Vol. 4123, Springer Verlag, 460-485.

[189] Codes with the identifiable parent property and the multiple–access channel, (with N. Cai) General Theory of Information Transfer and Combinatorics, Lecture Notes in Computer Science, Vol. 4123, Springer Verlag, 249-257.

[190] Estimating with randomized encoding the joint empirical distribution in a correlated source, (with Zhen Zhang),(Preliminary version: Worst case estimation of permutation invariant functions and identification via compressed data, (with Zhen Zhang), Preprint 97–005, SFB 343 "Diskrete Strukturen in der Mathematik", Universität Bielefeld) General Theory of Information Transfer and Combinatorics, Lecture Notes in Computer Science, Vol. 4123, Springer Verlag, 535-546.

[191] On attractive and friendly sets in sequence spaces, (with L.H. Khachatrian), General Theory of Information Transfer and Combinatorics, Lecture Notes in Computer Science, Vol. 4123, Springer Verlag, 955-970.

[192] Information theoretic models in language evolution, (with E. Arikan, L. Bäumer and C. Deppe), General Theory of Information Transfer and Combinatorics, Lecture Notes in Computer Science, Vol. 4123, Springer Verlag, 769-787. Accepted as Language Evolution and Information Theory in ISIT, Chicago June 27 - July 2, 2004.

[193] A fast suffix–sorting algorithm, (with B. Balkenhol, C. Deppe, and M. Fröhlich), General Theory of Information Transfer and Combinatorics, Lecture Notes in Computer Science, Vol. 4123, Springer Verlag, 719-734.

[194] On partitions of a rectangle into rectangles with restricted number of cross sections, (with A. Yudin), General Theory of Information Transfer and Combinatorics, Lecture Notes in Computer Science, Vol. 4123, Springer Verlag, 941-954.

[195] On concepts of performance parameters for channels, (Original version: Concepts of performance parameters for channels, Preprint 00–126, SFB 343 "Diskrete Strukturen in der Mathematik", Universität Bielefeld) General Theory of Information Transfer and Combinatorics, Lecture Notes in Computer Science, Vol. 4123, Springer Verlag, 639-663.

[196] Report on models of write–efficient memories with localized errors and defects, (with M.S. Pinsker), General Theory of Information Transfer and Combinatorics, Lecture Notes in Computer Science, Vol. 4123, Springer Verlag, 628-632.

[197] Correlation inequalities in function spaces, (with V. Blinovsky), General Theory of Information Transfer and Combinatorics, Lecture Notes in Computer Science, Vol. 4123, Springer Verlag, 572-577.

[198] Solution of Burnashev's problem and a sharpening of the Erdős/Ko/Rado Theorem, General Theory of Information Transfer and Combinatorics, Lecture Notes in Computer Science, Vol. 4123, Springer Verlag, 1006-1009.

[199] Transmission, identification and common randomness capacities for wire-tape channels with secure feedback from the decoder, (with Ning Cai), General Theory of Information Transfer and Combinatorics, Lecture Notes in Computer Science, Vol. 4123, Springer Verlag, 258-275.

[200] On set coverings in cartesian product spaces, General Theory of Information Transfer and Combinatorics, Lecture Notes in Computer Science, Vol. 4123, Springer Verlag, 926-937.

[201] Identification for sources, (with B. Balkenhol and C. Kleinewächter), General Theory of Information Transfer and Combinatorics, Lecture Notes in Computer Science, Vol. 4123, Springer Verlag, 51-61.

[202] Secrecy systems for identification via channels with additive-like instantaneous block encipherers, (with Ning Cai and Zhaozhi Zhang), General Theory of Information Transfer and Combinatorics, Lecture Notes in Computer Science, Vol. 4123, Springer Verlag, 285-292.

[203] Identification entropy, General Theory of Information Transfer and Combinatorics, Lecture Notes in Computer Science, Vol. 4123, Springer Verlag, 595-613.

[204] On logarithmically asymptotically optimal hypothesis testing for arbitrarily varying sources with side information, (with Evgueni Haroutunian and Ella Aloyan), General Theory of Information Transfer and Combinatorics, Lecture Notes in Computer Science, Vol. 4123, Springer Verlag, 547-552.

[205] On logarithmically asymptotically optimal testing of hypothesis and identification, (with Evgueni Haroutunian), General Theory of Information Transfer and Combinatorics, Lecture Notes in Computer Science, Vol. 4123, Springer Verlag, 553-571.

[206] Problems in network coding and error correcting codes, (with S. Riis), General Theory of Information Transfer and Combinatorics, Lecture Notes in Computer Science, Vol. 4123, Springer Verlag, 861-897, also appeared in NetCod the first Workshop on Network Coding, Theory, and Applications , April 2005, Trento, Italy.

[207] On edge-isoperimetric theorems for uniform hypergraphs, (with N. Cai), General Theory of Information Transfer and Combinatorics, Lecture Notes in Computer Science, Vol. 4123, Springer Verlag, 979-1005.

[208] An interpretation of identification entropy, (with N. Cai), IEEE Trans. Inf. Theory, Vol. 52, No. 9, 4198-4207.

[209] Construction of asymmetric connectors of depth two, (with H. Aydinian) Special Issue in Honor of Jacobus H. van Lint of J. Combinatorial Theory, Series A, Vol. 113, No. 8, 1614-1620.

[210] Maximal sets of integers not containing $k+1$ pairwise coprimes and having divisors from a specified set of primes, (with V. Blinovsky), Special Issue in Honor of Jacobus H. van Lint of J. Combinatorial Theory, Series A, Vol. 113, No. 8, 1621-1628.

[211] About the number of step functions with restrictions, (with V. Blinovsky), Probability Theory and Applications, Vol. 50, No. 4, 537-560.

[212] Another diametric theorem in Hamming spaces: optimal group anticodes, Proc. Information Theory Workshop, Punta del Este, Uruguay, March 13-17, 212-216.

[213] Intersection theorems under dimension constraints part I: the restricted case and part II: the unrestricted case, (with H. Aydinian and L.H. Khachatrian), J. Comb. Theory, Series A 113, 483-519.

[214] On q–ary codes correcting all unidirectional errors of a limited magnitude, (with H. Aydinian, L.H. Khachatrian, L.M. Tolhuizen), Abstract included in Proceedings of the International workshop on Algebraic and Combinatorial Coding Theory (ACCT), Kranevo, Bulgaria, June 19 - 25, 2004, Preprint in Arxiv, CS.IT-0607132, Codes, Polynomials, and Numbers, Selected work of Rom R. Varshamov and some of his students and colleauges dedicated to the memory of Academicaan Rom R. Varshamov, 41-61, 2009.

2007

[215] On the oblivious transfer capacity, (with I. Csiszár), ISIT, Proceedings of the IEEE International Symposium on Information Theory, 2061-2064.

[216] Classical capacity of classical-quantum arbitrarily varying channels, (with V. Blinovsky), IEEE Trans. Inf. Theory, Vol. 53, No. 2, 526-533.

[217] The final form of Tao's inequality relating conditional expectation and conditional mutual information, Advances in Mathematics of Communications, Vol. 1, No. 2, 239-242, 2007.

[218] Towards a General Theory of Information Transfer, Shannon Lecture at ISIT in Seattle 13th July 2006, IEEE Inform. Theory Society Newsletter, Vol. 57, No. 3, 6-28.

2008

[219] Error control codes for parallel asymmetric channels, (with H. Aydinian), IEEE Trans. Inf. Theory, Vol. 54, No. 2, 831 - 836.

[220] General theory of information transfer: updated, (Original version: General theory of information transfer, Preprint 97–118, SFB 343 "Diskrete Strukturen in der Mathematik", Universität Bielefeld) General Theory of Information Transfer and Combinatorics, Special Issue of Discrete Applied Mathematics, Vol. 156, No. 9, 1348-1388, [DOI: 10.1016/j.dam.2007.07.007].

[221] Searching with lies under error transition cost constraints, (with F. Cicalese and C. Deppe), General Theory of Information Transfer and Combinatorics, Special Issue of Discrete Applied Mathematics, Vol. 156, No. 9, 1444-1460, [DOI: 10.1016/j.dam.2007.04.033].

[222] T–Shift synchronization Codes", (with B. Balkenhol, C. Deppe, H. Mashurian and T. Partner), General Theory of Information Transfer and Combinatorics, Special Issue of Discrete Applied Mathematics, Vol. 156, No. 9, 1461-1468, [DOI: 10.1016/j.dam.2007.06.020].

[223] Rate–wise optimal non–sequential search strategies under a cardinality constraint on the tests, General Theory of Information Transfer and Combinatorics, Special Issue of Discrete Applied Mathematics, Vol. 156, No. 9, 1431-1443, [DOI: 10.1016/j.dam.2006.06.013].

[224] On the correlation of binary sequences, (with J. Cassaigne and A. Sárközy), General Theory of Information Transfer and Combinatorics, Special Issue of Discrete Applied Mathematics, Vol. 156, No. 9, 1478-1487, [DOI: 10.1016/j.dam.2006.11.021].

[225] Multiple packing in sum-type metric spaces, (with V. Blinovsky), General Theory of Information Transfer and Combinatorics, Special Issue of Discrete Applied Mathematics, Vol. 156, No. 9, 1469-1477, [DOI: 10.1016/j.dam.2007.07.012].

[226] Towards Combinatorial Algebraic Number Theory, (with V. Blinovsky), Lectures on Advances in Combinatorics, (with V. Blinovsky), Universitext, Springer, 275-284.

[227] A diametric theorem in $\mathbb{Z}_m^n$ for Lee and related distances, (with F.I. Solov'eva), Proceedings of the Second International Castle Meeting on Coding Theory and Applications, Lecture Notes in Computer Science, Vol. 5228, 1-10.

[228] On diagnosability of large multiprocessor networks, (with H. Aydinian), Discrete Applied Mathematics, Vol. 156, No. 18, 3464-3474.

2009

[229] Two batch search with lie cost, (with F. Cicalese, C. Deppe, and U. Vaccaro), IEEE Trans. Inf. Theory, Vol. 55, No. 4, 1433-1439.

[230] Interactive communication, diagnosis, and error control in networks, (with H. Aydinian), Algorithmics of Large and Complex Networks (Design, Analysis, and Simulation), J. Lerner, D. Wagner, and K. Zweig (Eds.), Lecture Notes in Computer Science, Vol. 5515, 197-226.

[231] On error control codes for random network coding, (with H. Aydinian), Proceedings of NetCod 09, Workshop on Network Coding, Theory, and Applications, Lausanne, 15-16 June 2009, 68 - 73.

2010

[232] Entanglement transmission under adversarially selected quantum noise, (with I. Bjelakovic, H. Boche, and J. Nötzel) DPG-Frühjahrstagung in Hannover, Preprint in Arxiv, arXiv:1004.5551v1.

[233] Every channel with time structure has a capacity sequence, Proceedings of the IEEE Information Theory Workshop in Dublin, August 30 - September 3, 2010.

2011

[234] On security of statistical databases, (with H. Aydinian), SIAM Journal on Discrete Mathematics, Vol. 25, No. 4, 1778-1791.

2012

[235] On generic erasure correcting sets and related problems, (with H. Aydinian), IEEE Trans. Inf. Theory, Vol. 58, No. 2, 501-508.

[236] Shadows under the word-subword relation, (with V. Lebedev), Problems of Information Transmission, Vol. 48, No. 1, 31-46.

[237] Finding one of D defective elements in some group testing models, (with C. Deppe and V. Lebedev), Problems of Information Transmission, Vol. 48, No. 2, 173-181.

2013

[238] Threshold and majority group testing, (with C. Deppe and V. Lebedev), LNCS Festschrifts, Vol. 7777.

[239] The Restricted Word Shadow Problem, (with V. Lebedev), LNCS Festschrifts, Vol. 7777.

[240] The oblivious transfer capacity, (with I. Csiszr), LNCS Festschrifts, Vol. 7777.

[241] New construction of error-tolerant pooling designs, (with H. Aydinian), LNCS Festschrifts, Vol. 7777.

to appear

[242] Quantum capacity under adversarial quantum noise: arbitrarily varying quantum channels (with I. Bjelakovic, H. Boche, and J. Ntzel) Communications in Mathematical Physics. (Preprint in Arxiv, arXiv:1010.0418v2).

Books

[243] Suchprobleme, (with I. Wegener), Teubner Verlag, Stuttgart, 1979.

[244] Zadatsi Poiska, (with I. Wegener), Russian Edition of "Suchprobleme" with an Appendix by M. Maljutov, MIR-Verlag, Moscow, 1982.

[245] Search Problems, (with I. Wegener), English Edition of "Suchprobleme" with Supplement of recent Literature, R.L. Graham, J.K. Leenstra, R.E. Tarjan (Eds.), Wiley–Interscience Series in Discrete Mathematics and Optimization, 1988.

[246] General Theory of Information Transfer and Combinatorics (assisted by L. Bäumer, N. Cai; in cooperation with H. Aydinian, V. Blinovsky, C. Deppe, and H. Mashurian), Lecture Notes in Computer Science, Springer-Verlag, Vol. 4123, 2006.

[247] Lectures on Advances in Combinatorics (with V. Blinovsky), Universitext, Springer-Verlag, 2008.

[248] General Theory of Information Transfer and Combinatorics (with L. Bäumer and N. Cai (Eds.)), Discrete Applied Mathematics, Volume 156, Issue 9, 2008.

Author Index